CW01498017

The UK Pesticide Guide 2023

Editor: M.A. Lainsbury, BSc(Hons)

BCPC

BCPC (British Crop Production Council) promotes the use of good science and technology in understanding and applying effective and sustainable crop production. Its key objectives are to identify developing issues in the science and practice of crop protection and production and provide informed, independent analysis; to publish definitive information to growers, advisors and other stakeholders; to organise conferences and symposia, and to stimulate interest and learning. BCPC is part of the NIAB Group, a registered Charity and a Company limited by Guarantee. Further details available from BCPC, 93 Lawrence Weaver Road, Cambridge, CB3 0LE.

Telephone: 01223 342495
email: publications@bcpc.org
Web: www.bcpc.org

ISBN: 978-1-9998966-7-6

Typeset and Printed in the United Kingdom by
Hobbs the Printers Ltd, Totton, Hampshire, SO40 3WX
Website: www.hobbs.uk.com

Contents

Editor's Note

This is the 36th edition of *The UK Pesticide Guide* and Brexit continues to impact on the entries listed in this Guide because Northern Ireland still has to comply with EU pesticide regulations while Great Britain can make its own decisions. This means that as parallel approval products approach their final use up date in GB, they are still legal to use in Northern Ireland and some parallel product companies have gained new MAPP numbers for their products that are only legal to use in Northern Ireland.

Some new EAMUs granted by Chemicals Regulation Division (CRD) are only valid in GB and some important products such as cinmethylin, the new blackgrass and ryegrass product from BASF, is only approved in GB and cannot yet be used in Northern Ireland because it is not yet approved by the EU. Mancozeb and its co-formulations have had approval renewed in GB but the EU has banned it so it cannot be used to protect the efficacy of the potato fungicides used now in Northern Ireland. These differences have been incorporated in this latest edition of the Guide so where apparent duplicate entries are listed, one might be for GB use with a different expiry date to the identical product listed for use in Northern Ireland only.

On a more positive front, the 2023 edition does have a number of new active ingredients, formulations and new products, The gains and losses of actives are summarized here with new actives shown in **bold**. Unusually, this year there are more gains than losses:

GAINS:

Herbicide gains are:
- **cinmethylin**
- mesotrione + pyridate
- mesosulfuron + propoxycarbazone
- napropamide + quinmerac

Fungicide gains are:
- **1,4-dimethyl naphthalene**
- mefentrifluconazole + pyraclostrobin
- myclobutanil (GB only)
- ***Trichoderma asperellum* (T34)**
- cyflufenamid + spiroxamine
- dithianon + potassium phosphonates
- flonicamid + prothioconazole

Insecticide & PGR full entry gains are:
- **cholecalciferol (rodenticide)**
- dodecadienol + tetradecenyl actetate
- ethylene
- mepiquat chloride + prohexadione calcium + pyraclostrobin
- **orange oil**
- **(Z)-11-hexadecenal**

LOSSES:

Fungicides demoted to PAR or lost are:
- *Aureobasidium pullans*
- azoxystrobin + cyproconazole
- benthiavalicarb-isopropyl + mancozeb
- *Coniothyrium minitans*
- dimethomorph + zoxamide
- imazalil + ipconazole
- mancozeb + zoxamide
- pyriofenone

Insecticides & PGRs demoted to PAR or lost are:
- flocoumafen (rodenticide)
- magnesium phosphide
- warfarin

Herbicides demoted to PAR or lost are:
aminopyralid + halauxifen
- clomazone + napropamide
- imazamox + metazachlor
- pyroxsulam

Herbicide Resistance Action Committee (HRAC) has changed the mode of action codes for herbicides from an alpha-numeric to a numeric code and this edition of the book contains the new numbers with the old alpha-numeric code in brackets. In future years we will just list the numeric code. The rise in energy prices, raw ingredients and labour costs must inevitably lead to an increase in product prices so it becomes even more essential to use pesticides in the most efficient way. We hope that the information in this book and in the online database will help with this aim.

Many products have been demoted from Section 2 to Section 3 Products also Registered (PAR) as their approvals expire. As always, this book is a snapshot of the products with a valid approval on 16 October 2022 when the text went for initial typesetting. If you cannot see a product in the main section of the book, please check the index because it may be listed in 'Products also Registered'. For those who must have the most up-to-date information I can recommend the on-line version of the UKPG – *ukpesticideguide.co.uk*. This is a searchable database with much more information than we can squeeze into the book.

We express our thanks to all those who have provided the information for the compilation of this Guide. Criticisms and suggestions for improvement are always welcome and if you notice any errors or omissions, please let me know via the email address below.

Martin Lainsbury, Editor
email: publications@bcpc.org

Disclaimer

Every effort has been made to ensure that the information in this book is correct at the time of going to press, but the Editor and the publishers do not accept liability for any error or omission in the content, or for any loss, damage or other accident arising from the use of the products listed herein. Omission of a product does not necessarily mean that it is not approved or that it is unavailable.

It is essential to follow the instructions on the approved label when handling, storing or using any crop protection product. Approved 'off-label' extensions of use now termed Extension of Authorisation for Minor Use (EAMUs) are undertaken entirely at the risk of the user.

The information in this publication has been selected from official sources, from the suppliers product labels and from the product manuals of the pesticides approved for use in the UK under EU Regulation 1107/2009.

Changes Since 2022 Edition

Pesticides and adjuvants added or deleted since the 2022 edition are listed below. Every effort has been made to ensure the accuracy of this list, but late notifications to the Editor are not included.

Products Added

Products new to this edition are shown here. In addition, products that were listed in the previous edition whose CRD/HSE registration number and/or supplier have changed are included.

Product	Reg. No.	Supplier	Product	Reg. No.	Supplier
1,4Sight (GB only)	20158	Dormfresh	Kanaster	20008	Syngenta
Agate EW (GB only)	19000	Nufarm UK	Karamate Dry Flo Neotec	19994	Agrii
AQ 10	19968	Fargro			
Atlantis Star	20011	Bayer CropScience	Leystar	19938	Corteva
			Luxinum Plus (GB only)	19978	BASF
Backrow Max	A0889	Interagro			
Blaster Pro	19937	Corteva	Maister WG	20033	Bayer CropScience
Byte	A0847	De Sangosse			
Calculate	19929	Agrii	Menno Florades	19998	Fargro
Caligula (GB only)	20108	Bayer CropScience	Meristo	20163	Syngenta
			Mirage (GB only)	18361	Adama
Callisto	19756	Syngenta	Monkey (GB only)	18997	Nufarm UK
Clayton Cocoon (GB only)	20058	Clayton	Monolith	19973	Bayer CropScience
Clayton Mondello SC	19859	Clayton	Movento	18435	Bayer CropScience
Clayton Pontoon	19972	Clayton	Nevada	19967	Corteva
Cleancrop Wilbur	19976	Agrii	Orius P (GB only)	18994	Nufarm UK
Consul (GB only)	19597	FMC Agro	Pastor Trio	19977	Corteva
Dingo	19959	Corteva	Poquet (GB only)	19980	Corteva
Dow Shield 400	19874	Corteva	Prevail	19957	Corteva
Dynetic	A0743	De Sangosse	Pyrafen	19913	BASF
Eco-Tac	A0867	Global Adjuvants	Quotient	19969	BASF
			Reflect (GB only)	18573	Adama
Expert Met	20031	Bayer CropScience	Renu	A0627	AgriMax
			Renu DS	A0926	AgriMax
Explicit (GB only)	18763	FMC Agro	Retainer	A0711	Intracrop
Falcon	20010	Adama	Rumo (GB only)	18797	FMC Agro
Gateway	A0881	FMC Agro	Selontra	UK20-1257	BASF
Giddyup	20223	Pangaea			
Grazon Pro	19875	Corteva	Slither	A0458	De Sangosse
Grazon Spot	19956	Corteva	SP058	A0793	De Sangosse
Ground Force	A0890	Global Adjuvants	Spitfire	19945	Corteva
			Spraymate	A0621	De Sangosse
Impact (GB only)	19592	FMC Agro	Spryte Aqua	A0857	Amega
Innovator	A0935	Intracrop	Spryte Aqua CF	A0859	AgriMax
Intercept	20226	Pangaea	Steward (GB only)	18792	FMC Agro
Jettano	19931	Bayer CropScience	Stockholm EC	19965	UPL Europe

Product	Reg. No.	Supplier
Sublime	19964	Bayer CropScience
Thistlex	19876	Corteva
Topguard (GB only)	19596	FMC Agro
Unicur	19987	Bayer CropScience

Product	Reg. No.	Supplier
Univoq (GB only)	19930	Corteva
Variano Duo	19943	Bayer CropScience
Velomax	A0831	De Sangosse

Products Deleted

The appearance of a product name in the following list may not necessarily mean that it is no longer available. In cases of doubt, refer to Section 3 or the supplier.

Product	Reg. No.	Supplier
Avocet	14829	Corteva
Banka	A0245	Interagro
Consul	19597	FMC Agro
Impact	19592	FMC Agro
Integral Pro	18510	BASF
Intracrop Impetus	A0647	Intracrop
Intracrop Salute AF	A0809	Intracrop
Intracrop Saturn	A0494	Intracrop
Kansas	19292	Adama
Kobra	A0618	Interagro
Kurdi	15917	AgChem Access

Product	Reg. No.	Supplier
Mirador Xtra	18542	Adama
Mohawk CS	18505	Sipcam
Mycotal (GB only)	16644	Koppert
Reflect	18573	Adama
Sakarat Warfarin Whole Wheat	UK17-1059	Killgerm
Saxon	19289	FMC Agro
Slippa	A0206	Interagro
Topguard	19596	FMC Agro
Wolverine	19181	BASF
Zigzag	A0896	FMC Agro

Introduction

Purpose

The primary aim of this book is to provide a practical handbook of pesticides, plant growth regulators and adjuvants that the farmer or grower can realistically and legally obtain in the UK and to identify the purposes for which they can be used. It is designed to help in the identification of products appropriate to a particular problem. In addition to uses recommended on product labels, details are provided of those uses which do not appear on product labels but which have been granted Extension of Authorisation for Minor Use (EAMUs). As well as identifying the products available, the book provides guidance on how to use them safely and effectively but without giving details of doses, volumes, spray schedules or approved tank mixtures. Sections 5 and 6 provide essential background information on a wide range of pesticide-related issues including: legislation, biodiversity, codes of practice, poisons and treatment of poisoning, products for use in special situations and weed and crop growth stage keys.

While we have tried to cover all other important factors, this book does not provide a full statement of product recommendations. Before using any pesticide product it is essential that the user should read the label carefully and comply strictly with the instructions it contains. **This Guide does not substitute for the product label.**

Scope

The *Guide* is confined to pesticides registered in the UK for professional use in arable agriculture, horticulture (including amenity use), forestry and areas in or near water. Within these fields of use there are about 2580 products in the UK with current approval. The *Guide* is a reference for virtually all of them given in two sections. Section 2 gives full details of the products notified to the editor as available on the market. Products are included in this section only if requested by the supplier and supported by evidence of approval. Section 3 gives brief details of all other registered products with extant approval.

Section 2 lists some 355 individual active ingredients and mixtures. For each entry a list is shown of the available approved products together with the name of the supplier from whom each may be obtained. All types of pesticide covered by Control of Pesticides Regulations or Plant Protection Products Regulations are included. This embraces acaricides, algicides, herbicides, fungicides, insecticides, lumbricides, molluscicides, nematicides and rodenticides together with plant growth regulators used as straw shorteners, sprout inhibitors and for various horticultural purposes. The total number of products included in section 2 (i.e. available in the market in 2023) is about 1156.

The *Guide* also gives information (in section 4) on around 129 authorised adjuvants which, although not active pesticides themselves, may be added to pesticide products to improve their effectiveness. The Guide does not include products solely approved for amateur, home and garden, domestic, food storage, public or animal health uses.

Sources of Information

The information in this edition has been drawn from these authoritative sources:

- approved labels and product manuals received from suppliers of pesticides up to October 2022.
- websites of the Chemicals Regulation Division (CRD) *www.hse.gov.uk/crd/index.htm* and the Health and Safety Executive (HSE), *www.hse.gov.uk*.

Criteria for Inclusion

To be included, a pesticide must meet the following conditions:

- it must have extant approval under UK pesticides legislation

- information on the approved uses must have been provided by the supplier
- it must be expected to be on the UK market during the currency of this edition.

When a company changes its name, whether by merger, takeover or joint venture, it is obliged to re-register its products in the name of the new company, and new MAPP numbers are normally assigned. Where stocks of the previously registered product remain legally available, both old and new identities are included in the *Guide* and remain until approval of the former lapses or stocks are notified as exhausted.

Products that have been withdrawn from the market and whose approval will finally lapse during 2023 are identified in each profile. After the indicated date, sale, storage or use of the product bearing that approval number becomes illegal. Where there is a direct replacement product, this is indicated.

The Voluntary Initiative

The Voluntary Initiative (VI) is a programme of measures, agreed by the crop protection industry and related organisations with Government, to minimise the environmental impacts of crop protection products. The programme has provided a framework of practices and principles to help Government achieve its objective of protection of water quality and enhancement of farmland biodiversity. Many of the environmental protection schemes launched under the VI represent current best practice.

The first five-year phase of the programme formally concluded in March 2006, but it is continuing with a rolling two-year review of proposals and targets; current proposals include an online crop protection management plan and greater involvement in catchment-sensitive farming (see below).

A key element of the VI has been the provision of environmental information on crop protection products. Members of CropLife UK have committed to do this by producing Environmental Information Sheets (EISs) for all their marketed professional products.

EISs reinforce and supplement information on a product label by giving specific environmental impact information in a standardised format. They highlight any situations where risk management is essential to ensure environmental protection. Their purpose is to provide user-friendly information to advisers (including those in the amenity sector), farmers and growers on the environmental impact of crop protection products, to allow planning with a better understanding of the practical implications.

Links are provided to give users rapid access to EISs on the VI website: www.voluntaryinitiative.org.uk.

Catchment-sensitive Farming

Catchment-sensitive Farming (CSF) is the government's response to climate change and the European Water Directive. Rainfall events are likely to become heavier, with a greater risk of soil, nutrients and pesticides ending up in the waterways. The CSF programme is investigating the effects of specific targeted advice on 20 catchment areas perceived as being at risk, and new entries in this *Guide* can help identify the pesticides that pose the greatest risk of water contamination (see Environmental Safety). See the website: www.gov.uk/catchment-sensitive-farming

Notes on Contents and Layout

The book consists of six main sections:

1 Crop/Pest Guide
2 Pesticide Profiles
3 Products Also Registered
4 Adjuvants
5 Useful Information
6 Appendices

1 Crop/Pest Guide

This section enables the user to identify which active ingredients are approved for a particular crop/pest combination. The crops are grouped as shown in the Crop Index. For convenience, some crops and pests have been grouped into generic units (for example, 'cereals' are covered as a group, not individually). Indications from the Crop/Pest Guide must always be checked against the specific entry in the pesticide profiles section because a product may not be approved on all crops in the group. Chemicals indicated as having uses in cereals, for example, may be approved for use only on winter wheat and winter barley, not for other cereals. Because of the difference in wording of product labels, it may sometimes be necessary to refer to broader categories of organism in addition to specific organisms (e.g. annual grasses and annual weeds in addition to annual meadow-grass).

2 Pesticide Profiles

Each active ingredient and mixture of ingredients has a separate numbered profile entry. The entries are arranged in alphabetical order using the common names approved by the British Standards Institution and used in *The Pesticide Manual* (19th Edition; BCPC, 2021) now available as an online subscription resource visit: *www.bcpc.org*. Where an active ingredient is available only in mixtures, this is stated. The ingredients of the mixtures are themselves ordered alphabetically, and the entries appear at the correct point for the first named active ingredient.

Below the profile title, a brief phrase describes the main use of the active ingredient(s). This is followed where appropriate by the mode of action code(s) as published by the Fungicide, Herbicide and Insecticide Resistance Action Committees.

Within each profile entry, a table lists the products approved and available on the market in the following style:

Product name	Main supplier	Active ingredient contents	Formulation type	Registration number
1 Liberator	Bayer CropScience	100:400g/l	SC	15206
2 Mertil	UPL Europe	200:400g/l	SC	18504

Many of the **product names** are registered trademarks, but no special indication of this is given. Individual products are indexed under their entry number in the Index of Proprietary Names at the back of the book. The main supplier indicates the marketing outlet through which the product may be purchased. Full addresses and contact details of suppliers are listed in Appendix 1. For mixtures, the **active ingredient contents** are given in the same order as they appear in the profile heading. The **formulation types** are listed in full in the Key to Abbreviations and Acronyms (Appendix 5). The **registration number** normally refers to the registration with the Chemicals Regulation Division (CRD). In cases where the products registered with the Health and Safety Executive are included, HSE numbers are quoted, e.g. H0462 and approved biocides are numbered UK12-XXXX.

Below the product table, a **Uses** section lists all approved uses notified by the suppliers to the editor by October 2022, giving both the principal target organisms and the recommended crops or situations (identified in ***bold italics***). Where there is an important condition of use, or where the approval is off-label, this is shown in parentheses as (*off-label*). Numbers in square brackets refer to the numbered products in the table above. Thus, in the example shown above, a use approved for Liberator (product 1) but not for Mertil (product 2) appears as:

- Annual grasses in ***winter oats*** (off-label) [1]

Any Extensions of Authorisation for Minor Uses (EAMUs), for products in the profile are detailed below the list of approved uses. Each EAMU has a separate entry and shows the crops to which it applies, the Notice of Approval number, the expiry date (if due to fall during the current edition), and the product reference number in square brackets. EAMUs may now appear on the product label but are undertaken entirely at the risk of the user.

Below the EAMU paragraph, **Notes** are listed under the following headings. Unless otherwise stated, any reference to dose is made in terms of product rather than active ingredient. Where a note refers to a particular product rather than to the entry generally, this is indicated by numbers in square brackets as described above.

Approval information	Information of a general nature about the approval status of the active ingredient or products in the profile is given here. Notes on approval for aerial or ULV application are given.
	Inclusion of the active ingredient in Annex I under EC Directive 1107/2009 is shown, as well as any acceptance from the British Beer and Pub Association (BBPA) for its use on barley or hops for brewing or malting.
	Where a product approval will finally expire in 2023, the expiry date is shown here.
Efficacy	This entry identifies factors important in making the most effective use of the treatment. Certain factors, such as the need to apply chemicals uniformly, the need to use appropriate volume rates, and the need to prevent settling out of the active ingredient in the spray tank, are assumed to be important for all treatments and are not emphasised in individual profiles.
Restrictions	Notes are included in this section where products are subject to the Poisons Law, and where the label warns about organophosphorus and/or anticholinesterase compounds. Factors important in optimising performance and minimising the risk of crop damage, including statutory conditions relating to the maximum permitted number of applications, are listed. Any restrictions on crop varieties that may not be sprayed are mentioned here.
Environmental safety	Where the label specifies an environmental hazard classification, it is noted together with the associated risk phrases. Any other special operator precautions and any conditions concerning withholding livestock from treated areas are specified. Where any of the products in the profile are subject to Category A, Category B or broadcast air-assisted buffer zone restrictions under the LERAP scheme, the relevant classification is shown. Where a LERAP requires a buffer zone greater than 5m, detail of the larger buffer zone is given here. Arthropod buffer zones and statutory drift reduction limitations are also listed here.
	Other environmental hazards are also noted here, including potential dangers to livestock, game, wildlife, bees and fish. The need to avoid drift onto neighbouring crops and the need to wash out equipment thoroughly after use are important with all pesticide treatments but may receive special mention here if of particular significance.
Crop-specific information	Instructions about the timing of application or cultivations that are specific to a particular crop, rather than generally applicable, are mentioned here. The latest permitted use and harvest intervals (if any) are shown for individual crops.

Following crops guidance | Any specific label instructions about what may be grown after a treated crop, whether harvested normally or after failure, are shown here.

Hazard classification and safety precautions | The label hazard classifications and precautions are shown using a series of letter and number codes which are explained in Appendix 4. Hazard warnings are now given in full, with the other precautions listed under the following subheadings:

- Risk phrases
- Operator protection
- Environmental protection
- Consumer protection
- Storage and disposal
- Treated seed
- Vertebrate/rodent control products
- Medical advice

This section is given for information only and should not be used for the purpose of making a COSHH assessment without reference to the actual label of the product to be used.

3 Products also Registered

Products with an extant approval for all or part of the year of the edition in which they appear are listed in this section if they have not been requested by their supplier or manufacturer for inclusion in Section 2. Details shown are the same (apart from formulation) as those in the product tables of Section 2 and the approval expiry date is shown. Not all products listed here will be available in the market, but this list, together with the products in Section 2, comprises a comprehensive listing of all approved products in the UK for uses within the scope of the *Guide* at the time of going to print.

4 Adjuvants

Adjuvants are listed in a table ordered alphabetically by product name. For each product, details are shown of the main supplier, the authorisation number and the mode of action (e.g. wetter, sticker, etc.) as shown on the label. A brief statement of the uses of the adjuvant is given. Protective clothing requirements and label precautions are listed using the codes from Appendix 4

5 Useful Information

This section summarises legislation covering approval, storage, sale and use of pesticides in the UK. There are brief articles on the broader issues concerning the use of crop protection chemicals, including resistance management. Lists are provided of products approved for use in or near water, in forestry, as seed treatments and for aerial application. Chemicals subject to the Poisons Laws are listed, and there is a summary of first aid measures if pesticide poisoning should be suspected. Finally, this section provides guidance on environmental protection issues and covers the protection of bees and the use of pesticides in or near water.

6 Appendices

Appendix 1 | Gives contact details of all companies listed as main suppliers in the pesticide profiles.

Where a supplier is no longer trading under that name (usually following a merger or takeover), but products under that name are still listed in the Guide because they are still available in the supply chain, a cross-reference indicates the new 'parent' company from which technical or commercial information can be obtained.

Appendix 2 | Gives contact details of useful contacts, including the National Poisons Information Service.

Appendix 3   Gives details of the keys to crop and weed growth stages, including the publication reference for each. The numeric codes are used in the descriptive sections of the pesticide profiles (Section 2).

Appendix 4   Shows the full text for code letters and numbers used in the pesticide profiles (Section 2) to indicate personal protective equipment requirements and label precautions.

Appendix 5   Shows the full text for the formulation abbreviations used in the pesticide profiles (Section 2). Other abbreviations and acronyms used in the Guide are also explained here.

Appendix 6   Provides full definitions of officially agreed descriptive phrases for crops or situations used in the pesticide profiles (Section 2) where misunderstandings can occur.

Appendix 7   Shows a list of useful reference publications, which amplify the summarised information provided in this section.

SECTION 1
CROP/PEST GUIDE

Crop/Pest Guide Index

Important note: The Crop/Pest Guide Index refers to pages on which the subject can be located.

Products that are not classed as pesticides (and therefore are not evaluated by Chemicals Regulation Division (CRD) do not have a MAPP No. They work by physical means and can be effective but are not governed by pesticide legislation and can be used in all appropriate crop/ situations. These products are variously effective against aphids, mealy bugs, mites, mosquitoes, spider mites, suckers, whitefly and powdery mildew. For details see section 291 (Physical Pest Control) on page 472.

Crop/Pest Guide

Important note: For convenience, some crops and pests or targets have been brought together into genetic groups in this guide, e.g. 'cereals', 'annual grasses'. It is essential to check the profile entry in Section 2 *and* the label to ensure that a product is approved for a specific crop/pest combination, e.g. winter wheat/blackgrass.

Arable and vegetable crops

Agricultural herbage - Grass

Crop control	Desiccation	glyphosate
Pests	Aphids	esfenvalerate
	Flies	esfenvalerate
	Slugs/snails	ferric phosphate
Plant growth regulation	Growth control	gibberellins
Weeds	Broad-leaved weeds	2,4-D, 2,4-D + MCPA, 2,4-D + dicamba, MCPA, amidosulfuron, aminopyralid, aminopyralid + triclopyr, clopyralid, clopyralid + florasulam + fluroxypyr, clopyralid + triclopyr, florasulam + fluroxypyr, fluroxypyr, fluroxypyr + triclopyr, glyphosate, glyphosate (*wiper application*), thifensulfuron-methyl
	Crops as weeds	florasulam + fluroxypyr
	Grass weeds	glyphosate
	Weeds, miscellaneous	MCPA, aminopyralid + triclopyr, clopyralid + florasulam + fluroxypyr, clopyralid + triclopyr, florasulam + fluroxypyr, glyphosate
	Woody weeds/scrub	clopyralid + triclopyr

Agricultural herbage - Herbage legumes

Diseases	Powdery mildew	physical pest control
Pests	Aphids	physical pest control
	Mealybugs	physical pest control
	Scale insects	physical pest control
	Spider mites	physical pest control
	Suckers	physical pest control
	Whiteflies	physical pest control
Plant growth regulation	Growth control	trinexapac-ethyl (*off-label*)
Weeds	Broad-leaved weeds	propyzamide, thifensulfuron-methyl (*off-label*)
	Grass weeds	clethodim (*off-label*), propyzamide

Agricultural herbage - Herbage seed

Diseases	Crown rust	bixafen + prothioconazole (*off-label*)
	Disease control	tebuconazole (*off-label*)
	Drechslera leaf spot	bixafen + prothioconazole (*off-label*)
	Leaf spot	prothioconazole + tebuconazole (*off-label*)
	Powdery mildew	bixafen + prothioconazole (*off-label*), prothioconazole + tebuconazole (*off-label*)

	Rhynchosporium	bixafen + prothioconazole (*off-label*), prothioconazole + tebuconazole (*off-label*)
	Rust	bixafen + prothioconazole (*off-label*), prothioconazole + tebuconazole (*off-label*), pyraclostrobin (*off-label*), tebuconazole (*off-label*), trifloxystrobin (*off-label*)
Pests	Aphids	deltamethrin (*off-label*), tau-fluvalinate (*off-label*)
	Flies	esfenvalerate
	Pests, miscellaneous	lambda-cyhalothrin (*off-label*)
Plant growth regulation	Growth control	prohexadione-calcium + trinexapac-ethyl (*off-label*), trinexapac-ethyl, trinexapac-ethyl (*off-label*)
Weeds	Broad-leaved weeds	MCPA, bifenox (*off-label*), clopyralid + florasulam + fluroxypyr, dicamba + MCPA + mecoprop-P, diflufenican (*off-label*), ethofumesate (*off-label*), florasulam (*off-label*), florasulam + fluroxypyr, florasulam + halauxifen-methyl (*off-label*), fluroxypyr, mecoprop-P, pendimethalin (*off-label*), propyzamide
	Crops as weeds	diflufenican (*off-label*), ethofumesate (*off-label*), florasulam + fluroxypyr
	Grass weeds	clodinafop-propargyl (*off-label*), diflufenican (*off-label*), ethofumesate (*off-label*), fenoxaprop-P-ethyl (*off-label*), pendimethalin (*off-label*), propyzamide
	Weeds, miscellaneous	MCPA, clopyralid + florasulam + fluroxypyr, diflufenican (*off-label*), florasulam + fluroxypyr

Brassicas - Brassica seed crops

Weeds	Broad-leaved weeds	propyzamide
	Grass weeds	propyzamide

Brassicas - Brassicas, general

Diseases	Alternaria	azoxystrobin + difenoconazole (*off-label*), cyprodinil + fludioxonil (*off-label*)
	Botrytis	Bacillus amyloliquefaciens strain MBI600 (*off-label*), boscalid + pyraclostrobin (*off-label*), cyprodinil + fludioxonil (*off-label*)
	Downy mildew	dimethomorph (*off-label*)
	Pythium	metalaxyl-M (*off-label*)
	Rhizoctonia	boscalid + pyraclostrobin (*off-label*), cyprodinil + fludioxonil (*off-label*)
	Ring spot	azoxystrobin + difenoconazole (*off-label*)
	Sclerotinia	Bacillus amyloliquefaciens strain MBI600 (*off-label*), boscalid + pyraclostrobin (*off-label*)
	Stem canker	cyprodinil + fludioxonil (*off-label*)
Pests	Aphids	deltamethrin (*off-label*)
	Beetles	deltamethrin (*off-label*)
	Caterpillars	deltamethrin (*off-label*), indoxacarb (*off-label*)
	Whiteflies	deltamethrin (*off-label*)
Weeds	Broad-leaved weeds	pyridate (*off-label*)

Brassicas - Fodder brassicas

Diseases	Alternaria	azoxystrobin, cyprodinil + fludioxonil (*off-label*), difenoconazole
	Anthracnose	Bacillus amyloliquefaciens D747 (*off-label*)

	Black leg	fludioxonil (*off-label*)
	Black rot	Bacillus amyloliquefaciens D747 (*off-label*)
	Botrytis	Bacillus amyloliquefaciens D747 (*off-label*), Bacillus subtilis (Strain QST 713) (*off-label*), cyprodinil + fludioxonil (*off-label*)
	Damping off	Bacillus amyloliquefaciens D747 (*off-label*), Bacillus subtilis (Strain QST 713) (*off-label*)
	Downy mildew	Bacillus amyloliquefaciens D747 (*off-label*), fluopicolide + propamocarb hydrochloride (*off-label*), metalaxyl-M (*off-label*)
	Fusarium	Bacillus amyloliquefaciens D747 (*off-label*)
	Powdery mildew	Bacillus amyloliquefaciens D747 (*off-label*), azoxystrobin + difenoconazole
	Pythium	Bacillus amyloliquefaciens D747 (*off-label*), Bacillus subtilis (Strain QST 713) (*off-label*), metalaxyl-M (*off-label*)
	Rhizoctonia	Bacillus amyloliquefaciens D747 (*off-label*), cyprodinil + fludioxonil (*off-label*)
	Ring spot	azoxystrobin, difenoconazole
	Sclerotinia	Bacillus amyloliquefaciens D747 (*off-label*)
	Spear rot	Bacillus amyloliquefaciens D747 (*off-label*)
	Stem canker	cyprodinil + fludioxonil (*off-label*)
	White blister	azoxystrobin, azoxystrobin + difenoconazole, boscalid + pyraclostrobin (*off-label*)
Pests	Aphids	Beauveria bassiana GHA (*off-label*), deltamethrin (*off-label*), esfenvalerate, fatty acids (*off-label*), maltodextrin (*off-label*), spirotetramat
	Beetles	deltamethrin (*off-label*)
	Caterpillars	Bacillus thuringiensis kurstaki, Bacillus thuringiensis kurstaki (*off-label*), deltamethrin (*off-label*), esfenvalerate, indoxacarb (*off-label*), spinosad (*off-label*)
	Flies	spinosad (*off-label*)
	Leaf miners	maltodextrin (*off-label*)
	Slugs/snails	ferric phosphate
	Thrips	Beauveria bassiana GHA (*off-label*), fatty acids (*off-label*)
	Whiteflies	Beauveria bassiana GHA (*off-label*), deltamethrin (*off-label*), maltodextrin (*off-label*), spirotetramat
Weeds	Broad-leaved weeds	clomazone (*off-label*), dimethenamid-p + metazachlor (*off-label*), metazachlor (*off-label*), napropamide, pendimethalin (*off-label*), pyridate (*off-label*)
	Crops as weeds	fluazifop-P-butyl
	Grass weeds	S-metolachlor (*off-label*), dimethenamid-p + metazachlor (*off-label*), fluazifop-P-butyl, metazachlor (*off-label*), napropamide, pendimethalin (*off-label*)

Brassicas - Leaf and flowerhead brassicas

Diseases	Alternaria	azoxystrobin, boscalid + pyraclostrobin, cyprodinil + fludioxonil (*off-label*), difenoconazole, difenoconazole + fluxapyroxad (*reduction*), prothioconazole, tebuconazole, tebuconazole + trifloxystrobin
	Anthracnose	Bacillus amyloliquefaciens D747 (*off-label*)

	Black leg	fludioxonil, fludioxonil (*off-label*)
	Black rot	Bacillus amyloliquefaciens D747 (*off-label*), azoxystrobin (*off-label*)
	Botrytis	Bacillus amyloliquefaciens D747 (*off-label*), Bacillus amyloliquefaciens strain MBI600 (*off-label*), Bacillus subtilis (Strain QST 713) (*off-label*), cyprodinil + fludioxonil (*off-label*)
	Damping off	Bacillus amyloliquefaciens D747 (*off-label*), Bacillus amyloliquefaciens strain MBI600 (*off-label*), Bacillus subtilis (Strain QST 713) (*off-label*)
	Downy mildew	Bacillus amyloliquefaciens D747 (*off-label*), fluopicolide + propamocarb hydrochloride (*off-label*), mandipropamid, mandipropamid (*off-label*), metalaxyl-M (*off-label*)
	Fungus diseases	fludioxonil
	Fusarium	Bacillus amyloliquefaciens D747 (*off-label*)
	Light leaf spot	prothioconazole, tebuconazole, tebuconazole + trifloxystrobin
	Phoma leaf spot	tebuconazole + trifloxystrobin
	Powdery mildew	Bacillus amyloliquefaciens D747 (*off-label*), azoxystrobin + difenoconazole, physical pest control, prothioconazole, tebuconazole, tebuconazole + trifloxystrobin
	Pythium	Bacillus amyloliquefaciens D747 (*off-label*), Bacillus amyloliquefaciens strain MBI600 (*off-label*), Bacillus subtilis (Strain QST 713) (*off-label*), metalaxyl-M, metalaxyl-M (*off-label*)
	Rhizoctonia	Bacillus amyloliquefaciens D747 (*off-label*), Bacillus amyloliquefaciens strain MBI600 (*off-label*), cyprodinil + fludioxonil (*off-label*)
	Ring spot	azoxystrobin, boscalid + pyraclostrobin, difenoconazole, difenoconazole + fluxapyroxad (*reduction*), prothioconazole, tebuconazole, tebuconazole + trifloxystrobin
	Sclerotinia	Bacillus amyloliquefaciens D747 (*off-label*), Bacillus amyloliquefaciens strain MBI600 (*off-label*)
	Spear rot	Bacillus amyloliquefaciens D747 (*off-label*)
	Stem canker	cyprodinil + fludioxonil (*off-label*), prothioconazole
	Storage rots	1-methylcyclopropene (*off-label*), metalaxyl-M (*off-label*)
	White blister	azoxystrobin, azoxystrobin (*off-label*), azoxystrobin + difenoconazole, boscalid + pyraclostrobin, boscalid + pyraclostrobin (*off-label*), boscalid + pyraclostrobin (*qualified minor use*), mancozeb + metalaxyl-M (*off-label*), metalaxyl-M (*off-label*), tebuconazole + trifloxystrobin
Pests	Aphids	Beauveria bassiana GHA (*off-label*), cypermethrin, deltamethrin (*off-label*), esfenvalerate, fatty acids (*off-label*), flonicamid, lambda-cyhalothrin, maltodextrin (*off-label*), physical pest control, pyrethrins, spirotetramat
	Beetles	deltamethrin (*off-label*)
	Birds/mammals	aluminium ammonium sulphate
	Caterpillars	Bacillus thuringiensis kurstaki, Bacillus thuringiensis kurstaki (*off-label*), cypermethrin, deltamethrin,

		deltamethrin (*off-label*), esfenvalerate, indoxacarb, indoxacarb (*off-label*), lambda-cyhalothrin, pyrethrins, spinosad, spinosad (*off-label*)
	Flies	cyantraniliprole, spinosad (*off-label*)
	Leaf miners	maltodextrin (*off-label*)
	Mealybugs	physical pest control
	Scale insects	physical pest control
	Slugs/snails	ferric phosphate
	Spider mites	physical pest control, pyrethrins
	Suckers	physical pest control
	Thrips	Beauveria bassiana GHA (*off-label*), fatty acids (*off-label*), pyrethrins
	Whiteflies	Beauveria bassiana GHA (*off-label*), cypermethrin, deltamethrin (*off-label*), lambda-cyhalothrin, maltodextrin (*off-label*), physical pest control, spirotetramat
Weeds	Broad-leaved weeds	clomazone (*off-label*), clopyralid, dimethenamid-p + metazachlor (*off-label*), dimethenamid-p + pendimethalin (*off-label*), metazachlor, metazachlor (*off-label*), napropamide, napropamide (*off-label*), pendimethalin, pendimethalin (*off-label*), propyzamide (*off-label*), pyridate, pyridate (*off-label*)
	Crops as weeds	clethodim (*off-label*), cycloxydim, cycloxydim (*off-label*), pendimethalin
	Grass weeds	S-metolachlor (*off-label*), clethodim (*off-label*), cycloxydim, cycloxydim (*off-label*), dimethenamid-p + metazachlor (*off-label*), dimethenamid-p + pendimethalin (*off-label*), metazachlor, metazachlor (*off-label*), napropamide, napropamide (*off-label*), pendimethalin, pendimethalin (*off-label*), propyzamide (*off-label*)
	Weeds, miscellaneous	napropamide (*off-label*)

Brassicas - Mustard

Crop control	Desiccation	glyphosate
Diseases	Alternaria	Bacillus amyloliquefaciens D747, azoxystrobin (*off-label*), cyprodinil + fludioxonil (*off-label*), difenoconazole (*off-label*)
	Anthracnose	Bacillus amyloliquefaciens D747 (*off-label*)
	Bacterial blight	Bacillus amyloliquefaciens D747 (*off-label*)
	Black rot	azoxystrobin (*off-label*)
	Black scurf and stem canker	Bacillus amyloliquefaciens strain MBI600 (*off-label*)
	Botrytis	Bacillus amyloliquefaciens D747 (*off-label*), Bacillus subtilis (Strain QST 713) (*off-label*), cyprodinil + fludioxonil (*off-label*), difenoconazole + fluxapyroxad (*off-label*)
	Cercospora leaf spot	Bacillus amyloliquefaciens D747 (*off-label*)
	Damping off	Bacillus amyloliquefaciens D747 (*off-label*)
	Disease control	boscalid (*off-label*), mandipropamid, prothioconazole (*off-label*), tebuconazole (*off-label*)

	Downy mildew	Bacillus amyloliquefaciens D747 (*off-label*), fluopicolide + propamocarb hydrochloride (*off-label*), mandipropamid
	Fusarium	Bacillus amyloliquefaciens D747, Bacillus amyloliquefaciens D747 (*off-label*)
	Powdery mildew	Bacillus amyloliquefaciens D747, Bacillus amyloliquefaciens D747 (*off-label*), difenoconazole (*off-label*), difenoconazole + fluxapyroxad (*off-label*)
	Pythium	Bacillus amyloliquefaciens D747, Bacillus amyloliquefaciens D747 (*off-label*)
	Rhizoctonia	Bacillus amyloliquefaciens D747, Bacillus amyloliquefaciens D747 (*off-label*), cyprodinil + fludioxonil (*off-label*), difenoconazole + fluxapyroxad (*off-label*)
	Rust	tebuconazole (*off-label*)
	Sclerotinia	Bacillus amyloliquefaciens D747 (*off-label*), azoxystrobin (*off-label*), difenoconazole (*off-label*), difenoconazole + fluxapyroxad (*off-label*)
	Stem canker	cyprodinil + fludioxonil (*off-label*)
	Stemphylium	Bacillus amyloliquefaciens D747 (*off-label*)
	White blister	azoxystrobin (*off-label*)
Pests	Aphids	Beauveria bassiana GHA (*off-label*), cypermethrin, fatty acids (*off-label*), tau-fluvalinate (*off-label*)
	Beetles	Bacillus amyloliquefaciens strain MBI600 (*off-label - reduction*), deltamethrin, indoxacarb (*off-label*)
	Caterpillars	spinosad (*off-label*)
	Pests, miscellaneous	cypermethrin, lambda-cyhalothrin (*off-label*)
	Slugs/snails	ferric phosphate
	Spider mites	Beauveria bassiana GHA (*off-label*), fatty acids (*off-label*)
	Thrips	Beauveria bassiana GHA (*off-label*), fatty acids (*off-label*), spinosad (*off-label*)
	Weevils	deltamethrin
	Whiteflies	Beauveria bassiana GHA (*off-label*)
Weeds	Broad-leaved weeds	bifenox (*off-label*), clopyralid + picloram (*off-label*), glyphosate, lenacil (*off-label*), metazachlor (*off-label*), napropamide (*off-label*)
	Crops as weeds	fluazifop-P-butyl, glyphosate, propaquizafop (*off-label*)
	Grass weeds	clethodim (*off-label*), fluazifop-P-butyl, glyphosate, metazachlor (*off-label*), napropamide (*off-label*), propaquizafop (*off-label*)
	Weeds, miscellaneous	glyphosate, lenacil (*off-label*), napropamide (*off-label*)

Brassicas - Root brassicas

Diseases	Alternaria	azoxystrobin, azoxystrobin (*off-label*), difenoconazole + fluxapyroxad (*off-label*), isopyrazam (*off-label*), prothioconazole, tebuconazole, tebuconazole + trifloxystrobin (*off-label*)
	Anthracnose	Bacillus amyloliquefaciens D747 (*off-label*)
	Bacterial blight	Bacillus amyloliquefaciens D747 (*off-label*)
	Black leg	fludioxonil (*off-label*)
	Black rot	Bacillus amyloliquefaciens D747 (*off-label*)

	Botrytis	Bacillus amyloliquefaciens D747 (*off-label*), Bacillus subtilis (Strain QST 713) (*off-label*), boscalid + pyraclostrobin (*off-label*)
	Cercospora leaf spot	isopyrazam (*off-label*)
	Damping off	Bacillus amyloliquefaciens D747 (*off-label*)
	Disease control	boscalid + pyraclostrobin (*off-label*)
	Downy mildew	Bacillus amyloliquefaciens D747 (*off-label*), azoxystrobin (*off-label*), dimethomorph (*off-label*), fluopicolide + propamocarb hydrochloride (*off-label*), mandipropamid, metalaxyl-M (*off-label*), propamocarb hydrochloride (*off-label*)
	Fungus diseases	azoxystrobin + difenoconazole (*off-label*)
	Fusarium	Bacillus amyloliquefaciens D747 (*off-label*)
	Leaf spot	tebuconazole + trifloxystrobin (*off-label*)
	Mycosphaerella	tebuconazole + trifloxystrobin (*off-label*)
	Powdery mildew	Bacillus amyloliquefaciens D747 (*off-label*), difenoconazole + fluxapyroxad (*off-label*), isopyrazam (*off-label*), physical pest control, prothioconazole, sulphur, tebuconazole, tebuconazole + trifloxystrobin (*off-label*)
	Pythium	Bacillus amyloliquefaciens D747 (*off-label*), metalaxyl-M, metalaxyl-M (*off-label*)
	Ramularia leaf spots	isopyrazam (*off-label*)
	Rhizoctonia	Bacillus amyloliquefaciens D747 (*off-label*), azoxystrobin (*off-label*), difenoconazole + fluxapyroxad (*off-label*)
	Ring spot	azoxystrobin
	Rust	boscalid + pyraclostrobin (*off-label*)
	Sclerotinia	Bacillus amyloliquefaciens D747 (*off-label*), boscalid + pyraclostrobin (*off-label*), cyprodinil + fludioxonil (*off-label*), difenoconazole + fluxapyroxad (*off-label*), isopyrazam (*off-label*), tebuconazole + trifloxystrobin (*off-label*)
	Septoria	isopyrazam (*off-label*), tebuconazole + trifloxystrobin (*off-label*)
	Spear rot	Bacillus amyloliquefaciens D747 (*off-label*)
	Stem canker	azoxystrobin + difenoconazole (*off-label*), isopyrazam (*off-label*)
	White blister	azoxystrobin (*off-label*), metalaxyl-M (*off-label*)
Pests	Aphids	Beauveria bassiana GHA (*off-label*), acetamiprid (*off-label*), cypermethrin, deltamethrin (*off-label*), esfenvalerate, fatty acids (*off-label*), flonicamid (*off-label*), maltodextrin (*off-label*), physical pest control, pirimicarb (*off-label*), pyrethrins, spirotetramat
	Beetles	deltamethrin (*off-label*)
	Caterpillars	Bacillus thuringiensis aizawai GC-91, Bacillus thuringiensis kurstaki, Bacillus thuringiensis kurstaki (*off-label*), chlorantraniliprole (*off-label*), deltamethrin, deltamethrin (*off-label*), esfenvalerate, indoxacarb (*off-label*), lambda-cyhalothrin (*off-label*), maltodextrin (*off-label*), pyrethrins, spinosad (*off-label*)
	Cutworms	Bacillus thuringiensis kurstaki (*off-label*), cypermethrin, deltamethrin (*off-label*)

	Flies	acetamiprid (*off-label*), chlorantraniliprole (*off-label*), cypermethrin, deltamethrin (*off-label*), lambda-cyhalothrin (*off-label*), maltodextrin (*off-label*), spinosad (*off-label*)
	Leaf miners	maltodextrin (*off-label*), spinosad (*off-label*)
	Mealybugs	physical pest control
	Pests, miscellaneous	cyantraniliprole, cypermethrin, lambda-cyhalothrin (*off-label*)
	Scale insects	physical pest control
	Slugs/snails	ferric phosphate
	Spider mites	physical pest control, pyrethrins
	Suckers	physical pest control
	Thrips	Beauveria bassiana GHA (*off-label*), deltamethrin (*off-label*), fatty acids (*off-label*), pyrethrins, spinosad (*off-label*)
	Weevils	deltamethrin (*off-label*)
	Whiteflies	Beauveria bassiana GHA (*off-label*), maltodextrin (*off-label*), physical pest control, spirotetramat
Plant growth regulation	Quality/yield control	sulphur
Weeds	Broad-leaved weeds	S-metolachlor (*off-label*), clomazone (*off-label*), clopyralid, dimethenamid-p + metazachlor (*off-label*), glyphosate, isoxaben (*off-label*), metamitron (*off-label*), metazachlor (*off-label*), napropamide (*off-label*), pelargonic acid (*off-label*), pendimethalin (*off-label*), prosulfocarb (*off-label*)
	Crops as weeds	cycloxydim, fluazifop-P-butyl, fluazifop-P-butyl (*stockfeed only*), glyphosate
	Grass weeds	S-metolachlor (*off-label*), clethodim (*off-label*), cycloxydim, cycloxydim (*off-label*), dimethenamid-p + metazachlor (*off-label*), fluazifop-P-butyl, fluazifop-P-butyl (*stockfeed only*), glyphosate, metamitron (*off-label*), metazachlor (*off-label*), napropamide (*off-label*), pelargonic acid (*off-label*), propaquizafop, prosulfocarb (*off-label*)
	Mosses	pelargonic acid (*off-label*)
	Weeds, miscellaneous	glyphosate, glyphosate (*off-label*), napropamide (*off-label*), pelargonic acid (*off-label*)

Brassicas - Salad greens

Diseases	Alternaria	azoxystrobin + difenoconazole (*off-label*), cyprodinil + fludioxonil (*off-label*), difenoconazole (*off-label*)
	Anthracnose	Bacillus amyloliquefaciens D747 (*off-label*)
	Bacterial blight	Bacillus amyloliquefaciens D747 (*off-label*)
	Black rot	Bacillus amyloliquefaciens D747 (*off-label*), azoxystrobin (*off-label*)
	Botrytis	Bacillus amyloliquefaciens D747 (*off-label*), Bacillus amyloliquefaciens strain MBI600 (*off-label*), Bacillus subtilis (Strain QST 713) (*off-label*), cerevisane (*off-label*), cyprodinil + fludioxonil (*off-label*), difenoconazole + fluxapyroxad (*off-label*), fenhexamid (*off-label - baby leaf production*), fluopyram + trifloxystrobin (*off-label*), pyrimethanil (*off-label*)
	Cercospora leaf spot	Bacillus amyloliquefaciens D747 (*off-label*)

	Damping off	Bacillus amyloliquefaciens D747 (*off-label*), Bacillus amyloliquefaciens strain MBI600 (*off-label*), Bacillus subtilis (Strain QST 713) (*off-label*)
	Disease control	mandipropamid
	Downy mildew	Bacillus amyloliquefaciens D747 (*off-label*), cerevisane (*off-label*), cos-oga (*off-label*), dimethomorph (*off-label*), fluopicolide + propamocarb hydrochloride (*off-label*), mancozeb + metalaxyl-M (*off-label*), mandipropamid, metalaxyl-M (*off-label*)
	Fusarium	Bacillus amyloliquefaciens D747 (*off-label*), Trichoderma afroharzianum (Strain T22), Trichoderma asperellum (Strain T34) (*off-label*)
	Leaf spot	difenoconazole (*off-label*)
	Powdery mildew	Ampelomyces quisqualis (Strain AQ10) (*off-label*), Bacillus amyloliquefaciens D747 (*off-label*), cerevisane (*off-label*), cos-oga (*off-label*), difenoconazole + fluxapyroxad (*off-label*), fluopyram + trifloxystrobin (*off-label*)
	Pythium	Bacillus amyloliquefaciens D747 (*off-label*), Bacillus amyloliquefaciens strain MBI600 (*off-label*), Trichoderma afroharzianum (Strain T22), Trichoderma asperellum (Strain T34) (*off-label*), metalaxyl-M (*off-label*)
	Rhizoctonia	Bacillus amyloliquefaciens D747 (*off-label*), Bacillus amyloliquefaciens strain MBI600 (*off-label*), Trichoderma afroharzianum (Strain T22), cyprodinil + fludioxonil (*off-label*), difenoconazole + fluxapyroxad (*off-label*)
	Ring spot	azoxystrobin + difenoconazole (*off-label*), difenoconazole (*off-label*)
	Sclerotinia	Bacillus amyloliquefaciens D747 (*off-label*), Bacillus amyloliquefaciens strain MBI600 (*off-label*), Trichoderma afroharzianum (Strain T22), difenoconazole + fluxapyroxad (*off-label*), fluopyram + trifloxystrobin (*off-label*)
	Seed-borne diseases	fludioxonil (*off-label*)
	Spear rot	Bacillus amyloliquefaciens D747 (*off-label*)
	Stem canker	cyprodinil + fludioxonil (*off-label*)
	Stemphylium	Bacillus amyloliquefaciens D747 (*off-label*)
	White blister	azoxystrobin (*off-label*), boscalid + pyraclostrobin (*off-label*)
Pests	Aphids	Beauveria bassiana GHA (*off-label*), acetamiprid (*off-label*), esfenvalerate, fatty acids (*off-label*), maltodextrin (*off-label*), pirimicarb (*off-label*)
	Beetles	cyantraniliprole (*off-label*)
	Caterpillars	Bacillus thuringiensis kurstaki, Bacillus thuringiensis kurstaki (*off-label*), cyantraniliprole (*off-label*), esfenvalerate, indoxacarb (*off-label*), spinosad, spinosad (*off-label*)
	Cutworms	Bacillus thuringiensis kurstaki (*off-label*)
	Leaf miners	maltodextrin (*off-label*)
	Pests, miscellaneous	lambda-cyhalothrin (*off-label*)
	Slugs/snails	ferric phosphate

	Spider mites	Beauveria bassiana GHA (*off-label*), Lecanicillium muscarium (*off-label*), abamectin (*off-label*), fatty acids (*off-label*)
	Thrips	Beauveria bassiana GHA (*off-label*), Lecanicillium muscarium (*off-label*), fatty acids (*off-label*), spinosad (*off-label*)
	Whiteflies	Beauveria bassiana GHA (*off-label*), Lecanicillium muscarium (*off-label*), maltodextrin (*off-label*)
Weeds	Broad-leaved weeds	S-metolachlor (*off-label*), clomazone (*off-label*), dimethenamid-p + metazachlor (*off-label*), dimethenamid-p + pendimethalin (*off-label*), lenacil (*off-label*), metazachlor (*off-label*), napropamide (*off-label*), pendimethalin (*off-label*), propyzamide (*off-label*), pyridate (*off-label*)
	Crops as weeds	cycloxydim (*off-label*), fluazifop-P-butyl
	Grass weeds	S-metolachlor (*off-label*), cycloxydim (*off-label*), dimethenamid-p + metazachlor (*off-label*), dimethenamid-p + pendimethalin (*off-label*), fluazifop-P-butyl, metazachlor (*off-label*), napropamide (*off-label*), propyzamide (*off-label*)
	Weeds, miscellaneous	lenacil (*off-label*), napropamide (*off-label*)

Cereals - Barley

Crop control	Desiccation	glyphosate
Diseases	Covered smut	fludioxonil (*seed treatment*), fludioxonil + fluxapyroxad + triticonazole, fludioxonil + sedaxane, fluopyram + prothioconazole + tebuconazole, prothioconazole + tebuconazole
	Disease control	bixafen + fluopyram + prothioconazole, boscalid, fluxapyroxad + metconazole
	Ear blight	prothioconazole + tebuconazole
	Eyespot	bixafen + prothioconazole (*reduction of incidence and severity*), cyprodinil, fluoxastrobin + prothioconazole (*reduction*), fluoxastrobin + prothioconazole + trifloxystrobin (*reduction*), prothioconazole, prothioconazole + spiroxamine, prothioconazole + spiroxamine + tebuconazole (*reduction*), prothioconazole + tebuconazole, prothioconazole + tebuconazole (*reduction*), prothioconazole + trifloxystrobin (*reduction in severity*), prothioconazole + trifloxystrobin (*reduction*)
	Foot rot	fludioxonil (*seed treatment*), fludioxonil + fluxapyroxad + triticonazole, imazalil + ipconazole (*useful protection*)
	Fusarium	fluopyram + prothioconazole + tebuconazole, prothioconazole + tebuconazole
	Late ear diseases	bixafen + prothioconazole, fluoxastrobin + prothioconazole, fluoxastrobin + prothioconazole + trifloxystrobin, prothioconazole, prothioconazole + spiroxamine + tebuconazole, prothioconazole + tebuconazole
	Leaf stripe	fludioxonil (*seed treatment - reduction*), fludioxonil + fluxapyroxad + triticonazole, fludioxonil + sedaxane (*moderate control*), fluopyram + prothioconazole +

	tebuconazole, imazalil + ipconazole, prothioconazole + tebuconazole
Loose smut	fludioxonil + fluxapyroxad + triticonazole, fluopyram + prothioconazole + tebuconazole, imazalil + ipconazole, prothioconazole + tebuconazole
Net blotch	azoxystrobin, benzovindiflupyr, benzovindiflupyr + prothioconazole, bixafen, bixafen + fluopyram + prothioconazole, bixafen + prothioconazole, cyprodinil, cyprodinil + isopyrazam, fluopyram + prothioconazole + tebuconazole (*seed borne*), fluoxastrobin + prothioconazole, fluoxastrobin + prothioconazole + trifloxystrobin, fluxapyroxad, fluxapyroxad + mefentrifluconazole (*reduction*), fluxapyroxad + metconazole, fluxapyroxad + pyraclostrobin, mefentrifluconazole (*reduction*), mefentrifluconazole + pyraclostrobin, metconazole (*reduction*), prothioconazole, prothioconazole + spiroxamine, prothioconazole + spiroxamine + tebuconazole, prothioconazole + tebuconazole, prothioconazole + trifloxystrobin, pyraclostrobin, tebuconazole, trifloxystrobin
Powdery mildew	azoxystrobin, azoxystrobin (*moderate control*), bixafen + fluopyram + prothioconazole, bixafen + prothioconazole, bixafen + prothioconazole + spiroxamine, cyflufenamid, cyflufenamid + spiroxamine, cyprodinil, cyprodinil + isopyrazam, fluoxastrobin + prothioconazole, fluoxastrobin + prothioconazole + trifloxystrobin, fluxapyroxad (*reduction*), fluxapyroxad + metconazole (*moderate control*), fluxapyroxad + pyraclostrobin (*moderate control*), mefentrifluconazole + pyraclostrobin, metconazole, metconazole (*moderate control*), metrafenone, proquinazid, prothioconazole, prothioconazole + spiroxamine, prothioconazole + spiroxamine + tebuconazole, prothioconazole + tebuconazole, prothioconazole + trifloxystrobin, sulphur, tebuconazole, tebuconazole (*moderate control*)
Ramularia leaf spots	benzovindiflupyr, benzovindiflupyr + prothioconazole, bixafen + fluopyram + prothioconazole, bixafen + prothioconazole, cyprodinil + isopyrazam, fluxapyroxad + mefentrifluconazole, fluxapyroxad + metconazole, fluxapyroxad + pyraclostrobin, mefentrifluconazole, mefentrifluconazole + pyraclostrobin
Rhynchosporium	azoxystrobin, azoxystrobin (*reduction*), benzovindiflupyr (*moderate control*), benzovindiflupyr + prothioconazole (*moderate control*), bixafen, bixafen + prothioconazole, cyprodinil, cyprodinil (*moderate control*), cyprodinil + isopyrazam, fluoxastrobin + prothioconazole, fluoxastrobin + prothioconazole + trifloxystrobin, fluxapyroxad, fluxapyroxad + mefentrifluconazole, fluxapyroxad + metconazole, fluxapyroxad + pyraclostrobin, folpet (*reduction*), mefentrifluconazole, mefentrifluconazole + pyraclostrobin (*moderate control*), metconazole, metconazole (*reduction*), prothioconazole, prothioconazole + spiroxamine, prothioconazole + spiroxamine + tebuconazole, prothioconazole + tebuconazole, prothioconazole + trifloxystrobin,

		pyraclostrobin (*moderate control*), tebuconazole, tebuconazole (*moderate control*), trifloxystrobin
	Rust	azoxystrobin, benzovindiflupyr, benzovindiflupyr + prothioconazole, bixafen, bixafen + fluopyram, bixafen + prothioconazole, bixafen + prothioconazole + spiroxamine, cyflufenamid + spiroxamine, cyprodinil + isopyrazam, fluoxastrobin + prothioconazole, fluoxastrobin + prothioconazole + trifloxystrobin, fluxapyroxad, fluxapyroxad (*moderate control*), fluxapyroxad + mefentrifluconazole, fluxapyroxad + metconazole, fluxapyroxad + pyraclostrobin, mefentrifluconazole, mefentrifluconazole + pyraclostrobin, metconazole, prothioconazole, prothioconazole + spiroxamine, prothioconazole + spiroxamine + tebuconazole, prothioconazole + tebuconazole, prothioconazole + trifloxystrobin, pyraclostrobin, tebuconazole, trifloxystrobin
	Seed-borne diseases	fludioxonil + fluxapyroxad + triticonazole, fludioxonil + tebuconazole, fluopyram + prothioconazole + tebuconazole, imazalil + ipconazole (*useful protection*), ipconazole
	Snow mould	fludioxonil (*seed treatment*), fludioxonil + fluxapyroxad + triticonazole, fludioxonil + sedaxane
	Soil-borne diseases	ipconazole
	Sooty moulds	prothioconazole + tebuconazole
	Take-all	azoxystrobin (*reduction*), fluoxastrobin + prothioconazole (*reduction*), silthiofam (*seed treatment*)
Pests	Aphids	cypermethrin, cypermethrin (*autumn sown*), deltamethrin, esfenvalerate, flonicamid, lambda-cyhalothrin, tau-fluvalinate
	Beetles	lambda-cyhalothrin
	Birds/mammals	aluminium ammonium sulphate
	Caterpillars	lambda-cyhalothrin
	Flies	cypermethrin, cypermethrin (*autumn sown*), cypermethrin (*qualified recommendation*), cypermethrin (*reduction*)
	Pests, miscellaneous	cypermethrin
	Slugs/snails	ferric phosphate
	Weevils	lambda-cyhalothrin
Plant growth regulation	Growth control	chlormequat, chlormequat + ethephon, ethephon, ethephon + mepiquat chloride, mepiquat chloride + prohexadione-calcium, prohexadione-calcium, prohexadione-calcium + trinexapac-ethyl, trinexapac-ethyl
	Quality/yield control	ethephon (*low lodging situations*), ethephon + mepiquat chloride (*low lodging situations*), mepiquat chloride + prohexadione-calcium, sulphur
Weeds	Broad-leaved weeds	2,4-D, 2,4-D + MCPA, MCPA, aclonifen, amidosulfuron, amidosulfuron + iodosulfuron-methyl-sodium, bifenox, carfentrazone-ethyl, carfentrazone-ethyl + mecoprop-P, chlorotoluron + diflufenican + pendimethalin, cinmethylin, clopyralid, clopyralid + florasulam, clopyralid + florasulam + fluroxypyr, clopyralid + fluroxypyr + MCPA, dicamba + MCPA +

mecoprop-P, dicamba + mecoprop-P, dichlorprop-P + MCPA + mecoprop-P, diflufenican, diflufenican + florasulam, diflufenican + flufenacet, diflufenican + flufenacet (*off-label*), diflufenican + flufenacet + metribuzin, diflufenican + metsulfuron-methyl, diflufenican + pendimethalin, florasulam, florasulam + fluroxypyr, florasulam + halauxifen-methyl, florasulam + tribenuron-methyl, florasulam + tribenuron-methyl (*MS to 20 cm*), florasulam + tribenuron-methyl (*MS to 6 lvs*), flufenacet, flufenacet + metribuzin, flufenacet + pendimethalin, flufenacet + pendimethalin (*off-label*), flufenacet + picolinafen, fluroxypyr, fluroxypyr + halauxifen-methyl, fluroxypyr + metsulfuron-methyl, fluroxypyr + metsulfuron-methyl + thifensulfuron-methyl, glyphosate, isoxaben, mecoprop-P, metsulfuron-methyl, metsulfuron-methyl + thifensulfuron-methyl, metsulfuron-methyl + tribenuron-methyl, pendimethalin, pendimethalin + picolinafen, pendimethalin + picolinafen (*off-label*), prosulfocarb, prosulfocarb (*off-label*), thifensulfuron-methyl, thifensulfuron-methyl + tribenuron-methyl, tribenuron-methyl

	Crops as weeds	amidosulfuron + iodosulfuron-methyl-sodium, chlorotoluron + diflufenican + pendimethalin, cinmethylin, clopyralid + florasulam, diflufenican + florasulam, diflufenican + flufenacet, florasulam, florasulam + fluroxypyr, florasulam + halauxifen-methyl, flufenacet + picolinafen, fluroxypyr, glyphosate, metsulfuron-methyl, metsulfuron-methyl + tribenuron-methyl, pendimethalin
	Grass weeds	aclonifen, chlorotoluron + diflufenican + pendimethalin, cinmethylin, diflufenican + flufenacet, diflufenican + flufenacet (*off-label*), diflufenican + flufenacet + metribuzin, diflufenican + pendimethalin, fenoxaprop-P-ethyl, flufenacet, flufenacet + metribuzin, flufenacet + pendimethalin, flufenacet + pendimethalin (*off-label*), flufenacet + picolinafen, glyphosate, pendimethalin, pendimethalin + picolinafen, pendimethalin + picolinafen (*off-label*), pinoxaden, prosulfocarb, prosulfocarb (*off-label*), tri-allate
	Weeds, miscellaneous	chlorotoluron + diflufenican + pendimethalin, clopyralid + florasulam, diflufenican + flufenacet + metribuzin, diflufenican + metsulfuron-methyl, florasulam, glyphosate, metsulfuron-methyl, metsulfuron-methyl + thifensulfuron-methyl

Cereals - Cereals, general

Crop control	Desiccation	glyphosate, glyphosate (*off-label*)
Diseases	Alternaria	bromuconazole + tebuconazole, difenoconazole (*off-label*)
	Blue mould	prothioconazole + tebuconazole
	Bunt	Pseudomonas chlororaphis MA 342, prothioconazole + tebuconazole
	Cladosporium	bromuconazole + tebuconazole, difenoconazole (*off-label*)
	Crown rust	benzovindiflupyr + prothioconazole

Disease control	benzovindiflupyr + prothioconazole, bixafen + fluopyram + prothioconazole, boscalid, fluxapyroxad + metconazole
Ear blight	fenpicoxamid + prothioconazole, metconazole (*reduction*)
Eyespot	boscalid, fluoxastrobin + prothioconazole + trifloxystrobin (*reduction*), fluxapyroxad + metconazole (*good reduction*), prochloraz (*moderate control*), prothioconazole, prothioconazole + trifloxystrobin
Foot rot	fludioxonil (*seed treatment*)
Fungus diseases	prothioconazole + tebuconazole (*qualified minor use*)
Fusarium	Pseudomonas chlororaphis MA 342, bromuconazole + tebuconazole, prothioconazole + tebuconazole
Late ear diseases	fluoxastrobin + prothioconazole + trifloxystrobin, prothioconazole, prothioconazole + trifloxystrobin
Loose smut	prothioconazole + tebuconazole
Powdery mildew	azoxystrobin (*moderate control*), bixafen + fluopyram + prothioconazole, bromuconazole + tebuconazole (*moderate control*), cyflufenamid + spiroxamine, fenpicoxamid + prothioconazole, fluoxastrobin + prothioconazole + trifloxystrobin, flutriafol, fluxapyroxad (*reduction*), fluxapyroxad + mefentrifluconazole (*moderate control*), fluxapyroxad + metconazole (*moderate control*), fluxapyroxad + pyraclostrobin (*moderate control*), mefentrifluconazole (*moderate control*), mefentrifluconazole + pyraclostrobin, mefentrifluconazole + pyraclostrobin (*moderate control*), metconazole (*moderate control*), physical pest control, prochloraz, prothioconazole, prothioconazole + trifloxystrobin, sulphur, tebuconazole
Rhynchosporium	azoxystrobin (*reduction*), benzovindiflupyr (*moderate control*), benzovindiflupyr + prothioconazole (*moderate control*), fenpicoxamid + prothioconazole, fluxapyroxad, fluxapyroxad + mefentrifluconazole (*moderate control*), fluxapyroxad + metconazole, fluxapyroxad + pyraclostrobin, mefentrifluconazole (*moderate control*), mefentrifluconazole + pyraclostrobin (*moderate control*), metconazole (*moderate control*), prochloraz (*moderate control*), tebuconazole
Rust	azoxystrobin, benzovindiflupyr, benzovindiflupyr + prothioconazole, bromuconazole + tebuconazole, cyflufenamid + spiroxamine, difenoconazole (*off-label*), fenpicoxamid, fenpicoxamid + prothioconazole, fluoxastrobin + prothioconazole + trifloxystrobin, flutriafol, fluxapyroxad, fluxapyroxad (*moderate control*), fluxapyroxad + mefentrifluconazole, fluxapyroxad + mefentrifluconazole (*moderate control*), fluxapyroxad + metconazole, fluxapyroxad + pyraclostrobin, mancozeb, mefentrifluconazole, mefentrifluconazole (*moderate control*), mefentrifluconazole + pyraclostrobin, metconazole, prothioconazole, prothioconazole + trifloxystrobin, pyraclostrobin (*off-label*), tebuconazole, trifloxystrobin (*off-label*)

	Seed-borne diseases	fludioxonil (*seed treatment*), silthiofam (*off-label*)
	Septoria	Pseudomonas chlororaphis MA 342, boscalid (*moderate control*), bromuconazole + tebuconazole (*moderate control*), difenoconazole (*off-label*), fenpicoxamid, fenpicoxamid + prothioconazole, fluoxastrobin + prothioconazole + trifloxystrobin, flutriafol, fluxapyroxad + mefentrifluconazole, fluxapyroxad + mefentrifluconazole (*moderate control*), mancozeb, mefentrifluconazole, mefentrifluconazole (*moderate control*), mefentrifluconazole + pyraclostrobin, metconazole, prothioconazole, prothioconazole + trifloxystrobin
	Sooty moulds	mancozeb
	Stripe smut	fludioxonil
	Take-all	azoxystrobin (*reduction*), silthiofam (*off-label*)
	Tan spot	fenpicoxamid, fenpicoxamid + prothioconazole, fluoxastrobin + prothioconazole + trifloxystrobin, mefentrifluconazole + pyraclostrobin, mefentrifluconazole + pyraclostrobin (*reduction*), prothioconazole
Pests	Aphids	cypermethrin, deltamethrin (*off-label*), flonicamid, lambda-cyhalothrin, physical pest control, tau-fluvalinate (*off-label*)
	Beetles	deltamethrin (*off-label*), lambda-cyhalothrin
	Birds/mammals	aluminium ammonium sulphate
	Caterpillars	lambda-cyhalothrin
	Flies	deltamethrin (*off-label*)
	Mealybugs	physical pest control
	Midges	deltamethrin (*off-label*)
	Pests, miscellaneous	cypermethrin, deltamethrin (*off-label*)
	Scale insects	physical pest control
	Slugs/snails	ferric phosphate
	Spider mites	physical pest control
	Suckers	physical pest control
	Weevils	lambda-cyhalothrin
	Whiteflies	physical pest control
Plant growth regulation	Growth control	chlormequat, ethephon, mepiquat chloride + prohexadione-calcium, prohexadione-calcium, trinexapac-ethyl, trinexapac-ethyl (*off-label*)
	Quality/yield control	mepiquat chloride + prohexadione-calcium, sulphur
Weeds	Broad-leaved weeds	carfentrazone-ethyl (*off-label*), chlorotoluron + diflufenican + pendimethalin, clopyralid + florasulam + fluroxypyr, clopyralid + fluroxypyr + MCPA, dicamba, dicamba + MCPA + mecoprop-P (*off-label*), dicamba + mecoprop-P (*off-label*), dichlorprop-P + MCPA + mecoprop-P, diflufenican (*off-label*), diflufenican + florasulam, diflufenican + flufenacet (*off-label*), florasulam, florasulam + fluroxypyr, florasulam + halauxifen-methyl, florasulam + pyroxsulam, flufenacet + picolinafen, fluroxypyr, fluroxypyr (*off-label*), fluroxypyr + halauxifen-methyl, fluroxypyr + metsulfuron-methyl, iodosulfuron-methyl-sodium + mesosulfuron-methyl (*off-label*), mesosulfuron-methyl + propoxycarbazone-sodium, metsulfuron-methyl + tribenuron-methyl (*off-label*),

		pendimethalin + picolinafen, prosulfocarb (*off-label*), prosulfuron (*off-label*)
	Crops as weeds	chlorotoluron + diflufenican + pendimethalin, diflufenican + florasulam, florasulam, florasulam + fluroxypyr, florasulam + halauxifen-methyl, florasulam + pyroxsulam, flufenacet + picolinafen, iodosulfuron-methyl-sodium + mesosulfuron-methyl (*off-label*), mesosulfuron-methyl + propoxycarbazone-sodium, metsulfuron-methyl + tribenuron-methyl (*off-label*), prosulfocarb (*off-label*)
	Grass weeds	chlorotoluron + diflufenican + pendimethalin, diflufenican (*off-label*), diflufenican + flufenacet (*off-label*), florasulam + pyroxsulam, flufenacet (*off-label*), flufenacet + picolinafen, glyphosate, iodosulfuron-methyl-sodium + mesosulfuron-methyl (*off-label*), mesosulfuron-methyl + propoxycarbazone-sodium, pendimethalin + picolinafen, prosulfocarb (*off-label*)
	Weeds, miscellaneous	chlorotoluron + diflufenican + pendimethalin, dicamba, dicamba + MCPA + mecoprop-P (*off-label*), glyphosate (*off-label*)

Cereals - Maize/sweetcorn

Diseases	Anthracnose	Bacillus amyloliquefaciens D747 (*off-label*)
	Bacterial blight	Bacillus amyloliquefaciens D747 (*off-label*)
	Botrytis	Bacillus amyloliquefaciens D747 (*off-label*), Bacillus subtilis (Strain QST 713) (*off-label*)
	Cercospora leaf spot	Bacillus amyloliquefaciens D747 (*off-label*)
	Disease control	azoxystrobin, pyraclostrobin
	Downy mildew	Bacillus amyloliquefaciens D747 (*off-label*)
	Eyespot	pyraclostrobin, pyraclostrobin (*off-label*)
	Foliar diseases	pyraclostrobin, pyraclostrobin (*moderate control*), pyraclostrobin (*off-label*)
	Fusarium	Bacillus amyloliquefaciens D747 (*off-label*)
	Powdery mildew	Bacillus amyloliquefaciens D747 (*off-label*)
	Pythium	Bacillus amyloliquefaciens D747 (*off-label*)
	Rhizoctonia	Bacillus amyloliquefaciens D747 (*off-label*)
	Sclerotinia	Bacillus amyloliquefaciens D747 (*off-label*)
	Stemphylium	Bacillus amyloliquefaciens D747 (*off-label*)
Pests	Aphids	cypermethrin, fatty acids (*off-label*), maltodextrin (*off-label*), spirotetramat (*off-label*)
	Caterpillars	Bacillus thuringiensis kurstaki (*off-label*), chlorantraniliprole, indoxacarb (*off-label*)
	Cutworms	Bacillus thuringiensis kurstaki (*off-label*)
	Flies	lambda-cyhalothrin (*off-label*)
	Pests, miscellaneous	cypermethrin
	Slugs/snails	ferric phosphate
	Thrips	fatty acids (*off-label*)
	Whiteflies	maltodextrin (*off-label*)
Weeds	Bindweeds	dicamba + prosulfuron
	Broad-leaved weeds	2,4-D, S-metolachlor, S-metolachlor (*moderately susceptible*), S-metolachlor (*off-label*), clopyralid, clopyralid + florasulam + fluroxypyr, dicamba + nicosulfuron, dicamba + prosulfuron, dicamba +

		prosulfuron (*seedlings only*), dimethenamid-p + pendimethalin, dimethenamid-p + pendimethalin (*off-label*), fluroxypyr, fluroxypyr (*off-label*), foramsulfuron + iodosulfuron-methyl-sodium, mesotrione, mesotrione + S-metolachlor, mesotrione + S-metolachlor (*pre-em only*), mesotrione + nicosulfuron, mesotrione + pyridate, nicosulfuron, nicosulfuron (*off-label*), pelargonic acid (*off-label*), pendimethalin, pendimethalin (*off-label*), prosulfuron, pyridate, pyridate (*off-label*), rimsulfuron
	Crops as weeds	fluroxypyr (*off-label*), mesotrione, mesotrione + nicosulfuron, nicosulfuron, nicosulfuron (*off-label*), pendimethalin, rimsulfuron
	Grass weeds	S-metolachlor, S-metolachlor (*off-label*), dicamba + nicosulfuron, dimethenamid-p + pendimethalin, dimethenamid-p + pendimethalin (*off-label*), foramsulfuron + iodosulfuron-methyl-sodium, mesotrione, mesotrione (*moderately susceptible*), mesotrione + S-metolachlor (*pre-em only*), mesotrione + nicosulfuron, mesotrione + pyridate, nicosulfuron, nicosulfuron (*off-label*), pelargonic acid (*off-label*), pendimethalin
	Mosses	pelargonic acid (*off-label*)
	Weed grasses	dicamba + nicosulfuron, foramsulfuron + iodosulfuron-methyl-sodium, mesotrione, mesotrione + S-metolachlor, mesotrione + nicosulfuron, nicosulfuron
	Weeds, miscellaneous	mesotrione, mesotrione + S-metolachlor, mesotrione + nicosulfuron, nicosulfuron (*off-label*), pelargonic acid (*off-label*), pyridate (*off-label*)

Cereals - Oats

Crop control	Desiccation	glyphosate
Diseases	Blue mould	prothioconazole + tebuconazole
	Bunt	prothioconazole + tebuconazole
	Covered smut	prothioconazole + tebuconazole
	Crown rust	azoxystrobin, benzovindiflupyr + prothioconazole, bixafen + fluopyram + prothioconazole, bixafen + prothioconazole, bixafen + prothioconazole + spiroxamine, fluoxastrobin + prothioconazole, fluxapyroxad, fluxapyroxad + mefentrifluconazole, fluxapyroxad + mefentrifluconazole (*moderate control*), fluxapyroxad + pyraclostrobin, mefentrifluconazole, mefentrifluconazole + pyraclostrobin, prothioconazole, prothioconazole + spiroxamine, prothioconazole + spiroxamine + tebuconazole, prothioconazole + tebuconazole, pyraclostrobin, tebuconazole, tebuconazole (*reduction*)
	Disease control	benzovindiflupyr + prothioconazole, bixafen + fluopyram + prothioconazole
	Eyespot	bixafen + prothioconazole (*reduction of incidence and severity*), fluoxastrobin + prothioconazole, fluoxastrobin + prothioconazole (*reduction of incidence and severity*), prothioconazole, prothioconazole + spiroxamine, prothioconazole + spiroxamine + tebuconazole, prothioconazole + tebuconazole

	Foot rot	difenoconazole + fludioxonil, fludioxonil (*seed treatment*)
	Fusarium	prothioconazole + tebuconazole
	Leaf spot	difenoconazole + fludioxonil, fludioxonil (*seed treatment*)
	Loose smut	fludioxonil + fluxapyroxad + triticonazole, fludioxonil + sedaxane, prothioconazole + tebuconazole
	Net blotch	fluxapyroxad
	Powdery mildew	azoxystrobin, azoxystrobin (*moderate control*), bixafen + fluopyram + prothioconazole, bixafen + prothioconazole, bixafen + prothioconazole + spiroxamine, cyflufenamid, fluoxastrobin + prothioconazole, fluxapyroxad (*reduction*), fluxapyroxad + mefentrifluconazole, fluxapyroxad + pyraclostrobin (*moderate control*), mefentrifluconazole, mefentrifluconazole + pyraclostrobin, metrafenone (*evidence of mildew control on oats is limited*), proquinazid, prothioconazole, prothioconazole + spiroxamine, prothioconazole + spiroxamine + tebuconazole, prothioconazole + tebuconazole, sulphur, tebuconazole
	Rhynchosporium	fluxapyroxad
	Rust	fluxapyroxad, fluxapyroxad (*moderate control*), fluxapyroxad + pyraclostrobin
	Seed-borne diseases	difenoconazole + fludioxonil, fludioxonil + fluxapyroxad + triticonazole, fludioxonil + tebuconazole
	Snow mould	fludioxonil (*seed treatment*)
Pests	Aphids	cypermethrin, deltamethrin, flonicamid, lambda-cyhalothrin
	Beetles	lambda-cyhalothrin
	Birds/mammals	aluminium ammonium sulphate
	Caterpillars	lambda-cyhalothrin
	Flies	cypermethrin
	Pests, miscellaneous	cypermethrin
	Slugs/snails	ferric phosphate
	Weevils	lambda-cyhalothrin
Plant growth regulation	Growth control	chlormequat, prohexadione-calcium, prohexadione-calcium + trinexapac-ethyl, trinexapac-ethyl
	Quality/yield control	sulphur
Weeds	Broad-leaved weeds	2,4-D, 2,4-D + MCPA, MCPA, amidosulfuron, carfentrazone-ethyl, carfentrazone-ethyl + mecoprop-P, clopyralid, clopyralid + florasulam, clopyralid + florasulam + fluroxypyr, clopyralid + fluroxypyr + MCPA, dicamba + MCPA + mecoprop-P, dicamba + mecoprop-P, dichlorprop-P + MCPA + mecoprop-P, diflufenican (*off-label*), diflufenican + flufenacet (*off-label*), florasulam, florasulam + fluroxypyr, florasulam + halauxifen-methyl, florasulam + tribenuron-methyl, fluroxypyr, fluroxypyr + halauxifen-methyl, glyphosate, isoxaben, mecoprop-P, metsulfuron-methyl, metsulfuron-methyl + thifensulfuron-methyl, metsulfuron-methyl + tribenuron-methyl,

	thifensulfuron-methyl + tribenuron-methyl, tribenuron-methyl
Crops as weeds	clopyralid + florasulam, diflufenican (*off-label*), florasulam, florasulam + halauxifen-methyl, fluroxypyr, glyphosate, metsulfuron-methyl, metsulfuron-methyl + tribenuron-methyl
Grass weeds	diflufenican (*off-label*), diflufenican + flufenacet (*off-label*), glyphosate
Weeds, miscellaneous	clopyralid + florasulam, diflufenican (*off-label*), florasulam, glyphosate, metsulfuron-methyl, metsulfuron-methyl + thifensulfuron-methyl

Cereals - Rye/triticale

Crop control	Desiccation	glyphosate
Diseases	Alternaria	bromuconazole + tebuconazole, difenoconazole (*off-label*)
	Blue mould	prothioconazole + tebuconazole
	Bunt	Pseudomonas chlororaphis MA 342, prothioconazole + tebuconazole
	Cladosporium	bromuconazole + tebuconazole, difenoconazole (*off-label*)
	Disease control	bixafen + fluopyram + prothioconazole, fluxapyroxad + metconazole, tebuconazole (*off-label*)
	Ear blight	fenpicoxamid + prothioconazole, metconazole (*reduction*)
	Eyespot	bixafen + prothioconazole (*reduction of incidence and severity*), bixafen + prothioconazole + spiroxamine, bixafen + prothioconazole + tebuconazole (*reduction of incidence and severity*), fluoxastrobin + prothioconazole (*reduction*), fluoxastrobin + prothioconazole + trifloxystrobin (*reduction*), fluxapyroxad + metconazole (*good reduction*), prochloraz (*moderate control*), prochloraz + tebuconazole, prothioconazole, prothioconazole + spiroxamine, prothioconazole + spiroxamine + tebuconazole (*reduction*), prothioconazole + tebuconazole, prothioconazole + tebuconazole (*reduction*), prothioconazole + trifloxystrobin
	Foot rot	fludioxonil (*seed treatment*), fludioxonil + fluxapyroxad + triticonazole
	Fusarium	Pseudomonas chlororaphis MA 342, bromuconazole + tebuconazole, prothioconazole + tebuconazole
	Late ear diseases	bixafen + fluoxastrobin + prothioconazole, bixafen + prothioconazole, bixafen + prothioconazole + spiroxamine, bixafen + prothioconazole + tebuconazole, fluoxastrobin + prothioconazole + trifloxystrobin, prochloraz + tebuconazole, prothioconazole, prothioconazole + spiroxamine, prothioconazole + spiroxamine + tebuconazole, prothioconazole + trifloxystrobin
	Loose smut	fludioxonil + fluxapyroxad + triticonazole, prothioconazole + tebuconazole
	Mycosphaerella	benzovindiflupyr
	Powdery mildew	azoxystrobin, azoxystrobin (*moderate control*), bixafen + fluopyram + prothioconazole, bixafen + fluoxastrobin + prothioconazole, bixafen +

prothioconazole, bixafen + prothioconazole + spiroxamine, bixafen + prothioconazole + tebuconazole, bromuconazole + tebuconazole (*moderate control*), cyflufenamid, cyflufenamid + spiroxamine, fenpicoxamid + prothioconazole, fluoxastrobin + prothioconazole, fluoxastrobin + prothioconazole + trifloxystrobin, flutriafol, fluxapyroxad (*reduction*), fluxapyroxad + mefentrifluconazole (*moderate control*), fluxapyroxad + metconazole (*moderate control*), fluxapyroxad + pyraclostrobin (*moderate control*), mefentrifluconazole (*moderate control*), mefentrifluconazole + pyraclostrobin (*moderate control*), metconazole (*moderate control*), prochloraz + tebuconazole, proquinazid, prothioconazole, prothioconazole + spiroxamine, prothioconazole + spiroxamine + tebuconazole, prothioconazole + tebuconazole, prothioconazole + trifloxystrobin, sulphur, tebuconazole

Rhynchosporium   azoxystrobin, azoxystrobin (*reduction*), benzovindiflupyr + prothioconazole (*moderate control*), bixafen + fluoxastrobin + prothioconazole, bixafen + prothioconazole, bixafen + prothioconazole + spiroxamine, bixafen + prothioconazole + tebuconazole, fluoxastrobin + prothioconazole, folpet (*reduction*), prochloraz (*moderate control*), prochloraz + tebuconazole, prothioconazole, prothioconazole + spiroxamine, prothioconazole + spiroxamine + tebuconazole, prothioconazole + tebuconazole, tebuconazole, tebuconazole (*moderate control*)

Rust   azoxystrobin, benzovindiflupyr, benzovindiflupyr + prothioconazole, bixafen + fluoxastrobin + prothioconazole, bixafen + prothioconazole, bixafen + prothioconazole + spiroxamine, bixafen + prothioconazole + tebuconazole, bromuconazole + tebuconazole, cyflufenamid + spiroxamine, difenoconazole (*off-label*), fenpicoxamid, fenpicoxamid + prothioconazole, fluoxastrobin + prothioconazole, fluoxastrobin + prothioconazole + trifloxystrobin, flutriafol, fluxapyroxad, fluxapyroxad (*moderate control*), fluxapyroxad + mefentrifluconazole, fluxapyroxad + metconazole, fluxapyroxad + pyraclostrobin, mancozeb, mefentrifluconazole, mefentrifluconazole + pyraclostrobin, metconazole, prochloraz + tebuconazole, prothioconazole, prothioconazole + spiroxamine, prothioconazole + spiroxamine + tebuconazole, prothioconazole + tebuconazole, prothioconazole + trifloxystrobin, pyraclostrobin (*off-label*), tebuconazole, tebuconazole (*off-label*), trifloxystrobin (*off-label*)

Seed-borne diseases   fludioxonil, fludioxonil + fluxapyroxad + triticonazole, fludioxonil + tebuconazole, silthiofam (*off-label*)

Septoria   Pseudomonas chlororaphis MA 342, benzovindiflupyr, benzovindiflupyr + prothioconazole, bixafen + fluoxastrobin + prothioconazole, bixafen + prothioconazole, bixafen + prothioconazole + spiroxamine, bixafen + prothioconazole + tebuconazole, bromuconazole + tebuconazole (*moderate control*), difenoconazole (*off-label*),

		fenpicoxamid, fenpicoxamid + prothioconazole, fluoxastrobin + prothioconazole + trifloxystrobin, flutriafol, fluxapyroxad, fluxapyroxad + mefentrifluconazole, fluxapyroxad + metconazole, fluxapyroxad + pyraclostrobin, folpet (*reduction*), mancozeb, mefentrifluconazole, mefentrifluconazole + pyraclostrobin, metconazole, prochloraz + tebuconazole, prothioconazole, prothioconazole + spiroxamine, prothioconazole + spiroxamine + tebuconazole, prothioconazole + trifloxystrobin, tebuconazole
	Snow mould	fludioxonil + sedaxane
	Sooty moulds	mancozeb
	Stripe smut	difenoconazole + fludioxonil, fludioxonil + fluxapyroxad + triticonazole, fludioxonil + sedaxane
	Take-all	azoxystrobin (*reduction in severity*), azoxystrobin (*reduction*), silthiofam (*off-label*)
	Tan spot	bixafen + fluoxastrobin + prothioconazole, bixafen + prothioconazole, bixafen + prothioconazole + spiroxamine, bixafen + prothioconazole + tebuconazole, fenpicoxamid, fluoxastrobin + prothioconazole + trifloxystrobin, prothioconazole, prothioconazole + spiroxamine
Pests	Aphids	cypermethrin, deltamethrin (*off-label*), flonicamid, lambda-cyhalothrin, tau-fluvalinate (*off-label*)
	Beetles	deltamethrin (*off-label*), lambda-cyhalothrin
	Birds/mammals	aluminium ammonium sulphate
	Caterpillars	lambda-cyhalothrin
	Flies	cypermethrin, deltamethrin (*off-label*)
	Midges	deltamethrin (*off-label*)
	Pests, miscellaneous	cypermethrin, deltamethrin (*off-label*), lambda-cyhalothrin (*off-label*)
	Slugs/snails	ferric phosphate
	Weevils	lambda-cyhalothrin
Plant growth regulation	Growth control	chlormequat, ethephon, ethephon + mepiquat chloride, mepiquat chloride + prohexadione-calcium, prohexadione-calcium, prohexadione-calcium + trinexapac-ethyl, prohexadione-calcium + trinexapac-ethyl (*off-label*), trinexapac-ethyl
	Quality/yield control	mepiquat chloride + prohexadione-calcium, sulphur
Weeds	Broad-leaved weeds	2,4-D, MCPA, amidosulfuron, amidosulfuron + iodosulfuron-methyl-sodium, bifenox, carfentrazone-ethyl, chlorotoluron + diflufenican + pendimethalin, clopyralid + florasulam, clopyralid + florasulam + fluroxypyr, clopyralid + fluroxypyr + MCPA, dicamba + MCPA + mecoprop-P (*off-label*), dicamba + mecoprop-P (*off-label*), dichlorprop-P + MCPA + mecoprop-P, diflufenican, diflufenican + florasulam, diflufenican + flufenacet, diflufenican + flufenacet (*off-label*), diflufenican + pendimethalin, florasulam, florasulam + fluroxypyr, florasulam + halauxifen-methyl, florasulam + pyroxsulam, florasulam + tribenuron-methyl, flufenacet, flufenacet + picolinafen, fluroxypyr, fluroxypyr + halauxifen-methyl, fluroxypyr + metsulfuron-methyl, glyphosate, iodosulfuron-methyl-sodium + mesosulfuron-methyl

		(*off-label*), iodosulfuron-methyl-sodium + mesosulfuron-methyl + thiencarbazone-methyl, isoxaben, mesosulfuron-methyl + propoxycarbazone-sodium, metsulfuron-methyl, metsulfuron-methyl + thifensulfuron-methyl, metsulfuron-methyl + tribenuron-methyl, pendimethalin, pendimethalin + picolinafen, prosulfocarb (*off-label*), pyroxsulam, thifensulfuron-methyl + tribenuron-methyl, tribenuron-methyl
	Crops as weeds	amidosulfuron + iodosulfuron-methyl-sodium, chlorotoluron + diflufenican + pendimethalin, clopyralid + florasulam, diflufenican + florasulam, diflufenican + flufenacet, florasulam, florasulam + fluroxypyr, florasulam + halauxifen-methyl, florasulam + pyroxsulam, flufenacet + picolinafen, iodosulfuron-methyl-sodium + mesosulfuron-methyl + thiencarbazone-methyl, mesosulfuron-methyl + propoxycarbazone-sodium, metsulfuron-methyl, metsulfuron-methyl + tribenuron-methyl, pendimethalin, pyroxsulam
	Grass weeds	chlorotoluron + diflufenican + pendimethalin, clodinafop-propargyl, diflufenican + flufenacet, diflufenican + flufenacet (*off-label*), diflufenican + pendimethalin, florasulam + pyroxsulam, flufenacet, flufenacet (*off-label*), flufenacet + picolinafen, iodosulfuron-methyl-sodium + mesosulfuron-methyl (*off-label*), iodosulfuron-methyl-sodium + mesosulfuron-methyl + thiencarbazone-methyl, mesosulfuron-methyl + propoxycarbazone-sodium, pendimethalin, pendimethalin + picolinafen, prosulfocarb (*off-label*), pyroxsulam, tri-allate (*off-label*)
	Weeds, miscellaneous	chlorotoluron + diflufenican + pendimethalin, clopyralid + florasulam, dicamba + MCPA + mecoprop-P (*off-label*), florasulam, metsulfuron-methyl, metsulfuron-methyl + thifensulfuron-methyl

Cereals - Undersown cereals

| Weeds | Broad-leaved weeds | 2,4-D, MCPA, florasulam + fluroxypyr |
| | Crops as weeds | florasulam + fluroxypyr |

Cereals - Wheat

Crop control	Desiccation	glyphosate
Diseases	Alternaria	bromuconazole + tebuconazole, tebuconazole
	Blue mould	prothioconazole + tebuconazole
	Bunt	Pseudomonas chlororaphis MA 342, difenoconazole + fludioxonil, fludioxonil (*seed treatment*), fludioxonil + fluxapyroxad + triticonazole, fludioxonil + sedaxane, imazalil + ipconazole, prothioconazole + tebuconazole
	Cladosporium	bromuconazole + tebuconazole, fluxapyroxad + metconazole (*moderate control*), tebuconazole
	Disease control	boscalid, fluxapyroxad + metconazole
	Ear blight	fenpicoxamid + prothioconazole, fludioxonil + sedaxane (*moderate control*), fluxapyroxad + metconazole (*good reduction*), metconazole (*reduction*), prothioconazole + tebuconazole, tebuconazole

Eyespot	bixafen + fluopyram + prothioconazole (*reduction of incidence and severity*), bixafen + prothioconazole (*reduction of incidence and severity*), bixafen + prothioconazole + spiroxamine, bixafen + prothioconazole + tebuconazole (*reduction of incidence and severity*), boscalid, cyprodinil, fluoxastrobin + prothioconazole (*reduction*), fluoxastrobin + prothioconazole + trifloxystrobin (*reduction*), fluxapyroxad + metconazole (*good reduction*), metrafenone (*reduction*), prochloraz (*moderate control*), prochloraz + tebuconazole, prothioconazole, prothioconazole + spiroxamine, prothioconazole + spiroxamine + tebuconazole (*reduction*), prothioconazole + tebuconazole, prothioconazole + tebuconazole (*reduction*), prothioconazole + trifloxystrobin
Foot rot	difenoconazole + fludioxonil, fludioxonil (*seed treatment*), fludioxonil + fluxapyroxad + triticonazole, fluoxastrobin + prothioconazole (*reduction*), imazalil + ipconazole (*useful protection*)
Fusarium	Pseudomonas chlororaphis MA 342, benzovindiflupyr + prothioconazole (*moderate control*), bromuconazole + tebuconazole, fludioxonil + sedaxane, prothioconazole + tebuconazole, tebuconazole
Late ear diseases	azoxystrobin, bixafen + fluopyram + prothioconazole (*reduction of incidence and severity*), bixafen + fluoxastrobin + prothioconazole, bixafen + prothioconazole, bixafen + prothioconazole + spiroxamine, bixafen + prothioconazole + tebuconazole, fluoxastrobin + prothioconazole, fluoxastrobin + prothioconazole + trifloxystrobin, prochloraz + tebuconazole, prothioconazole, prothioconazole + spiroxamine, prothioconazole + spiroxamine + tebuconazole, prothioconazole + tebuconazole, prothioconazole + trifloxystrobin, tebuconazole
Leaf spot	prochloraz
Loose smut	fludioxonil + fluxapyroxad + triticonazole, fludioxonil + sedaxane, prothioconazole + tebuconazole
Mycosphaerella	benzovindiflupyr, benzovindiflupyr + prothioconazole
Net blotch	metconazole (*reduction*)
Powdery mildew	azoxystrobin + difenoconazole, bixafen + fluopyram + prothioconazole, bixafen + fluoxastrobin + prothioconazole, bixafen + prothioconazole, bixafen + prothioconazole + spiroxamine, bixafen + prothioconazole + tebuconazole, bromuconazole + tebuconazole (*moderate control*), cyflufenamid, cyflufenamid + spiroxamine, cyprodinil (*moderate control*), fenpicoxamid + prothioconazole, fluoxastrobin + prothioconazole, fluoxastrobin + prothioconazole + trifloxystrobin, flutriafol, fluxapyroxad (*reduction*), fluxapyroxad + mefentrifluconazole (*moderate control*), fluxapyroxad + metconazole (*moderate control*), fluxapyroxad + pyraclostrobin (*moderate control*), laminarin, mefentrifluconazole (*moderate control*), mefentrifluconazole + pyraclostrobin (*moderate control*), metconazole (*moderate control*), metrafenone, prochloraz + tebuconazole,

proquinazid, prothioconazole, prothioconazole + spiroxamine, prothioconazole + spiroxamine + tebuconazole, prothioconazole + tebuconazole, prothioconazole + trifloxystrobin, sulphur, tebuconazole, tebuconazole (*moderate control*)

Rust   azoxystrobin, azoxystrobin + difenoconazole, benzovindiflupyr, benzovindiflupyr + prothioconazole, bixafen + fluopyram, bixafen + fluopyram + prothioconazole, bixafen + fluoxastrobin + prothioconazole, bixafen + prothioconazole, bixafen + prothioconazole + spiroxamine, bixafen + prothioconazole + tebuconazole, bromuconazole + tebuconazole, cyflufenamid + spiroxamine, difenoconazole, fenpicoxamid, fenpicoxamid + prothioconazole, fluoxastrobin + prothioconazole, fluoxastrobin + prothioconazole + trifloxystrobin, flutriafol, fluxapyroxad, fluxapyroxad (*moderate control*), fluxapyroxad + mefentrifluconazole, fluxapyroxad + metconazole, fluxapyroxad + pyraclostrobin, mancozeb, mancozeb (*useful control*), mefentrifluconazole, mefentrifluconazole + pyraclostrobin, metconazole, prochloraz + tebuconazole, prothioconazole, prothioconazole + spiroxamine, prothioconazole + spiroxamine + tebuconazole, prothioconazole + tebuconazole, prothioconazole + trifloxystrobin, pyraclostrobin, tebuconazole, trifloxystrobin

Seed-borne diseases   difenoconazole + fludioxonil, fludioxonil + fluxapyroxad + triticonazole, fludioxonil + tebuconazole, imazalil + ipconazole (*useful protection*), ipconazole

Septoria   Pseudomonas chlororaphis MA 342, azoxystrobin, benzovindiflupyr, benzovindiflupyr (*moderate control*), benzovindiflupyr + prothioconazole, benzovindiflupyr + prothioconazole (*moderate control*), bixafen + fluopyram, bixafen + fluopyram + prothioconazole, bixafen + fluoxastrobin + prothioconazole, bixafen + prothioconazole, bixafen + prothioconazole + spiroxamine, bixafen + prothioconazole + tebuconazole, boscalid (*moderate control*), bromuconazole + tebuconazole (*moderate control*), difenoconazole, difenoconazole + fludioxonil, fenpicoxamid, fenpicoxamid + prothioconazole, fludioxonil (*seed treatment*), fludioxonil + sedaxane, fluoxastrobin + prothioconazole, fluoxastrobin + prothioconazole + trifloxystrobin, flutriafol, fluxapyroxad, fluxapyroxad + mefentrifluconazole, fluxapyroxad + metconazole, fluxapyroxad + pyraclostrobin, fluxapyroxad + pyraclostrobin (*moderate control*), folpet (*reduction*), laminarin, mancozeb, mancozeb (*reduction*), mefentrifluconazole, mefentrifluconazole + pyraclostrobin, metconazole, prochloraz, prochloraz + tebuconazole, prothioconazole, prothioconazole + spiroxamine, prothioconazole + spiroxamine + tebuconazole, prothioconazole + tebuconazole, prothioconazole + trifloxystrobin, pyraclostrobin, tebuconazole, trifloxystrobin

	Sharp eyespot	fluoxastrobin + prothioconazole, fluoxastrobin + prothioconazole (*reduction*)
	Snow mould	difenoconazole + fludioxonil, fludioxonil (*seed treatment*), fludioxonil + sedaxane
	Soil-borne diseases	ipconazole
	Sooty moulds	fluoxastrobin + prothioconazole (*reduction*), mancozeb, prothioconazole + tebuconazole, tebuconazole
	Take-all	azoxystrobin (*reduction*), fluoxastrobin + prothioconazole (*reduction*), silthiofam (*seed treatment*)
	Tan spot	bixafen + fluopyram + prothioconazole, bixafen + fluoxastrobin + prothioconazole, bixafen + prothioconazole, bixafen + prothioconazole + spiroxamine, bixafen + prothioconazole + tebuconazole, fenpicoxamid, fenpicoxamid + prothioconazole, fluoxastrobin + prothioconazole, fluoxastrobin + prothioconazole + trifloxystrobin, fluxapyroxad (*reduction*), fluxapyroxad + metconazole (*good reduction*), fluxapyroxad + pyraclostrobin, mefentrifluconazole + pyraclostrobin (*reduction*), prothioconazole, prothioconazole + spiroxamine, prothioconazole + tebuconazole
Pests	Aphids	cypermethrin, cypermethrin (*autumn sown*), deltamethrin, esfenvalerate, flonicamid, lambda-cyhalothrin, tau-fluvalinate
	Beetles	lambda-cyhalothrin
	Birds/mammals	aluminium ammonium sulphate
	Caterpillars	lambda-cyhalothrin
	Flies	cypermethrin, cypermethrin (*autumn sown*), cypermethrin (*qualified recommendation*), cypermethrin (*reduction*), lambda-cyhalothrin
	Midges	lambda-cyhalothrin
	Pests, miscellaneous	cypermethrin
	Slugs/snails	ferric phosphate
	Suckers	lambda-cyhalothrin
	Weevils	lambda-cyhalothrin
Plant growth regulation	Growth control	chlormequat, chlormequat + ethephon, ethephon, ethephon + mepiquat chloride, mepiquat chloride + prohexadione-calcium, prohexadione-calcium, prohexadione-calcium + trinexapac-ethyl, trinexapac-ethyl
	Quality/yield control	ethephon (*low lodging situations*), ethephon + mepiquat chloride (*low lodging situations*), mepiquat chloride + prohexadione-calcium, sulphur
Weeds	Broad-leaved weeds	2,4-D, 2,4-D + MCPA, MCPA, aclonifen, amidosulfuron, amidosulfuron + iodosulfuron-methyl-sodium, amidosulfuron + iodosulfuron-methyl-sodium + mesosulfuron-methyl, bifenox, carfentrazone-ethyl, carfentrazone-ethyl + mecoprop-P, chlorotoluron + diflufenican + pendimethalin, cinmethylin, clopyralid, clopyralid + florasulam, clopyralid + florasulam + fluroxypyr, clopyralid + fluroxypyr + MCPA, dicamba + MCPA + mecoprop-P, dicamba + mecoprop-P, dichlorprop-P + MCPA + mecoprop-P, diflufenican, diflufenican + florasulam,

diflufenican + flufenacet, diflufenican + flufenacet (*off-label*), diflufenican + flufenacet + metribuzin, diflufenican + iodosulfuron-methyl-sodium + mesosulfuron-methyl, diflufenican + metsulfuron-methyl, diflufenican + pendimethalin, ethofumesate, florasulam, florasulam + fluroxypyr, florasulam + halauxifen-methyl, florasulam + pyroxsulam, florasulam + tribenuron-methyl, florasulam + tribenuron-methyl (*MS to 20 cm*), florasulam + tribenuron-methyl (*MS to 6 lvs*), flufenacet, flufenacet + metribuzin, flufenacet + pendimethalin, flufenacet + picolinafen, fluroxypyr, fluroxypyr + halauxifen-methyl, fluroxypyr + metsulfuron-methyl, fluroxypyr + metsulfuron-methyl + thifensulfuron-methyl, glyphosate, iodosulfuron-methyl-sodium + mesosulfuron-methyl, iodosulfuron-methyl-sodium + mesosulfuron-methyl + thiencarbazone-methyl, isoxaben, mecoprop-P, mesosulfuron-methyl + propoxycarbazone-sodium, metsulfuron-methyl, metsulfuron-methyl + thifensulfuron-methyl, metsulfuron-methyl + tribenuron-methyl, pendimethalin, pendimethalin (*off-label*), pendimethalin + picolinafen, pendimethalin + picolinafen (*off-label*), prosulfocarb, pyroxsulam, sulfosulfuron, thifensulfuron-methyl, thifensulfuron-methyl + tribenuron-methyl, tribenuron-methyl

Crops as weeds   amidosulfuron + iodosulfuron-methyl-sodium, chlorotoluron + diflufenican + pendimethalin, cinmethylin, clopyralid + florasulam, diflufenican + florasulam, diflufenican + flufenacet, diflufenican + iodosulfuron-methyl-sodium + mesosulfuron-methyl, florasulam, florasulam + fluroxypyr, florasulam + halauxifen-methyl, florasulam + pyroxsulam, flufenacet + picolinafen, fluroxypyr, glyphosate, iodosulfuron-methyl-sodium + mesosulfuron-methyl + thiencarbazone-methyl, mesosulfuron-methyl + propoxycarbazone-sodium, metsulfuron-methyl, metsulfuron-methyl + tribenuron-methyl, pendimethalin, pyroxsulam

Grass weeds   aclonifen, amidosulfuron + iodosulfuron-methyl-sodium + mesosulfuron-methyl, chlorotoluron + diflufenican + pendimethalin, cinmethylin, clodinafop-propargyl, clodinafop-propargyl + cloquintocet-mexyl, diflufenican + flufenacet, diflufenican + flufenacet (*off-label*), diflufenican + flufenacet + metribuzin, diflufenican + iodosulfuron-methyl-sodium + mesosulfuron-methyl, diflufenican + pendimethalin, ethofumesate, fenoxaprop-P-ethyl, florasulam + pyroxsulam, flufenacet, flufenacet + metribuzin, flufenacet + pendimethalin, flufenacet + picolinafen, glyphosate, iodosulfuron-methyl-sodium + mesosulfuron-methyl, iodosulfuron-methyl-sodium + mesosulfuron-methyl + thiencarbazone-methyl, mesosulfuron-methyl + propoxycarbazone-sodium, pendimethalin, pendimethalin (*off-label*), pendimethalin + picolinafen, pendimethalin + picolinafen (*off-label*), pinoxaden, prosulfocarb, pyroxsulam, sulfosulfuron, sulfosulfuron (*moderate control of barren brome*), sulfosulfuron (*moderate control only*), tri-allate

Weeds, miscellaneous	chlorotoluron + diflufenican + pendimethalin, clopyralid + florasulam, diflufenican + flufenacet + metribuzin, diflufenican + metsulfuron-methyl, florasulam, glyphosate, metsulfuron-methyl, metsulfuron-methyl + thifensulfuron-methyl

Edible fungi - Mushrooms

Diseases	Alternaria	Bacillus amyloliquefaciens D747
	Cobweb	metrafenone
	Fusarium	Bacillus amyloliquefaciens D747
	Pythium	Bacillus amyloliquefaciens D747
	Rhizoctonia	Bacillus amyloliquefaciens D747
Pests	Flies	fatty acids (*off-label*)

Fruiting vegetables - Aubergines

Diseases	Alternaria	Bacillus amyloliquefaciens D747
	Botrytis	Bacillus subtilis (Strain QST 713) (*off-label*), azoxystrobin (*off-label*)
	Disease control	azoxystrobin
	Downy mildew	cyazofamid (*off-label*)
	Foot rot	Trichoderma asperellum (Strain T34)
	Fusarium	Bacillus amyloliquefaciens D747
	Phytophthora	azoxystrobin (*off-label*), cyazofamid (*off-label*)
	Powdery mildew	Bacillus amyloliquefaciens D747, azoxystrobin (*off-label*), physical pest control
	Pythium	Bacillus amyloliquefaciens D747, Bacillus subtilis (Strain QST 713) (*off-label*)
	Rhizoctonia	Bacillus amyloliquefaciens D747
Pests	Aphids	physical pest control
	Mealybugs	physical pest control
	Scale insects	physical pest control
	Slugs/snails	ferric phosphate
	Spider mites	physical pest control
	Suckers	physical pest control
	Whiteflies	physical pest control

Fruiting vegetables - Cucurbits

Diseases	Alternaria	Bacillus amyloliquefaciens D747
	Anthracnose	Bacillus amyloliquefaciens D747 (*off-label*)
	Bacterial blight	Bacillus amyloliquefaciens D747 (*off-label*)
	Botrytis	Bacillus amyloliquefaciens D747 (*off-label*), Bacillus subtilis (Strain QST 713) (*off-label*), azoxystrobin (*off-label*), cyprodinil + fludioxonil (*off-label*), mepanipyrim (*off-label*)
	Cercospora leaf spot	Bacillus amyloliquefaciens D747 (*off-label*)
	Cladosporium	boscalid + pyraclostrobin (*off-label*)
	Disease control	cyflufenamid, mancozeb
	Downy mildew	Bacillus amyloliquefaciens D747 (*off-label*), azoxystrobin (*off-label*), cyazofamid (*off-label*)
	Fusarium	Bacillus amyloliquefaciens D747, Bacillus amyloliquefaciens D747 (*off-label*)

31

	Leaf spot	boscalid + pyraclostrobin (*off-label*)
	Phytophthora	Bacillus subtilis (Strain QST 713) (*off-label*), cyazofamid (*off-label*)
	Powdery mildew	Bacillus amyloliquefaciens D747, Bacillus amyloliquefaciens D747 (*off-label*), azoxystrobin (*off-label*), boscalid + pyraclostrobin (*off-label*), cyflufenamid, cyflufenamid (*off-label*), difenoconazole + fluxapyroxad (*off-label*), penconazole (*off-label*), physical pest control, proquinazid (*off-label*)
	Pythium	Bacillus amyloliquefaciens D747, Bacillus amyloliquefaciens D747 (*off-label*), Bacillus subtilis (Strain QST 713) (*off-label*)
	Rhizoctonia	Bacillus amyloliquefaciens D747, Bacillus amyloliquefaciens D747 (*off-label*)
	Sclerotinia	Bacillus amyloliquefaciens D747 (*off-label*)
	Stemphylium	Bacillus amyloliquefaciens D747 (*off-label*)
Pests	Aphids	fatty acids (*off-label*), flonicamid (*off-label*), maltodextrin (*off-label*), physical pest control
	Caterpillars	Bacillus thuringiensis kurstaki, spinosad (*off-label*)
	Leaf miners	spinosad (*off-label*)
	Mealybugs	physical pest control
	Scale insects	physical pest control
	Slugs/snails	ferric phosphate
	Spider mites	physical pest control
	Suckers	physical pest control
	Thrips	fatty acids (*off-label*), spinosad (*off-label*)
	Whiteflies	flonicamid (*off-label*), maltodextrin (*off-label*), physical pest control
Weeds	Broad-leaved weeds	clomazone (*off-label*), dimethenamid-p + pendimethalin (*off-label*), isoxaben (*off-label*), pelargonic acid (*off-label*), propyzamide (*off-label*)
	Grass weeds	dimethenamid-p + pendimethalin (*off-label*), pelargonic acid (*off-label*), propyzamide (*off-label*)
	Mosses	pelargonic acid (*off-label*)
	Weeds, miscellaneous	pelargonic acid (*off-label*)

Fruiting vegetables - Peppers

Diseases	Alternaria	Bacillus amyloliquefaciens D747
	Botrytis	Bacillus subtilis (Strain QST 713) (*off-label*)
	Disease control	azoxystrobin
	Foot rot	Trichoderma asperellum (Strain T34)
	Fusarium	Bacillus amyloliquefaciens D747
	Powdery mildew	Bacillus amyloliquefaciens D747, azoxystrobin (*off-label*), physical pest control
	Pythium	Bacillus amyloliquefaciens D747, Bacillus subtilis (Strain QST 713) (*off-label*)
	Rhizoctonia	Bacillus amyloliquefaciens D747
Pests	Aphids	physical pest control, pirimicarb (*off-label*)
	Mealybugs	physical pest control
	Scale insects	physical pest control

Spider mites	physical pest control
Suckers	physical pest control
Whiteflies	physical pest control

Fruiting vegetables - Tomatoes

Diseases	Alternaria	Bacillus amyloliquefaciens D747
	Botrytis	Bacillus subtilis (Strain QST 713) (*off-label*), azoxystrobin (*off-label*), fenhexamid (*off-label*)
	Disease control	azoxystrobin
	Downy mildew	cyazofamid (*off-label*)
	Foot rot	Trichoderma asperellum (Strain T34)
	Fungus diseases	1-methylcyclopropene
	Fusarium	Bacillus amyloliquefaciens D747
	Phytophthora	azoxystrobin (*off-label*), cyazofamid (*off-label*), mancozeb
	Powdery mildew	Bacillus amyloliquefaciens D747, azoxystrobin (*off-label*), physical pest control
	Pythium	Bacillus amyloliquefaciens D747, Bacillus subtilis (Strain QST 713) (*off-label*)
	Rhizoctonia	Bacillus amyloliquefaciens D747
Pests	Aphids	physical pest control
	Caterpillars	chlorantraniliprole (*off-label*)
	Mealybugs	physical pest control
	Scale insects	physical pest control
	Spider mites	physical pest control
	Suckers	physical pest control
	Whiteflies	physical pest control

Herb crops - Herbs

Crop control	Desiccation	glyphosate (*off-label*)
Diseases	Alternaria	Bacillus amyloliquefaciens D747, azoxystrobin (*off-label*), boscalid (*off-label*), boscalid + dimoxystrobin (*off-label*), boscalid + pyraclostrobin (*off-label*), cyprodinil + fludioxonil (*off-label*), fludioxonil (*off-label*)
	Anthracnose	Bacillus amyloliquefaciens D747 (*off-label*)
	Bacterial blight	Bacillus amyloliquefaciens D747 (*off-label*)
	Bacterial canker	Bacillus amyloliquefaciens D747 (*off-label*)
	Black rot	azoxystrobin (*off-label*)
	Botrytis	Bacillus amyloliquefaciens D747 (*off-label*), Bacillus amyloliquefaciens strain MBI600 (*off-label*), Bacillus subtilis (Strain QST 713) (*off-label*), boscalid (*off-label*), boscalid + pyraclostrobin (*off-label*), cerevisane (*off-label*), cyprodinil + fludioxonil (*off-label*), difenoconazole + fluxapyroxad (*off-label*), fenhexamid (*off-label*)
	Cercospora leaf spot	Bacillus amyloliquefaciens D747 (*off-label*)
	Damping off	Bacillus amyloliquefaciens D747 (*off-label*), Bacillus amyloliquefaciens strain MBI600 (*off-label*), Bacillus subtilis (Strain QST 713) (*off-label*), fosetyl-aluminium + propamocarb hydrochloride (*off-label*)
	Disease control	boscalid (*off-label*), difenoconazole (*off-label*), mandipropamid

	Downy mildew	Bacillus amyloliquefaciens D747 (*off-label*), cerevisane (*off-label*), cos-oga (*off-label*), dimethomorph (*off-label*), fluopicolide + propamocarb hydrochloride (*off-label*), fosetyl-aluminium + propamocarb hydrochloride (*off-label*), mancozeb + metalaxyl-M (*off-label*), mandipropamid, metalaxyl-M (*off-label*)
	Fire	boscalid (*off-label*), boscalid + pyraclostrobin (*off-label*)
	Fireblight	Bacillus amyloliquefaciens D747 (*off-label*)
	Fungus diseases	Bacillus amyloliquefaciens D747 (*off-label*)
	Fusarium	Bacillus amyloliquefaciens D747, Bacillus amyloliquefaciens D747 (*off-label*)
	Leaf spot	difenoconazole (*off-label*), fludioxonil (*off-label*)
	Phytophthora	difenoconazole (*off-label*)
	Powdery mildew	Bacillus amyloliquefaciens D747, Bacillus amyloliquefaciens D747 (*off-label*), azoxystrobin (*off-label*), azoxystrobin + difenoconazole, cerevisane (*off-label*), cos-oga (*off-label*), difenoconazole + fluxapyroxad (*off-label*), physical pest control, prothioconazole (*off-label*)
	Pythium	Bacillus amyloliquefaciens D747, Bacillus amyloliquefaciens D747 (*off-label*), Bacillus amyloliquefaciens strain MBI600 (*off-label*), metalaxyl-M (*off-label*)
	Rhizoctonia	Bacillus amyloliquefaciens D747, Bacillus amyloliquefaciens D747 (*off-label*), Bacillus amyloliquefaciens strain MBI600 (*off-label*), cyprodinil + fludioxonil (*off-label*), difenoconazole + fluxapyroxad (*off-label*)
	Rust	azoxystrobin + difenoconazole (*off-label*)
	Sclerotinia	Bacillus amyloliquefaciens D747 (*off-label*), Bacillus amyloliquefaciens strain MBI600 (*off-label*), boscalid (*off-label*), boscalid + dimoxystrobin (*off-label*), boscalid + pyraclostrobin (*off-label*), cyprodinil + fludioxonil (*off-label*), difenoconazole + fluxapyroxad (*off-label*), prothioconazole (*off-label*)
	Septoria	azoxystrobin + difenoconazole (*off-label*)
	Stem canker	cyprodinil + fludioxonil (*off-label*), prothioconazole (*off-label*)
	Stemphylium	Bacillus amyloliquefaciens D747 (*off-label*)
	White blister	azoxystrobin (*off-label*), boscalid + pyraclostrobin (*off-label*)
	White rot	tebuconazole (*off-label*)
Pests	Aphids	Beauveria bassiana GHA (*off-label*), acetamiprid (*off-label*), deltamethrin (*off-label*), fatty acids (*off-label*), flonicamid (*off-label*), maltodextrin (*off-label*), physical pest control, pirimicarb (*off-label*), pyrethrins
	Beetles	deltamethrin (*off-label*), lambda-cyhalothrin (*off-label*)
	Capsid bugs	deltamethrin (*off-label*)
	Caterpillars	Bacillus thuringiensis kurstaki, Bacillus thuringiensis kurstaki (*off-label*), deltamethrin (*off-label*), pyrethrins, spinosad (*off-label*)
	Cutworms	Bacillus thuringiensis kurstaki (*off-label*), deltamethrin (*off-label*), lambda-cyhalothrin (*off-label*)

	Flies	lambda-cyhalothrin (*off-label*), maltodextrin (*off-label*), spinosad (*off-label*)
	Leaf miners	deltamethrin (*off-label*), spinosad (*off-label*)
	Mealybugs	physical pest control
	Midges	lambda-cyhalothrin (*off-label*)
	Pests, miscellaneous	lambda-cyhalothrin (*off-label*)
	Scale insects	physical pest control
	Slugs/snails	ferric phosphate
	Spider mites	Beauveria bassiana GHA (*off-label*), fatty acids (*off-label*), physical pest control, pyrethrins
	Springtails	lambda-cyhalothrin (*off-label*)
	Suckers	physical pest control
	Thrips	Beauveria bassiana GHA (*off-label*), fatty acids (*off-label*), lambda-cyhalothrin (*off-label*), maltodextrin (*off-label*), pyrethrins, spinosad (*off-label*)
	Weevils	lambda-cyhalothrin (*off-label*)
	Whiteflies	Beauveria bassiana GHA (*off-label*), deltamethrin (*off-label*), physical pest control
	Wireworms	deltamethrin (*off-label*)
Weeds	Broad-leaved weeds	S-metolachlor (*off-label*), aclonifen (*off-label*), amidosulfuron (*off-label*), bentazone (*off-label*), clomazone (*off-label*), clopyralid + picloram (*off-label*), dimethenamid-p + metazachlor (*off-label*), dimethenamid-p + metazachlor + quinmerac (*off-label*), dimethenamid-p + pendimethalin (*off-label*), fluroxypyr (*off-label*), lenacil (*off-label*), metazachlor + quinmerac (*off-label*), napropamide (*off-label*), pendimethalin (*off-label*), phenmedipham (*off-label*), propyzamide (*off-label*), prosulfocarb (*off-label*)
	Crops as weeds	clethodim (*off-label*), cycloxydim (*off-label*), fluazifop-P-butyl, fluazifop-P-butyl (*off-label*), fluroxypyr (*off-label*), propaquizafop (*off-label*)
	Grass weeds	S-metolachlor (*off-label*), clethodim (*off-label*), cycloxydim (*off-label*), dimethenamid-p + metazachlor (*off-label*), dimethenamid-p + metazachlor + quinmerac (*off-label*), dimethenamid-p + pendimethalin (*off-label*), fluazifop-P-butyl, fluazifop-P-butyl (*off-label*), metazachlor + quinmerac (*off-label*), napropamide (*off-label*), pendimethalin (*off-label*), propaquizafop (*off-label*), propyzamide (*off-label*), prosulfocarb (*off-label*), tri-allate (*off-label*)
	Weeds, miscellaneous	fluroxypyr (*off-label*), lenacil (*off-label*), napropamide (*off-label*)

Leafy vegetables - Endives

Diseases	Alternaria	Bacillus amyloliquefaciens D747, cyprodinil + fludioxonil (*off-label*)
	Botrytis	Bacillus amyloliquefaciens strain MBI600 (*off-label*), Bacillus subtilis (Strain QST 713) (*off-label*), cerevisane (*off-label*), cyprodinil + fludioxonil (*off-label*), difenoconazole + fluxapyroxad (*off-label*), fenhexamid (*off-label*)
	Damping off	Bacillus amyloliquefaciens strain MBI600 (*off-label*)
	Disease control	mandipropamid

	Downy mildew	azoxystrobin, boscalid + pyraclostrobin (*off-label*), cerevisane (*off-label*), mandipropamid
	Fusarium	Bacillus amyloliquefaciens D747
	Powdery mildew	Bacillus amyloliquefaciens D747, cerevisane (*off-label*), difenoconazole + fluxapyroxad (*off-label*)
	Pythium	Bacillus amyloliquefaciens D747, Bacillus amyloliquefaciens strain MBI600 (*off-label*)
	Rhizoctonia	Bacillus amyloliquefaciens D747, Bacillus amyloliquefaciens strain MBI600 (*off-label*), cyprodinil + fludioxonil (*off-label*), difenoconazole + fluxapyroxad (*off-label*)
	Sclerotinia	Bacillus amyloliquefaciens strain MBI600 (*off-label*), difenoconazole + fluxapyroxad (*off-label*)
	Stem canker	cyprodinil + fludioxonil (*off-label*)
Pests	Aphids	deltamethrin (*off-label*), fatty acids (*off-label*)
	Beetles	deltamethrin (*off-label*)
	Caterpillars	Bacillus thuringiensis kurstaki, Bacillus thuringiensis kurstaki (*off-label*), deltamethrin (*off-label*), indoxacarb (*off-label*), spinosad (*off-label*)
	Pests, miscellaneous	lambda-cyhalothrin (*off-label*)
	Slugs/snails	ferric phosphate
	Spider mites	fatty acids (*off-label*)
	Thrips	fatty acids (*off-label*), spinosad (*off-label*)
Weeds	Broad-leaved weeds	S-metolachlor (*off-label*), pendimethalin (*off-label*), propyzamide (*off-label*), triflusulfuron-methyl (*off-label*)
	Crops as weeds	cycloxydim (*off-label*), fluazifop-P-butyl
	Grass weeds	S-metolachlor (*off-label*), cycloxydim (*off-label*), fluazifop-P-butyl, propyzamide (*off-label*)

Leafy vegetables - Lettuce

Diseases	Alternaria	Bacillus amyloliquefaciens D747, cyprodinil + fludioxonil (*off-label*)
	Bacterial canker	Bacillus amyloliquefaciens D747 (*off-label*)
	Black rot	azoxystrobin (*off-label*)
	Botrytis	Bacillus amyloliquefaciens D747 (*off-label*), Bacillus amyloliquefaciens strain MBI600 (*off-label*), Bacillus subtilis (Strain QST 713) (*off-label*), boscalid + pyraclostrobin, boscalid + pyraclostrobin (*off-label*), cerevisane (*off-label*), cyprodinil + fludioxonil (*off-label*), difenoconazole + fluxapyroxad (*off-label*), fenhexamid (*off-label*), fluopyram + trifloxystrobin (*off-label*)
	Bottom rot	boscalid + pyraclostrobin
	Damping off	Bacillus amyloliquefaciens strain MBI600 (*off-label*), Bacillus subtilis (Strain QST 713) (*off-label*)
	Disease control	boscalid + pyraclostrobin (*off-label*), mandipropamid
	Downy mildew	azoxystrobin, cerevisane (*off-label*), cos-oga (*off-label*), dimethomorph (*off-label*), fluopicolide + propamocarb hydrochloride (*off-label*), fosetyl-aluminium + propamocarb hydrochloride, mancozeb + metalaxyl-M (*off-label*), mandipropamid
	Fireblight	Bacillus amyloliquefaciens D747 (*off-label*)

	Fungus diseases	Bacillus amyloliquefaciens D747 (*off-label*), Bacillus subtilis (Strain QST 713) (*off-label*)
	Fusarium	Bacillus amyloliquefaciens D747
	Powdery mildew	Bacillus amyloliquefaciens D747, Bacillus amyloliquefaciens D747 (*off-label*), cerevisane (*off-label*), cos-oga (*off-label*), difenoconazole + fluxapyroxad (*off-label*), fluopyram + trifloxystrobin (*off-label*), physical pest control
	Pythium	Bacillus amyloliquefaciens D747, Bacillus amyloliquefaciens strain MBI600 (*off-label*), fosetyl-aluminium + propamocarb hydrochloride
	Rhizoctonia	Bacillus amyloliquefaciens D747, Bacillus amyloliquefaciens strain MBI600 (*off-label*), Bacillus subtilis (Strain QST 713) (*off-label*), cyprodinil + fludioxonil (*off-label*), difenoconazole + fluxapyroxad (*off-label*)
	Sclerotinia	Bacillus amyloliquefaciens D747 (*off-label*), Bacillus amyloliquefaciens strain MBI600 (*off-label*), difenoconazole + fluxapyroxad, difenoconazole + fluxapyroxad (*off-label*), fluopyram + trifloxystrobin (*off-label*)
	Soft rot	boscalid + pyraclostrobin
	Stem canker	cyprodinil + fludioxonil (*off-label*)
	White blister	azoxystrobin (*off-label*)
Pests	Aphids	Beauveria bassiana GHA (*off-label*), acetamiprid (*off-label*), deltamethrin, deltamethrin (*off-label*), fatty acids (*off-label*), lambda-cyhalothrin, physical pest control, pyrethrins, spirotetramat
	Beetles	deltamethrin, deltamethrin (*off-label*)
	Caterpillars	Bacillus thuringiensis kurstaki, Bacillus thuringiensis kurstaki (*off-label*), deltamethrin, deltamethrin (*off-label*), indoxacarb (*off-label*), pyrethrins, spinosad (*off-label*)
	Cutworms	Bacillus thuringiensis kurstaki (*off-label*), deltamethrin, lambda-cyhalothrin
	Leafhoppers	deltamethrin
	Mealybugs	physical pest control
	Pests, miscellaneous	cyantraniliprole, lambda-cyhalothrin (*off-label*)
	Scale insects	physical pest control
	Slugs/snails	ferric phosphate
	Spider mites	Beauveria bassiana GHA (*off-label*), fatty acids (*off-label*), physical pest control, pyrethrins
	Suckers	physical pest control
	Thrips	Beauveria bassiana GHA (*off-label*), deltamethrin, fatty acids (*off-label*), pyrethrins, spinosad (*off-label*)
	Whiteflies	Beauveria bassiana GHA (*off-label*), physical pest control
Weeds	Broad-leaved weeds	S-metolachlor (*off-label*), dimethenamid-p + pendimethalin (*off-label*), napropamide (*off-label*), pendimethalin (*off-label*), propyzamide, propyzamide (*off-label*)
	Crops as weeds	cycloxydim (*off-label*), fluazifop-P-butyl
	Grass weeds	S-metolachlor (*off-label*), cycloxydim (*off-label*), dimethenamid-p + pendimethalin (*off-label*), fluazifop-

P-butyl, napropamide (*off-label*), pendimethalin (*off-label*), propyzamide, propyzamide (*off-label*)

Weeds, miscellaneous   napropamide (*off-label*)

Leafy vegetables - Spinach

Diseases	Alternaria	Bacillus amyloliquefaciens D747, cyprodinil + fludioxonil (*off-label*)
	Anthracnose	Bacillus amyloliquefaciens D747 (*off-label*)
	Bacterial blight	Bacillus amyloliquefaciens D747 (*off-label*)
	Black rot	azoxystrobin (*off-label*)
	Botrytis	Bacillus amyloliquefaciens D747 (*off-label*), Bacillus amyloliquefaciens strain MBI600 (*off-label*), Bacillus subtilis (Strain QST 713) (*off-label*), cyprodinil + fludioxonil (*off-label*)
	Cercospora leaf spot	Bacillus amyloliquefaciens D747 (*off-label*)
	Damping off	Bacillus amyloliquefaciens D747 (*off-label*), Bacillus amyloliquefaciens strain MBI600 (*off-label*), Bacillus subtilis (Strain QST 713) (*off-label*)
	Disease control	mandipropamid
	Downy mildew	Bacillus amyloliquefaciens D747 (*off-label*), boscalid + pyraclostrobin (*off-label*), dimethomorph (*off-label*), mandipropamid, metalaxyl-M (*off-label*)
	Fusarium	Bacillus amyloliquefaciens D747, Bacillus amyloliquefaciens D747 (*off-label*)
	Powdery mildew	Bacillus amyloliquefaciens D747, Bacillus amyloliquefaciens D747 (*off-label*)
	Pythium	Bacillus amyloliquefaciens D747, Bacillus amyloliquefaciens D747 (*off-label*), Bacillus amyloliquefaciens strain MBI600 (*off-label*), metalaxyl-M (*off-label*)
	Rhizoctonia	Bacillus amyloliquefaciens D747, Bacillus amyloliquefaciens D747 (*off-label*), Bacillus amyloliquefaciens strain MBI600 (*off-label*), cyprodinil + fludioxonil (*off-label*)
	Sclerotinia	Bacillus amyloliquefaciens D747 (*off-label*), Bacillus amyloliquefaciens strain MBI600 (*off-label*)
	Stem canker	cyprodinil + fludioxonil (*off-label*)
	Stemphylium	Bacillus amyloliquefaciens D747 (*off-label*)
	White blister	azoxystrobin (*off-label*)
Pests	Aphids	Beauveria bassiana GHA (*off-label*), acetamiprid (*off-label*), fatty acids (*off-label*), maltodextrin (*off-label*)
	Caterpillars	Bacillus thuringiensis kurstaki, Bacillus thuringiensis kurstaki (*off-label*), spinosad (*off-label*)
	Slugs/snails	ferric phosphate
	Spider mites	Beauveria bassiana GHA (*off-label*), fatty acids (*off-label*)
	Thrips	Beauveria bassiana GHA (*off-label*), fatty acids (*off-label*), spinosad (*off-label*)
	Whiteflies	Beauveria bassiana GHA (*off-label*)
Weeds	Broad-leaved weeds	clopyralid (*off-label*), lenacil (*off-label*), phenmedipham (*off-label*)
	Crops as weeds	cycloxydim (*off-label*), fluazifop-P-butyl

| Grass weeds | cycloxydim (*off-label*), fluazifop-P-butyl |
| Weeds, miscellaneous | clopyralid (*off-label*), lenacil (*off-label*) |

Leafy vegetables - Watercress

Diseases	Alternaria	Bacillus amyloliquefaciens D747
	Anthracnose	Bacillus amyloliquefaciens D747 (*off-label*)
	Bacterial blight	Bacillus amyloliquefaciens D747 (*off-label*)
	Botrytis	Bacillus amyloliquefaciens D747 (*off-label*), Bacillus subtilis (Strain QST 713) (*off-label*)
	Cercospora leaf spot	Bacillus amyloliquefaciens D747 (*off-label*)
	Damping off	Bacillus amyloliquefaciens D747 (*off-label*), metalaxyl-M (*off-label*)
	Downy mildew	Bacillus amyloliquefaciens D747 (*off-label*), metalaxyl-M (*off-label*), propamocarb hydrochloride (*off-label*)
	Fusarium	Bacillus amyloliquefaciens D747, Bacillus amyloliquefaciens D747 (*off-label*), Trichoderma asperellum (Strain T34) (*off-label*)
	Phytophthora	propamocarb hydrochloride (*off-label*)
	Powdery mildew	Ampelomyces quisqualis (Strain AQ10) (*off-label*), Bacillus amyloliquefaciens D747, Bacillus amyloliquefaciens D747 (*off-label*)
	Pythium	Bacillus amyloliquefaciens D747, Bacillus amyloliquefaciens D747 (*off-label*), Trichoderma asperellum (Strain T34) (*off-label*), propamocarb hydrochloride (*off-label*)
	Rhizoctonia	Bacillus amyloliquefaciens D747, Bacillus amyloliquefaciens D747 (*off-label*)
	Root rot	propamocarb hydrochloride (*off-label*)
	Sclerotinia	Bacillus amyloliquefaciens D747 (*off-label*)
	Stemphylium	Bacillus amyloliquefaciens D747 (*off-label*)
Pests	Aphids	Beauveria bassiana GHA (*off-label*), fatty acids (*off-label*), maltodextrin (*off-label*), pyrethrins (*off-label*)
	Beetles	pyrethrins (*off-label*)
	Caterpillars	Bacillus thuringiensis kurstaki (*off-label*)
	Cutworms	Bacillus thuringiensis kurstaki (*off-label*)
	Spider mites	Beauveria bassiana GHA (*off-label*), fatty acids (*off-label*)
	Thrips	Beauveria bassiana GHA (*off-label*), fatty acids (*off-label*)
	Whiteflies	Beauveria bassiana GHA (*off-label*)
Weeds	Crops as weeds	cycloxydim (*off-label*)
	Grass weeds	cycloxydim (*off-label*)

Legumes - Beans (Phaseolus)

Diseases	Ascochyta	azoxystrobin, azoxystrobin (*useful control*)
	Botrytis	Bacillus subtilis (Strain QST 713) (*off-label*), azoxystrobin, azoxystrobin (*some control*), boscalid + pyraclostrobin (*off-label*), cyprodinil + fludioxonil (*moderate control*), difenoconazole + fluxapyroxad (*off-label*)
	Disease control	azoxystrobin

	Downy mildew	azoxystrobin, azoxystrobin (*reduction*)
	Fusarium	Trichoderma afroharzianum (Strain T22)
	Leaf spot	difenoconazole + fluxapyroxad (*off-label*)
	Mycosphaerella	azoxystrobin (*some control*)
	Phytophthora	difenoconazole + fluxapyroxad (*off-label*)
	Powdery mildew	physical pest control
	Pythium	Trichoderma afroharzianum (Strain T22)
	Rhizoctonia	Trichoderma afroharzianum (Strain T22)
	Rust	azoxystrobin, tebuconazole (*off-label*)
	Sclerotinia	Trichoderma afroharzianum (Strain T22), cyprodinil + fludioxonil
Pests	Aphids	cypermethrin, deltamethrin (*off-label*), fatty acids (*off-label*), flonicamid (*off-label*), physical pest control, pirimicarb (*off-label*), pyrethrins
	Birds/mammals	aluminium ammonium sulphate
	Caterpillars	Bacillus thuringiensis kurstaki, Bacillus thuringiensis kurstaki (*off-label*), deltamethrin (*off-label*), lambda-cyhalothrin (*off-label*), pyrethrins, spinosad (*off-label*)
	Cutworms	Bacillus thuringiensis kurstaki (*off-label*)
	Mealybugs	physical pest control
	Midges	deltamethrin (*off-label*)
	Pests, miscellaneous	cypermethrin, lambda-cyhalothrin (*off-label*)
	Scale insects	physical pest control
	Slugs/snails	ferric phosphate
	Spider mites	physical pest control, pyrethrins
	Suckers	physical pest control
	Thrips	deltamethrin (*off-label*), fatty acids (*off-label*), pyrethrins
	Weevils	deltamethrin (*off-label*)
	Whiteflies	physical pest control
Weeds	Broad-leaved weeds	bentazone, clomazone (*off-label*), pendimethalin, pendimethalin (*off-label*)
	Crops as weeds	clethodim (*off-label*), cycloxydim, cycloxydim (*off-label*), fluazifop-P-butyl
	Grass weeds	S-metolachlor (*off-label*), clethodim (*off-label*), cycloxydim, cycloxydim (*off-label*), fluazifop-P-butyl, pendimethalin (*off-label*), propaquizafop

Legumes - Beans (Vicia)

Crop control	Desiccation	glyphosate
Diseases	Botrytis	Bacillus subtilis (Strain QST 713) (*off-label*), benzovindiflupyr + prothioconazole (*useful control*), boscalid + pyraclostrobin (*off-label*), cyprodinil + fludioxonil (*moderate control*)
	Chocolate spot	azoxystrobin + tebuconazole, azoxystrobin + tebuconazole (*moderate control*), boscalid + pyraclostrobin (*moderate control*), tebuconazole, tebuconazole (*moderate control*)
	Disease control	azoxystrobin + tebuconazole, azoxystrobin + tebuconazole (*moderate control*), benzovindiflupyr + prothioconazole, metconazole (*off-label*), tebuconazole

	Downy mildew	metalaxyl-M (*off-label*)
	Rust	azoxystrobin, azoxystrobin + tebuconazole, benzovindiflupyr + prothioconazole, boscalid + pyraclostrobin, metconazole, tebuconazole, tebuconazole (*off-label*)
	Sclerotinia	cyprodinil + fludioxonil
Pests	Aphids	esfenvalerate, fatty acids (*off-label*), lambda-cyhalothrin, pirimicarb, pyrethrins
	Beetles	cyantraniliprole (*off-label*), lambda-cyhalothrin, lambda-cyhalothrin (*off-label*)
	Birds/mammals	aluminium ammonium sulphate
	Caterpillars	Bacillus thuringiensis kurstaki (*off-label*), lambda-cyhalothrin, pyrethrins
	Cutworms	Bacillus thuringiensis kurstaki (*off-label*)
	Slugs/snails	ferric phosphate
	Spider mites	pyrethrins
	Thrips	fatty acids (*off-label*), lambda-cyhalothrin (*off-label*), pyrethrins
	Weevils	cypermethrin, deltamethrin, esfenvalerate, lambda-cyhalothrin
Weeds	Broad-leaved weeds	S-metolachlor (*off-label*), bentazone, clomazone, clomazone (*off-label*), clomazone + pendimethalin, glyphosate, imazamox + pendimethalin, imazamox + pendimethalin (*off-label*), isoxaben (*off-label*), pendimethalin (*off-label*), propyzamide, prosulfocarb (*off-label*)
	Crops as weeds	clethodim (*off-label*), cycloxydim, cycloxydim (*off-label*), fluazifop-P-butyl, glyphosate, pendimethalin (*off-label*), propyzamide, quizalofop-P-ethyl, quizalofop-P-tefuryl
	Grass weeds	S-metolachlor (*off-label*), clethodim (*off-label*), cycloxydim, cycloxydim (*off-label*), fluazifop-P-butyl, glyphosate, pendimethalin (*off-label*), propaquizafop, propyzamide, prosulfocarb (*off-label*), quizalofop-P-ethyl, quizalofop-P-tefuryl
	Weeds, miscellaneous	S-metolachlor (*off-label*), glyphosate, pendimethalin (*off-label*)

Legumes - Forage legumes, general

Weeds	Broad-leaved weeds	pendimethalin (*off-label*)
	Grass weeds	pendimethalin (*off-label*)

Legumes - Lupins

Crop control	Desiccation	glyphosate (*off-label*)
Diseases	Ascochyta	metconazole (*qualified minor use*)
	Botrytis	Bacillus subtilis (Strain QST 713) (*off-label*), cyprodinil + fludioxonil (*off-label*), metconazole (*qualified minor use*)
	Powdery mildew	physical pest control
	Rust	azoxystrobin, metconazole (*qualified minor use*)
Pests	Aphids	cypermethrin, physical pest control
	Mealybugs	physical pest control
	Pests, miscellaneous	cypermethrin, lambda-cyhalothrin (*off-label*)

	Scale insects	physical pest control
	Slugs/snails	ferric phosphate
	Spider mites	physical pest control
	Suckers	physical pest control
	Whiteflies	physical pest control
Weeds	Broad-leaved weeds	clomazone (*off-label*), pendimethalin (*off-label*)
	Crops as weeds	fluazifop-P-butyl
	Grass weeds	fluazifop-P-butyl, pendimethalin (*off-label*), propaquizafop (*off-label*)
	Weeds, miscellaneous	glyphosate (*off-label*)

Legumes - Peas

Crop control	Desiccation	glyphosate
Diseases	Ascochyta	azoxystrobin, azoxystrobin (*useful control*), benzovindiflupyr + prothioconazole (*useful control*), boscalid + pyraclostrobin (*moderate control only*), cyprodinil + fludioxonil, difenoconazole + fluxapyroxad (*moderate control*), metconazole (*reduction*)
	Botrytis	Bacillus subtilis (Strain QST 713) (*off-label*), azoxystrobin (*some control*), cyprodinil + fludioxonil (*moderate control*), difenoconazole + fluxapyroxad (*off-label*), metconazole (*reduction*)
	Disease control	benzovindiflupyr + prothioconazole
	Downy mildew	azoxystrobin (*reduction*), mandipropamid
	Leaf spot	difenoconazole + fluxapyroxad (*off-label*)
	Mycosphaerella	azoxystrobin (*some control*), cyprodinil + fludioxonil, metconazole (*reduction*)
	Phytophthora	difenoconazole + fluxapyroxad (*off-label*)
	Powdery mildew	physical pest control, sulphur (*off-label*)
	Rust	benzovindiflupyr + prothioconazole (*moderate control*), metconazole
	Sclerotinia	cyprodinil + fludioxonil
Pests	Aphids	cypermethrin, deltamethrin (*off-label*), esfenvalerate, fatty acids (*off-label*), flonicamid, flonicamid (*off-label*), lambda-cyhalothrin, physical pest control, pirimicarb, pirimicarb (*off-label*), pyrethrins
	Beetles	lambda-cyhalothrin
	Birds/mammals	aluminium ammonium sulphate
	Caterpillars	Bacillus thuringiensis kurstaki, Bacillus thuringiensis kurstaki (*off-label*), cypermethrin, deltamethrin, deltamethrin (*off-label*), lambda-cyhalothrin, pyrethrins
	Cutworms	Bacillus thuringiensis kurstaki (*off-label*)
	Mealybugs	physical pest control
	Midges	deltamethrin, deltamethrin (*off-label*), lambda-cyhalothrin
	Pests, miscellaneous	cyantraniliprole, cypermethrin
	Scale insects	physical pest control
	Slugs/snails	ferric phosphate
	Spider mites	physical pest control, pyrethrins

	Suckers	physical pest control
	Thrips	deltamethrin (*off-label*), fatty acids (*off-label*), pyrethrins
	Weevils	cypermethrin, deltamethrin, deltamethrin (*off-label*), esfenvalerate, lambda-cyhalothrin
	Whiteflies	physical pest control
Weeds	Broad-leaved weeds	MCPB, MCPB (*off-label*), bentazone, bentazone (*off-label*), clomazone, clomazone (*off-label*), clomazone + pendimethalin, glyphosate, imazamox + pendimethalin, pendimethalin, pendimethalin (*off-label*), prosulfocarb (*off-label*)
	Crops as weeds	clethodim (*off-label*), cycloxydim, cycloxydim (*off-label*), fluazifop-P-butyl, glyphosate, pendimethalin, quizalofop-P-ethyl, quizalofop-P-tefuryl
	Grass weeds	S-metolachlor (*off-label*), clethodim (*off-label*), cycloxydim, cycloxydim (*off-label*), fluazifop-P-butyl, glyphosate, pendimethalin, pendimethalin (*off-label*), propaquizafop, quizalofop-P-ethyl, quizalofop-P-tefuryl
	Weeds, miscellaneous	bentazone (*off-label*), glyphosate, glyphosate (*off-label*)

Miscellaneous arable - Industrial crops

Crop control	Desiccation	glyphosate (*off-label*)
Diseases	Powdery mildew	physical pest control
Pests	Aphids	lambda-cyhalothrin (*off-label*), physical pest control
	Mealybugs	physical pest control
	Pests, miscellaneous	lambda-cyhalothrin (*off-label*)
	Scale insects	physical pest control
	Spider mites	physical pest control
	Suckers	physical pest control
	Whiteflies	physical pest control
Weeds	Broad-leaved weeds	2,4-D (*off-label*), florasulam + pyroxsulam (*off-label*), fluroxypyr (*off-label*), isoxaben (*off-label*), metsulfuron-methyl + thifensulfuron-methyl (*off-label*), metsulfuron-methyl + tribenuron-methyl (*off-label*), pendimethalin (*off-label*), prosulfocarb (*off-label*)
	Crops as weeds	florasulam + pyroxsulam (*off-label - from seed*), florasulam + pyroxsulam (*off-label*), fluroxypyr (*off-label*)
	Grass weeds	florasulam + pyroxsulam (*off-label*), prosulfocarb (*off-label*), tri-allate (*off-label*)
	Weeds, miscellaneous	clopyralid (*off-label*), glyphosate (*off-label*)

Miscellaneous arable - Miscellaneous arable crops

Crop control	Desiccation	glyphosate (*off-label*)
	Miscellaneous non-selective situations	benzoic acid
Diseases	Alternaria	Bacillus amyloliquefaciens D747
	Anthracnose	Bacillus amyloliquefaciens D747 (*off-label*)
	Bacterial blight	Bacillus amyloliquefaciens D747 (*off-label*)

43

	Botrytis	Bacillus amyloliquefaciens D747 (*off-label*), Bacillus subtilis (Strain QST 713) (*off-label*), azoxystrobin (*off-label*)
	Cercospora leaf spot	Bacillus amyloliquefaciens D747 (*off-label*)
	Didymella stem rot	Gliocladium catenulatum (*moderate control*)
	Disease control	boscalid (*off-label*), boscalid + pyraclostrobin (*off-label*), potassium bicarbonate (commodity substance)
	Downy mildew	Bacillus amyloliquefaciens D747 (*off-label*), azoxystrobin (*off-label*), boscalid + pyraclostrobin (*off-label*), cyazofamid (*off-label*)
	Fusarium	Bacillus amyloliquefaciens D747, Bacillus amyloliquefaciens D747 (*off-label*), Gliocladium catenulatum (*moderate control*)
	Leaf spot	difenoconazole (*off-label*)
	Phytophthora	Gliocladium catenulatum (*moderate control*), cyazofamid (*off-label*), difenoconazole (*off-label*)
	Powdery mildew	Bacillus amyloliquefaciens D747, Bacillus amyloliquefaciens D747 (*off-label*), azoxystrobin (*off-label*), cyflufenamid (*off-label*)
	Pythium	Bacillus amyloliquefaciens D747, Bacillus amyloliquefaciens D747 (*off-label*), Gliocladium catenulatum (*moderate control*)
	Rhizoctonia	Bacillus amyloliquefaciens D747, Bacillus amyloliquefaciens D747 (*off-label*), Gliocladium catenulatum (*moderate control*)
	Rust	difenoconazole (*off-label*)
	Sclerotinia	Bacillus amyloliquefaciens D747 (*off-label*)
	Stemphylium	Bacillus amyloliquefaciens D747 (*off-label*)
Pests	Aphids	fatty acids (*off-label*), flonicamid (*off-label*), lambda-cyhalothrin (*off-label*), maltodextrin, maltodextrin (*off-label*)
	Caterpillars	Bacillus thuringiensis kurstaki, Bacillus thuringiensis kurstaki (*off-label*)
	Pests, miscellaneous	lambda-cyhalothrin (*off-label*)
	Slugs/snails	ferric phosphate
	Spider mites	maltodextrin
	Thrips	fatty acids (*off-label*)
	Whiteflies	maltodextrin, maltodextrin (*off-label*)
Plant growth regulation	Growth control	chlormequat (*off-label*), trinexapac-ethyl (*off-label*)
Weeds	Broad-leaved weeds	amidosulfuron (*off-label*), bentazone (*off-label*), bifenox (*off-label*), carfentrazone-ethyl (*before planting*), clomazone (*off-label*), clopyralid (*off-label*), clopyralid + picloram (*off-label*), clopyralid + triclopyr (*off-label*), dicamba + MCPA + mecoprop-P (*off-label*), dicamba + mecoprop-P (*off-label*), diflufenican (*off-label*), diflufenican + flufenacet (*off-label*), dimethenamid-p + metazachlor (*off-label*), dimethenamid-p + pendimethalin (*off-label*), ethofumesate (*off-label*), flufenacet + pendimethalin (*off-label*), fluroxypyr (*off-label*), isoxaben (*off-label*), lenacil (*off-label*), metamitron (*off-label*), metazachlor (*off-label*), metazachlor + quinmerac (*off-label*), metsulfuron-methyl (*off-label*), napropamide (*off-label*), nicosulfuron (*off-label*), pelargonic acid (*off-*

		label), pendimethalin (*off-label*), pendimethalin + picolinafen (*off-label*), propyzamide (*off-label*), prosulfocarb (*off-label*), prosulfuron (*off-label*), thifensulfuron-methyl + tribenuron-methyl (*off-label*)
	Crops as weeds	carfentrazone-ethyl (*before planting*), clethodim (*off-label*), cycloxydim (*off-label*), diflufenican (*off-label*), fluroxypyr (*off-label*), nicosulfuron (*off-label*), pendimethalin (*off-label*), quizalofop-P-ethyl (*off-label*)
	Grass weeds	clethodim (*off-label*), cycloxydim (*off-label*), diflufenican (*off-label*), diflufenican + flufenacet (*off-label*), dimethenamid-p + metazachlor (*off-label*), dimethenamid-p + pendimethalin (*off-label*), ethofumesate (*off-label*), fluazifop-P-butyl (*off-label*), flufenacet + pendimethalin (*off-label*), metamitron (*off-label*), metazachlor (*off-label*), metazachlor + quinmerac (*off-label*), napropamide (*off-label*), nicosulfuron (*off-label*), pelargonic acid (*off-label*), pendimethalin (*off-label*), pendimethalin + picolinafen (*off-label*), propaquizafop (*off-label*), propyzamide (*off-label*), prosulfocarb (*off-label*), quizalofop-P-ethyl (*off-label*), tri-allate (*off-label*)
	Mosses	pelargonic acid (*off-label*)
	Weeds, miscellaneous	2,4-D + glyphosate, carfentrazone-ethyl, clopyralid (*off-label*), diflufenican (*off-label*), ethofumesate (*off-label*), glyphosate, glyphosate (*before planting*), glyphosate (*off-label*), glyphosate (*pre-sowing/ planting*), napropamide (*off-label*), nicosulfuron (*off-label*), pelargonic acid (*off-label*)

Miscellaneous arable - Miscellaneous arable situations

Weeds	Broad-leaved weeds	citronella oil
	Crops as weeds	glyphosate
	Grass weeds	glyphosate
	Weeds, miscellaneous	cycloxydim, glyphosate, metsulfuron-methyl, thifensulfuron-methyl

Miscellaneous field vegetables - All vegetables

Diseases	Anthracnose	Bacillus amyloliquefaciens D747 (*off-label*)
	Bacterial blight	Bacillus amyloliquefaciens D747 (*off-label*)
	Botrytis	Bacillus amyloliquefaciens D747 (*off-label*), Bacillus subtilis (Strain QST 713) (*off-label*)
	Cercospora leaf spot	Bacillus amyloliquefaciens D747 (*off-label*)
	Downy mildew	Bacillus amyloliquefaciens D747 (*off-label*)
	Fusarium	Bacillus amyloliquefaciens D747 (*off-label*)
	Powdery mildew	Bacillus amyloliquefaciens D747 (*off-label*)
	Pythium	Bacillus amyloliquefaciens D747 (*off-label*)
	Rhizoctonia	Bacillus amyloliquefaciens D747 (*off-label*)
	Sclerotinia	Bacillus amyloliquefaciens D747 (*off-label*)
	Stemphylium	Bacillus amyloliquefaciens D747 (*off-label*)
Weeds	Broad-leaved weeds	clomazone (*off-label*)

Oilseed crops - Linseed/flax

Crop control	Desiccation	glyphosate
Diseases	Alternaria	difenoconazole (*off-label*)

	Botrytis	tebuconazole (*reduction*)
	Damping off	Bacillus amyloliquefaciens strain MBI600 (*off-label*)
	Disease control	benzovindiflupyr + prothioconazole, boscalid (*off-label*), metconazole (*off-label*), prothioconazole (*off-label*), tebuconazole
	Foliar diseases	difenoconazole (*off-label*), difenoconazole + paclobutrazol (*off-label*), prothioconazole (*off-label*)
	Mycosphaerella	benzovindiflupyr + prothioconazole (*moderate control*)
	Powdery mildew	benzovindiflupyr + prothioconazole (*moderate control*), difenoconazole (*off-label*), difenoconazole + paclobutrazol (*off-label*), tebuconazole
	Pythium	Bacillus amyloliquefaciens strain MBI600 (*off-label*)
	Rhizoctonia	Bacillus amyloliquefaciens strain MBI600 (*off-label*)
	Rust	tebuconazole
	Sclerotinia	difenoconazole (*off-label*)
	Septoria	difenoconazole + paclobutrazol (*off-label*), prothioconazole (*off-label*)
Pests	Aphids	cypermethrin, lambda-cyhalothrin (*off-label*), tau-fluvalinate (*off-label*)
	Beetles	lambda-cyhalothrin (*off-label*)
	Pests, miscellaneous	cypermethrin, lambda-cyhalothrin (*off-label*)
	Slugs/snails	ferric phosphate
Weeds	Broad-leaved weeds	amidosulfuron, bentazone, bifenox (*off-label*), clopyralid, glyphosate, mesotrione (*off-label*), metazachlor (*off-label*), metsulfuron-methyl, napropamide (*off-label*), prosulfocarb (*off-label*)
	Crops as weeds	clethodim (*off-label*), cycloxydim, fluazifop-P-butyl, glyphosate, mesotrione (*off-label*), metsulfuron-methyl, quizalofop-P-ethyl, quizalofop-P-tefuryl
	Grass weeds	clethodim (*off-label*), cycloxydim, fluazifop-P-butyl, glyphosate, metazachlor (*off-label*), napropamide (*off-label*), propaquizafop, prosulfocarb (*off-label*), quizalofop-P-ethyl, quizalofop-P-tefuryl, tri-allate (*off-label*)
	Weed grasses	mesotrione (*off-label*)
	Weeds, miscellaneous	glyphosate, mesotrione (*off-label*), metsulfuron-methyl, napropamide (*off-label*)

Oilseed crops - Miscellaneous oilseeds

Crop control	Desiccation	glyphosate (*off-label*)
Diseases	Alternaria	azoxystrobin (*off-label*), boscalid + pyraclostrobin (*off-label*), difenoconazole (*off-label*)
	Botrytis	boscalid (*off-label*), boscalid + pyraclostrobin (*off-label*)
	Disease control	boscalid (*off-label*), tebuconazole (*off-label*)
	Downy mildew	mancozeb + metalaxyl-M (*off-label*)
	Fire	boscalid + pyraclostrobin (*off-label*)
	Powdery mildew	difenoconazole (*off-label*)
	Rust	tebuconazole (*off-label*)
	Sclerotinia	azoxystrobin (*off-label*), boscalid (*off-label*), difenoconazole (*off-label*)

Pests	Aphids	deltamethrin (*off-label*), lambda-cyhalothrin (*off-label*), tau-fluvalinate (*off-label*)
	Beetles	lambda-cyhalothrin (*off-label*)
	Caterpillars	deltamethrin (*off-label*)
	Pests, miscellaneous	deltamethrin (*off-label*), lambda-cyhalothrin (*off-label*)
Weeds	Broad-leaved weeds	bifenox (*off-label*), clomazone (*off-label*), clopyralid + picloram (*off-label*), fluroxypyr (*off-label*), metazachlor (*off-label*), napropamide (*off-label*), pendimethalin (*off-label*), prosulfocarb (*off-label*)
	Crops as weeds	clethodim (*off-label*), fluazifop-P-butyl, fluroxypyr (*off-label*), propaquizafop (*off-label*)
	Grass weeds	clethodim (*off-label*), fluazifop-P-butyl, metazachlor (*off-label*), napropamide (*off-label*), propaquizafop (*off-label*), prosulfocarb (*off-label*)
	Weeds, miscellaneous	glyphosate (*off-label*), napropamide (*off-label*)

Oilseed crops - Oilseed rape

Crop control	Desiccation	glyphosate
Diseases	Alternaria	azoxystrobin, boscalid, boscalid + dimoxystrobin (*moderate control*), boscalid + pyraclostrobin (*moderate control*), difenoconazole, metconazole, tebuconazole
	Black scurf and stem canker	Bacillus amyloliquefaciens strain MBI600 (*useful reduction*), bixafen + prothioconazole + tebuconazole, boscalid + dimoxystrobin (*moderate control*), boscalid + pyraclostrobin, difenoconazole, fluopyram + prothioconazole, mepiquat chloride + prohexadione-calcium + pyraclostrobin (*moderate control*), prothioconazole, prothioconazole + tebuconazole, tebuconazole
	Disease control	azoxystrobin + difenoconazole, azoxystrobin + tebuconazole, bixafen + prothioconazole, boscalid + dimoxystrobin, difenoconazole + paclobutrazol, fluopyram + prothioconazole
	Light leaf spot	bixafen + prothioconazole + tebuconazole, boscalid + dimoxystrobin (*reduction*), boscalid + pyraclostrobin (*moderate control*), difenoconazole, difenoconazole + paclobutrazol, fluopyram + prothioconazole, mepiquat chloride + prohexadione-calcium + pyraclostrobin (*moderate control*), metconazole, metconazole (*reduction*), prothioconazole, prothioconazole + tebuconazole, prothioconazole + tebuconazole (*moderate control only*), tebuconazole
	Phoma leaf spot	boscalid + pyraclostrobin, mepiquat chloride + prohexadione-calcium + pyraclostrobin, prothioconazole + tebuconazole, tebuconazole
	Powdery mildew	fluopyram + prothioconazole
	Ring spot	tebuconazole, tebuconazole (*reduction*)
	Sclerotinia	azoxystrobin, azoxystrobin + isopyrazam, azoxystrobin + tebuconazole (*moderate control*), bixafen + prothioconazole + tebuconazole, boscalid, boscalid + dimoxystrobin, boscalid + metconazole, boscalid + pyraclostrobin, fluopyram + prothioconazole, fluoxastrobin + tebuconazole,

		prothioconazole, prothioconazole + tebuconazole, tebuconazole
	Stem canker	bixafen + prothioconazole + tebuconazole, boscalid + dimoxystrobin (*moderate control*), difenoconazole + paclobutrazol, fluopyram + prothioconazole, metconazole (*reduction*), prothioconazole, prothioconazole + tebuconazole, tebuconazole
Pests	Aphids	acetamiprid, deltamethrin, flonicamid, lambda-cyhalothrin, tau-fluvalinate
	Beetles	Bacillus amyloliquefaciens strain MBI600 (*stimulation of plants own defenses*), cypermethrin, deltamethrin, indoxacarb, lambda-cyhalothrin, tau-fluvalinate
	Birds/mammals	aluminium ammonium sulphate
	Caterpillars	lambda-cyhalothrin
	Midges	cypermethrin, lambda-cyhalothrin
	Slugs/snails	ferric phosphate
	Weevils	cypermethrin, deltamethrin, lambda-cyhalothrin
Plant growth regulation	Growth control	mepiquat chloride + metconazole, mepiquat chloride + prohexadione-calcium + pyraclostrobin, metconazole, tebuconazole, trinexapac-ethyl
Weeds	Broad-leaved weeds	aminopyralid + metazachlor + picloram, aminopyralid + propyzamide, bifenox (*off-label*), clomazone, clopyralid, clopyralid + halauxifen-methyl, clopyralid + picloram, clopyralid + picloram (*off-label*), dimethachlor, dimethenamid-p + metazachlor, dimethenamid-p + metazachlor + quinmerac, dimethenamid-p + quinmerac, glyphosate, halauxifen-methyl + picloram, imazamox + quinmerac, imazamox + quinmerac (*cut-leaved*), metazachlor, metazachlor + quinmerac, napropamide, propyzamide
	Crops as weeds	clethodim, cycloxydim, fluazifop-P-butyl, glyphosate, imazamox + quinmerac, propyzamide, quizalofop-P-ethyl, quizalofop-P-tefuryl
	Grass weeds	aminopyralid + propyzamide, clethodim, cycloxydim, dimethachlor, dimethenamid-p + metazachlor + quinmerac, fluazifop-P-butyl, glyphosate, metazachlor, metazachlor + quinmerac, napropamide, propaquizafop, propyzamide, quizalofop-P-ethyl, quizalofop-P-tefuryl
	Weeds, miscellaneous	glyphosate

Oilseed crops - Soya

Diseases	Sclerotinia	azoxystrobin (*off-label*)
Pests	Aphids	pyrethrins
	Birds/mammals	aluminium ammonium sulphate
	Caterpillars	pyrethrins
	Flies	lambda-cyhalothrin (*off-label*)
	Slugs/snails	ferric phosphate
	Spider mites	pyrethrins
	Thrips	pyrethrins

Weeds	Broad-leaved weeds	bentazone, clomazone (*off-label*), flufenacet + metribuzin (*off-label*), imazamox + pendimethalin (*off-label*), pendimethalin (*off-label*), thifensulfuron-methyl (*off-label*)
	Crops as weeds	cycloxydim (*off-label*)
	Grass weeds	cycloxydim (*off-label*), flufenacet + metribuzin (*off-label*), pendimethalin (*off-label*)
	Weeds, miscellaneous	imazamox + pendimethalin (*off-label*)

Oilseed crops - Sunflowers

Diseases	Powdery mildew	physical pest control
Pests	Aphids	physical pest control
	Mealybugs	physical pest control
	Scale insects	physical pest control
	Slugs/snails	ferric phosphate
	Spider mites	physical pest control
	Suckers	physical pest control
	Whiteflies	physical pest control
Weeds	Broad-leaved weeds	aclonifen (*off-label*), pendimethalin
	Crops as weeds	aclonifen (*off-label*), fluazifop-P-butyl, pendimethalin
	Grass weeds	clethodim (*off-label*), fluazifop-P-butyl, pendimethalin

Root and tuber crops - Beet crops

Diseases	Alternaria	Bacillus amyloliquefaciens D747, cyprodinil + fludioxonil (*off-label*), difenoconazole + fluxapyroxad (*off-label*), fludioxonil (*off-label*), isopyrazam (*off-label*), prothioconazole (*off-label*)
	Anthracnose	Bacillus amyloliquefaciens D747 (*off-label*), fludioxonil (*moderate control*)
	Bacterial blight	Bacillus amyloliquefaciens D747 (*off-label*)
	Black leg	hymexazol (*seed treatment*)
	Black rot	azoxystrobin (*off-label*)
	Botrytis	Bacillus amyloliquefaciens D747 (*off-label*), Bacillus amyloliquefaciens strain MBI600 (*off-label*), Bacillus subtilis (Strain QST 713) (*off-label*), cerevisane (*off-label*), cyprodinil + fludioxonil (*off-label*)
	Cercospora leaf spot	Bacillus amyloliquefaciens D747 (*off-label*), fluopyram + prothioconazole, isopyrazam (*off-label*), prothioconazole (*off-label*)
	Damping off	Bacillus amyloliquefaciens D747 (*off-label*), Bacillus amyloliquefaciens strain MBI600 (*off-label*), Bacillus subtilis (Strain QST 713) (*off-label*)
	Disease control	azoxystrobin + difenoconazole, boscalid + pyraclostrobin (*off-label*), mandipropamid
	Downy mildew	Bacillus amyloliquefaciens D747 (*off-label*), boscalid + pyraclostrobin (*off-label*), cerevisane (*off-label*), dimethomorph (*off-label*), fluopicolide + propamocarb hydrochloride (*off-label*), fosetyl-aluminium + propamocarb hydrochloride (*off-label*), mandipropamid, metalaxyl-M (*off-label*)

	Fusarium	Bacillus amyloliquefaciens D747, Bacillus amyloliquefaciens D747 (*off-label*)
	Powdery mildew	Bacillus amyloliquefaciens D747, Bacillus amyloliquefaciens D747 (*off-label*), cerevisane (*off-label*), difenoconazole + fluxapyroxad (*off-label*), fluopyram + prothioconazole, flutriafol, isopyrazam (*off-label*), physical pest control, prothioconazole (*off-label*), sulphur
	Pythium	Bacillus amyloliquefaciens D747, Bacillus amyloliquefaciens D747 (*off-label*), Bacillus amyloliquefaciens strain MBI600 (*off-label*), metalaxyl-M, metalaxyl-M (*off-label*)
	Ramularia leaf spots	fluopyram + prothioconazole, isopyrazam (*off-label*)
	Rhizoctonia	Bacillus amyloliquefaciens D747, Bacillus amyloliquefaciens D747 (*off-label*), Bacillus amyloliquefaciens strain MBI600 (*off-label*), cyprodinil + fludioxonil (*off-label*), difenoconazole + fluxapyroxad (*off-label*)
	Root malformation disorder	azoxystrobin (*off-label*), metalaxyl-M (*off-label*)
	Rust	azoxystrobin, fluopyram + prothioconazole, flutriafol, mancozeb
	Sclerotinia	Bacillus amyloliquefaciens D747 (*off-label*), Bacillus amyloliquefaciens strain MBI600 (*off-label*), cyprodinil + fludioxonil (*off-label*), difenoconazole + fluxapyroxad (*off-label*), isopyrazam (*off-label*)
	Septoria	isopyrazam (*off-label*)
	Stem canker	cyprodinil + fludioxonil (*off-label*), isopyrazam (*off-label*), prothioconazole (*off-label*)
	Stemphylium	Bacillus amyloliquefaciens D747 (*off-label*)
	White blister	azoxystrobin (*off-label*)
Pests	Aphids	Beauveria bassiana GHA (*off-label*), acetamiprid, acetamiprid (*off-label*), deltamethrin (*off-label*), fatty acids (*off-label*), flonicamid, flonicamid (*off-label*), lambda-cyhalothrin, maltodextrin (*off-label*), physical pest control, pirimicarb (*off-label*), pyrethrins
	Beetles	deltamethrin, deltamethrin (*off-label*), lambda-cyhalothrin, tefluthrin (*seed treatment*)
	Birds/mammals	aluminium ammonium sulphate
	Caterpillars	Bacillus thuringiensis aizawai GC-91, Bacillus thuringiensis kurstaki, Bacillus thuringiensis kurstaki (*off-label*), chlorantraniliprole (*off-label*), deltamethrin (*off-label*), lambda-cyhalothrin, lambda-cyhalothrin (*off-label*), maltodextrin (*off-label*), pyrethrins, spinosad (*off-label*)
	Cutworms	Bacillus thuringiensis kurstaki (*off-label*), cypermethrin, lambda-cyhalothrin, lambda-cyhalothrin (*off-label*)
	Flies	chlorantraniliprole (*off-label*), deltamethrin (*off-label*), lambda-cyhalothrin, maltodextrin (*off-label*)
	Free-living nematodes	garlic extract (*off-label*)
	Leaf miners	lambda-cyhalothrin
	Mealybugs	physical pest control
	Millipedes	tefluthrin (*seed treatment*)

	Pests, miscellaneous	lambda-cyhalothrin (*off-label*)
	Scale insects	physical pest control
	Slugs/snails	ferric phosphate
	Spider mites	Beauveria bassiana GHA (*off-label*), fatty acids (*off-label*), physical pest control, pyrethrins
	Springtails	tefluthrin (*seed treatment*)
	Suckers	physical pest control
	Symphylids	tefluthrin (*seed treatment*)
	Thrips	Beauveria bassiana GHA (*off-label*), deltamethrin (*off-label*), fatty acids (*off-label*), pyrethrins, spinosad (*off-label*)
	Weevils	lambda-cyhalothrin
	Whiteflies	Beauveria bassiana GHA (*off-label*), physical pest control
Plant growth regulation	Quality/yield control	sulphur
Weeds	Broad-leaved weeds	S-metolachlor (*off-label*), clomazone (*off-label*), clopyralid, clopyralid (*off-label*), dimethenamid-p + quinmerac, ethofumesate, ethofumesate + metamitron, ethofumesate + phenmedipham, foramsulfuron + thiencarbazone-methyl, glyphosate, lenacil, lenacil (*off-label*), lenacil + triflusulfuron-methyl, lenacil + triflusulfuron-methyl (*off-label*), metamitron, metamitron + quinmerac, pelargonic acid (*off-label*), phenmedipham, phenmedipham (*off-label*), propyzamide, triflusulfuron-methyl, triflusulfuron-methyl (*off-label*)
	Crops as weeds	clethodim, cycloxydim, cycloxydim (*off-label*), fluazifop-P-butyl, foramsulfuron + thiencarbazone-methyl, glyphosate, glyphosate (*wiper application*), lenacil + triflusulfuron-methyl, lenacil + triflusulfuron-methyl (*off-label*), propaquizafop (*off-label*), quizalofop-P-ethyl, quizalofop-P-tefuryl, triflusulfuron-methyl (*off-label*)
	Grass weeds	S-metolachlor (*off-label*), clethodim, clethodim (*off-label*), cycloxydim, cycloxydim (*off-label*), ethofumesate, ethofumesate + metamitron, ethofumesate + phenmedipham, fluazifop-P-butyl, glyphosate, lenacil, metamitron, metamitron + quinmerac, pelargonic acid (*off-label*), propaquizafop, propaquizafop (*off-label*), propyzamide, quizalofop-P-ethyl, quizalofop-P-tefuryl
	Mosses	pelargonic acid (*off-label*)
	Weeds, miscellaneous	clopyralid (*off-label*), glyphosate, glyphosate (*off-label*), lenacil (*off-label*), pelargonic acid (*off-label*)

Root and tuber crops - Carrots/parsnips

Diseases	Alternaria	azoxystrobin, azoxystrobin (*off-label*), azoxystrobin + difenoconazole, boscalid + pyraclostrobin (*moderate control*), cyprodinil + fludioxonil (*moderate control*), difenoconazole + fluxapyroxad (*off-label*), fludioxonil, fludioxonil (*off-label*), isopyrazam, isopyrazam (*off-label*), prothioconazole, tebuconazole, tebuconazole + trifloxystrobin

	Botrytis	Bacillus subtilis (Strain QST 713) (*off-label*), cyprodinil + fludioxonil (*moderate control*)
	Cavity spot	Bacillus amyloliquefaciens D747 (*off-label*), Bacillus subtilis (Strain QST 713) (*off-label*), metalaxyl-M, metalaxyl-M (*off-label - reduction*), metalaxyl-M (*off-label*), metalaxyl-M (*reduction only*)
	Cercospora leaf spot	isopyrazam (*off-label*)
	Damping off	Bacillus subtilis (Strain QST 713) (*off-label*)
	Disease control	mancozeb
	Fungus diseases	azoxystrobin + difenoconazole (*off-label*)
	Leaf spot	fludioxonil (*off-label*), tebuconazole + trifloxystrobin (*off-label*)
	Powdery mildew	azoxystrobin, azoxystrobin (*off-label*), azoxystrobin + difenoconazole, boscalid + pyraclostrobin, difenoconazole + fluxapyroxad (*off-label*), isopyrazam, isopyrazam (*off-label*), prothioconazole, tebuconazole, tebuconazole + trifloxystrobin, tebuconazole + trifloxystrobin (*off-label*)
	Pythium	Bacillus subtilis (Strain QST 713) (*off-label*)
	Ramularia leaf spots	isopyrazam (*off-label*)
	Rhizoctonia	difenoconazole + fluxapyroxad (*off-label*)
	Rust	azoxystrobin
	Sclerotinia	Bacillus amyloliquefaciens D747 (*off-label*), boscalid + pyraclostrobin (*moderate control*), cyprodinil + fludioxonil (*moderate control*), cyprodinil + fludioxonil (*off-label*), difenoconazole + fluxapyroxad, difenoconazole + fluxapyroxad (*off-label*), isopyrazam (*off-label*), prothioconazole, tebuconazole, tebuconazole + trifloxystrobin, tebuconazole + trifloxystrobin (*off-label*)
	Septoria	isopyrazam (*off-label*)
	Stem canker	azoxystrobin + difenoconazole (*off-label*), isopyrazam (*off-label*)
Pests	Aphids	acetamiprid (*off-label*), cypermethrin, fatty acids (*off-label*), flonicamid (*off-label*), maltodextrin (*off-label*), pyrethrins, spirotetramat
	Birds/mammals	aluminium ammonium sulphate
	Caterpillars	chlorantraniliprole, maltodextrin (*off-label*), pyrethrins
	Cutworms	Bacillus thuringiensis kurstaki (*off-label*), deltamethrin (*off-label*), lambda-cyhalothrin
	Flies	chlorantraniliprole (*off-label*), cypermethrin, deltamethrin (*off-label*), lambda-cyhalothrin, maltodextrin (*off-label*)
	Free-living nematodes	fluopyram, fluopyram (*off-label*), garlic extract
	Pests, miscellaneous	cyantraniliprole, cypermethrin
	Slugs/snails	ferric phosphate
	Spider mites	pyrethrins
	Thrips	fatty acids (*off-label*), pyrethrins
Plant growth regulation	Growth control	maleic hydrazide (*off-label*)
Weeds	Broad-leaved weeds	aclonifen (*off-label*), clomazone, clomazone (*off-label*), clomazone + pendimethalin, diflufenican (*off-label*), isoxaben (*off-label*), metamitron (*off-label*), metribuzin

	(*off-label*), pelargonic acid (*off-label*), pendimethalin, pendimethalin (*off-label*), prosulfocarb (*off-label*)
Crops as weeds	clethodim (*off-label*), cycloxydim, diflufenican (*off-label*), fluazifop-P-butyl, metribuzin (*off-label*), pendimethalin
Grass weeds	clethodim (*off-label*), cycloxydim, fluazifop-P-butyl, metamitron (*off-label*), metribuzin (*off-label*), pelargonic acid (*off-label*), pendimethalin, propaquizafop, prosulfocarb (*off-label*)
Mosses	pelargonic acid (*off-label*)
Weeds, miscellaneous	clomazone (*off-label*), glyphosate (*off-label*), pelargonic acid (*off-label*)

Root and tuber crops - Miscellaneous root crops

Diseases	Alternaria	difenoconazole + fluxapyroxad (*off-label*), fludioxonil (*off-label*), isopyrazam (*off-label*), tebuconazole (*off-label*), tebuconazole + trifloxystrobin (*off-label*)
	Bacterial canker	Bacillus amyloliquefaciens D747 (*off-label*)
	Botrytis	Bacillus amyloliquefaciens D747 (*off-label*), Bacillus subtilis (Strain QST 713) (*off-label*), azoxystrobin + difenoconazole (*off-label*), cyprodinil + fludioxonil (*moderate control*)
	Cercospora leaf spot	isopyrazam (*off-label*)
	Damping off	Bacillus subtilis (Strain QST 713) (*off-label*)
	Disease control	tebuconazole (*off-label*)
	Fireblight	Bacillus amyloliquefaciens D747 (*off-label*)
	Fungus diseases	Bacillus amyloliquefaciens D747 (*off-label*), azoxystrobin + difenoconazole (*off-label*)
	Leaf spot	boscalid + pyraclostrobin (*off-label*), fludioxonil (*off-label*), tebuconazole + trifloxystrobin (*off-label*)
	Mycosphaerella	tebuconazole + trifloxystrobin (*off-label*)
	Phytophthora	difenoconazole (*off-label*)
	Powdery mildew	Bacillus amyloliquefaciens D747 (*off-label*), boscalid + pyraclostrobin (*off-label*), difenoconazole + fluxapyroxad (*off-label*), isopyrazam (*off-label*), physical pest control, tebuconazole (*off-label*), tebuconazole + trifloxystrobin (*off-label*)
	Ramularia leaf spots	isopyrazam (*off-label*)
	Rhizoctonia	difenoconazole + fluxapyroxad (*off-label*)
	Rust	azoxystrobin, boscalid + pyraclostrobin (*off-label*), tebuconazole (*off-label*)
	Sclerotinia	Bacillus amyloliquefaciens D747 (*off-label*), azoxystrobin (*off-label*), azoxystrobin + difenoconazole (*off-label*), boscalid + pyraclostrobin (*off-label*), cyprodinil + fludioxonil (*off-label*), difenoconazole + fluxapyroxad (*off-label*), isopyrazam (*off-label*), tebuconazole + trifloxystrobin (*off-label*)
	Septoria	Bacillus amyloliquefaciens D747 (*off-label*), isopyrazam (*off-label*), tebuconazole (*off-label*), tebuconazole + trifloxystrobin (*off-label*)
	Stem canker	azoxystrobin + difenoconazole (*off-label*), isopyrazam (*off-label*)

Pests	Aphids	acetamiprid (*off-label*), cypermethrin, deltamethrin (*off-label*), fatty acids (*off-label*), flonicamid (*off-label*), maltodextrin (*off-label*), physical pest control, pirimicarb (*off-label*), pyrethrins
	Beetles	deltamethrin (*off-label*)
	Caterpillars	deltamethrin (*off-label*), maltodextrin (*off-label*), pyrethrins, spinosad (*off-label*)
	Cutworms	Bacillus thuringiensis kurstaki (*off-label*)
	Flies	chlorantraniliprole (*off-label*), cypermethrin, lambda-cyhalothrin (*off-label*), maltodextrin (*off-label*)
	Leaf miners	spinosad (*off-label*)
	Mealybugs	physical pest control
	Pests, miscellaneous	cyantraniliprole, cypermethrin, lambda-cyhalothrin (*off-label*)
	Scale insects	physical pest control
	Slugs/snails	ferric phosphate
	Spider mites	physical pest control, pyrethrins
	Suckers	physical pest control
	Thrips	deltamethrin (*off-label*), fatty acids (*off-label*), pyrethrins, spinosad (*off-label*)
	Weevils	deltamethrin (*off-label*)
	Whiteflies	physical pest control
Weeds	Broad-leaved weeds	aclonifen (*off-label*), clomazone (*off-label*), metribuzin (*off-label*), napropamide (*off-label*), pendimethalin (*off-label*), propyzamide (*off-label*), prosulfocarb (*off-label*), triflusulfuron-methyl (*off-label*)
	Crops as weeds	fluazifop-P-butyl, metribuzin (*off-label*)
	Grass weeds	clethodim (*off-label*), cycloxydim (*off-label*), fluazifop-P-butyl, metribuzin (*off-label*), napropamide (*off-label*), propyzamide (*off-label*), prosulfocarb (*off-label*)
	Weeds, miscellaneous	glyphosate (*off-label*), napropamide (*off-label*), prosulfocarb (*off-label*)

Root and tuber crops - Potatoes

Crop control	Desiccation	carfentrazone-ethyl, pyraflufen-ethyl
Diseases	Alternaria	difenoconazole + fluxapyroxad (*moderate control*), difenoconazole + mandipropamid, fluopyram + prothioconazole, mancozeb
	Bacterial blight	Bacillus subtilis (Strain QST 713) (*off-label*)
	Black dot	azoxystrobin, azoxystrobin (*reduction*), fludioxonil (*some reduction*)
	Black scurf and stem canker	azoxystrobin, azoxystrobin (*reduction*), fludioxonil, flutolanil (*tuber treatment*)
	Blight	ametoctradin + dimethomorph (*reduction*), amisulbrom, mandipropamid (*protection*), oxathiapiprolin
	Disease control	Bacillus subtilis (Strain QST 713), azoxystrobin, azoxystrobin + fluazinam, difenoconazole, fluxapyroxad, mancozeb
	Dry rot	imazalil, thiabendazole (*tuber treatment - post-harvest*)

	Gangrene	imazalil, thiabendazole (*tuber treatment - post-harvest*)
	Helminthosporium seedling rot	Bacillus subtilis (Strain QST 713) (*off-label*)
	Phytophthora	ametoctradin, ametoctradin + dimethomorph, amisulbrom, azoxystrobin + fluazinam, benthiavalicarb + oxathiapiprolin, benthiavalicarb-isopropyl, boscalid + pyraclostrobin, cyazofamid, cymoxanil, cymoxanil + fluazinam, cymoxanil + mancozeb, cymoxanil + mandipropamid, cymoxanil + propamocarb, cymoxanil + zoxamide, difenoconazole + mandipropamid, dimethomorph, dimethomorph + fluazinam, dimethomorph + propamocarb , fluazinam, fluopicolide + propamocarb hydrochloride, mancozeb, mancozeb + metalaxyl-M, mancozeb + metalaxyl-M (*reduction*), mandipropamid, oxathiapiprolin
	Powdery mildew	physical pest control
	Powdery scab	fluazinam (*off-label*)
	Rhizoctonia	Bacillus subtilis (Strain QST 713) (*off-label*), Pseudomonas SP (DSMZ 13134) (*reduction*)
	Scab	fludioxonil (*off-label*)
	Silver scurf	fludioxonil (*reduction*), imazalil, thiabendazole (*tuber treatment - post-harvest*)
	Skin spot	imazalil, thiabendazole (*tuber treatment - post-harvest*)
Pests	Aphids	acetamiprid, esfenvalerate, flonicamid, lambda-cyhalothrin, physical pest control, spirotetramat
	Beetles	lambda-cyhalothrin
	Caterpillars	lambda-cyhalothrin
	Cutworms	cypermethrin
	Cyst nematodes	fluopyram, fosthiazate, garlic extract
	Free-living nematodes	fluopyram, fosthiazate (*reduction*)
	Mealybugs	physical pest control
	Scale insects	physical pest control
	Slugs/snails	ferric phosphate
	Spider mites	physical pest control
	Suckers	physical pest control
	Weevils	lambda-cyhalothrin
	Whiteflies	physical pest control
	Wireworms	fosthiazate (*reduction*)
Plant growth regulation	Growth control	1,4-dimethylnaphthalene, ethylene, maleic hydrazide, orange oil, spearmint oil
Weeds	Broad-leaved weeds	aclonifen, bentazone, carfentrazone-ethyl, clomazone, clomazone + metribuzin, clomazone + pendimethalin, flufenacet + metribuzin, glyphosate, metobromuron, metribuzin, pendimethalin, prosulfocarb, pyraflufen-ethyl, rimsulfuron
	Crops as weeds	carfentrazone-ethyl, cycloxydim, fluazifop-P-butyl, glyphosate, metribuzin, pendimethalin, quizalofop-P-ethyl, quizalofop-P-tefuryl, rimsulfuron
	Grass weeds	clomazone + metribuzin, cycloxydim, fluazifop-P-butyl, flufenacet + metribuzin, glyphosate, metobromuron, metribuzin, pendimethalin,

propaquizafop, prosulfocarb, quizalofop-P-ethyl, quizalofop-P-tefuryl

	Weeds, miscellaneous	carfentrazone-ethyl, glyphosate

Stem and bulb vegetables - Asparagus

Diseases	Ascochyta	azoxystrobin + difenoconazole (*off-label*)
	Botrytis	Bacillus subtilis (Strain QST 713) (*off-label*), boscalid + pyraclostrobin (*off-label*), cyprodinil + fludioxonil (*off-label*)
	Downy mildew	metalaxyl-M (*off-label*)
	Powdery mildew	physical pest control
	Rust	azoxystrobin, azoxystrobin + difenoconazole (*off-label*), boscalid + pyraclostrobin (*off-label*), difenoconazole (*off-label*)
	Stemphylium	azoxystrobin
Pests	Aphids	Beauveria bassiana GHA (*off-label*), acetamiprid (*off-label*), cypermethrin, fatty acids (*off-label*), lambda-cyhalothrin (*off-label*), maltodextrin (*off-label*), physical pest control, pyrethrins
	Beetles	acetamiprid (*off-label*), cyantraniliprole (*off-label*), cypermethrin (*off-label*), spinosad (*off-label*)
	Caterpillars	pyrethrins
	Flies	maltodextrin (*off-label*)
	Mealybugs	physical pest control
	Pests, miscellaneous	cypermethrin
	Scale insects	physical pest control
	Slugs/snails	ferric phosphate
	Spider mites	physical pest control, pyrethrins
	Suckers	physical pest control
	Thrips	Beauveria bassiana GHA (*off-label*), fatty acids (*off-label*), maltodextrin (*off-label*), pyrethrins, spinosad (*off-label*)
	Whiteflies	physical pest control
Weeds	Broad-leaved weeds	carfentrazone-ethyl (*off-label*), clomazone (*off-label*), flufenacet + metribuzin (*off-label*), isoxaben, metribuzin (*off-label - from seed*), metribuzin (*off-label*), pelargonic acid (*off-label*), pendimethalin (*off-label*), pyridate (*off-label*)
	Crops as weeds	carfentrazone-ethyl (*off-label*), fluazifop-P-butyl, metribuzin (*off-label*)
	Grass weeds	fluazifop-P-butyl, flufenacet + metribuzin (*off-label*), metribuzin (*off-label*), pelargonic acid (*off-label*)
	Weeds, miscellaneous	carfentrazone-ethyl (*off-label*), clomazone (*off-label*), glyphosate, glyphosate (*off-label*), metribuzin (*off-label*), pelargonic acid (*off-label*)

Stem and bulb vegetables - Celery/chicory

Diseases	Alternaria	Bacillus amyloliquefaciens D747, cyprodinil + fludioxonil (*off-label*)
	Bacterial canker	Bacillus amyloliquefaciens D747 (*off-label*)
	Botrytis	Bacillus amyloliquefaciens D747 (*off-label*), Bacillus subtilis (Strain QST 713) (*off-label*), azoxystrobin (*off-*

		label), azoxystrobin + difenoconazole (*off-label*), cyprodinil + fludioxonil (*off-label*), difenoconazole + fluxapyroxad (*off-label*)
	Damping off	Bacillus amyloliquefaciens strain MBI600 (*off-label*), Bacillus subtilis (Strain QST 713) (*off-label*)
	Disease control	difenoconazole (*off-label*), mandipropamid
	Downy mildew	azoxystrobin, mandipropamid
	Fireblight	Bacillus amyloliquefaciens D747 (*off-label*)
	Fungus diseases	Bacillus amyloliquefaciens D747 (*off-label*)
	Fusarium	Bacillus amyloliquefaciens D747
	Leaf spot	azoxystrobin (*off-label*), cyprodinil + fludioxonil (*off-label*), difenoconazole (*off-label*)
	Phytophthora	difenoconazole (*off-label*), fosetyl-aluminium + propamocarb hydrochloride (*off-label*)
	Powdery mildew	Ampelomyces quisqualis (Strain AQ10) (*off-label*), Bacillus amyloliquefaciens D747, Bacillus amyloliquefaciens D747 (*off-label*), difenoconazole + fluxapyroxad (*off-label*)
	Pythium	Bacillus amyloliquefaciens D747, Bacillus amyloliquefaciens strain MBI600 (*off-label*)
	Rhizoctonia	Bacillus amyloliquefaciens D747, Bacillus amyloliquefaciens strain MBI600 (*off-label*), azoxystrobin (*off-label*), difenoconazole + fluxapyroxad (*off-label*)
	Sclerotinia	Bacillus amyloliquefaciens D747 (*off-label*), azoxystrobin (*off-label*), azoxystrobin + difenoconazole (*off-label*), cyprodinil + fludioxonil (*off-label*), difenoconazole + fluxapyroxad (*off-label*)
	Septoria	tebuconazole + trifloxystrobin (*off-label*)
Pests	Aphids	Beauveria bassiana GHA (*off-label*), deltamethrin (*off-label*), fatty acids (*off-label*), maltodextrin (*off-label*), pirimicarb (*off-label*), spirotetramat (*off-label*)
	Beetles	deltamethrin (*off-label*)
	Capsid bugs	deltamethrin (*off-label*)
	Caterpillars	Bacillus thuringiensis kurstaki, Bacillus thuringiensis kurstaki (*off-label*), chlorantraniliprole (*off-label*), deltamethrin (*off-label*), lambda-cyhalothrin (*off-label*), spinosad (*off-label*)
	Cutworms	deltamethrin (*off-label*), lambda-cyhalothrin (*off-label*)
	Flies	chlorantraniliprole (*off-label*), lambda-cyhalothrin (*off-label*), maltodextrin (*off-label*), spinosad (*off-label*)
	Leaf miners	spinosad (*off-label*)
	Pests, miscellaneous	lambda-cyhalothrin (*off-label*)
	Slugs/snails	ferric phosphate
	Spider mites	Beauveria bassiana GHA (*off-label*), fatty acids (*off-label*)
	Thrips	Beauveria bassiana GHA (*off-label*), fatty acids (*off-label*), maltodextrin (*off-label*), spinosad (*off-label*)
	Whiteflies	Beauveria bassiana GHA (*off-label*)
Weeds	Broad-leaved weeds	S-metolachlor (*off-label*), aclonifen (*off-label*), clomazone (*off-label*), isoxaben (*off-label*), pelargonic acid (*off-label*), pendimethalin (*off-label*), propyzamide (*off-label*), prosulfocarb (*off-label*)

	Crops as weeds	cycloxydim (*off-label*), fluazifop-P-butyl
	Grass weeds	S-metolachlor (*off-label*), clethodim (*off-label*), cycloxydim (*off-label*), fluazifop-P-butyl, pelargonic acid (*off-label*), pendimethalin (*off-label*), propyzamide (*off-label*), prosulfocarb (*off-label*)
	Mosses	pelargonic acid (*off-label*)

Stem and bulb vegetables - Globe artichokes/cardoons

Diseases	Botrytis	Bacillus subtilis (Strain QST 713) (*off-label*)
	Disease control	difenoconazole (*off-label*)
	Leaf spot	difenoconazole (*off-label*)
	Phytophthora	difenoconazole (*off-label*)
Pests	Slugs/snails	ferric phosphate
Weeds	Crops as weeds	fluazifop-P-butyl
	Grass weeds	fluazifop-P-butyl

Stem and bulb vegetables - Onions/leeks/garlic

Diseases	Blight	fluoxastrobin + prothioconazole (*useful reduction*)
	Botrytis	Bacillus amyloliquefaciens D747 (*off-label*), Bacillus subtilis (Strain QST 713) (*off-label*), cyprodinil + fludioxonil (*off-label*), difenoconazole + fluxapyroxad (*off-label*), fludioxonil, fludioxonil (*off-label*), fluoxastrobin + prothioconazole (*useful reduction*)
	Damping off	Bacillus subtilis (Strain QST 713) (*off-label*)
	Disease control	benthiavalicarb + oxathiapiprolin, mancozeb
	Downy mildew	azoxystrobin (*moderate control*), azoxystrobin (*off-label*), azoxystrobin (*reduction*), benthiavalicarb + oxathiapiprolin, dimethomorph + pyraclostrobin, fluopicolide + propamocarb hydrochloride (*off-label*), fluoxastrobin + prothioconazole (*control*), mancozeb, mancozeb + metalaxyl-M (*off-label*), mancozeb + metalaxyl-M (*useful control*), metalaxyl-M (*off-label*)
	Fusarium	boscalid + pyraclostrobin (*off-label*)
	Leaf spot	difenoconazole + fluxapyroxad (*off-label*)
	Powdery mildew	physical pest control
	Purple blotch	azoxystrobin, azoxystrobin + difenoconazole (*moderate control*), difenoconazole + fluxapyroxad (*off-label*), prothioconazole
	Pythium	metalaxyl-M, metalaxyl-M (*off-label*)
	Rhynchosporium	difenoconazole + fluxapyroxad (*off-label*), prothioconazole (*useful reduction*)
	Rust	azoxystrobin, azoxystrobin + difenoconazole, benzovindiflupyr, difenoconazole + fluxapyroxad (*off-label*), prothioconazole, tebuconazole, tebuconazole + trifloxystrobin (*off-label*)
	Stemphylium	prothioconazole (*useful reduction*)
	White rot	Bacillus subtilis (Strain QST 713) (*off-label*), boscalid + pyraclostrobin (*off-label*), tebuconazole (*off-label*)
	White tip	ametoctradin, azoxystrobin, azoxystrobin + difenoconazole (*qualified minor use*), boscalid + pyraclostrobin (*off-label*), difenoconazole + fluxapyroxad (*off-label*)

Pests	Aphids	Beauveria bassiana GHA (*off-label*), fatty acids (*off-label*), lambda-cyhalothrin (*off-label*), maltodextrin (*off-label*), physical pest control, pyrethrins, spirotetramat
	Caterpillars	Bacillus thuringiensis kurstaki, Bacillus thuringiensis kurstaki (*off-label*), pyrethrins
	Cutworms	Bacillus thuringiensis kurstaki (*off-label*), deltamethrin (*off-label*)
	Flies	maltodextrin (*off-label*)
	Free-living nematodes	garlic extract (*off-label*)
	Mealybugs	physical pest control
	Pests, miscellaneous	cyantraniliprole, deltamethrin (*off-label*), lambda-cyhalothrin (*off-label*)
	Root-knot nematodes	garlic extract (*off-label*)
	Scale insects	physical pest control
	Slugs/snails	ferric phosphate
	Spider mites	physical pest control, pyrethrins
	Stem nematodes	garlic extract (*off-label*)
	Suckers	physical pest control
	Thrips	Beauveria bassiana GHA (*off-label*), fatty acids (*off-label*), lambda-cyhalothrin (*off-label*), maltodextrin (*off-label*), pyrethrins, spinosad
	Whiteflies	physical pest control, spirotetramat
Plant growth regulation	Growth control	maleic hydrazide, maleic hydrazide (*off-label*)
Weeds	Broad-leaved weeds	S-metolachlor (*off-label*), aclonifen (*off-label*), bentazone (*off-label*), clopyralid, clopyralid (*off-label*), dimethenamid-p + metazachlor (*off-label*), dimethenamid-p + pendimethalin (*off-label*), fluroxypyr (*off-label*), glyphosate, isoxaben (*off-label*), pelargonic acid (*off-label*), pendimethalin, pendimethalin (*off-label*), pendimethalin (*pre + post emergence treatment*), prosulfocarb (*off-label*), pyridate, pyridate (*off-label*)
	Crops as weeds	clethodim (*off-label*), clopyralid (*off-label*), cycloxydim, fluazifop-P-butyl, fluroxypyr (*off-label*), glyphosate, isoxaben (*off-label*), pendimethalin, prosulfocarb (*off-label*)
	Grass weeds	S-metolachlor (*off-label*), clethodim (*off-label*), cycloxydim, dimethenamid-p + metazachlor (*off-label*), dimethenamid-p + pendimethalin (*off-label*), fluazifop-P-butyl, glyphosate, pelargonic acid (*off-label*), pendimethalin, pendimethalin (*off-label*), propaquizafop, propaquizafop (*off-label*), prosulfocarb (*off-label*)
	Mosses	pelargonic acid (*off-label*)
	Weeds, miscellaneous	fluroxypyr (*off-label*), glyphosate, glyphosate (*off-label*), pelargonic acid (*off-label*)

Stem and bulb vegetables - Vegetables

Diseases	Alternaria	difenoconazole + fluxapyroxad (*off-label*)
	Botrytis	Bacillus subtilis (Strain QST 713) (*off-label*)
	Powdery mildew	difenoconazole + fluxapyroxad (*off-label*), penconazole (*off-label*)

	Rhizoctonia	difenoconazole + fluxapyroxad (*off-label*)
	Sclerotinia	difenoconazole + fluxapyroxad (*off-label*)
Pests	Aphids	acetamiprid (*off-label*), deltamethrin (*off-label*), fatty acids (*off-label*), maltodextrin (*off-label*), pyrethrins
	Beetles	deltamethrin (*off-label*)
	Caterpillars	Bacillus thuringiensis kurstaki, deltamethrin (*off-label*), maltodextrin (*off-label*), pyrethrins
	Flies	maltodextrin (*off-label*)
	Slugs/snails	ferric phosphate
	Spider mites	pyrethrins
	Thrips	deltamethrin (*off-label*), fatty acids (*off-label*), maltodextrin (*off-label*), pyrethrins
	Weevils	deltamethrin (*off-label*)
Weeds	Broad-leaved weeds	aclonifen (*off-label*)
	Crops as weeds	fluazifop-P-butyl
	Grass weeds	cycloxydim (*off-label*), fluazifop-P-butyl

Flowers and ornamentals

Flowers - Bedding plants, general

Diseases	Disease control	azoxystrobin
	Powdery mildew	bupirimate, physical pest control
Pests	Aphids	physical pest control
	Mealybugs	physical pest control
	Scale insects	physical pest control
	Spider mites	physical pest control
	Suckers	physical pest control
	Whiteflies	physical pest control
Plant growth regulation	Flowering control	paclobutrazol
	Growth control	paclobutrazol

Flowers - Bulbs/corms

Crop control	Desiccation	carfentrazone-ethyl (*off-label*)
Diseases	Basal stem rot	cyprodinil + fludioxonil (*off-label*)
	Botrytis	mancozeb
	Crown rot	metalaxyl-M (*off-label*)
	Downy mildew	metalaxyl-M (*off-label*)
	Fusarium	cyprodinil + fludioxonil (*off-label*)
	Powdery mildew	bupirimate, physical pest control
Pests	Aphids	physical pest control
	Mealybugs	physical pest control
	Scale insects	physical pest control
	Spider mites	physical pest control
	Suckers	physical pest control
	Whiteflies	physical pest control

Plant growth regulation	Growth control	benzyladenine + gibberellin
Weeds	Broad-leaved weeds	metobromuron (*off-label*)
	Grass weeds	S-metolachlor (*off-label*), metobromuron (*off-label*)

Flowers - Miscellaneous flowers

Diseases	Anthracnose	Bacillus amyloliquefaciens D747 (*off-label*)
	Bacterial blight	Bacillus amyloliquefaciens D747 (*off-label*)
	Botrytis	Bacillus amyloliquefaciens D747 (*off-label*), difenoconazole + fluxapyroxad (*off-label*)
	Cercospora leaf spot	Bacillus amyloliquefaciens D747 (*off-label*)
	Damping off	Bacillus amyloliquefaciens D747 (*off-label*), Bacillus amyloliquefaciens strain MBI600 (*off-label*)
	Disease control	tebuconazole (*off-label*)
	Downy mildew	Bacillus amyloliquefaciens D747 (*off-label*)
	Fusarium	Bacillus amyloliquefaciens D747 (*off-label*)
	Powdery mildew	Bacillus amyloliquefaciens D747 (*off-label*), bupirimate, difenoconazole + fluxapyroxad (*off-label*), physical pest control
	Pythium	Bacillus amyloliquefaciens D747 (*off-label*), Bacillus amyloliquefaciens strain MBI600 (*off-label*)
	Rhizoctonia	Bacillus amyloliquefaciens D747 (*off-label*), Bacillus amyloliquefaciens strain MBI600 (*off-label*), difenoconazole + fluxapyroxad (*off-label*)
	Rust	tebuconazole (*off-label*)
	Sclerotinia	Bacillus amyloliquefaciens D747 (*off-label*), difenoconazole + fluxapyroxad (*off-label*)
	Stemphylium	Bacillus amyloliquefaciens D747 (*off-label*)
Pests	Aphids	Beauveria bassiana GHA (*off-label*), fatty acids (*off-label*), physical pest control
	Cutworms	deltamethrin (*off-label*)
	Flies	deltamethrin (*off-label*), lambda-cyhalothrin (*off-label*)
	Mealybugs	physical pest control
	Pests, miscellaneous	lambda-cyhalothrin (*off-label*)
	Scale insects	physical pest control
	Slugs/snails	ferric phosphate
	Spider mites	Beauveria bassiana GHA (*off-label*), fatty acids (*off-label*), physical pest control
	Suckers	physical pest control
	Thrips	Beauveria bassiana GHA (*off-label*), fatty acids (*off-label*)
	Whiteflies	Beauveria bassiana GHA (*off-label*), physical pest control
Plant growth regulation	Flowering control	paclobutrazol
	Growth control	paclobutrazol
Weeds	Broad-leaved weeds	lenacil (*off-label*), metribuzin (*off-label*), pendimethalin (*off-label*), prosulfocarb (*off-label*)
	Crops as weeds	fluazifop-P-butyl, metribuzin (*off-label*)

| | Grass weeds | clethodim (*off-label*), fluazifop-P-butyl, metribuzin (*off-label*) |
| | Weeds, miscellaneous | lenacil (*off-label*) |

Flowers - Pot plants

Diseases	Powdery mildew	physical pest control
Pests	Aphids	physical pest control
	Mealybugs	physical pest control
	Scale insects	physical pest control
	Spider mites	physical pest control
	Suckers	physical pest control
	Whiteflies	physical pest control
Plant growth regulation	Flowering control	paclobutrazol
	Growth control	paclobutrazol

Flowers - Protected bulbs/corms

Diseases	Downy mildew	propamocarb hydrochloride (*off-label*)
	Fungus diseases	propamocarb hydrochloride (*off-label*)
	Phytophthora	propamocarb hydrochloride (*off-label*)
	Pythium	propamocarb hydrochloride (*off-label*)

Flowers - Protected flowers

Diseases	Crown rot	metalaxyl-M (*off-label*)
	Downy mildew	metalaxyl-M (*off-label*)
	Foot rot	Trichoderma asperellum (Strain T34)
	Fusarium	Trichoderma afroharzianum (Strain T22) (*off-label*)
	Powdery mildew	bupirimate
	Pythium	Trichoderma afroharzianum (Strain T22) (*off-label*)
	Rhizoctonia	Trichoderma afroharzianum (Strain T22) (*off-label*)
Pests	Aphids	fatty acids (*off-label*), pirimicarb (*off-label*)
	Spider mites	fatty acids (*off-label*)
	Thrips	fatty acids (*off-label*)
Plant growth regulation	Flowering control	benzyladenine

Miscellaneous flowers and ornamentals - Miscellaneous uses

Crop control	Desiccation	glyphosate
Diseases	Botrytis	boscalid + pyraclostrobin (*off-label*), isopyrazam (*off-label*)
	Disease control	boscalid + pyraclostrobin (*off-label*), potassium bicarbonate (commodity substance)
	Leaf spot	boscalid + pyraclostrobin (*off-label*)
	Powdery mildew	boscalid + pyraclostrobin (*off-label*), isopyrazam (*off-label*), physical pest control
	Rust	boscalid + pyraclostrobin (*off-label*)
	Sclerotinia	isopyrazam (*off-label*)
Pests	Aphids	physical pest control
	Caterpillars	Bacillus thuringiensis kurstaki (*off-label*)

	Mealybugs	physical pest control
	Scale insects	physical pest control
	Slugs/snails	ferric phosphate
	Spider mites	physical pest control
	Suckers	physical pest control
	Whiteflies	physical pest control
Weeds	Broad-leaved weeds	carfentrazone-ethyl (*before planting*), glyphosate, propyzamide
	Crops as weeds	carfentrazone-ethyl (*before planting*)
	Grass weeds	glyphosate, propyzamide
	Weeds, miscellaneous	carfentrazone-ethyl, glyphosate, glyphosate (*before planting*), glyphosate (*pre-sowing/planting*)

Ornamentals - Nursery stock

Diseases	Alternaria	Bacillus amyloliquefaciens strain FZB24 (*off-label*), prothioconazole (*off-label*)
	Anthracnose	Bacillus amyloliquefaciens D747 (*off-label*)
	Bacterial canker	Bacillus amyloliquefaciens D747 (*off-label*)
	Black root rot	azoxystrobin (*off-label*), tebuconazole + trifloxystrobin (*off-label*)
	Black spot	myclobutanil (*off-label*), sulphur (*off-label*)
	Botrytis	Bacillus amyloliquefaciens D747 (*off-label*), Bacillus amyloliquefaciens strain FZB24 (*off-label*), Bacillus pumilus QST2808 (*off-label*), Bacillus subtilis (Strain QST 713) (*off-label*), azoxystrobin (*off-label*), boscalid + pyraclostrobin (*off-label*), captan (*off-label*), cerevisane (*off-label*), cyprodinil + fludioxonil, fluopyram (*off-label*), fluxapyroxad (*off-label*), mepanipyrim (*off-label*), prohexadione-calcium (*off-label*), pyrimethanil (*off-label*)
	Cercospora leaf spot	Bacillus amyloliquefaciens D747 (*off-label*)
	Damping off	Bacillus amyloliquefaciens D747 (*off-label*)
	Disease control	fluoxastrobin + prothioconazole (*off-label*), mancozeb, metrafenone (*off-label*), prohexadione-calcium (*off-label*)
	Downy mildew	Bacillus amyloliquefaciens D747 (*off-label*), Bacillus amyloliquefaciens strain FZB24 (*off-label*), ametoctradin + dimethomorph (*off-label*), azoxystrobin (*off-label*), captan (*off-label*), cerevisane (*off-label*), cos-oga (*off-label*), cyazofamid (*off-label*), dimethomorph (*off-label*), fluopicolide + propamocarb hydrochloride (*off-label*), fosetyl-aluminium + propamocarb hydrochloride (*off-label*), mancozeb, mancozeb + metalaxyl-M (*off-label*), mandipropamid (*off-label*), propamocarb hydrochloride (*off-label*)
	Fireblight	Bacillus amyloliquefaciens D747 (*off-label*)
	Foliar diseases	trifloxystrobin (*off-label*)
	Fungus diseases	Bacillus amyloliquefaciens D747 (*off-label*), captan (*off-label*), propamocarb hydrochloride (*off-label*)
	Fusarium	Bacillus amyloliquefaciens D747 (*off-label*), Trichoderma afroharzianum (Strain T22), Trichoderma asperellum (Strain T34) (*off-label*), azoxystrobin (*off-label*)

	Leaf spot	azoxystrobin (*off-label*), difenoconazole (*off-label*), prothioconazole (*off-label*), tebuconazole (*off-label*)
	Phytophthora	difenoconazole (*off-label*), metalaxyl-M, propamocarb hydrochloride (*off-label*), tebuconazole (*off-label*)
	Powdery mildew	Bacillus amyloliquefaciens D747 (*off-label*), Bacillus amyloliquefaciens strain FZB24 (*off-label*), Bacillus pumilus QST2808 (*off-label*), azoxystrobin (*off-label*), bupirimate, cerevisane (*off-label*), cos-oga (*off-label*), cyflufenamid (*off-label*), fluopyram (*off-label*), fluxapyroxad (*off-label*), isopyrazam (*off-label*), kresoxim-methyl, mepanipyrim (*off-label*), metrafenone, myclobutanil (*off-label*), penconazole (*off-label*), physical pest control, prothioconazole (*off-label*), sulphur (*off-label*), tebuconazole (*off-label*)
	Pythium	Bacillus amyloliquefaciens D747 (*off-label*), Trichoderma afroharzianum (Strain T22), Trichoderma asperellum (Strain T34) (*off-label*), fosetyl-aluminium + propamocarb hydrochloride (*off-label*), metalaxyl-M, metalaxyl-M (*off-label*), propamocarb hydrochloride (*off-label*)
	Rhizoctonia	Bacillus amyloliquefaciens D747 (*off-label*), Trichoderma afroharzianum (Strain T22)
	Rust	azoxystrobin (*off-label*), difenoconazole (*off-label*), myclobutanil (*off-label*), sulphur (*off-label*), tebuconazole + trifloxystrobin (*off-label*)
	Scab	azoxystrobin (*off-label*), captan (*off-label*), sulphur (*off-label*)
	Sclerotinia	Bacillus amyloliquefaciens D747 (*off-label*), Bacillus amyloliquefaciens strain FZB24 (*off-label*), Trichoderma afroharzianum (Strain T22), prothioconazole (*off-label*)
	Stem canker	prothioconazole (*off-label*), tebuconazole (*off-label*)
	Stemphylium	Bacillus amyloliquefaciens D747 (*off-label*)
	White blister	azoxystrobin (*off-label*)
	White mould	Bacillus pumilus QST2808 (*off-label*)
Pests	Aphids	Beauveria bassiana GHA, Beauveria bassiana GHA (*off-label*), acetamiprid, cypermethrin, esfenvalerate, fatty acids (*off-label*), lambda-cyhalothrin (*off-label*), maltodextrin (*off-label*), physical pest control, pirimicarb (*off-label*), pyrethrins, spirotetramat (*off-label*), sulfoxaflor
	Beetles	Beauveria bassiana GHA
	Caterpillars	Bacillus thuringiensis aizawai GC-91, Bacillus thuringiensis kurstaki, indoxacarb, indoxacarb (*off-label*), pyrethrins
	Flies	Bacillus thuringiensis israelensis, strain AM65-52
	Free-living nematodes	fosthiazate (*off-label*)
	Gall, rust and leaf & bud mites	clofentezine (*off-label*), sulphur (*off-label*)
	Leaf miners	deltamethrin (*off-label*)
	Mealybugs	physical pest control
	Mites	Beauveria bassiana GHA (*off-label*), clofentezine (*off-label*), maltodextrin (*off-label*)

	Pests, miscellaneous	Bacillus thuringiensis israelensis, strain AM65-52, cypermethrin, esfenvalerate, indoxacarb (*off-label*), lambda-cyhalothrin (*off-label*), spinosad (*off-label*)
	Scale insects	physical pest control
	Slugs/snails	ferric phosphate
	Spider mites	Beauveria bassiana GHA (*off-label*), abamectin, bifenazate (*off-label*), clofentezine (*off-label*), cyflumetofen, etoxazole (*off-label*), fatty acids (*off-label*), physical pest control, pyrethrins
	Suckers	physical pest control
	Tarsonemid mites	Beauveria bassiana GHA (*off-label*)
	Thrips	Beauveria bassiana ATCC 74040 (*reduction*), Beauveria bassiana GHA (*off-label*), abamectin, azadirachtin, deltamethrin (*off-label*), pyrethrins, spinosad, spinosad (*off-label*), spirotetramat (*off-label*)
	Whiteflies	Beauveria bassiana ATCC 74040, Beauveria bassiana GHA, Beauveria bassiana GHA (*off-label*), Lecanicillium muscarium, acetamiprid, buprofezin, fatty acids (*off-label*), flonicamid (*off-label*), physical pest control, spirotetramat (*off-label*), sulfoxaflor
Plant growth regulation	Flowering control	paclobutrazol, paclobutrazol (*off-label*), sodium silver thiosulphate
	Growth control	4-indol-3-ylbutyric acid, benzyladenine, benzyladenine + gibberellin, benzyladenine + gibberellin (*off-label*), chlormequat (*off-label*), daminozide, ethephon, ethephon (*off-label*), ethephon + mepiquat chloride (*off-label*), gibberellins, gibberellins (*off-label*), mepiquat chloride + prohexadione-calcium (*off-label*), paclobutrazol, paclobutrazol (*off-label*), prohexadione-calcium (*off-label*), trinexapac-ethyl (*off-label*)
Weeds	Broad-leaved weeds	S-metolachlor (*off-label*), amidosulfuron (*off-label*), bentazone, bentazone (*off-label*), clomazone (*off-label*), clopyralid, clopyralid (*off-label*), clopyralid + picloram (*off-label*), diflufenican (*off-label*), dimethenamid-p + metazachlor (*off-label*), dimethenamid-p + pendimethalin (*off-label*), florasulam (*off-label*), flufenacet (*off-label*), fluroxypyr (*off-label*), imazamox + pendimethalin (*off-label*), isoxaben, lenacil (*off-label*), metamitron (*off-label*), metazachlor, metobromuron (*off-label*), metribuzin (*off-label*), napropamide (*off-label*), nicosulfuron (*off-label*), pelargonic acid, pendimethalin (*off-label*), phenmedipham (*off-label*), propyzamide, propyzamide (*off-label*), prosulfocarb (*off-label*), triflusulfuron-methyl (*off-label*)
	Crops as weeds	clethodim (*off-label*), diflufenican (*off-label*), florasulam (*off-label*), fluroxypyr (*off-label*), metribuzin (*off-label*), nicosulfuron (*off-label*)
	Grass weeds	S-metolachlor (*off-label*), clethodim (*off-label*), cycloxydim, cycloxydim (*off-label*), dimethenamid-p + metazachlor (*off-label*), dimethenamid-p + pendimethalin (*off-label*), flufenacet (*off-label*), metamitron (*off-label*), metazachlor, metobromuron (*off-label*), metribuzin (*off-label*), napropamide (*off-label*), nicosulfuron (*off-label*), propyzamide, propyzamide (*off-label*), prosulfocarb (*off-label*)
	Mosses	pelargonic acid

| | Weeds, miscellaneous | carfentrazone-ethyl (*off-label*), clopyralid (*off-label*), diflufenican (*off-label*), florasulam (*off-label*), glyphosate, glyphosate (*off-label - as a directed spray*), metobromuron (*off-label*), metribuzin (*off-label*), napropamide (*off-label*), nicosulfuron (*off-label*), pelargonic acid |

Ornamentals - Trees and shrubs

Crop control	Desiccation	glyphosate
Diseases	Botrytis	boscalid + pyraclostrobin (*off-label*)
	Disease control	mancozeb
	Leaf spot	boscalid + pyraclostrobin (*off-label*)
	Phytophthora	Bacillus subtilis (Strain QST 713) (*off-label*), metalaxyl-M (*off-label*)
	Powdery mildew	boscalid + pyraclostrobin (*off-label*)
	Rust	boscalid + pyraclostrobin (*off-label*)
Pests	Aphids	deltamethrin, lambda-cyhalothrin (*off-label*)
	Capsid bugs	deltamethrin
	Caterpillars	Bacillus thuringiensis aizawai GC-91, Bacillus thuringiensis kurstaki, deltamethrin
	Mealybugs	deltamethrin
	Pests, miscellaneous	indoxacarb (*off-label*), lambda-cyhalothrin (*off-label*)
	Scale insects	deltamethrin
	Slugs/snails	ferric phosphate
	Thrips	deltamethrin
	Whiteflies	deltamethrin
Plant growth regulation	Growth control	glyphosate
	Plant growth regulation, miscellaneous	glyphosate (*off-label*)
Weeds	Broad-leaved weeds	isoxaben, pelargonic acid, propyzamide
	Grass weeds	propyzamide
	Mosses	pelargonic acid
	Weeds, miscellaneous	2,4-D + glyphosate, diflufenican + iodosulfuron-methyl-sodium, flazasulfuron, glyphosate, glyphosate + sulfosulfuron, pelargonic acid, propyzamide
	Woody weeds/scrub	glyphosate

Forestry

Forest nurseries, general - Forest nurseries

Crop control	Chemical stripping/thinning	glyphosate
Diseases	Black spot	sulphur (*off-label*)
	Blight	copper oxychloride (*off-label*)
	Botrytis	boscalid + pyraclostrobin (*off-label*), cyprodinil + fludioxonil, fenhexamid (*off-label*), mepanipyrim (*off-label*), pyrimethanil (*off-label*)

	Damping off	fosetyl-aluminium + propamocarb hydrochloride (*off-label*)
	Disease control	cyprodinil (*off-label*), fluoxastrobin + prothioconazole (*off-label*), mancozeb (*off-label*), pyrimethanil (*off-label*)
	Downy mildew	fosetyl-aluminium + propamocarb hydrochloride (*off-label*), mancozeb + metalaxyl-M (*off-label*), propamocarb hydrochloride (*off-label*)
	Foot rot	Phlebiopsis gigantea
	Fungus diseases	propamocarb hydrochloride (*off-label*)
	Fusarium	Trichoderma asperellum (Strain T34) (*off-label*)
	Leaf spot	difenoconazole (*off-label*)
	Phytophthora	Bacillus subtilis (Strain QST 713) (*off-label*), difenoconazole (*off-label*), metalaxyl-M (*off-label*), propamocarb hydrochloride (*off-label*)
	Powdery mildew	proquinazid (*off-label*), sulphur (*off-label*)
	Pythium	Trichoderma asperellum (Strain T34) (*off-label*), propamocarb hydrochloride (*off-label*)
	Root rot	Phlebiopsis gigantea
	Rust	difenoconazole (*off-label*), sulphur (*off-label*)
	Scab	sulphur (*off-label*)
Pests	Aphids	acetamiprid (*off-label*), lambda-cyhalothrin (*off-label*)
	Beetles	acetamiprid (*off-label*)
	Caterpillars	Bacillus thuringiensis kurstaki (*off-label*)
	Cutworms	Bacillus thuringiensis kurstaki (*off-label*)
	Flies	acetamiprid (*off-label*)
	Gall, rust and leaf & bud mites	clofentezine (*off-label*), sulphur (*off-label*)
	Mites	clofentezine (*off-label*)
	Pests, miscellaneous	lambda-cyhalothrin (*off-label*)
	Spider mites	bifenazate (*off-label*), clofentezine (*off-label*)
	Thrips	Beauveria bassiana ATCC 74040 (*off-label - reduction*)
	Wasps	acetamiprid (*off-label*)
	Weevils	acetamiprid (*off-label*)
	Whiteflies	Beauveria bassiana ATCC 74040 (*off-label*)
Plant growth regulation	Growth control	ethephon + mepiquat chloride (*off-label*), trinexapac-ethyl (*off-label*)
Weeds	Broad-leaved weeds	S-metolachlor (*off-label*), amidosulfuron (*off-label*), amidosulfuron + iodosulfuron-methyl-sodium (*off-label*), bifenox (*off-label*), clomazone (*off-label*), clopyralid (*off-label*), clopyralid + picloram (*off-label*), diflufenican (*off-label*), dimethenamid-p + metazachlor (*off-label*), flufenacet (*off-label*), fluroxypyr (*off-label*), isoxaben, nicosulfuron (*off-label*), pelargonic acid, pendimethalin (*off-label*), pendimethalin + picolinafen (*off-label*), propyzamide, propyzamide (*off-label*), prosulfocarb (*off-label*)
	Crops as weeds	amidosulfuron + iodosulfuron-methyl-sodium (*off-label*), diflufenican (*off-label*), fluroxypyr (*off-label*), nicosulfuron (*off-label*)

	Grass weeds	S-metolachlor (*off-label*), cycloxydim, dimethenamid-p + metazachlor (*off-label*), flufenacet (*off-label*), nicosulfuron (*off-label*), pendimethalin + picolinafen (*off-label*), propaquizafop (*off-label*), propyzamide, propyzamide (*off-label*), prosulfocarb (*off-label*)
	Weeds, miscellaneous	clopyralid (*off-label*), diflufenican (*off-label*), glyphosate, nicosulfuron (*off-label*)
	Woody weeds/scrub	glyphosate

Forestry plantations, general - Forestry plantations

Crop control	Chemical stripping/ thinning	glyphosate
	Sucker/shoot control	glyphosate
Diseases	Foot rot	Phlebiopsis gigantea
	Root rot	Phlebiopsis gigantea
Pests	Aphids	maltodextrin (*off-label*)
	Caterpillars	Bacillus thuringiensis kurstaki (*off-label*)
	Mites	maltodextrin (*off-label*)
	Weevils	acetamiprid (*off-label*), cypermethrin
Plant growth regulation	Growth control	1-naphthylacetic acid (*off-label*), glyphosate
	Plant growth regulation, miscellaneous	glyphosate (*off-label*)
Weeds	Broad-leaved weeds	isoxaben, pendimethalin (*off-label*), propyzamide
	Grass weeds	cycloxydim, glyphosate, propaquizafop (*off-label*), propyzamide
	Weeds, miscellaneous	carfentrazone-ethyl (*off-label*), glyphosate, glyphosate (*off-label*)
	Woody weeds/scrub	glyphosate

Miscellaneous forestry situations - Cut logs/timber

Pests	Beetles	cypermethrin

Woodland on farms - Woodland

Crop control	Chemical stripping/ thinning	glyphosate
Diseases	Foot rot	Phlebiopsis gigantea
	Root rot	Phlebiopsis gigantea
Pests	Aphids	lambda-cyhalothrin (*off-label*)
	Pests, miscellaneous	lambda-cyhalothrin (*off-label*)
Plant growth regulation	Growth control	gibberellins
Weeds	Broad-leaved weeds	amidosulfuron (*off-label*), fluroxypyr (*off-label*), lenacil (*off-label*), pendimethalin (*off-label*), propyzamide
	Crops as weeds	fluazifop-P-butyl, fluroxypyr (*off-label*)
	Grass weeds	fluazifop-P-butyl, propyzamide
	Weeds, miscellaneous	clopyralid (*off-label*), glyphosate, glyphosate (*off-label*), lenacil (*off-label*)
	Woody weeds/scrub	glyphosate

Fruit and hops

All bush fruit - All currants

Diseases	Alternaria	Bacillus amyloliquefaciens D747
	Botrytis	Bacillus amyloliquefaciens D747 (*off-label*), Bacillus subtilis (Strain QST 713) (*off-label*), boscalid + pyraclostrobin, cerevisane (*off-label*), cyprodinil + fludioxonil (*qualified minor use recommendation*), fenhexamid, pyrimethanil (*off-label*)
	Cane blight	Bacillus amyloliquefaciens D747 (*off-label*)
	Cladosporium	Bacillus amyloliquefaciens D747 (*off-label*)
	Downy mildew	cerevisane (*off-label*)
	Fusarium	Bacillus amyloliquefaciens D747
	Leaf spot	boscalid + pyraclostrobin (*moderate control only*), dodine
	Phytophthora	Bacillus subtilis (Strain QST 713) (*off-label*)
	Powdery mildew	Bacillus amyloliquefaciens D747, Bacillus amyloliquefaciens D747 (*off-label*), boscalid + pyraclostrobin (*moderate control only*), bupirimate, cerevisane (*off-label*), kresoxim-methyl, kresoxim-methyl (*off-label*), myclobutanil, penconazole, proquinazid (*off-label*), sulphur
	Pythium	Bacillus amyloliquefaciens D747
	Rhizoctonia	Bacillus amyloliquefaciens D747
Pests	Aphids	fatty acids (*off-label*), maltodextrin (*off-label*), pyrethrins (*off-label*), spirotetramat (*off-label*)
	Beetles	fatty acids (*off-label*)
	Birds/mammals	aluminium ammonium sulphate
	Capsid bugs	deltamethrin (*off-label*), pyrethrins (*off-label*)
	Caterpillars	Bacillus thuringiensis kurstaki (*off-label*), chlorantraniliprole (*off-label*), deltamethrin (*off-label*), pyrethrins (*off-label*), spinosad (*off-label*)
	Cutworms	Bacillus thuringiensis kurstaki (*off-label*)
	Flies	lambda-cyhalothrin (*off-label*), pyrethrins (*off-label*)
	Gall, rust and leaf & bud mites	spirotetramat (*off-label*), sulphur
	Midges	lambda-cyhalothrin (*off-label*)
	Pests, miscellaneous	Bacillus thuringiensis kurstaki (*off-label*), indoxacarb (*off-label*), lambda-cyhalothrin (*off-label*), spinosad (*off-label*)
	Sawflies	deltamethrin (*off-label*), lambda-cyhalothrin (*off-label*), spinosad (*off-label*)
	Spider mites	Beauveria bassiana GHA (*off-label*), fatty acids (*off-label*), maltodextrin (*off-label*), pyrethrins (*off-label*)
	Thrips	deltamethrin (*off-label*), fatty acids (*off-label*), lambda-cyhalothrin (*off-label*)
	Weevils	fatty acids (*off-label*)
	Whiteflies	Beauveria bassiana GHA (*off-label*), fatty acids (*off-label*), maltodextrin (*off-label*)
Weeds	Broad-leaved weeds	carfentrazone-ethyl (*off-label*), flufenacet + metribuzin (*off-label*), isoxaben, napropamide, pendimethalin, pendimethalin (*off-label*), propyzamide

	Crops as weeds	fluazifop-P-butyl, pendimethalin
	Grass weeds	clethodim, clethodim (*off-label*), fluazifop-P-butyl, flufenacet + metribuzin (*off-label*), napropamide, pendimethalin, propyzamide
	Weeds, miscellaneous	carfentrazone-ethyl (*off-label*), glyphosate (*off-label*)

All bush fruit - All protected bush fruit

Diseases	Alternaria	Bacillus amyloliquefaciens strain MBI600
	Anthracnose	boscalid + pyraclostrobin (*off-label*), cyprodinil + fludioxonil (*off-label*)
	Botrytis	Bacillus amyloliquefaciens D747 (*off-label*), boscalid + pyraclostrobin (*off-label*), cerevisane (*off-label*), cyprodinil + fludioxonil (*off-label*), fenhexamid (*off-label*)
	Cane blight	Bacillus amyloliquefaciens D747 (*off-label*)
	Cladosporium	Bacillus amyloliquefaciens D747 (*off-label*)
	Disease control	Bacillus pumilus QST2808
	Downy mildew	cerevisane (*off-label*), mancozeb + metalaxyl-M (*off-label*)
	Fusarium	Bacillus amyloliquefaciens strain MBI600, Trichoderma afroharzianum (Strain T22) (*off-label*), Trichoderma asperellum (Strain T34) (*off-label*)
	Leaf spot	boscalid + pyraclostrobin (*off-label*)
	Phytophthora	fluazinam (*off-label*)
	Powdery mildew	Bacillus amyloliquefaciens D747 (*off-label*), Bacillus amyloliquefaciens strain MBI600, Bacillus pumilus QST2808, bupirimate, cerevisane (*off-label*)
	Pythium	Bacillus amyloliquefaciens strain MBI600, Trichoderma afroharzianum (Strain T22) (*off-label*), Trichoderma asperellum (Strain T34) (*off-label*)
	Rhizoctonia	Bacillus amyloliquefaciens strain MBI600, Trichoderma afroharzianum (Strain T22) (*off-label*)
	Rust	Bacillus pumilus QST2808
Pests	Aphids	fatty acids (*off-label*)
	Beetles	fatty acids (*off-label*)
	Caterpillars	Bacillus thuringiensis kurstaki (*off-label*), spinosad (*off-label*)
	Flies	spinosad (*off-label*)
	Pests, miscellaneous	indoxacarb (*off-label*)
	Spider mites	Beauveria bassiana GHA (*off-label*), Lecanicillium muscarium (*off-label*), fatty acids (*off-label*)
	Thrips	Lecanicillium muscarium (*off-label*), fatty acids (*off-label*)
	Weevils	fatty acids (*off-label*)
	Whiteflies	Beauveria bassiana GHA (*off-label*), Lecanicillium muscarium (*off-label*), fatty acids (*off-label*)
Weeds	Weeds, miscellaneous	carfentrazone-ethyl (*off-label*)

All bush fruit - All vines

Diseases	Bacterial blight	Trichoderma atroviride strain SC1
	Bacterial canker	Bacillus amyloliquefaciens D747 (*off-label*)
	Black rot	kresoxim-methyl (*off-label*)
	Botrytis	Bacillus amyloliquefaciens D747 (*off-label*), Bacillus amyloliquefaciens strain FZB24, Bacillus subtilis (Strain QST 713) (*off-label*), cerevisane, fenhexamid, fenpyrazamine, prohexadione-calcium (*off-label*)
	Disease control	mancozeb, potassium phosphonates
	Downy mildew	ametoctradin + dimethomorph (*off-label*), amisulbrom (*off-label*), cerevisane, copper oxychloride, cos-oga (*off-label*), cymoxanil (*off-label*)
	Fireblight	Bacillus amyloliquefaciens D747 (*off-label*)
	Fungus diseases	Bacillus amyloliquefaciens D747 (*off-label*), Trichoderma atroviride strain SC1
	Fusarium	Trichoderma afroharzianum (Strain T22) (*off-label*)
	Phytophthora	potassium phosphonates
	Powdery mildew	Bacillus amyloliquefaciens D747 (*off-label*), Bacillus amyloliquefaciens strain FZB24, cerevisane, cos-oga (*off-label*), cyflufenamid (*off-label*), fluxapyroxad (*off-label*), kresoxim-methyl (*off-label*), metrafenone (*off-label*), myclobutanil, penconazole, sulphur, tebuconazole + trifloxystrobin (*off-label*)
	Pythium	Trichoderma afroharzianum (Strain T22) (*off-label*), potassium phosphonates
	Rhizoctonia	Trichoderma afroharzianum (Strain T22) (*off-label*)
	Sclerotinia	Bacillus amyloliquefaciens D747 (*off-label*)
Pests	Aphids	fatty acids (*off-label*), spirotetramat (*off-label*)
	Beetles	fatty acids (*off-label*)
	Caterpillars	Bacillus thuringiensis kurstaki, Bacillus thuringiensis kurstaki (*off-label*)
	Pests, miscellaneous	indoxacarb (*off-label*)
	Scale insects	spirotetramat (*off-label*)
	Spider mites	Lecanicillium muscarium (*off-label*), fatty acids (*off-label*)
	Thrips	Lecanicillium muscarium (*off-label*)
	Whiteflies	Lecanicillium muscarium (*off-label*)
Plant growth regulation	Growth control	gibberellins
Weeds	Broad-leaved weeds	propyzamide (*off-label*)
	Crops as weeds	fluazifop-P-butyl
	Grass weeds	fluazifop-P-butyl, propyzamide (*off-label*)
	Weeds, miscellaneous	carfentrazone-ethyl (*off-label*), glyphosate

All bush fruit - Bilberries/blueberries/cranberries

Diseases	Alternaria	Bacillus amyloliquefaciens D747
	Anthracnose	boscalid + pyraclostrobin (*off-label*)
	Botrytis	Bacillus amyloliquefaciens D747 (*off-label*), Bacillus subtilis (Strain QST 713) (*off-label*), boscalid + pyraclostrobin (*off-label*), cerevisane (*off-label*), cyprodinil + fludioxonil (*qualified minor use*

		recommendation), fenhexamid (off-label), pyrimethanil (off-label)
	Cane blight	Bacillus amyloliquefaciens D747 (off-label)
	Cladosporium	Bacillus amyloliquefaciens D747 (off-label)
	Downy mildew	cerevisane (off-label)
	Fusarium	Bacillus amyloliquefaciens D747
	Leaf spot	boscalid + pyraclostrobin (off-label)
	Phytophthora	Bacillus subtilis (Strain QST 713) (off-label)
	Powdery mildew	Bacillus amyloliquefaciens D747, Bacillus amyloliquefaciens D747 (off-label), cerevisane (off-label), kresoxim-methyl (off-label)
	Pythium	Bacillus amyloliquefaciens D747
	Rhizoctonia	Bacillus amyloliquefaciens D747
Pests	Aphids	fatty acids (off-label), maltodextrin (off-label), pyrethrins (off-label), spirotetramat (off-label)
	Beetles	fatty acids (off-label)
	Birds/mammals	aluminium ammonium sulphate
	Capsid bugs	pyrethrins (off-label)
	Caterpillars	Bacillus thuringiensis kurstaki (off-label), chlorantraniliprole (off-label), pyrethrins (off-label), spinosad (off-label)
	Cutworms	Bacillus thuringiensis kurstaki (off-label)
	Flies	lambda-cyhalothrin (off-label), pyrethrins (off-label), spinosad (off-label)
	Gall, rust and leaf & bud mites	spirotetramat (off-label)
	Pests, miscellaneous	Bacillus thuringiensis kurstaki (off-label), indoxacarb (off-label)
	Spider mites	Beauveria bassiana GHA (off-label), fatty acids (off-label), maltodextrin (off-label), pyrethrins (off-label)
	Thrips	fatty acids (off-label), lambda-cyhalothrin (off-label)
	Weevils	fatty acids (off-label)
	Whiteflies	Beauveria bassiana GHA (off-label), fatty acids (off-label), maltodextrin (off-label)
Weeds	Broad-leaved weeds	carfentrazone-ethyl (off-label), flufenacet + metribuzin (off-label), napropamide (off-label), pendimethalin (off-label), propyzamide (off-label)
	Crops as weeds	fluazifop-P-butyl
	Grass weeds	clethodim, clethodim (off-label), fluazifop-P-butyl, flufenacet + metribuzin (off-label), napropamide (off-label), propyzamide (off-label)
	Weeds, miscellaneous	carfentrazone-ethyl (off-label), glyphosate (off-label)

All bush fruit - Bush fruit, general

Diseases	Botrytis	Bacillus amyloliquefaciens D747 (off-label), Bacillus subtilis (Strain QST 713) (off-label)
	Cane blight	Bacillus amyloliquefaciens D747 (off-label)
	Cladosporium	Bacillus amyloliquefaciens D747 (off-label)
	Powdery mildew	Bacillus amyloliquefaciens D747 (off-label)
Pests	Aphids	fatty acids (off-label), maltodextrin (off-label), pyrethrins (off-label)

	Beetles	fatty acids (*off-label*)
	Capsid bugs	pyrethrins (*off-label*)
	Caterpillars	pyrethrins (*off-label*)
	Flies	pyrethrins (*off-label*)
	Pests, miscellaneous	indoxacarb (*off-label*), spinosad (*off-label*)
	Spider mites	Beauveria bassiana GHA (*off-label*), fatty acids (*off-label*), maltodextrin (*off-label*), pyrethrins (*off-label*)
	Thrips	fatty acids (*off-label*), spinosad (*off-label*)
	Weevils	fatty acids (*off-label*)
	Whiteflies	Beauveria bassiana GHA (*off-label*), fatty acids (*off-label*), maltodextrin (*off-label*)
Weeds	Broad-leaved weeds	napropamide (*off-label*)
	Crops as weeds	fluazifop-P-butyl
	Grass weeds	fluazifop-P-butyl, napropamide (*off-label*)

All bush fruit - Gooseberries

Diseases	Alternaria	Bacillus amyloliquefaciens D747
	Anthracnose	boscalid + pyraclostrobin (*off-label*)
	Botrytis	Bacillus amyloliquefaciens D747 (*off-label*), Bacillus subtilis (Strain QST 713) (*off-label*), boscalid + pyraclostrobin (*off-label*), cerevisane (*off-label*), cyprodinil + fludioxonil (*qualified minor use recommendation*), fenhexamid, pyrimethanil (*off-label*)
	Cane blight	Bacillus amyloliquefaciens D747 (*off-label*)
	Cladosporium	Bacillus amyloliquefaciens D747 (*off-label*)
	Downy mildew	cerevisane (*off-label*)
	Fusarium	Bacillus amyloliquefaciens D747
	Leaf spot	boscalid + pyraclostrobin (*off-label*)
	Phytophthora	Bacillus subtilis (Strain QST 713) (*off-label*)
	Powdery mildew	Bacillus amyloliquefaciens D747, Bacillus amyloliquefaciens D747 (*off-label*), bupirimate, cerevisane (*off-label*), kresoxim-methyl (*off-label*), myclobutanil, proquinazid (*off-label*), sulphur
	Pythium	Bacillus amyloliquefaciens D747
	Rhizoctonia	Bacillus amyloliquefaciens D747
Pests	Aphids	fatty acids (*off-label*), maltodextrin (*off-label*), pyrethrins (*off-label*), spirotetramat (*off-label*)
	Beetles	fatty acids (*off-label*)
	Birds/mammals	aluminium ammonium sulphate
	Capsid bugs	deltamethrin (*off-label*), pyrethrins (*off-label*)
	Caterpillars	Bacillus thuringiensis kurstaki (*off-label*), chlorantraniliprole (*off-label*), deltamethrin (*off-label*), pyrethrins (*off-label*), spinosad (*off-label*)
	Cutworms	Bacillus thuringiensis kurstaki (*off-label*)
	Flies	pyrethrins (*off-label*), spinosad (*off-label*)
	Gall, rust and leaf & bud mites	spirotetramat (*off-label*)
	Midges	lambda-cyhalothrin (*off-label*)
	Pests, miscellaneous	Bacillus thuringiensis kurstaki (*off-label*), indoxacarb (*off-label*), lambda-cyhalothrin (*off-label*)

	Sawflies	deltamethrin (*off-label*), lambda-cyhalothrin (*off-label*)
	Spider mites	Beauveria bassiana GHA (*off-label*), fatty acids (*off-label*), maltodextrin (*off-label*), pyrethrins (*off-label*)
	Thrips	deltamethrin (*off-label*), fatty acids (*off-label*)
	Weevils	fatty acids (*off-label*)
	Whiteflies	Beauveria bassiana GHA (*off-label*), fatty acids (*off-label*), maltodextrin (*off-label*)
Weeds	Broad-leaved weeds	carfentrazone-ethyl (*off-label*), flufenacet + metribuzin (*off-label*), isoxaben, napropamide, pendimethalin, propyzamide
	Crops as weeds	fluazifop-P-butyl, pendimethalin
	Grass weeds	clethodim, clethodim (*off-label*), fluazifop-P-butyl, flufenacet + metribuzin (*off-label*), napropamide, pendimethalin, propyzamide
	Weeds, miscellaneous	carfentrazone-ethyl (*off-label*), glyphosate (*off-label*)

All bush fruit - Miscellaneous bush fruit

Diseases	Botrytis	cyprodinil + fludioxonil (*off-label*), pyrimethanil (*off-label*)
	Downy mildew	metalaxyl-M (*off-label*)
	Powdery mildew	boscalid (*off-label*), physical pest control, proquinazid (*off-label*)
Pests	Aphids	physical pest control
	Capsid bugs	lambda-cyhalothrin (*off-label*)
	Caterpillars	indoxacarb (*off-label*), spinosad (*off-label*)
	Flies	maltodextrin (*off-label*)
	Mealybugs	physical pest control
	Scale insects	physical pest control
	Spider mites	physical pest control
	Suckers	physical pest control
	Thrips	maltodextrin (*off-label*)
	Wasps	lambda-cyhalothrin (*off-label*)
	Whiteflies	physical pest control
Weeds	Weeds, miscellaneous	glyphosate (*off-label*)

Cane fruit - All outdoor cane fruit

Crop control	Desiccation	carfentrazone-ethyl (*off-label*)
Diseases	Alternaria	Bacillus amyloliquefaciens D747
	Botrytis	Bacillus amyloliquefaciens D747 (*off-label*), Bacillus subtilis (Strain QST 713) (*off-label*), cerevisane (*off-label*), fenhexamid, pyrimethanil, pyrimethanil (*off-label*), tebuconazole (*off-label*)
	Botrytis fruit rot	cyprodinil + fludioxonil
	Cane blight	Bacillus amyloliquefaciens D747 (*off-label*), boscalid + pyraclostrobin (*off-label*), tebuconazole (*off-label*)
	Cladosporium	Bacillus amyloliquefaciens D747 (*off-label*)
	Downy mildew	azoxystrobin (*off-label*), cerevisane (*off-label*), metalaxyl-M (*off-label*)
	Fusarium	Bacillus amyloliquefaciens D747

	Phytophthora	Bacillus subtilis (Strain QST 713) (*off-label*), fluazinam (*off-label*)
	Powdery mildew	Bacillus amyloliquefaciens D747, Bacillus amyloliquefaciens D747 (*off-label*), azoxystrobin (*off-label*), bupirimate, cerevisane (*off-label*), cos-oga (*off-label*), myclobutanil, tebuconazole (*off-label*)
	Purple blotch	boscalid + pyraclostrobin (*off-label*)
	Pythium	Bacillus amyloliquefaciens D747
	Rhizoctonia	Bacillus amyloliquefaciens D747
	Root rot	dimethomorph, dimethomorph (*moderate control*)
	Rust	tebuconazole (*off-label*)
Pests	Aphids	fatty acids (*off-label*), maltodextrin (*off-label*), pyrethrins (*off-label*)
	Beetles	deltamethrin, deltamethrin (*off-label*), fatty acids (*off-label*)
	Birds/mammals	aluminium ammonium sulphate
	Capsid bugs	lambda-cyhalothrin (*off-label*), pyrethrins (*off-label*)
	Caterpillars	Bacillus thuringiensis kurstaki, Bacillus thuringiensis kurstaki (*off-label*), deltamethrin (*off-label*), pyrethrins (*off-label*)
	Cutworms	Bacillus thuringiensis kurstaki (*off-label*)
	Flies	pyrethrins (*off-label*), spinosad (*off-label*)
	Pests, miscellaneous	Bacillus thuringiensis kurstaki (*off-label*), deltamethrin (*off-label*), indoxacarb (*off-label*), lambda-cyhalothrin (*off-label*), spinosad (*off-label*)
	Spider mites	Beauveria bassiana GHA (*off-label*), clofentezine (*off-label*), fatty acids (*off-label*), maltodextrin (*off-label*), pyrethrins (*off-label*)
	Thrips	fatty acids (*off-label*), maltodextrin (*off-label*), spinosad (*off-label*)
	Weevils	fatty acids (*off-label*), lambda-cyhalothrin (*off-label*)
	Whiteflies	Beauveria bassiana GHA (*off-label*), fatty acids (*off-label*), maltodextrin (*off-label*)
Weeds	Broad-leaved weeds	isoxaben, napropamide, napropamide (*off-label*), pendimethalin, propyzamide, propyzamide (*England only*)
	Crops as weeds	fluazifop-P-butyl, pendimethalin
	Grass weeds	clethodim, clethodim (*off-label*), fluazifop-P-butyl, napropamide, napropamide (*off-label*), pendimethalin, propyzamide, propyzamide (*England only*)
	Weeds, miscellaneous	carfentrazone-ethyl (*off-label*), glyphosate (*off-label*)

Cane fruit - All protected cane fruit

Crop control	Desiccation	carfentrazone-ethyl (*off-label*)
Diseases	Alternaria	Bacillus amyloliquefaciens strain MBI600
	Botrytis	Bacillus amyloliquefaciens D747 (*off-label*), cerevisane (*off-label*), pyrimethanil (*off-label*), tebuconazole (*off-label*)
	Cane blight	Bacillus amyloliquefaciens D747 (*off-label*), boscalid + pyraclostrobin (*off-label*), tebuconazole (*off-label*)
	Cladosporium	Bacillus amyloliquefaciens D747 (*off-label*)
	Disease control	Bacillus pumilus QST2808

	Downy mildew	azoxystrobin (*off-label*), cerevisane (*off-label*), mancozeb + metalaxyl-M (*off-label*)
	Fusarium	Bacillus amyloliquefaciens strain MBI600, Trichoderma afroharzianum (Strain T22) (*off-label*), Trichoderma asperellum (Strain T34) (*off-label*)
	Phytophthora	fluazinam (*off-label*)
	Powdery mildew	Bacillus amyloliquefaciens D747 (*off-label*), Bacillus amyloliquefaciens strain MBI600, Bacillus pumilus QST2808, azoxystrobin (*off-label*), bupirimate, cerevisane (*off-label*), cos-oga (*off-label*), tebuconazole (*off-label*)
	Purple blotch	boscalid + pyraclostrobin (*off-label*)
	Pythium	Bacillus amyloliquefaciens strain MBI600, Trichoderma afroharzianum (Strain T22) (*off-label*), Trichoderma asperellum (Strain T34) (*off-label*)
	Rhizoctonia	Bacillus amyloliquefaciens strain MBI600, Trichoderma afroharzianum (Strain T22) (*off-label*)
	Root rot	dimethomorph (*moderate control*)
	Rust	Bacillus pumilus QST2808, boscalid + pyraclostrobin (*off-label*)
Pests	Aphids	fatty acids (*off-label*), sulfoxaflor (*off-label*)
	Beetles	fatty acids (*off-label*)
	Pests, miscellaneous	indoxacarb (*off-label*)
	Spider mites	Beauveria bassiana GHA (*off-label*), Lecanicillium muscarium (*off-label*), abamectin (*off-label*), fatty acids (*off-label*)
	Thrips	Lecanicillium muscarium (*off-label*), fatty acids (*off-label*), spinosad (*off-label*)
	Weevils	fatty acids (*off-label*)
	Whiteflies	Beauveria bassiana GHA (*off-label*), Lecanicillium muscarium (*off-label*), fatty acids (*off-label*)
Weeds	Weeds, miscellaneous	carfentrazone-ethyl (*off-label*)

Hops, general - Hops

Crop control	Desiccation	pyraflufen-ethyl (*off-label*)
Diseases	Alternaria	cyprodinil + fludioxonil (*off-label*)
	Botrytis	Bacillus subtilis (Strain QST 713) (*off-label*), cyprodinil + fludioxonil (*off-label*), fenhexamid (*off-label*)
	Disease control	azoxystrobin, boscalid + pyraclostrobin (*off-label*), captan (*off-label*), cymoxanil + mancozeb (*off-label*)
	Downy mildew	ametoctradin + dimethomorph (*off-label*), cymoxanil (*off-label*), mandipropamid (*off-label*), metalaxyl-M (*off-label*)
	Fungus diseases	cyprodinil + fludioxonil (*off-label*)
	Penicillium rot	cyprodinil + fludioxonil (*off-label*)
	Powdery mildew	bupirimate (*off-label*), cos-oga (*off-label*), fluopyram (*off-label*), metrafenone (*off-label*), myclobutanil (*off-label*), penconazole (*off-label*), sulphur (*off-label*)
Pests	Aphids	acetamiprid (*off-label*), flonicamid (*off-label*), lambda-cyhalothrin (*off-label*), maltodextrin (*off-label*), spirotetramat (*off-label*)
	Beetles	lambda-cyhalothrin (*off-label*)

	Capsid bugs	lambda-cyhalothrin (*off-label*)
	Caterpillars	Bacillus thuringiensis kurstaki, Bacillus thuringiensis kurstaki (*off-label*), indoxacarb (*off-label*), lambda-cyhalothrin (*off-label*)
	Free-living nematodes	fosthiazate (*off-label*)
	Pests, miscellaneous	indoxacarb (*off-label*), lambda-cyhalothrin (*off-label*), spinosad (*off-label*)
	Spider mites	acequinocyl (*off-label*), clofentezine (*off-label*), hexythiazox, maltodextrin (*off-label*)
	Thrips	spinosad (*off-label*)
Plant growth regulation	Growth control	prohexadione-calcium (*off-label*)
Weeds	Broad-leaved weeds	bentazone (*off-label*), clomazone (*off-label*), imazamox + pendimethalin (*off-label*), isoxaben, metazachlor (*off-label*), pendimethalin (*off-label*), propyzamide (*off-label*), pyraflufen-ethyl (*off-label*)
	Crops as weeds	fluazifop-P-butyl
	Grass weeds	fluazifop-P-butyl, metazachlor (*off-label*), propyzamide (*off-label*)
	Weeds, miscellaneous	glyphosate (*off-label*)

Miscellaneous fruit situations - Fruit crops, general

Diseases	Bacterial canker	Bacillus amyloliquefaciens D747 (*off-label*)
	Botrytis	Bacillus amyloliquefaciens D747 (*off-label*), Bacillus subtilis (Strain QST 713) (*off-label*), fenhexamid (*off-label*), pyrimethanil (*off-label*)
	Damping off	fosetyl-aluminium + propamocarb hydrochloride (*off-label*)
	Disease control	boscalid + pyraclostrobin (*off-label*), pyrimethanil (*off-label*)
	Downy mildew	fosetyl-aluminium + propamocarb hydrochloride (*off-label*), metalaxyl-M (*off-label*)
	Fireblight	Bacillus amyloliquefaciens D747 (*off-label*)
	Fungus diseases	Bacillus amyloliquefaciens D747 (*off-label*)
	Powdery mildew	Bacillus amyloliquefaciens D747 (*off-label*), kresoxim-methyl (*reduction*), physical pest control, proquinazid (*off-label*)
	Scab	kresoxim-methyl
	Sclerotinia	Bacillus amyloliquefaciens D747 (*off-label*)
	Storage rots	ethylene
Pests	Aphids	acetamiprid (*off-label*), lambda-cyhalothrin (*off-label*), physical pest control
	Caterpillars	Bacillus thuringiensis kurstaki (*off-label*)
	Food/grain storage pests	ethylene
	Free-living nematodes	fosthiazate (*off-label*)
	Mealybugs	physical pest control
	Pests, miscellaneous	indoxacarb (*off-label*), lambda-cyhalothrin (*off-label*)
	Scale insects	physical pest control
	Spider mites	physical pest control

	Suckers	physical pest control
	Whiteflies	physical pest control
Plant growth regulation	Growth control	ethylene
Weeds	Broad-leaved weeds	imazamox + pendimethalin (*off-label*), nicosulfuron (*off-label*), pendimethalin (*off-label*), prosulfocarb (*off-label*)
	Grass weeds	nicosulfuron (*off-label*), prosulfocarb (*off-label*)
	Weeds, miscellaneous	glyphosate (*off-label*)

Miscellaneous fruit situations - Fruit nursery stock

Diseases	Alternaria	Bacillus amyloliquefaciens strain FZB24 (*off-label*), cyprodinil + fludioxonil (*off-label*)
	Botrytis	Bacillus amyloliquefaciens strain FZB24 (*off-label*), cyprodinil + fludioxonil (*off-label*)
	Disease control	captan (*off-label*)
	Downy mildew	Bacillus amyloliquefaciens strain FZB24 (*off-label*)
	Fungus diseases	cyprodinil + fludioxonil (*off-label*), potassium hydrogen carbonate (commodity substance)
	Leaf spot	difenoconazole (*off-label*)
	Penicillium rot	cyprodinil + fludioxonil (*off-label*)
	Phytophthora	difenoconazole (*off-label*)
	Powdery mildew	Bacillus amyloliquefaciens strain FZB24 (*off-label*), bupirimate (*off-label*), penconazole (*off-label*)
	Rust	difenoconazole (*off-label*)
	Sclerotinia	Bacillus amyloliquefaciens strain FZB24 (*off-label*)
	Sooty moulds	potassium hydrogen carbonate (commodity substance)
Pests	Aphids	Beauveria bassiana GHA (*off-label*), maltodextrin (*off-label*)
	Mites	Beauveria bassiana GHA (*off-label*), maltodextrin (*off-label*)
	Pests, miscellaneous	spinosad (*off-label*)
	Spider mites	Beauveria bassiana GHA (*off-label*)
	Tarsonemid mites	Beauveria bassiana GHA (*off-label*)
	Thrips	Beauveria bassiana GHA (*off-label*), spinosad (*off-label*)
	Whiteflies	Beauveria bassiana GHA (*off-label*)
Plant growth regulation	Growth control	prohexadione-calcium (*off-label*)
Weeds	Broad-leaved weeds	florasulam (*off-label*), lenacil (*off-label*), metazachlor
	Crops as weeds	clethodim (*off-label*), florasulam (*off-label*)
	Grass weeds	clethodim (*off-label*), metazachlor, tri-allate (*off-label*)
	Weeds, miscellaneous	florasulam (*off-label*)

Other fruit - All outdoor strawberries

Diseases	Alternaria	Bacillus amyloliquefaciens D747
	Anthracnose	azoxystrobin
	Bacterial canker	Bacillus amyloliquefaciens D747 (*off-label*)

	Black spot	boscalid + pyraclostrobin, cyprodinil + fludioxonil (*qualified minor use*)
	Botrytis	Bacillus amyloliquefaciens D747 (*off-label*), Bacillus subtilis (Strain QST 713) (*off-label*), boscalid + pyraclostrobin, cerevisane (*off-label*), cyprodinil + fludioxonil, fenhexamid, fenpyrazamine, mepanipyrim, pyrimethanil
	Crown rot	dimethomorph (*moderate control*)
	Didymella stem rot	Gliocladium catenulatum (*moderate control*)
	Disease control	azoxystrobin + difenoconazole, difenoconazole + fluxapyroxad
	Downy mildew	cerevisane (*off-label*)
	Fireblight	Bacillus amyloliquefaciens D747 (*off-label*)
	Fungus diseases	Bacillus amyloliquefaciens D747 (*off-label*)
	Fusarium	Bacillus amyloliquefaciens D747, Gliocladium catenulatum (*moderate control*)
	Phytophthora	Gliocladium catenulatum (*moderate control*)
	Powdery mildew	Bacillus amyloliquefaciens D747, Bacillus amyloliquefaciens D747 (*off-label*), azoxystrobin, boscalid + pyraclostrobin, bupirimate, cerevisane (*off-label*), cos-oga (*off-label*), cyflufenamid (*off-label*), kresoxim-methyl, myclobutanil, penconazole, physical pest control, sulphur
	Pythium	Bacillus amyloliquefaciens D747, Gliocladium catenulatum (*moderate control*)
	Rhizoctonia	Bacillus amyloliquefaciens D747, Gliocladium catenulatum (*moderate control*)
	Sclerotinia	Bacillus amyloliquefaciens D747 (*off-label*)
Pests	Aphids	fatty acids (*off-label*), maltodextrin (*off-label*), physical pest control, spirotetramat
	Beetles	fatty acids (*off-label*)
	Birds/mammals	aluminium ammonium sulphate
	Capsid bugs	indoxacarb (*off-label*), maltodextrin (*off-label*)
	Caterpillars	Bacillus thuringiensis kurstaki, deltamethrin (*off-label*)
	Flies	spinosad (*off-label*)
	Mealybugs	physical pest control
	Pests, miscellaneous	indoxacarb (*off-label*), lambda-cyhalothrin (*off-label*), spinosad (*off-label*)
	Scale insects	physical pest control
	Slugs/snails	ferric phosphate
	Spider mites	clofentezine (*off-label*), fatty acids (*off-label*), maltodextrin (*off-label*), physical pest control
	Suckers	physical pest control
	Tarsonemid mites	spirotetramat
	Thrips	Beauveria bassiana GHA (*off-label*), deltamethrin (*off-label*), fatty acids (*off-label*), spinosad (*off-label*)
	Weevils	cyantraniliprole (*off-label*), fatty acids (*off-label*)
	Whiteflies	Beauveria bassiana GHA (*off-label*), fatty acids (*off-label*), maltodextrin (*off-label*), physical pest control
Weeds	Broad-leaved weeds	carfentrazone-ethyl (*off-label*), clopyralid (*off-label*), dimethenamid-p + pendimethalin (*off-label*),

		isoxaben, metamitron (*off-label*), napropamide, pendimethalin, phenmedipham, propyzamide
	Crops as weeds	cycloxydim, pendimethalin
	Grass weeds	S-metolachlor (*off-label*), clethodim, clethodim (*off-label*), cycloxydim, dimethenamid-p + pendimethalin (*off-label*), metamitron (*off-label*), napropamide, pendimethalin, propyzamide
	Weeds, miscellaneous	carfentrazone-ethyl (*off-label*), clopyralid (*off-label*), glyphosate (*off-label*)

Other fruit - Protected miscellaneous fruit

Diseases	Alternaria	Bacillus amyloliquefaciens strain MBI600
	Anthracnose	azoxystrobin
	Black spot	boscalid + pyraclostrobin, cyprodinil + fludioxonil (*qualified minor use*)
	Botrytis	Bacillus amyloliquefaciens strain FZB24, Bacillus subtilis (Strain QST 713) (*reduction of damage to fruit*), boscalid + pyraclostrobin, cerevisane, cerevisane (*off-label*), cyprodinil + fludioxonil, fenpyrazamine, fluopyram + trifloxystrobin (*moderate control*), pyrimethanil, pyrimethanil (*moderate control*)
	Botrytis fruit rot	Bacillus subtilis (Strain QST 713) (*reduction of damage to fruit*)
	Crown rot	dimethomorph (*moderate control*)
	Damping off	fosetyl-aluminium + propamocarb hydrochloride (*off-label*)
	Disease control	Bacillus pumilus QST2808, azoxystrobin + difenoconazole, boscalid + pyraclostrobin (*off-label*), difenoconazole + fluxapyroxad
	Downy mildew	cerevisane (*off-label*), fosetyl-aluminium + propamocarb hydrochloride (*off-label*), mancozeb + metalaxyl-M (*off-label*)
	Fusarium	Bacillus amyloliquefaciens strain MBI600, Trichoderma afroharzianum (Strain T22) (*off-label*), Trichoderma asperellum (Strain T34) (*off-label*)
	Powdery mildew	Ampelomyces quisqualis (Strain AQ10), Bacillus amyloliquefaciens strain MBI600, Bacillus pumilus QST2808, azoxystrobin, boscalid + pyraclostrobin, bupirimate, cerevisane (*off-label*), cos-oga (*off-label*), fluopyram + trifloxystrobin, kresoxim-methyl, myclobutanil, penconazole, penconazole (*off-label*), proquinazid (*off-label*)
	Pythium	Bacillus amyloliquefaciens strain MBI600, Trichoderma afroharzianum (Strain T22) (*off-label*), Trichoderma asperellum (Strain T34) (*off-label*)
	Rhizoctonia	Bacillus amyloliquefaciens strain MBI600, Trichoderma afroharzianum (Strain T22) (*off-label*)
	Rust	Bacillus pumilus QST2808
Pests	Aphids	Beauveria bassiana GHA, acetamiprid (*off-label*), fatty acids, fatty acids (*off-label*), spirotetramat
	Beetles	Beauveria bassiana GHA, fatty acids (*off-label*)
	Caterpillars	Bacillus thuringiensis kurstaki (*off-label*), deltamethrin (*off-label*), indoxacarb (*off-label*), spinosad (*off-label*)

	Flies	deltamethrin (*off-label*)
	Leaf miners	abamectin (*off-label*)
	Pests, miscellaneous	indoxacarb (*off-label*), lambda-cyhalothrin (*off-label*), spinosad
	Spider mites	Lecanicillium muscarium (*off-label*), abamectin, abamectin (*off-label*), bifenazate, bifenazate (*off-label*), clofentezine (*off-label*), cyflumetofen, etoxazole (*off-label*), fatty acids, fatty acids (*off-label*)
	Tarsonemid mites	spirotetramat
	Thrips	Beauveria bassiana GHA (*off-label*), Lecanicillium muscarium (*off-label*), abamectin, abamectin (*off-label*), deltamethrin (*off-label*), fatty acids (*off-label*), spinosad
	Weevils	cyantraniliprole (*off-label*), fatty acids (*off-label*)
	Whiteflies	Beauveria bassiana GHA, Beauveria bassiana GHA (*off-label*), Lecanicillium muscarium, Lecanicillium muscarium (*off-label*), fatty acids, fatty acids (*off-label*)
Plant growth regulation	Growth control	gibberellins (*off-label*)
Weeds	Weeds, miscellaneous	carfentrazone-ethyl (*off-label*)

Other fruit - Rhubarb

Diseases	Botrytis	Bacillus subtilis (Strain QST 713) (*off-label*), cyprodinil + fludioxonil (*off-label*)
	Disease control	difenoconazole (*off-label*)
	Downy mildew	mancozeb + metalaxyl-M (*off-label*)
	Leaf spot	difenoconazole (*off-label*)
	Phytophthora	difenoconazole (*off-label*)
Pests	Aphids	deltamethrin (*off-label*), fatty acids (*off-label*), maltodextrin (*off-label*), pirimicarb (*off-label*)
	Beetles	deltamethrin (*off-label*)
	Capsid bugs	deltamethrin (*off-label*)
	Caterpillars	Bacillus thuringiensis kurstaki (*off-label*), deltamethrin (*off-label*)
	Cutworms	Bacillus thuringiensis kurstaki (*off-label*), deltamethrin (*off-label*)
	Flies	maltodextrin (*off-label*)
	Slugs/snails	ferric phosphate
	Thrips	fatty acids (*off-label*), maltodextrin (*off-label*)
Plant growth regulation	Growth control	gibberellins
Weeds	Broad-leaved weeds	aclonifen (*off-label*), clomazone (*off-label*), isoxaben (*off-label*), pelargonic acid (*off-label*), pendimethalin (*off-label*), propyzamide
	Crops as weeds	aclonifen (*off-label*), fluazifop-P-butyl
	Grass weeds	fluazifop-P-butyl, pelargonic acid (*off-label*), pendimethalin (*off-label*), propyzamide
	Weeds, miscellaneous	glyphosate (*off-label*), metribuzin (*off-label*), pelargonic acid (*off-label*)

Tree fruit - All nuts

Diseases	Botrytis	Bacillus subtilis (Strain QST 713) (*off-label*)
	Powdery mildew	tebuconazole (*off-label*)
	Stem canker	tebuconazole (*off-label*)
Pests	Aphids	fatty acids (*off-label*), lambda-cyhalothrin (*off-label*)
	Beetles	fatty acids (*off-label*)
	Caterpillars	Bacillus thuringiensis kurstaki (*off-label*)
	Pests, miscellaneous	indoxacarb (*off-label*), lambda-cyhalothrin (*off-label*)
	Spider mites	fatty acids (*off-label*)
Weeds	Broad-leaved weeds	fluroxypyr (*off-label*), isoxaben (*off-label*), nicosulfuron (*off-label*), pendimethalin (*off-label*), propyzamide (*off-label*), prosulfocarb (*off-label*)
	Crops as weeds	fluazifop-P-butyl, nicosulfuron (*off-label*)
	Grass weeds	fluazifop-P-butyl, nicosulfuron (*off-label*), propyzamide (*off-label*), prosulfocarb (*off-label*)
	Weeds, miscellaneous	glyphosate (*off-label*), nicosulfuron (*off-label*)

Tree fruit - All pome fruit

Crop control	Sucker/shoot control	glyphosate
Diseases	Alternaria	cyprodinil + fludioxonil, fludioxonil (*reduction*)
	Bacterial canker	Bacillus amyloliquefaciens D747 (*off-label*)
	Botrytis	Bacillus amyloliquefaciens D747 (*off-label*), Bacillus subtilis (Strain QST 713) (*off-label*), fludioxonil (*reduction*), fludioxonil + pyrimethanil, pyrimethanil (*off-label*)
	Botrytis fruit rot	cyprodinil + fludioxonil
	Disease control	captan (*off-label*), difenoconazole, fludioxonil, fludioxonil + pyrimethanil, fluopyram, mancozeb, trifloxystrobin
	Fireblight	Bacillus amyloliquefaciens D747 (*off-label*)
	Fungus diseases	1-methylcyclopropene, Bacillus amyloliquefaciens D747 (*off-label*), fludioxonil (*reduction*), potassium hydrogen carbonate (commodity substance)
	Fusarium	cyprodinil + fludioxonil
	Penicillium rot	cyprodinil + fludioxonil, fludioxonil (*reduction*)
	Phytophthora	mancozeb + metalaxyl-M (*off-label*)
	Powdery mildew	Bacillus amyloliquefaciens D747 (*off-label*), boscalid + pyraclostrobin, bupirimate, cyflufenamid, fluxapyroxad, kresoxim-methyl (*reduction*), myclobutanil, penconazole, penthiopyrad (*suppression*), physical pest control, proquinazid (*off-label*), sulphur (*off-label*), tebuconazole (*off-label*)
	Scab	boscalid + pyraclostrobin, captan, difenoconazole, dithianon, dithianon + potassium phosphonates, dodine, fluxapyroxad, kresoxim-methyl, mancozeb, myclobutanil, penthiopyrad, potassium phosphonates, pyrimethanil, pyrimethanil (*off-label*), sulphur (*off-label*)
	Sclerotinia	Bacillus amyloliquefaciens D747 (*off-label*)

	Sooty moulds	potassium hydrogen carbonate (commodity substance), potassium hydrogen carbonate (commodity substance) (*reduction*)
	Stem canker	tebuconazole (*off-label*)
	Storage rots	captan, cyprodinil + fludioxonil, fludioxonil + pyrimethanil
Pests	Aphids	acetamiprid, deltamethrin, fatty acids (*off-label*), flonicamid, lambda-cyhalothrin, maltodextrin (*off-label*), physical pest control, spirotetramat, spirotetramat (*off-label*)
	Beetles	fatty acids (*off-label*)
	Capsid bugs	deltamethrin
	Caterpillars	Bacillus thuringiensis kurstaki, Bacillus thuringiensis kurstaki (*off-label*), Cydia pomonella GV, chlorantraniliprole (*off-label*), deltamethrin, dodecadienol + tetradecenyl acetate + tetradecylacetate, indoxacarb, lambda-cyhalothrin, pyriproxyfen (*reduction*), spinosad
	Gall, rust and leaf & bud mites	maltodextrin (*off-label*)
	Mealybugs	physical pest control
	Mites	acequinocyl
	Pests, miscellaneous	indoxacarb (*off-label*), potassium hydrogen carbonate (commodity substance)
	Sawflies	deltamethrin
	Scale insects	maltodextrin (*off-label*), physical pest control, spirotetramat
	Spider mites	acequinocyl, clofentezine, cyflumetofen, fatty acids (*off-label*), hexythiazox, maltodextrin (*off-label*), physical pest control, tebufenpyrad
	Suckers	deltamethrin, lambda-cyhalothrin, physical pest control, potassium hydrogen carbonate (commodity substance) (*reduction*), pyriproxyfen (*reduction*)
	Whiteflies	acetamiprid, physical pest control
Plant growth regulation	Fruiting control	1-methylcyclopropene (*off-label - post harvest only*), 1-methylcyclopropene (*post-harvest use*), 1-naphthylacetic acid, benzyladenine, gibberellins, metamitron
	Growth control	benzyladenine, benzyladenine + gibberellin, ethephon (*off-label*), gibberellins, metamitron, prohexadione-calcium, prohexadione-calcium (*off-label*)
	Quality/yield control	gibberellins, gibberellins (*off-label*)
Weeds	Broad-leaved weeds	clopyralid (*off-label*), fluroxypyr (*off-label*), isoxaben, nicosulfuron (*off-label*), pendimethalin, pendimethalin (*off-label*), propyzamide, propyzamide (*off-label*), prosulfocarb (*off-label*)
	Crops as weeds	fluazifop-P-butyl, nicosulfuron (*off-label*), pendimethalin
	Grass weeds	fluazifop-P-butyl, glyphosate, nicosulfuron (*off-label*), pendimethalin, propyzamide, propyzamide (*off-label*), prosulfocarb (*off-label*)
	Weeds, miscellaneous	2,4-D + glyphosate, clopyralid (*off-label*), glyphosate, glyphosate (*off-label*), nicosulfuron (*off-label*)

Tree fruit - All stone fruit

Crop control	Sucker/shoot control	glyphosate
Diseases	Bacterial canker	Bacillus amyloliquefaciens D747 (*off-label*)
	Blossom wilt	boscalid + pyraclostrobin (*off-label*), cyprodinil + fludioxonil (*off-label*)
	Botrytis	1-methylcyclopropene (*off-label*), Bacillus amyloliquefaciens D747 (*off-label*), Bacillus subtilis (Strain QST 713) (*off-label*), cyprodinil + fludioxonil (*off-label*), fenhexamid (*off-label*)
	Disease control	myclobutanil
	Fireblight	Bacillus amyloliquefaciens D747 (*off-label*)
	Fungus diseases	1-methylcyclopropene, Bacillus amyloliquefaciens D747 (*off-label*)
	Powdery mildew	Bacillus amyloliquefaciens D747 (*off-label*), fluxapyroxad (*off-label*)
	Rust	myclobutanil (*off-label*)
	Scab	fluxapyroxad (*off-label*)
	Sclerotinia	Bacillus amyloliquefaciens D747 (*off-label*), boscalid + pyraclostrobin (*off-label*)
	Storage rots	1-methylcyclopropene (*off-label*)
Pests	Aphids	Beauveria bassiana GHA (*off-label*), acetamiprid, fatty acids (*off-label*), flonicamid, maltodextrin (*off-label*), spirotetramat, spirotetramat (*off-label*)
	Beetles	fatty acids (*off-label*)
	Caterpillars	Bacillus thuringiensis kurstaki (*off-label*), chlorantraniliprole (*off-label*), dodecadienol + tetradecenyl acetate + tetradecylacetate, indoxacarb (*off-label*)
	Flies	maltodextrin (*off-label*)
	Gall, rust and leaf & bud mites	Beauveria bassiana GHA (*off-label*), maltodextrin (*off-label*)
	Mites	acequinocyl
	Pests, miscellaneous	indoxacarb (*off-label*), lambda-cyhalothrin (*off-label*)
	Scale insects	maltodextrin (*off-label*)
	Spider mites	Beauveria bassiana GHA (*off-label*), acequinocyl, fatty acids (*off-label*), maltodextrin (*off-label*)
	Whiteflies	acetamiprid, maltodextrin (*off-label*)
Plant growth regulation	Fruiting control	gibberellins (*off-label*)
	Growth control	gibberellins (*off-label*), prohexadione-calcium (*off-label*)
Weeds	Broad-leaved weeds	isoxaben, nicosulfuron (*off-label*), pendimethalin, propyzamide, propyzamide (*off-label*), prosulfocarb (*off-label*)
	Crops as weeds	fluazifop-P-butyl, nicosulfuron (*off-label*), pendimethalin
	Grass weeds	fluazifop-P-butyl, nicosulfuron (*off-label*), pendimethalin, propyzamide, propyzamide (*off-label*), prosulfocarb (*off-label*)
	Weeds, miscellaneous	glyphosate, glyphosate (*off-label*), nicosulfuron (*off-label*)

Tree fruit - Miscellaneous top fruit

Diseases	Bacterial canker	Bacillus amyloliquefaciens D747 (*off-label*)
	Botrytis	Bacillus amyloliquefaciens D747 (*off-label*), Bacillus subtilis (Strain QST 713) (*off-label*)
	Fireblight	Bacillus amyloliquefaciens D747 (*off-label*)
	Fungus diseases	Bacillus amyloliquefaciens D747 (*off-label*)
	Powdery mildew	Bacillus amyloliquefaciens D747 (*off-label*)
	Sclerotinia	Bacillus amyloliquefaciens D747 (*off-label*)
Pests	Aphids	maltodextrin (*off-label*)
	Gall, rust and leaf & bud mites	maltodextrin (*off-label*)
	Pests, miscellaneous	indoxacarb (*off-label*)
	Scale insects	maltodextrin (*off-label*)
	Spider mites	cyflumetofen, maltodextrin (*off-label*)
Weeds	Broad-leaved weeds	nicosulfuron (*off-label*), prosulfocarb (*off-label*)
	Crops as weeds	nicosulfuron (*off-label*)
	Grass weeds	nicosulfuron (*off-label*), prosulfocarb (*off-label*)
	Weeds, miscellaneous	glyphosate (*off-label*), nicosulfuron (*off-label*)

Tree fruit - Protected fruit

Diseases	Alternaria	Bacillus amyloliquefaciens strain FZB24 (*off-label*)
	Botrytis	Bacillus amyloliquefaciens strain FZB24 (*off-label*)
	Downy mildew	Bacillus amyloliquefaciens strain FZB24 (*off-label*)
	Powdery mildew	Bacillus amyloliquefaciens strain FZB24 (*off-label*), penconazole (*off-label*)
	Sclerotinia	Bacillus amyloliquefaciens strain FZB24 (*off-label*)
Pests	Aphids	Beauveria bassiana GHA, Beauveria bassiana GHA (*off-label*), fatty acids (*off-label*)
	Beetles	Beauveria bassiana GHA, fatty acids (*off-label*)
	Caterpillars	Bacillus thuringiensis kurstaki (*off-label*)
	Leaf miners	abamectin (*off-label*)
	Mites	Beauveria bassiana GHA (*off-label*)
	Pests, miscellaneous	indoxacarb (*off-label*)
	Spider mites	Beauveria bassiana GHA (*off-label*), Lecanicillium muscarium (*off-label*), abamectin (*off-label*), fatty acids (*off-label*)
	Tarsonemid mites	Beauveria bassiana GHA (*off-label*)
	Thrips	Beauveria bassiana GHA (*off-label*), Lecanicillium muscarium (*off-label*), abamectin (*off-label*)
	Whiteflies	Beauveria bassiana GHA, Beauveria bassiana GHA (*off-label*), Lecanicillium muscarium (*off-label*)
Weeds	Broad-leaved weeds	florasulam (*off-label*)
	Crops as weeds	clethodim (*off-label*), florasulam (*off-label*)
	Grass weeds	clethodim (*off-label*)
	Weeds, miscellaneous	florasulam (*off-label*)

Grain/crop store uses

Stored produce - Food/produce storage

Crop control	Miscellaneous non-selective situations	benzoic acid
Pests	Food/grain storage pests	pirimiphos-methyl
	Pests, miscellaneous	cypermethrin, deltamethrin

Stored seed - Stored grain/rape/linseed

Pests	Food/grain storage pests	phenothrin + tetramethrin, pirimiphos-methyl
	Mites	pirimiphos-methyl
	Pests, miscellaneous	deltamethrin, pirimiphos-methyl

Grass

Turf/amenity grass - Amenity grassland

Diseases	Anthracnose	fludioxonil (*reduction*), tebuconazole + trifloxystrobin
	Dollar spot	tebuconazole + trifloxystrobin
	Drechslera leaf spot	fludioxonil (*useful levels of control*)
	Fusarium	fludioxonil, tebuconazole + trifloxystrobin
	Melting out	tebuconazole + trifloxystrobin
	Red thread	tebuconazole + trifloxystrobin
	Rust	tebuconazole + trifloxystrobin
	Seed-borne diseases	fludioxonil
Pests	Birds/mammals	aluminium ammonium sulphate
Plant growth regulation	Growth control	prohexadione-calcium, trinexapac-ethyl
Weeds	Broad-leaved weeds	2,4-D, 2,4-D + dicamba, 2,4-D + dicamba + fluroxypyr, 2,4-D + florasulam, aminopyralid + fluroxypyr, aminopyralid + triclopyr, citronella oil, clopyralid + florasulam + fluroxypyr, clopyralid + triclopyr, dicamba + MCPA + mecoprop-P, dicamba + MCPA + mecoprop-P (*moderately susceptible*), dicamba + mecoprop-P, florasulam + fluroxypyr, fluroxypyr + triclopyr, mecoprop-P
	Crops as weeds	2,4-D + dicamba + fluroxypyr, 2,4-D + florasulam, dicamba + MCPA + mecoprop-P (*moderately susceptible*), florasulam + fluroxypyr
	Mosses	carfentrazone-ethyl + mecoprop-P, ferrous sulphate, pelargonic acid
	Weeds, miscellaneous	2,4-D + glyphosate, aminopyralid + triclopyr, carfentrazone-ethyl + mecoprop-P, clopyralid + florasulam + fluroxypyr, dicamba + MCPA + mecoprop-P, florasulam + fluroxypyr, pelargonic acid
	Woody weeds/scrub	aminopyralid + triclopyr, clopyralid + triclopyr

Turf/amenity grass - Managed amenity turf

Diseases	Anthracnose	azoxystrobin, azoxystrobin (*moderate control*), fludioxonil (*reduction*), prochloraz (*off-label*), tebuconazole + trifloxystrobin
	Brown patch	azoxystrobin
	Crown rust	azoxystrobin
	Disease control	difenoconazole + fludioxonil
	Dollar spot	fluopyram + trifloxystrobin, fluxapyroxad (*off-label*), pyraclostrobin (*reduction*), pyraclostrobin (*useful reduction*), tebuconazole + trifloxystrobin
	Drechslera leaf spot	fludioxonil (*useful levels of control*)
	Fairy rings	azoxystrobin, azoxystrobin (*reduction*)
	Fusarium	azoxystrobin, fludioxonil, prochloraz (*off-label*), pyraclostrobin, pyraclostrobin (*moderate control*), tebuconazole + trifloxystrobin
	Leaf spot	prochloraz (*off-label*)
	Melting out	azoxystrobin, tebuconazole + trifloxystrobin
	Red thread	difenoconazole + fludioxonil, pyraclostrobin, tebuconazole + trifloxystrobin
	Rust	azoxystrobin, tebuconazole + trifloxystrobin
	Seed-borne diseases	fludioxonil, fluopyram + trifloxystrobin
	Take-all patch	azoxystrobin
Pests	Birds/mammals	aluminium ammonium sulphate
	Flies	esfenvalerate
	Pests, miscellaneous	garlic extract
	Slugs/snails	ferric phosphate
Plant growth regulation	Growth control	prohexadione-calcium, trinexapac-ethyl
Weeds	Broad-leaved weeds	2,4-D, 2,4-D + dicamba + MCPA + mecoprop-P, 2,4-D + dicamba + fluroxypyr, 2,4-D + florasulam, MCPA + mecoprop-P, clopyralid + florasulam + fluroxypyr, dicamba + MCPA + mecoprop-P, dicamba + MCPA + mecoprop-P (*moderately susceptible*), dicamba + mecoprop-P, ferrous sulphate + MCPA + mecoprop-P, florasulam + fluroxypyr, mecoprop-P
	Crops as weeds	2,4-D + dicamba + MCPA + mecoprop-P, 2,4-D + dicamba + fluroxypyr, 2,4-D + florasulam, dicamba + MCPA + mecoprop-P (*moderately susceptible*), florasulam + fluroxypyr
	Grass weeds	cycloxydim (*off-label*)
	Mosses	carfentrazone-ethyl + mecoprop-P, ferrous sulphate, ferrous sulphate + MCPA + mecoprop-P, pelargonic acid
	Weeds, miscellaneous	carfentrazone-ethyl + mecoprop-P, dicamba + MCPA + mecoprop-P, florasulam + fluroxypyr, glyphosate, glyphosate (*pre-establishment only*), pelargonic acid

Non-crop pest control

Farm buildings/yards - Farm buildings

Pests	Ants	cypermethrin + tetramethrin, pyrethrins
	Bedbugs	cypermethrin + tetramethrin
	Birds/mammals	brodifacoum, bromadiolone, cholecalciferol, difenacoum
	Caterpillars	cypermethrin + tetramethrin
	Cockroaches	cypermethrin + tetramethrin
	Flies	cypermethrin + tetramethrin, phenothrin + tetramethrin, pyrethrins
	Midges	pyrethrins
	Pests, miscellaneous	physical pest control
	Silverfish	cypermethrin + tetramethrin
	Wasps	cypermethrin + tetramethrin, phenothrin + tetramethrin, pyrethrins

Farm buildings/yards - Farmyards

Pests	Birds/mammals	brodifacoum, bromadiolone, difenacoum

Farmland pest control - Farmland situations

Pests	Birds/mammals	aluminium phosphide, brodifacoum, bromadiolone, difenacoum

Miscellaneous non-crop pest control - Manure/rubbish

Pests	Ants	cypermethrin + tetramethrin
	Bedbugs	cypermethrin + tetramethrin
	Caterpillars	cypermethrin + tetramethrin
	Cockroaches	cypermethrin + tetramethrin
	Flies	cypermethrin + tetramethrin
	Silverfish	cypermethrin + tetramethrin
	Wasps	cypermethrin + tetramethrin

Miscellaneous non-crop pest control - Miscellaneous pest control situations

Pests	Birds/mammals	carbon dioxide (commodity substance)

Protected salad and vegetable crops

Protected brassicas - Protected brassica vegetables

Diseases	Alternaria	Bacillus amyloliquefaciens strain MBI600, boscalid + pyraclostrobin (off-label)
	Black leg	fludioxonil (off-label)
	Botrytis	Bacillus amyloliquefaciens strain MBI600 (off-label)
	Damping off	fosetyl-aluminium + propamocarb hydrochloride, propamocarb hydrochloride

	Downy mildew	fluopicolide + propamocarb hydrochloride (*off-label*), fosetyl-aluminium + propamocarb hydrochloride, propamocarb hydrochloride
	Fungus diseases	propamocarb hydrochloride
	Fusarium	Bacillus amyloliquefaciens strain MBI600, Trichoderma afroharzianum (Strain T22), Trichoderma afroharzianum (Strain T22) (*off-label*), Trichoderma asperellum (Strain T34) (*off-label*)
	Mycosphaerella	boscalid + pyraclostrobin (*off-label*)
	Phytophthora	propamocarb hydrochloride
	Powdery mildew	Bacillus amyloliquefaciens strain MBI600
	Pythium	Bacillus amyloliquefaciens strain MBI600, Trichoderma afroharzianum (Strain T22), Trichoderma afroharzianum (Strain T22) (*off-label*), Trichoderma asperellum (Strain T34) (*off-label*), propamocarb hydrochloride
	Rhizoctonia	Bacillus amyloliquefaciens strain MBI600, Trichoderma afroharzianum (Strain T22), Trichoderma afroharzianum (Strain T22) (*off-label*)
	Ring spot	fluxapyroxad (*off-label*)
	Root rot	propamocarb hydrochloride
	Sclerotinia	Bacillus amyloliquefaciens strain MBI600 (*off-label*), Trichoderma afroharzianum (Strain T22)
	White blister	boscalid + pyraclostrobin (*off-label*)
Pests	Aphids	acetamiprid (*off-label*), fatty acids (*off-label*), pyrethrins
	Caterpillars	Bacillus thuringiensis kurstaki (*off-label*), indoxacarb (*off-label*), pyrethrins
	Spider mites	pyrethrins
	Thrips	fatty acids (*off-label*), pyrethrins

Protected brassicas - Protected salad brassicas

Diseases	Alternaria	Bacillus amyloliquefaciens strain MBI600
	Botrytis	Bacillus amyloliquefaciens strain MBI600 (*off-label*)
	Damping off	fosetyl-aluminium + propamocarb hydrochloride
	Downy mildew	fosetyl-aluminium + propamocarb hydrochloride
	Fusarium	Bacillus amyloliquefaciens strain MBI600, Trichoderma afroharzianum (Strain T22) (*off-label*), Trichoderma asperellum (Strain T34) (*off-label*)
	Powdery mildew	Bacillus amyloliquefaciens strain MBI600
	Pythium	Bacillus amyloliquefaciens strain MBI600, Trichoderma afroharzianum (Strain T22) (*off-label*), Trichoderma asperellum (Strain T34) (*off-label*)
	Rhizoctonia	Bacillus amyloliquefaciens strain MBI600, Trichoderma afroharzianum (Strain T22) (*off-label*)
	Sclerotinia	Bacillus amyloliquefaciens strain MBI600 (*off-label*)
Pests	Aphids	fatty acids (*off-label*), lambda-cyhalothrin (*off-label*)
	Caterpillars	Bacillus thuringiensis kurstaki (*off-label*)
	Pests, miscellaneous	lambda-cyhalothrin (*off-label*)
	Thrips	fatty acids (*off-label*)

Protected crops, general - All protected crops

Pests	Caterpillars	Bacillus thuringiensis kurstaki (*off-label*), indoxacarb (*off-label*)
	Pests, miscellaneous	indoxacarb (*off-label*)
	Thrips	Beauveria bassiana ATCC 74040 (*reduction*)
	Whiteflies	Beauveria bassiana ATCC 74040

Protected crops, general - Protected nuts

Pests	Aphids	fatty acids (*off-label*)
	Beetles	fatty acids (*off-label*)
	Pests, miscellaneous	indoxacarb (*off-label*)
	Spider mites	fatty acids (*off-label*)

Protected crops, general - Protected vegetables, general

Diseases	Anthracnose	Bacillus amyloliquefaciens D747 (*off-label*)
	Bacterial blight	Bacillus amyloliquefaciens D747 (*off-label*)
	Botrytis	Bacillus amyloliquefaciens D747 (*off-label*)
	Cercospora leaf spot	Bacillus amyloliquefaciens D747 (*off-label*)
	Downy mildew	Bacillus amyloliquefaciens D747 (*off-label*)
	Fusarium	Bacillus amyloliquefaciens D747 (*off-label*), Trichoderma afroharzianum (Strain T22) (*off-label*)
	Powdery mildew	Bacillus amyloliquefaciens D747 (*off-label*)
	Pythium	Bacillus amyloliquefaciens D747 (*off-label*), Trichoderma afroharzianum (Strain T22) (*off-label*)
	Rhizoctonia	Bacillus amyloliquefaciens D747 (*off-label*), Trichoderma afroharzianum (Strain T22) (*off-label*)
	Sclerotinia	Bacillus amyloliquefaciens D747 (*off-label*)
	Stemphylium	Bacillus amyloliquefaciens D747 (*off-label*)
Pests	Aphids	fatty acids (*off-label*)
	Pests, miscellaneous	indoxacarb (*off-label*)
	Spider mites	Lecanicillium muscarium (*off-label*)
	Thrips	Lecanicillium muscarium (*off-label*), fatty acids (*off-label*)
	Whiteflies	Lecanicillium muscarium (*off-label*)
Weeds	Broad-leaved weeds	nicosulfuron (*off-label*)
	Grass weeds	nicosulfuron (*off-label*)

Protected fruiting vegetables - Protected aubergines

Diseases	Alternaria	Bacillus amyloliquefaciens strain MBI600
	Botrytis	Bacillus subtilis (Strain QST 713) (*reduction of damage to fruit*), boscalid + pyraclostrobin (*off-label*), cerevisane, cyprodinil + fludioxonil (*off-label*), fenhexamid (*off-label*), fenpyrazamine, mepanipyrim (*off-label*), pyrimethanil (*off-label*)
	Botrytis fruit rot	Bacillus subtilis (Strain QST 713) (*reduction of damage to fruit*)
	Damping off	fosetyl-aluminium + propamocarb hydrochloride (*off-label*)
	Disease control	Bacillus pumilus QST2808
	Fusarium	Bacillus amyloliquefaciens strain MBI600, Trichoderma afroharzianum (Strain T22)

	Powdery mildew	Ampelomyces quisqualis (Strain AQ10), Bacillus amyloliquefaciens strain FZB24, Bacillus amyloliquefaciens strain MBI600, Bacillus pumilus QST2808, cos-oga, isopyrazam, mepanipyrim (*off-label*), proquinazid (*off-label*), sulphur (*off-label*)
	Pythium	Bacillus amyloliquefaciens strain MBI600, Trichoderma afroharzianum (Strain T22)
	Rhizoctonia	Bacillus amyloliquefaciens strain MBI600, Trichoderma afroharzianum (Strain T22)
	Rust	Bacillus pumilus QST2808
	Sclerotinia	Trichoderma afroharzianum (Strain T22)
Pests	Aphids	Beauveria bassiana GHA, acetamiprid, fatty acids (*off-label*), lambda-cyhalothrin (*off-label*), pirimicarb (*off-label*), potassium salts of fatty acids (*some activity*), pyrethrins (*off-label*), sulfoxaflor
	Beetles	Beauveria bassiana GHA, pyrethrins (*off-label*)
	Caterpillars	Bacillus thuringiensis kurstaki, cyantraniliprole (*off-label*), indoxacarb, pyrethrins (*off-label*)
	Leaf miners	deltamethrin (*off-label*), tetradecadienyl + trienyl acetate
	Mealybugs	flonicamid (*off-label*)
	Mites	etoxazole, sulphur (*off-label*)
	Pests, miscellaneous	lambda-cyhalothrin (*off-label*), pyrethrins
	Spider mites	abamectin, bifenazate (*off-label*), fatty acids (*off-label*), hexythiazox, potassium salts of fatty acids (*some activity*), pyrethrins (*off-label*)
	Thrips	abamectin, deltamethrin (*off-label*), pyrethrins (*off-label*), spinosad
	Whiteflies	Beauveria bassiana GHA, acetamiprid, cyantraniliprole (*off-label*), fatty acids (*off-label*), flonicamid (*off-label*), potassium salts of fatty acids (*moderate control*), pyrethrins (*off-label*), sulfoxaflor

Protected fruiting vegetables - Protected cucurbits

Diseases	Alternaria	Bacillus amyloliquefaciens strain MBI600
	Anthracnose	Bacillus amyloliquefaciens D747 (*off-label*)
	Bacterial blight	Bacillus amyloliquefaciens D747 (*off-label*)
	Botrytis	Bacillus amyloliquefaciens D747 (*off-label*), Bacillus subtilis (Strain QST 713) (*reduction of damage to fruit*), boscalid + pyraclostrobin (*off-label*), cyprodinil + fludioxonil (*off-label*), fenhexamid (*off-label*), fenpyrazamine, fluopyram + trifloxystrobin, imazalil (*off-label*), mepanipyrim (*off-label*)
	Botrytis fruit rot	Bacillus subtilis (Strain QST 713) (*reduction of damage to fruit*)
	Cercospora leaf spot	Bacillus amyloliquefaciens D747 (*off-label*)
	Cladosporium	boscalid + pyraclostrobin (*off-label*)
	Damping off	fosetyl-aluminium + propamocarb hydrochloride, fosetyl-aluminium + propamocarb hydrochloride (*off-label*)
	Disease control	Bacillus pumilus QST2808, difenoconazole + fluxapyroxad, isopyrazam

	Downy mildew	Bacillus amyloliquefaciens D747 (*off-label*), fosetyl-aluminium + propamocarb hydrochloride, metalaxyl-M (*off-label*)
	Fungus diseases	boscalid + pyraclostrobin (*off-label*)
	Fusarium	Bacillus amyloliquefaciens D747 (*off-label*), Bacillus amyloliquefaciens strain MBI600, Trichoderma afroharzianum (Strain T22), cyprodinil + fludioxonil (*off-label*), Trichoderma asperellum (Strain T34) (*off-label*)
	Powdery mildew	Ampelomyces quisqualis (Strain AQ10), Bacillus amyloliquefaciens D747 (*off-label*), Bacillus amyloliquefaciens strain FZB24, Bacillus amyloliquefaciens strain MBI600, Bacillus pumilus QST2808, boscalid + pyraclostrobin (*off-label*), bupirimate, cerevisane, cos-oga, fluopyram + trifloxystrobin, imazalil (*off-label*), isopyrazam, mepanipyrim (*off-label*), penconazole (*off-label*), proquinazid (*off-label*), sulphur (*off-label*)
	Pythium	Bacillus amyloliquefaciens D747 (*off-label*), Bacillus amyloliquefaciens strain MBI600, Trichoderma afroharzianum (Strain T22), fosetyl-aluminium + propamocarb hydrochloride, Trichoderma asperellum (Strain T34) (*off-label*)
	Rhizoctonia	Bacillus amyloliquefaciens D747 (*off-label*), Bacillus amyloliquefaciens strain MBI600, Trichoderma afroharzianum (Strain T22)
	Rust	Bacillus pumilus QST2808
	Sclerotinia	Bacillus amyloliquefaciens D747 (*off-label*), Trichoderma afroharzianum (Strain T22)
	Stemphylium	Bacillus amyloliquefaciens D747 (*off-label*)
Pests	Aphids	Beauveria bassiana GHA, acetamiprid, cyantraniliprole (*off-label*), deltamethrin, fatty acids, fatty acids (*off-label*), flonicamid (*off-label*), lambda-cyhalothrin (*off-label*), pirimicarb (*off-label*), potassium salts of fatty acids (*some activity*), pyrethrins (*off-label*), sulfoxaflor
	Beetles	Beauveria bassiana GHA, pyrethrins (*off-label*)
	Caterpillars	Bacillus thuringiensis kurstaki, Bacillus thuringiensis kurstaki (*off-label*), cyantraniliprole (*off-label*), deltamethrin, indoxacarb, pyrethrins (*off-label*), spinosad (*off-label*)
	Cutworms	Bacillus thuringiensis kurstaki (*off-label*)
	Leaf miners	cyantraniliprole (*off-label*), deltamethrin (*off-label*), spinosad (*off-label*), tetradecadienyl + trienyl acetate
	Mealybugs	deltamethrin
	Mites	sulphur (*off-label*)
	Pests, miscellaneous	lambda-cyhalothrin (*off-label*), pyrethrins
	Scale insects	deltamethrin
	Spider mites	Lecanicillium muscarium (*off-label*), abamectin, abamectin (*off-label*), bifenazate (*off-label*), fatty acids, fatty acids (*off-label*), hexythiazox, potassium salts of fatty acids (*some activity*), pyrethrins (*off-label*)
	Thrips	Lecanicillium muscarium (*off-label*), abamectin, deltamethrin, deltamethrin (*off-label*), fatty acids (*off-*

		label), pyrethrins (*off-label*), spinosad, spinosad (*off-label*)
	Whiteflies	Beauveria bassiana GHA, Lecanicillium muscarium, Lecanicillium muscarium (*off-label*), acetamiprid, cyantraniliprole (*off-label*), deltamethrin, fatty acids, fatty acids (*off-label*), flonicamid (*off-label*), potassium salts of fatty acids (*moderate control*), pyrethrins (*off-label*), pyriproxyfen (*reduction*), sulfoxaflor
Weeds	Broad-leaved weeds	clomazone (*off-label*)

Protected fruiting vegetables - Protected tomatoes

Crop control	Miscellaneous non-selective situations	Mild Pepino Mosaic Virus isolate VC1 + VX1
Diseases	Alternaria	Bacillus amyloliquefaciens strain FZB24, Bacillus amyloliquefaciens strain MBI600
	Botrytis	Bacillus subtilis (Strain QST 713) (*reduction of damage to fruit*), boscalid + pyraclostrobin (*off-label*), cerevisane, cyprodinil + fludioxonil (*off-label*), fenhexamid (*off-label*), fenpyrazamine, imazalil (*off-label*), mepanipyrim (*off-label*), pyrimethanil (*off-label*)
	Botrytis fruit rot	Bacillus subtilis (Strain QST 713) (*reduction of damage to fruit*)
	Damping off	propamocarb hydrochloride (*off-label*)
	Disease control	Bacillus pumilus QST2808, difenoconazole + fluxapyroxad
	Fusarium	Bacillus amyloliquefaciens strain MBI600, Trichoderma afroharzianum (Strain T22)
	Phytophthora	propamocarb hydrochloride (*off-label*)
	Powdery mildew	Ampelomyces quisqualis (Strain AQ10), Bacillus amyloliquefaciens strain FZB24, Bacillus amyloliquefaciens strain MBI600, Bacillus pumilus QST2808, bupirimate (*off-label*), cos-oga, cyflufenamid (*off-label*), imazalil (*off-label*), isopyrazam, mepanipyrim (*off-label*), penconazole (*off-label*), proquinazid (*off-label*), sulphur (*off-label*)
	Pythium	Bacillus amyloliquefaciens strain MBI600, Trichoderma afroharzianum (Strain T22), fosetyl-aluminium + propamocarb hydrochloride
	Rhizoctonia	Bacillus amyloliquefaciens strain MBI600, Trichoderma afroharzianum (Strain T22)
	Root rot	propamocarb hydrochloride (*off-label*)
	Rust	Bacillus pumilus QST2808
	Sclerotinia	Trichoderma afroharzianum (Strain T22)
Pests	Aphids	Beauveria bassiana GHA, acetamiprid, fatty acids, lambda-cyhalothrin (*off-label*), pirimicarb (*off-label*), potassium salts of fatty acids (*some activity*), pyrethrins, sulfoxaflor
	Beetles	Beauveria bassiana GHA
	Caterpillars	Bacillus thuringiensis kurstaki, cyantraniliprole (*off-label*), indoxacarb, pyrethrins
	Gall, rust and leaf & bud mites	sulphur (*off-label*)
	Leaf miners	deltamethrin (*off-label*), spinosad (*off-label*), tetradecadienyl + trienyl acetate

	Mealybugs	flonicamid (*off-label*)
	Mites	etoxazole, sulphur (*off-label*)
	Pests, miscellaneous	lambda-cyhalothrin (*off-label*), pyrethrins
	Spider mites	abamectin, abamectin (*off-label*), acequinocyl (*off-label*), bifenazate (*off-label*), fatty acids, hexythiazox, potassium salts of fatty acids (*some activity*), pyrethrins
	Thrips	abamectin, deltamethrin (*off-label*), pyrethrins, spinosad, spinosad (*off-label*)
	Whiteflies	Beauveria bassiana GHA, Lecanicillium muscarium, acetamiprid, cyantraniliprole (*off-label*), fatty acids, flonicamid (*off-label*), potassium salts of fatty acids (*moderate control*), pyriproxyfen (*reduction*), sulfoxaflor
Plant growth regulation	Growth control	ethephon (*off-label*)

Protected herb crops - Protected herbs

Diseases	Alternaria	Bacillus amyloliquefaciens strain MBI600, cyprodinil + fludioxonil (*off-label*), fludioxonil (*off-label*)
	Botrytis	Bacillus amyloliquefaciens strain MBI600 (*off-label*), Bacillus subtilis (Strain QST 713) (*reduction of damage to fruit*), cerevisane (*off-label*), cyprodinil + fludioxonil (*off-label*), difenoconazole + fluxapyroxad (*off-label*), fenhexamid (*off-label*), fluopyram + trifloxystrobin, fluopyram + trifloxystrobin (*off-label*), pyrimethanil (*off-label*)
	Botrytis fruit rot	Bacillus subtilis (Strain QST 713) (*reduction of damage to fruit*)
	Damping off	Bacillus amyloliquefaciens strain MBI600 (*off-label*), fosetyl-aluminium + propamocarb hydrochloride (*off-label*)
	Disease control	Bacillus pumilus QST2808, difenoconazole + fluxapyroxad, mandipropamid
	Downy mildew	cerevisane (*off-label*), cos-oga (*off-label*), dimethomorph (*off-label*), mancozeb + metalaxyl-M (*off-label*), mandipropamid, metalaxyl-M (*off-label*)
	Fusarium	Bacillus amyloliquefaciens strain MBI600, Trichoderma afroharzianum (Strain T22), Trichoderma afroharzianum (Strain T22) (*off-label*), Trichoderma asperellum (Strain T34) (*off-label*)
	Leaf spot	fludioxonil (*off-label*)
	Powdery mildew	Ampelomyces quisqualis (Strain AQ10), Ampelomyces quisqualis (Strain AQ10) (*off-label*), Bacillus amyloliquefaciens strain FZB24, Bacillus amyloliquefaciens strain MBI600, Bacillus pumilus QST2808, cerevisane (*off-label*), cos-oga, cos-oga (*off-label*), difenoconazole + fluxapyroxad (*off-label*), fluopyram + trifloxystrobin, fluopyram + trifloxystrobin (*off-label*), penconazole (*off-label*), sulphur (*off-label*)
	Pythium	Bacillus amyloliquefaciens strain MBI600, Bacillus amyloliquefaciens strain MBI600 (*off-label*), Trichoderma afroharzianum (Strain T22), Trichoderma afroharzianum (Strain T22) (*off-label*), Trichoderma asperellum (Strain T34) (*off-label*)

	Rhizoctonia	Bacillus amyloliquefaciens strain MBI600, Bacillus amyloliquefaciens strain MBI600 (*off-label*), Trichoderma afroharzianum (Strain T22), Trichoderma afroharzianum (Strain T22) (*off-label*), cyprodinil + fludioxonil (*off-label*), difenoconazole + fluxapyroxad (*off-label*)
	Rust	Bacillus pumilus QST2808
	Sclerotinia	Bacillus amyloliquefaciens strain MBI600 (*off-label*), Trichoderma afroharzianum (Strain T22), difenoconazole + fluxapyroxad (*off-label*), fluopyram + trifloxystrobin (*off-label*)
	Stem canker	cyprodinil + fludioxonil (*off-label*)
	White blister	boscalid + pyraclostrobin (*off-label*)
Pests	Aphids	Beauveria bassiana GHA, acetamiprid (*off-label*), fatty acids (*off-label*), flonicamid (*off-label*), pirimicarb (*off-label*), potassium salts of fatty acids (*some activity*), sulfoxaflor, sulfoxaflor (*off-label*)
	Beetles	Beauveria bassiana GHA
	Caterpillars	Bacillus thuringiensis kurstaki, Bacillus thuringiensis kurstaki (*off-label*), indoxacarb, indoxacarb (*off-label*), spinosad (*off-label*)
	Cutworms	Bacillus thuringiensis kurstaki (*off-label*)
	Leaf miners	tetradecadienyl + trienyl acetate
	Mites	sulphur (*off-label*)
	Pests, miscellaneous	pyrethrins
	Spider mites	Lecanicillium muscarium (*off-label*), abamectin, abamectin (*off-label*), bifenazate (*off-label*), fatty acids (*off-label*), hexythiazox, potassium salts of fatty acids (*some activity*)
	Thrips	Lecanicillium muscarium (*off-label*), abamectin, fatty acids (*off-label*), spinosad (*off-label*)
	Whiteflies	Beauveria bassiana GHA, Lecanicillium muscarium (*off-label*), fatty acids (*off-label*), potassium salts of fatty acids (*moderate control*), sulfoxaflor
Weeds	Broad-leaved weeds	napropamide (*off-label*), propyzamide (*off-label*)
	Grass weeds	napropamide (*off-label*), propyzamide (*off-label*)
	Weeds, miscellaneous	napropamide (*off-label*)

Protected leafy vegetables - Mustard and cress

Diseases	Alternaria	Bacillus amyloliquefaciens D747, cyprodinil + fludioxonil (*off-label*)
	Black rot	azoxystrobin (*off-label*)
	Botrytis	Bacillus amyloliquefaciens strain MBI600 (*off-label*), Bacillus subtilis (Strain QST 713) (*off-label*), cerevisane (*off-label*), cyprodinil + fludioxonil (*off-label*), difenoconazole + fluxapyroxad (*off-label*)
	Damping off	Bacillus amyloliquefaciens strain MBI600 (*off-label*)
	Disease control	mandipropamid
	Downy mildew	cerevisane (*off-label*), fluopicolide + propamocarb hydrochloride (*off-label*), mandipropamid
	Fusarium	Bacillus amyloliquefaciens D747

	Powdery mildew	Bacillus amyloliquefaciens D747, cerevisane (*off-label*), difenoconazole + fluxapyroxad (*off-label*)
	Pythium	Bacillus amyloliquefaciens D747, Bacillus amyloliquefaciens strain MBI600 (*off-label*)
	Rhizoctonia	Bacillus amyloliquefaciens D747, Bacillus amyloliquefaciens strain MBI600 (*off-label*), cyprodinil + fludioxonil (*off-label*), difenoconazole + fluxapyroxad (*off-label*)
	Sclerotinia	Bacillus amyloliquefaciens strain MBI600 (*off-label*), difenoconazole + fluxapyroxad (*off-label*)
	Stem canker	cyprodinil + fludioxonil (*off-label*)
	White blister	azoxystrobin (*off-label*)
Pests	Aphids	Beauveria bassiana GHA (*off-label*), acetamiprid (*off-label*), deltamethrin (*off-label*), fatty acids (*off-label*)
	Beetles	deltamethrin (*off-label*)
	Caterpillars	Bacillus thuringiensis kurstaki (*off-label*), deltamethrin (*off-label*), spinosad (*off-label*)
	Cutworms	Bacillus thuringiensis kurstaki (*off-label*)
	Spider mites	Beauveria bassiana GHA (*off-label*), fatty acids (*off-label*)
	Thrips	Beauveria bassiana GHA (*off-label*), fatty acids (*off-label*), spinosad (*off-label*)
	Whiteflies	Beauveria bassiana GHA (*off-label*)
Weeds	Broad-leaved weeds	pendimethalin (*off-label*), propyzamide (*off-label*)
	Crops as weeds	cycloxydim (*off-label*)
	Grass weeds	cycloxydim (*off-label*), propyzamide (*off-label*)

Protected leafy vegetables - Protected leafy vegetables

Diseases	Alternaria	Bacillus amyloliquefaciens strain MBI600, cyprodinil + fludioxonil (*off-label*)
	Botrytis	Bacillus amyloliquefaciens strain MBI600 (*off-label*), Bacillus subtilis (Strain QST 713) (*reduction of damage to fruit*), boscalid + pyraclostrobin, boscalid + pyraclostrobin (*off-label*), cerevisane (*off-label*), cyprodinil + fludioxonil (*off-label*), difenoconazole + fluxapyroxad (*off-label*), fluopyram + trifloxystrobin (*off-label*), pyrimethanil (*off-label*)
	Botrytis fruit rot	Bacillus subtilis (Strain QST 713) (*reduction of damage to fruit*)
	Bottom rot	boscalid + pyraclostrobin
	Disease control	Bacillus pumilus QST2808, mandipropamid
	Downy mildew	Bacillus amyloliquefaciens strain FZB24, azoxystrobin, boscalid + pyraclostrobin (*off-label*), cerevisane, cerevisane (*off-label*), cos-oga (*off-label*), dimethomorph (*off-label*), fosetyl-aluminium + propamocarb hydrochloride, mandipropamid
	Fusarium	Bacillus amyloliquefaciens strain MBI600, Trichoderma afroharzianum (Strain T22), Trichoderma afroharzianum (Strain T22) (*off-label*), Trichoderma asperellum (Strain T34) (*off-label*)
	Powdery mildew	Ampelomyces quisqualis (Strain AQ10) (*off-label*), Bacillus amyloliquefaciens strain MBI600, Bacillus pumilus QST2808, cerevisane (*off-label*), cos-oga (*off-*

		label), difenoconazole + fluxapyroxad (*off-label*), fluopyram + trifloxystrobin (*off-label*)
	Pythium	Bacillus amyloliquefaciens strain MBI600, Trichoderma afroharzianum (Strain T22), Trichoderma afroharzianum (Strain T22) (*off-label*), Trichoderma asperellum (Strain T34) (*off-label*), fosetyl-aluminium + propamocarb hydrochloride
	Rhizoctonia	Bacillus amyloliquefaciens strain MBI600, Trichoderma afroharzianum (Strain T22), boscalid + pyraclostrobin (*off-label*), cyprodinil + fludioxonil (*off-label*), difenoconazole + fluxapyroxad (*off-label*)
	Rust	Bacillus pumilus QST2808
	Sclerotinia	Bacillus amyloliquefaciens strain MBI600 (*off-label*), Trichoderma afroharzianum (Strain T22), boscalid + pyraclostrobin (*off-label*), difenoconazole + fluxapyroxad (*moderate control*), difenoconazole + fluxapyroxad (*off-label*), fluopyram + trifloxystrobin (*off-label*)
	Soft rot	boscalid + pyraclostrobin
	Stem canker	cyprodinil + fludioxonil (*off-label*)
Pests	Aphids	acetamiprid (*off-label*), fatty acids (*off-label*), lambda-cyhalothrin (*off-label*), pirimicarb (*off-label*), pyrethrins, sulfoxaflor (*off-label*)
	Caterpillars	Bacillus thuringiensis kurstaki (*off-label*), pyrethrins
	Pests, miscellaneous	lambda-cyhalothrin (*off-label*)
	Spider mites	abamectin (*off-label*), fatty acids (*off-label*), pyrethrins
	Thrips	fatty acids (*off-label*), pyrethrins
Weeds	Broad-leaved weeds	napropamide (*off-label*), propyzamide (*off-label*)
	Grass weeds	napropamide (*off-label*), propyzamide (*off-label*)
	Weeds, miscellaneous	napropamide (*off-label*)

Protected leafy vegetables - Protected spinach

Diseases	Alternaria	Bacillus amyloliquefaciens strain MBI600, cyprodinil + fludioxonil (*off-label*)
	Botrytis	Bacillus amyloliquefaciens strain MBI600 (*off-label*), boscalid + pyraclostrobin (*off-label*), cerevisane (*off-label*), cyprodinil + fludioxonil (*off-label*)
	Disease control	mandipropamid
	Downy mildew	cerevisane (*off-label*), mandipropamid, metalaxyl-M (*off-label*)
	Fusarium	Bacillus amyloliquefaciens strain MBI600, Trichoderma afroharzianum (Strain T22) (*off-label*), Trichoderma asperellum (Strain T34) (*off-label*)
	Powdery mildew	Ampelomyces quisqualis (Strain AQ10) (*off-label*), Bacillus amyloliquefaciens strain MBI600, cerevisane (*off-label*)
	Pythium	Bacillus amyloliquefaciens strain MBI600, Trichoderma afroharzianum (Strain T22) (*off-label*), Trichoderma asperellum (Strain T34) (*off-label*)
	Rhizoctonia	Bacillus amyloliquefaciens strain MBI600, Trichoderma afroharzianum (Strain T22) (*off-label*),

		boscalid + pyraclostrobin (*off-label*), cyprodinil + fludioxonil (*off-label*)
	Sclerotinia	Bacillus amyloliquefaciens strain MBI600 (*off-label*), boscalid + pyraclostrobin (*off-label*)
	Stem canker	cyprodinil + fludioxonil (*off-label*)
Pests	Aphids	fatty acids (*off-label*), sulfoxaflor (*off-label*)
	Caterpillars	Bacillus thuringiensis kurstaki (*off-label*)
	Spider mites	fatty acids (*off-label*)
	Thrips	fatty acids (*off-label*)

Protected legumes - Protected peas and beans

Diseases	Damping off	fosetyl-aluminium + propamocarb hydrochloride (*off-label*)
Pests	Aphids	fatty acids (*off-label*)
	Caterpillars	Bacillus thuringiensis kurstaki
	Thrips	fatty acids (*off-label*)

Protected root and tuber vegetables - Protected artichokes/cardoons

Pests	Aphids	fatty acids (*off-label*)
	Thrips	fatty acids (*off-label*)

Protected root and tuber vegetables - Protected carrots/parsnips/celeriac

Diseases	Alternaria	fludioxonil (*off-label*)
	Cavity spot	Bacillus amyloliquefaciens D747 (*off-label*)
	Leaf spot	fludioxonil (*off-label*)
	Sclerotinia	Bacillus amyloliquefaciens D747 (*off-label*)

Protected root and tuber vegetables - Protected root brassicas

Diseases	Alternaria	cyprodinil + fludioxonil (*off-label*)
	Anthracnose	Bacillus amyloliquefaciens D747 (*off-label*)
	Bacterial blight	Bacillus amyloliquefaciens D747 (*off-label*)
	Black leg	fludioxonil (*off-label*)
	Botrytis	Bacillus amyloliquefaciens D747 (*off-label*)
	Damping off	Bacillus amyloliquefaciens D747 (*off-label*), fosetyl-aluminium + propamocarb hydrochloride
	Disease control	boscalid + pyraclostrobin (*off-label*)
	Downy mildew	Bacillus amyloliquefaciens D747 (*off-label*), fosetyl-aluminium + propamocarb hydrochloride
	Fusarium	Bacillus amyloliquefaciens D747 (*off-label*)
	Leaf spot	cyprodinil + fludioxonil (*off-label*)
	Powdery mildew	Bacillus amyloliquefaciens D747 (*off-label*)
	Pythium	Bacillus amyloliquefaciens D747 (*off-label*)
	Rhizoctonia	Bacillus amyloliquefaciens D747 (*off-label*)
	Sclerotinia	Bacillus amyloliquefaciens D747 (*off-label*), cyprodinil + fludioxonil (*off-label*)

Protected stem and bulb vegetables - Protected asparagus

Diseases	Botrytis	cyprodinil + fludioxonil (*off-label*)
Pests	Aphids	Beauveria bassiana GHA (*off-label*), fatty acids (*off-label*), lambda-cyhalothrin (*off-label*)
	Thrips	Beauveria bassiana GHA (*off-label*), fatty acids (*off-label*)

Protected stem and bulb vegetables - Protected celery/chicory

Diseases	Botrytis	azoxystrobin (*off-label*)
	Disease control	Bacillus pumilus QST2808
	Fusarium	Trichoderma afroharzianum (Strain T22) (*off-label*)
	Leaf spot	azoxystrobin (*off-label*)
	Powdery mildew	Ampelomyces quisqualis (Strain AQ10) (*off-label*), Bacillus pumilus QST2808
	Pythium	Trichoderma afroharzianum (Strain T22) (*off-label*)
	Rhizoctonia	Trichoderma afroharzianum (Strain T22) (*off-label*), azoxystrobin (*off-label*)
	Rust	Bacillus pumilus QST2808
	Sclerotinia	azoxystrobin (*off-label*)
Pests	Aphids	Beauveria bassiana GHA (*off-label*), fatty acids (*off-label*), sulfoxaflor (*off-label*)
	Caterpillars	Bacillus thuringiensis kurstaki, Bacillus thuringiensis kurstaki (*off-label*), deltamethrin (*off-label*)
	Cutworms	Bacillus thuringiensis kurstaki (*off-label*)
	Spider mites	fatty acids (*off-label*)
	Thrips	Beauveria bassiana GHA (*off-label*), fatty acids (*off-label*)

Protected stem and bulb vegetables - Protected onions/leeks/garlic

Pests	Aphids	Beauveria bassiana GHA (*off-label*), fatty acids (*off-label*)
	Leaf miners	deltamethrin (*off-label*)
	Thrips	Beauveria bassiana GHA (*off-label*), fatty acids (*off-label*)

Total vegetation control

Aquatic situations, general - Aquatic situations

Pests	Aphids	Bacillus thuringiensis israelensis
	Flies	Bacillus thuringiensis israelensis
Plant growth regulation	Growth control	glyphosate
Weeds	Aquatic weeds	glyphosate
	Grass weeds	glyphosate
	Weeds, miscellaneous	glyphosate

Non-crop areas, general - Miscellaneous non-crop situations

Diseases	Botrytis	Bacillus subtilis (Strain QST 713) (*off-label*)
Weeds	Weeds, miscellaneous	2,4-D + glyphosate, diflufenican + iodosulfuron-methyl-sodium, flazasulfuron

Non-crop areas, general - Non-crop farm areas

Weeds	Broad-leaved weeds	metsulfuron-methyl, pyridate (*off-label*), tribenuron-methyl
	Crops as weeds	metsulfuron-methyl
	Grass weeds	glyphosate
	Weeds, miscellaneous	2,4-D + glyphosate, glyphosate, metsulfuron-methyl

Non-crop areas, general - Paths/roads etc

Weeds	Broad-leaved weeds	pelargonic acid
	Grass weeds	glyphosate
	Mosses	pelargonic acid, starch, protein, oil and water
		acetic acid, diflufenican + glyphosate, diflufenican + iodosulfuron-methyl-sodium, flazasulfuron, glyphosate, glyphosate + sulfosulfuron, pelargonic acid, starch, protein, oil and water

SECTION 2
PESTICIDE PROFILES

1   1,4-dimethylnaphthalene

A sprouting inhibitor for use in harvested potatoes

Products

1	1,4Sight (GB only)	Dormfresh	1000 g/l	HN	20158

Uses
* Sprout suppression in **Potatoes** [1];

Approval information
* 1,4-dimethylnaphthalene included in Annex I under EC Regulation 1107/2009

Efficacy guidance
* Must only be applied via hot fogging equipment.
* Allow a minimum of 28 days between applications
* Do not allow condensation to drip on potatoes since this can cause skin damage

Restrictions
* Do not use on seed potatoes
* Do not feed treated crops to livestock

Hazard classification and safety precautions
    Hazard Harmful if swallowed
    Transport code 9 [1]
    Packaging group III
    UN Number 3082
    Risk phrases H304, H319
    Operator protection A, D, H
    Environmental protection H410

2   abamectin

A selective acaricide and insecticide for use in ornamentals and other protected crops
IRAC mode of action code: 6

Products

1	Clayton Abba	Clayton	18 g/l	EC	18559
2	Dynamec	Syngenta	18 g/l	EC	18316
3	Smitten	Pan Agriculture	18 g/l	EC	18477

Uses
* Leaf miner in **Protected hops in propagation** *(off-label)* [2]; **Protected nursery fruit trees** *(off-label)* [2];
* Red spider mites in **Protected baby leaf crops** *(off-label)* [2]; **Protected blackberries** *(off-label)* [2]; **Protected cayenne peppers** *(off-label)* [2]; **Protected cherry tomatoes** *(off-label)* [2]; **Protected herbs (see appendix 6)** *(off-label)* [2]; **Protected lamb's lettuce** *(off-label)* [2]; **Protected loganberries** *(off-label)* [2]; **Protected raspberries** *(off-label)* [2]; **Protected Rubus hybrids** *(off-label)* [2];
* Two-spotted spider mite in **Protected aubergines** [1-3]; **Protected chilli peppers** [1-3]; **Protected hops in propagation** *(off-label)* [2]; **Protected nursery fruit trees** *(off-label)* [2]; **Protected ornamentals** [1-3]; **Protected peppers** [1-3]; **Protected strawberries** [1-3]; **Protected tomatoes** [1-3];
* Western flower thrips in **Protected aubergines** [1-3]; **Protected chilli peppers** [1-3]; **Protected hops in propagation** *(off-label)* [2]; **Protected nursery fruit trees** *(off-label)* [2]; **Protected ornamentals** [1-3]; **Protected peppers** [1-3]; **Protected strawberries** [1-3]; **Protected tomatoes** [1-3];

Extension of Authorisation for Minor Use (EAMUs)
* **Protected baby leaf crops** 20172832 [2]
* **Protected blackberries** 20172833 [2]

SEE SECTION 3 FOR PRODUCTS ALSO REGISTERED

- *Protected cayenne peppers* 20172831 [2]
- *Protected cherry tomatoes* 20172831 [2]
- *Protected herbs (see appendix 6)* 20172832 [2]
- *Protected hops in propagation* 20194251 [2]
- *Protected lamb's lettuce* 20172832 [2]
- *Protected loganberries* 20172833 [2]
- *Protected nursery fruit trees* 20194251 [2]
- *Protected raspberries* 20172833 [2]
- *Protected Rubus hybrids* 20172833 [2]

Approval information
- Abamectin included in Annex I under EC Regulation 1107/2009

Efficacy guidance
- Abamectin controls adults and immature stages of the two-spotted spider mite, the larval stages of leaf miners and the nymphs of Western Flower Thrips, plus a useful reduction in adults
- Treat at first sign of infestation. Repeat sprays may be required
- For effective control total cover of all plant surfaces is essential, but avoid run-off
- Target pests quickly become immobilised but 3-5 d may be required for maximum mortality
- Indoor applications should be made through a hydraulic nozzle applicator or a knapsack applicator. Outdoors suitable high volume hydraulic nozzle applicators should be used
- Limited data shows abamectin only slightly harmful to Anthocorid bugs and so compatible with biological control systems in which Anthocorid bugs are important

Restrictions
- Number of treatments 6 on protected tomatoes and cucumbers (only 4 of which can be made when flowers or fruit present); not restricted on flowers but rotation with other products advised
- Maximum concentration must not exceed 50 ml per 100 l water
- Do not mix with wetters, stickers or other adjuvants
- Do not use on ferns (*Adiantum* spp) or Shasta daisies
- Do not treat protected crops which are in flower or have set fruit between 1 Nov and 28 Feb
- Do not treat cherry tomatoes (but protected cherry tomatoes may be treated off-label)
- Consult manufacturer for list of plant varieties tested for safety
- There is insufficient evidence to support product compatibility with integrated and biological pest control programmes
- Unprotected persons must be kept out of treated areas until the spray has dried

Crop-specific information
- HI 3 d for protected edible crops
- On tomato or cucumber crops that are in flower, or have started to set fruit, treat only between 1 Mar and 31 Oct. Seedling tomatoes that have not started to flower or set fruit may be treated at any time
- Some spotting or staining may occur on carnation, kalanchoe and begonia foliage

Environmental safety
- Dangerous for the environment
- Very toxic to aquatic organisms
- High risk to bees. Do not apply to crops in flower or to those in which bees are actively foraging. Do not apply when flowering weeds are present
- Keep in original container, tightly closed, in a safe place, under lock and key
- Where bumble bees are used in tomatoes as pollinators, keep them out for 24 h after treatment

Hazard classification and safety precautions
Hazard Harmful, Dangerous for the environment, Harmful if swallowed, Very toxic to aquatic organisms
Transport code 9 [1-3]
Packaging group III
UN Number 3082
Risk phrases H319, H373
Operator protection U02a, U05a, U20a; A, C, D, H, K, M
Environmental protection E02a (until spray has dried), E12a, E12e, E15b, E34, E38, H410

FOR FULL CONDITIONS OF USE ALWAYS READ THE PRODUCT LABEL

Storage and disposal D01, D02, D05, D09b, D10c, D12a
Medical advice M04a

3   acequinocyl

A miticide for use on hops
IRAC mode of action code: 20b

Products

1	Kanemite SC	Certis Belchim B V	164 g/l	SC	19281

Uses
- Mites in **Apples** [1]; **Cherries** [1]; **Pears** [1]; **Plums** [1]; **Protected plums** [1];
- Red spider mites in **Protected tomatoes** *(off-label)* [1];
- Spider mites in **Apples** [1]; **Cherries** [1]; **Pears** [1]; **Plums** [1]; **Protected plums** [1];
- Two-spotted spider mite in **Hops** *(off-label)* [1];

Extension of Authorisation for Minor Use (EAMUs)
- **Hops** *20222409* [1]
- **Protected tomatoes** *20222071* [1]

Approval information
- Acequinocyl included in Annex I under EC Regulation 1107/2009

Efficacy guidance
- An acaricide with contact mode of action and some ingestion, adjust dose according to tree height and canopy density.
- Optimum timing is when 70% of the spider mites have hatched

Restrictions
- May be harmful to some biological control agents
- Alternate use with insecticides of a different mode of action to reduce the risk of the development of resistance
- Low drift spraying equipment must be operated according to the specific conditions stated in the official three star rating for that equipment as published on HSE Chemicals Regulation Division?s website. These operating conditions must be maintained until the operator is 30m from the top of the bank of any surface water bodies
- The maximum concentration of 0.09% (90 ml per 100 L of water) must not be exceeded
- Treated crops must not be handled for at least 4 days after treatment

Environmental safety
- Broadcast air-assisted LERAP [1] (20m pre BBCH70, 15m post BBCH 70)

Hazard classification and safety precautions
    Hazard Very toxic to aquatic organisms
    Transport code 9 [1]
    Packaging group III
    UN Number 3082
    Risk phrases H317, H373
    Operator protection U05a, U08, U20b, U24a; A, H
    Environmental protection E17b (20m pre BBCH70, 15m post BBCH 70), H410
    Storage and disposal D01, D02, D05, D07, D09a, D10b
    Medical advice M03

4   acetamiprid

A neonicotinoid insecticide for use in top fruit and horticulture
IRAC mode of action code: 4A

Products

1	Clayton Vault	Clayton	20% w/w	SG	18181

SEE SECTION 3 FOR PRODUCTS ALSO REGISTERED

Products – continued

2	Gazelle SG	Certis Belchim B V	20% w/w	SG	13725
3	Insyst	Certis Belchim B V	20% w/w	SP	19873
4	Pan Vulcan SG	Pan Agriculture	20% w/w	SG	19502
5	Vulcan	Pan Agriculture	20% w/w	SG	16689
6	Vulcan SP	Pan Agriculture	20% w/w	SG	18664

Uses

* Aphids in **Apples** [1-2,4-6]; **Asparagus** *(off-label)* [2]; **Carrots** *(off-label)* [2]; **Celeriac** *(off-label)* [2]; **Chard** *(off-label)* [2]; **Cherries** [1-2,4-6]; **Cress** *(off-label)* [2]; **Herbs (see appendix 6)** *(off-label)* [2]; **Hops** *(off-label)* [2]; **Horseradish** *(off-label)* [2]; **Jerusalem artichokes** *(off-label)* [2]; **Lamb's lettuce** *(off-label)* [2]; **Leaf brassicas** *(off-label)* [2]; **Lettuce** *(off-label)* [2]; **Ornamental plant production** [1-2,4-6]; **Parsley** *(off-label)* [2]; **Parsley root** *(off-label)* [2]; **Parsnips** *(off-label)* [2]; **Pears** [1-2,4-6]; **Plums** [1-2,4-6]; **Protected aubergines** [1-2,4-6]; **Protected cress** *(off-label)* [2]; **Protected forest nurseries** *(off-label)* [2]; **Protected herbs (see appendix 6)** *(off-label)* [2]; **Protected hops** *(off-label)* [2]; **Protected lamb's lettuce** *(off-label)* [2]; **Protected leaf brassicas** *(off-label)* [2]; **Protected ornamentals** [1-2,4-6]; **Protected parsley** *(off-label)* [2]; **Protected peppers** [1-2,4-6]; **Protected rocket** *(off-label)* [2]; **Protected soft fruit** *(off-label)* [2]; **Protected tomatoes** [1-2,4-6]; **Radishes** *(off-label)* [2]; **Red beet** *(off-label)* [2]; **Rocket** *(off-label)* [2]; **Seed potatoes** [3]; **Soft fruit** *(off-label)* [2]; **Spinach** *(off-label)* [2]; **Spinach beet** *(off-label)* [2]; **Spring oilseed rape** [3]; **Sugar beet** [3]; **Swedes** *(off-label)* [2]; **Top fruit** *(off-label)* [2]; **Turnips** *(off-label)* [2]; **Ware potatoes** [3]; **Winter oilseed rape** [3];
* Asparagus beetle in **Asparagus** *(off-label)* [2];
* Cabbage root fly in **Swedes** *(off-label)* [2]; **Turnips** *(off-label)* [2];
* Common oak thelaxid in **Protected forest nurseries** *(off-label)* [2];
* Gall wasp in **Protected forest nurseries** *(off-label)* [2];
* Great spruce bark beetle in **Protected forest nurseries** *(off-label)* [2];
* Green aphid in **Protected forest nurseries** *(off-label)* [2];
* Large pine weevil in **Forest** *(off-label)* [2];
* Pine weevil in **Forest nurseries** *(off-label)* [2];
* Whitefly in **Apples** [1-2,4-6]; **Cherries** [1-2,4-6]; **Ornamental plant production** [1-2,4-6]; **Pears** [1-2,4-6]; **Plums** [1-2,4-6]; **Protected aubergines** [1-2,4-6]; **Protected ornamentals** [1-2,4-6]; **Protected peppers** [1-2,4-6]; **Protected tomatoes** [1-2,4-6];
* Woolly aphid in **Protected forest nurseries** *(off-label)* [2];

Extension of Authorisation for Minor Use (EAMUs)

* **Asparagus** *20201841* [2]
* **Carrots** *20220382* [2]
* **Celeriac** *20220382* [2]
* **Chard** *20113144* [2]
* **Cress** *20192251* [2]
* **Forest** *20121068* [2]
* **Forest nurseries** *20200452* [2]
* **Herbs (see appendix 6)** *20192251* [2]
* **Hops** *20082857* [2]
* **Horseradish** *20220382* [2]
* **Jerusalem artichokes** *20220382* [2]
* **Lamb's lettuce** *20192250* [2]
* **Leaf brassicas** *20192250* [2]
* **Lettuce** *20192250* [2]
* **Parsley** *20192250* [2]
* **Parsley root** *20220382* [2]
* **Parsnips** *20220382* [2]
* **Protected cress** *20192251* [2]
* **Protected forest nurseries** *20162653* [2]
* **Protected herbs (see appendix 6)** *20192251* [2]
* **Protected hops** *20082857* [2]
* **Protected lamb's lettuce** *20192250* [2]
* **Protected leaf brassicas** *20192250* [2]
* **Protected parsley** *20192250* [2]

FOR FULL CONDITIONS OF USE ALWAYS READ THE PRODUCT LABEL

- *Protected rocket* *20192250* [2]
- *Protected soft fruit* *20082857* [2]
- *Radishes* *20220382* [2]
- *Red beet* *20220382* [2]
- *Rocket* *20192250* [2]
- *Soft fruit* *20082857* [2]
- *Spinach* *20192251* [2]
- *Spinach beet* *20113144* [2]
- *Swedes* *20220382* [2]
- *Top fruit* *20082857* [2]
- *Turnips* *20220382* [2]

Approval information
- Acetamiprid included in Annex I under EC Regulation 1107/2009

Efficacy guidance
- Best results obtained from application at the first sign of pest attack or when appropriate thresholds are reached
- Thorough coverage of foliage is essential to ensure best control. Acetamiprid has contact, systemic and translaminar activity

Restrictions
- Maximum number of treatments 1 per yr for cherries, ware potatoes; 2 per yr or crop for apples, pears, plums, ornamental plant production, protected crops, seed potatoes
- Do not use more than two applications of any neonicotinoid insecticide on any crop. Previous soil or seed treatment with a neonicotinoid counts as one such treatment
- 2 treatments are permitted on seed potatoes but must not be used consecutively [3]
- Broadcast air assisted sprayers must only be used where the operator?s normal working position is within a closed cab with a suitable in-cab filtration system* during application in protected situations. *Closed cabin meeting at least EN 15695 category [2]

Crop-specific information
- HI 3 d for protected crops; 14 d for apples, cherries, pears, plums, potatoes

Environmental safety
- Harmful to aquatic organisms
- Acetamiprid is slightly toxic to predatory mites and generally slightly toxic to other beneficials
- Aquatic buffer zone 12 m when used Under EAMU approval in sugar beet
- To protect non target insects/arthropods respect an unsprayed buffer zone of 5 m to non-crop land and do not apply after 31 July
- Broadcast air-assisted LERAP [1-2,4-6] (18 m); LERAP Category B [1-6]

Hazard classification and safety precautions
Hazard Dangerous for the environment, Harmful if swallowed
UN Number N/C
Risk phrases H361 [3]
Operator protection U02a, U19a, U20a [3], U20b [1-2,4-6]; A, D, H
Environmental protection E15b, E16a, E16b, E17b (18 m) [1-2,4-6], E38 [3], H410 [1-4,6], H412 [5]
Storage and disposal D05 [1-2,4-6], D09a, D10b, D12a [3]

5   acetic acid

A non-selective herbicide for non-crop situations

Products

1	New-Way Weed Spray	Headland Amenity	240 g/l	SL	15319

Uses
- General weed control in *Hard surfaces* [1]; *Natural surfaces not intended to bear vegetation* [1]; *Permeable surfaces overlying soil* [1];

SEE SECTION 3 FOR PRODUCTS ALSO REGISTERED

Approval information
- Acetic acid is included in Annex 1 under EC Regulation 1107/2009

Efficacy guidance
- Best results obtained from treatment of young tender weeds less than 10 cm high
- Treat in spring and repeat as necessary throughout the growing season
- Treat survivors as soon as fresh growth is seen
- Ensure complete coverage of foliage to the point of run-off
- Rainfall after treatment may reduce efficacy

Restrictions
- No restriction on number of treatments

Following crops guidance
- There is no residual activity in the soil and sowing or planting may take place as soon as treated weeds have died

Environmental safety
- Harmful to aquatic organisms
- Buffer zone requirement 1 m
- Applications must not be made via tractor mounted horizontal boom sprayers
- High risk to bees
- Do not apply to crops in flower or to those in which bees are actively foraging. Do not apply when flowering weeds are present
- Keep people and animals off treated dense weed patches until spray has dried. This is not necessary for areas with occasional low growing prostrate weeds such as on pathways

Hazard classification and safety precautions
Hazard Irritant
UN Number N/C
Risk phrases H315, H318
Operator protection U05a, U09a, U19a, U20b; A, C, D, H
Environmental protection E12a, E12e, E15a
Storage and disposal D01, D02, D09a, D12a

6   aclonifen

A pre-emergence herbicide for broad-leaf and grass weed control in potatoes
HRAC mode of action code: 32 (F3)

Products

1 Bandur	Bayer CropScience	600 g/l	SC	19679
2 Emerger	Bayer CropScience	600 g/l	SC	19056
3 Proclus	Bayer CropScience	600 g/l	SC	19387

Uses
- Annual dicotyledons in *Angelica* (off-label) [2]; *Caraway* (off-label) [2]; *Carrots* (off-label) [2]; *Celeriac* (off-label) [2]; *Celery (outdoor)* (off-label) [2]; *Celery leaves* (off-label) [2]; *Coriander* (off-label) [2]; *Dill* (off-label) [2]; *Fennel leaves* (off-label) [2]; *Garlic* (off-label) [2]; *Jerusalem artichokes* (off-label) [2]; *Leeks* (off-label) [2]; *Lovage* (off-label) [2]; *Onions* (off-label) [2]; *Parsley* (off-label) [2]; *Parsley root* (off-label) [2]; *Parsnips* (off-label) [2]; *Potatoes* [2]; *Salad burnet* (off-label) [2]; *Salad onions* (off-label) [2]; *Shallots* (off-label) [2]; *Sweet ciceley* (off-label) [2]; *Winter barley* [3]; *Winter wheat* [3];
- Black bindweed in *Rhubarb* (off-label) [2]; *Sunflowers* (off-label) [2];
- Blackgrass in *Winter barley* [1]; *Winter wheat* [1];
- Chickweed in *Rhubarb* (off-label) [2]; *Sunflowers* (off-label) [2];
- Cleavers in *Rhubarb* (off-label) [2]; *Sunflowers* (off-label) [2]; *Winter barley* [1]; *Winter wheat* [1];
- Crane's-bill in *Sunflowers* (off-label) [2];
- Fat hen in *Potatoes* [2]; *Rhubarb* (off-label) [2]; *Sunflowers* (off-label) [2]; *Winter barley* [1,3]; *Winter wheat* [1,3];

FOR FULL CONDITIONS OF USE ALWAYS READ THE PRODUCT LABEL

- Field pansy in **Winter barley** [1]; **Winter wheat** [1];
- Knotgrass in **Rhubarb** *(off-label)* [2]; **Sunflowers** *(off-label)* [2];
- Mayweeds in **Potatoes** [2]; **Sunflowers** *(off-label)* [2]; **Winter barley** [1,3]; **Winter wheat** [1,3];
- Redshank in **Rhubarb** *(off-label)* [2]; **Sunflowers** *(off-label)* [2];
- Shepherd's purse in **Rhubarb** *(off-label)* [2]; **Sunflowers** *(off-label)* [2];
- Small nettle in **Rhubarb** *(off-label)* [2]; **Sunflowers** *(off-label)* [2];
- Sowthistle in **Rhubarb** *(off-label)* [2]; **Sunflowers** *(off-label)* [2];
- Speedwells in **Rhubarb** *(off-label)* [2]; **Sunflowers** *(off-label)* [2];
- Volunteer oilseed rape in **Rhubarb** *(off-label)* [2]; **Sunflowers** *(off-label)* [2];

Extension of Authorisation for Minor Use (EAMUs)
- **Angelica** *20194485* [2]
- **Caraway** *20194485* [2]
- **Carrots** *20201101* [2]
- **Celeriac** *20201101* [2]
- **Celery (outdoor)** *20201100* [2]
- **Celery leaves** *20194485* [2]
- **Coriander** *20194485* [2]
- **Dill** *20194485* [2]
- **Fennel leaves** *20194485* [2]
- **Garlic** *20201102* [2]
- **Jerusalem artichokes** *20201101* [2]
- **Leeks** *20201100* [2]
- **Lovage** *20194485* [2]
- **Onions** *20201102* [2]
- **Parsley** *20194485* [2]
- **Parsley root** *20201101* [2]
- **Parsnips** *20201101* [2]
- **Rhubarb** *20222274* [2]
- **Salad burnet** *20194485* [2]
- **Salad onions** *20201100* [2]
- **Shallots** *20201102* [2]
- **Sunflowers** *20191615* [2]
- **Sweet ciceley** *20194485* [2]

Approval information
- Aclonifen included in Annex I under EC Regulation 1107/2009

Efficacy guidance
- DO NOT disrupt the soil surface after application as this will reduce the level of weed control provided

Restrictions
- Horizontal boom sprayers must be fitted with three star drift reduction technology for all uses
- Low drift spraying equipment must be operated according to the specific conditions stated in the official three star rating for that equipment as published on HSE Chemicals Regulation Division?s website. These operating conditions must be maintained until the operator is 30m from the top of the bank of any surface water bodies

Following crops guidance
- Oilseed rape, field pea, dwarf french bean, wheat, barley, oat, winter triticale, winter rye, sugar beet, maize, and sunflower can be established in the normal rotation or in the event of failure of the treated crop, provided that cultivation (using cultivator, disc harrow or similar) and thorough mixing of treated soil has been conducted to at least 10 cm depth.
- Radish and phacelia require a period of at least 120 days after application, or plough to at least 20 cm before drilling.

Environmental safety
- Aquatic buffer zone requirement 6 m
- LERAP Category B [1-3]

SEE SECTION 3 FOR PRODUCTS ALSO REGISTERED

Hazard classification and safety precautions
    Hazard Very toxic to aquatic organisms
    Transport code 9 [1-3]
    Packaging group III
    UN Number 3082
    Risk phrases H351
    Operator protection U05a, U20b; A
    Environmental protection E16a, H410
    Storage and disposal D01, D02, D05, D09a, D10b, D12c [1]

7 aluminium ammonium sulphate

An inorganic bird and animal repellent

Products

1	Curb Liquid Crop Spray	Sphere	83.3 g/l	LI	18643

Uses
- Animal repellent in *Amenity grassland* [1]; *Beans without pods (dry)* [1]; *Beans without pods (Fresh)* [1]; *Bilberries* [1]; *Blackberries* [1]; *Blackcurrants* [1]; *Blueberries* [1]; *Broad beans* [1]; *Broccoli* [1]; *Brussels sprouts* [1]; *Cabbages* [1]; *Calabrese* [1]; *Carrots* [1]; *Cauliflowers* [1]; *Combining peas* [1]; *Cranberries* [1]; *Durum wheat* [1]; *Dwarf beans* [1]; *French beans* [1]; *Gooseberries* [1]; *Loganberries* [1]; *Managed amenity turf* [1]; *Raspberries* [1]; *Redcurrants* [1]; *Rubus hybrids* [1]; *Runner beans* [1]; *Rye* [1]; *Soya beans* [1]; *Spring barley* [1]; *Spring field beans* [1]; *Spring oats* [1]; *Spring oilseed rape* [1]; *Spring wheat* [1]; *Strawberries* [1]; *Sugar beet* [1]; *Triticale* [1]; *Vining peas* [1]; *Winter barley* [1]; *Winter field beans* [1]; *Winter oats* [1]; *Winter oilseed rape* [1]; *Winter wheat* [1];
- Bird repellent in *Amenity grassland* [1]; *Beans without pods (dry)* [1]; *Beans without pods (Fresh)* [1]; *Bilberries* [1]; *Blackberries* [1]; *Blackcurrants* [1]; *Blueberries* [1]; *Broad beans* [1]; *Broccoli* [1]; *Brussels sprouts* [1]; *Cabbages* [1]; *Calabrese* [1]; *Carrots* [1]; *Cauliflowers* [1]; *Combining peas* [1]; *Cranberries* [1]; *Durum wheat* [1]; *Dwarf beans* [1]; *French beans* [1]; *Gooseberries* [1]; *Loganberries* [1]; *Managed amenity turf* [1]; *Raspberries* [1]; *Redcurrants* [1]; *Rubus hybrids* [1]; *Runner beans* [1]; *Rye* [1]; *Soya beans* [1]; *Spring barley* [1]; *Spring field beans* [1]; *Spring oats* [1]; *Spring oilseed rape* [1]; *Spring wheat* [1]; *Strawberries* [1]; *Sugar beet* [1]; *Triticale* [1]; *Vining peas* [1]; *Winter barley* [1]; *Winter field beans* [1]; *Winter oats* [1]; *Winter oilseed rape* [1]; *Winter wheat* [1];

Approval information
- aluminium ammonium sulphate is included in Annex 1 under EC Regulation 1107/2009

Efficacy guidance
- Apply as overall spray to growing crops before damage starts or mix powder with seed depending on type of protection required
- Spray deposit protects growth present at spraying but gives little protection to new growth
- Product must be sprayed onto dry foliage to be effective and must dry completely before dew or frost forms. In winter this may require some wind

Crop-specific information
- Latest use: no restriction

Hazard classification and safety precautions
    UN Number N/C
    Operator protection U05a, U20a; A
    Environmental protection E15a, E19b
    Storage and disposal D01, D02, D05, D11a, D12a
    Medical advice M03

FOR FULL CONDITIONS OF USE ALWAYS READ THE PRODUCT LABEL

8   aluminium phosphide

A phosphine generating compound used against vertebrates and grain store pests. Under new regulations any person wishing to purchase or use Aluminium Phosphide must be suitably qualified and licensed by January 2015. A new licensing body has been set up.
IRAC mode of action code: 24A

Products

1	Phostoxin	Rentokil	56% w/w	GE	17000
2	Talunex	Killgerm	56% w/w	GE	17001
3	Talunex	Killgerm	56% w/w	GE	UK16-1044
4	Talunex	Killgerm	56% w/w	GE	UK18-1118

Uses
- Moles in *All situations* [1-4]; *Farmland* [1];
- Rabbits in *All situations* [1-4]; *Farmland* [1];
- Rats in *All situations* [1-4]; *Farmland* [1];

Approval information
- Aluminium phosphide is included in Annex 1 under EC Regulation 1107/2009
- Accepted by BBPA for use in stores for malting barley

Efficacy guidance
- Product releases poisonous hydrogen phosphide gas in contact with moisture
- Place pellets in burrows or runs and seal hole by heeling in or covering with turf. Do not cover pellets with soil. Inspect daily and treat any new or re-opened holes [1, 2]

Restrictions
- Aluminium phosphide is subject to the Poisons Rules 1982 and the Poisons Act 1972. See Section 5 for more information
- Only to be used by operators instructed or trained in the use of aluminium phosphide and familiar with the precautionary measures to be taken. See label and HSE Guidance Notes for full precautions
- Only open container outdoors [1, 2] and for immediate use. Keep away from liquid or water as this causes immediate release of gas. Do not use in wet weather. Must not be placed or allowed to remain on the ground surface.
- Do not use within 10 m of human or animal habitation. Before application ensure that no humans or domestic animals are in adjacent buildings or structures. Allow a minimum airing-off period of 4 h before re-admission
- Before applying this product to a ship cargo hold, users must follow current guidance issued by the International Maritime Organisation?s Marine Safety Committee on the safe use of pesticides

Environmental safety
- Product liberates very toxic, highly flammable gas
- Dangerous to fish or other aquatic life. Do not contaminate surface waters or ditches with chemical or used container
- Prevent access by livestock, pets and other non-target mammals and birds to buildings under fumigation and ventilation
- Pellets must never be placed or allowed to remain on ground surface
- Do not use adjacent to watercourses
- Take particular care to avoid gassing non-target animals, especially those protected under the Wildlife and Countryside Act (e.g. badgers, polecat, reptiles, natterjack toads, most birds). Do not use in burrows where there is evidence of badger or fox activity, or when burrows might be occupied by birds
- Dust remaining after decomposition is harmless and of no environmental hazard
- Keep in original container, tightly closed, in a safe place, under lock and key
- Dispose of empty containers as directed on label

SEE SECTION 3 FOR PRODUCTS ALSO REGISTERED

Hazard classification and safety precautions

> **Hazard** Very toxic, Harmful, Highly flammable [1], Dangerous for the environment [1], In contact with water releases flammable gases which may ignite spontaneously, Fatal if swallowed, Toxic in contact with skin, Fatal if inhaled, Very toxic to aquatic organisms
> **Transport code** 4.1 [2-4], 4.3 [1]
> **Packaging group** I
> **UN Number** 1397
> **Risk phrases** H315, H318
> **Operator protection** U01, U05a, U07, U13, U19a, U20a; A, D, H
> **Environmental protection** E02a (4 h), E13b, E34
> **Storage and disposal** D01, D02, D07, D09b, D11b
> **Vertebrate/rodent control products** V04a
> **Medical advice** M04a

9   ametoctradin + dimethomorph

A systemic and protectant fungicide for potato blight control
FRAC mode of action code: 40 + 45

Products

1	Percos	BASF	300:225 g/l	SC	19700
2	Zampro DM	BASF	300:225 g/l	SC	19701

Uses

- Blight in **Potatoes** [1-2];
- Downy mildew in **Hops** *(off-label)* [1]; **Ornamental plant production** *(off-label)* [1]; **Protected ornamentals** *(off-label)* [1]; **Table grapes** *(off-label)* [1]; **Wine grapes** *(off-label)* [1];
- Tuber blight in **Potatoes** *(reduction)* [1-2];

Extension of Authorisation for Minor Use (EAMUs)

- **Hops** *20210964* [1]
- **Ornamental plant production** *20210962* [1]
- **Protected ornamentals** *20210962* [1]
- **Table grapes** *20210963* [1]
- **Wine grapes** *20210963* [1]

Approval information

- Ametoctradin and dimethomorph are included in Annex 1 under EC Regulation 1107/2009.

Efficacy guidance

- Application to dry foliage is rainfast within 1 hour of drying on the leaf.
- Will give effective control of phenylamide-resistant strains of blight

Crop-specific information

- HI 7 days for potatoes

Hazard classification and safety precautions

> **Hazard** Harmful, Dangerous for the environment, Harmful if swallowed
> **UN Number** N/C
> **Risk phrases** H317 [1]
> **Operator protection** U05a
> **Environmental protection** E15b, E34, E38, H410
> **Storage and disposal** D01, D02, D09a, D10c
> **Medical advice** M03, M05a

10   amidosulfuron

A post-emergence sulfonylurea herbicide for cleavers and other broad-leaved weed control in cereals
HRAC mode of action code: 2 (B)

Products

1 Eagle	Sumitomo	75% w/w	WG	18902
2 Squire Ultra	Sumitomo	75% w/w	WG	18906

Uses

- Annual dicotyledons in **Durum wheat** [1]; **Grassland** [2]; **Linseed** [1]; **Ornamental plant production** *(off-label)* [2]; **Spring barley** [1]; **Spring oats** [1]; **Spring rye** [1]; **Spring wheat** [1]; **Triticale** [1]; **Winter barley** [1]; **Winter oats** [1]; **Winter rye** [1]; **Winter wheat** [1];
- Charlock in **Corn Gromwell** *(off-label)* [1]; **Durum wheat** [1]; **Farm forestry** *(off-label)* [1]; **Forest nurseries** *(off-label)* [1]; **Game cover** *(off-label)* [1]; **Linseed** [1]; **Ornamental plant production** *(off-label)* [1]; **Spring barley** [1]; **Spring oats** [1]; **Spring rye** [1]; **Spring wheat** [1]; **Triticale** [1]; **Winter barley** [1]; **Winter oats** [1]; **Winter rye** [1]; **Winter wheat** [1];
- Cleavers in **Corn Gromwell** *(off-label)* [1]; **Durum wheat** [1]; **Farm forestry** *(off-label)* [1]; **Forest nurseries** *(off-label)* [1]; **Game cover** *(off-label)* [1]; **Linseed** [1]; **Ornamental plant production** *(off-label)* [1]; **Spring barley** [1]; **Spring oats** [1]; **Spring rye** [1]; **Spring wheat** [1]; **Triticale** [1]; **Winter barley** [1]; **Winter oats** [1]; **Winter rye** [1]; **Winter wheat** [1];
- Docks in **Grassland** [2];
- Forget-me-not in **Corn Gromwell** *(off-label)* [1]; **Durum wheat** [1]; **Farm forestry** *(off-label)* [1]; **Forest nurseries** *(off-label)* [1]; **Game cover** *(off-label)* [1]; **Linseed** [1]; **Ornamental plant production** *(off-label)* [1]; **Spring barley** [1]; **Spring oats** [1]; **Spring rye** [1]; **Spring wheat** [1]; **Triticale** [1]; **Winter barley** [1]; **Winter oats** [1]; **Winter rye** [1]; **Winter wheat** [1];
- Shepherd's purse in **Corn Gromwell** *(off-label)* [1]; **Durum wheat** [1]; **Farm forestry** *(off-label)* [1]; **Forest nurseries** *(off-label)* [1]; **Game cover** *(off-label)* [1]; **Linseed** [1]; **Ornamental plant production** *(off-label)* [1]; **Spring barley** [1]; **Spring oats** [1]; **Spring rye** [1]; **Spring wheat** [1]; **Triticale** [1]; **Winter barley** [1]; **Winter oats** [1]; **Winter rye** [1]; **Winter wheat** [1];

Extension of Authorisation for Minor Use (EAMUs)

- **Corn Gromwell** *20193096* [1]
- **Farm forestry** *20193093* [1]
- **Forest nurseries** *20193093* [1]
- **Game cover** *20193093* [1]
- **Ornamental plant production** *20193093* [1], *20201444* [2]

Approval information

- Accepted by BBPA for use on malting barley
- Amidosulfuron included in Annex I under EC Regulation 1107/2009

Efficacy guidance

- For best results apply in spring (from 1 Feb) in warm weather when soil moist and weeds growing actively. When used in grassland following cutting or grazing, docks should be allowed to regrow before treatment
- Weed kill is slow, especially under cool, dry conditions. Weeds may sometimes only be stunted but will have little or no competitive effect on crop
- May be used on all soil types unless certain sequences are used on linseed. See label
- Spray is rainfast after 1 h
- Cleavers controlled from emergence to flower bud stage. If present at application charlock (up to flower bud), shepherd's purse (up to flower bud) and field forget-me-not (up to 6 leaves) will also be controlled
- Amidosulfuron is a member of the ALS-inhibitor group of herbicides and products should be used in a planned Resistance Management strategy. See Section 5 for more information

Restrictions

- Maximum number of treatments 1 per crop [1]
- Use after 1 Feb and do not apply to rotational grass after 30 Jun, or to permanent grassland after 15 Oct [2]

SEE SECTION 3 FOR PRODUCTS ALSO REGISTERED

- Do not apply to crops undersown or due to be undersown with clover or alfalfa [1]
- Do not spray crops under stress, suffering drought, waterlogged, grazed, lacking nutrients or if soil compacted
- Do not spray if frost expected
- Do not roll or harrow within 1 wk of spraying
- Specific restrictions apply to use in sequence or tank mixture with other sulfonylurea or ALS-inhibiting herbicides. See label for details. There are no recommendations for mixtures with metsulfuron-methyl products on linseed
- Certain mixtures with fungicides are expressly forbidden. See label for details
- Livestock must be kept out of treated areas for at least 7 days following treatment and until poisonous weeds such as ragwort have died and become unpalatable [2]

Crop-specific information
- Latest use: before first spikelets just visible (GS 51) for cereals; before flower buds visible for linseed; 15 Oct for grassland
- Broadcast cereal crops should be sprayed post-emergence after plants have a well established root system

Following crops guidance
- If a treated crop fails cereals may be sown after 15 d and thorough cultivation
- After normal harvest of a treated crop only cereals, winter oilseed rape, mustard, turnips, winter field beans or vetches may be sown in the same year as treatment and these must be preceded by ploughing or thorough cultivation
- Only cereals may be sown within 12 mth of application to grassland
- Cereals or potatoes must be sown as the following crop after use of permitted mixtures or sequences with other sulfonylurea herbicides in cereals. Only cereals may be sown after the use of such sequences in linseed

Environmental safety
- Take care to wash out sprayers thoroughly. See label for details
- Avoid drift onto neighbouring broad-leaved plants or onto surface waters or ditches

Hazard classification and safety precautions
    Hazard Dangerous for the environment
    Transport code 9 [1-2]
    Packaging group III
    UN Number 3077
    Operator protection U20a [1], U20b [2]
    Environmental protection E07b (1 week) [2], E15a, E38, E41 [2], H410
    Storage and disposal D10a, D12a

11   amidosulfuron + iodosulfuron-methyl-sodium

A post-emergence sulfonylurea herbicide mixture for cereals
HRAC mode of action code: 2 + 2 (B + B)

See also iodosulfuron-methyl-sodium

Products
1 Chekker   Bayer CropScience   12.5:1.25% w/w   WG   16495

Uses
- Annual dicotyledons in *Forest nurseries (off-label)* [1]; *Spring barley* [1]; *Spring rye* [1]; *Spring wheat* [1]; *Triticale* [1]; *Winter barley* [1]; *Winter rye* [1]; *Winter wheat* [1];
- Chickweed in *Forest nurseries (off-label)* [1]; *Spring barley* [1]; *Spring rye* [1]; *Spring wheat* [1]; *Triticale* [1]; *Winter barley* [1]; *Winter rye* [1]; *Winter wheat* [1];
- Cleavers in *Forest nurseries (off-label)* [1]; *Spring barley* [1]; *Spring rye* [1]; *Spring wheat* [1]; *Triticale* [1]; *Winter barley* [1]; *Winter rye* [1]; *Winter wheat* [1];
- Mayweeds in *Forest nurseries (off-label)* [1]; *Spring barley* [1]; *Spring rye* [1]; *Spring wheat* [1]; *Triticale* [1]; *Winter barley* [1]; *Winter rye* [1]; *Winter wheat* [1];

FOR FULL CONDITIONS OF USE ALWAYS READ THE PRODUCT LABEL

- Volunteer oilseed rape in **Forest nurseries** *(off-label)* [1]; **Spring barley** [1]; **Spring rye** [1]; **Spring wheat** [1]; **Triticale** [1]; **Winter barley** [1]; **Winter rye** [1]; **Winter wheat** [1];

Extension of Authorisation for Minor Use (EAMUs)
- **Forest nurseries** *20142722* [1]

Approval information
- Amidosulfuron and iodosulfuron-methyl-sodium included in Annex I under EC Regulation 1107/2009
- Accepted by BBPA for use on malting barley

Efficacy guidance
- Best results obtained from treatment in warm weather when soil is moist and the weeds are growing actively
- Weeds must be present at application to be controlled
- Dry conditions resulting in moisture stress may reduce effectiveness
- Weed control is slow especially under cool dry conditions
- Occasionally weeds may only be stunted but they will normally have little or no competitive effect on the crop
- Amidosulfuron and iodosulfuron are members of the ALS-inhibitor group of herbicides and products should be used in a planned Resistance Management strategy. See Section 5 for more information

Restrictions
- Maximum number of treatments 1 per crop
- Must only be applied between 1 Feb in yr of harvest and specified latest time of application
- Do not apply to crops undersown or to be undersown with grass, clover or alfalfa
- Do not roll or harrow within 1 wk of spraying
- Do not spray crops under stress from any cause or if the soil is compacted
- Do not spray if rain or frost expected
- Do not apply in mixture or in sequence with any other ALS inhibitor

Crop-specific information
- Latest use: before first spikelet of inflorescence just visible (GS 51)
- Treat drilled crops after the 2-leaf stage; treat broadcast crops after the plants have a well-established root system
- Applications to spring barley may cause transient crop yellowing

Following crops guidance
- Cereals, winter oilseed rape and winter field beans may be sown in the same yr as treatment provided they are preceded by ploughing or thorough cultivation. Any crop may be sown in the spring of the yr following treatment
- A minimum of 3 mth must elapse between treatment and sowing winter oilseed rape

Environmental safety
- Dangerous for the environment
- Toxic to aquatic organisms
- Take extreme care to avoid damage by drift onto broad-leaved plants outside the target area or onto ponds, waterways and ditches
- Observe carefully label instructions for sprayer cleaning
- LERAP Category B [1]

Hazard classification and safety precautions
Hazard Irritant, Dangerous for the environment, Very toxic to aquatic organisms
Transport code 9 [1]
Packaging group III
UN Number 3077
Risk phrases H319
Operator protection U05a, U08, U11, U14, U15, U20b; A, C, H
Environmental protection E15a, E16a, E16b, E38, H410
Storage and disposal D01, D02, D10a, D12a

SEE SECTION 3 FOR PRODUCTS ALSO REGISTERED

12 amidosulfuron + iodosulfuron-methyl-sodium + mesosulfuron-methyl

A herbicide mixture for weed control in winter wheat
HRAC mode of action code: 2 + 2 + 2 (B + B + B)

See also amidosulfuron
*   amidosulfuron + iodosulfuron-methyl-sodium*
*   iodosulfuron-methyl-sodium*
*   mesosulfuron-methyl*

Products

1	Pacifica Plus	Bayer CropScience	0.5:0.1:0.3% w/w	WG	17272

Uses
- Annual dicotyledons in *Winter wheat* [1];
- Annual grasses in *Winter wheat* [1];
- Blackgrass in *Winter wheat* [1];

Approval information
- Amidosulfuron, iodosulfuron-methyl-sodium and mesosulfuron-methyl included in Annex I under EC Regulation 1107/2009
- Accepted by BBPA for use on malting barley

Efficacy guidance
- Apply as early as possible in spring and before stem extension stage of any grass weeds
- Monitor efficacy and investigate areas of poor control

Restrictions
- Do not apply in tank-mixture or sequence with any product containing any other sulfonylurea or ALS inhibiting herbicide
- Do not use on crops undersown with grasses, clover, legumes or any other broad-leaved crop
- Take care to avoid drift on to non-target plants

Following crops guidance
- Winter wheat and winter barley may be sown in the same year as harvest of a crop treated with 0.5 kg/ha. Winter oilseed rape may be sown after ploughing following a crop treated with 0.4 kg/ha. In the event of crop failure, sow only winter or spring wheat.

Environmental safety
- LERAP Category B [1]

Hazard classification and safety precautions
Hazard Very toxic to aquatic organisms
Transport code 9 [1]
Packaging group III
UN Number 3077
Risk phrases H317, H318
Operator protection U05a, U11, U20a; A, C, H
Environmental protection E15a, E16a, H410
Storage and disposal D01, D02, D05, D09a, D10b

13 aminopyralid

A pyridine carboxylic acid herbicide for weed control in grassland
HRAC mode of action code: 4 (O)

Products

1	Pro-Banish	Corteva	30 g/l	SL	18339

Uses
- Annual dicotyledons in *Grassland* [1];

FOR FULL CONDITIONS OF USE ALWAYS READ THE PRODUCT LABEL

Approval information
- Approvals of aminopyralid products have now been restored with extra advice stating that manure from animals fed on pasture treated with aminopyralid should not leave the farm.
- Aminopyralid is included in Annex 1 under 1107/2009.

Restrictions
- Livestock must be kept out of treated areas for at least 1 week following treatment and until poisonous weeds such as ragwort have died and become unpalatable.
- To protect groundwater do not apply to leys less than 1 year old.
- Users must have received adequate instruction, training and guidance in the safe and efficient use of the product and must take all reasonable precautions to protect the health of human beings, creatures and plants and safeguard the environment.
- This product must not be used on grassland grazed by animals other than cattle and sheep.
- The product must not be used on land where vegetation will be cut for animal feed, fodder or bedding nor for composting or mulching within one calendar year of treatment.

Following crops guidance
- In the event of failure of newly seeded treated grassland, grass may be re-seeded immediately or wheat may be sown provided 4 mth have elapsed since application
- Residues in plant tissues, including manure, may affect succeeding susceptible crops of peas, beans, other legumes, carrots, other Umbelliferae, potatoes, tomatoes, lettuce and other Compositae. These crops should not be sown within 3 mth of ploughing up treated grassland

Hazard classification and safety precautions
Hazard Dangerous for the environment
UN Number N/C
Operator protection U05a, U08, U11, U14, U23a; A, H
Environmental protection E06a (7 days), E07a, E15b, E38, H411
Consumer protection C01
Storage and disposal D01, D02, D05, D09a, D10b, D12a

14   aminopyralid + fluroxypyr

A foliar acting herbicide mixture for use in grassland
HRAC mode of action code: 4 + 4 (O + O)

See also fluroxypyr

Products
1 Synero	Corteva	30:100 g/l	EW	18578

Uses
- Buttercups in **Amenity grassland** [1];
- Chickweed in **Amenity grassland** [1];
- Dandelions in **Amenity grassland** [1];
- Docks in **Amenity grassland** [1];
- Stinging nettle in **Amenity grassland** [1];
- Thistles in **Amenity grassland** [1];

Approval information
- Aminopyralid and fluroxypyr are included in Annex I under EC Regulation 1107/2009.

Efficacy guidance
- For best results and to avoid crop check, grass and weeds must be growing actively
- Allow 2-3 wk after cutting for hay or silage for sufficient regrowth to occur before spraying and leave 7 d afterwards to allow maximum translocation
- Where there is a high reservoir of weed seed or a historically high weed population a programmed approach may be needed involving a second treatment in the following yr
- Control may be reduced if rain falls within 1 h of spraying

Restrictions
- Maximum number of treatments 1 per yr.

SEE SECTION 3 FOR PRODUCTS ALSO REGISTERED

- Do not apply to leys less than 1 year old.
- Do not use on grassland that will be used for animal feed, bedding, composting or mulching within 1 calender year of application.
- Do not use on grassland that will be grazed by animals other than cattle or sheep.
- Use of an antifoam is compulsory
- Do not use any treated plant material, or manure from animals fed on treated crops, for composting or mulching
- Do not use on crops grown for seed
- Manure from animals fed on pasture or silage treated with this herbicide should not leave the farm
- Do not use between 31st July and 1st March
- For applications made between 1st March and 31st May, 1 application may be made per calendar year
- For applications made between 1st June and 31st July, only 1 application may be made in a 2 year period
- Do not apply spot applications to more than 20% of the area: if infestation levels exceed this figure then a broadcast application should be made

Crop-specific information
- Treatment will kill clover
- Treatment may occasionally cause transient yellowing of the sward which is quickly outgrown
- Late treatments may lead to a slight transient leaning of grass that does not affect yield

Following crops guidance
- In the event of failure of newly seeded treated grassland, grass may be re-seeded immediately, or wheat may be sown provided 4 mth have elapsed since application
- Residues in plant tissues, including manure, may affect succeeding susceptible crops of peas, beans, other legumes, carrots, other Umbelliferae, potatoes, tomatoes, lettuce and other Compositae. These crops should not be sown within 3 mth of ploughing up treated grassland
- Only barley, wheat, oat, rye and maize can be grown in rotation following harvest of treated crops. Oilseed rape can only be planted after 1 year following application and all other rotational crops must not be planted until 2 years after application.

Environmental safety
- Dangerous for the environment
- Toxic to aquatic organisms
- Keep livestock out of treated areas for at least 7 d following treatment and until poisonous weeds, such as ragwort, have died down and become unpalatable
- Avoid damage by drift onto susceptible crops, non-target plants or waterways
- LERAP Category B [1]

Hazard classification and safety precautions
Hazard Irritant, Dangerous for the environment, Very toxic to aquatic organisms
Transport code 9 [1]
Packaging group III
UN Number 3082
Risk phrases H304, H315, H318, H336
Operator protection U05a, U08, U11, U14, U23a; A, C, H
Environmental protection E07a, E15b, E16a, E38, H410
Consumer protection C01
Storage and disposal D01, D02, D05, D09a, D10b, D12a

FOR FULL CONDITIONS OF USE ALWAYS READ THE PRODUCT LABEL

15   aminopyralid + metazachlor + picloram

A herbicide mixture for weed control in oilseed rape.
HRAC mode of action code: 15 + 4 (K3 + O)

See also metazachlor

Products
| 1 | Ralos | Corteva | 5.3:500:13.3 g/l | SC | 16737 |

Uses
- Chickweed in **Winter oilseed rape** [1];
- Dead nettle in **Winter oilseed rape** [1];
- Fat hen in **Winter oilseed rape** [1];
- Penny cress in **Winter oilseed rape** [1];
- Poppies in **Winter oilseed rape** [1];
- Scentless mayweed in **Winter oilseed rape** [1];

Approval information
- Aminopyralid, metazachlor and picloram are included in Annex 1 under 1107/2009.

Restrictions
- Applications shall be limited to a total dose of not more than 1.0 kg metazachlor/ha in a three year period on the same field.
- Do not use between 16th October and 31st July
- Livestock must be kept out of treated areas for at least 1 week following treatment and until poisonous weeds such as ragwort have died and become unpalatable
- Do not use on sands and very light soils or on soils with more than 10% organic matter.
- Straw from treated crops must not be baled and must stay on the field.
- Do not use treated plant material for composting or mulching.
- Do not use digestate from anaerobic digesters on crops such as peas, beans and other legumes, carrots and other Umbelliferae, potatoes, lettuce or other Compositae, glasshouse or protected crops.

Following crops guidance
- Only wheat, barley, oats, maize or oilseed rape should be planted within 4 months (120 days) of application. All other crops should wait until 12 months after application.
- In the event of crop failure, only spring wheat, spring barley, spring oats, maize or ryegrass should be planted after ploughing or thorough cultivation of the soil.

Environmental safety
- Metazachlor stewardship guidelines advise a maximum dose of 750 g.a.i/ha/annum. Applications to drained fields should be complete by 15th Oct but, if drains are flowing, complete applications by 1st Oct.
- LERAP Category B [1]

Hazard classification and safety precautions
Hazard Dangerous for the environment
Transport code 9 [1]
Packaging group III
UN Number 3082
Risk phrases H351
Operator protection U05a, U20a; A
Environmental protection E06c (1 week), E15b, E16a, E34, H410
Storage and disposal D01, D02, D09a, D10c
Medical advice M03

16   aminopyralid + propyzamide

A herbicide mixture for weed control in winter oilseed rape
HRAC mode of action code: 3 + 4 (K1 + O)

Products

1	AstroKerb	Corteva	6.3:500 g/l	SC	16184
2	Dymid	Corteva	5.3:500 g/l	SC	19187

Uses

- Annual dicotyledons in *Winter oilseed rape* [1-2];
- Annual grasses in *Winter oilseed rape* [1-2];
- Mayweeds in *Winter oilseed rape* [1-2];
- Poppies in *Winter oilseed rape* [1-2];

Approval information

- Aminopyralid and propyzamide are included in Annex 1 under EC Regulation 1107/2009.

Efficacy guidance

- Allow soil temperatures to drop before application to winter oilseed rape to optimise activity of propyzamide component

Restrictions

- Users must have received adequate instruction, training and guidance on the safe and efficient use of the product.
- Must not be used on land where vegetation will be cut for animal feed, fodder, bedding nor for composting or mulching within one calendar year of treatment.
- Do not remove oilseed rape straw from the field unless it is for burning for heat or electricity production. Do not use the oilseed rape straw for animal feed, animal bedding, composting or mulching.
- Livestock must be kept out of treated areas for at least 1 week following treatment and until poisonous weeds such as ragwort have died and become unpalatable

Crop-specific information

- Only cereals may follow application of aminopyralid + propyzamide and these should not be planted within 30 weeks of treatment.
- Aminopyralid residues in plant tissue which have not completely decayed may affect susceptible crops such as legumes (e.g. peas and beans), sugar and fodder beet, carrots and other Umbelliferae, potatoes, tomatoes, lettuce and other Compositae.

Following crops guidance

- May only be followed by spring or winter cereals. Treated land must be mouldboard ploughed to a depth of at least 15 cm before a following cereal crop is planted.
- Following crops (winter and spring wheat) should not be planted within 30 weeks of application [2]

Hazard classification and safety precautions

Hazard Harmful, Dangerous for the environment
Transport code 9 [1-2]
Packaging group III
UN Number 3082
Risk phrases H351
Operator protection U20c; A, H
Environmental protection E07b (1), E15b, E38, E39, H410
Storage and disposal D09a, D11a, D12a

17   aminopyralid + triclopyr

A foliar acting herbicide mixture for broad-leaved weed control in grassland
HRAC mode of action code: 4 + 4 (O + O)

See also triclopyr

Products

1	Forefront T	Corteva	30:240g/l	EW	15568
2	Garlon Ultra	Nomix Enviro	12:120 g/l	SL	19758
3	Icade	Corteva	12:120 g/l	SL	19069
4	Speedline Pro	Nomix Enviro	12:120 g/l	SL	19282

Uses

- Annual dicotyledons in **Amenity grassland** [2-4];
- Brambles in **Amenity grassland** [3-4];
- Broom in **Amenity grassland** [3-4];
- Buddleia in **Amenity grassland** [3-4];
- Buttercups in **Amenity grassland** [1]; **Grassland** [1];
- Common mugwort in **Amenity grassland** [3-4];
- Common nettle in **Amenity grassland** [1,3-4]; **Grassland** [1];
- Creeping thistle in **Amenity grassland** [3-4];
- Dandelions in **Amenity grassland** [1]; **Grassland** [1];
- Docks in **Amenity grassland** [1]; **Grassland** [1];
- Gorse in **Amenity grassland** [3-4];
- Hogweed in **Amenity grassland** [3-4];
- Japanese knotweed in **Amenity grassland** [3-4];
- Perennial dicotyledons in **Amenity grassland** [3-4];
- Rosebay willowherb in **Amenity grassland** [3-4];
- Thistles in **Amenity grassland** [1]; **Grassland** [1];

Approval information

- Aminopyralid and triclopyr are included in Annex I under EC Regulation 1107/2009.

Efficacy guidance

- For best results and to avoid crop check, grass and weeds must be growing actively
- Use adeqate water volume to ensure good weed coverage. Increase water volume if necessary where the weed population is high and where the grass is dense
- Allow 2-3 wk after cutting for hay or silage for sufficient regrowth to occur before spraying and leave 7 d afterwards to allow maximum translocation
- Where there is a high reservoir of weed seed or a historically high weed population a programmed approach may be needed involving a second treatment in the following yr
- Control may be reduced if rain falls within 1 h of spraying

Restrictions

- Maximum number of treatments 1 per yr.
- Do not apply to leys less than 1 year old.
- Do not apply by hand-held equipment [1]
- Do not use on grassland that will be used for animal feed, bedding, composting or mulching within 1 calendar year of application.
- Do not use on grassland that will be grazed by animals other than cattle or sheep.
- Do not use any treated plant material, or manure from animals fed on treated crops, for composting or mulching
- Do not use on crops grown for seed
- Manure from animals fed on pasture or silage from treated crops should not leave the farm
- Extreme care must be taken to avoid drift onto non-crop plants outside of the target area.
- For annual applications do not use between 31st July and 1st March; For applications made between 1st March and 1st June, 1 application may be made per calendar year; For applications made between 1st June and 1st August, only 1 application may be made in a 2 year period; For

applications made between 1st August and 31st August, only 1 application may be made in a 3 year period [3, 4]
- Do not apply spot applications to more than 20% of the area [3, 4]

Crop-specific information
- Latest use: 7 d before grazing or harvest for grassland
- Late applications may lead to transient leaning of grass which does not affect final yield

Following crops guidance
- Ensure that all plant remains of a treated crop have completely decayed before planting susceptible crops such as peas, beans and other legumes, sugar beet, carrots and other Umbelliferae, potatoes and tomatoes, lettuce and other Compositae
- Do not plant potatoes, sugar beet, vegetables, beans or other leguminous crops in the calendar yr following application

Environmental safety
- Dangerous for the environment
- Toxic to aquatic organisms
- Keep livestock out of treated areas for at least 7 d after treatment or until foliage of any poisonous weeds such as ragwort has died and become unpalatable
- To protect groundwater do not apply to grass leys less than 1 yr old
- Take extreme care to avoid drift onto susceptible crops, non-target plants or waterways. All conifers, especially pine and larch, are very sensitive and may be damaged by vapour drift in hot conditions
- LERAP Category B [1]

Hazard classification and safety precautions
Hazard Harmful, Irritant [1], Dangerous for the environment
Transport code 9 [1,4]
Packaging group III
UN Number 3082, N/C
Risk phrases H317 [1], H319, H373 [3]
Operator protection U05a, U08, U14 [1], U23a [1]; A, C, H, M
Environmental protection E06a (7 days) [1], E07c (7 days), E15b, E15c [1], E16a [1], E38, E39, H410 [2-4], H411 [1]
Consumer protection C01
Storage and disposal D01, D02, D05, D09a, D10b [1], D10c, D12a [1]
Medical advice M05a

18   amisulbrom

A fungicide for use in potatoes
FRAC mode of action code: 21

Products

1	Gachinko	Corteva	200 g/l	SC	18219
2	Shinkon	Gowan	200 g/l	SC	17498

Uses
- Downy mildew in **Wine grapes** *(off-label)* [2];
- Late blight in **Potatoes** [1-2];
- Tuber blight in **Potatoes** [1-2];

Extension of Authorisation for Minor Use (EAMUs)
- **Wine grapes** *20193119* [2]

Approval information
- Amisulbrom is included in Annex 1 under EC Regulation 1107/2009

Efficacy guidance
- Make no more than 3 consecutive applications before switching to a blight fungicide with a different mode of action to protect against the risk of resistance.
- Do not use if blight is already visible on 1% of leaves.

Crop-specific information
- Rainfast within 3 hours of application but apply to dry foliage.

Following crops guidance
- Plough before planting new crops in the same soil.

Environmental safety
- LERAP Category B [1-2]

Hazard classification and safety precautions
Hazard Dangerous for the environment, Very toxic to aquatic organisms
Transport code 9 [1-2]
Packaging group III
UN Number 3082
Risk phrases H351
Operator protection U05a, U15, U20b; A, C
Environmental protection E13b, E15a, E16a, E34, E38, H410
Storage and disposal D01, D02, D05, D09a, D10b, D12a

19   Ampelomyces quisqualis (Strain AQ10)

A biological fungicide that combats powdery mildew in fruit and ornamentals
FRAC mode of action code: BM 02

Products

1	AQ 10	Fargro	58% w/w	WG	19968

Uses
- Powdery mildew in *Protected aubergines* [1]; *Protected baby leaf crops* (off-label) [1]; *Protected chicory* (off-label) [1]; *Protected chilli peppers* [1]; *Protected courgettes* [1]; *Protected cress* (off-label) [1]; *Protected cucumbers* [1]; *Protected lamb's lettuce* (off-label) [1]; *Protected land cress* (off-label) [1]; *Protected lettuce* (off-label) [1]; *Protected melons* [1]; *Protected peppers* [1]; *Protected pumpkins* [1]; *Protected purslane* (off-label) [1]; *Protected red mustard* (off-label) [1]; *Protected rocket* (off-label) [1]; *Protected spinach* (off-label) [1]; *Protected spinach beet* (off-label) [1]; *Protected squashes* [1]; *Protected strawberries* [1]; *Protected tomatoes* [1]; *Protected watercress* (off-label) [1]; *Protected winter squash* [1]; *Protected witloof* (off-label) [1];

Extension of Authorisation for Minor Use (EAMUs)
- *Protected baby leaf crops* *20220659* [1]
- *Protected chicory* *20220659* [1]
- *Protected cress* *20220659* [1]
- *Protected lamb's lettuce* *20220659* [1]
- *Protected land cress* *20220659* [1]
- *Protected lettuce* *20220659* [1]
- *Protected purslane* *20220659* [1]
- *Protected red mustard* *20220659* [1]
- *Protected rocket* *20220659* [1]
- *Protected spinach* *20220659* [1]
- *Protected spinach beet* *20220659* [1]
- *Protected watercress* *20220659* [1]
- *Protected witloof* *20220659* [1]

Approval information
- Ampelomyces quisqualis is included in Annex 1 under EC Regulation 1107/2009

SEE SECTION 3 FOR PRODUCTS ALSO REGISTERED

SECTION 2

Restrictions
- Do not store below 4°C

Hazard classification and safety precautions
UN Number N/C
Operator protection U14, U20b; A, D, H
Environmental protection E15b, E34
Storage and disposal D01, D02, D05, D09a, D10a, D16, D20
Medical advice M03

20   azadirachtin

An insecticide for use on ornamentals under full enclosure acting as a feeding inhibitor and a growth disruptor
IRAC mode of action code: Unknown

Products

1	Azatin	Certis Belchim B V	217 g/l	EC	18301

Uses
- Onion thrips in **Protected ornamentals** [1];
- Western flower thrips in **Protected ornamentals** [1];

Approval information
- Azadirachtin included in Annex I under EC Regulation 1107/2009

Efficacy guidance
- A minimum interval of 7 days must be observed between applications within a block (i.e. 4 consecutive sprays represents a block) and 42 days between blocks
- Soil bound systems may not be treated until BBCH 40 or later
- Applications are carried out at a dose rate of 1.4 l/ha for ornamental plants grown under cover and 1.68 l/ha for roses grown on substrate under cover in 0.14% concentration and a minimum interval of 7 days between applications.
- The total number of applications per year is a maximum of 20 applications, divided over 5 blocks. 42 days between each block of 4 applications must be observed.

Restrictions
- Reasonable precautions must be taken to prevent access of birds, wild mammals and bees to treated crops
- To minimise airborne environmental exposure, vents, doors and other openings must be closed during and after application until the applied product has fully settled
- To protect ground water, a minimum interval of 39 days after final treatment must be observed before planting out of containerised forest nursery plants
- To protect the environment, plants grown in synthetic media treated with azadirachtin must not be planted out into soil.

Following crops guidance
- Because of the diversity of crops that may be treated, users are advised to carry out a small scale trial application to crops to ascertain safety to the variety being treated before large scale applications

Hazard classification and safety precautions
Transport code 9 [1]
Packaging group III
UN Number 3082
Risk phrases H319
Operator protection U11; A, C
Environmental protection E15b, E34, H410
Storage and disposal D01, D06e, D07, D09a

FOR FULL CONDITIONS OF USE ALWAYS READ THE PRODUCT LABEL

21   azoxystrobin

A systemic translaminar and protectant strobilurin fungicide for a wide range of crops
FRAC mode of action code: 11

See also azoxystrobin + difenoconazole
azoxystrobin + fluazinam
azoxystrobin + isopyrazam
azoxystrobin + tebuconazole

Products

1	Affix	UPL Europe	250 g/l	SC	18324
2	Amistar	Syngenta	250 g/l	SC	18039
3	Ampect	Syngenta	250 g/l	SC	19739
4	Azofin Plus	Clayton	250 g/l	SC	18552
5	Azoxystar	Life Scientific	250 g/l	SC	17407
6	Clayton Ozark	Clayton	250 g/l	SC	19392
7	Cleancrop Celeb	Agrii	250 g/l	SC	18040
8	Cleancrop Granada	Agrii	250 g/l	SC	19427
9	Conclude AZT 250SC	Certis Belchim B V	250 g/l	SC	18440
10	Cortina	Agrii	250 g/l	SC	19748
11	Heritage	Syngenta	50% w/w	WG	13536
12	Heritage Maxx	Syngenta	95 g/l	DC	18246
13	Hill-Star	Stefes	250 g/l	SC	18150
14	Lincano	BASF	250 g/l	SC	19318
15	Newton (N I Only)	Pan Amenity	250 g/l	SC	19683
16	Opal	Pan Agriculture	250 g/l	SC	19462
17	Oxe	Albaugh UK	250 g/l	SC	19796
18	Plazma	Nufarm UK	250 g/l	SC	19416
19	Profi Az 250 SC	Agrovista	250 g/l	SC	19404
20	Quadris	Syngenta	250 g/l	SC	19211
21	Sinstar	Agrii	250 g/l	SC	16852
22	Sinstar Pro	Agrii	250 g/l	SC	18523
23	Soar	Agform	250 g/l	SC	19525
24	Tazer	Nufarm UK	250 g/l	SC	15495
25	Valiant	FMC Agro	250 g/l	SC	18742
26	Zoxis	Arysta	250 g/l	SC	18438

Uses

- Alternaria in **Broccoli** [1-10,13-14,16-17,19-20,23-24,26]; **Brussels sprouts** [1-10,13-14,16-17,19-20,23-24,26]; **Cabbages** [1-10,13-14,16-17,19-20,23-24,26]; **Calabrese** [1-10,13-14,16-17,19-20,23-24,26]; **Carrots** [1-10,13-14,16-17,19-20,23,26]; **Cauliflowers** [1-10,13-14,16-17,19-20,23-24,26]; **Collards** [1-10,13-14,16-17,19-20,23-24,26]; **Crambe** *(off-label)* [2]; **Horseradish** *(off-label)* [2]; **Kale** [1-10,13-14,16-17,19-20,23-24,26]; **Mustard** *(off-label)* [2]; **Oriental cabbage** [24]; **Parsnips** *(off-label)* [2]; **Poppies** *(off-label)* [2]; **Poppies grown for seed production** *(off-label)* [2]; **Spring oilseed rape** [1-10,13-14,16-17,19-21,23-24,26]; **Swedes** [24]; **Turnips** [24]; **Winter oilseed rape** [1-10,13-14,16-17,19-21,23-24,26];
- Alternaria blight in **Carrots** [24];
- Anthracnose in **Managed amenity turf** *(moderate control)* [11-12,15,18]; **Protected strawberries** [2-10,13-14,16-17,19-20,23]; **Strawberries** [2-10,13-14,16-17,19-20,23];
- Ascochyta in **Beans without pods (Fresh)** [5]; **Combining peas** *(useful control)* [1-10,13-14,16-17,19-20,23,26]; **Dwarf beans** *(useful control)* [2-4,6-10,13-14,16-17,19-20,23]; **Edible podded peas** *(useful control)* [2-4,6-10,13-14,16-17,19-20,23]; **French beans** *(useful control)* [2-4,6-10,13-14,16-17,19-20,23]; **Mange-tout peas** *(useful control)* [2-4,6-10,13-14,16-17,19-20,23]; **Sugar snap peas** *(useful control)* [2-4,6-10,13-14,16-17,19-20,23]; **Vining peas** *(useful control)* [1-10,13-14,16-17,19-20,23,26];
- Black dot in **Potatoes** *(reduction)* [1-10,13-14,16-17,19-20,22-23,26];
- Black root rot in **Container-grown ornamentals** *(off-label)* [2]; **Ornamental plant production** *(off-label)* [2,5]; **Protected ornamentals** *(off-label)* [2,5];

- Black rot in *Baby leaf crops* (off-label) [2]; *Cress* (off-label) [2]; *Herbs (see appendix 6)* (off-label) [2]; *Lamb's lettuce* (off-label) [2]; *Land cress* (off-label) [2]; *Oriental cabbage* (off-label) [2]; *Purslane* (off-label) [2]; *Red mustard* (off-label) [2]; *Rocket* (off-label) [2]; *Spinach* (off-label) [2]; *Spinach beet* (off-label) [2];
- Black scurf in *Potatoes* (reduction) [5,22];
- Black scurf and stem canker in *Potatoes* [1-4,6-10,13-14,16-17,19-20,23,26];
- Botrytis in *Celery (outdoor)* (off-label) [2]; *Container-grown ornamentals* (off-label) [2]; *Ornamental plant production* (off-label) [2,5]; *Protected celery* (off-label) [2]; *Protected ornamentals* (off-label) [2,5];
- Brown patch in *Managed amenity turf* [11-12,15,18];
- Brown rust in *Rye* [24-25]; *Spring barley* [1-10,13-14,16-17,19-21,23-26]; *Spring rye* [1-10,13-14,16-17,19-20,23,26]; *Spring wheat* [1-10,13-14,16-17,19-21,23-26]; *Triticale* [1-10,13-14,16-17,19-20,23,26]; *Winter barley* [1-10,13-14,16-17,19-21,23-26]; *Winter rye* [1-10,13-14,16-17,19-20,23,26]; *Winter wheat* [1-10,13-14,16-17,19-21,23-26];
- Crown rust in *Managed amenity turf* [11,15,18]; *Spring oats* [1-10,13-14,16-17,19-20,23-26]; *Winter oats* [1-10,13-14,16-17,19-20,23-26];
- Dark leaf spot in *Spring oilseed rape* [24-25]; *Winter oilseed rape* [24-25];
- Disease control in *Aubergines* [24]; *China asters* [24]; *Dwarf French beans* [24]; *Forage maize* [24]; *Grain maize* [24]; *Hops* [24]; *Peppers* [24]; *Potatoes* [24]; *Tomatoes* [24];
- Downy mildew in *Beans without pods (Fresh)* [5]; *Blackberries* (off-label) [5]; *Bulb onions* (reduction) [1-10,13-14,16-17,19-20,23-24,26]; *Chicory* [5]; *Combining peas* (reduction) [1-4,6-10,13-14,16-17,19-20,23-24,26]; *Container-grown ornamentals* (off-label) [2]; *Courgettes* (off-label) [2]; *Cucumbers* (off-label) [2]; *Dwarf beans* (reduction) [2-4,6-10,13-14,16-17,19-20,23]; *Edible podded peas* (reduction) [2-4,6-10,13-14,16-17,19-20,23]; *Endives* [2-10,13-14,16-17,19-20,23]; *French beans* (reduction) [2-4,6-10,13-14,16-17,19-20,23]; *Garlic* (reduction) [2-10,13-14,16-17,19-20,23]; *Gherkins* (off-label) [2]; *Lettuce* [2-10,13-14,16-17,19-20,23]; *Mange-tout peas* (reduction) [2-4,6-10,13-14,16-17,19-20,23]; *Melons* (off-label) [2]; *Ornamental plant production* (off-label) [2,5]; *Protected blackberries* (off-label) [5]; *Protected endives* [2-10,13-14,16-17,19-20,23]; *Protected lettuce* [2-10,13-14,16-17,19-20,23]; *Protected ornamentals* (off-label) [2,5]; *Protected raspberries* (off-label) [5]; *Pumpkins* (off-label) [2]; *Radishes* (off-label) [2]; *Raspberries* (off-label) [5]; *Salad onions* (reduction) [2-10,13-14,16-17,19-20,23]; *Shallots* (reduction) [2-10,13-14,16-17,19-20,23]; *Sugar snap peas* (reduction) [2-4,6-10,13-14,16-17,19-20,23]; *Summer squash* (off-label) [2]; *Vining peas* (reduction) [1-4,6-10,13-14,16-17,19-20,23-24,26]; *Watermelons* (off-label) [2]; *Winter squash* (off-label) [2]; *Witloof* [5];
- Ear diseases in *Spring wheat* [24-25]; *Winter wheat* [24-25];
- Fairy rings in *Managed amenity turf* (reduction) [11-12,15,18];
- Fusarium in *Container-grown ornamentals* (off-label) [2]; *Ornamental plant production* (off-label) [2]; *Protected ornamentals* (off-label) [2];
- Fusarium patch in *Managed amenity turf* [11-12,15,18];
- Glume blotch in *Spring wheat* [1-10,13-14,16-17,19-21,23-26]; *Winter wheat* [1-10,13-14,16-17,19-21,23-26];
- Grey mould in *Aubergines* (off-label) [2]; *Beans without pods (Fresh)* [5]; *Combining peas* (some control) [1-4,6-10,13-14,16-17,19-20,23-24,26]; *Courgettes* (off-label) [2]; *Dwarf beans* (some control) [2-4,6-10,13-14,16-17,19-20,23]; *Edible podded peas* (some control) [2-4,6-10,13-14,16-17,19-20,23]; *French beans* (some control) [2-4,6-10,13-14,16-17,19-20,23]; *Mange-tout peas* (some control) [2-4,6-10,13-14,16-17,19-20,23]; *Melons* (off-label) [2]; *Pumpkins* (off-label) [2]; *Sugar snap peas* (some control) [2-4,6-10,13-14,16-17,19-20,23]; *Summer squash* (off-label) [2]; *Tomatoes* (off-label) [2]; *Vining peas* (some control) [1-4,6-10,13-14,16-17,19-20,23-24,26]; *Watermelons* (off-label) [2]; *Winter squash* (off-label) [2];
- Late blight in *Aubergines* (off-label) [2]; *Tomatoes* (off-label) [2];
- Late ear diseases in *Spring wheat* [1-10,13-14,16-17,19-21,23,26]; *Winter wheat* [1-10,13-14,16-17,19-21,23,26];
- Leaf and pod spot in *Combining peas* (useful control) [1-4,6-10,13-14,16-17,19-20,23-24,26]; *Dwarf beans* (useful control) [2-4,6-10,13-14,16-17,19-20,23]; *Edible podded peas* (useful control) [2-4,6-10,13-14,16-17,19-20,23]; *French beans* (useful control) [2-4,6-10,13-14,16-17,19-20,23]; *Mange-tout peas* (useful control) [2-4,6-10,13-14,16-17,19-20,23]; *Sugar snap peas* (useful control) [2-4,6-10,13-14,16-17,19-20,23]; *Vining peas* (useful control) [1-4,6-10,13-14,16-17,19-20,23-24,26];

FOR FULL CONDITIONS OF USE ALWAYS READ THE PRODUCT LABEL

- Leaf spot in *Celery (outdoor)* *(off-label)* [2]; *Container-grown ornamentals* *(off-label)* [2]; *Ornamental plant production* *(off-label)* [2,5]; *Protected celery* *(off-label)* [2]; *Protected ornamentals* *(off-label)* [2,5];
- Melting out in *Managed amenity turf* [11-12,15,18];
- Mycosphaerella in *Combining peas* *(some control)* [1-4,6-10,13-14,16-17,19-20,23-24,26]; *Dwarf beans* *(some control)* [2-4,6-10,13-14,16-17,19-20,23]; *Edible podded peas* *(some control)* [2-4,6-10,13-14,16-17,19-20,23]; *French beans* *(some control)* [2-4,6-10,13-14,16-17,19-20,23]; *Mange-tout peas* *(some control)* [2-4,6-10,13-14,16-17,19-20,23]; *Sugar snap peas* *(some control)* [2-4,6-10,13-14,16-17,19-20,23]; *Vining peas* *(some control)* [1-4,6-10,13-14,16-17,19-20,23-24,26];
- Needle blight in *Ornamental plant production* *(off-label)* [5]; *Protected ornamentals* *(off-label)* [5];
- Needle casts in *Container-grown ornamentals* *(off-label)* [2]; *Ornamental plant production* *(off-label)* [2,5]; *Protected ornamentals* *(off-label)* [2,5];
- Net blotch in *Spring barley* [1-10,13-14,16-17,19-21,23-26]; *Winter barley* [1-10,13-14,16-17,19-21,23-26];
- Powdery mildew in *Aubergines* *(off-label)* [2]; *Blackberries* *(off-label)* [2,5]; *Carrots* [1-10,13-14,16-17,19-20,23-24,26]; *Chillies* *(off-label)* [2]; *Container-grown ornamentals* *(off-label)* [2]; *Courgettes* *(off-label)* [2]; *Cucumbers* *(off-label)* [2]; *Gherkins* *(off-label)* [2]; *Loganberries* *(off-label)* [2]; *Melons* *(off-label)* [2]; *Ornamental plant production* *(off-label)* [2,5]; *Parsnips* *(off-label)* [2]; *Peppers* *(off-label)* [2]; *Poppies* *(off-label)* [2]; *Poppies grown for seed production* *(off-label)* [2]; *Protected blackberries* *(off-label)* [2,5]; *Protected loganberries* *(off-label)* [2]; *Protected ornamentals* *(off-label)* [2,5]; *Protected raspberries* *(off-label)* [2,5]; *Protected Rubus hybrids* *(off-label)* [2]; *Protected strawberries* [2-10,13-14,16-17,19-20,23]; *Pumpkins* *(off-label)* [2]; *Raspberries* *(off-label)* [2,5]; *Rubus hybrids* *(off-label)* [2]; *Rye* *(moderate control)* [24-25]; *Spring barley* *(moderate control)* [5,21,24-25]; *Spring oats* *(moderate control)* [5,25]; *Strawberries* [2-10,13-14,16-17,19-20,23]; *Summer squash* *(off-label)* [2]; *Tomatoes* *(off-label)* [2]; *Triticale* *(moderate control)* [5,24-25]; *Watermelons* *(off-label)* [2]; *Winter barley* *(moderate control)* [5,21,24-25]; *Winter oats* *(moderate control)* [5,25]; *Winter rye* [1-10,13-14,16-17,19-20,23,26]; *Winter squash* *(off-label)* [2];
- Purple blotch in *Leeks* [1-10,13-14,16-17,19-20,23-24,26];
- Rhizoctonia in *Celery (outdoor)* *(off-label)* [2]; *Protected celery* *(off-label)* [2]; *Radishes* *(off-label)* [2]; *Swedes* *(off-label)* [2]; *Turnips* *(off-label)* [2];
- Rhynchosporium in *Rye* *(reduction)* [24-25]; *Spring barley* *(reduction)* [1-10,13-14,16-17,19-21,23-26]; *Spring rye* [1-10,13-14,16-17,19-20,23,26]; *Triticale* *(reduction)* [1-10,13-14,16-17,19-20,23,26]; *Winter barley* *(reduction)* [1-10,13-14,16-17,19-21,23-26]; *Winter rye* [1-10,13-14,16-17,19-20,23,26];
- Ring spot in *Broccoli* [1-10,13-14,16-17,19-20,23-24,26]; *Brussels sprouts* [1-10,13-14,16-17,19-20,23-24,26]; *Cabbages* [1-10,13-14,16-17,19-20,23-24,26]; *Calabrese* [1-10,13-14,16-17,19-20,23-24,26]; *Cauliflowers* [1-10,13-14,16-17,19-20,23-24,26]; *Collards* [1-10,13-14,16-17,19-20,23-24,26]; *Kale* [1-10,13-14,16-17,19-20,23-24,26]; *Oriental cabbage* [24]; *Swedes* [24]; *Turnips* [24];
- Root malformation disorder in *Red beet* *(off-label)* [2];
- Rust in *Asparagus* [1-10,13-14,16-17,19-20,23-24,26]; *Broad beans* [2-10,13-14,16-17,19-20,23-24]; *Container-grown ornamentals* *(off-label)* [2]; *Dwarf French beans* [24]; *Garlic* [24]; *Leeks* [1-10,13-14,16-17,19-20,23-24,26]; *Lupins* [2-10,13-14,16-17,19-20,23]; *Managed amenity turf* [12]; *Ornamental plant production* *(off-label)* [2,5]; *Parsley root* [24]; *Parsnips* [24]; *Protected ornamentals* *(off-label)* [2,5]; *Red beet* [24]; *Salad onions* [24]; *Shallots* [24]; *Spring field beans* [1-10,13-14,16-17,19-20,23-24,26]; *Winter field beans* [1-10,13-14,16-17,19-20,23-24,26];
- Scab in *Container-grown ornamentals* *(off-label)* [2]; *Ornamental plant production* *(off-label)* [2,5]; *Protected ornamentals* *(off-label)* [2,5];
- Sclerotinia in *Borage for oilseed production* *(off-label)* [2]; *Celeriac* *(off-label)* [2]; *Celery (outdoor)* *(off-label)* [2]; *Crambe* *(off-label)* [2]; *Mustard* *(off-label)* [2]; *Protected celery* *(off-label)* [2]; *Soya beans* *(off-label)* [5]; *Spring oilseed rape* [24]; *Winter oilseed rape* [24];
- Sclerotinia stem rot in *Spring oilseed rape* [1-10,13-14,16-17,19-21,23,25-26]; *Winter oilseed rape* [1-10,13-14,16-17,19-21,23,25-26];
- Septoria leaf blotch in *Spring wheat* [5,21,24-25]; *Winter wheat* [1-10,13-14,16-17,19-21,23-26];

SEE SECTION 3 FOR PRODUCTS ALSO REGISTERED

- Stem canker in **Potatoes** *(reduction)* [5,22];
- Stemphylium in **Asparagus** [1-10,13-14,16-17,19-20,23-24,26];
- Take-all in **Rye** *(reduction)* [24-25]; **Spring barley** *(reduction)* [1-10,13-14,16-17,19-21,23-26]; **Spring rye** *(reduction in severity)* [1-4,6-10,13-14,16-17,19-20,23,26]; **Spring wheat** *(reduction)* [1-10,13-14,16-17,19-21,23-26]; **Triticale** *(reduction)* [1-4,6-10,13-14,16-17,19-20,23-26]; **Winter barley** *(reduction)* [1-10,13-14,16-17,19-21,23-26]; **Winter wheat** *(reduction)* [1-10,13-14,16-17,19-21,23-26];
- Take-all patch in **Managed amenity turf** [11-12,15,18];
- White blister in **Baby leaf crops** *(off-label)* [2]; **Broccoli** [1-10,13-14,16-17,19-20,23,26]; **Brussels sprouts** [1-10,13-14,16-17,19-20,23,26]; **Cabbages** [1-10,13-14,16-17,19-20,23,26]; **Calabrese** [1-10,13-14,16-17,19-20,23,26]; **Cauliflowers** [1-10,13-14,16-17,19-20,23,26]; **Collards** [1-10,13-14,16-17,19-20,23,26]; **Container-grown ornamentals** *(off-label)* [2]; **Cress** *(off-label)* [2]; **Herbs (see appendix 6)** *(off-label)* [2]; **Horseradish** *(off-label)* [2]; **Kale** [1-10,13-14,16-17,19-20,23,26]; **Lamb's lettuce** *(off-label)* [2]; **Land cress** *(off-label)* [2]; **Oriental cabbage** *(off-label)* [2]; **Ornamental plant production** *(off-label)* [2,5]; **Protected ornamentals** *(off-label)* [2,5]; **Purslane** *(off-label)* [2]; **Red mustard** *(off-label)* [2]; **Rocket** *(off-label)* [2]; **Spinach** *(off-label)* [2]; **Spinach beet** *(off-label)* [2];
- White tip in **Leeks** [1-10,13-14,16-17,19-20,23-24,26];
- Yellow rust in **Spring wheat** [1-10,13-14,16-17,19-21,23-26]; **Winter wheat** [1-10,13-14,16-17,19-21,23-26];

Extension of Authorisation for Minor Use (EAMUs)
- **Aubergines** *20170894* [2]
- **Baby leaf crops** *20172069* [2]
- **Blackberries** *20170895* [2], *20222133* [5]
- **Borage for oilseed production** *20210757* [2]
- **Celeriac** *20192198* [2]
- **Celery (outdoor)** *20170891* [2]
- **Chillies** *20170894* [2]
- **Container-grown ornamentals** *20183388* [2]
- **Courgettes** *20170893* [2], *20170894* [2]
- **Crambe** *20201908* [2]
- **Cress** *20172069* [2]
- **Cucumbers** *20170894* [2]
- **Gherkins** *20170894* [2]
- **Herbs (see appendix 6)** *20172069* [2]
- **Horseradish** *20192198* [2]
- **Lamb's lettuce** *20172069* [2]
- **Land cress** *20172069* [2]
- **Loganberries** *20170895* [2]
- **Melons** *20170893* [2], *20170894* [2]
- **Mustard** *20201403* [2]
- **Oriental cabbage** *20170889* [2]
- **Ornamental plant production** *20183388* [2], *20222139* [5]
- **Parsnips** *20192198* [2]
- **Peppers** *20170894* [2]
- **Poppies** *20180713* [2]
- **Poppies grown for seed production** *20180713* [2]
- **Protected blackberries** *20170895* [2], *20222133* [5]
- **Protected celery** *20170891* [2]
- **Protected loganberries** *20170895* [2]
- **Protected ornamentals** *20183388* [2], *20222139* [5]
- **Protected raspberries** *20170895* [2], *20222140* [5]
- **Protected Rubus hybrids** *20170895* [2]
- **Pumpkins** *20170893* [2], *20170894* [2]
- **Purslane** *20172069* [2]
- **Radishes** *20192198* [2]
- **Raspberries** *20170895* [2], *20222140* [5]
- **Red beet** *20192198* [2]

FOR FULL CONDITIONS OF USE ALWAYS READ THE PRODUCT LABEL

- **Red mustard** *20172069* [2]
- **Rocket** *20172069* [2]
- **Rubus hybrids** *20170895* [2]
- **Salad onions** *20170890* [2]
- **Soya beans** *20180786* [5]
- **Spinach** *20172069* [2]
- **Spinach beet** *20172069* [2]
- **Summer squash** *20170893* [2], *20170894* [2]
- **Swedes** *20192198* [2]
- **Tomatoes** *20170894* [2]
- **Turnips** *20192198* [2]
- **Watermelons** *20170893* [2], *20170894* [2]
- **Winter squash** *20170893* [2], *20170894* [2]

Approval information
- Azoxystrobin included in Annex I under EC Regulation 1107/2009
- Accepted by BBPA for use on malting barley

Efficacy guidance
- Best results obtained from use as a protectant or during early stages of disease establishment or when a predictive assessment indicates a risk of disease development
- Azoxystrobin inhibits fungal respiration and should always be used in mixture with fungicides with other modes of action
- Treatment under poor growing conditions may give less reliable results
- For good control of *Fusarium* patch in amenity turf and grass repeat treatment at minimum intervals of 2 wk
- Azoxystrobin is a member of the QoI cross resistance group. Product should be used preventatively and not relied on for its curative potential
- Use product in cereals as part of an Integrated Crop Management strategy incorporating other methods of control, including where appropriate other fungicides with a different mode of action. Do not apply more than two foliar applications of QoI containing products to any cereal crop
- There is a significant risk of widespread resistance occurring in *Septoria tritici* populations in UK. Failure to follow resistance management action may result in reduced levels of disease control
- On cereal crops product must always be used in mixture with another product, recommended for control of the same target disease, that contains a fungicide from a different cross resistance group and is applied at a dose that will give robust control
- Strains of barley powdery mildew resistant to QoIs are common in the UK

Restrictions
- Maximum number of treatments 1 per crop for potatoes; 2 per crop for brassicas, peas, cereals, oilseed rape; 4 per crop or yr for onions, carrots, leeks, amenity turf
- Maximum total dose ranges from 2-4 times the single full dose depending on crop and product. See labels for details
- On turf the maximum number of treatments is 4 per yr but they must not exceed one third of the total number of fungicide treatments applied
- Do not use where there is risk of spray drift onto neighbouring apple crops
- The same spray equipment should not be used to treat apples
- To reduce the risk of resistance developing on target diseases the total number of applications of products containing QoI fungicides made to any cereal crop must not exceed two
- To protect aquatic life, the maximum total dose applied to crops grown outdoors must not exceed 500 g a.i/ha per year
- To protect aquatic life, for uses on crops of broccoli, calabrese, Brussels sprouts, cabbage, cauliflower, collards, lettuce and kale, the maximum total dose applied must not exceed 500 g azoxystrobin per hectare per year
- Do not apply to protected or outdoor ornamentals at temperatures above 30?C or below 10?C [2]

Crop-specific information
- Latest use: at planting for potatoes; grain watery ripe (GS 71) for cereals; before senescence for asparagus

- HI: 10 d for carrots;14 d for broccoli, Brussels sprouts, bulb onions, cabbages, calabrese, cauliflowers, collards, kale, vining peas; 21 d for leeks, spring oilseed rape, winter oilseed rape; 36 d for combining peas, 35 d for field beans
- In cereals control of established infections can be improved by appropriate tank mixtures or application as part of a programme. Always use in mixture with another product from a different cross-resistance group
- In turf use product at full dose rate in a disease control programme, alternating with fungicides of different modes of action
- In potatoes when used as incorporated treatment apply overall to the entire area to be planted, incorporate to 15 cm and plant on the same day. In-furrow spray should be directed at the furrow and not the seed tubers
- Applications to brassica crops must only be made to a developed leaf canopy and not before growth stages specified on the label
- Heavy disease pressure in brassicae and oilseed rape may require a second treatment
- All crops should be treated when not under stress. Check leaf wax on peas if necessary
- Consult processor before treating any crops for processing
- Treat asparagus after the harvest season. Where a new bed is established do not treat within 3 wk of transplanting out the crowns
- Do not apply to turf when ground is frozen or during drought

Environmental safety
- Dangerous for the environment
- Very toxic to aquatic organisms
- Avoid spray drift onto surrounding areas or crops, especially apples, plums or privet
- Buffer zone requirement 5 m in winter wheat, winter barley and winter oilseed rape [21]
- Buffer zone requirement 10 m in blackberry, loganberry, raspberry and rubus hybrid [5, 2]
- Buffer zone requirement 6 m in bulb onion, carrots and leeks [24]
- LERAP Category B [1-21,23-26]

Hazard classification and safety precautions
Hazard Harmful [24], Dangerous for the environment, Harmful if swallowed [12], Harmful if inhaled [2-3,6,9-10,13-14,17,26], Very toxic to aquatic organisms [5,11,17,22,24]
Transport code 9 [1-26]
Packaging group III
UN Number 3077, 3082
Risk phrases H317 [26], R50 [25], R53a [25]
Operator protection U05a, U09a [1-10,13-26], U19a [1-10,13-26], U20b; A
Environmental protection E15a [1-10,13-23,26], E15b [11-12,24-25], E16a [1-21,23-26], E38, H410 [1-11,13-24,26], H411 [12]
Storage and disposal D01, D02, D03 [11-12], D05 [1-23,26], D09a, D10b [11-12], D10c [1-10,13-26], D12a
Medical advice M05a [11-12]

22   azoxystrobin + difenoconazole

A broad spectrum fungicide mixture for field crops
FRAC mode of action code: 11 + 3

See also difenoconazole

Products

1 Amistar Top	Syngenta	200:125 g/l	SC	18050
2 Angle	Syngenta	125:125 g/l	SC	19119
3 Priori Gold	Syngenta	125:125 g/l	SC	19226

Uses
- Alternaria in **Choi sum** *(off-label)* [1]; **Oriental brassicas** *(off-label)* [1];
- Alternaria blight in **Carrots** [1];
- Black canker in **Horseradish** *(off-label)* [1]; **Parsley root** *(off-label)* [1]; **Parsnips** *(off-label)* [1]; **Salsify** *(off-label)* [1];

FOR FULL CONDITIONS OF USE ALWAYS READ THE PRODUCT LABEL

- Botrytis in **Chicory** *(off-label)* [1]; **Chicory grown outside for forcing** *(off-label)* [1];
- Disease control in **Fodder beet** [2-3]; **Protected strawberries** [1]; **Spring oilseed rape** [2-3]; **Strawberries** [1]; **Sugar beet** [2-3]; **Winter oilseed rape** [2-3];
- Phoma in **Horseradish** *(off-label)* [1]; **Parsley root** *(off-label)* [1]; **Parsnips** *(off-label)* [1]; **Salsify** *(off-label)* [1];
- Powdery mildew in **Broccoli** [1]; **Brussels sprouts** [1]; **Cabbages** [1]; **Calabrese** [1]; **Carrots** [1]; **Cauliflowers** [1]; **Collards** [1]; **Kale** [1]; **Rocket** [1]; **Spring wheat** [2]; **Winter wheat** [2];
- Purple blotch in **Leeks** *(moderate control)* [1];
- Purple spot in **Asparagus** *(off-label)* [1];
- Ring spot in **Choi sum** *(off-label)* [1]; **Oriental brassicas** *(off-label)* [1];
- Rust in **Asparagus** *(off-label)* [1]; **Herbs (see appendix 6)** *(off-label)* [1]; **Leeks** [1]; **Spring wheat** [2]; **Winter wheat** [2];
- Sclerotinia in **Chicory** *(off-label)* [1]; **Chicory grown outside for forcing** *(off-label)* [1];
- Septoria seedling blight in **Herbs (see appendix 6)** *(off-label)* [1];
- White blister in **Broccoli** [1]; **Brussels sprouts** [1]; **Cabbages** [1]; **Calabrese** [1]; **Cauliflowers** [1]; **Collards** [1]; **Kale** [1];
- White tip in **Leeks** *(qualified minor use)* [1];

Extension of Authorisation for Minor Use (EAMUs)
- **Asparagus** *20171501* [1]
- **Chicory** *20171503* [1]
- **Chicory grown outside for forcing** *20171503* [1]
- **Choi sum** *20171502* [1]
- **Herbs (see appendix 6)** *20182671* [1]
- **Horseradish** *20171340* [1]
- **Oriental brassicas** *20171502* [1]
- **Parsley root** *20171340* [1]
- **Parsnips** *20171340* [1]
- **Salsify** *20171340* [1]

Approval information
- Azoxystrobin and difenoconazole included in Annex I under EC Regulation 1107/2009

Efficacy guidance
- Best results obtained from applications made in the earliest stages of disease development or as a protectant treatment following a disease risk assessment
- Ensure the crop is free from any stress caused by environmental or agronomic effects
- Azoxystrobin is a member of the QoI cross resistance group. Product should be used preventatively and not relied on for its curative potential
- Use as part of an Integrated Crop Management strategy incorporating other methods of control, including where appropriate other fungicides with a different mode of action. Do not apply more than two foliar applications of QoI containing products

Restrictions
- Maximum number of treatments 2 per crop
- Do not apply where there is a risk of spray drift onto neighbouring apple crops
- Consult processors before treating a crop destined for processing

Crop-specific information
- HI 14 d for carrots; 21 d for brassicas, leeks
- Minimum spray interval of 14 d must be observed on brassicas

Environmental safety
- Dangerous for the environment
- Very toxic to aquatic organisms
- LERAP Category B [1-3]

Hazard classification and safety precautions
Hazard Irritant, Dangerous for the environment, Harmful if swallowed, Harmful if inhaled
Transport code 9 [1-3]
Packaging group III
UN Number 3082

SEE SECTION 3 FOR PRODUCTS ALSO REGISTERED

Risk phrases H317 [1]
Operator protection U05a, U09a, U19a, U20b; A, H
Environmental protection E15b, E16a, E38, H410
Storage and disposal D01, D02, D05, D09a, D10c, D12a

23   azoxystrobin + fluazinam

A blight fungicide for use in potatoes
FRAC mode of action code: 11 + 29

Products

1	Vendetta	FMC Agro	150:375 g/l	SC	18709

Uses
- Blight in *Potatoes* [1];
- Leaf blight in *Potatoes* [1];

Approval information
- Azoxystrobin and fluazinam included in Annex I under EC Regulation 1107/2009

Efficacy guidance
- Applications must begin prior to blight development. The first application should be applied at the first blight warning or when local weather conditions are favourable for disease development, whichever is the sooner. In the absence of weather conducive to disease development, the first application should be made just before the crop meets in the rows

Restrictions
- Do not use more than 3 consecutive QoI-containing sprays
- Certain apple varieties are highly sensitive. As a precaution, do not apply when there is a risk of spray drift onto neighbouring apple crops. Spray equipment used for application should not be used to treat apples

Environmental safety
- Buffer zone requirement 7m [1]

Hazard classification and safety precautions
Hazard Very toxic to aquatic organisms
Transport code 9 [1]
Packaging group III
UN Number 3082
Risk phrases H317, H361
Operator protection A, H
Environmental protection H410
Storage and disposal D01, D02, D05

24   azoxystrobin + isopyrazam

A fungicide mixture for use in oilseed rape
FRAC mode of action code: 11 + 7

See also azoxystrobin

Products

1	Symetra (GB only)	Adama	200:125 g/l	SC	18556

Uses
- Sclerotinia stem rot in *Spring oilseed rape* [1]; *Winter oilseed rape* [1];

Approval information
- Azoxystrobin and isopyrazam are included in Appendix 1 under EC Regulation 1107/2009

Efficacy guidance
- Rainfast within 1 hour of application.

Restrictions
- Do not use application equipment used to apply Symetra on apples or damage will occur.
- Contains a member of the QoI cross resistance group and a member of the SDHI cross resistance group. Should be used preventatively and should not be relied on for its curative potential.

Environmental safety
- LERAP Category B [1]

Hazard classification and safety precautions

Hazard Toxic, Dangerous for the environment, Harmful if swallowed, Toxic if inhaled
Transport code 9 [1]
Packaging group III
UN Number 3082
Risk phrases H361
Operator protection U05a, U19a; A, H
Environmental protection E16a, E38, H410
Storage and disposal D01, D02, D12a

25   azoxystrobin + tebuconazole

A strobilurin/triazole fungicide mixture for disease control in oilseed rape
FRAC mode of action code: 11 + 3

Products

1 Custodia	Adama	120:200 g/l	SC	16393
2 Seraphin	Adama	120:200 g/l	SC	16248

Uses
- Chocolate spot in *Spring field beans* (moderate control) [1]; *Winter field beans* [1];
- Disease control in *Oilseed rape* [2]; *Spring field beans* [1-2]; *Winter field beans* (moderate control) [1-2];
- Rust in *Spring field beans* [2]; *Winter field beans* [2];
- Sclerotinia in *Spring oilseed rape* (moderate control) [1]; *Winter oilseed rape* (moderate control) [1];

Approval information
- Azoxystrobin and tebuconazole included in Annex I under EC Regulation 1107/2009

Efficacy guidance
- Azoxystrobin is a member of the QoI cross resistance group. Product should be used preventatively and not relied on for its curative potential
- Use product as part of an Integrated Crop Management strategy incorporating other methods of control, including where appropriate other fungicides with a different mode of action. Do not apply more than two foliar applications of QoI containing products to any crop

Restrictions
- Avoid drift on to neighbouring crops since damage may occur especially to broad-leaved plants
- Certain apple varieties are highly sensitive to azoxystrobin. As a precaution, azoxystrobin should not be applied when there is a risk of spray drift onto neighbouring apple crops. Spray equipment used to apply azoxystrobin to other crops should not be used to treat apples.
- Newer authorisations for tebuconazole products require application to cereals only after GS 30 and applications to oilseed rape and linseed after GS20 - check label

Environmental safety
- LERAP Category B [1-2]

Hazard classification and safety precautions

Hazard Harmful, Dangerous for the environment, Harmful if swallowed, Very toxic to aquatic organisms
Transport code 9 [1-2]
Packaging group III
UN Number 3082
Risk phrases H361
Operator protection U02a, U04a, U05a, U20b; A, H
Environmental protection E15b, E16a, E34, E38, H410
Storage and disposal D01, D02, D09a, D10b, D12a
Medical advice M03, M05a

26 Bacillus amyloliquefaciens D747

A fungicide for use on a range of horticultural crops
FRAC mode of action code: 44

Products

1	Amylo X WG	Certis Belchim B V	25% w/w	PO	17978

Uses

* Alternaria in *Aubergines* [1]; *Blackberries* [1]; *Blackcurrants* [1]; *Blueberries* [1]; *Chicory* [1]; *Chillies* [1]; *Courgettes* [1]; *Cress* [1]; *Cucumbers* [1]; *Endives* [1]; *Gooseberries* [1]; *Lamb's lettuce* [1]; *Land cress* [1]; *Lettuce* [1]; *Loganberries* [1]; *Melons* [1]; *Mushrooms* [1]; *Peppers* [1]; *Pumpkins* [1]; *Raspberries* [1]; *Red mustard* [1]; *Redcurrants* [1]; *Rocket* [1]; *Rubus hybrids* [1]; *Spinach* [1]; *Spinach beet* [1]; *Strawberries* [1]; *Summer squash* [1]; *Tomatoes* [1]; *Watercress* [1]; *Watermelons* [1]; *Winter squash* [1];
* Anthracnose in *Baby leaf crops* (off-label) [1]; *Broccoli* (off-label) [1]; *Brussels sprouts* (off-label) [1]; *Cabbages* (off-label) [1]; *Calabrese* (off-label) [1]; *Cauliflowers* (off-label) [1]; *Choi sum* (off-label) [1]; *Collards* (off-label) [1]; *Courgettes* (off-label) [1]; *Edible flowers* (off-label) [1]; *Gherkins* (off-label) [1]; *Herbs (see appendix 6)* (off-label) [1]; *Kale* (off-label) [1]; *Kohlrabi* (off-label) [1]; *Mustard* (off-label) [1]; *Okra* (off-label) [1]; *Oriental cabbage* (off-label) [1]; *Ornamental plant production* (off-label) [1]; *Parsley* (off-label) [1]; *Protected gherkins* (off-label) [1]; *Protected okra* (off-label) [1]; *Protected radishes* (off-label) [1]; *Protected swedes* (off-label) [1]; *Protected turnips* (off-label) [1]; *Pumpkins* (off-label) [1]; *Radishes* (off-label) [1]; *Spinach* (off-label) [1]; *Spinach beet* (off-label) [1]; *Summer squash* (off-label) [1]; *Swedes* (off-label) [1]; *Sweetcorn* (off-label) [1]; *Turnips* (off-label) [1]; *Watercress* (off-label) [1]; *Winter squash* (off-label) [1];
* Bacterial canker in *Apples* (off-label) [1]; *Apricots* (off-label) [1]; *Cherries* (off-label) [1]; *Chicory* (off-label) [1]; *Chicory root* (off-label) [1]; *Chinese dates* (off-label) [1]; *Crab apples* (off-label) [1]; *Lamb's lettuce* (off-label) [1]; *Lettuce* (off-label) [1]; *Medlar* (off-label) [1]; *Ornamental plant production* (off-label) [1]; *Peaches* (off-label) [1]; *Pears* (off-label) [1]; *Plums* (off-label) [1]; *Quinces* (off-label) [1]; *Rocket* (off-label) [1]; *Strawberries* (off-label) [1]; *Wine grapes* (off-label) [1]; *Witloof* (off-label) [1];
* Bacterial rot in *Protected radishes* (off-label) [1]; *Protected swedes* (off-label) [1]; *Protected turnips* (off-label) [1]; *Radishes* (off-label) [1]; *Swedes* (off-label) [1]; *Turnips* (off-label) [1];
* Black rot in *Broccoli* (off-label) [1]; *Brussels sprouts* (off-label) [1]; *Cabbages* (off-label) [1]; *Calabrese* (off-label) [1]; *Cauliflowers* (off-label) [1]; *Choi sum* (off-label) [1]; *Collards* (off-label) [1]; *Kale* (off-label) [1]; *Kohlrabi* (off-label) [1]; *Oriental cabbage* (off-label) [1];
* Botrytis in *Apples* (off-label) [1]; *Apricots* (off-label) [1]; *Baby leaf crops* (off-label) [1]; *Bilberries* (off-label) [1]; *Blackberries* (off-label) [1]; *Blackcurrants* (off-label) [1]; *Blueberries* (off-label) [1]; *Broccoli* (off-label) [1]; *Brussels sprouts* (off-label) [1]; *Cabbages* (off-label) [1]; *Calabrese* (off-label) [1]; *Cauliflowers* (off-label) [1]; *Cherries* (off-label) [1]; *Chicory* (off-label) [1]; *Chicory root* (off-label) [1]; *Chinese dates* (off-label) [1]; *Choi sum* (off-label) [1]; *Collards* (off-label) [1]; *Courgettes* (off-label) [1]; *Crab apples* (off-label) [1]; *Cranberries* (off-label) [1]; *Edible flowers* (off-label) [1]; *Gherkins* (off-label) [1]; *Gooseberries* (off-label) [1]; *Herbs (see appendix 6)* (off-label) [1]; *Kale* (off-label) [1]; *Kohlrabi* (off-label) [1]; *Lamb's lettuce* (off-label) [1]; *Lettuce* (off-label) [1]; *Loganberries* (off-label) [1]; *Medlar* (off-label) [1]; *Mustard* (off-label) [1]; *Okra* (off-label) [1]; *Oriental cabbage* (off-label) [1]; *Ornamental plant production* (off-label) [1]; *Parsley*

(off-label) [1]; **Peaches** *(off-label)* [1]; **Pears** *(off-label)* [1]; **Plums** *(off-label)* [1]; **Protected bilberries** *(off-label)* [1]; **Protected blackberries** *(off-label)* [1]; **Protected blackcurrants** *(off-label)* [1]; **Protected blueberry** *(off-label)* [1]; **Protected cranberries** *(off-label)* [1]; **Protected gherkins** *(off-label)* [1]; **Protected gooseberries** *(off-label)* [1]; **Protected loganberries** *(off-label)* [1]; **Protected okra** *(off-label)* [1]; **Protected radishes** *(off-label)* [1]; **Protected raspberries** *(off-label)* [1]; **Protected redcurrants** *(off-label)* [1]; **Protected rose hips** *(off-label)* [1]; **Protected Rubus hybrids** *(off-label)* [1]; **Protected swedes** *(off-label)* [1]; **Protected turnips** *(off-label)* [1]; **Pumpkins** *(off-label)* [1]; **Quinces** *(off-label)* [1]; **Radishes** *(off-label)* [1]; **Raspberries** *(off-label)* [1]; **Redcurrants** *(off-label)* [1]; **Rocket** *(off-label)* [1]; **Rose hips** *(off-label)* [1]; **Rubus hybrids** *(off-label)* [1]; **Spinach** *(off-label)* [1]; **Spinach beet** *(off-label)* [1]; **Strawberries** *(off-label)* [1]; **Summer squash** *(off-label)* [1]; **Swedes** *(off-label)* [1]; **Sweetcorn** *(off-label)* [1]; **Turnips** *(off-label)* [1]; **Watercress** *(off-label)* [1]; **Wine grapes** *(off-label)* [1]; **Winter squash** *(off-label)* [1]; **Witloof** *(off-label)* [1];

- Brown rot in **Apples** *(off-label)* [1]; **Apricots** *(off-label)* [1]; **Cherries** *(off-label)* [1]; **Chicory** *(off-label)* [1]; **Chicory root** *(off-label)* [1]; **Chinese dates** *(off-label)* [1]; **Crab apples** *(off-label)* [1]; **Lamb's lettuce** *(off-label)* [1]; **Lettuce** *(off-label)* [1]; **Medlar** *(off-label)* [1]; **Peaches** *(off-label)* [1]; **Pears** *(off-label)* [1]; **Plums** *(off-label)* [1]; **Quinces** *(off-label)* [1]; **Rocket** *(off-label)* [1]; **Strawberries** *(off-label)* [1]; **Wine grapes** *(off-label)* [1]; **Witloof** *(off-label)* [1];
- Cane blight in **Bilberries** *(off-label)* [1]; **Blackberries** *(off-label)* [1]; **Blackcurrants** *(off-label)* [1]; **Blueberries** *(off-label)* [1]; **Cranberries** *(off-label)* [1]; **Gooseberries** *(off-label)* [1]; **Loganberries** *(off-label)* [1]; **Protected bilberries** *(off-label)* [1]; **Protected blackberries** *(off-label)* [1]; **Protected blackcurrants** *(off-label)* [1]; **Protected blueberry** *(off-label)* [1]; **Protected cranberries** *(off-label)* [1]; **Protected gooseberries** *(off-label)* [1]; **Protected loganberries** *(off-label)* [1]; **Protected raspberries** *(off-label)* [1]; **Protected redcurrants** *(off-label)* [1]; **Protected rose hips** *(off-label)* [1]; **Protected Rubus hybrids** *(off-label)* [1]; **Raspberries** *(off-label)* [1]; **Redcurrants** *(off-label)* [1]; **Rose hips** *(off-label)* [1]; **Rubus hybrids** *(off-label)* [1];
- Cavity spot in **Carrots** *(off-label)* [1]; **Parsnips** *(off-label)* [1]; **Protected carrots** *(off-label)* [1]; **Protected parsnips** *(off-label)* [1];
- Cercospora leaf spot in **Baby leaf crops** *(off-label)* [1]; **Courgettes** *(off-label)* [1]; **Edible flowers** *(off-label)* [1]; **Gherkins** *(off-label)* [1]; **Herbs (see appendix 6)** *(off-label)* [1]; **Mustard** *(off-label)* [1]; **Okra** *(off-label)* [1]; **Ornamental plant production** *(off-label)* [1]; **Parsley** *(off-label)* [1]; **Protected gherkins** *(off-label)* [1]; **Protected okra** *(off-label)* [1]; **Pumpkins** *(off-label)* [1]; **Spinach** *(off-label)* [1]; **Spinach beet** *(off-label)* [1]; **Summer squash** *(off-label)* [1]; **Sweetcorn** *(off-label)* [1]; **Watercress** *(off-label)* [1]; **Winter squash** *(off-label)* [1];
- Cladosporium in **Bilberries** *(off-label)* [1]; **Blackberries** *(off-label)* [1]; **Blackcurrants** *(off-label)* [1]; **Blueberries** *(off-label)* [1]; **Cranberries** *(off-label)* [1]; **Gooseberries** *(off-label)* [1]; **Loganberries** *(off-label)* [1]; **Protected bilberries** *(off-label)* [1]; **Protected blackberries** *(off-label)* [1]; **Protected blackcurrants** *(off-label)* [1]; **Protected blueberry** *(off-label)* [1]; **Protected cranberries** *(off-label)* [1]; **Protected gooseberries** *(off-label)* [1]; **Protected loganberries** *(off-label)* [1]; **Protected raspberries** *(off-label)* [1]; **Protected redcurrants** *(off-label)* [1]; **Protected rose hips** *(off-label)* [1]; **Protected Rubus hybrids** *(off-label)* [1]; **Raspberries** *(off-label)* [1]; **Redcurrants** *(off-label)* [1]; **Rose hips** *(off-label)* [1]; **Rubus hybrids** *(off-label)* [1];
- Damping off in **Baby leaf crops** *(off-label)* [1]; **Broccoli** *(off-label)* [1]; **Brussels sprouts** *(off-label)* [1]; **Cabbages** *(off-label)* [1]; **Calabrese** *(off-label)* [1]; **Cauliflowers** *(off-label)* [1]; **Choi sum** *(off-label)* [1]; **Collards** *(off-label)* [1]; **Edible flowers** *(off-label)* [1]; **Herbs (see appendix 6)** *(off-label)* [1]; **Kale** *(off-label)* [1]; **Kohlrabi** *(off-label)* [1]; **Mustard** *(off-label)* [1]; **Oriental cabbage** *(off-label)* [1]; **Ornamental plant production** *(off-label)* [1]; **Parsley** *(off-label)* [1]; **Protected radishes** *(off-label)* [1]; **Protected swedes** *(off-label)* [1]; **Protected turnips** *(off-label)* [1]; **Radishes** *(off-label)* [1]; **Spinach** *(off-label)* [1]; **Spinach beet** *(off-label)* [1]; **Swedes** *(off-label)* [1]; **Turnips** *(off-label)* [1]; **Watercress** *(off-label)* [1];
- Downy mildew in **Baby leaf crops** *(off-label)* [1]; **Broccoli** *(off-label)* [1]; **Brussels sprouts** *(off-label)* [1]; **Cabbages** *(off-label)* [1]; **Calabrese** *(off-label)* [1]; **Cauliflowers** *(off-label)* [1]; **Choi sum** *(off-label)* [1]; **Collards** *(off-label)* [1]; **Courgettes** *(off-label)* [1]; **Edible flowers** *(off-label)* [1]; **Gherkins** *(off-label)* [1]; **Herbs (see appendix 6)** *(off-label)* [1]; **Kale** *(off-label)* [1]; **Kohlrabi** *(off-label)* [1]; **Mustard** *(off-label)* [1]; **Okra** *(off-label)* [1]; **Oriental cabbage** *(off-label)* [1]; **Ornamental plant production** *(off-label)* [1]; **Parsley** *(off-label)* [1]; **Protected gherkins** *(off-label)* [1]; **Protected okra** *(off-label)* [1]; **Protected radishes** *(off-label)* [1]; **Protected swedes** *(off-label)* [1]; **Protected turnips** *(off-label)* [1]; **Pumpkins** *(off-label)* [1]; **Radishes** *(off-label)* [1]; **Spinach** *(off-label)* [1]; **Spinach beet** *(off-label)* [1]; **Summer squash** *(off-label)* [1]; **Swedes** *(off-*

label) [1]; *Sweetcorn (off-label)* [1]; *Turnips (off-label)* [1]; *Watercress (off-label)* [1]; *Winter squash (off-label)* [1];

- Fireblight in *Apples (off-label)* [1]; *Apricots (off-label)* [1]; *Cherries (off-label)* [1]; *Chicory (off-label)* [1]; *Chicory root (off-label)* [1]; *Chinese dates (off-label)* [1]; *Crab apples (off-label)* [1]; *Lamb's lettuce (off-label)* [1]; *Lettuce (off-label)* [1]; *Medlar (off-label)* [1]; *Ornamental plant production (off-label)* [1]; *Peaches (off-label)* [1]; *Pears (off-label)* [1]; *Plums (off-label)* [1]; *Quinces (off-label)* [1]; *Rocket (off-label)* [1]; *Strawberries (off-label)* [1]; *Wine grapes (off-label)* [1]; *Witloof (off-label)* [1];

- Fusarium in *Aubergines* [1]; *Baby leaf crops (off-label)* [1]; *Blackberries* [1]; *Blackcurrants* [1]; *Blueberries* [1]; *Broccoli (off-label)* [1]; *Brussels sprouts (off-label)* [1]; *Cabbages (off-label)* [1]; *Calabrese (off-label)* [1]; *Cauliflowers (off-label)* [1]; *Chicory* [1]; *Chillies* [1]; *Choi sum (off-label)* [1]; *Collards (off-label)* [1]; *Courgettes (off-label)* [1]; *Cress* [1]; *Cucumbers* [1]; *Edible flowers (off-label)* [1]; *Endives (off-label)* [1]; *Gherkins (off-label)* [1]; *Gooseberries* [1]; *Herbs (see appendix 6) (off-label)* [1]; *Kale (off-label)* [1]; *Kohlrabi (off-label)* [1]; *Lamb's lettuce* [1]; *Land cress* [1]; *Lettuce* [1]; *Loganberries* [1]; *Melons* [1]; *Mushrooms* [1]; *Mustard (off-label)* [1]; *Okra (off-label)* [1]; *Oriental cabbage (off-label)* [1]; *Ornamental plant production (off-label)* [1]; *Parsley (off-label)* [1]; *Peppers* [1]; *Protected gherkins (off-label)* [1]; *Protected okra (off-label)* [1]; *Protected radishes (off-label)* [1]; *Protected swedes (off-label)* [1]; *Protected turnips (off-label)* [1]; *Pumpkins (off-label)* [1]; *Radishes (off-label)* [1]; *Raspberries* [1]; *Red mustard* [1]; *Redcurrants* [1]; *Rocket* [1]; *Rubus hybrids* [1]; *Spinach (off-label)* [1]; *Spinach beet (off-label)* [1]; *Strawberries* [1]; *Summer squash (off-label)* [1]; *Swedes (off-label)* [1]; *Sweetcorn (off-label)* [1]; *Tomatoes* [1]; *Turnips (off-label)* [1]; *Watercress (off-label)* [1]; *Watermelons* [1]; *Winter squash (off-label)* [1];

- Monilinia spp. in *Apples (off-label)* [1]; *Apricots (off-label)* [1]; *Cherries (off-label)* [1]; *Chicory (off-label)* [1]; *Chicory root (off-label)* [1]; *Chinese dates (off-label)* [1]; *Crab apples (off-label)* [1]; *Lamb's lettuce (off-label)* [1]; *Lettuce (off-label)* [1]; *Medlar (off-label)* [1]; *Ornamental plant production (off-label)* [1]; *Peaches (off-label)* [1]; *Pears (off-label)* [1]; *Plums (off-label)* [1]; *Quinces (off-label)* [1]; *Rocket (off-label)* [1]; *Strawberries (off-label)* [1]; *Wine grapes (off-label)* [1]; *Witloof (off-label)* [1];

- Neck rot in *Bulb onions (off-label)* [1]; *Shallots (off-label)* [1];

- Powdery mildew in *Apples (off-label)* [1]; *Apricots (off-label)* [1]; *Aubergines* [1]; *Baby leaf crops (off-label)* [1]; *Bilberries (off-label)* [1]; *Blackberries (off-label)* [1]; *Blackcurrants (off-label)* [1]; *Blueberries (off-label)* [1]; *Broccoli (off-label)* [1]; *Brussels sprouts (off-label)* [1]; *Cabbages (off-label)* [1]; *Calabrese (off-label)* [1]; *Cauliflowers (off-label)* [1]; *Cherries (off-label)* [1]; *Chicory (off-label)* [1]; *Chicory root (off-label)* [1]; *Chillies* [1]; *Chinese dates (off-label)* [1]; *Choi sum (off-label)* [1]; *Collards (off-label)* [1]; *Courgettes (off-label)* [1]; *Crab apples (off-label)* [1]; *Cranberries (off-label)* [1]; *Cress* [1]; *Cucumbers* [1]; *Edible flowers (off-label)* [1]; *Endives* [1]; *Gherkins (off-label)* [1]; *Gooseberries (off-label)* [1]; *Herbs (see appendix 6) (off-label)* [1]; *Kale (off-label)* [1]; *Kohlrabi (off-label)* [1]; *Lamb's lettuce (off-label)* [1]; *Land cress* [1]; *Lettuce (off-label)* [1]; *Loganberries (off-label)* [1]; *Medlar (off-label)* [1]; *Melons* [1]; *Mustard (off-label)* [1]; *Okra (off-label)* [1]; *Oriental cabbage (off-label)* [1]; *Ornamental plant production (off-label)* [1]; *Parsley (off-label)* [1]; *Peaches (off-label)* [1]; *Pears (off-label)* [1]; *Peppers* [1]; *Plums (off-label)* [1]; *Protected bilberries (off-label)* [1]; *Protected blackberries (off-label)* [1]; *Protected blackcurrants (off-label)* [1]; *Protected blueberry (off-label)* [1]; *Protected cranberries (off-label)* [1]; *Protected gherkins (off-label)* [1]; *Protected gooseberries (off-label)* [1]; *Protected loganberries (off-label)* [1]; *Protected okra (off-label)* [1]; *Protected radishes (off-label)* [1]; *Protected raspberries (off-label)* [1]; *Protected redcurrants (off-label)* [1]; *Protected rose hips (off-label)* [1]; *Protected Rubus hybrids (off-label)* [1]; *Protected swedes (off-label)* [1]; *Protected turnips (off-label)* [1]; *Pumpkins (off-label)* [1]; *Quinces (off-label)* [1]; *Radishes (off-label)* [1]; *Raspberries (off-label)* [1]; *Red mustard* [1]; *Redcurrants (off-label)* [1]; *Rocket (off-label)* [1]; *Rose hips (off-label)* [1]; *Rubus hybrids (off-label)* [1]; *Spinach (off-label)* [1]; *Spinach beet (off-label)* [1]; *Strawberries (off-label)* [1]; *Summer squash (off-label)* [1]; *Swedes (off-label)* [1]; *Sweetcorn (off-label)* [1]; *Tomatoes* [1]; *Turnips (off-label)* [1]; *Watercress (off-label)* [1]; *Watermelons* [1]; *Wine grapes (off-label)* [1]; *Winter squash (off-label)* [1]; *Witloof (off-label)* [1];

- Pseudomonas in *Baby leaf crops (off-label)* [1]; *Courgettes (off-label)* [1]; *Edible flowers (off-label)* [1]; *Gherkins (off-label)* [1]; *Herbs (see appendix 6) (off-label)* [1]; *Mustard (off-label)* [1]; *Okra (off-label)* [1]; *Parsley (off-label)* [1]; *Protected gherkins (off-label)* [1]; *Protected okra (off-label)* [1]; *Pumpkins (off-label)* [1]; *Spinach (off-label)* [1]; *Spinach beet (off-label)* [1]; *Summer squash* (off-label) [1]; *Sweetcorn* (off-label) [1]; *Watercress* (off-label) [1]; *Winter squash* (off-label) [1];

- Pythium in *Aubergines* [1]; *Baby leaf crops* (off-label) [1]; *Blackberries* [1]; *Blackcurrants* [1]; *Blueberries* [1]; *Broccoli* (off-label) [1]; *Brussels sprouts* (off-label) [1]; *Cabbages* (off-label) [1]; *Calabrese* (off-label) [1]; *Cauliflowers* (off-label) [1]; *Chicory* [1]; *Chillies* [1]; *Choi sum* (off-label) [1]; *Collards* (off-label) [1]; *Courgettes* (off-label) [1]; *Cress* [1]; *Cucumbers* [1]; *Edible flowers* (off-label) [1]; *Endives* [1]; *Gherkins* (off-label) [1]; *Gooseberries* [1]; *Herbs (see appendix 6)* (off-label) [1]; *Kale* (off-label) [1]; *Kohlrabi* (off-label) [1]; *Lamb's lettuce* [1]; *Land cress* [1]; *Lettuce* [1]; *Loganberries* [1]; *Melons* [1]; *Mushrooms* [1]; *Mustard* (off-label) [1]; *Okra* (off-label) [1]; *Oriental cabbage* (off-label) [1]; *Ornamental plant production* (off-label) [1]; *Parsley* (off-label) [1]; *Peppers* [1]; *Protected gherkins* (off-label) [1]; *Protected okra* (off-label) [1]; *Protected radishes* (off-label) [1]; *Protected swedes* (off-label) [1]; *Protected turnips* (off-label) [1]; *Pumpkins* (off-label) [1]; *Radishes* (off-label) [1]; *Raspberries* [1]; *Red mustard* [1]; *Redcurrants* [1]; *Rocket* [1]; *Rubus hybrids* [1]; *Spinach* (off-label) [1]; *Spinach beet* (off-label) [1]; *Strawberries* [1]; *Summer squash* (off-label) [1]; *Swedes* (off-label) [1]; *Sweetcorn* (off-label) [1]; *Tomatoes* [1]; *Turnips* (off-label) [1]; *Watercress* (off-label) [1]; *Watermelons* [1]; *Winter squash* (off-label) [1];

- Rhizoctonia in *Aubergines* [1]; *Baby leaf crops* (off-label) [1]; *Blackberries* [1]; *Blackcurrants* [1]; *Blueberries* [1]; *Broccoli* (off-label) [1]; *Brussels sprouts* (off-label) [1]; *Cabbages* (off-label) [1]; *Calabrese* (off-label) [1]; *Cauliflowers* (off-label) [1]; *Chicory* [1]; *Chillies* [1]; *Choi sum* (off-label) [1]; *Collards* (off-label) [1]; *Courgettes* (off-label) [1]; *Cress* [1]; *Cucumbers* [1]; *Edible flowers* (off-label) [1]; *Endives* [1]; *Gherkins* (off-label) [1]; *Gooseberries* [1]; *Herbs (see appendix 6)* (off-label) [1]; *Kale* (off-label) [1]; *Kohlrabi* (off-label) [1]; *Lamb's lettuce* [1]; *Land cress* [1]; *Lettuce* [1]; *Loganberries* [1]; *Melons* [1]; *Mushrooms* [1]; *Mustard* (off-label) [1]; *Okra* (off-label) [1]; *Oriental cabbage* (off-label) [1]; *Ornamental plant production* (off-label) [1]; *Parsley* (off-label) [1]; *Peppers* [1]; *Protected gherkins* (off-label) [1]; *Protected okra* (off-label) [1]; *Protected radishes* (off-label) [1]; *Protected swedes* (off-label) [1]; *Protected turnips* (off-label) [1]; *Pumpkins* (off-label) [1]; *Radishes* (off-label) [1]; *Raspberries* [1]; *Red mustard* [1]; *Redcurrants* [1]; *Rocket* [1]; *Rubus hybrids* [1]; *Spinach* (off-label) [1]; *Spinach beet* (off-label) [1]; *Strawberries* [1]; *Summer squash* (off-label) [1]; *Swedes* (off-label) [1]; *Sweetcorn* (off-label) [1]; *Tomatoes* [1]; *Turnips* (off-label) [1]; *Watercress* (off-label) [1]; *Watermelons* [1]; *Winter squash* (off-label) [1];

- Sclerotinia in *Baby leaf crops* (off-label) [1]; *Broccoli* (off-label) [1]; *Brussels sprouts* (off-label) [1]; *Cabbages* (off-label) [1]; *Calabrese* (off-label) [1]; *Carrots* (off-label) [1]; *Cauliflowers* (off-label) [1]; *Celeriac* (off-label) [1]; *Choi sum* (off-label) [1]; *Collards* (off-label) [1]; *Courgettes* (off-label) [1]; *Edible flowers* (off-label) [1]; *Gherkins* (off-label) [1]; *Herbs (see appendix 6)* (off-label) [1]; *Kale* (off-label) [1]; *Kohlrabi* (off-label) [1]; *Mustard* (off-label) [1]; *Okra* (off-label) [1]; *Oriental cabbage* (off-label) [1]; *Ornamental plant production* (off-label) [1]; *Parsley* (off-label) [1]; *Parsnips* (off-label) [1]; *Protected carrots* (off-label) [1]; *Protected gherkins* (off-label) [1]; *Protected okra* (off-label) [1]; *Protected parsnips* (off-label) [1]; *Protected radishes* (off-label) [1]; *Protected swedes* (off-label) [1]; *Protected turnips* (off-label) [1]; *Pumpkins* (off-label) [1]; *Radishes* (off-label) [1]; *Spinach* (off-label) [1]; *Spinach beet* (off-label) [1]; *Summer squash* (off-label) [1]; *Swedes* (off-label) [1]; *Sweetcorn* (off-label) [1]; *Turnips* (off-label) [1]; *Watercress* (off-label) [1]; *Winter squash* (off-label) [1];

- Sclerotinia rot in *Apples* (off-label) [1]; *Apricots* (off-label) [1]; *Cherries* (off-label) [1]; *Chicory* (off-label) [1]; *Chicory root* (off-label) [1]; *Chinese dates* (off-label) [1]; *Crab apples* (off-label) [1]; *Lamb's lettuce* (off-label) [1]; *Lettuce* (off-label) [1]; *Medlar* (off-label) [1]; *Peaches* (off-label) [1]; *Pears* (off-label) [1]; *Plums* (off-label) [1]; *Quinces* (off-label) [1]; *Rocket* (off-label) [1]; *Strawberries* (off-label) [1]; *Wine grapes* (off-label) [1]; *Witloof* (off-label) [1];

- Septoria seedling blight in *Celeriac* (off-label) [1];

- Spear rot in *Broccoli* (off-label) [1]; *Brussels sprouts* (off-label) [1]; *Cabbages* (off-label) [1]; *Calabrese* (off-label) [1]; *Cauliflowers* (off-label) [1]; *Choi sum* (off-label) [1]; *Collards* (off-label) [1]; *Kale* (off-label) [1]; *Kohlrabi* (off-label) [1]; *Oriental cabbage* (off-label) [1];

- Stemphylium in *Baby leaf crops* (off-label) [1]; *Courgettes* (off-label) [1]; *Edible flowers* (off-label) [1]; *Gherkins* (off-label) [1]; *Herbs (see appendix 6)* (off-label) [1]; *Mustard* (off-label) [1]; *Okra* (off-label) [1]; *Ornamental plant production* (off-label) [1]; *Parsley* (off-label) [1]; *Protected gherkins* (off-label) [1]; *Protected okra* (off-label) [1]; *Pumpkins* (off-label) [1]; *Spinach* (off-label) [1]; *Spinach beet* (off-label) [1]; *Summer squash* (off-label) [1]; *Sweetcorn* (off-label) [1]; *Watercress* (off-label) [1]; *Winter squash* (off-label) [1];

SEE SECTION 3 FOR PRODUCTS ALSO REGISTERED

- Xanthomonas in *Baby leaf crops* (off-label) [1]; *Courgettes* (off-label) [1]; *Edible flowers* (off-label) [1]; *Gherkins* (off-label) [1]; *Herbs (see appendix 6)* (off-label) [1]; *Mustard* (off-label) [1]; *Okra* (off-label) [1]; *Parsley* (off-label) [1]; *Protected gherkins* (off-label) [1]; *Protected okra* (off-label) [1]; *Pumpkins* (off-label) [1]; *Spinach* (off-label) [1]; *Spinach beet* (off-label) [1]; *Summer squash* (off-label) [1]; *Sweetcorn* (off-label) [1]; *Watercress* (off-label) [1]; *Winter squash* (off-label) [1];

Extension of Authorisation for Minor Use (EAMUs)

- *Apples* 20180469 [1]
- *Apricots* 20180469 [1]
- *Baby leaf crops* 20183116 [1]
- *Bilberries* 20193147 [1]
- *Blackberries* 20193147 [1]
- *Blackcurrants* 20193147 [1]
- *Blueberries* 20193147 [1]
- *Broccoli* 20190427 [1]
- *Brussels sprouts* 20190427 [1]
- *Bulb onions* 20181580 [1]
- *Cabbages* 20190427 [1], 20213083 [1]
- *Calabrese* 20190427 [1]
- *Carrots* 20194096 [1]
- *Cauliflowers* 20190427 [1]
- *Celeriac* 20194096 [1]
- *Cherries* 20180469 [1]
- *Chicory* 20180469 [1]
- *Chicory root* 20180469 [1]
- *Chinese dates* 20180469 [1]
- *Choi sum* 20190427 [1]
- *Collards* 20190427 [1]
- *Courgettes* 20192204 [1]
- *Crab apples* 20180469 [1]
- *Cranberries* 20193147 [1]
- *Edible flowers* 20183116 [1]
- *Gherkins* 20192204 [1]
- *Gooseberries* 20193147 [1]
- *Herbs (see appendix 6)* 20183116 [1]
- *Kale* 20190427 [1]
- *Kohlrabi* 20190427 [1]
- *Lamb's lettuce* 20180469 [1]
- *Lettuce* 20180469 [1]
- *Loganberries* 20193147 [1]
- *Medlar* 20180469 [1]
- *Mustard* 20183116 [1]
- *Okra* 20192204 [1]
- *Oriental cabbage* 20190427 [1]
- *Ornamental plant production* 20190428 [1]
- *Parsley* 20183116 [1]
- *Parsnips* 20194096 [1]
- *Peaches* 20180469 [1]
- *Pears* 20180469 [1]
- *Plums* 20180469 [1]
- *Protected bilberries* 20193147 [1]
- *Protected blackberries* 20193147 [1]
- *Protected blackcurrants* 20193147 [1]
- *Protected blueberry* 20193147 [1]
- *Protected carrots* 20194096 [1]
- *Protected cranberries* 20193147 [1]
- *Protected gherkins* 20192204 [1]
- *Protected gooseberries* 20193147 [1]

FOR FULL CONDITIONS OF USE ALWAYS READ THE PRODUCT LABEL

- *Protected loganberries* *20193147* [1]
- *Protected okra* *20192204* [1]
- *Protected parsnips* *20194096* [1]
- *Protected radishes* *20194096* [1]
- *Protected raspberries* *20193147* [1]
- *Protected redcurrants* *20193147* [1]
- *Protected rose hips* *20193147* [1]
- *Protected Rubus hybrids* *20193147* [1]
- *Protected swedes* *20194096* [1]
- *Protected turnips* *20194096* [1]
- *Pumpkins* *20192204* [1]
- *Quinces* *20180469* [1]
- *Radishes* *20194096* [1]
- *Raspberries* *20193147* [1]
- *Redcurrants* *20193147* [1]
- *Rocket* *20180469* [1]
- *Rose hips* *20193147* [1]
- *Rubus hybrids* *20193147* [1]
- *Shallots* *20181580* [1]
- *Spinach* *20183116* [1]
- *Spinach beet* *20183116* [1]
- *Strawberries* *20180469* [1]
- *Summer squash* *20192204* [1]
- *Swedes* *20194096* [1]
- *Sweetcorn* *20192586* [1]
- *Turnips* *20194096* [1]
- *Watercress* *20183116* [1]
- *Wine grapes* *20180469* [1]
- *Winter squash* *20192204* [1]
- *Witloof* *20180469* [1]

Approval information
- Bacillus amyloliquefaciens included in Annex I under EC Regulation 1107/2009

Restrictions
- Do not apply before BBCH 14 on leafy vegetables and BBCH 10 on pome fruit, stone fruit, vine and strawberry

Hazard classification and safety precautions
UN Number N/C
Operator protection U05b; A, D, H
Environmental protection E15b, E34
Storage and disposal D01, D02, D10c

27   Bacillus amyloliquefaciens strain FZB24

A fungicide for use on a range of horticultural crops
FRAC mode of action code: BM 02

Products

| 1 | Taegro | Fargro | 130 g/kg | WP | 19204 |

Uses
- Alternaria in **Nursery fruit trees** *(off-label)* [1]; **Ornamental plant production** *(off-label)* [1]; **Protected nursery fruit trees** *(off-label)* [1]; **Protected ornamentals** *(off-label)* [1];
- Botrytis in **Nursery fruit trees** *(off-label)* [1]; **Ornamental plant production** *(off-label)* [1]; **Protected nursery fruit trees** *(off-label)* [1]; **Protected ornamentals** *(off-label)* [1];
- Downy mildew in **Nursery fruit trees** *(off-label)* [1]; **Ornamental plant production** *(off-label)* [1]; **Protected lettuce** [1]; **Protected nursery fruit trees** *(off-label)* [1]; **Protected ornamentals** *(off-label)* [1];

SEE SECTION 3 FOR PRODUCTS ALSO REGISTERED

- Early blight in **Protected tomatoes** [1];
- Grey mould in **Protected strawberries** [1]; **Table grapes** [1]; **Wine grapes** [1];
- Powdery mildew in **Nursery fruit trees** *(off-label)* [1]; **Ornamental plant production** *(off-label)* [1]; **Protected aubergines** [1]; **Protected chilli peppers** [1]; **Protected courgettes** [1]; **Protected cucumbers** [1]; **Protected melons** [1]; **Protected nursery fruit trees** *(off-label)* [1]; **Protected ornamentals** *(off-label)* [1]; **Protected peppers** [1]; **Protected summer squash** [1]; **Protected tomatoes** [1]; **Protected watermelon** [1]; **Table grapes** [1]; **Wine grapes** [1];
- Sclerotinia in **Nursery fruit trees** *(off-label)* [1]; **Ornamental plant production** *(off-label)* [1]; **Protected nursery fruit trees** *(off-label)* [1]; **Protected ornamentals** *(off-label)* [1];

Extension of Authorisation for Minor Use (EAMUs)
- **Nursery fruit trees** *20222404* [1]
- **Ornamental plant production** *20222404* [1]
- **Protected nursery fruit trees** *20222404* [1]
- **Protected ornamentals** *20222404* [1]

Approval information
- Bacillus amyloliquefaciens FZB24 included in Annex I under EC Regulation 1107/2009

Efficacy guidance
- If the pH of the water is less than 5 or greater than 8, adjust to a pH within the range 5-8 before mixing
- Keep the mixture in constant agitation in the spray tank prior to application
- Apply content of entire suspension within a few hours of mixing, do not leave in direct sunlight
- Taegro is fit for organic programmes

Hazard classification and safety precautions
UN Number N/C
Operator protection U14, U15, U20a; A, D, H
Environmental protection E15b
Storage and disposal D02, D06c, D06d, D10a, D20

28 Bacillus amyloliquefaciens strain MBI600

A flowable seed treatment for use in winter oilseed rape
FRAC mode of action code: 44

Products
1	Integral Pro	BASF	6.85% w/v	FS	18510
2	Serifel	BASF	11% w/w	WP	19236

Uses
- Alternaria in **Protected aubergines** [2]; **Protected blackberries** [2]; **Protected blueberry** [2]; **Protected chilli peppers** [2]; **Protected choi sum** [2]; **Protected endives** [2]; **Protected lamb's lettuce** [2]; **Protected lettuce** [2]; **Protected loganberries** [2]; **Protected oriental cabbage** [2]; **Protected peppers** [2]; **Protected raspberries** [2]; **Protected rocket** [2]; **Protected Rubus hybrids** [2]; **Protected spinach** [2]; **Protected spinach beet** [2]; **Protected strawberries** [2]; **Protected tomatoes** [2];
- Botrytis in **Baby leaf crops** *(off-label)* [2]; **Beet leaves** *(off-label)* [2]; **Chard** *(off-label)* [2]; **Choi sum** *(off-label)* [2]; **Cress** *(off-label)* [2]; **Endives** *(off-label)* [2]; **Herbs (see appendix 6)** *(off-label)* [2]; **Lamb's lettuce** *(off-label)* [2]; **Lettuce** *(off-label)* [2]; **Oriental cabbage** *(off-label)* [2]; **Protected baby leaf crops** *(off-label)* [2]; **Protected beet leaves** *(off-label)* [2]; **Protected chard** *(off-label)* [2]; **Protected choi sum** *(off-label)* [2]; **Protected endives** *(off-label)* [2]; **Protected herbs (see appendix 6)** *(off-label)* [2]; **Protected lamb's lettuce** *(off-label)* [2]; **Protected lettuce** *(off-label)* [2]; **Protected oriental cabbage** *(off-label)* [2]; **Protected rocket** *(off-label)* [2]; **Protected spinach** *(off-label)* [2]; **Protected spinach beet** *(off-label)* [2]; **Rocket** *(off-label)* [2]; **Spinach** *(off-label)* [2]; **Spinach beet** *(off-label)* [2];
- Cabbage stem flea beetle in **Mustard** *(off-label - reduction)* [1]; **Winter oilseed rape** *(stimulation of plants own defenses)* [1];

- Damping off in **Angelica** *(off-label)* [1]; **Baby leaf crops** *(off-label)* [1]; **Balm** *(off-label)* [1]; **Basil** *(off-label)* [1]; **Bay** *(off-label)* [1]; **Caraway** *(off-label)* [1]; **Celery leaves** *(off-label)* [1]; **Chervil** *(off-label)* [1]; **Chives** *(off-label)* [1]; **Choi sum** *(off-label)* [1]; **Coriander** *(off-label)* [1]; **Cress** *(off-label)* [1]; **Curry plant** *(off-label)* [1]; **Dill** *(off-label)* [1]; **Edible flowers** *(off-label)* [1]; **Endives** *(off-label)* [1]; **Fennel leaves** *(off-label)* [1]; **Fenugreek** *(off-label)* [1]; **Hyssop** *(off-label)* [1]; **Lamb's lettuce** *(off-label)* [1]; **Land cress** *(off-label)* [1]; **Lavender** *(off-label)* [1]; **Lettuce** *(off-label)* [1]; **Linseed** *(off-label)* [1]; **Lovage** *(off-label)* [1]; **Marjoram** *(off-label)* [1]; **Mint** *(off-label)* [1]; **Nasturtium** *(off-label)* [1]; **Nettle** *(off-label)* [1]; **Oregano** *(off-label)* [1]; **Oriental cabbage** *(off-label)* [1]; **Parsley** *(off-label)* [1]; **Plantain** *(off-label)* [1]; **Protected angelica** *(off-label)* [1]; **Purslane** *(off-label)* [1]; **Rocket** *(off-label)* [1]; **Rosemary** *(off-label)* [1]; **Sage** *(off-label)* [1]; **Salad burnet** *(off-label)* [1]; **Savory** *(off-label)* [1]; **Sorrel** *(off-label)* [1]; **Spinach** *(off-label)* [1]; **Spinach beet** *(off-label)* [1]; **Sweet ciceley** *(off-label)* [1]; **Tarragon** *(off-label)* [1]; **Tatsoi** *(off-label)* [1]; **Thyme** *(off-label)* [1].
- Fusarium in **Protected aubergines** [2]; **Protected blackberries** [2]; **Protected blueberry** [2]; **Protected chilli peppers** [2]; **Protected choi sum** [2]; **Protected endives** [2]; **Protected lamb's lettuce** [2]; **Protected lettuce** [2]; **Protected loganberries** [2]; **Protected oriental cabbage** [2]; **Protected peppers** [2]; **Protected raspberries** [2]; **Protected rocket** [2]; **Protected Rubus hybrids** [2]; **Protected spinach** [2]; **Protected spinach beet** [2]; **Protected strawberries** [2]; **Protected tomatoes** [2];
- Powdery mildew in **Protected aubergines** [2]; **Protected blackberries** [2]; **Protected blueberry** [2]; **Protected chilli peppers** [2]; **Protected choi sum** [2]; **Protected endives** [2]; **Protected lamb's lettuce** [2]; **Protected lettuce** [2]; **Protected loganberries** [2]; **Protected oriental cabbage** [2]; **Protected peppers** [2]; **Protected raspberries** [2]; **Protected rocket** [2]; **Protected Rubus hybrids** [2]; **Protected spinach** [2]; **Protected spinach beet** [2]; **Protected strawberries** [2]; **Protected tomatoes** [2];
- Pythium in **Angelica** *(off-label)* [1]; **Baby leaf crops** *(off-label)* [1]; **Balm** *(off-label)* [1]; **Basil** *(off-label)* [1]; **Bay** *(off-label)* [1]; **Caraway** *(off-label)* [1]; **Celery leaves** *(off-label)* [1]; **Chervil** *(off-label)* [1]; **Chives** *(off-label)* [1]; **Choi sum** *(off-label)* [1]; **Coriander** *(off-label)* [1]; **Cress** *(off-label)* [1]; **Curry plant** *(off-label)* [1]; **Dill** *(off-label)* [1]; **Edible flowers** *(off-label)* [1]; **Endives** *(off-label)* [1]; **Fennel leaves** *(off-label)* [1]; **Fenugreek** *(off-label)* [1]; **Hyssop** *(off-label)* [1]; **Lamb's lettuce** *(off-label)* [1]; **Land cress** *(off-label)* [1]; **Lavender** *(off-label)* [1]; **Lettuce** *(off-label)* [1]; **Linseed** *(off-label)* [1]; **Lovage** *(off-label)* [1]; **Marjoram** *(off-label)* [1]; **Mint** *(off-label)* [1]; **Nasturtium** *(off-label)* [1]; **Nettle** *(off-label)* [1]; **Oregano** *(off-label)* [1]; **Oriental cabbage** *(off-label)* [1]; **Parsley** *(off-label)* [1]; **Plantain** *(off-label)* [1]; **Protected angelica** *(off-label)* [1]; **Protected aubergines** [2]; **Protected blackberries** [2]; **Protected blueberry** [2]; **Protected chilli peppers** [2]; **Protected choi sum** [2]; **Protected endives** [2]; **Protected lamb's lettuce** [2]; **Protected lettuce** [2]; **Protected loganberries** [2]; **Protected oriental cabbage** [2]; **Protected peppers** [2]; **Protected raspberries** [2]; **Protected rocket** [2]; **Protected Rubus hybrids** [2]; **Protected spinach** [2]; **Protected spinach beet** [2]; **Protected strawberries** [2]; **Protected tomatoes** [2]; **Purslane** *(off-label)* [1]; **Rocket** *(off-label)* [1]; **Rosemary** *(off-label)* [1]; **Sage** *(off-label)* [1]; **Salad burnet** *(off-label)* [1]; **Savory** *(off-label)* [1]; **Sorrel** *(off-label)* [1]; **Spinach** *(off-label)* [1]; **Spinach beet** *(off-label)* [1]; **Sweet ciceley** *(off-label)* [1]; **Tarragon** *(off-label)* [1]; **Tatsoi** *(off-label)* [1]; **Thyme** *(off-label)* [1];
- Rhizoctonia in **Angelica** *(off-label)* [1]; **Baby leaf crops** *(off-label)* [1]; **Balm** *(off-label)* [1]; **Basil** *(off-label)* [1]; **Bay** *(off-label)* [1]; **Caraway** *(off-label)* [1]; **Celery leaves** *(off-label)* [1]; **Chervil** *(off-label)* [1]; **Chives** *(off-label)* [1]; **Choi sum** *(off-label)* [1]; **Coriander** *(off-label)* [1]; **Cress** *(off-label)* [1]; **Curry plant** *(off-label)* [1]; **Dill** *(off-label)* [1]; **Edible flowers** *(off-label)* [1]; **Endives** *(off-label)* [1]; **Fennel leaves** *(off-label)* [1]; **Fenugreek** *(off-label)* [1]; **Hyssop** *(off-label)* [1]; **Lamb's lettuce** *(off-label)* [1]; **Land cress** *(off-label)* [1]; **Lavender** *(off-label)* [1]; **Lettuce** *(off-label)* [1]; **Linseed** *(off-label)* [1]; **Lovage** *(off-label)* [1]; **Marjoram** *(off-label)* [1]; **Mint** *(off-label)* [1]; **Nasturtium** *(off-label)* [1]; **Nettle** *(off-label)* [1]; **Oregano** *(off-label)* [1]; **Oriental cabbage** *(off-label)* [1]; **Parsley** *(off-label)* [1]; **Plantain** *(off-label)* [1]; **Protected angelica** *(off-label)* [1]; **Protected aubergines** [2]; **Protected blackberries** [2]; **Protected blueberry** [2]; **Protected chilli peppers** [2]; **Protected choi sum** [2]; **Protected endives** [2]; **Protected lamb's lettuce** [2]; **Protected lettuce** [2]; **Protected loganberries** [2]; **Protected oriental cabbage** [2]; **Protected peppers** [2]; **Protected raspberries** [2]; **Protected rocket** [2]; **Protected Rubus hybrids** [2]; **Protected spinach** [2]; **Protected spinach beet** [2]; **Protected strawberries** [2]; **Protected tomatoes** [2]; **Purslane** *(off-label)* [1]; **Rocket** *(off-label)* [1]; **Rosemary** *(off-label)* [1]; **Sage** *(off-label)* [1]; **Salad burnet** *(off-label)* [1]; **Savory** *(off-label)* [1]; **Sorrel** *(off-label)* [1]; **Spinach** *(off-label)* [1]; **Spinach**

SEE SECTION 3 FOR PRODUCTS ALSO REGISTERED

beet *(off-label)* [1]; **Sweet ciceley** *(off-label)* [1]; **Tarragon** *(off-label)* [1]; **Tatsoi** *(off-label)* [1]; **Thyme** *(off-label)* [1];

- Sclerotinia in **Baby leaf crops** *(off-label)* [2]; **Beet leaves** *(off-label)* [2]; **Chard** *(off-label)* [2]; **Choi sum** *(off-label)* [2]; **Cress** *(off-label)* [2]; **Endives** *(off-label)* [2]; **Herbs (see appendix 6)** *(off-label)* [2]; **Lamb's lettuce** *(off-label)* [2]; **Lettuce** *(off-label)* [2]; **Oriental cabbage** *(off-label)* [2]; **Protected baby leaf crops** *(off-label)* [2]; **Protected beet leaves** *(off-label)* [2]; **Protected chard** *(off-label)* [2]; **Protected choi sum** *(off-label)* [2]; **Protected endives** *(off-label)* [2]; **Protected herbs (see appendix 6)** *(off-label)* [2]; **Protected lamb's lettuce** *(off-label)* [2]; **Protected lettuce** *(off-label)* [2]; **Protected oriental cabbage** *(off-label)* [2]; **Protected rocket** *(off-label)* [2]; **Protected spinach** *(off-label)* [2]; **Protected spinach beet** *(off-label)* [2]; **Rocket** *(off-label)* [2]; **Spinach** *(off-label)* [2]; **Spinach beet** *(off-label)* [2];
- Stem canker in **Mustard** *(off-label)* [1]; **Winter oilseed rape** *(useful reduction)* [1];

Extension of Authorisation for Minor Use (EAMUs)
- **Angelica** *20210669* [1]
- **Baby leaf crops** *20210669* [1], *20212258* [2]
- **Balm** *20210669* [1]
- **Basil** *20210669* [1]
- **Bay** *20210669* [1]
- **Beet leaves** *20212249* [2]
- **Caraway** *20210669* [1]
- **Celery leaves** *20210669* [1]
- **Chard** *20212249* [2]
- **Chervil** *20210669* [1]
- **Chives** *20210669* [1]
- **Choi sum** *20210669* [1], *20212249* [2]
- **Coriander** *20210669* [1]
- **Cress** *20210669* [1], *20212246* [2], *20212258* [2]
- **Curry plant** *20210669* [1]
- **Dill** *20210669* [1]
- **Edible flowers** *20210669* [1]
- **Endives** *20210669* [1], *20212258* [2]
- **Fennel leaves** *20210669* [1]
- **Fenugreek** *20210669* [1]
- **Herbs (see appendix 6)** *20212258* [2]
- **Hyssop** *20210669* [1]
- **Lamb's lettuce** *20210669* [1], *20212258* [2]
- **Land cress** *20210669* [1]
- **Lavender** *20210669* [1]
- **Lettuce** *20210669* [1], *20212258* [2]
- **Linseed** *20210670* [1]
- **Lovage** *20210669* [1]
- **Marjoram** *20210669* [1]
- **Mint** *20210669* [1]
- **Mustard** *(reduction) 20202097* [1], *20202097* [1]
- **Nasturtium** *20210669* [1]
- **Nettle** *20210669* [1]
- **Oregano** *20210669* [1]
- **Oriental cabbage** *20210669* [1], *20212249* [2]
- **Parsley** *20210669* [1]
- **Plantain** *20210669* [1]
- **Protected angelica** *20210669* [1]
- **Protected baby leaf crops** *20212246* [2]
- **Protected beet leaves** *20212255* [2]
- **Protected chard** *20212255* [2]
- **Protected choi sum** *20212255* [2]
- **Protected endives** *20212246* [2]
- **Protected herbs (see appendix 6)** *20212246* [2]
- **Protected lamb's lettuce** *20212246* [2]

FOR FULL CONDITIONS OF USE ALWAYS READ THE PRODUCT LABEL

- *Protected lettuce* 20212246 [2]
- *Protected oriental cabbage* 20212255 [2]
- *Protected rocket* 20212246 [2]
- *Protected spinach* 20212255 [2]
- *Protected spinach beet* 20212255 [2]
- *Purslane* 20210669 [1]
- *Rocket* 20210669 [1], 20212258 [2]
- *Rosemary* 20210669 [1]
- *Sage* 20210669 [1]
- *Salad burnet* 20210669 [1]
- *Savory* 20210669 [1]
- *Sorrel* 20210669 [1]
- *Spinach* 20210669 [1], 20212249 [2]
- *Spinach beet* 20210669 [1], 20212249 [2]
- *Sweet ciceley* 20210669 [1]
- *Tarragon* 20210669 [1]
- *Tatsoi* 20210669 [1]
- *Thyme* 20210669 [1]

Approval information
- Bacillus amyloliquefaciens MBI600 included in Annex I under EC Regulation 1107/2009

Efficacy guidance
- It is important to achieve good even coverage on the seed to obtain reliable disease control.
- The germination capacity should be checked prior to use and in the event of any minor effects a small change in sowing rate may be made
- Some slight delay in emergence may occur, but this will be outgrown with no lasting effects

Restrictions
- Do not treat grain with a moisture content exceeding 16% and do not allow the moisture content of treated seed to exceed 16%.
- Do not apply to cracked, split or sprouted seed.
- Must be used within 12 months from the date of manufacture [2]

Hazard classification and safety precautions
UN Number N/C
Operator protection U05a [1], U05b [2]; A, D, H
Environmental protection E03 [1], E15b, E34
Storage and disposal D01, D02, D09a [1], D10c [2], D11a [1], D21 [1]
Treated seed S01 [1], S02 [1], S04b [1], S05 [1], S07 [1]

29   Bacillus pumilus QST2808

A bacterial fungicide use on protected fruit crops
FRAC mode of action code: 44

Products
1	Sonata	Bayer CropScience	1016.6 g/l	SC	19161

Uses
- Disease control in *Protected aubergines* [1]; *Protected blackberries* [1]; *Protected blackcurrants* [1]; *Protected chicory* [1]; *Protected chilli peppers* [1]; *Protected cucumbers* [1]; *Protected lamb's lettuce* [1]; *Protected peppers* [1]; *Protected raspberries* [1]; *Protected redcurrants* [1]; *Protected strawberries* [1]; *Protected tomatoes* [1];
- Mildew in *Protected aubergines* [1]; *Protected blackberries* [1]; *Protected blackcurrants* [1]; *Protected chicory* [1]; *Protected chilli peppers* [1]; *Protected cucumbers* [1]; *Protected lamb's lettuce* [1]; *Protected peppers* [1]; *Protected raspberries* [1]; *Protected redcurrants* [1]; *Protected strawberries* [1]; *Protected tomatoes* [1];
- Powdery mildew in *Ornamental plant production* (off-label) [1]; *Protected ornamentals* (off-label) [1];

SEE SECTION 3 FOR PRODUCTS ALSO REGISTERED

- Rust in *Protected aubergines* [1]; *Protected blackberries* [1]; *Protected blackcurrants* [1]; *Protected chicory* [1]; *Protected chilli peppers* [1]; *Protected cucumbers* [1]; *Protected lamb's lettuce* [1]; *Protected peppers* [1]; *Protected raspberries* [1]; *Protected redcurrants* [1]; *Protected strawberries* [1]; *Protected tomatoes* [1];
- Smoulder in *Ornamental plant production* (off-label) [1]; *Protected ornamentals* (off-label) [1];
- White mould in *Ornamental plant production* (off-label) [1]; *Protected ornamentals* (off-label) [1];

Extension of Authorisation for Minor Use (EAMUs)
- *Ornamental plant production* 20210710 [1]
- *Protected ornamentals* 20210710 [1]

Approval information
- Bacillus pumilus QST2808 included in Annex I under EC Regulation 1107/2009

Restrictions
- Can be stored at room temperature for two years. Storage at higher temperatures may reduce its shelf-life

Hazard classification and safety precautions
UN Number N/C
Operator protection A, D, H
Environmental protection E15b, E34
Storage and disposal D10a, D12c

30   Bacillus subtilis (Strain QST 713)

A bacterial control for disease control in ornamentals
FRAC mode of action code: 44

Products

1	Serenade ASO	Bayer CropScience	13.96 g/l	SC	16139

Uses
- Botrytis in *Almonds* (off-label) [1]; *Apples* (off-label) [1]; *Apricots* (off-label) [1]; *Asparagus* (off-label) [1]; *Aubergines* (off-label) [1]; *Baby leaf crops* (off-label) [1]; *Beans without pods (dry)* (off-label) [1]; *Beans without pods (Fresh)* (off-label) [1]; *Bilberries* (off-label) [1]; *Blackberries* (off-label) [1]; *Blackcurrants* (off-label) [1]; *Blueberries* (off-label) [1]; *Broad beans* (off-label) [1]; *Broccoli* (off-label) [1]; *Brussels sprouts* (off-label) [1]; *Bulb onions* (off-label) [1]; *Cabbages* (off-label) [1]; *Calabrese* (off-label) [1]; *Canary grass* (off-label) [1]; *Cardoons* (off-label) [1]; *Carrots* (off-label) [1]; *Cauliflowers* (off-label) [1]; *Celeriac* (off-label) [1]; *Celery (outdoor)* (off-label) [1]; *Cherries* (off-label) [1]; *Chestnuts* (off-label) [1]; *Chickpeas* (off-label) [1]; *Chicory* (off-label) [1]; *Chicory root* (off-label) [1]; *Chillies* (off-label) [1]; *Choi sum* (off-label) [1]; *Collards* (off-label) [1]; *Combining peas* (off-label) [1]; *Courgettes* (off-label) [1]; *Cranberries* (off-label) [1]; *Cress* (off-label) [1]; *Cucumbers* (off-label) [1]; *Dwarf beans* (off-label) [1]; *Edible podded peas* (off-label) [1]; *Elderberries* (off-label) [1]; *Endives* (off-label) [1]; *Figs* (off-label) [1]; *Florence fennel* (off-label) [1]; *French beans* (off-label) [1]; *Garlic* (off-label) [1]; *Gherkins* (off-label) [1]; *Globe artichoke* (off-label) [1]; *Gooseberries* (off-label) [1]; *Hazel nuts* (off-label) [1]; *Herbs (see appendix 6)* (off-label) [1]; *Hops* (off-label) [1]; *Horseradish* (off-label) [1]; *Jerusalem artichokes* (off-label) [1]; *Kale* (off-label) [1]; *Kohlrabi* (off-label) [1]; *Lamb's lettuce* (off-label) [1]; *Land clearance* (off-label) [1]; *Leeks* (off-label) [1]; *Lentils* (off-label) [1]; *Lettuce* (off-label) [1]; *Loganberries* (off-label) [1]; *Lupins* (off-label) [1]; *Medlar* (off-label) [1]; *Melons* (off-label) [1]; *Mulberries* (off-label) [1]; *Nectarines* (off-label) [1]; *Okra* (off-label) [1]; *Oriental cabbage* (off-label) [1]; *Ornamental plant production* (off-label) [1]; *Parsley root* (off-label) [1]; *Parsnips* (off-label) [1]; *Peaches* (off-label) [1]; *Pears* (off-label) [1]; *Peppers* (off-label) [1]; *Plums* (off-label) [1]; *Pumpkins* (off-label) [1]; *Quinces* (off-label) [1]; *Radishes* (off-label) [1]; *Raspberries* (off-label) [1]; *Red beet* (off-label) [1]; *Red mustard* (off-label) [1]; *Redcurrants* (off-label) [1]; *Rhubarb* (off-label) [1]; *Rocket* (off-label) [1]; *Rose hips* (off-label) [1]; *Rubus hybrids* (off-label) [1]; *Runner beans* (off-label) [1]; *Salad onions* (off-label) [1]; *Salsify* (off-label) [1]; *Shallots* (off-label) [1]; *Spinach* (off-label) [1]; *Spinach beet* (off-label) [1]; *Strawberries* (off-label) [1]; *Summer*

squash *(off-label)* [1]; **Swedes** *(off-label)* [1]; **Sweetcorn** *(off-label)* [1]; **Table grapes** *(off-label)* [1]; **Tomatoes (outdoor)** *(off-label)* [1]; **Turnips** *(off-label)* [1]; **Vining peas** *(off-label)* [1]; **Walnuts** *(off-label)* [1]; **Watercress** *(off-label)* [1]; **Watermelons** *(off-label)* [1]; **Wine grapes** *(off-label)* [1]; **Winter squash** *(off-label)* [1]; **Witloof** *(off-label)* [1];

- Botrytis fruit rot in **Protected aubergines** *(reduction of damage to fruit)* [1]; **Protected chilli peppers** *(reduction of damage to fruit)* [1]; **Protected lettuce** *(reduction of damage to fruit)* [1]; **Protected peppers** *(reduction of damage to fruit)* [1]; **Protected strawberries** *(reduction of damage to fruit)* [1]; **Protected tomatoes** *(reduction of damage to fruit)* [1];
- Butt rot in **Lettuce** *(off-label)* [1];
- Cavity spot in **Carrots** *(off-label)* [1]; **Parsnips** *(off-label)* [1];
- Damping off in **Baby leaf crops** *(off-label)* [1]; **Broccoli** *(off-label)* [1]; **Brussels sprouts** *(off-label)* [1]; **Bulb onions** *(off-label)* [1]; **Cabbages** *(off-label)* [1]; **Calabrese** *(off-label)* [1]; **Carrots** *(off-label)* [1]; **Cauliflowers** *(off-label)* [1]; **Celeriac** *(off-label)* [1]; **Celery (outdoor)** *(off-label)* [1]; **Chard** *(off-label)* [1]; **Collards** *(off-label)* [1]; **Garlic** *(off-label)* [1]; **Herbs (see appendix 6)** *(off-label)* [1]; **Kale** *(off-label)* [1]; **Leeks** *(off-label)* [1]; **Lettuce** *(off-label)* [1]; **Parsnips** *(off-label)* [1]; **Salad onions** *(off-label)* [1]; **Shallots** *(off-label)* [1]; **Spinach** *(off-label)* [1];
- Disease control in **Seed potatoes** [1];
- Grey mould in **Protected aubergines** *(reduction of damage to fruit)* [1]; **Protected chilli peppers** *(reduction of damage to fruit)* [1]; **Protected lettuce** *(reduction of damage to fruit)* [1]; **Protected peppers** *(reduction of damage to fruit)* [1]; **Protected strawberries** *(reduction of damage to fruit)* [1]; **Protected tomatoes** *(reduction of damage to fruit)* [1];
- Helminthosporium in **Potatoes** *(off-label)* [1];
- Phytophthora in **Amenity vegetation** *(off-label)* [1]; **Blackberries** *(off-label)* [1]; **Blackcurrants** *(off-label)* [1]; **Blueberries** *(off-label)* [1]; **Courgettes** *(off-label)* [1]; **Cranberries** *(off-label)* [1]; **Cucumbers** *(off-label)* [1]; **Forest nurseries** *(off-label)* [1]; **Gooseberries** *(off-label)* [1]; **Loganberries** *(off-label)* [1]; **Marrows** *(off-label)* [1]; **Pumpkins** *(off-label)* [1]; **Raspberries** *(off-label)* [1]; **Redcurrants** *(off-label)* [1]; **Rubus hybrids** *(off-label)* [1]; **Squashes** *(off-label)* [1];
- Pythium in **Aubergines** *(off-label)* [1]; **Broccoli** *(off-label)* [1]; **Brussels sprouts** *(off-label)* [1]; **Cabbages** *(off-label)* [1]; **Calabrese** *(off-label)* [1]; **Carrots** *(off-label)* [1]; **Cauliflowers** *(off-label)* [1]; **Chillies** *(off-label)* [1]; **Collards** *(off-label)* [1]; **Courgettes** *(off-label)* [1]; **Cucumbers** *(off-label)* [1]; **Kale** *(off-label)* [1]; **Marrows** *(off-label)* [1]; **Parsnips** *(off-label)* [1]; **Peppers** *(off-label)* [1]; **Pumpkins** *(off-label)* [1]; **Squashes** *(off-label)* [1]; **Tomatoes (outdoor)** *(off-label)* [1];
- Rhizoctonia in **Lettuce** *(off-label)* [1]; **Potatoes** *(off-label)* [1];
- Streptomyces in **Potatoes** *(off-label)* [1];
- White rot in **Bulb onions** *(off-label)* [1]; **Garlic** *(off-label)* [1]; **Leeks** *(off-label)* [1]; **Salad onions** *(off-label)* [1]; **Shallots** *(off-label)* [1];

Extension of Authorisation for Minor Use (EAMUs)
- **Almonds** *20182358* [1]
- **Amenity vegetation** *20130704* [1]
- **Apples** *20182360* [1]
- **Apricots** *20182354* [1]
- **Asparagus** *20182353* [1]
- **Aubergines** *20150306* [1], *20182343* [1]
- **Baby leaf crops** *20150306* [1], *20182363* [1]
- **Beans without pods (dry)** *20182352* [1]
- **Beans without pods (Fresh)** *20182365* [1]
- **Bilberries** *20182346* [1]
- **Blackberries** *20150306* [1], *20182336* [1]
- **Blackcurrants** *20150306* [1], *20182346* [1]
- **Blueberries** *20150306* [1], *20182346* [1]
- **Broad beans** *20182365* [1]
- **Broccoli** *20150306* [1], *20182357* [1]
- **Brussels sprouts** *20150306* [1], *20182357* [1]
- **Bulb onions** *20150306* [1], *20182351* [1]
- **Cabbages** *20150306* [1], *20182357* [1], *20182825* [1]
- **Calabrese** *20150306* [1], *20182357* [1]
- **Canary grass** *20182344* [1]
- **Cardoons** *20182353* [1]

SEE SECTION 3 FOR PRODUCTS ALSO REGISTERED

SECTION 2

- **Carrots** *20150306* [1], *20182359* [1]
- **Cauliflowers** *20150306* [1], *20182357* [1]
- **Celeriac** *20150306* [1], *20182359* [1]
- **Celery (outdoor)** *20150306* [1], *20182353* [1]
- **Chard** *20150306* [1]
- **Cherries** *20182354* [1]
- **Chestnuts** *20182358* [1]
- **Chickpeas** *20182352* [1]
- **Chicory** *20182363* [1]
- **Chicory root** *20182359* [1]
- **Chillies** *20150306* [1], *20182343* [1]
- **Choi sum** *20182357* [1]
- **Collards** *20150306* [1], *20182357* [1]
- **Combining peas** *20182352* [1]
- **Courgettes** *20150306* [1], *20182343* [1]
- **Cranberries** *20150306* [1], *20182346* [1]
- **Cress** *20182363* [1]
- **Cucumbers** *20150306* [1], *20182343* [1]
- **Dwarf beans** *20182365* [1]
- **Edible podded peas** *20182365* [1]
- **Elderberries** *20182346* [1]
- **Endives** *20182363* [1]
- **Figs** *20182341* [1]
- **Florence fennel** *20182353* [1]
- **Forest nurseries** *20130704* [1]
- **French beans** *20182365* [1]
- **Garlic** *20150306* [1], *20182351* [1]
- **Gherkins** *20182343* [1]
- **Globe artichoke** *20182353* [1]
- **Gooseberries** *20150306* [1], *20182346* [1]
- **Hazel nuts** *20182358* [1]
- **Herbs (see appendix 6)** *20150306* [1], *20182345* [1]
- **Hops** *20182355* [1]
- **Horseradish** *20182359* [1]
- **Jerusalem artichokes** *20182359* [1]
- **Kale** *20150306* [1], *20182357* [1]
- **Kohlrabi** *20182357* [1]
- **Lamb's lettuce** *20182363* [1]
- **Land clearance** *20182363* [1]
- **Leeks** *20150306* [1], *20182353* [1]
- **Lentils** *20182352* [1], *20182365* [1]
- **Lettuce** *20150306* [1], *20182363* [1]
- **Loganberries** *20150306* [1], *20182336* [1]
- **Lupins** *20182352* [1]
- **Marrows** *20150306* [1]
- **Medlar** *20182360* [1]
- **Melons** *20182343* [1]
- **Mulberries** *20182346* [1]
- **Nectarines** *20182354* [1]
- **Okra** *20182343* [1]
- **Oriental cabbage** *20182357* [1]
- **Ornamental plant production** *20182364* [1]
- **Parsley root** *20182359* [1]
- **Parsnips** *20150306* [1], *20182359* [1]
- **Peaches** *20182354* [1]
- **Pears** *20182360* [1]
- **Peppers** *20150306* [1], *20182343* [1]
- **Plums** *20182354* [1]
- **Potatoes** *20150306* [1]

FOR FULL CONDITIONS OF USE ALWAYS READ THE PRODUCT LABEL

- **Pumpkins** *20150306* [1], *20182343* [1]
- **Quinces** *20182360* [1]
- **Radishes** *20182359* [1]
- **Raspberries** *20150306* [1], *20182336* [1]
- **Red beet** *20182359* [1]
- **Red mustard** *20182363* [1]
- **Redcurrants** *20150306* [1], *20182346* [1]
- **Rhubarb** *20182353* [1]
- **Rocket** *20182363* [1]
- **Rose hips** *20182346* [1]
- **Rubus hybrids** *20150306* [1], *20182336* [1]
- **Runner beans** *20182365* [1]
- **Salad onions** *20150306* [1], *20182351* [1]
- **Salsify** *20182359* [1]
- **Shallots** *20150306* [1], *20182351* [1]
- **Spinach** *20150306* [1], *20182363* [1]
- **Spinach beet** *20182363* [1]
- **Squashes** *20150306* [1]
- **Strawberries** *20182356* [1]
- **Summer squash** *20182343* [1]
- **Swedes** *20182359* [1]
- **Sweetcorn** *20182343* [1]
- **Table grapes** *20182342* [1]
- **Tomatoes (outdoor)** *20150306* [1], *20182343* [1]
- **Turnips** *20182359* [1]
- **Vining peas** *20182365* [1]
- **Walnuts** *20182358* [1]
- **Watercress** *20182363* [1]
- **Watermelons** *20182343* [1]
- **Wine grapes** *20182342* [1]
- **Winter squash** *20182343* [1]
- **Witloof** *20182363* [1]

Approval information
- Bacillus subtilis QST713 included in Annex I under EC Regulation 1107/2009

Efficacy guidance
- Alternating applications with fungicides using a different mode of action is recommended for resistance management.
- Do not apply using irrigation equipment.
- For maximum effectiveness, start applications before disease development.
- Apply in a minimum water volume of 400 l/ha.

Restrictions
- Consult processor before using on crops grown for processing

Hazard classification and safety precautions
Hazard Irritant
UN Number N/C
Operator protection U05a, U11, U14, U20b; A, D, H
Environmental protection E12h, E15b
Storage and disposal D01, D02, D05, D10a, D16

31   Bacillus thuringiensis aizawai GC-91

An insecticide for use in amenity, ornamentals and some crops
IRAC mode of action code: 11

Products

	1 Agree 50 WG	Certis Belchim B V	50% w/w	WG	18965

SEE SECTION 3 FOR PRODUCTS ALSO REGISTERED

Uses
- Caterpillars in *Amenity vegetation* [1]; *Ornamental plant production* [1]; *Radishes* [1]; *Red beet* [1]; *Swedes* [1];

Approval information
- Bacillus thuringiensis aizawai GC-91 included in Annex I under EC Regulation 1107/2009

Hazard classification and safety precautions
UN Number N/C
Risk phrases H317
Operator protection U05a, U08, U14, U19a, U20a; A, D, H
Environmental protection E15b
Storage and disposal D01, D02, D05, D09a, D10b

32   Bacillus thuringiensis israelensis

A bacterial insecticide for control of larvae of chironomid midges
IRAC mode of action code: 11

Products
1	Bactimos SC	Resource Chemicals	123 g/l	SC	UK15-0947
2	Vectobac AS	Resource Chemicals	123 g/l	AS	UK19-1198

Uses
- Blackfly in *Open waters* [2];
- Filter fly in *Open waters* [1];
- Mosquitoes in *Open waters* [2];

Approval information
- Bacillus thuringiensis israelensis included in Annex I under EC Regulation 1107/2009

Hazard classification and safety precautions
UN Number N/C
Operator protection A, H

33   Bacillus thuringiensis israelensis, strain AM65-52

A bacterial insecticide for control of larvae of chironomid midges
IRAC mode of action code: 11

See also Bacillus thuringiensis israelensis

Products
1	Gnatrol SC	Resource Chemicals	123 g/l	SC	17998

Uses
- Insect pests in *Ornamental plant production* [1];
- Sciiard & Phorids in *Ornamental plant production* [1];

Approval information
- Bacillus thuringiensis israelensis strain AM65-52 included in Annex I under EC Regulation 1107/2009

Hazard classification and safety precautions
UN Number N/C

34   Bacillus thuringiensis kurstaki

A bacterial insecticide for control of caterpillars
IRAC mode of action code: 11

Products

1	Clayton Expel	Clayton	54% w/w	WG	18879
2	Dipel DF	Sumitomo	54% w/w	WG	18874
3	Lepinox Plus	Fargro	37.5% w/w	WP	16269

Uses

- Bramble shoot moth in *Bilberries* (off-label) [3]; *Blackberries* (off-label) [3]; *Blackcurrants* (off-label) [3]; *Blueberries* (off-label) [3]; *Cranberries* (off-label) [3]; *Gooseberries* (off-label) [3]; *Loganberries* (off-label) [3]; *Redcurrants* (off-label) [3]; *Rubus hybrids* (off-label) [3];
- Cabbage moth in *Baby leaf crops* (off-label) [3]; *Choi sum* (off-label) [3]; *Collards* (off-label) [3]; *Cress* (off-label) [3]; *Kohlrabi* (off-label) [3]; *Lamb's lettuce* (off-label) [3]; *Oriental cabbage* (off-label) [3]; *Protected chicory* (off-label) [3]; *Rocket* (off-label) [3]; *Watercress* (off-label) [3];
- Cabbage white butterfly in *Baby leaf crops* (off-label) [3]; *Choi sum* (off-label) [3]; *Collards* (off-label) [3]; *Cress* (off-label) [3]; *Kohlrabi* (off-label) [3]; *Lamb's lettuce* (off-label) [3]; *Oriental cabbage* (off-label) [3]; *Protected chicory* (off-label) [3]; *Rocket* (off-label) [3]; *Watercress* (off-label) [3];
- Caterpillars in *All edible seed crops grown outdoors* (off-label) [2]; *All non-edible seed crops grown outdoors* (off-label) [2]; *Amenity vegetation* [1-2]; *Apples* (off-label) [2-3]; *Apricots* (off-label) [2]; *Baby leaf crops* (off-label) [2]; *Broccoli* [1-3]; *Brussels sprouts* [2-3]; *Cabbages* [1-3]; *Calabrese* (off-label) [1-3]; *Cauliflowers* [1-2]; *Celery (outdoor)* (off-label) [2-3]; *Cherries* (off-label) [2]; *Chestnuts* (off-label) [2]; *Chicory* [3]; *Chinese cabbage* [3]; *Choi sum* (off-label) [2]; *Cob nuts* (off-label) [2]; *Collards* (off-label) [2]; *Combining peas* [1-3]; *Courgettes* [3]; *Dwarf beans* [3]; *Edible podded peas* [1-2]; *Endives* (off-label) [2-3]; *Fennel* (off-label) [2]; *Filberts* (off-label) [2]; *French beans* [3]; *Globe artichoke* [1-3]; *Hazel nuts* (off-label) [2]; *Herbs (see appendix 6)* (off-label) [2-3]; *Hops* [3]; *Hops in propagation* (off-label) [2]; *Kale* (off-label) [2-3]; *Kiwi fruit* (off-label) [2]; *Kohlrabi* (off-label) [2]; *Leeks* [1-2]; *Lettuce* (off-label) [2-3]; *Nectarines* (off-label) [2]; *Olives* (off-label) [2]; *Oriental cabbage* (off-label) [2]; *Ornamental plant production* [1-2]; *Peaches* (off-label) [2]; *Pears* (off-label) [2-3]; *Plums* (off-label) [2]; *Protected all edible seed crops* (off-label) [2]; *Protected all non-edible seed crops* (off-label) [2]; *Protected apricots* (off-label) [2]; *Protected aubergines* [1-3]; *Protected baby leaf crops* (off-label) [2]; *Protected broad beans* [1-2]; *Protected broccoli* (off-label) [2]; *Protected Brussels sprouts* (off-label) [2]; *Protected cabbages* (off-label) [2]; *Protected calabrese* (off-label) [2]; *Protected cauliflowers* (off-label) [2]; *Protected celery* (off-label) [2-3]; *Protected chestnuts* (off-label) [2]; *Protected chilli peppers* [1-2]; *Protected choi sum* (off-label) [2]; *Protected collards* (off-label) [2]; *Protected cucumbers* [1-2]; *Protected dwarf French beans* [1-2]; *Protected endives* (off-label) [2]; *Protected hazelnuts* (off-label) [2]; *Protected herbs (see appendix 6)* (off-label) [2]; *Protected kale* (off-label) [2]; *Protected kiwi fruit* (off-label) [2]; *Protected lettuce* (off-label) [2]; *Protected melons* [3]; *Protected nectarines* (off-label) [2]; *Protected olives* (off-label) [2]; *Protected oriental cabbage* (off-label) [2]; *Protected peaches* (off-label) [2]; *Protected peppers* [1-3]; *Protected plums* (off-label) [2]; *Protected quince* (off-label) [2]; *Protected rhubarb* (off-label) [2]; *Protected runner beans* [1-2]; *Protected spinach* (off-label) [2]; *Protected spinach beet* (off-label) [2]; *Protected tatsoi* (off-label) [2]; *Protected tomatoes* [1-3]; *Protected walnuts* (off-label) [2]; *Protected watercress* (off-label) [2]; *Protected watermelon* [3]; *Protected wine grapes* (off-label) [2]; *Pumpkins* [3]; *Quinces* (off-label) [2]; *Radishes* [3]; *Raspberries* [1-2]; *Rhubarb* (off-label) [2]; *Spinach* (off-label) [2-3]; *Spinach beet* (off-label) [2-3]; *Spring cabbage* (off-label) [2]; *Strawberries* [1-3]; *Tatsoi* (off-label) [2]; *Turnips* [3]; *Vining peas* [1-2]; *Walnuts* (off-label) [2]; *Watercress* (off-label) [2]; *Wine grapes* (off-label) [2-3]; *Winter squash* [3];
- Cherry fruit moth in *Apricots* (off-label) [3]; *Cherries* (off-label) [3]; *Nectarines* (off-label) [3]; *Peaches* (off-label) [3]; *Plums* (off-label) [3];
- Cutworms in *Baby leaf crops* (off-label) [3]; *Bilberries* (off-label) [3]; *Blackberries* (off-label) [3]; *Blackcurrants* (off-label) [3]; *Blueberries* (off-label) [3]; *Broad beans* (off-label) [3]; *Bulb onions* (off-label) [2]; *Carrots* (off-label) [2]; *Celeriac* (off-label) [2]; *Cranberries* (off-label) [3]; *Cress* (off-label) [3]; *Edible podded peas* (off-label) [3]; *Forest nurseries* (off-label) [3]; *Garlic* (off-label) [2];

Gooseberries (off-label) [3]; *Horseradish* (off-label) [2]; *Lamb's lettuce* (off-label) [3]; *Leeks* (off-label) [2-3]; *Loganberries* (off-label) [3]; *Mooli* (off-label) [2]; *Parsley root* (off-label) [2]; *Parsnips* (off-label) [2]; *Protected chicory* (off-label) [3]; *Protected chilli peppers* (off-label) [3]; *Protected courgettes* (off-label) [3]; *Protected dwarf French beans* (off-label) [3]; *Protected peppers* (off-label) [3]; *Protected summer squash* (off-label) [3]; *Radishes* (off-label) [2]; *Red beet* (off-label) [2]; *Redcurrants* (off-label) [3]; *Rhubarb* (off-label) [3]; *Rocket* (off-label) [3]; *Rubus hybrids* (off-label) [3]; *Runner beans* (off-label) [3]; *Salsify* (off-label) [2]; *Shallots* (off-label) [2]; *Swedes* (off-label) [2]; *Sweetcorn* (off-label) [3]; *Turnips* (off-label) [2]; *Watercress* (off-label) [3];

- Diamond-back moth in *Baby leaf crops* (off-label) [3]; *Broccoli* [3]; *Brussels sprouts* [3]; *Cabbages* [3]; *Calabrese* [3]; *Chinese cabbage* [3]; *Choi sum* (off-label) [3]; *Collards* (off-label) [3]; *Cress* (off-label) [3]; *Kohlrabi* (off-label) [3]; *Lamb's lettuce* (off-label) [3]; *Oriental cabbage* (off-label) [3]; *Protected chicory* (off-label) [3]; *Rocket* (off-label) [3]; *Turnips* [3]; *Watercress* (off-label) [3];
- Flax tortrix moth in *Broad beans* (off-label) [3]; *Edible podded peas* (off-label) [3]; *Protected dwarf French beans* (off-label) [3]; *Runner beans* (off-label) [3];
- Leek moth in *Leeks* (off-label) [3]; *Rhubarb* (off-label) [3];
- Light brown apple moth in *Apricots* (off-label) [3]; *Cherries* (off-label) [3]; *Nectarines* (off-label) [3]; *Peaches* (off-label) [3]; *Plums* (off-label) [3];
- Oak processionary moth in *Forest* (off-label) [2];
- Plum fruit moth in *Apricots* (off-label) [3]; *Cherries* (off-label) [3]; *Nectarines* (off-label) [3]; *Peaches* (off-label) [3]; *Plums* (off-label) [3];
- Plum tortrix moth in *Apricots* (off-label) [3]; *Cherries* (off-label) [3]; *Nectarines* (off-label) [3]; *Peaches* (off-label) [3]; *Plums* (off-label) [3];
- Plutella xylostella in *Broccoli* [3]; *Brussels sprouts* [3]; *Cabbages* [3]; *Calabrese* [3]; *Chinese cabbage* [3]; *Turnips* [3];
- Raspberry moth in *Bilberries* (off-label) [3]; *Blackberries* (off-label) [3]; *Blackcurrants* (off-label) [3]; *Blueberries* (off-label) [3]; *Cranberries* (off-label) [3]; *Gooseberries* (off-label) [3]; *Loganberries* (off-label) [3]; *Redcurrants* (off-label) [3]; *Rubus hybrids* (off-label) [3];
- Silver Y moth in *Baby leaf crops* (off-label) [3]; *Broad beans* (off-label) [2-3]; *Cress* (off-label) [3]; *Dwarf beans* (off-label) [2]; *Edible podded peas* (off-label) [3]; *French beans* (off-label) [2]; *Lamb's lettuce* (off-label) [3]; *Protected chicory* (off-label) [3]; *Protected chilli peppers* (off-label) [3]; *Protected courgettes* (off-label) [3]; *Protected dwarf French beans* (off-label) [3]; *Protected hemp* (off-label) [3]; *Protected peppers* (off-label) [3]; *Protected summer squash* (off-label) [3]; *Rocket* (off-label) [3]; *Runner beans* (off-label) [2-3]; *Sweetcorn* (off-label) [2-3]; *Watercress* (off-label) [3];
- Strawberry tortrix in *Bilberries* (off-label) [3]; *Blackberries* (off-label) [3]; *Blackcurrants* (off-label) [3]; *Blueberries* (off-label) [3]; *Cranberries* (off-label) [3]; *Gooseberries* (off-label) [3]; *Loganberries* (off-label) [3]; *Redcurrants* (off-label) [3]; *Rubus hybrids* (off-label) [3];
- Summer-fruit tortrix moth in *Apricots* (off-label) [3]; *Cherries* (off-label) [3]; *Nectarines* (off-label) [3]; *Peaches* (off-label) [3]; *Plums* (off-label) [3];
- Tomato moth in *Protected chilli peppers* (off-label) [3]; *Protected courgettes* (off-label) [3]; *Protected peppers* (off-label) [3]; *Protected summer squash* (off-label) [3]; *Sweetcorn* (off-label) [3];
- Tortrix moths in *Apples* [3]; *Pears* [3]; *Raspberries* [1-2]; *Strawberries* [3];
- Vapourer moth in *Forest nurseries* (off-label) [3];
- White ermine moth in *Forest nurseries* (off-label) [3];
- Winter moth in *Apricots* (off-label) [3]; *Bilberries* (off-label) [2]; *Blackcurrants* (off-label) [2]; *Blueberries* (off-label) [2]; *Cherries* (off-label) [3]; *Cranberries* (off-label) [2]; *Gooseberries* (off-label) [2]; *Nectarines* (off-label) [3]; *Peaches* (off-label) [3]; *Plums* (off-label) [3]; *Protected bilberries* (off-label) [2]; *Protected blackcurrants* (off-label) [2]; *Protected blueberry* (off-label) [2]; *Protected cranberries* (off-label) [2]; *Protected gooseberries* (off-label) [2]; *Protected redcurrants* (off-label) [2]; *Protected Vaccinium spp.* (off-label) [2]; *Redcurrants* (off-label) [2]; *Vaccinium spp.* (off-label) [2];

Extension of Authorisation for Minor Use (EAMUs)
- *All edible seed crops grown outdoors* 20193040 [2]
- *All non-edible seed crops grown outdoors* 20193040 [2]
- *Apples* 20193028 [2]
- *Apricots* 20193029 [2], 20142700 [3]

- **Baby leaf crops** *20193031* [2], *20142704* [3]
- **Bilberries** *20193041* [2], *20142706* [3]
- **Blackberries** *20142706* [3]
- **Blackcurrants** *20193041* [2], *20142706* [3]
- **Blueberries** *20193041* [2], *20142706* [3]
- **Broad beans** *20193034* [2], *20142702* [3]
- **Bulb onions** *20193030* [2]
- **Calabrese** *20193039* [2]
- **Carrots** *20193037* [2]
- **Celeriac** *20193037* [2]
- **Celery (outdoor)** *20193039* [2]
- **Cherries** *20193028* [2], *20142700* [3]
- **Chestnuts** *20193029* [2]
- **Choi sum** *20193039* [2], *20142701* [3]
- **Cob nuts** *20193029* [2]
- **Collards** *20193039* [2], *20142701* [3]
- **Cranberries** *20193041* [2], *20142706* [3]
- **Cress** *20142704* [3]
- **Dwarf beans** *20193034* [2]
- **Edible podded peas** *20142702* [3]
- **Endives** *20193031* [2]
- **Fennel** *20193039* [2]
- **Filberts** *20193029* [2]
- **Forest** *20193035* [2]
- **Forest nurseries** *20142705* [3]
- **French beans** *20193034* [2]
- **Garlic** *20193030* [2]
- **Gooseberries** *20193041* [2], *20142706* [3]
- **Hazel nuts** *20193029* [2]
- **Herbs (see appendix 6)** *20193031* [2]
- **Hops in propagation** *20193867* [2]
- **Horseradish** *20193037* [2]
- **Kale** *20193039* [2]
- **Kiwi fruit** *20193029* [2]
- **Kohlrabi** *20193039* [2], *20142701* [3]
- **Lamb's lettuce** *20142704* [3]
- **Leeks** *20193033* [2], *20142703* [3]
- **Lettuce** *20193031* [2]
- **Loganberries** *20142706* [3]
- **Mooli** *20193037* [2]
- **Nectarines** *20193029* [2], *20142700* [3]
- **Olives** *20193029* [2]
- **Oriental cabbage** *20193039* [2], *20142701* [3]
- **Parsley root** *20193037* [2]
- **Parsnips** *20193037* [2]
- **Peaches** *20193029* [2], *20142700* [3]
- **Pears** *20193028* [2]
- **Plums** *20193029* [2], *20142700* [3]
- **Protected all edible seed crops** *20193040* [2]
- **Protected all non-edible seed crops** *20193040* [2]
- **Protected apricots** *20193029* [2]
- **Protected baby leaf crops** *20193031* [2]
- **Protected bilberries** *20193041* [2]
- **Protected blackcurrants** *20193041* [2]
- **Protected blueberry** *20193041* [2]
- **Protected broccoli** *20193039* [2]
- **Protected Brussels sprouts** *20193039* [2]
- **Protected cabbages** *20193039* [2]
- **Protected calabrese** *20193039* [2]

SEE SECTION 3 FOR PRODUCTS ALSO REGISTERED

- *Protected cauliflowers* 20193039 [2]
- *Protected celery* 20193039 [2]
- *Protected chestnuts* 20193029 [2]
- *Protected chicory* 20142704 [3]
- *Protected chilli peppers* 20142707 [3]
- *Protected choi sum* 20193039 [2]
- *Protected collards* 20193039 [2]
- *Protected courgettes* 20142707 [3]
- *Protected cranberries* 20193041 [2]
- *Protected dwarf French beans* 20142702 [3]
- *Protected endives* 20193031 [2]
- *Protected gooseberries* 20193041 [2]
- *Protected hazelnuts* 20193029 [2]
- *Protected hemp* 20213350 [3]
- *Protected herbs (see appendix 6)* 20193031 [2]
- *Protected kale* 20193039 [2]
- *Protected kiwi fruit* 20193029 [2]
- *Protected lettuce* 20193031 [2]
- *Protected nectarines* 20193029 [2]
- *Protected olives* 20193029 [2]
- *Protected oriental cabbage* 20193039 [2]
- *Protected peaches* 20193029 [2]
- *Protected peppers* 20142707 [3]
- *Protected plums* 20193029 [2]
- *Protected quince* 20193029 [2]
- *Protected redcurrants* 20193041 [2]
- *Protected rhubarb* 20193039 [2]
- *Protected spinach* 20193031 [2]
- *Protected spinach beet* 20193031 [2]
- *Protected summer squash* 20142707 [3]
- *Protected tatsoi* 20193039 [2]
- *Protected Vaccinium spp.* 20193041 [2]
- *Protected walnuts* 20193029 [2]
- *Protected watercress* 20193032 [2]
- *Protected wine grapes* 20193029 [2]
- *Quinces* 20193029 [2]
- *Radishes* 20193037 [2]
- *Red beet* 20193036 [2]
- *Redcurrants* 20193041 [2], 20142706 [3]
- *Rhubarb* 20193039 [2], 20142703 [3]
- *Rocket* 20142704 [3]
- *Rubus hybrids* 20142706 [3]
- *Runner beans* 20193034 [2], 20142702 [3]
- *Salsify* 20193037 [2]
- *Shallots* 20193030 [2]
- *Spinach* 20193031 [2]
- *Spinach beet* 20193031 [2]
- *Spring cabbage* 20193039 [2]
- *Swedes* 20193037 [2]
- *Sweetcorn* 20193038 [2], 20142707 [3]
- *Tatsoi* 20193039 [2]
- *Turnips* 20193037 [2]
- *Vaccinium spp.* 20193041 [2]
- *Walnuts* 20193029 [2]
- *Watercress* 20193032 [2], 20142704 [3]
- *Wine grapes* 20193029 [2]

Approval information
- Bacillus thuringiensis kurstaki included in Annex I under EC Regulation 1107/2009

FOR FULL CONDITIONS OF USE ALWAYS READ THE PRODUCT LABEL

Efficacy guidance
- Pest control achieved by ingestion by caterpillars of the treated plant vegetation. Caterpillars cease feeding and die in 1-3 d
- Apply as soon as larvae appear on crop and repeat every 7-10 d until the end of the hatching period
- Good coverage is essential, especially of undersides of leaves. Spray onto dry foliage and do not apply if rain expected within 6 h

Restrictions
- Apply spray mixture as soon as possible after preparation

Crop-specific information
- HI zero

Environmental safety
- Store out of direct sunlight

Hazard classification and safety precautions
UN Number N/C
Operator protection U05a [1-2], U11 [3], U12 [3], U14 [3], U15 [1-2], U16b [3], U19a, U20b [3], U20c [1-2]; A, C, D, H
Environmental protection E15a [1-2], E15b [3], E34 [3]
Storage and disposal D01, D02, D05 [1-2], D09a, D10b [3], D11a [1-2]
Medical advice M03, M05a [3]

35   Beauveria bassiana ATCC 74040

A biological insecticide
IRAC mode of action code: UNF

Products

1	Naturalis-L	Fargro	7.16% w/w	OD	17526

Uses
- Thrips in *Protected edible crops* (reduction) [1]; *Protected forest nurseries* (off-label - reduction) [1]; *Protected ornamentals* (reduction) [1];
- Whitefly in *Protected edible crops* [1]; *Protected forest nurseries* (off-label) [1]; *Protected ornamentals* [1];

Extension of Authorisation for Minor Use (EAMUs)
- *Protected forest nurseries* (reduction) 20162195 [1], 20162195 [1]

Approval information
- Beauveria bassiana ATCC 74040 included in Annex 1 under EC Regulation 1107/2009

Hazard classification and safety precautions
UN Number N/C
Operator protection U05a, U11, U14, U15, U16b, U19a, U20b; A, D, H
Environmental protection E15b, E34
Storage and disposal D01, D02, D05, D09a, D10a
Medical advice M03, M04a

36   Beauveria bassiana GHA

It is an entomopathogenic fungus causing white muscardine disease. It can be used as a biological insecticide to control a number of pests such as termites, whitefly and some beetles
IRAC mode of action code: UNF

Products

1	Botanigard WP	Certis Belchim B V	22% w/w	WP	17054

SEE SECTION 3 FOR PRODUCTS ALSO REGISTERED

Uses

- Aphids in **Angelica** *(off-label)* [1]; **Apricots** *(off-label)* [1]; **Asparagus** *(off-label)* [1]; **Baby leaf crops** *(off-label)* [1]; **Balm** *(off-label)* [1]; **Basil** *(off-label)* [1]; **Bay** *(off-label)* [1]; **Caraway** *(off-label)* [1]; **Celery (outdoor)** *(off-label)* [1]; **Celery leaves** *(off-label)* [1]; **Cherries** *(off-label)* [1]; **Chervil** *(off-label)* [1]; **Chicory** *(off-label)* [1]; **Chives** *(off-label)* [1]; **Coriander** *(off-label)* [1]; **Cress** *(off-label)* [1]; **Dill** *(off-label)* [1]; **Edible flowers** *(off-label)* [1]; **Fennel leaves** *(off-label)* [1]; **Herbs (see appendix 6)** *(off-label)* [1]; **Hyssop** *(off-label)* [1]; **Lamb's lettuce** *(off-label)* [1]; **Land cress** *(off-label)* [1]; **Leeks** *(off-label)* [1]; **Lettuce** *(off-label)* [1]; **Lovage** *(off-label)* [1]; **Marjoram** *(off-label)* [1]; **Mint** *(off-label)* [1]; **Nectarines** *(off-label)* [1]; **Oregano** *(off-label)* [1]; **Peaches** *(off-label)* [1]; **Plums** *(off-label)* [1]; **Protected asparagus** *(off-label)* [1]; **Protected aubergines** [1]; **Protected celery** *(off-label)* [1]; **Protected chilli peppers** [1]; **Protected courgettes** [1]; **Protected cucumbers** [1]; **Protected leeks** *(off-label)* [1]; **Protected melons** [1]; **Protected nursery fruit trees** [1]; **Protected ornamentals** [1]; **Protected peppers** [1]; **Protected strawberries** [1]; **Protected summer squash** [1]; **Protected tomatoes** [1]; **Purslane** *(off-label)* [1]; **Red mustard** *(off-label)* [1]; **Rocket** *(off-label)* [1]; **Rosemary** *(off-label)* [1]; **Sage** *(off-label)* [1]; **Salad burnet** *(off-label)* [1]; **Savory** *(off-label)* [1]; **Spinach** *(off-label)* [1]; **Spinach beet** *(off-label)* [1]; **Sweet ciceley** *(off-label)* [1]; **Tarragon** *(off-label)* [1]; **Thyme** *(off-label)* [1]; **Watercress** *(off-label)* [1];
- Beetles in **Protected aubergines** [1]; **Protected chilli peppers** [1]; **Protected courgettes** [1]; **Protected cucumbers** [1]; **Protected melons** [1]; **Protected nursery fruit trees** [1]; **Protected ornamentals** [1]; **Protected peppers** [1]; **Protected strawberries** [1]; **Protected summer squash** [1]; **Protected tomatoes** [1];
- Broad mite in **Nursery fruit trees** *(off-label)* [1]; **Ornamental plant production** *(off-label)* [1]; **Protected nursery fruit trees** *(off-label)* [1]; **Protected ornamentals** *(off-label)* [1];
- Cabbage aphid in **Broccoli** *(off-label)* [1]; **Brussels sprouts** *(off-label)* [1]; **Cabbages** *(off-label)* [1]; **Calabrese** *(off-label)* [1]; **Cauliflowers** *(off-label)* [1]; **Choi sum** *(off-label)* [1]; **Collards** *(off-label)* [1]; **Kale** *(off-label)* [1]; **Kohlrabi** *(off-label)* [1]; **Oriental cabbage** *(off-label)* [1];
- Cyclamen mite in **Nursery fruit trees** *(off-label)* [1]; **Ornamental plant production** *(off-label)* [1]; **Protected nursery fruit trees** *(off-label)* [1]; **Protected ornamentals** *(off-label)* [1];
- Damson-hop aphid in **Apricots** *(off-label)* [1]; **Cherries** *(off-label)* [1]; **Nectarines** *(off-label)* [1]; **Peaches** *(off-label)* [1]; **Plums** *(off-label)* [1];
- Fruit tree red spider mite in **Apricots** *(off-label)* [1]; **Cherries** *(off-label)* [1]; **Nectarines** *(off-label)* [1]; **Peaches** *(off-label)* [1]; **Plums** *(off-label)* [1];
- Gall mite in **Apricots** *(off-label)* [1]; **Cherries** *(off-label)* [1]; **Nectarines** *(off-label)* [1]; **Peaches** *(off-label)* [1]; **Plums** *(off-label)* [1];
- Glasshouse red spider mite in **Nursery fruit trees** *(off-label)* [1]; **Ornamental plant production** *(off-label)* [1]; **Protected nursery fruit trees** *(off-label)* [1]; **Protected ornamentals** *(off-label)* [1];
- Leaf curling plum aphid in **Apricots** *(off-label)* [1]; **Cherries** *(off-label)* [1]; **Nectarines** *(off-label)* [1]; **Peaches** *(off-label)* [1]; **Plums** *(off-label)* [1];
- Lettuce root aphid in **Nursery fruit trees** *(off-label)* [1]; **Ornamental plant production** *(off-label)* [1]; **Protected nursery fruit trees** *(off-label)* [1]; **Protected ornamentals** *(off-label)* [1];
- Mealy aphid in **Broccoli** *(off-label)* [1]; **Brussels sprouts** *(off-label)* [1]; **Cabbages** *(off-label)* [1]; **Calabrese** *(off-label)* [1]; **Cauliflowers** *(off-label)* [1]; **Choi sum** *(off-label)* [1]; **Collards** *(off-label)* [1]; **Kale** *(off-label)* [1]; **Kohlrabi** *(off-label)* [1]; **Oriental cabbage** *(off-label)* [1];
- Mealy plum aphid in **Apricots** *(off-label)* [1]; **Cherries** *(off-label)* [1]; **Nectarines** *(off-label)* [1]; **Peaches** *(off-label)* [1]; **Plums** *(off-label)* [1];
- Onion thrips in **Broccoli** *(off-label)* [1]; **Brussels sprouts** *(off-label)* [1]; **Cabbages** *(off-label)* [1]; **Calabrese** *(off-label)* [1]; **Cauliflowers** *(off-label)* [1]; **Choi sum** *(off-label)* [1]; **Collards** *(off-label)* [1]; **Kale** *(off-label)* [1]; **Kohlrabi** *(off-label)* [1]; **Nursery fruit trees** *(off-label)* [1]; **Oriental cabbage** *(off-label)* [1]; **Ornamental plant production** *(off-label)* [1]; **Protected nursery fruit trees** *(off-label)* [1]; **Protected ornamentals** *(off-label)* [1];
- Peach-potato aphid in **Apricots** *(off-label)* [1]; **Broccoli** *(off-label)* [1]; **Brussels sprouts** *(off-label)* [1]; **Cabbages** *(off-label)* [1]; **Calabrese** *(off-label)* [1]; **Cauliflowers** *(off-label)* [1]; **Cherries** *(off-label)* [1]; **Choi sum** *(off-label)* [1]; **Collards** *(off-label)* [1]; **Kale** *(off-label)* [1]; **Kohlrabi** *(off-label)* [1]; **Nectarines** *(off-label)* [1]; **Oriental cabbage** *(off-label)* [1]; **Peaches** *(off-label)* [1]; **Plums** *(off-label)* [1];
- Red spider mites in **Nursery fruit trees** *(off-label)* [1]; **Ornamental plant production** *(off-label)* [1]; **Protected nursery fruit trees** *(off-label)* [1]; **Protected ornamentals** *(off-label)* [1];
- Rust mite in **Apricots** *(off-label)* [1]; **Cherries** *(off-label)* [1]; **Nectarines** *(off-label)* [1]; **Peaches** *(off-label)* [1]; **Plums** *(off-label)* [1];

FOR FULL CONDITIONS OF USE ALWAYS READ THE PRODUCT LABEL

- Thrips in **Angelica** *(off-label)* [1]; **Asparagus** *(off-label)* [1]; **Baby leaf crops** *(off-label)* [1]; **Balm** *(off-label)* [1]; **Basil** *(off-label)* [1]; **Bay** *(off-label)* [1]; **Bulb onions** *(off-label)* [1]; **Caraway** *(off-label)* [1]; **Celery (outdoor)** *(off-label)* [1]; **Celery leaves** *(off-label)* [1]; **Chervil** *(off-label)* [1]; **Chicory** *(off-label)* [1]; **Chives** *(off-label)* [1]; **Coriander** *(off-label)* [1]; **Cress** *(off-label)* [1]; **Dill** *(off-label)* [1]; **Edible flowers** *(off-label)* [1]; **Fennel leaves** *(off-label)* [1]; **Garlic** *(off-label)* [1]; **Herbs (see appendix 6)** *(off-label)* [1]; **Hyssop** *(off-label)* [1]; **Lamb's lettuce** *(off-label)* [1]; **Land cress** *(off-label)* [1]; **Leeks** *(off-label)* [1]; **Lettuce** *(off-label)* [1]; **Lovage** *(off-label)* [1]; **Marjoram** *(off-label)* [1]; **Mint** *(off-label)* [1]; **Nursery fruit trees** *(off-label)* [1]; **Oregano** *(off-label)* [1]; **Ornamental plant production** *(off-label)* [1]; **Protected asparagus** *(off-label)* [1]; **Protected celery** *(off-label)* [1]; **Protected leeks** *(off-label)* [1]; **Protected nursery fruit trees** *(off-label)* [1]; **Protected ornamentals** *(off-label)* [1]; **Protected strawberries** *(off-label)* [1]; **Purslane** *(off-label)* [1]; **Red mustard** *(off-label)* [1]; **Rocket** *(off-label)* [1]; **Rosemary** *(off-label)* [1]; **Sage** *(off-label)* [1]; **Salad burnet** *(off-label)* [1]; **Salad onions** *(off-label)* [1]; **Savory** *(off-label)* [1]; **Shallots** *(off-label)* [1]; **Spinach** *(off-label)* [1]; **Spinach beet** *(off-label)* [1]; **Strawberries** *(off-label)* [1]; **Sweet ciceley** *(off-label)* [1]; **Tarragon** *(off-label)* [1]; **Thyme** *(off-label)* [1]; **Watercress** *(off-label)* [1];
- Two-spotted spider mite in **Angelica** *(off-label)* [1]; **Apricots** *(off-label)* [1]; **Baby leaf crops** *(off-label)* [1]; **Balm** *(off-label)* [1]; **Basil** *(off-label)* [1]; **Bay** *(off-label)* [1]; **Bilberries** *(off-label)* [1]; **Blackberries** *(off-label)* [1]; **Blackcurrants** *(off-label)* [1]; **Blueberries** *(off-label)* [1]; **Caraway** *(off-label)* [1]; **Celery leaves** *(off-label)* [1]; **Cherries** *(off-label)* [1]; **Chervil** *(off-label)* [1]; **Chicory** *(off-label)* [1]; **Chives** *(off-label)* [1]; **Coriander** *(off-label)* [1]; **Cranberries** *(off-label)* [1]; **Cress** *(off-label)* [1]; **Dill** *(off-label)* [1]; **Edible flowers** *(off-label)* [1]; **Elderberries** *(off-label)* [1]; **Fennel leaves** *(off-label)* [1]; **Gooseberries** *(off-label)* [1]; **Herbs (see appendix 6)** *(off-label)* [1]; **Hyssop** *(off-label)* [1]; **Lamb's lettuce** *(off-label)* [1]; **Land cress** *(off-label)* [1]; **Lettuce** *(off-label)* [1]; **Loganberries** *(off-label)* [1]; **Lovage** *(off-label)* [1]; **Marjoram** *(off-label)* [1]; **Mint** *(off-label)* [1]; **Mulberries** *(off-label)* [1]; **Nectarines** *(off-label)* [1]; **Oregano** *(off-label)* [1]; **Peaches** *(off-label)* [1]; **Plums** *(off-label)* [1]; **Protected bilberries** *(off-label)* [1]; **Protected blackberries** *(off-label)* [1]; **Protected blackcurrants** *(off-label)* [1]; **Protected blueberry** *(off-label)* [1]; **Protected cranberries** *(off-label)* [1]; **Protected elderberries** *(off-label)* [1]; **Protected gooseberries** *(off-label)* [1]; **Protected loganberries** *(off-label)* [1]; **Protected mulberry** *(off-label)* [1]; **Protected raspberries** *(off-label)* [1]; **Protected redcurrants** *(off-label)* [1]; **Protected rose hips** *(off-label)* [1]; **Protected Rubus hybrids** *(off-label)* [1]; **Purslane** *(off-label)* [1]; **Raspberries** *(off-label)* [1]; **Red mustard** *(off-label)* [1]; **Redcurrants** *(off-label)* [1]; **Rocket** *(off-label)* [1]; **Rose hips** *(off-label)* [1]; **Rosemary** *(off-label)* [1]; **Rubus hybrids** *(off-label)* [1]; **Sage** *(off-label)* [1]; **Salad burnet** *(off-label)* [1]; **Savory** *(off-label)* [1]; **Spinach** *(off-label)* [1]; **Spinach beet** *(off-label)* [1]; **Sweet ciceley** *(off-label)* [1]; **Tarragon** *(off-label)* [1]; **Thyme** *(off-label)* [1]; **Watercress** *(off-label)* [1];
- Western flower thrips in **Nursery fruit trees** *(off-label)* [1]; **Ornamental plant production** *(off-label)* [1]; **Protected nursery fruit trees** *(off-label)* [1]; **Protected ornamentals** *(off-label)* [1];
- Whitefly in **Angelica** *(off-label)* [1]; **Baby leaf crops** *(off-label)* [1]; **Balm** *(off-label)* [1]; **Basil** *(off-label)* [1]; **Bay** *(off-label)* [1]; **Bilberries** *(off-label)* [1]; **Blackberries** *(off-label)* [1]; **Blackcurrants** *(off-label)* [1]; **Blueberries** *(off-label)* [1]; **Broccoli** *(off-label)* [1]; **Brussels sprouts** *(off-label)* [1]; **Cabbages** *(off-label)* [1]; **Calabrese** *(off-label)* [1]; **Caraway** *(off-label)* [1]; **Cauliflowers** *(off-label)* [1]; **Celery leaves** *(off-label)* [1]; **Chervil** *(off-label)* [1]; **Chicory** *(off-label)* [1]; **Chives** *(off-label)* [1]; **Choi sum** *(off-label)* [1]; **Collards** *(off-label)* [1]; **Coriander** *(off-label)* [1]; **Cranberries** *(off-label)* [1]; **Cress** *(off-label)* [1]; **Dill** *(off-label)* [1]; **Edible flowers** *(off-label)* [1]; **Elderberries** *(off-label)* [1]; **Fennel leaves** *(off-label)* [1]; **Gooseberries** *(off-label)* [1]; **Herbs (see appendix 6)** *(off-label)* [1]; **Hyssop** *(off-label)* [1]; **Kale** *(off-label)* [1]; **Kohlrabi** *(off-label)* [1]; **Lamb's lettuce** *(off-label)* [1]; **Land cress** *(off-label)* [1]; **Lettuce** *(off-label)* [1]; **Loganberries** *(off-label)* [1]; **Lovage** *(off-label)* [1]; **Marjoram** *(off-label)* [1]; **Mint** *(off-label)* [1]; **Mulberries** *(off-label)* [1]; **Nursery fruit trees** *(off-label)* [1]; **Oregano** *(off-label)* [1]; **Oriental cabbage** *(off-label)* [1]; **Ornamental plant production** *(off-label)* [1]; **Protected aubergines** [1]; **Protected bilberries** *(off-label)* [1]; **Protected blackberries** *(off-label)* [1]; **Protected blackcurrants** *(off-label)* [1]; **Protected blueberry** *(off-label)* [1]; **Protected chilli peppers** [1]; **Protected courgettes** [1]; **Protected cranberries** *(off-label)* [1]; **Protected cucumbers** [1]; **Protected elderberries** *(off-label)* [1]; **Protected gooseberries** *(off-label)* [1]; **Protected loganberries** *(off-label)* [1]; **Protected melons** [1]; **Protected mulberry** *(off-label)* [1]; **Protected nursery fruit trees** *(off-label)* [1]; **Protected ornamentals** *(off-label)* [1]; **Protected peppers** [1]; **Protected raspberries** *(off-label)* [1]; **Protected redcurrants** *(off-label)* [1]; **Protected rose hips** *(off-label)* [1]; **Protected Rubus**

SEE SECTION 3 FOR PRODUCTS ALSO REGISTERED

hybrids (off-label) [1]; **Protected strawberries** (off-label) [1]; **Protected summer squash** [1]; **Protected tomatoes** [1]; **Purslane** (off-label) [1]; **Raspberries** (off-label) [1]; **Red mustard** (off-label) [1]; **Redcurrants** (off-label) [1]; **Rocket** (off-label) [1]; **Rose hips** (off-label) [1]; **Rosemary** (off-label) [1]; **Rubus hybrids** (off-label) [1]; **Sage** (off-label) [1]; **Salad burnet** (off-label) [1]; **Savory** (off-label) [1]; **Spinach** (off-label) [1]; **Spinach beet** (off-label) [1]; **Strawberries** (off-label) [1]; **Sweet ciceley** (off-label) [1]; **Tarragon** (off-label) [1]; **Thyme** (off-label) [1]; **Watercress** (off-label) [1];

Extension of Authorisation for Minor Use (EAMUs)

- **Angelica** *20181792* [1]
- **Apricots** *20200791* [1]
- **Asparagus** *20201997* [1]
- **Baby leaf crops** *20181792* [1]
- **Balm** *20181792* [1]
- **Basil** *20181792* [1]
- **Bay** *20181792* [1]
- **Bilberries** *20181507* [1]
- **Blackberries** *20181507* [1]
- **Blackcurrants** *20181507* [1]
- **Blueberries** *20181507* [1]
- **Broccoli** *20221624* [1]
- **Brussels sprouts** *20221624* [1]
- **Bulb onions** *20201993* [1]
- **Cabbages** *20221624* [1]
- **Calabrese** *20221624* [1]
- **Caraway** *20181792* [1]
- **Cauliflowers** *20221624* [1]
- **Celery (outdoor)** *20201997* [1]
- **Celery leaves** *20181792* [1]
- **Cherries** *20200791* [1]
- **Chervil** *20181792* [1]
- **Chicory** *20181792* [1]
- **Chives** *20181792* [1]
- **Choi sum** *20221624* [1]
- **Collards** *20221624* [1]
- **Coriander** *20181792* [1]
- **Cranberries** *20181507* [1]
- **Cress** *20181792* [1]
- **Dill** *20181792* [1]
- **Edible flowers** *20181792* [1]
- **Elderberries** *20181507* [1]
- **Fennel leaves** *20181792* [1]
- **Garlic** *20201993* [1]
- **Gooseberries** *20181507* [1]
- **Herbs (see appendix 6)** *20181792* [1]
- **Hyssop** *20181792* [1]
- **Kale** *20221624* [1]
- **Kohlrabi** *20221624* [1]
- **Lamb's lettuce** *20181792* [1]
- **Land cress** *20181792* [1]
- **Leeks** *20201997* [1]
- **Lettuce** *20181792* [1]
- **Loganberries** *20181507* [1]
- **Lovage** *20181792* [1]
- **Marjoram** *20181792* [1]
- **Mint** *20181792* [1]
- **Mulberries** *20181507* [1]
- **Nectarines** *20200791* [1]
- **Nursery fruit trees** *20221644* [1]

FOR FULL CONDITIONS OF USE ALWAYS READ THE PRODUCT LABEL

- *Oregano* *20181792* [1]
- *Oriental cabbage* *20221624* [1]
- *Ornamental plant production* *20221644* [1]
- *Peaches* *20200791* [1]
- *Plums* *20200791* [1]
- *Protected asparagus* *20201997* [1]
- *Protected bilberries* *20181507* [1]
- *Protected blackberries* *20181507* [1]
- *Protected blackcurrants* *20181507* [1]
- *Protected blueberry* *20181507* [1]
- *Protected celery* *20201997* [1]
- *Protected cranberries* *20181507* [1]
- *Protected elderberries* *20181507* [1]
- *Protected gooseberries* *20181507* [1]
- *Protected leeks* *20201997* [1]
- *Protected loganberries* *20181507* [1]
- *Protected mulberry* *20181507* [1]
- *Protected nursery fruit trees* *20221644* [1]
- *Protected ornamentals* *20221644* [1]
- *Protected raspberries* *20181507* [1]
- *Protected redcurrants* *20181507* [1]
- *Protected rose hips* *20181507* [1]
- *Protected Rubus hybrids* *20181507* [1]
- *Protected strawberries* *20160110* [1]
- *Purslane* *20181792* [1]
- *Raspberries* *20181507* [1]
- *Red mustard* *20181792* [1]
- *Redcurrants* *20181507* [1]
- *Rocket* *20181792* [1]
- *Rose hips* *20181507* [1]
- *Rosemary* *20181792* [1]
- *Rubus hybrids* *20181507* [1]
- *Sage* *20181792* [1]
- *Salad burnet* *20181792* [1]
- *Salad onions* *20201993* [1]
- *Savory* *20181792* [1]
- *Shallots* *20201993* [1]
- *Spinach* *20181792* [1]
- *Spinach beet* *20181792* [1]
- *Strawberries* *20160110* [1]
- *Sweet ciceley* *20181792* [1]
- *Tarragon* *20181792* [1]
- *Thyme* *20181792* [1]
- *Watercress* *20181792* [1]

Approval information
- Beauveria bassiana GHA included in Annex 1 under EC Regulation 1107/2009

Environmental safety
- Buffer zone requirement 10 m for bilberries, blackberry, blackcurrants, blueberry, cranberry, elderberry, gooseberry, loganberry, rubus hybrid, raspberry, redcurrant, mulberry and rosehips, 15 m for apricots, cherries, nectarines, peaches and plums [1]

Hazard classification and safety precautions
UN Number N/C
Risk phrases H317, H334
Operator protection U05a, U11, U14, U15, U16b, U19a, U20b; A, D, H
Environmental protection E15b, E34
Storage and disposal D01, D02, D05, D09a, D10a
Medical advice M03, M04a

SEE SECTION 3 FOR PRODUCTS ALSO REGISTERED

37   bentazone

A post-emergence contact benzothiadiazinone herbicide
HRAC mode of action code: 6 (C3)

Products

1	Basagran SG	BASF	87% w/w	SG	08360
2	Benta 480 SL	Nufarm UK	480 g/l	SL	17355
3	Clayton Baritone	Clayton	87% w/w	SG	19568

Uses

- Annual dicotyledons in *Beans without pods (dry)* [1,3]; *Beans without pods (Fresh)* [1,3]; *Broad beans* [1-3]; *Bulb onions* (off-label) [1]; *Chives* (off-label) [1]; *Combining peas* [1-3]; *Dwarf beans* [1,3]; *Dwarf French beans* [2]; *French beans* [1,3]; *Garlic* (off-label) [1]; *Herbs (see appendix 6)* (off-label) [1]; *Leeks* (off-label) [1]; *Linseed* [1-3]; *Ornamental plant production* [1-3]; *Potatoes* [1-3]; *Runner beans* [1-3]; *Salad onions* (off-label) [1]; *Shallots* (off-label) [1]; *Soya beans* [1,3]; *Spring field beans* [1-3]; *Vining peas* [1-3]; *Winter field beans* [1-3];
- Black nightshade in *Edible podded peas* (off-label) [1];
- Charlock in *Edible podded peas* (off-label) [1];
- Chickweed in *Edible podded peas* (off-label) [1]; *Game cover* (off-label) [1]; *Hops* (off-label) [1]; *Ornamental plant production* (off-label) [1];
- Cleavers in *Game cover* (off-label) [1]; *Hops* (off-label) [1]; *Ornamental plant production* (off-label) [1];
- Common storksbill in *Leeks* (off-label) [1];
- Fool's parsley in *Leeks* (off-label) [1];
- Groundsel in *Game cover* (off-label) [1]; *Herbs (see appendix 6)* (off-label) [1]; *Hops* (off-label) [1]; *Ornamental plant production* (off-label) [1];
- Mayweeds in *Bulb onions* (off-label) [1]; *Edible podded peas* (off-label) [1]; *Game cover* (off-label) [1]; *Garlic* (off-label) [1]; *Herbs (see appendix 6)* (off-label) [1]; *Hops* (off-label) [1]; *Ornamental plant production* (off-label) [1]; *Shallots* (off-label) [1];
- Pale persicaria in *Edible podded peas* (off-label) [1];
- Penny cress in *Edible podded peas* (off-label) [1];
- Redshank in *Edible podded peas* (off-label) [1];
- Shepherd's purse in *Edible podded peas* (off-label) [1];

Extension of Authorisation for Minor Use (EAMUs)

- *Bulb onions* 20061631 [1]
- *Chives* 20102994 [1]
- *Edible podded peas* 20202707 [1]
- *Game cover* 20082819 [1]
- *Garlic* 20061631 [1]
- *Herbs (see appendix 6)* 20120744 [1]
- *Hops* 20082819 [1]
- *Leeks* 20210711 [1]
- *Ornamental plant production* 20082819 [1]
- *Salad onions* 20102994 [1]
- *Shallots* 20061631 [1]

Approval information

- Bentazone included in Annex I under EC Regulation 1107/2009

Efficacy guidance

- Most effective control obtained when weeds are growing actively and less than 5 cm high or across. Good spray cover is essential
- The addition of specified adjuvant oils is recommended for use on some crops to improve fat hen control. Do not use under hot or humid conditions. See label for details
- Split dose application may be made in all recommended crops except peas and generally gives better weed control. See label for details
- See label for guidance on water stewardship

FOR FULL CONDITIONS OF USE ALWAYS READ THE PRODUCT LABEL

SECTION 2

Restrictions

- Maximum number of treatments normally 2 per crop but check label
- Crops must be treated at correct stage of growth to avoid danger of scorch. See label for details
- Not all varieties of recommended crops are fully tolerant. Use only on tolerant varieties named in label. Do not use on forage or mange-tout varieties of peas
- Do not use on crops which have been affected by drought, waterlogging, frost or other stress conditions
- Do not apply insecticides within 7 d of treatment
- Leave 14 d after using a post-emergence grass herbicide and carry out a leaf wax test where relevant or 7 d where treatment precedes the grass herbicide
- Do not spray at temperatures above 21°C. Delay spraying until evening if necessary
- Do not apply if rain or frost expected, if unseasonably cold, if foliage wet or in drought
- A minimum of 6 h (preferably 12 h) free from rain is required after application
- May be used on selected varieties of maincrop and second early potatoes (see label for details), not on seed crops or first earlies
- Do not apply to narcissi during flower bud formation

Crop-specific information

- Latest use: before shoots exceed 15 cm high for potatoes and spring field beans (or 6-7 leaf pairs); 4 leaf pairs (6 pairs or 15 cm high with split dose) for broad beans; before flower buds visible for French, navy, runner and winter field beans, and linseed; before flower buds can be found enclosed in terminal shoot for peas
- Best results in narcissi obtained by using a suitable pre-emergence herbicide first
- Consult processor before using on crops for processing
- A satisfactory wax test must be carried out before use on peas
- Do not treat narcissi during flower bud formation

Environmental safety

- Dangerous for the environment
- Harmful to aquatic organisms
- Some pesticides pose a greater threat of contamination of water than others and bentazone is one of these pesticides. Take special care when applying bentazone near water and do not apply if heavy rain is forecast

Hazard classification and safety precautions

Hazard Harmful, Harmful if swallowed
Packaging group III
UN Number 2588, N/C
Risk phrases H317, H318 [1,3]
Operator protection U05a, U08, U11 [1,3], U14 [1,3], U19a, U20a [2], U20b [1,3]; A, C
Environmental protection E15a [2], E15b [1,3], E34 [1,3], H412 [2]
Storage and disposal D01, D02, D09a, D10b [1,3], D10c [2]
Medical advice M03 [1,3], M05a [1,3]

38 benthiavalicarb + oxathiapiprolin

A fungicide mixture for use in allium crops and potatoes
FRAC mode of action code: 40 + 49

See also benthiavalicarb-isopropyl

Products

1	Zorvec Endavia	Corteva	70:30 g/l	OD	19407

Uses

- Disease control in **Bulb onions** [1]; **Garlic** [1]; **Shallots** [1];
- Downy mildew in **Bulb onions** [1]; **Garlic** [1]; **Shallots** [1];
- Late blight in **Potatoes** [1];

Approval information

- Benthiavalicarb and oxathiapiprolin included in Annex I under EC Regulation 1107/2009

SEE SECTION 3 FOR PRODUCTS ALSO REGISTERED

Restrictions
- A minimum interval of 7 days must be observed between applications
- On potatoes, a minimum interval of 7 days must elapse between application and initial haulm destruction and/or harvest, whichever comes first

Hazard classification and safety precautions
>**Transport code** 9 [1]
>**Packaging group** III
>**UN Number** 3082
>**Risk phrases** H317, H351
>**Operator protection** A, H
>**Environmental protection** E15b, E34, H411
>**Storage and disposal** D01, D02, D05, D09a, D11a

39   benthiavalicarb-isopropyl

An amino acid amide carbamate fungicide for blight control in potatoes
FRAC mode of action code: 40

Products

1	Versilus	Certis Belchim B V	15% w/w	WG	18618

Uses
- Blight in **Potatoes** [1];

Approval information
- Benthiavalicarb-isopropyl included in Annex I under EC Regulation 1107/2009

Efficacy guidance
- Consult processor before using on potatoes grown for processing
- Apply at first local blight warning before infection is seen in the crop
- Works by contact activity so good coverage is essential

Restrictions
- No more than 3 applications of any CAA fungicide should be made consecutively

Hazard classification and safety precautions
>**UN Number** N/C
>**Risk phrases** H334, H351, H372
>**Operator protection** A, H
>**Environmental protection** E15b, H412
>**Storage and disposal** D01, D02, D05, D09a, D11a

40   benzoic acid

An organic horticultural disinfectant

Products

1	Menno Florades	Fargro	90 g/l	SL	13985
2	Menno Florades	Fargro	90 g/l	SL	19998

Uses
- Bacteria in **All protected non-edible crops** [1]; **Crop handling & storage structures** [2];
- Fungal spores in **All protected non-edible crops** [1]; **Crop handling & storage structures** [2];
- Viroids in **All protected non-edible crops** [1]; **Crop handling & storage structures** [2];
- Virus in **All protected non-edible crops** [1]; **Crop handling & storage structures** [2];

Approval information
- Benzoic acid is included in Annex 1 under EC Regulation 1107/2009

FOR FULL CONDITIONS OF USE ALWAYS READ THE PRODUCT LABEL

Hazard classification and safety precautions
> **Hazard** Corrosive, Highly flammable liquid and vapour [2], Flammable liquid and vapour [1]
> **Transport code** 3 [1-2]
> **Packaging group** III
> **UN Number** 1987
> **Risk phrases** H318, H336, H373 [2]
> **Operator protection** U05a, U11, U13, U14, U15; A, C, H
> **Environmental protection** E15a
> **Storage and disposal** D01, D02, D09a, D10a
> **Medical advice** M05a

41   benzovindiflupyr

An SDHI fungicide for disease control in cereals
FRAC mode of action code: 7

See also benzovindiflupyr + prothioconazole

Products

1	Ceravato Plus	Syngenta	100 g/l	EC	17865
2	Elatus Plus	Syngenta	100 g/l	EC	17841
3	Velogy Plus	Syngenta	100 g/l	EC	17866

Uses
- Brown rust in *Rye* [1-3]; *Spring barley* [1-3]; *Spring wheat* [1-3]; *Triticale* [1-3]; *Winter barley* [1-3]; *Winter wheat* [1-3];
- Glume blotch in *Spring wheat (moderate control)* [1-3]; *Winter wheat (moderate control)* [1-3];
- Mycosphaerella in *Spring wheat* [1-3]; *Triticale* [1-3]; *Winter wheat* [1-3];
- Net blotch in *Spring barley* [1-3]; *Winter barley* [1-3];
- Ramularia leaf spots in *Spring barley* [1-3]; *Winter barley* [1-3];
- Rhynchosporium in *Rye (moderate control)* [1-3]; *Spring barley (moderate control)* [1-3]; *Winter barley (moderate control)* [1-3];
- Rust in *Bulb onions* [2]; *Garlic* [2]; *Shallots* [2];
- Septoria leaf blotch in *Spring wheat* [1-3]; *Triticale* [1-3]; *Winter wheat* [1-3];
- Yellow rust in *Spring wheat* [1-3]; *Triticale* [1-3]; *Winter wheat* [1-3];

Approval information
- Benzovindiflupyr is included in Annex 1 under EC Regulation 1107/2009

Efficacy guidance
- Benzovindiflupyr works predominately via protective action

Restrictions
- Do not apply via hand-held equipment
- No more than 2 applications of SDHI fungicides should be applied to any cereal crop
- Application to bulb onion, garlic and shallot may only be made between BBCH 40-48
- Horizontal boom sprayers must be fitted with three star drift reduction technology for all uses
- Low drift spraying equipment must be operated according to the specific conditions stated in the official three star rating for that equipment as published on HSE Chemicals Regulation Division?s website. These operating conditions must be maintained until the operator is 30m from the top of the bank of any surface water bodies

Following crops guidance
- There are no restrictions on succeeding crops in a normal rotation

Environmental safety
- Buffer zone requirement 6 m [1, 2, 3]
- LERAP Category B [1-3]

Hazard classification and safety precautions
> **Hazard** Harmful if swallowed, Harmful if inhaled, Very toxic to aquatic organisms
> **Transport code** 9 [1-3]

SEE SECTION 3 FOR PRODUCTS ALSO REGISTERED

SECTION 2

Packaging group III
UN Number 3082
Risk phrases H317, H318, H335 [2]
Operator protection U09a, U10, U11, U19a, U20a; A, C, H
Environmental protection E15b, E16a, H410
Storage and disposal D01, D09a, D10c, D12a, D12b, D22

42   benzovindiflupyr + prothioconazole

An SDHI and triazole fungicide mixture for disease control in cereals
FRAC mode of action code: 3 + 7

See also benzovindiflupyr

Products

1	Elatus Era	Syngenta	75:150 g/l	EC	17889
2	Tacanza Era	Syngenta	75:150 g/l	EC	19217
3	Velogy Era	Syngenta	75:150 g/l	EC	18981

Uses

- Ascochyta in **Combining peas** *(useful control)* [1];
- Botrytis in **Spring field beans** *(useful control)* [1]; **Winter field beans** *(useful control)* [1];
- Brown rust in **Rye** [1-3]; **Spring barley** [1-3]; **Spring wheat** [1,3]; **Triticale** [1-3]; **Winter barley** [1-3]; **Winter wheat** [1,3];
- Crown rust in **Oats** [2]; **Spring oats** [1,3]; **Winter oats** [1,3];
- Disease control in **Combining peas** [2-3]; **Linseed** [2-3]; **Rye** [1,3]; **Spring field beans** [2-3]; **Spring oats** [3]; **Winter field beans** [2-3]; **Winter oats** [3];
- Fusarium in **Spring wheat** *(moderate control)* [1,3]; **Winter wheat** *(moderate control)* [1,3];
- Glume blotch in **Spring wheat** *(moderate control)* [1,3]; **Winter wheat** *(moderate control)* [1,3];
- Mycosphaerella in **Linseed** *(moderate control)* [1]; **Spring wheat** [1,3]; **Winter wheat** [1,3];
- Net blotch in **Spring barley** [1-3]; **Winter barley** [1-3];
- Powdery mildew in **Linseed** *(moderate control)* [1];
- Ramularia leaf spots in **Spring barley** [1-3]; **Winter barley** [1-3];
- Rhynchosporium in **Rye** *(moderate control)* [1-3]; **Spring barley** *(moderate control)* [1-3]; **Triticale** *(moderate control)* [1-3]; **Winter barley** *(moderate control)* [1-3];
- Rust in **Combining peas** *(moderate control)* [1]; **Spring field beans** [1]; **Winter field beans** [1];
- Septoria leaf blotch in **Spring wheat** [1,3]; **Triticale** [1-3]; **Winter wheat** [1,3];
- Yellow rust in **Spring wheat** [1,3]; **Winter wheat** [1,3];

Approval information

- Benzovindiflupyr and prothioconazole are included in Annex 1 under EC Regulation 1107/2009
- Accepted by BBPA for use on malting barley

Efficacy guidance

- Benzovindiflupyr is predominantly protectant while prothioconazole is a triazole with systemic and protectant activity. Use as a protectant treatment or in the earliest stages of disease development.

Restrictions

- The earliest time of application is GS31
- No more than 2 applications of SDHI fungicides should be applied to any cereal crop
- Do not apply via hand-held equipment
- Horizontal boom sprayers must be fitted with three star drift reduction technology for all uses
- Low drift spraying equipment must be operated according to the specific conditions stated in the official three star rating for that equipment as published on HSE Chemicals Regulation Division?s website. These operating conditions must be maintained until the operator is 30m from the top of the bank of any surface water bodies.
- The earliest time of application on cereals is GS31, on linseed/flax is GS32 and on pulses is GS51

Following crops guidance

- There are no restrictions on succeeding crops in a normal rotation

FOR FULL CONDITIONS OF USE ALWAYS READ THE PRODUCT LABEL

Environmental safety
- Buffer zone requirement 6m [1, 3, 2]
- LERAP Category B [1-3]

Hazard classification and safety precautions
   Hazard Harmful if swallowed, Harmful if inhaled
   Transport code 9 [1-3]
   Packaging group III
   UN Number 3082
   Risk phrases H315, H317, H318, H335
   Operator protection U09a, U10, U11, U19a, U20a; A, C, H
   Environmental protection E15b, E16a, H410
   Storage and disposal D01, D09a, D10c, D12a, D12b
   Medical advice M04a

43   benzyladenine

A cytokinin plant growth regulator for use in apples and pears or on named ornamentals

Products

1	Configure	Fine	20 g/l	SC	17523
2	Exilis	Fine	20 g/l	SL	15706
3	Globaryll 100	Certis Belchim B V	100 g/l	SL	16097
4	MaxCel	Sumitomo	20 g/l	SL	18878
5	Sitis	Fine	20 g/l	SL	19662

Uses
- Fruit thinning in *Apples* [2,4-5]; *Pears* [2,5];
- Growth regulation in *Apples* [3]; *Ornamental plant production* [1]; *Pears* [3];
- Increasing flowering in *Phalaenopsis spp.* [1]; *Schlumbergera spp.* [1]; *Sempervivium spp.* [1];

Approval information
- Benzyladenine is included in Annex 1 under EC Directive 1107/2009

Efficacy guidance
- A post-bloom thinner for apples and pears with excessive blossom [2, 3, 4]
- Apply when the temperature will exceed 15°C on day of application but note that temperatures above 28°C may result in excessive thinning [2, 3, 4]

Restrictions
- Use in pears is a Qualified Minor Use recommendation. Evidence of efficacy and crop safety in pears is limited to the pear variety 'Conference' [2]

Environmental safety
- Broadcast air-assisted LERAP [4] (5 m)

Hazard classification and safety precautions
   Hazard Harmful [3], Dangerous for the environment [3]
   UN Number N/C
   Risk phrases H317 [1], H318 [3], H361 [3]
   Operator protection U02a [2,4-5], U05a [2-5], U09a [2,4-5], U11 [3], U14 [3], U15 [3], U19a [2,4-5], U20b [2,4-5], U20c [3]; A, C, H
   Environmental protection E15a [3], E15b [1-2,4-5], E17b (5 m) [4], E38 [2-5], H412 [2-5]
   Storage and disposal D01 [1,3], D02, D03 [2,4-5], D05 [2,4-5], D09a, D10a [1], D10b, D12a [2,4-5]
   Medical advice M04a [3], M05a [2,4-5]

SEE SECTION 3 FOR PRODUCTS ALSO REGISTERED

44   benzyladenine + gibberellin

A plant growth regulator mixture for use on ornamental plants

See also benzyladenine
            gibberellins

Products

1	Chrysal BVB	Chrysal	19:19 g/l	SL	17780
2	Floralife Bulb 100	Fine	19:19 g/l	SL	17995
3	Promalin	Sumitomo	19:19 g/l	SL	18892

Uses
- Growth regulation in **Apples** [3]; **Ornamental plant production** [1]; **Pears** [3]; **Tulips** [2];
- Reduction of leaf yellowing in **Ornamental plant production** *(off-label)* [2];

Extension of Authorisation for Minor Use (EAMUs)
- **Ornamental plant production** *20221782* [2]

Approval information
- Benzyladenine and gibberellins are included in Annex 1 under EC Directive 1107/2009

Hazard classification and safety precautions
   UN Number N/C
   Operator protection A, H
   Environmental protection H412

45   bifenazate

A bifenazate acaricide for use on protected strawberries
IRAC mode of action code: 25

Products

1	Floramite 240 SC	UPL Europe	240 g/l	SC	19607
2	Pan Scarlet	Pan Agriculture	240 g/l	SC	19293

Uses
- Two-spotted spider mite in **Protected aubergines** *(off-label)* [1]; **Protected chilli peppers** *(off-label)* [1]; **Protected cucumbers** *(off-label)* [1]; **Protected forest nurseries** *(off-label)* [1]; **Protected hops** *(off-label)* [1]; **Protected ornamentals** *(off-label)* [1]; **Protected peppers** *(off-label)* [1]; **Protected strawberries** [1-2]; **Protected tomatoes** *(off-label)* [1];

Extension of Authorisation for Minor Use (EAMUs)
- **Protected aubergines** *20202718* [1]
- **Protected chilli peppers** *20202718* [1]
- **Protected cucumbers** *20202718* [1]
- **Protected forest nurseries** *20202716* [1]
- **Protected hops** *20202716* [1]
- **Protected ornamentals** *20202717* [1]
- **Protected peppers** *20202718* [1]
- **Protected tomatoes** *20202718* [1]

Approval information
- Bifenazate included in Annex I under EC Regulation 1107/2009

Efficacy guidance
- Bifenazate acts by contact knockdown after approximately 4 d and a subsequent period of residual control
- All mobile stages of mites are controlled with occasionally some ovicidal activity
- Best results obtained from a programme of two treatments started as soon as first spider mites are seen

FOR FULL CONDITIONS OF USE ALWAYS READ THE PRODUCT LABEL

- To minimise possible development of resistance use in a planned Resistance Management strategy. Alternate at least two products with different modes of action between treatment programmes of bifenazate. See Section 5 for more information

Restrictions
- Maximum number of treatments 2 per yr for protected strawberries
- After use in protected environments avoid re-entry for at least 2 h while full ventilation is carried out allowing spray to dry on leaves
- Personnel working among treated crops should wear suitable long-sleeved garments and gloves for 14 d after treatment
- Do not apply as a low volume spray

Crop-specific information
- HI: 7 d for protected strawberries
- Carry out small scale tolerance test on the strawberry cultivar before large scale use

Environmental safety
- Dangerous for the environment
- Toxic to aquatic organisms
- Harmful to non-target predatory mites. Avoid spray drift onto field margins, hedges, ditches, surface water and neighbouring crops

Hazard classification and safety precautions
Hazard Irritant, Dangerous for the environment
Transport code 9 [1-2]
Packaging group III
UN Number 3082
Risk phrases H317, H373
Operator protection U02a, U04a, U05a, U10, U14, U15, U19a, U20b; A, H
Environmental protection E15b, E38, H411
Storage and disposal D01, D02, D05, D09a, D10b, D12b
Medical advice M03

46   bifenox

A diphenyl ether herbicide for use in cereals and (off-label) in oilseed rape.
HRAC mode of action code: 14 (E)

Products
1 Clayton Belstone	Clayton	480 g/l	SC	14033
2 Fox	Adama	480 g/l	SC	11981

Uses
- Annual dicotyledons in *Canary flower (Echium spp.)* *(off-label)* [1-2]; *Evening primrose* *(off-label)* [2]; *Grass seed crops* *(off-label)* [1-2]; *Honesty* *(off-label)* [2]; *Linseed* *(off-label)* [2]; *Mustard* *(off-label)* [2]; *Oilseed rape* *(off-label)* [2]; *Triticale* [1]; *Winter barley* [1]; *Winter rye* [1]; *Winter wheat* [1];
- Charlock in *Grass seed crops* *(off-label)* [2];
- Cleavers in *Triticale* [2]; *Winter barley* [2]; *Winter rye* [2]; *Winter wheat* [2];
- Field pansy in *Forest nurseries* *(off-label)* [2]; *Game cover* *(off-label)* [2]; *Triticale* [2]; *Winter barley* [2]; *Winter rye* [2]; *Winter wheat* [2];
- Field speedwell in *Triticale* [2]; *Winter barley* [2]; *Winter rye* [2]; *Winter wheat* [2];
- Forget-me-not in *Forest nurseries* *(off-label)* [2]; *Game cover* *(off-label)* [2]; *Triticale* [2]; *Winter barley* [2]; *Winter rye* [2]; *Winter wheat* [2];
- Ivy-leaved speedwell in *Triticale* [2]; *Winter barley* [2]; *Winter rye* [2]; *Winter wheat* [2];
- Poppies in *Forest nurseries* *(off-label)* [2]; *Game cover* *(off-label)* [2]; *Triticale* [2]; *Winter barley* [2]; *Winter rye* [2]; *Winter wheat* [2];
- Red dead-nettle in *Forest nurseries* *(off-label)* [2]; *Game cover* *(off-label)* [2]; *Triticale* [2]; *Winter barley* [2]; *Winter rye* [2]; *Winter wheat* [2];
- Speedwells in *Forest nurseries* *(off-label)* [2]; *Game cover* *(off-label)* [2];

Extension of Authorisation for Minor Use (EAMUs)
- *Canary flower (Echium spp.) 20121422* [1], *20142318* [2]
- *Evening primrose 20142318* [2]
- *Forest nurseries 20202126* [2]
- *Game cover 20202125* [2]
- *Grass seed crops 20193238* [1], *20193257* [2]
- *Honesty 20142318* [2]
- *Linseed 20142318* [2]
- *Mustard 20142318* [2]
- *Oilseed rape 20142318* [2]

Approval information
- Accepted by BBPA for use on malting barley
- Bifenox included in Annex I under EC Regulation 1107/2009

Efficacy guidance
- Bifenox is absorbed by foliage and emerging roots of susceptible species
- Best results obtained when weeds are growing actively with adequate soil moisture

Restrictions
- Maximum number of treatments: one per crop
- Do not apply to crops suffering from stress from whatever cause
- Do not apply if the crop is wet or if rain or frost is expected
- Avoid drift onto broad-leaved plants outside the target area

Crop-specific information
- Latest use: before 2nd node detectable (GS 32) for all crops

Environmental safety
- Dangerous for the environment
- Very toxic to aquatic organisms
- Do not empty into drains

Hazard classification and safety precautions
Hazard Dangerous for the environment
Transport code 9 [1-2]
Packaging group III
UN Number 3082
Operator protection U05a, U08, U20b; A
Environmental protection E15a, E19b, E34, E38, H410
Storage and disposal D01, D02, D09a, D12a
Medical advice M03

47   bixafen

A succinate dehydrogenase inhibitor (SDHI) fungicide for disease control in barley
FRAC mode of action code: 7

Products
1 Inception Xpro   Bayer CropScience   125 g/l   EC   18312

Uses
- Brown rust in *Spring barley* [1]; *Winter barley* [1];
- Net blotch in *Spring barley* [1]; *Winter barley* [1];
- Rhynchosporium in *Spring barley* [1]; *Winter barley* [1];

Approval information
- Bixafen is included in Annex I under EC Regulation 1107/2009

Environmental safety
- LERAP Category B [1]

FOR FULL CONDITIONS OF USE ALWAYS READ THE PRODUCT LABEL

Hazard classification and safety precautions
    Hazard Harmful if swallowed
    Transport code 9 [1]
    Packaging group III
    UN Number 3082
    Risk phrases H304, H315, H319
    Operator protection U12, U14; A, C, H
    Environmental protection E16a, E34, H410
    Storage and disposal D01, D05, D10a

48   bixafen + fluopyram

A fungicide mixture for disease control in wheat
FRAC mode of action code: 7

Products

1	Jettano	Bayer CropScience	100:100 g/l	EC	19931
2	Silvron Xpro	Bayer CropScience	100:100 g/l	EC	19384

Uses
- Brown rust in *Spring barley* [1]; *Spring wheat* [2]; *Winter barley* [1]; *Winter wheat* [2];
- Septoria leaf blotch in *Spring wheat* [2]; *Winter wheat* [2];

Restrictions
- Do not apply prior to the beginning of stem elongation (BBCH30)

Environmental safety
- LERAP Category B [1-2]

Hazard classification and safety precautions
    Hazard Very toxic to aquatic organisms [2]
    Transport code 9 [1-2]
    Packaging group III
    UN Number 3082
    Risk phrases H304, H315, H317, H319, H335
    Operator protection U05a, U08, U11, U20a; A, C, H
    Environmental protection E16a, E34, H410
    Storage and disposal D01, D02, D05, D09a, D10b, D12c
    Medical advice M04c, M05b

49   bixafen + fluopyram + prothioconazole

A fungicide mixture for disease control in wheat
FRAC mode of action code: 7 + 3

See also bixafen + prothioconazole

Products

1	Ascra Xpro	Bayer CropScience	65:65:130 g/l	EC	17623

Uses
- Brown rust in *Spring wheat* [1]; *Winter wheat* [1];
- Crown rust in *Spring oats* [1]; *Winter oats* [1];
- Disease control in *Rye* [1]; *Spring barley* [1]; *Spring oats* [1]; *Triticale* [1]; *Winter barley* [1]; *Winter oats* [1];
- Ear diseases in *Spring wheat* (reduction of incidence and severity) [1]; *Winter wheat* (reduction of incidence and severity) [1];
- Eyespot in *Spring wheat* (reduction of incidence and severity) [1]; *Winter wheat* (reduction of incidence and severity) [1];
- Glume blotch in *Spring wheat* [1]; *Winter wheat* [1];
- Net blotch in *Spring barley* [1]; *Winter barley* [1];

SEE SECTION 3 FOR PRODUCTS ALSO REGISTERED

- Powdery mildew in *Rye* [1]; *Spring barley* [1]; *Spring oats* [1]; *Spring wheat* [1]; *Triticale* [1]; *Winter barley* [1]; *Winter oats* [1]; *Winter wheat* [1];
- Ramularia leaf spots in *Spring barley* [1]; *Winter barley* [1];
- Septoria leaf blotch in *Spring wheat* [1]; *Winter wheat* [1];
- Tan spot in *Spring wheat* [1]; *Winter wheat* [1];
- Yellow rust in *Spring wheat* [1]; *Winter wheat* [1];

Approval information
- Bixafen, fluopyram and prothioconazole are included in Annex 1 under EC Regulation 1107/2009.

Efficacy guidance
- Applications to upper leaves where *S. tritici* symptoms are present are likely to be less effective

Restrictions
- Bixafen is an SDH respiration inhibitor; Do not apply more than two foliar applications of products containing an SDH inhibitor to any cereal crop
- Resistance to some DMI fungicides has been identified in Septoria leaf blotch (*Mycosphaerella graminicola*) which may seriously affect the performance of some products. For further advice on resistance management in DMIs contact your agronomist or specialist advisor, and visit the FRAG-UK website.

Environmental safety
- LERAP Category B [1]

Hazard classification and safety precautions
Hazard Harmful if swallowed
Transport code 9 [1]
Packaging group III
UN Number 3082
Risk phrases H317, H318
Operator protection U05a, U09b, U11, U20a; A, C, H
Environmental protection E15b, E16a, E34, H410
Storage and disposal D01, D02, D05, D09a, D10b, D12b

50   bixafen + fluoxastrobin + prothioconazole

An SDHI + strobilurin + triazole mixture for disease control in cereals
FRAC mode of action code: 7 + 11 + 3

Products

1	Variano Xpro	Bayer CropScience	40:50:100 g/l	EC	15218

Uses
- Brown rust in *Spring wheat* [1]; *Triticale* [1]; *Winter rye* [1]; *Winter wheat* [1];
- Ear diseases in *Spring wheat* [1]; *Triticale* [1]; *Winter rye* [1]; *Winter wheat* [1];
- Glume blotch in *Spring wheat* [1]; *Triticale* [1]; *Winter rye* [1]; *Winter wheat* [1];
- Leaf blotch in *Triticale* [1]; *Winter rye* [1];
- Powdery mildew in *Spring wheat* [1]; *Triticale* [1]; *Winter rye* [1]; *Winter wheat* [1];
- Rhynchosporium in *Winter rye* [1];
- Tan spot in *Spring wheat* [1]; *Triticale* [1]; *Winter rye* [1]; *Winter wheat* [1];
- Yellow rust in *Spring wheat* [1]; *Triticale* [1]; *Winter rye* [1]; *Winter wheat* [1];

Approval information
- Bixafen, fluoxastrobin and prothioconazole are included in Annex 1 under EC Regulation 1107/2009.

Restrictions
- Bixafen is an SDH respiration inhibitor; Do not apply more than two foliar applications of products containing an SDH inhibitor to any cereal crop

Environmental safety
- LERAP Category B [1]

FOR FULL CONDITIONS OF USE ALWAYS READ THE PRODUCT LABEL

Hazard classification and safety precautions
   Hazard Irritant, Dangerous for the environment
   Transport code 9 [1]
   Packaging group III
   UN Number 3082
   Risk phrases H317, H319
   Operator protection U05b, U08, U14, U20b; A, H
   Environmental protection E15a, E16a, E16b, E34, E38, H410
   Storage and disposal D01, D02, D05, D09a, D10b, D12a
   Medical advice M03

51   bixafen + prothioconazole

Succinate dehydrogenase inhibitor (SDHI) + triazole fungicide mixture for disease control in cereals
FRAC mode of action code: 3 + 7

Products

1	Aviator 235 Xpro	Bayer CropScience	75:160 g/l	EC	15026
2	Siltra Xpro	Bayer CropScience	60:200 g/l	EC	15082

Uses
- Brown rust in **Grass seed crops** *(off-label)* [2]; **Spring barley** [2]; **Spring wheat** [1]; **Triticale** [1]; **Winter barley** [2]; **Winter rye** [1]; **Winter wheat** [1];
- Crown rust in **Grass seed crops** *(off-label)* [2]; **Spring oats** [2]; **Winter oats** [2];
- Disease control in **Spring oilseed rape** [1]; **Winter oilseed rape** [1];
- Drechslera leaf spot in **Grass seed crops** *(off-label)* [2];
- Ear diseases in **Spring barley** [2]; **Spring wheat** [1]; **Triticale** [1]; **Winter barley** [2]; **Winter wheat** [1];
- Eyespot in **Spring barley** *(reduction of incidence and severity)* [2]; **Spring oats** *(reduction of incidence and severity)* [2]; **Spring wheat** *(reduction of incidence and severity)* [1]; **Triticale** *(reduction of incidence and severity)* [1]; **Winter barley** *(reduction of incidence and severity)* [2]; **Winter oats** *(reduction of incidence and severity)* [2]; **Winter wheat** *(reduction of incidence and severity)* [1];
- Glume blotch in **Spring wheat** [1]; **Triticale** [1]; **Winter wheat** [1];
- Mildew in **Grass seed crops** *(off-label)* [2];
- Net blotch in **Spring barley** [2]; **Winter barley** [2];
- Powdery mildew in **Spring barley** [2]; **Spring oats** [2]; **Spring wheat** [1]; **Triticale** [1]; **Winter barley** [2]; **Winter oats** [2]; **Winter rye** [1]; **Winter wheat** [1];
- Ramularia leaf spots in **Spring barley** [2]; **Winter barley** [2];
- Rhynchosporium in **Grass seed crops** *(off-label)* [2]; **Spring barley** [2]; **Winter barley** [2]; **Winter rye** [1];
- Septoria leaf blotch in **Spring wheat** [1]; **Triticale** [1]; **Winter wheat** [1];
- Tan spot in **Spring wheat** [1]; **Triticale** [1]; **Winter wheat** [1];
- Yellow rust in **Spring barley** [2]; **Spring wheat** [1]; **Triticale** [1]; **Winter barley** [2]; **Winter wheat** [1];

Extension of Authorisation for Minor Use (EAMUs)
- **Grass seed crops** *20171034* [2]

Approval information
- Bixafen and prothioconazole are included in Annex I under EC Regulation 1107/2009.
- Accepted by BBPA for use on malting barley

Efficacy guidance
- Resistance to some DMI fungicides has been identified in Septoria leaf blotch (*Mycosphaerella graminicola*) which may seriously affect the performance of some products. For further advice on resistance management in DMIs contact your agronomist or specialist advisor, and visit the FRAG-UK website.

SEE SECTION 3 FOR PRODUCTS ALSO REGISTERED

Restrictions

- Bixafen is an SDH respiration inhibitor; Do not apply more than two foliar applications of products containing an SDH inhibitor to any cereal crop

Environmental safety

- LERAP Category B [1-2]

Hazard classification and safety precautions

Hazard Harmful, Dangerous for the environment, Harmful if swallowed [2]
Transport code 9 [1-2]
Packaging group III
UN Number 3082
Risk phrases H317 [2], H319
Operator protection U05a, U09a, U20b; A, C, H
Environmental protection E15b, E16a, E16b, E34, E38, H410
Storage and disposal D01, D02, D09a, D10b, D12a

52   bixafen + prothioconazole + spiroxamine

A broad-spectrum fungicide mixture for disease control in cereals
FRAC mode of action code: 3 + 5 + 7

Products

1	Boogie Xpro	Bayer CropScience	50:100:250 g/l	EC	15061
2	Boogie Xpro Plus	Bayer CropScience	55:110:250 g/l	EC	18835
3	Propel Xpro	Bayer CropScience	50:100:250 g/l	EC	19352

Uses

- Brown rust in *Spring barley* [3]; *Spring wheat* [1-2]; *Triticale* [1-2]; *Winter barley* [3]; *Winter rye* [1-2]; *Winter wheat* [1-2];
- Crown rust in *Spring oats* [3]; *Winter oats* [3];
- Ear diseases in *Spring wheat* [1-2]; *Triticale* [1-2]; *Winter rye* [1-2]; *Winter wheat* [1-2];
- Eyespot in *Spring wheat* [1-2]; *Triticale* [1-2]; *Winter rye* [1-2]; *Winter wheat* [1-2];
- Glume blotch in *Spring wheat* [1-2]; *Triticale* [1-2]; *Winter rye* [1-2]; *Winter wheat* [1-2];
- Powdery mildew in *Spring barley* [3]; *Spring oats* [3]; *Spring wheat* [1-2]; *Triticale* [1-2]; *Winter barley* [3]; *Winter oats* [3]; *Winter rye* [1-2]; *Winter wheat* [1-2];
- Rhynchosporium in *Triticale* [1-2]; *Winter rye* [1-2];
- Septoria leaf blotch in *Spring wheat* [1-2]; *Triticale* [1-2]; *Winter rye* [1-2]; *Winter wheat* [1-2];
- Tan spot in *Spring wheat* [1-2]; *Triticale* [1-2]; *Winter rye* [1-2]; *Winter wheat* [1-2];
- Yellow rust in *Spring wheat* [1-2]; *Triticale* [1-2]; *Winter rye* [1-2]; *Winter wheat* [1-2];

Approval information

- Bixafen, prothioconazole and spiroxamine included in Annex I under EC Regulation 1107/2009.

Efficacy guidance

- Resistance to some DMI fungicides has been identified in Septoria leaf blotch (*Mycosphaerella graminicola*) which may seriously affect the performance of some products. For further advice on resistance management in DMIs contact your agronomist or specialist advisor, and visit the FRAG-UK website.

Restrictions

- No more than two applications of SDH inhibitors must be applied to the same cereal crop.
- Maintain an interval of at least 21 days between applications.
- Horizontal boom sprayers must be fitted with three star drift reduction technology for all uses.
- Low drift spraying equipment must be operated according to the specific conditions stated in the official three star rating for that equipment as published on HSE Chemicals Regulation Division?s website. These operating conditions must be maintained until the operator is 30m from the top of the bank of any surface water bodies
- Only one application may be made before 30 April (BBCH 30-37) followed by a second application after 1 May (BBCH 37-69). Alternatively two applications can be made after 1 May (BBCH 37-69).

FOR FULL CONDITIONS OF USE ALWAYS READ THE PRODUCT LABEL

Environmental safety
- Maintain an aquatic buffer zone of 6 m
- LERAP Category B [1-3]

Hazard classification and safety precautions

Hazard Harmful, Dangerous for the environment, Harmful if swallowed, Harmful if inhaled, Very toxic to aquatic organisms

Transport code 9 [1-3]

Packaging group III

UN Number 3082

Risk phrases H318, H335, H361, H373

Operator protection U05a, U09b, U11, U20a, U26; A, C, H

Environmental protection E15b, E16a, E16b, E34, E38, H410

Storage and disposal D01, D02, D05, D09a, D10b, D12a

Medical advice M03

53   bixafen + prothioconazole + tebuconazole

A succinate dehydrogenase inhibitor (SDHI) + triazole fungicide mixture for disease control in cereals

FRAC mode of action code: 3 + 7

Products

1	Skyway 285 Xpro	Bayer CropScience	75:110:100 g/l	EC	15028
2	Sparticus Xpro	Bayer CropScience	75:110:90 g/l	EC	15162

Uses
- Brown rust in *Spring wheat* [1-2]; *Triticale* [1-2]; *Winter rye* [1-2]; *Winter wheat* [1-2];
- Ear diseases in *Spring wheat* [1-2]; *Triticale* [1-2]; *Winter wheat* [1-2];
- Eyespot in *Spring wheat* (reduction of incidence and severity) [1-2]; *Triticale* (reduction of incidence and severity) [1-2]; *Winter wheat* (reduction of incidence and severity) [1-2];
- Glume blotch in *Spring wheat* [1-2]; *Triticale* [1-2]; *Winter wheat* [1-2];
- Light leaf spot in *Winter oilseed rape* [1-2];
- Phoma in *Winter oilseed rape* [1-2];
- Powdery mildew in *Spring wheat* [1-2]; *Triticale* [1-2]; *Winter rye* [1-2]; *Winter wheat* [1-2];
- Rhynchosporium in *Winter rye* [1-2];
- Sclerotinia in *Winter oilseed rape* [1-2];
- Septoria leaf blotch in *Spring wheat* [1-2]; *Triticale* [1-2]; *Winter wheat* [1-2];
- Stem canker in *Winter oilseed rape* [1-2];
- Tan spot in *Spring wheat* [1-2]; *Triticale* [1-2]; *Winter wheat* [1-2];
- Yellow rust in *Spring wheat* [1-2]; *Triticale* [1-2]; *Winter wheat* [1-2];

Approval information
- Bixafen, prothioconazole and tebuconazole included in Annex I under EC Regulation 1107/2009.

Efficacy guidance
- Resistance to some DMI fungicides has been identified in Septoria leaf blotch (*Mycosphaerella graminicola*) which may seriously affect the performance of some products. For further advice on resistance management in DMIs contact your agronomist or specialist advisor, and visit the FRAG-UK website.

Restrictions
- Bixafen is an SDH respiration inhibitor; Do not apply more than two foliar applications of products containing an SDH inhibitor to any cereal crop
- Newer authorisations for tebuconazole products require application to cereals only after GS 30 and applications to oilseed rape and linseed after GS20 - check label

Environmental safety
- LERAP Category B [1-2]

SEE SECTION 3 FOR PRODUCTS ALSO REGISTERED

Hazard classification and safety precautions

    Hazard Harmful, Dangerous for the environment, Harmful if swallowed

    Transport code 9 [1-2]

    Packaging group III

    UN Number 3082

    Risk phrases H315, H317, H361

    Operator protection U05a, U09a, U20b; A, H

    Environmental protection E15b, E16a, E16b, E34, E38, H410

    Storage and disposal D01, D02, D09a, D10b, D12a

54   boscalid

A translocated and translaminar anilide fungicide
FRAC mode of action code: 7

Products

1	Entargo	BASF	500 g/l	SC	19312
2	Filan	BASF	50% w/w	WG	11449
3	Fulmar	AgChem Access	50% w/w	WG	15767

Uses

- Alternaria in *Spring oilseed rape* [2-3]; *Winter oilseed rape* [2-3];
- Botrytis in *Poppies* (off-label) [2]; *Poppies for morphine production* (off-label) [2]; *Poppies grown for seed production* (off-label) [2];
- Dark leaf and pod spot in *Poppies* (off-label) [2]; *Poppies grown for seed production* (off-label) [2];
- Disease control in *All edible seed crops grown outdoors* (off-label) [2]; *All non-edible seed crops grown outdoors* (off-label) [2]; *Borage for oilseed production* (off-label) [2]; *Canary flower (Echium spp.)* (off-label) [2]; *Corn Gromwell* (off-label) [2]; *Durum wheat* [1]; *Evening primrose* (off-label) [2]; *Honesty* (off-label) [2]; *Linseed* (off-label) [2]; *Mustard* (off-label) [2]; *Spelt* [1]; *Spring barley* [1]; *Spring wheat* [1]; *Winter barley* [1]; *Winter wheat* [1];
- Eyespot in *Durum wheat* [1]; *Spelt* [1]; *Spring wheat* [1]; *Winter wheat* [1];
- Poppy fire in *Poppies* (off-label) [2]; *Poppies grown for seed production* (off-label) [2];
- Powdery mildew in *Grapevines* (off-label) [2];
- Sclerotinia in *Poppies for morphine production* (off-label) [2];
- Sclerotinia stem rot in *Poppies* (off-label) [2]; *Poppies grown for seed production* (off-label) [2]; *Spring oilseed rape* [2-3]; *Winter oilseed rape* [2-3];
- Septoria leaf blotch in *Durum wheat* (moderate control) [1]; *Spelt* (moderate control) [1]; *Spring wheat* (moderate control) [1]; *Winter wheat* (moderate control) [1];

Extension of Authorisation for Minor Use (EAMUs)

- *All edible seed crops grown outdoors* 20093231 [2]
- *All non-edible seed crops grown outdoors* 20093231 [2]
- *Borage for oilseed production* 20131293 [2]
- *Canary flower (Echium spp.)* 20131293 [2]
- *Corn Gromwell* 20131293 [2]
- *Evening primrose* 20131293 [2]
- *Grapevines* 20131947 [2]
- *Honesty* 20131293 [2]
- *Linseed* 20131293 [2]
- *Mustard* 20131293 [2]
- *Poppies* 20180175 [2]
- *Poppies for morphine production* 20093222 [2]
- *Poppies grown for seed production* 20180175 [2]

Approval information

- Boscalid included in Annex I under EC Regulation 1107/2009
- Accepted by BBPA on malting barley before ear emergence only and for use on hops

FOR FULL CONDITIONS OF USE ALWAYS READ THE PRODUCT LABEL

Efficacy guidance
- For best results apply as a protectant spray before symptoms are visible
- Applications against *Sclerotinia* should be made in high disease risk situations at early to full flower
- *Sclerotinia* control may be reduced when high risk conditions occur after flowering, leading to secondary disease spread
- Applications against *Alternaria* should be made at full flowering. Later applications may result in reduced levels of control
- Ensure adequate spray penetration and good coverage
- Use with an effective tank-mix partner to reduce the risk of resistance developing

Restrictions
- Maximum total dose equivalent to two full dose treatments
- Do not treat crops to be used for seed production later than full flowering stage
- Avoid spray drift onto neighbouring crops
- Must not be applied before BBCH30 [1]
- Effects of baking have not been fully tested. Consult the grain merchant or processor before use [1]

Crop-specific information
- Latest use: up to and including when 50% of pods have reached final size (GS 75) for oilseed rape

Environmental safety
- Dangerous for the environment
- Toxic to aquatic organisms
- Product represents minimal hazard to bees when used as directed. However local bee-keepers should be notified if crops are to be sprayed when in flower

Hazard classification and safety precautions
Hazard Dangerous for the environment [2-3]
Transport code 9 [1-3]
Packaging group III
UN Number 3077, 3082
Risk phrases H317 [1]
Operator protection U05a, U20a [1], U20c [2-3]; A, H
Environmental protection E15a [2-3], E15b [1], E38 [2-3], H411
Storage and disposal D01, D02, D05 [2-3], D09a [2-3], D10c, D12a [2-3]
Medical advice M03 [1]

55 boscalid + dimoxystrobin

A fungicide mixture for disease control in oilseed rape
FRAC mode of action code: 7 + 11

See also boscalid

Products
1 Pictor BASF 200:200 g/l SC 16783

Uses
- Alternaria in **Corn Gromwell** *(off-label)* [1]; **Winter oilseed rape** *(moderate control)* [1];
- Dark leaf and pod spot in **Corn Gromwell** *(off-label)* [1];
- Dark leaf spot in **Winter oilseed rape** *(moderate control)* [1];
- Disease control in **Winter oilseed rape** [1];
- Light leaf spot in **Winter oilseed rape** *(reduction)* [1];
- Phoma in **Winter oilseed rape** *(moderate control)* [1];
- Sclerotinia in **Corn Gromwell** *(off-label)* [1]; **Winter oilseed rape** [1];
- Stem canker in **Winter oilseed rape** *(moderate control)* [1];

SEE SECTION 3 FOR PRODUCTS ALSO REGISTERED

Extension of Authorisation for Minor Use (EAMUs)
- *Corn Gromwell* *20151106* [1]

Approval information
- Boscalid and dimoxystrobin included in Annex I under EC Regulation 1107/2009

Efficacy guidance
- Treat light leaf spot and phoma/stem canker at first sign of infection in the spring. Treat sclerotinia at early - full flower in high disease risk situations. Treat altenaria from full flower to when 50% of pods have reached full size.

Environmental safety
- LERAP Category B [1]

Hazard classification and safety precautions
  Hazard Harmful if swallowed, Harmful if inhaled, Very toxic to aquatic organisms
  Transport code 9 [1]
  Packaging group III
  UN Number 3082
  Risk phrases H317, H351, H361
  Operator protection U05a; A, H
  Environmental protection E15b, E16a, E34, H410
  Storage and disposal D01, D02, D09a, D10c, D12b
  Medical advice M05a

56   boscalid + metconazole

An anilide and triazole fungicide mixture for disease control in oilseed rape
FRAC mode of action code: 3 + 7

Products

1	Highgate	BASF	133:60 g/l	SC	15251
2	Tectura	BASF	133:60 g/l	SC	15232

Uses
- Sclerotinia stem rot in *Oilseed rape* [1-2];

Approval information
- Boscalid and metconazole included in Annex I under EC Regulation 1107/2009

Following crops guidance
- Any crop can follow normally-harvested oilseed rape treated with [1] or [2]

Environmental safety
- LERAP Category B [1-2]

Hazard classification and safety precautions
  Hazard Harmful
  Transport code 9 [1-2]
  Packaging group III
  UN Number 3082
  Risk phrases H317 [2], H361
  Operator protection U05a, U20a; A, H
  Environmental protection E15b, E16a, E16b, E34, E38, H412
  Storage and disposal D01, D02, D05, D09a, D10c, D12a
  Medical advice M03, M05a

57   boscalid + pyraclostrobin

A protectant and systemic fungicide mixture
FRAC mode of action code: 7 + 11

See also pyraclostrobin

Products

1	Bellis	BASF	25.2:12.8% w/w	WG	12522
2	Darwin	Pan Agriculture	26.7:6.7% w/w	WG	19614
3	Pyrafen	BASF	150:250 g/l	SC	19913
4	Shepherd	BASF	150:250 g/l	SC	19550
5	Signum	BASF	26.7:6.7% w/w	WG	11450

Uses

- Alternaria in **Cabbages** [2,5]; **Calabrese** [2,5]; **Carrots** *(moderate control)* [2,5]; **Cauliflowers** [2,5]; **Poppies** *(off-label)* [5]; **Poppies for morphine production** *(off-label)* [5]; **Poppies grown for seed production** *(off-label)* [5]; **Protected oriental cabbage** *(off-label)* [5]; **Spring oilseed rape** *(moderate control)* [3-4]; **Winter oilseed rape** *(moderate control)* [3-4];
- American gooseberry mildew in **Blackcurrants** *(moderate control only)* [2,5]; **Redcurrants** *(moderate control only)* [5];
- Anthracnose in **Blueberries** *(off-label)* [5]; **Gooseberries** *(off-label)* [5]; **Protected blueberry** *(off-label)* [5]; **Protected gooseberries** *(off-label)* [5];
- Black spot in **Protected strawberries** [2,5]; **Strawberries** [2,5];
- Blight in **Potatoes** [5];
- Blossom wilt in **Cherries** *(off-label)* [5]; **Mirabelles** *(off-label)* [5]; **Plums** *(off-label)* [5]; **Protected cherries** *(off-label)* [5]; **Protected mirabelles** *(off-label)* [5]; **Protected plums** *(off-label)* [5];
- Botrytis in **Amenity vegetation** *(off-label)* [5]; **Asparagus** *(off-label)* [5]; **Beans without pods (Fresh)** *(off-label)* [5]; **Blueberries** *(off-label)* [5]; **Broad beans** *(off-label)* [5]; **Forest nurseries** *(off-label)* [5]; **Gooseberries** *(off-label)* [5]; **Interior landscapes** *(off-label)* [5]; **Lamb's lettuce** *(off-label)* [5]; **Ornamental plant production** *(off-label)* [5]; **Poppies** *(off-label)* [5]; **Poppies for morphine production** *(off-label)* [5]; **Poppies grown for seed production** *(off-label)* [5]; **Protected aubergines** *(off-label)* [5]; **Protected blueberry** *(off-label)* [5]; **Protected chard** *(off-label)* [5]; **Protected forest nurseries** *(off-label)* [5]; **Protected gooseberries** *(off-label)* [5]; **Protected lettuce** *(off-label)* [5]; **Protected ornamentals** *(off-label)* [5]; **Protected peppers** *(off-label)* [5]; **Protected spinach beet** *(off-label)* [5]; **Protected tomatoes** *(off-label)* [5]; **Swedes** *(off-label)* [5]; **Turnips** *(off-label)* [5];
- Bottom rot in **Lettuce** [2,5]; **Protected lettuce** [2,5];
- Brown rot in **Apricots** *(off-label)* [5]; **Nectarines** *(off-label)* [5]; **Peaches** *(off-label)* [5];
- Cane blight in **Blackberries** *(off-label)* [5]; **Protected blackberries** *(off-label)* [5]; **Raspberries** *(off-label)* [5];
- Chocolate spot in **Spring field beans** *(moderate control)* [2,5]; **Winter field beans** *(moderate control)* [2,5];
- Cladosporium in **Protected cucumbers** *(off-label)* [5];
- Cladosporium leaf blotch in **Courgettes** *(off-label)* [5]; **Summer squash** *(off-label)* [5];
- Colletotrichum orbiculare in **Courgettes** *(off-label)* [5]; **Summer squash** *(off-label)* [5];
- Disease control in **All edible crops (outdoor and protected)** *(off-label)* [5]; **All non-edible crops (outdoor)** *(off-label)* [5]; **All protected non-edible crops** *(off-label)* [5]; **Beetroot** *(off-label)* [5]; **Hops** *(off-label)* [1,5]; **Kohlrabi** *(off-label)* [5]; **Lamb's lettuce** *(off-label)* [5]; **Protected hops** *(off-label)* [5]; **Protected radishes** *(off-label)* [5]; **Protected soft fruit** *(off-label)* [5]; **Protected top fruit** *(off-label)* [5]; **Radishes** *(off-label)* [5]; **Red beet** *(off-label)* [5]; **Soft fruit** *(off-label)* [5]; **Top fruit** *(off-label)* [5];
- Downy mildew in **Bulls blood** *(off-label)* [5]; **Chard** *(off-label)* [5]; **Endives** *(off-label)* [5]; **Protected endives** *(off-label)* [5]; **Spinach** *(off-label)* [5]; **Spinach beet** *(off-label)* [5];
- Fusarium in **Bulb onion sets** *(off-label)* [5];
- Grey mould in **Blackcurrants** [2,5]; **Lettuce** [2,5]; **Protected lettuce** [2,5]; **Protected strawberries** [2,5]; **Redcurrants** [5]; **Strawberries** [2,5];
- Gummy stem blight in **Protected cucumbers** *(off-label)* [5];
- Leaf and pod spot in **Combining peas** *(moderate control only)* [2,5]; **Vining peas** *(moderate control only)* [2,5];

SEE SECTION 3 FOR PRODUCTS ALSO REGISTERED

SECTION 2

- Leaf spot in *Amenity vegetation* *(off-label)* [5]; *Black salsify* *(off-label)* [5]; *Blackcurrants* *(moderate control only)* [2,5]; *Blueberries* *(off-label)* [5]; *Gooseberries* *(off-label)* [5]; *Interior landscapes* *(off-label)* [5]; *Parsley root* *(off-label)* [5]; *Protected blueberry* *(off-label)* [5]; *Protected gooseberries* *(off-label)* [5]; *Redcurrants* *(moderate control only)* [5]; *Salsify* *(off-label)* [5];
- Light leaf spot in *Spring oilseed rape* *(moderate control)* [3-4]; *Winter oilseed rape* *(moderate control)* [3-4];
- Mycosphaerella in *Protected oriental cabbage* *(off-label)* [5];
- Phoma leaf spot in *Spring oilseed rape* [3-4]; *Winter oilseed rape* [3-4];
- Poppy fire in *Poppies* *(off-label)* [5]; *Poppies for morphine production* *(off-label)* [5]; *Poppies grown for seed production* *(off-label)* [5];
- Powdery mildew in *Amenity vegetation* *(off-label)* [5]; *Apples* [1]; *Black salsify* *(off-label)* [5]; *Carrots* [2,5]; *Interior landscapes* *(off-label)* [5]; *Parsley root* *(off-label)* [5]; *Protected cucumbers* *(off-label)* [5]; *Protected strawberries* [2,5]; *Pumpkins* *(off-label)* [5]; *Salsify* *(off-label)* [5]; *Strawberries* [2,5];
- Purple blotch in *Blackberries* *(off-label)* [5]; *Protected blackberries* *(off-label)* [5]; *Raspberries* *(off-label)* [5];
- Rhizoctonia in *Protected chard* *(off-label)* [5]; *Protected lettuce* *(off-label)* [5]; *Protected spinach beet* *(off-label)* [5];
- Ring spot in *Broccoli* [2,5]; *Brussels sprouts* [2,5]; *Cabbages* [2,5]; *Calabrese* [2,5]; *Cauliflowers* [2,5];
- Rust in *Amenity vegetation* *(off-label)* [5]; *Asparagus* *(off-label)* [5]; *Black salsify* *(off-label)* [5]; *Interior landscapes* *(off-label)* [5]; *Kohlrabi* *(off-label)* [5]; *Parsley root* *(off-label)* [5]; *Protected blackberries* *(off-label)* [5]; *Salsify* *(off-label)* [5]; *Spring field beans* [2,5]; *Winter field beans* [2,5];
- Scab in *Apples* [1]; *Pears* [1];
- Sclerotinia in *Black salsify* *(off-label)* [5]; *Carrots* *(moderate control)* [2,5]; *Celeriac* *(off-label)* [5]; *Horseradish* *(off-label)* [5]; *Parsley root* *(off-label)* [5]; *Poppies* *(off-label)* [5]; *Poppies grown for seed production* *(off-label)* [5]; *Protected chard* *(off-label)* [5]; *Protected lettuce* *(off-label)* [5]; *Protected spinach beet* *(off-label)* [5]; *Salsify* *(off-label)* [5]; *Spring oilseed rape* [3-4]; *Winter oilseed rape* [3-4];
- Soft rot in *Lettuce* [2,5]; *Protected lettuce* [2,5];
- Stem canker in *Spring oilseed rape* [3-4]; *Winter oilseed rape* [3-4];
- White blister in *Broccoli* [2,5]; *Brussels sprouts* [2,5]; *Cabbages* *(qualified minor use)* [2,5]; *Calabrese* *(qualified minor use)* [2,5]; *Chinese cabbage* *(off-label)* [5]; *Choi sum* *(off-label)* [5]; *Collards* *(off-label)* [5]; *Herbs (see appendix 6)* *(off-label)* [5]; *Kale* *(off-label)* [5]; *Leaf brassicas* *(off-label)* [5]; *Pak choi* *(off-label)* [5]; *Protected herbs (see appendix 6)* *(off-label)* [5]; *Protected leaf brassicas* *(off-label)* [5]; *Tatsoi* *(off-label)* [5];
- White rot in *Bulb onions* *(off-label)* [5]; *Garlic* *(off-label)* [5]; *Salad onions* *(off-label)* [5]; *Shallots* *(off-label)* [5];
- White rust in *Kohlrabi* *(off-label)* [5];
- White tip in *Leeks* *(off-label)* [5];

Extension of Authorisation for Minor Use (EAMUs)
- *All edible crops (outdoor and protected)* *20102111* [5]
- *All non-edible crops (outdoor)* *20102111* [5]
- *All protected non-edible crops* *20102111* [5]
- *Amenity vegetation* *20122317* [5]
- *Apricots* *20121721* [5]
- *Asparagus* *20102105* [5]
- *Beans without pods (Fresh)* *20121009* [5]
- *Beetroot* *20121717* [5]
- *Black salsify* *20121720* [5]
- *Blackberries* *20102110* [5]
- *Blueberries* *20121722* [5]
- *Broad beans* *20121009* [5]
- *Bulb onion sets* *20103122* [5]
- *Bulb onions* *20102108* [5]
- *Bulls blood* *20102136* [5]

- **Celeriac** *20141059* [5]
- **Chard** *20102136* [5]
- **Cherries** *20102109* [5]
- **Chinese cabbage** *20130285* [5]
- **Choi sum** *20130285* [5]
- **Collards** *20130285* [5]
- **Courgettes** *20142855* [5]
- **Endives** *20102136* [5]
- **Forest nurseries** *20102119* [5]
- **Garlic** *20102108* [5]
- **Gooseberries** *20121722* [5]
- **Herbs (see appendix 6)** *20102115* [5]
- **Hops** *20112732* [1], *20102111* [5]
- **Horseradish** *20093375* [5]
- **Interior landscapes** *20122317* [5]
- **Kale** *20130285* [5]
- **Kohlrabi** *20121719* [5]
- **Lamb's lettuce** *20121718* [5]
- **Leaf brassicas** *20102115* [5]
- **Leeks** *20102134* [5]
- **Mirabelles** *20102109* [5]
- **Nectarines** *20121721* [5]
- **Ornamental plant production** *20122141* [5]
- **Pak choi** *20130285* [5]
- **Parsley root** *20121720* [5]
- **Peaches** *20121721* [5]
- **Plums** *20102109* [5]
- **Poppies** *20180176* [5]
- **Poppies for morphine production** *20101233* [5]
- **Poppies grown for seed production** *20180176* [5]
- **Protected aubergines** *20120427* [5]
- **Protected blackberries** *20102102* [5]
- **Protected blueberry** *20121722* [5]
- **Protected chard** *20131807* [5], *20181744* [5]
- **Protected cherries** *20102109* [5]
- **Protected cucumbers** *20151178* [5]
- **Protected endives** *20102136* [5]
- **Protected forest nurseries** *20102119* [5]
- **Protected gooseberries** *20121722* [5]
- **Protected herbs (see appendix 6)** *20102115* [5]
- **Protected hops** *20102111* [5]
- **Protected leaf brassicas** *20102115* [5]
- **Protected lettuce** *20131807* [5]
- **Protected mirabelles** *20102109* [5]
- **Protected oriental cabbage** *20183055* [5]
- **Protected ornamentals** *20122141* [5]
- **Protected peppers** *20120427* [5]
- **Protected plums** *20102109* [5]
- **Protected radishes** *20121717* [5]
- **Protected soft fruit** *20102111* [5]
- **Protected spinach beet** *20131807* [5]
- **Protected tomatoes** *20120427* [5]
- **Protected top fruit** *20102111* [5]
- **Pumpkins** *20152651* [5]
- **Radishes** *20121717* [5]
- **Raspberries** *20102110* [5]
- **Red beet** *20121717* [5]
- **Salad onions** *20102107* [5]
- **Salsify** *20121720* [5]

SEE SECTION 3 FOR PRODUCTS ALSO REGISTERED

- ***Shallots*** *20102108* [5]
- ***Soft fruit*** *20102111* [5]
- ***Spinach*** *20102136* [5]
- ***Spinach beet*** *20102136* [5]
- ***Summer squash*** *20142855* [5]
- ***Swedes*** *20151961* [5]
- ***Tatsoi*** *20130285* [5]
- ***Top fruit*** *20102111* [5]
- ***Turnips*** *20151961* [5]

Approval information
- Boscalid and pyraclostrobin included in Annex I under EC Regulation 1107/2009
- Accepted by BBPA on malting barley before ear emergence only and for use on hops

Efficacy guidance
- On brassicas apply as a protectant spray or at the first sign of disease and repeat at 3-4 wk intervals depending on disease pressure [5]
- Ensure adequate spray penetration and coverage by increasing water volume in dense crops [5]
- For best results on strawberries apply as a protectant spray at the white bud stage. Applications should be made in sequence with other products as part of a fungicide spray programme during flowering at 7-10 day intervals [5]
- On carrots and field beans apply as a protectant spray or at the first sign of disease with a repeat treatment if needed, as directed on the label [5]
- On lettuce apply as a protectant spray 1-2 wk after planting [5]
- Optimum results on apples and pears obtained from a protectant treatment from bud burst [1]
- Application as the final 2 sprays on apples and pears gives reduction in storage rots [1]
- Pyraclostrobin is a member of the QoI cross-resistance group of fungicides and should be used in programmes with fungicides with a different mode of action

Restrictions
- Maximum total dose equivalent to three full dose treatments on brassicas; two full dose treatments on all other field crops [5]
- Maximum number of treatments (including other QoI treatments) on apples and pears 4 per yr if total number of applications is 12 or more, or 3 per yr if the total number is fewer than 12 [1]
- Do not use more than 2 consecutive treatments on apples and pears, and these must be separated by a minimum of 2 applications of a fungicide with a different mode of action [1]
- Use a maximum of three applications per yr on brassicas and no more than two per yr on other field crops [5]
- Do not use consecutive treatments; apply in alternation with fungicides from a different cross resistance group and effective against the target diseases [5]
- Do not apply more than 6 kg/ha to the same area of land per yr [5]
- Consult processor before use on crops for processing
- Applications to lettuce and protected salad crops and to crops grown under temporary structures or polytunnels must only be made between 1 April and 31 October [5]
- Horizontal boom sprayers must be fitted with three star drift reduction technology for all uses [4]
- Low drift spraying equipment must be operated according to the specific conditions stated in the official three star rating for that equipment as published on HSE Chemicals Regulation Division?s website. These operating conditions must be maintained until the operator is 30m from the top of the bank of any surface water bodies [4]
- Applications to winter oilseed rape must only be made between GS 14-75 - four leaves unfolded / 50% of pods have reached final size. Applications to spring oilseed rape must only be made between GS 25-75 - five side shoots detectable / 50% of pods have reached final size. Applications must not be made between 1 June and 31 October [4]

Crop-specific information
- HI 21 d for field beans; 14 d for brassica crops, carrots, lettuce; 7 d for apples, pears; 3 d for strawberries
- Any crop may follow a normally harvested or failed crop [3, 4]

Environmental safety
- Dangerous for the environment

FOR FULL CONDITIONS OF USE ALWAYS READ THE PRODUCT LABEL

- Very toxic to aquatic organisms
- Buffer zone requirement 12m [4]
- Broadcast air-assisted LERAP [1] (40 m); LERAP Category B

Hazard classification and safety precautions

Hazard Harmful, Dangerous for the environment, Harmful if swallowed, Harmful if inhaled [3-4], Very toxic to aquatic organisms
Transport code 9 [1-5]
Packaging group III
UN Number 3077, 3082
Risk phrases H315 [3-4], H317 [3-4], H335 [3-4]
Operator protection U05a, U13 [1], U14 [3-4], U20b [3-4], U20c [1-2,5]; A
Environmental protection E15a [1-2,5], E15b [3-4], E16a [2-5], E16b [3-4], E17b (40 m) [1], E34 [1,3-4], E38, H410
Storage and disposal D01, D02, D08 [1,3-4], D09a [2-5], D10c, D12a
Medical advice M03 [1,3-4], M05a

58   brodifacoum

An anticoagulant coumarin rodenticide

Products

1	Brodiag Blocks	Killgerm	0.0029% w/w	RB	UK15-0858
2	Brodiag Fresh Bait	Killgerm	0.0029% w/w	RB	UK15-0857
3	Brodiag Whole Wheat	Killgerm	0.0029% w/w	RB	UK17-1051
4	Pest Expert Formula B+	Killgerm	0.0029% w/w	RB	UK17-1051
5	Sakarat Brodikill Whole Wheat	Killgerm	0.0029% w/w	RB	UK17-1051

Uses

- Mice in **Farm buildings** [1-2,5]; **Farmyards** [1-2,5];
- Rats in **Farm buildings** [1-2,5]; **Farmyards** [1-2,5];

Approval information

- Brodifacoum is not included in Annex 1 under EC Regulation 1107/2009

Efficacy guidance

- Product is a self-contained control device which requires no bait handling
- Best results obtained by placing tubes where mice are active and will readily enter the device
- To acquire a lethal dose mice must pass through the tube a number of times
- Where a continuous source of infestation is present, or where sustained protection is needed, tubes can be installed permanently and replaced at monthly intervals
- For complete eradication other control methods should also be used bearing in mind the resistance status of the target population

Restrictions

- For use only by professional pest contractors
- Do not use outdoors. Products must be used in situations where baits are placed within a building or other enclosed structure, and the target is living or feeding predominantly within that building or structure
- Do not attempt to open or recharge the tubes

Environmental safety

- Prevent access to baits by children, birds and other animals
- Do not place tubes where food, feed or water could become contaminated
- Remove all remains of tubes after use and burn or bury
- Search for and burn or bury all rodent bodies. Do not place in refuse bins or on open rubbish tips
- Keep in original container, tightly closed, in a safe place, under lock and key

SEE SECTION 3 FOR PRODUCTS ALSO REGISTERED

Hazard classification and safety precautions
    Operator protection U13, U20b
    Storage and disposal D05, D07, D09a, D11a
    Vertebrate/rodent control products V01b, V02, V03b, V04b
    Medical advice M03

59   bromadiolone

An anti-coagulant coumarin-derivative rodenticide

Products

1	Bromag Fresh Bait	Killgerm	0.005% w/w	RB	UK13-0754
2	Bromag Wax Blocks	Killgerm	0.005% w/w	RB	UK13-0755

Uses
- Mice in **Farm buildings/yards** [1-2]; **Farmyards** [1-2];
- Rats in **Farm buildings/yards** [1-2]; **Farmyards** [1-2];

Approval information
- Bromadiolone is not included in Annex I under EC Directive 1107/2009

Efficacy guidance
- Ready-to-use baits are formulated on a mould-resistant, whole-wheat base
- Use in baiting programme. Place baits in protected situations, sufficient for continuous feeding between treatments
- Chemical is effective against warfarin- and coumatetralyl-resistant rats and mice and does not induce bait shyness
- Use bait bags where loose baiting inconvenient (eg behind ricks, silage clamps etc)
- The resistance status of the rodent population should be assessed when considering the choice of product to use. Resistance to bromadiolone in rats is now widespread in the UK.

Restrictions
- For use only by professional operators

Environmental safety
- Access to baits by children, birds and animals, particularly cats, dogs, pigs and poultry, must be prevented
- Baits must not be placed where food, feed or water could become contaminated
- Remains of bait and bait containers must be removed after treatment and burned or buried
- Rodent bodies must be searched for and burned or buried. They must not be placed in refuse bins or on rubbish tips
- Take extreme care to prevent domestic animals having access to the bait

Hazard classification and safety precautions
    Operator protection U13, U20b; A
    Storage and disposal D05, D07, D09a, D11a
    Vertebrate/rodent control products V01b, V02, V03b, V04b
    Medical advice M03

60   bromuconazole

A systemic triazole fungicide for cereals only available in mixtures
FRAC mode of action code: 3

61   bromuconazole + tebuconazole

A triazole mixture for disease control in wheat, rye and triticale
FRAC mode of action code: 3

See also tebuconazole

Products

1	Djembe	Sumitomo	167:107 g/l	EC	19223
2	Sakura	Sumitomo	167:107 g/l	EC	19224
3	Soleil	Sumitomo	167:107 g/l	EC	19222

Uses
- Alternaria in **Rye** [1-3]; **Spring wheat** [1-3]; **Triticale** [1-3]; **Winter wheat** [1-3];
- Brown rust in **Rye** [1-3]; **Spring wheat** [1-3]; **Triticale** [1-3]; **Winter wheat** [1-3];
- Cladosporium in **Rye** [1-3]; **Spring wheat** [1-3]; **Triticale** [1-3]; **Winter wheat** [1-3];
- Fusarium in **Rye** [1-3]; **Spring wheat** [1-3]; **Triticale** [1-3]; **Winter wheat** [1-3];
- Powdery mildew in **Rye** *(moderate control)* [1-3]; **Spring wheat** *(moderate control)* [1-3]; **Triticale** *(moderate control)* [1-3]; **Winter wheat** *(moderate control)* [1-3];
- Septoria leaf blotch in **Rye** *(moderate control)* [1-3]; **Spring wheat** *(moderate control)* [1-3]; **Triticale** *(moderate control)* [1-3]; **Winter wheat** *(moderate control)* [1-3];

Approval information
- Bromuconazole and tebuconazole included in Annex 1 under EC Directive 1107/2009

Restrictions
- Newer authorisations for tebuconazole products require application to cereals only after GS 30 and applications to oilseed rape and linseed after GS20 - check label

Following crops guidance
- Only sugar beet, cereals, oilseed rape, field beans, peas, potatoes, linseed and Italian rye-grass may be sown as the following crop. The effect on other crops has not been assessed.
- Some effects may be seen on sugar beet crops grown following a cereal crop but the sugar beet will usually recover completely.

Environmental safety
- LERAP Category B [1-3]

Hazard classification and safety precautions
Hazard Harmful, Dangerous for the environment, Very toxic to aquatic organisms
Transport code 9 [1-3]
Packaging group III
UN Number 3082
Risk phrases H304, H318, H336, H361
Operator protection U11, U19a, U19f, U20b; A, C, H
Environmental protection E16a, E34, E38, H410
Storage and disposal D01, D02, D05, D09a, D11a, D12a, D12b
Medical advice M03

62   bupirimate

A systemic aminopyrimidinol fungicide active against powdery mildew
FRAC mode of action code: 8

Products

1	Nimrod	Adama	250 g/l	EC	18522

Uses
- Powdery mildew in **Apples** [1]; **Begonias** [1]; **Blackcurrants** [1]; **Chrysanthemums** [1]; **Gooseberries** [1]; **Hops in propagation** *(off-label)* [1]; **Nursery fruit trees** *(off-label)* [1]; **Ornamental plant production** [1]; **Pears** [1]; **Protected begonias** [1]; **Protected**

SECTION 2

blackcurrants [1]; *Protected chrysanthemums* [1]; *Protected cucumbers* [1]; *Protected gooseberries* [1]; *Protected ornamentals* [1]; *Protected raspberries* [1]; *Protected redcurrants* [1]; *Protected roses* [1]; *Protected strawberries* [1]; *Protected tomatoes* (off-label) [1]; *Raspberries* [1]; *Redcurrants* [1]; *Roses* [1]; *Strawberries* [1];

Extension of Authorisation for Minor Use (EAMUs)
- *Hops in propagation* 20200606 [1]
- *Nursery fruit trees* 20200606 [1]
- *Protected tomatoes* 20192504 [1]

Approval information
- Bupirimate included in Annex 1 under EC Regulation 1107/2009
- Accepted by BBPA for use on hops

Efficacy guidance
- On apples during periods that favour disease development lower doses applied weekly give better results than higher rates fortnightly
- Not effective in protected crops against strains of mildew resistant to bupirimate

Restrictions
- Maximum number of treatments or maximum total dose depends on crop and dose (see label for details)
- Reasonable precautions must be taken to prevent access of birds, wild mammals and honey bees to treated crops

Crop-specific information
- HI: 1 d for apples, pears, strawberries; 2 d for cucurbits; 7 d for blackcurrants; 8 d for raspberries; 14 d for gooseberries, hops
- Apply before or at first signs of disease and repeat at 7-14 d intervals. Timing and maximum dose vary with crop. See label for details
- With apples, hops and ornamentals cultivars may vary in sensitivity to spray. See label for details
- If necessary to spray cucurbits in winter or early spring spray a few plants 10-14 d before spraying whole crop to test for likelihood of leaf spotting problem
- On roses some leaf puckering may occur on young soft growth in early spring or under low light intensity. Avoid use of high rates or wetter on such growth
- Never spray flowering begonias (or buds showing colour) as this can scorch petals
- Do not mix with other chemicals for application to begonias, cucumbers or gerberas

Environmental safety
- Dangerous for the environment
- Toxic to aquatic organisms
- Flammable
- Product has negligible effect on *Phytoseiulus* and *Encarsia* and may be used in conjunction with biological control of red spider mite

Hazard classification and safety precautions
Hazard Irritant, Flammable, Dangerous for the environment, Flammable liquid and vapour
Transport code 3 [1]
Packaging group III
UN Number 1993
Risk phrases H304, H319, H335, H351
Operator protection U05a, U20b; A, C
Environmental protection E15a, H410
Storage and disposal D01, D02, D09a, D10c
Medical advice M05b

63   buprofezin

A moulting inhibitor, thiadiazine insecticide for whitefly control
IRAC mode of action code: 16

Products

1 Applaud 25 SC   Certis Belchim B V   250 g/l   SL   17196

Uses
- Glasshouse whitefly in **Protected ornamentals** [1];
- Tobacco whitefly in **Protected ornamentals** [1];

Approval information
- Buprofezin included in Annex 1 under EC Regulation 1107/2009

Efficacy guidance
- Product has contact, residual and some vapour activity
- Whitefly most susceptible at larval stages but residual effect can also kill nymphs emerging from treated eggs and application to pupae reduces emergence
- Adult whitefly not directly affected. Resistant strains of tobacco whitefly are known and where present control likely to be reduced or ineffective

Restrictions
- Maximum number of treatments 8 per crop for tomatoes and cucumbers; 4 per crop on protected ornamentals; 2 per crop for aubergines and peppers
- Do not apply more than 2 sprays within a 65 d period on tomatoes, or within a 45 d period on cucumbers
- Do not treat *Dieffenbachia* or *Closmoplictrum*
- Do not apply to crops under stress
- Do not leave spray liquid in sprayer for long periods
- Do not apply as fog or mist

Crop-specific information
- HI for all crops 3 d
- In IPM programme apply as single application and allow at least 60 d before re-applying
- In All Chemical programme apply twice at 7-14 d interval and allow at least 60 d before re-applying
- See label for list of ornamentals successfully treated but small scale test advised to check varietal tolerance. This is especially important if spraying flowering ornamentals with buds showing colour

Environmental safety
- Product may be used either in IPM programme in association with *Encarsia formosa* or in All Chemical programme

Hazard classification and safety precautions
 Transport code 9 [1]
 Packaging group III
 UN Number 3082
 Operator protection U02a, U09a, U09c, U19a, U20a; A
 Environmental protection E15b, H411
 Storage and disposal D05, D09a, D10c, D12b
 Medical advice M03a

64   captan

A protectant phthalimide fungicide with horticultural uses
FRAC mode of action code: M4

Products

1 Captan 80 WDG   Adama   80% w/w   WG   16293

SEE SECTION 3 FOR PRODUCTS ALSO REGISTERED

Products – continued

2	Clayton Core	Clayton	80% w/w	WG	16934
3	Multicap	BelCrop	80% w/w	WG	17652
4	PP Captan 80 WG	Arysta	80% w/w	WG	16294

Uses

* Botrytis in *Ornamental plant production* (off-label) [1];
* Disease control in *Hops in propagation* (off-label) [1]; *Nursery fruit trees* (off-label) [1]; *Quinces* (off-label) [1];
* Downy mildew in *Ornamental plant production* (off-label) [1];
* Gloeosporium rot in *Apples* [1-4]; *Pears* [1];
* Scab in *Apples* [1-4]; *Ornamental plant production* (off-label) [1]; *Pears* [1,3-4];
* Shot-hole in *Ornamental plant production* (off-label) [1];

Extension of Authorisation for Minor Use (EAMUs)

* *Hops in propagation* 20142510 [1]
* *Nursery fruit trees* 20142510 [1]
* *Ornamental plant production* 20151919 [1], 20151920 [1]
* *Quinces* 20141276 [1]

Approval information

* Captan included in Annex I under EC Regulation 1107/2009

Restrictions

* Maximum number of treatments 12 per yr on apples and pears as pre-harvest sprays
* Do not use on apple cultivars Bramley, Monarch, Winston, King Edward, Spartan, Kidd's Orange or Red Delicious or on pear cultivar D'Anjou
* Do not mix with alkaline materials or oils
* Do not use on fruit for processing
* Powered visor respirator with hood and neck cape must be used when handling concentrate

Crop-specific information

* HI apples, pears 14 d; strawberries 7 d
* For control of scab apply at bud burst and repeat at 10-14 d intervals until danger of scab infection ceased
* For suppression of fruit storage rots apply from late Jul and repeat at 2-3 wk intervals
* For black spot control in roses apply after pruning with 3 further applications at 14 d intervals or spray when spots appear and repeat at 7-10 d intervals
* For grey mould in strawberries spray at first open flower and repeat every 7-10 d
* Do not leave diluted material for more than 2 h. Agitate well before and during spraying

Environmental safety

* Dangerous for the environment
* Very toxic to aquatic organisms
* LERAP Category B [1-4]

Hazard classification and safety precautions

Hazard Harmful, Dangerous for the environment, Very toxic to aquatic organisms [1-3]
Transport code 9 [1-4]
Packaging group III
UN Number 3077
Risk phrases H317, H318 [4], H319 [1-3], H320 [4], H351
Operator protection U05a, U09a, U11, U19a, U20c; A, D, E, H
Environmental protection E15a, E16a, H410 [4], H412 [1-3]
Storage and disposal D01, D02, D09a, D11a, D12a

65 carbon dioxide (commodity substance)

A gas for the control of trapped rodents and other vertebrates. Approval valid until 31/8/2025

Products

1	carbon dioxide	various	99.9%	GA	-

Uses
- Birds in **Traps** [1];
- Mice in **Traps** [1];
- Rats in **Traps** [1];

Approval information
- Carbon dioxide included in Annex 1 under EC Regulation 1107/2009

Efficacy guidance
- Use to destroy trapped rodent pests
- Use to control birds covered by general licences issued by the Agriculture and Environment Departments under Section 16(1) of the Wildlife and Countryside Act (1981) for the control of opportunistic bird species, where birds have been trapped or stupefied with alphachloralose/seconal

Restrictions
- Operators must wear self-contained breathing apparatus when carbon dioxide levels are greater than 0.5% v/v
- Operators must be suitably trained and competent

Environmental safety
- Unprotected persons and non-target animals must be excluded from the treatment enclosures and surrounding areas unless the carbon dioxide levels are below 0.5% v/v

Hazard classification and safety precautions
Operator protection G

66 carfentrazone-ethyl

A triazolinone contact herbicide
HRAC mode of action code: 14 (E)

Products

1	Aurora 40 WG	FMC Agro	40% w/w	WG	18703
2	Shark	FMC Agro	60 g/l	ME	18700
3	Spotlight Plus	FMC Agro	60 g/l	ME	18698

Uses
- Annual and perennial weeds in **All edible crops (outdoor)** [3]; **All non-edible crops (outdoor)** [3]; **Bilberries** (off-label) [2]; **Blackberries** (off-label) [2]; **Blackcurrants** (off-label) [2]; **Blueberries** (off-label) [2]; **Cranberries** (off-label) [2]; **Forestry plantations** (off-label) [2]; **Gooseberries** (off-label) [2]; **Loganberries** (off-label) [2]; **Ornamental plant production** (off-label) [2]; **Potatoes** [3]; **Protected bilberries** (off-label) [2]; **Protected blackberries** (off-label) [2]; **Protected blackcurrants** (off-label) [2]; **Protected blueberry** (off-label) [2]; **Protected cranberries** (off-label) [2]; **Protected forest** (off-label) [2]; **Protected gooseberries** (off-label) [2]; **Protected loganberries** (off-label) [2]; **Protected ornamentals** (off-label) [2]; **Protected raspberries** (off-label) [2]; **Protected redcurrants** (off-label) [2]; **Protected Rubus hybrids** (off-label) [2]; **Protected strawberries** (off-label) [2]; **Raspberries** (off-label) [2]; **Redcurrants** (off-label) [2]; **Ribes species** (off-label) [2]; **Rubus hybrids** (off-label) [2]; **Strawberries** (off-label) [2]; **Wine grapes** (off-label) [2];
- Annual dicotyledons in **All edible crops (outdoor)** (before planting) [2]; **All non-edible crops (outdoor)** (before planting) [2]; **Potatoes** [2]; **Rye** (off-label) [1]; **Spring barley** [1]; **Spring oats** [1]; **Spring wheat** [1]; **Triticale** [1]; **Winter barley** [1]; **Winter oats** [1]; **Winter wheat** [1];
- Black bindweed in **Asparagus** (off-label) [2]; **Potatoes** [2];

SEE SECTION 3 FOR PRODUCTS ALSO REGISTERED

SECTION 2

- Black nightshade in **Strawberries** *(off-label)* [2];
- Cleavers in **Asparagus** *(off-label)* [2]; **Potatoes** [2]; **Strawberries** *(off-label)* [2];
- Desiccation in **Blackberries** *(off-label)* [2]; **Loganberries** *(off-label)* [2]; **Narcissi** *(off-label)* [3]; **Potatoes** [3]; **Protected blackberries** *(off-label)* [2]; **Protected loganberries** *(off-label)* [2]; **Protected raspberries** *(off-label)* [2]; **Protected Rubus hybrids** *(off-label)* [2]; **Raspberries** *(off-label)* [2]; **Rubus hybrids** *(off-label)* [2];
- Fat hen in **Asparagus** *(off-label)* [2]; **Potatoes** [2];
- Field bindweed in **Asparagus** *(off-label)* [2];
- Field pansy in **Asparagus** *(off-label)* [2];
- Groundsel in **Asparagus** *(off-label)* [2];
- Hairy bittercress in **Strawberries** *(off-label)* [2];
- Ivy-leaved speedwell in **Potatoes** [2];
- Knotgrass in **Asparagus** *(off-label)* [2]; **Potatoes** [2]; **Strawberries** *(off-label)* [2];
- Polygonums in **Blackcurrants** *(off-label)* [2]; **Blueberries** *(off-label)* [2]; **Cranberries** *(off-label)* [2]; **Gooseberries** *(off-label)* [2]; **Redcurrants** *(off-label)* [2]; **Ribes species** *(off-label)* [2];
- Redshank in **Asparagus** *(off-label)* [2]; **Potatoes** [2]; **Strawberries** *(off-label)* [2];
- Shepherd's purse in **Asparagus** *(off-label)* [2];
- Small nettle in **Asparagus** *(off-label)* [2]; **Strawberries** *(off-label)* [2];
- Sowthistle in **Asparagus** *(off-label)* [2];
- Speedwells in **Asparagus** *(off-label)* [2]; **Strawberries** *(off-label)* [2];
- Thistles in **Blackcurrants** *(off-label)* [2]; **Blueberries** *(off-label)* [2]; **Cranberries** *(off-label)* [2]; **Gooseberries** *(off-label)* [2]; **Redcurrants** *(off-label)* [2]; **Ribes species** *(off-label)* [2];
- Volunteer oilseed rape in **All edible crops (outdoor)** *(before planting)* [2]; **All non-edible crops (outdoor)** *(before planting)* [2]; **Asparagus** *(off-label)* [2]; **Potatoes** [2];
- Willowherb in **Strawberries** *(off-label)* [2];

Extension of Authorisation for Minor Use (EAMUs)
- **Asparagus** *20190634* [2]
- **Bilberries** *20190633* [2]
- **Blackberries** *20190622* [2], *20190633* [2]
- **Blackcurrants** *20190627* [2], *20190633* [2]
- **Blueberries** *20190627* [2], *20190633* [2]
- **Cranberries** *20190627* [2], *20190633* [2]
- **Forestry plantations** *20190630* [2]
- **Gooseberries** *20190627* [2], *20190633* [2]
- **Loganberries** *20190622* [2], *20190633* [2]
- **Narcissi** *20190896* [3]
- **Ornamental plant production** *20190630* [2], *20190633* [2]
- **Protected bilberries** *20190633* [2]
- **Protected blackberries** *20190622* [2], *20190633* [2]
- **Protected blackcurrants** *20190633* [2]
- **Protected blueberry** *20190633* [2]
- **Protected cranberries** *20190633* [2]
- **Protected forest** *20190630* [2]
- **Protected gooseberries** *20190633* [2]
- **Protected loganberries** *20190622* [2], *20190633* [2]
- **Protected ornamentals** *20190630* [2], *20190633* [2]
- **Protected raspberries** *20190622* [2], *20190633* [2]
- **Protected redcurrants** *20190633* [2]
- **Protected Rubus hybrids** *20190622* [2], *20190633* [2]
- **Protected strawberries** *20190633* [2]
- **Raspberries** *20190622* [2], *20190633* [2]
- **Redcurrants** *20190627* [2], *20190633* [2]
- **Ribes species** *20190627* [2]
- **Rubus hybrids** *20190622* [2], *20190633* [2]
- **Rye** *20190752* [1]
- **Strawberries** *20190623* [2], *20190633* [2]
- **Wine grapes** *20190624* [2]

Approval information
- Carfentrazone-ethyl included in Annex I under EC Regulation 1107/2009
- Accepted by BBPA for use on malting barley and hops

Efficacy guidance
- Best weed control results achieved from good spray cover applied to small actively growing weeds
- Carfentrazone-ethyl acts by contact only; see label for optimum timing on specified weeds. Weeds emerging after application will not be controlled
- For weed control in cereals use as two spray programme with one application in autumn and one in spring
- Efficacy of haulm destruction will be reduced where flailed haulm covers the stems at application
- For potato crops with very dense vigorous haulm or where regrowth occurs following a single application a second application may be necessary to achieve satisfactory desiccation. A minimum interval between applications of 7 d should be observed to achieve optimum performance

Restrictions
- Maximum number of treatments 2 per crop for cereals (1 in Autumn and 1 in Spring); 1 per crop for weed control in potatoes; 1 per yr for pre-planting treatments
- Maximum total dose for potato haulm destruction equivalent to 1.6 full dose treatments
- Do not treat cereal crops under stress from drought, waterlogging, cold, pests, diseases, nutrient or lime deficiency or any factors reducing plant growth
- Do not treat cereals undersown with clover or other legumes
- Allow at least 2-3 wk between application and lifting potatoes to allow skins to set if potatoes are to be stored
- Follow label instructions for sprayer cleaning
- Do not apply through knapsack sprayers
- Contact processor before using a split dose on potatoes for processing
- Do not apply within 10 days of an application of iodosulfuron-methyl-sodium + mesosulfuron-methyl formulations

Crop-specific information
- Latest use: 1 mth before planting edible and non-edible crops; before 3rd node detectable (GS 33) on cereals
- HI 7 d for potatoes
- For weed control in cereals treat from 2 leaf stage
- When used as a treatment prior to planting a subsequent crop apply before weeds exceed maximum sizes indicated in the label
- If treated potato tubers are to be stored then allow at least 2-3 wk between the final application and lifting to allow skins to set

Following crops guidance
- No restrictions apply on the planting of succeeding crops 1 mth after application to potatoes for haulm destruction or as a pre-planting treatment, or 3 mth after application to cereals for weed control
- In the event of failure of a treated cereal crop, all cereals, ryegrass, maize, oilseed rape, peas, sunflowers, *Phacelia*, vetches, carrots or onions may be planted within 1 mth of treatment

Environmental safety
- Dangerous for the environment
- Very toxic to aquatic organisms
- Some non-target crops are sensitive. Avoid drift onto broad-leaved plants outside the treated area, or onto ponds waterways or ditches

Hazard classification and safety precautions
Hazard Irritant, Dangerous for the environment, Very toxic to aquatic organisms
Transport code 9 [1-3]
Packaging group III
UN Number 3077, 3082
Risk phrases H317 [2-3]
Operator protection U05a, U08 [1], U13 [1], U14, U20a [2-3]; A, H

SEE SECTION 3 FOR PRODUCTS ALSO REGISTERED

SECTION 2

Environmental protection E15a [1-2], E15b [3], E34 [1], E38, H410
Storage and disposal D01, D02, D09a, D10b [1], D10c [2-3], D12a
Medical advice M05a [1]

67   carfentrazone-ethyl + mecoprop-P

A foliar applied herbicide for cereals
HRAC mode of action code: 14 + 4 (E + O)

See also mecoprop-P

Products

1 Pan Glory	Pan Amenity	1.5:60% w/w	WG	14487
2 Platform S	FMC Agro	1.5:60% w/w	WG	18701

Uses

- Annual and perennial weeds in *Amenity grassland* [1]; *Managed amenity turf* [1];
- Charlock in *Spring barley* [2]; *Spring oats* [2]; *Spring wheat* [2]; *Winter barley* [2]; *Winter oats* [2]; *Winter wheat* [2];
- Chickweed in *Spring barley* [2]; *Spring oats* [2]; *Spring wheat* [2]; *Winter barley* [2]; *Winter oats* [2]; *Winter wheat* [2];
- Cleavers in *Spring barley* [2]; *Spring oats* [2]; *Spring wheat* [2]; *Winter barley* [2]; *Winter oats* [2]; *Winter wheat* [2];
- Field speedwell in *Spring barley* [2]; *Spring oats* [2]; *Spring wheat* [2]; *Winter barley* [2]; *Winter oats* [2]; *Winter wheat* [2];
- Ivy-leaved speedwell in *Spring barley* [2]; *Spring oats* [2]; *Spring wheat* [2]; *Winter barley* [2]; *Winter oats* [2]; *Winter wheat* [2];
- Moss in *Amenity grassland* [1]; *Managed amenity turf* [1];
- Red dead-nettle in *Spring barley* [2]; *Spring oats* [2]; *Spring wheat* [2]; *Winter barley* [2]; *Winter oats* [2]; *Winter wheat* [2];

Approval information

- Carfentrazone-ethyl and mecoprop-P included in Annex I under EC Regulation 1107/2009
- Accepted by BBPA for use on malting barley

Efficacy guidance

- Best results obtained when weeds have germinated and growing vigorously in warm moist conditions
- Treatment of large weeds and poor spray coverage may result in reduced weed control

Restrictions

- Maximum number of treatments 2 per crop. The total amount of mecoprop-P applied in a single yr must not exceed the maximum total dose approved for any single product for the crop per situation
- Do not treat crops suffering from stress from any cause
- Do not treat crops undersown or to be undersown
- Do not apply between 1st October and 1st March

Crop-specific information

- Latest use: before 3rd node detectable (GS 33)
- Can be used on all varieties of wheat and barley in autumn or spring from the beginning of tillering

Following crops guidance

- In the event of crop failure, any cereal, maize, oilseed rape, peas, vetches or sunflowers may be sown 1 mth after a spring treatment. Any crop may be planted 3 mth after treatment

Environmental safety

- Dangerous for the environment
- Very toxic to aquatic organisms

FOR FULL CONDITIONS OF USE ALWAYS READ THE PRODUCT LABEL

- Keep livestock out of treated areas for at least two weeks following treatment and until poisonous weeds, such as ragwort, have died down and become unpalatable
- Some pesticides pose a greater threat of contamination of water than others and mecoprop-P is one of these pesticides. Take special care when applying mecoprop-P near water and do not apply if heavy rain is forecast

Hazard classification and safety precautions

Hazard Harmful, Dangerous for the environment, Harmful if swallowed, Very toxic to aquatic organisms [2]

Transport code 9 [1-2]

Packaging group III

UN Number 3077

Risk phrases H317, H318

Operator protection U05a, U08 [2], U09a [1], U11, U13, U14, U20b; A, C, H, M

Environmental protection E07a, E15a [2], E15b [1], E34, E38, H410 [1], H411 [2]

Storage and disposal D01, D02, D09a, D10b, D12a [2]

Medical advice M05a

68   cerevisane

A systemic resistance inducer derived from yeast
FRAC mode of action code: P6

Products

| 1 | Romeo | Fargro | 94.1% w/w | WP | 19170 |

Uses

- Botrytis in **Baby leaf crops** *(off-label)* [1]; **Bilberries** *(off-label)* [1]; **Blackberries** *(off-label)* [1]; **Blackcurrants** *(off-label)* [1]; **Blueberries** *(off-label)* [1]; **Chives** *(off-label)* [1]; **Cranberries** *(off-label)* [1]; **Cress** *(off-label)* [1]; **Endives** *(off-label)* [1]; **Gooseberries** *(off-label)* [1]; **Herbs (see appendix 6)** *(off-label)* [1]; **Lamb's lettuce** *(off-label)* [1]; **Lettuce** *(off-label)* [1]; **Loganberries** *(off-label)* [1]; **Mint** *(off-label)* [1]; **Ornamental plant production** *(off-label)* [1]; **Parsley** *(off-label)* [1]; **Protected aubergines** [1]; **Protected baby leaf crops** *(off-label)* [1]; **Protected bilberries** *(off-label)* [1]; **Protected blackberries** *(off-label)* [1]; **Protected blackcurrants** *(off-label)* [1]; **Protected blueberries** *(off-label)* [1]; **Protected chives** *(off-label)* [1]; **Protected cranberries** *(off-label)* [1]; **Protected cress** *(off-label)* [1]; **Protected endives** *(off-label)* [1]; **Protected gooseberries** *(off-label)* [1]; **Protected herbs (see appendix 6)** *(off-label)* [1]; **Protected lamb's lettuce** *(off-label)* [1]; **Protected lettuce** *(off-label)* [1]; **Protected loganberries** *(off-label)* [1]; **Protected mint** *(off-label)* [1]; **Protected ornamentals** *(off-label)* [1]; **Protected parsley** *(off-label)* [1]; **Protected raspberries** *(off-label)* [1]; **Protected redcurrants** *(off-label)* [1]; **Protected Rubus hybrids** *(off-label)* [1]; **Protected spinach** *(off-label)* [1]; **Protected strawberries** *(off-label)* [1]; **Protected tomatoes** [1]; **Raspberries** *(off-label)* [1]; **Redcurrants** *(off-label)* [1]; **Rubus hybrids** *(off-label)* [1]; **Spinach** *(off-label)* [1]; **Strawberries** *(off-label)* [1];
- Botrytis bunch rot in **Table grapes** [1]; **Wine grapes** [1];
- Downy mildew in **Baby leaf crops** *(off-label)* [1]; **Bilberries** *(off-label)* [1]; **Blackberries** *(off-label)* [1]; **Blackcurrants** *(off-label)* [1]; **Blueberries** *(off-label)* [1]; **Chives** *(off-label)* [1]; **Cranberries** *(off-label)* [1]; **Cress** *(off-label)* [1]; **Endives** *(off-label)* [1]; **Gooseberries** *(off-label)* [1]; **Herbs (see appendix 6)** *(off-label)* [1]; **Lamb's lettuce** *(off-label)* [1]; **Lettuce** *(off-label)* [1]; **Loganberries** *(off-label)* [1]; **Mint** *(off-label)* [1]; **Ornamental plant production** *(off-label)* [1]; **Parsley** *(off-label)* [1]; **Protected baby leaf crops** *(off-label)* [1]; **Protected bilberries** *(off-label)* [1]; **Protected blackberries** *(off-label)* [1]; **Protected blackcurrants** *(off-label)* [1]; **Protected blueberries** *(off-label)* [1]; **Protected chives** *(off-label)* [1]; **Protected cranberries** *(off-label)* [1]; **Protected cress** *(off-label)* [1]; **Protected endives** *(off-label)* [1]; **Protected gooseberries** *(off-label)* [1]; **Protected herbs (see appendix 6)** *(off-label)* [1]; **Protected lamb's lettuce** *(off-label)* [1]; **Protected lettuce** *(off-label)* [1]; **Protected loganberries** *(off-label)* [1]; **Protected mint** *(off-label)* [1]; **Protected ornamentals** *(off-label)* [1]; **Protected parsley** *(off-label)* [1]; **Protected raspberries** *(off-label)* [1]; **Protected redcurrants** *(off-label)* [1]; **Protected Rubus hybrids** *(off-label)* [1]; **Protected spinach** *(off-label)* [1]; **Protected strawberries** *(off-label)* [1]; **Raspberries**

(off-label) [1]; ***Redcurrants*** *(off-label)* [1]; ***Rubus hybrids*** *(off-label)* [1]; ***Spinach*** *(off-label)* [1]; ***Strawberries*** *(off-label)* [1]; ***Table grapes*** [1]; ***Wine grapes*** [1];

- Powdery mildew in ***Baby leaf crops*** *(off-label)* [1]; ***Bilberries*** *(off-label)* [1]; ***Blackberries*** *(off-label)* [1]; ***Blackcurrants*** *(off-label)* [1]; ***Blueberries*** *(off-label)* [1]; ***Chives*** *(off-label)* [1]; ***Cranberries*** *(off-label)* [1]; ***Cress*** *(off-label)* [1]; ***Endives*** *(off-label)* [1]; ***Gooseberries*** *(off-label)* [1]; ***Herbs (see appendix 6)*** *(off-label)* [1]; ***Lamb's lettuce*** *(off-label)* [1]; ***Lettuce*** *(off-label)* [1]; ***Loganberries*** *(off-label)* [1]; ***Mint*** *(off-label)* [1]; ***Ornamental plant production*** *(off-label)* [1]; ***Parsley*** *(off-label)* [1]; ***Protected baby leaf crops*** *(off-label)* [1]; ***Protected bilberries*** *(off-label)* [1]; ***Protected blackberries*** *(off-label)* [1]; ***Protected blackcurrants*** *(off-label)* [1]; ***Protected blueberries*** *(off-label)* [1]; ***Protected chives*** *(off-label)* [1]; ***Protected courgettes*** [1]; ***Protected cranberries*** *(off-label)* [1]; ***Protected cress*** *(off-label)* [1]; ***Protected cucumbers*** [1]; ***Protected endives*** *(off-label)* [1]; ***Protected gherkins*** [1]; ***Protected gooseberries*** *(off-label)* [1]; ***Protected herbs (see appendix 6)*** *(off-label)* [1]; ***Protected lamb's lettuce*** *(off-label)* [1]; ***Protected lettuce*** *(off-label)* [1]; ***Protected loganberries*** *(off-label)* [1]; ***Protected melons*** [1]; ***Protected mint*** *(off-label)* [1]; ***Protected ornamentals*** *(off-label)* [1]; ***Protected parsley*** *(off-label)* [1]; ***Protected pumpkins*** [1]; ***Protected raspberries*** *(off-label)* [1]; ***Protected redcurrants*** *(off-label)* [1]; ***Protected Rubus hybrids*** *(off-label)* [1]; ***Protected spinach*** *(off-label)* [1]; ***Protected strawberries*** *(off-label)* [1]; ***Protected summer squash*** [1]; ***Protected watermelon*** [1]; ***Protected winter squash*** [1]; ***Raspberries*** *(off-label)* [1]; ***Redcurrants*** *(off-label)* [1]; ***Rubus hybrids*** *(off-label)* [1]; ***Spinach*** *(off-label)* [1]; ***Strawberries*** *(off-label)* [1]; ***Table grapes*** [1]; ***Wine grapes*** [1];

Extension of Authorisation for Minor Use (EAMUs)

- ***Baby leaf crops*** *20211443* [1]
- ***Bilberries*** *20211449* [1]
- ***Blackberries*** *20211449* [1]
- ***Blackcurrants*** *20211449* [1]
- ***Blueberries*** *20211449* [1]
- ***Chives*** *20211443* [1]
- ***Cranberries*** *20211449* [1]
- ***Cress*** *20211443* [1]
- ***Endives*** *20211443* [1]
- ***Gooseberries*** *20211449* [1]
- ***Herbs (see appendix 6)*** *20211443* [1]
- ***Lamb's lettuce*** *20211443* [1]
- ***Lettuce*** *20211443* [1]
- ***Loganberries*** *20211449* [1]
- ***Mint*** *20211443* [1]
- ***Ornamental plant production*** *20211455* [1]
- ***Parsley*** *20211443* [1]
- ***Protected baby leaf crops*** *20211443* [1]
- ***Protected bilberries*** *20211449* [1]
- ***Protected blackberries*** *20211449* [1]
- ***Protected blackcurrants*** *20211449* [1]
- ***Protected blueberries*** *20211449* [1]
- ***Protected chives*** *20211443* [1]
- ***Protected cranberries*** *20211449* [1]
- ***Protected cress*** *20211443* [1]
- ***Protected endives*** *20211443* [1]
- ***Protected gooseberries*** *20211449* [1]
- ***Protected herbs (see appendix 6)*** *20211443* [1]
- ***Protected lamb's lettuce*** *20211443* [1]
- ***Protected lettuce*** *20211443* [1]
- ***Protected loganberries*** *20211449* [1]
- ***Protected mint*** *20211443* [1]
- ***Protected ornamentals*** *20211455* [1]
- ***Protected parsley*** *20211443* [1]
- ***Protected raspberries*** *20211449* [1]
- ***Protected redcurrants*** *20211449* [1]

FOR FULL CONDITIONS OF USE ALWAYS READ THE PRODUCT LABEL

- *Protected **Rubus** hybrids* 20211449 [1]
- *Protected **spinach*** 20211443 [1]
- *Protected **strawberries*** 20211449 [1]
- ***Raspberries*** 20211449 [1]
- ***Redcurrants*** 20211449 [1]
- ***Rubus** hybrids* 20211449 [1]
- ***Spinach*** 20211443 [1]
- ***Strawberries*** 20211449 [1]

Approval information
- Cerevisane included in Annex I under EC Regulation 1107/2009

Restrictions
- When treating protected crops treatment must only be made under ?permanent protection? situations which provide full enclosure (including continuous top and side barriers down to below ground level) and which are present and maintained over a number of years.
- Reasonable precautions must be taken to prevent access of birds, wild mammals and honey bees to treated crops.
- To minimise airborne environmental exposure, vents, doors and other openings must be closed during and after application until the applied product has fully settled.

Hazard classification and safety precautions
  UN Number N/C
  Operator protection U19a; A, D, H
  Environmental protection E15b, E34
  Storage and disposal D01, D05, D09a, D10b, D12b

69   chlorantraniliprole

An ingested and contact insecticide for insect pest control in apples and pears and, in Ireland only, for colorado beetle control in potatoes.
IRAC mode of action code: 28

Products

1	Coragen	FMC Agro	200 g/l	SC	19498
2	Voliam	Syngenta	200 g/l	SC	19718

Uses
- Carnation tortrix moth in ***Bilberries*** *(off-label)* [1]; ***Blackcurrants*** *(off-label)* [1]; ***Blueberries*** *(off-label)* [1]; ***Gooseberries*** *(off-label)* [1]; ***Redcurrants*** *(off-label)* [1];
- Carrot fly in ***Celeriac*** *(off-label)* [1]; ***Celery (outdoor)*** *(off-label)* [1]; ***Horseradish*** *(off-label)* [1]; ***Parsley root*** *(off-label)* [1]; ***Parsnips*** *(off-label)* [1]; ***Red beet*** *(off-label)* [1]; ***Swedes*** *(off-label)* [1]; ***Turnips*** *(off-label)* [1];
- Caterpillars in ***Carrots*** [1-2]; ***Celery (outdoor)*** *(off-label)* [1]; ***Forage maize*** [1-2]; ***Grain maize*** [1-2]; ***Sweetcorn*** [1-2];
- Codling moth in ***Apples*** *(off-label)* [1]; ***Pears*** *(off-label)* [1];
- Fruitlet mining tortrix in ***Damsons*** *(off-label)* [1]; ***Plums*** *(off-label)* [1];
- Light brown apple moth in ***Bilberries*** *(off-label)* [1]; ***Blackcurrants*** *(off-label)* [1]; ***Blueberries*** *(off-label)* [1]; ***Damsons*** *(off-label)* [1]; ***Gooseberries*** *(off-label)* [1]; ***Plums*** *(off-label)* [1]; ***Redcurrants*** *(off-label)* [1];
- Plum fruit moth in ***Damsons*** *(off-label)* [1]; ***Plums*** *(off-label)* [1];
- Silver Y moth in ***Horseradish*** *(off-label)* [1]; ***Red beet*** *(off-label)* [1]; ***Swedes*** *(off-label)* [1]; ***Turnips*** *(off-label)* [1];
- South American Tomato Moth in ***Tomatoes*** *(off-label)* [1];
- Tortrix moths in ***Damsons*** *(off-label)* [1]; ***Plums*** *(off-label)* [1];
- Winter moth in ***Apples*** *(off-label)* [1]; ***Damsons*** *(off-label)* [1]; ***Pears*** *(off-label)* [1]; ***Plums*** *(off-label)* [1];

Extension of Authorisation for Minor Use (EAMUs)
- ***Apples*** 20220303 [1]

SEE SECTION 3 FOR PRODUCTS ALSO REGISTERED

- *Bilberries* *20220311* [1]
- *Blackcurrants* *20220311* [1]
- *Blueberries* *20220311* [1]
- *Celeriac* *20220800* [1]
- *Celery (outdoor)* *20220801* [1]
- *Damsons* *20220374* [1]
- *Gooseberries* *20220311* [1]
- *Horseradish* *20220799* [1]
- *Parsley root* *20220800* [1]
- *Parsnips* *20220800* [1]
- *Pears* *20220303* [1]
- *Plums* *20220374* [1]
- *Red beet* *20220799* [1]
- *Redcurrants* *20220311* [1]
- *Swedes* *20220799* [1]
- *Tomatoes* *20220335* [1]
- *Turnips* *20220799* [1]

Approval information
- Chlorantraniliprole is included in Annex 1 under EC Regulation 1107/2009

Efficacy guidance
- Can be used as part of an Integrated Pest Management programme.
- For best fruit protection in apples and pears, apply before egg hatch.
- Best applied early morning or late evening to avoid applications when bees may be present.

Restrictions
- Must not be applied before BBCH 51 when applying to maize and sweetcorn [1]

Crop-specific information
- Latest time of application is 14 days before harvest

Environmental safety
- To protect bees and pollinating insects, do not apply to crops when in flower. Do not apply when bees are actively foraging or when flowering plants are present.
- Broadcast air-assisted LERAP [1-2] (10 m); LERAP Category B [1-2]

Hazard classification and safety precautions
Hazard Dangerous for the environment, Very toxic to aquatic organisms
Transport code 9 [1-2]
Packaging group III
UN Number 3082
Environmental protection E12e, E15b, E16a, E17b (10 m), E22b, E38, H410
Storage and disposal D01, D02, D05, D09a, D12a

70   chlormequat

A plant-growth regulator for reducing stem growth and lodging

Products

1	3C Chlormequat 750	BASF	750 g/l	SL	16690
2	Agrovista 3 See 750	Agrovista	750 g/l	SL	15975
3	Clayton Everest	Clayton	620 g/l	SL	19624
4	CleanCrop Transist	Agrii	720 g/l	SL	16684
5	Stabilan 750	Nufarm UK	750 g/l	SL	09303
6	Stefes CCC 720	Stefes	720 g/l	SL	17731

Uses
- Growth regulation in *Canary seed* (off-label) [5]; *Protected ornamentals* (off-label) [5]; *Rye* [6]; *Spring barley* [6]; *Spring oats* [3-4,6]; *Spring wheat* [3-4,6]; *Triticale* [5-6]; *Winter barley* [3-4,6]; *Winter oats* [3-4,6]; *Winter wheat* [3-4,6];

FOR FULL CONDITIONS OF USE ALWAYS READ THE PRODUCT LABEL

- Lodging control in **Canary seed** *(off-label)* [5]; **Oats** [1]; **Rye** [1,5-6]; **Spring barley** [1,3,6]; **Spring oats** [2-6]; **Spring rye** [2]; **Spring wheat** [1-6]; **Triticale** [1-2,5-6]; **Winter barley** [1-6]; **Winter oats** [2-6]; **Winter rye** [2]; **Winter wheat** [1-6];

Extension of Authorisation for Minor Use (EAMUs)
- **Canary seed** *20180371* [5]
- **Protected ornamentals** *20171416* [5]

Approval information
- Chlormequat included in Annex 1 under EC Regulation 1107/2009
- Accepted by BBPA for use on malting barley

Efficacy guidance
- Most effective results on cereals normally achieved from application from Apr onwards, on wheat and rye from leaf sheath erect to first node detectable (GS 30-31), on oats at second node detectable (GS 32), on winter barley from mid-tillering to leaf sheath erect (GS 25-30). However, recommendations vary with product. See label for details
- Influence on growth varies with crop and growth stage. Risk of lodging reduced by application at early stem extension. Root development and yield can be improved by earlier treatment
- Results on barley can be variable
- In tank mixes with other pesticides on cereals optimum timing for herbicide action may differ from that for growth reduction. See label for details of tank mix recommendations
- Most products recommended for use on oats require addition of approved non-ionic wetter. Check label
- Some products are formulated with trace elements to help compensate for increased demand during rapid growth

Restrictions
- Maximum number of treatments or maximum total dose varies with crop and product and whether split dose treatments are recommended. Check labels
- Do not use on very late sown spring wheat or oats or on crops under stress
- Mixtures with liquid nitrogen fertilizers may cause scorch and are specifically excluded on some labels
- Do not use on soils of low fertility unless such crops regularly receive adequate dressings of nitrogen
- At least 6 h, preferably 24 h, required before rain for maximum effectiveness. Do not apply to wet crops
- Check labels for tank mixtures known to be incompatible
- Must only be applied between growth stages BBCH 31 to BBCH 39
- A maximum dose 1.33 L per hectare (750 g/l formulations) or 1.38 l/ha (720 g/l formulations) must not be exceeded when applied to winter wheat and spring barley before stem elongation (GS 30)
- No applications to be made before 1st February in the year of harvest [5, 2]

Crop-specific information
- Latest use varies with crop and product. See label for details
- May be used on cereals undersown with grass or clovers
- Ornamentals to be treated must be well established and growing vigorously. Do not treat in strong sunlight or when temperatures are likely to fall below 10°C
- Temporary yellow spotting may occur on poinsettias. It can be minimised by use of a non-ionic wetting agent - see label

Environmental safety
- Wash equipment thoroughly with water and wetting agent immediately after use and spray out. Spray out again before storing or using for another product. Traces can cause harm to susceptible crops sprayed later
- Do not use straw from treated cereals as horticultural growth medium or mulch

Hazard classification and safety precautions
   Hazard Harmful [1-2,4-6], Harmful if swallowed
   Transport code 8 [1-6]
   Packaging group III

SEE SECTION 3 FOR PRODUCTS ALSO REGISTERED

SECTION 2

UN Number 1760
Risk phrases H290, H312 [1-2,4-6]
Operator protection U05a, U08 [1-4,6], U09a [5], U19a, U20b; A, H
Environmental protection E15a, E34, H411 [6], H412 [2-5]
Storage and disposal D01, D02, D05 [1-2,5], D09a [3-6], D10a [5], D10b [1-4,6]
Medical advice M03, M05a [4-6]

71   chlormequat + ethephon

A plant growth regulator for use in cereals

See also ethephon

Products

1 Bogota	Adama	305:155 g/l	SL	19486
2 Chlormephon	Adama	305:155 g/l	SL	19489
3 Socom	Adama	305:155 g/l	SL	19483
4 Spatial Plus	Sumitomo	300:150 g/l	SL	18888
5 Vivax	Sumitomo	300:150 g/l	SL	18901

Uses

- Lodging control in **Spring barley** [1-5]; **Winter barley** [1-5]; **Winter wheat** [1-5];

Approval information

- Chlormequat and 2-chloroethylphosphonic acid (ethephon) included in Annex I under EC Regulation 1107/2009
- Accepted by BBPA for use on malting barley
- All products containing 2-chloroethylphosphonic acid carry the warning: '2-chloroethylphosphonic acid is an anticholinesterase organophosphate. Handle with care'.

Efficacy guidance

- Best results obtained when crops growing vigorously
- Recommended dose varies with growth stage. See labels for details and recommendations for use of sequential treatments

Restrictions

- 2-chloroethylphosphonic acid is an anticholinesterase organophosphorus compound. Do not use if under medical advice not to work with such compounds
- Maximum number of treatments 1 per crop; maximum total dose equivalent to one full dose treatment
- Product must always be used with specified non-ionic wetter - see labels
- Do not use on any crop in sequence with any other product containing 2-chloroethylphosphonic acid
- Do not spray when crop wet or rain imminent
- Do not spray during cold weather or periods of night frost, when soil is very dry, when crop diseased or suffering pest damage, nutrient deficiency or herbicide stress
- If used on seed crops grown for certification inform seed merchant beforehand
- Do not use on wheat variety Moulin or on any winter varieties sown in spring
- Do not use on spring barley variety Triumph
- Do not treat barley on soils with more than 10% organic matter
- Only crops growing under conditions of high fertility should be treated

Crop-specific information

- Latest use: before flag leaf ligule/collar just visible (GS 39) or 1st spikelet visible (GS 51) for wheat or barley at top dose; or before flag leaf sheath opening (GS 47) for winter wheat at reduced dose
- Apply before lodging has started

Environmental safety

- Harmful to fish or other aquatic life. Do not contaminate surface waters or ditches with chemical or used container
- Do not use straw from treated cereals as a horticultural growth medium or as a mulch

FOR FULL CONDITIONS OF USE ALWAYS READ THE PRODUCT LABEL

Hazard classification and safety precautions
Hazard Harmful, Toxic if swallowed [4-5], Harmful if swallowed [1-3]
Transport code 8 [1-5]
Packaging group III
UN Number 3265
Risk phrases H290 [4-5], H319 [4-5], H335 [1-3]
Operator protection U05a, U08, U14 [1-3], U19a, U20b; A, C, H
Environmental protection E15a, E34, E38, H412 [4-5]
Storage and disposal D01, D02, D09a, D10b, D12a [1-3]
Medical advice M01, M03, M05a

72   chlorotoluron

A contact and residual urea herbicide for cereals only available in mixtures
HRAC mode of action code: 5 (C2)

73   chlorotoluron + diflufenican + pendimethalin

A herbicide mixture for weed control in cereals
HRAC mode of action code: 5 + 12 + 3 (C2 + F1 + K1)

Products

1	Tower	Adama	250:40:300 g/l	SC	16586
2	Tribal	Adama	250:40:300 g/l	SC	17075

Uses
- Annual meadow grass in *Rye* [1-2]; *Spring barley* [1-2]; *Spring wheat* [1-2]; *Triticale* [1-2]; *Winter barley* [1-2]; *Winter wheat* [1-2];
- Charlock in *Rye* [1-2]; *Spring barley* [1-2]; *Spring wheat* [1-2]; *Triticale* [1-2]; *Winter barley* [1-2]; *Winter wheat* [1-2];
- Chickweed in *Rye* [1-2]; *Spring barley* [1-2]; *Spring wheat* [1-2]; *Triticale* [1-2]; *Winter barley* [1-2]; *Winter wheat* [1-2];
- Cleavers in *Rye* [1-2]; *Spring barley* [1-2]; *Spring wheat* [1-2]; *Triticale* [1-2]; *Winter barley* [1-2]; *Winter wheat* [1-2];
- Crane's-bill in *Rye* [1-2]; *Spring barley* [1-2]; *Spring wheat* [1-2]; *Triticale* [1-2]; *Winter barley* [1-2]; *Winter wheat* [1-2];
- Dead nettle in *Rye* [1-2]; *Spring barley* [1-2]; *Spring wheat* [1-2]; *Triticale* [1-2]; *Winter barley* [1-2]; *Winter wheat* [1-2];
- Field pansy in *Rye* [1-2]; *Spring barley* [1-2]; *Spring wheat* [1-2]; *Triticale* [1-2]; *Winter barley* [1-2]; *Winter wheat* [1-2];
- Forget-me-not in *Rye* [1-2]; *Spring barley* [1-2]; *Spring wheat* [1-2]; *Triticale* [1-2]; *Winter barley* [1-2]; *Winter wheat* [1-2];
- Fumitory in *Rye* [1-2]; *Spring barley* [1-2]; *Spring wheat* [1-2]; *Triticale* [1-2]; *Winter barley* [1-2]; *Winter wheat* [1-2];
- Loose silky bent in *Rye* [1-2]; *Spring barley* [1-2]; *Spring wheat* [1-2]; *Triticale* [1-2]; *Winter barley* [1-2]; *Winter wheat* [1-2];
- Mayweeds in *Rye* [1-2]; *Spring barley* [1-2]; *Spring wheat* [1-2]; *Triticale* [1-2]; *Winter barley* [1-2]; *Winter wheat* [1-2];
- Penny cress in *Rye* [1-2]; *Spring barley* [1-2]; *Spring wheat* [1-2]; *Triticale* [1-2]; *Winter barley* [1-2]; *Winter wheat* [1-2];
- Poppies in *Rye* [1-2]; *Spring barley* [1-2]; *Spring wheat* [1-2]; *Triticale* [1-2]; *Winter barley* [1-2]; *Winter wheat* [1-2];
- Runch in *Rye* [1-2]; *Spring barley* [1-2]; *Spring wheat* [1-2]; *Triticale* [1-2]; *Winter barley* [1-2]; *Winter wheat* [1-2];
- Shepherd's purse in *Rye* [1-2]; *Spring barley* [1-2]; *Spring wheat* [1-2]; *Triticale* [1-2]; *Winter barley* [1-2]; *Winter wheat* [1-2];
- Speedwells in *Rye* [1-2]; *Spring barley* [1-2]; *Spring wheat* [1-2]; *Triticale* [1-2]; *Winter barley* [1-2]; *Winter wheat* [1-2];

SEE SECTION 3 FOR PRODUCTS ALSO REGISTERED

- Volunteer oilseed rape in *Rye* [1-2]; ***Spring barley*** [1-2]; ***Spring wheat*** [1-2]; ***Triticale*** [1-2]; ***Winter barley*** [1-2]; ***Winter wheat*** [1-2];

Approval information
- Chlorotoluron, diflufenican and pendimethalin included in Annex I under EC Regulation 1107/2009

Efficacy guidance
- Minimum water volume for full dose (2.0 l/ha) is 200 l/ha while reducing the dose to 1.0 l/ha allows a lowest water volume of 100 l/ha.

Restrictions
- Maximum number of treatments 1 per crop
- Horizontal boom sprayers must be fitted with three star drift reduction technology for all uses
- Low drift spraying equipment must be operated according to the specific conditions stated in the official three star rating for that equipment as published on HSE Chemicals Regulation Division?s website. These operating conditions must be maintained until the operator is 30m from the top of the bank of any surface water bodies
- Use only on listed crop varieties when applying post-emergence to winter wheat. Do not apply to undersown crops
- Substances containing the chlorotoluron active agent may not be applied more than once per year on the same area
- Application must not be made after 31 October in the year of drilling
- Do not use if it is frosty
- DO NOT apply to soils with greater than 10% organic matter

Crop-specific information
- Following normal harvest, there are no restrictions. In the event of crop failure, plough before drilling the following crop in the spring. Only winter wheat can be re-drilled in the same autumn following crop failure

Environmental safety
- Buffer zone requirement 6 m
- LERAP Category B [1-2]

Hazard classification and safety precautions
Transport code 9 [1-2]
Packaging group III
UN Number 3082
Risk phrases H351, H361
Operator protection U02a, U04a, U20b; A
Environmental protection E15b, E16a, E34, E38, H410
Storage and disposal D01, D09a, D10b, D12a
Medical advice M03

74   cholecalciferol

A non-anticoagulant rodenticide

Products

1	Selontra	BASF	0.075% w/w	RB	UK20-1257

Uses
- Mice in ***Farm buildings*** [1];
- Rats in ***Farm buildings*** [1];

Hazard classification and safety precautions
UN Number N/C
Operator protection A, H
Storage and disposal D01, D02, D06c, D09a
Medical advice M03, M05a

FOR FULL CONDITIONS OF USE ALWAYS READ THE PRODUCT LABEL

75   cinmethylin

A grass and broad-leaf weed herbicide for use in winter wheat
HRAC mode of action code: 30

Products

1 Luxinum Plus (GB only)   BASF   714 g/l   EC   19978

Uses

- Annual meadow grass in *Spring barley* [1]; *Winter wheat* [1];
- Blackgrass in *Spring barley* [1]; *Winter wheat* [1];
- Loose silky bent in *Spring barley* [1]; *Winter wheat* [1];
- Poppies in *Spring barley* [1]; *Winter wheat* [1];
- Ryegrass in *Spring barley* [1]; *Winter wheat* [1];

Approval information

- Cinmethlin not currently included in Annex I under EC Regulation 1107/2009

Efficacy guidance

- Efficacy is most effective when applied pre-emergence but it can be applied post-emergence if absolutely necessary.
- Apply in mixture with other effective active ingredients such as florasulam and/or pendimethalin to protect activity
- Consolidate loose soils before application
- Ensure seed is covered by at least 3 cm of settled soil. Shallow-drilled crops should only be treated post-emergence
- BASF advise against pre-em application to winter wheat after 15th November due to the risk of poor seedbeds and waterlogged soils but this is not a label restriction

Restrictions

- Do not use on sands, very light, stony or gravelly soils
- Do not apply when heavy rain is forecast or when soils are waterlogged

Crop-specific information

- There are no restrictions on following crops after normal harvest. In the event of crop failure, plough to at least 15 cm and sow spring wheat or spring barley

Environmental safety

- Allow a 5 m buffer zone to un-cropped land

Hazard classification and safety precautions

Hazard Very toxic to aquatic organisms
Transport code 9 [1]
Packaging group III
UN Number 3082
Risk phrases H315, H317, H319, H371
Operator protection U05a, U20b; A, C, H
Environmental protection E15b, H410
Storage and disposal D01, D05, D09a, D10c
Medical advice M03

76   citronella oil

A natural plant extract herbicide

Products

1 Barrier H   Barrier   22.9% w/w   OD   17145

Uses

- Ragwort in *Amenity grassland* [1]; *Land temporarily removed from production* [1];

Approval information
- Plant oils such as citronella oil are included in Annex 1 under EC Regulation 1107/2009

Efficacy guidance
- Best results obtained from spot treatment of ragwort in the rosette stage, during dry still conditions
- Aerial growth of ragwort is rapidly destroyed. Longer term control depends on overall management strategy
- Check for regrowth after 28 d and re-apply as necessary

Crop-specific information
- Contact with grasses will result in transient scorch which is outgrown in good growing conditions

Environmental safety
- Apply away from bees
- Harmful to fish or other aquatic life. Do not contaminate surface waters or ditches with chemical or used container
- Keep livestock out of treated areas for at least 2 wk and until foliage of any poisonous weeds such as ragwort has died and become unpalatable

Hazard classification and safety precautions
Hazard Irritant
UN Number N/C
Risk phrases H317
Operator protection U05a, U20c; A, C
Environmental protection E07b (2 wk), E12g, E13c
Storage and disposal D01, D02, D05, D09a, D11a

77   clethodim

A post-emergence grass herbicide for use in listed broad-leaved crops
HRAC mode of action code: 1 (A)

Products

1	Balistik	Arysta	120 g/l	EC	18129
2	Centurion Max	Arysta	120 g/l	EC	17911
3	Clayton Gatso	Clayton	120 g/l	EC	18611
4	Cleancrop Signifier	Clayton	120 g/l	EC	18998
5	Select Prime	Arysta	120 g/l	EC	16304
6	VextaDim 240 EC	Agrii	240 g/l	EC	19516

Uses
- Annual grasses in *Balm* (off-label) [1-2,5]; *Blackberries* (off-label) [1-2,5]; *Blackcurrants* (off-label) [1-2,5]; *Blueberries* (off-label) [1-2,5]; *Broad beans* (off-label) [1-2,5]; *Brussels sprouts* (off-label) [1-2,5]; *Bulb onions* (off-label) [1-2,5]; *Cabbages* (off-label) [1-2,5]; *Carrots* (off-label) [1-2,5]; *Celeriac* (off-label) [1-2,5]; *Celery leaves* (off-label) [1-2,5]; *Chives* (off-label) [1-2,5]; *Corn Gromwell* (off-label) [5]; *Crambe* (off-label) [5]; *Cranberries* (off-label) [1-2,5]; *Dill* (off-label) [1-2,5]; *Dwarf beans* (off-label) [1-2,5]; *Edible podded peas* (off-label) [5]; *Fennel leaves* (off-label) [1-2,5]; *Fodder beet* (off-label) [1-2,5-6]; *French beans* (off-label) [1-2,5]; *Garlic* (off-label) [1-2,5]; *Gold-of-pleasure* (off-label) [1-2,5]; *Gooseberries* (off-label) [1-2,5]; *Horseradish* (off-label) [1-2,5]; *Leeks* (off-label) [1,5]; *Linseed* (off-label) [5]; *Lucerne* (off-label) [5]; *Mint* (off-label) [1-2,5]; *Mustard* (off-label) [1-2,5]; *Nursery fruit trees* (off-label) [2]; *Ornamental plant production* (off-label) [2]; *Parsley* (off-label) [1-2,5]; *Parsley root* (off-label) [1-2,5]; *Parsnips* (off-label) [1-2,5]; *Poppies* (off-label) [1-2,5]; *Protected nursery fruit trees* (off-label) [2]; *Raspberries* (off-label) [1-2,5]; *Red beet* (off-label) [1-2,5]; *Redcurrants* (off-label) [1-2,5]; *Runner beans* (off-label) [1-2,5]; *Salad onions* (off-label) [1,5]; *Salsify* (off-label) [1-2,5]; *Shallots* (off-label) [1-2,5]; *Strawberries* (off-label) [1-2,5]; *Sugar beet* [1-6]; *Sunflowers* (off-label) [1-2,5]; *Swedes* (off-label) [1-2,5]; *Turnips* (off-label) [1-2,5]; *Valerian root* (off-label) [1-2,5]; *Vining peas* (off-label) [1-2,5]; *Winter oilseed rape* [1-5];
- Annual meadow grass in *Angelica* (off-label) [2]; *Balm* (off-label) [1-2,5]; *Blackberries* (off-label) [1-2,5]; *Blackcurrants* (off-label) [1-2,5]; *Blueberries* (off-label) [1-2,5]; *Borage* (off-label) [1-2,5];

SECTION 2

Broad beans (off-label) [1-2,5]; *Brussels sprouts* (off-label) [1-2,5]; *Bulb onions* (off-label) [1-2,5]; *Cabbages* (off-label) [1-2,5]; *Caraway* (off-label) [2]; *Carrots* (off-label) [1-2,5]; *Celeriac* (off-label) [1-2,5]; *Celery leaves* (off-label) [1-2,5]; *Chives* (off-label) [1-2,5]; *Coriander* (off-label) [2]; *Corn Gromwell* (off-label) [1-2,5]; *Crambe* (off-label) [1-2,5]; *Cranberries* (off-label) [1-2,5]; *Dill* (off-label) [1-2,5]; *Dwarf beans* (off-label) [1-2,5]; *Edible flowers* (off-label) [2]; *Edible podded peas* (off-label) [1-2,5]; *Fennel leaves* (off-label) [1-2,5]; *Fodder beet* (off-label) [1-2,5-6]; *French beans* (off-label) [1-2,5]; *Garlic* (off-label) [1-2,5]; *Gold-of-pleasure* (off-label) [1-2,5]; *Gooseberries* (off-label) [1-2,5]; *Horseradish* (off-label) [1-2,5]; *Leeks* (off-label) [1-2,5]; *Linseed* (off-label) [1-2,5]; *Lovage* (off-label) [2]; *Lucerne* (off-label) [5]; *Meadowfoam (Limnanthes alba)* (off-label) [2]; *Mint* (off-label) [1-2,5]; *Mustard* (off-label) [1-2,5]; *Nursery fruit trees* (off-label) [2]; *Ornamental plant production* (off-label) [2]; *Parsley* (off-label) [1-2,5]; *Parsley root* (off-label) [1-2,5]; *Parsnips* (off-label) [1-2,5]; *Poppies* (off-label) [1-2,5]; *Protected nursery fruit trees* (off-label) [2]; *Raspberries* (off-label) [1-2,5]; *Red beet* (off-label) [1-2,5]; *Redcurrants* (off-label) [1-2,5]; *Runner beans* (off-label) [1-2,5]; *Salad burnet* (off-label) [2]; *Salad onions* (off-label) [1-2,5]; *Salsify* (off-label) [1-2,5]; *Shallots* (off-label) [1-2,5]; *Strawberries* (off-label) [1-2,5]; *Sugar beet* [1-6]; *Sunflowers* (off-label) [1-2,5]; *Swedes* (off-label) [1-2,5]; *Sweet ciceley* (off-label) [2]; *Turnips* (off-label) [1-2,5]; *Valerian root* (off-label) [1-2,5]; *Vining peas* (off-label) [1-2,5]; *Winter oilseed rape* [1-5];

- Blackgrass in *Angelica* (off-label) [2]; *Balm* (off-label) [1-2,5]; *Blackberries* (off-label) [1-2,5]; *Blackcurrants* (off-label) [1-2,5]; *Blueberries* (off-label) [1-2,5]; *Borage* (off-label) [1-2,5]; *Broad beans* (off-label) [1-2,5]; *Brussels sprouts* (off-label) [1-2,5]; *Bulb onions* (off-label) [1-2,5]; *Cabbages* (off-label) [1-2,5]; *Caraway* (off-label) [2]; *Carrots* (off-label) [1-2,5]; *Celeriac* (off-label) [1-2,5]; *Celery leaves* (off-label) [1-2,5]; *Chives* (off-label) [1-2,5]; *Coriander* (off-label) [2]; *Corn Gromwell* (off-label) [1-2,5]; *Crambe* (off-label) [1-2,5]; *Cranberries* (off-label) [1-2,5]; *Dill* (off-label) [1-2,5]; *Dwarf beans* (off-label) [1-2,5]; *Edible flowers* (off-label) [2]; *Edible podded peas* (off-label) [1-2,5]; *Fennel leaves* (off-label) [1-2,5]; *Fodder beet* (off-label) [1-2,5-6]; *French beans* (off-label) [1-2,5]; *Garlic* (off-label) [1-2,5]; *Gold-of-pleasure* (off-label) [1-2,5]; *Gooseberries* (off-label) [1-2,5]; *Horseradish* (off-label) [1-2,5]; *Leeks* (off-label) [1-2,5]; *Linseed* (off-label) [1-2,5]; *Lovage* (off-label) [2]; *Lucerne* (off-label) [5]; *Meadowfoam (Limnanthes alba)* (off-label) [2]; *Mint* (off-label) [1-2,5]; *Mustard* (off-label) [1-2,5]; *Nursery fruit trees* (off-label) [2]; *Ornamental plant production* (off-label) [2]; *Parsley* (off-label) [1-2,5]; *Parsley root* (off-label) [1-2,5]; *Parsnips* (off-label) [1-2,5]; *Poppies* (off-label) [1-2,5]; *Protected nursery fruit trees* (off-label) [2]; *Raspberries* (off-label) [1-2,5]; *Red beet* (off-label) [1-2,5]; *Redcurrants* (off-label) [1-2,5]; *Runner beans* (off-label) [1-2,5]; *Salad burnet* (off-label) [2]; *Salad onions* (off-label) [1-2,5]; *Salsify* (off-label) [1-2,5]; *Shallots* (off-label) [1-2,5]; *Strawberries* (off-label) [1-2,5]; *Sugar beet* [1-6]; *Sunflowers* (off-label) [1-2,5]; *Swedes* (off-label) [1-2,5]; *Sweet ciceley* (off-label) [2]; *Turnips* (off-label) [1-2,5]; *Valerian root* (off-label) [1-2,5]; *Vining peas* (off-label) [1-2,5]; *Winter oilseed rape* [1-5];

- Couch in *Angelica* (off-label) [2]; *Balm* (off-label) [1-2,5]; *Blackberries* (off-label) [1-2,5]; *Blackcurrants* (off-label) [1-2,5]; *Blueberries* (off-label) [1-2,5]; *Broad beans* (off-label) [1,5]; *Brussels sprouts* (off-label) [1-2,5]; *Bulb onions* (off-label) [1-2,5]; *Cabbages* (off-label) [1-2,5]; *Caraway* (off-label) [2]; *Carrots* (off-label) [1-2,5]; *Celeriac* (off-label) [1-2,5]; *Celery leaves* (off-label) [1-2,5]; *Chives* (off-label) [1-2,5]; *Coriander* (off-label) [2]; *Cranberries* (off-label) [1-2,5]; *Dill* (off-label) [1-2,5]; *Dwarf beans* (off-label) [1,5]; *Edible flowers* (off-label) [2]; *Fennel leaves* (off-label) [1-2,5]; *Fodder beet* (off-label) [1-2,5]; *French beans* (off-label) [1,5]; *Garlic* (off-label) [1-2,5]; *Gooseberries* (off-label) [1-2,5]; *Horseradish* (off-label) [1-2,5]; *Leeks* (off-label) [5]; *Lovage* (off-label) [2]; *Lucerne* (off-label) [5]; *Mint* (off-label) [1-2,5]; *Parsley* (off-label) [1-2,5]; *Parsley root* (off-label) [1-2,5]; *Parsnips* (off-label) [1-2,5]; *Raspberries* (off-label) [1-2,5]; *Red beet* (off-label) [1-2,5]; *Redcurrants* (off-label) [1-2,5]; *Runner beans* (off-label) [1,5]; *Salad burnet* (off-label) [2]; *Salad onions* (off-label) [5]; *Salsify* (off-label) [1-2,5]; *Shallots* (off-label) [1-2,5]; *Strawberries* (off-label) [1-2,5]; *Swedes* (off-label) [1-2,5]; *Sweet ciceley* (off-label) [2]; *Turnips* (off-label) [1-2,5]; *Valerian root* (off-label) [1-2,5]; *Vining peas* (off-label) [1,5];

- Italian ryegrass in *Meadowfoam (Limnanthes alba)* (off-label) [2];
- Loose silky bent in *Meadowfoam (Limnanthes alba)* (off-label) [2];
- Volunteer barley in *Borage* (off-label) [1-2,5]; *Corn Gromwell* (off-label) [1-2,5]; *Crambe* (off-label) [1-2,5]; *Fodder beet* [6]; *Linseed* (off-label) [1-2,5]; *Sugar beet* [1-6]; *Winter oilseed rape* [1-5];
- Volunteer cereals in *Broad beans* (off-label) [2,5]; *Brussels sprouts* (off-label) [2,5]; *Bulb onions* (off-label) [2,5]; *Cabbages* (off-label) [2,5]; *Carrots* (off-label) [2,5]; *Dwarf beans* (off-label) [2,5];

SEE SECTION 3 FOR PRODUCTS ALSO REGISTERED

French beans *(off-label)* [2,5]; **Garlic** *(off-label)* [2,5]; **Leeks** *(off-label)* [5]; **Meadowfoam (Limnanthes alba)** *(off-label)* [2]; **Nursery fruit trees** *(off-label)* [2]; **Ornamental plant production** *(off-label)* [2]; **Protected nursery fruit trees** *(off-label)* [2]; **Runner beans** *(off-label)* [2,5]; **Salad onions** *(off-label)* [5]; **Shallots** *(off-label)* [2,5]; **Vining peas** *(off-label)* [2,5];

- Volunteer wheat in **Borage** *(off-label)* [1-2,5]; **Corn Gromwell** *(off-label)* [1-2,5]; **Crambe** *(off-label)* [1-2,5]; **Fodder beet** [6]; **Linseed** *(off-label)* [1-2,5]; **Sugar beet** [1-6]; **Winter oilseed rape** [1-5];

- Wild oats in **Angelica** *(off-label)* [2]; **Balm** *(off-label)* [1-2,5]; **Blackberries** *(off-label)* [1-2,5]; **Blackcurrants** *(off-label)* [1-2,5]; **Blueberries** *(off-label)* [1-2,5]; **Borage** *(off-label)* [1-2,5]; **Broad beans** *(off-label)* [2,5]; **Brussels sprouts** *(off-label)* [2,5]; **Bulb onions** *(off-label)* [2,5]; **Cabbages** *(off-label)* [2,5]; **Caraway** *(off-label)* [2]; **Carrots** *(off-label)* [2,5]; **Celeriac** *(off-label)* [1-2,5]; **Celery leaves** *(off-label)* [1-2,5]; **Chives** *(off-label)* [1-2,5]; **Coriander** *(off-label)* [2]; **Corn Gromwell** *(off-label)* [1-2,5]; **Crambe** *(off-label)* [1-2,5]; **Cranberries** *(off-label)* [1-2,5]; **Dill** *(off-label)* [1-2,5]; **Dwarf beans** *(off-label)* [2,5]; **Edible flowers** *(off-label)* [2]; **Fennel leaves** *(off-label)* [1-2,5]; **Fodder beet** *(off-label)* [1-2,5]; **French beans** *(off-label)* [2,5]; **Garlic** *(off-label)* [2,5]; **Gold-of-pleasure** *(off-label)* [1-2,5]; **Gooseberries** *(off-label)* [1-2,5]; **Horseradish** *(off-label)* [1-2,5]; **Leeks** *(off-label)* [5]; **Linseed** *(off-label)* [1-2,5]; **Lovage** *(off-label)* [2]; **Lucerne** *(off-label)* [5]; **Meadowfoam (Limnanthes alba)** *(off-label)* [2]; **Mint** *(off-label)* [1-2,5]; **Mustard** *(off-label)* [1-2,5]; **Nursery fruit trees** *(off-label)* [2]; **Ornamental plant production** *(off-label)* [2]; **Parsley** *(off-label)* [1-2,5]; **Parsley root** *(off-label)* [1-2,5]; **Parsnips** *(off-label)* [1-2,5]; **Poppies** *(off-label)* [1-2,5]; **Protected nursery fruit trees** *(off-label)* [2]; **Raspberries** *(off-label)* [1-2,5]; **Red beet** *(off-label)* [1-2,5]; **Redcurrants** *(off-label)* [1-2,5]; **Runner beans** *(off-label)* [2,5]; **Salad burnet** *(off-label)* [2]; **Salad onions** *(off-label)* [5]; **Salsify** *(off-label)* [1-2,5]; **Shallots** *(off-label)* [2,5]; **Strawberries** *(off-label)* [1-2,5]; **Sunflowers** *(off-label)* [1-2,5]; **Swedes** *(off-label)* [1-2,5]; **Sweet ciceley** *(off-label)* [2]; **Turnips** *(off-label)* [1-2,5]; **Valerian root** *(off-label)* [1-2,5]; **Vining peas** *(off-label)* [2,5];

Extension of Authorisation for Minor Use (EAMUs)

- **Angelica** *20211032* [2]
- **Balm** *20192137* [1], *20193642* [2], *20192127* [5]
- **Blackberries** *20192136* [1], *20192128* [5]
- **Blackcurrants** *20192136* [1], *20192128* [5]
- **Blueberries** *20192136* [1], *20192128* [5]
- **Borage** *20193402* [1], *20193183* [2], *20193403* [5]
- **Broad beans** *20192141* [1], *20170460* [2], *20192134* [5]
- **Brussels sprouts** *20192141* [1], *20170460* [2], *20192134* [5]
- **Bulb onions** *20192141* [1], *20170460* [2], *20192134* [5]
- **Cabbages** *20192141* [1], *20170460* [2], *20192134* [5]
- **Caraway** *20211032* [2]
- **Carrots** *20192141* [1], *20170460* [2], *20192134* [5]
- **Celeriac** *20192138* [1], *20193641* [2], *20192126* [5]
- **Celery leaves** *20192137* [1], *20193642* [2], *20192127* [5]
- **Chives** *20192137* [1], *20193642* [2], *20192127* [5]
- **Coriander** *20211032* [2]
- **Corn Gromwell** *20193402* [1], *20193183* [2], *20192132* [5], *20193403* [5]
- **Crambe** *20192143* [1], *20170461* [2], *20192130* [5]
- **Cranberries** *20192136* [1], *20192128* [5]
- **Dill** *20192137* [1], *20193642* [2], *20192127* [5]
- **Dwarf beans** *20192141* [1], *20170460* [2], *20192134* [5]
- **Edible flowers** *20211032* [2]
- **Edible podded peas** *20192142* [1], *20193633* [2], *20192131* [5]
- **Fennel leaves** *20192137* [1], *20193642* [2], *20192127* [5]
- **Fodder beet** *20192138* [1], *20193641* [2], *20192126* [5]
- **French beans** *20192141* [1], *20170460* [2], *20192134* [5]
- **Garlic** *20192141* [1], *20170460* [2], *20192134* [5]
- **Gold-of-pleasure** *20192135* [1], *20193632* [2], *20192125* [5]
- **Gooseberries** *20192136* [1], *20192128* [5]
- **Horseradish** *20192138* [1], *20193641* [2], *20192126* [5]
- **Leeks** *20192140* [1], *20190792* [2], *20192133* [5]

- **Linseed** *20192144* [1], *20170458* [2], *20192129* [5]
- **Lovage** *20211032* [2]
- **Lucerne** *20192126* [5]
- **Meadowfoam (Limnanthes alba)** *20201990* [2]
- **Mint** *20192137* [1], *20193642* [2], *20192127* [5]
- **Mustard** *20192135* [1], *20193632* [2], *20192125* [5]
- **Nursery fruit trees** *20222336* [2]
- **Ornamental plant production** *20222336* [2]
- **Parsley** *20192137* [1], *20193642* [2], *20192127* [5]
- **Parsley root** *20192138* [1], *20193641* [2], *20192126* [5]
- **Parsnips** *20192138* [1], *20193641* [2], *20192126* [5]
- **Poppies** *20192135* [1], *20193632* [2], *20192125* [5]
- **Protected nursery fruit trees** *20222336* [2]
- **Raspberries** *20192136* [1], *20192128* [5]
- **Red beet** *20192138* [1], *20193641* [2], *20192126* [5]
- **Redcurrants** *20192136* [1], *20192128* [5]
- **Runner beans** *20192141* [1], *20170460* [2], *20192134* [5]
- **Salad burnet** *20211032* [2]
- **Salad onions** *20192140* [1], *20190792* [2], *20192133* [5]
- **Salsify** *20192138* [1], *20193641* [2], *20192126* [5]
- **Shallots** *20192141* [1], *20170460* [2], *20192134* [5]
- **Strawberries** *20192136* [1], *20192137* [1], *20193642* [2], *20192127* [5], *20192128* [5]
- **Sunflowers** *20192135* [1], *20193632* [2], *20192125* [5]
- **Swedes** *20192138* [1], *20193641* [2], *20192126* [5]
- **Sweet ciceley** *20211032* [2]
- **Turnips** *20192138* [1], *20193641* [2], *20192126* [5]
- **Valerian root** *20192137* [1], *20193642* [2], *20192127* [5]
- **Vining peas** *20192141* [1], *20170460* [2], *20192134* [5]

Approval information
- Clethodim included in Annex I under EC Regulation 1107/2009
- Products [3-4] due to expire 9/11/2023

Efficacy guidance
- Optimum efficacy is achieved from applications made when grasses are actively growing and have 3 lvs up to beginning of tillering
- Avoid late applications to winter oilseed rape in mixture with other products or adjuvants due to risk of damage to terminal shoot
- Use of an acidifying water conditioner can improve grass weed activity

Restrictions
- To avoid the build-up of resistance, do not apply products containing an ACCase inhibitor more than twice to any crop.
- Label requires a 14 day interval before and after application before other products are applied, particularly in oilseed rape.
- Do not apply in tank mix with other products.
- Do not apply to winter oilseed rape after end of October in year of sowing

Crop-specific information
- Consult processors before use on crops grown for processing.
- Do not apply to crops suffering from stress for any reason.

Following crops guidance
- Broad-leaved crops can be sown at any time after application but when planting cereals, maize or grasses, it is advisable to wait at least 4 weeks after application and to cultivate to at least 20 cm deep before sowing.

Hazard classification and safety precautions
    Hazard Harmful, Dangerous for the environment
    Transport code 9 [1-6]
    Packaging group III

SEE SECTION 3 FOR PRODUCTS ALSO REGISTERED

UN Number 3082
Risk phrases H304, H317 [6], H319 [6], H336
Operator protection U02a, U04a, U05a, U08, U19a, U20a; A, C, H
Environmental protection E15b, E38, H411 [1-5], H413 [6]
Storage and disposal D01, D02, D05, D09a, D10b, D12a
Medical advice M05b

78   clodinafop-propargyl

A foliar acting herbicide for annual grass weed control in wheat, triticale and rye
HRAC mode of action code: 1 (A)

Products

1	Kipota	Life Scientific	240 g/l	EC	17993
2	Ravena	FMC Agro	240 g/l	EC	18782
3	Sword	Nufarm UK	240 g/l	EC	18592
4	Topik	Adama	240 g/l	EC	18568
5	Viscount	Adama	240 g/l	EC	18567

Uses
- Blackgrass in *Durum wheat* [1-5]; *Grass seed crops* (off-label) [4]; *Spring rye* [1-5]; *Spring wheat* [1-5]; *Triticale* [1-5]; *Winter rye* [1-5]; *Winter wheat* [1-5].
- Rough-stalked meadow grass in *Durum wheat* [1-5]; *Grass seed crops* (off-label) [4]; *Spring rye* [1-5]; *Spring wheat* [1-5]; *Triticale* [1-5]; *Winter rye* [1-5]; *Winter wheat* [1-5];
- Wild oats in *Durum wheat* [1-5]; *Grass seed crops* (off-label) [4]; *Spring rye* [1-5]; *Spring wheat* [1-5]; *Triticale* [1-5]; *Winter rye* [1-5]; *Winter wheat* [1-5];

Extension of Authorisation for Minor Use (EAMUs)
- *Grass seed crops* 20201238 [4]

Approval information
- Clodinafop-propargyl included in Annex I under EC Regulation 1107/2009

Efficacy guidance
- Spray when majority of weeds have germinated but before competition reduces yield
- Products contain a herbicide safener (cloquintocet-mexyl) that improves crop tolerance to clodinafop-propargyl
- Optimum control achieved when all grass weeds emerged. Wait for delayed germination on dry or cloddy seedbed
- A mineral oil additive is recommended to give more consistent control of very high blackgrass populations or for late season treatments. See label for details
- Weed control not affected by soil type, organic matter or straw residues
- Control may be reduced if rain falls within 1 h of treatment
- Clodinafop-propargyl is an ACCase inhibitor herbicide. To avoid the build up of resistance do not apply products containing an ACCase inhibitor herbicide more than twice to any crop. In addition do not use any product containing clodinafop-propargyl in mixture or sequence with any other product containing the same ingredient
- Use these products as part of a resistance management strategy that includes cultural methods of control and does not use ACCase inhibitors as the sole chemical method of grass weed control
- Applying a second product containing an ACCase inhibitor to a crop will increase the risk of resistance development; only use a second ACCase inhibitor to control different weeds at a different timing
- Always follow WRAG guidelines for preventing and managing herbicide resistant weeds. See Section 5 for more information

Restrictions
- Maximum number of treatments 1 per crop
- Do not use on barley or oats
- Do not treat crops under stress or suffering from waterlogging, pest attack, disease or frost

FOR FULL CONDITIONS OF USE ALWAYS READ THE PRODUCT LABEL

- Do not treat crops undersown with grass mixtures
- Do not mix with products containing MCPA, mecoprop-P, 2,4-D or 2,4-DB
- MCPA, mecoprop, 2,4-D or 2,4-DB should not be applied within 21 d before, or 7 d after, treatment

Crop-specific information
- Latest use: before second node detectable stage (GS 32) for durum wheat, triticale, rye; before flag leaf sheath extending (GS 41) for wheat
- Spray in autumn, winter or spring from 1 true leaf stage (GS 11) to before second node detectable (GS 32) on durum wheat, rye, triticale; before flag leaf sheath extends (GS 41) on wheat

Following crops guidance
- Any broad leaved crop or cereal (except oats) may be sown after failure of a treated crop provided that at least 3 wk have elapsed between application and drilling a cereal
- After normal harvest of a treated crop any broad leaved crop or wheat, durum wheat, rye, triticale or barley should be sown. Oats and grass should not be sown until the following spring

Environmental safety
- Dangerous for the environment
- Very toxic to aquatic organisms

Hazard classification and safety precautions
Hazard Dangerous for the environment, Harmful if swallowed [3], Very toxic to aquatic organisms [3]
Transport code 9 [1-5]
Packaging group III
UN Number 3082
Risk phrases H304, H373
Operator protection U02a, U05a, U09a, U14 [2-3], U20a; A, C, H, K
Environmental protection E15a, E38, H410
Storage and disposal D01, D02, D05, D09a, D10c, D12a

79   clodinafop-propargyl + cloquintocet-mexyl

A grass herbicide for use in winter wheat and durum wheat
HRAC mode of action code: 1 (A)

Products
| 1 | Buguis | UPL Europe | 100:25 g/l | EC | 17151 |

Uses
- Annual grasses in **Durum wheat** [1]; **Winter wheat** [1];
- Blackgrass in **Durum wheat** [1]; **Winter wheat** [1];

Approval information
- Clodinafop-propargyl included in Annex I under EC Regulation 1107/2009 but cloquintocet-mexyl not yet listed

Efficacy guidance
- Reduced doses require addition of an adjuvant - see label

Restrictions
- Do not use on barley or oats.
- Do not spray crops under stress or crops suffering from waterlogging, pest attack, disease or frost
- Do not spray crops undersown with grass mixtures
- Rain within one hour after application may reduce grass weed control
- Avoid the use of hormone-containing herbicides in mixture or sequence

SEE SECTION 3 FOR PRODUCTS ALSO REGISTERED

Following crops guidance
- Activity is not affected by soil type, organic matter or straw residues. In the event of crop failure, any broad-leaved crop may be sown or after an interval of 3 weeks any cereal may be sown. Following normal harvest of a treated crop, any broad-leaved or cereal crop maybe sown.

Hazard classification and safety precautions
    Transport code 9 [1]
    Packaging group III
    UN Number 3082
    Risk phrases H304, H317, H319, H373
    Operator protection A, H
    Environmental protection H411

80   clofentezine

A selective ovicidal tetrazine acaricide for use in top fruit
IRAC mode of action code: 10A

Products

1 Apollo 50 SC	Adama	500 g/l	SC	17187
2 Clayton Sputnik	Clayton	500 g/l	SC	17344

Uses
- Conifer spinning mite in *Forest nurseries* (off-label) [1]; *Ornamental plant production* (off-label) [1];
- Red spider mites in *Apples* [1-2]; *Hops in propagation* (off-label) [1]; *Pears* [1-2];
- Rust mite in *Forest nurseries* (off-label) [1]; *Ornamental plant production* (off-label) [1];
- Spider mites in *Blackberries* (off-label) [1]; *Forest nurseries* (off-label) [1]; *Ornamental plant production* (off-label) [1]; *Protected strawberries* (off-label) [1]; *Raspberries* (off-label) [1]; *Strawberries* (off-label) [1];
- Two-spotted spider mite in *Hops in propagation* (off-label) [1];

Extension of Authorisation for Minor Use (EAMUs)
- *Blackberries* 20180620 [1]
- *Forest nurseries* 20162082 [1]
- *Hops in propagation* 20193790 [1]
- *Ornamental plant production* 20162082 [1]
- *Protected strawberries* 20180620 [1]
- *Raspberries* 20180620 [1]
- *Strawberries* 20180620 [1]

Approval information
- Clofentezine included in Annex I under EC Regulation 1107/2009

Efficacy guidance
- Acts on eggs and early motile stages of mites. For effective control total cover of plants is essential, particular care being needed to cover undersides of leaves

Restrictions
- Maximum number of treatments 1 per yr for apples, pears, cherries, plums
- The maximum spray concentration must not exceed 0.4 litres of product in 800 litres of water

Crop-specific information
- HI apples, pears 28 d; cherries, plums 8 wk
- For red spider mite control spray apples and pears between bud burst and pink bud, plums and cherries between white bud and first flower. Rust mite is also suppressed
- On established infestations apply in conjunction with an adult acaricide

Environmental safety
- Harmful to aquatic organisms
- Product safe on predatory mites, bees and other predatory insects

Hazard classification and safety precautions
UN Number N/C
Operator protection U05a, U08, U20b
Environmental protection E15a, H412
Storage and disposal D01, D02, D05, D09a, D11a

81   clomazone

An isoxazolidinone residual herbicide for oilseed rape, field beans, combining and vining peas
HRAC mode of action code: 13 (F4)

Products

1 Argosy	Sipcam	360 g/l	CS	19406
2 Beaufort	Sipcam	360 g/l	CS	19704
3 Blanco	Adama	360 g/l	CS	16704
4 Centium 360 CS	FMC Agro	360 g/l	CS	18719
5 Cirrus CS	FMC Agro	360 g/l	CS	18721
6 Cleancrop Chicane	FMC Agro	360 g/l	CS	18720
7 Cleancrop Covert	FMC Agro	360 g/l	CS	18722
8 Clomate	Albaugh UK	360 g/l	CS	15565
9 Gamit 36 CS	FMC Agro	360 g/l	CS	18718
10 Mohawk CS	Sipcam	360 g/l	CS	18505
11 Sirtaki CS	Sipcam	360 g/l	CS	18032
12 Upstage	UPL Europe	360 g/l	CS	18984

Uses

- Annual dicotyledons in *Asparagus* (off-label) [7]; *Baby leaf crops* (off-label) [7,9]; *Beans without pods (Fresh)* (off-label) [9]; *Borage* (off-label) [7,9]; *Broad beans* (off-label) [7,9]; *Broccoli* (off-label) [7,9]; *Brussels sprouts* (off-label) [7,9]; *Cabbages* (off-label) [7,9]; *Calabrese* (off-label) [7,9]; *Carrots* [8]; *Cauliflowers* (off-label) [7,9]; *Celeriac* (off-label) [7,9]; *Celery (outdoor)* (off-label) [7,9]; *Choi sum* (off-label) [7,9]; *Collards* (off-label) [7,9]; *Combining peas* [3,8]; *Courgettes* (off-label) [7]; *Edible podded peas* (off-label) [9]; *Fennel* (off-label) [7,9]; *Forest nurseries* (off-label) [4-5,7,9]; *French beans* (off-label) [7,9]; *Game cover* (off-label) [4-5,7,9]; *Herbs (see appendix 6)* (off-label) [7,9]; *Hops* (off-label) [4-5,7,9]; *Kale* (off-label) [7,9]; *Lupins* (off-label) [7,9]; *Oriental cabbage* (off-label) [7,9]; *Ornamental plant production* (off-label) [4-5,7,9]; *Poppies for morphine production* (off-label) [7,9]; *Potatoes* [3,8,12]; *Protected courgettes* (off-label) [7]; *Protected summer squash* (off-label) [7]; *Rhubarb* (off-label) [7,9]; *Runner beans* (off-label) [7,9]; *Soya beans* (off-label) [7,9]; *Spinach* (off-label) [7,9]; *Spring cabbage* (off-label) [7,9]; *Spring field beans* [3,8]; *Spring oilseed rape* [3]; *Summer squash* (off-label) [7]; *Swedes* (off-label) [7,9]; *Sweet potato* (off-label) [7,9]; *Vining peas* [3,8]; *Winter field beans* [3,8]; *Winter oilseed rape* [3,8,12];
- Chickweed in *Carrots* (off-label) [1-2,7,9-11]; *Combining peas* [1-2,4-6,10-11]; *Courgettes* (off-label) [9]; *Cucumbers* (off-label) [9]; *Gherkins* (off-label) [9]; *Melons* (off-label) [9]; *Okra* (off-label) [9]; *Potatoes* [1-2,7,9-11]; *Pumpkins* (off-label) [9]; *Spring field beans* [1-2,4-6,10-11]; *Spring oilseed rape* [1-2,4-6,10-11]; *Summer squash* (off-label) [9]; *Vining peas* [1-2,4-6,10-11]; *Watermelons* (off-label) [9]; *Winter field beans* [1-2,4-6,10-11]; *Winter oilseed rape* [1-2,4-6,10-11]; *Winter squash* (off-label) [9];
- Cleavers in *Carrots* (off-label) [1-2,7-11]; *Combining peas* [1-6,8,10-11]; *Courgettes* (off-label) [9]; *Cucumbers* (off-label) [9]; *Gherkins* (off-label) [9]; *Melons* (off-label) [9]; *Okra* (off-label) [9]; *Potatoes* [1-3,7-12]; *Pumpkins* (off-label) [9]; *Spring field beans* [1-6,8,10-11]; *Spring oilseed rape* [1-6,10-11]; *Summer squash* (off-label) [9]; *Vining peas* [1-6,8,10-11]; *Watermelons* (off-label) [9]; *Winter field beans* [1-6,8,10-11]; *Winter oilseed rape* [1-6,8,10-12]; *Winter squash* (off-label) [9];
- Fool's parsley in *Carrots* (off-label) [1-2,7,9-11]; *Combining peas* [1-2,4-6,10-11]; *Courgettes* (off-label) [9]; *Cucumbers* (off-label) [9]; *Gherkins* (off-label) [9]; *Melons* (off-label) [9]; *Okra* (off-label) [9]; *Potatoes* [1-2,7,9-11]; *Pumpkins* (off-label) [9]; *Spring field beans* [1-2,4-6,10-11]; *Spring oilseed rape* [1-2,4-6,10-11]; *Summer squash* (off-label) [9]; *Vining peas* [1-2,4-6,10-11]; *Watermelons* (off-label) [9]; *Winter squash* (off-label) [9];
- Groundsel in *Carrots* (off-label) [9];

SEE SECTION 3 FOR PRODUCTS ALSO REGISTERED

- Ivy-leaved speedwell in **Carrots** *(off-label)* [7,9]; **Courgettes** *(off-label)* [9]; **Cucumbers** *(off-label)* [9]; **Gherkins** *(off-label)* [9]; **Melons** *(off-label)* [9]; **Okra** *(off-label)* [9]; **Potatoes** [7]; **Pumpkins** *(off-label)* [9]; **Summer squash** *(off-label)* [9]; **Watermelons** *(off-label)* [9]; **Winter squash** *(off-label)* [9];
- Red dead-nettle in **Carrots** *(off-label)* [1-2,7,9-11]; **Combining peas** [1-2,4-6,10-11]; **Courgettes** *(off-label)* [9]; **Cucumbers** *(off-label)* [9]; **Gherkins** *(off-label)* [9]; **Melons** *(off-label)* [9]; **Okra** *(off-label)* [9]; **Potatoes** [1-2,7,9-11]; **Pumpkins** *(off-label)* [9]; **Spring field beans** [1-2,4-6,10-11]; **Spring oilseed rape** [1-2,4-6,10-11]; **Summer squash** *(off-label)* [9]; **Vining peas** [1-2,4-6,10-11]; **Watermelons** *(off-label)* [9]; **Winter field beans** [1-2,4-6,10-11]; **Winter oilseed rape** [1-2,4-6,10-11]; **Winter squash** *(off-label)* [9];
- Redshank in **Carrots** *(off-label)* [9];
- Runch in **Carrots** *(off-label)* [9];
- Shepherd's purse in **Carrots** [1-2,7,9-11]; **Combining peas** [1-2,4-6,10-11]; **Courgettes** *(off-label)* [9]; **Cucumbers** *(off-label)* [9]; **Gherkins** *(off-label)* [9]; **Melons** *(off-label)* [9]; **Okra** *(off-label)* [9]; **Potatoes** [1-2,7,9-11]; **Pumpkins** *(off-label)* [9]; **Spring field beans** [1-2,4-6,10-11]; **Spring oilseed rape** [1-2,4-6,10-11]; **Summer squash** *(off-label)* [9]; **Vining peas** [1-2,4-6,10-11]; **Watermelons** *(off-label)* [9]; **Winter field beans** [1-2,4-6,10-11]; **Winter oilseed rape** [1-2,4-6,10-11]; **Winter squash** *(off-label)* [9];
- Speedwells in **Carrots** [8]; **Combining peas** [3,8]; **Potatoes** [3,8,12]; **Spring field beans** [3,8]; **Spring oilseed rape** [3]; **Vining peas** [3,8]; **Winter field beans** [3,8]; **Winter oilseed rape** [3,8,12];
- Spring-germinating perennial weeds in **Asparagus** *(off-label)* [9];

Extension of Authorisation for Minor Use (EAMUs)
- **Asparagus** *20190601* [7], *20190802* [9]
- **Baby leaf crops** *20190613* [7], *20190778* [9]
- **Beans without pods (Fresh)** *20193047* [9]
- **Borage** *20190603* [7], *20190794* [9]
- **Broad beans** *20190606* [7], *20190776* [9]
- **Broccoli** *20190604* [7], *20190799* [9]
- **Brussels sprouts** *20190604* [7], *20190799* [9]
- **Cabbages** *20190604* [7], *20190799* [9]
- **Calabrese** *20190604* [7], *20190799* [9]
- **Carrots** *20210285* [9]
- **Cauliflowers** *20190604* [7], *20190799* [9]
- **Celeriac** *20190602* [7], *20190803* [9]
- **Celery (outdoor)** *20190611* [7], *20190772* [9]
- **Choi sum** *20190604* [7], *20190799* [9]
- **Collards** *20190604* [7], *20190799* [9]
- **Courgettes** *20190599* [7], *20190793* [9]
- **Cucumbers** *20190793* [9]
- **Edible podded peas** *20193047* [9]
- **Fennel** *20190609* [7], *20190771* [9]
- **Forest nurseries** *20190743* [4], *20190744* [5], *20190612* [7], *20190774* [9]
- **French beans** *20190600* [7], *20190798* [9]
- **Game cover** *20190743* [4], *20190744* [5], *20190612* [7], *20190774* [9]
- **Gherkins** *20190793* [9]
- **Herbs (see appendix 6)** *20190610* [7], *20190777* [9]
- **Hops** *20190743* [4], *20190744* [5], *20190612* [7], *20190774* [9]
- **Kale** *20190604* [7], *20190799* [9]
- **Lupins** *20190605* [7], *20190795* [9]
- **Melons** *20190793* [9]
- **Okra** *20190793* [9]
- **Oriental cabbage** *20190604* [7], *20190799* [9]
- **Ornamental plant production** *20190743* [4], *20190744* [5], *20190612* [7], *20190774* [9]
- **Poppies for morphine production** *20190607* [7], *20190796* [9]
- **Protected courgettes** *20190599* [7]
- **Protected summer squash** *20190599* [7]
- **Pumpkins** *20190793* [9]

FOR FULL CONDITIONS OF USE ALWAYS READ THE PRODUCT LABEL

- **Rhubarb** *20190614* [7], *20190797* [9]
- **Runner beans** *20183494* [7], *20190800* [9]
- **Soya beans** *20190597* [7], *20190801* [9]
- **Spinach** *20190613* [7], *20190778* [9]
- **Spring cabbage** *20190604* [7], *20190799* [9]
- **Summer squash** *20190599* [7], *20190793* [9]
- **Swedes** *20190596* [7], *20190775* [9]
- **Sweet potato** *20190608* [7], *20190773* [9]
- **Watermelons** *20190793* [9]
- **Winter squash** *20190793* [9]

Approval information
- Clomazone included in Annex I under EC Regulation 1107/2009

Efficacy guidance
- Best results obtained from application as soon as possible after sowing crop and before emergence of crop or weeds
- Uptake is via roots and shoots. Seedbeds should be firm, level and free from clods. Loose puffy seedbeds should be consolidated before spraying
- Efficacy is reduced on organic soils, on dry cloddy seedbeds and if prolonged dry weather follows application
- Clomazone acts by inhibiting synthesis of chlorophyll pigments. Susceptible weeds emerge but are chlorotic and die shortly afterwards
- Season-long control of weeds may not be achieved
- Always follow WRAG guidelines for preventing and managing herbicide resistant weeds. See Section 5 for more information

Restrictions
- Maximum number of treatments one per crop
- Crops must be covered by a minimum of 20 mm settled soil. Do not apply to broadcast crops. Direct-drilled crops should be harrowed across the slits to cover seed before spraying
- Do not use on compacted soils or soils of poor structure that may be liable to waterlogging
- Do not use on Sands or Very Light soils or those with more than 10% organic matter
- Do not treat two consecutive crops of carrots with clomazone in one calendar yr
- Consult manufacturer or your advisor before use on potato seed crops
- Do not overlap spray swaths. Crop plants emerged at time of treatment may be severely damaged
- Application must be made using a coarse spray quality.

Crop-specific information
- Latest use: pre-emergence of crop
- Severe, but normally transient, crop damage may occur in overlaps on field beans
- Some transient crop bleaching may occur under certain climatic conditions and can be severe where heavy rain follows application. This is normally rapidly outgrown and has no effect on final crop yield.

Following crops guidance
- Following normal harvest of a spring or autumn treated crop, cereals, oilseed rape, field beans, combining peas, potatoes, maize, turnips, linseed or sugar beet may be sown
- In the event of failure of an autumn treated crop, winter cereals or winter beans may be sown in the autumn if 6 wk have elapsed since treatment. In the spring following crop failure combining peas, field beans or potatoes may be sown if 6 wk have elapsed since treatment, and spring cereals, maize, turnips, onions, carrots or linseed may be sown if 7 mth have elapsed since treatment
- In the event of a failure of a spring treated crop a wide range of crops may be sown provided intervals of 6-9 wk have elapsed since treatment. See label for details
- Prior to resowing any listed replacement crop the soil should be ploughed and cultivated to 15 cm

SEE SECTION 3 FOR PRODUCTS ALSO REGISTERED

Environmental safety
- Take extreme care to avoid drift outside the target area, or on to ponds, waterways or ditches as considerable damage may occur. Apply using a coarse quality spray
- To protect non-target plants respect an untreated buffer zone of 10 meters to non-crop land for [3, 8, 11, 12] or 5 meters for [3]

Hazard classification and safety precautions
> **Transport code** 9 [3,12]
> **Packaging group** III
> **UN Number** N/C
> **Risk phrases** H315 [12], H319 [12]
> **Operator protection** U05a [1-2,4-11], U05b [3,12], U14 [1-2,4-11], U20b [1-2,4-11]; A, C, H
> **Environmental protection** E15b [1-2,4-11], E39 [8], H410 [1-2,10-12], H412 [3], H413 [9]
> **Storage and disposal** D01, D02, D09a [1-2,4-11], D10b [1-2,4-7,9-11], D10c [8], D12a [3,12]

82   clomazone + metribuzin

A mixture of isoxazolidinone and triazinone residual herbicides for use in potatoes
HRAC mode of action code: 13 + 5 (F4 + C1)

See also diflufenican + flufenacet + metribuzin
*   metribuzin*

Products

1	Metric	Certis Belchim B V	60:233 g/l	CC	16720

Uses
- Annual dicotyledons in **Potatoes** [1];
- Annual meadow grass in **Potatoes** [1];

Approval information
- Clomazone and metribuzin included in Annex I under EC Regulation 1107/2009

Restrictions
- For use on specified varieties of potato only
- Safety to daughter tubers has not been tested; consult manufacturer before treating seed crops
- Application must be made using a coarse spray quality.

Following crops guidance
- Before drilling or planting any succeeding crop, soil MUST be mouldboard ploughed to a depth of at least 15 cm (6?) taking care to ensure that the furrow slice is inverted. Ploughing should be carried out as soon as possible (preferably within 3-4 weeks) after lifting the potato crop, but certainly no later than the end of December.
- In the same year: Provided at least 16 weeks have elapsed after the application of the recommended rate cereals and winter beans may be grown as following crops.
- In the following year: Do not grow any vegetable brassica crop (including cauliflower, calabrese, Brussels sprout and cabbage), lettuce or radish on land treated in the previous year.
- Cereals, oilseed rape, field beans, combining peas, potatoes, maize, turnip, linseed and sugar beet may be sown from spring onwards in the year following use.

Environmental safety
- Buffer zone requirement 10 m [1]
- To protect non-target plants respect an untreated buffer zone of 5 meters to non-crop land
- LERAP Category B [1]

Hazard classification and safety precautions
> **Hazard** Dangerous for the environment, Very toxic to aquatic organisms
> **Transport code** 9 [1]
> **Packaging group** III
> **UN Number** 3082
> **Operator protection** U05a, U19a, U20b; A, H
> **Environmental protection** E15b, E16a, E16b, E34, E38, H410

FOR FULL CONDITIONS OF USE ALWAYS READ THE PRODUCT LABEL

Storage and disposal D01, D02, D09a, D10b, D12a
Medical advice M05a

83   clomazone + pendimethalin

A residual herbicide mixture for weed control in combining peas and field beans
HRAC mode of action code: 13 + 3 (F4 + K1)

Products

1	Stallion Sync Tec	FMC Agro	30:333 g/l	CS	18714

Uses

- Annual dicotyledons in *Carrots* [1]; *Combining peas* [1]; *Potatoes* [1]; *Spring field beans* [1]; *Vining peas* [1]; *Winter field beans* [1];
- Cleavers in *Carrots* [1]; *Combining peas* [1]; *Potatoes* [1]; *Spring field beans* [1]; *Vining peas* [1]; *Winter field beans* [1];

Approval information

- Clomazone and pendimethalin included in Annex I under EC Regulation 1107/2009

Restrictions

- Application must be made using a coarse spray quality.

Following crops guidance

- In the event of crop failure plough the soil to at least 25cm and see label for required elapsed time before planting. Note that sugar beet should not be sown until 12 months after treatment.

Environmental safety

- LERAP Category B [1]

Hazard classification and safety precautions

Hazard Dangerous for the environment, Very toxic to aquatic organisms
Transport code 9 [1]
Packaging group III
UN Number 3082
Risk phrases H317
Operator protection U02a, U05a, U08, U13, U14, U19a, U20b; A
Environmental protection E15b, E16a, E38, H410
Storage and disposal D01, D02, D05, D09a, D10b, D12a

84   clopyralid

A foliar translocated picolinic herbicide for a wide range of crops
HRAC mode of action code: 4 (O)

Products

1	Clayton Cocoon (GB only)	Clayton	400 g/l	SL	20058
2	Cliophar 400	Arysta	400 g/l	SL	15008
3	Dow Shield 400	Corteva	400 g/l	SL	19874
4	Vivendi 200	UPL Europe	200 g/l	SL	16966

Uses

- Annual Compositae weeds in *Chard* (off-label) [3]; *Farm forestry* (off-label) [3]; *Miscanthus* (off-label) [3]; *Spinach* (off-label) [3]; *Spinach beet* (off-label) [3];
- Annual dicotyledons in *Broccoli* [3]; *Brussels sprouts* [1-3]; *Bulb onions* [3]; *Cabbages* [3]; *Calabrese* [3]; *Cauliflowers* [3]; *Fodder beet* [1-4]; *Forage maize* [1-3]; *Grassland* [1-4]; *Linseed* [1-3]; *Mangels* [1-4]; *Ornamental plant production* [1-4]; *Red beet* [1-4]; *Spring barley* [1-4]; *Spring oats* [1-4]; *Spring oilseed rape* [1-4]; *Spring wheat* [1-4]; *Sugar beet* [1-4]; *Swedes* [1-4]; *Turnips* [1-4]; *Winter barley* [1-4]; *Winter oats* [1-4]; *Winter oilseed rape* [1-4]; *Winter wheat* [1-4];

SEE SECTION 3 FOR PRODUCTS ALSO REGISTERED

- Compositae weeds in **All edible seed crops grown outdoors** *(off-label)* [3]; **All non-edible seed crops grown outdoors** *(off-label)* [3]; **Apples** *(off-label)* [3]; **Forest nurseries** *(off-label)* [3]; **Game cover** *(off-label)* [3]; **Hemp grown for fibre production** *(off-label)* [3]; **Ornamental plant production** *(off-label)* [3]; **Pears** *(off-label)* [3]; **Strawberries** *(off-label)* [3];
- Corn marigold in **Broccoli** [3]; **Brussels sprouts** [1-3]; **Bulb onions** [3]; **Cabbages** [3]; **Calabrese** [3]; **Cauliflowers** [3]; **Fodder beet** [1-3]; **Forage maize** [1-3]; **Grassland** [1-3]; **Linseed** [1-3]; **Mangels** [1-3]; **Ornamental plant production** [1-3]; **Red beet** [1-3]; **Spring barley** [1-3]; **Spring oats** [1-3]; **Spring oilseed rape** [1-3]; **Spring wheat** [1-3]; **Sugar beet** [1-3]; **Swedes** [1-3]; **Turnips** [1-3]; **Winter barley** [1-3]; **Winter oats** [1-3]; **Winter oilseed rape** [1-3]; **Winter wheat** [1-3];
- Creeping thistle in **Apples** *(off-label)* [3]; **Broccoli** [3]; **Brussels sprouts** [1-3]; **Bulb onions** [3]; **Cabbages** [3]; **Calabrese** [3]; **Cauliflowers** [3]; **Fodder beet** [1-3]; **Forage maize** [1-3]; **Grassland** [1-3]; **Linseed** [1-3]; **Mangels** [1-3]; **Ornamental plant production** [1-3]; **Pears** *(off-label)* [3]; **Red beet** [1-3]; **Spring barley** [1-3]; **Spring oats** [1-3]; **Spring oilseed rape** [1-3]; **Spring wheat** [1-3]; **Sugar beet** [1-3]; **Swedes** [1-3]; **Turnips** [1-3]; **Winter barley** [1-3]; **Winter oats** [1-3]; **Winter oilseed rape** [1-3]; **Winter wheat** [1-3];
- Groundsel in **Chard** *(off-label)* [3]; **Forest nurseries** *(off-label)* [4]; **Game cover** *(off-label)* [2,4]; **Spinach** *(off-label)* [3]; **Spinach beet** *(off-label)* [3]; **Strawberries** *(off-label)* [3];
- Mayweeds in **Broccoli** [3]; **Brussels sprouts** [1-3]; **Bulb onions** [3]; **Cabbages** [3]; **Calabrese** [3]; **Cauliflowers** [3]; **Chard** *(off-label)* [3]; **Fodder beet** [1-4]; **Forage maize** [1-3]; **Forest nurseries** *(off-label)* [4]; **Game cover** *(off-label)* [2,4]; **Grassland** [1-4]; **Leeks** *(off-label)* [3]; **Linseed** [1-3]; **Mangels** [1-4]; **Ornamental plant production** [1-4]; **Red beet** [1-4]; **Salad onions** *(off-label)* [3]; **Spinach** *(off-label)* [3]; **Spinach beet** *(off-label)* [3]; **Spring barley** [1-4]; **Spring oats** [1-4]; **Spring oilseed rape** [1-4]; **Spring wheat** [1-4]; **Strawberries** *(off-label)* [3]; **Sugar beet** [1-4]; **Swedes** [1-4]; **Turnips** [1-4]; **Winter barley** [1-4]; **Winter oats** [1-4]; **Winter oilseed rape** [1-4]; **Winter wheat** [1-4];
- Sowthistle in **Forest nurseries** *(off-label)* [4]; **Game cover** *(off-label)* [4];
- Thistles in **Fodder beet** [4]; **Forest nurseries** *(off-label)* [3-4]; **Game cover** *(off-label)* [2,4]; **Grassland** [4]; **Leeks** *(off-label)* [3]; **Mangels** [4]; **Ornamental plant production** *(off-label)* [3-4]; **Red beet** [4]; **Salad onions** *(off-label)* [3]; **Spring barley** [4]; **Spring oats** [4]; **Spring oilseed rape** [4]; **Spring wheat** [4]; **Sugar beet** [4]; **Swedes** [4]; **Turnips** [4]; **Winter barley** [4]; **Winter oats** [4]; **Winter oilseed rape** [4]; **Winter wheat** [4];
- Volunteer potatoes in **Leeks** *(off-label)* [3]; **Salad onions** *(off-label)* [3];

Extension of Authorisation for Minor Use (EAMUs)
- **All edible seed crops grown outdoors** *20221091* [3]
- **All non-edible seed crops grown outdoors** *20221091* [3]
- **Apples** *20221090* [3]
- **Chard** *20221086* [3]
- **Farm forestry** *20221091* [3]
- **Forest nurseries** *20221089* [3], *20151529* [4]
- **Game cover** *20160786* [2], *20221091* [3], *20151529* [4]
- **Hemp grown for fibre production** *20221091* [3]
- **Leeks** *20221088* [3]
- **Miscanthus** *20221091* [3]
- **Ornamental plant production** *20221089* [3]
- **Pears** *20221090* [3]
- **Salad onions** *20221088* [3]
- **Spinach** *20221086* [3]
- **Spinach beet** *20221086* [3]
- **Strawberries** *20221087* [3]

Approval information
- Clopyralid included in Annex I under EC Regulation 1107/2009
- Accepted by BBPA for use on malting barley

Efficacy guidance
- Best results achieved by application to young actively growing weed seedlings. Treat creeping thistle at rosette stage and repeat 3-4 wk later as directed

FOR FULL CONDITIONS OF USE ALWAYS READ THE PRODUCT LABEL

- High activity on weeds of Compositae family. For most crops recommended for use in tank mixes. See label for details

Restrictions
- Maximum total dose varies between the equivalent of one and two full dose treatments, depending on the crop treated. See labels for details
- Do not apply to cereals later than the second node detectable stage (GS 32)
- Do not apply when crop damp or when rain expected within 6 h
- Do not use straw from treated cereals in compost or any other form for glasshouse crops. Straw may be used for strawing down strawberries
- Straw from treated grass seed crops or linseed should be baled and carted away. If incorporated do not plant winter beans in same year
- Do not use on onions at temperatures above 20°C or when under stress
- Do not treat maiden strawberries or runner beds or apply to early leaf growth during blossom period or within 4 wk of picking. Aug or early Sep sprays may reduce yield
- Do not use any treated plant material for composting or mulching, and do not use manure for composting from animals fed on treated crops
- Must not be used on grassland that will be cut for animal feed (i.e. fresh cut grass, silage, hay and haylage), fodder or bedding within 12 months of treatment.
- An interval of at least 12 months from the treatment date must elapse before making grass silage, hay or haylage.
- Manure from animals fed silage, hay or haylage produced from grassland treated in the previous 12 months must not leave the farm.
- Manure from animals grazed on treated grassland must not leave the farm. The product must not be used on grass land that will be grazed by horses and ponies.
- Do not use between the 31st August and 1st March
- Clopyralid is being found in ground and surface waters. Take care when applying to sloping soils near water.

Crop-specific information
- Latest use: 7 d before cutting grass for hay or silage; before 3rd node detectable (GS 33) for cereals; before flower buds visible from above for oilseed rape, linseed
- HI grassland 7 d; apples, pears, strawberries 4 wk; maize, sweetcorn, onions, Brussels sprouts, broccoli, cabbage, cauliflowers, calabrese, kale, fodder rape, oilseed rape, swedes, turnips, sugar beet, red beet, fodder beet, mangels, sage, honesty 6 wk
- Timing of application varies with weed problem, crop and other ingredients of tank mixes. See labels for details
- Apply as directed spray in woody ornamentals, avoiding leaves, buds and green stems. Do not apply in root zone of families Compositae or Leguminosae

Following crops guidance
- Do not plant susceptible autumn-sown crops in same year as treatment. Do not apply later than Jul where susceptible crops are to be planted in spring. See label for details

Environmental safety
- Harmful to aquatic organisms [1, 3]
- Wash spray equipment thoroughly with water and detergent immediately after use. Traces of product can damage susceptible plants sprayed later
- Keep livestock out of treated areas for at least 7 d and until foliage of any poisonous weeds such as ragwort has died and become unpalatable
- Some pesticides pose a greater threat of contamination of water than others and clopyralid is one of these pesticides. Take special care when applying clopyralid near water and do not apply if heavy rain is forecast

Hazard classification and safety precautions
UN Number N/C
Risk phrases R52 [2], R53a [2]
Operator protection U05a [4], U08, U19a, U20b; A, C
Environmental protection E07a [1-3], E15a, E34 [4], H410 [1,3]
Storage and disposal D01, D02 [4], D05, D09a, D10b, D12a [1-3]

SEE SECTION 3 FOR PRODUCTS ALSO REGISTERED

SECTION 2

85 clopyralid + florasulam

A translocated herbicide mixture for weed control in cereals
HRAC mode of action code: 4 + 2 (O + B)

Products

| 1 | Gartrel | Corteva | 300:25 g/l | SC | 16828 |

Uses

- Black bindweed in *Durum wheat* [1]; *Spring barley* [1]; *Spring oats* [1]; *Spring wheat* [1]; *Triticale* [1]; *Winter barley* [1]; *Winter rye* [1]; *Winter wheat* [1];
- Charlock in *Durum wheat* [1]; *Spring barley* [1]; *Spring oats* [1]; *Spring wheat* [1]; *Triticale* [1]; *Winter barley* [1]; *Winter rye* [1]; *Winter wheat* [1];
- Chickweed in *Durum wheat* [1]; *Spring barley* [1]; *Spring oats* [1]; *Spring wheat* [1]; *Triticale* [1]; *Winter barley* [1]; *Winter rye* [1]; *Winter wheat* [1];
- Cleavers in *Durum wheat* [1]; *Spring barley* [1]; *Spring oats* [1]; *Spring wheat* [1]; *Triticale* [1]; *Winter barley* [1]; *Winter rye* [1]; *Winter wheat* [1];
- Mayweeds in *Durum wheat* [1]; *Spring barley* [1]; *Spring oats* [1]; *Spring wheat* [1]; *Triticale* [1]; *Winter barley* [1]; *Winter rye* [1]; *Winter wheat* [1];
- Runch in *Durum wheat* [1]; *Spring barley* [1]; *Spring oats* [1]; *Spring wheat* [1]; *Triticale* [1]; *Winter barley* [1]; *Winter rye* [1]; *Winter wheat* [1];
- Shepherd's purse in *Durum wheat* [1]; *Spring barley* [1]; *Spring oats* [1]; *Spring wheat* [1]; *Triticale* [1]; *Winter barley* [1]; *Winter rye* [1]; *Winter wheat* [1];
- Volunteer oilseed rape in *Durum wheat* [1]; *Spring barley* [1]; *Spring oats* [1]; *Spring wheat* [1]; *Triticale* [1]; *Winter barley* [1]; *Winter rye* [1]; *Winter wheat* [1];
- Wild radish in *Durum wheat* [1]; *Spring barley* [1]; *Spring oats* [1]; *Spring wheat* [1]; *Triticale* [1]; *Winter barley* [1]; *Winter rye* [1]; *Winter wheat* [1];

Approval information

- Clopyralid and florasulam included in Annex I under EC Regulation 1107/2009

Efficacy guidance

- Best results obtained when weeds are small and actively growing
- Effectiveness may be reduced when soil is very dry
- Use adequate water volume to achieve complete spray coverage of the weeds
- Florasulam is a member of the ALS-inhibitor group of herbicides

Restrictions

- Do not roll or harrow for 7 d before or after application
- Do not use any treated plant material for composting or mulching
- Do not use manure from animals fed on treated crops for composting
- Specific restrictions apply to use in sequence or tank mixture with other sulfonylurea or ALS-inhibiting herbicides. See label for details
- The total amount of florasulam applied to a cereal crop must not exceed 7.5 g/ha
- Clopyralid is being found in ground and surface waters. Take care when applying to sloping soils near water.

Following crops guidance

- Where residues of a treated crop have not completely decayed by the time of planting a succeeding crop, avoid planting peas, beans and other legumes, carrots and other Umbelliferae, potatoes, lettuce and other Compositae, glasshouse and protected crops
- Where the product has been used in mixture with certain named products (see label) only cereals or grass may be sown in the autumn following harvest. Otherwise cereals, oilseed rape, grass or vegetable brassicas as transplants may be sown as a following crop in the same calendar yr as treatment. Oilseed rape may show some temporary reduction of vigour after a dry summer, but yields are not affected
- In addition to the above, field beans, linseed, peas, sugar beet, potatoes, maize, clover (for use in grass/clover mixtures) or carrots may be sown in the calendar yr following treatment
- In the event of failure of a treated crop in spring, only spring wheat, spring barley, spring oats, maize or ryegrass may be sown

Environmental safety
- LERAP Category B [1]

Hazard classification and safety precautions
Hazard Very toxic to aquatic organisms
Transport code 9 [1]
Packaging group III
UN Number 3082
Operator protection U05a, U19a, U20a
Environmental protection E15a, E16a, E34, H410
Storage and disposal D05, D09a, D10b, D12a
Medical advice M03

86   clopyralid + florasulam + fluroxypyr

A translocated herbicide mixture for cereals
HRAC mode of action code: 4 + 2 + 4 (O + B + O)

See also florasulam
fluroxypyr

Products

1	Dakota	Corteva	80:2.5:100 g/l	EC	19179
2	Dingo	Corteva	80:2.5:100 g/l	EC	19959
3	Galaxy	Corteva	80:2.5:100 g/l	EC	18952
4	Leystar	Corteva	80:2.5:100 g/l	EC	19938
5	Mogul	Pan Amenity	80:2.5:100 g/l	EC	19357
6	Pastor Trio	Corteva	80:2.5:100 g/l	EC	19977
7	Praxys	ICL (Everris) Ltd	80:2.5;100 g/l	EC	19820

Uses
- Annual dicotyledons in **Amenity grassland** [5,7]; **Durum wheat** [1]; **Forage maize** [2,4,6]; **Grass seed crops** [2,4,6]; **Grassland** [2,4,6]; **Lawns** [5]; **Managed amenity turf** [5,7]; **Newly sown grass leys** [2,4,6]; **Spelt** [1-4,6]; **Spring barley** [1-4,6]; **Spring oats** [1-4,6]; **Spring rye** [1-4,6]; **Spring wheat** [1-4,6]; **Triticale** [1-2,4,6]; **Winter barley** [1-4,6]; **Winter oats** [1-4,6]; **Winter rye** [1-4,6]; **Winter triticale** [3]; **Winter wheat** [1-4,6];
- Black medick in **Amenity grassland** [5,7];
- Bristly oxtongue in **Amenity grassland** [5,7]; **Grass seed crops** [2,4,6]; **Grassland** [2,4,6]; **Newly sown grass leys** [2,4,6];
- Buttercups in **Amenity grassland** [5,7]; **Grass seed crops** [2,4,6]; **Grassland** [2,4,6]; **Newly sown grass leys** [2,4,6];
- Chickweed in **Durum wheat** [1]; **Forage maize** [2,4,6]; **Spelt** [1-4,6]; **Spring barley** [1-4,6]; **Spring oats** [1-4,6]; **Spring rye** [1-4,6]; **Spring wheat** [1-4,6]; **Triticale** [1-2,4,6]; **Winter barley** [1-4,6]; **Winter oats** [1-4,6]; **Winter rye** [1-4,6]; **Winter triticale** [3]; **Winter wheat** [1-4,6];
- Cleavers in **Durum wheat** [1]; **Forage maize** [2,4,6]; **Spelt** [1-4,6]; **Spring barley** [1-4,6]; **Spring oats** [1-4,6]; **Spring rye** [1-4,6]; **Spring wheat** [1-4,6]; **Triticale** [1-2,4,6]; **Winter barley** [1-4,6]; **Winter oats** [1-4,6]; **Winter rye** [1-4,6]; **Winter triticale** [3]; **Winter wheat** [1-4,6];
- Common mouse-ear in **Amenity grassland** [5,7]; **Grass seed crops** [2,4,6]; **Grassland** [2,4,6]; **Newly sown grass leys** [2,4,6];
- Creeping thistle in **Durum wheat** [1]; **Forage maize** [2,4,6]; **Spelt** [1-2,4,6]; **Spring barley** [1-2,4,6]; **Spring oats** [1-2,4,6]; **Spring rye** [1-2,4,6]; **Spring wheat** [1-2,4,6]; **Triticale** [1-2,4,6]; **Winter barley** [1-2,4,6]; **Winter oats** [1-2,4,6]; **Winter rye** [1-2,4,6]; **Winter wheat** [1-2,4,6];
- Daisies in **Amenity grassland** [5,7]; **Grass seed crops** [2,4,6]; **Grassland** [2,4,6]; **Newly sown grass leys** [2,4,6];
- Dandelions in **Amenity grassland** [5,7]; **Grass seed crops** [2,4,6]; **Grassland** [2,4,6]; **Newly sown grass leys** [2,4,6];
- Mayweeds in **Durum wheat** [1]; **Forage maize** [2,4,6]; **Spelt** [1-2,4,6]; **Spring barley** [1-2,4,6]; **Spring oats** [1-2,4,6]; **Spring rye** [1-2,4,6]; **Spring wheat** [1-2,4,6]; **Triticale** [1-2,4,6]; **Winter barley** [1-2,4,6]; **Winter oats** [1-2,4,6]; **Winter rye** [1-2,4,6]; **Winter wheat** [1-2,4,6];

SEE SECTION 3 FOR PRODUCTS ALSO REGISTERED

- Plantains in **Amenity grassland** [5,7]; **Grass seed crops** [2,4,6]; **Grassland** [2,4,6]; **Newly sown grass leys** [2,4,6];
- Self-heal in **Amenity grassland** [5,7];
- Slender speedwell in **Amenity grassland** [5,7];

Approval information
- Clopyralid, florasulam and fluroxypyr are all included in Annex I under EC Regulation 1107/2009

Efficacy guidance
- Best results obtained when weeds are small and actively growing
- Effectiveness may be reduced when soil is very dry
- Use adequate water volume to achieve complete spray coverage of the weeds
- Florasulam is a member of the ALS-inhibitor group of herbicides

Restrictions
- Maximum number of treatments 1 per yr for all crops
- Do not spray when crops are under stress from any cause
- Do not apply through CDA applicators
- Do not roll or harrow for 7 d before or after application
- Do not use any treated plant material for composting or mulching
- Do not use manure from animals fed on treated crops for composting
- Specific restrictions apply to use in sequence or tank mixture with other sulfonylurea or ALS-inhibiting herbicides. See label for details
- Applications to cereals must only be made between 1st February and 30th June
- Application to any grass crop must not be made before 1st February [2, 4]
- To protect surface water, application to maize must not take place before 10th April [2, 4]
- Extreme care must be taken to avoid spray drift onto non-crop plants outside the target area
- For autumn planted crops a maximum total dose of 3.75 g florasulam must be observed for applications made between crop emergence in the year of planting and 1st Feb in the year of harvest. The total amount of florasulam applied to a cereal crop must not exceed 7.5 g/ha.
- Do not use any treated plant material for composting or mulching, and do not use manure for composting from animals fed on treated crops
- Must not be used on grassland that will be cut for animal feed (i.e. fresh cut grass, silage, hay and haylage), fodder or bedding within 12 months of treatment.
- An interval of at least 12 months from the treatment date must elapse before making grass silage, hay or haylage.
- Manure from animals fed silage, hay or haylage produced from grassland treated in the previous 12 months must not leave the farm.
- Manure from animals grazed on treated grassland must not leave the farm. The product must not be used on grass land that will be grazed by horses and ponies.
- Clopyralid is being found in ground and surface waters. Take care when applying to sloping soils near water.

Crop-specific information
- Latest use: before second node detectable for oats; before flag leaf detectable in wheat, barley rye, spelt and triticale [1,2, 3, 4]; before 30th Sept in established grassland and grass seed crops, before 31st Aug in newly sown leys and before 30th June in forage maize [2, 4]; latest timing not specified in amenity grassland, lawn or managed amenity turf [7]

Following crops guidance
- Where residues of a treated crop have not completely decayed by the time of planting a succeeding crop, avoid planting peas, beans and other legumes, carrots and other Umbelliferae, potatoes, lettuce and other Compositae, glasshouse and protected crops
- Where the product has been used in mixture with certain named products (see label) only cereals or grass may be sown in the autumn following harvest. Otherwise cereals, oilseed rape, grass or vegetable brassicas as transplants may be sown as a following crop in the same calendar yr as treatment. Oilseed rape may show some temporary reduction of vigour after a dry summer, but yields are not affected
- In addition to the above, field beans, linseed, peas, sugar beet, potatoes, maize, clover (for use in grass/clover mixtures) or carrots may be sown in the calendar yr following treatment

FOR FULL CONDITIONS OF USE ALWAYS READ THE PRODUCT LABEL

- In the event of failure of a treated crop in spring, only spring wheat, spring barley, spring oats, maize or ryegrass may be sown
- A period of at least 125 days must be observed prior to planting of succeeding crops [2, 4]

Environmental safety
- Dangerous for the environment
- Very toxic to aquatic organisms
- Take extreme care to avoid drift outside the target area
- Some pesticides pose a greater threat of contamination of water than others and clopyralid is one of these pesticides. Take special care when applying clopyralid near water and do not apply if heavy rain is forecast
- LERAP Category B [2-7]

Hazard classification and safety precautions
> **Hazard** Harmful, Dangerous for the environment, Harmful if inhaled
> **Transport code** 9 [1-7]
> **Packaging group** III
> **UN Number** 3082
> **Risk phrases** H304, H315, H317, H319
> **Operator protection** U05a, U11, U19a, U20a; A, C, H
> **Environmental protection** E15b, E16a [2-7], E34, E38, H410
> **Storage and disposal** D01, D02, D05 [5,7], D09a, D10b, D12a
> **Medical advice** M03, M05a [1-6]

87   clopyralid + fluroxypyr + MCPA

A translocated herbicide mixture for use in sports and amenity turf
HRAC mode of action code: 4 + 4 + 4 (O + O + O)

See also fluroxypyr
MCPA

Products

1	Kinvara	Barclay	28:72:233 g/l	ME	18436

Uses
- Annual dicotyledons in **Rye** [1]; **Spring barley** [1]; **Spring oats** [1]; **Spring wheat** [1]; **Triticale** [1]; **Winter barley** [1]; **Winter oats** [1]; **Winter wheat** [1];

Approval information
- Clopyralid, fluroxypyr and MCPA included in Annex I under EC Regulation 1107/2009

Efficacy guidance
- Best results achieved when weeds actively growing and turf grass competitive
- Treatment should normally be between Apr-Sep when the soil is moist
- Do not apply during drought unless irrigation is applied
- Allow 3 d before or after mowing established turf to ensure sufficient weed leaf surface present to allow uptake and movement

Restrictions
- Do not treat grass under stress from frost, drought, waterlogging, trace element deficiency, disease or pest attack
- Do not treat if night temperatures are low, when frost is imminent or during prolonged cold weather
- To protect groundwater applications must not be made between 31st August and 1st March
- Clopyralid is being found in ground and surface waters. Take care when applying to sloping soils near water.

Crop-specific information
- Treat young turf only in spring when at least 2 mth have elapsed since sowing
- Allow 5 d after mowing young turf before treatment

SEE SECTION 3 FOR PRODUCTS ALSO REGISTERED

- Product selective on a number of turf grass species (see label) but consultation or testing recommended before treatment of any cultivar

Following crops guidance
- A period of at least 125 days must be observed prior to planting succeeding crops.

Environmental safety
- Dangerous for the environment
- Very toxic to aquatic organisms
- Wash spray equipment thoroughly with water and detergent immediately after use. Traces of product can damage susceptible plants sprayed later
- Some pesticides pose a greater threat of contamination of water than others and clopyralid is one of these pesticides. Take special care when applying clopyralid near water and do not apply if heavy rain is forecast
- LERAP Category B [1]

Hazard classification and safety precautions
Hazard Irritant, Dangerous for the environment, Very toxic to aquatic organisms
Transport code 9 [1]
Packaging group III
UN Number 3082
Risk phrases H317, H319
Operator protection U05a, U08, U14, U19a, U20b; A, C, H
Environmental protection E15a, E16a, E38, H410
Storage and disposal D01, D02, D05, D09a, D10b, D12a

88   clopyralid + halauxifen-methyl

A herbicide mixture for broad-leaved weed control in oilseed rape
HRAC mode of action code: 4 + 4 (O + O)

Products

1	Korvetto	Corteva	120:5 g/l	EC	19111

Uses
- Annual dicotyledons in *Winter oilseed rape* [1];
- Cleavers in *Winter oilseed rape* [1];
- Mayweeds in *Winter oilseed rape* [1];
- Thistles in *Winter oilseed rape* [1];

Restrictions
- Do not apply between 1st June and end of February
- Do not apply before BBCH 30
- Clopyralid is being found in ground and surface waters. Take care when applying to sloping soils near water.

Following crops guidance
- Do not plant susceptible autumn crops such as winter beans in the same year as treatment
- Residues in plant tissues (including manure and digestate) that have not completely decayed may affect succeeding susceptible crops. If treated crop remains have not fully decayed by the time of planting following crops then avoid planting: peas, beans and other legumes; carrots and other Umbelliferae; potatoes; lettuce and other Compositae; glasshouse and protected crops
- Ploughing or thorough cultivation should be undertaken prior to planting leguminous crops (e.g. field beans and peas

Environmental safety
- Aquatic buffer zone requirement 14 m.
- LERAP Category B [1]

Hazard classification and safety precautions
Hazard Very toxic to aquatic organisms

FOR FULL CONDITIONS OF USE ALWAYS READ THE PRODUCT LABEL

Transport code 9 [1]
Packaging group III
UN Number 3082
Risk phrases H319, H335
Operator protection A, C, H
Environmental protection E16a, H410
Storage and disposal D01, D02, D05

89 clopyralid + picloram

A post-emergence herbicide mixture for oilseed rape
HRAC mode of action code: 4 + 4 (O + O)

See also picloram

Products

1	Clopic	UPL Europe	267:67 g/l	SL	17387
2	Galera	Corteva	267:67 g/l	SL	16413
3	Legara	AgChem Access	267:67 g/l	SL	16789

Uses

- Annual dicotyledons in **Corn Gromwell** *(off-label)* [2]; **Crambe** *(off-label)* [2]; **Forest nurseries** *(off-label)* [2]; **Game cover** *(off-label)* [2]; **Mustard** *(off-label)* [2]; **Ornamental plant production** *(off-label)* [2]; **Spring oilseed rape** *(off-label)* [2];
- Cleavers in **Corn Gromwell** *(off-label)* [2]; **Crambe** *(off-label)* [2]; **Forest nurseries** *(off-label)* [2]; **Game cover** *(off-label)* [2]; **Mustard** *(off-label)* [2]; **Ornamental plant production** *(off-label)* [2]; **Spring oilseed rape** *(off-label)* [2]; **Winter oilseed rape** [1-3];
- Mayweeds in **Corn Gromwell** *(off-label)* [2]; **Crambe** *(off-label)* [2]; **Forest nurseries** *(off-label)* [2]; **Game cover** *(off-label)* [2]; **Mustard** *(off-label)* [2]; **Ornamental plant production** *(off-label)* [2]; **Spring oilseed rape** *(off-label)* [2]; **Winter oilseed rape** [1-3];

Extension of Authorisation for Minor Use (EAMUs)

- **Corn Gromwell** *20210952* [2]
- **Crambe** *20210953* [2]
- **Forest nurseries** *20210951* [2]
- **Game cover** *20210951* [2]
- **Mustard** *20210950* [2]
- **Ornamental plant production** *20210949* [2]
- **Spring oilseed rape** *20210954* [2]

Approval information

- Clopyralid and picloram included in Annex I under EC Regulation 1107/2009

Efficacy guidance

- Best results obtained from treatment when weeds are small and actively growing
- Cleavers that germinate after treatment will not be controlled

Restrictions

- Maximum total dose equivalent to one full dose treatment
- Do not treat crops under stress from cold, drought, pest damage, nutrient deficiency or any other cause
- Do not roll or harrow for 7 d before or after spraying
- Do not apply through CDA applicators
- Extreme care must be taken to avoid spray drift onto non-crop plants outside of the target area
- Do not use any treated plant material for composting or mulching
- Do not use manure from animals fed on treated crops for composting
- Chop and incorporate all treated plant remains in early autumn, or as soon as possible after harvest, to release any residues into the soil. Ensure that all treated plant remains have completely decayed before planting susceptible crops
- Clopyralid is being found in ground and surface waters. Take care when applying to sloping soils near water.

SEE SECTION 3 FOR PRODUCTS ALSO REGISTERED

Crop-specific information
- Latest use: before flower buds visible above crop canopy for winter oilseed rape

Following crops guidance
- Wheat, barley, oats, maize, or oilseed rape may be sown 120 days after application, all other crops should only be sown 12 months after application [2]
- Ploughing or thorough cultivation should be carried out before planting leguminous crops
- Do not attempt to plant peas, beans, other legumes, carrots, other umbelliferous crops, potatoes, lettuce, other Compositae, or any glasshouse or protected crops if treated crop remains have not fully decayed by the time of planting
- In the event of failure of an autumn treated crop only oilseed rape, wheat, barley, oats, maize or ryegrass may be sown in the spring and only after ploughing or thorough cultivation

Environmental safety
- Dangerous for the environment
- Toxic to aquatic organisms
- Take extreme care to avoid drift onto crops and non-target plants outside the target area
- Some pesticides pose a greater threat of contamination of water than others and clopyralid is one of these pesticides. Take special care when applying clopyralid near water and do not apply if heavy rain is forecast

Hazard classification and safety precautions
Hazard Dangerous for the environment
UN Number N/C
Operator protection U05a; A, C
Environmental protection E15a, E34, E38 [1-2], H411 [2], H413 [1]
Storage and disposal D01, D02, D05, D07, D09a, D12a [1-2]

90 clopyralid + triclopyr

A perennial and woody weed herbicide for use in grassland
HRAC mode of action code: 4 + 4 (O + O)

See also triclopyr

Products

1	Blaster Pro	Corteva	60:240 g/l	EC	19937
2	Grazon Pro	Corteva	60:240 g/l	EC	19875
3	Grazon Spot	Corteva	60:240 g/l	EC	19956
4	Prevail	Corteva	200:200 g/l	SL	19957
5	Thistlex	Corteva	200:200 g/l	SL	19876

Uses
- Annual and perennial weeds in *Grassland* [2-3];
- Brambles in *Amenity grassland* [1]; *Grassland* [1-3];
- Broom in *Amenity grassland* [1]; *Grassland* [1-3];
- Creeping thistle in *Grassland* [2-5]; *Rotational grass* [4];
- Docks in *Amenity grassland* [1]; *Grassland* [1-3];
- Gorse in *Amenity grassland* [1]; *Grassland* [1-3];
- Perennial dicotyledons in *Amenity grassland* [1]; *Grassland* [1];
- Stinging nettle in *Amenity grassland* [1]; *Grassland* [1];
- Thistles in *Amenity grassland* [1]; *Game cover* *(off-label)* [5]; *Grassland* [1];

Extension of Authorisation for Minor Use (EAMUs)
- *Game cover* 20220276 [5]

Approval information
- Clopyralid and triclopyr included in Annex I under EC Regulation 1107/2009

Efficacy guidance
- Must be applied to actively growing weeds

FOR FULL CONDITIONS OF USE ALWAYS READ THE PRODUCT LABEL

- Correct timing crucial for good control. Spray stinging nettle before flowering, docks in rosette stage in spring, creeping thistle before flower stems 15-20 cm high, brambles, broom and gorse in Jun-Aug
- Allow 2-3 wk regrowth after grazing or mowing before spraying perennial weeds
- Where there is a large reservoir of weed seed in the soil further treatment in the following yr may be needed
- [5] available as a twin pack with Doxstar as Pas.Tor for control of docks, nettles and thistles in grassland

Restrictions
- Maximum number of treatments 1 per yr
- Only use on permanent pasture or rotational grassland established for at least 1 yr
- Do not apply where clover is an important constituent of sward
- Do not roll or harrow within 10 d before or 7 d after spraying
- Do not cut grass for 21 d before or 28 d after spraying
- Do not use any treated plant material for composting or mulching, and do not use manure for composting from animals fed on treated crops
- Must not be used on grassland that will be cut for animal feed (i.e. fresh cut grass, silage, hay and haylage), fodder or bedding within 12 months of treatment.
- An interval of at least 12 months from the treatment date must elapse before making grass silage, hay or haylage.
- Manure from animals fed silage, hay or haylage produced from grassland treated in the previous 12 months must not leave the farm.
- Manure from animals grazed on treated grassland must not leave the farm. The product must not be used on grass land that will be grazed by horses and ponies.
- Do not apply by hand-held rotary atomiser equipment
- Do not allow drift onto other crops, amenity plantings or gardens, ponds, lakes or water courses. All conifers, especially pine and larch, are very sensitive
- Maximum concentration must not exceed 60 mls product per 10 litres of water (6 ml product per litre of water) [1, 2]
- Do not apply onto or around manure or other compost heaps [1]
- Clopyralid is being found in ground and surface waters. Take care when applying to sloping soils near water.

Crop-specific information
- Latest use: 7 d before grazing or cutting grass
- Some transient yellowing of treated swards may occur but is quickly outgrown

Following crops guidance
- Residues in plant tissues which have not completely decayed may affect succeeding susceptible crops such as peas, beans, other legumes, carrots, parsnips, potatoes, tomatoes, lettuce, glasshouse and protected crops
- Do not plant susceptible autumn-sown crops (eg winter beans) in same year as treatment and allow at least 9 mth from treatment before planting a susceptible crop in the following yr
- Do not direct drill kale, swedes, turnips, grass or grass mixtures within 6 wk of spraying
- Do not spray after end Jul where susceptible crops are to be planted in the next spring

Environmental safety
- Dangerous for the environment
- Very toxic to aquatic organisms
- Keep livestock out of treated areas for at least 7 d after spraying and until foliage of any poisonous weeds such as ragwort or buttercup has died down and become unpalatable
- Some pesticides pose a greater threat of contamination of water than others and clopyralid is one of these pesticides. Take special care when applying clopyralid near water and do not apply if heavy rain is forecast
- LERAP Category B [1-5]

Hazard classification and safety precautions
Hazard Harmful [1-3], Irritant [1,4-5], Dangerous for the environment, Flammable liquid and vapour [1-3], Very toxic to aquatic organisms [1]
Transport code 3 [1-3]

SEE SECTION 3 FOR PRODUCTS ALSO REGISTERED

Packaging group III
UN Number 1993, N/C
Risk phrases H304 [1-3], H315 [1-3], H317 [1-3], H318 [4-5], H319 [1-3], H335, H336 [1-3], H373 [1,4-5]
Operator protection U02a, U05a, U08 [1-3], U11, U14 [1-3], U15 [1,4-5], U19a [1-3], U20b, U23a [4-5], U23b [1-3]; A, C, H, M
Environmental protection E07a [1-3], E07c (7 days) [4-5], E15a [1-3], E16a, E23 [1-3], E34 [1-3], E38 [1-3], H410 [1,4-5], H411 [2-3]
Consumer protection C01 [1-3]
Storage and disposal D01, D02, D05, D09a, D10b, D12a
Medical advice M03, M05b [1]

91   copper oxychloride

A protectant copper fungicide and bactericide
FRAC mode of action code: M1

Products

1	Cuprokylt	Certis Belchim B V	87% w/w (copper)	WP	17079

Uses
* Dothistroma needle blight in *Forest nurseries* *(off-label)* [1];
* Downy mildew in *Table grapes* [1]; *Wine grapes* [1];

Extension of Authorisation for Minor Use (EAMUs)
* *Forest nurseries 20152416* [1]

Approval information
* Copper oxychloride included in Annex 1 under EC Regulation 1107/2009

Efficacy guidance
* Spray crops at high volume when foliage dry but avoid run off. Do not spray if rain expected soon
* Spray interval commonly 10-14 d but varies with crop, see label for details

Environmental safety
* Dangerous for the environment
* Very toxic to aquatic organisms
* Keep all livestock out of treated areas for at least 3 wks
* Buffer zone requirement 18 m in potatoes and ornamentals when using a horizontal sprayer, 50m when using broadcast air-assisted sprayer on crops up to 1.2m height
* LERAP Category B [1]

Hazard classification and safety precautions
Hazard Dangerous for the environment, Harmful if swallowed, Harmful if inhaled, Very toxic to aquatic organisms
Transport code 9 [1]
Packaging group III
UN Number 3077
Operator protection U20a; A, D, H
Environmental protection E06a (3 wk), E13c, E16a, E17a (50 m), E34, E38, H410, H411
Storage and disposal D01, D09a, D10b

92   cos-oga

An oligosaccharidic elicitor for powdery mildew control in protected crops
FRAC mode of action code: Unclassified

Products

1	Fytosave	Gowan	12.5 g/l	SC	18433

Uses

- Downy mildew in *Baby leaf crops* (off-label) [1]; *Herbs (see appendix 6)* (off-label) [1]; *Lettuce* (off-label) [1]; *Ornamental plant production* (off-label) [1]; *Protected baby leaf crops* (off-label) [1]; *Protected herbs (see appendix 6)* (off-label) [1]; *Protected lettuce* (off-label) [1]; *Protected ornamentals* (off-label) [1]; *Table grapes* (off-label) [1]; *Wine grapes* (off-label) [1];
- Powdery mildew in *Baby leaf crops* (off-label) [1]; *Blackberries* (off-label) [1]; *Herbs (see appendix 6)* (off-label) [1]; *Hops* (off-label) [1]; *Lettuce* (off-label) [1]; *Loganberries* (off-label) [1]; *Ornamental plant production* (off-label) [1]; *Protected aubergines* [1]; *Protected baby leaf crops* (off-label) [1]; *Protected blackberries* (off-label) [1]; *Protected chilli peppers* [1]; *Protected courgettes* [1]; *Protected cucumbers* [1]; *Protected herbs (see appendix 6)* (off-label) [1]; *Protected lettuce* (off-label) [1]; *Protected loganberries* (off-label) [1]; *Protected melons* [1]; *Protected ornamentals* (off-label) [1]; *Protected peppers* [1]; *Protected pumpkins* [1]; *Protected raspberries* (off-label) [1]; *Protected Rubus hybrids* (off-label) [1]; *Protected strawberries* (off-label) [1]; *Protected summer squash* [1]; *Protected tomatoes* [1]; *Protected winter squash* [1]; *Raspberries* (off-label) [1]; *Rubus hybrids* (off-label) [1]; *Strawberries* (off-label) [1]; *Table grapes* (off-label) [1]; *Wine grapes* (off-label) [1];

Extension of Authorisation for Minor Use (EAMUs)

- *Baby leaf crops* 20191911 [1]
- *Blackberries* 20191910 [1]
- *Herbs (see appendix 6)* 20191911 [1]
- *Hops* 20191910 [1]
- *Lettuce* 20191911 [1]
- *Loganberries* 20191910 [1]
- *Ornamental plant production* 20191911 [1]
- *Protected baby leaf crops* 20191911 [1]
- *Protected blackberries* 20191910 [1]
- *Protected herbs (see appendix 6)* 20191911 [1]
- *Protected lettuce* 20191911 [1]
- *Protected loganberries* 20191910 [1]
- *Protected ornamentals* 20191911 [1]
- *Protected raspberries* 20191910 [1]
- *Protected Rubus hybrids* 20191910 [1]
- *Protected strawberries* 20191910 [1]
- *Raspberries* 20191910 [1]
- *Rubus hybrids* 20191910 [1]
- *Strawberries* 20191910 [1]
- *Table grapes* 20191911 [1]
- *Wine grapes* 20191911 [1]

Approval information

- Cos-oga included in Annex I under EC Regulation 1107/2009

Efficacy guidance

- A minimum interval of 7 days must be observed between applications
- To minimise airborne environmental exposure, vents, doors and other openings must be closed during and after application until the applied product has fully settled

Restrictions

- Reasonable precautions must be taken to prevent access of birds, wild mammals and honey bees to treated crops

Hazard classification and safety precautions

UN Number N/C
Operator protection U05a
Storage and disposal D01, D12a

SEE SECTION 3 FOR PRODUCTS ALSO REGISTERED

93   cyantraniliprole

An ingested and contact insecticide for insect pest control in brassica crops
IRAC mode of action code: 28

Products

1	Minecto One	Syngenta	40% w/w	WG	18649
2	Verimark 20 SC	FMC Agro	200 g/l	SC	18756

Uses

- Aphids in **Protected cucumbers** *(off-label)* [2];
- Asparagus beetle in **Asparagus** *(off-label)* [1];
- Bruchid beetle in **Broad beans** *(off-label)* [1];
- Cabbage root fly in **Broccoli** [1-2]; **Brussels sprouts** [1-2]; **Cabbages** [1-2]; **Calabrese** [1]; **Cauliflowers** [1-2];
- Caterpillars in **Choi sum** *(off-label)* [1]; **Protected cucumbers** *(off-label)* [2];
- Flea beetle in **Choi sum** *(off-label)* [1];
- Insect pests in **Bulb onions** [1]; **Carrots** [1]; **Celeriac** [1]; **Edible podded peas** [1]; **Garlic** [1]; **Horseradish** [1]; **Lettuce** [1]; **Parsley root** [1]; **Parsnips** [1]; **Salad onions** [1]; **Salsify** [1]; **Shallots** [1]; **Swedes** [1]; **Turnips** [1]; **Vining peas** [1];
- Leaf miner in **Protected cucumbers** *(off-label)* [2];
- South American Tomato Moth in **Protected aubergines** *(off-label)* [2]; **Protected tomatoes** *(off-label)* [2];
- Strawberry blossom weevil in **Protected strawberries** *(off-label)* [2]; **Strawberries** *(off-label)* [2];
- Tomato moth in **Protected aubergines** *(off-label)* [2]; **Protected tomatoes** *(off-label)* [2];
- Whitefly in **Protected aubergines** *(off-label)* [2]; **Protected cucumbers** *(off-label)* [2]; **Protected tomatoes** *(off-label)* [2];

Extension of Authorisation for Minor Use (EAMUs)

- **Asparagus** *20210278* [1]
- **Broad beans** *20201664* [1]
- **Choi sum** *20194392* [1]
- **Protected aubergines** *20210959* [2]
- **Protected cucumbers** *20220094* [2]
- **Protected strawberries** *20192766* [2]
- **Protected tomatoes** *20210959* [2]
- **Strawberries** *20192766* [2]

Approval information

- Cyantraniliprole is included in Annex 1 under EC Regulation 1107/2009

Efficacy guidance

- May be used as part of an Integrated Pest Management (IPM) programme

Restrictions

- To protect bees and pollinating insects avoid application to crops when in flower.
- Consult processor when the product is to be applied to crops grown for processing
- The maximum total dose of cyantraniliprole per crop must not be exceeded in any calendar year. Any land treated with cyantraniliprole at the maximum total dose must not be treated with any other cyantraniliprole containing products in the same calendar year, including either foliar applications in the growing crop or drench treatments to transplants applied pre-planting.
- Must not be applied in each calendar year after 31 August in brassica crops

Environmental safety

- Aquatic buffer zone 5 m [1]
- LERAP Category B [1]

Hazard classification and safety precautions

Hazard Very toxic to aquatic organisms
Transport code 9 [1-2]
Packaging group III
UN Number 3077, 3082

FOR FULL CONDITIONS OF USE ALWAYS READ THE PRODUCT LABEL

Operator protection U05a, U09a, U19a, U20a; A, H
Environmental protection E12c [1], E12e, E15b, E16a [1], H410
Storage and disposal D01, D02, D09a, D12a, D22
Medical advice M03

94   cyazofamid

A cyanoimidazole sulfonamide protectant fungicide for potatoes
FRAC mode of action code: 21

Products

1	Ranman Top	Certis Belchim B V	160 g/l	SC	14753

Uses

- Blight in **Potatoes** [1];
- Downy mildew in **Aubergines** *(off-label)* [1]; **Courgettes** *(off-label)* [1]; **Cucumbers** *(off-label)* [1]; **Gherkins** *(off-label)* [1]; **Ornamental plant production** *(off-label)* [1]; **Protected ornamentals** *(off-label)* [1]; **Pumpkins** *(off-label)* [1]; **Summer squash** *(off-label)* [1]; **Tomatoes** *(off-label)* [1]; **Winter squash** *(off-label)* [1];
- Late blight in **Aubergines** *(off-label)* [1]; **Courgettes** *(off-label)* [1]; **Cucumbers** *(off-label)* [1]; **Gherkins** *(off-label)* [1]; **Pumpkins** *(off-label)* [1]; **Summer squash** *(off-label)* [1]; **Tomatoes** *(off-label)* [1]; **Winter squash** *(off-label)* [1];

Extension of Authorisation for Minor Use (EAMUs)

- **Aubergines** *20180489* [1]
- **Courgettes** *20180489* [1]
- **Cucumbers** *20180489* [1]
- **Gherkins** *20180489* [1]
- **Ornamental plant production** *20201205* [1]
- **Protected ornamentals** *20201205* [1]
- **Pumpkins** *20180489* [1]
- **Summer squash** *20180489* [1]
- **Tomatoes** *20180489* [1]
- **Winter squash** *20180489* [1]

Approval information

- Cyazofamid included in Annex I under EC Regulation 1107/2009

Efficacy guidance

- Apply as a protectant treatment before blight enters the crop and repeat every 5 -10 d depending on severity of disease pressure
- Commence spray programme immediately the risk of blight in the locality occurs, usually when the crop meets along the rows
- Product must always be used with organosilicone adjuvant provided in the twin pack
- To minimise the chance of development of resistance no more than three applications should be made consecutively (out of a permissible total of six) in the blight control programme. For more information on Resistance Management see Section 5

Restrictions

- Maximum number of treatments 6 per crop (no more than three of which should be consecutive)
- Mixed product must not be allowed to stand overnight
- Consult processor before using on crops intended for processing

Crop-specific information

- HI 7 d

Environmental safety

- Dangerous for the environment
- Very toxic to aquatic organisms
- Do not empty into drains

Hazard classification and safety precautions

Hazard Irritant, Dangerous for the environment
Transport code 9 [1]
Packaging group III
UN Number 3082
Risk phrases H315
Operator protection U02a, U05a, U11, U15, U20b; A, C
Environmental protection E15b, E19b, E38, H410
Storage and disposal D01, D02, D05, D09a, D10c, D12a, D12b
Medical advice M03

95   cycloxydim

A translocated post-emergence cyclohexanedione oxime herbicide for grass weed control
HRAC mode of action code: 1 (A)

Products

1	Laser	BASF	200 g/l	EC	17339

Uses

- Agrostis spp. in *Managed amenity turf* (off-label) [1];
- Annual grasses in *Angelica* (off-label) [1]; *Baby leaf crops* (off-label) [1]; *Beans without pods (Fresh)* (off-label) [1]; *Broad beans* (off-label) [1]; *Broccoli* (off-label) [1]; *Calabrese* (off-label) [1]; *Caraway* (off-label) [1]; *Carrots* [1]; *Cauliflowers* [1]; *Celeriac* (off-label) [1]; *Celery leaves* (off-label) [1]; *Christmas trees* [1]; *Combining peas* [1]; *Coriander* (off-label) [1]; *Cress* (off-label) [1]; *Dill* (off-label) [1]; *Dwarf beans* [1]; *Early potatoes* [1]; *Edible podded peas* (off-label) [1]; *Endives* (off-label) [1]; *Fennel leaves* (off-label) [1]; *Fodder beet* [1]; *Forest* [1]; *Forest nurseries* [1]; *Herbs (see appendix 6)* (off-label) [1]; *Horseradish* (off-label) [1]; *Jerusalem artichokes* (off-label) [1]; *Lamb's lettuce* (off-label) [1]; *Land cress* (off-label) [1]; *Leeks* [1]; *Lentils* (off-label) [1]; *Lettuce* (off-label) [1]; *Linseed* [1]; *Lovage* (off-label) [1]; *Maincrop potatoes* [1]; *Mangels* [1]; *Ornamental plant production* (off-label) [1]; *Purslane* (off-label) [1]; *Red beet* (off-label) [1]; *Rocket* (off-label) [1]; *Runner beans* (off-label) [1]; *Salad burnet* (off-label) [1]; *Salad onions* [1]; *Salsify* (off-label) [1]; *Soya beans* (off-label) [1]; *Spinach* (off-label) [1]; *Spinach beet* (off-label) [1]; *Spring field beans* [1]; *Spring oilseed rape* [1]; *Strawberries* [1]; *Sugar beet* [1]; *Swedes* [1]; *Sweet ciceley* (off-label) [1]; *Turnips* (off-label) [1]; *Vining peas* [1]; *Watercress* (off-label) [1]; *Winter field beans* [1]; *Winter oilseed rape* [1];
- Bent grasses in *Managed amenity turf* (off-label) [1];
- Black bent in *All edible seed crops grown outdoors* (off-label) [1]; *All non-edible seed crops grown outdoors* (off-label) [1]; *Carrots* [1]; *Cauliflowers* [1]; *Combining peas* [1]; *Dwarf beans* [1]; *Early potatoes* [1]; *Fodder beet* [1]; *Leeks* [1]; *Linseed* [1]; *Maincrop potatoes* [1]; *Mangels* [1]; *Salad onions* [1]; *Spring field beans* [1]; *Spring oilseed rape* [1]; *Strawberries* [1]; *Sugar beet* [1]; *Swedes* [1]; *Vining peas* [1]; *Winter field beans* [1]; *Winter oilseed rape* [1];
- Blackgrass in *All edible seed crops grown outdoors* (off-label) [1]; *All non-edible seed crops grown outdoors* (off-label) [1]; *Carrots* [1]; *Cauliflowers* [1]; *Combining peas* [1]; *Dwarf beans* [1]; *Early potatoes* [1]; *Fodder beet* [1]; *Leeks* [1]; *Linseed* [1]; *Maincrop potatoes* [1]; *Mangels* [1]; *Salad onions* [1]; *Spring field beans* [1]; *Spring oilseed rape* [1]; *Strawberries* [1]; *Sugar beet* [1]; *Swedes* [1]; *Vining peas* [1]; *Winter field beans* [1]; *Winter oilseed rape* [1];
- Brome grasses in *All edible seed crops grown outdoors* (off-label) [1]; *All non-edible seed crops grown outdoors* (off-label) [1];
- Canary grass in *All edible seed crops grown outdoors* (off-label) [1]; *All non-edible seed crops grown outdoors* (off-label) [1];
- Couch in *All edible seed crops grown outdoors* (off-label) [1]; *All non-edible seed crops grown outdoors* (off-label) [1]; *Carrots* [1]; *Cauliflowers* [1]; *Combining peas* [1]; *Dwarf beans* [1]; *Early potatoes* [1]; *Fodder beet* [1]; *Leeks* [1]; *Linseed* [1]; *Maincrop potatoes* [1]; *Mangels* [1]; *Salad onions* [1]; *Spring field beans* [1]; *Spring oilseed rape* [1]; *Strawberries* [1]; *Sugar beet* [1]; *Swedes* [1]; *Vining peas* [1]; *Winter field beans* [1]; *Winter oilseed rape* [1];

FOR FULL CONDITIONS OF USE ALWAYS READ THE PRODUCT LABEL

- Creeping bent in *All edible seed crops grown outdoors* (off-label) [1]; *All non-edible seed crops grown outdoors* (off-label) [1]; *Carrots* [1]; *Cauliflowers* [1]; *Combining peas* [1]; *Dwarf beans* [1]; *Early potatoes* [1]; *Fodder beet* [1]; *Leeks* [1]; *Linseed* [1]; *Maincrop potatoes* [1]; *Mangels* [1]; *Salad onions* [1]; *Spring field beans* [1]; *Spring oilseed rape* [1]; *Strawberries* [1]; *Sugar beet* [1]; *Swedes* [1]; *Vining peas* [1]; *Winter field beans* [1]; *Winter oilseed rape* [1];
- Green cover in *Land temporarily removed from production* [1];
- Italian ryegrass in *All edible seed crops grown outdoors* (off-label) [1]; *All non-edible seed crops grown outdoors* (off-label) [1];
- Loose silky bent in *All edible seed crops grown outdoors* (off-label) [1]; *All non-edible seed crops grown outdoors* (off-label) [1];
- Onion couch in *Carrots* [1]; *Cauliflowers* [1]; *Combining peas* [1]; *Dwarf beans* [1]; *Early potatoes* [1]; *Fodder beet* [1]; *Leeks* [1]; *Linseed* [1]; *Maincrop potatoes* [1]; *Mangels* [1]; *Salad onions* [1]; *Spring field beans* [1]; *Spring oilseed rape* [1]; *Strawberries* [1]; *Sugar beet* [1]; *Swedes* [1]; *Vining peas* [1]; *Winter field beans* [1]; *Winter oilseed rape* [1];
- Perennial grasses in *Angelica* (off-label) [1]; *Beans without pods (Fresh)* (off-label) [1]; *Broad beans* (off-label) [1]; *Caraway* (off-label) [1]; *Celeriac* (off-label) [1]; *Celery leaves* (off-label) [1]; *Christmas trees* [1]; *Coriander* (off-label) [1]; *Cress* (off-label) [1]; *Dill* (off-label) [1]; *Edible podded peas* (off-label) [1]; *Endives* (off-label) [1]; *Fennel leaves* (off-label) [1]; *Forest* [1]; *Forest nurseries* [1]; *Horseradish* (off-label) [1]; *Jerusalem artichokes* (off-label) [1]; *Lamb's lettuce* (off-label) [1]; *Land cress* (off-label) [1]; *Lentils* (off-label) [1]; *Lettuce* (off-label) [1]; *Lovage* (off-label) [1]; *Ornamental plant production* (off-label) [1]; *Purslane* (off-label) [1]; *Red beet* (off-label) [1]; *Rocket* (off-label) [1]; *Runner beans* (off-label) [1]; *Salad burnet* (off-label) [1]; *Salsify* (off-label) [1]; *Soya beans* (off-label) [1]; *Spinach* (off-label) [1]; *Spinach beet* (off-label) [1]; *Sweet ciceley* (off-label) [1]; *Turnips* (off-label) [1];
- Perennial ryegrass in *All edible seed crops grown outdoors* (off-label) [1]; *All non-edible seed crops grown outdoors* (off-label) [1];
- Volunteer cereals in *All edible seed crops grown outdoors* (off-label) [1]; *All non-edible seed crops grown outdoors* (off-label) [1]; *Angelica* (off-label) [1]; *Baby leaf crops* (off-label) [1]; *Beans without pods (Fresh)* (off-label) [1]; *Broad beans* (off-label) [1]; *Broccoli* (off-label) [1]; *Calabrese* (off-label) [1]; *Caraway* (off-label) [1]; *Carrots* [1]; *Cauliflowers* [1]; *Celery leaves* (off-label) [1]; *Combining peas* [1]; *Coriander* (off-label) [1]; *Cress* (off-label) [1]; *Dill* (off-label) [1]; *Dwarf beans* [1]; *Early potatoes* [1]; *Edible podded peas* (off-label) [1]; *Endives* (off-label) [1]; *Fennel leaves* (off-label) [1]; *Fodder beet* [1]; *Herbs (see appendix 6)* (off-label) [1]; *Lamb's lettuce* (off-label) [1]; *Land cress* (off-label) [1]; *Leeks* [1]; *Lentils* (off-label) [1]; *Lettuce* (off-label) [1]; *Linseed* [1]; *Lovage* (off-label) [1]; *Maincrop potatoes* [1]; *Mangels* [1]; *Purslane* (off-label) [1]; *Rocket* (off-label) [1]; *Runner beans* (off-label) [1]; *Salad burnet* (off-label) [1]; *Salad onions* [1]; *Soya beans* (off-label) [1]; *Spinach* (off-label) [1]; *Spinach beet* (off-label) [1]; *Spring field beans* [1]; *Spring oilseed rape* [1]; *Strawberries* [1]; *Sugar beet* [1]; *Swedes* [1]; *Sweet ciceley* (off-label) [1]; *Vining peas* [1]; *Watercress* (off-label) [1]; *Winter field beans* [1]; *Winter oilseed rape* [1];
- Wild oats in *All edible seed crops grown outdoors* (off-label) [1]; *All non-edible seed crops grown outdoors* (off-label) [1]; *Carrots* [1]; *Cauliflowers* [1]; *Combining peas* [1]; *Dwarf beans* [1]; *Early potatoes* [1]; *Fodder beet* [1]; *Leeks* [1]; *Linseed* [1]; *Maincrop potatoes* [1]; *Mangels* [1]; *Salad onions* [1]; *Spring field beans* [1]; *Spring oilseed rape* [1]; *Strawberries* [1]; *Sugar beet* [1]; *Swedes* [1]; *Vining peas* [1]; *Winter field beans* [1]; *Winter oilseed rape* [1];

Extension of Authorisation for Minor Use (EAMUs)
- *All edible seed crops grown outdoors* 20193890 [1]
- *All non-edible seed crops grown outdoors* 20193890 [1]
- *Angelica* 20171266 [1]
- *Baby leaf crops* 20171266 [1]
- *Beans without pods (Fresh)* 20171265 [1]
- *Broad beans* 20171265 [1]
- *Broccoli* 20171267 [1]
- *Calabrese* 20171267 [1]
- *Caraway* 20171266 [1]
- *Celeriac* 20170438 [1]

- **Celery leaves** *20171266* [1]
- **Coriander** *20171266* [1]
- **Cress** *20171266* [1]
- **Dill** *20171266* [1]
- **Edible podded peas** *20171265* [1]
- **Endives** *20171266* [1]
- **Fennel leaves** *20171266* [1]
- **Herbs (see appendix 6)** *20171266* [1]
- **Horseradish** *20170438* [1]
- **Jerusalem artichokes** *20170438* [1]
- **Lamb's lettuce** *20171266* [1]
- **Land cress** *20171266* [1]
- **Lentils** *20171265* [1]
- **Lettuce** *20171266* [1]
- **Lovage** *20171266* [1]
- **Managed amenity turf** *20171715* [1]
- **Ornamental plant production** *20170437* [1]
- **Purslane** *20171266* [1]
- **Red beet** *20170438* [1]
- **Rocket** *20171266* [1]
- **Runner beans** *20171265* [1]
- **Salad burnet** *20171266* [1]
- **Salsify** *20170438* [1]
- **Soya beans** *20171265* [1]
- **Spinach** *20171266* [1]
- **Spinach beet** *20171266* [1]
- **Sweet ciceley** *20171266* [1]
- **Turnips** *20170438* [1]
- **Watercress** *20171266* [1]

Approval information
- Cycloxydim included in Annex 1 under EC Regulation 1107/2009

Efficacy guidance
- Best results achieved when weeds small and have not begun to compete with crop. Effectiveness reduced by drought, cool conditions or stress. Weeds emerging after application are not controlled
- Foliage death usually complete after 3-4 wk but longer under cool conditions, especially late treatments to winter oilseed rape
- Perennial grasses should have sufficient foliage to absorb spray and should not be cultivated for at least 14 d after treatment
- On established couch pre-planting cultivation recommended to fragment rhizomes and encourage uniform emergence
- Split applications to volunteer wheat and barley at GS 12-14 will often give adequate control in winter oilseed rape. See label for details
- Apply to dry foliage when rain not expected for at least 2 h
- Cycloxydim is an ACCase inhibitor herbicide. To avoid the build up of resistance do not apply products containing an ACCase inhibitor herbicide more than twice to any crop. In addition do not use any product containing cycloxydim in mixture or sequence with any other product containing the same ingredient
- Use these products as part of a resistance management strategy that includes cultural methods of control and does not use ACCase inhibitors as the sole chemical method of grass weed control
- Applying a second product containing an ACCase inhibitor to a crop will increase the risk of resistance development; only use a second ACCase inhibitor to control different weeds at a different timing
- Always follow WRAG guidelines for preventing and managing herbicide resistant weeds. See Section 5 for more information

FOR FULL CONDITIONS OF USE ALWAYS READ THE PRODUCT LABEL

Restrictions
- Maximum number of treatments 1 per crop or yr in most situations. See label
- Must be used with authorised adjuvant oil. See label
- Do not apply to crops damaged or stressed by adverse weather, pest or disease attack or other pesticide treatment
- Prevent drift onto other crops, especially cereals and grass

Crop-specific information
- HI cabbage, cauliflower, calabrese, salad onions 4 wk; peas, dwarf beans 5 wk; bulb onions, carrots, parsnips, strawberries 6 wk; sugar and fodder beet, leeks, mangels, potatoes, field beans, swedes, Brussels sprouts, winter field beans 8 wk; oilseed rape, soya beans, linseed 12 wk
- Recommended time of application varies with crop. See label for details
- On peas a crystal violet wax test should be done if leaf wax likely to have been affected by weather conditions or other chemical treatment. The wax test is essential if other products are to be sprayed before or after treatment
- May be used on ornamental bulbs when crop 5-10 cm tall. Product has been used on tulips, narcissi, hyacinths and irises but some subjects may be more sensitive and growers advised to check tolerance on small number of plants before treating the rest of the crop
- May be applied to land temporarily removed from production where the green cover is made up predominantly of tolerant crops listed on label. Use on industrial crops of linseed and oilseed rape on land temporarily removed from production also permitted.

Following crops guidance
- Guideline intervals for sowing succeeding crops after failed treated crop: field beans, peas, sugar beet, rape, kale, swedes, radish, white clover, lucerne 1 wk; dwarf French beans 4 wk; wheat, barley, maize 8 wk
- Oats should not be sown after failure of a treated crop

Environmental safety
- Dangerous for the environment
- Toxic to aquatic organisms
- Harmful to fish or other aquatic life. Do not contaminate surface waters or ditches with chemical or used container

Hazard classification and safety precautions
  Hazard Harmful, Dangerous for the environment
  UN Number N/C
  Risk phrases H304, H315, H318, H336, H361
  Operator protection U05a, U08, U20b; A, C, H
  Environmental protection E15a, E38, H411
  Storage and disposal D01, D02, D09a, D10c, D12a
  Medical advice M05b

96   Cydia pomonella GV

An baculovirus insecticide for codling moth control

Products

1	Carpovirusine	Arysta	1.0 x10^13 GV/l	SC	15243
2	Carpovirusine EVO 2	Arysta	1.0 x10^13 GV/l	SC	17565

Uses
- Codling moth in **Apples** [1-2]; **Pears** [1-2];

Approval information
- Cydia pomonella included in Annex I under EC Regulation 1107/2009

Efficacy guidance
- Optimum results require a precise spraying schedule commencing soon after egg laying and before first hatch

SEE SECTION 3 FOR PRODUCTS ALSO REGISTERED

SECTION 2

- First generation of codling moth normally occurs in first 2 wk of Jun in UK
- Apply in sufficient water to achieve good coverage of whole tree
- Normally 3 applications at intervals of 8 sunny days are needed for each codling moth generation
- Fruit damage is reduced significantly in the first yr and the population of codling moth in the following year will be reduced

Restrictions
- Maximum number of treatments 3 per season
- Do not tank mix with copper or any products with a pH lower than 5 or higher than 8
- Do not use as the exclusive measure for codling moth control. Adopt an anti-resistance strategy. See Section 5 for more information
- Consult processor before use on crops for processing

Crop-specific information
- HI for apples and pears 1 day

Environmental safety
- Product has no adverse effect on predatory insects and is fully compatible with IPM programmes

Hazard classification and safety precautions
UN Number N/C
Risk phrases H317
Operator protection U05a, U14, U16b, U19a; A, D, H
Environmental protection E15b, E34
Storage and disposal D01, D02, D09a, D10b, D12a
Medical advice M05a

97   cyflufenamid

An amidoxime fungicide for cereals
FRAC mode of action code: U6

Products

1	CF-50	Pan Agriculture	50 g/l	EW	18487
2	Clayton Midas	Clayton	50 g/l	EW	18362
3	Clayton Roulette	Clayton	50 g/l	EW	18017
4	Cosine	Certis Belchim B V	50 g/l	EW	16404
5	Cyflamid	Certis Belchim B V	50 g/l	EW	12403
6	Flinch	Pan Agriculture	50 g/l	EW	18464
7	Inari	Certis Belchim B V	50 g/l	EW	19366
8	Laquna	Pan Agriculture	50 g/l	EW	19232
9	Takumi SC	Certis Belchim B V	100 g/l	SC	16000
10	Vegas	Certis Belchim B V	50 g/l	EW	15238

Uses
- Disease control in **Courgettes** [9]; **Cucumbers** [9]; **Melons** [9]; **Summer squash** [9];
- Powdery mildew in **Apples** [1,4]; **Canary seed** (off-label) [5]; **Courgettes** [9]; **Cucumbers** [9]; **Durum wheat** [2-3,5-8,10]; **Melons** [9]; **Ornamental plant production** (off-label) [9]; **Pears** [1,4]; **Protected tomatoes** (off-label) [9]; **Pumpkins** (off-label) [9]; **Spring barley** [2-3,5-8,10]; **Spring oats** [2-3,5-8,10]; **Spring wheat** [2-3,5-8,10]; **Strawberries** (off-label) [9]; **Summer squash** [9]; **Triticale** [2-3,5-8,10]; **Wine grapes** (off-label) [4]; **Winter barley** [2-3,5-8,10]; **Winter oats** [2-3,5-8,10]; **Winter rye** [2-3,5-8,10]; **Winter squash** (off-label) [9]; **Winter wheat** [2-3,5-8,10];

Extension of Authorisation for Minor Use (EAMUs)
- **Canary seed** 20201579 [5]
- **Ornamental plant production** 20131294 [9]
- **Protected tomatoes** 20140800 [9]
- **Pumpkins** 20162915 [9]
- **Strawberries** 20162055 [9]

FOR FULL CONDITIONS OF USE ALWAYS READ THE PRODUCT LABEL

- **Wine grapes** *20170846* [4]
- **Winter squash** *20162915* [9]

Approval information
- Cyflufenamid included in Annex 1 under EC Regulation 1107/2009
- Accepted by BBPA on malting barley

Efficacy guidance
- Best results obtained from treatment at the first visible signs of infection by powdery mildew
- Sustained disease pressure may require a second treatment
- Disease spectrum may be broadened by appropriate tank mixtures. See label
- Product is rainfast within 1 h
- Must be used as part of an Integrated Crop Management programme that includes alternating use, or mixture, with fungicides with a different mode of action effective against powdery mildew

Restrictions
- Maximum number of treatments 2 per crop on all recommended cereals
- Apply only in the spring
- Do not apply to crops under stress from drought, waterlogging, cold, pests or diseases, lime or nutrient deficiency or other factors affecting crop growth

Crop-specific information
- Latest use: before start of flowering of cereal crops

Following crops guidance
- No restrictions

Environmental safety
- Dangerous for the environment
- Very toxic to aquatic organisms
- Avoid drift onto ponds, waterways or ditches

Hazard classification and safety precautions
Hazard Harmful [1,4,10], Dangerous for the environment, Harmful if swallowed [1,4]
Transport code 9 [1-8,10]
Packaging group III
UN Number 3082, N/C
Risk phrases R22b [10]
Operator protection U05a; A
Environmental protection E15b, E17a (5 m) [1,4], E34, E38 [1,4], H411
Storage and disposal D01, D02, D05, D10c, D12b
Medical advice M05b [1,4,10]

98   cyflufenamid + spiroxamine

A fungicide mixture for use in cereals
FRAC mode of action code: 5 + U6

See also cyflufenamid

Products
1	Vegas Plus	Certis Belchim B V	12.5:312.5 g/l	EC	19743

Uses
- Powdery mildew in **Durum wheat** [1]; **Rye** [1]; **Spelt** [1]; **Spring barley** [1]; **Spring wheat** [1]; **Triticale** [1]; **Winter barley** [1]; **Winter wheat** [1];
- Rust in **Durum wheat** [1]; **Rye** [1]; **Spelt** [1]; **Spring barley** [1]; **Spring wheat** [1]; **Triticale** [1]; **Winter barley** [1]; **Winter wheat** [1];

Approval information
- Cyflufenamid and spiroxamine included in Annex I under EC Regulation 1107/2009

SEE SECTION 3 FOR PRODUCTS ALSO REGISTERED

Restrictions
- Do not apply by hand-held equipment.
- Non-returnable containers must not be re-used for any purpose

Environmental safety
- Aquatic buffer zone requirement 12m
- LERAP Category B [1]

Hazard classification and safety precautions
    Hazard Harmful if swallowed, Harmful if inhaled
    Transport code 9 [1]
    Packaging group III
    UN Number 3082
    Risk phrases H315, H317, H318, H335, H361, H373
    Operator protection A, C, H
    Environmental protection E16a, H410
    Storage and disposal D01, D02, D05

99   cyflumetofen

An acaricide for spider mite control in apples, protected strawberries and protected ornamentals.
IRAC mode of action code: 25

Products

1	Nealta	BASF	200 g/l	SC	19411

Uses
- Red spider mites in *Apples* [1]; *Medlar* [1]; *Pears* [1]; *Protected ornamentals* [1]; *Protected strawberries* [1]; *Quinces* [1];
- Spider mites in *Apples* [1]; *Medlar* [1]; *Pears* [1]; *Protected ornamentals* [1]; *Protected strawberries* [1]; *Quinces* [1];
- Two-spotted spider mite in *Apples* [1]; *Medlar* [1]; *Pears* [1]; *Protected ornamentals* [1]; *Protected strawberries* [1]; *Quinces* [1];

Approval information
- Cyflumetofen included in Annex 1 under EC Regulation 1107/2009

Restrictions
- A minimum interval of 2 years must be observed between applications
- Approval is limited to use on roses when applying to protected ornamentals

Environmental safety
- LERAP Category B [1]

Hazard classification and safety precautions
    Transport code 9 [1]
    Packaging group III
    UN Number 3082
    Risk phrases H317, H351
    Environmental protection E12c, E12e, E16a, E17a (15m), H411

100   cymoxanil

A cyanoacetamide oxime fungicide for potatoes
FRAC mode of action code: 27

See also cyazofamid + cymoxanil

Products

1	Cymbal 45	Certis Belchim B V	45% w/w	WG	16647
2	Drum	Certis Belchim B V	45% w/w	WG	16750

Products – continued

3 Option	Corteva	60% w/w	WG	16959
4 Sacron WG	UPL Europe	45% w/w	WG	16433
5 Sipcam C50 WG	Sipcam	50% w/w	WG	16743

Uses

- Blight in *Early potatoes* [5]; *Potatoes* [1-5];
- Downy mildew in *Hops* *(off-label)* [3]; *Wine grapes* *(off-label)* [3];

Extension of Authorisation for Minor Use (EAMUs)

- *Hops 20161101* [3]
- *Wine grapes* *20160542* [3]

Approval information

- Cymoxanil included in Annex I under EC Regulation 1107/2009
- Accepted by BBPA for use on hops

Efficacy guidance

- Product to be used in mixture with specified mixture partners in order to combine systemic and protective activity. See labels for details
- Commence spray programme as soon as weather conditions favourable for disease development occur and before infection appears. At latest the first treatment should be made as the foliage meets along the rows
- Repeat treatments at 7-14 day intervals according to disease incidence and weather conditions
- Treat irrigated crops as soon as possible after irrigation and repeat at 10 d intervals

Restrictions

- Minimum spray interval 7 d
- Do not allow packs to become wet during storage and do not keep opened packs from one season to another
- Cymoxanil must be used in tank mixture with other fungicides
- The maximum concentration must not exceed 220 g of product per 300 litre water [4]

Crop-specific information

- HI for potatoes 1 d [3], 7 d [2, 5, 1], 14 d [4]
- Consult processor before treating potatoes grown for processing

Environmental safety

- Dangerous for the environment
- Very toxic to aquatic organisms
- LERAP Category B [4]

Hazard classification and safety precautions

Hazard Harmful [1,3-5], Irritant [2], Dangerous for the environment [1-4], Harmful if swallowed [3-5], Very toxic to aquatic organisms [3]

Transport code 9 [1-4]

Packaging group III

UN Number 3077, N/C

Risk phrases H317, H319 [1-2], H361 [1-2,4], H373 [1-2,4]

Operator protection U05a, U08 [3,5], U09a [1-2,4], U14 [1-2,4-5], U19a [3], U20a [1], U20b; A, D, H

Environmental protection E13a [3], E13c [5], E15a [1-2,4], E16a [4], E34 [3], E38 [1-4], H410 [1-3], H411 [4-5]

Storage and disposal D01, D02, D09a, D11a, D12a [1-4]

Medical advice M03 [3,5], M05a [1,4]

101   cymoxanil + fluazinam

A fungicide mixture for blight control in potatoes
FRAC mode of action code: 27 + 29

Products

1	Kunshi	Certis Belchim B V	25:37.5% w/w	WG	16991

Uses
- Blight in **Potatoes** [1];
- Late blight in **Potatoes** [1];
- Phytophthora in **Potatoes** [1];

Approval information
- Cymoxanil and fluazinam included in Annex 1 under EC Regulation 1107/2009

Restrictions
- Use as part of a programme with fungicides acting via a different mode of action to reduce the risk of resistance
- Use low drift spraying equipment up to 30m from the top of the bank of any surface water bodies [1]

Following crops guidance
- No restrictions on following crops

Environmental safety
- Buffer zone requirement 6 m [1]
- To protect non target insects/arthropods respect an unsprayed buffer zone of 5 m to non-crop land
- Horizontal boom sprayers must be fitted with three star drift reduction technology for all uses [1]

Hazard classification and safety precautions
Hazard Very toxic to aquatic organisms
Transport code 9 [1]
Packaging group III
UN Number 3077
Risk phrases H317, H319, H361, H373
Environmental protection H410

102   cymoxanil + mancozeb

A protectant and systemic fungicide for potato blight control only approved in GB
FRAC mode of action code: 27 + M3

See also mancozeb

Products

1	Curzate M WG	Corteva	4.5:68% w/w	WG	11901
2	Nautile DG	UPL Europe	5:68% w/w	WG	16653
3	Profilux	Certis Belchim B V	4.5:68% w/w	WG	16125
4	Rhapsody	Corteva	4.5:68% w/w	WG	11958
5	Video	UPL Europe	5:68% w/w	WG	16685

Uses
- Blight in **Potatoes** [1-5];
- Disease control in **Hops in propagation** *(off-label)* [1];
- Late blight in **Potatoes** [3];

Extension of Authorisation for Minor Use (EAMUs)
- **Hops in propagation** *20212140* [1]

Approval information
- Cymoxanil and mancozeb included in Annex I under EC Regulation 1107/2009
- Approval expired in Northern Ireland 4/1/2022. Do not use in Northern Ireland

Efficacy guidance
- Apply immediately after blight warning or as soon as local conditions dictate and repeat at 7-14 d intervals until haulm dies down or is burnt off
- Spray interval should not be more than 7-10 d in irrigated crops (see product label). Apply treatment after irrigation
- To minimise the likelihood of development of resistance these products should be used in a planned Resistance Management strategy. See Section 5 for more information

Restrictions
- Not specified for most products
- Do not apply at less than 7 d intervals
- Do not allow packs to become wet during storage
- Use low drift spraying equipment up to 30m from the top of the bank of any surface water bodies [2, 5]
- Horizontal boom sprayers must be fitted with three star drift reduction technology for all uses [5, 2]
- The maximum concentration must not exceed 2 kg of product per 300 litres water [5]

Crop-specific information
- Destroy and remove any haulm that remains after harvest of early varieties to reduce blight pressure on neighbouring maincrop potatoes

Following crops guidance
- HI 7 days [3]

Environmental safety
- Dangerous for the environment
- Very toxic to aquatic organisms
- Keep product away from fire or sparks
- Buffer zone requirement 6m [2, 5]
- LERAP Category B [1-4]

Hazard classification and safety precautions
Hazard Harmful [1,3-4], Irritant [2,5], Dangerous for the environment, Very toxic to aquatic organisms
Transport code 9 [1-5]
Packaging group III
UN Number 3077
Risk phrases H317, H361
Operator protection U05a, U08, U11 [2,5], U14, U19a, U20b; A, C, D, H
Environmental protection E13c [1,3-4], E15a [1,3-5], E16a [1-4], E38, H410 [1-2,4-5], H411 [3]
Storage and disposal D01, D02, D06b [2,5], D09a, D11a [1,3-4], D12a [2,5]
Medical advice M05a [5]

103   cymoxanil + mandipropamid

A fungicide mixture for use in potatoes
FRAC mode of action code: 27 + 40

See also cymoxanil

Products
1 Carial Flex	Syngenta	18:25% w/w	WG	16629

Uses
- Blight in **Potatoes** [1];
- Late blight in **Potatoes** [1];

SEE SECTION 3 FOR PRODUCTS ALSO REGISTERED

Approval information
- Cymoxanil and mandipropamid are included in Annex 1 under EC Regulation 1107/2009

Efficacy guidance
- Where possible, use an alternating strategy using fungicides from different mode of action groups.
- Where CAA fungicides are applied as a mixture (co-formulated or as a tank mix) up to six applications (or max. of 50% of the total number of applications) may be made per crop or season.
- No more than 3 applications of any CAA fungicide should be made consecutively

Restrictions
- Application may only be made between 1 May and 31 August

Hazard classification and safety precautions
   Hazard Harmful if swallowed
   Transport code 9 [1]
   Packaging group III
   UN Number 3077
   Risk phrases H361, H373
   Operator protection U05a, U08, U20b; A, H
   Environmental protection E15b, E34, E38, H410
   Storage and disposal D01, D02, D09a, D10c, D12a, D19
   Medical advice M03, M05a

104   cymoxanil + propamocarb

A cyanoacetamide oxime fungicide + a carbamate fungicide for potatoes
FRAC mode of action code: 27 + 28

Products

1 Axidor	Arysta	50:400 g/l	SC	16830
2 Proxanil	Arysta	50:400 g/l	SC	16664

Uses
- Blight in *Potatoes* [1-2];

Approval information
- Cymoxanil and propamocarb hydrochloride included in Annex I under EC Regulation 1107/2009

Restrictions
- Must be applied using 300-350 l water/ha
- The maximum in use concentration must not exceed 3.33 g propamacarb and 0.417 g cymoxanil/litre

Hazard classification and safety precautions
   Hazard Irritant
   Transport code 8 [1-2]
   Packaging group III
   UN Number 3265
   Risk phrases H290, H317, H361
   Operator protection U05a, U08, U14, U20b; A, H
   Environmental protection E15a, E38, H411
   Storage and disposal D01, D02, D09a, D10a

105 cymoxanil + zoxamide

A protectant and systemic fungicide for blight control in potatoes
FRAC mode of action code: 27 + 22

See also cymoxanil
dimethomorph + zoxamide

Products

1	Lieto	Sipcam	33:33% w/w	WG	16703

Uses
* Blight in **Potatoes** [1];

Approval information
* Cymoxanil and zoxamide included in Annex I under EC Regulation 1107/2009

Environmental safety
* Buffer zone requirement 6 m

Hazard classification and safety precautions
Hazard Harmful, Dangerous for the environment, Harmful if swallowed
Transport code 9 [1]
Packaging group III
UN Number 3077
Risk phrases H317
Operator protection U05a, U11, U14, U15, U19a; A, C, H
Environmental protection E38, H410
Storage and disposal D01, D02, D12a, D12b
Medical advice M05a

106 cypermethrin

A contact and stomach acting pyrethroid insecticide
IRAC mode of action code: 3

Products

1	Cythrin 500 EC	Arysta	500 g/l	EC	16993
2	Cythrin Max EC	Arysta	500 g/l	EC	17138
3	Forester	Arysta	100 g/l	EW	13164
4	Signal 300 ES	Arysta	300 g/l	ES	15949
5	Signal 300 ES	UPL Europe	300 g/l	ES	19631
6	Talisma EC	Arysta	80 g/l	EC	16541
7	Talisma UL	Arysta	20 g/l	UL	16542

Uses
* Aphids in **Asparagus** [1-2]; **Broccoli** [1-2]; **Brussels sprouts** [1-2]; **Cabbages** [1-2]; **Calabrese** [1-2]; **Carrots** [1-2]; **Cauliflowers** [1-2]; **Celeriac** [1-2]; **Dwarf beans** [1-2]; **Edible podded peas** [1-2]; **Forage maize** [1-2]; **French beans** [1-2]; **Grain maize** [1-2]; **Horseradish** [1-2]; **Linseed** [1-2]; **Lupins** [1-2]; **Mustard** [1-2]; **Ornamental plant production** [1-2]; **Parsley root** [1-2]; **Parsnips** [1-2]; **Radishes** [1-2]; **Runner beans** [1-2]; **Salsify** [1-2]; **Spelt** [1-2]; **Spring barley** *(autumn sown)* [1-2]; **Spring oats** [1-2]; **Spring rye** [1-2]; **Spring wheat** *(autumn sown)* [1-2]; **Triticale** [1-2]; **Vining peas** [1-2]; **Winter barley** [1-2]; **Winter oats** [1-2]; **Winter rye** [1-2]; **Winter wheat** [1-2];
* Asparagus beetle in **Asparagus** *(off-label)* [1-2];
* Barley yellow dwarf virus vectors in **Spring barley** *(autumn sown)* [1-2]; **Spring oats** [1-2]; **Spring rye** [1-2]; **Spring wheat** *(autumn sown)* [1-2]; **Triticale** [1-2]; **Winter barley** [1-2]; **Winter oats** [1-2]; **Winter rye** [1-2]; **Winter wheat** [1-2];
* Beetles in **Cut logs** [3];
* Bladder pod midge in **Spring oilseed rape** [1-2]; **Winter oilseed rape** [1-2];
* Cabbage stem flea beetle in **Winter oilseed rape** [1-2];

SEE SECTION 3 FOR PRODUCTS ALSO REGISTERED

- Carrot fly in *Carrots* [1-2]; *Celeriac* [1-2]; *Horseradish* [1-2]; *Parsley root* [1-2]; *Parsnips* [1-2];
- Caterpillars in *Broccoli* [1-2]; *Brussels sprouts* [1-2]; *Cabbages* [1-2]; *Calabrese* [1-2]; *Cauliflowers* [1-2];
- Cutworms in *Fodder beet* [1-2]; *Mangels* [1-2]; *Potatoes* [1-2]; *Red beet* [1-2]; *Sugar beet* [1-2]; *Swedes* [1-2]; *Turnips* [1-2];
- Frit fly in *Spring barley* (qualified recommendation) [4-5]; *Spring wheat* (qualified recommendation) [4-5]; *Winter barley* (qualified recommendation) [4-5]; *Winter wheat* (qualified recommendation) [4-5];
- Insect pests in *Asparagus* [1-2]; *Carrots* [1-2]; *Celeriac* [1-2]; *Crop handling & storage structures* [6]; *Dwarf beans* [1-2]; *Edible podded peas* [1-2]; *French beans* [1-2]; *Grain maize* [1-2]; *Horseradish* [1-2]; *Linseed* [1-2]; *Lupins* [1-2]; *Mustard* [1-2]; *Ornamental plant production* [1-2]; *Parsley root* [1-2]; *Parsnips* [1-2]; *Radishes* [1-2]; *Rice* [6-7]; *Runner beans* [1-2]; *Rye* [7]; *Salsify* [1-2]; *Spelt* [1-2,6-7]; *Spring barley* [6-7]; *Spring oats* [6-7]; *Spring rye* [6]; *Spring wheat* [6-7]; *Triticale* [6-7]; *Winter barley* [6-7]; *Winter oats* [6]; *Winter rye* [6]; *Winter wheat* [6-7];
- Large pine weevil in *Forest* [3];
- Pea and bean weevil in *Combining peas* [1-2]; *Spring field beans* [1-2]; *Vining peas* [1-2]; *Winter field beans* [1-2];
- Pea moth in *Vining peas* [1-2];
- Pollen beetle in *Spring oilseed rape* [1-2]; *Winter oilseed rape* [1-2];
- Rape winter stem weevil in *Winter oilseed rape* [1-2];
- Seed weevil in *Spring oilseed rape* [1-2]; *Winter oilseed rape* [1-2];
- Wheat bulb fly in *Spring barley* (reduction) [4-5]; *Spring wheat* (reduction) [4-5]; *Winter barley* (reduction) [4-5]; *Winter wheat* (reduction) [4-5];
- Whitefly in *Broccoli* [1-2]; *Brussels sprouts* [1-2]; *Cabbages* [1-2]; *Calabrese* [1-2]; *Cauliflowers* [1-2];
- Yellow cereal fly in *Spring barley* (autumn sown) [1-2]; *Spring oats* [1-2]; *Spring rye* [1-2]; *Spring wheat* (autumn sown) [1-2]; *Triticale* [1-2]; *Winter barley* [1-2]; *Winter oats* [1-2]; *Winter rye* [1-2]; *Winter wheat* [1-2];

Extension of Authorisation for Minor Use (EAMUs)
- *Asparagus* 20222131 [1], 20170903 [2]

Approval information
- Cypermethrin included in Annex I under EC Regulation 1107/2009
- Accepted by BBPA for use on malting barley and hops

Efficacy guidance
- Products combine rapid action, good persistence, and high activity on Lepidoptera.
- As effect is mainly via contact, good coverage is essential for effective action. Spray volume should be increased on dense crops
- A repeat spray after 10-14 d is needed for some pests of outdoor crops
- Rates and timing of sprays vary with crop and pest. See label for details

Restrictions
- Maximum number of treatments varies with crop and product. See label or approval notice for details [3]
- Horizontal boom sprayers must be fitted with three star drift reduction technology for all uses
- Apply by knapsack or hand-held sprayer [3]
- Some new products advise using low drift spraying equipment up to 30m from the top of the bank of any surface water bodies - check label before use.
- Must only be applied to cereal seed sown in the autumn/winter [4]

Crop-specific information
- HI vining peas 7 d; other crops 0 d
- Test spray sample of new or unusual ornamentals or trees before committing whole batches [3]
- Post-planting forestry applications should be made before damage is seen or at the onset of damage during the first 2 yrs after transplanting [3]

Environmental safety
- Dangerous for the environment

FOR FULL CONDITIONS OF USE ALWAYS READ THE PRODUCT LABEL

- Very toxic to aquatic organisms
- High risk to bees. Do not apply to crops in flower, or to those in which bees are actively foraging, except as directed. Do not apply when flowering weeds are present
- Give local beekeepers warning when using in flowering crops
- Do not spray cereals after 31 Mar within 6 m of the edge of the growing crop [3]
- Buffer zone requirement 18 m [1, 2]
- To protect non target insects/arthropods respect an unsprayed buffer zone of 5 m to non-crop land [1]
- Horizontal boom sprayers must be fitted with three star drift reduction technology for all uses [1]
- Use low drift spraying equipment up to 30m from the top of the bank of any surface water bodies [1]
- Broadcast air-assisted LERAP [2] (18 m); LERAP Category A [1-2]

SECTION 2

Hazard classification and safety precautions

Hazard Harmful [1-5], Flammable [1-2], Dangerous for the environment [1-5], Flammable liquid and vapour [2], Harmful if swallowed [1,3-6], Harmful if inhaled [2], Very toxic to aquatic organisms [6-7], Toxic to aquatic organisms [4-5]

Transport code 3 [1-2], 9 [3-7]

Packaging group III

UN Number 1993, 3082

Risk phrases H304 [2], H315 [1-3], H317 [1,3,6], H318 [2], H320 [1], H335 [1-2,4-5], H336 [1-2], H373 [1]

Operator protection U02a [3], U04a [3], U05a, U08 [1-2,4-5], U10 [3], U14 [3-5], U16b [3], U19a [1-3,6-7], U20b [1-2,4-5]; A, C, D, H, M

Environmental protection E12c, E12f, E13a [1-2], E15a [3], E15b [4-5], E16c [1-2], E16d [1-2], E16h [1-2], E17b (18 m) [2], E34 [1-5], E38 [3-5], H410

Storage and disposal D01, D02, D05 [1-3,6], D09a [1-6], D09b [7], D10a [4-5], D10b [1-3], D12a [1-3,6]

Treated seed S01 [4-5], S02 [4-5], S03 [4-5], S04b [4-5], S05 [4-5], S06a [4-5], S07 [4-5]

Medical advice M03 [3-5], M05b [1-2]

107   cypermethrin + tetramethrin

A pyrethroid insecticide mixture for control of crawling and flying insects in a variety of situations
IRAC mode of action code: 3

See also cypermethrin

Products

1	Vazor Cypermax Plus	Killgerm	10% + 5% w/w	ME	H9988

Uses

- Ants in *Agricultural premises* [1]; *Manure heaps* [1]; *Poultry houses* [1]; *Refuse tips* [1];
- Bedbugs in *Agricultural premises* [1]; *Manure heaps* [1]; *Poultry houses* [1]; *Refuse tips* [1];
- Cockroaches in *Agricultural premises* [1]; *Manure heaps* [1]; *Poultry houses* [1]; *Refuse tips* [1];
- Mosquitoes in *Agricultural premises* [1]; *Manure heaps* [1]; *Poultry houses* [1]; *Refuse tips* [1];
- Moths in *Agricultural premises* [1]; *Manure heaps* [1]; *Poultry houses* [1]; *Refuse tips* [1];
- Silverfish in *Agricultural premises* [1]; *Manure heaps* [1]; *Poultry houses* [1]; *Refuse tips* [1];
- Wasps in *Agricultural premises* [1]; *Manure heaps* [1]; *Poultry houses* [1]; *Refuse tips* [1];

Approval information

- Cypermethrin is included but tetramethrin not included in Annex 1 under EC Regulation 1107/2009

Hazard classification and safety precautions

Transport code 9 [1]

Packaging group III

UN Number 3082

SEE SECTION 3 FOR PRODUCTS ALSO REGISTERED

Operator protection U20b; A, C, D, H
Environmental protection E05c, E12b, H410
Storage and disposal D09a, D12a

108 cyprodinil

An anilinopyrimidine systemic broad spectrum fungicide for cereals
FRAC mode of action code: 9

See also cyproconazole + cyprodinil

Products

1	Coracle	AgChem Access	300 g/l	EC	15866
2	Kayak	Syngenta	300 g/l	EC	14847
3	Unix	Syngenta	75% w/w	WG	14846

Uses

- Disease control in *Forest nurseries* (off-label) [2];
- Eyespot in *Spring barley* [1-3]; *Spring wheat* [3]; *Winter barley* [1-3]; *Winter wheat* [3];
- Net blotch in *Spring barley* [1-3]; *Winter barley* [1-3];
- Powdery mildew in *Spring barley* [1-3]; *Spring wheat* (moderate control) [3]; *Winter barley* [1-3]; *Winter wheat* (moderate control) [3];
- Rhynchosporium in *Spring barley* (moderate control) [1-3]; *Winter barley* (moderate control) [1-3];

Extension of Authorisation for Minor Use (EAMUs)

- *Forest nurseries* 20122056 [2]

Approval information

- Cyprodinil included in Annex I under EC Regulation 1107/2009
- Accepted by BBPA for use on malting barley

Efficacy guidance

- Best results obtained from treatment at early stages of disease development
- For best control of eyespot spray before or during the period of stem extension in spring. Control may be reduced if very dry conditions follow treatment

Restrictions

- Maximum number of treatments 2 per crop on wheat and barley [2]

Crop-specific information

- Latest use: before first spikelet of inflorescence just visible stage for barley, before early milk stage for wheat [3]; up to and including first awns visible stage (GS49) in barley [2]

Environmental safety

- Dangerous for the environment
- Very toxic to aquatic organisms
- LERAP Category B [1-3]

Hazard classification and safety precautions

Hazard Irritant [1-2], Dangerous for the environment, Very toxic to aquatic organisms [1-2]
Transport code 9 [1-3]
Packaging group III
UN Number 3077, 3082
Risk phrases H317 [1-2], R50 [3], R53a [3]
Operator protection U05a, U09a [1-2], U20a; A, H
Environmental protection E15a [3], E15b [1-2], E16a, E34 [1-2], E38, H410 [1-2]
Storage and disposal D01, D02, D05, D07, D09a, D10c [1-2], D12a

109   cyprodinil + fludioxonil

A broad-spectrum fungicide mixture for fruit crops, legumes, some vegetables, forest nurseries and ornamentals
FRAC mode of action code: 9 + 12

See also fludioxonil

Products

1	Switch	Syngenta	37.5:25 % w/w	WG	15129

Uses

- Alternaria in **Apples** [1]; **Brassica leaves and sprouts** *(off-label)* [1]; **Broccoli** *(off-label)* [1]; **Brussels sprouts** *(off-label)* [1]; **Cabbages** *(off-label)* [1]; **Calabrese** *(off-label)* [1]; **Cauliflowers** *(off-label)* [1]; **Celery (outdoor)** *(off-label)* [1]; **Collards** *(off-label)* [1]; **Crab apples** [1]; **Cress** *(off-label)* [1]; **Endives** *(off-label)* [1]; **Herbs (see appendix 6)** *(off-label)* [1]; **Hops in propagation** *(off-label)* [1]; **Kale** *(off-label)* [1]; **Lamb's lettuce** *(off-label)* [1]; **Lettuce** *(off-label)* [1]; **Nursery fruit trees** *(off-label)* [1]; **Pears** [1]; **Protected baby leaf crops** *(off-label)* [1]; **Protected cress** *(off-label)* [1]; **Protected endives** *(off-label)* [1]; **Protected herbs (see appendix 6)** *(off-label)* [1]; **Protected lamb's lettuce** *(off-label)* [1]; **Protected lettuce** *(off-label)* [1]; **Protected radishes** *(off-label)* [1]; **Protected red mustard** *(off-label)* [1]; **Protected rocket** *(off-label)* [1]; **Protected sorrel** *(off-label)* [1]; **Protected spinach** *(off-label)* [1]; **Protected spinach beet** *(off-label)* [1]; **Quinces** [1]; **Red mustard** *(off-label)* [1]; **Rocket** *(off-label)* [1]; **Sorrel** *(off-label)* [1]; **Spinach** *(off-label)* [1]; **Spinach beet** *(off-label)* [1]; **Spring cabbage** *(off-label)* [1];
- Alternaria blight in **Carrots** *(moderate control)* [1];
- Anthracnose in **Protected blueberry** *(off-label)* [1];
- Ascochyta in **Mange-tout peas** [1]; **Sugar snap peas** [1]; **Vining peas** [1];
- Basal stem rot in **Ornamental bulbs** *(off-label)* [1];
- Black spot in **Protected strawberries** *(qualified minor use)* [1]; **Strawberries** *(qualified minor use)* [1];
- Blossom wilt in **Apricots** *(off-label)* [1]; **Cherries** *(off-label)* [1]; **Peaches** *(off-label)* [1]; **Plums** *(off-label)* [1];
- Botrytis in **Apricots** *(off-label)* [1]; **Asparagus** *(off-label)* [1]; **Brassica leaves and sprouts** *(off-label)* [1]; **Broad beans** *(moderate control)* [1]; **Broad beans (dry-harvested)** *(moderate control)* [1]; **Broccoli** *(off-label)* [1]; **Brussels sprouts** *(off-label)* [1]; **Bulb onions** *(off-label)* [1]; **Cabbages** *(off-label)* [1]; **Calabrese** *(off-label)* [1]; **Carrots** *(moderate control)* [1]; **Cauliflowers** *(off-label)* [1]; **Celeriac** *(moderate control)* [1]; **Cherries** *(off-label)* [1]; **Chicory** *(off-label)* [1]; **Collards** *(off-label)* [1]; **Combining peas** *(moderate control)* [1]; **Cress** *(off-label)* [1]; **Endives** *(off-label)* [1]; **Forest nurseries** [1]; **French beans (dry-harvested)** *(moderate control)* [1]; **Garlic** *(off-label)* [1]; **Grapevines** *(off-label)* [1]; **Green beans** *(moderate control)* [1]; **Herbs (see appendix 6)** *(off-label)* [1]; **Hops in propagation** *(off-label)* [1]; **Kale** *(off-label)* [1]; **Lamb's lettuce** *(off-label)* [1]; **Lettuce** *(off-label)* [1]; **Lupins** *(off-label)* [1]; **Mange-tout peas** *(moderate control)* [1]; **Nursery fruit trees** *(off-label)* [1]; **Ornamental plant production** [1]; **Peaches** *(off-label)* [1]; **Plums** *(off-label)* [1]; **Protected asparagus** *(off-label)* [1]; **Protected aubergines** *(off-label)* [1]; **Protected baby leaf crops** *(off-label)* [1]; **Protected blueberry** *(off-label)* [1]; **Protected courgettes** *(off-label)* [1]; **Protected cress** *(off-label)* [1]; **Protected cucumbers** *(off-label)* [1]; **Protected endives** *(off-label)* [1]; **Protected forest nurseries** [1]; **Protected gherkins** *(off-label)* [1]; **Protected herbs (see appendix 6)** *(off-label)* [1]; **Protected lamb's lettuce** *(off-label)* [1]; **Protected lettuce** *(off-label)* [1]; **Protected ornamentals** [1]; **Protected peppers** *(off-label)* [1]; **Protected pumpkins** *(off-label)* [1]; **Protected red mustard** *(off-label)* [1]; **Protected rocket** *(off-label)* [1]; **Protected sorrel** *(off-label)* [1]; **Protected spinach** *(off-label)* [1]; **Protected spinach beet** *(off-label)* [1]; **Protected strawberries** [1]; **Protected summer squash** *(off-label)* [1]; **Protected tomatoes** *(off-label)* [1]; **Pumpkins** *(off-label)* [1]; **Red mustard** *(off-label)* [1]; **Rhubarb** *(off-label)* [1]; **Rocket** *(off-label)* [1]; **Runner beans** *(moderate control)* [1]; **Salad onions** *(off-label)* [1]; **Shallots** *(off-label)* [1]; **Sorrel** *(off-label)* [1]; **Spinach** *(off-label)* [1]; **Spinach beet** *(off-label)* [1]; **Spring cabbage** *(off-label)* [1]; **Strawberries** [1]; **Sugar snap peas** *(moderate control)* [1]; **Vining peas** *(moderate control)* [1];
- Botrytis fruit rot in **Apples** [1]; **Blackberries** [1]; **Crab apples** [1]; **Pears** [1]; **Quinces** [1]; **Raspberries** [1];

- Botrytis root rot in *Bilberries* *(qualified minor use recommendation)* [1]; *Blackcurrants* *(qualified minor use recommendation)* [1]; *Blueberries* *(qualified minor use recommendation)* [1]; *Cranberries* *(qualified minor use recommendation)* [1]; *Gooseberries* *(qualified minor use recommendation)* [1]; *Redcurrants* *(qualified minor use recommendation)* [1]; *Whitecurrants* *(qualified minor use recommendation)* [1];
- Fusarium in *Protected cucumbers* *(off-label)* [1];
- Fusarium bulb rot in *Ornamental bulbs* *(off-label)* [1];
- Fusarium diseases in *Apples* [1]; *Crab apples* [1]; *Pears* [1]; *Quinces* [1];
- Gloeosporium rot in *Apples* [1]; *Crab apples* [1]; *Pears* [1]; *Quinces* [1];
- Leaf spot in *Celery (outdoor)* *(off-label)* [1]; *Protected radishes* *(off-label)* [1];
- Monilinia spp. in *Hops in propagation* *(off-label)* [1]; *Nursery fruit trees* *(off-label)* [1];
- Mycosphaerella in *Mange-tout peas* [1]; *Sugar snap peas* [1]; *Vining peas* [1];
- Penicillium rot in *Apples* [1]; *Crab apples* [1]; *Hops in propagation* *(off-label)* [1]; *Nursery fruit trees* *(off-label)* [1]; *Pears* [1]; *Quinces* [1];
- Phoma in *Brassica leaves and sprouts* *(off-label)* [1]; *Broccoli* *(off-label)* [1]; *Brussels sprouts* *(off-label)* [1]; *Cabbages* *(off-label)* [1]; *Calabrese* *(off-label)* [1]; *Cauliflowers* *(off-label)* [1]; *Collards* *(off-label)* [1]; *Cress* *(off-label)* [1]; *Endives* *(off-label)* [1]; *Herbs (see appendix 6)* *(off-label)* [1]; *Kale* *(off-label)* [1]; *Lamb's lettuce* *(off-label)* [1]; *Lettuce* *(off-label)* [1]; *Protected baby leaf crops* *(off-label)* [1]; *Protected cress* *(off-label)* [1]; *Protected endives* *(off-label)* [1]; *Protected herbs (see appendix 6)* *(off-label)* [1]; *Protected lamb's lettuce* *(off-label)* [1]; *Protected lettuce* *(off-label)* [1]; *Protected red mustard* *(off-label)* [1]; *Protected rocket* *(off-label)* [1]; *Protected sorrel* *(off-label)* [1]; *Protected spinach* *(off-label)* [1]; *Protected spinach beet* *(off-label)* [1]; *Red mustard* *(off-label)* [1]; *Rocket* *(off-label)* [1]; *Sorrel* *(off-label)* [1]; *Spinach* *(off-label)* [1]; *Spinach beet* *(off-label)* [1]; *Spring cabbage* *(off-label)* [1];
- Rhizoctonia in *Brassica leaves and sprouts* *(off-label)* [1]; *Broccoli* *(off-label)* [1]; *Brussels sprouts* *(off-label)* [1]; *Cabbages* *(off-label)* [1]; *Calabrese* *(off-label)* [1]; *Cauliflowers* *(off-label)* [1]; *Collards* *(off-label)* [1]; *Cress* *(off-label)* [1]; *Endives* *(off-label)* [1]; *Herbs (see appendix 6)* *(off-label)* [1]; *Kale* *(off-label)* [1]; *Lamb's lettuce* *(off-label)* [1]; *Lettuce* *(off-label)* [1]; *Protected baby leaf crops* *(off-label)* [1]; *Protected cress* *(off-label)* [1]; *Protected endives* *(off-label)* [1]; *Protected herbs (see appendix 6)* *(off-label)* [1]; *Protected lamb's lettuce* *(off-label)* [1]; *Protected lettuce* *(off-label)* [1]; *Protected red mustard* *(off-label)* [1]; *Protected rocket* *(off-label)* [1]; *Protected sorrel* *(off-label)* [1]; *Protected spinach* *(off-label)* [1]; *Protected spinach beet* *(off-label)* [1]; *Red mustard* *(off-label)* [1]; *Rocket* *(off-label)* [1]; *Sorrel* *(off-label)* [1]; *Spinach* *(off-label)* [1]; *Spinach beet* *(off-label)* [1]; *Spring cabbage* *(off-label)* [1];
- Sclerotinia in *Beetroot* *(off-label)* [1]; *Broad beans* [1]; *Broad beans (dry-harvested)* [1]; *Carrots* *(moderate control)* [1]; *Celery (outdoor)* *(off-label)* [1]; *Combining peas* [1]; *French beans (dry-harvested)* [1]; *Ginger* *(off-label)* [1]; *Green beans* [1]; *Herbs for medicinal uses (see Appendix 6)* *(off-label)* [1]; *Horseradish* *(off-label)* [1]; *Mange-tout peas* [1]; *Parsley root* *(off-label)* [1]; *Parsnips* *(off-label)* [1]; *Protected radishes* *(off-label)* [1]; *Runner beans* [1]; *Salsify* *(off-label)* [1]; *Spice roots* *(off-label)* [1]; *Sugar snap peas* [1]; *Turmeric* *(off-label)* [1]; *Vining peas* [1];

Extension of Authorisation for Minor Use (EAMUs)
- *Apricots* 20103092 [1]
- *Asparagus* 20103173 [1]
- *Beetroot* 20103087 [1]
- *Brassica leaves and sprouts* 20150085 [1]
- *Broccoli* 20150085 [1]
- *Brussels sprouts* 20150085 [1]
- *Bulb onions* 20122285 [1]
- *Cabbages* 20150085 [1]
- *Calabrese* 20150085 [1]
- *Cauliflowers* 20150085 [1]
- *Celery (outdoor)* 20150865 [1]
- *Cherries* 20103092 [1]
- *Chicory* 20122285 [1]
- *Collards* 20150085 [1]
- *Cress* 20150085 [1]
- *Endives* 20150085 [1]

FOR FULL CONDITIONS OF USE ALWAYS READ THE PRODUCT LABEL

- **Garlic** *20122285* [1]
- **Ginger** *20103087* [1]
- **Grapevines** *20112098* [1]
- **Herbs (see appendix 6)** *20150085* [1]
- **Herbs for medicinal uses (see Appendix 6)** *20103087* [1]
- **Hops in propagation** *20194254* [1]
- **Horseradish** *20103087* [1]
- **Kale** *20150085* [1]
- **Lamb's lettuce** *20150085* [1]
- **Lettuce** *20150085* [1]
- **Lupins** *20103170* [1]
- **Nursery fruit trees** *20194254* [1]
- **Ornamental bulbs** *20152274* [1]
- **Parsley root** *20103087* [1]
- **Parsnips** *20103087* [1]
- **Peaches** *20103092* [1]
- **Plums** *20103092* [1]
- **Protected asparagus** *20170202* [1]
- **Protected aubergines** *20103172* [1]
- **Protected baby leaf crops** *20150085* [1]
- **Protected blueberry** *20160648* [1]
- **Protected courgettes** *20170202* [1]
- **Protected cress** *20150085* [1]
- **Protected cucumbers** *20103171* [1]
- **Protected endives** *20150085* [1]
- **Protected gherkins** *20170202* [1]
- **Protected herbs (see appendix 6)** *20150085* [1]
- **Protected lamb's lettuce** *20150085* [1]
- **Protected lettuce** *20150085* [1]
- **Protected peppers** *20103172* [1]
- **Protected pumpkins** *20170202* [1]
- **Protected radishes** *20150865* [1]
- **Protected red mustard** *20150085* [1]
- **Protected rocket** *20150085* [1]
- **Protected sorrel** *20150085* [1]
- **Protected spinach** *20150085* [1]
- **Protected spinach beet** *20150085* [1]
- **Protected summer squash** *20170202* [1]
- **Protected tomatoes** *20110302* [1]
- **Pumpkins** *20170202* [1]
- **Red mustard** *20150085* [1]
- **Rhubarb** *20211446* [1]
- **Rocket** *20150085* [1]
- **Salad onions** *20103174* [1]
- **Salsify** *20103087* [1]
- **Shallots** *20122285* [1]
- **Sorrel** *20150085* [1]
- **Spice roots** *20103087* [1]
- **Spinach** *20150085* [1]
- **Spinach beet** *20150085* [1]
- **Spring cabbage** *20150085* [1]
- **Turmeric** *20103087* [1]

Approval information
- Cyprodinil and fludioxonil included in Annex I under EC Regulation 1107/2009

Efficacy guidance
- Best results obtained from application at the earliest stage of disease development or as a protective treatment following disease risk assessment
- Subsequent treatments may follow after a minimum of 10 d if disease pressure remains high

SEE SECTION 3 FOR PRODUCTS ALSO REGISTERED

- To minimise the likelihood of development of resistance product should be used in a planned Resistance Management strategy. See Section 5 for more information

Restrictions
- Maximum number of treatments 2 per crop
- Consult processor before treating any crop for processing
- Spray equipment must only be used where the operator?s normal working position is within a closed cab on a tractor or on a self-propelled sprayer when making air-assisted applications to top fruit.

Crop-specific information
- HI broad beans, green beans, mange-tout peas, runner beans, sugar snap peas, vining peas 14 d; protected strawberries, strawberries 3 d
- First signs of disease infection likely to be seen from early flowering for all crops
- Ensure peas are free from any stress before treatment. If necessary check wax with a crystal violet test

Environmental safety
- Dangerous for the environment
- Very toxic to aquatic organisms
- Do not allow direct spray from broadcast air-assisted sprayers to fall within 10 m of the top of the bank of any static or flowing waterbody when treating blackberry, raspberry, blueberry, bilberry, cranberry, redcurrant, whitecurrant, blackcurrant, gooseberry, ornamental plant production and forest nursery; increase this to 30 m when treating apple, crab apple, pear and quince and not within 5 m of a ditch which is dry at the time of application
- LERAP Category B [1]

Hazard classification and safety precautions
    Hazard Dangerous for the environment, Very toxic to aquatic organisms
    Transport code 9 [1]
    Packaging group III
    UN Number 3077
    Risk phrases H317
    Operator protection U05a, U20a; A, H
    Environmental protection E15b, E16a, E16b, E38, H410
    Storage and disposal D01, D02, D05, D07, D09a, D10c, D12a

110   cyprodinil + isopyrazam

A fungicide mixture for disease control in barley
FRAC mode of action code: 7 + 9

See also cyprodinil

Products

1 Bontima	Adama	187.5:62.5 g/l	EC	18566
2 Cebara	Adama	187.5:62.5 g/l	EC	18569
3 Concorde	Adama	150:62.5 g/l	EC	18570

Uses
- Brown rust in *Spring barley* [1-3]; *Winter barley* [1-3];
- Net blotch in *Spring barley* [1-3]; *Winter barley* [1-3];
- Powdery mildew in *Spring barley* [1-3]; *Winter barley* [1-3];
- Ramularia leaf spots in *Spring barley* [1-3]; *Winter barley* [1-3];
- Rhynchosporium in *Spring barley* [1-3]; *Winter barley* [1-3];

Approval information
- Cyprodinil and isopyrazam are included in Annex 1 under EC Regulation 1107/2009
- Accepted by BBPA for use on malting barley before ear emergence only

FOR FULL CONDITIONS OF USE ALWAYS READ THE PRODUCT LABEL

Restrictions
- Isopyrazam is an SDH respiration inhibitor; Do not apply more than two foliar applications of products containing an SDH inhibitor to any cereal crop

Crop-specific information
- There are no restrictions on succeeding crops in a normal rotation

Environmental safety
- LERAP Category B [1-3]

Hazard classification and safety precautions
Hazard Harmful, Dangerous for the environment, Harmful if inhaled, Very toxic to aquatic organisms
Transport code 9 [1-3]
Packaging group III
UN Number 3082
Risk phrases H304, H351 [3], H361
Operator protection U05a, U09a, U11 [1-2], U20b; A, C, H
Environmental protection E15b, E16a, E16b, E38, H410
Storage and disposal D01, D02, D05, D09a, D10c, D11a, D12a
Medical advice M05b

111   2,4-D

A translocated phenoxycarboxylic acid herbicide for cereals, grass and amenity use
HRAC mode of action code: 4 (O)

See also 2,4-D + dicamba
2,4-D + dicamba + fluroxypyr
2,4-D + dicamba + MCPA + mecoprop-P
2,4-D + florasulam
2,4-D + glyphosate
2,4-D + MCPA

Products
1	Darby	Barclay	500 g/l	SL	19780
2	Depitone Ultra	Nufarm UK	600 g/l	EC	17615
3	Depitox 500	Nufarm UK	500 g/l	SL	17597
4	HY-D Super	Agrichem	500 g/l	SL	13198
5	Stapler	Barclay	500 g/l	SL	19781

Uses
- Annual dicotyledons in **Amenity grassland** [1-5]; **Forage maize** [2]; **Grain maize** [2]; **Grassland** [1-3,5]; **Managed amenity turf** [1-5]; **Miscanthus** *(off-label)* [4]; **Permanent grassland** [4]; **Spring barley** [1-5]; **Spring rye** [1-3,5]; **Spring wheat** [1-5]; **Undersown barley** [1,3-5]; **Undersown oats** [1,3-5]; **Undersown rye** [1,3-5]; **Undersown wheat** [1,3-5]; **Winter barley** [1-5]; **Winter oats** [1-5]; **Winter rye** [1-5]; **Winter wheat** [1-5];
- Perennial dicotyledons in **Amenity grassland** [1-5]; **Forage maize** [2]; **Grain maize** [2]; **Grassland** [1-3,5]; **Managed amenity turf** [1-5]; **Permanent grassland** [4]; **Spring barley** [1-5]; **Spring rye** [1-3,5]; **Spring wheat** [1-5]; **Undersown barley** [1,3-5]; **Undersown oats** [1,3-5]; **Undersown rye** [1,3-5]; **Undersown wheat** [1,3-5]; **Winter barley** [1-5]; **Winter oats** [1-5]; **Winter rye** [1-5]; **Winter wheat** [1-5];

Extension of Authorisation for Minor Use (EAMUs)
- **Miscanthus** *20082914* [4]

Approval information
- 2,4-D included in Annex I under EC Regulation 1107/2009
- Accepted by BBPA for use on malting barley

SEE SECTION 3 FOR PRODUCTS ALSO REGISTERED

Efficacy guidance
- Best results achieved by spraying weeds in seedling to young plant stage when growing actively in a strongly competing crop
- Most effective stage for spraying perennials varies with species. See label for details

Restrictions
- Maximum number of treatments normally 1 per crop and in forestry, 1 per yr in grassland and 2 or 3 per yr in amenity turf. Check individual labels
- Do not use on newly sown leys containing clover
- Do not spray grass seed crops after ear emergence
- Do not spray within 6 mth of laying turf or sowing fine grass
- Do not dump surplus herbicide in water or ditch bottoms
- Do not plant conifers until at least 1 mth after treatment
- Do not spray crops stressed by cold weather or drought or if frost expected
- Do not roll or harrow within 7 d before or after spraying
- Do not spray if rain falling or imminent
- Do not mow or roll turf or amenity grass 4 d before or after spraying. The first 4 mowings after treatment must be composted for at least 6 mth before use
- Do not mow grassland or graze for at least 10 d after spraying

Crop-specific information
- Latest use: before 1st node detectable in cereals; end Aug for conifer plantations; before established grassland 25 cm high
- Spray winter cereals in spring when leaf-sheath erect but before first node detectable (GS 31), spring cereals from 5-leaf stage to before first node detectable (GS 15-31)
- Cereals undersown with grass and/or clover, but not with lucerne, may be treated
- Selective treatment of resistant conifers can be made in Aug when growth ceased and plants hardened off, spray must be directed if applied earlier. See label for details

Following crops guidance
- Do not use shortly before or after sowing any crop
- Do not plant succeeding crops within 3 mth
- Do not direct drill brassicas or grass/clover mixtures within 3 wk of application

Environmental safety
- Dangerous for the environment
- Dangerous to aquatic higher plants. Do not contaminate surface waters or ditches with chemical or used container
- 2,4-D is active at low concentrations. Take extreme care to avoid drift onto neighbouring crops, especially beet crops, brassicas, most market garden crops including lettuce and tomatoes under glass, pears and vines
- Keep livestock out of treated areas for at least 2 wk following treatment and until poisonous weeds, such as ragwort, have died down and become unpalatable
- Water containing the herbicide must not be used for irrigation purposes within 3 wk of treatment or until the concentration in water is below 0.05 ppm

Hazard classification and safety precautions
Hazard Harmful, Irritant [1-3,5], Dangerous for the environment [4], Harmful if swallowed [1,3-5]
Transport code 9 [1-3,5]
Packaging group III
UN Number 3082, N/C
Risk phrases H315 [2], H317 [2,4], H318 [1,3-5]
Operator protection U05a, U08, U11, U14 [4], U15 [4], U20b; A, C, H, M
Environmental protection E07a, E15a [1-3,5], E15b [4], E34, E38 [4], H410 [1-3,5], H411 [4]
Storage and disposal D01, D02, D05, D09a, D10a [4], D10b [1-3,5], D12a [4]
Medical advice M03, M05a [4]

112   2,4-D + dicamba

A translocated herbicide for use on turf
HRAC mode of action code: 4 + 4 (O + O)

See also dicamba

Products

1 Magneto	Nufarm UK	344:120 g/l	SL	17444
2 Thrust	Nufarm UK	344:120 g/l	SL	15408

Uses
* Annual dicotyledons in **Amenity grassland** [1-2]; **Grassland** [1-2];
* Perennial dicotyledons in **Amenity grassland** [1-2]; **Grassland** [1-2];

Approval information
* 2,4-D and dicamba included in Annex I under EC Regulation 1107/2009
* Accepted by BBPA for use on malting barley

Efficacy guidance
* Best results achieved by application when weeds growing actively in spring or early summer (later with irrigation and feeding)
* More resistant weeds may need repeat treatment after 3 wk
* Improved control of some weeds can be obtained by use of specifed oil adjuvant [2, 1]
* Do not use during drought conditions

Restrictions
* Maximum number of treatments 2 per yr on amenity grass and established grassland [2, 1]
* Do not treat newly sown or turfed areas or grass less than 1 yr old
* Do not treat grass crops intended for seed production [2, 1]
* Do not treat grass suffering from drought, disease or other adverse factors [2, 1]
* Do not roll or harrow for 7 d before or after treatment [2, 1]
* Do not apply when grassland is flowering [2, 1]
* Avoid spray drift onto cultivated crops or ornamentals
* Do not graze grass for at least seven days after spraying [2, 1]
* Do not mow or roll four days before or after application. The first four mowings after treatment must be composted for at least six months before use [2, 1]

Crop-specific information
* Latest use: before grass 25 cm high for amenity grass, established grassland [2]
* The first four mowings after treatment must be composted for at least 6 mth before use

Environmental safety
* Dangerous for the environment [2]
* Toxic to aquatic organisms
* Keep livestock out of treated areas for at least two weeks following treatment and until poisonous weeds, such as ragwort, have died down and become unpalatable [2, 1]

Hazard classification and safety precautions
Hazard Harmful, Dangerous for the environment, Harmful if swallowed
Transport code 9 [1-2]
Packaging group III
UN Number 3082
Risk phrases H317, H318
Operator protection U05a, U11; A, C, H, M
Environmental protection E07a, E15a, E22c [1], E34, E38, H410 [2], H411 [1]
Storage and disposal D01, D02, D05, D09a, D10a, D12a
Medical advice M03

SECTION 2

113   2,4-D + dicamba + fluroxypyr

A translocated and contact herbicide mixture for amenity turf
HRAC mode of action code: 4 + 4 + 4 (O + O + O)

See also dicamba
fluroxypyr

Products

1 Enstar	Headland Amenity	285:52.5:105 g/l	EC	18911
2 Holster XL	Barclay	285:52.5:105 g/l	EC	16407
3 Mascot Crossbar	Rigby Taylor	285:52.5:105 g/l	EC	16454
4 Speedster	Pan Amenity	285:52.5:105 g/l	EC	17025

Uses

- Annual dicotyledons in *Amenity grassland* [1-4]; *Managed amenity turf* [1-4];
- Buttercups in *Amenity grassland* [1-2]; *Managed amenity turf* [1-2];
- Clover in *Amenity grassland* [1-2]; *Managed amenity turf* [1-2];
- Daisies in *Amenity grassland* [1-2]; *Managed amenity turf* [1-2];
- Dandelions in *Amenity grassland* [1-2]; *Managed amenity turf* [1-2];
- Yarrow in *Amenity grassland* [1-2]; *Managed amenity turf* [1-2];

Approval information

- 2,4-D, dicamba and fluroxypyr included in Annex I under EC Regulation 1107/2009

Efficacy guidance

- Apply when weeds actively growing (normally between Apr and Sep) and when soil is moist
- Best results obtained from treatment in spring or early summer before weeds begin to flower
- Do not apply if turf is wet or if rainfall expected within 4 h of treatment. Both circumstances will reduce weed control

Restrictions

- Maximum number of treatments 2 per yr
- Do not mow for 3 d before or after treatment
- Avoid overlapping or overdosing, especially on newly sown turf
- Do not spray in drought conditions or if turf under stress from frost, waterlogging, trace element deficiency, pest or disease attack

Crop-specific information

- Latest use: normally Sep for managed amenity turf
- New turf may be treated in spring provided at least 2 mth have elapsed since sowing

Environmental safety

- Dangerous for the environment
- Toxic to aquatic organisms
- Keep livestock out of treated areas for at least two weeks following treatment and until poisonous weeds, such as ragwort, have died down and become unpalatable
- LERAP Category B [1-4]

Hazard classification and safety precautions

Hazard Harmful, Dangerous for the environment, Harmful if swallowed, Very toxic to aquatic organisms [1-3]
Transport code 9 [1-4]
Packaging group III
UN Number 3082
Risk phrases H304 [1-3], H315, H319 [1-3], H320 [4], H336, H371 [4]
Operator protection U05a, U08, U11, U14, U15, U19a, U20b; A, C, H, M
Environmental protection E07a, E15a, E16a, E34, E38, H410 [1-3], H411 [4]
Storage and disposal D01, D02, D09a, D10a
Medical advice M03, M05a, M05b

FOR FULL CONDITIONS OF USE ALWAYS READ THE PRODUCT LABEL

114   2,4-D + dicamba + MCPA + mecoprop-P

A translocated herbicide mixture for grassland
HRAC mode of action code: 4 + 4 + 4 + 4 (O + O + O + O)

See also dicamba
MCPA
mecoprop-P

Products

1	Enforcer	ICL (Everris) Ltd	70:20:70:42 g/l	SL	17274
2	Longbow	Bayer CropScience	70:20:70:42 g/l	SL	16528

Uses
- Buttercups in *Managed amenity turf* [1-2];
- Clovers in *Managed amenity turf* [1-2];
- Daisies in *Managed amenity turf* [1-2];
- Dandelions in *Managed amenity turf* [1-2];
- Self-heal in *Managed amenity turf* [1-2];

Approval information
- 2,4-D, dicamba, MCPA and mecoprop-P included in Annex I under EC Regulation 1107/2009

Environmental safety
- Some pesticides pose a greater threat of contamination of water than others and mecoprop-P is one of these pesticides. Take special care when applying mecoprop-P near water and do not apply if heavy rain is forecast

Hazard classification and safety precautions
  UN Number N/C
  Risk phrases H319
  Operator protection A, C, H, M
  Environmental protection E07a, E15a, H410
  Storage and disposal D05, D09a, D10a, D12a

115   2,4-D + florasulam

A post-emergence herbicide mixture for control of broad leaved weeds in managed amenity turf
HRAC mode of action code: 4 + 2 (O + B)

See also florasulam

Products

1	Esteron T	ICL (Everris) Ltd	300:6.25 g/l	ME	17432
2	Junction	Rigby Taylor	452:6.25 g/l	ME	12493

Uses
- Clover in *Amenity grassland* [1-2]; *Managed amenity turf* [1-2];
- Daisies in *Amenity grassland* [1-2]; *Managed amenity turf* [1-2];
- Dandelions in *Amenity grassland* [1-2]; *Managed amenity turf* [1-2];
- Plantains in *Amenity grassland* [1-2]; *Managed amenity turf* [1-2];
- Sticky mouse-ear in *Amenity grassland* [1-2]; *Managed amenity turf* [1-2];

Approval information
- 2,4-D and florasulam included in Annex I under EC Regulation 1107/2009

Efficacy guidance
- Best results obtained from treatment of actively growing weeds between Mar and Oct when soil is moist
- Do not apply when rain is imminent or during periods of drought unless irrigation is applied

SEE SECTION 3 FOR PRODUCTS ALSO REGISTERED

SECTION 2

Restrictions
- Maximum number of treatments on managed amenity turf: 1 per yr

Crop-specific information
- Avoid mowing turf 3 d before and after spraying
- Ensure newly sown turf has established before treating. Turf sown in late summer or autumn should not be sprayed until growth is resumed in the following spring

Following crops guidance
- An interval of 4 wk must elapse between application and re-seeding turf

Environmental safety
- Dangerous for the environment
- Toxic to aquatic organisms

Hazard classification and safety precautions
> **Hazard** Harmful, Dangerous for the environment, Harmful if swallowed
> **Transport code** 9 [1-2]
> **Packaging group** III
> **UN Number** 3082
> **Risk phrases** H317
> **Operator protection** U02a, U05a, U08, U14, U20a; A, H
> **Environmental protection** E15a, E34, E38, H410
> **Storage and disposal** D01, D02, D09a, D10b, D12a

116   2,4-D + glyphosate

A total herbicide for weed control in crops and on railway ballast
HRAC mode of action code: 4 + 9 (O + G)

See also glyphosate

Products

1	Kyleo	Nufarm UK	160:240 g/l	SL	16567

Uses
- Annual and perennial weeds in *All edible crops (outdoor)* [1]; *All non-edible seed crops grown outdoors* [1]; *Amenity grassland* [1]; *Amenity vegetation* [1]; *Apple orchards* [1]; *Green cover on land temporarily removed from production* [1]; *Pear orchards* [1]; *Railway tracks* [1];

Approval information
- 2,4-D and glyphosate included in Annex I under EC Regulation 1107/2009

Restrictions
- Do not use under polythene or glass
- Application of lime, fertiliser, farmyard manure and pesticides should be delayed until 5 days after application of Kyleo. After application large quantities of decaying foliage, stolons, roots or rhizomes should be dispersed or buried by thorough cultivation before crop drilling
- Do not use in or alongside hedges
- Do not apply by hand-held equipment
- Windfall (fruit fall to ground) must not be used as food or feeding stuff

Environmental safety
- LERAP Category B [1]

Hazard classification and safety precautions
> **Transport code** 9 [1]
> **Packaging group** III
> **UN Number** 3082
> **Risk phrases** H317, H319
> **Operator protection** U05a, U19a; A, C, H

Environmental protection E15a, E16a, E34, E38, H410
Storage and disposal D01, D09a, D10a, D12a, D12b

117   2,4-D + MCPA

A translocated herbicide mixture for cereals and grass
HRAC mode of action code: 4 + 4 (O + O)

See also MCPA

Products

1	Cirran 360	Nufarm UK	360:315 g/l	SL	17346
2	Lupo	Nufarm UK	360:315 g/l	SL	14931
3	PastureMaster	Nufarm UK	360:315 g/l	SL	17994
4	Polo	Nufarm UK	360:315 g/l	SL	19599

Uses
- Annual dicotyledons in **Grassland** [1-4]; **Spring barley** [1-4]; **Spring wheat** [1-4]; **Winter barley** [1-4]; **Winter oats** [1-4]; **Winter wheat** [1-4];
- Perennial dicotyledons in **Grassland** [1-3]; **Spring barley** [1-4]; **Spring wheat** [1-4]; **Winter barley** [1-4]; **Winter oats** [1-4]; **Winter wheat** [1-4];

Approval information
- 2,4-D and MCPA included in Annex I under EC Regulation 1107/2009
- Accepted by BBPA for use on malting barley

Efficacy guidance
- Best results achieved by spraying weeds in seedling to young plant stage when growing actively in a strongly competing crop
- Most effective stage for spraying perennials varies with species. See label for details

Restrictions
- Maximum number of treatments 1 per crop in cereals and 1 per year in grassland [2]
- Do not spray if rain falling or imminent
- Do not cut grass or graze for at least 10 d after spraying
- Do not use on newly sown leys containing clover or other legumes
- Do not spray crops stressed by cold weather or drought or if frost expected
- Do not use shortly before or after sowing any crop
- Do not roll or harrow within 7 d before or after spraying
- Do not apply before end of February in the year of harvest
- Do not apply by hand-held equipment
- Livestock must be kept out of treated areas for at least 2 weeks following treatment and until poisonous weeds such as ragwort have died and become unpalatable.

Crop-specific information
- Spray winter cereals in spring when leaf-sheath erect but before first node detectable (GS 31), spring cereals from 5-leaf stage to before first node detectable (GS 15-31)
- Latest use: before first node detectable (GS 31) in cereals;

Environmental safety
- Dangerous for the environment
- Toxic to aquatic organisms
- Keep livestock out of treated areas for at least two weeks following treatment and until poisonous weeds, such as ragwort, have died down and become unpalatable
- 2,4-D and MCPA are active at low concentrations. Take extreme care to avoid drift onto neighbouring crops, especially beet crops, brassicas, most market garden crops including lettuce and tomatoes under glass, pears and vines

Hazard classification and safety precautions
Hazard Harmful, Dangerous for the environment, Very toxic to aquatic organisms [4]
Transport code 9 [1-4]

SEE SECTION 3 FOR PRODUCTS ALSO REGISTERED

Packaging group III
UN Number 3082
Risk phrases H318
Operator protection U05a, U08, U11, U14, U15 [4], U20b; A, C, H, M
Environmental protection E07a, E15a, E34, E38 [1-3], H410 [1,3], H411 [4]
Storage and disposal D01, D02, D05, D09a, D10a, D12a [1-3]
Medical advice M03, M05a

118   daminozide

A hydrazide plant growth regulator for use in certain ornamentals

Products

1	B-Nine SG	UPL Europe	85% w/w	SG	19611
2	Dazide Enhance	Fine	85% w/w	SG	16092

Uses
- Growth regulation in **Protected ornamentals** [1-2];

Approval information
- Daminozide included in Annex I under EC Regulation 1107/2009

Efficacy guidance
- Best results obtained by applying during a cool part of the day to well-watered but dry plants
- Use a compressed air or power sprayer, which must be thoroughly cleansed before and after use
- Response to treatment differs widely depending on variety, stage of growth and physiological condition of the plant. It is recommended that any new variety is first tested on a small scale to observe if adverse effects occur

Restrictions
- Apply only to turgid, well watered plants. Do not water for 24 h after spraying
- Do not mix with other spray chemicals unless specifically recommended
- Do not store product in metal containers

Crop-specific information
- Do not use on chrysanthemum variety Fandango
- See label for guidance on a range of bedding plant species

Hazard classification and safety precautions
  UN Number N/C
  Operator protection U02a [2], U05a, U08 [2], U20b [2]; A, H
  Environmental protection E15b
  Storage and disposal D01, D02 [1], D09a, D10b [2], D12a [2]
  Medical advice M05a [2]

119   deltamethrin

A pyrethroid insecticide with contact and residual activity
IRAC mode of action code: 3

Products

1	Decis Protech	Bayer CropScience	15 g/l	EW	16160
2	Grainstore	Pan Agriculture	25 g/l	EC	14932
3	Grainstore 25EC	Pan Agriculture	25.4 g/l	EC	19218
4	Grainstore ULV	Pan Agriculture	0.69% w/v	UL	19195
5	K-Obiol EC 25	Bayer CropScience	25 g/l	EC	13573
6	K-Obiol ULV 6	Bayer CropScience	0.69% w/v	UL	13572

Uses
- American serpentine leaf miner in **Protected ornamentals** *(off-label)* [1];

FOR FULL CONDITIONS OF USE ALWAYS READ THE PRODUCT LABEL

- Aphids in *Amenity vegetation* [1]; *Apples* [1]; *Borecole* *(off-label)* [1]; *Celeriac* *(off-label)* [1]; *Celery (outdoor)* *(off-label)* [1]; *Collards* *(off-label)* [1]; *Cress* *(off-label)* [1]; *Endives* *(off-label)* [1]; *Evening primrose* *(off-label)* [1]; *Fennel* *(off-label)* [1]; *Grass seed crops* *(off-label)* [1]; *Herbs (see appendix 6)* *(off-label)* [1]; *Jerusalem artichokes* *(off-label)* [1]; *Kale* [1]; *Lamb's lettuce* *(off-label)* [1]; *Land cress* *(off-label)* [1]; *Leaves and shoots* *(off-label)* [1]; *Lettuce* [1]; *Oriental cabbage* *(off-label)* [1]; *Protected peppers* [1]; *Radishes* *(off-label)* [1]; *Red beet* *(off-label)* [1]; *Rhubarb* *(off-label)* [1]; *Rye* *(off-label)* [1]; *Salsify* *(off-label)* [1]; *Spring barley* [1]; *Spring greens* *(off-label)* [1]; *Spring oats* [1]; *Spring wheat* [1]; *Triticale* *(off-label)* [1]; *Winter barley* [1]; *Winter oats* [1]; *Winter wheat* [1];
- Apple sucker in *Apples* [1];
- Barley yellow dwarf virus vectors in *Winter barley* [1]; *Winter wheat* [1];
- Beet flea beetle in *Red beet* *(off-label)* [1];
- Beet fly in *Red beet* *(off-label)* [1];
- Beet virus yellows vectors in *Winter oilseed rape* [1];
- Black bean aphid in *Dwarf beans* *(off-label)* [1]; *Edible podded peas* *(off-label)* [1]; *French beans* *(off-label)* [1]; *Runner beans* *(off-label)* [1];
- Bramble shoot moth in *Blackberries* *(off-label)* [1];
- Cabbage seed weevil in *Spring mustard* [1]; *Spring oilseed rape* [1]; *Winter mustard* [1]; *Winter oilseed rape* [1];
- Cabbage stem flea beetle in *Winter oilseed rape* [1];
- Cabbage stem weevil in *Spring mustard* [1]; *Spring oilseed rape* [1]; *Winter mustard* [1]; *Winter oilseed rape* [1];
- Capsids in *Amenity vegetation* [1]; *Apples* [1]; *Blackcurrants* *(off-label)* [1]; *Celery (outdoor)* *(off-label)* [1]; *Fennel* *(off-label)* [1]; *Gooseberries* *(off-label)* [1]; *Redcurrants* *(off-label)* [1]; *Rhubarb* *(off-label)* [1];
- Carrot fly in *Carrots* *(off-label)* [1]; *Horseradish* *(off-label)* [1]; *Mallow (Althaea spp.)* *(off-label)* [1]; *Parsnips* *(off-label)* [1];
- Caterpillars in *Amenity vegetation* [1]; *Apples* [1]; *Blackberries* *(off-label)* [1]; *Blackcurrants* *(off-label)* [1]; *Borecole* *(off-label)* [1]; *Brussels sprouts* [1]; *Cabbages* [1]; *Cauliflowers* [1]; *Celeriac* *(off-label)* [1]; *Celery (outdoor)* *(off-label)* [1]; *Collards* *(off-label)* [1]; *Cress* *(off-label)* [1]; *Endives* *(off-label)* [1]; *Evening primrose* *(off-label)* [1]; *Fennel* *(off-label)* [1]; *Gooseberries* *(off-label)* [1]; *Herbs (see appendix 6)* *(off-label)* [1]; *Jerusalem artichokes* *(off-label)* [1]; *Kale* *(off-label)* [1]; *Lamb's lettuce* *(off-label)* [1]; *Land cress* *(off-label)* [1]; *Leaves and shoots* *(off-label)* [1]; *Lettuce* [1]; *Oriental cabbage* *(off-label)* [1]; *Protected celery* *(off-label)* [1]; *Protected peppers* [1]; *Protected rhubarb* *(off-label)* [1]; *Radishes* *(off-label)* [1]; *Red beet* *(off-label)* [1]; *Redcurrants* *(off-label)* [1]; *Rhubarb* *(off-label)* [1]; *Salsify* *(off-label)* [1]; *Spring greens* *(off-label)* [1]; *Strawberries* *(off-label)* [1]; *Swedes* [1]; *Turnips* [1];
- Cereal flea beetle in *Rye* *(off-label)* [1]; *Triticale* *(off-label)* [1];
- Codling moth in *Apples* [1];
- Cutworms in *Carrots* *(off-label)* [1]; *Celery (outdoor)* *(off-label)* [1]; *Fennel* *(off-label)* [1]; *Herbs (see appendix 6)* *(off-label)* [1]; *Horseradish* *(off-label)* [1]; *Leeks* *(off-label)* [1]; *Lettuce* [1]; *Mallow (Althaea spp.)* *(off-label)* [1]; *Parsnips* *(off-label)* [1]; *Rhubarb* *(off-label)* [1];
- Flea beetle in *Borecole* *(off-label)* [1]; *Celery (outdoor)* *(off-label)* [1]; *Collards* *(off-label)* [1]; *Cress* *(off-label)* [1]; *Endives* *(off-label)* [1]; *Fennel* *(off-label)* [1]; *Herbs (see appendix 6)* *(off-label)* [1]; *Kale* *(off-label)* [1]; *Lamb's lettuce* *(off-label)* [1]; *Land cress* *(off-label)* [1]; *Leaves and shoots* *(off-label)* [1]; *Lettuce* [1]; *Oriental cabbage* *(off-label)* [1]; *Radishes* *(off-label)* [1]; *Rhubarb* *(off-label)* [1]; *Salsify* *(off-label)* [1]; *Spring greens* *(off-label)* [1]; *Sugar beet* [1];
- Gall midge in *Dwarf beans* *(off-label)* [1]; *Edible podded peas* *(off-label)* [1]; *French beans* *(off-label)* [1]; *Runner beans* *(off-label)* [1];
- Gout fly in *Rye* *(off-label)* [1]; *Triticale* *(off-label)* [1];
- Insect pests in *Crop handling & storage structures* [5]; *Evening primrose* *(off-label)* [1]; *Grain* *(off-label)* [6]; *Grain stores* [2-3,5]; *Stored grain* [2-6]; *Stored pulses* [2-6];
- Leaf beetle in *Celeriac* *(off-label)* [1]; *Jerusalem artichokes* *(off-label)* [1]; *Radishes* *(off-label)* [1]; *Salsify* *(off-label)* [1];
- Leaf miner in *Herbs (see appendix 6)* *(off-label)* [1]; *Protected aubergines* *(off-label)* [1]; *Protected courgettes* *(off-label)* [1]; *Protected cucumbers* *(off-label)* [1]; *Protected gherkins* *(off-label)* [1]; *Protected spring onions* *(off-label)* [1]; *Protected summer squash* *(off-label)* [1]; *Protected tomatoes* *(off-label)* [1];
- Leaf moth in *Leeks* *(off-label)* [1];

SEE SECTION 3 FOR PRODUCTS ALSO REGISTERED

- Leafhoppers in **Lettuce** [1];
- Mealybugs in **Amenity vegetation** [1]; **Protected peppers** [1];
- Pea and bean weevil in **Broad beans** [1]; **Combining peas** [1]; **Dwarf beans** *(off-label)* [1]; **Edible podded peas** *(off-label)* [1]; **French beans** *(off-label)* [1]; **Runner beans** *(off-label)* [1]; **Spring field beans** [1]; **Vining peas** [1]; **Winter field beans** [1];
- Pea aphid in **Dwarf beans** *(off-label)* [1]; **Edible podded peas** *(off-label)* [1]; **French beans** *(off-label)* [1]; **Runner beans** *(off-label)* [1];
- Pea midge in **Combining peas** [1]; **Vining peas** [1];
- Pea moth in **Combining peas** [1]; **Dwarf beans** *(off-label)* [1]; **Edible podded peas** *(off-label)* [1]; **French beans** *(off-label)* [1]; **Runner beans** *(off-label)* [1]; **Vining peas** [1];
- Peach-potato aphid in **Dwarf beans** *(off-label)* [1]; **Edible podded peas** *(off-label)* [1]; **French beans** *(off-label)* [1]; **Runner beans** *(off-label)* [1];
- Pear sucker in **Pears** [1];
- Pollen beetle in **Lettuce** [1]; **Spring mustard** [1]; **Spring oilseed rape** [1]; **Winter mustard** [1]; **Winter oilseed rape** [1];
- Raspberry beetle in **Blackberries** *(off-label)* [1]; **Raspberries** [1];
- Saddle gall midge in **Rye** *(off-label)* [1]; **Triticale** *(off-label)* [1];
- Sawflies in **Apples** [1]; **Blackcurrants** *(off-label)* [1]; **Gooseberries** *(off-label)* [1]; **Redcurrants** *(off-label)* [1];
- Scale insects in **Amenity vegetation** [1]; **Protected peppers** [1];
- Silver Y moth in **Dwarf beans** *(off-label)* [1]; **Edible podded peas** *(off-label)* [1]; **French beans** *(off-label)* [1]; **Runner beans** *(off-label)* [1];
- Spotted wing drosophila in **Protected strawberries** *(off-label)* [1];
- Thrips in **Amenity vegetation** [1]; **Blackcurrants** *(off-label)* [1]; **Celeriac** *(off-label)* [1]; **Dwarf beans** *(off-label)* [1]; **Edible podded peas** *(off-label)* [1]; **French beans** *(off-label)* [1]; **Gooseberries** *(off-label)* [1]; **Jerusalem artichokes** *(off-label)* [1]; **Lettuce** [1]; **Protected peppers** [1]; **Radishes** *(off-label)* [1]; **Red beet** *(off-label)* [1]; **Redcurrants** *(off-label)* [1]; **Runner beans** *(off-label)* [1]; **Salsify** *(off-label)* [1]; **Strawberries** *(off-label)* [1];
- Tortrix moths in **Apples** [1];
- Weevils in **Celeriac** *(off-label)* [1]; **Jerusalem artichokes** *(off-label)* [1]; **Radishes** *(off-label)* [1]; **Salsify** *(off-label)* [1];
- Western flower thrips in **Protected aubergines** *(off-label)* [1]; **Protected courgettes** *(off-label)* [1]; **Protected cucumbers** *(off-label)* [1]; **Protected gherkins** *(off-label)* [1]; **Protected ornamentals** *(off-label)* [1]; **Protected peppers** [1]; **Protected strawberries** *(off-label)* [1]; **Protected summer squash** *(off-label)* [1]; **Protected tomatoes** *(off-label)* [1];
- Whitefly in **Amenity vegetation** [1]; **Borecole** *(off-label)* [1]; **Collards** *(off-label)* [1]; **Herbs (see appendix 6)** *(off-label)* [1]; **Kale** *(off-label)* [1]; **Oriental cabbage** *(off-label)* [1]; **Protected peppers** [1]; **Spring greens** *(off-label)* [1];
- Wireworm in **Herbs (see appendix 6)** *(off-label)* [1];
- Yellow cereal fly in **Rye** *(off-label)* [1]; **Triticale** *(off-label)* [1];

Extension of Authorisation for Minor Use (EAMUs)
- **Blackberries** *20131673* [1]
- **Blackcurrants** *20190635* [1]
- **Borecole** *20201050* [1]
- **Carrots** *20131672* [1]
- **Celeriac** *20190679* [1]
- **Celery (outdoor)** *20201053* [1]
- **Collards** *20201050* [1]
- **Cress** *20131670* [1]
- **Dwarf beans** *20190678* [1]
- **Edible podded peas** *20190678* [1]
- **Endives** *20131670* [1]
- **Evening primrose** *20142584* [1]
- **Fennel** *20201053* [1]
- **French beans** *20190678* [1]
- **Gooseberries** *20190635* [1]
- **Grain** *20112491* [6]
- **Grass seed crops** *20131671* [1]

- *Herbs (see appendix 6)* 20191272 [1]
- *Horseradish* 20131672 [1]
- *Jerusalem artichokes* 20190679 [1]
- *Kale* 20201050 [1]
- *Lamb's lettuce* 20131670 [1]
- *Land cress* 20131670 [1]
- *Leaves and shoots* 20131670 [1]
- *Leeks* 20131665 [1]
- *Mallow (Althaea spp.)* 20131672 [1]
- *Oriental cabbage* 20201050 [1]
- *Parsnips* 20131672 [1]
- *Protected aubergines* 20201051 [1]
- *Protected celery* 20142585 [1]
- *Protected courgettes* 20201052 [1]
- *Protected cucumbers* 20201052 [1]
- *Protected gherkins* 20201052 [1]
- *Protected ornamentals* 20131662 [1]
- *Protected rhubarb* 20142585 [1]
- *Protected spring onions* 20131664 [1]
- *Protected strawberries* 20211296 [1]
- *Protected summer squash* 20201052 [1]
- *Protected tomatoes* 20201051 [1]
- *Radishes* 20190679 [1]
- *Red beet* 20190679 [1]
- *Redcurrants* 20190635 [1]
- *Rhubarb* 20201053 [1]
- *Runner beans* 20190678 [1]
- *Rye* 20193280 [1]
- *Salsify* 20190679 [1]
- *Spring greens* 20201050 [1]
- *Strawberries* 20190636 [1]
- *Triticale* 20193280 [1]

Approval information
- Deltamethrin included in Annex I under EC Regulation 1107/2009
- Accepted by BBPA for use on malting barley and hops

Efficacy guidance
- A contact and stomach poison with 3-4 wk persistence, particularly effective on caterpillars and sucking insects
- Normally applied at first signs of damage with follow-up treatments where necessary at 10-14 d intervals. Rates, timing and recommended combinations with other pesticides vary with crop and pest. See label for details
- Spray is rainfast within 1 h
- May be applied in frosty weather provided foliage not covered in ice
- Temperatures above 35°C may reduce effectiveness or persistence

Restrictions
- Maximum number of treatments varies with crop and pest, 4 per crop for wheat and barley, only 1 application between 1 Apr and 31 Aug. See label for other crops
- Do not apply more than 1 aphicide treatment to cereals in summer
- Do not spray crops suffering from drought or other physical stress
- Consult processer before treating crops for processing
- Do not apply to a cereal crop if any product containing a pyrethroid insecticide or dimethoate has been applied to that crop after the start of ear emergence (GS 51)
- Do not spray cereals after 31 Mar in the year of harvest within 6 m of the outside edge of the crop
- Reduced volume spraying must not be used on cereals after 31 Mar in yr of harvest

SEE SECTION 3 FOR PRODUCTS ALSO REGISTERED

Crop-specific information
- Latest use: early dough (GS 83) for barley, oats, wheat; before flowering for mustard, oilseed rape; before 31 Mar for grass seed crops [1]

Environmental safety
- Dangerous for the environment
- Toxic to aquatic organisms
- Flammable [1]
- Extremely dangerous to fish or other aquatic life. Do not contaminate surface waters or ditches with chemical or used container
- Dangerous to bees. Do not apply to crops in flower or to those in which bees are actively foraging. Do not apply when flowering weeds are present
- Do not apply in tank-mixture with a triazole-containing fungicide when bees are likely to be actively foraging in the crop
- To protect non target insects/arthropods respect an unsprayed buffer zone of 5 m to non-crop land [1]
- Buffer zone requirement 7 m [1]
- Broadcast air-assisted LERAP 30m in raspberries, 50m in apples and pears [1] ; LERAP category B [1]
- Broadcast air-assisted LERAP [2,1] (30 or 50m); LERAP Category A [1-3]

Hazard classification and safety precautions
Hazard Harmful [2-3], Flammable [2-3], Dangerous for the environment [1-3], Harmful if swallowed [2-3], Harmful if inhaled [2-3], Very toxic to aquatic organisms [1]
Transport code 3 [1-3], 9 [4-6]
Packaging group III
UN Number 1993, 3082
Risk phrases H315 [2-3], H318 [2-3], H371 [2-3]
Operator protection U04a [2-6], U05a [2-6], U08 [2-3], U09a, U14 [1], U19a, U20b [1-3], U20c; A, C, D, H, M
Environmental protection E12a [2-3], E12f (cereals, oilseed rape, peas, beans - see label for guidance) [1-3], E15a [2-3], E15b [1,4-6], E16c [1-3], E16d [2-3], E17a (18 m) [3], E17b (18 m) [2], E17b (30 or 50m) [1], E22c [1], E34, E38 [1-3], H410 [1], H411 [2-3]
Storage and disposal D01, D02, D05 [2-3], D09a, D10c, D12a [1-3]
Medical advice M03 [2-6], M05b [2-3]

120   dicamba

A translocated benzoic herbicide available in mixtures or formulated alone for weed control in maize
HRAC mode of action code: 4 (O)

See also 2,4-D + dicamba
2,4-D + dicamba + fluroxypyr
2,4-D + dicamba + MCPA + mecoprop-P
dicamba + mecoprop-P
dicamba + nicosulfuron
dicamba + prosulfuron

Products

1 Oceal	Rotam	70% w/w	SG	15618

Uses
- Amaranthus in *Maize* [1];
- Fat hen in *Maize* [1];
- Field pansy in *Maize* [1];
- Maple-leaved goosefoot in *Maize* [1];

Approval information
- Dicamba included in Annex I under EC Regulation 1107/2009

FOR FULL CONDITIONS OF USE ALWAYS READ THE PRODUCT LABEL

Restrictions
- Do not treat maize during cold or frosty conditions, during periods of high day/night temperature variations or during periods of very high temperatures.
- Avoid drift into greenhouses or onto other agricultural or horticultural crops, amenity plantings or gardens. No to be used in glasshouses. Do not allow spray applications to come into contact with desired broad-leaved trees. Beets, all brassicae (including oilseed rape), lettuce, peas, tomatoes, potatoes, all crops and ornamentals are particularly susceptible to dicamba and may be damaged by spray drift.
- Application may only be made between 1 May and 30 September

Following crops guidance
- In the event of crop failure, only redrill with maize but any crop may be sown after a normal harvest date

Hazard classification and safety precautions
Hazard Dangerous for the environment
Transport code 9 [1]
Packaging group III
UN Number 3077
Operator protection U20b
Environmental protection E15b, E38, H411
Storage and disposal D01, D02, D10c, D12b

121   dicamba + MCPA + mecoprop-P

A translocated herbicide for cereals, grassland, amenity grass and orchards
HRAC mode of action code: 4 + 4 + 4 (O + O + O)

See also MCPA
mecoprop-P

Products

1	Hyprone P	UPL Europe	16:101:92 g/l	SL	17418
2	Hysward-P	Agrichem	16:101:92 g/l	SL	17609
3	Hysward-p	Agrichem	16:101:92 g/l	SL	19767
4	Mircam Plus	Nufarm UK	19.5:245:43.3 g/l	SL	15868
5	Relay Turf Elite	Nufarm UK	19.5:245:43.3 g/l	SL	19633
6	T2 Green Pro	Nufarm UK	19.5:245:43.3 g/l	SL	16366
7	Turfmaster	Nufarm UK	19.5:245:43.3 g/l	SL	16344

Uses
- Annual dicotyledons in *Amenity grassland* [2-7]; *Canary seed* (off-label) [4]; *Grass seed crops* [4-5]; *Managed amenity turf* [2-7]; *Spring barley* [1,4]; *Spring oats* [1,4]; *Spring wheat* [1,4]; *Winter barley* [1,4]; *Winter oats* [1,4]; *Winter wheat* [1,4];
- Black bindweed in *Amenity grassland* [6-7]; *Managed amenity turf* [6-7];
- Charlock in *Amenity grassland* [6-7]; *Managed amenity turf* [6-7];
- Chickweed in *Amenity grassland* [6-7]; *Managed amenity turf* [6-7];
- Clover in *Amenity grassland* (moderately susceptible) [6-7]; *Managed amenity turf* (moderately susceptible) [6-7];
- Common orache in *Amenity grassland* [6-7]; *Managed amenity turf* [6-7]; *Rye* (off-label) [4]; *Triticale* (off-label) [4];
- Dandelions in *Amenity grassland* (moderately susceptible) [6-7]; *Managed amenity turf* (moderately susceptible) [6-7];
- Docks in *Amenity grassland* [2-3]; *Managed amenity turf* [2-3];
- Fat hen in *Amenity grassland* [6-7]; *Managed amenity turf* [6-7];
- Forget-me-not in *Amenity grassland* [6-7]; *Managed amenity turf* [6-7]; *Rye* (off-label) [4]; *Triticale* (off-label) [4];
- Fumitory in *Amenity grassland* [6-7]; *Managed amenity turf* [6-7];
- Groundsel in *Amenity grassland* [6-7]; *Managed amenity turf* [6-7];
- Knotgrass in *Amenity grassland* [6-7]; *Managed amenity turf* [6-7];

SEE SECTION 3 FOR PRODUCTS ALSO REGISTERED

- Mayweeds in *Amenity grassland* [6-7]; *Managed amenity turf* [6-7];
- Pale persicaria in *Amenity grassland* [6-7]; *Managed amenity turf* [6-7]; *Rye* (off-label) [4]; *Triticale* (off-label) [4];
- Penny cress in *Amenity grassland* [6-7]; *Managed amenity turf* [6-7];
- Perennial dicotyledons in *Amenity grassland* [2-7]; *Grass seed crops* [4-5]; *Managed amenity turf* [2-7]; *Spring barley* [1,4]; *Spring oats* [1,4]; *Spring wheat* [1,4]; *Winter barley* [1,4]; *Winter oats* [1,4]; *Winter wheat* [1,4];
- Plantains in *Amenity grassland* (moderately susceptible) [6-7]; *Managed amenity turf* (moderately susceptible) [6-7];
- Poppies in *Amenity grassland* [6-7]; *Managed amenity turf* [6-7]; *Rye* (off-label) [4]; *Triticale* (off-label) [4];
- Redshank in *Amenity grassland* [6-7]; *Managed amenity turf* [6-7];
- Shepherd's purse in *Amenity grassland* [6-7]; *Managed amenity turf* [6-7];
- Wild radish in *Amenity grassland* [6-7]; *Managed amenity turf* [6-7];

Extension of Authorisation for Minor Use (EAMUs)
- *Canary seed* 20160549 [4]
- *Rye* 20180355 [4]
- *Triticale* 20180355 [4]

Approval information
- Dicamba, MCPA and mecoprop-P included in Annex I under EC Regulation 1107/2009
- Accepted by BBPA for use on malting barley

Efficacy guidance
- Treatment should be made when weeds growing actively. Weeds hardened by winter weather may be less susceptible
- For best results apply in fine warm weather, preferably when soil is moist. Do not spray if rain expected within 6 h or in drought
- Application of fertilizer 1-2 wk before spraying aids weed control in turf
- Where a second treatment later in the season is needed in amenity situations and on grass allow 4-6 wk between applications to permit sufficient foliage regrowth for uptake

Restrictions
- Maximum number of treatments (including other mecoprop-P products) or maximum total dose varies with crop and product. See label for details. The total amount of mecoprop-P applied in a single yr must not exceed the maximum total dose approved for any single product for the crop/situation
- Do not apply to cereals after the first node is detectable (GS 31), or to grass under stress from drought or cold weather
- Applications to cereals must not be made between 1st October and 1st March [4]
- Do not spray cereals undersown with clovers or legumes, to be undersown with grass or legumes or grassland where clovers or other legumes are important
- Do not spray leys established less than 18 mth or orchards established less than 3 yr
- Do not roll or harrow within 7 d before or after treatment, or graze for at least 7 d afterwards (longer if poisonous weeds present)
- Do not use on turf or grass in year of establishment. Allow 6-8 wk after treatment before seeding bare patches
- The first 4 mowings after use should not be used for mulching unless composted for 6 mth
- Turf should not be mown for 24 h before or after treatment (3-4 d for closely mown turf)
- Avoid drift onto all broad-leaved plants outside the target area

Crop-specific information
- Latest use: before first node detectable (GS 31) for cereals; 5-6 wk before head emergence for grass seed crops; mid-Oct for established grass
- HI 7-14 d before cutting or grazing for leys, permanent pasture
- Apply to winter cereals from the leaf sheath erect stage (GS 30), and to spring cereals from the 5 expanded leaf stage (GS 15)
- Spray grass seed crops 4-6 wk before flower heads begin to emerge (Timothy 6 wk)
- Turf containing bulbs may be treated once the foliage has died down completely [2]

FOR FULL CONDITIONS OF USE ALWAYS READ THE PRODUCT LABEL

Environmental safety
- Harmful to aquatic organisms
- Keep livestock out of treated areas for at least 2 wk following treatment and until poisonous weeds, such as ragwort, have died down and become unpalatable
- Harmful to fish or other aquatic life. Do not contaminate surface waters or ditches with chemical or used container
- Some pesticides pose a greater threat of contamination of water than others and mecoprop-P is one of these pesticides. Take special care when applying mecoprop-P near water and do not apply if heavy rain is forecast

Hazard classification and safety precautions
Hazard Harmful [6-7], Irritant [1-5], Dangerous for the environment [4-7], Very toxic to aquatic organisms [7]
Transport code 9 [4-7]
Packaging group III
UN Number 3082, N/C
Risk phrases H315 [1-3], H318
Operator protection U05a, U08, U11, U15 [4-5], U19a [2-3], U20b; A, C, H, M
Environmental protection E07a, E13c [1-3], E15a [4-5], E15b [6,7], E34 [4-7], E38 [4-5], H410 [4-7], H411 [1-3]
Storage and disposal D01, D02, D05, D09a, D10b, D10c [1-3], D12a [4-5]
Medical advice M03 [4-7], M05a [1-3]

122   dicamba + mecoprop-P

A translocated post-emergence herbicide for cereals and grassland
HRAC mode of action code: 4 + 4 (O + O)

See also mecoprop-P

Products
1	High Load Mircam	Nufarm UK	80:600 g/l	SL	11930
2	Hyban P	Agrichem	18.7:150 g/l	SL	16799
3	Hygrass P	Agrichem	18.7:150 g/l	SL	19771
4	Prompt	Nufarm UK	84:600 g/l	SL	19604
5	Quickfire	Headland Amenity	18.7:150 g/l	SL	17245
6	Saxon	Nufarm UK	84:600 g/l	SL	19632

Uses
- Annual dicotyledons in *Amenity grassland* [1,4,6]; *Game cover* (off-label) [1]; *Managed amenity turf* [2-3,5]; *Rye* (off-label) [1]; *Spring barley* [1-2,4,6]; *Spring oats* [1-2,4,6]; *Spring wheat* [1-2,4,6]; *Triticale* (off-label) [1]; *Winter barley* [1-2,4,6]; *Winter oats* [1-2,4,6]; *Winter wheat* [1-2,4,6];
- Chickweed in *Amenity grassland* [1]; *Game cover* (off-label) [1]; *Managed amenity turf* [2]; *Rye* (off-label) [1]; *Spring barley* [1-2,4,6]; *Spring oats* [1-2,4,6]; *Spring wheat* [1-2,4,6]; *Triticale* (off-label) [1]; *Winter barley* [1-2,4,6]; *Winter oats* [1-2,4,6]; *Winter wheat* [1-2,4,6];
- Cleavers in *Amenity grassland* [1]; *Game cover* (off-label) [1]; *Managed amenity turf* [2]; *Rye* (off-label) [1]; *Spring barley* [1-2,4,6]; *Spring oats* [1-2,4,6]; *Spring wheat* [1-2,4,6]; *Triticale* (off-label) [1]; *Winter barley* [1-2,4,6]; *Winter oats* [1-2,4,6]; *Winter wheat* [1-2,4,6];
- Docks in *Managed amenity turf* [3,5];
- Mayweeds in *Amenity grassland* [1]; *Managed amenity turf* [2]; *Rye* (off-label) [1]; *Spring barley* [1-2,4,6]; *Spring oats* [1-2,4,6]; *Spring wheat* [1-2,4,6]; *Triticale* (off-label) [1]; *Winter barley* [1-2,4,6]; *Winter oats* [1-2,4,6]; *Winter wheat* [1-2,4,6];
- Perennial dicotyledons in *Amenity grassland* [4,6]; *Managed amenity turf* [2-3,5]; *Spring barley* [2]; *Spring oats* [2]; *Spring wheat* [2]; *Winter barley* [2]; *Winter oats* [2]; *Winter wheat* [2];
- Plantains in *Managed amenity turf* [2]; *Spring barley* [2]; *Spring oats* [2]; *Spring wheat* [2]; *Winter barley* [2]; *Winter oats* [2]; *Winter wheat* [2];

- Polygonums in **Amenity grassland** [1]; **Managed amenity turf** [2]; **Rye** *(off-label)* [1]; **Spring barley** [1-2,4,6]; **Spring oats** [1-2,4,6]; **Spring wheat** [1-2,4,6]; **Triticale** *(off-label)* [1]; **Winter barley** [1-2,4,6]; **Winter oats** [1-2,4,6]; **Winter wheat** [1-2,4,6];
- Thistles in **Managed amenity turf** [3,5];

Extension of Authorisation for Minor Use (EAMUs)
- **Game cover** *20194255* [1]
- **Rye** *20180470* [1]
- **Triticale** *20180470* [1]

Approval information
- Dicamba and mecoprop-P included in Annex I under EC Regulation 1107/2009
- Accepted by BBPA for use on malting barley
- M19604 expires 30/9/2023

Efficacy guidance
- Best results by application in warm, moist weather when weeds are actively growing

Restrictions
- Maximum number of treatments 1 per crop for cereals and 1 or 2 per yr on grass depending on label. The total amount of mecoprop-P applied in a single yr must not exceed the maximum total dose approved for any single product for the crop/situation
- Applications to cereals must not be made between 1st October and 1st March
- Do not spray in cold or frosty conditions
- Do not spray if rain expected within 6 h
- Do not treat undersown grass until tillering begins
- Do not spray cereals undersown with clover or legume mixtures
- Do not roll or harrow within 7 d before or after spraying
- Do not treat crops suffering from stress from any cause
- Use product immediately following dilution; do not allow diluted product to stand before use
- Avoid treatment when drift may damage neighbouring susceptible crops
- Do not use on new grass for at least 6 months after establishment [5]
- Do not cut grass for at least 1 day after treatment
- Do not apply via hand-held equipment [1]

Crop-specific information
- Latest use: before 1st node detectable for cereals; 7 d before cutting or 14 d before grazing grass
- Apply to winter sown crops from 5 expanded leaf stage (GS 15)
- Apply to spring sown cereals from 5 expanded leaf stage but before first node is detectable (GS 15-31)
- Treat grassland just before perennial weeds flower
- Transient crop prostration may occur after spraying but recovery is rapid

Following crops guidance
- The total amount of mecoprop-P applied in a single year must not exceed the maximum total dose approved for that crop for any single product

Environmental safety
- Dangerous for the environment [4, 1]
- Toxic to aquatic organisms
- Harmful to fish or other aquatic life. Do not contaminate surface waters or ditches with chemical or used container
- Keep livestock out of treated areas for at least 2 wk and until foliage of poisonous weeds such as ragwort has died and become unpalatable
- Some pesticides pose a greater threat of contamination of water than others and mecoprop-P is one of these pesticides. Take special care when applying mecoprop-P near water and do not apply if heavy rain is forecast

Hazard classification and safety precautions
Hazard Harmful [1,4,6], Irritant [2-3,5], Dangerous for the environment [1,4,6], Harmful if swallowed [1,4,6], Very toxic to aquatic organisms [4]
Transport code 9 [1,4]

FOR FULL CONDITIONS OF USE ALWAYS READ THE PRODUCT LABEL

Packaging group III
UN Number 3082, N/C
Risk phrases H315 [1,4-6], H318
Operator protection U05a, U08 [2,4,6], U09a [1,3,5], U11, U15 [2-3,5], U19a [3,5], U20b; A, C, H, M
Environmental protection E07a [2-6], E13c [1-3,5], E15a [4,6], E34, E38 [2-6], H410 [1,4], H412 [2-3,5-6]
Storage and disposal D01, D02, D05, D09a, D10b [4,6], D10c [2-3,5], D12a [2-6]
Medical advice M03, M05a [2-3,5]

123   dicamba + nicosulfuron

A herbicide mixture for weed control in forage and grain maize
HRAC mode of action code: 4 + 2 (O + B)

Products

1	Kaltor	Rotam	60:15% w/w	SG	18749
2	Kingsley	Rotam	60:15% w/w	SG	19372

Uses
- Annual dicotyledons in *Forage maize* [1-2]; *Grain maize* [1-2];
- Annual meadow grass in *Forage maize* [1-2];
- Chickweed in *Forage maize* [1-2];
- Cockspur grass in *Forage maize* [1-2];
- Docks in *Forage maize* [1-2]; *Grain maize* [1-2];
- Fat hen in *Forage maize* [1-2];
- Redshank in *Forage maize* [1-2];

Approval information
- Dicamba and nicosulfuron included in Annex I under EC Regulation 1107/2009

Efficacy guidance
- To avoid damage to crops other than maize, immediately after spraying thoroughly clean all spray equipment as advised on the label, including inside and outside the lid

Restrictions
- To avoid the buildup of resistance do not apply this or any other product containing an ALS inhibitor herbicide with claims for control of grass-weeds more than once to any crop
- Application may only be made between 1 May and 30 September
- Do not apply to crops suffering from herbicide damage or stress caused by pests, nutrition defects or weather.
- Do not spray when cold or frosty conditions are prevalent nor during periods of high temperature
- Do not spray maize subjected to substantial day and night temperature fluctuations.
- Avoid overlapping spray swaths as considerable crop damage may occur which may not grow out and could lead to yield reductions.
- Do not mix with foliar or liquid fertilisers.
- In the case of particularly sensitive varieties, e.g. Abraxas, Fjord, Rival and Nancis, some crop damage may be caused from which recovery may not be complete.
- Do not apply in sequence or in tank-mixture with a product containing any other sulfonylurea.
- Do not mix with adjuvants except those specified on the label

Following crops guidance
- Following normal harvest or in the event of crop failure maize or sunflowers may be sown immediately after ploughing, winter wheat may be sown 5 months after harrowing and the following year sunflower, barley, wheat, oilseed rape, maize may be sown after harrowing.

Environmental safety
- LERAP Category B [1-2]

Hazard classification and safety precautions
Transport code 9 [1-2]

SEE SECTION 3 FOR PRODUCTS ALSO REGISTERED

Packaging group III
UN Number 3077
Risk phrases H318
Operator protection U05a, U08, U20b; A, C, H
Environmental protection E15b, E16a, E39, H410
Storage and disposal D01, D02, D09a, D10c, D11a

124   dicamba + prosulfuron

A herbicide mixture for weed control in forage and grain maize
HRAC mode of action code: 4 + 2 (O + B)

Products

1 Rosan	FMC Agro	50:5% w/w	WG	18778

Uses
- Annual dicotyledons in **Forage maize** [1]; **Grain maize** [1];
- Bindweeds in **Forage maize** [1]; **Grain maize** [1];
- Docks in **Forage maize** *(seedlings only)* [1]; **Grain maize** *(seedlings only)* [1];

Approval information
- Dicamba and prosulfuron included in Annex I under EC Regulation 1107/2009

Efficacy guidance
- Always apply in mixture with a non-ionic adjuvant.
- Do not apply in mixture with organo-phosphate insecticides

Following crops guidance
- Following normal harvest wheat, barley, rye, triticale and perennial ryegrass may be sown in the autumn, wheat, barley, rye, triticale, combining peas, maize, field beans, forage kale, broccoli and cauliflower may be sown the following spring but sugar beet, sunflowers or lucerne are not recommended.

Environmental safety
- LERAP Category B [1]

Hazard classification and safety precautions
Hazard Dangerous for the environment, Very toxic to aquatic organisms
Transport code 9 [1]
Packaging group III
UN Number 3077
Operator protection U02a, U05a, U14, U20b; A, C, H
Environmental protection E15b, E16a, E38, H410
Storage and disposal D01, D02, D09a, D10c, D12a

125   dichlorprop-P + MCPA + mecoprop-P

A translocated herbicide mixture for winter and spring cereals
HRAC mode of action code: 4 + 4 + 4 (O + O + O)

See also MCPA
*   mecoprop-P*

Products

1 Duplosan Super	Nufarm UK	310:160:130 g/l	SL	18231
2 Isomec Ultra	Nufarm UK	310:160:130 g/l	SL	16033

Uses
- Annual dicotyledons in **Durum wheat** [1-2]; **Rye** [1-2]; **Spelt** [1-2]; **Spring barley** [1-2]; **Spring oats** [1-2]; **Spring wheat** [1-2]; **Triticale** [1-2]; **Winter barley** [1-2]; **Winter oats** [1-2]; **Winter wheat** [1-2];

FOR FULL CONDITIONS OF USE ALWAYS READ THE PRODUCT LABEL

- Chickweed in **Durum wheat** [1-2]; **Rye** [1-2]; **Spelt** [1-2]; **Spring barley** [1-2]; **Spring oats** [1-2]; **Spring wheat** [1-2]; **Triticale** [1-2]; **Winter barley** [1-2]; **Winter oats** [1-2]; **Winter wheat** [1-2];
- Cleavers in **Durum wheat** [1-2]; **Rye** [1-2]; **Spelt** [1-2]; **Spring barley** [1-2]; **Spring oats** [1-2]; **Spring wheat** [1-2]; **Triticale** [1-2]; **Winter barley** [1-2]; **Winter oats** [1-2]; **Winter wheat** [1-2];
- Field pansy in **Durum wheat** [1-2]; **Rye** [1-2]; **Spelt** [1-2]; **Spring barley** [1-2]; **Spring oats** [1-2]; **Spring wheat** [1-2]; **Triticale** [1-2]; **Winter barley** [1-2]; **Winter oats** [1-2]; **Winter wheat** [1-2];
- Mayweeds in **Durum wheat** [1-2]; **Rye** [1-2]; **Spelt** [1-2]; **Spring barley** [1-2]; **Spring oats** [1-2]; **Spring wheat** [1-2]; **Triticale** [1-2]; **Winter barley** [1-2]; **Winter oats** [1-2]; **Winter wheat** [1-2];
- Poppies in **Durum wheat** [1-2]; **Rye** [1-2]; **Spelt** [1-2]; **Spring barley** [1-2]; **Spring oats** [1-2]; **Spring wheat** [1-2]; **Triticale** [1-2]; **Winter barley** [1-2]; **Winter oats** [1-2]; **Winter wheat** [1-2];

Approval information
- Dichlorprop-P, MCPA and mecoprop-P included in Annex I under EC Regulation 1107/2009
- Accepted by BBPA for use on malting barley

Efficacy guidance
- Best results obtained if application is made while majority of weeds are at seedling stage but not if temperatures are too low
- Optimum results achieved by spraying when temperature is above 10°C. If temperatures are lower delay spraying until growth becomes more active

Restrictions
- Maximum number of treatments 1 per crop
- Do not spray in windy conditions where spray drift may cause damage to neighbouring crops, especially sugar beet, oilseed rape, peas, turnips and most horticultural crops including lettuce and tomatoes under glass
- Do not apply before 1st March in year of application.

Crop-specific information
- Latest use: before second node detectable (GS 32) for all crops

Environmental safety
- Harmful to aquatic organisms
- Harmful to fish or other aquatic life. Do not contaminate surface waters or ditches with chemical or used container
- Some pesticides pose a greater threat of contamination of water than others and mecoprop-P is one of these pesticides. Take special care when applying mecoprop-P near water and do not apply if heavy rain is forecast

Hazard classification and safety precautions
Hazard Harmful, Dangerous for the environment, Harmful if swallowed
Transport code 9 [1-2]
Packaging group III
UN Number 3082
Risk phrases H317, H318, R51
Operator protection U05a, U08, U11, U14, U15, U20b; A, C, H, M
Environmental protection E13c, E34, E38, H410
Storage and disposal D01, D02, D05, D09a, D10c, D12a
Medical advice M05a

126 difenacoum

An anticoagulant coumarin rodenticide

Products

1	Difenag Wax Bait	Killgerm	0.005% w/w	RB	UK14-0851
2	Ratak Cut Wheat	Killgerm	0.005% w/w	RB	UK12-0313
3	Sakarat D Wax Bait	Killgerm	0.005% w/w	RB	UK12-0370

SEE SECTION 3 FOR PRODUCTS ALSO REGISTERED

Uses
- Mice in **Farm buildings** [1-3]; **Farmyards** [1-3];
- Rats in **Farm buildings** [1-2]; **Farmyards** [1-2];

Approval information
- Difenacoum included in Annex I under EC Regulation 1107/2009

Efficacy guidance
- Difenacoum is a chronic poison and rodents need to feed several times before accumulating a lethal dose. Effective against rodents resistant to other commonly used anticoagulants
- Best results achieved by placing baits at points between nesting and feeding places, at entry points, in holes and where droppings are seen
- A minimum of five baiting points normally required for a small infestation; more than 40 for a large infestation
- Inspect bait sites frequently and top up as long as there is evidence of feeding
- Maintain a few baiting points to guard against reinfestation after a successful control campaign
- Resistance to difenacoum in rats is now widespread in the UK. Take local professional advice before relying on these products to control infestations.

Restrictions
- Only for use by farmers, horticulturists and other professional users
- When working in rodent infested areas wear synthetic rubber/PVC gloves to protect against rodent-borne diseases

Environmental safety
- Harmful to wildlife
- Cover baits by placing in bait boxes, drain pipes or under boards to prevent access by children, animals or birds
- Products contain human taste deterrent

Hazard classification and safety precautions
UN Number N/C
Operator protection U13, U20b
Storage and disposal D09a, D11a
Vertebrate/rodent control products V01b [3], V02 [3], V03b [3], V04b [3]
Medical advice M03

127   difenoconazole

A diphenyl-ether triazole protectant and curative fungicide
FRAC mode of action code: 3

See also azoxystrobin + difenoconazole
benzovindiflupyr + difenoconazole

Products
1	Alternet	BelCrop	250 g/l	EC	17689
2	Difcor 250 EC	Certis Belchim B V	250 g/l	EC	13917
3	Difenostar	Life Scientific	250 g/l	EC	19118
4	Difference	Certis Belchim B V	250 g/l	EC	16129
5	Kix	Certis Belchim B V	250 g/l	EC	17424
6	Narita	Certis Belchim B V	250 g/l	EC	16210
7	Piccolo	Agrii	250 g/l	EC	19415
8	Plover	Syngenta	250 g/l	EC	17288
9	Septuna	Novastar	250 g/l	EC	17390
10	Slick	Syngenta	250 g/l	EC	17331

Uses
- Alternaria in **Borage for oilseed production** *(off-label)* [2]; **Broccoli** [2-3,7-9]; **Brussels sprouts** [2-3,7-9]; **Cabbages** [2-3,7-9]; **Calabrese** [2-3,7-9]; **Canary flower (Echium spp.)** *(off-label)* [2]; **Cauliflowers** [2-3,7-9]; **Choi sum** *(off-label)* [8]; **Collards** [3,7-9]; **Evening primrose** *(off-label)*

[2]; **Honesty** *(off-label)* [2]; **Kale** [3,7-9]; **Linseed** *(off-label)* [2-3]; **Mustard** *(off-label)* [2]; **Rye** *(off-label)* [2]; **Spring oilseed rape** [2-3,7-10]; **Triticale** *(off-label)* [2]; **Winter oilseed rape** [2-3,7-10];

- Blight in **All edible seed crops grown outdoors** *(off-label)* [8]; **All non-edible seed crops grown outdoors** *(off-label)* [8]; **Forest nurseries** *(off-label)* [2]; **Nursery fruit trees** *(off-label)* [8]; **Ornamental plant production** *(off-label)* [8];
- Brown rust in **Rye** *(off-label)* [2]; **Triticale** *(off-label)* [2]; **Winter wheat** [2];
- Cladosporium in **Rye** *(off-label)* [2]; **Triticale** *(off-label)* [2];
- Disease control in **Apples** [4]; **Pears** [4]; **Potatoes** [1,5-6];
- Kabatiella lini in **Linseed** *(off-label)* [3];
- Late blight in **Cardoons** *(off-label)* [8]; **Celeriac** *(off-label)* [8]; **Celery (outdoor)** *(off-label)* [8]; **Florence fennel** *(off-label)* [8]; **Rhubarb** *(off-label)* [8];
- Leaf blight in **Cardoons** *(off-label)* [8]; **Celery (outdoor)** *(off-label)* [8]; **Florence fennel** *(off-label)* [8]; **Rhubarb** *(off-label)* [8];
- Leaf spot in **All edible seed crops grown outdoors** *(off-label)* [8]; **All non-edible seed crops grown outdoors** *(off-label)* [8]; **Cardoons** *(off-label)* [8]; **Celery (outdoor)** *(off-label)* [8]; **Choi sum** *(off-label)* [8]; **Florence fennel** *(off-label)* [8]; **Forest nurseries** *(off-label)* [2]; **Nursery fruit trees** *(off-label)* [8]; **Ornamental plant production** *(off-label)* [8]; **Rhubarb** *(off-label)* [8];
- Light leaf spot in **Spring oilseed rape** [2-3,7-9]; **Winter oilseed rape** [2-3,7-9];
- Powdery mildew in **Borage for oilseed production** *(off-label)* [2]; **Canary flower (Echium spp.)** *(off-label)* [2]; **Evening primrose** *(off-label)* [2]; **Honesty** *(off-label)* [2]; **Linseed** *(off-label)* [2]; **Mustard** *(off-label)* [2];
- Ring spot in **Broccoli** [2-3,7-9]; **Brussels sprouts** [2-3,7-9]; **Cabbages** [2-3,7-9]; **Calabrese** [2-3,7-9]; **Cauliflowers** [2-3,7-9]; **Choi sum** *(off-label)* [8]; **Collards** [3,7-9]; **Kale** [3,7-9];
- Rust in **All edible seed crops grown outdoors** *(off-label)* [8]; **All non-edible seed crops grown outdoors** *(off-label)* [8]; **Asparagus** *(off-label)* [8]; **Forest nurseries** *(off-label)* [2]; **Nursery fruit trees** *(off-label)* [8]; **Ornamental plant production** *(off-label)* [8];
- Scab in **Apples** [4]; **Pears** [4];
- Sclerotinia in **Borage for oilseed production** *(off-label)* [2]; **Canary flower (Echium spp.)** *(off-label)* [2]; **Evening primrose** *(off-label)* [2]; **Honesty** *(off-label)* [2]; **Linseed** *(off-label)* [2-3]; **Mustard** *(off-label)* [2];
- Septoria leaf blotch in **Rye** *(off-label)* [2]; **Triticale** *(off-label)* [2]; **Winter wheat** [2];
- Stem canker in **Spring oilseed rape** [2-3,7-10]; **Winter oilseed rape** [2-3,7-10];

Extension of Authorisation for Minor Use (EAMUs)
- **All edible seed crops grown outdoors** *20194256* [8]
- **All non-edible seed crops grown outdoors** *20194256* [8]
- **Asparagus** *20171713* [8]
- **Borage for oilseed production** *20140211* [2]
- **Canary flower (Echium spp.)** *20140211* [2]
- **Cardoons** *20192713* [8]
- **Celeriac** *20182953* [8]
- **Celery (outdoor)** *20192713* [8]
- **Choi sum** *20171714* [8]
- **Evening primrose** *20140211* [2]
- **Florence fennel** *20192713* [8]
- **Forest nurseries** *20140205* [2]
- **Honesty** *20140211* [2]
- **Linseed** *20140211* [2], *20212263* [3]
- **Mustard** *20140211* [2]
- **Nursery fruit trees** *20194256* [8]
- **Ornamental plant production** *20194256* [8]
- **Rhubarb** *20192713* [8]
- **Rye** *20193253* [2]
- **Triticale** *20193253* [2]

Approval information
- Difenoconazole included in Annex I under EC Regulation 1107/2009

SEE SECTION 3 FOR PRODUCTS ALSO REGISTERED

Efficacy guidance

- In brassicas a 3-spray programme should be used starting at the first sign of disease and repeated at 14-21 d intervals
- Product is fully rainfast 2 h after application
- For most effective control of Septoria, apply as part of a programme of sprays which includes a suitable flag leaf treatment
- Difenoconazole is a DMI fungicide. Resistance to some DMI fungicides has been identified in Septoria leaf blotch which may seriously affect performance of some products. For further advice contact a specialist advisor and visit the Fungicide Resistance Action Group (FRAG)-UK website

Restrictions

- Maximum number of treatments 3 per crop for brassicas; 2 per crop for oilseed rape; 1 per crop for wheat [2, 4], 4 for potatoes [5, 6]
- Maximum total dose equivalent to 3 full dose treatments on brassicas; 2 full dose treatments on oilseed rape; 1 full dose treatment on wheat [8]
- Apply to wheat any time from ear fully emerged stage but before early milk-ripe stage (GS 59-73)
- Maintain an interval of at least 14 days between applications to Brassica crops and 10 - 14 days in potatoes

Crop-specific information

- Latest use: before grain early milk-ripe stage (GS 73) for cereals; end of flowering for oilseed rape
- HI brassicas 14 or 21 d
- Treat oilseed rape in autumn from 4 expanded true leaf stage (GS 1,4). A repeat spray may be made in spring at the beginning of stem extension (GS 2,0) if visible symptoms develop

Environmental safety

- Dangerous for the environment
- Very toxic to aquatic organisms
- Buffer zone requirement 6 m [1]
- Broadcast air-assisted LERAP [4] (20m); LERAP Category B [1-3,5-10]

Hazard classification and safety precautions

Hazard Harmful [4], Irritant [2], Dangerous for the environment, Harmful if swallowed [2,4-6], Very toxic to aquatic organisms [3]
Transport code 9 [1-10]
Packaging group III
UN Number 3082
Risk phrases H304, H315 [1], H318 [1], H319 [9], H336 [1]
Operator protection U05a, U09a [1-3,5-9], U11 [2,4], U20b [1-9]; A, C, H
Environmental protection E15a [1-3,5-8], E15b [4], E16a [1-3,5-10], E17b (20m) [4], E38, H410 [1,3,7-10], H411 [2,4-6]
Storage and disposal D01, D02, D05 [1,3,5-9], D07 [1,3,5-9], D09a [1,3,5-9], D10b [2], D10c [1,3,5-9], D12a [1-3,5-10]
Medical advice M05b [4]

128   difenoconazole + fludioxonil

A triazole + phenylpyrrole seed treatment for use in cereals
FRAC mode of action code: 3 + 12

See also fludioxonil

Products

1	Celest Extra	Syngenta	25:25 g/l	FS	16630
2	Difend Extra	Certis Belchim B V	25:25 g/l	FS	17739
3	Instrata Elite	Syngenta	80.3:80.3 g/l	SC	17976

Uses

- Bunt in *Winter wheat* [1];
- Disease control in *Managed amenity turf* [3];

FOR FULL CONDITIONS OF USE ALWAYS READ THE PRODUCT LABEL

- Fusarium foot rot and seedling blight in **Winter oats** [1]; **Winter wheat** [1];
- Microdochium nivale in **Winter oats** [1]; **Winter wheat** [1];
- Pyrenophora leaf spot in **Winter oats** [1];
- Red thread in **Managed amenity turf** [3];
- Seed-borne diseases in **Winter wheat** [1-2];
- Seedling blight and foot rot in **Winter wheat** [1];
- Septoria seedling blight in **Winter wheat** [1];
- Snow mould in **Winter wheat** [1];
- Stripe smut in **Winter rye** [1];

Approval information
- Difenoconazole and fludioxonil included in Annex I under EC Regulation 1107/2009

Efficacy guidance
- Effective against benzimidazole-resistant and benzimidazole-sensitive strains of *Microdochium nivale*

Crop-specific information
- Under adverse environmental or soil conditions, seed rates should be increased to compensate for a slight drop in germination capacity. Flow rates of treated seed should be checked before drilling commences.

Environmental safety
- LERAP Category B

Hazard classification and safety precautions
Hazard Harmful [1], Dangerous for the environment [1-2], Harmful if inhaled [1], Very toxic to aquatic organisms [1]
Transport code 9 [1-3]
Packaging group III
UN Number 3082
Operator protection U05a [1-2], U20c [1]; A, H
Environmental protection E15a [1], E16a [3], E38 [1-2], H410
Storage and disposal D01, D02, D05 [2-3], D09a [1], D11a [1], D12a [1-2]
Treated seed S01 [1-2], S02 [1-2], S03 [1-2], S04a [1-2], S04b [1], S05 [1-2], S06a [1-2], S07 [1-2], S08 [1-2], S09 [1]
Medical advice M05a [1]

129   difenoconazole + fluxapyroxad

A fungicide mixture for disease control in a range of horticultural crops
FRAC mode of action code: 3 + 7

Products

1 Atticus	Pan Agriculture	50:75 g/l	SC	19268
2 Charm	BASF	50:75 g/l	SC	18396
3 Perseus	BASF	50:75 g/l	SC	18397

Uses
- Alternaria in **Broccoli** (reduction) [3]; **Calabrese** (reduction) [3]; **Celeriac** (off-label) [3]; **Horseradish** (off-label) [3]; **Jerusalem artichokes** (off-label) [3]; **Parsnips** (off-label) [3]; **Potatoes** (moderate control) [3]; **Radishes** (off-label) [3]; **Red beet** (off-label) [3]; **Swedes** (off-label) [3]; **Turnips** (off-label) [3];
- Blight in **Beans with pods** (off-label) [3]; **Dwarf beans** (off-label) [3]; **Edible podded peas** (off-label) [3]; **French beans** (off-label) [3]; **Runner beans** (off-label) [3];
- Botrytis in **Beans with pods** (off-label) [3]; **Celery leaves** (off-label) [3]; **Chives** (off-label) [3]; **Cress** (off-label) [3]; **Dwarf beans** (off-label) [3]; **Edible flowers** (off-label) [3]; **Edible podded peas** (off-label) [3]; **Endives** (off-label) [3]; **French beans** (off-label) [3]; **Herbs (see appendix 6)** (off-label) [3]; **Lamb's lettuce** (off-label) [3]; **Leeks** (off-label) [3]; **Parsley** (off-label) [3]; **Protected baby leaf crops** (off-label) [3]; **Protected cress** (off-label) [3]; **Protected endives** (off-label) [3]; **Protected lamb's lettuce** (off-label) [3]; **Protected land cress** (off-label) [3]; **Protected red**

SEE SECTION 3 FOR PRODUCTS ALSO REGISTERED

mustard *(off-label)* [3]; **Protected rocket** *(off-label)* [3]; **Purslane** *(off-label)* [3]; **Red mustard** *(off-label)* [3]; **Runner beans** *(off-label)* [3]; **Salad onions** *(off-label)* [3]; **Sweet ciceley** *(off-label)* [3];

- Disease control in **Protected chilli peppers** [1-2]; **Protected cucumbers** [1-2]; **Protected peppers** [1-2]; **Protected strawberries** [1-2]; **Protected tomatoes** [1-2]; **Strawberries** [1-2];
- Leaf and pod spot in **Vining peas** *(moderate control)* [3];
- Leaf blotch in **Leeks** *(off-label)* [3]; **Salad onions** *(off-label)* [3];
- Leaf spot in **Beans with pods** *(off-label)* [3]; **Dwarf beans** *(off-label)* [3]; **Edible podded peas** *(off-label)* [3]; **French beans** *(off-label)* [3]; **Leeks** *(off-label)* [3]; **Runner beans** *(off-label)* [3]; **Salad onions** *(off-label)* [3];
- Powdery mildew in **Celeriac** *(off-label)* [3]; **Celery leaves** *(off-label)* [3]; **Chives** *(off-label)* [3]; **Courgettes** *(off-label)* [3]; **Cress** *(off-label)* [3]; **Edible flowers** *(off-label)* [3]; **Endives** *(off-label)* [3]; **Gherkins** *(off-label)* [3]; **Herbs (see appendix 6)** *(off-label)* [3]; **Horseradish** *(off-label)* [3]; **Jerusalem artichokes** *(off-label)* [3]; **Lamb's lettuce** *(off-label)* [3]; **Parsley** *(off-label)* [3]; **Parsnips** *(off-label)* [3]; **Protected baby leaf crops** *(off-label)* [3]; **Protected cress** *(off-label)* [3]; **Protected endives** *(off-label)* [3]; **Protected lamb's lettuce** *(off-label)* [3]; **Protected land cress** *(off-label)* [3]; **Protected red mustard** *(off-label)* [3]; **Protected rocket** *(off-label)* [3]; **Pumpkins** *(off-label)* [3]; **Purslane** *(off-label)* [3]; **Radishes** *(off-label)* [3]; **Red beet** *(off-label)* [3]; **Red mustard** *(off-label)* [3]; **Squashes** *(off-label)* [3]; **Swedes** *(off-label)* [3]; **Sweet ciceley** *(off-label)* [3]; **Turnips** *(off-label)* [3];
- Purple blotch in **Leeks** *(off-label)* [3]; **Salad onions** *(off-label)* [3];
- Rhizoctonia in **Celeriac** *(off-label)* [3]; **Celery leaves** *(off-label)* [3]; **Chives** *(off-label)* [3]; **Cress** *(off-label)* [3]; **Edible flowers** *(off-label)* [3]; **Endives** *(off-label)* [3]; **Herbs (see appendix 6)** *(off-label)* [3]; **Horseradish** *(off-label)* [3]; **Jerusalem artichokes** *(off-label)* [3]; **Lamb's lettuce** *(off-label)* [3]; **Parsley** *(off-label)* [3]; **Parsnips** *(off-label)* [3]; **Protected baby leaf crops** *(off-label)* [3]; **Protected cress** *(off-label)* [3]; **Protected endives** *(off-label)* [3]; **Protected lamb's lettuce** *(off-label)* [3]; **Protected land cress** *(off-label)* [3]; **Protected red mustard** *(off-label)* [3]; **Protected rocket** *(off-label)* [3]; **Purslane** *(off-label)* [3]; **Radishes** *(off-label)* [3]; **Red beet** *(off-label)* [3]; **Red mustard** *(off-label)* [3]; **Swedes** *(off-label)* [3]; **Sweet ciceley** *(off-label)* [3]; **Turnips** *(off-label)* [3];
- Ring spot in **Broccoli** *(reduction)* [3]; **Calabrese** *(reduction)* [3];
- Rust in **Leeks** *(off-label)* [3]; **Salad onions** *(off-label)* [3];
- Sclerotinia in **Carrots** [3]; **Celeriac** *(off-label)* [3]; **Celery leaves** *(off-label)* [3]; **Chives** *(off-label)* [3]; **Cress** *(off-label)* [3]; **Edible flowers** *(off-label)* [3]; **Endives** *(off-label)* [3]; **Herbs (see appendix 6)** *(off-label)* [3]; **Horseradish** *(off-label)* [3]; **Jerusalem artichokes** *(off-label)* [3]; **Lamb's lettuce** *(off-label)* [3]; **Lettuce** *(off-label)* [3]; **Parsley** *(off-label)* [3]; **Parsnips** *(off-label)* [3]; **Protected baby leaf crops** *(off-label)* [3]; **Protected cress** *(off-label)* [3]; **Protected endives** *(off-label)* [3]; **Protected lamb's lettuce** *(off-label)* [3]; **Protected land cress** *(off-label)* [3]; **Protected lettuce** *(moderate control)* [3]; **Protected red mustard** *(off-label)* [3]; **Protected rocket** *(off-label)* [3]; **Purslane** *(off-label)* [3]; **Radishes** *(off-label)* [3]; **Red beet** *(off-label)* [3]; **Red mustard** *(off-label)* [3]; **Swedes** *(off-label)* [3]; **Sweet ciceley** *(off-label)* [3]; **Turnips** *(off-label)* [3];
- White tip in **Leeks** *(off-label)* [3]; **Salad onions** *(off-label)* [3];

Extension of Authorisation for Minor Use (EAMUs)
- **Beans with pods** *20193285* [3]
- **Celeriac** *20193426* [3]
- **Celery leaves** *20220275* [3]
- **Chives** *20220275* [3]
- **Courgettes** *20194197* [3]
- **Cress** *20220275* [3]
- **Dwarf beans** *20193285* [3]
- **Edible flowers** *20220275* [3]
- **Edible podded peas** *20193285* [3]
- **Endives** *20220275* [3]
- **French beans** *20193285* [3]
- **Gherkins** *20194197* [3]
- **Herbs (see appendix 6)** *20220275* [3]
- **Horseradish** *20193426* [3]
- **Jerusalem artichokes** *20193426* [3]
- **Lamb's lettuce** *20220275* [3]

FOR FULL CONDITIONS OF USE ALWAYS READ THE PRODUCT LABEL

- *Leeks* 20220822 [3]
- *Parsley* 20220275 [3]
- *Parsnips* 20193426 [3]
- *Protected baby leaf crops* 20220275 [3]
- *Protected cress* 20220275 [3]
- *Protected endives* 20220275 [3]
- *Protected lamb's lettuce* 20220275 [3]
- *Protected land cress* 20220275 [3]
- *Protected red mustard* 20220275 [3]
- *Protected rocket* 20220275 [3]
- *Pumpkins* 20194197 [3]
- *Purslane* 20220275 [3]
- *Radishes* 20193426 [3]
- *Red beet* 20193426 [3]
- *Red mustard* 20220275 [3]
- *Runner beans* 20193285 [3]
- *Salad onions* 20220822 [3]
- *Squashes* 20194197 [3]
- *Swedes* 20193426 [3]
- *Sweet ciceley* 20220275 [3]
- *Turnips* 20193426 [3]

Approval information
- Difenoconazole and fluxapyroxad included in Annex I under EC Regulation 1107/2009

Restrictions
- Do not use in protected tomato and pepper crops between 1st October and 1st March [2]
- Consult processors before using Perseus on vining peas or potatoes.
- Do not use on crops intended for fermentation.
- Do not use on protected crops intended for production of seed or propagation material.
- Avoid overlapping sprays.
- The total number of applications of SDHI containing products should make up no more than 50% of the total number of fungicides applied per season.
- No more than two consecutive applications of SHDI fungicides should be applied per crop.

Environmental safety
- LERAP Category B [1-3]

Hazard classification and safety precautions
Transport code 9 [1-3]
Packaging group III
UN Number 3082
Risk phrases H319, H351
Operator protection U11 [3], U20b; A, C
Environmental protection E15b [1-2], E16a, E34, H410
Storage and disposal D01, D02, D09a, D10c

130   difenoconazole + mandipropamid

A triazole/mandelamide fungicide mixture for blight control in potatoes
FRAC mode of action code: 3 + 40

Products
1	Amphore Plus	Syngenta	250:250 g/l	SL	16327
2	Carial Star	Syngenta	250:250 g/l	SL	16323

Uses
- Alternaria blight in *Potatoes* [1-2];
- Blight in *Potatoes* [1-2];
- Early blight in *Potatoes* [1-2];

SEE SECTION 3 FOR PRODUCTS ALSO REGISTERED

Approval information
- Difenoconazole and mandipropamid included in Annex I under EC Regulation 1107/2009

Efficacy guidance
- Rainfast within 15 minutes
- Spray programme must start before blight enters the crop

Crop-specific information
- Can be used on all varieties of potato, including seed potatoes

Environmental safety
- LERAP Category B [1-2]

Hazard classification and safety precautions
Transport code 9 [1-2]
Packaging group III
UN Number 3082
Operator protection U05a, U20b; A, H
Environmental protection E15b, E16a, E22c, E34, E38, H410
Storage and disposal D01, D09a, D10c
Medical advice M03

131 difenoconazole + paclobutrazol

A triazole fungicide mixture for growth regulation and disease control in oilseed rape
FRAC mode of action code: 3

Products

1	Toprex	Syngenta	250:125 g/l	SC	16456

Uses
- Disease control in **Winter oilseed rape** [1];
- Kabatiella lini in **Linseed** *(off-label)* [1];
- Light leaf spot in **Winter oilseed rape** [1];
- Mildew in **Linseed** *(off-label)* [1];
- Phoma in **Winter oilseed rape** [1];
- Septoria leaf blotch in **Linseed** *(off-label)* [1];

Extension of Authorisation for Minor Use (EAMUs)
- **Linseed** 20151633 [1]

Approval information
- Difenoconazole and paclobutrazol included in Annex I under EC Regulation 1107/2009

Efficacy guidance
- Contains a growth regulator for height reduction and lodging control and a fungicide for disease control. It should only be used when both growth regulation and disease control are required. If this is not the case, use appropriate alternative products at the required timing

Environmental safety
- LERAP Category B [1]

Hazard classification and safety precautions
Hazard Dangerous for the environment, Very toxic to aquatic organisms
Transport code 9 [1]
Packaging group III
UN Number 3082
Risk phrases H361
Operator protection U05a; A, H
Environmental protection E16a, E38, H410
Storage and disposal D01, D02, D12a
Medical advice M05a

FOR FULL CONDITIONS OF USE ALWAYS READ THE PRODUCT LABEL

132   diflufenican

A shoot absorbed pyridinecarboxamide herbicide for winter cereals
HRAC mode of action code: 12 (F1)

See also chlorotoluron + diflufenican
chlorotoluron + diflufenican + pendimethalin
clodinafop-propargyl + diflufenican
clopyralid + diflufenican + MCPA

Products

1 Cachet	Albaugh UK	500 g/l	SC	19733
2 Clayton El Nino	Clayton	500 g/l	SC	17337
3 Hurricane SC	Adama	500 g/l	SC	16027
4 Quotient	BASF	500 g/l	SC	19969
5 Sempra XL	UPL Europe	500 g/l	SC	17447
6 Whip	Agform	500 g/l	SC	19540

Uses

- Annual dicotyledons in *Game cover* (off-label) [3]; *Grass seed crops* (off-label) [3]; *Oats* (off-label) [3];
- Annual meadow grass in *Game cover* (off-label) [3]; *Grass seed crops* (off-label) [3,5]; *Oat seed crops* (off-label) [5]; *Oats* (off-label) [3];
- Black bindweed in *Grass seed crops* (off-label) [5]; *Oat seed crops* (off-label) [5];
- Charlock in *Forest nurseries* (off-label) [5]; *Game cover* (off-label) [5]; *Grass seed crops* (off-label) [5]; *Oat seed crops* (off-label) [5]; *Ornamental plant production* (off-label) [5];
- Chickweed in *Carrots* (off-label) [3]; *Forest nurseries* (off-label) [5]; *Game cover* (off-label) [5]; *Grass seed crops* (off-label) [5]; *Oat seed crops* (off-label) [5]; *Ornamental plant production* (off-label) [5];
- Cleavers in *Durum wheat* [1-3,5-6]; *Spring barley* [1-6]; *Spring rye* [3]; *Spring wheat* [1-3,5-6]; *Triticale* [1-6]; *Winter barley* [1-6]; *Winter rye* [1-6]; *Winter wheat* [1-6];
- Corn spurrey in *Forest nurseries* (off-label) [5]; *Game cover* (off-label) [5]; *Grass seed crops* (off-label) [5]; *Oat seed crops* (off-label) [5]; *Ornamental plant production* (off-label) [5];
- Dead nettle in *Forest nurseries* (off-label) [5]; *Game cover* (off-label) [5]; *Grass seed crops* (off-label) [5]; *Oat seed crops* (off-label) [5]; *Ornamental plant production* (off-label) [5];
- Field pansy in *Carrots* (off-label) [3]; *Durum wheat* [1-3,5-6]; *Forest nurseries* (off-label) [5]; *Game cover* (off-label) [5]; *Grass seed crops* (off-label) [5]; *Oat seed crops* (off-label) [5]; *Ornamental plant production* (off-label) [5]; *Spring barley* [1-6]; *Spring rye* [3]; *Spring wheat* [1-3,5-6]; *Triticale* [1-6]; *Winter barley* [1-6]; *Winter rye* [1-6]; *Winter wheat* [1-6];
- Field speedwell in *Durum wheat* [1-3,5-6]; *Spring barley* [1-6]; *Spring rye* [3]; *Spring wheat* [1-3,5-6]; *Triticale* [1-6]; *Winter barley* [1-6]; *Winter rye* [1-6]; *Winter wheat* [1-6];
- Flixweed in *Forest nurseries* (off-label) [5]; *Game cover* (off-label) [5]; *Grass seed crops* (off-label) [5]; *Oat seed crops* (off-label) [5]; *Ornamental plant production* (off-label) [5];
- Forget-me-not in *Forest nurseries* (off-label) [5]; *Game cover* (off-label) [5]; *Ornamental plant production* (off-label) [5];
- Ivy-leaved speedwell in *Durum wheat* [1-3,5-6]; *Spring barley* [1-6]; *Spring rye* [3]; *Spring wheat* [1-3,5-6]; *Triticale* [1-6]; *Winter barley* [1-6]; *Winter rye* [1-6]; *Winter wheat* [1-6];
- Knotgrass in *Grass seed crops* (off-label) [5]; *Oat seed crops* (off-label) [5];
- Mayweeds in *Durum wheat* [1-3,5-6]; *Spring barley* [1-6]; *Spring rye* [3]; *Spring wheat* [1-3,5-6]; *Triticale* [1-6]; *Winter barley* [1-6]; *Winter rye* [1-6]; *Winter wheat* [1-6];
- Mouse-ear chickweed in *Forest nurseries* (off-label) [5]; *Game cover* (off-label) [5]; *Grass seed crops* (off-label) [5]; *Oat seed crops* (off-label) [5]; *Ornamental plant production* (off-label) [5];
- Nipplewort in *Grass seed crops* (off-label) [5]; *Oat seed crops* (off-label) [5];
- Parsley-piert in *Forest nurseries* (off-label) [5]; *Game cover* (off-label) [5]; *Grass seed crops* (off-label) [5]; *Oat seed crops* (off-label) [5]; *Ornamental plant production* (off-label) [5];
- Poppies in *Durum wheat* [1-3,5-6]; *Forest nurseries* (off-label) [5]; *Game cover* (off-label) [5]; *Grass seed crops* (off-label) [5]; *Oat seed crops* (off-label) [5]; *Ornamental plant production* (off-label) [5]; *Spring barley* [1-6]; *Spring rye* [3]; *Spring wheat* [1-3,5-6]; *Triticale* [1-6]; *Winter barley* [1-6]; *Winter rye* [1-6]; *Winter wheat* [1-6];

SEE SECTION 3 FOR PRODUCTS ALSO REGISTERED

- Red dead-nettle in *Durum wheat* [1-3,5-6]; *Spring barley* [1-6]; *Spring rye* [3]; *Spring wheat* [1-3,5-6]; *Triticale* [1-6]; *Winter barley* [1-6]; *Winter rye* [1-6]; *Winter wheat* [1-6];
- Runch in *Forest nurseries* *(off-label)* [5]; *Game cover* *(off-label)* [5]; *Grass seed crops* *(off-label)* [5]; *Oat seed crops* *(off-label)* [5]; *Ornamental plant production* *(off-label)* [5];
- Shepherd's purse in *Carrots* *(off-label)* [3]; *Forest nurseries* *(off-label)* [5]; *Game cover* *(off-label)* [5]; *Ornamental plant production* *(off-label)* [5];
- Sowthistle in *Grass seed crops* *(off-label)* [5]; *Oat seed crops* *(off-label)* [5];
- Speedwells in *Carrots* *(off-label)* [3]; *Forest nurseries* *(off-label)* [5]; *Game cover* *(off-label)* [5]; *Grass seed crops* *(off-label)* [5]; *Oat seed crops* *(off-label)* [5]; *Ornamental plant production* *(off-label)* [5];
- Treacle mustard in *Forest nurseries* *(off-label)* [5]; *Game cover* *(off-label)* [5]; *Grass seed crops* *(off-label)* [5]; *Oat seed crops* *(off-label)* [5]; *Ornamental plant production* *(off-label)* [5];
- Volunteer oilseed rape in *Carrots* *(off-label)* [3]; *Forest nurseries* *(off-label)* [5]; *Game cover* *(off-label)* [5]; *Grass seed crops* *(off-label)* [5]; *Oat seed crops* *(off-label)* [5]; *Ornamental plant production* *(off-label)* [5];
- Wild radish in *Forest nurseries* *(off-label)* [5]; *Game cover* *(off-label)* [5]; *Grass seed crops* *(off-label)* [5]; *Oat seed crops* *(off-label)* [5]; *Ornamental plant production* *(off-label)* [5];

Extension of Authorisation for Minor Use (EAMUs)
- *Carrots* 20190180 [3]
- *Forest nurseries* 20170374 [5]
- *Game cover* 20183440 [3], 20170374 [5]
- *Grass seed crops* 20141324 [3], 20162935 [5]
- *Oat seed crops* 20162935 [5]
- *Oats* 20212217 [3]
- *Ornamental plant production* 20170374 [5]

Approval information
- Diflufenican included in Annex I under EC Regulation 1107/2009
- Accepted by BBPA for use on malting barley

Efficacy guidance
- Best results achieved from treatment of small actively growing weeds in early autumn or spring
- Good weed control depends on efficient burial of trash or straw before or during seedbed preparation
- Loose or fluffy seedbeds should be rolled before application
- The final seedbed should be moist, fine and firm with clods no bigger than fist size
- Ensure good even spray coverage and increase spray volume for post-emergence treatments where the crop or weed foliage is dense
- Activity may be slow under cool conditions and final level of weed control may take some time to appear
- Where cleavers are a particular problem a separate specific herbicide treatment may be required
- Efficacy may be impaired on soils with a Kd factor greater than 6
- Always follow WRAG guidelines for preventing and managing herbicide resistant weeds. See Section 5 for more information

Restrictions
- Do not treat broadcast crops [3]
- Do not roll treated crops or harrow at any time after treatment
- Do not apply to soils with more than 10% organic matter or on Sands, or very stony or gravelly soils
- Do not treat after a period of cold frosty weather
- Horizontal boom sprayers must be fitted with three star drift reduction technology for all uses
- Must not be applied via hand-held equipment
- Low drift spraying equipment must be used for 30 m from the top of the bank of any surface water bodies

Crop-specific information
- Latest use: before end of tillering (GS 29) for wheat and barley [3], pre crop emergence for triticale and winter rye [3]
- Treat only named varieties of rye or triticale [3]

FOR FULL CONDITIONS OF USE ALWAYS READ THE PRODUCT LABEL

Following crops guidance
- Labels vary slightly but in general ploughing to 150 mm and thoroughly mixing the soil is recommended before drilling or planting any succeeding crops either after crop failure or after normal harvest
- In the event of crop failure only winter wheat or winter barley may be re-drilled immediately after ploughing. Spring crops of wheat, barley, oilseed rape, peas, field beans, sugar beet [3], potatoes, carrots, edible brassicas or onions may be sown provided an interval of 12 wk has elapsed after ploughing
- After normal harvest of a treated crop winter cereals, oilseed rape, field beans, leaf brassicas, sugar beet seed crops and winter onions may be drilled in the following autumn. Other crops listed above for crop failure may be sown in the spring after normal harvest
- Successive treatments with any products containing diflufenican can lead to soil build-up and inversion ploughing to 150 mm must precede sowing any following non-cereal crop. Even where ploughing occurs some crops may be damaged

Environmental safety
- Dangerous for the environment
- Very toxic to aquatic organisms
- Buffer zone requirement 6m [1, 2, 3, 5, 6]
- Buffer zone requirement 7 m for spring barley, 8m for winter barley, winter rye, triticale and winter wheat [4]
- Buffer zone requirement 12 m on crops grown for game cover or ornamental plant production [3]
- LERAP Category B [1-6]

Hazard classification and safety precautions
Hazard Dangerous for the environment, Very toxic to aquatic organisms [1,6]
Transport code 9 [1-6]
Packaging group III
UN Number 3082
Operator protection U08, U13, U15, U19a, U20a; A, C
Environmental protection E16a, E16i [1-2], E34, E38, H410
Storage and disposal D01, D02, D05, D09a, D10b, D12a
Medical advice M05a

133   diflufenican + florasulam

A herbicide mixture for weed control in cereals
HRAC mode of action code: 2 + 12 (B + F1)

See also diflufenican
florasulam

Products

1 Bow	Nufarm UK	500:50 g/l	SC	19054
2 Lector Delta	Nufarm UK	500:50 g/l	SC	19055

Uses
- Annual dicotyledons in *Rye* [2]; *Spring barley* [1-2]; *Triticale* [1-2]; *Winter barley* [1-2]; *Winter rye* [1]; *Winter wheat* [1-2];
- Black bindweed in *Rye* [2]; *Spring barley* [1-2]; *Triticale* [1-2]; *Winter barley* [1-2]; *Winter rye* [1]; *Winter wheat* [1-2];
- Charlock in *Rye* [2]; *Spring barley* [1-2]; *Triticale* [1-2]; *Winter barley* [1-2]; *Winter rye* [1]; *Winter wheat* [1-2];
- Chickweed in *Rye* [2]; *Spring barley* [1-2]; *Triticale* [1-2]; *Winter barley* [1-2]; *Winter rye* [1]; *Winter wheat* [1-2];
- Cleavers in *Rye* [2]; *Spring barley* [1-2]; *Triticale* [1-2]; *Winter barley* [1-2]; *Winter rye* [1]; *Winter wheat* [1-2];
- Field pansy in *Rye* [2]; *Spring barley* [1-2]; *Triticale* [1-2]; *Winter barley* [1-2]; *Winter rye* [1]; *Winter wheat* [1-2];

SEE SECTION 3 FOR PRODUCTS ALSO REGISTERED

- Forget-me-not in *Rye* [2]; *Spring barley* [1-2]; *Triticale* [1-2]; *Winter barley* [1-2]; *Winter rye* [1]; *Winter wheat* [1-2];
- Hemp-nettle in *Rye* [2]; *Spring barley* [1-2]; *Triticale* [1-2]; *Winter barley* [1-2]; *Winter rye* [1]; *Winter wheat* [1-2];
- Mayweeds in *Rye* [2]; *Spring barley* [1-2]; *Triticale* [1-2]; *Winter barley* [1-2]; *Winter rye* [1]; *Winter wheat* [1-2];
- Penny cress in *Rye* [2]; *Spring barley* [1-2]; *Triticale* [1-2]; *Winter barley* [1-2]; *Winter rye* [1]; *Winter wheat* [1-2];
- Shepherd's purse in *Rye* [2]; *Spring barley* [1-2]; *Triticale* [1-2]; *Winter barley* [1-2]; *Winter rye* [1]; *Winter wheat* [1-2];
- Volunteer oilseed rape in *Rye* [2]; *Spring barley* [1-2]; *Triticale* [1-2]; *Winter barley* [1-2]; *Winter rye* [1]; *Winter wheat* [1-2];

Approval information
- Diflufenican and florasulam included in Annex I under EC Regulation 1107/2009

Restrictions
- Following application to grass seed crops, the treated grass must not be fed to livestock.
- The total amount of florasulam applied to a cereal crop must not exceed 7.5 g.a.i/ha; the maximum amount of florasulam applied in the Autumn must not exceed 3.75g/ha
- Only one other product with an ALS inhibitor mode of action may be applied to a treated cereal crop. However, a further application or another product containing florasulam may also be made providing the maximum total dose of florasulam is not exceeded. A joint application with one of the listed ALS products can be made to the same cereal crop (see label).
- To avoid subsequent injury to crops other than cereals all spraying equipment must be thoroughly cleaned both inside and outside using All Clear Extra.

Following crops guidance
- In the event of crop failure in the spring, cultivate to 20 cms and then only plant spring wheat, spring barley or spring oats.
- Crops that can be sown in the same calendar year following treatment are cereals, oilseed rape, field beans, grass, peas, sugar beet, potatoes, maize and vegetable brassicas as transplants.

Environmental safety
- Buffer zone requirement 6 m for spring barley, 5 m for winter barley, winter wheat, winter rye and triticale
- LERAP Category B [1-2]

Hazard classification and safety precautions
Hazard Dangerous for the environment, Very toxic to aquatic organisms
Transport code 9 [1-2]
Packaging group III
UN Number 3082
Operator protection U05a [1]; A
Environmental protection E15b, E16a, E38, H410
Storage and disposal D01, D02, D03 [2], D05, D06a, D09a, D10c, D12a

134   diflufenican + flufenacet

A contact and residual herbicide mixture for cereals
HRAC mode of action code: 12 + 15 (F1 + K3)

See also flufenacet

Products

1	Ambush	Albaugh UK	100:400 g/l	SC	19647
2	Ascent	Albaugh UK	100:400 g/l	SC	19644
3	Calculate	Agrii	100:400 g/l	SC	19929
4	Clayton Aspect XL	Clayton	100:400 g/l	SC	19721
5	Clayton Facet XL	Clayton	100:400 g/l	SC	19724
6	Clayton Sabre XL	Clayton	100:400 g/l	SC	19725

FOR FULL CONDITIONS OF USE ALWAYS READ THE PRODUCT LABEL

Products – continued

7	Cleancrop Jab	Agrii	100:400 g/l	SC	19731
8	Cleancrop Wilbur	Agrii	200:400 g/l	SC	19976
9	Dephend	FMC Agro	100:400 g/l	SC	18739
10	Eridant	Albaugh UK	100:400 g/l	SC	19723
11	Firestarter	Life Scientific	100:400 g/l	SC	18422
12	Firestorm	Certis Belchim B V	100:400 g/l	SC	17631
13	Fosburi Super	Bayer CropScience	100:400 g/l	SC	19709
14	Giddo	Bayer CropScience	100:400 g/l	SC	19128
15	Golding	Rotam	100:400 g/l	SC	18654
16	Herold	Adama	200:400 g/l	SC	16195
17	Hopper	Agform	100:400 g/l	SC	19579
18	Isolator	Agrii	100:400 g/l	SC	19762
19	Liberator	Bayer CropScience	100:400 g/l	SC	15206
20	Mertil	UPL Europe	200:400 g/l	SC	18504
21	Nucleus	FMC Agro	100:400 g/l	SC	19026
22	Pincer	Albaugh UK	100:400 g/l	SC	19649
23	Regatta	Bayer CropScience	100:400 g/l	SC	15353
24	Reliance	UPL Europe	200:400 g/l	SC	18372
25	Terrane	FMC Agro	100:400 g/l	SC	18738
26	Terrane Total	FMC Agro	200:400 g/l	SC	19370

SECTION 2

Uses

- Annual dicotyledons in **Hemp grown for fibre production** *(off-label)* [19]; **Rye** *(off-label)* [13-14,16,19]; **Spring barley** *(off-label)* [16]; **Spring wheat** *(off-label)* [16]; **Triticale** *(off-label)* [13-14,16,19]; **Winter oats** *(off-label)* [16];
- Annual meadow grass in **Hemp grown for fibre production** *(off-label)* [19]; **Rye** *(off-label)* [13-14,16,19]; **Spring barley** *(off-label)* [4-7,11-14,16,18-19,23]; **Spring wheat** *(off-label)* [4-7,11-14,16,18-19,23]; **Triticale** *(off-label)* [9,13-14,16,19-21,24-26]; **Winter barley** [1-26]; **Winter oats** *(off-label)* [16]; **Winter rye** [9,20-21,24-26]; **Winter wheat** [1-26];
- Blackgrass in **Hemp grown for fibre production** *(off-label)* [19]; **Rye** *(off-label)* [13-14,19,23]; **Spring barley** [4-7,11-14,18-19,23]; **Spring wheat** [4-7,11-14,18-19,23]; **Triticale** *(off-label)* [9,13-14,19-21,23-26]; **Winter barley** [1-26]; **Winter oats** *(off-label)* [19,23]; **Winter rye** [9,20-21,24-26]; **Winter wheat** [1-26];
- Chickweed in **Spring barley** [4-7,11-14,18-19,23]; **Spring wheat** [4-7,11-14,18-19,23]; **Triticale** [9,20-21,24-26]; **Winter barley** [1-26]; **Winter rye** [9,20-21,24-26]; **Winter wheat** [1-26];
- Field pansy in **Spring barley** [4-7,11-14,18-19,23]; **Spring wheat** [4-7,11-14,18-19,23]; **Triticale** [9,20-21,24-26]; **Winter barley** [1-26]; **Winter rye** [9,20-21,24-26]; **Winter wheat** [1-26];
- Field speedwell in **Spring barley** [4-7,11-14,18-19,23]; **Spring wheat** [4-7,11-14,18-19,23]; **Triticale** [9,20-21,24-26]; **Winter barley** [1-26]; **Winter rye** [9,20-21,24-26]; **Winter wheat** [1-26];
- Groundsel in **Spring barley** [13]; **Spring wheat** [13]; **Winter barley** [13]; **Winter wheat** [13];
- Ivy-leaved speedwell in **Spring barley** [4-7,11-14,18-19,23]; **Spring wheat** [4-7,11-14,18-19,23]; **Triticale** [9,20-21,24-26]; **Winter barley** [1-26]; **Winter rye** [9,20-21,24-26]; **Winter wheat** [1-26];
- Mayweeds in **Spring barley** [4-7,11-14,18-19,23]; **Spring wheat** [4-7,11-14,18-19,23]; **Triticale** [9,20-21,24-26]; **Winter barley** [1-26]; **Winter rye** [9,20-21,24-26]; **Winter wheat** [1-26];
- Red dead-nettle in **Spring barley** [4-7,11-14,18-19,23]; **Spring wheat** [4-7,11-14,18-19,23]; **Winter barley** [1-7,10-15,17-19,22-23]; **Winter wheat** [1-7,10-15,17-19,22-23];
- Shepherd's purse in **Triticale** [9,20-21,24-26]; **Winter barley** [8-9,16,20-21,24-26]; **Winter rye** [9,20-21,24-26]; **Winter wheat** [8-9,16,20-21,24-26];
- Volunteer oilseed rape in **Triticale** [9,20-21,24-26]; **Winter barley** [8-9,16,20-21,24-26]; **Winter rye** [9,20-21,24-26]; **Winter wheat** [8-9,16,20-21,24-26];

Extension of Authorisation for Minor Use (EAMUs)

- **Hemp grown for fibre production** *20193868* [19]
- **Rye** *20212394* [13], *20193349* [14], *20193259* [16], *20193348* [19], *20212400* [23]
- **Spring barley** *20212911* [16]
- **Spring wheat** *20212911* [16]

SEE SECTION 3 FOR PRODUCTS ALSO REGISTERED

- **Triticale** *20212394* [13], *20193349* [14], *20193259* [16], *20193348* [19], *20212400* [23]
- **Winter oats** *20212218* [16], *20212219* [19], *20212220* [23]

Approval information
- Diflufenican and flufenacet included in Annex I under EC Regulation 1107/2009
- Accepted by BBPA for use on malting barley

Efficacy guidance
- Best results obtained when there is moist soil at and after application and rain falls within 7 d
- Residual control may be reduced under prolonged dry conditions
- Activity may be slow under cool conditions and final level of weed control may take some time to appear
- Good weed control depends on burying any trash or straw before or during seedbed preparation
- Established perennial grasses and broad-leaved weeds will not be controlled
- Do not use as a stand-alone treatment for blackgrass control. Always follow WRAG guidelines for preventing and managing herbicide resistant weeds. Section 5 for more information

Restrictions
- Do not treat undersown cereals or those to be undersown
- Do not use on waterlogged soils or soils prone to waterlogging
- Do not use on Sands or Very Light soils, or very stony or gravelly soils, or on soils containing more than 10% organic matter
- Do not treat broadcast crops and treat shallow-drilled crops post-emergence only
- Do not incorporate into the soil or disturb the soil after application by rolling or harrowing
- Avoid treating crops under stress from whatever cause and avoid treating during periods of prolonged or severe frosts
- Where the second application of a sequence is made after 31st December, or where the total dose exceeds 0.6 L/ha, the first application must be made before GS13 of the crop and a minimum interval of 6 weeks must be observed between applications [14, 26].
- Horizontal boom sprayers must be fitted with three star drift reduction technology for all uses
- Must not be applied via hand-held equipment
- Low drift spraying equipment must be operated according to the specific conditions stated in the official three star rating for that equipment as published on HSE website. These operating conditions must be maintained until the operator is 30m from the top of the bank of any surface water bodies

Crop-specific information
- Latest use: before 31 Dec in yr of sowing and before 3rd tiller stage (GS 23) for wheat or 4th tiller stage (GS 24) for barley
- For pre-emergence treatments the seed should be covered with a minimum of 32 mm settled soil

Following crops guidance
- In the event of crop failure wheat, barley or potatoes may be sown provided the soil is ploughed to 15 cm, and a minimum of 12 weeks elapse between treatment and sowing spring wheat or spring barley
- After normal harvest wheat, barley or potatoes may be sown without special cultivations. Soil must be ploughed or cultivated to 15 cm before sowing oilseed rape, field beans, peas, sugar beet, carrots, onions or edible brassicae
- Successive treatments with any products containing diflufenican can lead to soil build-up and inversion ploughing to 150 mm must precede sowing any following non-cereal crop. Even where ploughing occurs some crops may be damaged

Environmental safety
- Dangerous for the environment
- Very toxic to aquatic organisms
- Risk to non-target insects or other arthropods. Avoid spraying within 6 m of the field boundary to reduce the effects on non-target insects or other arthropods
- Buffer zone requirement 6 m [9, 21, 25]
- Buffer zone requirement 12 m [20, 24]
- LERAP Category B [1-26]

FOR FULL CONDITIONS OF USE ALWAYS READ THE PRODUCT LABEL

Hazard classification and safety precautions

Hazard Harmful, Dangerous for the environment, Harmful if swallowed [1-8,10-15,17-20,22-24], Very toxic to aquatic organisms [1-8,10-15,17-20,22-24,26]

Transport code 9 [1-26]

Packaging group III

UN Number 3082

Risk phrases H317 [3,8,18,20,24], H373 [1-15,17-26], R22a [16], R43 [16,25], R48 [16], R50 [16], R53a [16]

Operator protection U05a, U14, U23a [9,21]; A, C, H

Environmental protection E15a [15,23], E15b [8-9,16,20-21,24-26], E16a, E22c [1-14,16-22,24-26], E34, E38, H410 [1-15,17-26]

Storage and disposal D01, D02, D09a, D10b, D12a [1-12,14-26], D12c [13]

Medical advice M03, M05a

135   diflufenican + flufenacet + metribuzin

A residual herbicide mixture for weed control in winter cereals
HRAC mode of action code: 12 + 15 + 5 (F1 + K3 + C1)

See also diflufenican + flufenacet
diflufenican + metribuzin
flufenacet + metribuzin

Products

1	Alternator Met	Bayer CropScience	60:240:70 g/l	SC	18267
2	Octavian Met	Bayer CropScience	90:240:70 g/l	SC	18266

Uses

- Annual dicotyledons in **Winter barley** [1-2]; **Winter wheat** [1-2];
- Annual meadow grass in **Winter barley** [1-2]; **Winter wheat** [1-2];
- Chickweed in **Winter barley** [1]; **Winter wheat** [1];
- Common field speedwell in **Winter barley** [1]; **Winter wheat** [1];
- Field pansy in **Winter barley** [1]; **Winter wheat** [1];
- Red dead-nettle in **Winter barley** [1]; **Winter wheat** [1];
- Scented mayweed in **Winter barley** [1]; **Winter wheat** [1];
- Wild radish in **Winter barley** [1]; **Winter wheat** [1];

Approval information

- Diflufenican, flufenacet and metribuzin included in Annex I under EC Regulation 1107/2009

Following crops guidance

- Following normal harvest wheat and barley can be sown without cultivation but thorough mixing of the top 15 cm of soil is necessary before sowing oilseed rape, mustard or edible Brassicae. The following spring after normal harvest wheat, barley, oilseed rape, sugarbeet, mustard, peas, sunflowers or maize can be sown without special cultivations.
- In the event of crop failure, allow at least 5 months from application before sowing wheat or barley and at least 6 months before sowing maize or sunflowers.
- Be aware of the potential for diflufenican to build up in the soil from successive applications over the crop rotation.

Environmental safety

- Buffer zone requirement 6 m
- Horizontal boom sprayers must be fitted with three star drift reduction technology for all uses
- Low drift spraying equipment must be used until the operator is 30m from the top of the bank of any surface water bodies
- LERAP Category B [1-2]

Hazard classification and safety precautions

Hazard Harmful, Dangerous for the environment, Very toxic to aquatic organisms

Transport code 9 [1-2]

Packaging group III

SEE SECTION 3 FOR PRODUCTS ALSO REGISTERED

UN Number 3082
Risk phrases H373
Operator protection U05a, U08, U13, U14, U19a, U20b; A, H
Environmental protection E15a, E16a, E38, H410
Storage and disposal D01, D02, D09a, D11a, D12a
Medical advice M03

136 diflufenican + glyphosate

A foliar non-selective herbicide mixture
HRAC mode of action code: 12 + 9 (F1 + G)

See also glyphosate

Products

1	Pistol	BASF	40:250 g/l	SC	18854
2	Pistol Rail	BASF	40:250 g/l	SC	18853
3	Proshield	ICL (Everris) Ltd	40:250 g/l	SC	17525

Uses
- Annual and perennial weeds in **Hard surfaces** [1-2]; **Natural surfaces not intended to bear vegetation** [1,3]; **Permeable surfaces overlying soil** [1,3];

Approval information
- Diflufenican and glyphosate included in Annex I under EC Regulation 1107/2009

Efficacy guidance
- Treat when weeds actively growing from Mar to end Sep and before they begin to senesce
- Performance may be reduced if application is made to plants growing under stress, such as drought or water-logging
- Pre-emergence activity is reduced on soils containing more than 10% organic matter or where organic debris has collected
- Perennial weeds such as docks, perennial sowthistle and willowherb are best treated before flowering or setting seed
- Perennial weeds emerging from established rootstocks after treatment will not be controlled
- A rainfree period of at least 6 h (preferably 24 h) should follow spraying for optimum control
- For optimum control do not cultivate or rake after treatment

Restrictions
- Maximum number of treatments 1 per yr
- May only be used on porous surfaces overlying soil. Must not be used if an impermeable membrane lies between the porous surface and the soil, and must not be used on any non-porous man-made surfaces
- Do not add any wetting agent or adjuvant oil
- Do not spray in windy weather
- Use a knapsack sprayer giving a coarse spray via anti-drift nozzles for application
- Low drift spraying equipment must be operated according to the specific conditions stated in the official three star rating for that equipment as published on HSE Chemicals Regulation Division?s website. These operating conditions must be maintained until the operator is 30m from the top of the bank of any surface water bodies

Following crops guidance
- A period of at least 6 mth must be allowed after treatment of sites that are to be cleared and grubbed before sowing and planting. Soil should be ploughed or dug first to ensure thorough mixing and dilution of any herbicide residues

Environmental safety
- Dangerous for the environment
- Very toxic to aquatic organisms
- Avoid drift onto non-target plants
- Heavy rain after application may wash product onto sensitive areas such as newly sown grass or areas about to be planted

FOR FULL CONDITIONS OF USE ALWAYS READ THE PRODUCT LABEL

- Buffer zone requirement 2 m [3, 1] for hand-held sprayers, 6 m for horizontal boom sprayers [1]
- LERAP Category B [1,3]

Hazard classification and safety precautions
  Hazard Dangerous for the environment, Very toxic to aquatic organisms
  Transport code 9 [1-3]
  Packaging group III
  UN Number 3082
  Operator protection U20b; A, C, H
  Environmental protection E15a, E16a [1,3], E16b, E38, H410
  Storage and disposal D01, D02, D09a, D11a, D12a

137   diflufenican + iodosulfuron-methyl-sodium

A residual herbicide mixture for use on hard ballast areas
HRAC mode of action code: 12 + 2 (F1 + B)

See also diflufenican + iodosulfuron-methyl-sodium + mesosulfuron-methyl

Products

1	LockStar	ICL (Everris) Ltd	36:1% w/w	SG	19101

Uses
- Annual and perennial weeds in **Amenity vegetation** [1]; **Hard surfaces** [1]; **Natural surfaces not intended to bear vegetation** [1]; **Permeable surfaces overlying soil** [1]; **Railway tracks** [1];

Approval information
- Diflufenican and iodosulfuron-methyl-sodium included in Annex I under EC Regulation 1107/2009

Restrictions
- Horizontal boom sprayers must be fitted with three star drift reduction technology for all uses
- Low drift spraying equipment must be used for 30 m from the top of the bank of any surface water bodies

Environmental safety
- Aquatic buffer zone requirement 6 m, 2 m for hand-held applications
- LERAP Category B [1]

Hazard classification and safety precautions
  Hazard Very toxic to aquatic organisms
  Transport code 9 [1]
  Packaging group III
  UN Number 3077
  Risk phrases H319
  Operator protection A, C, H
  Environmental protection E16a, H410

138   diflufenican + iodosulfuron-methyl-sodium + mesosulfuron-methyl

A contact and residual herbicide mixture containing sulfonyl ureas for winter wheat
HRAC mode of action code: 12 + 2 + 2 (F1 + B + B)

See also iodosulfuron-methyl-sodium
mesosulfuron-methyl

Products

1	Hamlet	Bayer CropScience	50:2.5:7.5 g/l	OD	17370
2	Othello	Bayer CropScience	50:2.5:7.5 g/l	OD	16149

SEE SECTION 3 FOR PRODUCTS ALSO REGISTERED

SECTION 2

Uses

- Annual dicotyledons in **Winter wheat** [1-2];
- Annual meadow grass in **Winter wheat** [1-2];
- Cleavers in **Winter wheat** [1-2];
- Mayweeds in **Winter wheat** [1-2];
- Rough-stalked meadow grass in **Winter wheat** [1-2];
- Volunteer oilseed rape in **Winter wheat** [1-2];

Approval information

- Diflufenican, iodosulfuron-methyl-sodium and mesosulfuron-methyl included in Annex I under EC Regulation 1107/2009

Efficacy guidance

- Optimum control obtained when all weeds are emerged at spraying. Activity is primarily via foliar uptake and good spray coverage of the target weeds is essential
- Translocation occurs readily within the target weeds and growth is inhibited within hours of treatment but symptoms may not be apparent for up to 4 wk, depending on weed species, timing of treatment and weather conditions
- Iodosulfuron-methyl and mesosulfuron-methyl are both members of the ALS-inhibitor group of herbicides. To avoid the build up of resistance do not use any product containing an ALS-inhibitor herbicide with claims for control of grass weeds more than once on any crop
- Use this product as part of a Resistance Management Strategy that includes cultural methods of control and does not use ALS inhibitors as the sole chemical method of weed control in successive crops. See Section 5 for more information

Restrictions

- Maximum number of treatments 1 per crop
- Do not use on crops undersown with grasses, clover or other legumes or any other broad-leaved crop
- Do not use where annual grass weeds other than annual meadow grass and rough meadow grass are present
- Do not use as a stand-alone treatment for control of common chickweed or common poppy. Only use mixtures with non ALS-inhibitor herbicides for these weeds
- Do not use as the sole means of weed control in successive crops
- Specific restrictions apply to use in sequence or tank mixture with other sulfonylurea or ALS-inhibiting herbicides. See label for details
- Do not apply to crops under stress from any cause
- Do not apply when rain is imminent or during periods of frosty weather
- Specified adjuvant must be used. See label

Crop-specific information

- Latest use: before 2nd node detectable for winter wheat
- Transitory crop effects may occur, particularly on overlaps and after late season/spring applications. Recovery is normally complete and yield not affected

Following crops guidance

- In the event of crop failure winter or spring wheat may be drilled after normal cultivation and an interval of 6 wk
- Winter wheat, winter barley or winter oilseed rape may be drilled in the autumn following normal harvest of a treated crop. Spring wheat, spring barley, spring oilseed rape or sugar beet may be drilled in the following spring
- Where the product has been applied in sequence with a permitted ALS-inhibitor herbicide (see label) only winter or spring wheat or barley, or sugar beet may be sown as following crops
- Successive treatments with any products containing diflufenican can lead to soil build-up and inversion ploughing to 150 mm must precede sowing any following non-cereal crop. Even where ploughing occurs some crops may be damaged

Environmental safety

- Dangerous for the environment
- Very toxic to aquatic organisms
- Take extreme care to avoid drift outside the target area
- LERAP Category B [1-2]

FOR FULL CONDITIONS OF USE ALWAYS READ THE PRODUCT LABEL

Hazard classification and safety precautions

Hazard Irritant, Dangerous for the environment, Very toxic to aquatic organisms
Transport code 9 [1-2]
Packaging group III
UN Number 3082
Risk phrases H319
Operator protection U05a, U11, U20b; A, C
Environmental protection E15b, E16a, E38, H410
Storage and disposal D01, D02, D05, D09a, D10a, D12a

139   diflufenican + metsulfuron-methyl

A contact and residual herbicide mixture containing sulfonyl ureas for cereals
HRAC mode of action code: 12 + 2 (F1 + B)

Products

1	Pelican Delta	Nufarm UK	60:6% w/w	WG	18513

Uses

- Annual dicotyledons in *Spring barley* [1]; *Winter barley* [1]; *Winter wheat* [1];
- Charlock in *Spring barley* [1]; *Winter barley* [1]; *Winter wheat* [1];
- Corn spurrey in *Spring barley* [1]; *Winter barley* [1]; *Winter wheat* [1];
- Docks in *Spring barley* [1]; *Winter barley* [1]; *Winter wheat* [1];
- Hemp-nettle in *Spring barley* [1]; *Winter barley* [1]; *Winter wheat* [1];
- Mayweeds in *Spring barley* [1]; *Winter barley* [1]; *Winter wheat* [1];
- Pale persicaria in *Spring barley* [1]; *Winter barley* [1]; *Winter wheat* [1];
- Parsley-piert in *Spring barley* [1]; *Winter barley* [1]; *Winter wheat* [1];
- Redshank in *Spring barley* [1]; *Winter barley* [1]; *Winter wheat* [1];
- Scarlet pimpernel in *Spring barley* [1]; *Winter barley* [1]; *Winter wheat* [1];
- Shepherd's purse in *Spring barley* [1]; *Winter barley* [1]; *Winter wheat* [1];

Approval information

- Diflufenican and metsulfuron-methyl included in Annex I under EC Regulation 1107/2009

Efficacy guidance

- Do not apply to frosted crops or scorch may occur

Restrictions

- Do not use on crops grown for seed
- Do not apply in mixture or sequence with any other ALS herbicide
- This product must only be applied from 1st February up to the specified latest time of application

Crop-specific information

- Do not use on soils containing more than 10% organic matter

Following crops guidance

- In the event of crop failure, only sow wheat within 3 months of application

Environmental safety

- Buffer zone requirement 7m [1]
- LERAP Category B [1]

Hazard classification and safety precautions

Hazard Irritant, Dangerous for the environment, Very toxic to aquatic organisms
Transport code 9 [1]
Packaging group III
UN Number 3077
Risk phrases H319
Operator protection U05a, U08, U11, U19a, U20b; A, C
Environmental protection E15a, E16a, E34, E38, H410
Storage and disposal D01, D02, D05, D09a, D10b, D12a

SEE SECTION 3 FOR PRODUCTS ALSO REGISTERED

140 diflufenican + pendimethalin

A mixture of pyridinecarboxamide and dinitroanaline herbicides for grass and broad-leaved weed control in cereals
HRAC mode of action code: 12 + 3 (F1 + K1)

See also pendimethalin

Products

1 Bulldog	Sipcam	15.6:313 g/l	SC	17538
2 Omaha 2	Adama	40:400 g/l	SC	16846

Uses
- Annual dicotyledons in *Spring barley* [2]; *Spring wheat* [2]; *Triticale* [1-2]; *Winter barley* [1-2]; *Winter rye* [1-2]; *Winter wheat* [1-2];
- Annual grasses in *Spring barley* [2]; *Spring wheat* [2]; *Triticale* [2]; *Winter barley* [2]; *Winter rye* [2]; *Winter wheat* [2];
- Annual meadow grass in *Triticale* [1]; *Winter barley* [1]; *Winter rye* [1]; *Winter wheat* [1];

Approval information
- Diflufenican and pendimethalin included in Annex I under EC Regulation 1107/2009

Efficacy guidance
- Do not apply pre-emergence to winter crops drilled after 20th November
- Do not treat broadcast crops

Restrictions
- Horizontal boom sprayers must be fitted with three star drift reduction technology for all uses
- Low drift spraying equipment must be operated according to the specific conditions stated in the official three star rating for that equipment as published on HSE Chemicals Regulation Division?s website. These operating conditions must be maintained until the operator is 30m from the top of the bank of any surface water bodies

Following crops guidance
- Plough to 150 mm and thoroughly mix the soil before planting any following crop

Environmental safety
- Buffer zone requirement 6 m

Hazard classification and safety precautions
Hazard Dangerous for the environment, Very toxic to aquatic organisms
Transport code 9 [1-2]
Packaging group III
UN Number 3082
Risk phrases H334 [1]
Operator protection U05a, U08, U20a; A
Environmental protection E38, H410
Storage and disposal D01, D02, D09a, D10c, D12a

141 dimethachlor

A residual anilide herbicide for use in oilseed rape
HRAC mode of action code: 15 (K3)

Products

1 Teridox	Syngenta	500 g/l	EC	15876

Uses
- Annual dicotyledons in *Winter oilseed rape* [1];
- Chickweed in *Winter oilseed rape* [1];
- Docks in *Winter oilseed rape* [1];
- Groundsel in *Winter oilseed rape* [1];

FOR FULL CONDITIONS OF USE ALWAYS READ THE PRODUCT LABEL

- Loose silky bent in **Winter oilseed rape** [1];
- Mayweeds in **Winter oilseed rape** [1];
- Red dead-nettle in **Winter oilseed rape** [1];
- Scarlet pimpernel in **Winter oilseed rape** [1];

Approval information
- Dimethachlor included in Annex 1 under EC Regulation 1107/2009

Efficacy guidance
- Optimum weed control requires moist soils.
- Heavy rain within a few days of application may lead to poor weed control

Restrictions
- Do not use on sands or very light soils, stony soils, organic soil, broadcast crops or late-drilled crops.
- Apply after drilling but before crop germination which can occur within 48 hours of drilling.

Crop-specific information
- High transpiration rates or persistent wet weather during the first weeks after treatment may reduce crop vigour and plant stand.
- Crops must be covered by at least 20mm of settled soil before application.
- Do not cultivate after application.

Following crops guidance
- In the event of autumn crop failure, winter oilseed rape may be re-drilled after cultivating the soil to at least 15 cms.
- If re-drilling the following spring, spring oilseed rape, field beans, peas, maize or potatoes may be sown. If planting spring wheat, barley, oats, spring linseed, sunflowers, sugar beet or grass crops the land should be ploughed before sowing.

Environmental safety
- Buffer zone requirement 10m [1]
- LERAP Category B [1]

Hazard classification and safety precautions
  Hazard Irritant, Dangerous for the environment, Very toxic to aquatic organisms
  Transport code 9 [1]
  Packaging group III
  UN Number 3082
  Risk phrases H304, H315, H317
  Operator protection U05a, U09a, U19a, U20c; A, C, H
  Environmental protection E15b, E16a, E40b, H410
  Storage and disposal D01, D02, D05, D09a, D10c, D12a
  Medical advice M05a

142   dimethenamid-p

A chloroacetamide herbicide available only in mixtures
HRAC mode of action code: 15 (K3)

See also clomazone + dimethenamid-p + metazachlor

143   dimethenamid-p + metazachlor

A soil acting herbicide mixture for oilseed rape
HRAC mode of action code: 15 + 15 (K3 + K3)

See also metazachlor

Products

1	Springbok	BASF	200:200 g/l	EC	16786

Uses

- Annual dicotyledons in **Broccoli** *(off-label)* [1]; **Brussels sprouts** *(off-label)* [1]; **Cabbages** *(off-label)* [1]; **Calabrese** *(off-label)* [1]; **Cauliflowers** *(off-label)* [1]; **Chinese leaf** *(off-label)* [1]; **Chives** *(off-label)* [1]; **Choi sum** *(off-label)* [1]; **Collards** *(off-label)* [1]; **Forest nurseries** *(off-label)* [1]; **Game cover** *(off-label)* [1]; **Kale** *(off-label)* [1]; **Kohlrabi** *(off-label)* [1]; **Leeks** *(off-label)* [1]; **Oriental cabbage** *(off-label)* [1]; **Ornamental plant production** *(off-label)* [1]; **Salad onions** *(off-label)* [1]; **Swedes** *(off-label)* [1]; **Tatsoi** *(off-label)* [1]; **Turnips** *(off-label)* [1]; **Winter oilseed rape** [1];
- Annual grasses in **Broccoli** *(off-label)* [1]; **Brussels sprouts** *(off-label)* [1]; **Cabbages** *(off-label)* [1]; **Calabrese** *(off-label)* [1]; **Cauliflowers** *(off-label)* [1]; **Chinese leaf** *(off-label)* [1]; **Chives** *(off-label)* [1]; **Choi sum** *(off-label)* [1]; **Collards** *(off-label)* [1]; **Forest nurseries** *(off-label)* [1]; **Game cover** *(off-label)* [1]; **Kale** *(off-label)* [1]; **Kohlrabi** *(off-label)* [1]; **Leeks** *(off-label)* [1]; **Oriental cabbage** *(off-label)* [1]; **Ornamental plant production** *(off-label)* [1]; **Salad onions** *(off-label)* [1]; **Swedes** *(off-label)* [1]; **Tatsoi** *(off-label)* [1]; **Turnips** *(off-label)* [1];
- Annual meadow grass in **Broccoli** *(off-label)* [1]; **Brussels sprouts** *(off-label)* [1]; **Cabbages** *(off-label)* [1]; **Calabrese** *(off-label)* [1]; **Cauliflowers** *(off-label)* [1]; **Chinese leaf** *(off-label)* [1]; **Chives** *(off-label)* [1]; **Choi sum** *(off-label)* [1]; **Collards** *(off-label)* [1]; **Kale** *(off-label)* [1]; **Kohlrabi** *(off-label)* [1]; **Leeks** *(off-label)* [1]; **Oriental cabbage** *(off-label)* [1]; **Ornamental plant production** *(off-label)* [1]; **Salad onions** *(off-label)* [1]; **Swedes** *(off-label)* [1]; **Tatsoi** *(off-label)* [1]; **Turnips** *(off-label)* [1];
- Chickweed in **Ornamental plant production** *(off-label)* [1]; **Winter oilseed rape** [1];
- Cleavers in **Ornamental plant production** *(off-label)* [1]; **Winter oilseed rape** [1];
- Common storksbill in **Winter oilseed rape** [1];
- Crane's-bill in **Chives** *(off-label)* [1]; **Forest nurseries** *(off-label)* [1]; **Game cover** *(off-label)* [1]; **Leeks** *(off-label)* [1]; **Ornamental plant production** *(off-label)* [1]; **Salad onions** *(off-label)* [1]; **Swedes** *(off-label)* [1]; **Turnips** *(off-label)* [1];
- Fat hen in **Broccoli** *(off-label)* [1]; **Brussels sprouts** *(off-label)* [1]; **Cabbages** *(off-label)* [1]; **Calabrese** *(off-label)* [1]; **Cauliflowers** *(off-label)* [1]; **Chinese leaf** *(off-label)* [1]; **Choi sum** *(off-label)* [1]; **Collards** *(off-label)* [1]; **Kale** *(off-label)* [1]; **Kohlrabi** *(off-label)* [1]; **Oriental cabbage** *(off-label)* [1]; **Tatsoi** *(off-label)* [1];
- Field speedwell in **Winter oilseed rape** [1];
- Groundsel in **Broccoli** *(off-label)* [1]; **Brussels sprouts** *(off-label)* [1]; **Cabbages** *(off-label)* [1]; **Calabrese** *(off-label)* [1]; **Cauliflowers** *(off-label)* [1]; **Chinese leaf** *(off-label)* [1]; **Choi sum** *(off-label)* [1]; **Collards** *(off-label)* [1]; **Kale** *(off-label)* [1]; **Kohlrabi** *(off-label)* [1]; **Oriental cabbage** *(off-label)* [1]; **Ornamental plant production** *(off-label)* [1]; **Tatsoi** *(off-label)* [1];
- Mayweeds in **Broccoli** *(off-label)* [1]; **Brussels sprouts** *(off-label)* [1]; **Cabbages** *(off-label)* [1]; **Calabrese** *(off-label)* [1]; **Cauliflowers** *(off-label)* [1]; **Chinese leaf** *(off-label)* [1]; **Choi sum** *(off-label)* [1]; **Collards** *(off-label)* [1]; **Kale** *(off-label)* [1]; **Kohlrabi** *(off-label)* [1]; **Oriental cabbage** *(off-label)* [1]; **Ornamental plant production** *(off-label)* [1]; **Tatsoi** *(off-label)* [1];
- Poppies in **Winter oilseed rape** [1];
- Redshank in **Broccoli** *(off-label)* [1]; **Brussels sprouts** *(off-label)* [1]; **Cabbages** *(off-label)* [1]; **Calabrese** *(off-label)* [1]; **Cauliflowers** *(off-label)* [1]; **Chinese leaf** *(off-label)* [1]; **Choi sum** *(off-label)* [1]; **Collards** *(off-label)* [1]; **Kale** *(off-label)* [1]; **Kohlrabi** *(off-label)* [1]; **Oriental cabbage** *(off-label)* [1]; **Tatsoi** *(off-label)* [1];
- Scented mayweed in **Winter oilseed rape** [1];
- Shepherd's purse in **Broccoli** *(off-label)* [1]; **Brussels sprouts** *(off-label)* [1]; **Cabbages** *(off-label)* [1]; **Calabrese** *(off-label)* [1]; **Cauliflowers** *(off-label)* [1]; **Chinese leaf** *(off-label)* [1]; **Chives** *(off-label)* [1]; **Choi sum** *(off-label)* [1]; **Collards** *(off-label)* [1]; **Forest nurseries** *(off-label)* [1]; **Game cover** *(off-label)* [1]; **Kale** *(off-label)* [1]; **Kohlrabi** *(off-label)* [1]; **Leeks** *(off-label)* [1]; **Oriental cabbage** *(off-label)* [1]; **Ornamental plant production** *(off-label)* [1]; **Salad onions** *(off-label)* [1]; **Swedes** *(off-label)* [1]; **Tatsoi** *(off-label)* [1]; **Turnips** *(off-label)* [1]; **Winter oilseed rape** [1];
- Small nettle in **Broccoli** *(off-label)* [1]; **Brussels sprouts** *(off-label)* [1]; **Cabbages** *(off-label)* [1]; **Calabrese** *(off-label)* [1]; **Cauliflowers** *(off-label)* [1]; **Chinese leaf** *(off-label)* [1]; **Choi sum** *(off-label)* [1]; **Collards** *(off-label)* [1]; **Kale** *(off-label)* [1]; **Kohlrabi** *(off-label)* [1]; **Oriental cabbage** *(off-label)* [1]; **Ornamental plant production** *(off-label)* [1]; **Tatsoi** *(off-label)* [1];
- Sowthistle in **Ornamental plant production** *(off-label)* [1];

Extension of Authorisation for Minor Use (EAMUs)

- **Broccoli** *20151540* [1]
- **Brussels sprouts** *20151540* [1]

- *Cabbages* 20151540 [1]
- *Calabrese* 20151540 [1]
- *Cauliflowers* 20151540 [1]
- *Chinese leaf* 20151540 [1]
- *Chives* 20143008 [1]
- *Choi sum* 20151540 [1]
- *Collards* 20151540 [1]
- *Forest nurseries* 20143006 [1]
- *Game cover* 20143006 [1]
- *Kale* 20151540 [1]
- *Kohlrabi* 20151540 [1]
- *Leeks* 20143008 [1]
- *Oriental cabbage* 20151540 [1]
- *Ornamental plant production* 20143006 [1], 20152108 [1]
- *Salad onions* 20143008 [1]
- *Swedes* 20143007 [1]
- *Tatsoi* 20151540 [1]
- *Turnips* 20143007 [1]

Approval information
- Dimethenamid-p and metazachlor included in Annex I under EC Regulation 1107/2009

Efficacy guidance
- Best results obtained from treatments to fine, firm and moist seedbeds
- Apply pre- or post-emergence of the crop and ideally before weed emergence
- Residual weed control may be reduced under prolonged dry conditions
- Weeds germinating from depth may not be controlled
- Dimethenamid-p is more active than metazachlor under dry soil conditions

Restrictions
- Maximum total dose on winter oilseed rape equivalent to one full dose treatment
- Do not disturb soil after application
- Do not treat broadcast crops until they have attained two fully expanded cotyledons
- Do not use on Sands, Very Light soils, or soils containing 10% organic matter
- Do not apply when heavy rain forecast or on soils waterlogged or prone to waterlogging
- Do not treat crops suffering from stress from any cause
- Do not use pre-emergence when crop seed has started to germinate or if not covered with 15 mm of soil
- Applications shall be limited to a total dose of not more than 1.0 kg metazachlor/ha in a three year period on the same field
- Do not apply in mixture with phosphate liquid fertilisers
- Metazachlor stewardship requires that all autumn applications should be made before the end of September to reduce the risk to water

Crop-specific information
- Latest use: before 7th true leaf for winter oilseed rape
- All varieties of winter oilseed rape may be treated

Following crops guidance
- Any crop may follow a normally harvested treated winter oilseed rape crop. Ploughing is not essential before a following cereal crop but is required for all other crops
- In the event of failure of a treated crop winter wheat (excluding durum) or winter barley may be drilled in the same autumn, and any cereal (excluding durum wheat), spring oilseed rape, peas or field beans may be sown in the following spring. Ploughing to at least 150 mm should precede planting in all cases

Environmental safety
- Dangerous for the environment
- Very toxic to aquatic organisms
- Take extreme care to avoid spray drift onto non-crop plants outside the target area

SEE SECTION 3 FOR PRODUCTS ALSO REGISTERED

- Some pesticides pose a greater threat of contamination of water than others and metazachlor is one of these pesticides. Take special care when applying metazachlor near water and do not apply if heavy rain is forecast
- Metazachlor stewardship guidelines advise a maximum dose of 750 g.a.i/ha/annum. Applications to drained fields should be complete by 15th Oct but, if drains are flowing, complete applications by 1st Oct.
- LERAP Category B [1]

Hazard classification and safety precautions

Hazard Harmful, Dangerous for the environment, Harmful if swallowed, Harmful if inhaled, Very toxic to aquatic organisms
Transport code 9 [1]
Packaging group III
UN Number 3082
Risk phrases H304, H317, H319, H351
Operator protection U05a, U08, U20c; A, C, H
Environmental protection E15a, E16a, E34, E38
Storage and disposal D01, D02, D09a, D10c, D12a
Medical advice M03, M05a

144   dimethenamid-p + metazachlor + quinmerac

A soil-acting herbicide mixture for use in oilseed rape
HRAC mode of action code: 15 + 15 + 4 (K3 + K3 + O)

See also metazachlor
quinmerac

Products

1	Banastar	BASF	100:300:100 g/l	SE	16834
2	Katamaran Turbo	BASF	200:200:100 g/l	SE	16921
3	Shadow	BASF	200:200:100 g/l	SE	16804

Uses

- Annual dicotyledons in *Winter oilseed rape* [1-3];
- Annual grasses in *Winter oilseed rape* [2-3];
- Annual meadow grass in *Winter oilseed rape* [1];
- Blackgrass in *Winter corn gromwell* (off-label) [2];
- Chickweed in *Winter corn gromwell* (off-label) [2];
- Cleavers in *Winter corn gromwell* (off-label) [2]; *Winter oilseed rape* [1-3];
- Crane's-bill in *Winter corn gromwell* (off-label) [2]; *Winter oilseed rape* [1];
- Fat hen in *Winter corn gromwell* (off-label) [2];
- Field pansy in *Winter corn gromwell* (off-label) [2];
- Forget-me-not in *Winter corn gromwell* (off-label) [2];
- Groundsel in *Winter corn gromwell* (off-label) [2];
- Ivy-leaved speedwell in *Winter corn gromwell* (off-label) [2];
- Mayweeds in *Winter corn gromwell* (off-label) [2];
- Poppies in *Winter corn gromwell* (off-label) [2]; *Winter oilseed rape* [1];
- Red dead-nettle in *Winter corn gromwell* (off-label) [2];
- Shepherd's purse in *Winter corn gromwell* (off-label) [2];
- Sowthistle in *Winter corn gromwell* (off-label) [2];
- Speedwells in *Winter oilseed rape* [1];

Extension of Authorisation for Minor Use (EAMUs)

- *Winter corn gromwell* 20152318 [2]

Approval information

- Dimethenamid-p, metazachlor and quinmerac included in Annex I under EC Regulation 1107/2009

SECTION 2

Efficacy guidance

- Dimethenamid-p is more active than metazachlor under dry soil conditions

Restrictions

- Applications shall be limited to a total dose of not more than 1.0 kg metazachlor/ha in a three year period on the same field
- Do not apply in mixture with phosphate liquid fertilisers
- Metazachlor stewardship requires that all autumn applications should be made before the end of September to reduce the risk to water. Quinmerac is being found in ground and surface waters. Take care when applying to sloping soils near water.

Following crops guidance

- For all situations following or rotational crops must not be planted until four months after application

Environmental safety

- Some pesticides pose a greater threat of contamination of water than others and metazachlor is one of these pesticides. Take special care when applying metazachlor near water and do not apply if heavy rain is forecast.
- Metazachlor stewardship guidelines advise a maximum dose of 750 g.a.i/ha/annum. Applications to drained fields should be complete by 15th Oct but, if drains are flowing, complete applications by 1st Oct.
- LERAP Category B [1-3]

Hazard classification and safety precautions

Hazard Harmful [2-3], Dangerous for the environment [2-3], Very toxic to aquatic organisms [2-3]
Transport code 9 [1-3]
Packaging group III
UN Number 3082
Risk phrases H317, H351
Operator protection U05a, U08, U14 [2-3], U20a [1], U20c [2-3]; A, H
Environmental protection E07a [2-3], E07d [1], E15b, E16a, E34, E38 [2-3], H410
Storage and disposal D01, D02, D05 [1], D09a, D10c, D12a [2-3]
Medical advice M03 [2-3], M05a [2-3]

145   dimethenamid-p + pendimethalin

A chloracetamide and dinitroaniline herbicide mixture for weed control in maize crops
HRAC mode of action code: 3 + 15 (K1 + K3)

Products

1	Clayton Launch	Clayton	212.5:250 g/l	EC	16277
2	Dime	Pan Agriculture	212.5:250 g/l	EC	17398
3	Dime	Pan Agriculture	212.5:250 g/l	EC	18388
4	Dime	Pan Agriculture	212.5:250 g/l	EC	18661
5	Dime	Pan Agriculture	212.5:250 g/l	EC	18825
6	Wing - P	BASF	212.5:250 g/l	EC	15425

Uses

- Annual dicotyledons in *Baby leaf crops* (off-label) [6]; *Broccoli* (off-label) [6]; *Brussels sprouts* (off-label) [6]; *Bulb onion sets* (off-label) [6]; *Bulb onions* (off-label) [6]; *Cabbages* (off-label) [6]; *Calabrese* (off-label) [6]; *Cauliflowers* (off-label) [6]; *Chives* (off-label) [6]; *Courgettes* (off-label) [6]; *Forage maize* [1-6]; *Forage maize (under plastic mulches)* (off-label) [6]; *Game cover* (off-label) [6]; *Garlic* (off-label) [6]; *Grain maize* [1-6]; *Herbs (see appendix 6)* (off-label) [6]; *Leeks* (off-label) [6]; *Lettuce* (off-label) [6]; *Ornamental plant production* (off-label) [6]; *Pumpkins* (off-label) [6]; *Salad onions* (off-label) [6]; *Shallots* (off-label) [6]; *Strawberries* (off-label) [6]; *Summer squash* (off-label) [6]; *Sweetcorn* (off-label) [6]; *Sweetcorn under plastic mulches* (off-label) [6]; *Winter squash* (off-label) [6];
- Annual grasses in *Baby leaf crops* (off-label) [6]; *Broccoli* (off-label) [6]; *Brussels sprouts* (off-label) [6]; *Bulb onion sets* (off-label) [6]; *Bulb onions* (off-label) [6]; *Cabbages* (off-label) [6]; *Calabrese* (off-label) [6]; *Cauliflowers* (off-label) [6]; *Chives* (off-label) [6]; *Forage maize (under*

plastic mulches) *(off-label)* [6]; **Garlic** *(off-label)* [6]; **Herbs (see appendix 6)** *(off-label)* [6]; **Leeks** *(off-label)* [6]; **Lettuce** *(off-label)* [6]; **Ornamental plant production** *(off-label)* [6]; **Salad onions** *(off-label)* [6]; **Shallots** *(off-label)* [6]; **Strawberries** *(off-label)* [6];

- Annual meadow grass in **Broccoli** *(off-label)* [6]; **Brussels sprouts** *(off-label)* [6]; **Calabrese** *(off-label)* [6]; **Cauliflowers** *(off-label)* [6]; **Courgettes** *(off-label)* [6]; **Forage maize** [1-6]; **Game cover** *(off-label)* [6]; **Grain maize** [1-6]; **Pumpkins** *(off-label)* [6]; **Summer squash** *(off-label)* [6]; **Sweetcorn** *(off-label)* [6]; **Sweetcorn under plastic mulches** *(off-label)* [6]; **Winter squash** *(off-label)* [6];
- Wild oats in **Ornamental plant production** *(off-label)* [6];

Extension of Authorisation for Minor Use (EAMUs)
- **Baby leaf crops** *20170810* [6]
- **Broccoli** *20131656* [6]
- **Brussels sprouts** *20131656* [6]
- **Bulb onion sets** *20122250* [6]
- **Bulb onions** *20122250* [6]
- **Cabbages** *20122251* [6]
- **Calabrese** *20131656* [6]
- **Cauliflowers** *20131656* [6]
- **Chives** *20122249* [6]
- **Courgettes** *20180619* [6]
- **Forage maize (under plastic mulches)** *20150101* [6]
- **Game cover** *20213087* [6]
- **Garlic** *20122250* [6]
- **Herbs (see appendix 6)** *20170810* [6]
- **Leeks** *20122248* [6]
- **Lettuce** *20170810* [6]
- **Ornamental plant production** *20130253* [6]
- **Pumpkins** *20180619* [6]
- **Salad onions** *20122249* [6]
- **Shallots** *20122250* [6]
- **Strawberries** *20160933* [6]
- **Summer squash** *20180619* [6]
- **Sweetcorn** *20180917* [6]
- **Sweetcorn under plastic mulches** *20180917* [6]
- **Winter squash** *20180619* [6]

Approval information
- Dimethenamid-p and pendimethalin are included in Annex 1 under EC Regulation 1107/2009

Efficacy guidance
- Do not use on soil types with more than 10% organic matter
- Best results obtained when rain falls within 7 days of application

Crop-specific information
- Use on maize crops grown under plastic mulches is by EAMU only; do not use in greenhouses, covers or other forms of protection
- Risk of crop damage if heavy rain falls soon after application to stoney or gravelly soils
- Seed should be covered with at least 5 cm of settled soil

Environmental safety
- LERAP Category B [1-6]

Hazard classification and safety precautions
Hazard Harmful, Dangerous for the environment, Harmful if swallowed [2-6], Very toxic to aquatic organisms [3-6]

Transport code 9 [1-6]

Packaging group III

UN Number 3082

Risk phrases H304 [3-6], H315 [2-6], H317 [2-6], H371 [2], R22a [1], R22b [1], R38 [1], R43 [1], R50 [1], R53a [1]

FOR FULL CONDITIONS OF USE ALWAYS READ THE PRODUCT LABEL

Operator protection U05a, U14; A, H
Environmental protection E15b, E16a, E34, E38, H410
Storage and disposal D01, D02, D09a, D10c, D12a
Medical advice M03, M05a

146   dimethenamid-p + quinmerac

A herbicide mixture for weed control in oilseed rape
HRAC mode of action code: 15 + 4 (K3 + O)

See also dimethenamid-p + metazachlor + quinmerac
quinmerac

Products

1	Tanaris	BASF	333 + 167 g/l	SE	17173
2	Topkat	BASF	333 + 167 g/l	SE	17356

Uses

- Annual dicotyledons in *Fodder beet* [1-2]; *Sugar beet* [1-2]; *Winter oilseed rape* [1-2];
- Cleavers in *Fodder beet* [1-2]; *Sugar beet* [1-2]; *Winter oilseed rape* [1-2];
- Crane's-bill in *Fodder beet* [1-2]; *Sugar beet* [1-2]; *Winter oilseed rape* [1-2];
- Dead nettle in *Fodder beet* [1-2]; *Sugar beet* [1-2]; *Winter oilseed rape* [1-2];
- Mayweeds in *Fodder beet* [1-2]; *Sugar beet* [1-2]; *Winter oilseed rape* [1-2];
- Poppies in *Fodder beet* [1-2]; *Sugar beet* [1-2]; *Winter oilseed rape* [1-2];
- Shepherd's purse in *Fodder beet* [1-2]; *Sugar beet* [1-2]; *Winter oilseed rape* [1-2];
- Sowthistle in *Fodder beet* [1-2]; *Sugar beet* [1-2]; *Winter oilseed rape* [1-2];
- Speedwells in *Fodder beet* [1-2]; *Sugar beet* [1-2]; *Winter oilseed rape* [1-2];

Approval information

- Dimethenamid-p and quinmerac included in Annex I under EC Regulation 1107/2009

Restrictions

- Quinmerac is being found in ground and surface waters. Take care when applying to sloping soils near water.

Environmental safety

- LERAP Category B [1-2]

Hazard classification and safety precautions

Hazard Very toxic to aquatic organisms
Transport code 9 [1-2]
Packaging group III
UN Number 3082
Risk phrases H317, H319
Operator protection U05a, U08, U20a; A, C, H
Environmental protection E15b, E16a, E34, H410
Storage and disposal D01, D02, D09a, D10c

147   dimethomorph

A cinnamic acid fungicide with translaminar activity
FRAC mode of action code: 40

See also ametoctradin + dimethomorph

Products

1	Clayton Macaw	Clayton	50% w/w	WP	19488
2	Coronam	Pan Agriculture	50% w/w	WP	18366
3	Coronam	Pan Agriculture	50% w/w	WP	19510
4	Dimix 500 SC	Arysta	500 g/l	SC	18459
5	Paraat	BASF	50% w/w	WP	15445

SEE SECTION 3 FOR PRODUCTS ALSO REGISTERED

Uses

- Blight in **Potatoes** [4];
- Crown rot in **Protected strawberries** *(moderate control)* [1-3,5]; **Strawberries** *(moderate control)* [1-3,5];
- Downy mildew in **Brassica leaves and sprouts** *(off-label)* [5]; **Chard** *(off-label)* [5]; **Herbs (see appendix 6)** *(off-label)* [5]; **Lamb's lettuce** *(off-label)* [5]; **Lettuce** *(off-label)* [5]; **Protected baby leaf crops** *(off-label)* [5]; **Protected herbs (see appendix 6)** *(off-label)* [5]; **Protected lamb's lettuce** *(off-label)* [5]; **Protected lettuce** *(off-label)* [5]; **Protected ornamentals** *(off-label)* [5]; **Radishes** *(off-label)* [5]; **Rocket** *(off-label)* [5]; **Seedling brassicas** *(off-label)* [5]; **Spinach** *(off-label)* [5];
- Root rot in **Blackberries** [1-3,5]; **Protected blackberries** *(moderate control)* [1-3,5]; **Protected raspberries** *(moderate control)* [1-3,5]; **Raspberries** *(moderate control)* [1-3,5];

Extension of Authorisation for Minor Use (EAMUs)

- **Brassica leaves and sprouts** *20122244* [5]
- **Chard** *20122244* [5]
- **Herbs (see appendix 6)** *20122244* [5]
- **Lamb's lettuce** *20122244* [5]
- **Lettuce** *20122244* [5]
- **Protected baby leaf crops** *20112584* [5]
- **Protected herbs (see appendix 6)** *20112584* [5]
- **Protected lamb's lettuce** *20112584* [5]
- **Protected lettuce** *20112584* [5]
- **Protected ornamentals** *20112585* [5]
- **Radishes** *20160190* [5]
- **Rocket** *20122244* [5]
- **Seedling brassicas** *20130274* [5]
- **Spinach** *20122244* [5]

Approval information

- Dimethomorph included in Annex I under EC Regulation 1107/2009

Restrictions

- Do not sow clover until the spring following application.
- Dimethomorph is a carboxylic acid amide fungicide. No more than three consecutive applications of CAA fungicides should be made.

Crop-specific information

- After application, allow 10 months before sowing clover [5]

Following crops guidance

- Cereal crops may be planted as part of a normal rotation following use. 120 days must elapse after the application before any other crops are planted.

Environmental safety

- LERAP Category B [1-3,5]

Hazard classification and safety precautions

Hazard Harmful if swallowed [4]
Transport code 9 [1-5]
Packaging group III
UN Number 3077, 3082
Operator protection U05a, U09a [4], U20b [4]; A, D, H
Environmental protection E15b, E16a [1-3,5], E38, H410 [4], H411 [1,5]
Storage and disposal D01, D02, D09a, D10a [1-3,5], D10c [4], D12a

FOR FULL CONDITIONS OF USE ALWAYS READ THE PRODUCT LABEL

148   dimethomorph + fluazinam

A mixture of cinnamic acid and dinitroaniline fungicides for blight control in potatoes
FRAC mode of action code: 29 + 40

See also fluazinam

Products

1	Hubble	Adama	200:200 g/l	SC	16089

Uses
- Blight in **Potatoes** [1];
- Late blight in **Potatoes** [1];

Approval information
- Dimethomorph and fluazinam included in Annex I under EC Regulation 1107/2009

Environmental safety
- Buffer zone requirement 6m [1]
- LERAP Category B [1]

Hazard classification and safety precautions
   Hazard Harmful, Dangerous for the environment
   Transport code 9 [1]
   Packaging group III
   UN Number 3082
   Risk phrases H361
   Operator protection U02a, U04a, U05a, U08, U14, U15, U20a; A, H
   Environmental protection E15b, E16a, E34, E38, H410
   Storage and disposal D01, D02, D05, D09a, D12a
   Medical advice M03

149   dimethomorph + propamocarb

A fungicide mixture for blight control in potatoes
FRAC mode of action code: 28 + 40

See also dimethomorph
   propamocarb hydrochloride

Products

1	Diprospero	Arysta	90 + 500 g/l	SC	19188

Uses
- Blight in **Potatoes** [1];
- Late blight in **Potatoes** [1];

Approval information
- Dimethomorph and propamocarb hydrochloride included in Annex I under EC Regulation 1107/2009

Efficacy guidance
- Spraying should commence when the crop meets within the row or following the first blight infection period (whichever is the earlier)
- Do not apply Diprospero if rainfall is imminent

Restrictions
- Up to 5 applications may be made, making up no more than half of the total number of intended late blight sprays per season.
- No more than three consecutive applications of dimethomorph should be made in any crop

SEE SECTION 3 FOR PRODUCTS ALSO REGISTERED

SECTION 2

Hazard classification and safety precautions
    Transport code 9 [1]
    Packaging group III
    UN Number 3082
    Operator protection U05a, U08, U20a; A, H
    Environmental protection E15b, H411
    Storage and disposal D09a, D10b

150   dimethomorph + pyraclostrobin

A fungicide mixture for disease control in onions, garlic and shallots
FRAC mode of action code: 11 + 40

Products

1	Cassiopeia	BASF	72:40 g/l	EC	16522

Uses
- Downy mildew in **Bulb onions** [1]; **Garlic** [1]; **Shallots** [1];

Approval information
- Dimethomorph and pyraclostrobin included in Annex I under EC Regulation 1107/2009

Efficacy guidance
- Use before the disease is established in the crop and use in a programme with other effective fungicides that have a different mode of action to reduce the risk of resistance

Environmental safety
- LERAP Category B [1]

Hazard classification and safety precautions
    Hazard Harmful, Dangerous for the environment, Harmful if swallowed, Harmful if inhaled
    Transport code 9 [1]
    Packaging group III
    UN Number 3082
    Risk phrases H304, H315
    Operator protection U05a; A, H
    Environmental protection E16a, E38, H410
    Storage and disposal D01, D02, D12a, D12b
    Medical advice M05b

151   dimoxystrobin

A protectant strobilurin fungicide for cereals available only in mixtures
FRAC mode of action code: 11

See also boscalid + dimoxystrobin

152   dithianon

A protectant and eradicant dicarbonitrile fungicide for scab control
FRAC mode of action code: M9

See also dithianon + potassium phosphonates

Products

1	Alcoban	Certis Belchim B V	70% w/w	WG	18151
2	Cleancrop Malling	Agrii	70% w/w	WG	19822
3	Diozinos	UPL Europe	70% w/w	WG	19615

FOR FULL CONDITIONS OF USE ALWAYS READ THE PRODUCT LABEL

Uses
- Scab in **Apples** [1-3]; **Pears** [1-3];

Approval information
- Dithianon included in Annex I under EC Regulation 1107/2009

Efficacy guidance
- Apply at bud-burst and repeat every 7-14 d until danger of scab infection ceases
- Application at high rate within 48 h of a Mills period prevents new infection
- Spray programme also reduces summer infection with apple canker

Restrictions
- Maximum number of treatments on apples and pears 6 per crop
- Do not use on Golden Delicious apples after green cluster
- Do not mix with lime sulphur or highly alkaline products

Crop-specific information
- HI 4 wk for apples, pears

Environmental safety
- Dangerous for the environment
- Very toxic to aquatic organisms

Hazard classification and safety precautions
Hazard Dangerous for the environment, Toxic if swallowed, Very toxic to aquatic organisms
Transport code 9 [1-3]
Packaging group III
UN Number 3077
Risk phrases H317, H318, H351
Operator protection U05a, U11; A, D, H
Environmental protection E15a, E17a (30 m) [1], E34, E38, H410
Storage and disposal D01, D02, D05, D12a

153 dithianon + potassium phosphonates

A fungicide mixture for use in apple and pears
FRAC mode of action code: M9 + P07

See also dithianon

Products

1 Clayton Prolan	Clayton	125:561 g/l	SC	19618
2 Delan Pro	BASF	125:561 g/l	SC	17374

Uses
- Scab in **Apples** [1-2]; **Pears** [1-2];

Approval information
- Dithianon and potassium phosphonate included in Annex I under EC Regulation 1107/2009

Efficacy guidance
- Apply at bud-burst and repeat every 7-14 d until danger of scab infection ceases
- Application at high rate within 48 h of a Mills period prevents new infection
- Spray programme also reduces summer infection with apple canker

Restrictions
- When mixed with products containing carbonate or bicarbonate carbon dioxide may be released and foaming may occur.

Environmental safety
- Dangerous for the environment
- Very toxic to aquatic organisms
- LERAP Category B [1-2]

SEE SECTION 3 FOR PRODUCTS ALSO REGISTERED

Hazard classification and safety precautions
 Hazard Very toxic to aquatic organisms
 Transport code 9 [1-2]
 Packaging group III
 UN Number 3082
 Risk phrases H317, H319, H351
 Operator protection U19a, U20a; A, C, H
 Environmental protection E16a, E17a (30 m), H410
 Storage and disposal D01, D02, D05

154   dodecadienol + tetradecenyl acetate + tetradecylacetate

An insect pheromone mixture

Products

1	RAK 3+4	BASF	3.8:4.1:1.9 % w/w	VP	17824

Uses
- Codling moth in **Apples** [1]; **Pears** [1];
- Fruit tree tortrix moth in **Apples** [1]; **Cherries** [1]; **Pears** [1];

Approval information
- Dodecadienol, tetradecenyl acetate and tetradecylacetate are included in Annex 1 under EC Regulation 1107/2009

Efficacy guidance
- Use in conjunction with an insecticide if damage threshold is exceeded during treatment.

Restrictions
- Do not use in orchards less than 1 ha in area
- Efficacy impaired if high density of Tortrix moths in surrounding untreated area

Hazard classification and safety precautions
 UN Number N/C
 Risk phrases H315, H317
 Operator protection U05a, U20b; A
 Environmental protection E15b, E34, H411
 Storage and disposal D01, D02, D09a

155   dodine

A protectant and eradicant guanidine fungicide
FRAC mode of action code: M7

Products

1	Syllit 400 SC	Arysta	400 g/l	SC	13363

Uses
- Leaf spot in **Blackcurrants** [1];
- Scab in **Apples** [1]; **Pears** [1];

Approval information
- Dodine included in Annex I under EC Regulation 1107/2009

Efficacy guidance
- Apply protective spray on apples and pears at bud-burst and at 10-14 d intervals until late Jun to early Jul
- Apply post-infection spray within 36 h of rain responsible for initiating infection. Where scab already present spray prevents production of spores
- On blackcurrants commence spraying at early grape stage and repeat at 2-3 wk intervals, and at least once after picking

FOR FULL CONDITIONS OF USE ALWAYS READ THE PRODUCT LABEL

Restrictions
- Do not apply in very cold weather (under 5°C) or under slow drying conditions to pears or dessert apples during bloom or immediately after petal fall
- Do not mix with lime sulphur or tetradifon
- Consult processors before use on crops grown for processing

Crop-specific information
- Latest use: early Jul for culinary apples; pre-blossom for dessert apples and pears

Environmental safety
- Dangerous for the environment
- Very toxic to aquatic organisms

Hazard classification and safety precautions
Hazard Toxic if inhaled
Transport code 6.1 [1]
Packaging group III
UN Number 2902
Risk phrases H315, H318
Operator protection U09c, U11; A, C, H
Environmental protection H410
Storage and disposal D01, D02, D09a, D12a

---

156   esfenvalerate

A contact and ingested pyrethroid insecticide
IRAC mode of action code: 3

Products
1	Gocha	Sumitomo	25 g/l	EC	18623
2	Kingpin	Sumitomo	25 g/l	EC	18626
3	Sumi-Alpha	Sumitomo	25 g/l	EC	18637
4	Sven	Sumitomo	25 g/l	EC	18627

Uses
- Aphids in *Broccoli* [1-4]; *Brussels sprouts* [1-4]; *Cabbages* [1-4]; *Calabrese* [1-4]; *Cauliflowers* [1-4]; *Chinese cabbage* [3-4]; *Combining peas* [1-4]; *Edible podded peas* [1-4]; *Grassland* [3]; *Kale* [1-4]; *Kohlrabi* [1-4]; *Oriental cabbage* [1-2]; *Ornamental plant production* [3]; *Potatoes* [1-4]; *Protected ornamentals* [3]; *Spring barley* [1-4]; *Spring field beans* [1-4]; *Spring wheat* [1-4]; *Vining peas* [1-4]; *Winter barley* [1-4]; *Winter field beans* [1-4]; *Winter wheat* [1-4];
- Bibionids in *Grass seed crops* [4]; *Grassland* [1-2,4]; *Managed amenity turf* [1-2,4];
- Caterpillars in *Broccoli* [1-2,4]; *Brussels sprouts* [1-2,4]; *Cabbages* [1-2,4]; *Calabrese* [1-2,4]; *Cauliflowers* [1-2,4]; *Chinese cabbage* [4]; *Kale* [1-2,4]; *Kohlrabi* [1-2,4]; *Oriental cabbage* [1-2];
- Insect pests in *Ornamental plant production* [1-2,4]; *Protected ornamentals* [4];
- Pea and bean weevil in *Combining peas* [1-2,4]; *Edible podded peas* [1-2,4]; *Spring field beans* [1-2,4]; *Vining peas* [1-2,4]; *Winter field beans* [1-2,4];

Approval information
- Esfenvalerate included in Annex I under EC Regulation 1107/2009
- Accepted by BBPA for use on malting barley

Efficacy guidance
- For best reduction of spread of barley yellow dwarf virus winter sown crops at high risk (e.g. after grass or in areas with history of BYDV) should be treated when aphids first seen or by mid-Oct. Otherwise treat in late Oct-early Nov
- High risk winter sown crops will need a second treatment
- Spring sown crops should be treated from the 2-3 leaf stage if aphids are found colonising in the crop and a further application may be needed before the first node stage if they reinfest
- Product also recommended between onset of flowering and milky ripe stages (GS 61-73) for control of summer cereal aphids

SEE SECTION 3 FOR PRODUCTS ALSO REGISTERED

SECTION 2

Restrictions
- Maximum number of treatments 3 per crop of which 2 may be in autumn
- Do not use if another pyrethroid or dimethoate has been applied to crop after start of ear emergence (GS 51)
- Do not leave spray solution standing in spray tank

Crop-specific information
- Latest use: 31 Mar in yr of harvest (winter use dose); early milk stage for barley (summer use); late milk stage for wheat (summer use)

Environmental safety
- Dangerous for the environment
- Very toxic to aquatic organisms
- High risk to non-target insects or other arthropods. Do not spray within 5 m of the field boundary
- Give local beekeepers warning when using in flowering crops
- Flammable
- Store product in dark away from direct sunlight
- LERAP Category A [1-4]

Hazard classification and safety precautions
Hazard Harmful, Flammable, Dangerous for the environment, Flammable liquid and vapour, Harmful if swallowed, Harmful if inhaled, Very toxic to aquatic organisms
Transport code 3 [1-4]
Packaging group III
UN Number 1993
Risk phrases H304, H317, H318, H373
Operator protection U04a, U05a, U08, U11, U14, U19a, U20b; A, C, H
Environmental protection E15a, E16c, E16d, E22a, E34, E38, H410
Storage and disposal D01, D02, D09a, D10c, D12a
Medical advice M03

157   ethephon

A plant growth regulator for cereals and various horticultural crops

See also chlormequat + ethephon
chlormequat + ethephon + imazaquin
chlormequat + ethephon + mepiquat chloride

Products

1	Cerone	Nufarm UK	480 g/l	SL	18903
2	Chrysal Plus	Chrysal	480 g/l	SL	17847
3	Ephon Top	Nufarm UK	660 g/l	SL	18872
4	Ipanema	Adama	480 g/l	SL	19484
5	Padawan	Adama	480 g/l	SL	19492
6	Telsee	Adama	480 g/l	SL	19490
7	Tephon	Clayton	480 g/l	SL	19652

Uses
- Growth regulation in *Apples* (off-label) [1]; *Durum wheat* [3]; *Ornamental plant production* (off-label) [1-2]; *Rye* [3]; *Spelt* [3]; *Spring barley* [3]; *Spring wheat* [3]; *Triticale* [3]; *Winter barley* [3]; *Winter wheat* [3];
- Growth retardation in *Protected tomatoes* (off-label) [1];
- Increasing yield in *Winter barley* (low lodging situations) [7]; *Winter wheat* (low lodging situations) [7];
- Lodging control in *Spring barley* [1,4-7]; *Triticale* [1,7]; *Winter barley* [1,4-7]; *Winter rye* [1,7]; *Winter wheat* [1,4-7];

Extension of Authorisation for Minor Use (EAMUs)
- *Apples* 20193088 [1]
- *Ornamental plant production* 20193091 [1]
- *Protected tomatoes* 20193092 [1]

Approval information
- Ethephon (2-chloroethylphosphonic acid) included in Annex I under EC Regulation 1107/2009
- All products containing ethephon carry the warning: 'ethephon is an anticholinesterase organophosphate. Handle with care'
- Accepted by BBPA for use on malting barley

Efficacy guidance
- Best results achieved on crops growing vigorously under conditions of high fertility
- Optimum timing varies between crops and products. See labels for details
- Do not spray crops when wet or if rain imminent

Restrictions
- Ethephon (2-chloroethylphosphonic acid) is an anticholinesterase organophosphorus compound. Do not use if under medical advice not to work with such compounds
- Maximum number of treatments 1 per crop or yr
- Do not spray crops suffering from stress caused by any factor, during cold weather or period of night frost nor when soil very dry
- Do not apply to cereals within 10 d of herbicide or liquid fertilizer application
- Do not spray wheat or triticale where the leaf sheaths have split and the ear is visible

Crop-specific information
- Latest use: before 1st spikelet visible (GS 51) for spring barley, winter barley, winter rye; before flag leaf sheath opening (GS 47) for triticale, winter wheat
- HI cider apples, tomatoes 5 d

Environmental safety
- Harmful to aquatic organisms
- Avoid accidental deposits on painted objects such as cars, trucks, aircraft

Hazard classification and safety precautions
Hazard Harmful [1,7], Irritant [4-6]
Transport code 8 [1-7]
Packaging group III
UN Number 3265
Risk phrases H290, H312 [3], H314 [3], H315, H318 [1-2,4-7], H335 [3]
Operator protection U05a [1,4-6], U08 [1,4-6], U11 [1,4-6], U13 [1,4-6], U19c [1,4-6], U20b [1,4-7]; A, C, H
Environmental protection E13c [1,4-6], E15a [7], E38 [1,4-6], H411 [1-3,7]
Storage and disposal D01, D02, D05 [2-3], D08 [7], D09a [1,4-7], D10b [1,4-6], D10c [7], D12a [1,4-6]
Medical advice M01 [1,4-7], M05a [7]

158   ethephon + mepiquat chloride

A plant growth regulator for use in cereals

See also mepiquat chloride

Products

1 Clayton Proud	Clayton	155:305 g/l	SL	17067
2 Terpal	BASF	155:305 g/l	SL	16463

Uses
- Growth regulation in *Forest nurseries* (off-label) [2]; *Ornamental plant production* (off-label) [2];
- Increasing yield in *Winter barley* (low lodging situations) [1-2]; *Winter wheat* (low lodging situations) [1-2];

SEE SECTION 3 FOR PRODUCTS ALSO REGISTERED

- Lodging control in **Spring barley** [1-2]; **Triticale** [1-2]; **Winter barley** [1-2]; **Winter rye** [1-2]; **Winter wheat** [1-2];

Extension of Authorisation for Minor Use (EAMUs)
- **Forest nurseries** *20142725* [2]
- **Ornamental plant production** *20180151* [2]

Approval information
- Ethephon and mepiquat chloride included in Annex I under EC Regulation 1107/2009
- All products containing ethephon (2-chloroethylphosphonic acid) carry the warning: '2-chloroethylphosphonic acid is an anticholinesterase organophosphate. Handle with care'
- Accepted by BBPA for use on malting barley

Efficacy guidance
- Best results achieved on crops growing vigorously under conditions of high fertility
- Recommended dose and timing vary with crop, cultivar, growing conditions, previous treatment and desired degree of lodging control. See label for details
- May be applied to crops undersown with grass or clovers
- Do not apply to crops if wet or rain expected as efficacy will be impaired

Restrictions
- Ethephon is an anticholinesterase organophosphorus compound. Do not use if under medical advice not to work with such compounds
- Maximum number of treatments 2 per crop
- Add an authorised non-ionic wetter to spray solution. See label for recommended product and rate
- Do not treat crops damaged by herbicides or stressed by drought, waterlogging etc
- Do not treat crops on soils of low fertility unless adequately fertilized
- Do not apply to winter cultivars sown in spring or treat winter barley, triticale or winter rye on soils with more than 10% organic matter (winter wheat may be treated)
- Do not apply at temperatures above 21°C

Crop-specific information
- Latest use: before ear visible (GS 49) for winter barley, spring barley, winter wheat and triticale; flag leaf just visible (GS 37) for winter rye
- Late tillering may be increased with crops subject to moisture stress and may reduce quality of malting barley

Environmental safety
- Do not use straw from treated cereals as a mulch or growing medium

Hazard classification and safety precautions
Hazard Harmful, Harmful if swallowed [2]
Transport code 8 [1-2]
Packaging group III
UN Number 3265
Risk phrases H290 [2]
Operator protection U20b; A, C
Environmental protection E15a, H413 [2]
Storage and disposal D01, D02, D08, D09a, D10c
Medical advice M01, M05a

159   ethofumesate

A benzofuran herbicide for grass weed control in various crops
HRAC mode of action code: 15 (N)

See also ethofumesate + metamitron
      ethofumesate + phenmedipham

Products

1 Efeckt	UPL Europe	500 g/l	SC	19328
2 Ethofol	UPL Europe	500 g/l	SC	19338
3 Oblix 500	UPL Europe	500 g/l	SC	19304
4 Xerton	UPL Europe	417 g/l	SC	17335

Uses

- Annual dicotyledons in *Fodder beet* [1-3]; *Grass seed crops* *(off-label)* [1,3]; *Mangels* [1-3]; *Red beet* [1-3]; *Sugar beet* [1-3]; *Winter wheat* [4];
- Annual grasses in *Fodder beet* [1,3]; *Grass seed crops* *(off-label)* [3]; *Mangels* [1,3]; *Red beet* [1,3]; *Sugar beet* [1,3]; *Winter wheat* [4];
- Annual meadow grass in *Grass seed crops* *(off-label)* [1]; *Meadowfoam (Limnanthes alba)* *(off-label)* [1];
- Black bindweed in *Meadowfoam (Limnanthes alba)* *(off-label)* [1];
- Blackgrass in *Fodder beet* [2]; *Grass seed crops* *(off-label)* [1]; *Mangels* [2]; *Meadowfoam (Limnanthes alba)* *(off-label)* [1]; *Red beet* [2]; *Sugar beet* [2]; *Winter wheat* [4];
- Charlock in *Meadowfoam (Limnanthes alba)* *(off-label)* [1];
- Chickweed in *Meadowfoam (Limnanthes alba)* *(off-label)* [1];
- Cleavers in *Fodder beet* [2]; *Mangels* [2]; *Meadowfoam (Limnanthes alba)* *(off-label)* [1]; *Red beet* [2]; *Sugar beet* [2];
- Fat hen in *Meadowfoam (Limnanthes alba)* *(off-label)* [1];
- Mayweeds in *Meadowfoam (Limnanthes alba)* *(off-label)* [1];
- Runch in *Meadowfoam (Limnanthes alba)* *(off-label)* [1];
- Volunteer cereals in *Grass seed crops* *(off-label)* [1];

Extension of Authorisation for Minor Use (EAMUs)

- *Grass seed crops* *20211812* [1], *20212214* [3]
- *Meadowfoam (Limnanthes alba)* *20210008* [1]

Approval information

- Ethofumesate included in Annex I under EC Regulation 1107/2009

Efficacy guidance

- Most products may be applied pre- or post-emergence of crop or weeds but some restricted to pre-emergence or post-emergence use only. Check label
- Some products recommended for use only in mixtures. Check label
- Volunteer cereals not well controlled pre-emergence, weed grasses should be sprayed before fully tillered

Restrictions

- The maximum total dose must not exceed 1.0 kg/ha ethofumesate in any three year period
- Do not use on Sands or Heavy soils, Very Light soils containing a high percentage of stones, or soils with more than 5-10% organic matter (percentage varies according to label)
- No food or feed crops except sugar beet, fodder beet, mangel and red beet may be grown within 120 days of treatment with ethofumesate
- Do not use on swards reseeded without ploughing
- Clovers will be killed or severely checked

Crop-specific information

- Latest use: before crops meet across rows for beet crops and mangels; not specified for other crops
- Apply in beet crops in tank mixes with other pre- or post-emergence herbicides. Recommendations vary for different mixtures. See label for details
- Safe timing on beet crops varies with other ingredient of tank mix. See label for details

SEE SECTION 3 FOR PRODUCTS ALSO REGISTERED

Following crops guidance
- In the event of failure of a treated crop only sugar beet, red beet, fodder beet and mangels may be redrilled within 120 days of application
- Any crop may be sown 120 days after application of mixtures in beet crops following ploughing, 5 mth after application in grass crops

Environmental safety
- Dangerous for the environment
- Toxic to aquatic organisms
- Do not empty into drains

Hazard classification and safety precautions
    Hazard Irritant [2]
    Transport code 9 [1-2]
    Packaging group III
    UN Number 3082, N/C
    Risk phrases H317 [4]
    Operator protection U05a [1,3-4], U09a [2], U14 [2], U19a [2], U20b [2]; A, H
    Environmental protection E15b, E23 [1,3-4], E34 [1,3-4], E38 [2-4], H410 [1-3], H411 [4]
    Storage and disposal D01, D02, D05 [2], D09a, D10a [1,3-4], D10c [2], D12a [1,3-4]
    Medical advice M05a [2]

160   ethofumesate + metamitron

A contact and residual herbicide mixture for beet crops
HRAC mode of action code: 15 + 5 (N + C1)

See also metamitron

Products

1	Oblix MT	UPL Europe	150:350 g/l	SC	18857
2	Torero	Adama	150:350 g/l	SC	18625
3	Volcano	UPL Europe	150:350 g/l	SC	18926

Uses
- Annual dicotyledons in **Fodder beet** [1-3]; **Mangels** [1-3]; **Sugar beet** [1-3];
- Annual meadow grass in **Fodder beet** [1-3]; **Mangels** [1-3]; **Sugar beet** [1-3];

Approval information
- Ethofumesate and metamitron included in Annex I under EC Regulation 1107/2009

Efficacy guidance
- Best results obtained from a series of treatments applied as an overall fine spray commencing when earliest germinating weeds are no larger than fully expanded cotyledon and the majority of the crop at fully expanded cotyledon
- Apply subsequent sprays as each new flush of weeds reaches early cotyledon and continue until weed emergence ceases
- Product may be used on all soil types but residual activity may be reduced on those with more than 5% organic matter

Restrictions
- Maximum total dose equivalent to three full dose treatments
- The maximum total dose must not exceed 1.0 kg/ha ethofumesate in any three year period

Crop-specific information
- Latest use: before crop leaves meet between rows
- Crop tolerance may be reduced by stress caused by growing conditions, effects of pests, disease or other pesticides, nutrient deficiency etc

Following crops guidance
- Beet crops may be sown at any time after treatment. Any other crop may be sown after mouldboard ploughing to 15 cm and a minimum interval of 3 mth after treatment

Environmental safety
- Dangerous for the environment
- Very toxic to aquatic organisms
- Do not empty into drains

Hazard classification and safety precautions

Hazard Harmful [1], Dangerous for the environment, Harmful if swallowed [1,3]
Transport code 9 [1-3]
Packaging group III
UN Number 3082
Operator protection U05a, U08 [2-3], U14 [2-3], U15 [2-3], U19a, U20a; A, C, H
Environmental protection E13c [2-3], E15a [1], E19b, E34 [2-3], E38, H410 [1,3], H411 [2]
Storage and disposal D01, D02, D05 [2-3], D09a, D10b, D12a
Medical advice M05a [1]

161   ethofumesate + phenmedipham

A contact and residual herbicide for use in beet crops
HRAC mode of action code: 15 + 5 (N + C1)

See also phenmedipham

Products

1	Betanal Tandem	Bayer CropScience	190:200 g/l	SC	19257
2	Magic Tandem	Bayer CropScience	190:200 g/l	SL	17358
3	Powertwin	Adama	200:200 g/l	SC	14004

Uses
- Annual dicotyledons in **Fodder beet** [1-3]; **Mangels** [1-3]; **Sugar beet** [1-3];
- Annual meadow grass in **Fodder beet** [1-3]; **Mangels** [1-3]; **Sugar beet** [1-3];
- Blackgrass in **Fodder beet** [1-3]; **Mangels** [1-3]; **Sugar beet** [1-3];

Approval information
- Ethofumesate and phenmedipham included in Annex I under EC Regulation 1107/2009

Efficacy guidance
- Best results achieved by repeat applications to cotyledon stage weeds. Larger susceptible weeds not killed by first treatment usually checked and controlled by second application
- Apply on all soil types at 5-10 d intervals
- On soils with more than 5-10% organic matter residual activity may be reduced

Restrictions
- Maximum number of treatments normally 3 per crop or maximum total dose equivalent to three full dose treatments, or less - see labels for details
- Do not spray wet foliage or if rain imminent
- Spray in evening if daytime temperatures above 21°C expected
- Avoid or delay treatment if frost expected within 7 days
- Avoid or delay treating crops under stress from wind damage, manganese or lime deficiency, pest or disease attack etc.
- The maximum total dose must not exceed 1.0 kg/ha ethofumesate in any three year period

Crop-specific information
- Latest use: before crop foliage meets in the rows
- Check from which recovery may not be complete may occur if treatment made during conditions of sharp diurnal temperature fluctuation

Following crops guidance
- Beet crops may be sown at any time after treatment. Any other crop may be sown after mouldboard ploughing to 15 cm and a minimum interval of 3 mth after treatment

Environmental safety
- Dangerous for the environment

SEE SECTION 3 FOR PRODUCTS ALSO REGISTERED

- Toxic to aquatic organisms
- Do not empty into drains
- Extra care necessary to avoid drift because product is recommended for use as a fine spray

Hazard classification and safety precautions
    Hazard Irritant [1,3], Dangerous for the environment [1,3]
    Transport code 9 [1-3]
    Packaging group III
    UN Number 3082
    Risk phrases H319 [1], R43 [3], R51 [3], R53a [3]
    Operator protection U05a [1,3], U08 [1,3], U14 [1,3], U19a [1,3], U20a [1,3]; A, H
    Environmental protection E15a [1,3], E19b [1,3], E34 [1,3], H410 [1], H412 [2]
    Storage and disposal D01 [1,3], D02 [1,3], D05 [1,3], D09a [1,3], D10b [1,3], D12a [1,3]
    Medical advice M05a [1,3]

162   ethylene

A gas used for fruit ripening and potato storage approved until 31st Aug 2025

Products

1 Banarg	BOC	40 g/Kg	GA	18190
2 Biofresh Safestore	Biofresh	99.9% w/w	GA	15729

Uses
- Growth regulation in **Banana** [1]; **Citrus fruits** [1];
- Sprout suppression in **Potatoes** [2];
- Storage pests in **Banana** [1]; **Citrus fruits** [1];
- Storage rots in **Banana** [1]; **Citrus fruits** [1];

Approval information
- Ethylene is included in Annex 1 under EC Regulation 1107/2009

Efficacy guidance
- For use in stored fruit or potatoes after harvest

Restrictions
- Handling and release of ethylene must only be undertaken by operators suitably trained and competent to carry out the work
- Operators must vacate treated areas immediately after ethylene introduction
- Unprotected persons must be excluded from the treated areas until atmospheres have been thoroughly ventilated for 15 minutes minimum before re-entry
- Ambient atmospheric ethylene concentration must not exceed 1000 ppm. Suitable self-contained breathing apparatus must be worn in atmospheres containing ethylene in excess of 1000 ppm
- A minimum 3 d post treatment period is required before removal of treated crop from storage
- Ethylene treatment must only be undertaken in fully enclosed storage areas that are air tight with appropriate air circulation and venting facilities

Hazard classification and safety precautions
    Hazard Extremely flammable [2], Extremely flammable gas [2]
    Transport code 2 [2]
    UN Number , 1962
    Risk phrases H280, H336 [2]
    Operator protection U19a [2]; A, D, H
    Environmental protection E38 [2], H412 [2]
    Storage and disposal D06e [2], D06f [2], D06g [2]
    Medical advice M04a [2]

163   etoxazole

A mite growth inhibitor
IRAC mode of action code: 10B

Products

1	Borneo	Sumitomo	110 g/l	SC	18873
2	Clayton Java	Clayton	110 g/l	SC	15155

Uses

* Mites in **Protected aubergines** [1-2]; **Protected tomatoes** [1-2];
* Spider mites in **Protected ornamentals** *(off-label)* [2]; **Protected strawberries** *(off-label)* [2];
* Two-spotted spider mite in **Protected ornamentals** *(off-label)* [1-2]; **Protected strawberries** *(off-label)* [1-2];

Extension of Authorisation for Minor Use (EAMUs)

* **Protected ornamentals** *20193043* [1], *20220191* [1], *20111544* [2]
* **Protected strawberries** *20193046* [1], *20220192* [1], *20111545* [2]

Approval information

* Etoxazole included in Annex I under EC Regulation 1107/2009

Restrictions

* A maximum individual dose of 500 ml/ha must not be exceeded

Environmental safety

* Keep in original container, tightly closed, in a safe place, under lock and key [2, 1]

Hazard classification and safety precautions

Hazard Dangerous for the environment
Transport code 9 [1-2]
Packaging group III
UN Number 3082
Operator protection U05a, U20b; A
Environmental protection E15a, E34, H410
Storage and disposal D01, D02, D09b, D11a, D12b

164   fatty acids

A soap concentrate insecticide and acaricide

Products

1	Flipper	Bayer CropScience	480 g/l	EW	19154

Uses

* Aphids in **Almonds** *(off-label)* [1]; **Apples** *(off-label)* [1]; **Apricots** *(off-label)* [1]; **Asparagus** *(off-label)* [1]; **Baby leaf crops** *(off-label)* [1]; **Beans without pods (Fresh)** *(off-label)* [1]; **Bilberries** *(off-label)* [1]; **Blackberries** *(off-label)* [1]; **Blackcurrants** *(off-label)* [1]; **Blueberries** *(off-label)* [1]; **Broad beans** *(off-label)* [1]; **Broccoli** *(off-label)* [1]; **Brussels sprouts** *(off-label)* [1]; **Bulb onions** *(off-label)* [1]; **Cabbages** *(off-label)* [1]; **Calabrese** *(off-label)* [1]; **Carrots** *(off-label)* [1]; **Cauliflowers** *(off-label)* [1]; **Celeriac** *(off-label)* [1]; **Celery (outdoor)** *(off-label)* [1]; **Celery leaves** *(off-label)* [1]; **Cherries** *(off-label)* [1]; **Chestnuts** *(off-label)* [1]; **Chives** *(off-label)* [1]; **Choi sum** *(off-label)* [1]; **Collards** *(off-label)* [1]; **Courgettes** *(off-label)* [1]; **Cranberries** *(off-label)* [1]; **Cress** *(off-label)* [1]; **Dwarf beans** *(off-label)* [1]; **Edible flowers** *(off-label)* [1]; **Edible podded peas** *(off-label)* [1]; **Elderberries** *(off-label)* [1]; **Endives** *(off-label)* [1]; **Florence fennel** *(off-label)* [1]; **French beans** *(off-label)* [1]; **Garlic** *(off-label)* [1]; **Globe artichoke** *(off-label)* [1]; **Gooseberries** *(off-label)* [1]; **Hazel nuts** *(off-label)* [1]; **Herbs (see appendix 6)** *(off-label)* [1]; **Herbs for medicinal uses (see Appendix 6)** *(off-label)* [1]; **Horseradish** *(off-label)* [1]; **Jerusalem artichokes** *(off-label)* [1]; **Kale** *(off-label)* [1]; **Kohlrabi** *(off-label)* [1]; **Lamb's lettuce** *(off-label)* [1]; **Land cress** *(off-label)* [1]; **Leeks** *(off-label)* [1]; **Lentils** *(off-label)* [1]; **Lettuce** *(off-label)* [1]; **Loganberries** *(off-label)* [1]; **Mulberries** *(off-label)* [1]; **Nectarines** *(off-label)* [1]; **Oriental**

cabbage (off-label) [1]; **Ornamental plant production** (off-label) [1]; **Parsley root** (off-label) [1]; **Parsnips** (off-label) [1]; **Peaches** (off-label) [1]; **Pears** (off-label) [1]; **Plums** (off-label) [1]; **Protected almonds** (off-label) [1]; **Protected apple** (off-label) [1]; **Protected apricots** (off-label) [1]; **Protected asparagus** (off-label) [1]; **Protected aubergines** (off-label) [1]; **Protected baby leaf crops** (off-label) [1]; **Protected beans without pods** (off-label) [1]; **Protected bilberries** (off-label) [1]; **Protected blackberries** (off-label) [1]; **Protected blackcurrants** (off-label) [1]; **Protected blueberry** (off-label) [1]; **Protected broad beans** (off-label) [1]; **Protected broccoli** (off-label) [1]; **Protected Brussels sprouts** (off-label) [1]; **Protected cabbages** (off-label) [1]; **Protected calabrese** (off-label) [1]; **Protected cauliflowers** (off-label) [1]; **Protected celery** (off-label) [1]; **Protected cherries** (off-label) [1]; **Protected chestnuts** (off-label) [1]; **Protected chilli peppers** (off-label) [1]; **Protected chives** (off-label) [1]; **Protected choi sum** (off-label) [1]; **Protected collards** (off-label) [1]; **Protected courgettes** (off-label) [1]; **Protected cranberries** (off-label) [1]; **Protected cucumbers** [1]; **Protected dwarf French beans** (off-label) [1]; **Protected edible flowers** (off-label) [1]; **Protected edible podded peas** (off-label) [1]; **Protected elderberries** (off-label) [1]; **Protected endives** (off-label) [1]; **Protected Florence fennel** (off-label) [1]; **Protected globe artichokes** (off-label) [1]; **Protected gooseberries** (off-label) [1]; **Protected hazelnuts** (off-label) [1]; **Protected herbs (see appendix 6)** (off-label) [1]; **Protected kale** (off-label) [1]; **Protected kohlrabi** (off-label) [1]; **Protected lamb's lettuce** (off-label) [1]; **Protected land cress** (off-label) [1]; **Protected leeks** (off-label) [1]; **Protected lentils** (off-label) [1]; **Protected lettuce** (off-label) [1]; **Protected loganberries** (off-label) [1]; **Protected mulberry** (off-label) [1]; **Protected nectarines** (off-label) [1]; **Protected oriental cabbage** (off-label) [1]; **Protected ornamentals** (off-label) [1]; **Protected peaches** (off-label) [1]; **Protected pear** (off-label) [1]; **Protected peppers** (off-label) [1]; **Protected plums** (off-label) [1]; **Protected pumpkins** (off-label) [1]; **Protected quince** (off-label) [1]; **Protected raspberries** (off-label) [1]; **Protected red mustard** (off-label) [1]; **Protected redcurrants** (off-label) [1]; **Protected rhubarb** (off-label) [1]; **Protected rocket** (off-label) [1]; **Protected rose hips** (off-label) [1]; **Protected Rubus hybrids** (off-label) [1]; **Protected runner beans** (off-label) [1]; **Protected spinach** (off-label) [1]; **Protected spinach beet** (off-label) [1]; **Protected strawberries** (off-label) [1]; **Protected summer squash** (off-label) [1]; **Protected sweetcorn** (off-label) [1]; **Protected tomatoes** [1]; **Protected vining peas** (off-label) [1]; **Protected walnuts** (off-label) [1]; **Protected watercress** (off-label) [1]; **Protected wine grapes** (off-label) [1]; **Protected winter squash** (off-label) [1]; **Pumpkins** (off-label) [1]; **Quinces** (off-label) [1]; **Radishes** (off-label) [1]; **Raspberries** (off-label) [1]; **Red beet** (off-label) [1]; **Red mustard** (off-label) [1]; **Redcurrants** (off-label) [1]; **Rhubarb** (off-label) [1]; **Rocket** (off-label) [1]; **Rose hips** (off-label) [1]; **Rubus hybrids** (off-label) [1]; **Runner beans** (off-label) [1]; **Salad onions** (off-label) [1]; **Salsify** (off-label) [1]; **Shallots** (off-label) [1]; **Spinach** (off-label) [1]; **Spinach beet** (off-label) [1]; **Strawberries** (off-label) [1]; **Summer squash** (off-label) [1]; **Swedes** (off-label) [1]; **Sweetcorn** (off-label) [1]; **Turnips** (off-label) [1]; **Vining peas** (off-label) [1]; **Walnuts** (off-label) [1]; **Watercress** (off-label) [1]; **Wine grapes** (off-label) [1]; **Winter squash** (off-label) [1];

- Blossom weevil in **Almonds** (off-label) [1]; **Apples** (off-label) [1]; **Apricots** (off-label) [1]; **Cherries** (off-label) [1]; **Chestnuts** (off-label) [1]; **Hazel nuts** (off-label) [1]; **Nectarines** (off-label) [1]; **Peaches** (off-label) [1]; **Pears** (off-label) [1]; **Plums** (off-label) [1]; **Protected almonds** (off-label) [1]; **Protected apple** (off-label) [1]; **Protected apricots** (off-label) [1]; **Protected cherries** (off-label) [1]; **Protected chestnuts** (off-label) [1]; **Protected hazelnuts** (off-label) [1]; **Protected nectarines** (off-label) [1]; **Protected peaches** (off-label) [1]; **Protected pear** (off-label) [1]; **Protected plums** (off-label) [1]; **Protected quince** (off-label) [1]; **Protected walnuts** (off-label) [1]; **Protected wine grapes** (off-label) [1]; **Quinces** (off-label) [1]; **Walnuts** (off-label) [1]; **Wine grapes** (off-label) [1];

- Cabbage aphid in **Asparagus** (off-label) [1]; **Beans without pods (Fresh)** (off-label) [1]; **Broad beans** (off-label) [1]; **Broccoli** (off-label) [1]; **Brussels sprouts** (off-label) [1]; **Bulb onions** (off-label) [1]; **Cabbages** (off-label) [1]; **Calabrese** (off-label) [1]; **Carrots** (off-label) [1]; **Cauliflowers** (off-label) [1]; **Celeriac** (off-label) [1]; **Celery (outdoor)** (off-label) [1]; **Choi sum** (off-label) [1]; **Collards** (off-label) [1]; **Courgettes** (off-label) [1]; **Dwarf beans** (off-label) [1]; **Edible podded peas** (off-label) [1]; **Florence fennel** (off-label) [1]; **French beans** (off-label) [1]; **Garlic** (off-label) [1]; **Globe artichoke** (off-label) [1]; **Horseradish** (off-label) [1]; **Jerusalem artichokes** (off-label) [1]; **Kale** (off-label) [1]; **Kohlrabi** (off-label) [1]; **Leeks** (off-label) [1]; **Lentils** (off-label) [1]; **Oriental cabbage** (off-label) [1]; **Parsley root** (off-label) [1]; **Parsnips** (off-label) [1]; **Protected asparagus** (off-label) [1]; **Protected beans without pods** (off-label) [1]; **Protected broad beans** (off-label) [1]; **Protected broccoli** (off-label) [1]; **Protected Brussels sprouts** (off-label) [1]; **Protected** *cabbages* (off-label) [1]; **Protected calabrese** (off-label) [1]; **Protected cauliflowers** (off-label) [1]; **Protected celery** (off-label) [1]; **Protected choi sum** (off-label) [1]; **Protected collards** (off-label) [1]; **Protected courgettes** (off-label) [1]; **Protected dwarf French beans** (off-label) [1]; **Protected edible podded peas** (off-label) [1]; **Protected Florence fennel** (off-label) [1]; **Protected globe artichokes** (off-label) [1]; **Protected kale** (off-label) [1]; **Protected kohlrabi** (off-label) [1]; **Protected leeks** (off-label) [1]; **Protected lentils** (off-label) [1]; **Protected oriental cabbage** (off-label) [1]; **Protected pumpkins** (off-label) [1]; **Protected rhubarb** (off-label) [1]; **Protected runner beans** (off-label) [1]; **Protected summer squash** (off-label) [1]; **Protected sweetcorn** (off-label) [1]; **Protected vining peas** (off-label) [1]; **Protected winter squash** (off-label) [1]; **Pumpkins** (off-label) [1]; **Radishes** (off-label) [1]; **Red beet** (off-label) [1]; **Rhubarb** (off-label) [1]; **Runner beans** (off-label) [1]; **Salad onions** (off-label) [1]; **Salsify** (off-label) [1]; **Shallots** (off-label) [1]; **Summer squash** (off-label) [1]; **Swedes** (off-label) [1]; **Sweetcorn** (off-label) [1]; **Turnips** (off-label) [1]; **Vining peas** (off-label) [1]; **Winter squash** (off-label) [1];

- Leaf hoppers in **Bilberries** (off-label) [1]; **Blackberries** (off-label) [1]; **Blackcurrants** (off-label) [1]; **Blueberries** (off-label) [1]; **Cranberries** (off-label) [1]; **Elderberries** (off-label) [1]; **Gooseberries** (off-label) [1]; **Loganberries** (off-label) [1]; **Mulberries** (off-label) [1]; **Protected bilberries** (off-label) [1]; **Protected blackberries** (off-label) [1]; **Protected blackcurrants** (off-label) [1]; **Protected blueberry** (off-label) [1]; **Protected cranberries** (off-label) [1]; **Protected elderberries** (off-label) [1]; **Protected gooseberries** (off-label) [1]; **Protected loganberries** (off-label) [1]; **Protected mulberry** (off-label) [1]; **Protected raspberries** (off-label) [1]; **Protected redcurrants** (off-label) [1]; **Protected rose hips** (off-label) [1]; **Protected Rubus hybrids** (off-label) [1]; **Protected strawberries** (off-label) [1]; **Raspberries** (off-label) [1]; **Redcurrants** (off-label) [1]; **Rose hips** (off-label) [1]; **Rubus hybrids** (off-label) [1]; **Strawberries** (off-label) [1];
- Phorid flies in **Mushrooms** (off-label) [1];
- Sciarid flies in **Mushrooms** (off-label) [1];
- Spider mites in **Baby leaf crops** (off-label) [1]; **Celery leaves** (off-label) [1]; **Chives** (off-label) [1]; **Cress** (off-label) [1]; **Edible flowers** (off-label) [1]; **Endives** (off-label) [1]; **Herbs (see appendix 6)** (off-label) [1]; **Herbs for medicinal uses (see Appendix 6)** (off-label) [1]; **Lamb's lettuce** (off-label) [1]; **Land cress** (off-label) [1]; **Lettuce** (off-label) [1]; **Ornamental plant production** (off-label) [1]; **Protected aubergines** (off-label) [1]; **Protected baby leaf crops** (off-label) [1]; **Protected celery** (off-label) [1]; **Protected chilli peppers** (off-label) [1]; **Protected chives** (off-label) [1]; **Protected cucumbers** [1]; **Protected edible flowers** (off-label) [1]; **Protected endives** (off-label) [1]; **Protected herbs (see appendix 6)** (off-label) [1]; **Protected lamb's lettuce** (off-label) [1]; **Protected land cress** (off-label) [1]; **Protected lettuce** (off-label) [1]; **Protected ornamentals** (off-label) [1]; **Protected peppers** (off-label) [1]; **Protected red mustard** (off-label) [1]; **Protected rocket** (off-label) [1]; **Protected spinach** (off-label) [1]; **Protected spinach beet** (off-label) [1]; **Protected strawberries** [1]; **Protected tomatoes** [1]; **Protected watercress** (off-label) [1]; **Red mustard** (off-label) [1]; **Rocket** (off-label) [1]; **Spinach** (off-label) [1]; **Spinach beet** (off-label) [1]; **Watercress** (off-label) [1];
- Strawberry blossom weevil in **Bilberries** (off-label) [1]; **Blackberries** (off-label) [1]; **Blackcurrants** (off-label) [1]; **Blueberries** (off-label) [1]; **Cranberries** (off-label) [1]; **Elderberries** (off-label) [1]; **Gooseberries** (off-label) [1]; **Loganberries** (off-label) [1]; **Mulberries** (off-label) [1]; **Protected bilberries** (off-label) [1]; **Protected blackberries** (off-label) [1]; **Protected blackcurrants** (off-label) [1]; **Protected blueberry** (off-label) [1]; **Protected cranberries** (off-label) [1]; **Protected elderberries** (off-label) [1]; **Protected gooseberries** (off-label) [1]; **Protected loganberries** (off-label) [1]; **Protected mulberry** (off-label) [1]; **Protected raspberries** (off-label) [1]; **Protected redcurrants** (off-label) [1]; **Protected rose hips** (off-label) [1]; **Protected Rubus hybrids** (off-label) [1]; **Protected strawberries** (off-label) [1]; **Raspberries** (off-label) [1]; **Redcurrants** (off-label) [1]; **Rose hips** (off-label) [1]; **Rubus hybrids** (off-label) [1]; **Strawberries** (off-label) [1];
- Thrips in **Asparagus** (off-label) [1]; **Baby leaf crops** (off-label) [1]; **Beans without pods (Fresh)** (off-label) [1]; **Bilberries** (off-label) [1]; **Blackberries** (off-label) [1]; **Blackcurrants** (off-label) [1]; **Blueberries** (off-label) [1]; **Broad beans** (off-label) [1]; **Broccoli** (off-label) [1]; **Brussels sprouts** (off-label) [1]; **Bulb onions** (off-label) [1]; **Cabbages** (off-label) [1]; **Calabrese** (off-label) [1]; **Carrots** (off-label) [1]; **Cauliflowers** (off-label) [1]; **Celeriac** (off-label) [1]; **Celery (outdoor)** (off-label) [1]; **Celery leaves** (off-label) [1]; **Chives** (off-label) [1]; **Choi sum** (off-label) [1]; **Collards** (off-label) [1]; **Courgettes** (off-label) [1]; **Cranberries** (off-label) [1]; **Cress** (off-label) [1]; **Dwarf beans** (off-label) [1]; **Edible flowers** (off-label) [1]; **Edible podded peas** (off-label) [1]; **Elderberries** (off-label) [1]; **Endives** (off-label) [1]; **Florence fennel** (off-label) [1]; **French beans** (off-label) [1]; **Garlic** (off-label) [1]; **Globe artichoke** (off-label) [1]; **Gooseberries** (off-label) [1];

SEE SECTION 3 FOR PRODUCTS ALSO REGISTERED

Herbs (see appendix 6) *(off-label)* [1]; *Herbs for medicinal uses (see Appendix 6)* *(off-label)* [1]; *Horseradish* *(off-label)* [1]; *Jerusalem artichokes* *(off-label)* [1]; *Kale* *(off-label)* [1]; *Kohlrabi* *(off-label)* [1]; *Lamb's lettuce* *(off-label)* [1]; *Land cress* *(off-label)* [1]; *Leeks* *(off-label)* [1]; *Lentils* *(off-label)* [1]; *Lettuce* *(off-label)* [1]; *Loganberries* *(off-label)* [1]; *Mulberries* *(off-label)* [1]; *Oriental cabbage* *(off-label)* [1]; *Parsley root* *(off-label)* [1]; *Parsnips* *(off-label)* [1]; *Protected asparagus* *(off-label)* [1]; *Protected baby leaf crops* *(off-label)* [1]; *Protected beans without pods* *(off-label)* [1]; *Protected bilberries* *(off-label)* [1]; *Protected blackberries* *(off-label)* [1]; *Protected blackcurrants* *(off-label)* [1]; *Protected blueberry* *(off-label)* [1]; *Protected broad beans* *(off-label)* [1]; *Protected broccoli* *(off-label)* [1]; *Protected Brussels sprouts* *(off-label)* [1]; *Protected cabbages* *(off-label)* [1]; *Protected calabrese* *(off-label)* [1]; *Protected cauliflowers* *(off-label)* [1]; *Protected celery* *(off-label)* [1]; *Protected chives* *(off-label)* [1]; *Protected choi sum* *(off-label)* [1]; *Protected collards* *(off-label)* [1]; *Protected courgettes* *(off-label)* [1]; *Protected cranberries* *(off-label)* [1]; *Protected dwarf French beans* *(off-label)* [1]; *Protected edible flowers* *(off-label)* [1]; *Protected edible podded peas* *(off-label)* [1]; *Protected elderberries* *(off-label)* [1]; *Protected endives* *(off-label)* [1]; *Protected Florence fennel* *(off-label)* [1]; *Protected globe artichokes* *(off-label)* [1]; *Protected gooseberries* *(off-label)* [1]; *Protected herbs (see appendix 6)* *(off-label)* [1]; *Protected kale* *(off-label)* [1]; *Protected kohlrabi* *(off-label)* [1]; *Protected lamb's lettuce* *(off-label)* [1]; *Protected land cress* *(off-label)* [1]; *Protected leeks* *(off-label)* [1]; *Protected lentils* *(off-label)* [1]; *Protected lettuce* *(off-label)* [1]; *Protected loganberries* *(off-label)* [1]; *Protected mulberry* *(off-label)* [1]; *Protected oriental cabbage* *(off-label)* [1]; *Protected pumpkins* *(off-label)* [1]; *Protected raspberries* *(off-label)* [1]; *Protected red mustard* *(off-label)* [1]; *Protected redcurrants* *(off-label)* [1]; *Protected rhubarb* *(off-label)* [1]; *Protected rocket* *(off-label)* [1]; *Protected rose hips* *(off-label)* [1]; *Protected Rubus hybrids* *(off-label)* [1]; *Protected runner beans* *(off-label)* [1]; *Protected spinach* *(off-label)* [1]; *Protected spinach beet* *(off-label)* [1]; *Protected strawberries* *(off-label)* [1]; *Protected summer squash* *(off-label)* [1]; *Protected sweetcorn* *(off-label)* [1]; *Protected vining peas* *(off-label)* [1]; *Protected watercress* *(off-label)* [1]; *Protected winter squash* *(off-label)* [1]; *Pumpkins* *(off-label)* [1]; *Radishes* *(off-label)* [1]; *Raspberries* *(off-label)* [1]; *Red beet* *(off-label)* [1]; *Red mustard* *(off-label)* [1]; *Redcurrants* *(off-label)* [1]; *Rhubarb* *(off-label)* [1]; *Rocket* *(off-label)* [1]; *Rose hips* *(off-label)* [1]; *Rubus hybrids* *(off-label)* [1]; *Runner beans* *(off-label)* [1]; *Salad onions* *(off-label)* [1]; *Salsify* *(off-label)* [1]; *Shallots* *(off-label)* [1]; *Spinach* *(off-label)* [1]; *Spinach beet* *(off-label)* [1]; *Strawberries* *(off-label)* [1]; *Summer squash* *(off-label)* [1]; *Swedes* *(off-label)* [1]; *Sweetcorn* *(off-label)* [1]; *Turnips* *(off-label)* [1]; *Vining peas* *(off-label)* [1]; *Watercress* *(off-label)* [1]; *Winter squash* *(off-label)* [1];

- Two-spotted spider mite in *Almonds* *(off-label)* [1]; *Apples* *(off-label)* [1]; *Apricots* *(off-label)* [1]; *Bilberries* *(off-label)* [1]; *Blackberries* *(off-label)* [1]; *Blackcurrants* *(off-label)* [1]; *Blueberries* *(off-label)* [1]; *Cherries* *(off-label)* [1]; *Chestnuts* *(off-label)* [1]; *Cranberries* *(off-label)* [1]; *Elderberries* *(off-label)* [1]; *Gooseberries* *(off-label)* [1]; *Hazel nuts* *(off-label)* [1]; *Loganberries* *(off-label)* [1]; *Mulberries* *(off-label)* [1]; *Nectarines* *(off-label)* [1]; *Peaches* *(off-label)* [1]; *Pears* *(off-label)* [1]; *Plums* *(off-label)* [1]; *Protected almonds* *(off-label)* [1]; *Protected apple* *(off-label)* [1]; *Protected apricots* *(off-label)* [1]; *Protected bilberries* *(off-label)* [1]; *Protected blackberries* *(off-label)* [1]; *Protected blackcurrants* *(off-label)* [1]; *Protected blueberry* *(off-label)* [1]; *Protected cherries* *(off-label)* [1]; *Protected chestnuts* *(off-label)* [1]; *Protected cranberries* *(off-label)* [1]; *Protected elderberries* *(off-label)* [1]; *Protected gooseberries* *(off-label)* [1]; *Protected hazelnuts* *(off-label)* [1]; *Protected loganberries* *(off-label)* [1]; *Protected mulberry* *(off-label)* [1]; *Protected nectarines* *(off-label)* [1]; *Protected peaches* *(off-label)* [1]; *Protected pear* *(off-label)* [1]; *Protected plums* *(off-label)* [1]; *Protected quince* *(off-label)* [1]; *Protected raspberries* *(off-label)* [1]; *Protected redcurrants* *(off-label)* [1]; *Protected rose hips* *(off-label)* [1]; *Protected Rubus hybrids* *(off-label)* [1]; *Protected strawberries* *(off-label)* [1]; *Protected walnuts* *(off-label)* [1]; *Protected wine grapes* *(off-label)* [1]; *Quinces* *(off-label)* [1]; *Raspberries* *(off-label)* [1]; *Redcurrants* *(off-label)* [1]; *Rose hips* *(off-label)* [1]; *Rubus hybrids* *(off-label)* [1]; *Strawberries* *(off-label)* [1]; *Walnuts* *(off-label)* [1]; *Wine grapes* *(off-label)* [1];
- Western flower thrips in *Baby leaf crops* *(off-label)* [1]; *Bilberries* *(off-label)* [1]; *Blackberries* *(off-label)* [1]; *Blackcurrants* *(off-label)* [1]; *Blueberries* *(off-label)* [1]; *Celery leaves* *(off-label)* [1]; *Chives* *(off-label)* [1]; *Cranberries* *(off-label)* [1]; *Cress* *(off-label)* [1]; *Edible flowers* *(off-label)* [1]; *Elderberries* *(off-label)* [1]; *Endives* *(off-label)* [1]; *Gooseberries* *(off-label)* [1]; *Herbs (see appendix 6)* *(off-label)* [1]; *Herbs for medicinal uses (see Appendix 6)* *(off-label)* [1]; *Lamb's lettuce* *(off-label)* [1]; *Land cress* *(off-label)* [1]; *Lettuce* *(off-label)* [1]; *Loganberries* *(off-label)*

[1]; *Mulberries* (off-label) [1]; *Protected baby leaf crops* (off-label) [1]; *Protected bilberries* (off-label) [1]; *Protected blackberries* (off-label) [1]; *Protected blackcurrants* (off-label) [1]; *Protected blueberry* (off-label) [1]; *Protected celery* (off-label) [1]; *Protected chives* (off-label) [1]; *Protected cranberries* (off-label) [1]; *Protected edible flowers* (off-label) [1]; *Protected elderberries* (off-label) [1]; *Protected endives* (off-label) [1]; *Protected gooseberries* (off-label) [1]; *Protected herbs (see appendix 6)* (off-label) [1]; *Protected lamb's lettuce* (off-label) [1]; *Protected land cress* (off-label) [1]; *Protected lettuce* (off-label) [1]; *Protected loganberries* (off-label) [1]; *Protected mulberry* (off-label) [1]; *Protected raspberries* (off-label) [1]; *Protected red mustard* (off-label) [1]; *Protected redcurrants* (off-label) [1]; *Protected rocket* (off-label) [1]; *Protected rose hips* (off-label) [1]; *Protected Rubus hybrids* (off-label) [1]; *Protected spinach* (off-label) [1]; *Protected spinach beet* (off-label) [1]; *Protected strawberries* (off-label) [1]; *Protected watercress* (off-label) [1]; *Raspberries* (off-label) [1]; *Red mustard* (off-label) [1]; *Redcurrants* (off-label) [1]; *Rocket* (off-label) [1]; *Rose hips* (off-label) [1]; *Rubus hybrids* (off-label) [1]; *Spinach* (off-label) [1]; *Spinach beet* (off-label) [1]; *Strawberries* (off-label) [1]; *Watercress* (off-label) [1];

- Whitefly in *Bilberries* (off-label) [1]; *Blackberries* (off-label) [1]; *Blackcurrants* (off-label) [1]; *Blueberries* (off-label) [1]; *Cranberries* (off-label) [1]; *Elderberries* (off-label) [1]; *Gooseberries* (off-label) [1]; *Loganberries* (off-label) [1]; *Mulberries* (off-label) [1]; *Ornamental plant production* (off-label) [1]; *Protected aubergines* (off-label) [1]; *Protected bilberries* (off-label) [1]; *Protected blackberries* (off-label) [1]; *Protected blackcurrants* (off-label) [1]; *Protected blueberry* (off-label) [1]; *Protected chilli peppers* (off-label) [1]; *Protected cranberries* (off-label) [1]; *Protected cucumbers* [1]; *Protected elderberries* (off-label) [1]; *Protected gooseberries* (off-label) [1]; *Protected loganberries* (off-label) [1]; *Protected mulberry* (off-label) [1]; *Protected ornamentals* (off-label) [1]; *Protected peppers* (off-label) [1]; *Protected raspberries* (off-label) [1]; *Protected redcurrants* (off-label) [1]; *Protected rose hips* (off-label) [1]; *Protected Rubus hybrids* (off-label) [1]; *Protected strawberries* (off-label) [1]; *Protected tomatoes* [1]; *Raspberries* (off-label) [1]; *Redcurrants* (off-label) [1]; *Rose hips* (off-label) [1]; *Rubus hybrids* (off-label) [1]; *Strawberries* (off-label) [1];

Extension of Authorisation for Minor Use (EAMUs)

- *Almonds* 20193419 [1]
- *Apples* 20193419 [1]
- *Apricots* 20193419 [1]
- *Asparagus* 20200103 [1]
- *Baby leaf crops* 20193416 [1]
- *Beans without pods (Fresh)* 20200103 [1]
- *Bilberries* 20193418 [1]
- *Blackberries* 20193418 [1]
- *Blackcurrants* 20193418 [1]
- *Blueberries* 20193418 [1]
- *Broad beans* 20200103 [1]
- *Broccoli* 20200103 [1]
- *Brussels sprouts* 20200103 [1]
- *Bulb onions* 20200103 [1]
- *Cabbages* 20200103 [1]
- *Calabrese* 20200103 [1]
- *Carrots* 20200103 [1]
- *Cauliflowers* 20200103 [1]
- *Celeriac* 20200103 [1]
- *Celery (outdoor)* 20200103 [1]
- *Celery leaves* 20193416 [1]
- *Cherries* 20193419 [1]
- *Chestnuts* 20193419 [1]
- *Chives* 20193416 [1]
- *Choi sum* 20200103 [1]
- *Collards* 20200103 [1]
- *Courgettes* 20200103 [1]
- *Cranberries* 20193418 [1]
- *Cress* 20193416 [1]

SEE SECTION 3 FOR PRODUCTS ALSO REGISTERED

- **Dwarf beans** *20200103* [1]
- **Edible flowers** *20193416* [1]
- **Edible podded peas** *20200103* [1]
- **Elderberries** *20193418* [1]
- **Endives** *20193416* [1]
- **Florence fennel** *20200103* [1]
- **French beans** *20200103* [1]
- **Garlic** *20200103* [1]
- **Globe artichoke** *20200103* [1]
- **Gooseberries** *20193418* [1]
- **Hazel nuts** *20193419* [1]
- **Herbs (see appendix 6)** *20193416* [1]
- **Herbs for medicinal uses (see Appendix 6)** *20193416* [1]
- **Horseradish** *20200103* [1]
- **Jerusalem artichokes** *20200103* [1]
- **Kale** *20200103* [1]
- **Kohlrabi** *20200103* [1]
- **Lamb's lettuce** *20193416* [1]
- **Land cress** *20193416* [1]
- **Leeks** *20200103* [1]
- **Lentils** *20200103* [1]
- **Lettuce** *20193416* [1]
- **Loganberries** *20193418* [1]
- **Mulberries** *20193418* [1]
- **Mushrooms** *20211802* [1]
- **Nectarines** *20193419* [1]
- **Oriental cabbage** *20200103* [1]
- **Ornamental plant production** *20201415* [1]
- **Parsley root** *20200103* [1]
- **Parsnips** *20200103* [1]
- **Peaches** *20193419* [1]
- **Pears** *20193419* [1]
- **Plums** *20193419* [1]
- **Protected almonds** *20193419* [1]
- **Protected apple** *20193419* [1]
- **Protected apricots** *20193419* [1]
- **Protected asparagus** *20200103* [1]
- **Protected aubergines** *20193172* [1]
- **Protected baby leaf crops** *20193416* [1]
- **Protected beans without pods** *20200103* [1]
- **Protected bilberries** *20193418* [1]
- **Protected blackberries** *20193418* [1]
- **Protected blackcurrants** *20193418* [1]
- **Protected blueberry** *20193418* [1]
- **Protected broad beans** *20200103* [1]
- **Protected broccoli** *20200103* [1]
- **Protected Brussels sprouts** *20200103* [1]
- **Protected cabbages** *20200103* [1]
- **Protected calabrese** *20200103* [1]
- **Protected cauliflowers** *20200103* [1]
- **Protected celery** *20193416* [1], *20200103* [1]
- **Protected cherries** *20193419* [1]
- **Protected chestnuts** *20193419* [1]
- **Protected chilli peppers** *20193172* [1]
- **Protected chives** *20193416* [1]
- **Protected choi sum** *20200103* [1]
- **Protected collards** *20200103* [1]
- **Protected courgettes** *20200103* [1]
- **Protected cranberries** *20193418* [1]

- *Protected dwarf French beans* 20200103 [1]
- *Protected edible flowers* 20193416 [1]
- *Protected edible podded peas* 20200103 [1]
- *Protected elderberries* 20193418 [1]
- *Protected endives* 20193416 [1]
- *Protected Florence fennel* 20200103 [1]
- *Protected globe artichokes* 20200103 [1]
- *Protected gooseberries* 20193418 [1]
- *Protected hazelnuts* 20193419 [1]
- *Protected herbs (see appendix 6)* 20193416 [1]
- *Protected kale* 20200103 [1]
- *Protected kohlrabi* 20200103 [1]
- *Protected lamb's lettuce* 20193416 [1]
- *Protected land cress* 20193416 [1]
- *Protected leeks* 20200103 [1]
- *Protected lentils* 20200103 [1]
- *Protected lettuce* 20193416 [1]
- *Protected loganberries* 20193418 [1]
- *Protected mulberry* 20193418 [1]
- *Protected nectarines* 20193419 [1]
- *Protected oriental cabbage* 20200103 [1]
- *Protected ornamentals* 20193171 [1], 20201415 [1]
- *Protected peaches* 20193419 [1]
- *Protected pear* 20193419 [1]
- *Protected peppers* 20193172 [1]
- *Protected plums* 20193419 [1]
- *Protected pumpkins* 20200103 [1]
- *Protected quince* 20193419 [1]
- *Protected raspberries* 20193418 [1]
- *Protected red mustard* 20193416 [1]
- *Protected redcurrants* 20193418 [1]
- *Protected rhubarb* 20200103 [1]
- *Protected rocket* 20193416 [1]
- *Protected rose hips* 20193418 [1]
- *Protected Rubus hybrids* 20193418 [1]
- *Protected runner beans* 20200103 [1]
- *Protected spinach* 20193416 [1]
- *Protected spinach beet* 20193416 [1]
- *Protected strawberries* 20193418 [1]
- *Protected summer squash* 20200103 [1]
- *Protected sweetcorn* 20200103 [1]
- *Protected vining peas* 20200103 [1]
- *Protected walnuts* 20193419 [1]
- *Protected watercress* 20193416 [1]
- *Protected wine grapes* 20193419 [1]
- *Protected winter squash* 20200103 [1]
- *Pumpkins* 20200103 [1]
- *Quinces* 20193419 [1]
- *Radishes* 20200103 [1]
- *Raspberries* 20193418 [1]
- *Red beet* 20200103 [1]
- *Red mustard* 20193416 [1]
- *Redcurrants* 20193418 [1]
- *Rhubarb* 20200103 [1]
- *Rocket* 20193416 [1]
- *Rose hips* 20193418 [1]
- *Rubus hybrids* 20193418 [1]
- *Runner beans* 20200103 [1]
- *Salad onions* 20200103 [1]

SEE SECTION 3 FOR PRODUCTS ALSO REGISTERED

- **Salsify** *20200103* [1]
- **Shallots** *20200103* [1]
- **Spinach** *20193416* [1]
- **Spinach beet** *20193416* [1]
- **Strawberries** *20193418* [1]
- **Summer squash** *20200103* [1]
- **Swedes** *20200103* [1]
- **Sweetcorn** *20200103* [1]
- **Turnips** *20200103* [1]
- **Vining peas** *20200103* [1]
- **Walnuts** *20193419* [1]
- **Watercress** *20193416* [1]
- **Wine grapes** *20193419* [1]
- **Winter squash** *20200103* [1]

Approval information
- Fatty acids C7 - C20 are included in Annex 1 under EC Regulation 1107/2009

Efficacy guidance
- Use only soft or rain water for diluting spray
- Pests must be sprayed directly to achieve any control. Spray all plant parts thoroughly to run off
- For glasshouse use apply when insects first seen and repeat as necessary. For scale insects apply several applications at weekly intervals after egg hatch
- To control whitefly spray when required and use biological control after 12 h

Restrictions
- Do not use on new transplants, newly rooted cuttings or plants under stress
- Do not use on specified susceptible shrubs. See label for details

Crop-specific information
- HI zero

Environmental safety
- Harmful to fish or other aquatic life. Do not contaminate surface waters or ditches with chemical or used container
- Buffer zone requirement 10 m, 10 m for ornamental plant production between 50cm and 1m in height and ornamental plant production over 1m in height [1]

Hazard classification and safety precautions
Hazard Irritant
UN Number N/C
Risk phrases H315, H319, H335
Operator protection A, C, H
Environmental protection H412

165 fenhexamid

A protectant hydroxyanilide fungicide for soft fruit and a range of horticultural crops
FRAC mode of action code: 17

Products
1 Teldor   Bayer CropScience   50% w/w   WG   11229

Uses
- Botrytis in **Bilberries** *(off-label)* [1]; **Blackberries** [1]; **Blackcurrants** [1]; **Blueberries** *(off-label)* [1]; **Cherries** *(off-label)* [1]; **Frise** *(off-label)* [1]; **Gooseberries** [1]; **Herbs (see appendix 6)** *(off-label)* [1]; **Hops** *(off-label)* [1]; **Lamb's lettuce** *(off-label)* [1]; **Leaf brassicas** *(off-label - baby leaf production)* [1]; **Loganberries** [1]; **Plums** *(off-label)* [1]; **Protected aubergines** *(off-label)* [1]; **Protected blueberry** *(off-label)* [1]; **Protected chilli peppers** *(off-label)* [1]; **Protected courgettes** *(off-label)* [1]; **Protected cucumbers** *(off-label)* [1]; **Protected forest nurseries** *(off-label)* [1]; **Protected gherkins** *(off-label)* [1]; **Protected peppers** *(off-label)* [1]; **Protected**

squashes (off-label) [1]; **Protected tomatoes** (off-label) [1]; **Raspberries** [1]; **Redcurrants** [1]; **Rocket** (off-label) [1]; **Rubus hybrids** [1]; **Scarole** (off-label) [1]; **Soft fruit** (off-label) [1]; **Strawberries** [1]; **Tomatoes (outdoor)** (off-label) [1]; **Whitecurrants** [1]; **Wine grapes** [1];
- Grey mould in **Protected aubergines** (off-label) [1]; **Protected chilli peppers** (off-label) [1]; **Protected courgettes** (off-label) [1]; **Protected cucumbers** (off-label) [1]; **Protected gherkins** (off-label) [1]; **Protected peppers** (off-label) [1]; **Protected squashes** (off-label) [1]; **Protected tomatoes** (off-label) [1];

Extension of Authorisation for Minor Use (EAMUs)
- **Bilberries** 20161214 [1]
- **Blueberries** 20161214 [1]
- **Cherries** 20031866 [1]
- **Frise** 20082062 [1]
- **Herbs (see appendix 6)** 20082062 [1]
- **Hops** 20082926 [1]
- **Lamb's lettuce** 20082062 [1]
- **Leaf brassicas** (baby leaf production) 20082062 [1]
- **Plums** 20031866 [1]
- **Protected aubergines** 20042087 [1]
- **Protected blueberry** 20161214 [1]
- **Protected chilli peppers** 20042086 [1]
- **Protected courgettes** 20042085 [1]
- **Protected cucumbers** 20042085 [1]
- **Protected forest nurseries** 20082926 [1]
- **Protected gherkins** 20042085 [1]
- **Protected peppers** 20042086 [1]
- **Protected squashes** 20042085 [1]
- **Protected tomatoes** 20042087 [1]
- **Rocket** 20082062 [1]
- **Scarole** 20082062 [1]
- **Soft fruit** 20082926 [1]
- **Tomatoes (outdoor)** 20042399 [1]

Approval information
- Fenhexamid included in Annex I under EC Regulation 1107/2009

Efficacy guidance
- Use as part of a programme of sprays throughout the flowering period to achieve effective control of Botrytis
- To minimise possibility of development of resistance, no more than two sprays of the product may be applied consecutively. Other fungicides from a different chemical group should then be used for at least two consecutive sprays. If only two applications are made on grapevines, only one may include fenhexamid
- Complete spray cover of all flowers and fruitlets throughout the blossom period is essential for successful control of Botrytis
- Spray programmes should normally start at the start of flowering

Restrictions
- Maximum number of treatments 2 per yr on grapevines; 4 per yr on other listed crops but no more than 2 sprays may be applied consecutively

Crop-specific information
- HI 1 d for strawberries, raspberries, loganberries, blackberries, Rubus hybrids; 3 d for cherries; 7 d for blackcurrants, redcurrants, whitecurrants, gooseberries; 21d for outdoor grapes

Environmental safety
- Dangerous for the environment
- Harmful to fish or other aquatic life. Do not contaminate surface waters or ditches with chemical or used container

SEE SECTION 3 FOR PRODUCTS ALSO REGISTERED

Hazard classification and safety precautions
   Hazard Dangerous for the environment
   Transport code 9 [1]
   Packaging group III
   UN Number 3077
   Operator protection U08, U19a, U20b
   Environmental protection E13c, E38, H411
   Storage and disposal D05, D09a, D11a, D12a

166 fenoxaprop-P-ethyl

An aryloxyphenoxypropionate herbicide for use in wheat
HRAC mode of action code: 1 (A)

See also diclofop-methyl + fenoxaprop-P-ethyl

Products

1 Foxtrot	FMC Agro	69 g/l	EW	18808
2 Oskar	FMC Agro	69 g/l	EW	18696

Uses
- Blackgrass in *Grass seed crops* (off-label) [1-2]; *Spring barley* [1-2]; *Spring wheat* [1-2]; *Winter barley* [1-2]; *Winter wheat* [1-2];
- Canary grass in *Spring barley* [1-2]; *Spring wheat* [1-2]; *Winter barley* [1-2]; *Winter wheat* [1-2];
- Rough-stalked meadow grass in *Grass seed crops* (off-label) [1-2]; *Spring barley* [1-2]; *Spring wheat* [1-2]; *Winter barley* [1-2]; *Winter wheat* [1-2];
- Wild oats in *Spring barley* [1-2]; *Spring wheat* [1-2]; *Winter barley* [1-2]; *Winter wheat* [1-2];

Extension of Authorisation for Minor Use (EAMUs)
- *Grass seed crops* 20190835 [1], 20190880 [2]

Approval information
- Fenoxaprop-P-ethyl included in Annex I under EC Regulation 1107/2009

Efficacy guidance
- Treat weeds from 2 fully expanded leaves up to flag leaf ligule just visible; for awned canary-grass and rough meadow-grass from 2 leaves to the end of tillering
- A second application may be made in spring where susceptible weeds emerge after an autumn application
- Spray is rainfast 1 h after application
- Dry conditions resulting in moisture stress may reduce effectiveness
- Fenoxaprop-P-ethyl is an ACCase inhibitor herbicide. To avoid the build up of resistance do not apply products containing an ACCase inhibitor more than twice to any crop. In addition do not use any product containing fenoxaprop-P-ethyl in mixture or sequence with any other product containing the same ingredient
- Use these products as part of a resistance management strategy that includes cultural methods of control and does not use ACCase inhibitors as the sole chemical method of grass weed control
- Applying a second product containing an ACCase inhibitor to a crop will increase the risk of resistance development; only use a second ACCase inhibitor to control different weeds at a different timing
- Always follow WRAG guidelines for preventing and managing herbicide resistant weeds. See Section 5 for more information

Restrictions
- Maximum total dose equivalent to one or two full dose treatments depending on product used
- Do not apply to barley, durum wheat, undersown crops or crops to be undersown
- Do not roll or harrow within 1 wk of spraying
- Do not spray crops under stress, suffering from drought, waterlogging or nutrient deficiency or those grazed or if soil compacted

FOR FULL CONDITIONS OF USE ALWAYS READ THE PRODUCT LABEL

- Avoid spraying immediately before or after a sudden drop in temperature or a period of warm days/cold nights
- Do not mix with hormone weedkillers

Crop-specific information
- Latest use: before flag leaf sheath extending (GS 41)
- Treat from crop emergence to flag leaf fully emerged (GS 41).
- Product may be sprayed in frosty weather provided crop hardened off but do not spray wet foliage or leaves covered with ice
- Broadcast crops should be sprayed post-emergence after plants have developed well-established root system

Environmental safety
- Dangerous for the environment
- Very toxic to aquatic organisms

Hazard classification and safety precautions
    Hazard Irritant, Dangerous for the environment
    Transport code 9 [1-2]
    Packaging group III
    UN Number 3082
    Risk phrases H315, H317 [1]
    Operator protection U05a, U08, U14, U19a, U20b; A, C, H
    Environmental protection E15a, E34, E38, H411
    Storage and disposal D01, D02, D05, D09a, D10b

167   fenpicoxamid

A novel cereal fungicide for septoria control in cereals with activity on rusts and other diseases
FRAC mode of action code: 21

Products

1	Apaveq	Corteva	50 g/l	EC	19710
2	Aquino	Corteva	50 g/l	EC	19711
3	Inconiq	Corteva	50 g/l	EC	19712
4	Peacoq	Corteva	50 g/l	EC	19714
5	Peqtiga	Corteva	50 g/l	EC	19681
6	Poquet (GB only)	Corteva	50 g/l	EC	19980
7	Questar	Corteva	50 g/l	EC	19713

Uses
- Brown rust in *Durum wheat* [1-7]; *Rye* [1-7]; *Spelt* [1-7]; *Spring wheat* [1-7]; *Triticale* [1-7]; *Winter wheat* [1-7];
- Septoria leaf blotch in *Durum wheat* [1-7]; *Rye* [1-7]; *Spelt* [1-7]; *Spring wheat* [1-7]; *Triticale* [1-7]; *Winter wheat* [1-7];
- Tan spot in *Durum wheat* [1-7]; *Rye* [1-7]; *Spelt* [1-7]; *Spring wheat* [1-7]; *Triticale* [1-7]; *Winter wheat* [1-7];
- Yellow rust in *Durum wheat* [1-7]; *Rye* [1-7]; *Spelt* [1-7]; *Spring wheat* [1-7]; *Triticale* [1-7]; *Winter wheat* [1-7];

Approval information
- Fenpicoxamide included in Annex I under EC Regulation 1107/2009

Efficacy guidance
- Apply from the start of stem extension (BBCH 30)

Restrictions
- Only one Qil fungicide application is allowed per crop in each season.

SEE SECTION 3 FOR PRODUCTS ALSO REGISTERED

- Horizontal boom sprayers must be fitted with three star drift reduction technology for all uses and use low drift spraying equipment up to 30m from the top of the bank of any surface water bodies. Use of these nozzles in the main part of the field does not affect efficacy.

Following crops guidance
- There are no following crop restrictions

Environmental safety
- Buffer zone requirement 12 m.
- Dangerous to bees. To protect bees and pollinating insects do not apply to crop plants when in flower. Do not use where bees are actively foraging. Do not apply when flowering weeds are present
- LERAP Category B [1-7]

Hazard classification and safety precautions
    Transport code 9 [1-7]
    Packaging group III
    UN Number 3082
    Risk phrases H315, H318, H335
    Operator protection U05a, U20a; A, C, H
    Environmental protection E12c, E15b, E16a, E34, H410
    Storage and disposal D01, D02, D09a, D10c
    Medical advice M03

168   fenpicoxamid + prothioconazole

A fungicide mixture for disease control in cereals
FRAC mode of action code: 3 + 21

Products

1	Univoq (GB only)	Corteva	50:100 g/l	EC	19930

Uses
- Brown rust in *Durum wheat* [1]; *Rye* [1]; *Spelt* [1]; *Spring wheat* [1]; *Triticale* [1]; *Winter wheat* [1];
- Fusarium ear blight in *Durum wheat* [1]; *Rye* [1]; *Spelt* [1]; *Spring wheat* [1]; *Triticale* [1]; *Winter wheat* [1];
- Powdery mildew in *Durum wheat* [1]; *Rye* [1]; *Spelt* [1]; *Spring wheat* [1]; *Triticale* [1]; *Winter wheat* [1];
- Rhynchosporium in *Rye* [1];
- Septoria leaf blotch in *Durum wheat* [1]; *Spelt* [1]; *Spring wheat* [1]; *Triticale* [1]; *Winter wheat* [1];
- Tan spot in *Durum wheat* [1]; *Spelt* [1]; *Spring wheat* [1]; *Winter wheat* [1];
- Yellow rust in *Durum wheat* [1]; *Rye* [1]; *Spelt* [1]; *Spring wheat* [1]; *Triticale* [1]; *Winter wheat* [1];

Approval information
- Fenpicoxamide and prothioconazole included in Annex I under EC Regulation 1107/2009

Efficacy guidance
- Apply from the start of stem extension (BBCH 30)

Restrictions
- Only one QiI fungicide application is allowed per crop in each season.
- Horizontal boom sprayers must be fitted with three star drift reduction technology for all uses and use low drift spraying equipment up to 30m from the top of the bank of any surface water bodies. Use of these nozzles in the main part of the field does not affect efficacy.

Following crops guidance
- There are no following crop restrictions

Environmental safety
- Buffer zone requirement 12 m.
- Dangerous to bees. To protect bees and pollinating insects do not apply to crop plants when in flower. Do not use where bees are actively foraging. Do not apply when flowering weeds are present
- LERAP Category B [1]

Hazard classification and safety precautions
> **Transport code** 9 [1]
> **Packaging group** III
> **UN Number** 3082
> **Risk phrases** H315, H318
> **Operator protection** U05a, U20a; A, C, H
> **Environmental protection** E12c, E15b, E16a, E34, H410
> **Storage and disposal** D01, D02, D09a, D10c
> **Medical advice** M03

169   fenpyrazamine

A botriticide for protected crops
FRAC mode of action code: 17

Products

| 1 | Empire | Pan Agriculture | 50% w/w | WG | 17357 |

Uses
- Botrytis in **Protected aubergines** [1]; **Protected courgettes** [1]; **Protected cucumbers** [1]; **Protected peppers** [1]; **Protected strawberries** [1]; **Protected tomatoes** [1]; **Strawberries** [1]; **Wine grapes** [1];

Approval information
- Fenpyrazamine is included in Annex I under EC Regulation 1107/2009

Restrictions
- Consult processors before use on crops grown for processing
- No more than a third of the intended botryticide applications made per crop, per year, should contain 3-keto reductase (FRAC code 17) fungicides

Hazard classification and safety precautions
> **Hazard** Dangerous for the environment
> **Transport code** 9 [1]
> **Packaging group** III
> **UN Number** 3077
> **Operator protection** U05a, U11, U13, U14, U15, U20b
> **Environmental protection** E15b, H410
> **Storage and disposal** D01, D02, D09a, D11a, D12a, D12b
> **Medical advice** M03

170   ferric phosphate

A molluscicide bait for controlling slugs and snails

Products

1	Carakol Iron	Adama	2.94% w/w	RB	19511
2	Chicane	Certis Belchim B V	2.11% w/w	GB	19059
3	Daxxos	Certis Belchim B V	2.11% w/w	GB	18809
4	Enzo Iron	Adama	2.94% w/w	RB	19501
5	Epitaph	Certis Belchim B V	2.11% w/w	GB	19412
6	Fe-est	De Sangosse	2.42% w/w	RB	18517
7	Fe-Lyn	De Sangosse	2.42% w/w	RB	18518

SEE SECTION 3 FOR PRODUCTS ALSO REGISTERED

SECTION 2

Products – continued

8	Ferrex	Sipcam	2.45% w/w	GB	19138
9	Ferrimax Pro	De Sangosse	2.42% w/w	RB	18519
10	Gusto Iron	Adama	2.94% w/w	RB	19499
11	Ironflexx	Certis Belchim B V	2.11% w/w	GB	18555
12	Ironmax Pro	De Sangosse	2.42% w/w	RB	18839
13	Iroxx	Certis Belchim B V	2.11% w/w	GB	16640
14	Luminare	Certis Belchim B V	2.11% w/w	GB	19060
15	Menorexx	Certis Belchim B V	2.11% w/w	GB	18842
16	Minixx	Certis Belchim B V	2.11% w/w	GB	19068
17	Sigon	De Sangosse	2.4% w/w	RB	19014
18	Sluxx HP	Certis Belchim B V	2.97% w/w	GB	16571
19	Spinner	Certis Belchim B V	2.11% w/w	GB	19067
20	X-Ecute	De Sangosse	2.42% w/w	RB	18536

Uses

- Slugs and snails in *All edible crops (outdoor and protected)* [5-7,9,11-20]; *All non-edible crops (outdoor)* [2-3,5-7,9,11-20]; *All protected non-edible crops* [2-3,5,11,13,18-19]; *Amenity vegetation* [2-3,5,11,13-16,18-19]; *Asparagus* [8]; *Aubergines* [8]; *Baby leaf crops* [1,4,10]; *Basil* [8]; *Beans without pods (dry)* [8]; *Beans without pods (Fresh)* [1,4,8,10]; *Borage* [1,4,10]; *Broad beans* [1,4,8,10]; *Broccoli* [1,4,8,10]; *Brussels sprouts* [1,4,8,10]; *Bulb onions* [1,4,8,10]; *Cabbages* [1,4,8,10]; *Calabrese* [1,4,8,10]; *Cardoons* [8]; *Carrots* [1,4,6-7,9-10,12,17,20]; *Cauliflowers* [1,4,8,10]; *Celeriac* [1,4,6-7,9-10,12,17,20]; *Celery (outdoor)* [1,4,8,10]; *Chickpeas* [8]; *Chicory* [1,4,10]; *Chives* [8]; *Choi sum* [8]; *Collards* [8]; *Combining peas* [1,4,8,10]; *Coriander* [8]; *Courgettes* [8]; *Durum wheat* [1,4,8,10]; *Dwarf beans* [1,4,8,10]; *Edible flowers* [8]; *Edible podded peas* [8]; *Endives* [1,4,10]; *Fennel leaves* [1,4,10]; *Florence fennel* [8]; *Fodder beet* [1,4,6-10,12,17,20]; *Forage maize* [1,4,8,10]; *French beans* [1,4,8,10]; *Garlic* [1,4,8,10]; *Gherkins* [8]; *Globe artichoke* [1,4,8,10]; *Grain maize* [1,4,8,10]; *Grassland* [6-7,9,12,17,20]; *Herbs (see appendix 6)* [12,17,20]; *Kale* [1,4,8,10]; *Kohlrabi* [8]; *Lamb's lettuce* [1,4,10]; *Land cress* [1,4,10]; *Leeks* [1,4,8,10]; *Lettuce* [1,4,8,10]; *Linseed* [1,4,8,10]; *Lupins* [8]; *Managed amenity turf* [6-7,9,12,17,20]; *Melons* [8]; *Mint* [8]; *Mustard* [1,4,10]; *Newly sown grass leys* [8]; *Oriental cabbage* [8]; *Ornamental plant production* [6-9,12,17,20]; *Parsley* [8]; *Poppies* [1,4,8,10]; *Potatoes* [1,4,6-7,9-10,12,17,20]; *Pumpkins* [8]; *Purslane* [1,4,10]; *Radishes* [1,4,6-7,9-10,12,17,20]; *Red beet* [1,4,6-7,9-10,12,17,20]; *Red mustard* [1,4,10]; *Rhubarb* [8]; *Rocket* [1,4,8,10]; *Rosemary* [8]; *Runner beans* [1,4,8,10]; *Rye* [1,4,8,10]; *Salad onions* [1,4,8,10]; *Seakale* [8]; *Shallots* [8]; *Soya beans* [1,4,8,10]; *Spinach* [8]; *Spinach beet* [1,4,8,10]; *Spring barley* [1,4,8,10]; *Spring field beans* [1,4,8,10]; *Spring oats* [1,4,8,10]; *Spring oilseed rape* [1,4,8,10]; *Spring wheat* [1,4,8,10]; *Strawberries* [8]; *Sugar beet* [1,4,6-10,12,17,20]; *Summer squash* [8]; *Sunflowers* [1,4,8,10]; *Swedes* [1,4,10,12,17]; *Sweetcorn* [1,4,8,10]; *Thyme* [8]; *Triticale* [1,4,8,10]; *Turnips* [1,4,6-7,9-10,12,17,20]; *Vining peas* [1,4,8,10]; *Winter barley* [1,4,8,10]; *Winter field beans* [1,4,8,10]; *Winter oats* [1,4,8,10]; *Winter oilseed rape* [1,4,8,10]; *Winter squash* [8]; *Winter wheat* [1,4,8,10]; *Witloof* [1,4,10];

Approval information

- Ferric phosphate included in Annex 1 under EC Regulation 1107/2009
- Accepted by BBPA for use on malting barley

Efficacy guidance

- Treat as soon as damage first seen preferably in early evening. Repeat as necessary to maintain control
- Best results obtained from moist soaked granules. This will occur naturally on moist soils or in humid conditions
- Ferric phosphate does not cause excessive slime secretion and has no requirement to collect moribund slugs from the soil surface
- Active ingredient is degraded by micro-organisms to beneficial plant nutrients

Restrictions

- Maximum total dose equivalent to four full dose treatments
- The product must not be broadcast near water

FOR FULL CONDITIONS OF USE ALWAYS READ THE PRODUCT LABEL

Crop-specific information
- Latest use not specified for any crop

Hazard classification and safety precautions
   UN Number N/C
   Risk phrases H319 [8]
   Operator protection U05a, U20a; A, H
   Environmental protection E15a [2-3,5,8,11,13-16,19], E34
   Storage and disposal D01, D09a

171   ferrous sulphate

A herbicide/fertilizer combination for moss control in turf

Products

1	Ferromex Mosskiller Concentrate	Omex	381.15 g/l	SL	16896
2	Greenmaster Autumn	ICL (Everris) Ltd	16.3% w/w	GR	16762
3	Greentec Mosskiller Pro	Headland Amenity	16.78% w/w	GR	16728
4	Rigby Taylor Turf Moss Killer	Rigby Taylor	381.15 g/l	SL	17219

Uses
- Moss in *Amenity grassland* [1,3-4]; *Managed amenity turf* [1-4];

Approval information
- Ferrous sulphate included in Annex 1 under EC Regulation 1107/2009

Efficacy guidance
- For best results apply when turf is actively growing and the soil is moist
- Fertilizer component of most products encourages strong root growth and tillering
- Mow 3 d before treatment and do not mow for 3-4 d afterwards
- Water after 2 d if no rain
- Rake out dead moss thoroughly 7-14 d after treatment. Re-treatment may be necessary for heavy infestations

Restrictions
- Maximum number of treatments - see labels
- Do not apply during drought or when heavy rain expected
- Do not apply in frosty weather or when the ground is frozen
- Do not walk on treated areas until well watered

Crop-specific information
- If spilt on paving, concrete, clothes etc brush off immediately to avoid discolouration
- Observe label restrictions for interval before cutting after treatment

Environmental safety
- Harmful to fish or other aquatic life. Do not contaminate surface waters or ditches with chemical or used container

Hazard classification and safety precautions
   Hazard Harmful
   UN Number N/C
   Risk phrases H315 [3], H318 [3], H319 [2]
   Operator protection U20b [1,3], U20c [2,4]; A
   Environmental protection E13c [2], E15a [3-4], E15b [1]
   Storage and disposal D01 [1,3], D09a, D11a [2-4], D12a [1]

SECTION 2

172   ferrous sulphate + MCPA + mecoprop-P

A translocated herbicide and moss killer mixture
HRAC mode of action code: 4 + 4 (O + O)

See also MCPA
*   mecoprop-P*

Products

1	Landscaper Pro Feed Weed Mosskiller	ICL (Everris) Ltd	16.3:0.49:0.29% w/w	GR	17549
2	Landscaper Pro Triple Action	ICL (Everris) Ltd	16.3:0.49:0.29% w/w	GR	17913
3	Renovator Pro	ICL (Everris) Ltd	16.3:0.49:0.29 % w/w	GR	16803

Uses
* Annual dicotyledons in *Managed amenity turf* [1-3];
* Moss in *Managed amenity turf* [1-3];
* Perennial dicotyledons in *Managed amenity turf* [1-3];

Approval information
* Ferrous sulphate, MCPA and mecoprop-P included in Annex I under EC Regulation 1107/2009

Efficacy guidance
* Apply from Apr to Sep when weeds are growing
* For best results apply when light showers or heavy dews are expected
* Apply with a suitable calibrated fertilizer distributor
* Retreatment may be necessary after 6 wk if weeds or moss persist
* For best control of moss scarify vigorously after 2 wk to remove dead moss
* Where regrowth of moss or weeds occurs a repeat treatment may be made after 6 wk
* Avoid treatment of wet grass or during drought. If no rain falls within 48 h water in thoroughly

Restrictions
* Maximum number of treatments 3 per yr
* Do not treat new turf until established for 6 mth
* The first 4 mowings after treatment should not be used to mulch cultivated plants unless composted at least 6 mth
* Avoid walking on treated areas until it has rained or they have been watered
* Do not re-seed or turf within 8 wk of last treatment
* Do not cut grass for at least 3 d before and at least 4 d after treatment
* Do not apply during freezing conditions or when rain imminent

Environmental safety
* Keep livestock out of treated areas for up to two weeks following treatment and until poisonous weeds, such as ragwort, have died down and become unpalatable
* Harmful to fish or other aquatic life. Do not contaminate surface waters or ditches with chemical or used container
* Do not empty into drains
* Some pesticides pose a greater threat of contamination of water than others and mecoprop-P is one of these pesticides. Take special care when applying mecoprop-P near water and do not apply if heavy rain is forecast

Hazard classification and safety precautions
UN Number N/C
Risk phrases H319
Operator protection U20b; A, H
Environmental protection E07a, E13c, E19b
Storage and disposal D01, D09a, D12a
Medical advice M05a

173   flazasulfuron

A sulfonylurea herbicide for non-crop use
HRAC mode of action code: 2 (B)

Products

1	Chikara Weed Control	Certis Belchim B V	25% w/w	WG	14189
2	Clayton Apt	Clayton	25% w/w	WG	15157
3	Katana	Certis Belchim B V	25% w/w	WG	16162
4	Pacaya	ProKlass	25% w/w	WG	16797
5	Paradise	Pan Agriculture	25% w/w	WG	14504
6	Paradise	Pan Agriculture	25% w/w	WG	16829
7	Paradise	Pan Agriculture	25% w/w	WG	18851

Uses

- Annual and perennial weeds in **Amenity vegetation** [1,3,7]; **Hard surfaces** [1,3,7]; **Natural surfaces not intended to bear vegetation** [1-7]; **Permeable surfaces overlying soil** [1-7]; **Railway tracks** [2,4-6];

Approval information

- Flazasulfuron included in Annex 1 under EC Regulation 1107/2009

Restrictions

- Extreme care must be taken to avoid spray drift onto non-crop plants outside of the target area
- Direct spray from the train sprayer must not be allowed to fall within 5m of the top of the bank of a static or flowing water body. Do not allow direct overspray of static or flowing surface waters
- Must not be applied to any non-porous man made surfaces

Environmental safety

- LERAP Category B [1-7]

Hazard classification and safety precautions

Hazard Dangerous for the environment, Very toxic to aquatic organisms [1-4,7]
Transport code 9 [1-7]
Packaging group III
UN Number 3077
Risk phrases R50 [5], R53a [5]
Operator protection U02a, U05a, U08 [1,3-4,7], U09a [2,5-6], U20b; A, H, M
Environmental protection E15a, E16a, E16b, E38, H410 [1-4,6-7]
Storage and disposal D01, D02, D09a, D11a, D12a

174   flonicamid

A selective feeding blocker aphicide
IRAC mode of action code: 9C

Products

1	Mainman	Certis Belchim B V	50% w/w	WG	13123
2	Teppeki	Certis Belchim B V	50% w/w	WG	12402

Uses

- Aphids in **Apples** [1]; **Brussels sprouts** [2]; **Cabbages** [2]; **Canary seed** *(off-label)* [2]; **Cherries** [1]; **Fodder beet** [2]; **Herbs (see appendix 6)** *(off-label)* [1]; **Pears** [1]; **Potatoes** [2]; **Protected chilli peppers** *(off-label)* [1]; **Protected peppers** *(off-label)* [1]; **Rye** [2]; **Spring barley** [2]; **Spring oats** [2]; **Sugar beet** [2]; **Triticale** [2]; **Vining peas** [2]; **Winter oilseed rape** [2]; **Winter wheat** [2];
- Black bean aphid in **Dwarf French beans** *(off-label)* [2]; **Edible podded peas** *(off-label)* [2]; **Runner beans** *(off-label)* [2];
- Carrot willow aphid in **Carrots** *(off-label)* [2]; **Celeriac** *(off-label)* [2]; **Parsnips** *(off-label)* [2]; **Red beet** *(off-label)* [2]; **Swedes** *(off-label)* [2]; **Turnips** *(off-label)* [2];
- Damson-hop aphid in **Hops** *(off-label)* [1];

SEE SECTION 3 FOR PRODUCTS ALSO REGISTERED

SECTION 2

- Glasshouse whitefly in *Courgettes* *(off-label)* [1]; *Cucumbers* *(off-label)* [1]; *Protected courgettes* *(off-label)* [1]; *Protected cucumbers* *(off-label)* [1]; *Protected summer squash* *(off-label)* [1]; *Summer squash* *(off-label)* [1];
- Mealybugs in *Protected aubergines* *(off-label)* [1]; *Protected tomatoes* *(off-label)* [1];
- Parsnip aphid in *Carrots* *(off-label)* [2]; *Celeriac* *(off-label)* [2]; *Parsnips* *(off-label)* [2]; *Red beet* *(off-label)* [2]; *Swedes* *(off-label)* [2]; *Turnips* *(off-label)* [2];
- Pea aphid in *Dwarf French beans* *(off-label)* [2]; *Edible podded peas* *(off-label)* [2]; *Runner beans* *(off-label)* [2];
- Peach-potato aphid in *Carrots* *(off-label)* [2]; *Celeriac* *(off-label)* [2]; *Courgettes* *(off-label)* [1]; *Cucumbers* *(off-label)* [1]; *Dwarf French beans* *(off-label)* [2]; *Edible podded peas* *(off-label)* [2]; *Parsnips* *(off-label)* [2]; *Protected courgettes* *(off-label)* [1]; *Protected cucumbers* *(off-label)* [1]; *Protected summer squash* *(off-label)* [1]; *Red beet* *(off-label)* [2]; *Runner beans* *(off-label)* [2]; *Summer squash* *(off-label)* [1]; *Swedes* *(off-label)* [2]; *Turnips* *(off-label)* [2];
- Tobacco whitefly in *Ornamental plant production* *(off-label)* [1];
- Whitefly in *Ornamental plant production* *(off-label)* [1]; *Protected aubergines* *(off-label)* [1]; *Protected tomatoes* *(off-label)* [1];

Extension of Authorisation for Minor Use (EAMUs)
- *Canary seed* 20170673 [2]
- *Carrots* 20210772 [2]
- *Celeriac* 20210772 [2]
- *Courgettes* 20142923 [1]
- *Cucumbers* 20142923 [1]
- *Dwarf French beans* 20202697 [2]
- *Edible podded peas* 20202697 [2]
- *Herbs (see appendix 6)* 20172096 [1]
- *Hops* 20142225 [1]
- *Ornamental plant production* 20130045 [1]
- *Parsnips* 20210772 [2]
- *Protected aubergines* 20160191 [1]
- *Protected chilli peppers* 20191139 [1]
- *Protected courgettes* 20142923 [1]
- *Protected cucumbers* 20142923 [1]
- *Protected peppers* 20191139 [1]
- *Protected summer squash* 20142923 [1]
- *Protected tomatoes* 20160191 [1]
- *Red beet* 20210772 [2]
- *Runner beans* 20202697 [2]
- *Summer squash* 20142923 [1]
- *Swedes* 20210772 [2]
- *Turnips* 20210772 [2]

Approval information
- Flonicamid included in Annex 1 under EC Regulation 1107/2009
- Accepted by BBPA for use on hops

Efficacy guidance
- Apply when warning systems forecast significant aphid infestations
- Persistence of action is 21 d
- Currently has activity against pyrethroid-resistant aphids with no cross-resistance to carbamates or neonicotinoids

Restrictions
- Maximum number of treatments 2 per crop for potatoes, winter wheat [2], 3 per year on apples and pears [1]
- Must not be applied to winter wheat before 50% ear emerged stage (GS 53)
- Use a maximum of two consecutive applications of [1] in apples and pears. If further treatments are required, use an insecticide with a different mode of action before applying the final application of [1]

FOR FULL CONDITIONS OF USE ALWAYS READ THE PRODUCT LABEL

- Fodder beet must not be grazed by livestock or harvested for animal consumption until at least 60 days following the last application
- Do not use where bees are actively foraging. Do not apply when flowering weeds are present
- Must not be applied in tank mix with oil based adjuvants when used on ware potatoes

Crop-specific information
- HI: potatoes 14 d; winter wheat 28 d;
- HI: apples and pears 21 days [1]

Environmental safety
- Dangerous for the environment
- Harmful to aquatic organisms
- Dangerous to bees. To protect bees and pollinating insects do not apply to crop plants when in flower. Do not use where bees are actively foraging. Do not apply when flowering weeds are present

Hazard classification and safety precautions
Hazard Dangerous for the environment [2]
UN Number N/C
Risk phrases H319
Operator protection U05a, U13 [1], U14 [1], U15 [1]; A, C, E, H, M
Environmental protection E12c, E15a
Storage and disposal D01, D02, D03 [1], D07 [2], D08 [1], D12b

175　florasulam

A triazolopyrimidine herbicide for cereals
HRAC mode of action code: 2 (B)

See also 2,4-D + florasulam
clopyralid + florasulam
clopyralid + florasulam + fluroxypyr
diflufenican + florasulam

Products

1	Barton WG	Corteva	25% w/w	WG	13284
2	Boxer	Corteva	50 g/l	SC	19536
3	Lector	Nufarm UK	50 g/l	SC	18728
4	Paramount	Nufarm UK	50 g/l	SC	18727
5	Solstice	Nufarm UK	50 g/l	SC	18752
6	Sumir	Life Scientific	50 g/l	SC	18473

Uses
- Annual dicotyledons in *Durum wheat* [2]; *Grass seed crops* (off-label) [3-4]; *Nursery fruit trees* (off-label) [4-5]; *Ornamental plant production* (off-label) [4-5]; *Protected nursery fruit trees* (off-label) [4-5]; *Protected ornamentals* (off-label) [4-5]; *Spelt* [2]; *Spring barley* [1-2,4-6]; *Spring durum wheat* [2]; *Spring oats* [1-2,4-6]; *Spring wheat* [1-2,4-6]; *Triticale* [4-5]; *Winter barley* [1-2,4-6]; *Winter oats* [1-2,4-6]; *Winter rye* [4-5]; *Winter wheat* [1-2,4-6];
- Charlock in *Nursery fruit trees* (off-label) [3-5]; *Ornamental plant production* (off-label) [3-5]; *Protected nursery fruit trees* (off-label) [3-5]; *Protected ornamentals* (off-label) [4-5];
- Chickweed in *Durum wheat* [2]; *Nursery fruit trees* (off-label) [3-5]; *Ornamental plant production* (off-label) [3-5]; *Protected nursery fruit trees* (off-label) [3-5]; *Protected ornamentals* (off-label) [4-5]; *Spelt* [2]; *Spring barley* [1-6]; *Spring durum wheat* [2]; *Spring oats* [1-6]; *Spring wheat* [1-6]; *Triticale* [3-5]; *Winter barley* [1-6]; *Winter oats* [1-6]; *Winter rye* [3-5]; *Winter wheat* [1-6];
- Cleavers in *Durum wheat* [2]; *Nursery fruit trees* (off-label) [3-5]; *Ornamental plant production* (off-label) [3-5]; *Protected nursery fruit trees* (off-label) [3-5]; *Protected ornamentals* (off-label) [4-5]; *Spelt* [2]; *Spring barley* [1-6]; *Spring durum wheat* [2]; *Spring oats* [1-6]; *Spring wheat* [1-6]; *Triticale* [3-5]; *Winter barley* [1-6]; *Winter oats* [1-6]; *Winter rye* [3-5]; *Winter wheat* [1-6];

SEE SECTION 3 FOR PRODUCTS ALSO REGISTERED

- Groundsel in **Nursery fruit trees** *(off-label)* [3,5]; **Ornamental plant production** *(off-label)* [3,5]; **Protected nursery fruit trees** *(off-label)* [3,5]; **Protected ornamentals** *(off-label)* [5];
- Hedge mustard in **Nursery fruit trees** *(off-label)* [4]; **Ornamental plant production** *(off-label)* [4]; **Protected nursery fruit trees** *(off-label)* [4]; **Protected ornamentals** *(off-label)* [4]; **Spring barley** [3]; **Spring oats** [3]; **Spring wheat** [3]; **Triticale** [3]; **Winter barley** [3]; **Winter oats** [3]; **Winter rye** [3]; **Winter wheat** [3];
- Mayweeds in **Durum wheat** [2]; **Nursery fruit trees** *(off-label)* [3-5]; **Ornamental plant production** *(off-label)* [3-5]; **Protected nursery fruit trees** *(off-label)* [3-5]; **Protected ornamentals** *(off-label)* [4-5]; **Spelt** [2]; **Spring barley** [1-2,4-6]; **Spring durum wheat** [2]; **Spring oats** [1-2,4-6]; **Spring wheat** [1-2,4-6]; **Triticale** [4-5]; **Winter barley** [1-2,4-6]; **Winter oats** [1-2,4-6]; **Winter rye** [4-5]; **Winter wheat** [1-2,4-6];
- Polygonums in **Nursery fruit trees** *(off-label)* [3]; **Ornamental plant production** *(off-label)* [3]; **Protected nursery fruit trees** *(off-label)* [3];
- Runch in **Nursery fruit trees** *(off-label)* [3-5]; **Ornamental plant production** *(off-label)* [3-5]; **Protected nursery fruit trees** *(off-label)* [3-5]; **Protected ornamentals** *(off-label)* [4-5]; **Spring barley** [3]; **Spring oats** [3]; **Spring wheat** [3]; **Triticale** [3]; **Winter barley** [3]; **Winter oats** [3]; **Winter rye** [3]; **Winter wheat** [3];
- Scented mayweed in **Spring barley** [3]; **Spring oats** [3]; **Spring wheat** [3]; **Triticale** [3]; **Winter barley** [3]; **Winter oats** [3]; **Winter rye** [3]; **Winter wheat** [3];
- Scentless mayweed in **Spring barley** [3]; **Spring oats** [3]; **Spring wheat** [3]; **Triticale** [3]; **Winter barley** [3]; **Winter oats** [3]; **Winter rye** [3]; **Winter wheat** [3];
- Shepherd's purse in **Nursery fruit trees** *(off-label)* [3-5]; **Ornamental plant production** *(off-label)* [3-5]; **Protected nursery fruit trees** *(off-label)* [3-5]; **Protected ornamentals** *(off-label)* [4-5]; **Spring barley** [3]; **Spring oats** [3]; **Spring wheat** [3]; **Triticale** [3]; **Winter barley** [3]; **Winter oats** [3]; **Winter rye** [3]; **Winter wheat** [3];
- Sowthistle in **Nursery fruit trees** *(off-label)* [3,5]; **Ornamental plant production** *(off-label)* [3,5]; **Protected nursery fruit trees** *(off-label)* [3,5]; **Protected ornamentals** *(off-label)* [5];
- Volunteer oilseed rape in **Durum wheat** [2]; **Nursery fruit trees** *(off-label)* [3-5]; **Ornamental plant production** *(off-label)* [3-5]; **Protected nursery fruit trees** *(off-label)* [3-5]; **Protected ornamentals** *(off-label)* [4-5]; **Spelt** [2]; **Spring barley** [1-6]; **Spring durum wheat** [2]; **Spring oats** [1-6]; **Spring wheat** [1-6]; **Triticale** [3-5]; **Winter barley** [1-6]; **Winter oats** [1-6]; **Winter rye** [3-5]; **Winter wheat** [1-6];
- Wild radish in **Spring barley** [3]; **Spring oats** [3]; **Spring wheat** [3]; **Triticale** [3]; **Winter barley** [3]; **Winter oats** [3]; **Winter rye** [3]; **Winter wheat** [3];

Extension of Authorisation for Minor Use (EAMUs)
- **Grass seed crops** *20183488* [3], *20183493* [4]
- **Nursery fruit trees** *20220880* [3], *20222804* [4], *20222805* [5]
- **Ornamental plant production** *20220880* [3], *20222804* [4], *20222805* [5]
- **Protected nursery fruit trees** *20220880* [3], *20222804* [4], *20222805* [5]
- **Protected ornamentals** *20222804* [4], *20222805* [5]

Approval information
- Florasulam included in Annex I under EC Regulation 1107/2009
- Accepted by BBPA for use on malting barley

Efficacy guidance
- Best results obtained from treatment of small actively growing weeds in good conditions
- Apply in autumn or spring once crop has 3 leaves
- Product is mainly absorbed by leaves of weeds and is effective on all soil types
- Florasulam is a member of the ALS-inhibitor group of herbicides

Restrictions
- The total amount of florasulam applied to a cereal crop must not exceed 7.5 g.a.i/ha
- For autumn planted crops a maximum total dose of 3.75 g of florasulam must be observed for applications made between crop emergence in the year of planting and 1st January in the year of harvest
- Do not roll or harrow within 7 d before or after application
- Do not spray when crops under stress from cold, drought, pest damage, nutrient deficiency or any other cause

FOR FULL CONDITIONS OF USE ALWAYS READ THE PRODUCT LABEL

- Specific restrictions apply to use in sequence or tank mixture with other sulfonylurea or ALS-inhibiting herbicides. See label for details

Crop-specific information
- Latest use: up to and including flag leaf just visible stage for all crops

Following crops guidance
- Where the product has been used in mixture with certain named products (see label) only cereals or grass may be sown in the autumn following harvest
- Unless otherwise restricted cereals, oilseed rape, field beans, grass or vegetable brassicas as transplants may be sown as a following crop in the same calendar yr as treatment. Oilseed rape may show some temporary reduction of vigour after a dry summer, but yields are not affected
- In addition to the above, linseed, peas, sugar beet, potatoes, maize, clover (for use in grass/clover mixtures) or carrots may be sown in the calendar yr following treatment
- In the event of failure of a treated crop in spring only spring wheat, spring barley, spring oats, maize or ryegrass may be sown

Environmental safety
- Dangerous for the environment
- Very toxic to aquatic organisms
- See label for detailed instructions on tank cleaning

Hazard classification and safety precautions
Hazard Dangerous for the environment, Very toxic to aquatic organisms [6]
Transport code 9 [1-6]
Packaging group III
UN Number 3077, 3082
Risk phrases H319 [1]
Operator protection U05a
Environmental protection E15a [6], E15b [1,3-5], E34 [1-2,6], E38, H410
Storage and disposal D01, D02, D05, D06a [3-5], D09a, D10b [1-2,6], D10c [3-5], D12a [1-2,6], D12b [3-5]

176   florasulam + fluroxypyr

A post-emergence herbicide mixture for cereals or grass
HRAC mode of action code: 2 + 4 (B + O)

See also fluroxypyr

Products

1	Cabadex	Headland Amenity	2.5:100 g/l	SE	19745
2	Celadon	ICL (Everris) Ltd	2.5:100 g/l	SE	19833
3	Cleave	Adama	2.5:100 g/l	SE	16774
4	Envy	Corteva	2.5:100 g/l	SE	17901
5	Flurostar XL	Certis Belchim B V	2.5:100 g/l	SE	18617
6	Hunter	Corteva	2.5:100 g/l	SE	19776
7	Nevada	Corteva	5:100 g/l	SE	19967
8	Sickle	Corteva	2.5:100 g/l	SE	17923
9	Slalom	Corteva	2.5:100 g/l	SE	19777
10	Spitfire	Corteva	5:100 g/l	SE	19945
11	Starane XL	Corteva	2.5:100 g/l	SE	19775
12	Trafalgar	Pan Amenity	2.5:100 g/l	SE	14888

Uses
- Annual and perennial weeds in **Amenity grassland** [1-2]; **Grass seed crops** [8]; **Grassland** [8]; **Lawns** [1-2]; **Managed amenity turf** [1-2]; **Newly sown grass leys** [8];
- Annual dicotyledons in **Amenity grassland** [12]; **Lawns** [12]; **Managed amenity turf** [12]; **Oats** [7,10]; **Rye** [10]; **Spring barley** [3,5-7,9-11]; **Spring oats** [3,5-6,9,11]; **Spring wheat** [3,5-7,9-11]; **Triticale** [3,7,10]; **Undersown barley** [10]; **Undersown oats** [10]; **Undersown rye** [10];

SEE SECTION 3 FOR PRODUCTS ALSO REGISTERED

Undersown spring cereals [10]; *Undersown triticale* [10]; *Undersown wheat* [10]; *Winter barley* [3,5-7,9-11]; *Winter oats* [3,5-6,9,11]; *Winter rye* [3,7]; *Winter wheat* [3,5-7,9-11];

- Black bindweed in *Oats* [7,10]; *Rye* [10]; *Spring barley* [7,10]; *Spring wheat* [7,10]; *Triticale* [7,10]; *Undersown barley* [10]; *Undersown oats* [10]; *Undersown rye* [10]; *Undersown spring cereals* [10]; *Undersown triticale* [10]; *Undersown wheat* [10]; *Winter barley* [7,10]; *Winter rye* [7]; *Winter wheat* [7,10];
- Black nightshade in *Oats* [7,10]; *Rye* [10]; *Spring barley* [7,10]; *Spring wheat* [7,10]; *Triticale* [7,10]; *Undersown barley* [10]; *Undersown oats* [10]; *Undersown rye* [10]; *Undersown spring cereals* [10]; *Undersown triticale* [10]; *Undersown wheat* [10]; *Winter barley* [7,10]; *Winter rye* [7]; *Winter wheat* [7,10];
- Buttercups in *Amenity grassland* [1-2]; *Grass seed crops* [8]; *Grassland* [8]; *Lawns* [1-2]; *Managed amenity turf* [1-2]; *Newly sown grass leys* [8];
- Charlock in *Grass seed crops* [4]; *Grassland* [4]; *Newly sown grass leys* [4]; *Oats* [7,10]; *Rye* [10]; *Spring barley* [7,10]; *Spring wheat* [7,10]; *Triticale* [7,10]; *Undersown barley* [10]; *Undersown oats* [10]; *Undersown rye* [10]; *Undersown spring cereals* [10]; *Undersown triticale* [10]; *Undersown wheat* [10]; *Winter barley* [7,10]; *Winter rye* [7]; *Winter wheat* [7,10];
- Chickweed in *Amenity grassland* [12]; *Grass seed crops* [4]; *Grassland* [4]; *Lawns* [12]; *Managed amenity turf* [12]; *Newly sown grass leys* [4]; *Oats* [7,10]; *Rye* [10]; *Spring barley* [3,5-7,9-11]; *Spring oats* [3,5-6,9,11]; *Spring wheat* [3,5-7,9-11]; *Triticale* [3,7,10]; *Undersown barley* [10]; *Undersown oats* [10]; *Undersown rye* [10]; *Undersown spring cereals* [10]; *Undersown triticale* [10]; *Undersown wheat* [10]; *Winter barley* [3,5-7,9-11]; *Winter oats* [3,5-6,9,11]; *Winter rye* [3,7]; *Winter wheat* [3,5-7,9-11];
- Cleavers in *Grass seed crops* [4]; *Grassland* [4]; *Newly sown grass leys* [4]; *Oats* [7,10]; *Rye* [10]; *Spring barley* [3,5-7,9-11]; *Spring oats* [3,5-6,9,11]; *Spring wheat* [3,5-7,9-11]; *Triticale* [3,7,10]; *Undersown barley* [10]; *Undersown oats* [10]; *Undersown rye* [10]; *Undersown spring cereals* [10]; *Undersown triticale* [10]; *Undersown wheat* [10]; *Winter barley* [3,5-7,9-11]; *Winter oats* [3,5-6,9,11]; *Winter rye* [3,7]; *Winter wheat* [3,5-7,9-11];
- Clover in *Amenity grassland* [1-2]; *Grass seed crops* [8]; *Grassland* [8]; *Lawns* [1-2]; *Managed amenity turf* [1-2]; *Newly sown grass leys* [8];
- Common mouse-ear in *Grassland* [4];
- Creeping buttercup in *Grassland* [4];
- Daisies in *Amenity grassland* [1-2]; *Grass seed crops* [8]; *Grassland* [4,8]; *Lawns* [1-2]; *Managed amenity turf* [1-2]; *Newly sown grass leys* [8];
- Dandelions in *Amenity grassland* [1-2]; *Grass seed crops* [8]; *Grassland* [4,8]; *Lawns* [1-2]; *Managed amenity turf* [1-2]; *Newly sown grass leys* [8];
- Forget-me-not in *Grass seed crops* [4]; *Grassland* [4]; *Newly sown grass leys* [4]; *Oats* [7,10]; *Rye* [10]; *Spring barley* [7,10]; *Spring wheat* [7,10]; *Triticale* [7,10]; *Undersown barley* [10]; *Undersown oats* [10]; *Undersown rye* [10]; *Undersown spring cereals* [10]; *Undersown triticale* [10]; *Undersown wheat* [10]; *Winter barley* [7,10]; *Winter rye* [7]; *Winter wheat* [7,10];
- Knotgrass in *Grass seed crops* [4]; *Grassland* [4]; *Newly sown grass leys* [4]; *Oats* [7,10]; *Rye* [10]; *Spring barley* [7,10]; *Spring wheat* [7,10]; *Triticale* [7,10]; *Undersown barley* [10]; *Undersown oats* [10]; *Undersown rye* [10]; *Undersown spring cereals* [10]; *Undersown triticale* [10]; *Undersown wheat* [10]; *Winter barley* [7,10]; *Winter rye* [7]; *Winter wheat* [7,10];
- Mayweeds in *Grass seed crops* [4]; *Grassland* [4]; *Newly sown grass leys* [4]; *Oats* [7,10]; *Rye* [10]; *Spring barley* [3,5-7,9-11]; *Spring oats* [3,5-6,9,11]; *Spring wheat* [3,5-7,9-11]; *Triticale* [3,7,10]; *Undersown barley* [10]; *Undersown oats* [10]; *Undersown rye* [10]; *Undersown spring cereals* [10]; *Undersown triticale* [10]; *Undersown wheat* [10]; *Winter barley* [3,5-7,9-11]; *Winter oats* [3,5-6,9,11]; *Winter rye* [3,7]; *Winter wheat* [3,5-7,9-11];
- Plantains in *Amenity grassland* [1-2]; *Grass seed crops* [8]; *Grassland* [8]; *Lawns* [1-2]; *Managed amenity turf* [1-2]; *Newly sown grass leys* [8];
- Poppies in *Grass seed crops* [4]; *Grassland* [4]; *Newly sown grass leys* [4]; *Oats* [7,10]; *Rye* [10]; *Spring barley* [7,10]; *Spring wheat* [7,10]; *Triticale* [7,10]; *Undersown barley* [10]; *Undersown oats* [10]; *Undersown rye* [10]; *Undersown spring cereals* [10]; *Undersown triticale* [10]; *Undersown wheat* [10]; *Winter barley* [7,10]; *Winter rye* [7]; *Winter wheat* [7,10];

- Redshank in *Oats* [7,10]; *Rye* [10]; *Spring barley* [7,10]; *Spring wheat* [7,10]; *Triticale* [7,10]; *Undersown barley* [10]; *Undersown oats* [10]; *Undersown rye* [10]; *Undersown spring cereals* [10]; *Undersown triticale* [10]; *Undersown wheat* [10]; *Winter barley* [7,10]; *Winter rye* [7]; *Winter wheat* [7,10];
- Shepherd's purse in *Grass seed crops* [4]; *Grassland* [4]; *Newly sown grass leys* [4];
- Volunteer beans in *Oats* [7,10]; *Rye* [10]; *Spring barley* [7,10]; *Spring wheat* [7,10]; *Triticale* [7,10]; *Undersown barley* [10]; *Undersown oats* [10]; *Undersown rye* [10]; *Undersown spring cereals* [10]; *Undersown triticale* [10]; *Undersown wheat* [10]; *Winter barley* [7,10]; *Winter rye* [7]; *Winter wheat* [7,10];
- Volunteer oilseed rape in *Grass seed crops* [4]; *Grassland* [4]; *Newly sown grass leys* [4]; *Oats* [7,10]; *Rye* [10]; *Spring barley* [7,10]; *Spring wheat* [7,10]; *Triticale* [7,10]; *Undersown barley* [10]; *Undersown oats* [10]; *Undersown rye* [10]; *Undersown spring cereals* [10]; *Undersown triticale* [10]; *Undersown wheat* [10]; *Winter barley* [7,10]; *Winter rye* [7]; *Winter wheat* [7,10];

Approval information
- Florasulam and fluroxypyr included in Annex I under EC Regulation 1107/2009
- Accepted by BBPA for use on malting barley

Efficacy guidance
- Best results obtained when weeds are small and growing actively
- Products are mainly absorbed through weed foliage. Cleavers emerging after application will not be controlled
- Florasulam is a member of the ALS-inhibitor group of herbicides

Restrictions
- Maximum total dose equivalent to one full dose treatment
- Do not roll or harrow 7 d before or after application
- Do not spray when crops are under stress from cold, drought, pest damage or nutrient deficiency
- Do not apply through CDA applicators
- Specific restrictions apply to use in sequence or tank mixture with other sulfonylurea or ALS-inhibiting herbicides. See label for details
- The total amount of florasulam applied to a cereal crop must not exceed 7.5 g.a.i/ha
- For autumn planted crops a maximum total dose of 3.75 g of florasulam must be observed for applications made between crop emergence in the year of planting and February 1st in the year of harvest

Crop-specific information
- Latest use: before flag leaf sheath extended (before GS 41) for spring barley and spring wheat; before flag leaf sheath opening (before GS 47) for winter barley and winter wheat; before second node detectable (before GS 32) for winter oats

Following crops guidance
- Cereals, oilseed rape, field beans or grass may follow treated crops in the same yr. Oilseed rape may suffer temporary vigour reduction after a dry summer
- In addition to the above, linseed, peas, sugar beet, potatoes, maize or clover may be sown in the calendar yr following treatment
- In the event of failure of a treated crop in the spring, only spring cereals, maize or ryegrass may be planted
- May also be used on crops undersown with grass [7, 10]

Environmental safety
- Dangerous for the environment
- Toxic to aquatic organisms
- Take extreme care to avoid drift onto non-target crops or plants
- Aquatic buffer zone of 5m [8]
- LERAP Category B [2-10]

Hazard classification and safety precautions
Hazard Irritant, Flammable [10], Dangerous for the environment, Flammable liquid and vapour [7,10], Very toxic to aquatic organisms [1,5]
Transport code 3 [7,10], 9 [1-6,8-9,11-12]

SEE SECTION 3 FOR PRODUCTS ALSO REGISTERED

Packaging group III
UN Number 1993, 3082
Risk phrases H304 [4,7-8,10], H315, H317 [1-11], H318 [5], H319 [1-4,6-11], H320 [12], H335 [1-2,4-11], H336
Operator protection U05a, U08 [1-2,4], U09a [7,10], U11, U14 [1-2,4,7,10], U19a, U20b [1-2,4,7,10]; A, C, H
Environmental protection E07c (7 days) [4,8], E15a [3,5-12], E15b [1-2], E16a [2-5,7-10], E34, E38, H410 [1-11], H411 [12]
Storage and disposal D01, D02, D05 [4,12], D09a, D10c, D12a
Medical advice M05a [7,10,12]

177   florasulam + halauxifen-methyl

A herbicide mixture for control of broad-leaved weeds in cereals
HRAC mode of action code: 2 + 4 (B + O)

See also florasulam
florasulam + fluroxypyr

Products
1 Zypar   Corteva   5:6.25 g/l   OD   17938

Uses
- Annual dicotyledons in *Durum wheat* [1]; *Rye* [1]; *Spelt* [1]; *Spring barley* [1]; *Spring wheat* [1]; *Triticale* [1]; *Winter barley* [1]; *Winter oats* [1]; *Winter triticale* [1]; *Winter wheat* [1];
- Black bindweed in *Grass seed crops* (off-label) [1];
- Chickweed in *Durum wheat* [1]; *Grass seed crops* (off-label) [1]; *Rye* [1]; *Spelt* [1]; *Spring barley* [1]; *Spring wheat* [1]; *Triticale* [1]; *Winter barley* [1]; *Winter oats* [1]; *Winter triticale* [1]; *Winter wheat* [1];
- Cleavers in *Durum wheat* [1]; *Grass seed crops* (off-label) [1]; *Rye* [1]; *Spelt* [1]; *Spring barley* [1]; *Spring wheat* [1]; *Triticale* [1]; *Winter barley* [1]; *Winter oats* [1]; *Winter triticale* [1]; *Winter wheat* [1];
- Crane's-bill in *Durum wheat* [1]; *Grass seed crops* (off-label) [1]; *Rye* [1]; *Spelt* [1]; *Spring barley* [1]; *Spring wheat* [1]; *Triticale* [1]; *Winter barley* [1]; *Winter oats* [1]; *Winter triticale* [1]; *Winter wheat* [1];
- Fat hen in *Durum wheat* [1]; *Grass seed crops* (off-label) [1]; *Rye* [1]; *Spelt* [1]; *Spring barley* [1]; *Spring wheat* [1]; *Triticale* [1]; *Winter barley* [1]; *Winter oats* [1]; *Winter triticale* [1]; *Winter wheat* [1];
- Fumitory in *Durum wheat* [1]; *Grass seed crops* (off-label) [1]; *Rye* [1]; *Spelt* [1]; *Spring barley* [1]; *Spring wheat* [1]; *Triticale* [1]; *Winter barley* [1]; *Winter oats* [1]; *Winter triticale* [1]; *Winter wheat* [1];
- Groundsel in *Durum wheat* [1]; *Rye* [1]; *Spelt* [1]; *Spring barley* [1]; *Spring wheat* [1]; *Triticale* [1]; *Winter barley* [1]; *Winter oats* [1]; *Winter triticale* [1]; *Winter wheat* [1];
- Hemp-nettle in *Grass seed crops* (off-label) [1];
- Ivy-leaved speedwell in *Grass seed crops* (off-label) [1];
- Mayweeds in *Durum wheat* [1]; *Grass seed crops* (off-label) [1]; *Rye* [1]; *Spelt* [1]; *Spring barley* [1]; *Spring wheat* [1]; *Triticale* [1]; *Winter barley* [1]; *Winter oats* [1]; *Winter triticale* [1]; *Winter wheat* [1];
- Poppies in *Durum wheat* [1]; *Grass seed crops* (off-label) [1]; *Rye* [1]; *Spelt* [1]; *Spring barley* [1]; *Spring wheat* [1]; *Triticale* [1]; *Winter barley* [1]; *Winter oats* [1]; *Winter triticale* [1]; *Winter wheat* [1];
- Red dead-nettle in *Grass seed crops* (off-label) [1];
- Shepherd's purse in *Grass seed crops* (off-label) [1];
- Volunteer oilseed rape in *Durum wheat* [1]; *Rye* [1]; *Spelt* [1]; *Spring barley* [1]; *Spring wheat* [1]; *Triticale* [1]; *Winter barley* [1]; *Winter oats* [1]; *Winter triticale* [1]; *Winter wheat* [1];

Extension of Authorisation for Minor Use (EAMUs)
- *Grass seed crops* 20181557 [1]

FOR FULL CONDITIONS OF USE ALWAYS READ THE PRODUCT LABEL

Approval information
- Florasulam and halauxifen-methyl included in Annex I under EC Regulation 1107/2009

Efficacy guidance
- Rainfast 1 hour after application
- Can also be used on cereals undersown with grass

Restrictions
- One application per season
- The total amount of halauxifen-methyl applied to a winter cereal must not exceed 13.5 g a.e/ha per season: 7.5 g a.e/ha in the autumn followed by 6 g a.e/ha in the spring, with a minimal interval of 3 months between both applications of products which contain halauxifen-methyl.
- Do not apply more than 0.75 litre per hectare to any crop before the 15th February
- For autumn planted crops a maximum total dose of 3.75 g/ha of florasulam must be observed for applications made between crop emergence in the year of planting and February 15th in the year of harvest. The total amount of florasulam applied to a cereal crop must not exceed 7.5 g/ha
- To avoid subsequent injury to crops other than cereals (wheat, durum wheat, spelt, barley, rye, triticale), all spraying equipment must be thoroughly cleaned both inside and out using All Clear Extra spray cleaner at 0.5 % v/v or bleach containing 5 % hypochlorite.

Following crops guidance
- There are no restrictions for sowing any succeeding crop after the cereal harvest but for sensitive species such as clover, lentils or sunflower ploughing is recommended prior to drilling.
- Where crop failure after an autumn application occurs, sowing the following spring crops is possible: 1 month after application (no cultivation restrictions): spring wheat, spring barley, maize, ryegrass; 3 months after application (after ploughing): spring oilseed rape, field beans, peas, sunflower
- Where crop failure after a spring application occurs, it is possible to sow spring wheat and spring barley 1 month after application with no need to cultivate while maize can be sown 2 months after application after ploughing.

Environmental safety
- LERAP Category B [1]

Hazard classification and safety precautions
Transport code 9 [1]
Packaging group III
UN Number 3082
Risk phrases H315, H317, H319
Operator protection U05a, U08, U19a, U20c; A, C, H
Environmental protection E15b, E16a, E34, H410
Storage and disposal D09a, D10c

178 florasulam + pyroxsulam

A mixture of two triazolopirimidine sulfonamides for winter wheat
HRAC mode of action code: 2 (B)

See also pyroxsulam

Products

1	Broadway Star	Corteva	1.4:7.1% w/w	WG	18273
2	Capri Duo	Corteva	1.4:7.1% w/w	WG	19142
3	Palio	Corteva	1.4:7.1% w/w	WG	18349

Uses
- Blackgrass in **Miscanthus** (off-label) [1];
- Charlock in **Miscanthus** (off-label) [1]; **Rye** [2]; **Spring wheat** [1-3]; **Triticale** [1-3]; **Winter rye** [1,3]; **Winter wheat** [1-3];
- Chickweed in **Miscanthus** (off-label) [1]; **Rye** [2]; **Spring wheat** [1-3]; **Triticale** [1-3]; **Winter rye** [1,3]; **Winter wheat** [1-3];

SEE SECTION 3 FOR PRODUCTS ALSO REGISTERED

SECTION 2

- Cleavers in *Miscanthus* *(off-label)* [1]; *Rye* [2]; *Spring wheat* [1-3]; *Triticale* [1-3]; *Winter rye* [1,3]; *Winter wheat* [1-3];
- Field pansy in *Miscanthus* *(off-label)* [1]; *Rye* [2]; *Spring wheat* [1-3]; *Triticale* [1-3]; *Winter rye* [1,3]; *Winter wheat* [1-3];
- Field speedwell in *Rye* [2]; *Spring wheat* [1-3]; *Triticale* [1-3]; *Winter rye* [1,3]; *Winter wheat* [1-3];
- Geranium species in *Miscanthus* *(off-label)* [1]; *Rye* [2]; *Spring wheat* [1-3]; *Triticale* [1-3]; *Winter rye* [1,3]; *Winter wheat* [1-3];
- Ivy-leaved speedwell in *Rye* [2]; *Spring wheat* [1-3]; *Triticale* [1-3]; *Winter rye* [1,3]; *Winter wheat* [1-3];
- Mayweeds in *Miscanthus* *(off-label)* [1]; *Rye* [2]; *Spring wheat* [1-3]; *Triticale* [1-3]; *Winter rye* [1,3]; *Winter wheat* [1-3];
- Poppies in *Miscanthus* *(off-label)* [1]; *Rye* [2]; *Spring wheat* [1-3]; *Triticale* [1-3]; *Winter rye* [1,3]; *Winter wheat* [1-3];
- Ryegrass in *Miscanthus* *(off-label - from seed)* [1]; *Rye* [2]; *Spring wheat* [1-3]; *Triticale* [1-3]; *Winter rye* [1,3]; *Winter wheat* [1-3];
- Speedwells in *Miscanthus* *(off-label)* [1];
- Sterile brome in *Miscanthus* *(off-label)* [1]; *Rye* [2]; *Spring wheat* [1-3]; *Triticale* [1-3]; *Winter rye* [1,3]; *Winter wheat* [1-3];
- Volunteer oilseed rape in *Miscanthus* *(off-label)* [1]; *Rye* [2]; *Spring wheat* [1-3]; *Triticale* [1-3]; *Winter rye* [1,3]; *Winter wheat* [1-3];
- Wild oats in *Miscanthus* *(off-label)* [1]; *Rye* [2]; *Spring wheat* [1-3]; *Triticale* [1-3]; *Winter rye* [1,3]; *Winter wheat* [1-3];

Extension of Authorisation for Minor Use (EAMUs)
- *Miscanthus* *(from seed)* *20193887* [1], *20193887* [1]

Approval information
- Florasulam and pyroxsulam included in Annex I under EC Regulation 1107/2009.

Efficacy guidance
- Rainfast within 1 hour of application
- Requires an authorised adjuvant at application and recommended for use in a programme with herbicides employing a different mode of action.

Restrictions
- For autumn planted crops a maximum total dose of 3.75 g.a.i/ha of florasulam must be observed for applications made between crop emergence in the year of planting and February 1st in the year of harvest.
- The total amount of florasulam applied to a cereal crop must not exceed 7.5 g.a.i/ha
- Do not apply this or any other ALS inhibitor herbicide with claims for grass weed control more than once to any crop.

Crop-specific information
- Crop injury may occur if applied in tank mixture with plant growth regulators - allow a minimum interval of 7 days.
- Crop injury may occur if applied in tank mixture with OP insecticides, MCPB or dicamba - allow a minimum interval of 14 days

Following crops guidance
- Crop failure before 1st Feb - plough and allow 6 weeks to elapse and then drill spring wheat, spring barley, grass or maize.
- Crop failure after 1st Feb - plough and allow 6 weeks to elapse before drilling grass or maize.

Environmental safety
- Take extreme care to avoid drift on to susceptible crops, non-target plants or waterways

Hazard classification and safety precautions
Hazard Dangerous for the environment
Transport code 9 [1-3]
Packaging group III
UN Number 3077

FOR FULL CONDITIONS OF USE ALWAYS READ THE PRODUCT LABEL

Operator protection U05a, U14, U15, U20a
Environmental protection E15b, E34, E38, E39, H410
Storage and disposal D01, D02, D09a, D12a

179   florasulam + tribenuron-methyl

SECTION 2

A herbicide mixture for weed control in winter & spring cereals
HRAC mode of action code: 2 + 2 (B + B)

See also tribenuron-methyl

Products

1	Bolt	Nufarm UK	20:60% w/w	WG	18753
2	Flame Duo	Albaugh UK	10.4:25% w/w	SG	19097
3	Paramount Max	Nufarm UK	20:60% w/w	WG	18751

Uses

*   Annual dicotyledons in **Spring barley** [1-3]; **Spring oats** [1,3]; **Triticale** [1,3]; **Winter barley** [1-3]; **Winter oats** [1,3]; **Winter rye** [1,3]; **Winter wheat** [1-3];
*   Charlock in **Spring barley** [2]; **Winter barley** [2]; **Winter wheat** [2];
*   Chickweed in **Spring barley** [2]; **Winter barley** [2]; **Winter wheat** [2];
*   Cleavers in **Spring barley** *(MS to 20 cm)* [2]; **Winter barley** *(MS to 20 cm)* [2]; **Winter wheat** *(MS to 20 cm)* [2];
*   Fat hen in **Spring barley** [2]; **Winter barley** [2]; **Winter wheat** [2];
*   Poppies in **Spring barley** [2]; **Winter barley** [2]; **Winter wheat** [2];
*   Scented mayweed in **Spring barley** [2]; **Winter barley** [2]; **Winter wheat** [2];
*   Scentless mayweed in **Spring barley** *(MS to 6 lvs)* [2]; **Winter barley** *(MS to 6 lvs)* [2]; **Winter wheat** *(MS to 6 lvs)* [2];
*   Shepherd's purse in **Spring barley** *(MS to 6 lvs)* [2]; **Winter barley** *(MS to 6 lvs)* [2]; **Winter wheat** *(MS to 6 lvs)* [2];

Approval information

*   Florasulam and tribenuron-methyl included in Annex I under EC Regulation 1107/2009

Restrictions

*   For autumn planted crops a maximum total dose of 3.75 g.a.i/ha of florasulam must be observed for applications made between crop emergence in the year of planting and February 1st in the year of harvest.
*   The total amount of florasulam applied to a cereal crop must not exceed 7.5 g.a.i/ha
*   Only one other product with an ALS inhibitor mode of action may be applied to a cereal crop treated with [1] or [3]. However, a further application of [1], [3] or another product containing florasulam may also be made providing the maximum total dose of florasulam is not exceeded. [1] or [3] may be applied in joint application to the same cereal crop with one of the listed ALS products (see label).
*   To avoid subsequent injury to crops other than cereals all spraying equipment must be thoroughly cleaned both inside and outside using a cleaner such as All Clear Extra.
*   Applications must not be made to winter cereals before 1 March in the year of harvest, and before 5 tillers (GS25); Applications must not be made to spring barley before 15 March, and before 3 tillers (GS23) [2].
*   Applications must not be made before 15 March in the year of harvest; the maximum total dose of florasulam applied to a cereal crop must not exceed 7.5 g/ha and the maximum total dose of tribenuron applied to a cereal crop must not exceed 15 g/ha.

Following crops guidance

*   In the event of crop failure in the spring after application only cereals may be planted.
*   Only cereals, oilseed rape, field beans and grass can be sown in the same year as a treated crop is harvested. Vigour reductions may be seen in the following crops of oilseed rape after a dry summer. This will be outgrown and will not result in yield loss.

- Only cereals, oilseed rape, field beans, grass, linseed, peas, sugar beet, potatoes, maize, clover (for use in grass/clover mixtures), carrots and vegetable brassicas as transplants can be sown in the calendar year after treatment.

Environmental safety
- LERAP Category B [1,3]

Hazard classification and safety precautions
    Hazard Very toxic to aquatic organisms [1,3]
    Transport code 9 [1-3]
    Packaging group III
    UN Number 3077
    Operator protection U05a, U20b [2]; A, H
    Environmental protection E15b [1-2], E16a [1,3], E34, H410
    Storage and disposal D01, D02, D05, D06e [1], D07 [1,3], D09a [2], D10c
    Medical advice M03 [2]

180   fluazifop-P-butyl

A phenoxypropionic acid grass herbicide for broadleaved crops
HRAC mode of action code: 1 (A)

Products

1	Clayton Maximus	Clayton	125 g/l	EC	19299
2	Fusilade Forte	Nufarm UK	150 g/l	EC	19019
3	Fusilade Max	Nufarm UK	125 g/l	EC	19013

Uses
- Annual grasses in *Almonds* [1,3]; *Apples* [1,3]; *Apricots* [1,3]; *Asparagus* [1,3]; *Baby leaf crops* [1,3]; *Beans without pods (dry)* [1,3]; *Bilberries* [1,3]; *Blackberries* [1,3]; *Blackcurrants* [1,3]; *Blueberries* [1,3]; *Broad beans* [1,3]; *Bulb onions* [1,3]; *Cardoons* [1,3]; *Carrots* [1,3]; *Celeriac* [1,3]; *Celery (outdoor)* [1,3]; *Chestnuts* [1,3]; *Chicory root* [1,3]; *Combining peas* [1-3]; *Cranberries* [1,3]; *Edible podded peas* [1,3]; *Elderberries* [1,3]; *Endives* [1,3]; *Farm forestry* [1,3]; *Florence fennel* [1,3]; *Fodder beet* [1-3]; *Fodder rape* [1-3]; *Garlic* [1,3]; *Ginger* [1,3]; *Ginseng root* [1,3]; *Globe artichoke* [1,3]; *Gooseberries* [1,3]; *Hazel nuts* [1,3]; *Herbs (see appendix 6)* [1,3]; *Hops* [1,3]; *Horseradish* [1,3]; *Lamb's lettuce* [1,3]; *Land cress* [1,3]; *Lettuce* [1,3]; *Linseed* [1-3]; *Liquorice* [1,3]; *Loganberries* [1,3]; *Lupins* [1-3]; *Mallow (Althaea spp.)* [1,3]; *Mangels* [1,3]; *Mulberries* [1,3]; *Mustard* [2]; *Parsley root* [1,3]; *Parsnips* [1,3]; *Pears* [1,3]; *Plums* [1,3]; *Poppies* [1-3]; *Potatoes* [1,3]; *Purslane* [1,3]; *Quinces* [1,3]; *Radishes* [1,3]; *Raspberries* [1,3]; *Red beet* [1,3]; *Red mustard* [1,3]; *Redcurrants* [1,3]; *Rhubarb* [1,3]; *Rocket* [1,3]; *Rose hips* [1,3]; *Rubus hybrids* [1,3]; *Safflower* [1,3]; *Salsify* [1,3]; *Shallots* [1,3]; *Spinach* [1,3]; *Spinach beet* [1,3]; *Spring field beans* [1-3]; *Spring oilseed rape* [1-3]; *Sugar beet* [1-3]; *Sunflowers* [1,3]; *Swedes (stockfeed only)* [1,3]; *Table grapes* [1,3]; *Turmeric* [1,3]; *Turnips (stockfeed only)* [1,3]; *Valerian root* [1,3]; *Vining peas* [1,3]; *Walnuts* [1,3]; *Wine grapes* [1,3]; *Winter field beans* [1-3]; *Winter oilseed rape* [1-3];
- Black bent in *All edible seed crops grown outdoors (off-label)* [3]; *All non-edible seed crops grown outdoors (off-label)* [3]; *Corn Gromwell (off-label)* [3];
- Blackgrass in *Corn Gromwell (off-label)* [3]; *Spring field beans* [1-3]; *Sunflowers* [1,3]; *Winter field beans* [1-3];
- Brome grasses in *Corn Gromwell (off-label)* [3];
- Couch in *All edible seed crops grown outdoors (off-label)* [3]; *All non-edible seed crops grown outdoors (off-label)* [3]; *Corn Gromwell (off-label)* [3];
- Creeping bent in *All edible seed crops grown outdoors (off-label)* [3]; *All non-edible seed crops grown outdoors (off-label)* [3]; *Corn Gromwell (off-label)* [3];
- Italian ryegrass in *Corn Gromwell (off-label)* [3];
- Perennial grasses in *Almonds* [1,3]; *Apples* [1,3]; *Apricots* [1,3]; *Asparagus* [1,3]; *Baby leaf crops* [1,3]; *Beans without pods (dry)* [1,3]; *Bilberries* [1,3]; *Blackcurrants* [1,3]; *Blueberries* [1,3]; *Broad beans* [1,3]; *Bulb onions* [1,3]; *Cardoons* [1,3]; *Carrots* [1,3]; *Celeriac* [1,3]; *Celery (outdoor)* [1,3]; *Chestnuts* [1,3]; *Chicory root* [1,3]; *Combining peas* [1-

3]; *Cranberries* [1,3]; *Edible podded peas* [1,3]; *Elderberries* [1,3]; *Endives* [1,3]; *Farm forestry* [1,3]; *Florence fennel* [1,3]; *Fodder beet* [1-3]; *Fodder rape* [1-3]; *Garlic* [1,3]; *Ginger* [1,3]; *Ginseng root* [1,3]; *Globe artichoke* [1,3]; *Gooseberries* [1,3]; *Hazel nuts* [1,3]; *Herbs (see appendix 6)* [1,3]; *Hops* [1,3]; *Horseradish* [1,3]; *Lamb's lettuce* [1,3]; *Land cress* [1,3]; *Lettuce* [1,3]; *Linseed* [1-3]; *Liquorice* [1,3]; *Loganberries* [1,3]; *Lupins* [1-3]; *Mallow (Althaea spp.)* [1,3]; *Mangels* [1,3]; *Mulberries* [1,3]; *Mustard* [2]; *Parsley root* [1,3]; *Parsnips* [1,3]; *Pears* [1,3]; *Plums* [1,3]; *Poppies* [1-3]; *Potatoes* [1,3]; *Purslane* [1,3]; *Quinces* [1,3]; *Radishes* [1,3]; *Raspberries* [1,3]; *Red beet* [1,3]; *Red mustard* [1,3]; *Redcurrants* [1,3]; *Rhubarb* [1,3]; *Rocket* [1,3]; *Rose hips* [1,3]; *Rubus hybrids* [1,3]; *Safflower* [1,3]; *Salsify* [1,3]; *Shallots* [1,3]; *Spinach* [1,3]; *Spinach beet* [1,3]; *Spring field beans* [1-3]; *Spring oilseed rape* [1-3]; *Sugar beet* [1-3]; *Sunflowers* [1,3]; *Swedes* (stockfeed only) [1,3]; *Table grapes* [1,3]; *Turmeric* [1,3]; *Turnips* (stockfeed only) [1,3]; *Valerian root* [1,3]; *Vining peas* [1,3]; *Walnuts* [1,3]; *Wine grapes* [1,3]; *Winter field beans* [1-3]; *Winter oilseed rape* [1-3]

- Volunteer cereals in *Almonds* [1,3]; *Apples* [1,3]; *Apricots* [1,3]; *Asparagus* [1,3]; *Baby leaf crops* [1,3]; *Beans without pods (dry)* [1,3]; *Bilberries* [1,3]; *Blackberries* [1,3]; *Blackcurrants* [1,3]; *Blueberries* [1,3]; *Broad beans* [1,3]; *Bulb onions* [1,3]; *Cardoons* [1,3]; *Carrots* [1,3]; *Celeriac* [1,3]; *Celery (outdoor)* [1,3]; *Chestnuts* [1,3]; *Chicory root* [1,3]; *Combining peas* [1-3]; *Corn Gromwell* (off-label) [3]; *Cranberries* [1,3]; *Edible podded peas* [1,3]; *Elderberries* [1,3]; *Endives* [1,3]; *Farm forestry* [1,3]; *Florence fennel* [1,3]; *Fodder beet* [1-3]; *Fodder rape* [1-3]; *Garlic* [1,3]; *Ginger* [1,3]; *Ginseng root* [1,3]; *Globe artichoke* [1,3]; *Gooseberries* [1,3]; *Hazel nuts* [1,3]; *Herbs (see appendix 6)* [1,3]; *Hops* [1,3]; *Horseradish* [1,3]; *Lamb's lettuce* [1,3]; *Land cress* [1,3]; *Lettuce* [1,3]; *Linseed* [1-3]; *Liquorice* [1,3]; *Loganberries* [1,3]; *Lupins* [1-3]; *Mallow (Althaea spp.)* [1,3]; *Mangels* [1,3]; *Mulberries* [1,3]; *Mustard* [2]; *Parsley root* [1,3]; *Parsnips* [1,3]; *Pears* [1,3]; *Plums* [1,3]; *Poppies* [1-3]; *Potatoes* [1,3]; *Purslane* [1,3]; *Quinces* [1,3]; *Radishes* [1,3]; *Raspberries* [1,3]; *Red beet* [1,3]; *Red mustard* [1,3]; *Redcurrants* [1,3]; *Rhubarb* [1,3]; *Rocket* [1,3]; *Rose hips* [1,3]; *Rubus hybrids* [1,3]; *Safflower* [1,3]; *Salsify* [1,3]; *Shallots* [1,3]; *Spinach* [1,3]; *Spinach beet* [1,3]; *Spring field beans* [1-3]; *Spring oilseed rape* [1-3]; *Sugar beet* [1-3]; *Sunflowers* [1,3]; *Swedes* (stockfeed only) [1,3]; *Table grapes* [1,3]; *Turmeric* [1,3]; *Turnips* (stockfeed only) [1,3]; *Valerian root* [1,3]; *Vining peas* [1,3]; *Walnuts* [1,3]; *Wine grapes* [1,3]; *Winter field beans* [1-3]; *Winter oilseed rape* [1-3]

- Wild oats in *Almonds* [1,3]; *Apples* [1,3]; *Apricots* [1,3]; *Asparagus* [1,3]; *Baby leaf crops* [1,3]; *Beans without pods (dry)* [1,3]; *Bilberries* [1,3]; *Blackberries* [1,3]; *Blackcurrants* [1,3]; *Blueberries* [1,3]; *Broad beans* [1,3]; *Bulb onions* [1,3]; *Cardoons* [1,3]; *Carrots* [1,3]; *Celeriac* [1,3]; *Celery (outdoor)* [1,3]; *Chestnuts* [1,3]; *Chicory root* [1,3]; *Combining peas* [1-3]; *Corn Gromwell* (off-label) [3]; *Cranberries* [1,3]; *Edible podded peas* [1,3]; *Elderberries* [1,3]; *Endives* [1,3]; *Farm forestry* [1,3]; *Florence fennel* [1,3]; *Fodder beet* [1-3]; *Fodder rape* [1-3]; *Garlic* [1,3]; *Ginger* [1,3]; *Ginseng root* [1,3]; *Globe artichoke* [1,3]; *Gooseberries* [1,3]; *Hazel nuts* [1,3]; *Herbs (see appendix 6)* [1,3]; *Hops* [1,3]; *Horseradish* [1,3]; *Lamb's lettuce* [1,3]; *Land cress* [1,3]; *Lettuce* [1,3]; *Linseed* [1-3]; *Liquorice* [1,3]; *Loganberries* [1,3]; *Lupins* [1-3]; *Mallow (Althaea spp.)* [1,3]; *Mangels* [1,3]; *Mulberries* [1,3]; *Mustard* [2]; *Parsley root* [1,3]; *Parsnips* [1,3]; *Pears* [1,3]; *Plums* [1,3]; *Poppies* [1-3]; *Potatoes* [1,3]; *Purslane* [1,3]; *Quinces* [1,3]; *Radishes* [1,3]; *Raspberries* [1,3]; *Red beet* [1,3]; *Red mustard* [1,3]; *Redcurrants* [1,3]; *Rhubarb* [1,3]; *Rocket* [1,3]; *Rose hips* [1,3]; *Rubus hybrids* [1,3]; *Safflower* [1,3]; *Salsify* [1,3]; *Shallots* [1,3]; *Spinach* [1,3]; *Spinach beet* [1,3]; *Spring field beans* [1-3]; *Spring oilseed rape* [1-3]; *Sugar beet* [1-3]; *Sunflowers* [1,3]; *Swedes* (stockfeed only) [1,3]; *Table grapes* [1,3]; *Turmeric* [1,3]; *Turnips* (stockfeed only) [1,3]; *Valerian root* [1,3]; *Vining peas* [1,3]; *Walnuts* [1,3]; *Wine grapes* [1,3]; *Winter field beans* [1-3]; *Winter oilseed rape* [1-3];

Extension of Authorisation for Minor Use (EAMUs)

- *All edible seed crops grown outdoors* 20200607 [3]
- *All non-edible seed crops grown outdoors* 20200607 [3]
- *Corn Gromwell* 20212033 [3]

Approval information

- Fluazifop-P-butyl included in Annex I under EC Regulation 1107/2009
- Accepted by BBPA for use on hops

SEE SECTION 3 FOR PRODUCTS ALSO REGISTERED

Efficacy guidance

- Best results achieved by application when weed growth active under warm conditions with adequate soil moisture.
- Spray weeds from 2-expanded leaf stage to fully tillered, couch from 4 leaves when majority of shoots have emerged, with a second application if necessary
- Control may be reduced under dry conditions. Do not cultivate for 2 wk after spraying couch
- Annual meadow grass is not controlled
- May also be used to remove grass cover crops
- Fluazifop-P-butyl is an ACCase inhibitor herbicide. To avoid the build up of resistance do not apply products containing an ACCase inhibitor herbicide more than twice to any crop. In addition do not use any product containing fluazifop-P-butyl in mixture or sequence with any other product containing the same ingredient
- Use these products as part of a resistance management strategy that includes cultural methods of control and does not use ACCase inhibitors as the sole chemical method of grass weed control

Restrictions

- Maximum number of treatments 1 per crop or yr for all crops
- Do not sow cereals or grass crops for at least 8 wk after application of high rate or 2 wk after low rate
- Do not apply through CDA sprayer, with hand-held equipment or from air
- Avoid treatment before spring growth has hardened or when buds opening
- Do not treat bush and cane fruit or hops between flowering and harvest
- Consult processors before treating crops intended for processing
- Oilseed rape, linseed and flax for industrial use must not be harvested for human or animal consumption nor grazed
- Do not use for forestry establishment on land not previously under arable cultivation or improved grassland
- Treated vegetation in field margins, land temporarily removed from production etc, must not be grazed or harvested for human or animal consumption and unprotected persons must be kept out of treated areas for at least 24 h
- When applying to chicory root the maximum concentration of spray solution must not exceed 3 L product in 122.5 litres water/ha

Crop-specific information

- Latest use: before 50% ground cover for swedes, turnips; before 5 leaf stage for spring oilseed rape; before flowering for blackcurrants, gooseberries, hops, raspberries, strawberries; before flower buds visible for field beans, peas, linseed, flax, winter oilseed rape; 2 wk before sowing cereals or grass for field margins, land temporarily removed from production
- HI beet crops, kale, carrots 8 wk; onions 4 wk; oilseed rape for industrial use 2 wk
- Apply to sugar and fodder beet from 1-true leaf to 50% ground cover
- Apply to winter oilseed rape from 1-true leaf to established plant stage
- Apply to spring oilseed rape from 1-true leaf but before 5-true leaves
- Apply in fruit crops after harvest. See label for timing details on other crops
- Before using on onions or peas use crystal violet test to check that leaf wax is sufficient

Environmental safety

- Dangerous for the environment
- Very toxic to aquatic organisms

Hazard classification and safety precautions

Hazard Harmful, Dangerous for the environment, Very toxic to aquatic organisms [1,3]
Transport code 9 [1-3]
Packaging group III
UN Number 3082
Risk phrases H315 [1], H317 [2], H361
Operator protection U05a, U08, U20b; A, C, H, M
Environmental protection E15a, E38 [2-3], H410
Storage and disposal D01, D02, D05, D09a, D10b [1], D10c [2-3], D12a
Medical advice M05b [1]

FOR FULL CONDITIONS OF USE ALWAYS READ THE PRODUCT LABEL

181   fluazinam

A dinitroaniline fungicide for use in potatoes
FRAC mode of action code: 29

*See also azoxystrobin + fluazinam
      cymoxanil + fluazinam
      dimethomorph + fluazinam*

Products

1	Cleancrop Alicante	Agrii	500 g/l	SC	19821
2	Fluazinova	Barclay	500 g/l	SC	17625
3	Nando 500SC	Nufarm UK	500 g/l	SC	16388
4	Shirlan	Certis Belchim B V	500 g/l	SC	18406
5	Tizca	FMC Agro	500 g/l	SC	18813
6	Volley	Adama	500 g/l	SC	16451

Uses

- Blight in **Potatoes** ;
- Late blight in **Potatoes** [1-2];
- Powdery scab in **Potatoes grown for seed** *(off-label)* ; **Seed potatoes** *(off-label)* [5];
- Raspberry root rot in **Blackberries** *(off-label)* [5]; **Loganberries** *(off-label)* [5]; **Protected blackberries** *(off-label)* [5]; **Protected loganberries** *(off-label)* [5]; **Protected raspberries** *(off-label)* [5]; **Protected Ribes hybrids** *(off-label)* [5]; **Raspberries** *(off-label)* [5]; **Rubus hybrids** *(off-label)* [5];

Extension of Authorisation for Minor Use (EAMUs)

- **Blackberries** *20192066* [5]
- **Loganberries** *20192066* [5]
- **Potatoes grown for seed** *20180200* [3], *20192051* [4]
- **Protected blackberries** *20192066* [5]
- **Protected loganberries** *20192066* [5]
- **Protected raspberries** *20192066* [5]
- **Protected Ribes hybrids** *20192066* [5]
- **Raspberries** *20192066* [5]
- **Rubus hybrids** *20192066* [5]
- **Seed potatoes** *20190877* [5]

Approval information

- Fluazinam included in Annex I under EC Regulation 1107/2009

Efficacy guidance

- Commence treatment at the first blight risk warning (before blight enters the crop). Products are rainfast within 1 h
- In the absence of a warning, treatment should start before foliage of adjacent plants meets in the rows
- Spray at 5-14 d intervals depending on severity of risk (see label)
- Ensure complete coverage of the foliage and stems, increasing volume as haulm growth progresses, in dense crops and if blight risk increases

Restrictions

- Horizontal boom sprayers must be fitted with three star drift reduction technology for all uses
- Do not use with hand-held sprayers
- Low drift spraying equipment must be operated according to the specific conditions stated in the official three star rating for that equipment as published on HSE Chemicals Regulation Division?s website. These operating conditions must be maintained until the operator is 30m from the top of the bank of any surface water bodies

Crop-specific information

- HI 0 - 10 d for potatoes. Check label
- Ensure complete kill of potato haulm before lifting and do not lift crops for storage while there is any green tissue left on the leaves or stem bases

SEE SECTION 3 FOR PRODUCTS ALSO REGISTERED

Environmental safety
- Dangerous for the environment
- Very toxic to aquatic organisms
- Buffer zone requirement 6m [2, 6]
- Buffer zone requirement 7m [4, 5]
- Buffer zone requirement 8m [3]
- LERAP Category B [1-6]

Hazard classification and safety precautions

Hazard Harmful [3], Irritant [2,4-5], Dangerous for the environment [2-6], Very toxic to aquatic organisms [1,4,6]

Transport code 9 [1-6]

Packaging group III

UN Number 3082

Risk phrases H315 [3], H317 [2-5], H361

Operator protection U02a [1-4,6], U04a [1-4,6], U05a [2-5], U08 [1-4,6], U11 [5], U14 [2-5], U15 [2-4], U20a [2-4], U20b [1,5-6]; A, C, H

Environmental protection E15a, E16a, E16b [2-4,6], E34 [1-4,6], E38 [2-6], H410

Storage and disposal D01 [2-5], D02 [2-5], D05 [2-5], D09a, D10a [1,6], D10c [2-5], D12a [2-6]

Medical advice M03 [1-4,6], M05a [2-4]

---

182   fludioxonil

A phenylpyrrole fungicide seed treatment for wheat and barley
FRAC mode of action code: 12

See also cymoxanil + fludioxonil + metalaxyl-M
       cyprodinil + fludioxonil
       difenoconazole + fludioxonil
       difenoconazole + fludioxonil + tebuconazole

Products

1	Beret Gold	Syngenta	25 g/l	FS	16430
2	Geoxe	Syngenta	50% w/w	WG	16596
3	Maxim 100FS	Syngenta	100 g/l	FS	15683
4	Maxim 480FS	Syngenta	480 g/l	FS	16725
5	Medallion TL	Syngenta	125 g/l	SC	15287
6	Prepper	Certis Belchim B V	25 g/l	FS	19466

Uses
- Alternaria in **Apples** *(reduction)* [2]; **Carrots** [4]; **Celeriac** *(off-label)* [4]; **Crab apples** *(reduction)* [2]; **Fennel** *(off-label)* [4]; **Parsley root** *(off-label)* [4]; **Parsnips** *(off-label)* [4]; **Pears** *(reduction)* [2]; **Protected celeriac** *(off-label)* [4]; **Protected fennel** *(off-label)* [4]; **Protected parsley root** *(off-label)* [4]; **Protected parsnips** *(off-label)* [4]; **Quinces** *(reduction)* [2]; **Red beet** *(off-label)* [4];
- Anthracnose in **Amenity grassland** *(reduction)* [5]; **Managed amenity turf** *(reduction)* [5]; **Spinach** *(moderate control)* [4];
- Black dot in **Potatoes** *(some reduction)* [3];
- Black leg in **Broccoli** *(off-label)* [4]; **Brussels sprouts** *(off-label)* [4]; **Cabbages** [4]; **Cauliflowers** *(off-label)* [4]; **Collards** *(off-label)* [4]; **Kale** *(off-label)* [4]; **Kohlrabi** *(off-label)* [4]; **Oriental cabbage** *(off-label)* [4]; **Protected broccoli** *(off-label)* [4]; **Protected Brussels sprouts** *(off-label)* [4]; **Protected cauliflowers** *(off-label)* [4]; **Protected collards** *(off-label)* [4]; **Protected kale** *(off-label)* [4]; **Protected kohlrabi** *(off-label)* [4]; **Protected oriental cabbage** *(off-label)* [4]; **Protected radishes** *(off-label)* [4]; **Radishes** *(off-label)* [4];
- Black scurf in **Potatoes** [3];
- Botrytis in **Apples** *(reduction)* [2]; **Bulb onions** [4]; **Crab apples** *(reduction)* [2]; **Pears** *(reduction)* [2]; **Quinces** *(reduction)* [2];
- Bunt in **Spring wheat** *(seed treatment)* [1,6]; **Winter wheat** *(seed treatment)* [1,6];
- Covered smut in **Spring barley** *(seed treatment)* [1,6]; **Winter barley** *(seed treatment)* [1,6];
- Disease control in **Apples** [2]; **Crab apples** [2]; **Pears** [2]; **Quinces** [2];

- Drechslera leaf spot in **Amenity grassland** *(useful levels of control)* [5]; **Managed amenity turf** *(useful levels of control)* [5];
- Fusarium foot rot and seedling blight in **Rye** *(seed treatment)* [1,6]; **Spring barley** *(seed treatment)* [1,6]; **Spring oats** *(seed treatment)* [1,6]; **Spring wheat** *(seed treatment)* [1,6]; **Triticale** *(seed treatment)* [1,6]; **Winter barley** *(seed treatment)* [1,6]; **Winter oats** *(seed treatment)* [1,6]; **Winter wheat** *(seed treatment)* [1,6];
- Fusarium patch in **Amenity grassland** [5]; **Managed amenity turf** [5];
- Leaf spot in **Celeriac** *(off-label)* [4]; **Fennel** *(off-label)* [4]; **Parsley root** *(off-label)* [4]; **Parsnips** *(off-label)* [4]; **Protected celeriac** *(off-label)* [4]; **Protected fennel** *(off-label)* [4]; **Protected parsley root** *(off-label)* [4]; **Protected parsnips** *(off-label)* [4];
- Leaf stripe in **Spring barley** *(seed treatment - reduction)* [1,6]; **Winter barley** *(seed treatment - reduction)* [1,6];
- Leptosphariae maculans in **Cabbages** [4];
- Microdochium nivale in **Amenity grassland** [5]; **Managed amenity turf** [5]; **Rye** *(seed treatment)* [1,6]; **Triticale** [1,6];
- Monilinia spp. in **Apples** *(reduction)* [2]; **Crab apples** *(reduction)* [2]; **Pears** *(reduction)* [2]; **Quinces** *(reduction)* [2];
- Neck rot in **Salad onions** *(off-label)* [4]; **Shallots** *(off-label)* [4];
- Nectria spp in **Apples** *(reduction)* [2]; **Crab apples** *(reduction)* [2]; **Pears** *(reduction)* [2]; **Quinces** *(reduction)* [2];
- Penicillium rot in **Apples** *(reduction)* [2]; **Crab apples** *(reduction)* [2]; **Pears** *(reduction)* [2]; **Quinces** *(reduction)* [2];
- Phlyctema vagabunda in **Apples** *(reduction)* [2]; **Crab apples** *(reduction)* [2]; **Pears** *(reduction)* [2]; **Quinces** *(reduction)* [2];
- Pyrenophora leaf spot in **Spring oats** *(seed treatment)* [1,6]; **Winter oats** *(seed treatment)* [1,6];
- Scab in **Seed potatoes** *(off-label)* [3];
- Seed-borne diseases in **Baby leaf crops** *(off-label)* [4];
- Septoria seedling blight in **Spring wheat** *(seed treatment)* [1,6]; **Winter wheat** *(seed treatment)* [1,6];
- Silver scurf in **Potatoes** *(reduction)* [3];
- Snow mould in **Spring barley** *(seed treatment)* [1,6]; **Spring oats** *(seed treatment)* [1,6]; **Spring wheat** *(seed treatment)* [1,6]; **Winter barley** *(seed treatment)* [1,6]; **Winter oats** *(seed treatment)* [1,6]; **Winter wheat** *(seed treatment)* [1,6];
- Stripe smut in **Rye** [1,6];

Extension of Authorisation for Minor Use (EAMUs)
- **Baby leaf crops** *20142846* [4]
- **Broccoli** *20161150* [4]
- **Brussels sprouts** *20161150* [4]
- **Cauliflowers** *20161150* [4]
- **Celeriac** *20161149* [4]
- **Collards** *20161150* [4]
- **Fennel** *20161148* [4]
- **Kale** *20161150* [4]
- **Kohlrabi** *20161150* [4]
- **Oriental cabbage** *20161150* [4]
- **Parsley root** *20161149* [4]
- **Parsnips** *20152903* [4], *20161148* [4]
- **Protected broccoli** *20161150* [4]
- **Protected Brussels sprouts** *20161150* [4]
- **Protected cauliflowers** *20161150* [4]
- **Protected celeriac** *20161149* [4]
- **Protected collards** *20161150* [4]
- **Protected fennel** *20161148* [4]
- **Protected kale** *20161150* [4]
- **Protected kohlrabi** *20161150* [4]
- **Protected oriental cabbage** *20161150* [4]
- **Protected parsley root** *20161149* [4]
- **Protected parsnips** *20161148* [4]

SEE SECTION 3 FOR PRODUCTS ALSO REGISTERED

- **Protected radishes** *20161149* [4]
- **Radishes** *20161149* [4]
- **Red beet** *20152903* [4]
- **Salad onions** *20160461* [4]
- **Seed potatoes** *20160736* [3]
- **Shallots** *20160461* [4]

Approval information
- Fludioxonil included in Annex I under EC Regulation 1107/2009
- Accepted by BBPA for use on malting barley

Efficacy guidance
- Apply direct to seed using conventional seed treatment equipment. Continuous flow treaters should be calibrated using product before use
- Effective against benzimidazole-resistant strains of *Microdochium nivale*

Restrictions
- Maximum number of treatments 1 per seed batch
- Do not apply to cracked, split or sprouted seed
- Sow treated seed within 6 mth

Crop-specific information
- Latest use: before drilling
- Product may reduce flow rate of seed through drill. Recalibrate with treated seed before drilling

Environmental safety
- Dangerous for the environment
- Toxic to aquatic organisms
- Do not use treated seed as food or feed
- Treated seed harmful to game and wildlife
- LERAP Category B [2,5]

Hazard classification and safety precautions
Hazard Irritant [2], Dangerous for the environment, Very toxic to aquatic organisms [2]
Transport code 9 [1-6]
Packaging group III
UN Number 3077, 3082
Risk phrases H317 [2,6], H318 [6], R50 [4], R53a [4]
Operator protection U05a, U14 [2], U19a [2], U20a [2], U20b [1,3,5-6]; A, H
Environmental protection E03 [1,6], E15a [1,6], E15b [2-5], E16a [2,5], E16b [5], E34 [3-5], E38, H410 [2], H411 [1,3,5-6]
Storage and disposal D01, D02, D05 [1,3-4,6], D09a, D10c [2-5], D11a [1,3-4,6], D12a
Treated seed S01 [3], S02 [1,3,6], S03 [3], S04d [4], S05 [1,3,6], S07 [1,6]

183   fludioxonil + fluxapyroxad + triticonazole

A fungicide seed treatment for cereals
FRAC mode of action code: 3 + 7 + 12

Products

1	Kinto Plus	BASF	33:33:33 g/l	FS	19506

Uses
- Bunt in *Winter wheat* [1];
- Covered smut in *Winter barley* [1];
- Leaf stripe in *Winter barley* [1];
- Loose smut in *Winter barley* [1]; *Winter oats* [1]; *Winter rye* [1]; *Winter triticale* [1]; *Winter wheat* [1];
- Seed-borne diseases in *Winter barley* [1]; *Winter oats* [1]; *Winter rye* [1]; *Winter triticale* [1]; *Winter wheat* [1];

FOR FULL CONDITIONS OF USE ALWAYS READ THE PRODUCT LABEL

- Seedling blight and foot rot in *Winter barley* [1]; *Winter rye* [1]; *Winter triticale* [1]; *Winter wheat* [1];
- Snow rot in *Winter barley* [1];
- Stripe smut in *Winter rye* [1];

Approval information
- Fludioxonil, fluxapyroxad and triticonazole included in Annex I under EC Regulation 1107/2009

Efficacy guidance
- Seed should be drilled to a depth of 40 mm into a well-prepared seed bed

Restrictions
- Treated seed should not be broadcast. To protect birds and mammals treated seed should not be left on the soil surface. Bury or remove spillages
- If seed is present on the soil surface, or if spills have occurred, then, if conditions are appropriate, the field should be harrowed and then rolled to ensure good incorporation
- In crops grown for seed the control of leaf stripe may not be sufficient to prevent higher levels of leaf stripe in daughter crops
- Do not treat grain with a moisture content higher than 16%

Hazard classification and safety precautions
Transport code 9 [1]
Packaging group III
UN Number 3082
Risk phrases H315, H317, H362
Operator protection U05a, U20a; A, H
Environmental protection E15b, H410
Storage and disposal D01, D02, D09a, D11a, D21
Treated seed S01, S02, S03, S04b, S05, S08

184 fludioxonil + pyrimethanil

A fungicide mixture for use in apples and pears
FRAC mode of action code: 9 + 12

Products

1	Pomax	Certis Belchim B V	133:336 g/l	SC	18244

Uses
- Botrytis in *Apples* [1]; *Pears* [1];
- Disease control in *Apples* [1]; *Pears* [1];
- Gloeosporium in *Apples* [1];

Restrictions
- Do not apply by hand-held equipment

Environmental safety
- Broadcast air-assisted LERAP [1] (20 m)

Hazard classification and safety precautions
Hazard Very toxic to aquatic organisms
Transport code 9 [1]
Packaging group III
UN Number 3082
Operator protection A, H
Environmental protection E17b (20 m), E34, H410
Storage and disposal D01, D02, D05, D09a, D10a

185   fludioxonil + sedaxane

A fungicide seed treatment for cereals
FRAC mode of action code: 12 + 7

See also sedaxane

Products
1 Vibrance Duo   Syngenta   25:25 g/l   FS   17838

Uses
* Bunt in *Spring wheat* [1]; *Winter wheat* [1];
* Covered smut in *Spring barley* [1]; *Winter barley* [1];
* Fusarium diseases in *Spring wheat* [1]; *Winter wheat* [1];
* Fusarium ear blight in *Spring wheat* *(moderate control)* [1]; *Winter wheat* *(moderate control)* [1];
* Leaf stripe in *Spring barley* *(moderate control)* [1]; *Winter barley* *(moderate control)* [1];
* Loose smut in *Spring oats* [1]; *Spring wheat* [1]; *Winter wheat* [1];
* Septoria leaf blotch in *Spring wheat* [1]; *Winter wheat* [1];
* Snow mould in *Spring barley* [1]; *Spring wheat* [1]; *Triticale* [1]; *Winter barley* [1]; *Winter rye* [1]; *Winter wheat* [1];
* Stripe smut in *Winter rye* [1];

Approval information
* Fludioxonil and sedaxane included in Annex 1 under EC Regulation 1107/2009

Hazard classification and safety precautions
Hazard Harmful if inhaled
Transport code 9 [1]
Packaging group III
UN Number 3082
Risk phrases H317
Operator protection U05a, U19a; A, H
Environmental protection E15b, E34, H410
Storage and disposal D01, D02, D05, D09a, D10c, D11a, D12a
Medical advice M03

186   fludioxonil + tebuconazole

A seed treatment for use in cereals
FRAC mode of action code: 3 + 12

See also fludioxonil
tebuconazole

Products
1 Fountain   Certis Belchim B V   50:10 g/l   FS   17708

Uses
* Seed-borne diseases in *Triticale* [1]; *Winter barley* [1]; *Winter oats* [1]; *Winter rye* [1]; *Winter wheat* [1];

Approval information
* Fludioxonil and tebuconazole included in Annex I under EC Regulation 1107/2009

Efficacy guidance
* Do not broadcast treated seed. Ensure that it is covered by at least 40mm settled soil.

Hazard classification and safety precautions
Transport code 9 [1]
Packaging group III
UN Number 3082

FOR FULL CONDITIONS OF USE ALWAYS READ THE PRODUCT LABEL

Operator protection A, H
Environmental protection H410

187 flufenacet

A broad spectrum oxyacetamide herbicide for weed control in winter cereals
HRAC mode of action code: 15 (K3)

See also diflufenican + flufenacet
diflufenican + flufenacet + flurtamone
diflufenican + flufenacet + metribuzin
flufenacet + isoxaflutole
flufenacet + metribuzin
flufenacet + pendimethalin

Products

1	Clayton Tacit	Clayton	480 g/l	SL	19073
2	Fence	Albaugh UK	480 g/l	SL	17393
3	Firecloud	Certis Belchim B V	500 g/l	SL	18175
4	Glosset SC	Certis Belchim B V	600 g/l	SC	19265
5	Iconic	Adama	500 g/l	SC	19094
6	Macho	Albaugh UK	480 g/l	SL	18122
7	Osprey	Albaugh UK	480 g/l	SL	18657
8	Starfire	Certis Belchim B V	500 g/l	SL	18179
9	Steeple	Albaugh UK	480 g/l	SL	18082
10	Sunfire	Certis Belchim B V	500 g/l	SL	16745
11	System 50	Certis Belchim B V	500 g/l	SL	16612

Uses

- Annual dicotyledons in **Ornamental plant production** *(off-label)* [10]; **Spring barley** [3,8,10-11]; **Spring wheat** [3,8,10-11]; **Winter barley** [1-11]; **Winter rye** [4]; **Winter triticale** [4]; **Winter wheat** [1-11];
- Annual meadow grass in **Ornamental plant production** *(off-label)* [10]; **Protected forest nurseries** *(off-label)* [10]; **Rye** *(off-label)* [3,8,10-11]; **Triticale** *(off-label)* [3,8,10-11];
- Blackgrass in **Ornamental plant production** *(off-label)* [10]; **Rye** *(off-label)* [3,8,10-11]; **Spring barley** [3,8,10-11]; **Spring wheat** [3,8,10-11]; **Triticale** *(off-label)* [3,8,10-11]; **Winter barley** [1-11]; **Winter rye** [4]; **Winter triticale** [4]; **Winter wheat** [1-11];
- Chickweed in **Protected forest nurseries** *(off-label)* [10];
- Field pansy in **Protected forest nurseries** *(off-label)* [10];
- Penny cress in **Protected forest nurseries** *(off-label)* [10];
- Scented mayweed in **Protected forest nurseries** *(off-label)* [10];
- Shepherd's purse in **Protected forest nurseries** *(off-label)* [10];
- Sowthistle in **Protected forest nurseries** *(off-label)* [10];
- Speedwells in **Protected forest nurseries** *(off-label)* [10];

Extension of Authorisation for Minor Use (EAMUs)

- *Ornamental plant production 20171065* [10]
- *Protected forest nurseries 20170951* [10]
- *Rye 20193068* [3], *20193070* [8], *20170952* [10], *20162019* [11]
- *Triticale 20193068* [3], *20193070* [8], *20170952* [10], *20162019* [11]

Approval information

- Flufenacet included in Annex I under EC Regulation 1107/2009

Efficacy guidance

- Reports of reduced sensitivity in blackgrass and isolated instances of Enhanced Metabolic Resistance to flufenacet have been confirmed in a paper presented in Pest Management Science, March 2019

Restrictions
- Do not apply by hand-held equipment
- Do not handle treated crops for at least 2 days after treatment

Environmental safety
- Buffer zone requirement 8 m [4]
- LERAP Category B [1-11]

Hazard classification and safety precautions

Hazard Harmful, Dangerous for the environment, Harmful if swallowed, Very toxic to aquatic organisms [5,8,10-11]
Transport code 9 [1-11]
Packaging group III
UN Number 3082
Risk phrases H317 [3-5,8,10-11], H373
Operator protection U05a [5], U09a [5], U19a [5]; A, H
Environmental protection E16a, H410
Storage and disposal D01 [1-9,11], D02 [1-9,11], D05 [3-4,8,11], D09a [3-4,8], D10a [8], D10b [3-4], D11a [8], D12a [3-4], D12b [5]
Medical advice M05a [5]

188   flufenacet + metribuzin

A herbicide mixture for potatoes
HRAC mode of action code: 15 + 5 (K3 + C1)

See also metribuzin

Products

1	Artist	Bayer CropScience	24:17.5% w/w	WP	17049
2	Expert Met	Bayer CropScience	42:14% w/w	WP	20031

Uses
- Annual dicotyledons in *Asparagus* (off-label) [1]; *Bilberries* (off-label) [1]; *Blackcurrants* (off-label) [1]; *Cranberries* (off-label) [1]; *Early potatoes* [1]; *Gooseberries* (off-label) [1]; *Maincrop potatoes* [1]; *Redcurrants* (off-label) [1]; *Ribes species* (off-label) [1]; *Soya beans* (off-label) [1]; *Winter barley* [2]; *Winter wheat* [2];
- Annual grasses in *Asparagus* (off-label) [1]; *Bilberries* (off-label) [1]; *Blackcurrants* (off-label) [1]; *Cranberries* (off-label) [1]; *Gooseberries* (off-label) [1]; *Redcurrants* (off-label) [1]; *Ribes species* (off-label) [1]; *Soya beans* (off-label) [1];
- Annual meadow grass in *Early potatoes* [1]; *Maincrop potatoes* [1]; *Winter barley* [2]; *Winter wheat* [2];
- Black bindweed in *Soya beans* (off-label) [1];

Extension of Authorisation for Minor Use (EAMUs)
- *Asparagus* 20201340 [1]
- *Bilberries* 20152968 [1]
- *Blackcurrants* 20152968 [1]
- *Cranberries* 20152968 [1]
- *Gooseberries* 20152968 [1]
- *Redcurrants* 20152968 [1]
- *Ribes species* 20152968 [1]
- *Soya beans* 20171098 [1]

Approval information
- Flufenacet and metribuzin included in Annex I under EC Regulation 1107/2009

Efficacy guidance
- Product acts through root uptake and needs sufficient soil moisture at and shortly after application
- Effectiveness is reduced under dry soil conditions

FOR FULL CONDITIONS OF USE ALWAYS READ THE PRODUCT LABEL

- Residual activity is reduced on mineral soils with a high organic matter content and on peaty or organic soils
- Ensure application is made evenly to both sides of potato ridges
- Perennial weeds are not controlled

Restrictions
- Maximum total dose equivalent to one full dose treatment
- Potatoes must be sprayed before emergence of crop and weeds
- See label for list of tolerant varieties. Do not treat Maris Piper grown on Sands or Very Light soils
- Do not use on Sands
- On stony or gravelly soils there is risk of crop damage especially if heavy rain falls soon after application

Crop-specific information
- Latest use: before potato crop emergence
- Consult processor before use on crops for processing

Following crops guidance
- Before drilling or planting any succeeding crop soil must be mouldboard ploughed to at least 15 cm as soon as possible after lifting and no later than end Dec
- In W Cornwall on soils with more than 5% organic matter treated early potatoes may be followed by summer planted brassica crops 14 wk after treatment and after mouldboard ploughing. Elsewhere cereals or winter beans may be grown in the same year if at least 16 wk have elapsed since treatment
- In the yr following treatment any crop may be grown except lettuce or radish, or vegetable brassica crops on silt soils in Lincs

Environmental safety
- Dangerous for the environment
- Very toxic to aquatic organisms
- Take care to avoid spray drift onto neighbouring crops, especially lettuce or brassicas
- LERAP Category B [1-2]

Hazard classification and safety precautions
    Hazard Harmful, Dangerous for the environment, Harmful if swallowed
    Transport code 9 [1-2]
    Packaging group III
    UN Number 3077
    Risk phrases H317 [1], H319 [2], H373
    Operator protection U05a, U08, U13, U14, U19a, U20b; A, C, D, H
    Environmental protection E15a, E16a, E38, H410
    Storage and disposal D01, D02, D09a, D11a, D12a
    Medical advice M03

189   flufenacet + pendimethalin

A broad spectrum residual and contact herbicide mixture for winter cereals
HRAC mode of action code: 15 + 3 (K3 + K1)

See also pendimethalin

Products

1	Confluence	Pan Agriculture	60:300 g/l	EC	17199
2	Crystal	BASF	60:300 g/l	EC	13914
3	Shooter	BASF	60:300 g/l	EC	14106
4	Trooper	BASF	60:300 g/l	EC	13924

Uses
- Annual dicotyledons in **Game cover** *(off-label)* [2]; **Spring barley** *(off-label)* [2]; **Winter barley** [1-4]; **Winter wheat** [1-4];
- Annual grasses in **Spring barley** *(off-label)* [2]; **Winter barley** [1-4]; **Winter wheat** [1-4];

SEE SECTION 3 FOR PRODUCTS ALSO REGISTERED

- Annual meadow grass in *Game cover* *(off-label)* [2]; *Winter barley* [1-4]; *Winter wheat* [1-4];
- Blackgrass in *Game cover* *(off-label)* [2]; *Winter barley* [1-4]; *Winter wheat* [1-4];
- Chickweed in *Winter barley* [1-4]; *Winter wheat* [1-4];
- Corn marigold in *Winter barley* [1-4]; *Winter wheat* [1-4];
- Field speedwell in *Winter barley* [1-4]; *Winter wheat* [1-4];
- Ivy-leaved speedwell in *Winter barley* [1-4]; *Winter wheat* [1-4];

Extension of Authorisation for Minor Use (EAMUs)
- *Game cover* *20090450* [2]
- *Spring barley* *20212921* [2]

Approval information
- Flufenacet and pendimethalin included in Annex I under EC Regulation 1107/2009
- Accepted by BBPA for use on malting barley

Efficacy guidance
- Best results achieved when applied from pre-emergence of weeds up to 2 leaf stage but post emergence treatment is not recommended on clay soils
- Product requires some soil moisture to be activated ideally from rain within 7 d of application. Prolonged dry conditions may reduce residual control
- Product is slow acting and final level of weed control may take some time to appear
- For effective weed control seed bed preparations should ensure even incorporation of any trash, straw and ash to 15 cm
- Efficacy may be reduced on soils with more than 6% organic matter
- Always follow WRAG guidelines for preventing and managing herbicide resistant weeds. See Section 5 for more information

Restrictions
- Maximum total dose equivalent to one full dose treatment
- For pre-emergence treatments seed should be covered with at least 32 mm settled soil. Shallow drilled crops should be treated post-emergence only
- Do not treat undersown crops
- Avoid spraying during periods of prolonged or severe frosts
- Do not use on stony or gravelly soils or those with more than 10% organic matter
- Pre-emergence treatment may only be used on crops drilled before 30 Nov. All crops must be treated before 31 Dec in yr of planting
- Concentrated or diluted product may stain clothing or skin

Crop-specific information
- Latest use: before third tiller stage (GS 23) and before 31 Dec in yr of planting
- Very wet weather before and after treatment may result in loss of crop vigour and reduced yield, particularly where soils become waterlogged

Following crops guidance
- Any crop may follow a failed or normally harvested treated crop provided ploughing to at least 15 cm is carried out beforehand

Environmental safety
- Dangerous for the environment
- Very toxic to aquatic organisms
- Risk to certain non-target insects or other arthropods - avoid spraying within 6 m of field boundary
- Some products supplied in small volume returnable packs. Follow instructions for use
- LERAP Category B [1-4]

Hazard classification and safety precautions
Hazard Harmful, Dangerous for the environment, Harmful if swallowed, Very toxic to aquatic organisms [2-4]
Transport code 3 [1-4]
Packaging group III
UN Number 1993, 3082
Risk phrases H304 [2-4], H315, H351 [1], H370 [1]

FOR FULL CONDITIONS OF USE ALWAYS READ THE PRODUCT LABEL

Operator protection U02a, U05a, U14 [3], U20b [3], U20c [1-2,4]; A, C, H
Environmental protection E15a [1-2,4], E15b [3], E16a, E16b, E22b, E34, E38, H410
Storage and disposal D01, D02, D08 [3], D09a, D10a [1-2,4], D10c [3], D12a
Medical advice M03, M05b [1-2,4]

190   flufenacet + picolinafen

A herbicide mixture for weed control in winter cereals
HRAC mode of action code: 15 + 12 (K3 + F1)

Products

1 Pontos	BASF	240:100 g/l	SC	17811
2 Quirinus	BASF	240:50 g/l	SC	17711

Uses

- Annual dicotyledons in *Rye* [1-2]; *Triticale* [1-2]; *Winter barley* [1-2]; *Winter wheat* [1-2];
- Annual grasses in *Rye* [1-2]; *Triticale* [1-2]; *Winter barley* [1-2]; *Winter wheat* [1-2];
- Annual meadow grass in *Rye* [1-2]; *Triticale* [1-2]; *Winter barley* [1-2]; *Winter wheat* [1-2];
- Charlock in *Rye* [1-2]; *Triticale* [1-2]; *Winter barley* [1-2]; *Winter wheat* [1-2];
- Chickweed in *Rye* [1-2]; *Triticale* [1-2]; *Winter barley* [1-2]; *Winter wheat* [1-2];
- Field pansy in *Rye* [1-2]; *Triticale* [1-2]; *Winter barley* [1-2]; *Winter wheat* [1-2];
- Loose silky bent in *Rye* [1-2]; *Triticale* [1-2]; *Winter barley* [1-2]; *Winter wheat* [1-2];
- Mayweeds in *Rye* [1-2]; *Triticale* [1-2]; *Winter barley* [1-2]; *Winter wheat* [1-2];
- Shepherd's purse in *Rye* [1-2]; *Triticale* [1-2]; *Winter barley* [1-2]; *Winter wheat* [1-2];
- Speedwells in *Rye* [1-2]; *Triticale* [1-2]; *Winter barley* [1-2]; *Winter wheat* [1-2];
- Volunteer oilseed rape in *Rye* [1-2]; *Triticale* [1-2]; *Winter barley* [1-2]; *Winter wheat* [1-2];

Approval information

- Flufenacet and picolinafen are included in Annex I under EC Regulation 1107/2009

Efficacy guidance

- Can be used on all varieties of winter crops of wheat, barley, rye and triticale

Restrictions

- Always follow WRAG guidelines for preventing and managing herbicide resistant weeds
- Use low drift spraying equipment up to 30m from the top of the bank of any surface water bodies
- Horizontal boom sprayers must be fitted with three star drift reduction technology for all uses

Following crops guidance

- There are no restrictions on following crops after the normal harvest.
- In the event of crop failure, winter wheat can be re-sown in the same autumn provided soil is cultivated to a minimum depth of 15cm. Any of the following crops may be sown provided there has been a minimum of 60 days after the application and the soil is cultivated to a minimum depth of 15cm; legumes, maize, sugar beet and sunflower. Oilseed rape can be re-sown after 90 days following a pre-emergence application or 60 days following a post emergence application and the soil is cultivated to a minimum depth of 15cm. Spring barley can be re-drilled 120 days following application and the soil is cultivated to a minimum depth of 15cm.

Environmental safety

- Buffer zone requirement 6 m [1, 2]
- LERAP Category B [1-2]

Hazard classification and safety precautions

Transport code 9 [1-2]
Packaging group III
UN Number 3082
Risk phrases H373
Operator protection U05a, U19a; A, H
Environmental protection E16a, E34, H410
Storage and disposal D01, D02, D05, D09a, D10c, D12b
Medical advice M03

SECTION 2

SEE SECTION 3 FOR PRODUCTS ALSO REGISTERED

191   fluopicolide

An benzamide fungicide available only in mixtures
FRAC mode of action code: 43

192   fluopicolide + propamocarb hydrochloride

A protectant and systemic fungicide mixture for potato blight
FRAC mode of action code: 43 + 28

See also propamocarb hydrochloride

Products

1	Infinito	Bayer CropScience	62.5:625 g/l	SC	16335

Uses
*   Blight in **Potatoes** [1];
*   Downy mildew in **Baby leaf crops** *(off-label)* [1]; **Broccoli** *(off-label)* [1]; **Brussels sprouts** *(off-label)* [1]; **Bulb onions** *(off-label)* [1]; **Cabbages** *(off-label)* [1]; **Calabrese** *(off-label)* [1]; **Cauliflowers** *(off-label)* [1]; **Collards** *(off-label)* [1]; **Cress** *(off-label)* [1]; **Garlic** *(off-label)* [1]; **Herbs (see appendix 6)** *(off-label)* [1]; **Kale** *(off-label)* [1]; **Lamb's lettuce** *(off-label)* [1]; **Land cress** *(off-label)* [1]; **Leeks** *(off-label)* [1]; **Lettuce** *(off-label)* [1]; **Ornamental plant production** *(off-label)* [1]; **Protected broccoli** *(off-label)* [1]; **Protected Brussels sprouts** *(off-label)* [1]; **Protected cabbages** *(off-label)* [1]; **Protected calabrese** *(off-label)* [1]; **Protected cauliflowers** *(off-label)* [1]; **Protected collards** *(off-label)* [1]; **Protected kale** *(off-label)* [1]; **Radishes** *(off-label)* [1]; **Red mustard** *(off-label)* [1]; **Rocket** *(off-label)* [1]; **Salad onions** *(off-label)* [1]; **Shallots** *(off-label)* [1]; **Spinach** *(off-label)* [1];

Extension of Authorisation for Minor Use (EAMUs)
*   **Baby leaf crops** *20172431* [1]
*   **Broccoli** *20152557* [1]
*   **Brussels sprouts** *20152557* [1]
*   **Bulb onions** *20161552* [1]
*   **Cabbages** *20152557* [1]
*   **Calabrese** *20152557* [1]
*   **Cauliflowers** *20152557* [1]
*   **Collards** *20152557* [1]
*   **Cress** *20172431* [1]
*   **Garlic** *20161552* [1]
*   **Herbs (see appendix 6)** *20172431* [1]
*   **Kale** *20152557* [1]
*   **Lamb's lettuce** *20172431* [1]
*   **Land cress** *20172431* [1]
*   **Leeks** *20161552* [1]
*   **Lettuce** *20172431* [1]
*   **Ornamental plant production** *20142251* [1]
*   **Protected broccoli** *20152557* [1]
*   **Protected Brussels sprouts** *20152557* [1]
*   **Protected cabbages** *20152557* [1]
*   **Protected calabrese** *20152557* [1]
*   **Protected cauliflowers** *20152557* [1]
*   **Protected collards** *20152557* [1]
*   **Protected kale** *20152557* [1]
*   **Radishes** *20172432* [1]
*   **Red mustard** *20172431* [1]
*   **Rocket** *20172431* [1]
*   **Salad onions** *20161552* [1]
*   **Shallots** *20161552* [1]
*   **Spinach** *20172431* [1]

FOR FULL CONDITIONS OF USE ALWAYS READ THE PRODUCT LABEL

Approval information
- Fluopicolide and propamocarb hydrochloride included in Annex I under EC Regulation 1107/2009

Efficacy guidance
- Commence spray programme before infection appears as soon as weather conditions favourable for disease development occur. At latest the first treatment should be made as the foliage meets along the rows
- Repeat treatments at 7-10 day intervals according to disease incidence and weather conditions
- Reduce spray interval if conditions are conducive to the spread of blight
- Spray as soon as possible after irrigation
- Increase water volume in dense crops
- When used from full canopy development to haulm desiccation as part of a full blight protection programme tubers will be protected from late blight after harvest and tuber blight incidence will be reduced
- To reduce the development of resistance product should be used in single or block applications with fungicides from a different cross-resistance group

Restrictions
- Maximum total dose on potatoes equivalent to four full dose treatments
- Do not apply if rainfall or irrigation is imminent. Product is rainfast in 1 h provided spray has dried on leaf
- Do not apply as a curative treatment when blight is present in the crop
- Do not apply more than 3 consecutive treatments of the product
- To protect groundwater do not apply more than 400 g/ha fluopicolide in any three year period.

Crop-specific information
- HI 7 d for potatoes
- All varieties of potatoes, including seed crops, may be treated

Environmental safety
- Dangerous for the environment
- Toxic to aquatic organisms

Hazard classification and safety precautions
Hazard Irritant, Dangerous for the environment, Very toxic to aquatic organisms
Transport code 9 [1]
Packaging group III
UN Number 3082
Risk phrases H317
Operator protection U05a, U08, U11, U19a, U20a; A, H
Environmental protection E15a, E38, H410
Storage and disposal D01, D02, D09a, D10b, D12a

193   fluopyram

An SDHI fungicide for disease control in fruit trees
FRAC mode of action code: 7

See also bixafen + fluopyram + prothioconazole

Products

1	Luna Privilege	Bayer CropScience	500 g/l	SC	18393
2	Velum Prime	Bayer CropScience	400 g/l	SC	18880

Uses
- Botrytis in **Ornamental plant production** *(off-label)* [1]; **Protected ornamentals** *(off-label)* [1];
- Disease control in **Apples** [1]; **Pears** [1];
- Free-living nematodes in **Potatoes** [2];
- Globodera species in **Potatoes** [2];
- Nematodes in **Carrots** [2]; **Parsnips** *(off-label)* [2];
- Potato cyst nematode in **Potatoes** [2];

SEE SECTION 3 FOR PRODUCTS ALSO REGISTERED

SECTION 2

- Powdery mildew in *Hops* *(off-label)* [1]; *Ornamental plant production* *(off-label)* [1]; *Protected ornamentals* *(off-label)* [1];

Extension of Authorisation for Minor Use (EAMUs)
- *Hops* *20192567* [1]
- *Ornamental plant production* *20210289* [1]
- *Parsnips* *20210288* [2]
- *Protected ornamentals* *20210289* [1]

Approval information
- Fluopyram is included in Annex 1 under EC Regulation 1107/2009

Efficacy guidance
- Apply as a medium or coarse spray [2]

Restrictions
- Apply as an in-furrow application but note that it is important to direct spray into the planting furrow and not onto the seed tuber
- Can be used on all potato varieties but crops grown under cover are treated at growers risk only.
- If the crop is intended for processing, consult the processor before the use
- Where [2] has been applied, the first foliar fungicide application made to the crop must not be a member of the SDHI group

Crop-specific information
- Cardoons, celeries, Florence fennels and crops belonging to the category ?other stem vegetables? cannot be allowed as succeeding crops [2]

Environmental safety
- Broadcast air-assisted LERAP [1] (10m)

Hazard classification and safety precautions
 Transport code 9 [1-2]
 Packaging group III
 UN Number 3082
 Operator protection U05a, U09a [1], U20a [1], U20b [2]; A, C, H
 Environmental protection E15a [1], E15b [2], E17b (10m) [1], E34, H411
 Storage and disposal D01, D02, D05 [2], D09a, D10a [1], D10b [2], D12b [2], D22 [2]
 Medical advice M03

194 fluopyram + prothioconazole

A foliar fungicide for use in oilseed rape
FRAC mode of action code: 3 + 7

Products

1	Caligula	Bayer CropScience	125:125 g/l	SE	19297
2	Caligula (GB only)	Bayer CropScience	125:125 g/l	SE	20108
3	Propulse	Bayer CropScience	125:125 g/l	SE	17837
4	Recital	Bayer CropScience	125:125 g/l	SE	17909

Uses
- Alternaria in *Potatoes* [1-2];
- Cercospora leaf spot in *Sugar beet* [2];
- Disease control in *Spring oilseed rape* ; *Winter oilseed rape* [3-4];
- Early blight in *Potatoes* [1-2];
- Light leaf spot in *Winter oilseed rape* [3];
- Phoma in *Winter oilseed rape* [3];
- Powdery mildew in *Sugar beet* [2]; *Winter oilseed rape* [3];
- Ramularia leaf spots in *Sugar beet* [2];
- Rust in *Sugar beet* [2];

FOR FULL CONDITIONS OF USE ALWAYS READ THE PRODUCT LABEL

- Sclerotinia in **Winter oilseed rape** [3];
- Stem canker in **Winter oilseed rape** [3];

Approval information
- Fluopyram and prothioconazole included in Annex 1 under EC Regulation 1107/2009.

Restrictions
- A minimum interval of 14 days should be observed between applications
- Do not apply to sugar beet before 1st September [2]
- Do not feed beet tops to livestock [2]
- Manually remove bolters by hand before application [2]
- Do not enter field for 48 hours after application [2]

Crop-specific information
- A minimum interval of 365 days must elapse before planting stem vegetables such as cardoons, celery and Florence fennel as succeeding crops

Environmental safety
- LERAP Category B [1-4]

Hazard classification and safety precautions
Hazard Harmful [4], Dangerous for the environment, Very toxic to aquatic organisms [1,3-4]
Transport code 9 [1-4]
Packaging group III
UN Number 3082
Operator protection U05a, U09b, U19a, U20c; A, H
Environmental protection E15b, E16a, E34, E38, H410 [1,3-4], H411 [2]
Storage and disposal D01, D02, D05, D09a, D10b, D12a
Medical advice M03

195   fluopyram + prothioconazole + tebuconazole

An SDHI and triazole mixture for seed treatment in winter barley
FRAC mode of action code: 3 + 7

Products
1	Raxil Star	Bayer CropScience	20:100:60 g/l	FS	17805

Uses
- Covered smut in **Winter barley** [1];
- Fusarium in **Winter barley** [1];
- Leaf stripe in **Winter barley** [1];
- Loose smut in **Winter barley** [1];
- Microdochium nivale in **Winter barley** [1];
- Net blotch in **Winter barley** *(seed borne)* [1];

Approval information
- Fluopyram, prothioconazole and tebuconazole included in Annex 1 under EC Regulation 1107/2009
- Accepted by BBPA for use on malting barley

Efficacy guidance
- Seed treatment products must be applied by manufacturer's recommended treatment application equipment
- Treated cereal seed should preferably be drilled in the same season
- Follow-up treatments will be needed later in the season to give protection against air-borne and splash-borne diseases
- Seed should be drilled to a depth of 40 mm into a well prepared and firm seedbed. If seed is present on the soil surface, or spills have occurred, then, if conditions are appropriate, the field should be harrowed then rolled to ensure good incorporation

SEE SECTION 3 FOR PRODUCTS ALSO REGISTERED

SECTION 2

- Do not use on seed with more than 16% moisture content or on sprouted, cracked, skinned or otherwise damaged seed

Hazard classification and safety precautions
 Hazard Harmful, Dangerous for the environment
 Transport code 9 [1]
 Packaging group III
 UN Number 3082
 Risk phrases H361
 Operator protection U05a, U12b, U20b, U20e; A, C, D, H
 Environmental protection E15b, E34, E38, H410
 Storage and disposal D01, D02, D05, D09a, D10d, D12a, D14
 Treated seed S01, S02, S04b, S05, S07, S08

196   fluopyram + trifloxystrobin

A fungicide mixture for disease control in protected strawberries
FRAC mode of action code: 7 + 11

See also fluopyram
*   trifloxystrobin*

Products

1	Exteris Stressgard	Bayer CropScience	12.5:12.5 g/l	SC	17825
2	Luna Sensation	Bayer CropScience	250:250 g/l	SC	15793
3	Robin	Pan Agriculture	250:250 g/l	SC	18915
4	Robin	Pan Agriculture	250:250 g/l	SC	19209

Uses
- Botrytis in *Lettuce* (off-label) [2]; *Protected baby leaf crops* (off-label) [2]; *Protected lamb's lettuce* (off-label) [2]; *Protected land cress* (off-label) [2]; *Protected lettuce* (off-label) [2]; *Protected purslane* (off-label) [2]; *Protected red mustard* (off-label) [2]; *Protected rocket* (off-label) [2];
- Dollar spot in *Managed amenity turf* [1];
- Grey mould in *Protected chilli peppers* [2-4]; *Protected peppers* [2-4]; *Protected strawberries* (moderate control) [2-4];
- Microdochium nivale in *Managed amenity turf* [1];
- Powdery mildew in *Lettuce* (off-label) [2]; *Protected baby leaf crops* (off-label) [2]; *Protected chilli peppers* [2-4]; *Protected lamb's lettuce* (off-label) [2]; *Protected land cress* (off-label) [2]; *Protected lettuce* (off-label) [2]; *Protected peppers* [2-4]; *Protected purslane* (off-label) [2]; *Protected red mustard* (off-label) [2]; *Protected rocket* (off-label) [2]; *Protected strawberries* [2-4];
- Sclerotinia in *Lettuce* (off-label) [2]; *Protected baby leaf crops* (off-label) [2]; *Protected lamb's lettuce* (off-label) [2]; *Protected land cress* (off-label) [2]; *Protected lettuce* (off-label) [2]; *Protected purslane* (off-label) [2]; *Protected red mustard* (off-label) [2]; *Protected rocket* (off-label) [2];

Extension of Authorisation for Minor Use (EAMUs)
- *Lettuce* 20171179 [2]
- *Protected baby leaf crops* 20220888 [2]
- *Protected lamb's lettuce* 20220888 [2]
- *Protected land cress* 20220889 [2]
- *Protected lettuce* 20171179 [2]
- *Protected purslane* 20220888 [2]
- *Protected red mustard* 20220889 [2]
- *Protected rocket* 20220888 [2]

Approval information
- Fluopyram and trifloxystrobin are included in Annex 1 under EC Regulation 1107/2009

FOR FULL CONDITIONS OF USE ALWAYS READ THE PRODUCT LABEL

Restrictions
- If more than two QoI products are to be used on the crop then the FRAC advice on treatment must be followed.
- Cardoons, celeries, Florence fennels and crops belonging to the category "other stem vegetables" cannot be allowed as succeeding crops

Environmental safety
- Buffer zone requirement 14 m on protected strawberries
- LERAP Category B [1-4]

Hazard classification and safety precautions
   Hazard Harmful if swallowed [2-4], Very toxic to aquatic organisms
   Transport code 9 [1-4]
   Packaging group III
   UN Number 3082
   Risk phrases H317 [1]
   Operator protection U05a, U09a [2-4], U20a [2-4]; A, H
   Environmental protection E15a [2-4], E15b [1], E16a, E34, H410
   Storage and disposal D01, D02, D09a, D10a
   Medical advice M03

197   fluoxastrobin

A protectant stobilurin fungicide available in mixtures
FRAC mode of action code: 11

See also bixafen + fluoxastrobin + prothioconazole

198   fluoxastrobin + prothioconazole

A strobilurin and triazole fungicide mixture for cereals
FRAC mode of action code: 11 + 3

See also prothioconazole

Products

1	Fandango	Bayer CropScience	100:100 g/l	EC	17318
2	Firefly 155	Bayer CropScience	45:110 g/l	EC	14818
3	Maestro	Bayer CropScience	100:100 g/l	EC	18300
4	Sublime	Bayer CropScience	50:100 g/l	EC	19964
5	Unicur	Bayer CropScience	100:100 g/l	EC	17402
6	Unicur	Bayer CropScience	100:100 g/l	EC	19987
7	Variano Duo	Bayer CropScience	50:100 g/l	EC	19943

Uses
- Botrytis in *Bulb onions* (useful reduction) ; *Shallots* (useful reduction) [5-6];
- Botrytis squamosa in *Bulb onions* (useful reduction) [5-6]; *Shallots* (useful reduction) [5-6];
- Brown rust in *Spring barley* [1,3]; *Spring wheat* [1-4,7]; *Winter barley* [1,3]; *Winter rye* [1-4,7]; *Winter wheat* [1-4,7];
- Crown rust in *Spring oats* [1,3]; *Winter oats* [1,3];
- Disease control in *Forest nurseries* (off-label) [1]; *Ornamental plant production* (off-label) [1];
- Downy mildew in *Bulb onions* (control) [5-6]; *Shallots* (control) [5-6];
- Eyespot in *Spring barley* (reduction) [1,3]; *Spring oats* (reduction of incidence and severity) [1-3]; *Spring wheat* (reduction) [1-4,7]; *Winter barley* (reduction) [1,3]; *Winter oats* (reduction of incidence and severity) [1-3]; *Winter rye* (reduction) [1-4,7]; *Winter wheat* (reduction) [1-4,7];
- Fusarium root rot in *Spring wheat* (reduction) [1,3]; *Winter wheat* (reduction) [1,3];
- Glume blotch in *Spring wheat* [1-4,7]; *Winter wheat* [1-4,7];
- Late ear diseases in *Spring barley* [1,3]; *Spring wheat* [1,3]; *Winter barley* [1,3]; *Winter wheat* [1,3];
- Net blotch in *Spring barley* [1,3]; *Winter barley* [1,3];

- Powdery mildew in **Spring barley** [1,3]; **Spring oats** [1,3]; **Spring wheat** [1-4,7]; **Winter barley** [1,3]; **Winter oats** [1,3]; **Winter rye** [1-4,7]; **Winter wheat** [1-4,7];
- Rhynchosporium in **Spring barley** [1,3]; **Winter barley** [1,3]; **Winter rye** [1-4,7];
- Septoria leaf blotch in **Spring wheat** [1-4,7]; **Winter wheat** [1-4,7];
- Sharp eyespot in **Spring wheat** *(reduction)* [1,3]; **Winter wheat** [1,3];
- Sooty moulds in **Spring wheat** *(reduction)* [1,3]; **Winter wheat** *(reduction)* [1,3];
- Take-all in **Spring wheat** *(reduction)* [1,3]; **Winter barley** *(reduction)* [1,3]; **Winter wheat** *(reduction)* [1,3];
- Tan spot in **Spring wheat** [1,3]; **Winter wheat** [1,3];
- Yellow rust in **Spring wheat** [1-4,7]; **Winter wheat** [1-4,7];

Extension of Authorisation for Minor Use (EAMUs)
- **Forest nurseries** *20202547* [1]
- **Ornamental plant production** *20202547* [1]

Approval information
- Fluoxastrobin and prothioconazole included in Annex I under EC Regulation 1107/2009
- Accepted by BBPA for use on malting barley

Efficacy guidance
- Best results on foliar diseases obtained from treatment at early stages of disease development. Further treatment may be needed if disease attack is prolonged
- Foliar applications to established infections of any disease are likely to be less effective
- Best control of cereal ear diseases obtained by treatment during ear emergence
- Fluoxastrobin is a member of the QoI cross resistance group. Foliar product should be used preventatively and not relied on for its curative potential
- Use product as part of an Integrated Crop Management strategy incorporating other methods of control, including where appropriate other fungicides with a different mode of action. Do not apply more than two foliar applications of QoI containing products to any cereal crop
- There is a significant risk of widespread resistance occurring in *Septoria tritici* populations in UK. Failure to follow resistance management action may result in reduced levels of disease control
- Strains of wheat and barley powdery mildew resistant to QoIs are common in the UK. Control of wheat mildew can only be relied on from the triazole component
- Prothioconazole is a DMI fungicide. Resistance to some DMI fungicides has been identified in Septoria leaf blotch which may seriously affect performance of some products. For further advice contact a specialist advisor and visit the Fungicide Resistance Action Group (FRAG)-UK website
- Where specific control of wheat mildew is required this should be achieved through a programme of measures including products recommended for the control of mildew that contain a fungicide from a different cross-resistance group and applied at a dose that will give robust control

Restrictions
- Maximum total dose of foliar sprays equivalent to two full dose treatments for the crop
- Do not apply by hand-held equipment, e.g Knapsack Sprayer [5, 6]

Crop-specific information
- Latest use: before grain milky ripe for spray treatments on wheat and rye; beginning of flowering for barley, oats
- Some transient leaf chlorosis may occur after treatment of wheat or barley but this has not been found to affect yield

Environmental safety
- Dangerous for the environment
- Toxic to aquatic organisms
- Risk to non-target insects or other arthropods. Avoid spraying within 6 m of the field boundary to reduce the effects on non-target insects or other arthropods
- LERAP Category B [1-7]

Hazard classification and safety precautions
Hazard Dangerous for the environment, Very toxic to aquatic organisms
Transport code 9 [1-7]

FOR FULL CONDITIONS OF USE ALWAYS READ THE PRODUCT LABEL

Packaging group III
UN Number 3082
Risk phrases H318 [1,3]
Operator protection U05a, U09a [2,4,7], U09b [1,3,5-6], U20a [1,3,5-6], U20b [2,4,7]; A, H
Environmental protection E15a, E16a, E22c [1,3,5-6], E34, E38, H410
Storage and disposal D01, D05, D09a, D10b, D12a
Medical advice M03

199   fluoxastrobin + prothioconazole + trifloxystrobin

A triazole and strobilurin fungicide mixture for cereals
FRAC mode of action code: 11 + 3 + 11

See also prothioconazole
*   trifloxystrobin*

Products

1	Haven	Bayer CropScience	75:150:75 g/l	EC	18403
2	Jaunt	Bayer CropScience	75:150:75 g/l	EC	12350

Uses

- Brown rust in *Durum wheat* [1-2]; *Rye* [1-2]; *Spring barley* [1-2]; *Spring wheat* [1-2]; *Triticale* [1-2]; *Winter barley* [1-2]; *Winter wheat* [1-2];
- Eyespot in *Durum wheat* *(reduction)* [1-2]; *Rye* *(reduction)* [1-2]; *Spring barley* *(reduction)* [1-2]; *Spring wheat* *(reduction)* [1-2]; *Triticale* *(reduction)* [1-2]; *Winter barley* *(reduction)* [1-2]; *Winter wheat* *(reduction)* [1-2];
- Glume blotch in *Durum wheat* [1-2]; *Rye* [1-2]; *Spring wheat* [1-2]; *Triticale* [1-2]; *Winter wheat* [1-2];
- Late ear diseases in *Durum wheat* [1-2]; *Rye* [1-2]; *Spring barley* [1-2]; *Spring wheat* [1-2]; *Triticale* [1-2]; *Winter barley* [1-2]; *Winter wheat* [1-2];
- Net blotch in *Spring barley* [1-2]; *Winter barley* [1-2];
- Powdery mildew in *Durum wheat* [1-2]; *Rye* [1-2]; *Spring barley* [1-2]; *Spring wheat* [1-2]; *Triticale* [1-2]; *Winter barley* [1-2]; *Winter wheat* [1-2];
- Rhynchosporium in *Spring barley* [1-2]; *Winter barley* [1-2];
- Septoria leaf blotch in *Durum wheat* [1-2]; *Rye* [1-2]; *Spring wheat* [1-2]; *Triticale* [1-2]; *Winter wheat* [1-2];
- Tan spot in *Durum wheat* [1-2]; *Rye* [1-2]; *Spring wheat* [1-2]; *Triticale* [1-2]; *Winter wheat* [1-2];
- Yellow rust in *Durum wheat* [1-2]; *Rye* [1-2]; *Spring wheat* [1-2]; *Triticale* [1-2]; *Winter wheat* [1-2];

Approval information

- Fluoxastrobin, prothioconazole and trifloxystrobin included in Annex I under EC Regulation 1107/2009
- Accepted by BBPA for use on malting barley

Efficacy guidance

- Best results obtained from treatment at early stages of disease development. Further treatment may be needed if disease attack is prolonged
- Applications to established infections of any disease are likely to be less effective
- Best control of cereal ear diseases obtained by treatment during ear emergence
- Fluoxastrobin and trifloxystrobin are members of the QoI cross resistance group. Product should be used preventatively and not relied on for its curative potential
- Use product as part of an Integrated Crop Management strategy incorporating other methods of control, including where appropriate other fungicides with a different mode of action. Do not apply more than two foliar applications of QoI containing products to any cereal crop
- There is a significant risk of widespread resistance occurring in *Septoria tritici* populations in UK. Failure to follow resistance management action may result in reduced levels of disease control
- Strains of wheat and barley powdery mildew resistant to QoIs are common in the UK. Control of wheat mildew can only be relied on from the triazole component

SEE SECTION 3 FOR PRODUCTS ALSO REGISTERED

- Where specific control of wheat mildew is required this should be achieved through a programme of measures including products recommended for the control of mildew that contain a fungicide from a different cross-resistance group and applied at a dose that will give robust control
- Prothioconazole is a DMI fungicide. Resistance to some DMI fungicides has been identified in Septoria leaf blotch which may seriously affect performance of some products. For further advice contact a specialist advisor and visit the Fungicide Resistance Action Group (FRAG)-UK website

Restrictions
- Maximum total dose equivalent to two full dose treatments

Crop-specific information
- Latest use: before grain milky ripe for winter wheat; up to beginning of anthesis (GS 61) for barley

Environmental safety
- Dangerous for the environment
- Very toxic to aquatic organisms
- Risk to non-target insects or other arthropods. Avoid spraying within 6 m of the field boundary to reduce the effects on non-target insects or other arthropods
- LERAP Category B [1-2]

Hazard classification and safety precautions
Hazard Irritant, Dangerous for the environment, Very toxic to aquatic organisms
Transport code 9 [1-2]
Packaging group III
UN Number 3082
Operator protection U05a, U09b, U19a, U20b; A, C, H
Environmental protection E15a, E16a, E22c, E34, E38, H410
Storage and disposal D01, D02, D05, D09a, D10b, D12a
Medical advice M03

200   fluoxastrobin + tebuconazole

A fungicide mixture for disease control in oilseed rape
FRAC mode of action code: 11 + 3

See also fluoxastrobin + prothioconazole
fluoxastrobin + prothioconazole + tebuconazole

Products

1	Evito T	Arysta	180:250 g/l	SC	18671

Uses
- Sclerotinia in **Spring oilseed rape** [1]; **Winter oilseed rape** [1];

Approval information
- Fluoxastrobin and tebuconazole included in Annex I under EC Regulation 1107/2009

Restrictions
- Do not apply before 10% of flowers on main raceme open, main raceme elongating (GS61) or 1 May if this occurs after GS61 in the year of harvest

Environmental safety
- LERAP Category B [1]

Hazard classification and safety precautions
Hazard Very toxic to aquatic organisms
Transport code 9 [1]
Packaging group III
UN Number 3082

FOR FULL CONDITIONS OF USE ALWAYS READ THE PRODUCT LABEL

Risk phrases H319, H361
Operator protection U05a, U11, U20b; A, H
Environmental protection E16a, H410
Medical advice M04a

201   fluroxypyr

A post-emergence pyridinecarboxylic acid herbicide
HRAC mode of action code: 4 (O)

See also 2,4-D + dicamba + fluroxypyr
aminopyralid + fluroxypyr
clopyralid + florasulam + fluroxypyr
clopyralid + fluroxypyr + MCPA
clopyralid + fluroxypyr + triclopyr
florasulam + fluroxypyr

Products

1	Arbiter	Barclay	200 g/l	EC	18326
2	Clayton Flurry	Clayton	200 g/l	EC	19736
3	Cleancrop Gallifrey 3	Corteva	333 g/l	EC	17399
4	Flurostar 200	Globachem	200 g/l	EC	17438
5	Gal-gone	Certis Belchim B V	200 g/l	EC	17505
6	Hoist	Agform	200 g/l	EC	19542
7	Hudson 200	Barclay	200 g/l	EC	17749
8	Hurler	Barclay	200 g/l	EC	17715
9	Minstrel	UPL Europe	200 g/l	EC	13745
10	Moraine	Adama	200 g/l	EC	19608
11	Starane Hi-Load HL	Corteva	333 g/l	EC	16557
12	Tandus	Nufarm UK	200 g/l	EC	18071
13	Tensira	Albaugh UK	200 g/l	EC	19590
14	Tomahawk	Nufarm UK	200 g/l	EC	19782

Uses

- Annual dicotyledons in **Bulb onions** *(off-label)* [11]; **Durum wheat** [1-8,10-12,14]; **Forage maize** [1-14]; **Garlic** *(off-label)* [11]; **Grass seed crops** [3,11]; **Grassland** [1-14]; **Leeks** *(off-label)* [11]; **Rye** [2,5,10,12,14]; **Salad onions** *(off-label)* [11]; **Shallots** *(off-label)* [11]; **Spelt** [3,11]; **Spring barley** [1-2,4-10,12-14]; **Spring oats** [1-14]; **Spring rye** [1,3-4,6-8,11]; **Spring wheat** [1-14]; **Triticale** [1-14]; **Winter barley** [1-14]; **Winter oats** [1-14]; **Winter rye** [1,3-4,6-9,11,13]; **Winter wheat** [1-14];
- Black bindweed in **Bulb onions** *(off-label)* [3,11]; **Durum wheat** [1-8,10-12,14]; **Forage maize** [1-11,13-14]; **Garlic** *(off-label)* [3,11]; **Grass seed crops** [3,11]; **Grassland** [1-11,13-14]; **Leeks** *(off-label)* [3,11]; **Poppies** *(off-label)* [11]; **Poppies grown for seed production** *(off-label)* [11]; **Rye** [2,5,10,12,14]; **Salad onions** *(off-label)* [3,11]; **Shallots** *(off-label)* [3,11]; **Spelt** [3,11]; **Spring barley** [1-14]; **Spring oats** [1-14]; **Spring rye** [1,3-4,6-8,11]; **Spring wheat** [1-14]; **Triticale** [1-14]; **Winter barley** [1-14]; **Winter oats** [1-14]; **Winter rye** [1,3-4,6-9,11,13]; **Winter wheat** [1-14];
- Black nightshade in **Forage maize** [3];
- Chickweed in **Almonds** *(off-label)* [3,11]; **Apples** *(off-label)* [3,11]; **Bulb onions** *(off-label)* [3,11]; **Canary seed** *(off-label)* [11]; **Chestnuts** *(off-label)* [3,11]; **Durum wheat** [1-8,10-12,14]; **Farm forestry** *(off-label)* [11]; **Forage maize** [1-14]; **Forest nurseries** *(off-label)* [11]; **Game cover** *(off-label)* [11]; **Garlic** *(off-label)* [3,11]; **Grass seed crops** [3,11]; **Grassland** [1-14]; **Hazel nuts** *(off-label)* [3,11]; **Leeks** *(off-label)* [3,11]; **Millet** *(off-label)* [11]; **Miscanthus** *(off-label)* [11]; **Ornamental plant production** *(off-label)* [11]; **Pears** *(off-label)* [3,11]; **Poppies** *(off-label)* [11]; **Poppies for morphine production** *(off-label)* [3]; **Poppies grown for seed production** *(off-label)* [11]; **Rye** [2,5,10,12,14]; **Salad onions** *(off-label)* [3,11]; **Shallots** *(off-label)* [3,11]; **Spelt** [3,11]; **Spring barley** [1-14]; **Spring oats** [1-14]; **Spring rye** [1,3-4,6-8,11]; **Spring wheat** [1-14]; **Sweetcorn** *(off-label)* [3,11]; **Triticale** [1-14]; **Walnuts** *(off-label)* [3,11]; **Winter barley** [1-14]; **Winter oats** [1-14]; **Winter rye** [1,3-4,6-9,11,13]; **Winter wheat** [1-14];

SEE SECTION 3 FOR PRODUCTS ALSO REGISTERED

- Cleavers in *Almonds* (off-label) [3,11]; *Apples* (off-label) [3,11]; *Bulb onions* (off-label) [3,11]; *Canary seed* (off-label) [11]; *Chestnuts* (off-label) [3,11]; *Durum wheat* [1-8,10-12,14]; *Farm forestry* (off-label) [11]; *Forage maize* [1-11,13-14]; *Forest nurseries* (off-label) [11]; *Game cover* (off-label) [11]; *Garlic* (off-label) [3,11]; *Grass seed crops* [3,11]; *Grassland* [1-11,13-14]; *Hazel nuts* (off-label) [3,11]; *Leeks* (off-label) [3,11]; *Millet* (off-label) [11]; *Miscanthus* (off-label) [11]; *Ornamental plant production* (off-label) [11]; *Pears* (off-label) [3,11]; *Poppies* (off-label) [11]; *Poppies for morphine production* (off-label) [3,11]; *Poppies grown for seed production* (off-label) [11]; *Rye* [2,5,10,12,14]; *Salad onions* (off-label) [3,11]; *Shallots* (off-label) [3,11]; *Spelt* [3,11]; *Spring barley* [1-14]; *Spring oats* [1-14]; *Spring rye* [1,3-4,6-8,11]; *Spring wheat* [1-14]; *Sweetcorn* (off-label) [3,11]; *Triticale* [1-14]; *Walnuts* (off-label) [3,11]; *Winter barley* [1-14]; *Winter oats* [1-14]; *Winter rye* [1,3-4,6-9,11,13]; *Winter wheat* [1-14];
- Corn spurrey in *Bulb onions* (off-label) [3,11]; *Garlic* (off-label) [3,11]; *Leeks* (off-label) [3,11]; *Salad onions* (off-label) [3,11]; *Shallots* (off-label) [3,11];
- Dead nettle in *Poppies* (off-label) [11];
- Docks in *Durum wheat* [1-2,4-8,10,12,14]; *Forage maize* [1-2,4-10,13-14]; *Grassland* [1-2,4-8,10,13-14]; *Rye* [2,5,10,12,14]; *Spring barley* [1-2,4-10,12-14]; *Spring oats* [1-2,4-10,12-14]; *Spring rye* [1,4,6-8]; *Spring wheat* [1-2,4-10,12-14]; *Triticale* [1-2,4-10,12-14]; *Winter barley* [1-2,4-10,12-14]; *Winter oats* [1-2,4-10,12-14]; *Winter rye* [1,4,6-9,13]; *Winter wheat* [1-2,4-10,12-14];
- Forget-me-not in *Bulb onions* (off-label) [3,11]; *Durum wheat* [1-8,10-12,14]; *Forage maize* [1-11,13-14]; *Garlic* (off-label) [3,11]; *Grass seed crops* [3,11]; *Grassland* [1-11,13-14]; *Leeks* (off-label) [3,11]; *Poppies* (off-label) [11]; *Poppies grown for seed production* (off-label) [11]; *Rye* [2,5,10,12,14]; *Salad onions* (off-label) [3,11]; *Shallots* (off-label) [3,11]; *Spelt* [3,11]; *Spring barley* [1-14]; *Spring oats* [1-14]; *Spring rye* [1,3-4,6-8,11]; *Spring wheat* [1-14]; *Triticale* [1-14]; *Winter barley* [1-14]; *Winter oats* [1-14]; *Winter rye* [1,3-4,6-9,11,13]; *Winter wheat* [1-14];
- Fumitory in *Bulb onions* (off-label) [3,11]; *Garlic* (off-label) [3,11]; *Leeks* (off-label) [3,11]; *Poppies* (off-label) [11]; *Poppies grown for seed production* (off-label) [11]; *Salad onions* (off-label) [3,11]; *Shallots* (off-label) [3,11];
- Groundsel in *Bulb onions* (off-label) [3,11]; *Garlic* (off-label) [3,11]; *Leeks* (off-label) [3,11]; *Poppies* (off-label) [11]; *Poppies grown for seed production* (off-label) [11]; *Salad onions* (off-label) [3,11]; *Shallots* (off-label) [3,11];
- Hemp-nettle in *Almonds* (off-label) [3,11]; *Apples* (off-label) [3,11]; *Bulb onions* (off-label) [3,11]; *Chestnuts* (off-label) [3,11]; *Durum wheat* [1-8,10-12,14]; *Farm forestry* (off-label) [11]; *Forage maize* [1-11,13-14]; *Forest nurseries* (off-label) [11]; *Game cover* (off-label) [11]; *Garlic* (off-label) [3,11]; *Grass seed crops* [3,11]; *Grassland* [1-11,13-14]; *Hazel nuts* (off-label) [3,11]; *Leeks* (off-label) [3,11]; *Miscanthus* (off-label) [11]; *Ornamental plant production* (off-label) [11]; *Pears* (off-label) [3,11]; *Poppies* (off-label) [11]; *Poppies grown for seed production* (off-label) [11]; *Rye* [2,5,10,12,14]; *Salad onions* (off-label) [3,11]; *Shallots* (off-label) [3,11]; *Spelt* [3,11]; *Spring barley* [1-14]; *Spring oats* [1-14]; *Spring rye* [1,3-4,6-8,11]; *Spring wheat* [1-14]; *Sweetcorn* (off-label) [3,11]; *Triticale* [1-14]; *Walnuts* (off-label) [3,11]; *Winter barley* [1-14]; *Winter oats* [1-14]; *Winter rye* [1,3-4,6-9,11,13]; *Winter wheat* [1-14];
- Henbit in *Poppies grown for seed production* (off-label) [11];
- Ivy-leaved speedwell in *Poppies* (off-label) [11]; *Poppies grown for seed production* (off-label) [11];
- Knotgrass in *Bulb onions* (off-label) [3,11]; *Garlic* (off-label) [3,11]; *Leeks* (off-label) [3,11]; *Poppies* (off-label) [11]; *Poppies grown for seed production* (off-label) [11]; *Salad onions* (off-label) [3,11]; *Shallots* (off-label) [3,11];
- Mayweeds in *Bulb onions* (off-label) [3,11]; *Garlic* (off-label) [3,11]; *Leeks* (off-label) [3,11]; *Poppies* (off-label) [11]; *Poppies grown for seed production* (off-label) [11]; *Salad onions* (off-label) [3,11]; *Shallots* (off-label) [3,11];
- Pale persicaria in *Bulb onions* (off-label) [3,11]; *Garlic* (off-label) [3,11]; *Leeks* (off-label) [3,11]; *Poppies* (off-label) [11]; *Poppies grown for seed production* (off-label) [11]; *Salad onions* (off-label) [3,11]; *Shallots* (off-label) [3,11];
- Poppies in *Poppies* (off-label) [11]; *Poppies for morphine production* (off-label) [3,11]; *Poppies grown for seed production* (off-label) [11];
- Red dead-nettle in *Bulb onions* (off-label) [3,11]; *Garlic* (off-label) [3,11]; *Leeks* (off-label) [3,11]; *Salad onions* (off-label) [3,11]; *Shallots* (off-label) [3,11];

FOR FULL CONDITIONS OF USE ALWAYS READ THE PRODUCT LABEL

SECTION 2

- Redshank in **Bulb onions** *(off-label)* [3,11]; **Garlic** *(off-label)* [3,11]; **Leeks** *(off-label)* [3,11]; **Poppies** *(off-label)* [11]; **Poppies grown for seed production** *(off-label)* [11]; **Salad onions** *(off-label)* [3,11]; **Shallots** *(off-label)* [3,11];
- Speedwells in **Bulb onions** *(off-label)* [3]; **Garlic** *(off-label)* [3]; **Leeks** *(off-label)* [3]; **Poppies grown for seed production** *(off-label)* [11]; **Salad onions** *(off-label)* [3]; **Shallots** *(off-label)* [3];
- Volunteer potatoes in **Bulb onions** *(off-label)* [3,11]; **Farm forestry** *(off-label)* [11]; **Forest nurseries** *(off-label)* [11]; **Game cover** *(off-label)* [11]; **Garlic** *(off-label)* [3,11]; **Leeks** *(off-label)* [3,11]; **Miscanthus** *(off-label)* [11]; **Ornamental plant production** *(off-label)* [11]; **Poppies** *(off-label)* [11]; **Poppies for morphine production** *(off-label)* [3,11]; **Poppies grown for seed production** *(off-label)* [11]; **Salad onions** *(off-label)* [3,11]; **Shallots** *(off-label)* [3,11]; **Spring barley** [12]; **Spring oats** [12]; **Spring wheat** [12]; **Sweetcorn** *(off-label)* [3,11]; **Winter barley** [1-2,4-10,12-14]; **Winter wheat** [1-2,4-10,12-14];

Extension of Authorisation for Minor Use (EAMUs)
- **Almonds** *20210956* [3], *20210944* [11]
- **Apples** *20210956* [3], *20210944* [11]
- **Bulb onions** *20210957* [3], *20210946* [11]
- **Canary seed** *20210945* [11]
- **Chestnuts** *20210956* [3], *20210944* [11]
- **Farm forestry** *20210942* [11]
- **Forest nurseries** *20210942* [11]
- **Game cover** *20210942* [11]
- **Garlic** *20210957* [3], *20210946* [11]
- **Hazel nuts** *20210956* [3], *20210944* [11]
- **Leeks** *20210957* [3], *20210946* [11]
- **Millet** *20210762* [11]
- **Miscanthus** *20210942* [11]
- **Ornamental plant production** *20210942* [11]
- **Pears** *20210956* [3], *20210944* [11]
- **Poppies** *20180145* [11]
- **Poppies for morphine production** *20210955* [3], *20210947* [11]
- **Poppies grown for seed production** *20210948* [11]
- **Salad onions** *20210957* [3], *20210946* [11]
- **Shallots** *20210957* [3], *20210946* [11]
- **Sweetcorn** *20210958* [3], *20210943* [11]
- **Walnuts** *20210956* [3], *20210944* [11]

Approval information
- Fluroxypyr included in Annex I under EC Regulation 1107/2009
- Accepted by BBPA for use on malting barley

Efficacy guidance
- Best results achieved under good growing conditions in a strongly competing crop
- A number of tank mixtures with other herbicides are recommended for use in autumn and spring to extend range of species controlled. See label for details
- Spray is rainfast in 1 h

Restrictions
- Maximum number of treatments 1 per crop or yr or maximum total dose equivalent to one full dose treatment
- Do not apply in any tank-mix on triticale or forage maize
- Do not use on crops undersown with clovers or other legumes
- Do not treat crops suffering stress caused by any factor
- Do not roll or harrow for 7 d before or after treatment
- Do not spray if frost imminent
- Straw from treated crops must not be returned directly to the soil but must be removed and used only for livestock bedding
- A maximum total dose of 0.75 l/ha (200 g/l formulations) or 0.45 l/ha (333 g/l formulations) must be observed for applications made to cereals between crop emergence in the year of planting and 1st February in the year of harvest

SEE SECTION 3 FOR PRODUCTS ALSO REGISTERED

- When the product is applied as a spot treatment to permanent grassland, rotational grass and newly sown grass leys the maximum concentration must not exceed 30 ml product per 10 litres water

Crop-specific information
- Latest use: before flag leaf sheath opening (GS 47) for winter wheat and barley; before flag leaf sheath extending (GS 41) for spring wheat and barley; before second node detectable (GS 32) for oats, rye, triticale and durum wheat; before 7 leaves unfolded and before buttress roots appear for maize
- Apply to new leys from 3 expanded leaf stage
- Timing varies in tank mixtures. See label for details
- Crops undersown with grass may be sprayed provided grasses are tillering

Following crops guidance
- Clovers, peas, beans and other legumes must not be sown for 12 mth following treatment at the highest dose
- In the event of crop failure, spring cereals, spring oilseed rape, maize, onions, poppies and new leys may be sown 5 weeks after application with no requirements for soil cultivation [11]

Environmental safety
- Dangerous for the environment
- Very toxic to aquatic organisms
- Flammable
- Keep livestock out of treated areas for at least 3 d following treatment and until poisonous weeds, such as ragwort, have died down and become unpalatable
- Wash spray equipment thoroughly with water and detergent immediately after use. Traces of product can damage susceptible plants sprayed later
- LERAP Category B [2-6,9-11,13-14]

Hazard classification and safety precautions
Hazard Harmful [1-2,4-6,8-10,12-14], Flammable [1-2,4-6,8-10,13-14], Dangerous for the environment [1-6,8-14], Flammable liquid and vapour [1-2,6-9,12-13], Very toxic to aquatic organisms [4-5,9,13]
Transport code 3 [1-4,6,8-11,13-14], 9 [5,7,12]
Packaging group III
UN Number 1993, 3082
Risk phrases H304 [1-2,4-10,12-14], H315 [1-2,4-8,12], H317 [1-3,6-8,11-12], H318 [6,10,14], H319 [1-5,7-9,11-13], H335 [1-3,7-9,11-13], H336 [1-2,4-10,12-14]
Operator protection U04c [3,11], U05a [1,3-6,8,11], U08, U11 [3-6,11], U12 [3,11], U14 [1,4-6,8,12], U19a, U20b [1-11,13-14]; A, C, H
Environmental protection E07a [2,4-5,7,9-10,12-14], E07c (14 d) [1,8], E07f [6], E15a [2,7,9-10,13-14], E15b [1,3-6,8,11-12], E16a [2-6,9-11,13-14], E34 [1-6,8,10-11,13-14], E38 [1-2,4-10,12-14], H410 [1-9,11-13], H412 [10,14]
Storage and disposal D01 [1,3-6,8,11], D02 [1,3-6,8,11], D05 [1-6,8,10-11,13-14], D09a, D10a [1,8], D10b [2-7,9-14], D12a [1-2,4-10,12-14], D12b [3,11]
Treated seed S06a [12]
Medical advice M03 [1,4-6,8], M05b [1-2,4-10,12-14]

202   fluroxypyr + halauxifen-methyl

A herbicide mixture for use in cereals
HRAC mode of action code: 4 (O)

Products

1	Pixxaro EC	Corteva	280:12 g/l	EC	17545
2	Whorl	Corteva	280:12 g/l	EC	17819

Uses
- Chickweed in *Durum wheat* [1-2]; *Rye* [1-2]; *Spelt* [1-2]; *Spring barley* [1-2]; *Spring wheat* [1-2]; *Triticale* [1-2]; *Winter barley* [1-2]; *Winter oats* [1-2]; *Winter wheat* [1-2];

FOR FULL CONDITIONS OF USE ALWAYS READ THE PRODUCT LABEL

- Cleavers in **Durum wheat** [1-2]; **Rye** [1-2]; **Spelt** [1-2]; **Spring barley** [1-2]; **Spring wheat** [1-2]; **Triticale** [1-2]; **Winter barley** [1-2]; **Winter oats** [1-2]; **Winter wheat** [1-2];
- Crane's-bill in **Durum wheat** [1-2]; **Rye** [1-2]; **Spelt** [1-2]; **Spring barley** [1-2]; **Spring wheat** [1-2]; **Triticale** [1-2]; **Winter barley** [1-2]; **Winter oats** [1-2]; **Winter wheat** [1-2];
- Fat hen in **Durum wheat** [1-2]; **Rye** [1-2]; **Spelt** [1-2]; **Spring barley** [1-2]; **Spring wheat** [1-2]; **Triticale** [1-2]; **Winter barley** [1-2]; **Winter oats** [1-2]; **Winter wheat** [1-2];
- Fumitory in **Durum wheat** [1-2]; **Rye** [1-2]; **Spelt** [1-2]; **Spring barley** [1-2]; **Spring wheat** [1-2]; **Triticale** [1-2]; **Winter barley** [1-2]; **Winter oats** [1-2]; **Winter wheat** [1-2];
- Poppies in **Durum wheat** [1-2]; **Rye** [1-2]; **Spelt** [1-2]; **Spring barley** [1-2]; **Spring wheat** [1-2]; **Triticale** [1-2]; **Winter barley** [1-2]; **Winter oats** [1-2]; **Winter wheat** [1-2];

Approval information
- Fluroxypyr and halauxifen-methyl included in Annex I under EC Regulation 1107/2009

Efficacy guidance
- Rainfast one hour after application
- Addition of an adjuvant gives improved reliability against poppy, chickweed and volunteer potatoes

Restrictions
- The total amount of halauxifen-methyl applied to a winter cereal must not exceed 13.5 g.a.e/ha per season: 7.5 g.a.e/ha in the autumn followed by 6 g.a.e/ha in the spring, with a minimal interval of 3 months between both applications of products which contain halauxifen-methyl
- Do not use on spring oats
- Must only be applied to winter cereals between 1st February and 30th June, must only be applied to spring cereals between 1st March and 30th June.

Following crops guidance
- After an application of 0.5 l/ha up to BBCH 45 the following crops may be sown in the summer/ autumn of the same year after cereal harvest: white mustard, pea, broad bean, field beans, winter oilseed rape, phacelia, ryegrass, winter sown cereals (wheat, barley, rye, triticale, oats, spelt,) and vegetable transplants. Ploughing is recommended prior to drilling maize, alfalfa, soybean and clover. All crops can be sown in the spring following an application of 0.5 l/ha up to BBCH 45 in the previous calendar year.
- In the event of crop failure spring wheat, spring barley, spring oats and ryegrass may be sown one month after application with no cultivation. Maize, spring oilseed rape, peas, broad beans and field beans may be sown two months after application and ploughing. Three months after application sorghum may be sown.

Environmental safety
- LERAP Category B [1-2]

Hazard classification and safety precautions
Transport code 9 [1-2]
Packaging group III
UN Number 3082
Risk phrases H317, H319, H335
Operator protection U05a, U08, U19a, U20c; A, C, H
Environmental protection E15b, E16a, E34, H410
Storage and disposal D09a, D10c

203   fluroxypyr + metsulfuron-methyl

A foliar-applied herbicide mixture for use in wheat and barley
HRAC mode of action code: 4 + 2 (O + B)

Products

1	Croupier OD	Certis Belchim B V	225:9 g/l	OD	18457

SEE SECTION 3 FOR PRODUCTS ALSO REGISTERED

SECTION 2

Uses
- Annual dicotyledons in **Durum wheat** [1]; **Rye** [1]; **Spring barley** [1]; **Spring wheat** [1]; **Triticale** [1]; **Winter barley** [1]; **Winter wheat** [1];

Approval information
- Fluroxypyr and metsulfuron-methyl included in Annex I under EC Regulation 1107/2009

Efficacy guidance
- See label for ALS herbicides allowed in mixture
- Weed control may be reduced in dry soil conditions

Restrictions
- Do not apply to any cereal crop in tank mixture or sequence with any product containing ALS inhibiting or sulfonylurea herbicides, except as directed for specified products.
- Apply to dry foliage, when rain is not imminent.
- Transient chlorosis may occur following treatment during periods of rapid crop growth.
- Do not apply within 7 days of rolling the crop.
- Consult contract agents before using on crops grown for seed.
- For winter cereals, do not apply before March 15 in the year of harvest

Following crops guidance
- Only cereals, oilseed rape, field beans or grass may be sown in the same calendar year following a treated cereal crop.
- If a crop fails for any reason, only sow wheat within 3 months of the date of treatment.
- Only wheat, barley, rye and triticale can follow a cereal crop treated with a tank mixture of [1] and amidosulfuron e.g. MAPP 18902 or 16491.

Environmental safety
- LERAP Category B [1]

Hazard classification and safety precautions
Hazard Irritant, Dangerous for the environment
Transport code 9 [1]
Packaging group III
UN Number 3077
Operator protection U05a, U09a, U19a, U20b; A, C, H
Environmental protection E15b, E16a, E16b, E34, E38, H410
Storage and disposal D01, D02, D09a, D11a, D12a
Medical advice M05a

204   fluroxypyr + metsulfuron-methyl + thifensulfuron-methyl

A herbicide mixture for cereals
HRAC mode of action code: 4 + 2 + 2 (O + B + B)

See also fluroxypyr + metsulfuron-methyl
thifensulfuron-methyl

Products
1	Omnera LQM	FMC Agro	135:5:30 g/l	OD	18758
2	Provalia LQM	FMC Agro	135:5:30 g/l	OD	18798

Uses
- Annual dicotyledons in **Spring barley** [1]; **Spring wheat** [1]; **Winter barley** [1-2]; **Winter wheat** [1-2];
- Charlock in **Winter barley** [2]; **Winter wheat** [2];
- Chickweed in **Winter barley** [2]; **Winter wheat** [2];
- Cleavers in **Winter barley** [2]; **Winter wheat** [2];
- Field pansy in **Winter barley** [2]; **Winter wheat** [2];
- Fumitory in **Winter barley** [2]; **Winter wheat** [2];
- Mayweeds in **Winter barley** [2]; **Winter wheat** [2];
- Poppies in **Winter barley** [2]; **Winter wheat** [2];

- Red dead-nettle in *Winter barley* [2]; *Winter wheat* [2];
- Shepherd's purse in *Winter barley* [2]; *Winter wheat* [2];

Approval information
- Fluroxypyr, metsulfuron-methyl and thifensulfuron-methyl included in Annex I under EC Regulation 1107/2009

Efficacy guidance
- Apply in a minimum of 150 litres of water per hectare.

Restrictions
- Do not apply to any crop suffering from stress as a result of drought, waterlogging, low temperatures, pest or disease attack, nutrient or lime deficiency or other factors reducing crop growth.
- Do not use on cereal crops undersown with grasses, clover or other legumes or any other broad-leaved crop.
- Take special care to avoid damage by drift onto broad-leaved plants outside the target area, or onto ponds, waterways or ditches. Thorough cleansing of equipment is also very important.
- Do not apply within 7 days of rolling the crop.

Following crops guidance
- Cereals, oilseed rape, field beans or grass may be sown in the same calendar year as harvest of a cereal crop. In case of crop failure for any reason, sow only wheat within three months of application. Before sowing, soil should be ploughed and cultivated to a depth of at least 15cm.

Environmental safety
- LERAP Category B [1-2]

Hazard classification and safety precautions
Transport code 9 [1-2]
Packaging group III
UN Number 3082
Risk phrases H317
Operator protection U05b [2], U09a [2], U11 [2], U20a [2]; A, H
Environmental protection E15b [2], E16a, H410
Storage and disposal D01 [2], D02 [2], D09a [2], D10c [2], D12a [2]

205　fluroxypyr + triclopyr

A foliar acting herbicide for docks in grassland
HRAC mode of action code: 4 + 4 (O + O)

See also triclopyr

Products

1	Doxstar Pro	Corteva	150:150 g/l	EC	15664
2	Pivotal	Corteva	150:150 g/l	EC	16943

Uses
- Docks in *Amenity grassland* [1-2]; *Grassland* [1-2];

Approval information
- Fluroxypyr and triclopyr included in Annex I under EC Regulation 1107/2009

Efficacy guidance
- Seedling docks in established grass only are controlled up to 50 mm diameter. Apply in spring or autumn or, at lower dose, in spring and autumn on docks up to 200 mm. A second application in the subsequent yr may be needed
- Allow 2-3 wk after cutting or grazing to allow sufficient regrowth of docks to occur before spraying
- Control may be reduced if rain falls within 2 h of application
- To allow maximum translocation to the roots of docks do not cut grass for 28 d after spraying

SEE SECTION 3 FOR PRODUCTS ALSO REGISTERED

- [1] available as a twin pack with Thistlex as Pas.Tor for control of docks, nettles and thistles in grassland

Restrictions
- Maximum total dose equivalent to one full dose treatment
- Do not roll or harrow for 10 d before or 7 d after spraying
- Do not spray in drought, very hot or very cold weather

Crop-specific information
- Latest use: 7 d before grazing or harvest of grass
- Clover will be killed or severely checked by treatment

Following crops guidance
- Do not sow kale, turnips, swedes or grass mixtures containing clover by direct drilling or minimum cultivation techniques within 6 wk of application

Environmental safety
- Dangerous for the environment
- Toxic to aquatic organisms
- Keep livestock out of treated areas for at least 7 d following treatment and until poisonous weeds, such as ragwort, have died down and become unpalatable
- Do not allow drift to come into contact with crops, amenity plantings, gardens, ponds, lakes or watercourses
- Wash spray equipment thoroughly with water and detergent immediately after use. Traces of product can damage susceptible plants sprayed later
- LERAP Category B [1-2]

Hazard classification and safety precautions
> **Hazard** Irritant, Dangerous for the environment
> **Transport code** 3 [1-2]
> **Packaging group** III
> **UN Number** 1993
> **Risk phrases** H317, H373
> **Operator protection** U14; A
> **Environmental protection** E16a, E38, H410
> **Storage and disposal** D01, D02, D05, D12a
> **Medical advice** M05a

206 flutolanil

An carboxamide fungicide for treatment of potato seed tubers
FRAC mode of action code: 7

Products

1 Rhino	Certis Belchim B V	460 g/l	SC	14311
2 Rhino DSG	Certis Belchim B V	6% w/w	DS	19730

Uses
- Black scurf in **Potatoes** *(tuber treatment)* [1-2];
- Stem canker in **Potatoes** *(tuber treatment)* [1-2];

Approval information
- Flutolanil included in Annex I under EC Regulation 1107/2009

Efficacy guidance
- Apply to clean tubers before chitting, prior to planting, or at planting
- Apply flowable concentrate through canopied, hydraulic or spinning disc equipment (with or without electrostatics) mounted on a rolling conveyor or table [1]
- Flowable concentrate may be diluted with water up to 2.0 l per tonne to improve tuber coverage. Disease in areas not covered by spray will not be controlled [1]
- Dry powder may be applied via an on-planter applicator [2]

FOR FULL CONDITIONS OF USE ALWAYS READ THE PRODUCT LABEL

Restrictions
- Maximum number of treatments 1 per batch of seed tubers
- Check with processor before use on crops for processing

Crop-specific information
- Latest use: at planting
- Seed tubers should be of good quality and free from bacterial rots, physical damage or virus infection, and should not be sprouted to such an extent that mechanical damage to the shoots will occur during treatment or planting

Environmental safety
- Harmful to aquatic organisms

Hazard classification and safety precautions
Hazard Irritant, May be harmful if swallowed [1]
Transport code 9 [1]
Packaging group III
UN Number 3082, N/C
Risk phrases H317 [1], H319, R53a [2]
Operator protection U04a, U05a, U14 [1], U19a, U20a; A, C, H
Environmental protection E15a, E34, E38, H411 [1], H412 [2]
Storage and disposal D01, D02, D05 [1], D09a, D10a [1], D11a [2], D12a
Treated seed S01, S02, S03, S04a, S05
Medical advice M03

207   flutriafol

A broad-spectrum triazole fungicide for cereals
FRAC mode of action code: 3

See also fludioxonil + flutriafol

Products

1 Consul (GB only)	FMC Agro	125 g/l	SC	19597
2 Impact (GB only)	FMC Agro	125 g/l	SC	19592
3 Topguard (GB only)	FMC Agro	125 g/l	SC	19596

Uses
- Brown rust in *Rye* [2]; *Spring wheat* [2]; *Triticale* [2]; *Winter rye* [1,3]; *Winter triticale* [1,3]; *Winter wheat* [1-3];
- Powdery mildew in *Rye* [2]; *Spring wheat* [2]; *Sugar beet* [1-3]; *Triticale* [2]; *Winter rye* [1,3]; *Winter triticale* [1,3]; *Winter wheat* [1-3];
- Rust in *Sugar beet* [1-3];
- Septoria leaf blotch in *Rye* [2]; *Spring wheat* [2]; *Triticale* [2]; *Winter rye* [1,3]; *Winter triticale* [1,3]; *Winter wheat* [1-3];
- Yellow rust in *Rye* [2]; *Spring wheat* [2]; *Triticale* [2]; *Winter rye* [1,3]; *Winter triticale* [1,3]; *Winter wheat* [1-3];

Approval information
- Flutriafol included in Annex I under EC Regulation 1107/2009
- Accepted by BBPA for use on malting barley

Efficacy guidance
- Best results obtained from treatment in early stages of disease development. See label for detailed guidance on spray timing for specific diseases and the need for repeat treatments
- Tank mix options available on the label to broaden activity spectrum
- Good spray coverage essential for optimum performance
- Flutriafol is a DMI fungicide. Resistance to some DMI fungicides has been identified in Septoria leaf blotch which may seriously affect performance of some products. For further advice contact a specialist advisor and visit the Fungicide Resistance Action Group (FRAG)-UK website

SEE SECTION 3 FOR PRODUCTS ALSO REGISTERED

SECTION 2

Restrictions
- Maximum number of treatments 2 per crop (including other products containing flutriafol)
- Approval expired 30/11/2022 in Northern Ireland

Crop-specific information
- Latest use: before early grain milky ripe stage (GS 73)
- Flag leaf tip scorch on wheat caused by stress may be increased by fungicide treatment

Environmental safety
- Harmful to aquatic organisms
- Harmful to fish or other aquatic life. Do not contaminate surface waters or ditches with chemical or used container

Hazard classification and safety precautions
Hazard Harmful
Transport code 9 [2]
Packaging group III
UN Number 3082, N/C
Risk phrases H317
Operator protection U05a, U09a, U19a, U20b; A, H
Environmental protection E13c, E38, H411
Storage and disposal D01, D02, D09a, D10c

208 fluxapyroxad

Also known as Xemium, it is an SDHI fungicide for use in cereals.
FRAC mode of action code: 7

See also difenoconazole + fluxapyroxad
fluxapyroxad + metconazole

Products

1	Allstar	BASF	300 g/l	SC	19717
2	Bugle	BASF	59.4 g/l	EC	17821
3	Honesty	BASF	300 g/l	SC	19666
4	Imperis XE	BASF	62.5 g/l	EC	19310
5	Imtrex	BASF	62.5 g/l	EC	17108
6	Sercadis	BASF	300 g/l	SC	19716

Uses
- Botrytis in **Ornamental plant production** *(off-label)* [6]; **Protected ornamentals** *(off-label)* [6];
- Brown rust in **Durum wheat** [2,4-5]; **Rye** [2,4-5]; **Spring barley** [2,4-5]; **Spring oats** [4]; **Spring wheat** [2,4-5]; **Triticale** [2,4-5]; **Winter barley** [2,4-5]; **Winter oats** [4]; **Winter wheat** [2,4-5];
- Crown rust in **Spring oats** [2]; **Winter oats** [2,5];
- Disease control in **Potatoes** [1]; **Seed potatoes** [3];
- Dollar spot in **Managed amenity turf** *(off-label)* [6];
- Net blotch in **Spring barley** [2,4-5]; **Spring oats** [4]; **Winter barley** [2,4-5]; **Winter oats** [4];
- Powdery mildew in **Apples** [6]; **Durum wheat** *(reduction)* [4-5]; **Nectarines** *(off-label)* [6]; **Ornamental plant production** *(off-label)* [6]; **Peaches** *(off-label)* [6]; **Pears** [6]; **Protected ornamentals** *(off-label)* [6]; **Rye** *(reduction)* [4-5]; **Spring wheat** *(reduction)* [4-5]; **Triticale** *(reduction)* [4-5]; **Wine grapes** *(off-label)* [6]; **Winter barley** *(reduction)* [4-5]; **Winter oats** *(reduction)* [5]; **Winter wheat** *(reduction)* [4-5];
- Rhynchosporium in **Rye** [2,4-5]; **Spring barley** [2,4-5]; **Spring oats** [4]; **Winter barley** [2,4-5]; **Winter oats** [4];
- Ring spot in **Protected oriental cabbage** *(off-label)* [6];
- Scab in **Apples** [6]; **Nectarines** *(off-label)* [6]; **Peaches** *(off-label)* [6]; **Pears** [6];
- Septoria leaf blotch in **Durum wheat** [2,4-5]; **Spring wheat** [2,4-5]; **Triticale** [2,4-5]; **Winter wheat** [2,4-5];
- Smoulder in **Ornamental plant production** *(off-label)* [6]; **Protected ornamentals** *(off-label)* [6];
- Tan spot in **Durum wheat** *(reduction)* [4-5]; **Spring wheat** *(reduction)* [4-5]; **Winter wheat** *(reduction)* [4-5];

FOR FULL CONDITIONS OF USE ALWAYS READ THE PRODUCT LABEL

- Yellow rust in **Durum wheat** *(moderate control)* [2,4-5]; **Rye** *(moderate control)* [2,4-5]; **Spring barley** *(moderate control)* [2,4-5]; **Spring oats** *(moderate control)* [4]; **Spring wheat** *(moderate control)* [2,4-5]; **Triticale** *(moderate control)* [2,4-5]; **Winter barley** *(moderate control)* [2,4-5]; **Winter oats** *(moderate control)* [4]; **Winter wheat** *(moderate control)* [2,4-5];

Extension of Authorisation for Minor Use (EAMUs)
- *Managed amenity turf* 20222649 [6]
- *Nectarines* 20211302 [6]
- *Ornamental plant production* 20211303 [6]
- *Peaches* 20211302 [6]
- *Protected oriental cabbage* 20220749 [6]
- *Protected ornamentals* 20211303 [6]
- *Wine grapes* 20211301 [6]

Approval information
- Fluxapyroxad included in Annex I under EC Regulation 1107/2009
- Accepted by BBPA for use on malting barley up to GS 45 only (max. dose, 1 litre/ha)

Restrictions
- Do not apply more that two foliar applications of products containing SDHI fungicides to any cereal crop.
- Do not apply by hand-held equipment

Following crops guidance
- Only cereals, cabbages, carrots, chicory, clover, dwarf french beans, field beans, leeks, lettuce, linseed, maize, oats, oilseed rape, onions, peas, potatoes, radishes, ryegrass, soyabean, spinach, sunflowers or sugar beet may be sown as following crops after treatment

Hazard classification and safety precautions
Hazard Harmful [4-6], Dangerous for the environment [2,4-6], Harmful if inhaled [2,4-5], Very toxic to aquatic organisms [1,3,6]
Transport code 9 [1-6]
Packaging group III
UN Number 3082
Risk phrases H317 [3], H317 (15 m) [6], H319 [2,4-5], H351 [1-2,4-5], H362 [3,6]
Operator protection U05a [2,4-6], U20b [2,4-6], U23a [1,3]; A, H
Environmental protection E15b [2,4-6], E17a (15 m) [6], E34 [2,4-6], E38 [2,4-6], H410 [1,3,6], H411 [2,4-5]
Storage and disposal D01 [2,4-6], D02 [2,4-6], D05 [2,4-6], D09a [2,4-6], D10c [2,4-6], D12a [2,4-6]
Medical advice M05a [2,4-6]

209   fluxapyroxad + mefentrifluconazole

A fungicide mixture fr disease control in cereals
FRAC mode of action code: 3 + 7

See also mefentrifluconazole

Products

1	Aderya XE	BASF	63.3:70 g/l	EC	19302
2	Diadem XE	BASF	47.5:100 g/l	EC	19526
3	Lentyma XE	BASF	66.7:70 g/l	EC	19301
4	Mivyto XE	BASF	47.5:100 g/l	EC	19259
5	Provyto XE	BASF	47.5:100 g/l	EC	19585
6	Revystar XE	BASF	47.5:100 g/l	EC	19250
7	Verydor XE	BASF	52.5:100 g/l	EC	19251

Uses
- Brown rust in **Durum wheat** [1-7]; **Rye** *(moderate control)* [1-7]; **Spelt** [1-7]; **Spring barley** [1-7]; **Spring wheat** [1-7]; **Triticale** [1-7]; **Winter barley** [1-7]; **Winter wheat** [1-7];

SEE SECTION 3 FOR PRODUCTS ALSO REGISTERED

- Crown rust in **Spring oats** *(moderate control)* [1-7]; **Winter oats** *(moderate control)* [1-7];
- Net blotch in **Spring barley** *(reduction)* [1-7]; **Winter barley** *(reduction)* [1-7];
- Powdery mildew in **Durum wheat** *(moderate control)* [1-7]; **Rye** *(moderate control)* [1-7]; **Spelt** *(moderate control)* [1-7]; **Spring oats** [2,4-7]; **Spring wheat** *(moderate control)* [1-7]; **Triticale** *(moderate control)* [1-7]; **Winter oats** [2,4-7]; **Winter wheat** *(moderate control)* [1-7];
- Ramularia leaf spots in **Spring barley** [1-7]; **Winter barley** [1-7];
- Rhynchosporium in **Rye** *(moderate control)* [1-7]; **Spring barley** [1-7]; **Winter barley** [1-7];
- Septoria leaf blotch in **Durum wheat** [1-7]; **Rye** *(moderate control)* [1-7]; **Spelt** [1-7]; **Spring wheat** [1-7]; **Triticale** [1-7]; **Winter wheat** [1-7];
- Yellow rust in **Durum wheat** [1-7]; **Rye** *(moderate control)* [1-7]; **Spelt** [1-7]; **Spring barley** [1-7]; **Spring wheat** [1-7]; **Triticale** [1-7]; **Winter barley** [1-7]; **Winter wheat** [1-7];

Approval information
- Fluxapyroxad and mefentrifluconazole included in Annex I under EC Regulation 1107/2009

Efficacy guidance
- While it is a systemic fungicide with protectant and curative properties, it should be used preventatively and should not be relied upon for its curative potential.
- There is a significant risk of widespread resistance occurring in Septoria tritici and Ramularia populations in the UK. Failure to follow resistance management action may result in reduced levels of disease control

Restrictions
- Application must not be made before beginning of stem elongation (GS30)
- Do not apply more than two foliar applications of products containing complex II inhibitors (SDHIs) to any cereal crop

Environmental safety
- LERAP Category B [1-7]

Hazard classification and safety precautions
    Hazard Harmful if swallowed [2,4-7], Harmful if inhaled
    Transport code 9 [1-7]
    Packaging group III
    UN Number 3082
    Risk phrases H315, H317, H318, H335, H351
    Operator protection U05a, U11 [6], U20c; A, C, H
    Environmental protection E15b, E16a, H411
    Storage and disposal D01, D02, D05, D08, D09a, D10c
    Medical advice M03

210   fluxapyroxad + metconazole

An SDHI and triazole mixture for disease control in cereals
FRAC mode of action code: 3 + 7

Products

1	Clayton Tardis	Clayton	62.5:45 g/l	EC	17459
2	Vastimo	FMC Agro	62.5:45 g/l	EC	19549
3	Wolverine	FMC Agro	62.5:45 g/l	EC	19374

Uses
- Brown rust in **Durum wheat** [1-3]; **Rye** [1-3]; **Spring barley** [1-3]; **Spring wheat** [1-3]; **Triticale** [1-3]; **Winter barley** [1-3]; **Winter wheat** [1-3];
- Cladosporium in **Durum wheat** *(moderate control)* [1-3]; **Spring wheat** *(moderate control)* [1-3]; **Winter wheat** *(moderate control)* [1-3];
- Disease control in **Durum wheat** [1-3]; **Rye** [1-3]; **Spring barley** [1-3]; **Spring wheat** [1-3]; **Triticale** [1-3]; **Winter barley** [1-3]; **Winter wheat** [1-3];
- Eyespot in **Durum wheat** *(good reduction)* [1-3]; **Rye** *(good reduction)* [1-3]; **Spring wheat** *(good reduction)* [1-3]; **Triticale** *(good reduction)* [1-3]; **Winter wheat** *(good reduction)* [1-3];

- Fusarium ear blight in **Durum wheat** *(good reduction)* [1-3]; **Spring wheat** *(good reduction)* [1-3]; **Winter wheat** *(good reduction)* [1-3];
- Net blotch in **Spring barley** [1-3]; **Winter barley** [1-3];
- Powdery mildew in **Durum wheat** *(moderate control)* [1-3]; **Rye** *(moderate control)* [1-3]; **Spring barley** *(moderate control)* [1-3]; **Spring wheat** *(moderate control)* [1-3]; **Triticale** *(moderate control)* [1-3]; **Winter barley** *(moderate control)* [1-3]; **Winter wheat** *(moderate control)* [1-3];
- Ramularia leaf spots in **Spring barley** [1-3]; **Winter barley** [1-3];
- Rhynchosporium in **Rye** [1-3]; **Spring barley** [1-3]; **Winter barley** [1-3];
- Septoria leaf blotch in **Durum wheat** [1-3]; **Spring wheat** [1-3]; **Triticale** [1-3]; **Winter wheat** [1-3];
- Tan spot in **Durum wheat** *(good reduction)* [1-3]; **Spring wheat** *(good reduction)* [1-3]; **Winter wheat** *(good reduction)* [1-3];
- Yellow rust in **Durum wheat** [1-3]; **Rye** [1-3]; **Spring barley** [1-3]; **Spring wheat** [1-3]; **Triticale** [1-3]; **Winter barley** [1-3]; **Winter wheat** [1-3];

Approval information
- Fluxapyroxad and metconazole included in Annex I under EC Regulation 1107/2009

Efficacy guidance
- Metconazole is a DMI fungicide. Resistance to some DMI fungicides has been identified in Septoria leaf blotch which may seriously affect performance of some products. For further advice contact a specialist advisor and visit the Fungicide Resistance Action Group (FRAG)-UK website

Restrictions
- Do not apply more that two foliar applications of products containing SDHI fungicides to any cereal crop.
- To protect birds, only one application is allowed on cereals before GS 29 (end of tillering)

Following crops guidance
- Only cereals, cabbages, carrots, clover, dwarf french beans, field beans, lettuce, maize, oats, oilseed rape, onions, peas, potatoes, ryegrass, sunflowers or sugar beet may be sown as following crops after treatment

Environmental safety
- LERAP Category B [1-3]

Hazard classification and safety precautions
  Hazard Harmful, Dangerous for the environment, Harmful if inhaled, Very toxic to aquatic organisms
  Transport code 9 [1-3]
  Packaging group III
  UN Number 3082
  Risk phrases H317, H319, H351, H361
  Operator protection U05a, U20b; A, C, H
  Environmental protection E15b, E16a, E34, E38, H410
  Storage and disposal D01, D02, D09a, D10c, D12a
  Medical advice M03

211   fluxapyroxad + pyraclostrobin

A fungicide mixture for disease control in cereals
FRAC mode of action code: 7 + 11

Products

1	Priaxor EC	BASF	75:150 g/l	EC	17371
2	Serpent	BASF	75:150 g/l	EC	17427
3	Syrex	BASF	75:150 g/l	EC	18908
4	Tevos	BASF	75:142.5 g/l	EC	19789

SEE SECTION 3 FOR PRODUCTS ALSO REGISTERED

Uses

- Brown rust in *Durum wheat* [1]; *Rye* [1]; *Spring barley* [1-4]; *Spring wheat* [1]; *Triticale* [1]; *Winter barley* [1-4]; *Winter wheat* [1];
- Crown rust in *Spring oats* [1]; *Winter oats* [1];
- Glume blotch in *Durum wheat* *(moderate control)* [1]; *Spring wheat* *(moderate control)* [1]; *Winter wheat* *(moderate control)* [1];
- Net blotch in *Spring barley* [1-4]; *Winter barley* [1-4];
- Powdery mildew in *Durum wheat* *(moderate control)* [1]; *Rye* *(moderate control)* [1]; *Spring barley* *(moderate control)* [1-4]; *Spring oats* *(moderate control)* [1]; *Spring wheat* *(moderate control)* [1]; *Triticale* *(moderate control)* [1]; *Winter barley* *(moderate control)* [1-4]; *Winter oats* *(moderate control)* [1]; *Winter wheat* *(moderate control)* [1];
- Ramularia leaf spots in *Spring barley* [1-4]; *Winter barley* [1-4];
- Rhynchosporium in *Rye* [1]; *Spring barley* [1-4]; *Winter barley* [1-4];
- Septoria leaf blotch in *Durum wheat* [1]; *Spring wheat* [1]; *Triticale* [1]; *Winter wheat* [1];
- Tan spot in *Durum wheat* [1]; *Spring wheat* [1]; *Winter wheat* [1];
- Yellow rust in *Durum wheat* [1,3]; *Rye* [1,3-4]; *Spring barley* [1-4]; *Spring oats* [3-4]; *Spring wheat* [1,3]; *Triticale* [1,3-4]; *Winter barley* [1-4]; *Winter oats* [3-4]; *Winter wheat* [1,3];

Approval information

- Fluxapyroxad and pyraclostrobin included in Annex I under EC Regulation 1107/2009

Environmental safety

- LERAP Category B [1-4]

Hazard classification and safety precautions

Hazard Harmful if swallowed, Harmful if inhaled, Very toxic to aquatic organisms
Transport code 9 [1-4]
Packaging group III
UN Number 3082
Risk phrases H335 [3-4], H351 [1-3], H362 [4]
Operator protection U05a, U20a; A, H
Environmental protection E15b, E16a, H410
Storage and disposal D01, D02, D10c

212   folpet

A multi-site protectant fungicide for disease control in wheat and barley
FRAC mode of action code: M4

See also epoxiconazole + folpet

Products

1	Arizona	Adama	500 g/l	SC	15318
2	Clayton Canyon	Clayton	500 g/l	SC	18414
3	Colorado	Adama	500 g/l	SC	19346
4	Mojave	Adama	500 g/l	SC	19349
5	Phoenix	Adama	500 g/l	SC	15259

Uses

- Rhynchosporium in *Spring barley* *(reduction)* [1-5]; *Triticale* *(reduction)* [5]; *Winter barley* *(reduction)* [1-5];
- Septoria leaf blotch in *Spring wheat* *(reduction)* [1-5]; *Triticale* *(reduction)* [1,3-5]; *Winter wheat* *(reduction)* [1-5];

Approval information

- Folpet included in Annex I under EC Regulation 1107/2009
- Accepted by BBPA for use on cereals

Efficacy guidance

- Folpet is a protectant fungicide and so the first application must be made before the disease becomes established in the crop. A second application should be timed to protect new growth.

- Folpet only has contact activity and so good coverage of the target foliage is essential for good activity.

Environmental safety
- LERAP Category B [1-5]

Hazard classification and safety precautions
   Hazard Harmful, Dangerous for the environment, Very toxic to aquatic organisms
   Transport code 9 [1-5]
   Packaging group III
   UN Number 3082
   Risk phrases H317, H351, R40 [2]
   Operator protection U05a, U08, U11, U14, U15, U20a; A, H
   Environmental protection E15b, E16a, E34, E38, H410
   Storage and disposal D01, D02, D09a, D12a
   Medical advice M03

213   foramsulfuron + iodosulfuron-methyl-sodium

A sulfonylurea herbicide mixture for weed control in forage and grain maize.
HRAC mode of action code: 2 (B)

See also iodosulfuron-methyl-sodium

Products

1	Maister WG	Bayer CropScience	0.3:0.01% w/w	SG	20033

Uses
- Annual dicotyledons in *Forage maize* [1]; *Grain maize* [1];
- Annual meadow grass in *Forage maize* [1]; *Grain maize* [1];
- Black nightshade in *Forage maize* [1]; *Grain maize* [1];
- Chickweed in *Forage maize* [1]; *Grain maize* [1];
- Cockspur grass in *Forage maize* [1]; *Grain maize* [1];
- Couch in *Forage maize* [1]; *Grain maize* [1];
- Fat hen in *Forage maize* [1]; *Grain maize* [1];
- Knotgrass in *Forage maize* [1]; *Grain maize* [1];
- Mayweeds in *Forage maize* [1]; *Grain maize* [1];
- Shepherd's purse in *Forage maize* [1]; *Grain maize* [1];

Approval information
- Foramsulfuron and iodosulfuron-methyl-sodium included in Annex 1 under EC Regulation 1107/2009

Efficacy guidance
- Do not spray if rain is imminent or if temperature is above 25°C.
- Do not use in a non-CRD approved tank-mixture or sequence with another ALS or sulfonyl urea herbicide.
- Do not use on crops suffering from stress.
- Do not use on crops undersown with grass or a broad-leaved crop.

Environmental safety
- Buffer zone requirement 11m [1]
- LERAP Category B [1]

Hazard classification and safety precautions
   Hazard Irritant, Dangerous for the environment, Very toxic to aquatic organisms
   Transport code 9 [1]
   Packaging group III
   UN Number 3077
   Risk phrases H317, H351
   Operator protection U05a, U14; A, D, H

SECTION 2

SEE SECTION 3 FOR PRODUCTS ALSO REGISTERED

Environmental protection E15b, E16a, E38, H410
Storage and disposal D01, D02, D10a, D12a

214   foramsulfuron + thiencarbazone-methyl

Herbicide mixture for weed control in ALS-resistant sugar beet varieties
HRAC mode of action code: 2 (B)

Products

1	Conviso One	Bayer CropScience	50:30 g/l	OD	19036

Uses
- Annual dicotyledons in *Fodder beet* [1]; *Sugar beet* [1];
- Volunteer sugar beet in *Fodder beet* [1]; *Sugar beet* [1];

Approval information
- Foramsulfuron and thiencarbazone-methyl included in Annex I under EC Regulation 1107/2009

Efficacy guidance
- Application when the temperature is above 25°C under conditions of high light intensity and low water supply may cause phytotoxic symptoms to the crop - wait until the cool of the evening.

Restrictions
- Must only be used on Conviso Smart sugar beet hybrids

Following crops guidance
- After normal harvest plough or cultivate to 20cms before sowing wheat in the same year as application or before sowing wheat, barley, maize, sunflowers, peas or ryegrass the following spring.
- In the event of crop failure, Conviso Smart beet can be re-sown or, after ploughing or cultivating to 20 cms, maize may be sown 1 month after application and wheat may be sown 4 months after application.

Environmental safety
- Buffer zone requirement 10m

Hazard classification and safety precautions
Hazard Harmful if inhaled, Very toxic to aquatic organisms
Transport code 9 [1]
Packaging group III
UN Number 3082
Risk phrases H304, H315, H317, H318
Operator protection U05a, U12, U14, U20a; A, C, H
Environmental protection H410
Storage and disposal D05, D09a, D10b
Medical advice M03

215   fosetyl-aluminium + propamocarb hydrochloride

A systemic and protectant fungicide mixture for use in horticulture
FRAC mode of action code: 33 + 28

See also propamocarb hydrochloride

Products

1	Previcur Energy	Bayer CropScience	310:530 g/l	SL	15367

Uses
- Damping off in *Herbs (see appendix 6)* *(off-label)* [1]; *Protected aubergines* *(off-label)* [1]; *Protected broccoli* [1]; *Protected Brussels sprouts* [1]; *Protected cabbages* [1]; *Protected calabrese* [1]; *Protected cauliflowers* [1]; *Protected chilli peppers* *(off-label)* [1]; *Protected*

FOR FULL CONDITIONS OF USE ALWAYS READ THE PRODUCT LABEL

Chinese cabbage [1]; *Protected collards* [1]; *Protected courgettes* *(off-label)* [1]; *Protected forest nurseries* *(off-label)* [1]; *Protected gherkins* *(off-label)* [1]; *Protected hops* *(off-label)* [1]; *Protected kale* [1]; *Protected marrows* *(off-label)* [1]; *Protected melons* [1]; *Protected pumpkins* *(off-label)* [1]; *Protected radishes* [1]; *Protected soft fruit* *(off-label)* [1]; *Protected squashes* *(off-label)* [1]; *Protected sweet peppers* *(off-label)* [1]; *Protected top fruit* *(off-label)* [1]; *Protected watermelon* *(off-label)* [1];

- Downy mildew in *Herbs (see appendix 6)* *(off-label)* [1]; *Lettuce* [1]; *Ornamental plant production* *(off-label)* [1]; *Protected broccoli* [1]; *Protected Brussels sprouts* [1]; *Protected cabbages* [1]; *Protected calabrese* [1]; *Protected cauliflowers* [1]; *Protected Chinese cabbage* [1]; *Protected collards* [1]; *Protected forest nurseries* *(off-label)* [1]; *Protected hops* *(off-label)* [1]; *Protected kale* [1]; *Protected lettuce* [1]; *Protected melons* [1]; *Protected ornamentals* *(off-label)* [1]; *Protected radishes* [1]; *Protected soft fruit* *(off-label)* [1]; *Protected top fruit* *(off-label)* [1]; *Spinach* *(off-label)* [1];
- Phytophthora root rot in *Chicory* *(off-label)* [1]; *Witloof* *(off-label)* [1];
- Pythium in *Lettuce* [1]; *Protected cucumbers* [1]; *Protected lettuce* [1]; *Protected ornamentals* *(off-label)* [1]; *Protected tomatoes* [1];

Extension of Authorisation for Minor Use (EAMUs)
- *Chicory* 20193825 [1]
- *Herbs (see appendix 6)* 20130983 [1]
- *Ornamental plant production* 20131845 [1]
- *Protected aubergines* 20111553 [1]
- *Protected chilli peppers* 20111553 [1]
- *Protected courgettes* 20111556 [1]
- *Protected forest nurseries* 20122045 [1]
- *Protected gherkins* 20111556 [1]
- *Protected hops* 20122045 [1]
- *Protected marrows* 20111556 [1]
- *Protected ornamentals* 20111557 [1], 20131845 [1]
- *Protected pumpkins* 20111556 [1]
- *Protected soft fruit* 20122045 [1]
- *Protected squashes* 20111556 [1]
- *Protected sweet peppers* 20111553 [1]
- *Protected top fruit* 20122045 [1]
- *Protected watermelon* 20111556 [1]
- *Spinach* 20112452 [1]
- *Witloof* 20193825 [1]

Approval information
- Fosetyl-aluminium and propamocarb hydrochloride included in Annex I under EC Regulation 1107/2009

Hazard classification and safety precautions
Hazard Irritant
UN Number N/C
Risk phrases H317
Operator protection U04a, U05a, U08, U14, U19a, U20b; A, H, M
Environmental protection E15a
Storage and disposal D09a, D11a
Medical advice M03

216   fosthiazate

An organophosphorus contact nematicide for potatoes
IRAC mode of action code: 1B

Products

1	Nemathorin 10G	Syngenta	10% w/w	FG	11003

SEE SECTION 3 FOR PRODUCTS ALSO REGISTERED

Uses

- Nematodes in **Hops** *(off-label)* [1]; **Ornamental plant production** *(off-label)* [1]; **Soft fruit** *(off-label)* [1];
- Potato cyst nematode in **Potatoes** [1];
- Spraing vectors in **Potatoes** *(reduction)* [1];
- Wireworm in **Potatoes** *(reduction)* [1];

Extension of Authorisation for Minor Use (EAMUs)

- **Hops** *20082912* [1]
- **Ornamental plant production** *20082912* [1]
- **Soft fruit** *20082912* [1]

Approval information

- Fosthiazate included in Annex I under EC Regulation 1107/2009
- In 2006 CRD required that all products containing this active ingredient should carry the following warning in the main area of the container label: "Fosthiazate is an anticholinesterase organophosphate. Handle with care"

Efficacy guidance

- Application best achieved using equipment such as Horstine Farmery Microband Applicator, Matco or Stocks Micrometer applicators together with a rear mounted powered rotary cultivator
- Granules must not become wet or damp before use.
- For optimum efficacy incorporate evenly in to top 20 cm of soil.
- Apply as close as possible to time of planting.

Restrictions

- Contains an anticholinesterase organophosphorus compound. Do not use if under medical advice not to work with such compounds
- Maximum number of treatments 1 per crop
- Product must only be applied using tractor-mounted/drawn direct placement machinery. Do not use air assisted broadcast machinery other than that referenced on the product label
- Do not allow granules to stand overnight in the application hopper
- Do not apply more than once every four years on the same area of land
- Do not use on crops to be harvested less than 17 wk after treatment
- Consult before using on crops intended for processing

Crop-specific information

- Latest use: at planting
- HI 17 wk for potatoes

Environmental safety

- Dangerous for the environment
- Toxic to aquatic organisms
- Dangerous to game, wild birds and animals
- Dangerous to livestock. Keep all livestock out of treated areas for at least 13 wk
- Incorporation to 10-15 cm and ridging up of treated soil must be carried out immediately after application. Powered rotary cultivators are preferred implements for incorporation but discs, power, spring tine or Dutch harrows may be used provided two passes are made at right angles
- To protect birds and wild mammals remove spillages
- Failure completely to bury granules immediately after application is hazardous to wildlife
- To protect groundwater do not apply any product containing fosthiazate more than once every four yr

Hazard classification and safety precautions

Hazard Harmful, Dangerous for the environment, Toxic if swallowed, Very toxic to aquatic organisms
Transport code 9 [1]
Packaging group III
UN Number 3077
Operator protection U02a, U04a, U05a, U09a, U13, U14, U19a, U20a; A, E, G, H, K, M
Environmental protection E06b (13 wk), E15b, E34, E38, H411

FOR FULL CONDITIONS OF USE ALWAYS READ THE PRODUCT LABEL

Storage and disposal D01, D02, D05, D09a, D11a, D14
Medical advice M01, M03, M05a

217   garlic extract

A naturally occuring nematicide and animal repellent

Products

1	EGC Liquid	Rigby Taylor	999 g/l	SL	17852
2	EGCA Granules	Rigby Taylor	45% w/w	GR	17233
3	NEMguard A PCN Granules	Certis Belchim B V	45% w/w	GR	17922
4	NEMguard DE	Certis Belchim B V	45% w/w	GR	16749

Uses
- Free-living nematodes in *Fodder beet* (off-label) [4]; *Red beet* (off-label) [4];
- Insect pests in *Managed amenity turf* [1-2];
- Nematodes in *Carrots* [4]; *Parsnips* [4];
- Potato cyst nematode in *Potatoes* [3];
- Root nematodes in *Bulb onions* (off-label) [4]; *Garlic* (off-label) [4]; *Leeks* (off-label) [4]; *Shallots* (off-label) [4];
- Root-knot nematodes in *Bulb onions* (off-label) [4]; *Garlic* (off-label) [4]; *Leeks* (off-label) [4]; *Shallots* (off-label) [4];
- Stem nematodes in *Bulb onions* (off-label) [4]; *Garlic* (off-label) [4]; *Leeks* (off-label) [4]; *Shallots* (off-label) [4];

Extension of Authorisation for Minor Use (EAMUs)
- *Bulb onions* 20222454 [4], 20222456 [4]
- *Fodder beet* 20222453 [4], 20222455 [4]
- *Garlic* 20222454 [4], 20222456 [4]
- *Leeks* 20222454 [4], 20222456 [4]
- *Red beet* 20222453 [4], 20222455 [4]
- *Shallots* 20222454 [4], 20222456 [4]

Approval information
- Garlic extract included in Annex 1 under EC Regulation 1107/2009

Efficacy guidance
- Efficacy against free living nematodes requires adequate soil moisture

Environmental safety
- LERAP Category B [1]

Hazard classification and safety precautions
   Hazard Irritant [4], Dangerous for the environment [4]
   UN Number , N/C
   Risk phrases H315, H320 [4]
   Operator protection U11 [4], U14 [4]; A, C, H
   Environmental protection E16a [1], E38 [4], H411 [4]

218   gibberellins

A plant growth regulator for use in top fruit and grassland

Products

1	Florgib Tablet	Fine	20.4% w/w	ST	18285
2	Gibb 3	Certis Belchim B V	10% w/w	TB	17013
3	Gibb Plus	Certis Belchim B V	10 g/l	SL	17251
4	Novagib	Fine	10 g/l	SL	18341
5	Regulex 10 SG	Sumitomo	10% w/w	SG	18890

SEE SECTION 3 FOR PRODUCTS ALSO REGISTERED

Products – continued

| 6 | Smartgrass | Sumitomo | 40% w/w | SG | 18883 |

Uses

- Fruit retention in **Cherries** *(off-label)* [1];
- Growth regulation in **Cherries** *(off-label)* [2]; **Grassland** [6]; **Ornamental plant production** *(off-label)* [1-2,4]; **Pears** [1]; **Protected cherries** *(off-label)* [2]; **Protected rhubarb** *(off-label)* [2]; **Rhubarb** [1]; **Wine grapes** [1];
- Improved germination in **Nothofagus** [5];
- Increasing fruit set in **Pears** [2];
- Reducing fruit russeting in **Apples** [3-4]; **Pears** *(off-label)* [3-4]; **Quinces** *(off-label)* [3-4];
- Russet reduction in **Apples** [5]; **Pears** [5];

Extension of Authorisation for Minor Use (EAMUs)

- **Cherries** *20192566* [1], *20192041* [2]
- **Ornamental plant production** *20194261* [2], *20200608* [4]
- **Pears** *20171152* [3], *20211161* [4]
- **Protected cherries** *20192041* [2]
- **Protected rhubarb** *20162836* [2]
- **Quinces** *20171152* [3], *20211161* [4]

Approval information

- Gibberellins included in Annex 1 under EC Regulation 1107/2009

Efficacy guidance

- For optimum results on apples spray under humid, slow drying conditions and ensure good spray cover
- Treat apples and pears immediately after completion of petal fall and repeat as directed on the label
- Fruit set in pears can be improved when blossom is spare, setting is poor or where frost has killed many flowers [2]
- The maximum concentration must not exceed 6 g of product per 1 litre water [1]
- Tablets must be placed whole into the sprayer. They must not be broken-up or crumbled [1]

Restrictions

- Maximum total dose varies with crop and product. Check label
- Prepared spray solutions are unstable. Do not leave in the sprayer during meal breaks or overnight
- Return bloom may be reduced in yr following treatment
- Consult processor before treating crops grown for processing [2]
- Avoid storage at temperatures above 32°C [2]
- Do not exceed 2.5 g per 5 litres when used as a seed soak in Nothofagus [5]
- Do not apply by hand-held equipment [4]
- Livestock must be kept out of treated areas for at least 14 days after treatment [6]

Crop-specific information

- HI zero for apples [5, 3]
- Apply to apples at completion of petal fall and repeat 3 or 4 times at 7-10 d intervals. Number of sprays and spray interval depend on weather conditions and dose (see labels)
- Good results achieved on apple varieties Cox's Orange Pippin, Discovery, Golden Delicious and Karmijn. For other cultivars test on a small number of trees
- Split treatment on pears allows second spray to be omitted if pollinating conditions become very favourable [2]
- Pear variety Conference usually responds well to treatment; Doyenne du Comice can be variable [2]

Hazard classification and safety precautions

    UN Number N/C
    Operator protection U05a [5], U08 [2], U20b [4], U20c [2-3,5]; A
    Environmental protection E15a [2-5]
    Storage and disposal D01 [4-5], D02 [5], D03 [4], D05 [4], D07 [4], D09a [2-5], D10b [3-5]

FOR FULL CONDITIONS OF USE ALWAYS READ THE PRODUCT LABEL

219   Gliocladium catenulatum

A fungus that acts as an antagonist against other fungi
FRAC mode of action code: 44

Products

1	Prestop	ICL (Everris) Ltd	32% w/w	WP	17223

Uses

- Didymella in *All protected edible crops* (*moderate control*) [1]; *All protected non-edible crops* (*moderate control*) [1]; *Strawberries* (*moderate control*) [1];
- Fusarium in *All protected edible crops* (*moderate control*) [1]; *All protected non-edible crops* (*moderate control*) [1]; *Strawberries* (*moderate control*) [1];
- Phytophthora in *All protected edible crops* (*moderate control*) [1]; *All protected non-edible crops* (*moderate control*) [1]; *Strawberries* (*moderate control*) [1];
- Pythium in *All protected edible crops* (*moderate control*) [1]; *All protected non-edible crops* (*moderate control*) [1]; *Strawberries* (*moderate control*) [1];
- Rhizoctonia in *All protected edible crops* (*moderate control*) [1]; *All protected non-edible crops* (*moderate control*) [1]; *Strawberries* (*moderate control*) [1];

Approval information

- Gliocladium catenulatum included in Annex 1 under EC Regulation 1107/2009

Restrictions

- When used via broadcast air-assisted sprayers, spray equipment must only be used where the operator?s normal working position is within a closed cab on a tractor or on a self-propelled sprayer

Hazard classification and safety precautions

UN Number N/C
Operator protection U05a, U10, U14, U15, U20a; A, D, H
Environmental protection E15b, E17a (10 m), E34
Storage and disposal D01, D02, D09a, D10a
Medical advice M03

220   glyphosate

A translocated non-residual glycine derivative herbicide
HRAC mode of action code: 9 (G)

See also 2,4-D + glyphosate
diflufenican + glyphosate
glyphosate + sulfosulfuron

Products

1	Amega Duo	Nufarm UK	540 g/l	SL	13358
2	Ardee XL	Barclay	360 g/l	SL	17515
3	Azural	Monsanto	360 g/l	SL	16239
4	Barbarian XL	Barclay	360 g/l	SL	17665
5	Barclay Gallup Biograde 360	Barclay	360 g/l	SL	17612
6	Barclay Gallup Biograde Amenity	Barclay	360 g/l	SL	17674
7	Barclay Gallup Hi-Aktiv	Barclay	490 g/l	SL	17646
8	Barclay Glyde 144	Barclay	144 g/l	SL	15362
9	Boom Efekt	Albaugh UK	360 g/l	SL	17588
10	Cleancrop Corral 2	Agrii	360 g/l	SL	18023
11	Cleancrop Tungsten Ultra	Agrii	450 g/l	SL	18297
12	Clinic UP	Nufarm UK	360 g/l	SL	17893
13	Clipper	Adama	360 g/l	SL	14820

SEE SECTION 3 FOR PRODUCTS ALSO REGISTERED

Products – continued

14	Credit	Nufarm UK	540 g/l	SL	16775
15	Crestler	Nufarm UK	540 g/l	SL	16555
16	Discman Biograde	Barclay	216 g/l	SL	12856
17	Ecoplug Max	Monsanto	72 % w/w	GR	17581
18	Fettle	Agrii	360 g/l	SL	19558
19	Gallup Hi-Aktiv Amenity	Barclay	490 g/l	SL	17681
20	Gallup XL	Barclay	360 g/l	SL	17663
21	Garryowen XL	Barclay	360 g/l	SL	17508
22	Giddyup	Pangaea	360 g/l	SL	20223
23	Glypho-Rapid 450	Barclay	450 g/l	SL	17647
24	Hilite	Nomix Enviro	20.1% w/w	RH	18352
25	Intercept	Pangaea	450 g/l	SL	20226
26	Landmaster 360 TF	Albaugh UK	360 g/l	SL	17683
27	Liaison	Monsanto	360 g/l	SL	17089
28	Mascot Hi-Aktiv Amenity	Rigby Taylor	490 g/l	SL	17696
29	Mentor	Monsanto	360 g/l	SL	16508
30	Monsanto Amenity Glyphosate	Monsanto	360 g/l	SL	16382
31	Monsanto Amenity Glyphosate XL	Monsanto	360 g/l	SL	17997
32	Motif	Monsanto	360 g/l	SL	16509
33	Nomix Conqueror Amenity	Nomix Enviro	144 g/l	RC	18369
34	Ovation	Monsanto	360 g/l	SL	17476
35	Rattler	Nufarm UK	540 g/l	SL	15522
36	Rodeo	Monsanto	360 g/l	SL	16242
37	Rosate 360 TF	Albaugh UK	360 g/l	SL	17682
38	Roundup Biactive	Monsanto	360 g/l	SL	10320
39	Roundup Biactive GL	Monsanto	360 g/l	SL	17348
40	Roundup Energy	Monsanto	450 g/l	SL	12945
41	Roundup Flex	Monsanto	480 g/l	SL	15541
42	Roundup Metro XL	Monsanto	360 g/l	SL	17684
43	Roundup POWERMAX	Monsanto	72% w/w	SG	16373
44	Roundup ProActive	Monsanto	360 g/l	SL	17380
45	Roundup ProVantage	Monsanto	480 g/l	SL	15534
46	Roundup Sonic	Monsanto	450 g/l	SL	17152
47	Roundup Star	Monsanto	450 g/l	SL	15470
48	Roundup Ultimate	Monsanto	450 g/l	SL	12774
49	Roundup Vista Plus	Monsanto	450 g/l	SL	18002
50	Rustler Pro-Green	ChemSource	360 g/l	SL	15798
51	Samurai	Monsanto	360 g/l	SL	16238
52	Scorpion	Monsanto	360 g/l	SL	17516
53	Snapper	Nufarm UK	550 g/l	SL	19037
54	Tanker	Nufarm UK	540 g/l	SL	15016
55	Touchdown Quattro	Syngenta	360 g/l	SL	10608
56	Trustee Amenity	Barclay	450 g/l	SL	17697

Uses

- Annual and perennial weeds in **All edible crops (outdoor and protected)** [14-15,23,54]; **All edible crops (outdoor)** *(before planting)* [1-2,4-5,7,10-13,18,20-21,26,35,37,39,41-42,44-45,50,53,55]; **All edible crops (stubble)** *(before planting)* [2,4-5,7,18,20-21,39,44]; **All non-edible crops (outdoor)** *(before planting)* [1-2,4-5,7,10-15,18,20-21,23,26,35,37,39,41-42,44-45,50,53-55]; **All non-edible crops (stubble)** *(before planting)* [2,4-5,7,18,20-21,39,44]; **Almonds** *(off-label)* [38-40,43]; **Amenity vegetation** [1,3,6,8,11,13-16,19,24,27-36,39,41-47,49-54,56]; **Apple orchards** [1-2,4-5,7,9-15,18,20-21,23,26,35,37,40-42,45-50,53-54]; **Apples** *(off-label)* [33,38-40,43]; **Apricots** *(off-label)* [38-40,43]; **Asparagus** *(off-label)* [2,4-5,7,12-14,18,20-21,26,37-39,41-45,50]; **Beetroot** *(off-label)* [40-41,43]; **Bilberries** *(off-label)* [38,40,43]; **Blackberries** *(off-label)* [40,43]; **Blackcurrants** *(off-label)* [38,40,43]; **Blueberries** *(off-label)*

[38,40,43]; *Buckwheat (off-label)* [40]; *Bulb onions* [1,12-15,35,39,41,44-45,53-54]; *Canary seed (off-label)* [40]; *Carrots (off-label)* [40-41,43]; *Cherries (off-label)* [1-5,7,9-15,18,20-22,25-27,29-30,32-54]; *Chestnuts (off-label)* [38-40,43]; *Christmas trees (off-label)* [44]; *Cob nuts (off-label)* [38]; *Combining peas* [1,12,14-15,23,35,39,41,44-45,49,53-55]; *Crab apples (off-label)* [39]; *Cranberries (off-label)* [38,40,43]; *Damsons* [15,38,40,42,48,50,54]; *Durum wheat* [1,12,14-15,23,35,39,41,44-45,53-54]; *Edible podded peas (off-label)* [40-41]; *Enclosed waters* [3,6,10-11,19,22-23,25,27-29,32-34,36,43,46-47,49,51-52,56]; *Farm forestry (off-label)* [3,10-11,22,25,27,29,32,34,36,38,40,43,46-47,49,51-52]; *Forest* [1-8,10-16,18-30,32-37,39,41-47,49-54,56]; *Forest nurseries* [1,13,30-31,53]; *Forestry plantations* [1,14-15,35,53-54]; *Game cover (off-label)* [38,40,43]; *Garlic (off-label)* [40-41,43]; *Gooseberries (off-label)* [38,40,43]; *Grapevines (off-label)* [38]; *Grassland* [1,3,12,14-15,22-25,27,29-30,32,34-36,44,46-47,49,51-54]; *Green cover on land temporarily removed from production* [1-5,7,12,14-15,22-25,27,29-30,32,34-36,39,41,43,45,51-54]; *Hard surfaces* [1-8,10-14,16,18-37,39,41-47,49-53,56]; *Hazel nuts (off-label)* [38-40,43]; *Hemp (off-label)* [38,40,43]; *Hops (off-label)* [38,40,43]; *Horseradish (off-label)* [40-41,43]; *Kentish cobnuts (off-label)* [39,43]; *Land immediately adjacent to aquatic areas* [2-8,10-11,16,19,22-23,25-29,32-34,36-39,42-44,46-47,49,51-52,56]; *Land not intended to bear vegetation* [38]; *Leeks (off-label)* [1,3,12-15,22-23,25,27,29-30,32,34-36,39-41,43-45,51-54]; *Linseed* [1,3,12,14-15,22-23,25,27,29-30,32,34-36,39,41,43-45,51-55]; *Loganberries (off-label)* [40,43]; *Lupins (off-label)* [40-41,43]; *Managed amenity turf (pre-establishment only)* [24,33]; *Medlar (off-label)* [40,43]; *Millet (off-label)* [40]; *Miscanthus (off-label)* [38,40-41,43]; *Mustard* [1,12,14-15,23,35,39,41,44-45,53-55]; *Natural surfaces not intended to bear vegetation* [1-8,10-16,18-37,39,41-47,49-54,56]; *Nectarines (off-label)* [38,40,43]; *Oilseed rape* [3,14,22,25,27,29-30,32,34,36,43,51-52]; *Onions (off-label)* [23,40-41,43]; *Open waters* [3,6,10-11,19,22-23,25,27-29,32-34,36,46-47,49,51-52,56]; *Ornamental plant production (off-label - as a directed spray)* [12,24,33]; *Parsley root (off-label)* [40-41,43]; *Parsnips (off-label)* [40-41,43]; *Peaches (off-label)* [38-40,43]; *Pear orchards* [1-2,4-5,7,9-15,18,20-21,23,26,35,37,40-42,45-50,53-54]; *Pears (off-label)* [33,38-40,43]; *Permanent grassland* [38-40,48]; *Permeable surfaces overlying soil* [1-8,10-16,18-37,39,41-47,49-54,56]; *Plums (off-label)* [1-5,7,9-15,18,20-22,25-27,29-30,32-54]; *Poppies for morphine production (off-label)* [41]; *Potatoes* [39,41,43-45]; *Quinces (off-label)* [38-40,43]; *Quinoa (off-label)* [40,43]; *Raspberries (off-label)* [40,43]; *Redcurrants (off-label)* [38,40,43]; *Rhubarb (off-label)* [38,43]; *Rotational grass* [39]; *Rubus hybrids (off-label)* [40,43]; *Rye (off-label)* [40-41,43]; *Salad onions (off-label)* [40-41,43]; *Salsify (off-label)* [40-41,43]; *Shallots (off-label)* [40-41,43]; *Soft fruit (off-label)* [40]; *Sorghum (off-label)* [40]; *Spring barley* [1,12,14-15,23,35,39,41,44-45,53-55]; *Spring field beans* [1,12,14-15,23,35,39,41,44-45,53-55]; *Spring oats* [1,12,14-15,23,35,39,41,44-45,53-54]; *Spring oilseed rape* [1,9,12,14-15,23,35,39,41,44-45,53-55]; *Spring wheat* [1,12,14-15,23,35,39,41,44-45,53-55]; *Strawberries (off-label)* [40,43]; *Stubbles* [3,11,22,25-27,29-30,32,34,36-38,40,42-43,46-52,55]; *Sugar beet* [1,3,12-15,22-23,25,27,29-30,32,34-36,39,41,43-45,51-54]; *Swedes (off-label)* [1,3,12-15,22-23,25,27,29-30,32,34-36,39-41,43-45,51-54]; *Table grapes* [43]; *Top fruit (off-label)* [40]; *Turnips (off-label)* [1,3,12-15,22-23,25,27,29-30,32,34-36,39-41,43-45,51-54]; *Vaccinium spp. (off-label)* [38,43]; *Vining peas (off-label)* [1,3,12-15,27,29-30,32,34-36,39-41,43-45,51-54]; *Walnuts (off-label)* [38-40,43]; *Whitecurrants (off-label)* [38,43]; *Wine grapes* [43]; *Winter barley* [1,12,14-15,23,35,39,41,44-45,53-55]; *Winter field beans* [1,12,14-15,23,35,39,41,44-45,53-54]; *Winter oats* [1,12,14-15,23,35,39,41,44-45,53-54]; *Winter oilseed rape* [1,9,12,14-15,23,35,39,41,44-45,53-55]; *Winter wheat* [1,12,14-15,23,35,39,41,44-45,53-55]; *Woad (off-label)* [38,40,43];

- Annual dicotyledons in *All edible crops (stubble)* [10]; *All non-edible crops (stubble)* [10]; *Bulb onions* [2-5,7,10-11,18,20-22,25-27,29-30,32,34,36-38,40,42-43,46-52]; *Combining peas* [2,4-5,7,10-11,13,18,20-21,26,37,42,50]; *Durum wheat* [2,4-5,7,10-11,13,18,20-21,26,30,37-38,40,42,46-50]; *Grassland* [44,46-47,49]; *Leeks* [2,4-5,7,10-11,18,20-21,26,37-38,40,42,46-50]; *Linseed* [2,4-5,7,10-11,13,18,20-21,26,37-38,40,42,46-50]; *Mustard* [2,4-5,7,10-11,13,18,20-21,26,37-38,40,42,46-50]; *Peas* [38]; *Permanent grassland* [38,40,48]; *Potatoes* [46-47,49]; *Spring barley* [2,4-5,7,10-11,13,18,20-21,26,37-38,40,42,46-50]; *Spring field beans* [2,4-5,7,10-11,13,18,20-21,26,37-38,40,42,46-50]; *Spring oats* [2,4-5,7,10-11,13,18,20-21,26,37-38,40,42,46-50]; *Spring oilseed rape* [2,4-5,7,10-11,13,18,20-21,26,37-38,40,42,46-50]; *Spring wheat* [2,4-5,7,9-11,13,18,20-21,26,37-38,40,42,46-50]; *Sugar beet* [2,4-5,7,10-11,18,20-21,26,37-38,40,42,46-50]; *Swedes* [2,4-5,7,10-11,18,20-21,26,37-38,40,42,46-50]; *Triticale* [9]; *Turnips* [2,4-5,7,10-11,18,20-21,26,37-38,40,46-49]; *Vining peas* [2,4-5,7,18,20-21,26,37,40,42,46-50]; *Winter barley* [2,4-5,7,9,11,13,18,20-21,26,37-38,40,42,46-50]; *Winter field beans* [2,4-5,7,10-11,13,18,20-21,26,37-38,40,42,46-50]; *Winter oats* [2,4-5,7,10-11,13,18,20-21,26,37-38,40,42,46-50]; *Winter oilseed rape* [2,4-5,7,10-11,13,18,20-21,26,37-38,40,42,46-50]; *Winter rye* [9]; *Winter wheat* [2,4-5,7,9-11,13,18,20-21,26,37-38,40,42,46-50];

- Annual grasses in *Bulb onions* [3,22,25,27,29-30,32,34,36,38,40,43,46-49,51-52]; *Durum wheat* [30,38,40,46-49]; *Grassland* [44,46-47,49]; *Leeks* [38,40,46-49]; *Linseed* [38,40,46-49]; *Mustard* [38,40,46-49]; *Peas* [38]; *Permanent grassland* [38,40,48]; *Potatoes* [46-47,49]; *Spring barley* [38,40,46-49]; *Spring field beans* [38,40,46-49]; *Spring oats* [38,40,46-49]; *Spring oilseed rape* [38,40,46-49]; *Spring wheat* [38,40,46-49]; *Sugar beet* [38,40,46-49]; *Swedes* [38,40,46-49]; *Turnips* [38,40,46-49]; *Vining peas* [40,46-49]; *Winter barley* [38,40,46-49]; *Winter field beans* [38,40,46-49]; *Winter oats* [38,40,46-49]; *Winter oilseed rape* [38,40,46-49]; *Winter wheat* [38,40,46-49];

- Aquatic weeds in *Enclosed waters* [3,19,22,25,27,29,32,34,36,38-39,41,43-45,51-52,56]; *Land immediately adjacent to aquatic areas* [2-5,7,19,22,25-27,29,32,34,36-39,41-45,51-52,56]; *Open waters* [3,19,22,25,27,29,32,34,36,38-39,41,44-45,51-52,56];

- Biennial weeds in *Beetroot* (off-label) [41]; *Carrots* (off-label) [41]; *Garlic* (off-label) [41]; *Horseradish* (off-label) [41]; *Leeks* (off-label) [41]; *Onions* (off-label) [41]; *Parsley root* (off-label) [41]; *Parsnips* (off-label) [41]; *Salad onions* (off-label) [41]; *Salsify* (off-label) [41]; *Shallots* (off-label) [41]; *Swedes* (off-label) [41]; *Turnips* (off-label) [41];

- Bolters in *Sugar beet* (wiper application) [38,40,48];

- Bracken in *Amenity vegetation* [13]; *Farm forestry* [10-11]; *Forest* [2,4-5,7,10-11,13,18,20-21,26,37,41-42,45,50]; *Forest nurseries* [13];

- Chemical thinning in *Farm forestry* [3,10-11,22,25,27,29,32,34,36,46-47,49,51-52]; *Forest* [2-5,7,10-11,18,20-22,25-27,29-32,34,36-37,41-42,44-47,49-52]; *Forest nurseries* [30];

- Couch in *Combining peas* [3,22,25,27,29-30,32,34,36,38,40,43,46-48,51-52,55]; *Durum wheat* [3,22,25,27,29-30,32,34,36,38,40,43,46-49,51-52]; *Linseed* [38,40,46-49,55]; *Mustard* [3,22,25,27,29-30,32,34,36,38,40,43,46-49,51-52,55]; *Oats* [3,22,25,27,29-30,32,34,36,43,51-52]; *Spring barley* [3,22,25,27,29-30,32,34,36,38,40,43,46-49,51-52,55]; *Spring field beans* [3,22,25,27,29-30,32,34,36,38,40,43,46-49,51-52,55]; *Spring oats* [38,40,46-49]; *Spring oilseed rape* [38,40,46-49,55]; *Spring wheat* [3,22,25,27,29-30,32,34,36,38,40,43,46-49,51-52,55]; *Stubbles* [38,40,46-49,55]; *Winter barley* [3,22,25,27,29-30,32,34,36,38,40,43,46-49,51-52,55]; *Winter field beans* [3,22,25,27,29-30,32,34,36,38,40,43,46-49,51-52,55]; *Winter oats* [38,40,46-49]; *Winter oilseed rape* [38,40,46-49,55]; *Winter wheat* [3,22,25,27,29-30,32,34,36,38,40,43,46-49,51-52,55];

- Desiccation in *Amenity vegetation* [6]; *Borage* (off-label) [41]; *Buckwheat* (off-label) [40]; *Canary flower (Echium spp.)* (off-label) [41]; *Canary seed* (off-label) [40]; *Combining peas* [39,41,45,49]; *Corn Gromwell* (off-label) [41]; *Durum wheat* [41,45]; *Hemp* (off-label) [40]; *Honesty* (off-label) [41]; *Linseed* [41,45]; *Lupins* (off-label) [40-41,43]; *Millet* (off-label) [40]; *Mustard* [41,45]; *Perilla* (off-label) [41]; *Poppies for morphine production* (off-label) [11,38,40-41]; *Quinoa* (off-label) [40]; *Rye* (off-label) [40-41,43]; *Sorghum* (off-label) [40]; *Spring barley* [41,45]; *Spring field beans* [41,45]; *Spring oats* [41,45]; *Spring oilseed rape* [41,45]; *Spring wheat* [41,45]; *Winter barley* [41,45]; *Winter field beans* [41,45]; *Winter oats* [41,45]; *Winter oilseed rape* [41,45]; *Winter wheat* [41,45];

- Destruction of crops in *All edible crops (outdoor and protected)* [9,14,54]; *All edible crops (outdoor)* [3,11,22,25,27,29-30,32,34-36,39,43-44,51-53]; *All edible crops (stubble)* [9,39,44]; *All non-edible crops (outdoor)* [3,9,11,14,22,25,27,29-30,32,34-36,39,43-44,51-53]; *All non-edible crops (stubble)* [9,39,44]; *Amenity vegetation* [16,44]; *Grassland* [1,14,35,41,45,53-54]; *Green cover on land temporarily removed from production* [3,22,25,27,29-30,32,34,36,39,43,51-52]; *Hard surfaces* [44]; *Natural surfaces not intended to bear vegetation* [44]; *Permanent grassland* [39]; *Permeable surfaces overlying soil* [44]; *Rotational grass* [39];

- Grass weeds in *All edible crops (outdoor and protected)* [23]; *All non-edible crops (outdoor)* [23]; *Apple orchards* [23]; *Combining peas* [23]; *Durum wheat* [23]; *Enclosed waters* [23]; *Forest* [16,23]; *Grassland* [23]; *Green cover on land temporarily removed from production* [23]; *Hard surfaces* [16,23]; *Land immediately adjacent to aquatic areas* [16,23]; *Leeks* [23]; *Linseed* [23]; *Mustard* [23]; *Natural surfaces not intended to bear vegetation* [16,23]; *Onions* [23]; *Open waters* [23]; *Pear orchards* [23]; *Permeable surfaces overlying soil* [16,23]; *Spring barley* [23]; *Spring field beans* [23]; *Spring oats* [23]; *Spring oilseed rape* [23]; *Spring wheat* [23]; *Sugar beet* [23]; *Swedes* [23]; *Turnips* [23]; *Winter barley* [23]; *Winter field beans* [23]; *Winter oats* [23]; *Winter oilseed rape* [23]; *Winter wheat* [23];

FOR FULL CONDITIONS OF USE ALWAYS READ THE PRODUCT LABEL

- Green cover in **Land immediately adjacent to aquatic areas** [18,20-21]; **Land temporarily removed from production** [13,18,20-21,26,37-38,40,42,44,46-50,55];
- Growth suppression in **Tree stumps** [12,35,53];
- Harvest management/desiccation in **All edible crops (stubble)** [10]; **All non-edible crops (stubble)** [10]; **Combining peas** [1-5,7,10-14,18,20-22,25-27,29-30,32,34-38,40,42-43,46-48,50-54]; **Durum wheat** [1-5,7,10-14,18,20-22,25-27,29-30,32,34-38,40,42-43,46-54]; **Grassland** [3,22,25,27,29-30,32,34,36,51-52]; **Linseed** [1-5,7,10-14,18,20-22,25-27,29-30,32,34-38,40,42-43,46-54]; **Mustard** [1-5,7,10-14,18,20-22,25-27,29-30,32,34-38,40,42-43,46-55]; **Oats** [3,22,25,27,29-30,32,34,36,43,51-52]; **Oilseed rape** [3,22,25,27,29-30,32,34,36,43,51-52]; **Spring barley** [1-5,7,10-14,18,20-22,25-27,29-30,32,34-38,40,42-43,46-54]; **Spring field beans** [2-5,7,10-11,13,18,20-22,25-27,29-30,32,34-38,40,42-43,46-53]; **Spring oats** [1-2,4-5,7,10-14,18,20-21,26,35,37-38,40,42,46-50,53-54]; **Spring oilseed rape** [1-2,4-5,7,10-14,18,20-21,26,35,37-38,40,42,46-50,53-55]; **Spring wheat** [1-5,7,9-14,18,20-22,25-27,29-30,32,34-38,40,42-43,46-54]; **Triticale** [9]; **Winter barley** [1-5,7,9,11-14,18,20-22,25-27,29-30,32,34-38,40,42-43,46-54]; **Winter field beans** [2-5,7,10-11,13,18,20-22,25-27,29-30,32,34-38,40,42-43,46-53]; **Winter oats** [1-2,4-5,7,10-14,18,20-21,26,35,37-38,40,42,46-50,53-54]; **Winter oilseed rape** [1-2,4-5,7,10-14,18,20-21,26,35,37-38,40,42,46-50,53-55]; **Winter rye** [9]; **Winter wheat** [1-5,7,9-14,18,20-22,25-27,29-30,32,34-38,40,42-43,46-54];
- Heather in **Farm forestry** [10-11]; **Forest** [2,4-5,7,10-11,18,20-21,26,37,42,50];
- Perennial dicotyledons in **Grassland** *(wiper application)* [46-47,49]; **Permanent grassland** *(wiper application)* [38,40,48];
- Pre-harvest desiccation in **Combining peas** [55]; **Linseed** [55]; **Spring barley** [55]; **Spring field beans** [55]; **Spring wheat** [55]; **Winter barley** [55]; **Winter field beans** [55]; **Winter wheat** [55];
- Reeds in **Aquatic areas** [38]; **Enclosed waters** [41,45]; **Land immediately adjacent to aquatic areas** [41,45]; **Open waters** [41,45];
- Re-growth suppression in **Tree stumps** [44];
- Rhododendrons in **Farm forestry** [10-11]; **Forest** [2,4-5,7,10-11,18,20-21,26,37,42,50];
- Rushes in **Aquatic areas** [38]; **Enclosed waters** [41,45]; **Land immediately adjacent to aquatic areas** [41,45]; **Open waters** [41,45];
- Sedges in **Aquatic areas** [38]; **Enclosed waters** [41,45]; **Land immediately adjacent to aquatic areas** [41,45]; **Open waters** [41,45];
- Sprout suppression in **Forest** [39]; **Tree stumps** [19,41,45-47,49];
- Sucker control in **Apple orchards** [3,22,25,27,29-30,32,34,36,38,40,43-44,46-49,51-52]; **Cherries** [3,22,25,27,29-30,32,34,36,38,40,43-44,46-49,51-52]; **Damsons** [38,40,48]; **Forest** [39]; **Pear orchards** [3,22,25,27,29-30,32,34,36,38,40,43-44,46-49,51-52]; **Plums** [3,22,25,27,29-30,32,34,36,38,40,43-44,46-49,51-52];
- Sucker inhibition in **Enclosed waters** [17]; **Open waters** [17]; **Tree stumps** [1,3,14,17,22,25,27,29-30,32,34,36,51-52,54];
- Sward destruction in **Amenity vegetation** [19,28,56]; **Grassland** [2,4-5,7,10-11,13,18,20-21,26,37,42,44,46-47,49]; **Permanent grassland** [38,40,48,50,55]; **Rotational grass** [50,55];
- Total vegetation control in **All edible crops (outdoor)** *(pre-sowing/planting)* [40,46-49]; **All non-edible crops (outdoor)** *(pre-sowing/planting)* [40,46-49]; **Land not intended to bear vegetation** [55];
- Trees in **Amenity vegetation** *(off-label)* [17]; **Forest** *(off-label)* [17];
- Volunteer cereals in **Bulb onions** [3,22,25,27,29-30,32,34,36,38,40,43,46-49,51-52]; **Durum wheat** [30,38,40,46-49]; **Leeks** [38,40,46-49]; **Linseed** [38,40,46-49]; **Mustard** [38,40,46-49]; **Peas** [38]; **Potatoes** [46-47,49]; **Spring barley** [38,40,46-49]; **Spring field beans** [38,40,46-49]; **Spring oats** [38,40,46-49]; **Spring oilseed rape** [38,40,46-49]; **Spring wheat** [38,40,46-49]; **Stubbles** [38,40,46-49,55]; **Sugar beet** [38,40,46-49]; **Swedes** [38,40,46-49]; **Turnips** [38,40,46-49]; **Vining peas** [40,46-49]; **Winter barley** [38,40,46-49]; **Winter field beans** [38,40,46-49]; **Winter oats** [38,40,46-49]; **Winter oilseed rape** [38,40,46-49]; **Winter wheat** [38,40,46-49];
- Volunteer potatoes in **Stubbles** [38,40,46-49,55];
- Waterlilies in **Enclosed waters** [41,45]; **Land immediately adjacent to aquatic areas** [41,45]; **Open waters** [41,45];
- Woody weeds in **Farm forestry** [10-11]; **Forest** [2,4-5,7,10-11,18,20-21,26,37,42,50];

SECTION 2

SEE SECTION 3 FOR PRODUCTS ALSO REGISTERED

Extension of Authorisation for Minor Use (EAMUs)

- **Almonds** *20221348* [38], *20170234* [39], *20082888* [40], *20141300* [43]
- **Amenity vegetation** *20170195* [17]
- **Apples** *20221348* [38], *20170234* [39], *20082888* [40], *20141300* [43]
- **Apricots** *20221348* [38], *20170234* [39], *20082888* [40], *20141300* [43]
- **Asparagus** *20221350* [38], *20171123* [39]
- **Beetroot** *20130354* [40], *20132528* [41], *20141305* [43]
- **Bilberries** *20221352* [38], *20082888* [40], *20141300* [43], *20141316* [43]
- **Blackberries** *20082888* [40], *20141300* [43]
- **Blackcurrants** *20221352* [38], *20082888* [40], *20141300* [43], *20141316* [43]
- **Blueberries** *20221352* [38], *20082888* [40], *20141300* [43], *20141316* [43]
- **Borage** *20211583* [41]
- **Buckwheat** *20071839* [40]
- **Canary flower (Echium spp.)** *20211583* [41]
- **Canary seed** *20071839* [40]
- **Carrots** *20130354* [40], *20132528* [41], *20141305* [43]
- **Cherries** *20221348* [38], *20170234* [39], *20082888* [40], *20141300* [43]
- **Chestnuts** *20221348* [38], *20170234* [39], *20082888* [40], *20141300* [43], *20141317* [43]
- **Christmas trees** *20152945* [44]
- **Cob nuts** *20221348* [38]
- **Corn Gromwell** *20211583* [41]
- **Crab apples** *20170234* [39]
- **Cranberries** *20221352* [38], *20082888* [40], *20141300* [43], *20141316* [43]
- **Edible podded peas** *20141672* [40], *20141671* [41]
- **Farm forestry** *20221353* [38], *20082888* [40], *20141300* [43]
- **Forest** *20170195* [17]
- **Game cover** *20221353* [38], *20082888* [40], *20141300* [43]
- **Garlic** *20130354* [40], *20132528* [41], *20141305* [43]
- **Gooseberries** *20221352* [38], *20082888* [40], *20141300* [43], *20141316* [43]
- **Grapevines** *20221355* [38]
- **Hazel nuts** *20221348* [38], *20170234* [39], *20082888* [40], *20141300* [43], *20141317* [43]
- **Hemp** *20221353* [38], *20071840* [40], *20141300* [43], *20141321* [43]
- **Honesty** *20211583* [41]
- **Hops** *20221353* [38], *20082888* [40], *20141300* [43]
- **Horseradish** *20130354* [40], *20132528* [41], *20141305* [43]
- **Kentish cobnuts** *20170234* [39], *20141317* [43]
- **Leeks** *20130354* [40], *20132528* [41], *20141305* [43]
- **Loganberries** *20082888* [40], *20141300* [43]
- **Lupins** *20071279* [40], *20131667* [41], *20141306* [43]
- **Medlar** *20082888* [40], *20141300* [43]
- **Millet** *20071839* [40]
- **Miscanthus** *20221353* [38], *20082888* [40], *20132131* [41], *20141300* [43]
- **Nectarines** *20221348* [38], *20082888* [40], *20141300* [43]
- **Onions** *20130354* [40], *20132528* [41], *20141305* [43]
- **Ornamental plant production** (as a directed spray) *20172202* [12]
- **Parsley root** *20130354* [40], *20132528* [41], *20141305* [43]
- **Parsnips** *20130354* [40], *20132528* [41], *20141305* [43]
- **Peaches** *20221348* [38], *20170234* [39], *20082888* [40], *20141300* [43]
- **Pears** *20221348* [38], *20170234* [39], *20082888* [40], *20141300* [43]
- **Perilla** *20211583* [41]
- **Plums** *20221348* [38], *20170234* [39], *20082888* [40], *20141300* [43]
- **Poppies for morphine production** *20210830* [11], *20221351* [38], *20081058* [40], *20131668* [41]
- **Quinces** *20221348* [38], *20170234* [39], *20082888* [40], *20141300* [43]
- **Quinoa** *20071840* [40], *20141321* [43]
- **Raspberries** *20082888* [40], *20141300* [43]
- **Redcurrants** *20221352* [38], *20082888* [40], *20141300* [43], *20141316* [43]
- **Rhubarb** *20221349* [38], *20141320* [43]
- **Rubus hybrids** *20082888* [40], *20141300* [43]

FOR FULL CONDITIONS OF USE ALWAYS READ THE PRODUCT LABEL

- **Rye** *20111701* [40], *20131666* [41], *20141307* [43]
- **Salad onions** *20130354* [40], *20132528* [41], *20141305* [43]
- **Salsify** *20130354* [40], *20132528* [41], *20141305* [43]
- **Shallots** *20130354* [40], *20132528* [41], *20141305* [43]
- **Soft fruit** *20082888* [40]
- **Sorghum** *20071839* [40]
- **Strawberries** *20082888* [40], *20141300* [43]
- **Swedes** *20130354* [40], *20132528* [41], *20141305* [43]
- **Top fruit** *20082888* [40]
- **Turnips** *20130354* [40], *20132528* [41], *20141305* [43]
- **Vaccinium spp.** *20221352* [38], *20141316* [43]
- **Vining peas** *20141672* [40], *20141671* [41]
- **Walnuts** *20221348* [38], *20170234* [39], *20082888* [40], *20141300* [43], *20141317* [43]
- **Whitecurrants** *20221352* [38], *20141316* [43]
- **Woad** *20221353* [38], *20082888* [40], *20141300* [43]

Approval information
- Glyphosate included in Annex I under EC Regulation 1107/2009
- Accepted by BBPA for use on malting barley and hops

Efficacy guidance
- For best results apply to actively growing weeds with enough leaf to absorb chemical
- For most products a rainfree period of at least 6 h (preferably 24 h) should follow spraying
- Adjuvants are obligatory for some products and recommended for some uses with others. See labels
- Mixtures with other pesticides or fertilizers may lead to reduced control.
- Products are formulated as isopropylamine, ammonium, potassium, or trimesium salts of glyphosate and may vary in the details of efficacy claims. See individual product labels
- If using to treat hard surfaces, ensure spraying takes place only when weeds are actively growing (normally March to October) and is confined only to visible weeds including those in the 30cm swath covering the kerb edge and road gulley ? do not overspray drains.
- For use with hand held rotary atomisers, the spray droplet spectra produced must be of a minimum Volume Median Diameter (VMD) of 200 microns [24]
- With wiper application weeds should be at least 10 cm taller than crop
- Annual weed grasses should have at least 5 cm of leaf and annual broad-leaved weeds at least 2 expanded true leaves
- Perennial grass weeds should have 4-5 new leaves and be at least 10 cm long when treated. Perennial broad-leaved weeds should be treated at or near flowering but before onset of senescence
- Volunteer potatoes and polygonums are not controlled by harvest-aid rates
- Bracken must be treated at full frond expansion
- Fruit tree suckers best treated in late spring
- Chemical thinning treatment can be applied as stump spray or stem injection
- In order to allow translocation, do not cultivate before spraying and do not apply other pesticides, lime, fertilizer or farmyard manure within 5 d of treatment
- Recommended intervals after treatment and before cultivation vary. See labels
- When used on managed amenity turf application must only be carried out pre-establishment of the crop

Restrictions
- Maximum total dose per crop or season normally equivalent to one full dose treatment on field and edible crops and no restriction for non-crop uses. However some older labels indicate a maximum number of treatments. Check for details
- Do not treat cereals grown for seed or undersown crops
- Consult grain merchant before treating crops grown on contract or intended for malting
- Do not use treated straw as a mulch or growing medium for horticultural crops
- For use in nursery stock, shrubberies, orchards, grapevines and tree nuts care must be taken to avoid contact with the trees. Do not use in orchards established less than 2 yr and keep off low-lying branches
- Certain conifers may be sprayed overall in dormant season. See label for details

SEE SECTION 3 FOR PRODUCTS ALSO REGISTERED

- Use a tree guard when spraying in established forestry plantations
- Do not spray root suckers in orchards in late summer or autumn
- Do not use under glass or polythene as damage to crops may result
- Do not mix, store or apply in galvanised or unlined mild steel containers or spray tanks
- Do not leave diluted chemical in spray tanks for long periods and make sure that tanks are well vented
- The maximum concentration of glyphosate in water must not exceed 0.12 ppm or such lower concentration as the appropriate regulatory body may require

Crop-specific information
- Harvest intervals: 4 wk for blackcurrants, blueberries; 14 d for linseed, oilseed rape; 8 d for mustard; 5-7 d for all other edible crops. Check label for exact details
- Latest use: for most products 2-14 d before cultivating, drilling or planting a crop in treated land; after harvest (post-leaf fall) but before bud formation in the following season for nuts and most fruit and vegetable crops; before fruit set for grapevines. See labels for details

Following crops guidance
- Decaying remains of plants killed by spraying must be dispersed before direct drilling
- Crops may be drilled 48 h after application. Trees and shrubs may be planted 7 d after application.

Environmental safety
- Products differ in their hazard and environmental safety classification. See labels
- Do not dump surplus herbicide in water or ditch bottoms or empty into drains
- Check label for maximum permitted concentration in treated water
- The Environment Agency or Local River Purification Authority must be consulted before use in or near water
- Take extreme care to avoid drift and possible damage to neighbouring crops or plants
- Treated poisonous plants must be removed before grazing or conserving
- Do not use in covered areas such as greenhouses or under polythene
- For field edge treatment direct spray away from hedge bottoms
- Some products require livestock to be excluded from treated areas and do not permit treated forage to be used for hay, silage or bedding. Check label for details

Hazard classification and safety precautions
Hazard Irritant [2,4,10-13,17-22,25-26,30-31,34-35,37,42,53], Dangerous for the environment [2-4,10-13,17-22,24-27,29-37,41-42,45,51-53,55], Harmful if inhaled [42]

Transport code 9 [2-4,10,13,15,17-18,20-22,24-27,29-32,34-37,42-43,45,51-53]

Packaging group III

UN Number 3077, 3082, N/C

Risk phrases H318 [15,35,42,53], H319 [2-3,21,27,29,31-32,34,36,46-47,49,52], R36 [51], R41 [30], R51 [28,30,51], R52 [40,46,48], R53a [28,30,40-41,46,48,51]

Operator protection U02a [1,5-8,12,14-16,19,22-25,30-31,33-35,43,50,53-54,56], U05a [1-2,4,10-11,13-15,17-18,20-21,24,26,28,33,35,37,42-43,53-55], U08 [5-7,13,17,19,43,50,56], U09a [8,16,23-24,33], U11 [2-4,10-13,17-18,20-22,25-32,34,36-37,42,51-52], U12 [35,53], U15 [35,53], U19a [5-8,13,16,19,23-24,33,43,50,56], U20a [2,4,10-11,16-18,20-21,23,26,28,37-39,42,44,55], U20b [1,3,5-9,12-15,19,22,24-25,27,29-36,40-41,43,45-54,56]; A, C, H, M

Environmental protection E06d [55], E07d [54], E08b [55], E13c [7,24,33,56], E15a [38-39], E15b [1-6,8-12,14-27,29-37,40-55], E19a [7,56], E34 [14-15,54], E36a [14-15,54], E38 [2-4,10-12,17-18,20-22,25-27,29-32,34-37,41-43,45,51-53,55], H410 [24,35], H411 [2-3,10-11,15,21-22,25,27,29,32,39,42,55], H412 [14,17,47,53], H413 [1,12,45,54]

Storage and disposal D01 [1-6,8-11,13-21,23-24,26-29,32-33,35-55], D02 [1-6,8-11,13-21,23-24,26-29,32-33,35-55], D05 [1-8,10-12,14-27,29-37,39-42,45-56], D09a [2-13,16-27,29-53,55-56], D10a [5-6,19,50], D10b [7,13,23,56], D10c [1-4,9-12,14-15,18,20-22,25-27,29-32,34-42,44-49,51-55], D11a [8,16,24,33,43], D12a [2-4,10-12,17-18,20-22,24-27,29-34,36-37,41-42,45,51-52,55], D14 [14-15,54]

Medical advice M03, M05a [3,12-13,22,25,27-32,34,36,51-52]

221   glyphosate + sulfosulfuron

A translocated non-residual glycine derivative herbicide + a sulfonyl urea herbicide
HRAC mode of action code: 9 + 2 (G + B)

See also sulfosulfuron

Products
1 Nomix Dual   Nomix Enviro   120:2.22 g/l   RC   18351

Uses
• Annual and perennial weeds in **Amenity vegetation** [1]; **Hard surfaces** [1]; **Natural surfaces not intended to bear vegetation** [1]; **Permeable surfaces overlying soil** [1];

Approval information
• Glyphosate and sulfosulfuron included in Annex I under EC Regulation 1107/2009

Efficacy guidance
• For use with hand held rotary atomisers, the spray droplet spectra produced must be of a minimum Volume Median Diameter (VMD) of 200 microns

Hazard classification and safety precautions
Hazard Dangerous for the environment
Transport code 9 [1]
Packaging group III
UN Number 3082
Operator protection U02a, U09a, U19a, U20b; A, H, M
Environmental protection E15b, H410
Storage and disposal D01, D05, D09a, D11a

222   halauxifen-methyl

A herbicide for use in cereal crops only available in mixtures
HRAC mode of action code: 4 (O)

See also aminopyralid + halauxifen-methyl
         clopyralid + halauxifen-methyl
         florasulam + halauxifen-methyl
         fluroxypyr + halauxifen-methyl

223   halauxifen-methyl + picloram

A broad-leaved contact herbicide mixture for weed control in oilseed rape
HRAC mode of action code: 4 + 4 (O + O)

Products
1 Belkar   Corteva   10:48 g/l   EC   18615

Uses
• Annual dicotyledons in **Winter oilseed rape** [1];
• Cleavers in **Winter oilseed rape** [1];
• Crane's-bill in **Winter oilseed rape** [1];
• Fumitory in **Winter oilseed rape** [1];
• Poppies in **Winter oilseed rape** [1];
• Shepherd's purse in **Winter oilseed rape** [1];

Approval information
• Halauxifen-methyl and picloram included in Annex 1 under EC Regulation 1107/2009

SEE SECTION 3 FOR PRODUCTS ALSO REGISTERED

SECTION 2

Efficacy guidance
- Use in cold or warm (from 2 to 25°C), humid or dry conditions. In severe drought conditions there can be a slight reduction in efficacy.
- Rainfast 1 hour after application

Restrictions
- DO NOT apply 0.25 l/ha before 1st September or BBCH 12 (2 true leaves)
- DO NOT apply 0.5 l/ha before 15th September or BBCH 16 (6 true leaves)

Following crops guidance
- After normal harvest or in the case of crop failure wheat, barley, oats, maize or oilseed rape can be planted 120 days after application. Allow at least 12 months before planting other crops. Ploughing or thorough cultivation is necessary before planting legumes such as peas or beans.

Environmental safety
- LERAP Category B [1]

Hazard classification and safety precautions
Hazard Very toxic to aquatic organisms
Transport code 9 [1]
Packaging group III
UN Number 3082
Risk phrases H319, H335
Operator protection U05a, U08, U10, U19a, U20a; A, C, H
Environmental protection E15b, E16a, E23, E34, H410
Storage and disposal D09a, D10c

224   hexythiazox

A miticide for spider mite control in protected horticultural crops
IRAC mode of action code: 10A

Products
1 Nissorun SC   Certis Belchim B V   250 g/l   SC   19263

Uses
- Spider mites in *Apples* [1]; *Hops* [1]; *Pears* [1]; *Protected aubergines* [1]; *Protected chilli peppers* [1]; *Protected courgettes* [1]; *Protected cucumbers* [1]; *Protected melons* [1]; *Protected peppers* [1]; *Protected pumpkins* [1]; *Protected summer squash* [1]; *Protected tomatoes* [1]; *Protected watermelon* [1]; *Protected winter squash* [1];

Approval information
- Hexythiazox included in Annex I under EC Regulation 1107/2009

Efficacy guidance
- Applications should be made at first sign of spider mite attack. Where high infestation of spider mites occur alternating with active ingredients controlling the adult stages of spider mites should be considered. As there is limited effect on adults the product will not control adults which are in diapause stage
- Has contact and stomach action with good translaminar activity. Has ovicidal, larvicidal and nymphicidal activity. Not active against adults, but eggs laid by treated females are non-viable. The amount of water and spray quality or spraying equipment should allow a thorough wetting of all plant parts to achieve maximum coverage including on the underside of the leaves
- Can be used in Integrated Pest Management strategies. However, it may be harmful to some biological control agents, including the predatory mite Typhlodromus pyri
- Thoroughly clean sprayer and lines with water after use

Restrictions
- The effects on fermentation processes in hops and cider/perry production have not been fully tested

FOR FULL CONDITIONS OF USE ALWAYS READ THE PRODUCT LABEL

Environmental safety
- Broadcast air-assisted LERAP [1] (30m (hops or 15 m (pome fruit post BBCH 70)); LERAP Category B [1]

Hazard classification and safety precautions
    Risk phrases H319
    Operator protection A, C, H
    Environmental protection E16a, E17b (30m (hops or 15 m (pome fruit post BBCH 70)), H411
    Storage and disposal D01, D02, D05

SECTION 2

225   hymexazol

A systemic heteroaromatic fungicide for pelleting sugar beet seed
FRAC mode of action code: 32

Products

1	Tachigaren 70 WP	Mitsui	70% w/w	WP	17977

Uses
- Black leg in **Sugar beet** *(seed treatment)* [1];

Approval information
- Hymexazol included in Annex 1 under EC Regulation 1107/2009

Efficacy guidance
- Incorporate into pelleted seed using suitable seed pelleting machinery

Restrictions
- Maximum number of treatments 1 per batch of seed
- Do not use treated seed as food or feed

Crop-specific information
- Latest use: before planting sugar beet seed

Environmental safety
- Harmful to aquatic organisms
- Harmful to fish or other aquatic life. Do not contaminate surface waters or ditches with chemical or used container
- Treated seed harmful to game and wildlife

Hazard classification and safety precautions
    Hazard Flammable solid
    Transport code 3 [1]
    Packaging group III
    UN Number 1325
    Risk phrases H317, H318, H361
    Operator protection U05a, U11, U20b; A, C, F
    Environmental protection E03, E13c, H411
    Storage and disposal D01, D02, D09a, D11a
    Treated seed S01, S02, S03, S04a, S05

226   imazalil

A systemic and protectant imidazole fungicide
FRAC mode of action code: 3

See also guazatine + imazalil

Products

1	Gavel	Certis Belchim B V	100 g/l	SL	17586

SEE SECTION 3 FOR PRODUCTS ALSO REGISTERED

Uses
- Botrytis in **Protected courgettes** *(off-label)* [1]; **Protected cucumbers** *(off-label)* [1]; **Protected tomatoes** *(off-label)* [1];
- Dry rot in **Seed potatoes** [1];
- Gangrene in **Seed potatoes** [1];
- Powdery mildew in **Protected courgettes** *(off-label)* [1]; **Protected cucumbers** *(off-label)* [1]; **Protected tomatoes** *(off-label)* [1];
- Silver scurf in **Seed potatoes** [1];
- Skin spot in **Seed potatoes** [1];

Extension of Authorisation for Minor Use (EAMUs)
- **Protected courgettes** *20220343* [1]
- **Protected cucumbers** *20220343* [1]
- **Protected tomatoes** *20220343* [1]

Approval information
- Imazalil included in Annex I under EC Regulation 1107/2009

Efficacy guidance
- For best control of skin and wound diseases of ware potatoes treat as soon as possible after harvest, preferably within 7-10 d, before any wounds have healed

Restrictions
- Maximum number of treatments 1 per batch of ware tubers; 2 per batch of seed tubers;
- Consult processor before treating potatoes for processing

Crop-specific information
- Latest use: during storage and before chitting for seed potatoes
- Apply to clean soil-free potatoes post-harvest before putting into store, or at first grading. A further treatment may be applied in early spring before planting
- Apply through canopied hydraulic or spinning disc equipment preferably diluted with up to two litres water per tonne of potatoes to obtain maximum skin cover and penetration
- Use on ware potatoes subject to discharges of imazalil from potato washing plants being within emission limits set by the UK monitoring authority

Environmental safety
- Dangerous for the environment
- Toxic to aquatic organisms
- Do not empty into drains
- Personal protective equipment requirements may vary for each pack size. Check label

Hazard classification and safety precautions
Hazard Irritant
UN Number N/C
Risk phrases H318, H351
Operator protection U04a, U05a, U11, U14, U19a, U20a; A, C, H
Environmental protection E15a, E19b, E34, H410
Storage and disposal D01, D02, D05, D09a, D10c, D12a
Treated seed S01, S02, S03, S04a, S05, S06a

227 imazamox

An imidazolinone contact and residual herbicide available only in mixtures
HRAC mode of action code: 2 (B)

See also imazamox + pendimethalin
imazamox + quinmerac

228   imazamox + pendimethalin

A pre-emergence broad-spectrum herbicide mixture for legumes
HRAC mode of action code: 2 + 3 (B + K1)

See also pendimethalin

Products

1 Nirvana   BASF   16.7:250 g/l   EC   14256

Uses

- Annual dicotyledons in **Broad beans** *(off-label)* [1]; **Combining peas** [1]; **Hops** *(off-label)* [1]; **Ornamental plant production** *(off-label)* [1]; **Soft fruit** *(off-label)* [1]; **Spring field beans** [1]; **Top fruit** *(off-label)* [1]; **Vining peas** [1]; **Winter field beans** [1];
- Black bindweed in **Soya beans** *(off-label)* [1];
- Charlock in **Soya beans** *(off-label)* [1];
- Chickweed in **Soya beans** *(off-label)* [1];
- Common orache in **Soya beans** *(off-label)* [1];
- Fat hen in **Soya beans** *(off-label)* [1];
- Field speedwell in **Soya beans** *(off-label)* [1];
- Fumitory in **Soya beans** *(off-label)* [1];
- Poppies in **Soya beans** *(off-label)* [1];

Extension of Authorisation for Minor Use (EAMUs)

- **Broad beans** *20092891* [1]
- **Hops** *20092894* [1]
- **Ornamental plant production** *20092894* [1]
- **Soft fruit** *20092894* [1]
- **Soya beans** *20152425* [1]
- **Top fruit** *20092894* [1]

Approval information

- Imazamox and pendimethalin included in Annex I under EC Regulation 1107/2009

Efficacy guidance

- Best results obtained from applications to fine firm seedbeds in the presence of adequate moisture
- Weed control may be reduced on cloddy seedbeds and on soils with over 6% organic matter
- Residual control may be reduced under prolonged dry conditions

Restrictions

- Maximum number of treatments 1 per crop
- Seed must be drilled to at least 2.5 cm of settled soil
- Do not use on soils containing more than 10% organic matter
- Do not apply to soils that are waterlogged or are prone to waterlogging
- Do not apply if heavy rain is forecast
- Do not soil incorporate the product or disturb the soil after application
- Consult processors before use on crops destined for processing
- To avoid the build-up of resistance do not apply this or any other product containing an ALS inhibitor herbicide with claims for control of grass-weeds more than once to any crop

Crop-specific information

- Latest use: pre-crop emergence for all crops
- Inadequately covered seed may result in cupping of the leaves after application from which recovery is normally complete
- Crop damage may occur on stony or gravelly soils especially if heavy rain follows treatment
- Winter oilseed rape and other brassica crops should not be drilled as the following crop.

Following crops guidance

- Winter wheat or winter barley may be drilled as a following crop provided 3 mth have elapsed since treatment and the land has been at least cultivated by a non-inversion technique such as discing

- Winter oilseed rape or other brassica crops should not be drilled as following crops
- A minimum of 12 mth must elapse between treatment and sowing red beet, sugar beet or spinach

Environmental safety
- Dangerous for the environment
- Very toxic to aquatic organisms
- LERAP Category B [1]

Hazard classification and safety precautions
    Hazard Irritant, Dangerous for the environment, Very toxic to aquatic organisms
    Transport code 9 [1]
    Packaging group III
    UN Number 3082
    Risk phrases H317, R38
    Operator protection U05a, U14, U20b; A, H
    Environmental protection E15b, E16a, E38, H410
    Storage and disposal D01, D02, D09a, D10c, D12a
    Medical advice M03

229   imazamox + quinmerac

A herbicide mixture for weed control in CLEARFIELD oilseed rape only
HRAC mode of action code: 2 + 4 (B + O)

Products

1	Cleravo	BASF	35:250 g/l	SC	16877

Uses
- Annual dicotyledons in *Spring oilseed rape* [1]; *Winter oilseed rape* [1];
- Charlock in *Spring oilseed rape* [1]; *Winter oilseed rape* [1];
- Cleavers in *Spring oilseed rape* [1]; *Winter oilseed rape* [1];
- Crane's-bill in *Spring oilseed rape* *(cut-leaved)* [1]; *Winter oilseed rape* *(cut-leaved)* [1];
- Dead nettle in *Spring oilseed rape* [1]; *Winter oilseed rape* [1];
- Flixweed in *Spring oilseed rape* [1]; *Winter oilseed rape* [1];
- Volunteer oilseed rape in *Spring oilseed rape* [1]; *Winter oilseed rape* [1];

Approval information
- Imazamox and quinmerac included in Annex I under EC Regulation 1107/2009

Efficacy guidance
- Transient crop scorch may occur if applied under frosty conditions.
- Soil moisture is required for efficacy via root uptake.
- Maximum activity is from application before or shortly after weed emergence. Uptake is via cotyledons, roots and shoots.

Restrictions
- Must only be applied to CLEARFIELD oilseed rape hybrids.
- Extreme care must be taken to avoid spray drift onto non-target plants
- Must only be applied 1 year out of every 3 years
- Do not apply if heavy rain is forecast.
- Do not use on sands, very light soils or soils with more than 10% organic matter.
- Use on stony soils may result in a reduction in plant vigour or crop stand.
- To avoid the build-up of resistance do not apply this or any other product containing an ALS inhibitor herbicide with claims for control of grass-weeds more than once to any crop.
- Quinmerac is being found in ground and surface waters. Take care when applying to sloping soils near water.

Crop-specific information
- In the event of crop failure, do not re-drill with non CLEARFIELD oilseed rape or with sugar beet. After ploughing Clearfield rape, maize or peas may be sown 4 weeks after application, wheat,

barley or oats 8 weeks after application, field beans 10 weeks after application or sunflower 15 weeks after application.

Hazard classification and safety precautions
Transport code 9 [1]
Packaging group III
UN Number 3082
Risk phrases H317
Operator protection U05a, U20a
Environmental protection E15b, E38, H410
Storage and disposal D01, D02, D09a, D10c, D12b

230   4-indol-3-ylbutyric acid

A plant growth regulator promoting the rooting of cuttings

Products

1	Chryzoplus Grey 0.8%	Fargro	0.8% w/w	DP	17569
2	Chryzopon Rose 0.1%	Fargro	0.1% w/w	DP	17566
3	Chryzotek Beige 0.4%	Fargro	0.4% w/w	DP	17568
4	Chryzotop Green 0.25%	Fargro	0.25% w/w	DP	17567
5	Rhizopon AA Powder (0.5%)	Fargro	0.5% w/w	DP	17570
6	Rhizopon AA Powder (1%)	Fargro	1% w/w	DP	17571
7	Rhizopon AA Powder (2%)	Fargro	2% w/w	DP	17572
8	Rhizopon AA Tablets	Fargro	50 mg a.i.	WT	17573

Uses
* Rooting of cuttings in ***Ornamental plant production*** [1-8];

Approval information
* 4-indol-3-ylbutyric acid included in Annex 1 under EC Regulation 1107/2009

Efficacy guidance
* Dip base of cuttings into powder immediately before planting
* Powders or solutions of different concentration are required for different types of cutting. Lowest concentration for softwood, intermediate for semi-ripe, highest for hardwood
* See label for details of concentration and timing recommended for different species
* Use of planting holes recommended for powder formulations to ensure product is not removed on insertion of cutting. Cuttings should be watered in if necessary

Restrictions
* Maximum number of treatments 1 per situation
* Use of too strong a powder or solution may cause injury to cuttings
* No unused moistened powder should be returned to container

Crop-specific information
* Latest use: before cutting insertion for ornamental specimens

Hazard classification and safety precautions
UN Number N/C
Operator protection U14 [1-4], U15 [1-4], U19a, U20a [1-4,6-8], U20b [5]; A, H
Environmental protection E15a
Storage and disposal D09a, D11a

SEE SECTION 3 FOR PRODUCTS ALSO REGISTERED

231   indoxacarb

An oxadiazine insecticide for caterpillar control in a range of crops
IRAC mode of action code: 22A

Products

1	Explicit (GB only)	FMC Agro	30% w/w	WG	18763
2	Rumo (GB only)	FMC Agro	30% w/w	WG	18797
3	Steward (GB only)	FMC Agro	30% w/w	WG	18792

Uses

- Cabbage leafroller in **Choi sum** *(off-label)* [1-2]; **Collards** *(off-label)* [1-2]; **Kale** *(off-label)* [1-2]; **Kohlrabi** *(off-label)* [1-2]; **Oriental cabbage** *(off-label)* [1-2]; **Protected baby leaf crops** *(off-label)* [2]; **Protected herbs (see appendix 6)** *(off-label)* [2]; **Protected kohlrabi** *(off-label)* [2]; **Spring greens** *(off-label)* [1-2]; **Tatsoi** *(off-label)* [1-2];
- Cabbage moth in **Choi sum** *(off-label)* [1-2]; **Collards** *(off-label)* [1-2]; **Kale** *(off-label)* [1-2]; **Kohlrabi** *(off-label)* [1-2]; **Oriental cabbage** *(off-label)* [1-2]; **Protected baby leaf crops** *(off-label)* [1-2]; **Protected herbs (see appendix 6)** *(off-label)* [1-2]; **Protected kohlrabi** *(off-label)* [1-2]; **Spring greens** *(off-label)* [1-2]; **Tatsoi** *(off-label)* [1-2];
- Cabbage white butterfly in **Choi sum** *(off-label)* [1-2]; **Collards** *(off-label)* [1-2]; **Kale** *(off-label)* [1-2]; **Kohlrabi** *(off-label)* [1-2]; **Oriental cabbage** *(off-label)* [1-2]; **Protected baby leaf crops** *(off-label)* [1-2]; **Protected herbs (see appendix 6)** *(off-label)* [1-2]; **Protected kohlrabi** *(off-label)* [1-2]; **Spring greens** *(off-label)* [1-2]; **Tatsoi** *(off-label)* [1-2];
- Capsids in **Strawberries** *(off-label)* [1];
- Caterpillars in **Apples** [1,3]; **Broccoli** [1-3]; **Cabbages** [1-3]; **Calabrese** [2]; **Cauliflowers** [1-3]; **Hops** *(off-label)* [2]; **Ornamental plant production** *(off-label)* [2]; **Pears** [1,3]; **Protected all edible seed crops** *(off-label)* [2]; **Protected all non-edible seed crops** *(off-label)* [2]; **Protected aubergines** [1-3]; **Protected chilli peppers** [2]; **Protected courgettes** [1-3]; **Protected cucumbers** [1-3]; **Protected hops** *(off-label)* [2]; **Protected marrows** [1,3]; **Protected melons** [1-3]; **Protected ornamentals** *(off-label)* [1-3]; **Protected peppers** [1-3]; **Protected pumpkins** [1-3]; **Protected squashes** [1,3]; **Protected summer squash** [2]; **Protected tomatoes** [1-3]; **Protected winter squash** [2]; **Sweetcorn** *(off-label)* [1-2];
- Diamond-back moth in **Brussels sprouts** *(off-label)* [1-2]; **Choi sum** *(off-label)* [1-2]; **Collards** *(off-label)* [1-2]; **Endives** *(off-label)* [1-2]; **Kale** *(off-label)* [1-2]; **Kohlrabi** *(off-label)* [1-2]; **Lettuce** *(off-label)* [1-2]; **Oriental cabbage** *(off-label)* [1-2]; **Protected baby leaf crops** *(off-label)* [1-2]; **Protected herbs (see appendix 6)** *(off-label)* [1-2]; **Protected kohlrabi** *(off-label)* [1-2]; **Spring greens** *(off-label)* [1-2]; **Sweetcorn** *(off-label)* [1-2]; **Tatsoi** *(off-label)* [1-2];
- Flax tortrix moth in **Endives** *(off-label)* [1-2]; **Lettuce** *(off-label)* [1-2];
- Garden pebble moth in **Choi sum** *(off-label)* [1-2]; **Collards** *(off-label)* [1-2]; **Kale** *(off-label)* [1-2]; **Kohlrabi** *(off-label)* [1-2]; **Oriental cabbage** *(off-label)* [1-2]; **Protected baby leaf crops** *(off-label)* [1-2]; **Protected herbs (see appendix 6)** *(off-label)* [1-2]; **Protected kohlrabi** *(off-label)* [1-2]; **Spring greens** *(off-label)* [1-2]; **Tatsoi** *(off-label)* [1-2];
- Ghost moth in **Endives** *(off-label)* [1-2]; **Lettuce** *(off-label)* [1-2];
- Grape berry moth in **Grapevines** *(off-label)* [1];
- Insect pests in **Almonds** *(off-label)* [1]; **Apricots** *(off-label)* [1]; **Bilberries** *(off-label)* [1]; **Blackberries** *(off-label)* [1]; **Blackcurrants** *(off-label)* [1]; **Blueberries** *(off-label)* [1]; **Cherries** *(off-label)* [1]; **Chestnuts** *(off-label)* [1]; **Cranberries** *(off-label)* [1]; **Elderberries** *(off-label)* [1]; **Figs** *(off-label)* [1]; **Gooseberries** *(off-label)* [1]; **Hazel nuts** *(off-label)* [1]; **Hops** *(off-label)* [1]; **Kiwi fruit** *(off-label)* [1]; **Loganberries** *(off-label)* [1]; **Medlar** *(off-label)* [1]; **Mulberries** *(off-label)* [1]; **Nectarines** *(off-label)* [1]; **Ornamental plant production** *(off-label)* [1]; **Peaches** *(off-label)* [1]; **Plums** *(off-label)* [1]; **Protected all edible seed crops** *(off-label)* [1]; **Protected all non-edible seed crops** *(off-label)* [1]; **Protected almonds** *(off-label)* [1]; **Protected apricots** *(off-label)* [1]; **Protected bilberries** *(off-label)* [1]; **Protected blackberries** *(off-label)* [1]; **Protected blackcurrants** *(off-label)* [1]; **Protected blueberries** *(off-label)* [1]; **Protected cherries** *(off-label)* [1]; **Protected chestnuts** *(off-label)* [1]; **Protected cranberries** *(off-label)* [1]; **Protected elderberries** *(off-label)* [1]; **Protected figs** *(off-label)* [1]; **Protected gooseberries** *(off-label)* [1]; **Protected hazelnuts** *(off-label)* [1]; **Protected hops** *(off-label)* [1]; **Protected kiwi fruit** *(off-label)* [1]; **Protected loganberries** *(off-label)* [1]; **Protected medlars** *(off-label)* [1]; **Protected mulberry** *(off-label)* [1]; **Protected nectarines** *(off-label)* [1]; **Protected ornamentals** *(off-label)*

[1]; **Protected peaches** *(off-label)* [1]; **Protected plums** *(off-label)* [1]; **Protected quince** *(off-label)* [1]; **Protected raspberries** *(off-label)* [1]; **Protected redcurrants** *(off-label)* [1]; **Protected rose hips** *(off-label)* [1]; **Protected Rubus hybrids** *(off-label)* [1]; **Protected strawberries** *(off-label)* [1]; **Protected table grapes** *(off-label)* [1]; **Protected table olives** *(off-label)* [1]; **Protected walnuts** *(off-label)* [1]; **Protected wine grapes** *(off-label)* [1]; **Quinces** *(off-label)* [1]; **Raspberries** *(off-label)* [1]; **Redcurrants** *(off-label)* [1]; **Rose hips** *(off-label)* [1]; **Rubus hybrids** *(off-label)* [1]; **Strawberries** *(off-label)* [1]; **Table grapes** *(off-label)* [1]; **Table olives** *(off-label)* [1]; **Walnuts** *(off-label)* [1]; **Wine grapes** *(off-label)* [1];

- Light brown apple moth in **Cherries** *(off-label)* [1];
- Peach moth in **Apricots** *(off-label)* [1]; **Nectarines** *(off-label)* [1]; **Peaches** *(off-label)* [1];
- Pollen beetle in **Mustard** *(off-label)* [1]; **Spring oilseed rape** [1-3]; **Winter oilseed rape** [1-3];
- Silver Y moth in **Brussels sprouts** *(off-label)* [1-2]; **Choi sum** *(off-label)* [1]; **Collards** *(off-label)* [1]; **Endives** *(off-label)* [1-2]; **Kale** *(off-label)* [1]; **Lettuce** *(off-label)* [1-2]; **Oriental cabbage** *(off-label)* [1]; **Protected baby leaf crops** *(off-label)* [1-2]; **Protected herbs (see appendix 6)** *(off-label)* [1-2]; **Protected kohlrabi** *(off-label)* [1-2]; **Spring greens** *(off-label)* [1]; **Tatsoi** *(off-label)* [1];
- Summer-fruit tortrix moth in **Cherries** *(off-label)* [1];
- Swift moth in **Endives** *(off-label)* [1-2]; **Lettuce** *(off-label)* [1-2];
- Turnip moth in **Endives** *(off-label)* [1-2]; **Lettuce** *(off-label)* [1-2];
- Winter moth in **Apricots** *(off-label)* [1]; **Nectarines** *(off-label)* [1]; **Peaches** *(off-label)* [1];

Extension of Authorisation for Minor Use (EAMUs)
- **Almonds** *20221280* [1]
- **Apricots** *20221280* [1], *20221283* [1]
- **Bilberries** *20221280* [1]
- **Blackberries** *20221279* [1], *20221280* [1]
- **Blackcurrants** *20221280* [1]
- **Blueberries** *20221279* [1], *20221280* [1]
- **Brussels sprouts** *20221277* [1], *20221292* [2]
- **Cherries** *20221275* [1], *20221280* [1]
- **Chestnuts** *20221280* [1]
- **Choi sum** *20221285* [1], *20221287* [2]
- **Collards** *20221285* [1], *20221287* [2]
- **Cranberries** *20221280* [1]
- **Elderberries** *20221280* [1]
- **Endives** *20221276* [1], *20221288* [2]
- **Figs** *20221280* [1]
- **Gooseberries** *20221280* [1]
- **Grapevines** *20221286* [1]
- **Hazel nuts** *20221280* [1]
- **Hops** *20221280* [1], *20221291* [2]
- **Kale** *20221281* [1], *20221285* [1], *20221287* [2]
- **Kiwi fruit** *20221280* [1]
- **Kohlrabi** *20221281* [1], *20221287* [2]
- **Lettuce** *20221276* [1], *20221288* [2]
- **Loganberries** *20221280* [1]
- **Medlar** *20221280* [1]
- **Mulberries** *20221280* [1]
- **Mustard** *20221278* [1]
- **Nectarines** *20221280* [1], *20221283* [1]
- **Oriental cabbage** *20221281* [1], *20221285* [1], *20221287* [2]
- **Ornamental plant production** *20221280* [1], *20221291* [2]
- **Peaches** *20221280* [1], *20221283* [1]
- **Plums** *20221280* [1]
- **Protected all edible seed crops** *20221280* [1], *20221291* [2]
- **Protected all non-edible seed crops** *20221280* [1], *20221291* [2]
- **Protected almonds** *20221280* [1]
- **Protected apricots** *20221280* [1]
- **Protected baby leaf crops** *20221284* [1], *20221290* [2]

SEE SECTION 3 FOR PRODUCTS ALSO REGISTERED

- **Protected bilberries** *20221280* [1]
- **Protected blackberries** *20221279* [1], *20221280* [1]
- **Protected blackcurrants** *20221280* [1]
- **Protected blueberries** *20221280* [1]
- **Protected cherries** *20221280* [1]
- **Protected chestnuts** *20221280* [1]
- **Protected cranberries** *20221280* [1]
- **Protected elderberries** *20221280* [1]
- **Protected figs** *20221280* [1]
- **Protected gooseberries** *20221280* [1]
- **Protected hazelnuts** *20221280* [1]
- **Protected herbs (see appendix 6)** *20221284* [1], *20221290* [2]
- **Protected hops** *20221280* [1], *20221291* [2]
- **Protected kiwi fruit** *20221280* [1]
- **Protected kohlrabi** *20221284* [1], *20221290* [2]
- **Protected loganberries** *20221280* [1]
- **Protected medlars** *20221280* [1]
- **Protected mulberry** *20221280* [1]
- **Protected nectarines** *20221280* [1]
- **Protected ornamentals** *20221280* [1], *20221291* [2]
- **Protected peaches** *20221280* [1]
- **Protected plums** *20221280* [1]
- **Protected quince** *20221280* [1]
- **Protected raspberries** *20221279* [1], *20221280* [1]
- **Protected redcurrants** *20221280* [1]
- **Protected rose hips** *20221280* [1]
- **Protected Rubus hybrids** *20221280* [1]
- **Protected strawberries** *20221280* [1]
- **Protected table grapes** *20221280* [1]
- **Protected table olives** *20221280* [1]
- **Protected walnuts** *20221280* [1]
- **Protected wine grapes** *20221280* [1]
- **Quinces** *20221280* [1]
- **Raspberries** *20221279* [1], *20221280* [1]
- **Redcurrants** *20221280* [1]
- **Rose hips** *20221280* [1]
- **Rubus hybrids** *20221280* [1]
- **Spring greens** *20221285* [1], *20221287* [2]
- **Strawberries** *20221280* [1], *20221283* [1]
- **Sweetcorn** *20221282* [1], *20221289* [2]
- **Table grapes** *20221280* [1]
- **Table olives** *20221280* [1]
- **Tatsoi** *20221285* [1], *20221287* [2]
- **Walnuts** *20221280* [1]
- **Wine grapes** *20221280* [1]

Approval information
- Indoxacarb included in Annex I under EC Regulation 1107/2009

Efficacy guidance
- Best results in brassica and protected crops obtained from treatment when first caterpillars are detected, or when damage first seen, or 7-10 d after trapping first adults in pheromone traps
- In apples and pears apply at egg-hatch
- Subsequent treatments in all crops may be applied at 8-14 d intervals
- Indoxacarb acts by ingestion and contact. Only larval stages are controlled but there is some ovicidal action against some species

Restrictions
- Maximum number of treatments 3 per crop or yr for apples, pears, brassica crops; 6 per yr for protected crops and 1 per crop for oilseed rape

FOR FULL CONDITIONS OF USE ALWAYS READ THE PRODUCT LABEL

- Do not apply to any crop suffering from stress from any cause
- When applying to protected crops the maximum concentration must not exceed 12.5 g of product per 100 litre water

Crop-specific information
- HI: brassica crops, protected crops 1 d; apples, pears 7 d

Following crops guidance
- When treating protected crops, observe the maximum concentration permitted.

Environmental safety
- Dangerous for the environment
- Toxic to aquatic organisms
- In accordance with good agricultural practice apply in early morning or late evening when bees are less active
- Broadcast air-assisted LERAP [1,3] (15m); LERAP Category B [1-2]

Hazard classification and safety precautions
Hazard Harmful, Dangerous for the environment, Harmful if swallowed
Transport code 9 [1-3]
Packaging group III
UN Number 3077
Risk phrases H371
Operator protection U04a, U05a; A, D
Environmental protection E12f, E15b, E16a [1-2], E17b (15m) [1,3], E34, E38, H410
Storage and disposal D01, D02, D09a, D12a
Medical advice M03

232   iodosulfuron-methyl-sodium + mesosulfuron-methyl

A sulfonyl urea herbicide mixture for winter wheat
HRAC mode of action code: 2 + 2 (B + B)

See also mesosulfuron-methyl

Products

1	Atlantis OD	Bayer CropScience	2:10 g/l	OD	18100
2	Cintac	Life Scientific	1.0:3.0% w/w	WG	18222
3	Clayton Metropolis	Clayton	2:10 g/l	OD	18401
4	Hatra	Bayer CropScience	2:10 g/l	OD	16190
5	Horus	Bayer CropScience	2:10 g/l	OD	16216
6	Niantic	Life Scientific	0.6:3.0% w/w	WG	18217

Uses
- Annual meadow grass in *Rye* *(off-label)* [1-2,4-6]; *Triticale* *(off-label)* [1,4-5]; *Winter wheat* [1-6];
- Blackgrass in *Rye* *(off-label)* [1-2,4-6]; *Triticale* *(off-label)* [1,4-5]; *Winter wheat* [1-6];
- Brome grasses in *Rye* *(off-label)* [2];
- Charlock in *Rye* *(off-label)* [2];
- Chickweed in *Rye* *(off-label)* [1-2,4-6]; *Triticale* *(off-label)* [1,4-5]; *Winter wheat* [1-6];
- Italian ryegrass in *Rye* *(off-label)* [1-2,4-6]; *Triticale* *(off-label)* [1,4-5]; *Winter wheat* [1-6];
- Mayweeds in *Rye* *(off-label)* [1-2,4-6]; *Triticale* *(off-label)* [1,4-5]; *Winter wheat* [1-6];
- Perennial ryegrass in *Rye* *(off-label)* [1,4-5]; *Triticale* *(off-label)* [1,4-5]; *Winter wheat* [1-6];
- Rough-stalked meadow grass in *Rye* *(off-label)* [1-2,4-6]; *Triticale* *(off-label)* [1,4-5]; *Winter wheat* [1-6];
- Ryegrass in *Rye* *(off-label)* [6];
- Volunteer oilseed rape in *Rye* *(off-label)* [2];
- Wild oats in *Rye* *(off-label)* [1-2,4-6]; *Triticale* *(off-label)* [1,4-5]; *Winter wheat* [1-6];

Extension of Authorisation for Minor Use (EAMUs)
- *Rye* 20200860 [1], 20202518 [2], 20200861 [4], 20200862 [5], 20202517 [6]
- *Triticale* 20200860 [1], 20200861 [4], 20200862 [5]

SEE SECTION 3 FOR PRODUCTS ALSO REGISTERED

Approval information

- Iodosulfuron-methyl-sodium and mesosulfuron-methyl included in Annex I under EC Regulation 1107/2009

Efficacy guidance

- Optimum grass weed control obtained when all grass weeds are emerged at spraying. Activity is primarily via foliar uptake and good spray coverage of the target weeds is essential
- Translocation occurs readily within the target weeds and growth is inhibited within hours of treatment but symptoms may not be apparent for up to 4 wk, depending on weed species, timing of treatment and weather conditions
- Residual activity is important for best results and is optimised by treatment on fine moist seedbeds. Avoid application under very dry conditions
- Residual efficacy may be reduced by high soil temperatures and cloddy seedbeds
- Iodosulfuron-methyl and mesosulfuron-methyl are both members of the ALS-inhibitor group of herbicides. To avoid the build up of resistance do not use any product containing an ALS-inhibitor herbicide with claims for control of grass weeds more than once on any crop
- Use these products as part of a Resistance Management Strategy that includes cultural methods of control and does not use ALS inhibitors as the sole chemical method of grass weed control. See Section 5 for more information

Restrictions

- Maximum number of treatments 1 per crop with a maximum total dose equivalent to one full dose treatment
- Do not use on crops undersown with grasses, clover or other legumes or any other broad-leaved crop
- Do not use as a stand-alone treatment for blackgrass, ryegrass or chickweed control
- To avoid the build up of resistance do not apply this or any other product containing an ALS inhibitor herbicide with claims for control of grass-weeds more than once to any crop [6]
- Do not use as the sole means of weed control in successive crops
- Do not use in mixture or in sequence with any other ALS-inhibitor herbicide except those (if any) specified on the label
- Do not apply earlier than 1 Feb in the yr of harvest [2]
- Do not apply to crops under stress from any cause
- Do not apply when rain is imminent or during periods of frosty weather
- Specified adjuvant must be used. See label

Crop-specific information

- Latest use: flag leaf just visible (GS 39)
- Winter wheat may be treated from the two-leaf stage of the crop
- Safety to crops grown for seed not established
- Avoid application before a severe frost or transient yellowing of the crop may occur.

Following crops guidance

- In the event of crop failure sow only winter wheat in the same cropping season
- Only winter wheat or winter barley (or winter oilseed rape) may be sown in the year of harvest of a treated crop but ploughing before drilling oilseed rape is advisable, especially in a season where applications are made later than usual.
- Spring wheat, spring barley, sugar beet or spring oilseed rape may be drilled in the following spring. Plough before drilling oilseed rape

Environmental safety

- Dangerous for the environment
- Very toxic to aquatic organisms
- Dangerous to fish or other aquatic life. Do not contaminate surface waters or ditches with chemical or used container
- Take extreme care to avoid drift onto plants outside the target area or on to ponds, waterways or ditches
- LERAP Category B [1-6]

Hazard classification and safety precautions

Hazard Irritant, Dangerous for the environment, Very toxic to aquatic organisms [1-5]
Transport code 9 [1-6]

FOR FULL CONDITIONS OF USE ALWAYS READ THE PRODUCT LABEL

Packaging group III
UN Number 3077, 3082
Risk phrases H315 [6], H317 [6], H318 [2,6], H319 [1,3-5]
Operator protection U05a, U11, U20b; A, C, H
Environmental protection E15a, E16a, E38, H410 [1-5], H411 [6]
Storage and disposal D01, D02, D05, D09a, D10a, D12a

233 iodosulfuron-methyl-sodium + mesosulfuron-methyl + thiencarbazone-methyl

A sulfonyl urea mixture for weed control in winter cereals
HRAC mode of action code: 2 (B)

See also iodosulfuron-methyl-sodium + mesosulfuron-methyl
iodosulfuron-methyl-sodium + propoxycarbazone-sodium

Products

1	Atlantis Star	Bayer CropScience	0.9:4.5:2.25% w/w	WG	20011
2	Proverb	Bayer CropScience	0.9:4.5:2.25% w/w	WG	19584

Uses

- Annual meadow grass in *Winter triticale* [1-2]; *Winter wheat* [1-2];
- Blackgrass in *Winter triticale* [1-2]; *Winter wheat* [1-2];
- Brome grasses in *Winter triticale* [1-2]; *Winter wheat* [1-2];
- Chickweed in *Winter triticale* [1-2]; *Winter wheat* [1-2];
- Cleavers in *Winter triticale* [1-2]; *Winter wheat* [1-2];
- Loose silky bent in *Winter triticale* [1-2]; *Winter wheat* [1-2];
- Ryegrass in *Winter triticale* [1-2]; *Winter wheat* [1-2];
- Wild oats in *Winter triticale* [1-2]; *Winter wheat* [1-2];

Approval information

- Iodosulfuron-methyl-sodium, mesosulfuron-methyl and thiencarbazone-methyl are included in Annex I under EC Regulation 1107/2009

Restrictions

- Must only be applied between 1 February in the year of harvest and the specified latest time of application

Environmental safety

- LERAP Category B [1-2]

Hazard classification and safety precautions
Transport code 9 [1-2]
Packaging group III
UN Number 3077
Risk phrases H318
Operator protection U05a, U08, U11; A, C, H
Environmental protection E16a, H410
Storage and disposal D01, D02, D05, D08 [2], D10a, D12c

234 ipconazole

A triazole fungicide for disease control in cereals
FRAC mode of action code: 3

See also imazalil + ipconazole

Products

1	Rancona 15 ME	UPL Europe	15.1 g/l	MS	19629

Uses

- Seed-borne diseases in *Spring barley* [1]; *Spring wheat* [1]; *Winter barley* [1]; *Winter wheat* [1];
- Soil-borne diseases in *Spring barley* [1]; *Spring wheat* [1]; *Winter barley* [1]; *Winter wheat* [1];

Approval information

- Ipconazole included in Annex I under EC Regulation 1107/2009
- Accepted by BBPA for use on malting barley

Hazard classification and safety precautions

Transport code 9 [1]
Packaging group III
UN Number 3082
Risk phrases H360
Operator protection U05a, U09a, U20b; A, H
Environmental protection E03, E15b, E34, H410
Storage and disposal D01, D02, D09a, D10a
Treated seed S01, S02, S03, S04d, S05, S07
Medical advice M03

235   isopyrazam

An SDHI fungicide available for disease control in barley.
FRAC mode of action code: 7

See also azoxystrobin + isopyrazam
cyprodinil + isopyrazam

Products

1 Reflect (GB only)	Adama	125 g/l	EC	18573

Uses

- Alternaria in *Celeriac* (off-label) [1]; *Parsnips* (off-label) [1]; *Red beet* (off-label) [1]; *Swedes* (off-label) [1]; *Turnips* (off-label) [1];
- Alternaria blight in *Carrots* [1];
- Canker in *Celeriac* (off-label) [1]; *Parsnips* (off-label) [1]; *Red beet* (off-label) [1]; *Swedes* (off-label) [1]; *Turnips* (off-label) [1];
- Cercospora leaf spot in *Celeriac* (off-label) [1]; *Parsnips* (off-label) [1]; *Red beet* (off-label) [1]; *Swedes* (off-label) [1]; *Turnips* (off-label) [1];
- Disease control in *Protected melons* [1];
- Grey mould in *Protected hemp* (off-label) [1];
- Powdery mildew in *Carrots* [1]; *Celeriac* (off-label) [1]; *Ornamental plant production* (off-label) [1]; *Parsnips* (off-label) [1]; *Protected aubergines* [1]; *Protected courgettes* [1]; *Protected cucumbers* [1]; *Protected hemp* (off-label) [1]; *Protected melons* [1]; *Protected ornamentals* (off-label) [1]; *Protected summer squash* [1]; *Protected tomatoes* [1]; *Red beet* (off-label) [1]; *Swedes* (off-label) [1]; *Turnips* (off-label) [1];
- Ramularia leaf spots in *Celeriac* (off-label) [1]; *Parsnips* (off-label) [1]; *Red beet* (off-label) [1]; *Swedes* (off-label) [1]; *Turnips* (off-label) [1];
- Sclerotinia in *Celeriac* (off-label) [1]; *Parsnips* (off-label) [1]; *Protected hemp* (off-label) [1]; *Red beet* (off-label) [1]; *Swedes* (off-label) [1]; *Turnips* (off-label) [1];
- Septoria leaf blotch in *Celeriac* (off-label) [1]; *Parsnips* (off-label) [1]; *Red beet* (off-label) [1]; *Swedes* (off-label) [1]; *Turnips* (off-label) [1];

Extension of Authorisation for Minor Use (EAMUs)

- *Celeriac* 20221190 [1]
- *Ornamental plant production* 20221187 [1]
- *Parsnips* 20221190 [1]
- *Protected hemp* 20221188 [1]
- *Protected ornamentals* 20221187 [1]
- *Red beet* 20221190 [1]

FOR FULL CONDITIONS OF USE ALWAYS READ THE PRODUCT LABEL

- **Swedes** *20221190* [1]
- **Turnips** *20221190* [1]

Approval information
- Isopyrazam is included in Appendix 1 under EC Regulation 1107/2009

Efficacy guidance
- Use only in mixture with fungicides with a different mode of action that are effective against the target diseases.

Restrictions
- Isopyrazam is an SDH respiration inhibitor; Do not apply more than two foliar applications of products containing an SDH inhibitor to any cereal crop

Crop-specific information
- There are no restrictions on succeeding crops in a normal rotation

Environmental safety
- LERAP Category B [1]

Hazard classification and safety precautions
 Hazard Harmful if swallowed
 Transport code 9 [1]
 Packaging group III
 UN Number 3082
 Risk phrases H319, H351, H361
 Operator protection U05a, U09a, U19a, U20e; A, C, H
 Environmental protection E16a, E34, H410
 Storage and disposal D01, D02, D05, D09a, D10c, D11a
 Medical advice M03

236   isoxaben

A soil-acting benzamide herbicide for use in cereals, grass and fruit
HRAC mode of action code: 29 (L)

Products

1	Flexidor	Landseer	500 g/l	SC	18042
2	Pan Isoxaben 500	Pan Amenity	500 g/l	SC	18419

Uses
- Annual dicotyledons in **Almonds** *(off-label)* [1]; **Amenity vegetation** [1-2]; **Apples** [1-2]; **Asparagus** [1-2]; **Blackberries** [1-2]; **Blackcurrants** [1-2]; **Broad beans** *(off-label)* [1]; **Bulb onions** *(off-label)* [1]; **Carrots** *(off-label)* [1]; **Cherries** [1-2]; **Chestnuts** *(off-label)* [1]; **Chicory** *(off-label)* [1]; **Courgettes** *(off-label)* [1]; **Forest** [1-2]; **Forest nurseries** [1-2]; **Gooseberries** [1-2]; **Hazel nuts** *(off-label)* [1]; **Hops** [1-2]; **Horseradish** *(off-label)* [1]; **Leeks** *(off-label)* [1]; **Miscanthus** *(off-label)* [1]; **Ornamental plant production** [1-2]; **Parsnips** *(off-label)* [1]; **Pears** [1-2]; **Plums** [1-2]; **Pumpkins** *(off-label)* [1]; **Raspberries** [1-2]; **Redcurrants** [1-2]; **Rhubarb** *(off-label)* [1]; **Shallots** *(off-label)* [1]; **Strawberries** [1-2]; **Summer squash** *(off-label)* [1]; **Triticale** [1-2]; **Walnuts** *(off-label)* [1]; **Winter barley** [1-2]; **Winter oats** [1-2]; **Winter rye** [1-2]; **Winter squash** *(off-label)* [1]; **Winter wheat** [1-2]; **Witloof** *(off-label)* [1];
- Chickweed in **Pumpkins** *(off-label)* [1]; **Rhubarb** *(off-label)* [1]; **Winter squash** *(off-label)* [1];
- Volunteer oilseed rape in **Bulb onions** *(off-label)* [1]; **Leeks** *(off-label)* [1]; **Shallots** *(off-label)* [1];

Extension of Authorisation for Minor Use (EAMUs)
- **Almonds** *20180011* [1], *20210727* [1]
- **Broad beans** *20193020* [1], *20210728* [1]
- **Bulb onions** *20181145* [1], *20210724* [1]
- **Carrots** *20180020* [1], *20210730* [1]
- **Chestnuts** *20180011* [1], *20210727* [1]
- **Chicory** *20193019* [1], *20210726* [1]

SEE SECTION 3 FOR PRODUCTS ALSO REGISTERED

- ***Courgettes*** *20180012* [1], *20210731* [1]
- ***Hazel nuts*** *20180011* [1], *20210727* [1]
- ***Horseradish*** *20180020* [1], *20210730* [1]
- ***Leeks*** *20181145* [1], *20210724* [1]
- ***Miscanthus*** *20193907* [1], *20210725* [1]
- ***Parsnips*** *20180020* [1], *20210730* [1]
- ***Pumpkins*** *20200883* [1], *20210723* [1]
- ***Rhubarb*** *20192598* [1], *20210729* [1]
- ***Shallots*** *20181145* [1], *20210724* [1]
- ***Summer squash*** *20180012* [1], *20210731* [1]
- ***Walnuts*** *20180011* [1], *20210727* [1]
- ***Winter squash*** *20200883* [1], *20210723* [1]
- ***Witloof*** *20193019* [1], *20210726* [1]

Approval information
- Isoxaben included in Annex 1 under EC Regulation 1107/2009
- Accepted by BBPA for use on hops

Efficacy guidance
- When used alone apply pre-weed emergence
- Effectiveness is reduced in dry conditions. Weed seeds germinating at depth are not controlled
- Activity reduced on soils with more than 10% organic matter. Do not use on peaty soils
- Various tank mixtures are recommended for early post-weed emergence treatment (especially for grass weeds). See label for details

Restrictions
- Maximum number of treatments 1 per crop for all edible crops; 2 per yr on amenity vegetation and non-edible crops

Crop-specific information
- Latest use: before 1 Apr in yr of harvest for edible crops

Following crops guidance
- See label for details of crops which may be sown in the event of failure of a treated crop

Environmental safety
- Keep all livestock out of treated areas for at least 50 d
- Buffer zone requirement 20 m for rhubarb [1]

Hazard classification and safety precautions
    Hazard Very toxic to aquatic organisms
    Transport code 9 [1-2]
    Packaging group III
    UN Number 3082
    Operator protection U05a, U20a; A, C, H, M
    Environmental protection E06a (50 days), E15b, H410
    Storage and disposal D01, D02, D09a, D11a

237   kresoxim-methyl

A protectant strobilurin fungicide for apples
FRAC mode of action code: 11

Products

1	Stroby WG	BASF	50% w/w	WG	17316

Uses
- American gooseberry mildew in ***Blueberries*** *(off-label)* [1]; ***Cranberries*** *(off-label)* [1]; ***Gooseberries*** *(off-label)* [1]; ***Whitecurrants*** *(off-label)* [1];
- Black rot in ***Wine grapes*** *(off-label)* [1];

- Powdery mildew in **Apple for cider making** *(reduction)* [1]; **Apples** *(reduction)* [1]; **Blackcurrants** [1]; **Blueberries** *(off-label)* [1]; **Cranberries** *(off-label)* [1]; **Gooseberries** *(off-label)* [1]; **Ornamental plant production** [1]; **Protected strawberries** [1]; **Redcurrants** [1]; **Strawberries** [1]; **Whitecurrants** *(off-label)* [1]; **Wine grapes** *(off-label)* [1];
- Scab in **Apple for cider making** [1]; **Apples** [1];

Extension of Authorisation for Minor Use (EAMUs)
- **Blueberries** *20160089* [1]
- **Cranberries** *20160089* [1]
- **Gooseberries** *20160089* [1]
- **Whitecurrants** *20160089* [1]
- **Wine grapes** *20160960* [1]

Approval information
- Kresoxim-methyl included in Annex I under EC Regulation 1107/2009
- Approved for use in ULV systems

Efficacy guidance
- Activity is protectant. Best results achieved from treatments prior to disease development. See label for timing details on each crop. Treatments should be repeated at 10-14 d intervals but note limitations below
- To minimise the likelihood of development of resistance to strobilurin fungicides these products should be used in a planned Resistance Management strategy. See Section 5 for more information
- Product may be applied in ultra low volumes (ULV) but disease control may be reduced

Restrictions
- Maximum number of treatments 4 per yr on apples; 3 per yr on other crops. See notes in Efficacy about limitations on consecutive treatments
- Consult before using on crops intended for processing

Crop-specific information
- HI 14 d for blackcurrants, protected strawberries, strawberries; 35 d for apples
- On apples do not spray product more than twice consecutively and separate each block of two consecutive treatments with at least two applications from a different cross-resistance group. For all other crops do not apply consecutively and use a maximum of once in every three fungicide sprays
- Product should not be used as final spray of the season on apples

Environmental safety
- Dangerous for the environment
- Very toxic to aquatic organisms
- Harmless to ladybirds and predatory mites
- Harmless to honey bees and may be applied during flowering. Nevertheless local beekeepers should be notified when treatment of orchards in flower is to occur

Hazard classification and safety precautions
Hazard Harmful, Dangerous for the environment, Very toxic to aquatic organisms
Transport code 9 [1]
Packaging group III
UN Number 3077
Risk phrases H351
Operator protection U05a, U20b
Environmental protection E15a, E38, H410
Storage and disposal D01, D02, D08, D09a, D10c, D12a
Medical advice M05a

SEE SECTION 3 FOR PRODUCTS ALSO REGISTERED

238   lambda-cyhalothrin

A quick-acting contact and ingested pyrethroid insecticide
IRAC mode of action code: 3

Products

1	Audax	Sipcam	100 g/l	CS	19697
2	Balliol	Sipcam	100 g/l	CS	19343
3	Clayton Sparta	Clayton	50 g/l	EC	13457
4	Cleancrop Argent	Agrii	100 g/l	CS	18286
5	CleanCrop Corsair	Agrii	50 g/l	EC	14124
6	Colt 10 CS	FMC Agro	100 g/l	CS	18708
7	Coucal	Syngenta	50 g/l	CS	19751
8	Dalda 5	Globachem	50 g/l	EC	13688
9	Hallmark with Zeon Technology	Syngenta	100 g/l	CS	12629
10	Karis 10 CS	FMC Agro	100 g/l	CS	18707
11	Kendo	Syngenta	50 g/l	CS	15562
12	Kung Fu	Syngenta	50 g/l	CS	18974
13	Kusti	Syngenta	50 g/l	CS	16656
14	Lambdastar	Life Scientific	100 g/l	CS	17406
15	Ninja 5CS	Syngenta	50 g/l	CS	16417
16	Sceptre	Adama	2.5% w/w	GR	18868
17	Seal Z	AgChem Access	100 g/l	CS	14201
18	Sparviero	Sipcam	100 g/l	CS	15687
19	Stealth	Syngenta	100 g/l	CS	14551
20	Warrior	Syngenta	100 g/l	CS	13857

Uses

- Aphids in *All edible seed crops grown outdoors* (off-label) [9]; *All non-edible seed crops grown outdoors* (off-label) [9]; *Asparagus* (off-label) [9]; *Broccoli* [6,10]; *Brussels sprouts* [6,10]; *Bulb onions* (off-label) [9]; *Cabbages* [6]; *Calabrese* [6,10]; *Cauliflowers* [6]; *Chestnuts* (off-label) [9]; *Combining peas* [3,5,8,13]; *Crambe* (off-label) [9]; *Durum wheat* [1-2,4,6-12,15-17,19-20]; *Edible podded peas* [8]; *Farm forestry* (off-label) [9]; *Forest nurseries* (off-label) [9]; *Garlic* (off-label) [9]; *Hazel nuts* (off-label) [9]; *Hops* (off-label) [9]; *Leeks* (off-label) [9]; *Lettuce* [8]; *Miscanthus* (off-label) [9]; *Ornamental plant production* (off-label) [9]; *Pears* [8]; *Potatoes* [1-15,17-20]; *Protected asparagus* (off-label) [9]; *Protected aubergines* (off-label) [9]; *Protected forest nurseries* (off-label) [9]; *Protected ornamentals* (off-label) [9]; *Protected pak choi* (off-label) [9]; *Protected peppers* (off-label) [9]; *Protected tatsoi* (off-label) [9]; *Protected tomatoes* (off-label) [9]; *Rye* [8]; *Salad onions* (off-label) [9]; *Shallots* (off-label) [9]; *Soft fruit* (off-label) [9]; *Spring barley* [1-2,4,6-12,14-15,17-20]; *Spring field beans* [8]; *Spring oats* [1-2,4,6-12,14-15,17-20]; *Spring oilseed rape* [8]; *Spring wheat* [1-15,17-20]; *Sugar beet* [6,8,10]; *Top fruit* (off-label) [9]; *Triticale* [8]; *Vining peas* [3,5,8,13]; *Walnuts* (off-label) [9]; *Willow (Short rotation coppice)* (off-label) [9]; *Winter barley* [1-2,4,6-12,14-20]; *Winter field beans* [8]; *Winter oats* [1-2,4,6-12,14-20]; *Winter oilseed rape* [8]; *Winter wheat* [1-20];
- Barley yellow dwarf virus vectors in *Durum wheat* [1-2,4,6-7,9-12,15-17,19-20]; *Spring barley* [1-2,20]; *Spring oats* [20]; *Spring wheat* [1-2,20]; *Winter barley* [1-2,4,6-7,9-12,14-20]; *Winter oats* [1-2,4,6-7,9-12,14-20]; *Winter wheat* [1-2,4,6-7,9-12,14-20];
- Bean seed fly in *Soya beans* (off-label) [14];
- Beet leaf miner in *Sugar beet* [1-2,4,6-7,9-12,14-15,17-20];
- Beet virus yellows vectors in *Spring oilseed rape* [1-2,4,6-7,9-12,14-15,17-20]; *Winter oilseed rape* [1-2,4,6-7,9-12,14-20];
- Beetles in *Combining peas* [8]; *Durum wheat* [8]; *Edible podded peas* [8]; *Potatoes* [8]; *Rye* [8]; *Spring barley* [8]; *Spring field beans* [8]; *Spring oats* [8]; *Spring oilseed rape* [8]; *Spring wheat* [8]; *Sugar beet* [8]; *Triticale* [8]; *Vining peas* [8]; *Winter barley* [8]; *Winter field beans* [8]; *Winter oats* [8]; *Winter oilseed rape* [8]; *Winter wheat* [8];
- Black bean aphid in *Asparagus* (off-label) [9]; *Protected asparagus* (off-label) [9];
- Brassica pod midge in *Spring oilseed rape* [3,5,13]; *Winter oilseed rape* [3,5,13];

FOR FULL CONDITIONS OF USE ALWAYS READ THE PRODUCT LABEL

- Cabbage seed weevil in *Spring oilseed rape* [1-7,9-15,17-19]; *Winter oilseed rape* [1-7,9-15,17-19];
- Cabbage stem flea beetle in *Spring oilseed rape* [1-2,4,6-7,9-12,14-15,17-20]; *Winter oilseed rape* [1-2,4,6-7,9-12,14-20];
- Capsids in *Blackberries* (off-label) [9]; *Dewberries* (off-label) [9]; *Grapevines* (off-label) [9]; *Raspberries* (off-label) [9]; *Rubus hybrids* (off-label) [9];
- Carrot fly in *Carrots* [20]; *Celery (outdoor)* (off-label) [9]; *Fennel* (off-label) [9]; *Horseradish* (off-label) [9]; *Mallow (Althaea spp.)* (off-label) [9]; *Parsley root* (off-label) [9]; *Parsnips* [20];
- Caterpillars in *Beetroot* (off-label) [9]; *Broccoli* [1-2,4,7,9,11-12,14-15,17,19-20]; *Brussels sprouts* [1-2,4,7,9,11-12,14-15,17,19-20]; *Cabbages* [1-2,4,7,9,11-12,14-15,17,19-20]; *Calabrese* [1-2,4,7,9,11-12,14-15,17,19-20]; *Cauliflowers* [1-2,4,7,9,11-12,14-15,17,19-20]; *Celery (outdoor)* (off-label) [9]; *Combining peas* [8]; *Durum wheat* [8]; *Edible podded peas* [8]; *French beans* (off-label) [9]; *Navy beans* (off-label) [9]; *Pears* [8]; *Potatoes* [8]; *Rye* [8]; *Spring barley* [8]; *Spring field beans* [8]; *Spring oats* [8]; *Spring oilseed rape* [8]; *Spring wheat* [8]; *Sugar beet* [8]; *Swedes* (off-label) [9]; *Triticale* [8]; *Turnips* (off-label) [9]; *Vining peas* [8]; *Winter barley* [8]; *Winter field beans* [8]; *Winter oats* [8]; *Winter oilseed rape* [8]; *Winter wheat* [8];
- Clay-coloured weevil in *Blackberries* (off-label) [9]; *Dewberries* (off-label) [9]; *Raspberries* (off-label) [9]; *Rubus hybrids* (off-label) [9];
- Common green capsid in *Hops* (off-label) [9];
- Cutworms in *Beetroot* (off-label) [9]; *Carrots* [1-2,4,6-7,9-12,14-15,17-19]; *Chicory* (off-label) [9]; *Fennel* (off-label) [9]; *Lettuce* [1-2,4,6-7,9-12,14-15,17-19]; *Outdoor lettuce* [20]; *Parsnips* [4,6-7,9-12,14-15,17-19]; *Sugar beet* [1-7,9-15,17-20];
- Flax flea beetle in *Spring linseed* (off-label) [5]; *Winter linseed* (off-label) [5];
- Flea beetle in *Corn Gromwell* (off-label) [9]; *Hops* (off-label) [9]; *Poppies* (off-label) [9]; *Poppies for morphine production* (off-label) [9]; *Poppies grown for seed production* (off-label) [9]; *Spring linseed* (off-label) [5]; *Spring oilseed rape* [1-7,9-15,17-20]; *Sugar beet* [1-7,9-15,17-20]; *Winter linseed* (off-label) [5]; *Winter oilseed rape* [1-7,9-20];
- Frit fly in *Sweetcorn* (off-label) [9];
- Grain aphid in *Durum wheat* [20]; *Spring barley* [20]; *Spring oats* [20]; *Spring wheat* [20]; *Winter barley* [20]; *Winter oats* [20]; *Winter wheat* [20];
- Insect pests in *All edible seed crops grown outdoors* (off-label) [9]; *All non-edible seed crops grown outdoors* (off-label) [9]; *Beetroot* (off-label) [9]; *Blackberries* (off-label) [9]; *Blackcurrants* (off-label) [9]; *Borage for oilseed production* (off-label) [9]; *Bulb onions* (off-label) [9]; *Canary flower (Echium spp.)* (off-label) [9]; *Celeriac* (off-label) [9]; *Celery (outdoor)* (off-label) [9]; *Cherries* (off-label) [9]; *Chestnuts* (off-label) [9]; *Crambe* (off-label) [9]; *Dewberries* (off-label) [9]; *Evening primrose* (off-label) [9]; *Farm forestry* (off-label) [9]; *Fodder beet* (off-label) [9]; *Forest nurseries* (off-label) [9]; *Frise* (off-label) [9]; *Garlic* (off-label) [9]; *Gooseberries* (off-label) [9]; *Grass seed crops* (off-label) [9]; *Hazel nuts* (off-label) [9]; *Herbs (see appendix 6)* (off-label) [9]; *Honesty* (off-label) [9]; *Hops* (off-label) [9]; *Horseradish* (off-label) [9]; *Lamb's lettuce* (off-label) [9]; *Leaf brassicas* (off-label) [9]; *Leeks* (off-label) [9]; *Lupins* (off-label) [9]; *Mallow (Althaea spp.)* (off-label) [9]; *Mirabelles* (off-label) [9]; *Miscanthus* (off-label) [9]; *Mustard* (off-label) [9]; *Navy beans* (off-label) [9]; *Ornamental plant production* (off-label) [9]; *Parsley root* (off-label) [9]; *Plums* (off-label) [9]; *Poppies for morphine production* (off-label) [9]; *Protected aubergines* (off-label) [9]; *Protected forest nurseries* (off-label) [9]; *Protected ornamentals* (off-label) [9]; *Protected pak choi* (off-label) [9]; *Protected peppers* (off-label) [9]; *Protected strawberries* (off-label) [9]; *Protected tatsoi* (off-label) [9]; *Protected tomatoes* (off-label) [9]; *Radishes* (off-label) [9]; *Raspberries* (off-label) [9]; *Redcurrants* (off-label) [9]; *Rubus hybrids* (off-label) [9]; *Runner beans* (off-label) [9]; *Salad onions* (off-label) [9]; *Scarole* (off-label) [9]; *Shallots* (off-label) [9]; *Soft fruit* (off-label) [9]; *Spring linseed* (off-label) [9]; *Spring rye* (off-label) [9]; *Strawberries* (off-label) [9]; *Top fruit* (off-label) [9]; *Triticale* (off-label) [9]; *Walnuts* (off-label) [9]; *Whitecurrants* (off-label) [9]; *Willow (Short rotation coppice)* (off-label) [9]; *Winter linseed* (off-label) [9]; *Winter rye* (off-label) [9];
- Leaf midge in *Blackcurrants* (off-label) [9]; *Gooseberries* (off-label) [9]; *Redcurrants* (off-label) [9]; *Whitecurrants* (off-label) [9];
- Leaf miner in *Sugar beet* [3,5,13];
- Mangold fly in *Sugar beet* [6,10,20];
- Orange blossom midge in *Spring wheat* [20]; *Winter wheat* [20];

SEE SECTION 3 FOR PRODUCTS ALSO REGISTERED

- Pea and bean weevil in *Combining peas* [3-7,9-15,17-20]; *Edible podded peas* [1-2,4,6-7,9-12,14-15,17-20]; *Peas* [1-2]; *Spring field beans* [1-7,9-15,17-20]; *Vining peas* [3-7,9-15,17-20]; *Winter field beans* [1-7,9-15,17-20];
- Pea aphid in *Combining peas* [4,6-7,9-12,14-15,17-20]; *Edible podded peas* [1-2,4,6-7,9-12,14-15,17-20]; *Peas* [1-2]; *Vining peas* [4,6-7,9-12,14-15,17-20];
- Pea midge in *Combining peas* [4,7,9,11-12,14-15,17-19]; *Edible podded peas* [1-2,4,7,9,11-12,14-15,17-19]; *Peas* [1-2]; *Vining peas* [4,7,9,11-12,14-15,17-19];
- Pea moth in *Combining peas* [3-7,9-15,17-20]; *Edible podded peas* [1-2,4,6-7,9-12,14-15,17-20]; *Peas* [1-2]; *Vining peas* [9-15,17-20];
- Peach-potato aphid in *Asparagus* (off-label) [9]; *Protected asparagus* (off-label) [9]; *Spring linseed* (off-label) [5]; *Winter linseed* (off-label) [5];
- Pear midge in *Combining peas* [6,10,20]; *Edible podded peas* [6,10,20]; *Vining peas* [6,10,20];
- Pear sucker in *Pears* [1-7,9-15,17-20]; *Winter wheat* [3,5,13];
- Pod midge in *Poppies* (off-label) [9]; *Poppies grown for seed production* (off-label) [9]; *Spring oilseed rape* [1-2,4,6-7,9-12,14-15,17-20]; *Winter oilseed rape* [1-2,4,6-7,9-12,14-15,17-20];
- Pollen beetle in *Corn Gromwell* (off-label) [9]; *Poppies* (off-label) [9]; *Poppies for morphine production* (off-label) [9]; *Poppies grown for seed production* (off-label) [9]; *Spring oilseed rape* [1-7,9-15,17-20]; *Winter oilseed rape* [1-7,9-15,17-20];
- Rosy rustic moth in *Hops* (off-label) [9];
- Sawflies in *Blackcurrants* (off-label) [9]; *Gooseberries* (off-label) [9]; *Redcurrants* (off-label) [9]; *Whitecurrants* (off-label) [9];
- Seed beetle in *Broad beans* (off-label) [9];
- Seed weevil in *Poppies* (off-label) [9]; *Poppies grown for seed production* (off-label) [9]; *Spring oilseed rape* [20]; *Winter oilseed rape* [20];
- Silver Y moth in *French beans* (off-label) [9]; *Runner beans* (off-label) [9];
- Spotted wing drosophila in *Bilberries* (off-label) [9]; *Blueberries* (off-label) [9]; *Cranberries* (off-label) [9]; *Ribes hybrids* (off-label) [9];
- Springtails in *Poppies* (off-label) [9]; *Poppies grown for seed production* (off-label) [9];
- Thrips in *Bilberries* (off-label) [9]; *Blueberries* (off-label) [9]; *Broad beans* (off-label) [9]; *Bulb onions* (off-label) [9]; *Cranberries* (off-label) [9]; *Garlic* (off-label) [9]; *Leeks* (off-label) [9]; *Poppies* (off-label) [9]; *Poppies grown for seed production* (off-label) [9]; *Ribes hybrids* (off-label) [9]; *Salad onions* (off-label) [9]; *Shallots* (off-label) [9];
- Wasps in *Grapevines* (off-label) [9];
- Weevils in *Combining peas* [8]; *Durum wheat* [8]; *Edible podded peas* [8]; *Potatoes* [8]; *Rye* [8]; *Spring barley* [8]; *Spring field beans* [8]; *Spring oats* [8]; *Spring oilseed rape* [8]; *Spring wheat* [8]; *Sugar beet* [8]; *Triticale* [8]; *Vining peas* [8]; *Winter barley* [8]; *Winter field beans* [8]; *Winter oats* [8]; *Winter oilseed rape* [8]; *Winter wheat* [8];
- Wheat-blossom midge in *Winter wheat* [4,6-7,9-12,14-15,17-19];
- Whitefly in *Broccoli* [1-2,4,6-7,9-12,14-15,17,19-20]; *Brussels sprouts* [1-2,4,6-7,9-12,14-15,17,19-20]; *Cabbages* [1-2,4,6-7,9,11-12,14-15,17,19-20]; *Calabrese* [1-2,4,6-7,9-12,14-15,17,19-20]; *Cauliflowers* [1-2,4,6-7,9,11-12,14-15,17,19-20];
- Yellow cereal fly in *Winter wheat* [1-7,9-15,17-20];

Extension of Authorisation for Minor Use (EAMUs)
- *All edible seed crops grown outdoors* *20082944* [9]
- *All non-edible seed crops grown outdoors* *20082944* [9]
- *Asparagus* *20182878* [9]
- *Beetroot* *20060743* [9]
- *Bilberries* *20140521* [9]
- *Blackberries* *20060728* [9]
- *Blackcurrants* *20060727* [9]
- *Blueberries* *20140521* [9]
- *Borage for oilseed production* *20060634* [9]
- *Broad beans* *20060753* [9]
- *Bulb onions* *20190109* [9]
- *Canary flower (Echium spp.)* *20060634* [9]
- *Celeriac* *20060731* [9]
- *Celery (outdoor)* *20060744* [9]
- *Cherries* *20131273* [9]

FOR FULL CONDITIONS OF USE ALWAYS READ THE PRODUCT LABEL

- **Chestnuts** *20060742* [9]
- **Chicory** *20060740* [9]
- **Corn Gromwell** *20121047* [9]
- **Crambe** *20081046* [9]
- **Cranberries** *20140521* [9]
- **Dewberries** *20060728* [9]
- **Evening primrose** *20060634* [9]
- **Farm forestry** *20082944* [9]
- **Fennel** *20060733* [9]
- **Fodder beet** *20060637* [9]
- **Forest nurseries** *20082944* [9]
- **French beans** *20060739* [9]
- **Frise** *20190108* [9]
- **Garlic** *20190109* [9]
- **Gooseberries** *20060727* [9]
- **Grapevines** *20060266* [9]
- **Grass seed crops** *20060624* [9]
- **Hazel nuts** *20060742* [9]
- **Herbs (see appendix 6)** *20190108* [9]
- **Honesty** *20060634* [9]
- **Hops** *20082944* [9], *20131719* [9]
- **Horseradish** *20071301* [9]
- **Lamb's lettuce** *20190108* [9]
- **Leaf brassicas** *20190108* [9]
- **Leeks** *20190109* [9]
- **Lupins** *20060635* [9]
- **Mallow (Althaea spp.)** *20071301* [9]
- **Mirabelles** *20131273* [9]
- **Miscanthus** *20082944* [9]
- **Mustard** *20060634* [9]
- **Navy beans** *20060739* [9]
- **Ornamental plant production** *20082944* [9]
- **Parsley root** *20071301* [9]
- **Plums** *20131273* [9]
- **Poppies** *20180013* [9]
- **Poppies for morphine production** *20060749* [9]
- **Poppies grown for seed production** *20180013* [9]
- **Protected asparagus** *20182878* [9]
- **Protected aubergines** *20121994* [9]
- **Protected forest nurseries** *20082944* [9]
- **Protected ornamentals** *20082944* [9]
- **Protected pak choi** *20081263* [9]
- **Protected peppers** *20121994* [9]
- **Protected strawberries** *20111705* [9]
- **Protected tatsoi** *20081263* [9]
- **Protected tomatoes** *20121994* [9]
- **Radishes** *20060731* [9]
- **Raspberries** *20060728* [9]
- **Redcurrants** *20060727* [9]
- **Ribes hybrids** *20140521* [9]
- **Rubus hybrids** *20060728* [9]
- **Runner beans** *20060739* [9]
- **Salad onions** *20190109* [9]
- **Scarole** *20190108* [9]
- **Shallots** *20190109* [9]
- **Soft fruit** *20082944* [9]
- **Soya beans** *20180785* [14]
- **Spring linseed** *20142296* [5], *20060634* [9]
- **Spring rye** *20060624* [9]

SEE SECTION 3 FOR PRODUCTS ALSO REGISTERED

- ***Strawberries*** *20111705* [9]
- ***Swedes*** *20101856* [9]
- ***Sweetcorn*** *20060732* [9]
- ***Top fruit*** *20082944* [9]
- ***Triticale*** *20060624* [9]
- ***Turnips*** *20101856* [9]
- ***Walnuts*** *20060742* [9]
- ***Whitecurrants*** *20060727* [9]
- ***Willow (Short rotation coppice)*** *20060748* [9]
- ***Winter linseed*** *20142296* [5], *20060634* [9]
- ***Winter rye*** *20060624* [9]

Approval information
- Lambda-cyhalothrin included in Annex I under EC Regulation 1107/2009
- Accepted by BBPA for use on malting barley and hops

Efficacy guidance
- Best results normally obtained from treatment when pest attack first seen. See label for detailed recommendations on each crop
- Timing for control of barley yellow dwarf virus vectors depends on specialist assessment of the level of risk in the area
- Repeat applications recommended in some crops where prolonged attack occurs, up to maximum total dose. See label for details
- Where strains of aphids resistant to lambda-cyhalothrin occur control is unlikely to be satisfactory
- Addition of wetter recommended for control of certain pests in brassicas and oilseed rape
- Use of sufficient water volume to ensure thorough crop penetration recommended for optimum results
- Use of drop-legged sprayer gives improved results in crops such as Brussels sprouts

Restrictions
- Maximum number of applications or maximum total dose per crop varies - see labels
- Do not apply to a cereal crop if any product containing a pyrethroid insecticide or dimethoate has been applied to the crop after the start of ear emergence (GS 51)
- Do not spray cereals in the spring/summer (i.e. after 1 Apr) within 6 m of edge of crop

Crop-specific information
- Latest use before late milk stage on cereals; before end of flowering for winter oilseed rape
- HI 3 d for radishes, red beet; 7 d for lettuce [9]; 14 d for carrots and parsnips; 25 d for peas, field beans; 6 wk for spring oilseed rape; 8 wk for sugar beet

Environmental safety
- Dangerous for the environment
- Very toxic to aquatic organisms
- Flammable [3]
- To protect non-target arthropods respect an untreated buffer zone of 5 m to non-crop land
- Since there is a high risk to non-target insects or other anthropods, do not spray cereals in spring/summer i.e. after 1st April within 6m of the field boundary [16]
- Dangerous to bees
- Broadcast air-assisted LERAP [1-2,4,6-9,11-15,17,19-20,3,5,18] (38 m); LERAP Category A [3,5,13,18,20] ; LERAP Category B [1-2,4,6-12,14-17,19-20]

Hazard classification and safety precautions
Hazard Harmful, Corrosive [8], Flammable [3,5,13], Dangerous for the environment, Flammable liquid and vapour [3], Toxic if swallowed [8], Harmful if swallowed [1-4,6-7,9-15,17-20], Harmful if inhaled [1-4,6,8-10,13-20], Very toxic to aquatic organisms [1-4,6-7,9-15,17-20]
Transport code 3 [3,5], 8 [8,5], 9 [1-2,4,6-7,9-20]
Packaging group II, III
UN Number 1760, 1993, 3072, 3082
Risk phrases H304 [3,8], H314 [8], H317 [4,7,9,11-13,15,17-20], H319 [3], H336 [3,8], R20 [5], R21 [5], R22a [5], R22b [5], R36 [5], R37 [5], R38 [5], R50 [5], R53a [5], R67 [5]

FOR FULL CONDITIONS OF USE ALWAYS READ THE PRODUCT LABEL

Operator protection U02a [1-2,4,6-12,14-20], U04a [8], U05a, U08 [1-15,17-20], U09a [16], U11 [8,16], U14 [1-2,4,6-7,9-12,14-20], U19a [3,5,8,13], U20a [3,5,8,13], U20b [1-2,4,6-7,9-12,14-20]; A, C, H, J, K, M
Environmental protection E12a [8], E12c [3-7,9,11-20], E15a [1-2], E15b [5,8,13,16,20], E16a [1-2,4,6-12,14-17,19-20], E16b [1-2,6], E16c [3,5,13,18], E16d [20], E17b (25 m) [1-2,4,6-9,11-15,17,19-20], E17b (38 m) [3,5,18], E22b [4,6-9,11-12,14-16,19-20], E34 [1-15,17-20], E38 [4,6-12,14-20], H410 [1-4,6-20]
Storage and disposal D01, D02, D05 [1-3,5,13], D07 [3,5,13], D09a [1-15,17-20], D10b [3,5,8,13], D10c [1-2,4,6-7,9-12,14-15,17-20], D12a [1-7,9-20]
Medical advice M03 [1-2,4,6-7,9-12,14-20], M04a [8], M05a [1-2], M05b [5,8,13]

239 laminarin

A fungicide for winter wheat

Products

1	Iodus	UPL Europe	37 g/l	SL	19163
2	Vacciplant	UPL Europe	37 g/l	SL	13260

Uses
- Powdery mildew in **Winter wheat** [1-2];
- Septoria leaf blotch in **Winter wheat** [1-2];

Approval information
- Laminarin included in Annex 1 under EC Regulation 1107/2009

Hazard classification and safety precautions
UN Number N/C
Risk phrases H290, H312, H314, H335
Operator protection A, C, H
Environmental protection E15b, H411
Storage and disposal D01, D02, D05, D09a, D10b

240 Lecanicillium muscarium

A fungal pathogen of whitefly now reclassified as Akanthomyces muscarius
IRAC mode of action code: UNF

Products

1	Mycotal	Koppert	4.8% w/w	WP	16644

Uses
- Spider mites in **Leaf brassicas** (off-label) [1]; **Protected bilberries** (off-label) [1]; **Protected blackberries** (off-label) [1]; **Protected blackcurrants** (off-label) [1]; **Protected blueberry** (off-label) [1]; **Protected cayenne peppers** (off-label) [1]; **Protected cranberries** (off-label) [1]; **Protected gooseberries** (off-label) [1]; **Protected herbs (see appendix 6)** (off-label) [1]; **Protected hops** (off-label) [1]; **Protected loganberries** (off-label) [1]; **Protected nursery fruit trees** (off-label) [1]; **Protected raspberries** (off-label) [1]; **Protected redcurrants** (off-label) [1]; **Protected Ribes hybrids** (off-label) [1]; **Protected Rubus hybrids** (off-label) [1]; **Protected table grapes** (off-label) [1]; **Protected wine grapes** (off-label) [1];
- Thrips in **Leaf brassicas** (off-label) [1]; **Protected bilberries** (off-label) [1]; **Protected blackcurrants** (off-label) [1]; **Protected blueberry** (off-label) [1]; **Protected cayenne peppers** (off-label) [1]; **Protected cranberries** (off-label) [1]; **Protected gooseberries** (off-label) [1]; **Protected herbs (see appendix 6)** (off-label) [1]; **Protected hops** (off-label) [1]; **Protected loganberries** (off-label) [1]; **Protected nursery fruit trees** (off-label) [1]; **Protected raspberries** (off-label) [1]; **Protected redcurrants** (off-label) [1]; **Protected Ribes hybrids** (off-label) [1]; **Protected Rubus hybrids** (off-label) [1]; **Protected table grapes** (off-label) [1]; **Protected wine grapes** (off-label) [1];

- Whitefly in **Leaf brassicas** *(off-label)* [1]; **Protected bilberries** *(off-label)* [1]; **Protected blackberries** *(off-label)* [1]; **Protected blackcurrants** *(off-label)* [1]; **Protected blueberry** *(off-label)* [1]; **Protected cayenne peppers** *(off-label)* [1]; **Protected cranberries** *(off-label)* [1]; **Protected cucumbers** [1]; **Protected gooseberries** *(off-label)* [1]; **Protected herbs (see appendix 6)** *(off-label)* [1]; **Protected hops** *(off-label)* [1]; **Protected loganberries** *(off-label)* [1]; **Protected nursery fruit trees** *(off-label)* [1]; **Protected ornamentals** [1]; **Protected peppers** [1]; **Protected raspberries** *(off-label)* [1]; **Protected redcurrants** *(off-label)* [1]; **Protected Ribes hybrids** *(off-label)* [1]; **Protected Rubus hybrids** *(off-label)* [1]; **Protected strawberries** [1]; **Protected table grapes** *(off-label)* [1]; **Protected tomatoes** [1]; **Protected wine grapes** *(off-label)* [1];

Extension of Authorisation for Minor Use (EAMUs)

- **Leaf brassicas** *20142679* [1]
- **Protected bilberries** *20142685* [1]
- **Protected blackberries** *20142684* [1]
- **Protected blackcurrants** *20142685* [1]
- **Protected blueberry** *20142685* [1]
- **Protected cayenne peppers** *20142680* [1]
- **Protected cranberries** *20142685* [1]
- **Protected gooseberries** *20142685* [1]
- **Protected herbs (see appendix 6)** *20142679* [1]
- **Protected hops** *20142681* [1]
- **Protected loganberries** *20142684* [1]
- **Protected nursery fruit trees** *20142681* [1]
- **Protected raspberries** *20142684* [1]
- **Protected redcurrants** *20142685* [1]
- **Protected Ribes hybrids** *20142685* [1]
- **Protected Rubus hybrids** *20142684* [1]
- **Protected table grapes** *20142685* [1]
- **Protected wine grapes** *20142685* [1]

Approval information

- *Lecanicillium* included in Annex I under EC Regulation 1107/2009

Efficacy guidance

- *Lecanicillium muscarium* is a pathogenic fungus that infects the target pests and destroys them
- Apply spore powder as spray as part of biological control programme keeping the spray liquid well agitated
- Treat before infestations build to high levels and repeat as directed on the label
- Spray during late afternoon and early evening directing spray onto underside of leaves and to growing points
- Best results require minimum 80% relative humidity (70% if applied with the adjuvant Addit) and 18°C within the crop canopy

Restrictions

- Never use in tank mixture
- Do not use a fungicide within 3 d of treatment. Pesticides containing captan, imazalil, or prochloraz may not be used on the same crop
- Keep in a refrigerated store at 2-6°C but do not freeze

Environmental safety

- Products have negligible effects on commercially available natural predators or parasites but consult manufacturer before using with a particular biological control agent for the first time

Hazard classification and safety precautions

UN Number N/C
Operator protection U19a, U20b; A, D, H
Environmental protection E15a, E15b
Storage and disposal D09a, D11a

FOR FULL CONDITIONS OF USE ALWAYS READ THE PRODUCT LABEL

241   lenacil

A residual, soil-acting uracil herbicide for beet crops
HRAC mode of action code: 5 (C1)

Products

1 Lenazar Flo 500 SC	BelCrop	500 g/l	SC	18124
2 Venzar 500 SC	FMC Agro	500 g/l	SC	18799

Uses

- Annual dicotyledons in *Game cover* *(off-label)* [2]; *Nursery fruit trees* *(off-label)* [2]; *Ornamental plant production* *(off-label)* [2]; *Sugar beet* [1-2];
- Annual meadow grass in *Sugar beet* [1-2];
- Black bindweed in *Baby leaf crops* *(off-label)* [2]; *Edible flowers* *(off-label)* [2]; *Farm woodland* *(off-label)* [2]; *Fodder beet* *(off-label)* [2]; *Herbs (see appendix 6)* *(off-label)* [2]; *Mangels* *(off-label)* [2]; *Red beet* *(off-label)* [2]; *Red mustard* *(off-label)* [2]; *Spinach* *(off-label)* [2]; *Spinach beet* *(off-label)* [2];
- Brassica spp. in *Baby leaf crops* *(off-label)* [2]; *Edible flowers* *(off-label)* [2]; *Farm woodland* *(off-label)* [2]; *Fodder beet* *(off-label)* [2]; *Herbs (see appendix 6)* *(off-label)* [2]; *Mangels* *(off-label)* [2]; *Red beet* *(off-label)* [2]; *Red mustard* *(off-label)* [2]; *Spinach* *(off-label)* [2]; *Spinach beet* *(off-label)* [2];
- Polygonums in *Baby leaf crops* *(off-label)* [2]; *Edible flowers* *(off-label)* [2]; *Farm woodland* *(off-label)* [2]; *Fodder beet* *(off-label)* [2]; *Herbs (see appendix 6)* *(off-label)* [2]; *Mangels* *(off-label)* [2]; *Red beet* *(off-label)* [2]; *Red mustard* *(off-label)* [2]; *Spinach* *(off-label)* [2]; *Spinach beet* *(off-label)* [2];

Extension of Authorisation for Minor Use (EAMUs)

- *Baby leaf crops* *20191357* [2]
- *Edible flowers* *20191357* [2]
- *Farm woodland* *20190731* [2]
- *Fodder beet* *20192585* [2]
- *Game cover* *20194263* [2]
- *Herbs (see appendix 6)* *20191357* [2]
- *Mangels* *20192585* [2]
- *Nursery fruit trees* *20194263* [2]
- *Ornamental plant production* *20194263* [2]
- *Red beet* *20192585* [2]
- *Red mustard* *20191357* [2]
- *Spinach* *20191357* [2]
- *Spinach beet* *20210329* [2]

Approval information

- Lenacil included in Annex I under EC Regulation 1107/2009

Efficacy guidance

- Best results, especially from pre-emergence treatments, achieved on fine, even, firm and moist soils free from clods. Continuing presence of moisture from rain or irrigation gives improved residual control of later germinating weeds. Effectiveness may be reduced by dry conditions
- On beet crops may be used pre- or post-emergence, alone or in mixture to broaden weed spectrum
- Apply overall or as band spray to beet crops pre-drilling incorporated, pre- or post-emergence
- All labels have limitations on soil types that may be treated. Residual activity reduced on soils with high OM content

Restrictions

- Do not use any other residual herbicide within 3 mth of the initial application to fruit or ornamental crops
- Do not treat crops under stress from drought, low temperatures, nutrient deficiency, pest or disease attack, or waterlogging
- Must only be applied after BBCH10 growth stage
- May only be applied once in every three years

SEE SECTION 3 FOR PRODUCTS ALSO REGISTERED

SECTION 2

Crop-specific information

- Latest use: pre-emergence for red beet, fodder beet, spinach, spinach beet, mangels; before leaves meet over rows when used on these crops post-emergence; 24 h after planting new strawberry runners or before flowering for established strawberry crops, blackcurrants, gooseberries, raspberries
- Heavy rain after application to beet crops may cause damage especially if followed by very hot weather
- Reduction in beet stand may occur where crop emergence or vigour is impaired by soil capping or pest attack
- Strawberry runner beds to be treated should be level without depressions around the roots
- New soft fruit cuttings should be planted at least 15 cm deep and firmed before treatment
- Check varietal tolerance of ornamentals before large scale treatment

Following crops guidance

- Succeeding crops should not be planted or sown for at least 4 mth (6 mth on organic soils) after treatment following ploughing to at least 150 mm.
- Only beet crops, mangels or strawberries may be sown within 4 months of treatment

Environmental safety

- Dangerous for the environment
- Very toxic to aquatic organisms
- LERAP Category B [1-2]

Hazard classification and safety precautions

Hazard Dangerous for the environment
Transport code 9 [1-2]
Packaging group III
UN Number 3082
Risk phrases H351
Operator protection U05a, U08, U19a, U20a; A, C
Environmental protection E13c, E16a, E38, H410
Storage and disposal D01, D02, D05, D09a, D10a, D12a

242   Ienacil + triflusulfuron-methyl

A foliar and residual herbicide mixture for sugar beet
HRAC mode of action code: 5 + 2 (C1 + B)

See also triflusulfuron-methyl

Products

1	Debut Plus	FMC Agro	71.4:5.4% w/w	WG	18765
2	Safari Lite WSB	FMC Agro	71.4:5.4% w/w	ZZ	18784

Uses

- Annual dicotyledons in **Sugar beet** [1-2];
- Black bindweed in **Fodder beet** *(off-label)* [2];
- Charlock in **Fodder beet** *(off-label)* [2];
- Chickweed in **Fodder beet** *(off-label)* [2];
- Cleavers in **Fodder beet** *(off-label)* [2];
- Fat hen in **Fodder beet** *(off-label)* [2];
- Field pansy in **Fodder beet** *(off-label)* [2];
- Fool's parsley in **Fodder beet** *(off-label)* [2];
- Fumitory in **Fodder beet** *(off-label)* [2];
- Knotgrass in **Fodder beet** *(off-label)* [2];
- Mayweeds in **Fodder beet** *(off-label)* [2];
- Red dead-nettle in **Fodder beet** *(off-label)* [2];
- Redshank in **Fodder beet** *(off-label)* [2];
- Scentless mayweed in **Fodder beet** *(off-label)* [2];
- Small nettle in **Fodder beet** *(off-label)* [2];
- Volunteer oilseed rape in **Fodder beet** *(off-label)* [2]; **Sugar beet** [1-2];

FOR FULL CONDITIONS OF USE ALWAYS READ THE PRODUCT LABEL

Extension of Authorisation for Minor Use (EAMUs)
- *Fodder beet* 20190808 [2]

Approval information
- Lenacil and triflusulfuron-methyl included in Annex I under EC Regulation 1107/2009

Efficacy guidance
- Best results obtained when weeds are small and growing actively
- Product recommended for use in a programme of treatments in tank mixture with a suitable herbicide partner to broaden the weed spectrum
- Ensure good spray cover of weeds. Apply when first weeds have emerged provided crop has reached cotyledon stage
- Susceptible plants cease to grow almost immediately after treatment and symptoms can be seen 5-10 d later
- Weed control may be reduced in very dry soil conditions
- Triflusulfuron-methyl is a member of the ALS-inhibitor group of herbicides

Restrictions
- Maximum number of treatments on sugar beet 2 per crop and do not apply more than 4 applications of any product containing triflusulfuron-methyl
- Do not apply to crops suffering from stress caused by drought, water-logging, low temperatures, pest or disease attack, nutrient deficiency or any other factors affecting crop growth
- Do not use on Sands, stony or gravelly soils or on soils with more than 10% organic matter
- Do not apply when temperature above or likely to exceed 21˚C on day of spraying or under conditions of high light intensity
- Must only be applied after BBCH10 growth stage
- The maximum total dose of lenacil must not exceed 450 g lenacil /ha in any three-year period in the same field.
- The maximum total dose of triflusulfuron methyl must not exceed 45 g triflusulfuron methyl /ha in any three-year period in the same field.

Crop-specific information
- Latest use: before crop leaves meet between rows for sugar beet

Following crops guidance
- Only cereals may be sown in the same calendar yr as a treated sugar beet crop. Any crop may be sown in the following spring
- In the event of crop failure sow only sugar beet within 4 mth of treatment
- A minimum interval of 6 months must be observed prior to the planting of a succeeding cereal crop

Environmental safety
- Dangerous for the environment
- Very toxic to aquatic organisms
- Take extreme care to avoid drift onto broad-leaved plants outside the target area or onto surface waters or ditches, or land intended for cropping
- Spraying equipment should not be drained or flushed onto land planted, or to be planted, with trees or crops other than cereals and should be thoroughly cleansed after use - see label for instructions
- LERAP Category B [1-2]

Hazard classification and safety precautions
> **Hazard** Irritant, Dangerous for the environment
> **Transport code** 9 [1-2]
> **Packaging group** III
> **UN Number** 3077
> **Risk phrases** H351
> **Operator protection** U05a, U08, U11, U19a, U20b, U22a; A, C
> **Environmental protection** E15a, E16a, E38, H410
> **Storage and disposal** D01, D02, D09a, D12a

SEE SECTION 3 FOR PRODUCTS ALSO REGISTERED

243   maleic hydrazide

A pyridazine plant growth regulator suppressing sprout and bud growth

Products

1 Crown MH	Certis Belchim B V	270 g/l	SL	19495
2 Fazor	UPL Europe	60% w/w	SG	19605
3 Magna SL	Certis Belchim B V	270 g/l	SL	19496

Uses

* Growth regulation in **Bulb onions** [1-2]; **Carrots** *(off-label)* [1]; **Garlic** *(off-label)* [1]; **Parsnips** *(off-label)* [1]; **Potatoes** [1-2]; **Shallots** *(off-label)* [1];
* Sprout suppression in **Bulb onions** [3]; **Carrots** *(off-label)* [2]; **Garlic** *(off-label)* [2]; **Parsnips** *(off-label)* [2]; **Potatoes** [3]; **Shallots** *(off-label)* [2];

Extension of Authorisation for Minor Use (EAMUs)

* **Carrots** *20201913* [1], *20202714* [2]
* **Garlic** *20201657* [1], *20202715* [2]
* **Parsnips** *20201913* [1], *20202714* [2]
* **Shallots** *20201657* [1], *20202715* [2]

Approval information

* Maleic hydrazide included in Annex I under EC Regulation 1107/2009

Efficacy guidance

* Uniform coverage and dry weather necessary for effective results
* Accurate timing essential for good results on potatoes but rain or irrigation within 24 h may reduce effectiveness on onions and potatoes
* When used for suppression of volunteer potatoes treatment may also give some suppression of sprouting in store but separate treatment will be necessary if sprouting occurs

Restrictions

* Do not apply in drought or when crops are suffering from pest, disease or herbicide damage.
* Do not treat potatoes within 3 wk of applying a haulm desiccant or if temperatures above 26°C
* Consult processor before use on potato crops for processing
* Do not apply to seed potatoes or first year onion sets

Crop-specific information

* Latest use: 3 wk before haulm destruction for potatoes; before 50% necking for onions
* HI onions 1 wk; potatoes 3 wk
* Apply to onions at 10% necking and not later than 50% necking stage when the tops are still green
* Only treat onions in good condition and properly cured, and do not treat more than 2 wk before maturing. Treated onions may be stored until Mar but must then be removed to avoid browning
* Apply to second early or maincrop potatoes at least 3 wk before haulm destruction
* Only treat potatoes of good keeping quality; not on seed, first earlies or crops grown under polythene

Environmental safety

* Do not use treated water for irrigation purposes within 3 wk of treatment or until concentration in water falls below 0.02 ppm
* Maximum permitted concentration in water 2 ppm
* Do not dump surplus product in water or ditch bottoms
* Avoid drift onto nearby vegetables, flowers or other garden plants

Hazard classification and safety precautions

Transport code 9 [2]
Packaging group III
UN Number 3077, N/C
Operator protection U08 [1-2], U20b [1-2]; A, D, H
Environmental protection E15a [1-2], H411
Storage and disposal D09a [1-2], D11a [1-2]

FOR FULL CONDITIONS OF USE ALWAYS READ THE PRODUCT LABEL

244 maltodextrin

A polysaccharide used as a food additive and with activity against red spider mites
FRAC mode of action code: 12 + IRAC 3

Products

1	Eradicoat Max	Certis Belchim B V	476 g/l	SL	18852
2	Majestik	Certis Belchim B V	598 g/l	SL	17240

Uses

- Aphids in *All protected edible crops* [1-2]; *All protected non-edible crops* [1-2]; *Apples* (off-label) [1]; *Apricots* (off-label) [1]; *Asparagus* (off-label) [1]; *Baby leaf crops* (off-label) [1]; *Bilberries* (off-label) [1]; *Blackberries* (off-label) [1]; *Blackcurrants* (off-label) [1]; *Blueberries* (off-label) [1]; *Bulb onions* (off-label) [1]; *Carrots* (off-label) [1]; *Celeriac* (off-label) [1]; *Celery (outdoor)* (off-label) [1]; *Cherries* (off-label) [1]; *Christmas trees* (off-label) [1]; *Courgettes* (off-label) [1]; *Cranberries* (off-label) [1]; *Elderberries* (off-label) [1]; *Florence fennel* (off-label) [1]; *Garlic* (off-label) [1]; *Globe artichoke* (off-label) [1]; *Gooseberries* (off-label) [1]; *Herbs (see appendix 6)* (off-label) [1]; *Horseradish* (off-label) [1]; *Jerusalem artichokes* (off-label) [1]; *Leeks* (off-label) [1]; *Loganberries* (off-label) [1]; *Medlar* (off-label) [1]; *Mulberries* (off-label) [1]; *Nectarines* (off-label) [1]; *Nursery fruit trees* (off-label) [1]; *Ornamental plant production* (off-label) [1]; *Parsley root* (off-label) [1]; *Parsnips* (off-label) [1]; *Peaches* (off-label) [1]; *Plums* (off-label) [1]; *Pumpkins* (off-label) [1]; *Quinces* (off-label) [1]; *Radishes* (off-label) [1]; *Raspberries* (off-label) [1]; *Red beet* (off-label) [1]; *Redcurrants* (off-label) [1]; *Rhubarb* (off-label) [1]; *Rose hips* (off-label) [1]; *Rubus hybrids* (off-label) [1]; *Salad onions* (off-label) [1]; *Salsify* (off-label) [1]; *Shallots* (off-label) [1]; *Spinach* (off-label) [1]; *Spinach beet* (off-label) [1]; *Strawberries* (off-label) [1]; *Summer squash* (off-label) [1]; *Swedes* (off-label) [1]; *Sweet potato* (off-label) [1]; *Sweetcorn* (off-label) [1]; *Turnips* (off-label) [1]; *Watercress* (off-label) [1]; *Winter squash* (off-label) [1];
- Bean seed fly in *Asparagus* (off-label) [1]; *Celery (outdoor)* (off-label) [1]; *Florence fennel* (off-label) [1]; *Globe artichoke* (off-label) [1]; *Leeks* (off-label) [1]; *Rhubarb* (off-label) [1];
- Brassica leaf miner in *Broccoli* (off-label) [1]; *Brussels sprouts* (off-label) [1]; *Cabbages* (off-label) [1]; *Calabrese* (off-label) [1]; *Cauliflowers* (off-label) [1]; *Choi sum* (off-label) [1]; *Collards* (off-label) [1]; *Kale* (off-label) [1]; *Kohlrabi* (off-label) [1]; *Oriental cabbage* (off-label) [1];
- Cabbage aphid in *Broccoli* (off-label) [1]; *Brussels sprouts* (off-label) [1]; *Cabbages* (off-label) [1]; *Calabrese* (off-label) [1]; *Cauliflowers* (off-label) [1]; *Choi sum* (off-label) [1]; *Collards* (off-label) [1]; *Kale* (off-label) [1]; *Kohlrabi* (off-label) [1]; *Oriental cabbage* (off-label) [1]; *Swedes* (off-label) [1]; *Turnips* (off-label) [1];
- Cabbage root fly in *Carrots* (off-label) [1]; *Celeriac* (off-label) [1]; *Horseradish* (off-label) [1]; *Parsley root* (off-label) [1]; *Parsnips* (off-label) [1]; *Radishes* (off-label) [1]; *Swedes* (off-label) [1]; *Turnips* (off-label) [1];
- Carrot fly in *Red beet* (off-label) [1];
- Cherry black fly in *Cherries* (off-label) [1];
- Common green capsid in *Strawberries* (off-label) [1];
- Damson-hop aphid in *Hops* (off-label) [1]; *Plums* (off-label) [1];
- Diamond-back moth in *Swedes* (off-label) [1]; *Turnips* (off-label) [1];
- Large white butterfly in *Swedes* (off-label) [1]; *Turnips* (off-label) [1];
- Leaf curling aphid in *Apples* (off-label) [1]; *Medlar* (off-label) [1]; *Plums* (off-label) [1]; *Quinces* (off-label) [1];
- Mealy plum aphid in *Plums* (off-label) [1];
- Mites in *Christmas trees* (off-label) [1]; *Nursery fruit trees* (off-label) [1]; *Ornamental plant production* (off-label) [1];
- Mussel scale in *Apples* (off-label) [1]; *Apricots* (off-label) [1]; *Cherries* (off-label) [1]; *Medlar* (off-label) [1]; *Nectarines* (off-label) [1]; *Peaches* (off-label) [1]; *Plums* (off-label) [1]; *Quinces* (off-label) [1];
- Nut scale in *Apples* (off-label) [1]; *Apricots* (off-label) [1]; *Cherries* (off-label) [1]; *Medlar* (off-label) [1]; *Nectarines* (off-label) [1]; *Peaches* (off-label) [1]; *Plums* (off-label) [1]; *Quinces* (off-label) [1];
- Parsnip aphid in *Carrots* (off-label) [1]; *Parsnips* (off-label) [1];

SEE SECTION 3 FOR PRODUCTS ALSO REGISTERED

SECTION 2

- Peach-potato aphid in **Broccoli** *(off-label)* [1]; **Brussels sprouts** *(off-label)* [1]; **Cabbages** *(off-label)* [1]; **Calabrese** *(off-label)* [1]; **Carrots** *(off-label)* [1]; **Cauliflowers** *(off-label)* [1]; **Celeriac** *(off-label)* [1]; **Choi sum** *(off-label)* [1]; **Collards** *(off-label)* [1]; **Horseradish** *(off-label)* [1]; **Jerusalem artichokes** *(off-label)* [1]; **Kale** *(off-label)* [1]; **Kohlrabi** *(off-label)* [1]; **Oriental cabbage** *(off-label)* [1]; **Parsley root** *(off-label)* [1]; **Parsnips** *(off-label)* [1]; **Radishes** *(off-label)* [1]; **Red beet** *(off-label)* [1]; **Salsify** *(off-label)* [1]; **Sweet potato** *(off-label)* [1];
- Potato aphid in **Strawberries** *(off-label)* [1];
- Red spider mites in **Apples** *(off-label)* [1]; **Apricots** *(off-label)* [1]; **Cherries** *(off-label)* [1]; **Medlar** *(off-label)* [1]; **Nectarines** *(off-label)* [1]; **Peaches** *(off-label)* [1]; **Plums** *(off-label)* [1]; **Quinces** *(off-label)* [1];
- Rosy apple aphid in **Apples** *(off-label)* [1]; **Medlar** *(off-label)* [1]; **Quinces** *(off-label)* [1];
- Rust mite in **Apples** *(off-label)* [1]; **Medlar** *(off-label)* [1]; **Plums** *(off-label)* [1]; **Quinces** *(off-label)* [1];
- Silver Y moth in **Carrots** *(off-label)* [1]; **Jerusalem artichokes** *(off-label)* [1]; **Parsnips** *(off-label)* [1]; **Red beet** *(off-label)* [1]; **Salsify** *(off-label)* [1]; **Swedes** *(off-label)* [1]; **Sweet potato** *(off-label)* [1]; **Turnips** *(off-label)* [1];
- Small white butterfly in **Swedes** *(off-label)* [1]; **Turnips** *(off-label)* [1];
- Spider mites in **All protected edible crops** [1-2]; **All protected non-edible crops** [1-2]; **Apples** *(off-label)* [1]; **Apricots** *(off-label)* [1]; **Bilberries** *(off-label)* [1]; **Blackberries** *(off-label)* [1]; **Blackcurrants** *(off-label)* [1]; **Blueberries** *(off-label)* [1]; **Cherries** *(off-label)* [1]; **Cranberries** *(off-label)* [1]; **Elderberries** *(off-label)* [1]; **Gooseberries** *(off-label)* [1]; **Loganberries** *(off-label)* [1]; **Medlar** *(off-label)* [1]; **Mulberries** *(off-label)* [1]; **Nectarines** *(off-label)* [1]; **Peaches** *(off-label)* [1]; **Plums** *(off-label)* [1]; **Quinces** *(off-label)* [1]; **Raspberries** *(off-label)* [1]; **Redcurrants** *(off-label)* [1]; **Rose hips** *(off-label)* [1]; **Rubus hybrids** *(off-label)* [1]; **Strawberries** *(off-label)* [1];
- Spotted wing drosophila in **Grapevines** *(off-label)* [1];
- Tarnished plant bug in **Strawberries** *(off-label)* [1];
- Thrips in **Asparagus** *(off-label)* [1]; **Blackberries** *(off-label)* [1]; **Bulb onions** *(off-label)* [1]; **Celery (outdoor)** *(off-label)* [1]; **Florence fennel** *(off-label)* [1]; **Garlic** *(off-label)* [1]; **Globe artichoke** *(off-label)* [1]; **Grapevines** *(off-label)* [1]; **Leeks** *(off-label)* [1]; **Loganberries** *(off-label)* [1]; **Raspberries** *(off-label)* [1]; **Rhubarb** *(off-label)* [1]; **Rubus hybrids** *(off-label)* [1]; **Salad onions** *(off-label)* [1]; **Shallots** *(off-label)* [1];
- Turnip moth in **Red beet** *(off-label)* [1]; **Swedes** *(off-label)* [1]; **Turnips** *(off-label)* [1];
- Two-spotted spider mite in **Hops** *(off-label)* [1];
- Whitefly in **All protected edible crops** [1-2]; **All protected non-edible crops** [1-2]; **Apricots** *(off-label)* [1]; **Bilberries** *(off-label)* [1]; **Blackberries** *(off-label)* [1]; **Blackcurrants** *(off-label)* [1]; **Blueberries** *(off-label)* [1]; **Broccoli** *(off-label)* [1]; **Brussels sprouts** *(off-label)* [1]; **Cabbages** *(off-label)* [1]; **Calabrese** *(off-label)* [1]; **Cauliflowers** *(off-label)* [1]; **Cherries** *(off-label)* [1]; **Choi sum** *(off-label)* [1]; **Collards** *(off-label)* [1]; **Courgettes** *(off-label)* [1]; **Cranberries** *(off-label)* [1]; **Elderberries** *(off-label)* [1]; **Gooseberries** *(off-label)* [1]; **Kale** *(off-label)* [1]; **Kohlrabi** *(off-label)* [1]; **Loganberries** *(off-label)* [1]; **Mulberries** *(off-label)* [1]; **Nectarines** *(off-label)* [1]; **Oriental cabbage** *(off-label)* [1]; **Peaches** *(off-label)* [1]; **Plums** *(off-label)* [1]; **Pumpkins** *(off-label)* [1]; **Raspberries** *(off-label)* [1]; **Redcurrants** *(off-label)* [1]; **Rose hips** *(off-label)* [1]; **Rubus hybrids** *(off-label)* [1]; **Strawberries** *(off-label)* [1]; **Summer squash** *(off-label)* [1]; **Swedes** *(off-label)* [1]; **Sweetcorn** *(off-label)* [1]; **Turnips** *(off-label)* [1]; **Winter squash** *(off-label)* [1];
- Willow parsnip aphid in **Carrots** *(off-label)* [1]; **Parsnips** *(off-label)* [1];
- Willow-carrot aphid in **Carrots** *(off-label)* [1]; **Parsnips** *(off-label)* [1];

Extension of Authorisation for Minor Use (EAMUs)
- **Apples** *20221835* [1]
- **Apricots** *20221833* [1]
- **Asparagus** *20221841* [1]
- **Baby leaf crops** *20221839* [1]
- **Bilberries** *20222138* [1]
- **Blackberries** *20221834* [1]
- **Blackcurrants** *20222138* [1]
- **Blueberries** *20222138* [1]
- **Broccoli** *20221844* [1]
- **Brussels sprouts** *20221844* [1]
- **Bulb onions** *20221837* [1]

FOR FULL CONDITIONS OF USE ALWAYS READ THE PRODUCT LABEL

- **Cabbages** *20221844* [1]
- **Calabrese** *20221844* [1]
- **Carrots** *20221836* [1]
- **Cauliflowers** *20221844* [1]
- **Celeriac** *20221836* [1]
- **Celery (outdoor)** *20221841* [1]
- **Cherries** *20221833* [1]
- **Choi sum** *20221844* [1]
- **Christmas trees** *20221840* [1]
- **Collards** *20221844* [1]
- **Courgettes** *20221842* [1]
- **Cranberries** *20222138* [1]
- **Elderberries** *20222138* [1]
- **Florence fennel** *20221841* [1]
- **Garlic** *20221837* [1]
- **Globe artichoke** *20221841* [1]
- **Gooseberries** *20222138* [1]
- **Grapevines** *20221837* [1]
- **Herbs (see appendix 6)** *20221839* [1]
- **Hops** *20221843* [1]
- **Horseradish** *20221836* [1]
- **Jerusalem artichokes** *20221836* [1]
- **Kale** *20221844* [1]
- **Kohlrabi** *20221844* [1]
- **Leeks** *20221841* [1]
- **Loganberries** *20221834* [1]
- **Medlar** *20221835* [1]
- **Mulberries** *20222138* [1]
- **Nectarines** *20221833* [1]
- **Nursery fruit trees** *20221840* [1]
- **Oriental cabbage** *20221844* [1]
- **Ornamental plant production** *20221840* [1]
- **Parsley root** *20221836* [1]
- **Parsnips** *20221836* [1]
- **Peaches** *20221833* [1]
- **Plums** *20221833* [1]
- **Pumpkins** *20221842* [1]
- **Quinces** *20221835* [1]
- **Radishes** *20221836* [1]
- **Raspberries** *20221834* [1]
- **Red beet** *20221836* [1]
- **Redcurrants** *20222138* [1]
- **Rhubarb** *20221841* [1]
- **Rose hips** *20222138* [1]
- **Rubus hybrids** *20221834* [1]
- **Salad onions** *20221837* [1]
- **Salsify** *20221836* [1]
- **Shallots** *20221837* [1]
- **Spinach** *20221839* [1]
- **Spinach beet** *20221839* [1]
- **Strawberries** *20221845* [1]
- **Summer squash** *20221842* [1]
- **Swedes** *20221836* [1]
- **Sweet potato** *20221836* [1]
- **Sweetcorn** *20221842* [1]
- **Turnips** *20221836* [1]
- **Watercress** *20221839* [1]
- **Winter squash** *20221842* [1]

SEE SECTION 3 FOR PRODUCTS ALSO REGISTERED

Approval information
- Maltodextrin is included in Annex 1 under EC Regulation 1107/2009

Restrictions
- Reasonable precautions must be taken to prevent access of birds, wild mammals and honey bees to treated crops
- The maximum concentration must not exceed 25 ml of product per 1 litre of water [2]

Hazard classification and safety precautions
Hazard Irritant
UN Number N/C
Risk phrases H318 [1], H319 [2]
Operator protection U05a, U09a, U14, U19a, U20b; A, C, H
Environmental protection E12a [2], E12c [2], E12e
Storage and disposal D01, D02, D09a, D10b, D12a

245   mancozeb

A protective dithiocarbamate fungicide for potatoes and other crops only approved in GB
FRAC mode of action code: M3

See also cymoxanil + mancozeb
mancozeb + metalaxyl-M

Products

1	Agria Mancozeb 75WDG	Agria SA	75% w/w	WG	16807
2	Cleancrop Feudal	Agrii	75% w/w	WG	17894
3	Karamate Dry Flo Neotec	Agrii	75% w/w	WG	19994
4	Laminator 75 WG	Sumitomo	75% w/w	WG	19022
5	Manfil WP Plus	Agrii	80% w/w	WP	17939
6	Manzate 75 WG	UPL Europe	75% w/w	WG	15052
7	Penncozeb 80 WP	UPL Europe	80% w/w	WP	16953
8	Penncozeb WDG	UPL Europe	75% w/w	WG	16885
9	Unizeb Gold	UPL Europe	500 g/l	SC	18186

Uses
- Blight in **Potatoes** [1-2,4-8]; **Tomatoes (outdoor)** [4];
- Botrytis in **Flower bulbs** [6];
- Brown rust in **Durum wheat** *(useful control)* [5]; **Spelt** [9]; **Spring wheat** *(useful control)* [5,9]; **Triticale** [9]; **Winter wheat** *(useful control)* [5,9];
- Disease control in **Amenity vegetation** [3]; **Apples** [4]; **Bulb onions** [3]; **Carrots** [2-3]; **Courgettes** [3]; **Forest nurseries** *(off-label)* [8]; **Ornamental plant production** [3,8]; **Parsnips** [2-3]; **Potatoes** [1]; **Shallots** [3]; **Table grapes** [4]; **Wine grapes** [3-4];
- Downy mildew in **Bulb onions** [2,6-8]; **Ornamental plant production** [8]; **Shallots** [2];
- Early blight in **Potatoes** [5];
- Rust in **Red beet** [9];
- Scab in **Apples** [3,6-8]; **Pears** [3];
- Septoria leaf blotch in **Durum wheat** *(reduction)* [5]; **Spelt** [9]; **Spring wheat** *(reduction)* [2,5-9]; **Triticale** [9]; **Winter wheat** *(reduction)* [2,5-9];
- Sooty moulds in **Spelt** [9]; **Spring wheat** [9]; **Triticale** [9]; **Winter wheat** [9];

Extension of Authorisation for Minor Use (EAMUs)
- **Forest nurseries** *20212167* [8]

Approval information
- Mancozeb included in Annex I under EC Regulation 1107/2009
- Approval expired in Northern Ireland 4/1/2022. Do not use in Northern Ireland

FOR FULL CONDITIONS OF USE ALWAYS READ THE PRODUCT LABEL

Efficacy guidance

- Mancozeb is a protectant fungicide and will give moderate control, suppression or reduction of the cereal diseases listed if treated before they are established but in many cases mixture with carbendazim is essential to achieve satisfactory results. See labels for details
- May be recommended for suppression or control of mildew in cereals depending on product and tank mix. See label for details

Restrictions

- Maximum number of treatments varies with crop and product used - check labels for details
- Check labels for minimum interval that must elapse between treatments
- Avoid treating wet cereal crops or those suffering from drought or other stress
- Keep dry formulations away from fire and sparks
- Use dry formulations immediately. Do not store
- Avoid spraying within 5m of the field boundary to reduce effects on non-target insects or other arthropods [9]

Crop-specific information

- Latest use: before early milk stage (GS 73) for cereals;
- HI potatoes 7 d; apples, blackcurrants 28 d
- Apply to potatoes before haulm meets across rows (usually mid-Jun) or at earlier blight warning, and repeat every 7-14 d depending on conditions and product used (see label)
- May be used on potatoes up to desiccation of haulm
- Apply to cereals from 4-leaf stage to before early milk stage (GS 71). Recommendations vary, see labels for details

Environmental safety

- Dangerous for the environment
- Very toxic to aquatic organisms
- Harmful to fish or other aquatic life. Do not contaminate surface waters or ditches with chemical or used container
- Do not empty into drains
- Buffer zone for application via a broadcast sprayer is 10m for grapes, 40m for apples [4]
- LERAP Category B [1-9]

Hazard classification and safety precautions

Hazard Harmful [4,7-8], Irritant [1-3,5-6,9], Dangerous for the environment, Very toxic to aquatic organisms [1-3,5]

Transport code 9 [1-9]

Packaging group III

UN Number 3077, 3082

Risk phrases H317 [1-8], H319 [2-4], H335 [6], H360 [8], H361 [1-5,7,9]

Operator protection U05a [2-6,8-9], U05b [1], U08 [9], U14, U15 [1], U19a [1,7-8], U20b [2-6,9]; A, D, H

Environmental protection E15a [2-6,9], E15b [7-8], E16a, E17a (40 m) [4], E19b [8], E34 [9], E38, H410 [4,6-9], H411 [2-3,5]

Storage and disposal D01, D02, D05 [2-6,9], D07 [1], D09a [8-9], D09b [1], D11a [9], D12a

Medical advice M05a [1,7-8]

246   mancozeb + metalaxyl-M

A systemic and protectant fungicide mixture only approved in GB
FRAC mode of action code: M3 + 4

See also metalaxyl-M

Products

1	Fubol Gold WG	Syngenta	64:4% w/w	WG	14605

Uses

- Blight in **Potatoes** [1];

SEE SECTION 3 FOR PRODUCTS ALSO REGISTERED

- Downy mildew in **Baby leaf crops** *(off-label)* [1]; **Bulb onions** *(useful control)* [1]; **Herbs (see appendix 6)** *(off-label)* [1]; **Lettuce** *(off-label)* [1]; **Poppies for morphine production** *(off-label)* [1]; **Protected bilberries** *(off-label)* [1]; **Protected blackberries** *(off-label)* [1]; **Protected blackcurrants** *(off-label)* [1]; **Protected blueberries** *(off-label)* [1]; **Protected cranberries** *(off-label)* [1]; **Protected forest nurseries** *(off-label)* [1]; **Protected gooseberries** *(off-label)* [1]; **Protected herbs (see appendix 6)** *(off-label)* [1]; **Protected hops** *(off-label)* [1]; **Protected loganberries** *(off-label)* [1]; **Protected ornamentals** *(off-label)* [1]; **Protected raspberries** *(off-label)* [1]; **Protected redcurrants** *(off-label)* [1]; **Protected Rubus hybrids** *(off-label)* [1]; **Protected strawberries** *(off-label)* [1]; **Rhubarb** *(off-label)* [1]; **Salad onions** *(off-label)* [1]; **Shallots** *(useful control)* [1];
- Phytophthora fruit rot in **Apples** *(off-label)* [1];
- Pink rot in **Potatoes** *(reduction)* [1];
- White blister in **Cabbages** *(off-label)* [1];

Extension of Authorisation for Minor Use (EAMUs)
- **Apples** *20212119* [1]
- **Baby leaf crops** *20212125* [1]
- **Cabbages** *20212117* [1]
- **Herbs (see appendix 6)** *20212125* [1]
- **Lettuce** *20212125* [1]
- **Poppies for morphine production** *20212127* [1]
- **Protected bilberries** *20212133* [1]
- **Protected blackberries** *20212133* [1]
- **Protected blackcurrants** *20212133* [1]
- **Protected blueberries** *20212133* [1]
- **Protected cranberries** *20212133* [1]
- **Protected forest nurseries** *20212133* [1]
- **Protected gooseberries** *20212133* [1]
- **Protected herbs (see appendix 6)** *20212131* [1]
- **Protected hops** *20212133* [1]
- **Protected loganberries** *20212133* [1]
- **Protected ornamentals** *20212133* [1]
- **Protected raspberries** *20212133* [1]
- **Protected redcurrants** *20212133* [1]
- **Protected Rubus hybrids** *20212133* [1]
- **Protected strawberries** *20212133* [1]
- **Rhubarb** *20212121* [1]
- **Salad onions** *20212123* [1]

Approval information
- Mancozeb and metalaxyl-M included in Annex I under EC Regulation 1107/2009
- Approval expired in Northern Ireland 4/1/2022. Do not use in Northern Ireland

Efficacy guidance
- Commence potato blight programme before risk of infection occurs as crops begin to meet along the rows and repeat every 7-14 d according to blight risk. Do not exceed a 14 d interval between sprays
- If infection risk conditions occur earlier than the above growth stage commence spraying potatoes immediately
- Complete the potato blight programme using a protectant fungicide starting no later than 10 d after the last phenylamide spray. At least 2 such sprays should be applied
- To minimise the likelihood of development of resistance these products should be used in a planned Resistance Management strategy. See Section 5 for more information

Crop-specific information
- Latest use: before end of active potato haulm growth or before end Aug, whichever is earlier
- HI 7 d for potatoes
- After treating early potatoes destroy and remove any remaining haulm after harvest to minimise blight pressure on neighbouring maincrop potatoes

FOR FULL CONDITIONS OF USE ALWAYS READ THE PRODUCT LABEL

SECTION 2

Environmental safety
- Dangerous for the environment
- Very toxic to aquatic organisms
- Do not harvest crops for human consumption for at least 7 d after final application
- LERAP Category B [1]

Hazard classification and safety precautions
Hazard Harmful, Dangerous for the environment, Harmful if inhaled
Transport code 9 [1]
Packaging group III
UN Number 3077
Risk phrases H317, H351, H361, H373
Operator protection U05a, U08, U20b; A
Environmental protection E15a, E16a, E34, E38, H410
Consumer protection C02a (7 d)
Storage and disposal D01, D02, D05, D09a, D11a, D12a

247   mandipropamid

A mandelamide fungicide for the control of potato blight
FRAC mode of action code: 40

See also cymoxanil + mandipropamid
difenoconazole + mandipropamid

Products
1	Revus	Syngenta	250 g/l	SC	17443

Uses
- Blight in **Potatoes** [1];
- Disease control in **Baby leaf crops** [1]; **Celery leaves** [1]; **Chives** [1]; **Cress** [1]; **Endives** [1]; **Herbs (see appendix 6)** [1]; **Lamb's lettuce** [1]; **Land cress** [1]; **Lettuce** [1]; **Protected baby leaf crops** [1]; **Protected chives** [1]; **Protected cress** [1]; **Protected endives** [1]; **Protected herbs (see appendix 6)** [1]; **Protected lamb's lettuce** [1]; **Protected land cress** [1]; **Protected lettuce** [1]; **Protected red mustard** [1]; **Protected spinach** [1]; **Protected spinach beet** [1]; **Red mustard** [1]; **Spinach** [1]; **Spinach beet** [1];
- Downy mildew in **Baby leaf crops** [1]; **Broccoli** *(off-label)* [1]; **Brussels sprouts** *(off-label)* [1]; **Calabrese** *(off-label)* [1]; **Cauliflowers** *(off-label)* [1]; **Celery leaves** [1]; **Chives** [1]; **Cress** [1]; **Endives** [1]; **Herbs (see appendix 6)** [1]; **Hops** *(off-label)* [1]; **Lamb's lettuce** [1]; **Land cress** [1]; **Lettuce** [1]; **Ornamental plant production** *(off-label)* [1]; **Protected baby leaf crops** [1]; **Protected chives** [1]; **Protected cress** [1]; **Protected endives** [1]; **Protected herbs (see appendix 6)** [1]; **Protected lamb's lettuce** [1]; **Protected land cress** [1]; **Protected lettuce** [1]; **Protected ornamentals** *(off-label)* [1]; **Protected red mustard** [1]; **Protected spinach** [1]; **Protected spinach beet** [1]; **Radishes** [1]; **Red mustard** [1]; **Spinach** [1]; **Spinach beet** [1]; **Vining peas** [1];
- Tuber blight in **Potatoes** *(protection)* [1];

Extension of Authorisation for Minor Use (EAMUs)
- **Broccoli** *20212915* [1]
- **Brussels sprouts** *20212915* [1]
- **Calabrese** *20212915* [1]
- **Cauliflowers** *20212915* [1]
- **Hops** *20162024* [1]
- **Ornamental plant production** *20162763* [1]
- **Protected ornamentals** *20162763* [1]

Approval information
- Mandipropamid is included in Annex 1 under EC Regulation 1107/2009

Efficacy guidance

- Mandipropamid acts preventatively by preventing spore germination and inhibiting mycelial growth during incubation. Apply immediately after blight warning or as soon as local conditions favour disease development but before blight enters the crop
- Spray at 7-10 d intervals reducing the interval as blight risk increases
- Spray programme should include a complete haulm desiccant to prevent tuber infection at harvest
- Eliminate other potential infection sources
- To minimise the likelihood of development of resistance this product should be used in a planned Resistance Management strategy. See Section 5 for more information
- See label for details of tank mixtures that may be used as part of a resistance management strategy
- Rainfast in potatoes within 15 minutes of application

Restrictions

- Maximum number of treatments 4 per crop. Do not apply more than 3 treatments of this, or any other fungicide in the same resistance category, consecutively
- For outdoor crops and protected crops that are grown under a temporary cover, the maximum total dose must not exceed 1.2 L product/ha per year on any single area of land [1]
- Baby leaf crops must be harvested no later than the 8 true leaf stage [1]

Crop-specific information

- HI 3 d for potatoes

Environmental safety

- Buffer zone requirement 10 m in hops [1]

Hazard classification and safety precautions

UN Number N/C
Operator protection U05a, U20b; A, H
Environmental protection E15b, E34, E38, H411
Storage and disposal D01, D02, D05, D09a, D10c, D12a
Medical advice M03, M05a

248   MCPA

A translocated phenoxycarboxylic acid herbicide for cereals and grassland
HRAC mode of action code: 4 (O)

See also 2,4-D + dicamba + MCPA + mecoprop-P
2,4-D + MCPA
clopyralid + fluroxypyr + MCPA
dicamba + MCPA + mecoprop-P
dichlorprop-P + MCPA + mecoprop-P
ferrous sulphate + MCPA + mecoprop-P

Products

1	Agritox	Nufarm UK	500 g/l	SL	14894
2	Easel	Nufarm UK	750 g/l	SL	15548
3	HY-MCPA	Agrichem	500 g/l	SL	14927
4	HY-MCPA	Agrichem	500 g/l	SL	19769
5	Larke	Nufarm UK	750 g/l	SL	14914

Uses

- Annual and perennial weeds in *Grass seed crops* [1-2,5]; *Grassland* [1-2,5];
- Annual dicotyledons in *Grass seed crops* [3-4]; *Grassland* [3-4]; *Spring barley* [1-5]; *Spring oats* [1-5]; *Spring rye* [1-2,5]; *Spring wheat* [1-5]; *Undersown barley* [1-5]; *Undersown oats* [1-5]; *Undersown rye* [1-5]; *Undersown wheat* [1-5]; *Winter barley* [1-5]; *Winter oats* [1-5]; *Winter rye* [1-2,5]; *Winter wheat* [1-5];
- Charlock in *Spring barley* [1-5]; *Spring oats* [1-5]; *Spring rye* [1-2,5]; *Spring wheat* [1-5]; *Winter barley* [1-5]; *Winter oats* [1-5]; *Winter rye* [1-2,5]; *Winter wheat* [1-5];

FOR FULL CONDITIONS OF USE ALWAYS READ THE PRODUCT LABEL

- Fat hen in *Spring barley* [1-5]; *Spring oats* [1-5]; *Spring rye* [1-2,5]; *Spring wheat* [1-5]; *Winter barley* [1-5]; *Winter oats* [1-5]; *Winter rye* [1-2,5]; *Winter wheat* [1-5];
- Hemp-nettle in *Spring barley* [1-5]; *Spring oats* [1-5]; *Spring rye* [1-2,5]; *Spring wheat* [1-5]; *Winter barley* [1-5]; *Winter oats* [1-5]; *Winter rye* [1-2,5]; *Winter wheat* [1-5];
- Perennial dicotyledons in *Spring barley* [1-5]; *Spring oats* [1-5]; *Spring rye* [1-2,5]; *Spring wheat* [1-5]; *Winter barley* [1-5]; *Winter oats* [1-5]; *Winter rye* [1-2,5]; *Winter wheat* [1-5];
- Wild radish in *Spring barley* [1-5]; *Spring oats* [1-5]; *Spring rye* [1-2,5]; *Spring wheat* [1-5]; *Winter barley* [1-5]; *Winter oats* [1-5]; *Winter rye* [1-2,5]; *Winter wheat* [1-5];

Approval information
- MCPA included in Annex I under EC Regulation 1107/2009
- Accepted by BBPA for use on malting barley

Efficacy guidance
- Best results achieved by application to weeds in seedling to young plant stage under good growing conditions when crop growing actively
- Spray perennial weeds in grassland before flowering. Most susceptible growth stage varies between species. See label for details
- Do not spray during cold weather, drought, if rain or frost expected or if crop wet

Restrictions
- Maximum number of treatments normally 1 per crop or yr except grass (2 per yr) for some products. See label
- Do not treat grass within 3 mth of germination and preferably not in the first yr of a direct sown ley or after reseeding
- Do not use on cereals before undersowing
- Do not roll, harrow or graze for a few days before or after spraying; see label
- Do not use on grassland where clovers are an important part of the sward
- Do not use on any crop suffering from stress or herbicide damage
- Avoid spray drift onto nearby susceptible crops
- Do not apply by hand-held equipment
- Livestock must be kept out of treated areas until poisonous weeds such as ragwort have died and become unpalatable
- This product must not be applied before the end of February in the year of harvest
- Do not apply in volumes less than 200 litres of water per hectare

Crop-specific information
- Latest use: before 1st node detectable (GS 31) for cereals; 4-6 wk before heading for grass seed crops; before crop 15-25 cm high for linseed
- Apply to winter cereals in spring from fully tillered, leaf sheath erect stage to before first node detectable (GS 31)
- Apply to spring barley and wheat from 5-leaves unfolded (GS 15), to oats from 1-leaf unfolded (GS 11) to before first node detectable (GS 31)
- Apply to cereals undersown with grass after grass has 2-3 leaves unfolded
- Recommendations for crops undersown with legumes vary. Red clover may withstand low doses after 2-trifoliate leaf stage, especially if shielded by taller weeds but white clover is more sensitive. See label for details
- Apply to grass seed crops from 2-3 leaf stage to 5 wk before head emergence
- Temporary wilting may occur on linseed but without long term effects

Following crops guidance
- Do not direct drill brassicas or legumes within 6 wk of spraying grassland

Environmental safety
- Harmful to aquatic organisms
- MCPA is active at low concentrations. Take extreme care to avoid drift onto neighbouring crops, especially beet crops, brassicas, most market garden crops including lettuce and tomatoes under glass, pears and vines
- Keep livestock out of treated areas for at least 2 wk and until foliage of poisonous weeds such as ragwort has died and become unpalatable
- LERAP Category B [1-5]

SEE SECTION 3 FOR PRODUCTS ALSO REGISTERED

Hazard classification and safety precautions

Hazard Harmful, Dangerous for the environment [2-4], Harmful if swallowed
Transport code 9 [1-2,5]
Packaging group III
UN Number 3082, N/C
Risk phrases H318, H335 [3]
Operator protection U05a, U08, U11, U12 [2], U20b; A, C
Environmental protection E06a (2 wk) [3-4], E07a [1-2,5], E15a [1,3-5], E15b [2], E16a, E34, E38 [2], H410 [1-4]
Storage and disposal D01, D02, D05, D09a, D10a, D12a [2]
Medical advice M03, M05a

249 MCPA + mecoprop-P

A translocated selective herbicide for amenity grass
HRAC mode of action code: 4 + 4 (O + O)

See also mecoprop-P

Products

1	Cleanrun Pro	ICL (Everris) Ltd	0.49:0.29% w/w	GR	15828

Uses

* Annual dicotyledons in *Managed amenity turf* [1];
* Perennial dicotyledons in *Managed amenity turf* [1];

Approval information

* MCPA and mecoprop-P included in Annex I under EC Regulation 1107/2009

Efficacy guidance

* Apply from Apr to Sep, when weeds growing actively and have large leaf area available for chemical absorption

Restrictions

* The total amount of mecoprop-P applied in a single year must not exceed the maximum total dose approved for any single product for use on turf
* Avoid contact with cultivated plants
* Do not use first 4 mowings as compost or mulch unless composted for 6 mth
* Do not treat newly sown or turfed areas for at least 6 mth
* Do not reseed bare patches for 8 wk after treatment
* Do not apply when heavy rain expected or during prolonged drought. Irrigate after 1-2 d unless rain has fallen
* Do not mow within 2-3 d of treatment
* Treat areas planted with bulbs only after the foliage has died down
* Avoid walking on treated areas until it has rained or irrigation has been applied

Crop-specific information

* Granules contain NPK fertilizer to encourage grass growth

Environmental safety

* Take extreme care to avoid drift onto neighbouring crops, especially beet crops, brassicas, most market garden crops including lettuce and tomatoes under glass, pears and vines
* Harmful to fish or other aquatic life. Do not contaminate surface waters or ditches with chemical or used container
* Keep livestock out of treated areas for at least 2 wk and until foliage of any poisonous weeds such as ragwort has died and become unpalatable
* Some pesticides pose a greater threat of contamination of water than others and mecoprop-P is one of these pesticides. Take special care when applying mecoprop-P near water and do not apply if heavy rain is forecast

Hazard classification and safety precautions

UN Number N/C

Operator protection U20b; A, C, H, M
Environmental protection E07a, E13c, E19b
Storage and disposal D01, D09a, D12a
Medical advice M05a

250   MCPB

A translocated phenoxycarboxylic acid herbicide
HRAC mode of action code: 4 (O)

Products

1	Tropotox	Nufarm UK	400 g/l	SL	18309

Uses
- Annual dicotyledons in **Combining peas** *(off-label)* [1]; **Vining peas** *(off-label)* [1];
- Docks in **Combining peas** [1]; **Vining peas** [1];
- Polygonums in **Combining peas** *(off-label)* [1]; **Vining peas** *(off-label)* [1];
- Thistles in **Combining peas** [1]; **Vining peas** [1];

Extension of Authorisation for Minor Use (EAMUs)
- **Combining peas** *20210024* [1]
- **Vining peas** *20210024* [1]

Approval information
- MCPB included in Annex I under EC Regulation 1107/2009

Efficacy guidance
- Best results achieved by spraying young seedling weeds in good growing conditions
- Best results on perennials by spraying before flowering
- Effectiveness may be reduced by rain within 12 h, by very cold or dry conditions

Restrictions
- Maximum number of treatments 1 per crop or yr.
- Do not roll or harrow for 7-10 d before or after treatment (check label)

Crop-specific information
- Latest use: first node detectable stage (GS 31) for cereals; before flower buds appear in terminal leaf (GS 201) for peas; before flower buds form for clover
- Apply to undersown cereals from 2-leaves unfolded to first node detectable (GS 12-31), and after first trifoliate leaf stage of clover
- Red clover seedlings may be temporarily damaged but later growth is normal
- Apply to white clover seed crops in Mar to early Apr, not after mid-May, and allow 3 wk before cutting and closing up for seed
- Apply to peas from 3-6 leaf stage but before flower bud detectable (GS 103-201). Consult PGRO (see Appendix 2) or label for information on susceptibility of cultivars.
- Do not use on leguminous crops not mentioned on the label
- Apply to cane and bush fruit after harvest and after shoot growth ceased but before weeds are damaged by frost, usually in late Aug or Sep; direct spray onto weeds as far as possible

Environmental safety
- Harmful to aquatic organisms
- Harmful to fish or other aquatic life. Do not contaminate surface waters or ditches with chemical or used container
- Keep livestock out of treated areas until foliage of any poisonous weeds such as ragwort has died and become unpalatable
- Take extreme care to avoid drift onto neighbouring sensitive crops

Hazard classification and safety precautions
Hazard Harmful, Dangerous for the environment, Harmful if swallowed
Transport code 9 [1]
Packaging group III

SEE SECTION 3 FOR PRODUCTS ALSO REGISTERED

UN Number 3082
Risk phrases H315, H318
Operator protection U05a, U08, U19a, U20b; A, C
Environmental protection E07a, E15a, E34, E38, H411
Storage and disposal D01, D02, D09a, D10b
Medical advice M03, M05a

251   mecoprop-P

A translocated phenoxycarboxylic acid herbicide for cereals and grassland
HRAC mode of action code: 4 (O)

See also 2,4-D + dicamba + MCPA + mecoprop-P
carfentrazone-ethyl + mecoprop-P
dicamba + MCPA + mecoprop-P
dicamba + mecoprop-P
dichlorprop-P + MCPA + mecoprop-P
ferrous sulphate + MCPA + mecoprop-P
MCPA + mecoprop-P

Products

1	Compitox Plus	Nufarm UK	600 g/l	SL	18375
2	Duplosan Kv	Nufarm UK	600 g/l	SL	18374

Uses
- Annual dicotyledons in *Amenity grassland* [1-2]; *Grass seed crops* [1-2]; *Managed amenity turf* [1-2]; *Spring barley* [1-2]; *Spring oats* [1-2]; *Spring wheat* [1-2]; *Winter barley* [1-2]; *Winter oats* [1-2]; *Winter wheat* [1-2];
- Chickweed in *Amenity grassland* [1-2]; *Grass seed crops* [1-2]; *Managed amenity turf* [1-2]; *Spring barley* [1-2]; *Spring oats* [1-2]; *Spring wheat* [1-2]; *Winter barley* [1-2]; *Winter oats* [1-2]; *Winter wheat* [1-2];
- Cleavers in *Amenity grassland* [1-2]; *Grass seed crops* [1-2]; *Managed amenity turf* [1-2]; *Spring barley* [1-2]; *Spring oats* [1-2]; *Spring wheat* [1-2]; *Winter barley* [1-2]; *Winter oats* [1-2]; *Winter wheat* [1-2];
- Perennial dicotyledons in *Amenity grassland* [1-2]; *Grass seed crops* [1-2]; *Managed amenity turf* [1-2]; *Spring barley* [1-2]; *Spring oats* [1-2]; *Spring wheat* [1-2]; *Winter barley* [1-2]; *Winter oats* [1-2]; *Winter wheat* [1-2];

Approval information
- Mecoprop-P included in Annex I under EC Regulation 1107/2009
- Accepted by BBPA for use on malting barley

Efficacy guidance
- Best results achieved by application to seedling weeds which have not been frost hardened, when soil warm and moist and expected to remain so for several days

Restrictions
- Maximum number of treatments normally 1 per crop for spring cereals and 1 per yr for newly sown grass; 2 per crop or yr for winter cereals and grass crops. Check labels for details
- The total amount of mecoprop-P applied in a single year must not exceed the maximum total dose approved for any single product for the crop/situation
- Do not spray cereals undersown with clovers or legumes or to be undersown with legumes or grasses
- Do not spray grass seed crops within 5 wk of seed head emergence
- Do not spray crops suffering from herbicide damage or physical stress
- Do not spray during cold weather, periods of drought, if rain or frost expected or if crop wet
- Do not roll or harrow for 7 d before or after treatment
- Applications to cereals must not be made between 1 October and 1 March

Crop-specific information
- Latest use: generally before 1st node detectable (GS 31) for spring cereals and before 3rd node detectable (GS 33) for winter cereals, but individual labels vary; 5 wk before emergence of seed head for grass seed crops
- Spray winter cereals from 1 leaf stage in autumn up to and including first node detectable in spring (GS 10-31) or up to second node detectable (GS 32) if necessary. Apply to spring cereals from first fully expanded leaf stage (GS 11) but before first node detectable (GS 31)
- Spray cereals undersown with grass after grass starts to tiller
- Spray newly sown grass leys when grasses have at least 3 fully expanded leaves and have begun to tiller. Any clovers will be damaged

Environmental safety
- Harmful to aquatic organisms
- Harmful to fish or other aquatic life. Do not contaminate surface waters or ditches with chemical or used container
- Keep livestock out of treated areas for at least 2 wk and until foliage of any poisonous weeds, such as ragwort, has died and become unpalatable
- Take extreme care to avoid drift onto neighbouring crops, especially beet crops, brassicas, most market garden crops including lettuce and tomatoes under glass, pears and vines
- Some pesticides pose a greater threat of contamination of water than others and mecoprop-P is one of these pesticides. Take special care when applying mecoprop-P near water and do not apply if heavy rain is forecast

Hazard classification and safety precautions
Hazard Harmful, Dangerous for the environment, Harmful if swallowed
Transport code 9 [1-2]
Packaging group III
UN Number 3082
Risk phrases H315, H318
Operator protection U05a, U08, U11, U20b; A, C, H
Environmental protection E15b, E34, E38, H410
Storage and disposal D01, D02, D09a, D10c
Medical advice M03

252   mefentrifluconazole

A triazole fungicide for use in cereals
FRAC mode of action code: 3

See also fluxapyroxad + mefentrifluconazole

Products
1	Lenvyor Duo	BASF	97 g/l	EC	19255
2	Myresa	BASF	97 g/l	EC	19153

Uses
- Brown rust in **Durum wheat** [1-2]; **Rye** *(moderate control)* [1-2]; **Spelt** [1-2]; **Spring barley** [1-2]; **Spring wheat** [1-2]; **Triticale** [1-2]; **Winter barley** [1-2]; **Winter wheat** [1-2];
- Crown rust in **Spring oats** [1-2]; **Winter oats** [1-2];
- Net blotch in **Spring barley** *(reduction)* [1-2]; **Winter barley** *(reduction)* [1-2];
- Powdery mildew in **Durum wheat** *(moderate control)* [1-2]; **Rye** *(moderate control)* [1-2]; **Spelt** *(moderate control)* [1-2]; **Spring oats** [1-2]; **Spring wheat** *(moderate control)* [1-2]; **Triticale** *(moderate control)* [1-2]; **Winter oats** [1-2]; **Winter wheat** *(moderate control)* [1-2];
- Ramularia leaf spots in **Spring barley** [1-2]; **Winter barley** [1-2];
- Rhynchosporium in **Rye** *(moderate control)* [1-2]; **Spring barley** [1-2]; **Winter barley** [1-2];
- Septoria leaf blotch in **Durum wheat** [1-2]; **Rye** *(moderate control)* [1-2]; **Spelt** [1-2]; **Spring wheat** [1-2]; **Triticale** [1-2]; **Winter wheat** [1-2];
- Yellow rust in **Durum wheat** [1-2]; **Rye** *(moderate control)* [1-2]; **Spelt** [1-2]; **Spring barley** [1-2]; **Spring wheat** [1-2]; **Triticale** [1-2]; **Winter barley** [1-2]; **Winter wheat** [1-2];

SEE SECTION 3 FOR PRODUCTS ALSO REGISTERED

SECTION 2

Approval information
- Mefentrifluconazole included in Annex I under EC Regulation 1107/2009

Efficacy guidance
- For best results apply as a protectant treatment. Applications can be made from beginning of stem elongation BBCH 30 up to and including flowering anthesis complete (BBCH 69) in wheat, barley, rye, triticale and oats.

Environmental safety
- LERAP Category B [1-2]

Hazard classification and safety precautions

Hazard Harmful if inhaled, Very toxic to aquatic organisms
Transport code 9 [1-2]
Packaging group III
UN Number 3082
Risk phrases H315, H317, H319, H335
Operator protection U05a, U20c; A, C, H
Environmental protection E15b, E16a, H410, H411
Storage and disposal D01, D02, D05, D08, D09a, D10c
Medical advice M03

253   mefentrifluconazole + pyraclostrobin

A fungicide mixture for disease control in cereal crops
FRAC mode of action code: 3 + 11

See also mefentrifluconazole
pyraclostrobin

Products

1	Rylox	BASF	100:95 g/l	EC	19229

Uses
- Brown rust in *Durum wheat* [1]; *Rye* [1]; *Spelt* [1]; *Spring barley* [1]; *Spring wheat* [1]; *Triticale* [1]; *Winter barley* [1]; *Winter wheat* [1];
- Crown rust in *Spring oats* [1]; *Winter oats* [1];
- Net blotch in *Spring barley* [1]; *Winter barley* [1];
- Powdery mildew in *Durum wheat* *(moderate control)* [1]; *Rye* [1]; *Spelt* *(moderate control)* [1]; *Spring barley* [1]; *Spring oats* [1]; *Spring wheat* *(moderate control)* [1]; *Triticale* *(moderate control)* [1]; *Winter barley* [1]; *Winter oats* [1]; *Winter wheat* *(moderate control)* [1];
- Ramularia leaf spots in *Spring barley* [1]; *Winter barley* [1];
- Rhynchosporium in *Rye* *(moderate control)* [1]; *Spring barley* *(moderate control)* [1]; *Winter barley* *(moderate control)* [1];
- Septoria leaf blotch in *Durum wheat* [1]; *Rye* [1]; *Spelt* [1]; *Spring wheat* [1]; *Triticale* [1]; *Winter wheat* [1];
- Tan spot in *Durum wheat* *(reduction)* [1]; *Rye* [1]; *Spelt* *(reduction)* [1]; *Spring wheat* *(reduction)* [1]; *Winter wheat* *(reduction)* [1];
- Yellow rust in *Durum wheat* [1]; *Rye* [1]; *Spelt* [1]; *Spring barley* [1]; *Spring wheat* [1]; *Triticale* [1]; *Winter barley* [1]; *Winter wheat* [1];

Approval information
- Mefentrifluconazole and pyraclostrobin included in Annex I under EC Regulation 1107/2009

Efficacy guidance
- For best results apply at the start of foliar disease attack
- Pyraclostrobin is a member of the QoI cross resistance group. Product should be used preventatively and not relied on for its curative potential
- Use product as part of an Integrated Crop Management strategy incorporating other methods of control, including where appropriate other fungicides with a different mode of action. Do not apply more than two foliar applications of QoI containing products to any cereal crop

FOR FULL CONDITIONS OF USE ALWAYS READ THE PRODUCT LABEL

Restrictions
- Allow an interval of at least 21 days between applications
- Do not apply using hand-held equipment

Environmental safety
- Buffer zone requirement 6 m
- LERAP Category A [1]

Hazard classification and safety precautions
> **Hazard** Harmful if swallowed, Harmful if inhaled, Very toxic to aquatic organisms
> **Transport code** 9 [1]
> **Packaging group** III
> **UN Number** 3082
> **Risk phrases** H315, H317, H318, H335, H351
> **Operator protection** U05a, U20b; A, C, H
> **Environmental protection** E16c, E34, H410
> **Storage and disposal** D01, D02, D05, D09a, D10c
> **Medical advice** M03

254   mepanipyrim

An anilinopyrimidine fungicide for use in horticulture
FRAC mode of action code: 9

Products

1	Frupica SC	Certis Belchim B V	450 g/l	SC	12067

Uses
- Botrytis in *Courgettes* (off-label) [1]; *Forest nurseries* (off-label) [1]; *Ornamental plant production* (off-label) [1]; *Protected aubergines* (off-label) [1]; *Protected courgettes* (off-label) [1]; *Protected cucumbers* (off-label) [1]; *Protected ornamentals* (off-label) [1]; *Protected tomatoes* (off-label) [1]; *Strawberries* [1];
- Powdery mildew in *Ornamental plant production* (off-label) [1]; *Protected aubergines* (off-label) [1]; *Protected courgettes* (off-label) [1]; *Protected cucumbers* (off-label) [1]; *Protected ornamentals* (off-label) [1]; *Protected tomatoes* (off-label) [1];

Extension of Authorisation for Minor Use (EAMUs)
- *Courgettes* 20093235 [1]
- *Forest nurseries* 20082853 [1]
- *Ornamental plant production* 20191294 [1]
- *Protected aubergines* 20191523 [1]
- *Protected courgettes* 20191523 [1]
- *Protected cucumbers* 20191523 [1]
- *Protected ornamentals* 20191294 [1]
- *Protected tomatoes* 20191523 [1]

Approval information
- Mepanipyrim included in Annex I under EC Regulation 1107/2009

Efficacy guidance
- Product is protectant and should be applied as a preventative spray when conditions favourable for Botrytis development occur
- To maintain Botrytis control use as part of a programme with other fungicides that control the disease
- To minimise the possibility of development of resistance adopt resistance management procedures by using products from different chemical groups as part of a mixed spray programme

Restrictions
- Maximum number of treatments 2 per crop (including other anilinopyrimidine products)

SEE SECTION 3 FOR PRODUCTS ALSO REGISTERED

- Consult processor before use on crops for processing
- Use spray mixture immediately after preparation

Crop-specific information
- HI 3 d

Environmental safety
- Dangerous for the environment
- Very toxic to aquatic organisms
- LERAP Category B [1]

Hazard classification and safety precautions
>   **Hazard** Dangerous for the environment
>   **Transport code** 9 [1]
>   **Packaging group** III
>   **UN Number** 3082
>   **Risk phrases** H351
>   **Operator protection** U20c; A, H
>   **Environmental protection** E16a, E16b, E34, H410
>   **Storage and disposal** D01, D02, D05, D10c, D11a, D12b

255   mepiquat chloride

A quaternary ammonium plant growth regulator available only in mixtures

See also chlormequat + mepiquat chloride
           ethephon + mepiquat chloride
           mepiquat chloride + metconazole
           mepiquat chloride + prohexadione-calcium

256   mepiquat chloride + metconazole

A PGR mixture for growth control in winter oilseed rape

Products

1	Caryx	BASF	210:30 g/l	SL	16100

Uses
- Growth regulation in **Winter oilseed rape** [1];
- Lodging control in **Winter oilseed rape** [1];

Approval information
- Metconazole and mepiquat chloride included in Annex I under EC Regulation 1107/2009

Efficacy guidance
- Apply in 200 - 400 l/ha water. Use at lower water volumes has not been evaluated.
- Apply from the beginning of stem extension once the crop is actively growing.

Following crops guidance
- Beans, cabbage, carrots, cereals, clover, lettuce, linseed, maize, oilseed rape, onions, peas, potatoes, ryegrass, sugar beet or sunflowers may be sown as a following crop but the effect on other crops has not been tested.

Environmental safety
- LERAP Category B [1]

Hazard classification and safety precautions
>   **Hazard** Harmful, Dangerous for the environment, Harmful if swallowed, Harmful if inhaled
>   **Transport code** 9 [1]
>   **Packaging group** III
>   **UN Number** 3082

Risk phrases H317, H318
Operator protection U02a, U05a, U11, U14, U15; A, C, H
Environmental protection E15b, E16a, E38, H411
Storage and disposal D01, D02, D05, D09a, D10c, D19
Medical advice M03, M05a

257 mepiquat chloride + prohexadione-calcium

A growth regulator mixture for cereals

See also prohexadione-calcium

Products

1	Canopy	BASF	300:50 g/l	SC	16314

Uses
- Growth regulation in **Ornamental plant production** *(off-label)* [1]; **Protected ornamentals** *(off-label)* [1];
- Increasing yield in **Oats** [1]; **Spring barley** [1]; **Triticale** [1]; **Winter barley** [1]; **Winter rye** [1]; **Winter wheat** [1];
- Lodging control in **Oats** [1]; **Spring barley** [1]; **Triticale** [1]; **Winter barley** [1]; **Winter rye** [1]; **Winter wheat** [1];

Extension of Authorisation for Minor Use (EAMUs)
- **Ornamental plant production** *20194484* [1]
- **Protected ornamentals** *20194484* [1]

Approval information
- Mepiquat chloride and prohexadione-calcium included in Annex I under EC Regulation 1107/2009
- Accepted by BBPA for use on malting barley

Efficacy guidance
- Best results obtained from treatments applied to healthy crops from the beginning of stem extension

Restrictions
- Maximum total dose equivalent to one full dose treatment on all crops
- Do not apply to any crop suffering from physical stress caused by waterlogging, drought or other conditions
- Do not treat on soils with a substantial moisture deficit
- Consult grain merchant or processor before use on crops for bread making or brewing. Effects on these processes have not been tested

Crop-specific information
- Latest use: before flag leaf fully emerged on wheat and barley

Following crops guidance
- Any crop may follow a normally harvested treated crop. Ploughing is not essential.

Environmental safety
- Harmful to aquatic organisms
- Avoid spray drift onto neighbouring crops

Hazard classification and safety precautions
Hazard Harmful, Harmful if swallowed
UN Number N/C
Risk phrases H319
Operator protection U05a, U20b; A
Environmental protection E15a, E34, E38, H412
Storage and disposal D01, D02, D05, D09a, D10c
Medical advice M05a

SECTION 2

258   mepiquat chloride + prohexadione-calcium + pyraclostrobin

A growth regulator and fungicide mixture for use in oilseed rape

Products

1 Architect	BASF	150:25:100 g/l	SE	19728

Uses
- Growth regulation in *Winter oilseed rape* [1];
- Light leaf spot in *Winter oilseed rape* (moderate control) [1];
- Phoma leaf spot in *Winter oilseed rape* [1];
- Stem canker in *Winter oilseed rape* (moderate control) [1];

Approval information
- Mepiquat chloride, prohexadione-calcium and pyraclostrobin included in Annex I under EC Regulation 1107/2009

Crop-specific information
- Any crop may follow normally harvested or failed crops

Environmental safety
- Buffer zone requirement 9 m
- LERAP Category B [1]

Hazard classification and safety precautions
  Hazard Harmful, Harmful if swallowed, Very toxic to aquatic organisms
  Transport code 9 [1]
  Packaging group III
  UN Number 3082
  Risk phrases H315, H317
  Operator protection U05a, U20b; A, H
  Environmental protection E15a, E16a, E34, E38, H410
  Storage and disposal D01, D02, D05, D09a, D10c
  Medical advice M05a

259   mesosulfuron-methyl

A sulfonyl urea herbicide for cereals available only in mixtures
HRAC mode of action code: 2 (B)

See also amidosulfuron + iodosulfuron-methyl-sodium + mesosulfuron-methyl
diflufenican + iodosulfuron-methyl-sodium + mesosulfuron-methyl
iodosulfuron-methyl-sodium + mesosulfuron-methyl
iodosulfuron-methyl-sodium + mesosulfuron-methyl + thiencarbazone-methyl

260   mesosulfuron-methyl + propoxycarbazone-sodium

A herbicide mixture for use in winter cereals
HRAC mode of action code: 2 + 2 (B + B)

Products

1 Monolith	Bayer CropScience	4.5:6.75% w/w	WG	19973

Uses
- Annual dicotyledons in *Durum wheat* [1]; *Spelt* [1]; *Triticale* [1]; *Winter rye* [1]; *Winter wheat* [1];
- Annual meadow grass in *Durum wheat* [1]; *Spelt* [1]; *Triticale* [1]; *Winter rye* [1]; *Winter wheat* [1];
- Blackgrass in *Durum wheat* [1]; *Spelt* [1]; *Triticale* [1]; *Winter rye* [1]; *Winter wheat* [1];
- Brome grasses in *Durum wheat* [1]; *Spelt* [1]; *Triticale* [1]; *Winter rye* [1]; *Winter wheat* [1];

- Chickweed in **Durum wheat** [1]; **Spelt** [1]; **Triticale** [1]; **Winter rye** [1]; **Winter wheat** [1];
- Loose silky bent in **Durum wheat** [1]; **Spelt** [1]; **Triticale** [1]; **Winter rye** [1]; **Winter wheat** [1];
- Mayweeds in **Durum wheat** [1]; **Spelt** [1]; **Triticale** [1]; **Winter rye** [1]; **Winter wheat** [1];
- Ryegrass in **Durum wheat** [1]; **Spelt** [1]; **Triticale** [1]; **Winter rye** [1]; **Winter wheat** [1];
- Wild oats in **Durum wheat** [1]; **Spelt** [1]; **Triticale** [1]; **Winter rye** [1]; **Winter wheat** [1];

Approval information
- Mesosulfuron-methyl and propoxycarbazone-sodium included in Annex I under EC Regulation 1107/2009

Efficacy guidance
- Apply as early as possible and before GS 29 of grass weeds
- Always use in mixture with an authorised adjuvant such as Biopower (ADJ: 0617) at a rate of 1 l/ha

Restrictions
- This product must only be applied between 1 February in the year of harvest and the specified latest time of application.
- To avoid the build-up of resistance do not apply this or any other product containing an ALS herbicide with claims of control of grass-weeds more than once to any crop.
- This product must not be applied via hand-held equipment.
- DO NOT use on crops undersown with grasses, clover or other legumes or any other broad-leaved crop.
- Do not use on cereal crops grown for seed as effects on germination have not been established.
- Do not use as the sole means of grass weed or broad-leaved weed control in successive crops

Crop-specific information
- Clean the spray tank with clean water and add a liquid sprayer cleaner specifically formulated for sulfonylurea herbicides after use.

Following crops guidance
- Winter wheat, winter barley, winter oilseed rape, mustard, lupins and phacelia may be sown in the year of harvest to succeed a treated cereal crop. Spray overlaps in the treated cereal crop should be avoided in order to reduce the risk of localised adverse effects on following crops of winter oilseed rape and mustard.
- Spring wheat, spring barley, maize, spring oilseed rape, sugar beet, Italian rye-grass, peas and sunflowers may be drilled in the spring following harvest of a treated cereal crop.

Environmental safety
- LERAP Category B [1]

Hazard classification and safety precautions
Hazard Very toxic to aquatic organisms
Transport code 9 [1]
Packaging group III
UN Number 3077
Risk phrases H319
Operator protection U05a, U11, U20b; A, C, H
Environmental protection E15a, E16a, H410
Storage and disposal D01, D02, D05, D09a, D10a, D12a, D12b

261   mesotrione

A foliar applied triketone herbicide for maize
HRAC mode of action code: 27 (F2)

Products

1 Barracuda	Albaugh UK	100 g/l	SC	18348
2 Basilico	Life Scientific	100 g/l	SC	18028
3 Callisto	Syngenta	100 g/l	SC	19756
4 Clayton Cob	Clayton	100 g/l	SC	18971

SECTION 2

SEE SECTION 3 FOR PRODUCTS ALSO REGISTERED

Products – continued

5	Clayton Goldcob	Clayton	100 g/l	SC	19247
6	Clayton Tassel	Clayton	100 g/l	SC	19369
7	Daneva	Rotam	100 g/l	SC	18029
8	Evolya	Syngenta	50% w/w	WG	17490
9	Meristo	Syngenta	100 g/l	SC	20163
10	Raikiri	Sipcam	100 g/l	SC	17670
11	Temsa SC	Certis Belchim B V	100 g/l	SC	18261

Uses

- Annual dicotyledons in *Forage maize* [1-7,10-11]; *Grain maize* [7];
- Annual meadow grass in *Grain maize* [1-6,10-11];
- Barnyard grass in *Forage maize* *(moderately susceptible)* [8]; *Grain maize* *(moderately susceptible)* [8];
- Black bindweed in *Forage maize* [8]; *Grain maize* [8];
- Black nightshade in *Forage maize* [8-9]; *Grain maize* [8-9]; *Linseed* *(off-label)* [2];
- Charlock in *Forage maize* [9]; *Grain maize* [9]; *Linseed* *(off-label)* [2];
- Chickweed in *Forage maize* [8-9]; *Grain maize* [8-9]; *Linseed* *(off-label)* [2];
- Cockspur grass in *Forage maize* [9]; *Grain maize* [9]; *Linseed* *(off-label)* [2];
- Common amaranth in *Forage maize* [8-9]; *Grain maize* [8-9]; *Linseed* *(off-label)* [2];
- Fat hen in *Forage maize* [8-9]; *Grain maize* [8-9]; *Linseed* *(off-label)* [2];
- Field pansy in *Forage maize* [8-9]; *Grain maize* [8-9]; *Linseed* *(off-label)* [2];
- Gallant soldier in *Forage maize* [8]; *Grain maize* [8];
- Maple-leaved goosefoot in *Forage maize* [8]; *Grain maize* [8];
- Pale persicaria in *Forage maize* [8]; *Grain maize* [8];
- Red dead-nettle in *Forage maize* [8]; *Grain maize* [8];
- Redshank in *Forage maize* [8-9]; *Grain maize* [8-9]; *Linseed* *(off-label)* [2];
- Volunteer oilseed rape in *Forage maize* [1-11]; *Grain maize* [1-11]; *Linseed* *(off-label)* [2];

Extension of Authorisation for Minor Use (EAMUs)

- *Linseed* 20202705 [2]

Approval information

- Mesotrione included in Annex I under EC Regulation 1107/2009

Efficacy guidance

- Best results obtained from treatment of young actively growing weed seedlings in the presence of adequate soil moisture
- Treatment in poor growing conditions or in dry soil may give less reliable control
- Activity is mostly by foliar uptake with some soil uptake
- To minimise the possible development of resistance where continuous maize is grown the product should not be used for more than two consecutive seasons

Restrictions

- Maximum number of treatments 1 per crop of forage or grain maize
- Do not use on seed crops or on sweetcorn varieties
- Do not spray when crop foliage wet or when excessive rainfall is expected to follow application
- Do not treat crops suffering from stress from cold or drought conditions, or when wide temperature fluctuations are anticipated
- Extreme care must be taken to avoid spray drift onto plants outside of the target area
- Must not be applied via hand-held equipment [1]
- Do not handle treated crops for at least 2 days after treatment [1]

Crop-specific information

- Latest use: 8 leaves unfolded stage (GS 18) for forage maize
- Treatment under adverse conditions may cause mild to moderate chlorosis. The effect is transient and does not affect yield

Following crops guidance

- Winter wheat, durum wheat, winter barley or ryegrass may follow a normally harvested treated crop of maize. Oilseed rape may be sown provided it is preceded by deep ploughing to more than 15 cm

FOR FULL CONDITIONS OF USE ALWAYS READ THE PRODUCT LABEL

- In the spring following application only forage maize, ryegrass, spring wheat or spring barley may be sown
- In the event of crop failure maize may be re-seeded immediately. Some slight crop effects may be seen soon after emergence but these are normally transient

Environmental safety
- Dangerous for the environment
- Very toxic to aquatic organisms
- Take extreme care to avoid drift onto all plants outside the target area
- LERAP Category B [1-6,8-11]

Hazard classification and safety precautions
    Hazard Irritant, Dangerous for the environment [1-7,9-11], Very toxic to aquatic organisms [1-10]
    Transport code 9 [4-9]
    Packaging group III
    UN Number 3077, 3082, N/C
    Risk phrases H317 [1,6,10-11], H318 [1,6-7,10-11], H319 [2-5,9], H361 [3,9]
    Operator protection U05a, U09a, U14 [8], U20b; A, C, H
    Environmental protection E15b [1-7,9-11], E16a [1-6,8-11], E16b [1-6,9-11], E38 [1-7,9-11], H410
    Storage and disposal D01, D02, D05, D09a, D10c, D11a, D12a [1-7,9-11]

262   mesotrione + nicosulfuron

A herbicide for use in forage and grain maize
HRAC mode of action code: 2 + 27 (B + F2)

Products

| 1 | Choriste | Syngenta | 75:30 g/l | OD | 16899 |
| 2 | Elumis | Syngenta | 75:30 g/l | OD | 15800 |

Uses
- Amaranthus in *Forage maize* [1-2]; *Grain maize* [1-2];
- Annual dicotyledons in *Forage maize* [1-2]; *Grain maize* [1-2];
- Annual meadow grass in *Forage maize* [1-2]; *Grain maize* [1-2];
- Black bindweed in *Forage maize* [1-2]; *Grain maize* [1-2];
- Black nightshade in *Forage maize* [1-2]; *Grain maize* [1-2];
- Chickweed in *Forage maize* [1-2]; *Grain maize* [1-2];
- Cleavers in *Forage maize* [1-2]; *Grain maize* [1-2];
- Cockspur grass in *Forage maize* [1-2]; *Grain maize* [1-2];
- Fat hen in *Forage maize* [1-2]; *Grain maize* [1-2];
- Field pansy in *Forage maize* [1-2]; *Grain maize* [1-2];
- Field speedwell in *Forage maize* [1-2]; *Grain maize* [1-2];
- Fumitory in *Forage maize* [1-2]; *Grain maize* [1-2];
- Hemp-nettle in *Forage maize* [1-2]; *Grain maize* [1-2];
- Mayweeds in *Forage maize* [1-2]; *Grain maize* [1-2];
- Pale persicaria in *Forage maize* [1-2]; *Grain maize* [1-2];
- Red dead-nettle in *Forage maize* [1-2]; *Grain maize* [1-2];
- Redshank in *Forage maize* [1-2]; *Grain maize* [1-2];
- Ryegrass in *Forage maize* [1-2]; *Grain maize* [1-2];
- Shepherd's purse in *Forage maize* [1-2]; *Grain maize* [1-2];
- Volunteer oilseed rape in *Forage maize* [1-2]; *Grain maize* [1-2];

Approval information
- Mesotrione and nicosulfuron included in Annex 1 under EC Regulation 1107/2009

Efficacy guidance
- Optimum efficacy requires soils to be moist at application.
- To minimise the possible development of resistance where continuous maize is grown the product should not be used for more than two consecutive seasons.

SEE SECTION 3 FOR PRODUCTS ALSO REGISTERED

Following crops guidance
- In the event of crop failure, only forage or grain maize may be sown as a replacement after ploughing.
- After normal harvest wheat or barley may be sown the same autumn after ploughing to 15 cm and after 4 months has elapsed since application. The following spring maize, wheat, barley or ryegrass may be planted after ploughing but other crops are not recommended.

Environmental safety
- LERAP Category B [1-2]

Hazard classification and safety precautions
Hazard Irritant, Dangerous for the environment, Very toxic to aquatic organisms
Transport code 9 [1-2]
Packaging group III
UN Number 3082
Operator protection U02a, U05a, U14, U19a, U20a; A, H
Environmental protection E15b, E16a, E23, H410
Storage and disposal D01, D02, D05, D09a, D10c, D12a
Medical advice M05a, M05d

263   mesotrione + pyridate

A residual herbicide mixture for use in forage and grain maize

See also mesotrione
        pyridate

Products

1	Botiga	Certis Belchim B V	90:300 g/l	OD	19759

Uses
- Annual dicotyledons in *Forage maize* [1]; *Grain maize* [1];
- Annual grasses in *Forage maize* [1]; *Grain maize* [1];

Approval information
- Mesotrione and pyridate included in Annex I under EC Regulation 1107/2009

Hazard classification and safety precautions
Hazard Harmful if swallowed
Transport code 9 [1]
Packaging group III
UN Number 3082
Risk phrases H315, H317, H318, H361
Operator protection A, C, H
Environmental protection H410

264   mesotrione + S-metolachlor

A herbicide mixture for weed control in forage and grain maize.
HRAC mode of action code: 27 + 15 (F2 + K3)

Products

1	Camix	Syngenta	60:500 g/l	SE	17722
2	Clarido	Syngenta	60:500 g/l	SE	17891

Uses
- Black nightshade in *Forage maize* [1-2]; *Grain maize* [1-2];
- Bristle grasses in *Forage maize* (pre-em only) [1-2]; *Grain maize* (pre-em only) [1-2];
- Chickweed in *Forage maize* [1-2]; *Grain maize* [1-2];
- Cockspur grass in *Forage maize* [1-2]; *Grain maize* [1-2];
- Common amaranth in *Forage maize* [1-2]; *Grain maize* [1-2];

FOR FULL CONDITIONS OF USE ALWAYS READ THE PRODUCT LABEL

- Fat hen in **Forage maize** [1-2]; **Grain maize** [1-2];
- Redshank in **Forage maize** *(pre-em only)* [1-2]; **Grain maize** *(pre-em only)* [1-2];

Approval information
- Mesotrione and S-metolachlor included in Annex I under EC Regulation 1107/2009

Efficacy guidance
- Application must be made to healthy maize, preferably in good growing conditions, when the vegetation is dry.
- Do not use during periods of frosty weather, when frost is imminent, or onto crops under stress from frost, water-logging, insect attack or drought.
- Special care should be taken to avoid damage to plants outside the target area by drift or drift onto land intended for cropping.
- Some phytotoxicity may be seen after application. This effect is usually transient and does not affect yield.
- Consult processors before use on grain maize for processing.

Restrictions
- Ensure spraying equipment is thoroughly washed out according to specific instructions after use. Do not allow washings-out to drain onto land intended for cropping or growing crops.

Following crops guidance
- Winter wheat (including durum wheat), winter barley and rye grass can follow (after shallow tillage) a maize crop. Deep ploughing (greater than 15cm) followed by cultivation is necessary before drilling oilseed rape.
- Forage maize and grain maize, ryegrass, spring wheat and spring barley may be sown in the spring following application. Do not sow any other crop at this time.
- In case of maize failure after application, only grain and forage maize can be considered as alternative crops in the same spring (ploughing is recommended prior to reseeding to avoid slight and transitory crop effects soon after emergence).

Environmental safety
- LERAP Category B [1-2]

Hazard classification and safety precautions
Transport code 9 [1-2]
Packaging group III
UN Number 3082
Risk phrases H317
Operator protection U02a, U14, U20c; A, H
Environmental protection E16a, H410
Storage and disposal D01, D02, D05, D09a, D10c

265   metalaxyl-M

A phenylamide systemic fungicide
FRAC mode of action code: 4

See also mancozeb + metalaxyl-M

Products
1	Apron XL	Syngenta	339.2 g/l	ES	14654
2	Caveo	Pan Agriculture	465.2 g/l	SL	16361
3	Clayton Tine	Clayton	465 g/l	SL	14072
4	Floreo	Pan Agriculture	465.2 g/l	SL	16358
5	SL 567A	Syngenta	465.2 g/l	SL	12380
6	Subdue	Fargro	465 g/l	SL	12503

Uses
- Cavity spot in **Carrots** *(reduction only)* [2-3,5]; **Parsnips** *(off-label - reduction)* [3,5];
- Crown rot in **Protected water lilies** *(off-label)* [5]; **Water lilies** *(off-label)* [5];

SEE SECTION 3 FOR PRODUCTS ALSO REGISTERED

- Damping off in **Watercress** *(off-label)* [5];
- Downy mildew in **Asparagus** *(off-label)* [5]; **Baby leaf crops** *(off-label)* [5]; **Blackberries** *(off-label)* [5]; **Broad beans** *(off-label)* [5]; **Broccoli** *(off-label)* [5]; **Brussels sprouts** *(off-label)* [5]; **Cabbages** *(off-label)* [5]; **Calabrese** *(off-label)* [5]; **Cauliflowers** *(off-label)* [5]; **Collards** *(off-label)* [5]; **Grapevines** *(off-label)* [5]; **Herbs (see appendix 6)** *(off-label)* [5]; **Hops** *(off-label)* [5]; **Horseradish** *(off-label)* [5]; **Kale** *(off-label)* [5]; **Protected baby leaf crops** *(off-label)* [5]; **Protected cucumbers** *(off-label)* [5]; **Protected herbs (see appendix 6)** *(off-label)* [5]; **Protected soft fruit** *(off-label)* [5]; **Protected spinach** *(off-label)* [5]; **Protected spinach beet** *(off-label)* [5]; **Protected water lilies** *(off-label)* [5]; **Raspberries** *(off-label)* [5]; **Rubus hybrids** *(off-label)* [5]; **Salad onions** *(off-label)* [5]; **Soft fruit** *(off-label)* [5]; **Spinach** *(off-label)* [5]; **Spinach beet** *(off-label)* [5]; **Spring field beans** *(off-label)* [5]; **Water lilies** *(off-label)* [5]; **Watercress** *(off-label)* [5]; **Winter field beans** *(off-label)* [5];
- Phytophthora in **Amenity vegetation** *(off-label)* [6]; **Forest nurseries** *(off-label)* [6]; **Ornamental plant production** [4]; **Protected ornamentals** [4];
- Phytophthora root rot in **Ornamental plant production** [6]; **Protected ornamentals** [6];
- Pythium in **Beetroot** *(off-label)* [1]; **Brassica leaves and sprouts** *(off-label)* [1]; **Broccoli** [1]; **Brussels sprouts** [1]; **Bulb onions** [1]; **Cabbages** [1]; **Calabrese** [1]; **Cauliflowers** [1]; **Chard** *(off-label)* [1]; **Choi sum** *(off-label)* [1]; **Collards** *(off-label)* [1]; **Herbs (see appendix 6)** *(off-label)* [1]; **Kale** *(off-label)* [1]; **Kohlrabi** [1]; **Oriental cabbage** [1]; **Ornamental plant production** *(off-label)* [1,4,6]; **Protected ornamentals** [4,6]; **Radishes** *(off-label)* [1]; **Shallots** *(off-label)* [1]; **Spinach** [1];
- Root malformation disorder in **Red beet** *(off-label)* [5];
- Storage rots in **Cabbages** *(off-label)* [5];
- White blister in **Brussels sprouts** *(off-label)* [5]; **Cabbages** *(off-label)* [5]; **Cauliflowers** *(off-label)* [5]; **Horseradish** *(off-label)* [5];

Extension of Authorisation for Minor Use (EAMUs)

- **Amenity vegetation** *20120383* [6]
- **Asparagus** *20051502* [5]
- **Baby leaf crops** *20051507* [5]
- **Beetroot** *20211025* [1]
- **Blackberries** *20072195* [5]
- **Brassica leaves and sprouts** *20211026* [1]
- **Broad beans** *20201899* [5]
- **Broccoli** *20112502* [5]
- **Brussels sprouts** *20201898* [5]
- **Cabbages** *20062117* [5], *20201898* [5]
- **Calabrese** *20112502* [5]
- **Cauliflowers** *20201898* [5]
- **Chard** *20211026* [1]
- **Choi sum** *20211027* [1]
- **Collards** *20211027* [1], *20112048* [5]
- **Forest nurseries** *20120383* [6]
- **Grapevines** *20051504* [5]
- **Herbs (see appendix 6)** *20211025* [1], *20051507* [5]
- **Hops** *20051500* [5]
- **Horseradish** *20051499* [5], *20061040* [5]
- **Kale** *20211027* [1], *20112048* [5]
- **Ornamental plant production** *20211025* [1]
- **Parsnips** *(reduction)* *20111033* [3], *20051508* [5]
- **Protected baby leaf crops** *20051507* [5]
- **Protected cucumbers** *20051503* [5]
- **Protected herbs (see appendix 6)** *20051507* [5]
- **Protected soft fruit** *20082937* [5]
- **Protected spinach** *20051507* [5]
- **Protected spinach beet** *20051507* [5]
- **Protected water lilies** *20051501* [5]
- **Radishes** *20211025* [1]
- **Raspberries** *20072195* [5]

FOR FULL CONDITIONS OF USE ALWAYS READ THE PRODUCT LABEL

- **Red beet** *20051307* [5]
- **Rubus hybrids** *20072195* [5]
- **Salad onions** *20072194* [5]
- **Shallots** *20211024* [1]
- **Soft fruit** *20082937* [5]
- **Spinach** *20051507* [5]
- **Spinach beet** *20051507* [5]
- **Spring field beans** *20130917* [5]
- **Water lilies** *20051501* [5]
- **Watercress** *20072193* [5]
- **Winter field beans** *20130917* [5]

Approval information
- Metalaxyl-M included in Annex I under EC Regulation 1107/2009
- Accepted by BBPA for use on hops

Efficacy guidance
- Best results achieved when applied to damp soil or potting media
- Treatments to ornamentals should be followed immediately by irrigation to wash any residues from the leaves and allow penetration to the rooting zone [6]
- Efficacy may be reduced in prolonged dry weather [5]
- Results may not be satisfactory on soils with high organic matter content [5]
- Control of cavity spot on carrots overwintered in the ground or lifted in winter may be lower than expected [5]
- Use in an integrated pest management strategy and, where appropriate, alternate with products from different chemical groups
- Product should ideally be used preventatively and the number of phenylamide applications should be limited to 1-2 consecutive treatments
- Always follow FRAG guidelines for preventing and managing fungicide resistance. See Section 5 for more information

Restrictions
- Maximum number of treatments on ornamentals 1 per situation for media treatment. See label for details of drench treatment of protected ornamentals [6]
- Maximum total dose on carrots equivalent to one full dose treatment [5]
- Do not use where carrots have been grown on the same site within the previous eight yrs [5]
- Consult before use on crops intended for processing [5]
- Do not re-use potting media from treated plants for subsequent crops [6]
- Disinfect pots thoroughly prior to re-use [6]

Crop-specific information
- Latest use: 6 wk after drilling for carrots [5]
- Because of the large number of species and ornamental cultivars susceptibility should be checked before large scale treatment [6]
- Treatment of *Viburnum* and *Prunus* species not recommended [6]

Environmental safety
- Harmful to aquatic organisms
- Limited evidence suggests that metalaxyl-M is not harmful to soil dwelling predatory mites

Hazard classification and safety precautions
Hazard Harmful, Harmful if swallowed
UN Number N/C
Risk phrases H319, H335 [6]
Operator protection U02a, U04a [2-6], U05a, U10 [2-6], U19a [2-6], U20b; A, C, D, H
Environmental protection E15b, E34, E38, H412 [2-6]
Storage and disposal D01, D02, D05, D07 [3,5-6], D09a, D10c [2-6], D11a [1], D12a
Treated seed S02 [1], S04d [1], S05 [1], S07 [1], S08 [1], S09 [1]
Medical advice M03, M05a [1]

SEE SECTION 3 FOR PRODUCTS ALSO REGISTERED

266   metamitron

A contact and residual triazinone herbicide for use in beet crops and a fruit thinner for apples and pears.
HRAC mode of action code: 5 (C1)

See also ethofumesate + metamitron

Products

1	Beetron 700	AgChem Access	700 g/l	SC	17458
2	Bettix Flo	UPL Europe	700 g/l	SC	16559
3	Bettix WG	UPL Europe	70% w/w	WG	16496
4	Brevis	Adama	15% w/w	SG	17479
5	Clayton Mondello	Clayton	700 g/l	SC	19288
6	Clayton Mondello SC	Clayton	700 g/l	SC	19859
7	Clayton Neutron	Clayton	700 g/l	SC	19563
8	Cleancrop Savant	Agrii	700 g/l	SC	18399
9	Defiant SC	UPL Europe	700 g/l	SC	16531
10	Goltix 70 SC	Adama	700 g/l	SC	16638
11	Mitron 700 SC	BelCrop	700 g/l	SC	16908
12	Target Flo	UPL Europe	700 g/l	SC	19159

Uses

* Annual dicotyledons in *Fodder beet* [1-3,5-12]; *Game cover* (off-label) [10]; *Horseradish* (off-label) [10]; *Mangels* [1-3,5-12]; *Parsnips* (off-label) [10]; *Red beet* [1-3,5-12]; *Strawberries* (off-label) [10]; *Sugar beet* [1-3,5-12];
* Annual grasses in *Fodder beet* [1-3,5,9,11-12]; *Mangels* [1-3,5,9,11-12]; *Red beet* [1-3,5,9,11-12]; *Sugar beet* [1-3,5,9,11-12];
* Annual meadow grass in *Fodder beet* [1-3,5-12]; *Game cover* (off-label) [10]; *Horseradish* (off-label) [10]; *Mangels* [1-3,5-12]; *Ornamental plant production* (off-label) [10]; *Parsnips* (off-label) [10]; *Protected ornamentals* (off-label) [10]; *Red beet* [1-3,5-12]; *Strawberries* (off-label) [10]; *Sugar beet* [1-3,5-12];
* Fat hen in *Fodder beet* [1-3,5,9,11-12]; *Game cover* (off-label) [10]; *Mangels* [1-3,5,9,11-12]; *Ornamental plant production* (off-label) [10]; *Protected ornamentals* (off-label) [10]; *Red beet* [1-3,5,9,11-12]; *Sugar beet* [1-3,5,9,11-12];
* Fruit thinning in *Apples* [4]; *Pears* [4];
* Groundsel in *Game cover* (off-label) [10]; *Ornamental plant production* (off-label) [10]; *Protected ornamentals* (off-label) [10];
* Growth regulation in *Apples* [4]; *Pears* [4];
* Mayweeds in *Game cover* (off-label) [10]; *Ornamental plant production* (off-label) [10]; *Protected ornamentals* (off-label) [10];
* Red dead-nettle in *Ornamental plant production* (off-label) [10]; *Protected ornamentals* (off-label) [10];
* Scarlet pimpernel in *Ornamental plant production* (off-label) [10]; *Protected ornamentals* (off-label) [10];
* Small nettle in *Ornamental plant production* (off-label) [10]; *Protected ornamentals* (off-label) [10];

Extension of Authorisation for Minor Use (EAMUs)

* *Game cover* 20194264 [10]
* *Horseradish* 20142918 [10]
* *Ornamental plant production* 20151175 [10]
* *Parsnips* 20142918 [10]
* *Protected ornamentals* 20151175 [10]
* *Strawberries* 20142919 [10]

Approval information

* Metamitron included in Annex I under EC Regulation 1107/2009

Efficacy guidance
- May be used pre-emergence alone or post-emergence in tank mixture or with an authorised adjuvant oil
- Low dose programme (LDP). Apply a series of low-dose post-weed emergence sprays, including adjuvant oil, timing each treatment according to weed emergence and size. See label for details and for recommended tank mixes and sequential treatments. On mineral soils the LDP should be preceded by pre-drilling or pre-emergence treatment
- Traditional application. Apply either pre-drilling before final cultivation with incorporation to 8-10 cm, or pre-crop emergence at or soon after drilling into firm, moist seedbed to emerged weeds from cotyledon to first true leaf stage
- On emerged weeds at or beyond 2-leaf stage addition of adjuvant oil advised
- For control of wild oats and certain other weeds, tank mixes with other herbicides or sequential treatments are recommended. See label for details
- When used for fruit thinning in apples and pears, apply post blossom if fruit set is excessive. Apply in temperatures between 10 and 25°C [4]

Restrictions
- Maximum total dose equivalent to three full dose treatments for most products. Check label
- Using traditional method post-crop emergence on mineral soils do not apply before first true leaves have reached 1 cm long
- Fodder beet and mangels must not be grazed by livestock or harvested for animal consumption until at least 103 days following the last application

Crop-specific information
- Latest use: before crop foliage meets across rows for beet crops
- HI herbs 6 wk
- HI for apples and pears 60 days
- Crop tolerance may be reduced by stress caused by growing conditions, effects of pests, disease or other pesticides, nutrient deficiency etc

Following crops guidance
- Only sugar beet, fodder beet or mangels may be drilled within 4 mth after treatment. Winter cereals may be sown in same season after ploughing, provided 16 wk passed since last treatment

Environmental safety
- Dangerous for the environment
- Very toxic to aquatic organisms
- Dangerous to fish or other aquatic life. Do not contaminate surface waters or ditches with chemical or used container
- Do not empty into drains

Hazard classification and safety precautions
Hazard Harmful [1,3,6-11], Dangerous for the environment, Harmful if swallowed [1-4,6-8,10-11], Very toxic to aquatic organisms
Transport code 9 [1-12]
Packaging group III
UN Number 3077, 3082
Risk phrases H317 [2,6,10], H318 [4], R22a [9], R43 [9], R51 [9], R53a [9]
Operator protection U02a [4], U04a [4], U05a [3-4,6,8,11], U08 [6,8,11], U09a [2,5,9,12], U11 [4], U14 [2,5-6,8-9,11-12], U19a [3,6,8,11], U20a [11], U20b [1,3-4,6,8], U20c [2,5,9,12]; A, C, H
Environmental protection E13b [2-3,5,9,12], E15a [1,6,8,11], E15b [4], E19b [1], E34 [1,4-6,8], E38 [2,5-12], H410 [1,5-8,10,12], H411 [2-4,11]
Storage and disposal D01, D02, D05 [2,4-5,9,12], D09a [1-6,8-9,11-12], D10c [2,5,9,12], D11a [1,3-4,6,8], D12a
Medical advice M03 [1-3,5-6,8-9,12], M05a [1-3,5,7,9-12]

SEE SECTION 3 FOR PRODUCTS ALSO REGISTERED

SECTION 2

267   metamitron + quinmerac

A herbicide mixture for weed control in beet crops
HRAC mode of action code: 5 + 4 (C1 + O)

See also metamitron

Products

| 1 Goltix Titan | Adama | 525:40 g/l | SC | 17301 |

Uses
- Annual dicotyledons in **Fodder beet** [1]; **Sugar beet** [1];
- Annual meadow grass in **Fodder beet** [1]; **Sugar beet** [1];
- Cleavers in **Fodder beet** [1]; **Sugar beet** [1];

Approval information
- Metamitron and quinmerac included in Annex I under EC Regulation 1107/2009

Efficacy guidance
- Do not handle treated plants for at least 2 days after treatment

Restrictions
- Maintain an interval of at least 5 days between applications
- Quinmerac is being found in ground and surface waters. Take care when applying to sloping soils near water.

Crop-specific information
- Beet crops may be sown at any time following application. Allow 16 weeks from application before sowing winter cereals in the same season. After normal harvest, any crop can be grown in the following spring. Deep and thorough cultivation is necessary before sowing or planting succeeding crops. Mouldboard ploughing to a minimum depth of 15 cm is recommended before seedbed preparation.

Hazard classification and safety precautions
Transport code 9 [1]
Packaging group III
UN Number 3082
Operator protection U02a, U04a; A, H
Environmental protection E15b, H411
Storage and disposal D01, D09a, D10b, D12a

268   metazachlor

A residual anilide herbicide for use in brassicas, nurseries and forestry
HRAC mode of action code: 15 (K3)

See also aminopyralid + metazachlor + picloram
       dimethenamid-p + metazachlor
       dimethenamid-p + metazachlor + quinmerac
       imazamox + metazachlor

Products

1 Butisan S	BASF	500 g/l	SC	16569
2 Rapsan Solo	Certis Belchim B V	500 g/l	SC	18516
3 Stalwart	UPL Europe	500 g/l	SC	17405
4 Sultan 50 SC	Adama	500 g/l	SC	16680

Uses
- Annual dicotyledons in **Borage for oilseed production** *(off-label)* [1,4]; **Broccoli** [1,3-4]; **Brussels sprouts** [1-4]; **Cabbages** [1,3-4]; **Calabrese** [1,3-4]; **Canary flower (Echium spp.)** *(off-label)* [1,4]; **Cauliflowers** [1,3-4]; **Chinese cabbage** *(off-label)* [1]; **Choi sum** *(off-label)* [1]; **Collards** *(off-label)* [1]; **Evening primrose** *(off-label)* [1,4]; **Game cover** *(off-label)* [1]; **Honesty**

(off-label) [1,4]; **Hops in propagation** *(off-label)* [1,4]; **Kale** *(off-label)* [1]; **Kohlrabi** *(off-label)* [1]; **Linseed** *(off-label)* [1]; **Mustard** *(off-label)* [1,4]; **Nursery fruit trees** [1,3-4]; **Oriental cabbage** *(off-label)* [1]; **Ornamental plant production** [1-4]; **Pak choi** *(off-label)* [1]; **Spring linseed** *(off-label)* [4]; **Spring oilseed rape** [1-4]; **Swedes** *(off-label)* [1]; **Tatsoi** *(off-label)* [1]; **Turnips** *(off-label)* [1]; **Winter linseed** *(off-label)* [4]; **Winter oilseed rape** [1-4];

- Annual meadow grass in **Borage for oilseed production** *(off-label)* [1,4]; **Broccoli** [1,3-4]; **Brussels sprouts** [1-4]; **Cabbages** [1,3-4]; **Calabrese** [1,3-4]; **Canary flower (Echium spp.)** *(off-label)* [1,4]; **Cauliflowers** [1,3-4]; **Chinese cabbage** *(off-label)* [1]; **Choi sum** *(off-label)* [1]; **Collards** *(off-label)* [1]; **Evening primrose** *(off-label)* [1,4]; **Game cover** *(off-label)* [1]; **Honesty** *(off-label)* [1,4]; **Hops in propagation** *(off-label)* [1,4]; **Kale** *(off-label)* [1]; **Kohlrabi** *(off-label)* [1]; **Linseed** *(off-label)* [1]; **Mustard** *(off-label)* [1,4]; **Nursery fruit trees** [1,3-4]; **Oriental cabbage** *(off-label)* [1]; **Ornamental plant production** [1-4]; **Pak choi** *(off-label)* [1]; **Spring linseed** *(off-label)* [4]; **Spring oilseed rape** [1-4]; **Swedes** *(off-label)* [1]; **Tatsoi** *(off-label)* [1]; **Turnips** *(off-label)* [1]; **Winter linseed** *(off-label)* [4]; **Winter oilseed rape** [1-4];
- Blackgrass in **Broccoli** [1,3-4]; **Brussels sprouts** [1-4]; **Cabbages** [1,3-4]; **Calabrese** [1,3-4]; **Cauliflowers** [1,3-4]; **Nursery fruit trees** [1,3-4]; **Ornamental plant production** [1-4]; **Spring oilseed rape** [1-4]; **Winter oilseed rape** [1-4];
- Chickweed in **Choi sum** *(off-label)* [4]; **Collards** *(off-label)* [4]; **Kale** *(off-label)* [4]; **Kohlrabi** *(off-label)* [4]; **Oriental cabbage** *(off-label)* [4]; **Tatsoi** *(off-label)* [4];
- Groundsel in **Borage for oilseed production** *(off-label)* [1]; **Canary flower (Echium spp.)** *(off-label)* [1-2]; **Chinese cabbage** *(off-label)* [1]; **Choi sum** *(off-label)* [1-2,4]; **Collards** *(off-label)* [1-2,4]; **Evening primrose** *(off-label)* [1-2]; **Game cover** *(off-label)* [1]; **Honesty** *(off-label)* [1]; **Hops in propagation** *(off-label)* [1]; **Kale** *(off-label)* [1-2,4]; **Kohlrabi** *(off-label)* [1-2,4]; **Linseed** *(off-label)* [1-2]; **Mustard** *(off-label)* [1-2]; **Oriental cabbage** *(off-label)* [1-2,4]; **Pak choi** *(off-label)* [1]; **Swedes** *(off-label)* [1-2]; **Tatsoi** *(off-label)* [1-2,4]; **Turnips** *(off-label)* [1-2]; **Winter borage** *(off-label)* [2];
- Mayweeds in **Borage for oilseed production** *(off-label)* [1]; **Canary flower (Echium spp.)** *(off-label)* [1-2]; **Chinese cabbage** *(off-label)* [1]; **Choi sum** *(off-label)* [1-2,4]; **Collards** *(off-label)* [1-2,4]; **Evening primrose** *(off-label)* [1-2]; **Game cover** *(off-label)* [1]; **Honesty** *(off-label)* [1]; **Hops in propagation** *(off-label)* [1]; **Kale** *(off-label)* [1-2,4]; **Kohlrabi** *(off-label)* [1-2,4]; **Linseed** *(off-label)* [1-2]; **Mustard** *(off-label)* [1-2]; **Oriental cabbage** *(off-label)* [1-2,4]; **Pak choi** *(off-label)* [1]; **Swedes** *(off-label)* [1-2]; **Tatsoi** *(off-label)* [1-2,4]; **Turnips** *(off-label)* [1-2]; **Winter borage** *(off-label)* [2];
- Shepherd's purse in **Canary flower (Echium spp.)** *(off-label)* [1-2]; **Chinese cabbage** *(off-label)* [1]; **Choi sum** *(off-label)* [2,4]; **Collards** *(off-label)* [2,4]; **Evening primrose** *(off-label)* [2]; **Honesty** *(off-label)* [1]; **Kale** *(off-label)* [2,4]; **Kohlrabi** *(off-label)* [2,4]; **Linseed** *(off-label)* [2]; **Mustard** *(off-label)* [2]; **Oriental cabbage** *(off-label)* [2,4]; **Pak choi** *(off-label)* [1]; **Swedes** *(off-label)* [1-2]; **Tatsoi** *(off-label)* [2,4]; **Turnips** *(off-label)* [1-2]; **Winter borage** *(off-label)* [2];

Extension of Authorisation for Minor Use (EAMUs)

- **Borage for oilseed production** *20211133* [1], *20142381* [4]
- **Canary flower (Echium spp.)** *20141053* [1], *20181945* [2], *20142381* [4]
- **Chinese cabbage** *20141052* [1]
- **Choi sum** *20211134* [1], *20211154* [2], *20161213* [4]
- **Collards** *20211134* [1], *20211154* [2], *20161213* [4]
- **Evening primrose** *20211133* [1], *20211156* [2], *20142381* [4]
- **Game cover** *20211853* [1]
- **Honesty** *20141053* [1], *20142381* [4]
- **Hops in propagation** *20211853* [1], *20193865* [4]
- **Kale** *20211134* [1], *20211154* [2], *20161213* [4]
- **Kohlrabi** *20211134* [1], *20211154* [2], *20161213* [4]
- **Linseed** *20211133* [1], *20211156* [2]
- **Mustard** *20211133* [1], *20211156* [2], *20142381* [4]
- **Oriental cabbage** *20211134* [1], *20211154* [2], *20161213* [4]
- **Pak choi** *20141052* [1]
- **Spring linseed** *20142381* [4]
- **Swedes** *20211136* [1], *20211155* [2]
- **Tatsoi** *20211134* [1], *20211154* [2], *20161213* [4]
- **Turnips** *20211136* [1], *20211155* [2]

SEE SECTION 3 FOR PRODUCTS ALSO REGISTERED

- **Winter borage** *20211156* [2]
- **Winter linseed** *20142381* [4]

Approval information
- Metazachlor is included in Annex I under EC Regulation 1107/2009

Efficacy guidance
- Activity is dependent on root uptake. For pre-emergence use apply to firm, moist, clod-free seedbed
- Some weeds (chickweed, mayweed, blackgrass etc) susceptible up to 2- or 4-leaf stage. Moderate control of cleavers achieved provided weeds not emerged and adequate soil moisture present
- Split pre- and post-emergence treatments recommended for certain weeds in winter oilseed rape on light and/or stony soils
- Effectiveness is reduced on soils with more than 10% organic matter
- Always follow WRAG guidelines for preventing and managing herbicide resistant weeds. See Section 5 for more information

Restrictions
- Maximum number of treatments 1 per crop for spring oilseed rape, swedes, turnips and brassicas; 2 per crop for winter oilseed rape (split dose treatment); 3 per yr for ornamentals, nursery stock, nursery fruit trees, forestry and farm forestry
- Do not use on sand, very light or poorly drained soils
- Do not treat protected crops or spray overall on ornamentals with soft foliage
- Do not spray crops suffering from wilting, pest or disease
- Do not spray broadcast crops or if a period of heavy rain forecast
- When used on nursery fruit trees any fruit harvested within 1 yr of treatment must be destroyed
- Applications shall be limited to a total dose of not more than 1.0 kg metazachlor/ha in a three year period on the same field
- Do not apply in mixture with phosphate liquid fertilisers
- Metazachlor stewardship requires that all autumn applications should be made before the end of September to reduce the risk to water

Crop-specific information
- Latest use: pre-emergence for swedes and turnips; before 10 leaf stage for spring oilseed rape; before end of Jan for winter oilseed rape
- HI brassicas 6 wk
- On winter oilseed rape may be applied pre-emergence from drilling until seed chits, post-emergence after fully expanded cotyledon stage (GS 1,0) or by split dose technique depending on soil and weeds. See label for details
- On spring oilseed rape may also be used pre-weed-emergence from cotyledon to 10-leaf stage of crop (GS 1,0-1,10)
- With pre-emergence treatment ensure seed covered by 15 mm of well consolidated soil. Harrow across slits of direct-drilled crops
- Ensure brassica transplants have roots well covered and are well established. Direct drilled brassicas should not be treated before 3 leaf stage
- In ornamentals and hardy nursery stock apply after plants established and hardened off as a directed spray or, on some subjects, as an overall spray. See label for list of tolerant subjects. Do not treat plants in containers

Following crops guidance
- Any crop can follow normally harvested treated winter oilseed rape. See label for details of crops which may be planted after spring treatment and in event of crop failure

Environmental safety
- Dangerous for the environment
- Very toxic to aquatic organisms
- Keep livestock out of treated areas until foliage of any poisonous weeds such as ragwort has died and become unpalatable
- Keep livestock out of treated areas of swede and turnip for at least 5 wk following treatment

FOR FULL CONDITIONS OF USE ALWAYS READ THE PRODUCT LABEL

- Some pesticides pose a greater threat of contamination of water than others and metazachlor is one of these pesticides. Take special care when applying metazachlor near water and do not apply if heavy rain is forecast.
- Metazachlor stewardship guidelines advise a maximum dose of 750 g.a.i/ha/annum. Applications to drained fields should be complete by 15th Oct but, if drains are flowing, complete applications by 1st Oct.
- LERAP Category B [1-4]

Hazard classification and safety precautions

Hazard Harmful, Dangerous for the environment, Harmful if swallowed, Very toxic to aquatic organisms [3]

Transport code 9 [1-4]

Packaging group III

UN Number 3082

Risk phrases H317, H351

Operator protection U05a, U08, U14, U15 [3-4], U19a, U20b; A, C, H, M

Environmental protection E06a (5 wk for swedes, turnips) [3], E07a [1-3], E15a, E16a, E34, E38 [1-2], H410

Storage and disposal D01, D02, D05 [3-4], D09a, D10c, D12a [1-2], D12b [3-4], D19 [1-2]

Medical advice M03, M05a [1-2]

269   metazachlor + quinmerac

A residual herbicide mixture for oilseed rape
HRAC mode of action code: 15 + 4 (K3 + O)

See also quinmerac

Products

1	Katamaran	BASF	375:125 g/l	SC	16766
2	Legion	Adama	375:125 g/l	SC	16249
3	Naspar Extra	Certis Belchim B V	375:125 g/l	SC	17038

Uses

- Annual dicotyledons in *Game cover* (off-label) [1]; *Spring oilseed rape* [1,3]; *Winter oilseed rape* [1-3];
- Annual meadow grass in *Game cover* (off-label) [1]; *Spring oilseed rape* [1,3]; *Winter oilseed rape* [1-3];
- Blackgrass in *Corn Gromwell* (off-label) [2]; *Game cover* (off-label) [1]; *Spring oilseed rape* [1,3]; *Winter oilseed rape* [1-3];
- Chickweed in *Corn Gromwell* (off-label) [2];
- Cleavers in *Corn Gromwell* (off-label) [2]; *Spring oilseed rape* [1,3]; *Winter oilseed rape* [1-3];
- Mayweeds in *Corn Gromwell* (off-label) [2]; *Spring oilseed rape* [3]; *Winter oilseed rape* [2-3];
- Poppies in *Corn Gromwell* (off-label) [2]; *Spring oilseed rape* [1,3]; *Winter oilseed rape* [1-3];
- Speedwells in *Corn Gromwell* (off-label) [2];

Extension of Authorisation for Minor Use (EAMUs)

- *Corn Gromwell* 20162047 [2]
- *Game cover* 20193866 [1]

Approval information

- Metazachlor and quinmerac included in Annex I under EC Regulation 1107/2009

Efficacy guidance

- Activity is dependent on root uptake. Pre-emergence treatments should be applied to firm moist seedbeds. Applications to dry soil do not become effective until after rain has fallen
- Maximum activity achieved from treatment before weed emergence for some species
- Weed control may be reduced if excessive rain falls shortly after application especially on light soils
- May be used on all soil types except Sands, Very Light Soils, and soils containing more than 10% organic matter. Crop vigour and/or plant stand may be reduced on brashy and stony soils

SEE SECTION 3 FOR PRODUCTS ALSO REGISTERED

Restrictions
- Maximum total dose equivalent to one full dose treatment
- Damage may occur in waterlogged conditions. Do not use on poorly drained soils
- Do not treat stressed crops. In frosty conditions transient scorch may occur
- Applications shall be limited to a total dose of not more than 1.0 kg metazachlor/ha in a three year period on the same field
- Do not apply in mixture with phosphate liquid fertilisers
- Metazachlor stewardship requires that all autumn applications should be made before the end of September to reduce the risk to water. Quinmerac is being found in ground and surface waters. Take care when applying to sloping soils near water.

Crop-specific information
- Latest use: end Jan in yr of harvest
- To ensure crop safety it is essential that crop seed is well covered with soil to 15 mm. Loose or puffy seedbeds must be consolidated before treatment. Do not use on broadcast crops
- Crop vigour and possibly plant stand may be reduced if excessive rain falls shortly after treatment especially on light soils

Following crops guidance
- In the event of crop failure after use, wheat or barley may be sown in the autumn after ploughing to 15 cm. Spring cereals or brassicas may be planted after ploughing in the spring

Environmental safety
- Dangerous for the environment
- Very toxic to aquatic organisms
- Keep livestock out of treated areas until foliage of any poisonous weeds such as ragwort has died and become unpalatable
- To reduce risk of movement to water do not apply to dry soil or if heavy rain is forecast. On clay soils create a fine consolidated seedbed
- Some pesticides pose a greater threat of contamination of water than others and metazachlor is one of these pesticides. Take special care when applying metazachlor near water and do not apply if heavy rain is forecast.
- Metazachlor stewardship guidelines advise a maximum dose of 750 g.a.i/ha/annum. Applications to drained fields should be complete by 15th Oct but, if drains are flowing, complete applications by 1st Oct.
- LERAP Category B [1-3]

Hazard classification and safety precautions
Hazard Irritant, Dangerous for the environment, Very toxic to aquatic organisms [1,3]
Transport code 9 [1-3]
Packaging group III
UN Number 3082
Risk phrases H317 [1,3], H351
Operator protection U05a, U08, U14, U19a, U20b; A
Environmental protection E07a, E15a, E16a, E38, H410
Storage and disposal D01, D02, D08, D09a, D10c, D12a
Medical advice M05a

270   metconazole

A triazole fungicide for cereals and oilseed rape
FRAC mode of action code: 3

See also boscalid + metconazole
fluxapyroxad + metconazole
mepiquat chloride + metconazole

Products

1	Artina EC	Certis Belchim B V	90 g/l	EC	18680
2	Conatra 60	Corteva	60 g/l	EC	19294
3	Gringo	AgChem Access	60 g/l	EC	15774

FOR FULL CONDITIONS OF USE ALWAYS READ THE PRODUCT LABEL

Products – continued

4	Juventus	BASF	90 g/l	EC	15528
5	Metfin 90	Clayton	90 g/l	EC	18910
6	Redstar	AgChem Access	90 g/l	SL	15938
7	Simveris	BASF	90 g/l	EC	19619
8	Sirena	Certis Belchim B V	60 g/l	EC	17103
9	Sunorg Pro	BASF	90 g/l	EC	15433

Uses

- Alternaria in **Spring oilseed rape** [1-9]; **Winter oilseed rape** [1-9];
- Ascochyta in **Combining peas** *(reduction)* [1-9]; **Lupins** *(qualified minor use)* [1-9]; **Vining peas** *(reduction)* [1-9];
- Botrytis in **Combining peas** *(reduction)* [1-9]; **Lupins** *(qualified minor use)* [1-9]; **Vining peas** *(reduction)* [1-9];
- Brown rust in **Durum wheat** [1-9]; **Rye** [1-9]; **Spring barley** [1-9]; **Spring wheat** [1-2,4-7,9]; **Triticale** [1-9]; **Winter barley** [1-9]; **Winter wheat** [1-9];
- Disease control in **Broad beans** *(off-label)* [9]; **Spring linseed** *(off-label)* [4,9]; **Winter linseed** *(off-label)* [4,9];
- Fusarium ear blight in **Durum wheat** *(reduction)* [2-5,7-9]; **Rye** *(reduction)* [1-3,5-6,8-9]; **Spring wheat** *(reduction)* [1-2,4-7,9]; **Triticale** *(reduction)* [1-9]; **Winter wheat** *(reduction)* [1-9];
- Growth regulation in **Winter oilseed rape** [9];
- Light leaf spot in **Spring oilseed rape** *(reduction)* [1-9]; **Winter oilseed rape** *(reduction)* [1-9];
- Mycosphaerella in **Combining peas** *(reduction)* [1-9]; **Vining peas** *(reduction)* [1-9];
- Net blotch in **Durum wheat** *(reduction)* [1]; **Spring barley** *(reduction)* [1-9]; **Winter barley** *(reduction)* [1-9];
- Phoma in **Spring oilseed rape** *(reduction)* [1-9]; **Winter oilseed rape** *(reduction)* [1-9];
- Powdery mildew in **Durum wheat** *(moderate control)* [1-9]; **Rye** *(moderate control)* [1-9]; **Spring barley** *(moderate control)* [1-9]; **Spring wheat** *(moderate control)* [1-2,4-7,9]; **Triticale** *(moderate control)* [1-9]; **Winter barley** *(moderate control)* [1-9]; **Winter wheat** *(moderate control)* [1-9];
- Rhynchosporium in **Rye** *(moderate control)* [4,7]; **Spring barley** *(reduction)* [1-9]; **Winter barley** *(reduction)* [1-9];
- Rust in **Combining peas** [1-9]; **Lupins** *(qualified minor use)* [1-9]; **Spring field beans** [1-9]; **Vining peas** [1-9]; **Winter field beans** [1-9];
- Septoria leaf blotch in **Durum wheat** [2-5,7-9]; **Rye** [1-3,5-6,8-9]; **Spring wheat** [1-2,4-7,9]; **Triticale** [1-9]; **Winter wheat** [1-9];
- Yellow rust in **Durum wheat** [2-5,7-9]; **Rye** [1-9]; **Spring barley** [4,7]; **Spring wheat** [1-2,4-7,9]; **Triticale** [1-9]; **Winter barley** [4,7]; **Winter wheat** [1-9];

Extension of Authorisation for Minor Use (EAMUs)

- **Broad beans** *20111928* [9]
- **Spring linseed** *20130387* [4], *20111929* [9]
- **Winter linseed** *20130387* [4], *20111929* [9]

Approval information

- Metconazole included in Annex I under EC Regulation 1107/2009
- Accepted by BBPA for use on malting barley

Efficacy guidance

- Best results from application to healthy, vigorous crops when disease starts to develop
- Good spray cover of the target is essential for best results. Spray volume should be increased to improve spray penetration into dense crops
- Metconazole is a DMI fungicide. Resistance to some DMI fungicides has been identified in Septoria leaf blotch which may seriously affect performance of some products. For further advice contact a specialist advisor and visit the Fungicide Resistance Action Group (FRAG)-UK website

Restrictions

- Maximum total dose equivalent to two full dose treatments on all crops
- Do not apply to oilseed rape crops that are damaged or stressed from previous treatments, adverse weather, nutrient deficiency or pest attack. Spring application may lead to reduction of crop height

SEE SECTION 3 FOR PRODUCTS ALSO REGISTERED

SECTION 2

- The addition of adjuvants is neither advised nor necessary and can lead to enhanced growth regulatory effects on stressed crops of oilseed rape
- Ensure sprayer is free from residues of previous treatments that may harm the crop, especially oilseed rape. Use of a detergent cleaner is advised before and after use
- Do not apply with pyrethroid insecticides on oilseed rape at flowering
- Maintain an interval of at least 14 days between applications to oilseed rape, peas, beans and lupins [4], 21 days between applications to cereals [4]
- To protect birds, only one application is allowed on cereals before end of tillering (GS29)

Crop-specific information
- Latest use: up to and including milky ripe stage (GS 71) for cereals; 10% pods at final size for oilseed rape
- HI: 14 d for peas, field beans, lupins
- Spring treatments on oilseed rape can reduce the height of the crop
- Treatment for Septoria leaf spot in wheat should be made before second node detectable stage and when weather favouring development of the disease has occurred. If conditions continue to favour disease development a follow-up treatment may be needed
- Treat mildew infections in cereals before 3% infection on any green leaf. A specific mildewicide will improve control of established infections
- Treat yellow rust on wheat before 1% infection on any leaf or as preventive treatment after GS 39
- For brown rust in cereals spray susceptible varieties before any of top 3 leaves has more than 2% infection
- Peas should be treated at the start of flowering and repeat 3-4 wk later if required
- Field beans and lupins should be treated at petal fall and repeat 3-4 wk later if required

Following crops guidance
- Only cereals, oilseed rape, sugar beet, linseed, maize, clover, beans, peas, carrots, potatoes or onions may be sown as following crops after treatment

Environmental safety
- Dangerous for the environment
- Very toxic to aquatic organisms
- Avoid treatment close to field boundary, even if permitted by LERAP assessment, to reduce effects on non-target insects or other arthropods
- LERAP Category B [1-9]

Hazard classification and safety precautions
Hazard Harmful [2-5,7-8], Irritant [1,6,9], Flammable [2-3,5,8], Dangerous for the environment, Flammable liquid and vapour [2-3,8], Very toxic to aquatic organisms [3]
Transport code 3 [2-3,5,8], 9 [1,4,6-7,9]
Packaging group III
UN Number 1993, 3082
Risk phrases H304 [2-3,8], H315 [2-3,8], H317 [2-3,8], H318 [2-3,8], H319 [1,4-7,9], H335 [2-3,8], H361, H373 [1,4-6,9]
Operator protection U02a, U05a, U11 [2-3,5,8], U14 [2-3,5,8], U20c; A, C, H, M
Environmental protection E15a, E16a, E16b [2-5,7-8], E34, E38, H410 [2-3,8], H411 [1,4-6,9], H412 [7]
Storage and disposal D01, D02, D05, D06c [1,4,6-7,9], D09a, D10a [1-3,5-6,8-9], D10c [4,7], D12a
Medical advice M03 [4,7], M05a [4,7], M05b [2-3,5,8]

271   1-methylcyclopropene

An inhibitor of ethylene production for use in stored apples

Products

1	SmartFresh	Landseer	3.3% w/w	SP	11799
2	SmartFresh ProTabs	Landseer	2% w/w	TB	16546
3	SmartFresh SmartTabs	Landseer	0.63% w/w	TB	12684

FOR FULL CONDITIONS OF USE ALWAYS READ THE PRODUCT LABEL

Uses
- Botrytis in **Plums** *(off-label)* [3];
- Ethylene inhibition in **Apples** *(post-harvest use)* [1]; **Pears** *(off-label - post harvest only)* [1];
- Scald in **Apples** *(post-harvest use)* [1]; **Pears** *(off-label - post harvest only)* [1];
- Slow down of ripening in **Apples** [2]; **Pears** [2]; **Plums** [2]; **Tomatoes post harvest** [2-3];
- Storage rots in **Broccoli** *(off-label)* [1]; **Brussels sprouts** *(off-label)* [1]; **Cabbages** *(off-label)* [1]; **Calabrese** *(off-label)* [1]; **Cauliflowers** *(off-label)* [1]; **Plums** *(off-label)* [3];

Extension of Authorisation for Minor Use (EAMUs)
- **Broccoli** *20111502* [1]
- **Brussels sprouts** *20111503* [1]
- **Cabbages** *20111503* [1]
- **Calabrese** *20111502* [1]
- **Cauliflowers** *20111502* [1]
- **Pears** *(post harvest only) 20102424* [1]
- **Plums** *20072234* [3]

Approval information
- Methylcyclopropene included in Annex 1 under EC Regulation 1107/2009

Efficacy guidance
- Best results obtained from treatment of fruit in good condition and of proper quality for long-term storage
- Effects may be reduced in fruit that is in poor condition or ripe prior to storage or harvested late
- Product acts by releasing vapour into store when mixed with water
- Apply as soon as possible after harvest
- Treatment controls superficial scald and maintains fruit firmness and acid content for 3-6 mth in normal air, and 6-9 mth in controlled atmosphere storage
- Ethylene production recommences after removal from storage

Restrictions
- Maximum number of treatments 1 per batch of apples
- Must only be used by suitably trained and competent persons in fumigation operations
- Keep unprotected persons out of treated areas during the 24-hour treatment process
- Consult processors before treatment of fruit destined for processing or cider making
- Do not apply in mixture with other products
- Ventilate all areas thoroughly with all refrigeration fans operating at maximum power for at least 15 min before re-entry

Crop-specific information
- Latest use: 7 d after harvest of apples
- Product tested on Granny Smith, Gala, Jonagold, Bramley and Cox. Consult distributor or supplier before treating other varieties

Environmental safety
- Unprotected persons must be kept out of treated stores during the 24 h treatment period
- Prior to application ensure that the store can be properly and promptly sealed

Hazard classification and safety precautions
Hazard Irritant [2-3]
UN Number N/C
Risk phrases H319 [2-3]
Operator protection U01 [2-3], U05a, U11 [2-3], U12 [2-3]; C
Environmental protection E02a (24 h) [1], E02a (12 h) [2-3], E15a [1], E15b [2-3], E34, E38 [2], H412 [2]
Consumer protection C12 [1]
Storage and disposal D01, D02, D07 [1], D14

SECTION 2

272   metobromuron

A pre-emergence herbicide for weed control in potatoes
HRAC mode of action code: 5 (C2)

Products

1	Inigo	Certis Belchim B V	500 g/l	SC	16933
2	Praxim	Certis Belchim B V	500 g/l	SC	16871
3	Soleto	Certis Belchim B V	500 g/l	SC	16935

Uses
- Annual dicotyledons in **Ornamental bulbs** *(off-label)* [1-3]; **Ornamental plant production** *(off-label)* [2]; **Potatoes** [1-3];
- Annual grasses in **Ornamental bulbs** *(off-label)* [1-3]; **Ornamental plant production** *(off-label)* [2];
- Annual meadow grass in **Ornamental plant production** *(off-label)* [2]; **Potatoes** [1-3];
- Annual mercury in **Ornamental plant production** *(off-label)* [2];
- Charlock in **Potatoes** [1-3];
- Chickweed in **Potatoes** [1-3];
- Fat hen in **Potatoes** [1-3];
- Groundsel in **Potatoes** [1-3];
- Mayweeds in **Potatoes** [1-3];
- Shepherd's purse in **Potatoes** [1-3];

Extension of Authorisation for Minor Use (EAMUs)
- **Ornamental bulbs** *20183121* [1], *20183073* [2], *20183127* [3]
- **Ornamental plant production** *20222647* [2]

Approval information
- Metobromuron included in Annex I under EC Regulation 1107/2009

Restrictions
- Heavy rain after application may cause transient discolouration of the crop.

Crop-specific information
- May be used on all varieties of potato

Following crops guidance
- Following normal harvest of a treated crop Brassicas (oilseed rape, turnip, brassica vegetables) and sugar beet may be sown, providing the soil has been ploughed. No restrictions apply when other crops are sown.
- In the event of crop failure maize, beans, peas and carrots may be sown after ploughing. It is not recommended to sow Brassicas (oilseed rape, turnip, brassica vegetables) or sugar beet as a replacement crop after crop failure. Potatoes can be re-planted after minimal cultivation.

Hazard classification and safety precautions
Hazard Very toxic to aquatic organisms
Transport code 9 [1-3]
Packaging group III
UN Number 3082
Risk phrases H351, H373
Operator protection U05a, U09a, U19a, U20a; A, H, M
Environmental protection E15b, E34, E38, H410
Storage and disposal D01, D02, D05, D09a, D12b
Medical advice M05a

273   metrafenone

A benzophenone protectant and curative fungicide for cereals
FRAC mode of action code: U8

Products

1	Flexity	BASF	300 g/l	SC	11775
2	Vivando	BASF	500 g/l	SC	18026

Uses

- Cobweb in **Mushroom boxes** [2];
- Disease control/foliar feed in **Ornamental plant production** *(off-label)* [1];
- Eyespot in **Spring wheat** *(reduction)* [1]; **Winter wheat** *(reduction)* [1];
- Powdery mildew in **Hops** *(off-label)* [2]; **Ornamental plant production** [1]; **Spring barley** [1]; **Spring oats** *(evidence of mildew control on oats is limited)* [1]; **Spring wheat** [1]; **Wine grapes** *(off-label)* [2]; **Winter barley** [1]; **Winter oats** *(evidence of mildew control on oats is limited)* [1]; **Winter wheat** [1];

Extension of Authorisation for Minor Use (EAMUs)

- **Hops** *20192992* [2]
- **Ornamental plant production** *20082850* [1]
- **Wine grapes** *20192990* [2]

Approval information

- Metrafenone included in Annex I under EC Regulation 1107/2009
- Accepted by BBPA for use on malting barley

Efficacy guidance

- Best results obtained from treatment at the start of foliar disease attack
- Activity against mildew in wheat is mainly protectant with moderate curative control in the latent phase; activity is entirely protectant in barley
- Useful reduction of eyespot in wheat is obtained if treatment applied at GS 30-32
- Should be used as part of a resistance management strategy that includes mixtures or sequences effective against mildew and non-chemical methods

Restrictions

- Maximum number of treatments 2 per crop
- Avoid the use of sequential applications of metrafenone unless it is used in tank mixture with other products active against powdery mildew employing a different mode of action

Crop-specific information

- Latest use: beginning of flowering (GS 31) for wheat, barley, oats

Following crops guidance

- Cereals, oilseed rape, sugar beet, linseed, maize, clover, field beans, peas, turnips, carrots, cauliflowers, onions, lettuce or potatoes may follow a treated cereal crop

Environmental safety

- Dangerous for the environment
- Toxic to aquatic organisms

Hazard classification and safety precautions

Hazard Irritant, Dangerous for the environment
Transport code 9 [2]
Packaging group III
UN Number 3082, N/C
Risk phrases H317 [1], H319 [1]
Operator protection U05a, U20b; A, H
Environmental protection E15a, E34, H411
Storage and disposal D01, D02, D05, D09a, D10c
Medical advice M03

SECTION 2

274   metribuzin

A contact and residual triazinone herbicide for use in potatoes
HRAC mode of action code: 5 (C1)

See also clomazone + metribuzin
diflufenican + flufenacet + metribuzin
flufenacet + metribuzin

Products

1	Clayton Mizen	Clayton	70% w/w	WG	19207
2	Clayton Mizuna	Clayton	70% w/w	WG	19659
3	Sencorex Flow	Nufarm UK	600 g/l	SL	16167
4	Sencorex Flow	Sumitomo	600 g/l	SL	18895
5	Shotput	Adama	70% w/w	WG	15968

Uses

*   Annual and perennial weeds in **Asparagus** *(off-label)* [4];
*   Annual dicotyledons in **Asparagus** *(off-label)* [3,5]; **Carrots** *(off-label)* [3,5]; **Celeriac** *(off-label)* [3]; **Early potatoes** [1-5]; **Maincrop potatoes** [1-5]; **Mallow (Althaea spp.)** *(off-label)* [3,5]; **Ornamental plant production** *(off-label)* [3-4]; **Parsnips** *(off-label)* [3,5]; **Protected ornamentals** *(off-label)* [4]; **Sweet potato** *(off-label)* [3-5];
*   Annual grasses in **Asparagus** *(off-label)* [3,5]; **Carrots** *(off-label)* [3,5]; **Celeriac** *(off-label)* [3]; **Early potatoes** [1-5]; **Maincrop potatoes** [1-5]; **Mallow (Althaea spp.)** *(off-label)* [3,5]; **Ornamental plant production** *(off-label)* [3]; **Parsnips** *(off-label)* [3,5]; **Sweet potato** *(off-label)* [3,5];
*   Annual meadow grass in **Carrots** *(off-label)* [5]; **Maincrop potatoes** [1-2,5]; **Mallow (Althaea spp.)** *(off-label)* [5]; **Ornamental plant production** *(off-label)* [3]; **Parsnips** *(off-label)* [5]; **Protected ornamentals** *(off-label)* [3]; **Sweet potato** *(off-label)* [5];
*   Charlock in **Ornamental plant production** *(off-label)* [3]; **Protected ornamentals** *(off-label)* [3];
*   Cleavers in **Ornamental plant production** *(off-label)* [3]; **Protected ornamentals** *(off-label)* [3];
*   Common orache in **Ornamental plant production** *(off-label)* [3]; **Protected ornamentals** *(off-label)* [3];
*   Fat hen in **Ornamental plant production** *(off-label)* [3]; **Protected ornamentals** *(off-label)* [3];
*   Field pansy in **Ornamental plant production** *(off-label)* [3]; **Protected ornamentals** *(off-label)* [3];
*   Fool's parsley in **Asparagus** *(off-label)* [3]; **Carrots** *(off-label)* [3-5]; **Celeriac** *(off-label)* [3-4]; **Mallow (Althaea spp.)** *(off-label)* [3-5]; **Parsnips** *(off-label)* [3-5]; **Sweet potato** *(off-label)* [3];
*   Groundsel in **Ornamental plant production** *(off-label)* [3]; **Protected ornamentals** *(off-label)* [3];
*   Himalayan balsam in **Rhubarb** *(off-label)* [3-4];
*   Knotgrass in **Ornamental plant production** *(off-label)* [3]; **Protected ornamentals** *(off-label)* [3];
*   Mayweeds in **Ornamental plant production** *(off-label)* [3]; **Protected ornamentals** *(off-label)* [3];
*   Pale persicaria in **Ornamental plant production** *(off-label)* [3]; **Protected ornamentals** *(off-label)* [3];
*   Perennial dicotyledons in **Asparagus** *(off-label - from seed)* [5];
*   Redshank in **Ornamental plant production** *(off-label)* [3]; **Protected ornamentals** *(off-label)* [3];
*   Shepherd's purse in **Ornamental plant production** *(off-label)* [3]; **Protected ornamentals** *(off-label)* [3];
*   Small nettle in **Ornamental plant production** *(off-label)* [3]; **Protected ornamentals** *(off-label)* [3];
*   Volunteer oilseed rape in **Asparagus** *(off-label)* [3]; **Carrots** *(off-label)* [3]; **Celeriac** *(off-label)* [3]; **Early potatoes** [1-5]; **Maincrop potatoes** [1-5]; **Mallow (Althaea spp.)** *(off-label)* [3]; **Ornamental plant production** *(off-label)* [3]; **Parsnips** *(off-label)* [3]; **Protected ornamentals** *(off-label)* [3]; **Sweet potato** *(off-label)* [3];
*   Wild mignonette in **Asparagus** *(off-label)* [3]; **Carrots** *(off-label)* [3-5]; **Celeriac** *(off-label)* [3-4]; **Mallow (Althaea spp.)** *(off-label)* [3-5]; **Parsnips** *(off-label)* [3-5]; **Sweet potato** *(off-label)* [3];

- Willowherb in **Ornamental plant production** *(off-label)* [3]; **Protected ornamentals** *(off-label)* [3];

Extension of Authorisation for Minor Use (EAMUs)
- **Asparagus** *20150917* [3], *20193109* [4], *(from seed) 20142321* [5], *20142321* [5]
- **Carrots** *20150916* [3], *20193102* [4], *20142320* [5]
- **Celeriac** *20150916* [3], *20193102* [4]
- **Mallow (Althaea spp.)** *20150916* [3], *20193102* [4], *20142320* [5]
- **Ornamental plant production** *20131867* [3], *20171732* [3], *20193099* [4], *20193103* [4]
- **Parsnips** *20150916* [3], *20193102* [4], *20142320* [5]
- **Protected ornamentals** *20171732* [3], *20193099* [4]
- **Rhubarb** *20152218* [3], *20193110* [4]
- **Sweet potato** *20150915* [3], *20193106* [4], *20142252* [5]

Approval information
- Metribuzin included in Annex I under EC Regulation 1107/2009

Efficacy guidance
- Best results achieved on weeds at cotyledon to 1-leaf stage
- May be applied pre- or post-emergence of crop on named maincrop varieties and cv. Marfona; pre-emergence only on named early varieties
- Apply to moist soil with well-rounded ridges and few clods
- Activity reduced by dry conditions and on soils with high organic matter content
- On fen and moss soils pre-planting incorporation to 10-15 cm gives increased activity. Incorporate thoroughly and evenly
- With named maincrop and second early potato varieties on soils with more than 10% organic matter shallow pre- or post-planting incorporation may be used. See label for details
- Effective control using a programme of reduced doses is made possible by using a spray of smaller droplets, thus improving retention

Restrictions
- Maximum total dose equivalent to one full dose treatment on early potato varieties and one and a third full doses on maincrop varieties
- Only certain varieties may be treated. Apply pre-emergence only on named first earlies, pre- or post-emergence on named second earlies. On named maincrop varieties apply pre-emergence (except for certain varieties on Sands or Very Light soils) or post-emergence. See label
- All post-emergence treatments must be carried out before longest shoots reach 15 cm
- Do not cultivate after treatment
- Some recommended varieties may be sensitive to post-emergence treatment if crop under stress

Crop-specific information
- Latest use: pre-crop emergence for named early potato varieties; before most advanced shoots have reached 15 cm for post-emergence treatment of potatoes
- On stony or gravelly soils there is risk of crop damage, especially if heavy rain falls soon after application
- When days are hot and sunny delay spraying until evening

Following crops guidance
- Ryegrass, cereals or winter beans may be sown in same season provided at least 16 wk elapsed after treatment and ground ploughed to 15 cm and thoroughly cultivated as soon as possible after harvest and no later than end Dec
- In W Cornwall on soil with more than 5% organic matter early potatoes treated as recommended may be followed by summer planted brassica crops provided the soil has been ploughed, spring rainfall has been normal and at least 14 wk have elapsed since treatment
- Do not grow any vegetable brassicas, lettuces or radishes on land treated the previous yr. Other crops may be sown normally in spring of next yr

Environmental safety
- Dangerous for the environment
- Very toxic to aquatic organisms

SEE SECTION 3 FOR PRODUCTS ALSO REGISTERED

SECTION 2

- Do not empty into drains
- LERAP Category B [1-5]

Hazard classification and safety precautions

Hazard Harmful, Dangerous for the environment, Very toxic to aquatic organisms [1-2,4-5]
Transport code 9 [1-5]
Packaging group III
UN Number 3077, 3082
Risk phrases H373 [4]
Operator protection U05a, U08, U13, U14 [1-2,5], U19a, U20a [3-4], U20b [1-2,5]; A, D, H
Environmental protection E15a [1-2,5], E16a, E16b [1-2,5], E19b [1-2,5], E38 [3-4], H410
Storage and disposal D01, D02, D09a, D11a, D12a
Medical advice M03 [3-4], M05a

275   metsulfuron-methyl

A contact and residual sulfonylurea herbicide used in cereals, linseed and set-aside
HRAC mode of action code: 2 (B)

See also diflufenican + metsulfuron-methyl
fluroxypyr + metsulfuron-methyl
fluroxypyr + metsulfuron-methyl + thifensulfuron-methyl

Products

1	Accurate	Nufarm UK	20% w/w	WG	19237
2	Alias SX	FMC Agro	20% w/w	SG	18668
3	Answer SX	FMC Agro	20% w/w	SG	18684
4	Cleancrop Mondial	Agrii	20% w/w	SG	18598
5	Daytona	Nufarm UK	20% w/w	SG	19601
6	Deft Premium	Rotam	20% w/w	SG	18525
7	Finy	UPL Europe	20% w/w	WG	18274
8	Gropper SX	FMC Agro	20% w/w	SG	18685
9	Jubilee SX	FMC Agro	20% w/w	SG	18686
10	Lens	UPL Europe	20% w/w	SG	19386
11	Lorate	FMC Agro	20% w/w	SG	18687
12	Savvy Premium	Rotam	20% w/w	SG	18461
13	Simba SX	FMC Agro	20% w/w	SG	18791

Uses

- Amsinkia in *Green cover on land temporarily removed from production* [9]; *Linseed* [9]; *Spring barley* [9]; *Spring oats* [9]; *Spring wheat* [9]; *Triticale* [9]; *Winter barley* [9]; *Winter oats* [9]; *Winter wheat* [9];
- Annual dicotyledons in *Game cover (off-label)* [4]; *Green cover on land temporarily removed from production* [7,9]; *Linseed* [2-4,6,8-9,11-13]; *Spring barley* [1-13]; *Spring oats* [1-13]; *Spring wheat* [1-13]; *Triticale* [1-13]; *Winter barley* [1-13]; *Winter oats* [1-13]; *Winter wheat* [1-13];
- Charlock in *Green cover on land temporarily removed from production* [9]; *Linseed* [6,9]; *Spring barley* [5-6,9]; *Spring oats* [5-6,9]; *Spring wheat* [5-6,9]; *Triticale* [5-6,9]; *Winter barley* [5-6,9]; *Winter oats* [5-6,9]; *Winter wheat* [5-6,9];
- Chickweed in *Green cover on land temporarily removed from production* [9]; *Linseed* [2-4,6,8-9,11-13]; *Spring barley* [1-6,8,13]; *Spring oats* [1-6,8,13]; *Spring wheat* [1-6,8,13]; *Triticale* [1-6,8,13]; *Winter barley* [1-6,8,13]; *Winter oats* [1-6,8,13]; *Winter wheat* [1-6,8,13];
- Docks in *Green cover on land temporarily removed from production* [9]; *Linseed* [9]; *Spring barley* [9]; *Spring oats* [9]; *Spring wheat* [9]; *Triticale* [9]; *Winter barley* [9]; *Winter oats* [9]; *Winter wheat* [9];
- Green cover in *Land temporarily removed from production* [1-6,8,10-13];
- Groundsel in *Green cover on land temporarily removed from production* [9]; *Linseed* [9]; *Spring barley* [9]; *Spring oats* [9]; *Spring wheat* [9]; *Triticale* [9]; *Winter barley* [9]; *Winter oats* [9]; *Winter wheat* [9];

FOR FULL CONDITIONS OF USE ALWAYS READ THE PRODUCT LABEL

- Hemp-nettle in *Green cover on land temporarily removed from production* [9]; *Linseed* [9]; *Spring barley* [9]; *Spring oats* [9]; *Spring wheat* [9]; *Triticale* [9]; *Winter barley* [9]; *Winter oats* [9]; *Winter wheat* [9];
- Mayweeds in *Green cover on land temporarily removed from production* [9]; *Linseed* [2-4,6,8-9,11-13]; *Spring barley* [1-6,8-13]; *Spring oats* [1-6,8-13]; *Spring wheat* [1-6,8-13]; *Triticale* [1-6,8-13]; *Winter barley* [1-6,8-13]; *Winter oats* [1-6,8-13]; *Winter wheat* [1-6,8-13];
- Mouse-ear chickweed in *Green cover on land temporarily removed from production* [9]; *Linseed* [9]; *Spring barley* [9]; *Spring oats* [9]; *Spring wheat* [9]; *Triticale* [9]; *Winter barley* [9]; *Winter oats* [9]; *Winter wheat* [9];
- Scarlet pimpernel in *Green cover on land temporarily removed from production* [9]; *Linseed* [9]; *Spring barley* [9]; *Spring oats* [9]; *Spring wheat* [9]; *Triticale* [9]; *Winter barley* [9]; *Winter oats* [9]; *Winter wheat* [9];
- Venus looking-glass in *Green cover on land temporarily removed from production* [9]; *Linseed* [9]; *Spring barley* [9]; *Spring oats* [9]; *Spring wheat* [9]; *Triticale* [9]; *Winter barley* [9]; *Winter oats* [9]; *Winter wheat* [9];
- Volunteer oilseed rape in *Green cover on land temporarily removed from production* [9]; *Linseed* [9]; *Spring barley* [9]; *Spring oats* [9]; *Spring wheat* [9]; *Triticale* [9]; *Winter barley* [9]; *Winter oats* [9]; *Winter wheat* [9];
- Volunteer sugar beet in *Green cover on land temporarily removed from production* [9]; *Linseed* [9]; *Spring barley* [9]; *Spring oats* [9]; *Spring wheat* [9]; *Triticale* [9]; *Winter barley* [9]; *Winter oats* [9]; *Winter wheat* [9];

Extension of Authorisation for Minor Use (EAMUs)
- *Game cover 20183075* [4]

Approval information
- Metsulfuron-methyl included in Annex I under EC Regulation 1107/2009
- Accepted by BBPA for use on malting barley

Efficacy guidance
- Best results achieved on small, actively growing weeds up to 6 true leaf stage. Good spray cover is important
- Weed control may be reduced in dry soil conditions
- Commonly used in tank-mixture on wheat and barley with other cereal herbicides to improve control of resistant dicotyledons (cleavers, fumitory, ivy-leaved speedwell), larger weeds and grasses. See label for recommended mixtures
- Metsulfuron-methyl is a member of the ALS-inhibitor group of herbicides and products should be used in a planned Resistance Management strategy. See Section 5 for more information

Restrictions
- Maximum number of treatments 1 per crop or per yr (for set-aside)
- Do not apply within 7 d of rolling
- Do not use on cereal crops undersown with grass or legumes
- Do not use in tank mixture on oats, triticale or linseed. On linseed allow at least 7 d before or after other treatments
- Consult contract agents before use on a cereal crop for seed
- Do not use on any crop suffering stress from drought, waterlogging, frost, deficiency, pest or disease attack
- Specific restrictions apply to use in sequence or tank mixture with other sulfonylurea or ALS-inhibiting herbicides. See label for details
- Spraying equipment should not be drained or flushed onto land planted, or to be planted, with trees or crops other than cereals and should be thoroughly cleansed after use - see label for instructions
- Must only be applied from 1 February in the year of harvest until the specified latest time of application and must only be applied to green cover on land not being used for crop production where a full green cover is established [12, 7]
- For winter cereals; application should only be made after 15 March and before specified latest time of application [6, 1, 10]
- For spring cereals and linseed; this product must only be applied from 1 April in the year of harvest until the specified latest time of application; For winter cereals application should only be

made after 15 March and before specified latest time of application; For set-aside this product must only be used after 1 April and before the specified latest time of application [13, 1, 10]

Crop-specific information

*   Latest use: before flag leaf sheath extending stage for cereals (GS 41); before flower buds visible or crop 30 cm tall for linseed; before 1 Aug in yr of treatment for land not being used for crop production
*   Apply after 1 Feb to wheat, oats and triticale from 2-leaf (GS 12), and to barley from 3-leaf (GS 13) until flag-leaf sheath extending (GS 41)
*   On linseed allow at least 7 d (10 d if crop is growing poorly or under stress) before or after other treatments
*   Use in set-aside when a full green cover is established and made up predominantly of grassland, wheat, barley, oats or triticale. Do not use on seedling grasses

Following crops guidance

*   Only cereals, oilseed rape, field beans or grass may be sown in same calendar year after treating cereals with the product alone. Other restrictions apply to tank mixtures. See label for details
*   Only cereals should be planted within 16 mth of applying to a linseed crop or set-aside
*   In the event of failure of a treated crop sow only wheat within 3 mth after treatment
*   A minimum period of 120 days must be observed prior to planting of succeeding crops

Environmental safety

*   Dangerous for the environment
*   Very toxic to aquatic organisms
*   Take extreme care to avoid damage by drift onto broad-leaved plants outside the target area, onto surface waters or ditches or onto land intended for cropping
*   A range of broad leaved species will be fully or partially controlled when used in land temporarily removed from production, hence product may not be suitable where wild flower borders or other forms of conservation headland are being developed
*   Before use on land temporarily removed from production as part of grant-aided scheme, ensure compliance with the management rules
*   Green cover on land temporarily removed from production must not be grazed by livestock or harvested for human or animal consumption or used for animal bedding
*   LERAP Category B [1-13]

Hazard classification and safety precautions

Hazard Dangerous for the environment, Very toxic to aquatic organisms [1-4,8-9,11,13]
Transport code 9 [1-13]
Packaging group III
UN Number 3077
Risk phrases H315 [12], H318 [5-6,12], H319 [1]
Operator protection U04b [5-6], U05a [7], U11 [5-6], U20b [5-6]; A, C, H
Environmental protection E15a [12], E15b [1-11,13], E16a, E16b [3-4], E23 [7], E38, H410
Storage and disposal D01 [7,10], D02 [7,10], D09a [3-7,12], D10b [5-6,12], D11a [3-4,7], D12a [1-6,8-13], D12b [5-7]

276   metsulfuron-methyl + thifensulfuron-methyl

A contact residual and translocated sulfonylurea herbicide mixture for use in cereals
HRAC mode of action code: 2 + 2 (B + B)

See also thifensulfuron-methyl

Products

1 Accurate Extra	Nufarm UK	7:68% w/w	WG	18608
2 Avro SX	FMC Agro	2.9:42.9 % w/w	SG	18771
3 Chimera SX	FMC Agro	2.9:42.9% w/w	SG	18823
4 Concert SX	FMC Agro	4:40% w/w	SG	18806
5 Ergon	Rotam	6.8:68.2% w/w	WG	19580
6 Finish SX	FMC Agro	6.7:33.3% w/w	SG	18762

FOR FULL CONDITIONS OF USE ALWAYS READ THE PRODUCT LABEL

Products – continued

7	Harmony M SX	FMC Agro	4:40% w/w	SG	18824
8	Mozaic SX	FMC Agro	4:40% w/w	SG	18759
9	Pennant	FMC Agro	4:40% w/w	SG	18779
10	Presite SX	FMC Agro	6.7:33.3% w/w	SG	18776
11	Refine Max SX	FMC Agro	6.7:33.3% w/w	SG	18785

Uses

- Annual dicotyledons in **Miscanthus** *(off-label)* [3,7]; **Spring barley** [1-11]; **Spring oats** [5]; **Spring wheat** [1-11]; **Triticale** [5]; **Winter barley** [2-3,5-6,10-11]; **Winter oats** [6,10-11]; **Winter rye** [5]; **Winter wheat** [1-11];
- Black bindweed in **Spring barley** [2]; **Spring wheat** [2]; **Winter barley** [2]; **Winter wheat** [2];
- Charlock in **Spring barley** [2,5]; **Spring oats** [5]; **Spring wheat** [2,5]; **Triticale** [5]; **Winter barley** [2,5]; **Winter rye** [5]; **Winter wheat** [2,5];
- Chickweed in **Spring barley** [2-11]; **Spring oats** [5]; **Spring wheat** [2-11]; **Triticale** [5]; **Winter barley** [2-3,5-6,10-11]; **Winter oats** [6,10-11]; **Winter rye** [5]; **Winter wheat** [2-11];
- Creeping thistle in **Spring barley** [2-3]; **Spring wheat** [2-3]; **Winter barley** [2-3]; **Winter wheat** [2-3];
- Fat hen in **Spring barley** [2]; **Spring wheat** [2]; **Winter barley** [2]; **Winter wheat** [2];
- Field pansy in **Spring barley** [2,6,10-11]; **Spring wheat** [2,6,10-11]; **Winter barley** [2,6,10-11]; **Winter oats** [6,10-11]; **Winter wheat** [2,6,10-11];
- Field speedwell in **Spring barley** [2,6,10-11]; **Spring wheat** [2,6,10-11]; **Winter barley** [2,6,10-11]; **Winter oats** [6,10-11]; **Winter wheat** [2,6,10-11];
- Forget-me-not in **Spring barley** [2]; **Spring wheat** [2]; **Winter barley** [2]; **Winter wheat** [2];
- Hemp-nettle in **Spring barley** [2,5]; **Spring oats** [5]; **Spring wheat** [2,5]; **Triticale** [5]; **Winter barley** [2,5]; **Winter rye** [5]; **Winter wheat** [2,5];
- Ivy-leaved speedwell in **Spring barley** [10-11]; **Spring wheat** [6,10-11]; **Winter barley** [6,10-11]; **Winter oats** [6,10-11]; **Winter wheat** [6,10-11];
- Knotgrass in **Spring barley** [2,6,10-11]; **Spring wheat** [2,6,10-11]; **Winter barley** [2,6,10-11]; **Winter oats** [6,10-11]; **Winter wheat** [2,6,10-11];
- Mayweeds in **Spring barley** [2-11]; **Spring oats** [5]; **Spring wheat** [2-11]; **Triticale** [5]; **Winter barley** [2-3,5-6,10-11]; **Winter oats** [6,10-11]; **Winter rye** [5]; **Winter wheat** [2-11];
- Parsley-piert in **Spring barley** [5]; **Spring oats** [5]; **Spring wheat** [5]; **Triticale** [5]; **Winter barley** [5]; **Winter rye** [5]; **Winter wheat** [5];
- Polygonums in **Spring barley** [3-4,7-9]; **Spring wheat** [3-4,7-9]; **Winter barley** [3]; **Winter wheat** [3-4,7-9];
- Poppies in **Spring barley** [2]; **Spring wheat** [2]; **Winter barley** [2]; **Winter wheat** [2];
- Red dead-nettle in **Spring barley** [2-5,7-9]; **Spring oats** [5]; **Spring wheat** [2-5,7-9]; **Triticale** [5]; **Winter barley** [2-3,5]; **Winter rye** [5]; **Winter wheat** [2-5,7-9];
- Redshank in **Spring barley** [2,5]; **Spring oats** [5]; **Spring wheat** [2,5]; **Triticale** [5]; **Winter barley** [2,5]; **Winter rye** [5]; **Winter wheat** [2,5];
- Shepherd's purse in **Spring barley** [2-5,7-9]; **Spring oats** [5]; **Spring wheat** [2-5,7-9]; **Triticale** [5]; **Winter barley** [2-3,5]; **Winter rye** [5]; **Winter wheat** [2-5,7-9];

Extension of Authorisation for Minor Use (EAMUs)

- **Miscanthus** *20183485* [3], *20190866* [7]

Approval information

- Metsulfuron-methyl and thifensulfuron-methyl included in Annex I under EC Regulation 1107/2009
- Accepted by BBPA for use on malting barley

Efficacy guidance

- Best results by application to small, actively growing weeds up to 6-true leaf stage
- Ensure good spray cover
- Susceptible weeds stop growing almost immediately but symptoms may not be visible for about 2 wk
- Effectiveness may be reduced by heavy rain or if soil conditions very dry
- Metsulfuron-methyl and thifensulfuron-methyl are members of the ALS-inhibitor group of herbicides and products should be used in a planned Resistance Management strategy. See Section 5 for more information

SEE SECTION 3 FOR PRODUCTS ALSO REGISTERED

Restrictions
- Maximum number of treatments 1 per crop
- Products may only be used after 1 Feb
- Do not use on any crop suffering stress from drought, waterlogging, frost, deficiency, pest or disease attack or any other cause
- Do not use on crops undersown with grasses, clover or legumes, or any other broad leaved crop
- Specific restrictions apply to use in sequence or tank mixture with other sulfonylurea or ALS-inhibiting herbicides. See label for details
- Horizontal boom sprayers must be fitted with three star drift reduction technology for all uses. [5]
- Do not apply within 7 d of rolling
- Consult contract agents before use on a cereal crop grown for seed

Crop-specific information
- Latest use: before flag leaf sheath extending (GS 39) for barley and wheat, before 2nd node (GS32) for winter oats

Following crops guidance
- Only cereals, oilseed rape, field beans or grass may be sown in same calendar year after treatment
- Additional constraints apply after use of certain tank mixtures. See label
- In the event of crop failure sow only winter wheat within 3 mth after treatment and after ploughing and cultivating to a depth of at least 15 cm

Environmental safety
- Dangerous for the environment
- Very toxic to aquatic organisms
- Take extreme care to avoid damage by drift onto broad-leaved plants outside the target area, or onto ponds, waterways or ditches
- Spraying equipment should not be drained or flushed onto land planted, or to be planted, with trees or crops other than cereals and should be thoroughly cleansed after use - see label for instructions
- Buffer zone requirement 6 m.
- LERAP Category B [1-11]

Hazard classification and safety precautions
Hazard Dangerous for the environment, Very toxic to aquatic organisms [2-4,6-11]
Transport code 9 [1-11]
Packaging group III
UN Number 3077
Operator protection U05a [6,10-11], U08 [2-4,6-11], U19a [2-11], U20b [2-4,6-11], U20c [5]
Environmental protection E15a [10-11], E15b [1,4-5,7-9], E16a, E38 [1-4,6-11], H410
Storage and disposal D01 [6,10-11], D02 [6,10-11], D09a, D10b [5], D11a [1,4,6-11], D12a [1-4,6-11]

277   metsulfuron-methyl + tribenuron-methyl

A sulfonylurea herbicide mixture for cereals
HRAC mode of action code: 2 + 2 (B + B)

See also tribenuron-methyl

Products

1 Ally Max SX	FMC Agro	14.3:14.3% w/w	SG	18768
2 Biplay SX	FMC Agro	11.1:22.2% w/w	SG	18800
3 Boudha	Rotam	25:25% w/w	WG	19537
4 DP911 SX	FMC Agro	11.1:22.2% w/w	SG	18764
5 Traton SX	FMC Agro	11.1:22.2% w/w	SG	18793

Uses

- Annual dicotyledons in *Miscanthus* *(off-label)* [5]; *Rye* *(off-label)* [1]; *Spring barley* [1-5]; *Spring oats* [1,3]; *Spring wheat* [1-5]; *Triticale* [1-5]; *Winter barley* [1-5]; *Winter oats* [1-5]; *Winter rye* [3]; *Winter wheat* [1-5];
- Chickweed in *Spring barley* [1-3,5]; *Spring oats* [1,3]; *Spring wheat* [1-3,5]; *Triticale* [1-3,5]; *Winter barley* [1-3,5]; *Winter oats* [1-3,5]; *Winter rye* [3]; *Winter wheat* [1-3,5];
- Field pansy in *Spring barley* [1-2,5]; *Spring wheat* [1-2,5]; *Triticale* [1-2,5]; *Winter barley* [1-2,5]; *Winter oats* [1-2,5]; *Winter wheat* [1-2,5];
- Field speedwell in *Spring oats* [1];
- Hemp-nettle in *Spring barley* [1-2,5]; *Spring oats* [1]; *Spring wheat* [1-2,5]; *Triticale* [1-2,5]; *Winter barley* [1-2,5]; *Winter oats* [1-2,5]; *Winter wheat* [1-2,5];
- Mayweeds in *Spring barley* [1-3,5]; *Spring oats* [1,3]; *Spring wheat* [1-3,5]; *Triticale* [1-3,5]; *Winter barley* [1-3,5]; *Winter oats* [1-3,5]; *Winter rye* [3]; *Winter wheat* [1-3,5];
- Poppies in *Spring barley* [3]; *Spring oats* [3]; *Spring wheat* [3]; *Triticale* [3]; *Winter barley* [3]; *Winter oats* [3]; *Winter rye* [3]; *Winter wheat* [3];
- Red dead-nettle in *Spring barley* [1-3,5]; *Spring oats* [1,3]; *Spring wheat* [1-3,5]; *Triticale* [1-3,5]; *Winter barley* [1-3,5]; *Winter oats* [1-3,5]; *Winter rye* [3]; *Winter wheat* [1-3,5];
- Shepherd's purse in *Spring barley* [3]; *Spring oats* [3]; *Spring wheat* [3]; *Triticale* [3]; *Winter barley* [3]; *Winter oats* [3]; *Winter rye* [3]; *Winter wheat* [3];
- Volunteer oilseed rape in *Rye* *(off-label)* [1]; *Spring barley* [1,3]; *Spring oats* [1,3]; *Spring wheat* [1,3]; *Triticale* [1,3]; *Winter barley* [1,3]; *Winter oats* [1,3]; *Winter rye* [3]; *Winter wheat* [1,3];
- Volunteer sugar beet in *Spring barley* [1]; *Spring oats* [1]; *Spring wheat* [1]; *Triticale* [1]; *Winter barley* [1]; *Winter oats* [1]; *Winter wheat* [1];

Extension of Authorisation for Minor Use (EAMUs)

- *Miscanthus* *20193913* [5]
- *Rye* *20190562* [1]

Approval information

- Metsulfuron-methyl and tribenuron-methyl included in Annex I under EC Regulation 1107/2009
- Accepted by BBPA for use on malting barley

Efficacy guidance

- Best results obtained when applied to small actively growing weeds
- Product acts by foliar and root uptake. Good spray cover essential but performance may be reduced when soil conditions are very dry and residual effects may be reduced by heavy rain
- Weed growth inhibited within hours of treatment and many show marked colour changes as they die back. Full effects may not be apparent for up to 4 wk
- Metsulfuron-methyl and tribenuron-methyl are members of the ALS-inhibitor group of herbicides and products should be used in a planned Resistance Management strategy. See Section 5 for more information

Restrictions

- Maximum number of treatments 1 per crop
- Product must only be used after 1 Feb and after crop has three leaves
- Do not apply to a crop suffering from drought, waterlogging, low temperatures, pest or disease attack, nutrient deficiency, soil compaction or any other stress
- Do not use on crops undersown with grasses, clover or other legumes
- Do not apply within 7 d of rolling
- Specific restrictions apply to use in sequence or tank mixture with other sulfonylurea or ALS-inhibiting herbicides. See label for details

Crop-specific information

- Latest use: before flag leaf sheath extending

Following crops guidance

- Only cereals, field beans, grass or oilseed rape may be sown in the same calendar yr as harvest of a treated crop
- In the event of failure of a treated crop only winter wheat may be sown within 3 mth of treatment, and only after ploughing and cultivation to 15 cm minimum

SECTION 2

SEE SECTION 3 FOR PRODUCTS ALSO REGISTERED

Environmental safety
- Dangerous for the environment
- Very toxic to aquatic organisms
- Some non-target crops are highly sensitive. Take extreme care to avoid drift outside the target area, or onto ponds, waterways or ditches
- Spraying equipment should be thoroughly cleaned in accordance with manufacturer's instructions
- LERAP Category B [2,4-5]

Hazard classification and safety precautions
Hazard Irritant [1-2,4-5], Dangerous for the environment, Very toxic to aquatic organisms [1-2,4-5]
Transport code 9 [1-5]
Packaging group III
UN Number 3077
Operator protection U08, U19a, U20b [1-2,4-5]; A, H
Environmental protection E15a [1-2,4-5], E15b [3], E16a [2,4-5], E38 [1-3,5], E39 [3], H410
Storage and disposal D01, D02, D09a, D11a, D12a

278   Mild Pepino Mosaic Virus isolate VC1 + VX1

For use in tomatoes grown under permanent protection

Products

1	V10	Valto	10 - 50 mg/litre	LI	19124

Uses
- Virus in **Protected tomatoes** [1];

Approval information
- Pepino mosaic virus isolate VX1 and VC1 included in Annex 1 under EC Regulation 1107/2009

Efficacy guidance
- Plants must be free of Pepino Mosaic Virus at the time of treatment

Hazard classification and safety precautions
UN Number N/C
Operator protection D, H

279   myclobutanil

A systemic, protectant and curative triazole fungicide
FRAC mode of action code: 3

Products

1	Systhane 20 EW (GB Only)	Landseer	200 g/l	EW	19160

Uses
- American gooseberry mildew in **Blackberries** [1]; **Blackcurrants** [1]; **Gooseberries** [1]; **Redcurrants** [1];
- Black spot in **Protected ornamentals** *(off-label)* [1];
- Disease control in **Cherries** [1];
- Plum rust in **Plums** *(off-label)* [1];
- Powdery mildew in **Apples** [1]; **Hops** *(off-label)* [1]; **Pears** [1]; **Protected ornamentals** *(off-label)* [1]; **Protected strawberries** [1]; **Strawberries** [1]; **Wine grapes** [1];
- Rust in **Protected ornamentals** *(off-label)* [1];
- Scab in **Apples** [1]; **Pears** [1];

Extension of Authorisation for Minor Use (EAMUs)
- **Hops** *20210748* [1], *20220141* [1]

FOR FULL CONDITIONS OF USE ALWAYS READ THE PRODUCT LABEL

- **Plums** *20210749* [1], *20220139* [1]
- **Protected ornamentals** *20210747* [1], *20220140* [1]

Approval information
- Accepted by BBPA for use on hops
- Myclobutanil included in Annex 1 under EC Regulation 1107/2009

Efficacy guidance
- Best results achieved when used as part of routine preventive spray programme from bud burst to end of flowering in apples and pears and from just before the signs of mildew infection in blackcurrants and gooseberries
- In strawberries commence spraying at, or just prior to, first flower. Post-harvest sprays may be required on mildew-susceptible varieties where mildew is present and likely to be damaging
- Spray at 7-14 d intervals depending on disease pressure and dose applied
- For improved scab control on apples in post-blossom period tank-mix with mancozeb or captan
- Apply alone from mid-Jun for control of secondary mildew on apples and pears
- On roses spray at first signs of disease and repeat every 2 wk. In high risk areas spray when leaves emerge in spring, repeat 1 wk later and then continue normal programme
- Treatment of managed amenity turf may be carried out at any time of year before, or at, the first sign of disease
- Myclobutanil products should be used in conjunction with other fungicides with a different mode of action to reduce the possibility of resistance developing

Restrictions
- Do not handle treated grapevines for at least 12 days after treatment
- Only 1 application may be made to apple, pear and cherry during August

Crop-specific information
- HI 28 d (grapevines); 21 d (cherries, mirabelles); 14 d (apples, pears, blackcurrants, gooseberries, hops); 3 d (plums, protected blackberries, protected raspberries, protected Rubus hybrids, strawberries)
- Product may be applied to newly sown turf after the two-leaf stage but in view of the large number of turf grass cultivars a safety test on a small area is recommended before large scale treatment

Environmental safety
- Dangerous for the environment
- Buffer zone requirement 20 m when applied to apple, pear and cherry, within 10m when applied to blackcurrant, hops, red currant, gooseberry, raspberry and blackberry when using air-assisted sprayers

Hazard classification and safety precautions
Hazard Harmful, Dangerous for the environment
Transport code 9 [1]
Packaging group III
UN Number 3082
Risk phrases H319, H361, H373
Operator protection U08, U15, U20a; A, C, H
Environmental protection E15a, E17a (20 m), H411
Storage and disposal D01, D02, D05, D10c, D12a
Medical advice M05b

280   1-naphthylacetic acid

A plant growth regulator for fruit thinning

Products

1 Fixor	BelCrop	100 g/l	SL	17428

Uses
- Fruit thinning in *Apples* [1];
- Growth regulation in *Christmas trees* *(off-label)* [1];

Extension of Authorisation for Minor Use (EAMUs)
- *Christmas trees* *20181213* [1]

Approval information
- 1-naphthylacetic acid included in Annex I under EC Regulation 1107/2009

Efficacy guidance
- One application should be made when the fruit have a diameter between 8 and 12 mm (BBCH 71)
- Dose rate: Max 150 ml product/ha in 1000 litres water
- Adapt the spray volume to the tree size and canopy density to ensure a thorough fruit and leaf coverage

Hazard classification and safety precautions
UN Number N/C
Risk phrases H318, H361
Operator protection U20c
Environmental protection E15a
Storage and disposal D05, D09a, D11a

281   napropamide

A soil applied alkanamide herbicide for oilseed rape, fruit and woody ornamentals
HRAC mode of action code: 15 (K3)

See also clomazone + napropamide

Products

1	AC 650	UPL Europe	450 g/l	SC	19399
2	Devrinol	UPL Europe	450 g/l	SC	19358

Uses
- Amaranthus in *Baby leaf crops* *(off-label)* [2]; *Celeriac* *(off-label)* [2]; *Choi sum* *(off-label)* [2]; *Collards* *(off-label)* [2]; *Herbs (see appendix 6)* *(off-label)* [2]; *Horseradish* *(off-label)* [2]; *Lamb's lettuce* *(off-label)* [2]; *Oriental cabbage* *(off-label)* [2]; *Protected baby leaf crops* *(off-label)* [2]; *Protected herbs (see appendix 6)* *(off-label)* [2]; *Protected lamb's lettuce* *(off-label)* [2]; *Protected red mustard* *(off-label)* [2]; *Protected rocket* *(off-label)* [2]; *Red mustard* *(off-label)* [2]; *Rocket* *(off-label)* [2]; *Swedes* *(off-label)* [2]; *Turnips* *(off-label)* [2];
- Annual dicotyledons in *Blackcurrants* [2]; *Broccoli* [2]; *Brussels sprouts* [2]; *Cabbages* [2]; *Calabrese* [2]; *Cauliflowers* [2]; *Gooseberries* [2]; *Kale* [2]; *Raspberries* [2]; *Redcurrants* [2]; *Strawberries* [2]; *Winter oilseed rape* [1-2];
- Annual grasses in *Blackcurrants* [2]; *Broccoli* [2]; *Brussels sprouts* [2]; *Cabbages* [2]; *Calabrese* [2]; *Cauliflowers* [2]; *Gooseberries* [2]; *Kale* [2]; *Raspberries* [2]; *Redcurrants* [2]; *Strawberries* [2]; *Winter oilseed rape* [1-2];
- Annual meadow grass in *Baby leaf crops* *(off-label)* [2]; *Bilberries* *(off-label)* [2]; *Blackberries* *(off-label)* [2]; *Blueberries* *(off-label)* [2]; *Celeriac* *(off-label)* [2]; *Choi sum* *(off-label)* [2]; *Collards* *(off-label)* [2]; *Cranberries* *(off-label)* [2]; *Elderberries* *(off-label)* [2]; *Evening primrose* *(off-label)* [2]; *Game cover* *(off-label)* [1-2]; *Herbs (see appendix 6)* *(off-label)* [2]; *Horseradish* *(off-label)* [2]; *Lamb's lettuce* *(off-label)* [2]; *Linseed* *(off-label)* [2]; *Loganberries* *(off-label)* [2]; *Mulberries* *(off-label)* [2]; *Mustard* *(off-label)* [2]; *Oriental cabbage* *(off-label)* [2]; *Ornamental plant production* *(off-label)* [1-2]; *Protected baby leaf crops* *(off-label)* [2]; *Protected herbs (see appendix 6)* *(off-label)* [2]; *Protected lamb's lettuce* *(off-label)* [2]; *Protected red mustard* *(off-label)* [2]; *Protected rocket* *(off-label)* [2]; *Red mustard* *(off-label)* [2]; *Rocket* *(off-label)* [2]; *Rose hips* *(off-label)* [2]; *Rubus hybrids* *(off-label)* [2]; *Swedes* *(off-label)* [2]; *Turnips* *(off-label)* [2]; *Winter borage* *(off-label)* [2]; *Winter oilseed rape* [2];
- Barnyard grass in *Baby leaf crops* *(off-label)* [2]; *Celeriac* *(off-label)* [2]; *Choi sum* *(off-label)* [2]; *Collards* *(off-label)* [2]; *Herbs (see appendix 6)* *(off-label)* [2]; *Horseradish* *(off-label)* [2]; *Lamb's*

SECTION 2

lettuce (off-label) [2]; **Oriental cabbage** (off-label) [2]; **Protected baby leaf crops** (off-label) [2]; **Protected herbs (see appendix 6)** (off-label) [2]; **Protected lamb's lettuce** (off-label) [2]; **Protected red mustard** (off-label) [2]; **Protected rocket** (off-label) [2]; **Red mustard** (off-label) [2]; **Rocket** (off-label) [2]; **Swedes** (off-label) [2]; **Turnips** (off-label) [2];

- Black bindweed in **Baby leaf crops** (off-label) [2]; **Celeriac** (off-label) [2]; **Choi sum** (off-label) [2]; **Collards** (off-label) [2]; **Herbs (see appendix 6)** (off-label) [2]; **Horseradish** (off-label) [2]; **Lamb's lettuce** (off-label) [2]; **Oriental cabbage** (off-label) [2]; **Protected baby leaf crops** (off-label) [2]; **Protected herbs (see appendix 6)** (off-label) [2]; **Protected lamb's lettuce** (off-label) [2]; **Protected red mustard** (off-label) [2]; **Protected rocket** (off-label) [2]; **Red mustard** (off-label) [2]; **Rocket** (off-label) [2]; **Swedes** (off-label) [2]; **Turnips** (off-label) [2];

- Blackgrass in **Evening primrose** (off-label) [2]; **Game cover** (off-label) [2]; **Linseed** (off-label) [2]; **Mustard** (off-label) [2]; **Winter borage** (off-label) [2]; **Winter oilseed rape** [2];

- Chickweed in **Baby leaf crops** (off-label) [2]; **Bilberries** (off-label) [2]; **Blackberries** (off-label) [2]; **Blueberries** (off-label) [2]; **Celeriac** (off-label) [2]; **Choi sum** (off-label) [2]; **Collards** (off-label) [2]; **Cranberries** (off-label) [2]; **Elderberries** (off-label) [2]; **Evening primrose** (off-label) [2]; **Game cover** (off-label) [2]; **Herbs (see appendix 6)** (off-label) [2]; **Horseradish** (off-label) [2]; **Lamb's lettuce** (off-label) [2]; **Linseed** (off-label) [2]; **Loganberries** (off-label) [2]; **Mulberries** (off-label) [2]; **Mustard** (off-label) [2]; **Oriental cabbage** (off-label) [2]; **Ornamental plant production** (off-label) [2]; **Protected baby leaf crops** (off-label) [2]; **Protected herbs (see appendix 6)** (off-label) [2]; **Protected lamb's lettuce** (off-label) [2]; **Protected red mustard** (off-label) [2]; **Protected rocket** (off-label) [2]; **Red mustard** (off-label) [2]; **Rocket** (off-label) [2]; **Rose hips** (off-label) [2]; **Rubus hybrids** (off-label) [2]; **Swedes** (off-label) [2]; **Turnips** (off-label) [2]; **Winter borage** (off-label) [2];

- Cleavers in **Baby leaf crops** (off-label) [2]; **Blackcurrants** [2]; **Broccoli** [2]; **Brussels sprouts** [2]; **Cabbages** [2]; **Calabrese** [2]; **Cauliflowers** [2]; **Celeriac** (off-label) [2]; **Choi sum** (off-label) [2]; **Collards** (off-label) [2]; **Gooseberries** [2]; **Herbs (see appendix 6)** (off-label) [2]; **Horseradish** (off-label) [2]; **Kale** [2]; **Lamb's lettuce** (off-label) [2]; **Oriental cabbage** (off-label) [2]; **Protected baby leaf crops** (off-label) [2]; **Protected herbs (see appendix 6)** (off-label) [2]; **Protected lamb's lettuce** (off-label) [2]; **Protected red mustard** (off-label) [2]; **Protected rocket** (off-label) [2]; **Raspberries** [2]; **Red mustard** (off-label) [2]; **Redcurrants** [2]; **Rocket** (off-label) [2]; **Strawberries** [2]; **Swedes** (off-label) [2]; **Turnips** (off-label) [2];

- Common field speedwell in **Baby leaf crops** (off-label) [2]; **Celeriac** (off-label) [2]; **Choi sum** (off-label) [2]; **Collards** (off-label) [2]; **Evening primrose** (off-label) [2]; **Game cover** (off-label) [1-2]; **Herbs (see appendix 6)** (off-label) [2]; **Horseradish** (off-label) [2]; **Lamb's lettuce** (off-label) [2]; **Linseed** (off-label) [2]; **Mustard** (off-label) [2]; **Oriental cabbage** (off-label) [2]; **Ornamental plant production** (off-label) [1]; **Protected baby leaf crops** (off-label) [2]; **Protected herbs (see appendix 6)** (off-label) [2]; **Protected lamb's lettuce** (off-label) [2]; **Protected red mustard** (off-label) [2]; **Protected rocket** (off-label) [2]; **Red mustard** (off-label) [2]; **Rocket** (off-label) [2]; **Swedes** (off-label) [2]; **Turnips** (off-label) [2]; **Winter borage** (off-label) [2];

- Common purslane in **Baby leaf crops** (off-label) [2]; **Celeriac** (off-label) [2]; **Choi sum** (off-label) [2]; **Collards** (off-label) [2]; **Herbs (see appendix 6)** (off-label) [2]; **Horseradish** (off-label) [2]; **Lamb's lettuce** (off-label) [2]; **Oriental cabbage** (off-label) [2]; **Protected baby leaf crops** (off-label) [2]; **Protected herbs (see appendix 6)** (off-label) [2]; **Protected lamb's lettuce** (off-label) [2]; **Protected red mustard** (off-label) [2]; **Protected rocket** (off-label) [2]; **Red mustard** (off-label) [2]; **Rocket** (off-label) [2]; **Swedes** (off-label) [2]; **Turnips** (off-label) [2];

- Fat hen in **Baby leaf crops** (off-label) [2]; **Celeriac** (off-label) [2]; **Choi sum** (off-label) [2]; **Collards** (off-label) [2]; **Evening primrose** (off-label) [2]; **Game cover** (off-label) [1-2]; **Herbs (see appendix 6)** (off-label) [2]; **Horseradish** (off-label) [2]; **Lamb's lettuce** (off-label) [2]; **Linseed** (off-label) [2]; **Mustard** (off-label) [2]; **Oriental cabbage** (off-label) [2]; **Ornamental plant production** (off-label) [1]; **Protected baby leaf crops** (off-label) [2]; **Protected herbs (see appendix 6)** (off-label) [2]; **Protected lamb's lettuce** (off-label) [2]; **Protected red mustard** (off-label) [2]; **Protected rocket** (off-label) [2]; **Red mustard** (off-label) [2]; **Rocket** (off-label) [2]; **Swedes** (off-label) [2]; **Turnips** (off-label) [2]; **Winter borage** (off-label) [2];

- Field pansy in **Evening primrose** (off-label) [2]; **Game cover** (off-label) [2]; **Linseed** (off-label) [2]; **Mustard** (off-label) [2]; **Winter borage** (off-label) [2];

- Forget-me-not in **Evening primrose** (off-label) [2]; **Game cover** (off-label) [2]; **Linseed** (off-label) [2]; **Mustard** (off-label) [2]; **Winter borage** (off-label) [2];

- Fumitory in **Evening primrose** (off-label) [2]; **Game cover** (off-label) [2]; **Linseed** (off-label) [2]; **Mustard** (off-label) [2]; **Winter borage** (off-label) [2];

SEE SECTION 3 FOR PRODUCTS ALSO REGISTERED

- Groundsel in **Baby leaf crops** *(off-label)* [2]; **Blackcurrants** [2]; **Broccoli** [2]; **Brussels sprouts** [2]; **Cabbages** [2]; **Calabrese** [2]; **Cauliflowers** [2]; **Celeriac** *(off-label)* [2]; **Choi sum** *(off-label)* [2]; **Collards** *(off-label)* [2]; **Gooseberries** [2]; **Herbs (see appendix 6)** *(off-label)* [2]; **Horseradish** *(off-label)* [2]; **Kale** [2]; **Lamb's lettuce** *(off-label)* [2]; **Oriental cabbage** *(off-label)* [2]; **Protected baby leaf crops** *(off-label)* [2]; **Protected herbs (see appendix 6)** *(off-label)* [2]; **Protected lamb's lettuce** *(off-label)* [2]; **Protected red mustard** *(off-label)* [2]; **Protected rocket** *(off-label)* [2]; **Raspberries** [2]; **Red mustard** *(off-label)* [2]; **Redcurrants** [2]; **Rocket** *(off-label)* [2]; **Strawberries** [2]; **Swedes** *(off-label)* [2]; **Turnips** *(off-label)* [2];
- Ivy-leaved speedwell in **Evening primrose** *(off-label)* [2]; **Game cover** *(off-label)* [2]; **Linseed** *(off-label)* [2]; **Mustard** *(off-label)* [2]; **Winter borage** *(off-label)* [2];
- Knotgrass in **Baby leaf crops** *(off-label)* [2]; **Celeriac** *(off-label)* [2]; **Choi sum** *(off-label)* [2]; **Collards** *(off-label)* [2]; **Herbs (see appendix 6)** *(off-label)* [2]; **Horseradish** *(off-label)* [2]; **Lamb's lettuce** *(off-label)* [2]; **Oriental cabbage** *(off-label)* [2]; **Protected baby leaf crops** *(off-label)* [2]; **Protected herbs (see appendix 6)** *(off-label)* [2]; **Protected lamb's lettuce** *(off-label)* [2]; **Protected red mustard** *(off-label)* [2]; **Protected rocket** *(off-label)* [2]; **Red mustard** *(off-label)* [2]; **Rocket** *(off-label)* [2]; **Swedes** *(off-label)* [2]; **Turnips** *(off-label)* [2];
- Mayweeds in **Baby leaf crops** *(off-label)* [2]; **Bilberries** *(off-label)* [2]; **Blackberries** *(off-label)* [2]; **Blueberries** *(off-label)* [2]; **Celeriac** *(off-label)* [2]; **Choi sum** *(off-label)* [2]; **Collards** *(off-label)* [2]; **Cranberries** *(off-label)* [2]; **Elderberries** *(off-label)* [2]; **Evening primrose** *(off-label)* [2]; **Game cover** *(off-label)* [2]; **Herbs (see appendix 6)** *(off-label)* [2]; **Horseradish** *(off-label)* [2]; **Lamb's lettuce** *(off-label)* [2]; **Linseed** *(off-label)* [2]; **Loganberries** *(off-label)* [2]; **Mulberries** *(off-label)* [2]; **Mustard** *(off-label)* [2]; **Oriental cabbage** *(off-label)* [2]; **Ornamental plant production** *(off-label)* [2]; **Protected baby leaf crops** *(off-label)* [2]; **Protected herbs (see appendix 6)** *(off-label)* [2]; **Protected lamb's lettuce** *(off-label)* [2]; **Protected red mustard** *(off-label)* [2]; **Protected rocket** *(off-label)* [2]; **Red mustard** *(off-label)* [2]; **Rocket** *(off-label)* [2]; **Rose hips** *(off-label)* [2]; **Rubus hybrids** *(off-label)* [2]; **Swedes** *(off-label)* [2]; **Turnips** *(off-label)* [2]; **Winter borage** *(off-label)* [2];
- Pale persicaria in **Evening primrose** *(off-label)* [2]; **Game cover** *(off-label)* [2]; **Linseed** *(off-label)* [2]; **Mustard** *(off-label)* [2]; **Winter borage** *(off-label)* [2];
- Poppies in **Evening primrose** *(off-label)* [2]; **Game cover** *(off-label)* [1-2]; **Linseed** *(off-label)* [2]; **Mustard** *(off-label)* [2]; **Ornamental plant production** *(off-label)* [1]; **Winter borage** *(off-label)* [2]; **Winter oilseed rape** [1];
- Prickly poppy in **Evening primrose** *(off-label)* [2]; **Game cover** *(off-label)* [1-2]; **Linseed** *(off-label)* [2]; **Mustard** *(off-label)* [2]; **Ornamental plant production** *(off-label)* [1]; **Winter borage** *(off-label)* [2];
- Redshank in **Baby leaf crops** *(off-label)* [2]; **Celeriac** *(off-label)* [2]; **Choi sum** *(off-label)* [2]; **Collards** *(off-label)* [2]; **Evening primrose** *(off-label)* [2]; **Game cover** *(off-label)* [2]; **Herbs (see appendix 6)** *(off-label)* [2]; **Horseradish** *(off-label)* [2]; **Lamb's lettuce** *(off-label)* [2]; **Linseed** *(off-label)* [2]; **Mustard** *(off-label)* [2]; **Oriental cabbage** *(off-label)* [2]; **Protected baby leaf crops** *(off-label)* [2]; **Protected herbs (see appendix 6)** *(off-label)* [2]; **Protected lamb's lettuce** *(off-label)* [2]; **Protected red mustard** *(off-label)* [2]; **Protected rocket** *(off-label)* [2]; **Red mustard** *(off-label)* [2]; **Rocket** *(off-label)* [2]; **Swedes** *(off-label)* [2]; **Turnips** *(off-label)* [2]; **Winter borage** *(off-label)* [2];
- Shepherd's purse in **Baby leaf crops** *(off-label)* [2]; **Celeriac** *(off-label)* [2]; **Choi sum** *(off-label)* [2]; **Collards** *(off-label)* [2]; **Herbs (see appendix 6)** *(off-label)* [2]; **Horseradish** *(off-label)* [2]; **Lamb's lettuce** *(off-label)* [2]; **Oriental cabbage** *(off-label)* [2]; **Protected baby leaf crops** *(off-label)* [2]; **Protected herbs (see appendix 6)** *(off-label)* [2]; **Protected lamb's lettuce** *(off-label)* [2]; **Protected red mustard** *(off-label)* [2]; **Protected rocket** *(off-label)* [2]; **Red mustard** *(off-label)* [2]; **Rocket** *(off-label)* [2]; **Swedes** *(off-label)* [2]; **Turnips** *(off-label)* [2];
- Small nettle in **Baby leaf crops** *(off-label)* [2]; **Celeriac** *(off-label)* [2]; **Choi sum** *(off-label)* [2]; **Collards** *(off-label)* [2]; **Herbs (see appendix 6)** *(off-label)* [2]; **Horseradish** *(off-label)* [2]; **Lamb's lettuce** *(off-label)* [2]; **Oriental cabbage** *(off-label)* [2]; **Protected baby leaf crops** *(off-label)* [2]; **Protected herbs (see appendix 6)** *(off-label)* [2]; **Protected lamb's lettuce** *(off-label)* [2]; **Protected red mustard** *(off-label)* [2]; **Protected rocket** *(off-label)* [2]; **Red mustard** *(off-label)* [2]; **Rocket** *(off-label)* [2]; **Swedes** *(off-label)* [2]; **Turnips** *(off-label)* [2];

Extension of Authorisation for Minor Use (EAMUs)
- **Baby leaf crops** 20200647 [2]
- **Bilberries** 20200170 [2]

- **Blackberries** *20200169* [2]
- **Blueberries** *20200170* [2]
- **Celeriac** *20200648* [2]
- **Choi sum** *20200646* [2]
- **Collards** *20200646* [2]
- **Cranberries** *20200170* [2]
- **Elderberries** *20200170* [2]
- **Evening primrose** *20200650* [2]
- **Game cover** *20200643* [1], *20200644* [2], *20200649* [2]
- **Herbs (see appendix 6)** *20200645* [2]
- **Horseradish** *20200648* [2]
- **Lamb's lettuce** *20200647* [2]
- **Linseed** *20200650* [2]
- **Loganberries** *20200169* [2]
- **Mulberries** *20200170* [2]
- **Mustard** *20200650* [2]
- **Oriental cabbage** *20200646* [2]
- **Ornamental plant production** *20200643* [1], *20200168* [2]
- **Protected baby leaf crops** *20200647* [2]
- **Protected herbs (see appendix 6)** *20200645* [2]
- **Protected lamb's lettuce** *20200647* [2]
- **Protected red mustard** *20200647* [2]
- **Protected rocket** *20200647* [2]
- **Red mustard** *20200647* [2]
- **Rocket** *20200647* [2]
- **Rose hips** *20200170* [2]
- **Rubus hybrids** *20200169* [2]
- **Swedes** *20200648* [2]
- **Turnips** *20200648* [2]
- **Winter borage** *20200650* [2]

Approval information
- Napropamide included in Annex I under EC Regulation 1107/2009

Efficacy guidance
- Best results obtained from treatment pre-emergence of weeds but product may be used in conjunction with contact herbicides for control of emerged weeds. Otherwise remove existing weeds before application
- Weed control may be reduced where spray is mixed too deeply in the soil
- Manure, crop debris or other organic matter may reduce weed control
- Napropamide broken down by sunlight, so application during such conditions not recommended. Most crops recommended for treatment between Nov and end-Feb
- Post-emergence use of specific grass weedkilller recommended where volunteer cereals are a serious problem

Restrictions
- Maximum number of treatments 1 per crop or yr
- Do not use on Sands
- Do not use on soils with more than 10% organic matter
- Consult processors before use on any crop for processing
- Applications to strawberries must only be made to plants where the mature foliage has previously been removed (topped).
- Application to strawberries, established blackcurrants, redcurrants, gooseberries and raspberries must be made after 1st November and before the end of February

Crop-specific information
- Latest use: before transplanting for brassicas; pre-emergence for winter oilseed rape
- Where minimal cultivation used to establish oilseed rape, tank-mixture may be applied directly to stubble and mixed into top 25 mm as part of surface cultivations [1]
- Apply up to 14 d prior to drilling winter oilseed rape [1]

SEE SECTION 3 FOR PRODUCTS ALSO REGISTERED

Following crops guidance
- After use in oilseed rape only oilseed rape, swedes, fodder turnips, brassicas or potatoes should be sown within 12 mth of application
- Soil should be mould-board ploughed to a depth of at least 200 mm before drilling or planting any following crop

Environmental safety
- Dangerous for the environment
- Toxic to aquatic organisms
- LERAP Category B [1-2]

Hazard classification and safety precautions
Hazard Irritant, Dangerous for the environment, Very toxic to aquatic organisms
Transport code 9 [1-2]
Packaging group III
UN Number 3082
Operator protection U05a, U09a, U11, U20b; A, C, H
Environmental protection E15a [2], E16a, E34, E38, H410
Storage and disposal D01, D02, D05, D09a, D10c, D12a [1], D12b [2]

282 nicosulfuron

A sulfonylurea herbicide for maize
HRAC mode of action code: 2 (B)

See also dicamba + nicosulfuron
mesotrione + nicosulfuron

Products

1 Accent	Corteva	75% w/w	WG	15392
2 Bandera	Rotam	40.8 g/l	OD	16576
3 Crew	Nufarm UK	40 g/l	OD	15770
4 Entail	FMC Agro	240 g/l	OD	18740
5 Fornet 6 OD	Certis Belchim B V	60 g/l	OD	19520
6 Milagro 240 OD	Syngenta	240 g/l	OD	15957
7 Nico Pro 4 SC	Certis Belchim B V	40 g/l	OD	16424
8 Primero	Rotam	40.8 g/l	OD	16444
9 Samson Extra 6%	Certis Belchim B V	60 g/l	OD	19473
10 Templier	Rotam	75% w/w	WG	16866

Uses
- Amaranthus in *Forage maize* [7]; *Grain maize* [3];
- Annual dicotyledons in *Almonds* (off-label) [5]; *Apples* (off-label) [5]; *Apricots* (off-label) [5]; *Cherries* (off-label) [5]; *Chestnuts* (off-label) [5]; *Forage maize* [1-2,4-6,8-10]; *Forest nurseries* (off-label) [5,7]; *Game cover* (off-label) [5,7]; *Grain maize* [2,4-6,8-10]; *Hazel nuts* (off-label) [5]; *Medlar* (off-label) [5]; *Nectarines* (off-label) [5]; *Ornamental plant production* (off-label) [5,7]; *Peaches* (off-label) [5]; *Pears* (off-label) [5]; *Plums* (off-label) [5]; *Protected sweetcorn* (off-label) [9]; *Quinces* (off-label) [5]; *Sweetcorn* (off-label) [9]; *Top fruit* (off-label) [7]; *Walnuts* (off-label) [5];
- Annual grasses in *Forest nurseries* (off-label) [7]; *Game cover* (off-label) [7]; *Ornamental plant production* (off-label) [7]; *Protected sweetcorn* (off-label) [9]; *Sweetcorn* (off-label) [9]; *Top fruit* (off-label) [7];
- Annual meadow grass in *Almonds* (off-label) [5,9]; *Apples* (off-label) [5,9]; *Apricots* (off-label) [5,9]; *Cherries* (off-label) [5,9]; *Chestnuts* (off-label) [5,9]; *Forage maize* [1-3,5,7-8,10]; *Forest nurseries* (off-label) [5,7,9]; *Game cover* (off-label) [5,7,9]; *Grain maize* [2-3,5,8,10]; *Hazel nuts* (off-label) [5,9]; *Medlar* (off-label) [5,9]; *Nectarines* (off-label) [5,9]; *Ornamental plant production* (off-label) [5,7,9]; *Peaches* (off-label) [5,9]; *Pears* (off-label) [5,9]; *Plums* (off-label) [5,9]; *Quinces* (off-label) [5,9]; *Sweetcorn* (off-label) [5,7]; *Sweetcorn under plastic mulches* (off-label) [5,7]; *Top fruit* (off-label) [7]; *Walnuts* (off-label) [5,9];
- Black bindweed in *Sweetcorn* (off-label) [5]; *Sweetcorn under plastic mulches* (off-label) [5];

FOR FULL CONDITIONS OF USE ALWAYS READ THE PRODUCT LABEL

- Black nightshade in *Sweetcorn* *(off-label)* [5]; *Sweetcorn under plastic mulches* *(off-label)* [5];
- Blackgrass in *Forage maize* [1,10]; *Grain maize* [10];
- Chickweed in *Sweetcorn* *(off-label)* [5]; *Sweetcorn under plastic mulches* *(off-label)* [5];
- Cockspur grass in *Forage maize* [1,10]; *Grain maize* [10];
- Common amaranth in *Almonds* *(off-label)* [9]; *Apples* *(off-label)* [9]; *Apricots* *(off-label)* [9]; *Cherries* *(off-label)* [9]; *Chestnuts* *(off-label)* [9]; *Forest nurseries* *(off-label)* [9]; *Game cover* *(off-label)* [9]; *Hazel nuts* *(off-label)* [9]; *Medlar* *(off-label)* [9]; *Nectarines* *(off-label)* [9]; *Ornamental plant production* *(off-label)* [9]; *Peaches* *(off-label)* [9]; *Pears* *(off-label)* [9]; *Plums* *(off-label)* [9]; *Quinces* *(off-label)* [9]; *Sweetcorn* *(off-label)* [5,7]; *Sweetcorn under plastic mulches* *(off-label)* [5,7]; *Walnuts* *(off-label)* [9];
- Common orache in *Sweetcorn* *(off-label)* [5]; *Sweetcorn under plastic mulches* *(off-label)* [5];
- Couch in *Forage maize* [1,10]; *Grain maize* [10];
- Fat hen in *Sweetcorn* *(off-label)* [5]; *Sweetcorn under plastic mulches* *(off-label)* [5];
- Grass weeds in *Forage maize* [4,6,9]; *Grain maize* [4,6,9];
- Groundsel in *Almonds* *(off-label)* [9]; *Apples* *(off-label)* [9]; *Apricots* *(off-label)* [9]; *Cherries* *(off-label)* [9]; *Chestnuts* *(off-label)* [9]; *Forage maize* [3,7]; *Forest nurseries* *(off-label)* [9]; *Game cover* *(off-label)* [9]; *Grain maize* [3]; *Hazel nuts* *(off-label)* [9]; *Medlar* *(off-label)* [9]; *Nectarines* *(off-label)* [9]; *Ornamental plant production* *(off-label)* [9]; *Peaches* *(off-label)* [9]; *Pears* *(off-label)* [9]; *Plums* *(off-label)* [9]; *Quinces* *(off-label)* [9]; *Sweetcorn* *(off-label)* [7]; *Sweetcorn under plastic mulches* *(off-label)* [7]; *Walnuts* *(off-label)* [9];
- Mayweeds in *Forage maize* [3,7]; *Grain maize* [3]; *Sweetcorn* *(off-label)* [5]; *Sweetcorn under plastic mulches* *(off-label)* [5];
- Pale persicaria in *Sweetcorn* *(off-label)* [5]; *Sweetcorn under plastic mulches* *(off-label)* [5];
- Redshank in *Sweetcorn* *(off-label)* [5]; *Sweetcorn under plastic mulches* *(off-label)* [5];
- Ryegrass in *Almonds* *(off-label)* [5,9]; *Apples* *(off-label)* [5,9]; *Apricots* *(off-label)* [5,9]; *Cherries* *(off-label)* [5,9]; *Chestnuts* *(off-label)* [5,9]; *Forage maize* [1,3,7,10]; *Forest nurseries* *(off-label)* [5,9]; *Game cover* *(off-label)* [5,9]; *Grain maize* [3,10]; *Hazel nuts* *(off-label)* [5,9]; *Medlar* *(off-label)* [5,9]; *Nectarines* *(off-label)* [5,9]; *Ornamental plant production* *(off-label)* [5,9]; *Peaches* *(off-label)* [5,9]; *Pears* *(off-label)* [5,9]; *Plums* *(off-label)* [5,9]; *Quinces* *(off-label)* [5,9]; *Sweetcorn* *(off-label)* [5,7]; *Sweetcorn under plastic mulches* *(off-label)* [5,7]; *Walnuts* *(off-label)* [5,9];
- Shepherd's purse in *Almonds* *(off-label)* [9]; *Apples* *(off-label)* [9]; *Apricots* *(off-label)* [9]; *Cherries* *(off-label)* [9]; *Chestnuts* *(off-label)* [9]; *Forage maize* [3,7]; *Forest nurseries* *(off-label)* [9]; *Game cover* *(off-label)* [9]; *Grain maize* [3]; *Hazel nuts* *(off-label)* [9]; *Medlar* *(off-label)* [9]; *Nectarines* *(off-label)* [9]; *Ornamental plant production* *(off-label)* [9]; *Peaches* *(off-label)* [9]; *Pears* *(off-label)* [9]; *Plums* *(off-label)* [9]; *Quinces* *(off-label)* [9]; *Sweetcorn* *(off-label)* [5,7]; *Sweetcorn under plastic mulches* *(off-label)* [5,7]; *Walnuts* *(off-label)* [9];
- Volunteer cereals in *Forage maize* [1,10]; *Grain maize* [10];
- Wild oats in *Forage maize* [1,10]; *Grain maize* [10];

Extension of Authorisation for Minor Use (EAMUs)
- *Almonds* *20211442* [5], *20211440* [9]
- *Apples* *20211442* [5], *20211440* [9]
- *Apricots* *20211442* [5], *20211440* [9]
- *Cherries* *20211442* [5], *20211440* [9]
- *Chestnuts* *20211442* [5], *20211440* [9]
- *Forest nurseries* *20211442* [5], *20140752* [7], *20211440* [9]
- *Game cover* *20211442* [5], *20140752* [7], *20211440* [9]
- *Hazel nuts* *20211442* [5], *20211440* [9]
- *Medlar* *20211442* [5], *20211440* [9]
- *Nectarines* *20211442* [5], *20211440* [9]
- *Ornamental plant production* *20211442* [5], *20140752* [7], *20211440* [9]
- *Peaches* *20211442* [5], *20211440* [9]
- *Pears* *20211442* [5], *20211440* [9]
- *Plums* *20211442* [5], *20211440* [9]
- *Protected sweetcorn* *20210833* [9]
- *Quinces* *20211442* [5], *20211440* [9]
- *Sweetcorn* *20211441* [5], *20140941* [7], *20210833* [9]
- *Sweetcorn under plastic mulches* *20211441* [5], *20140941* [7]

SEE SECTION 3 FOR PRODUCTS ALSO REGISTERED

- **Top fruit** *20140752* [7]
- **Walnuts** *20211442* [5], *20211440* [9]

Approval information
- Nicosulfuron included in Annex I under EC Regulation 1107/2009

Efficacy guidance
- Product should be applied post-emergence between 2 and 8 crop leaf stage and to emerged weeds from the 2-leaf stage
- Product acts mainly by foliar activity. Ensure good spray cover of the weeds
- Nicosulfuron is a member of the ALS-inhibitor group of herbicides. To avoid the build up of resistance do not use any product containing an ALS-inhibitor herbicide with claims for control of grass weeds more than once on any crop
- Use these products as part of a resistance management strategy that includes cultural methods of control and does not use ALS inhibitors as the sole chemical method of grass weed control

Restrictions
- Maximum number of treatments 1 per crop
- Do not use if an organophosphorus soil insecticide has been used on the same crop
- Do not mix with foliar or liquid fertilisers or specified herbicides. See label for details
- Do not apply in mixture, or sequence, with any other sulfonyl-urea containing product
- Do not treat crops under stress
- Do not apply if rainfall is forecast to occur within 6 h of application
- To avoid the build up of resistance do not apply this or any other product containing an ALS inhibitor herbicide with claims for control of grass-weeds more than once to any crop

Crop-specific information
- Latest use: up to and including 8 true leaves for maize
- Some transient yellowing may be seen from 1-2 wk after treatment
- Only healthy maize crops growing in good field conditions should be treated

Following crops guidance
- Winter wheat and winter barley can be sown in normal crop rotation and all other crops can be sown the following spring. In the event of crop failure only maize can be sown after ploughing [1]
- In normal crop rotation, after ploughing winter wheat, winter barley, winter rye and triticale can be sown and all other crops can be sown the following spring. In the event of crop failure maize or soybeans can be sown after ploughing [5, 7, 9]

Environmental safety
- Dangerous for the environment
- Very toxic to aquatic organisms
- LERAP Category B [1-10]

Hazard classification and safety precautions
Hazard Harmful [5,9], Irritant [2-4,6-7], Dangerous for the environment, Harmful if inhaled [3], Very toxic to aquatic organisms [1-2,5-6,8-9]
Transport code 9 [1-10]
Packaging group III
UN Number 3077, 3082
Risk phrases H315 [3-4,6-7,9], H317 [3-6,9], H319 [5]
Operator protection U02a [4,6,9], U05a [1-7,9], U10 [4-6,9], U11 [4-6,9], U14 [2-7,9], U15 [2-3,7], U19a [2-7,9], U20b [4,6,9]; A, C, H
Environmental protection E15b [1-4,6-7,9], E16a, E16b [4,6,9], E34 [1-3,7], E38 [1-7,9-10], E40b [1], H410
Storage and disposal D01 [1-9], D02 [1-9], D06a [5], D09a [1-3,7], D09b [8], D10c [2-4,6-7,9], D11a [1], D12a [1,4-6,9-10], D12b [2-3,5,7]
Medical advice M03 [2-3,7], M05a [4-6,9]

283 orange oil

An insect repellant oil
IRAC mode of action code: Unknown

Products

1 Argos	Arysta	843.2 g/l	GE	19799

Uses
- Sprout suppression in **Potatoes** [1];

Approval information
- Orange oil included in Annex I under EC Regulation 1107/2009

Efficacy guidance
- Can be used as a sprout suppressant when applied to potatoes as a hot fog (1)
- Start treatment with the first symptoms of sprouting on dry potatoes
- Up to 9 repeat applications may be necessary and regular inspection for the onset of sprouting should be made to ensure timely application
- DO NOT fog potatoes with a high level of skin spot
- Potatoes should be clean, dry and free from signs of disease. Ensure skin is sufficiently cured before treatment. Daily crop inspection is essential to check curing and onset of sprouting

Restrictions
- Keep out of store during treatment and do not enter treated areas for at least 24 hours
- after treatment (1)
- Do not use on seed potatoes [1]
- To protect aquatic organisms, respect a period of 24 hours between the last application and the industrial rinsing of potatoes [1]
- Potatoes can be removed from the store a minimum of 48 hours after treatment [1]

Hazard classification and safety precautions
Hazard Flammable liquid and vapour, Very toxic to aquatic organisms
UN Number N/C
Risk phrases H304, H315, H317
Operator protection U05a, U09a, U19a, U19c, U20a; A, C, H
Environmental protection E15b, E34, H410
Storage and disposal D02, D06b, D09a, D12c

284 oxathiapiprolin

A blight fungicide for use in potatoes
FRAC mode of action code: 49

Products

1 Zorvec Enicade	Corteva	100 g/l	OD	18370

Uses
- Blight in **Potatoes** [1];
- Late blight in **Potatoes** [1];
- Tuber blight in **Potatoes** [1];

Approval information
- Oxathiapiprolin included in Annex I under EC Regulation 1107/2009

Efficacy guidance
- Sold in twinpacks with either Gachinko (amisulbrom) or Rhapsody (cymoxanil + mancozeb) and in co-formulation with benthiavalicarb to combat the risk of resistance

Restrictions
- Do not apply by hand-held equipment.

SEE SECTION 3 FOR PRODUCTS ALSO REGISTERED

Hazard classification and safety precautions
> **Transport code** 9 [1]
> **Packaging group** III
> **UN Number** 3082
> **Risk phrases** H317
> **Operator protection** H
> **Environmental protection** H411

285   paclobutrazol

A triazole plant growth regulator for ornamentals
FRAC mode of action code: 3

See also difenoconazole + paclobutrazol

Products

1	Bonzi	Syngenta	4 g/l	SC	17576
2	Pirouette	Fine	4 g/l	SC	17203

Uses
- Growth regulation in *Azaleas* [2]; *Bedding plants* [2]; *Kalanchoes* [2]; *Mini roses* [2]; *Pelargoniums* [2]; *Poinsettias* [2]; *Protected ornamentals* [1];
- Increasing flowering in *Azaleas* [2]; *Bedding plants* [2]; *Kalanchoes* [2]; *Mini roses* [2]; *Ornamental plant production* *(off-label)* [2]; *Pelargoniums* [2]; *Poinsettias* [2];
- Stem shortening in *Ornamental plant production* *(off-label)* [2];

Extension of Authorisation for Minor Use (EAMUs)
- *Ornamental plant production 20171269* [2]

Approval information
- Paclobutrazol included in Annex I under EC Regulation 1107/2009

Efficacy guidance
- Chemical is active via both foliage and root uptake. For best results apply in dull weather when relative humidity not high

Restrictions
- The maximum concentration must not exceed 250ml of product per 10 litre water
- Treatment must only be made under 'permanent protection' situations which provide full enclosure (including continuous top and side barriers down to below ground level) and which are present and maintained over a number of years
- Reasonable precautions must be taken to prevent access of birds, wild mammals and honey bees to treated crops
- Must not be used on ornamentals in compost where the compost will be re-used for cultivation of edible crops for human or animal consumption

Following crops guidance
- Chemical has residual soil activity which can affect growth of following crops. Do not use treated pots or soil for other crops

Environmental safety
- Harmful to aquatic organisms
- Do not use on food crops [2]

Hazard classification and safety precautions
> **Transport code** 9 [2]
> **Packaging group** III
> **UN Number** 3082, N/C
> **Operator protection** U05a, U08, U20a; A, H
> **Environmental protection** E13c [2], E15a, H411
> **Consumer protection** C01
> **Storage and disposal** D01, D02, D05, D09a, D10b [2], D10c [1]

286   pelargonic acid

A naturally occuring 9 carbon acid terminating in a carboxylic acid
HRAC mode of action code: Not classified

Products

1	Finalsan	Certis Belchim B V	186.7 g/l	EC	13102
2	Katoun Gold	Certis Belchim B V	500 g/l	EC	17879

Uses

- Algae in **Amenity grassland** [1]; **Amenity vegetation** [1]; **Beetroot** *(off-label)* [1]; **Bulb onions** *(off-label)* [1]; **Carrots** *(off-label)* [1]; **Courgettes** *(off-label)* [1]; **Garlic** *(off-label)* [1]; **Managed amenity turf** [1]; **Natural surfaces not intended to bear vegetation** [1]; **Ornamental plant production** [1]; **Parsnips** *(off-label)* [1]; **Permeable surfaces overlying soil** [1]; **Pumpkins** *(off-label)* [1]; **Salad onions** *(off-label)* [1]; **Shallots** *(off-label)* [1]; **Summer squash** *(off-label)* [1]; **Swedes** *(off-label)* [1]; **Sweetcorn** *(off-label)* [1]; **Turnips** *(off-label)* [1]; **Winter squash** *(off-label)* [1];
- Annual and perennial weeds in **Amenity grassland** [1]; **Amenity vegetation** [1]; **Managed amenity turf** [1]; **Natural surfaces not intended to bear vegetation** [1]; **Ornamental plant production** [1]; **Permeable surfaces overlying soil** [1];
- Annual dicotyledons in **Amenity vegetation** [2]; **Beetroot** *(off-label)* [1]; **Bulb onions** *(off-label)* [1]; **Carrots** *(off-label)* [1]; **Courgettes** *(off-label)* [1]; **Forest nurseries** [2]; **Garlic** *(off-label)* [1]; **Natural surfaces not intended to bear vegetation** [2]; **Ornamental plant production** [2]; **Parsnips** *(off-label)* [1]; **Permeable surfaces overlying soil** [2]; **Pumpkins** *(off-label)* [1]; **Salad onions** *(off-label)* [1]; **Shallots** *(off-label)* [1]; **Summer squash** *(off-label)* [1]; **Swedes** *(off-label)* [1]; **Sweetcorn** *(off-label)* [1]; **Turnips** *(off-label)* [1]; **Winter squash** *(off-label)* [1];
- Annual meadow grass in **Beetroot** *(off-label)* [1]; **Bulb onions** *(off-label)* [1]; **Carrots** *(off-label)* [1]; **Celery (outdoor)** *(off-label)* [1]; **Courgettes** *(off-label)* [1]; **Garlic** *(off-label)* [1]; **Leeks** *(off-label)* [1]; **Parsnips** *(off-label)* [1]; **Pumpkins** *(off-label)* [1]; **Salad onions** *(off-label)* [1]; **Shallots** *(off-label)* [1]; **Summer squash** *(off-label)* [1]; **Swedes** *(off-label)* [1]; **Sweetcorn** *(off-label)* [1]; **Turnips** *(off-label)* [1]; **Winter squash** *(off-label)* [1];
- Black nightshade in **Asparagus** *(off-label)* [1]; **Rhubarb** *(off-label)* [1];
- Burdock in **Asparagus** *(off-label)* [1]; **Rhubarb** *(off-label)* [1];
- Buttercups in **Asparagus** *(off-label)* [1]; **Beetroot** *(off-label)* [1]; **Bulb onions** *(off-label)* [1]; **Carrots** *(off-label)* [1]; **Courgettes** *(off-label)* [1]; **Garlic** *(off-label)* [1]; **Parsnips** *(off-label)* [1]; **Pumpkins** *(off-label)* [1]; **Rhubarb** *(off-label)* [1]; **Salad onions** *(off-label)* [1]; **Shallots** *(off-label)* [1]; **Summer squash** *(off-label)* [1]; **Swedes** *(off-label)* [1]; **Sweetcorn** *(off-label)* [1]; **Turnips** *(off-label)* [1]; **Winter squash** *(off-label)* [1];
- Canadian fleabane in **Asparagus** *(off-label)* [1]; **Rhubarb** *(off-label)* [1];
- Chickweed in **Celery (outdoor)** *(off-label)* [1]; **Leeks** *(off-label)* [1];
- Cleavers in **Asparagus** *(off-label)* [1]; **Rhubarb** *(off-label)* [1];
- Couch in **Asparagus** *(off-label)* [1]; **Rhubarb** *(off-label)* [1];
- Creeping thistle in **Asparagus** *(off-label)* [1]; **Beetroot** *(off-label)* [1]; **Bulb onions** *(off-label)* [1]; **Carrots** *(off-label)* [1]; **Courgettes** *(off-label)* [1]; **Garlic** *(off-label)* [1]; **Parsnips** *(off-label)* [1]; **Pumpkins** *(off-label)* [1]; **Rhubarb** *(off-label)* [1]; **Salad onions** *(off-label)* [1]; **Shallots** *(off-label)* [1]; **Summer squash** *(off-label)* [1]; **Swedes** *(off-label)* [1]; **Sweetcorn** *(off-label)* [1]; **Turnips** *(off-label)* [1]; **Winter squash** *(off-label)* [1];
- Field bindweed in **Asparagus** *(off-label)* [1]; **Rhubarb** *(off-label)* [1];
- Mayweeds in **Celery (outdoor)** *(off-label)* [1]; **Leeks** *(off-label)* [1];
- Moss in **Amenity grassland** [1]; **Amenity vegetation** [1]; **Beetroot** *(off-label)* [1]; **Bulb onions** *(off-label)* [1]; **Carrots** *(off-label)* [1]; **Celery (outdoor)** *(off-label)* [1]; **Courgettes** *(off-label)* [1]; **Garlic** *(off-label)* [1]; **Leeks** *(off-label)* [1]; **Managed amenity turf** [1]; **Natural surfaces not intended to bear vegetation** [1]; **Ornamental plant production** [1]; **Parsnips** *(off-label)* [1]; **Permeable surfaces overlying soil** [1]; **Pumpkins** *(off-label)* [1]; **Salad onions** *(off-label)* [1]; **Shallots** *(off-label)* [1]; **Summer squash** *(off-label)* [1]; **Swedes** *(off-label)* [1]; **Sweetcorn** *(off-label)* [1]; **Turnips** *(off-label)* [1]; **Winter squash** *(off-label)* [1];
- Mugwort in **Asparagus** *(off-label)* [1]; **Rhubarb** *(off-label)* [1];
- Polygonums in **Celery (outdoor)** *(off-label)* [1]; **Leeks** *(off-label)* [1];
- Ragwort in **Asparagus** *(off-label)* [1]; **Rhubarb** *(off-label)* [1];

SEE SECTION 3 FOR PRODUCTS ALSO REGISTERED

- Small nettle in *Asparagus (off-label)* [1]; *Rhubarb (off-label)* [1];
- Speedwells in *Celery (outdoor) (off-label)* [1]; *Leeks (off-label)* [1];

Extension of Authorisation for Minor Use (EAMUs)
- *Asparagus 20191786* [1]
- *Beetroot 20201665* [1]
- *Bulb onions 20201665* [1]
- *Carrots 20201665* [1]
- *Celery (outdoor) 20191786* [1]
- *Courgettes 20201609* [1]
- *Garlic 20201665* [1]
- *Leeks 20191786* [1]
- *Parsnips 20201665* [1]
- *Pumpkins 20201609* [1]
- *Rhubarb 20191786* [1]
- *Salad onions 20201665* [1]
- *Shallots 20201665* [1]
- *Summer squash 20201609* [1]
- *Swedes 20201665* [1]
- *Sweetcorn 20201609* [1]
- *Turnips 20201665* [1]
- *Winter squash 20201609* [1]

Approval information
- Pelargonic acid included in Annex 1 under EC Regulation 1107/2009

Environmental safety
- LERAP Category B [1]

Hazard classification and safety precautions
    Hazard Irritant [1]
    UN Number N/C
    Risk phrases H319
    Operator protection U05a [1], U11 [1], U15 [1], U16b [1], U19a [1]; A, C, P
    Environmental protection E12c [2], E12e [2], E15a [1], E16a [1], E16b [1], E19b [1]
    Storage and disposal D01, D02, D05 [1]
    Medical advice M05a [1]

287   penconazole

A protectant triazole fungicide with antisporulant activity
FRAC mode of action code: 3

Products

1	Pan Penco	Pan Agriculture	100 g/l	EC	18180
2	Topas	Syngenta	100 g/l	EC	16765

Uses
- Powdery mildew in *Apples* [1-2]; *Blackcurrants* [1-2]; *Courgettes (off-label)* [2]; *Crab apples* [1-2]; *Cucumbers (off-label)* [2]; *Gherkins (off-label)* [2]; *Globe artichoke (off-label)* [2]; *Hops in propagation (off-label)* [2]; *Nursery fruit trees (off-label)* [2]; *Ornamental plant production (off-label)* [2]; *Pears* [1-2]; *Protected chilli peppers (off-label)* [2]; *Protected courgettes (off-label)* [2]; *Protected cucumbers (off-label)* [2]; *Protected gherkins (off-label)* [2]; *Protected hops (off-label)* [2]; *Protected melons (off-label)* [2]; *Protected nursery fruit trees (off-label)* [2]; *Protected ornamentals (off-label)* [2]; *Protected peppers (off-label)* [2]; *Protected pumpkins (off-label)* [2]; *Protected strawberries* [1-2]; *Protected summer squash (off-label)* [2]; *Protected tomatoes (off-label)* [2]; *Protected watermelon (off-label)* [2]; *Protected winter squash (off-label)* [2]; *Redcurrants* [1-2]; *Strawberries* [1-2]; *Summer squash (off-label)* [2]; *Table grapes* [1-2]; *Wine grapes* [1-2];

Extension of Authorisation for Minor Use (EAMUs)
- *Courgettes* *20152336* [2]
- *Cucumbers* *20152336* [2]
- *Gherkins* *20152336* [2]
- *Globe artichoke* *20152901* [2]
- *Hops in propagation* *20152338* [2]
- *Nursery fruit trees* *20152338* [2]
- *Ornamental plant production* *20190169* [2]
- *Protected chilli peppers* *20152335* [2]
- *Protected courgettes* *20152336* [2]
- *Protected cucumbers* *20152336* [2]
- *Protected gherkins* *20152336* [2]
- *Protected hops* *20152338* [2]
- *Protected melons* *20152337* [2]
- *Protected nursery fruit trees* *20152338* [2]
- *Protected ornamentals* *20190169* [2]
- *Protected peppers* *20152335* [2]
- *Protected pumpkins* *20152337* [2]
- *Protected summer squash* *20152336* [2]
- *Protected tomatoes* *20152335* [2]
- *Protected watermelon* *20152337* [2]
- *Protected winter squash* *20152337* [2]
- *Summer squash* *20152336* [2]

Approval information
- Accepted by BBPA for use on hops
- Penconazole included in Annex I under EC Regulation 1107/2009

Efficacy guidance
- Use as a protectant fungicide by treating at the earliest signs of disease
- Treat crops every 10-14 d (every 7-10 d in warm, humid weather) at first sign of infection or as a protective spray ensuring complete coverage. See label for details of timing
- Increase dose and volume with growth of hops but do not exceed 2000 l/ha. Little or no activity will be seen on established powdery mildew
- Antisporulant activity reduces development of secondary mildew in apples

Restrictions
- Maximum number of treatments 10 per yr for apples, 6 per yr for hops, 4 per yr for blackcurrants
- Check for varietal susceptibility in roses. Some defoliation may occur after repeat applications on Dearest

Crop-specific information
- HI apples, hops 14 d; currants, bilberries, blueberries, cranberries, gooseberries 4 wk

Environmental safety
- Dangerous for the environment
- Toxic to aquatic organisms

Hazard classification and safety precautions
    Hazard Irritant, Dangerous for the environment
    Transport code 3 [1-2]
    Packaging group III
    UN Number 1915
    Risk phrases H319, H361
    Operator protection U02a, U05a, U08, U20b; A, C
    Environmental protection E15a, E38, H411
    Storage and disposal D01, D02, D05, D09a, D10c, D12a
    Medical advice M05b

SEE SECTION 3 FOR PRODUCTS ALSO REGISTERED

288 pendimethalin

A residual dinitroaniline herbicide for cereals and other crops
HRAC mode of action code: 3 (K1)

See also chlorotoluron + diflufenican + pendimethalin
clomazone + pendimethalin
diflufenican + pendimethalin
dimethenamid-p + pendimethalin
flufenacet + pendimethalin
imazamox + pendimethalin

Products

1	Anthem	Adama	400 g/l	SC	15761
2	Claymore	BASF	400 g/l	SC	13441
3	Cleancrop National	Adama	400 g/l	SC	17094
4	Most Micro	Sipcam	365 g/l	CS	16063
5	Pendifin 400 SC	Clayton	400 g/l	SC	18132
6	Stomp Aqua	BASF	455 g/l	CS	14664
7	Torres	BASF	400 g/l	SC	19017

Uses

* Annual dicotyledons in **Almonds** *(off-label)* [6]; **Apple orchards** [2,7]; **Apples** [1,3,5-6]; **Asparagus** *(off-label)* [6]; **Bilberries** *(off-label)* [6]; **Blackberries** [1-3,5-7]; **Blackcurrants** [1-3,5-7]; **Blueberries** *(off-label)* [6]; **Bog myrtle** *(off-label)* [6]; **Broad beans** *(off-label)* [5-6]; **Broccoli** *(off-label)* [1-3,5-7]; **Brussels sprouts** *(off-label)* [1-3,5-7]; **Bulb onion sets** *(off-label)* [1]; **Bulb onions** *(pre + post emergence treatment)* [1-3,5-7]; **Cabbages** *(off-label)* [1-3,5-7]; **Calabrese** *(off-label)* [1-3,5-7]; **Carrots** *(off-label)* [1-3,5-7]; **Cauliflowers** *(off-label)* [1-3,5-7]; **Celery (outdoor)** *(off-label)* [6]; **Cherries** [1-3,5-7]; **Chervil** *(off-label)* [6]; **Chestnuts** *(off-label)* [6]; **Collards** *(off-label)* [6]; **Combining peas** [1-3,5-7]; **Coriander** *(off-label)* [6]; **Cranberries** *(off-label)* [6]; **Cress** *(off-label)* [6]; **Dill** *(off-label)* [6]; **Durum wheat** [1-7]; **Dwarf beans** *(off-label)* [4]; **Edible podded peas** *(off-label)* [2,4,6]; **Evening primrose** *(off-label)* [6]; **Farm forestry** *(off-label)* [6]; **Fennel** *(off-label)* [6]; **Forage maize** [1-3,5-7]; **Forage maize (under plastic mulches)** [1,3,6]; **Forage soya bean** *(off-label)* [4,6]; **Forest** *(off-label)* [6]; **Forest nurseries** *(off-label)* [6]; **French beans** *(off-label)* [4,6]; **Frise** *(off-label)* [6]; **Game cover** *(off-label)* [1,3,6]; **Garlic** *(off-label)* [1,6]; **Gooseberries** [1-3,5-7]; **Grain maize** [1,3,6]; **Grain maize (under plastic mulches)** [1,3,6]; **Grass seed crops** *(off-label)* [6]; **Hazel nuts** *(off-label)* [6]; **Herbs (see appendix 6)** *(off-label)* [6]; **Hops** *(off-label)* [6]; **Horseradish** *(off-label)* [6]; **Kale** *(off-label)* [6]; **Lamb's lettuce** *(off-label)* [6]; **Leaf brassicas** *(off-label)* [6]; **Leeks** *(off-label)* [1-3,5-7]; **Lettuce** *(off-label)* [6]; **Loganberries** [1-3,5-7]; **Lupins** *(off-label)* [2,4,6]; **Mallow (Althaea spp.)** *(off-label)* [6]; **Miscanthus** *(off-label)* [6]; **Navy beans** [6]; **Ornamental plant production** *(off-label)* [6]; **Parsley** *(off-label)* [6]; **Parsley root** *(off-label)* [6]; **Parsnips** *(off-label)* [1-3,5-7]; **Pear orchards** [2,7]; **Pears** [1,3,5-6]; **Plums** [1-3,5-7]; **Potatoes** [1-7]; **Protected forest nurseries** *(off-label)* [6]; **Quinces** *(off-label)* [6]; **Radicchio** *(off-label)* [6]; **Raspberries** [1-3,5-7]; **Redcurrants** *(off-label)* [1,3,5-7]; **Rhubarb** *(off-label)* [6]; **Rubus hybrids** [1-3,5-7]; **Runner beans** *(off-label)* [2,4,6]; **Salad onions** *(off-label)* [6]; **Scarole** *(off-label)* [6]; **Shallots** *(off-label)* [1,6]; **Soya beans** *(off-label)* [4,6]; **Spring barley** [1-7]; **Spring field beans** *(off-label)* [2,4,6]; **Spring wheat** *(off-label)* [1,6]; **Strawberries** [1-3,5-7]; **Sunflowers** [1-7]; **Sweetcorn** *(off-label)* [6]; **Sweetcorn under plastic mulches** *(off-label)* [6]; **Top fruit** *(off-label)* [6]; **Triticale** [1-7]; **Vining peas** *(off-label)* [2,4,6]; **Walnuts** *(off-label)* [6]; **Whitecurrants** *(off-label)* [6]; **Winter barley** [1-7]; **Winter field beans** *(off-label)* [2,4,6]; **Winter rye** [1-7]; **Winter wheat** [1-7];
* Annual grasses in **Apples** [1,3,5-6]; **Blackberries** [1,3,5-6]; **Blackcurrants** [1,3,5-6]; **Bog myrtle** *(off-label)* [6]; **Broad beans** *(off-label)* [5]; **Broccoli** *(off-label)* [1,3,5-6]; **Brussels sprouts** *(off-label)* [1,3,5-6]; **Bulb onion sets** *(off-label)* [1]; **Bulb onions** *(off-label)* [1,3,5-6]; **Cabbages** [1,3,5-6]; **Calabrese** *(off-label)* [1,3,5-6]; **Carrots** [1,3,5-6]; **Cauliflowers** *(off-label)* [1,3,5-6]; **Celery (outdoor)** *(off-label)* [6]; **Cherries** [1,3,5-6]; **Collards** *(off-label)* [6]; **Combining peas** [1,3,5-6]; **Durum wheat** [1,3-6]; **Dwarf beans** *(off-label)* [4]; **Edible podded peas** *(off-label)* [2,6]; **Fennel** *(off-label)* [6]; **Forage maize** [1,3,5-6]; **Forage maize (under plastic mulches)** [1,3,6]; **Forage soya bean** *(off-label)* [4]; **French beans** *(off-label)* [4]; **Game cover** *(off-label)* [1,3]; **Garlic** *(off-label)* [1]; **Gooseberries** [1,3,5-6]; **Grain maize** [1,3,6]; **Grain maize (under plastic mulches)**

[1,3,6]; **Grass seed crops** *(off-label)* [6]; **Kale** *(off-label)* [6]; **Leeks** [1,3,5-6]; **Loganberries** [1,3,5-6]; **Lupins** *(off-label)* [2,6]; **Parsnips** [1,3,5-6]; **Pears** [1,3,5-6]; **Plums** [1,3,5-6]; **Potatoes** [1,3-6]; **Raspberries** [1,3,5-6]; **Redcurrants** [1,3,5]; **Rhubarb** *(off-label)* [6]; **Rubus hybrids** [1,3,5-6]; **Runner beans** *(off-label)* [2,6]; **Shallots** *(off-label)* [1]; **Spring barley** [1,3-6]; **Spring field beans** *(off-label)* [2,6]; **Spring wheat** *(off-label)* [1,6]; **Strawberries** [1,3,5-6]; **Sunflowers** [1,3-6]; **Triticale** [1,3-6]; **Vining peas** *(off-label)* [2,6]; **Winter barley** [1,3-6]; **Winter field beans** *(off-label)* [2,6]; **Winter rye** [1,3-6]; **Winter wheat** [1,3-6];

- Annual meadow grass in **Apple orchards** [2,7]; **Blackberries** [2,7]; **Blackcurrants** [2,7]; **Broad beans** *(off-label)* [4]; **Broccoli** [2,7]; **Brussels sprouts** [2,7]; **Bulb onion sets** *(off-label)* [1]; **Bulb onions** *(off-label)* [1-2,7]; **Cabbages** *(off-label)* [2,6-7]; **Calabrese** [2,7]; **Carrots** [2,7]; **Cauliflowers** [2,7]; **Cherries** [2,7]; **Combining peas** [2,7]; **Durum wheat** [2,4,7]; **Dwarf beans** *(off-label)* [4]; **Edible podded peas** *(off-label)* [4,6]; **Forage maize** [2,7]; **Forage soya bean** *(off-label)* [4,6]; **French beans** *(off-label)* [4]; **Garlic** *(off-label)* [1]; **Gooseberries** [2,7]; **Leeks** [2,7]; **Lettuce** *(off-label)* [6]; **Loganberries** [2,7]; **Lupins** *(off-label)* [4,6]; **Parsnips** [2,7]; **Pear orchards** [2,7]; **Plums** [2,7]; **Potatoes** [2,4,7]; **Raspberries** [2,7]; **Redcurrants** [7]; **Rubus hybrids** [2,7]; **Runner beans** *(off-label)* [4,6]; **Shallots** *(off-label)* [1]; **Soya beans** *(off-label)* [4,6]; **Spring barley** [2,4,7]; **Spring field beans** *(off-label)* [4,6]; **Spring wheat** *(off-label)* [6]; **Strawberries** [2,7]; **Sunflowers** [2,4,7]; **Triticale** [2,4,7]; **Vining peas** *(off-label)* [4,6]; **Winter barley** [2,4,7]; **Winter field beans** *(off-label)* [4,6]; **Winter rye** [2,4,7]; **Winter wheat** [2,4,7];

- Blackgrass in **Apple orchards** [2,7]; **Blackberries** [2,7]; **Blackcurrants** [2,7]; **Broad beans** *(off-label)* [1]; **Broccoli** [2,7]; **Brussels sprouts** [2,7]; **Bulb onions** [2,7]; **Cabbages** [2,7]; **Calabrese** [2,7]; **Carrots** [2,7]; **Cauliflowers** [2,7]; **Cherries** [2,7]; **Combining peas** [2,7]; **Durum wheat** [2,7]; **Forage soya bean** *(off-label)* [6]; **Game cover** *(off-label)* [1]; **Gooseberries** [2,7]; **Leeks** [2,7]; **Loganberries** [2,7]; **Parsnips** [2,7]; **Pear orchards** [2,7]; **Plums** [2,7]; **Raspberries** [2,7]; **Redcurrants** [7]; **Rubus hybrids** [2,7]; **Soya beans** *(off-label)* [4,6]; **Spring barley** [2,7]; **Spring field beans** *(off-label)* [1]; **Spring wheat** *(off-label)* [1]; **Strawberries** [2,7]; **Sunflowers** [2,7]; **Triticale** [2,7]; **Winter barley** [2,7]; **Winter field beans** *(off-label)* [1]; **Winter rye** [2,7]; **Winter wheat** [2,7];

- Chickweed in **Broad beans** *(off-label)* [4]; **Celeriac** *(off-label)* [6]; **Lettuce** *(off-label)* [6];
- Cleavers in **Broad beans** *(off-label)* [1,3]; **Spring field beans** *(off-label)* [1,3]; **Winter field beans** *(off-label)* [1,3];
- Common orache in **Broad beans** *(off-label)* [4];
- Corn marigold in **Broad beans** *(off-label)* [4];
- Dead nettle in **Broad beans** *(off-label)* [4];
- Fat hen in **Broad beans** *(off-label)* [4]; **Celeriac** *(off-label)* [6];
- Field pansy in **Broad beans** *(off-label)* [4];
- Forget-me-not in **Broad beans** *(off-label)* [4];
- Fumitory in **Broad beans** *(off-label)* [4];
- Groundsel in **Celeriac** *(off-label)* [6];
- Hemp-nettle in **Broad beans** *(off-label)* [4];
- Mayweeds in **Celeriac** *(off-label)* [6];
- Parsley-piert in **Broad beans** *(off-label)* [4];
- Polygonums in **Celeriac** *(off-label)* [6];
- Poppies in **Broad beans** *(off-label)* [4];
- Red dead-nettle in **Lettuce** *(off-label)* [6];
- Rough-stalked meadow grass in **Apple orchards** [2,7]; **Blackberries** [2,7]; **Blackcurrants** [2,7]; **Broad beans** *(off-label)* [4]; **Broccoli** [2,7]; **Brussels sprouts** [2,7]; **Bulb onions** [2,7]; **Cabbages** [2,7]; **Calabrese** [2,7]; **Carrots** [2,7]; **Cauliflowers** [2,7]; **Cherries** [2,7]; **Combining peas** [2,7]; **Durum wheat** [2,7]; **Forage maize** [2,7]; **Gooseberries** [2,7]; **Leeks** [2,7]; **Loganberries** [2,7]; **Parsnips** [2,7]; **Pear orchards** [2,7]; **Plums** [2,7]; **Potatoes** [2,7]; **Raspberries** [2,7]; **Redcurrants** [7]; **Rubus hybrids** [2,7]; **Spring barley** [2,7]; **Strawberries** [2,7]; **Sunflowers** [2,7]; **Triticale** [2,7]; **Winter barley** [2,7]; **Winter rye** [2,7]; **Winter wheat** [2,7];
- Scarlet pimpernel in **Broad beans** *(off-label)* [4];
- Shepherd's purse in **Broad beans** *(off-label)* [4];
- Small nettle in **Broad beans** *(off-label)* [4]; **Celeriac** *(off-label)* [6];
- Sowthistle in **Broad beans** *(off-label)* [4]; **Celeriac** *(off-label)* [6];
- Speedwells in **Broad beans** *(off-label)* [4]; **Lettuce** *(off-label)* [6];
- Volunteer oilseed rape in **Apple orchards** [2,7]; **Apples** [1,3,5]; **Blackberries** [1-3,5,7]; **Blackcurrants** [1-3,5,7]; **Broad beans** *(off-label)* [5]; **Broccoli** [1-3,5,7]; **Brussels sprouts** [1-

3,5,7]; **Bulb onions** [1-3,5,7]; **Cabbages** [1-3,5,7]; **Calabrese** [1-3,5,7]; **Carrots** [1-3,5,7];
Cauliflowers [1-3,5,7]; **Cherries** [1-3,5,7]; **Combining peas** [1-3,5,7]; **Durum wheat** [1-3,5,7];
Forage maize [1-3,5,7]; **Forage maize (under plastic mulches)** [1,3]; **Game cover** (off-label)
[3]; **Gooseberries** [1-3,5,7]; **Grain maize** [1,3]; **Grain maize (under plastic mulches)** [1,3];
Leeks [1-3,5,7]; **Loganberries** [1-3,5,7]; **Parsnips** [1-3,5,7]; **Pear orchards** [2,7]; **Pears** [1,3,5];
Plums [1-3,5,7]; **Potatoes** [1-3,5,7]; **Raspberries** [1-3,5,7]; **Redcurrants** [1,3,5,7]; **Rubus
hybrids** [1-3,5,7]; **Spring barley** [2,7]; **Strawberries** [2,7]; **Sunflowers** [2,7]; **Triticale** [1-3,5,7];
Winter barley [1-3,5,7]; **Winter rye** [1-3,5,7]; **Winter wheat** [1-3,5,7];
- Wild oats in **Durum wheat** [2,7]; **Triticale** [2,7]; **Winter barley** [2,7]; **Winter rye** [2,7]; **Winter
wheat** [2,7];

Extension of Authorisation for Minor Use (EAMUs)
- **Almonds** *20092916* [6]
- **Asparagus** *20092920* [6]
- **Bilberries** *20092907* [6]
- **Blueberries** *20092907* [6]
- **Bog myrtle** *20120323* [6]
- **Broad beans** *20213004* [1], *20213006* [3], *20140230* [4], *20210610* [5], *20092922* [6]
- **Broccoli** *20112451* [6]
- **Brussels sprouts** *20112451* [6]
- **Bulb onion sets** *20171415* [1]
- **Bulb onions** *20171415* [1], *20092914* [6]
- **Cabbages** *20100650* [6]
- **Calabrese** *20112451* [6]
- **Carrots** *20093526* [6]
- **Cauliflowers** *20112451* [6]
- **Celeriac** *20131625* [6]
- **Celery (outdoor)** *20111287* [6]
- **Chervil** *20092913* [6]
- **Chestnuts** *20092916* [6]
- **Collards** *20112451* [6]
- **Coriander** *20092913* [6]
- **Cranberries** *20092907* [6]
- **Cress** *20092921* [6]
- **Dill** *20092913* [6]
- **Dwarf beans** *20140232* [4]
- **Edible podded peas** *20212993* [2], *20212947* [4], *20212997* [6]
- **Evening primrose** *20092915* [6]
- **Farm forestry** *20092918* [6]
- **Fennel** *20111287* [6]
- **Forage soya bean** *20140231* [4], *20121634* [6]
- **Forest** *20092918* [6]
- **Forest nurseries** *20092919* [6]
- **French beans** *20140232* [4], *20092908* [6]
- **Frise** *20092921* [6]
- **Game cover** *20141260* [1], *20152496* [3], *20092919* [6]
- **Garlic** *20171415* [1], *20092914* [6]
- **Grass seed crops** *20092923* [6]
- **Hazel nuts** *20092916* [6]
- **Herbs (see appendix 6)** *20092921* [6]
- **Hops** *20092919* [6]
- **Horseradish** *20093526* [6]
- **Kale** *20112451* [6]
- **Lamb's lettuce** *20092921* [6]
- **Leaf brassicas** *20092921* [6]
- **Leeks** *20092914* [6]
- **Lettuce** *20092921* [6], *20170372* [6], *20170375* [6]
- **Lupins** *20212993* [2], *20212947* [4], *20212997* [6]
- **Mallow (Althaea spp.)** *20092921* [6]

FOR FULL CONDITIONS OF USE ALWAYS READ THE PRODUCT LABEL

- *Miscanthus* *20092919* [6]
- *Navy beans* *20092908* [6]
- *Ornamental plant production* *20092919* [6]
- *Parsley* *20092913* [6]
- *Parsley root* *20092911* [6], *20093526* [6]
- *Parsnips* *20093526* [6]
- *Protected forest nurseries* *20092919* [6]
- *Quinces* *20092917* [6]
- *Radicchio* *20092921* [6]
- *Redcurrants* *20092907* [6]
- *Rhubarb* *20111287* [6]
- *Runner beans* *20212993* [2], *20212947* [4], *20212997* [6]
- *Salad onions* *20092914* [6]
- *Scarole* *20092921* [6]
- *Shallots* *20171415* [1], *20092914* [6]
- *Soya beans* *20140231* [4], *20121634* [6]
- *Spring field beans* *20213003* [1], *20212992* [2], *20213005* [3], *20212946* [4], *20212996* [6]
- *Spring wheat* *20212924* [1], *20212923* [6]
- *Sweetcorn* *20092912* [6]
- *Sweetcorn under plastic mulches* *20092912* [6]
- *Top fruit* *20092919* [6]
- *Vining peas* *20212993* [2], *20212947* [4], *20212997* [6]
- *Walnuts* *20092916* [6]
- *Whitecurrants* *20092907* [6]
- *Winter field beans* *20213003* [1], *20212992* [2], *20213005* [3], *20212946* [4], *20212996* [6]

Approval information
- Pendimethalin included in Annex I under EC Regulation 1107/2009
- Accepted by BBPA for use on malting barley and hops

Efficacy guidance
- Apply as soon as possible after drilling. Weeds are controlled as they germinate and emerged weeds will not be controlled by use of the product alone
- For effective blackgrass control apply not more than 2 d after final cultivation and before weed seeds germinate
- Best results by application to fine firm, moist, clod-free seedbeds when rain follows treatment. Effectiveness reduced by prolonged dry weather after treatment
- Effectiveness reduced on soils with more than 6% organic matter. Do not use where organic matter exceeds 10%
- Any trash, ash or straw should be incorporated evenly during seedbed preparation
- Do not disturb soil after treatment
- Apply to potatoes as soon as possible after planting and ridging in tank-mix with metribuzin but note that this will restrict the varieties that may be treated
- Always follow WRAG guidelines for preventing and managing herbicide resistant weeds. See Section 5 for more information

Restrictions
- Maximum number of treatments 1 per crop or yr
- Maximum total dose equivalent to one full dose treatment on most crops; 2 full dose treatments on leeks
- May be applied pre-emergence of cereal crops sown before 30 Nov provided seed covered by at least 32 mm soil, or post-emergence to early tillering stage (GS 23)
- Do not undersow treated crops
- Do not use on crops suffering stress due to disease, drought, waterlogging, poor seedbed conditions or chemical treatment or on soils where water may accumulate
- Do not apply with hand-held equipment [1]
- Note timing restriction when used on spring wheat [6]: The latest time of application for pre-emergence use on spring wheat sown in the Autumn and Winter seasons is March 30th in England and Wales and 16 April in Scotland, in the year of harvest. For post emergence use the latest time of sowing of spring wheat is before 1st February in the year of harvest

SEE SECTION 3 FOR PRODUCTS ALSO REGISTERED

Crop-specific information

- Latest use: pre-emergence for spring barley, carrots, lettuce, fodder maize, parsnips, parsley, combining peas, potatoes, onions and leeks; before transplanting for brassicas; before leaf sheaths erect for winter cereals; before bud burst for blackcurrants, gooseberries, cane fruit, hops; before flower trusses emerge for strawberries; 14 d after transplanting for leaf herbs
- Do not use on spring barley after end Mar (mid-Apr in Scotland on some labels) because dry conditions likely. Do not apply to dry seedbeds in spring unless rain imminent
- Apply to combining peas as soon as possible after sowing. Do not spray if plumule less than 13 mm below soil surface
- Apply to potatoes up to 7 d before first shoot emerges
- Apply to drilled crops as soon as possible after drilling but before crop and weed emergence
- Apply in top fruit, bush fruit and hops from autumn to early spring when crop dormant
- In cane fruit apply to weed free soil from autumn to early spring, immediately after planting new crops and after cutting out canes in established crops
- Apply in strawberries from autumn to early spring (not before Oct on newly planted bed). Do not apply pre-planting or during flower initiation period (post-harvest to mid-Sep)
- Apply pre-emergence in drilled onions or leeks, not on Sands, Very Light, organic or peaty soils or when heavy rain forecast
- Apply to brassicas after final plant-bed cultivation but before transplanting. Avoid unnecessary soil disturbance after application and take care not to introduce treated soil into the root zone when transplanting. Follow transplanting with specified post-planting treatments - see label
- Do not use on protected crops or in greenhouses

Following crops guidance

- Before ryegrass is drilled after a very dry season plough or cultivate to at least 15 cm. If treated spring crops are to be followed by crops other than cereals, plough or cultivate to at least 15 cm
- In the event of crop failure land must be ploughed or thoroughly cultivated to at least 15 cm. See label for minimum intervals that should elapse between treatment and sowing a range of replacement crops

Environmental safety

- Dangerous for the environment
- Very toxic to aquatic organisms
- Dangerous to fish or other aquatic life. Do not contaminate surface waters or ditches with chemical or used container
- LERAP Category B [1-7]

Hazard classification and safety precautions

Hazard Irritant [4], Dangerous for the environment, Very toxic to aquatic organisms [2,5]
Transport code 9 [1-7]
Packaging group III
UN Number 3082
Risk phrases H317 [4]
Operator protection U02a [1-3,5-7], U05a, U08, U13 [1-3,5-7], U14 [2,4,6-7], U15 [4], U19a [1-3,5-7], U20b [1-5,7]; A, H
Environmental protection E15b, E16a, E16b [1,3,5], E38, E39 [6], H410 [1-3,5-7], H412 [4]
Storage and disposal D01, D02, D05 [1,3,5-6], D08 [2,7], D09a, D10b [1-3,5-7], D10c [4], D12a [1-3,5-7]
Medical advice M05a [1-3,5,7]

289   pendimethalin + picolinafen

A post-emergence broad-spectrum herbicide mixture for winter cereals
HRAC mode of action code: 3 + 12 (K1 + F1)

Products

1 Chronicle	BASF	320:16 g/l	SC	15394
2 Orient	BASF	330:7.5 g/l	SC	12541
3 Parade	BASF	320:16 g/l	SC	17246
4 PicoMax	BASF	320:16 g/l	SC	13456

FOR FULL CONDITIONS OF USE ALWAYS READ THE PRODUCT LABEL

Products – continued

5	Picona	BASF	320:16 g/l	SC	13428
6	PicoPro	BASF	320:16 g/l	SC	13454
7	PicoStomp	BASF	320:16 g/l	SC	13455

Uses

- Annual dicotyledons in **Forest nurseries** *(off-label)* [4,7]; **Game cover** *(off-label)* [5]; **Rye** [1]; **Spring barley** *(off-label)* [1,3-7]; **Spring wheat** *(off-label)* [1,3-7]; **Triticale** [1,3-7]; **Winter barley** [1-7]; **Winter rye** [3-7]; **Winter wheat** [1-7];
- Annual meadow grass in **Forest nurseries** *(off-label)* [4,7]; **Game cover** *(off-label)* [5]; **Rye** [1]; **Spring barley** *(off-label)* [1,3-7]; **Spring wheat** *(off-label)* [1,3-7]; **Triticale** [1,3-7]; **Winter barley** [1-7]; **Winter rye** [3-7]; **Winter wheat** [1-7];
- Chickweed in **Rye** [1]; **Spring barley** [1,3-7]; **Spring wheat** [1,3-7]; **Triticale** [1,3-7]; **Winter barley** [1-7]; **Winter rye** [3-7]; **Winter wheat** [1-7];
- Cleavers in **Rye** [1]; **Spring barley** [1,3-7]; **Spring wheat** [1,3-7]; **Triticale** [1,3-7]; **Winter barley** [1-7]; **Winter rye** [3-7]; **Winter wheat** [1-7];
- Field speedwell in **Rye** [1]; **Spring barley** [1,3-7]; **Spring wheat** [1,3-7]; **Triticale** [1,3-7]; **Winter barley** [1-7]; **Winter rye** [3-7]; **Winter wheat** [1-7];
- Ivy-leaved speedwell in **Rye** [1]; **Spring barley** [1,3-7]; **Spring wheat** [1,3-7]; **Triticale** [1,3-7]; **Winter barley** [1-7]; **Winter rye** [3-7]; **Winter wheat** [1-7];
- Loose silky bent in **Spring barley** [5];
- Rough-stalked meadow grass in **Forest nurseries** *(off-label)* [4,7]; **Game cover** *(off-label)* [5]; **Rye** [1]; **Spring barley** [1,3-7]; **Spring wheat** [1,3-7]; **Triticale** [1,3-7]; **Winter barley** [1-7]; **Winter rye** [3-7]; **Winter wheat** [1-7];

Extension of Authorisation for Minor Use (EAMUs)

- **Forest nurseries** *20082863* [4], *20082865* [7]
- **Game cover** *20082864* [5]
- **Spring barley** *20212920* [1]
- **Spring wheat** *20212920* [1]

Approval information

- Pendimethalin and picolinafen included in Annex I under EC Regulation 1107/2009

Efficacy guidance

- Best results obtained on crops growing in a fine, firm tilth and when rain falls within 7 d of application
- Loose or cloddy seed beds must be consolidated prior to application otherwise reduced weed control may occur
- Residual weed control may be reduced on soils with more than 6% organic matter or under prolonged dry conditions
- Always follow WRAG guidelines for preventing and managing herbicide resistant weeds. See Section 5 for more information

Restrictions

- Maximum number of treatments 1 per crop
- Do not treat undersown cereals or those to be undersown
- Do not roll emerged crops before treatment nor autumn treated crops until the following spring
- Do not use on stony or gravelly soils, on soils that are waterlogged or prone to waterlogging, or on soils with more than 10% organic matter
- Do not treat crops under stress from any cause
- Do not apply pre-emergence to crops drilled after 30 Nov [2]
- Consult processor before treating crops for processing [2]

Crop-specific information

- Latest use: before pseudo-stem erect stage (GS 30)
- Transient bleaching may occur after treatment but it does not lead to yield loss
- Do not use pre-emergence on crops drilled after 30th Nov [4]
- For pre-emergence applications, seed should be covered with a minimum of 3.2 cm settled soil [5, 6]

SEE SECTION 3 FOR PRODUCTS ALSO REGISTERED

Following crops guidance
- After normal harvest any crop except ryegrass can be sown. Cultivation or ploughing to at least 15 cm is required before sowing ryegrass.
- In the event of crop failure following autumn application spring wheat, spring barley or winter field beans can be sown after the land is ploughed and at least 8 weeks have elapsed since treatment. The following spring oilseed rape, peas or spring field beans can be sown after non-inversion tillage to at least 5 cm but maize should not be planted until at least 5 months have elapsed since application.
- In the event of crop failure following spring application spring wheat, spring barley, peas, spring field beans, maize or spring oilseed rape can be sown after the land is ploughed to at least 15 cm and at least 8 weeks have elapsed since application.

Environmental safety
- Dangerous for the environment
- Very toxic to aquatic organisms
- Product binds strongly to soil minimising likelihood of movement into groundwater
- LERAP Category B [1-7]

Hazard classification and safety precautions
> **Hazard** Dangerous for the environment, Very toxic to aquatic organisms
> **Transport code** 9 [1-7]
> **Packaging group** III
> **UN Number** 3082
> **Operator protection** U02a [1-3,5-7], U05a, U20c; A, H
> **Environmental protection** E15a [2], E15b [1,3-7], E16a, E34, E38, H410
> **Storage and disposal** D01, D02, D05, D09a, D10c, D12a
> **Medical advice** M03

290 penthiopyrad

An SDHI fungicide for disease control in apples and pears
FRAC mode of action code: 7

Products

1	Fontelis	Corteva	200 g/l	SC	17465

Uses
- Powdery mildew in **Apples** *(suppression)* [1]; **Pears** *(suppression)* [1];
- Scab in **Apples** [1]; **Pears** [1];

Approval information
- Penthiopyrad is included in Annex 1 under EC Regulation 1107/2009

Efficacy guidance
- Always use in mixture with another product from a different fungicide resistance group that is used for control of the same target disease.
- Best results achieved from applications made before the disease is established in the crop

Restrictions
- Apply no more than 2 consecutive sprays of penthiopyrad or other SDHI fungicides
- Avoid application in either frosty or hot, sunny weather conditions [1]
- Do not apply to any crop suffering from stress as a result of drought, water-logging, low temperatures, nutrient or lime deficiency or other factors reducing crop growth.

Environmental safety
- Broadcast air-assisted LERAP 40m [1]
- LERAP Category B [1]

Hazard classification and safety precautions
> **Hazard** Very toxic to aquatic organisms
> **Transport code** 9 [1]
> **Packaging group** III

FOR FULL CONDITIONS OF USE ALWAYS READ THE PRODUCT LABEL

UN Number 3082
Operator protection A
Environmental protection E15b, E16a, E17a (40 m), E34, H410
Storage and disposal D05, D09a, D12a

291   phenmedipham

A contact phenyl carbamate herbicide for beet crops and strawberries
HRAC mode of action code: 5 (C1)

See also ethofumesate + phenmedipham

Products

1	Beetup Flo	UPL Europe	160 g/l	SC	14328
2	Betasana SC	UPL Europe	160 g/l	SC	14209
3	Corzal SC	UPL Europe	160 g/l	SC	17751
4	Jupiter	UPL Europe	160 g/l	SC	19396
5	Shrapnel	Adama	320 g/l	SC	18234

Uses

* Annual dicotyledons in **Chard** *(off-label)* ; **Fodder beet** [1-5]; **Herbs (see appendix 6)** *(off-label)* [2-3]; **Mangels** [1-5]; **Red beet** [1-5]; **Rocket** *(off-label)* [2-3]; **Seakale** *(off-label)* [2-3]; **Spinach** *(off-label)* [2-3]; **Spinach beet** *(off-label)* [2-3]; **Strawberries** [5]; **Sugar beet** [1-5];
* Black bindweed in **Ornamental plant production** *(off-label)* [2-3];
* Chickweed in **Fodder beet** [1-2]; **Mangels** [1-2]; **Ornamental plant production** *(off-label)* [2-3]; **Red beet** [1-2]; **Sugar beet** [1-2];
* Fat hen in **Ornamental plant production** *(off-label)* [2-3];
* Field pansy in **Ornamental plant production** *(off-label)* [2-3];
* Knotgrass in **Ornamental plant production** *(off-label)* [2-3];
* Penny cress in **Ornamental plant production** *(off-label)* [2-3];
* Scentless mayweed in **Ornamental plant production** *(off-label)* [2-3];

Extension of Authorisation for Minor Use (EAMUs)

* **Chard** *20220014* [2], *20172731* [3]
* **Herbs (see appendix 6)** *20220014* [2], *20172731* [3]
* **Ornamental plant production** *20152050* [2], *20180376* [3]
* **Rocket** *20220014* [2], *20172731* [3]
* **Seakale** *20220014* [2], *20172731* [3]
* **Spinach** *20220014* [2], *20172731* [3]
* **Spinach beet** *20220014* [2], *20172731* [3]

Approval information

* Phenmedipham included in Annex I under EC Regulation 1107/2009

Efficacy guidance

* Best results achieved by application to young seedling weeds, preferably cotyledon stage, under good growing conditions when low doses are effective
* If using a low-dose programme 2-3 repeat applications at 7-10 d intervals are recommended on mineral soils, 3-5 applications may be needed on organic soils
* Addition of adjuvant oil may improve effectiveness on some weeds
* Various tank-mixtures with other beet herbicides recommended. See label for details
* Use of certain pre-emergence herbicides is recommended in combination with post-emergence treatment. See label for details

Restrictions

* Maximum number of treatments and maximum total dose varies with crop and product used. See label for details
* At high temperatures (above 21°C) reduce rate and spray after 5 pm
* Do not apply immediately after frost or if frost expected
* Do not spray wet foliage or if rain imminent

SEE SECTION 3 FOR PRODUCTS ALSO REGISTERED

- Do not spray crops stressed by wind damage, nutrient deficiency, pest or disease attack etc. Do not roll or harrow for 7 d before or after treatment
- Do not use on strawberries under cloches or polythene tunnels
- Consult processor before use on crops for processing
- Do not apply to crops with wet leaves

Crop-specific information
- Latest use: before crop leaves meet between rows for beet crops; before flowering for strawberries
- Apply to beet crops at any stage as low dose/low volume spray or from fully developed cotyledon stage with full rate. Apply to red beet after fully developed cotyledon stage
- Apply to strawberries at any time when weeds in susceptible stage, except in period from start of flowering to picking

Following crops guidance
- Beet crops may follow at any time after a treated crop. 3 mth must elapse from treatment before any other crop is sown and must be preceded by mould-board ploughing to 15 cm

Environmental safety
- Dangerous for the environment
- Very toxic to aquatic organisms
- Harmful to fish or other aquatic life. Do not contaminate surface waters or ditches with chemical or used container
- LERAP Category B [1-4]

Hazard classification and safety precautions
Hazard Irritant [1-2,5], Dangerous for the environment, Very toxic to aquatic organisms [5]
Transport code 9 [1-5]
Packaging group III
UN Number 3082
Risk phrases H317 [1-4], H318 [5], H319 [1-4]
Operator protection U05a [3-5], U08, U11 [1-2], U14 [1-2,5], U19a [1-2,5], U20a [5], U20c [1-2]; A, C, H
Environmental protection E15a [5], E15b [1-4], E16a [1-4], E16b [3-4], E19b [5], E34 [3-5], E38 [3-5], H410 [2,5], H411 [1,3-4]
Storage and disposal D01, D02 [1-4], D05 [1-2,5], D09a, D10b [3-5], D10c [1-2], D12a
Medical advice M05a [5]

292   phenothrin

A non-systemic pyrethroid insecticide available only in mixtures
IRAC mode of action code: 3A

293   phenothrin + tetramethrin

A pyrethroid insecticide mixture for control of flying insects
IRAC mode of action code: 3 + 3

See also tetramethrin

Products
1	Killgerm ULV 1500	Killgerm	4.8:2.4% w/w	UL	H6961
2	Vazor ULV 500	Killgerm	4.8:0.8% w/w	UL	H10814
3	Vazor ULV 500	Killgerm	4.8:0.8% w/w	UL	H4647

Uses
- Flies in *Agricultural premises* [1-3];
- Grain storage mite in *Grain stores* [1-3];
- Mosquitoes in *Agricultural premises* [1-3];
- Wasps in *Agricultural premises* [1-3];

FOR FULL CONDITIONS OF USE ALWAYS READ THE PRODUCT LABEL

Approval information
- Product approved for ULV application. See label for details
- Phenothrin and tetramethrin are not included in Annex 1 under EC Regulation 1107/2009

Efficacy guidance
- Close doors and windows and spray in all directions for 3-5 sec. Keep room closed for at least 10 min

Restrictions
- For use only by professional operators
- Do not use space sprays containing pyrethrins or pyrethroid more than once per week in intensive or controlled environment animal houses in order to avoid development of resistance. If necessary, use a different control method or product

Crop-specific information
- May be used in the presence of poultry and livestock

Environmental safety
- Dangerous for the environment
- Toxic to aquatic organisms
- Do not apply directly to livestock/poultry
- Remove exposed milk and collect eggs before application. Protect milk machinery and containers from contamination

Hazard classification and safety precautions
Hazard Harmful, Dangerous for the environment, Harmful if swallowed
Transport code 9 [1-3]
Packaging group III
UN Number 3082
Operator protection U02b, U09b, U14, U19a, U20a, U20b; A, C, D, E, H
Environmental protection E05a, E15a, E38, H411
Consumer protection C06, C07, C08, C09, C11, C12
Storage and disposal D01, D05, D06a, D09a, D10a, D12a
Medical advice M05b

294 Phlebiopsis gigantea

A fungal protectant against root and butt rot in conifers

Products
1 PG Suspension	Forest Research	0.5% w/w	SC	17009

Uses
- Fomes root and butt rot in *Farm forestry* [1]; *Forest* [1]; *Forest nurseries* [1];
- Heterobasidion annosum in *Farm forestry* [1]; *Forest* [1]; *Forest nurseries* [1];

Approval information
- Phlebiopsis gigantea included in Annex 1 under EC Regulation 1107/2009

Efficacy guidance
- Store sachets in the original box in a cool dry place between 2°C and 15°C out of the sun.
- Use the diluted solution within 24 hours of mixing. Ensure complete coverage of stumps for optimum efficacy.

Hazard classification and safety precautions
UN Number N/C
Operator protection U20c; A, D, H
Environmental protection E15a
Storage and disposal D11a

295 physical pest control

Products that work by physical action only

Products

1	SB Plant Invigorator	Fargro	1 - 3% v/v	SL	-
2	Vazor DE	Killgerm	100% w/w	PO	-

Uses

- Aphids in *All edible crops (outdoor and protected)* [1]; *Apples* [1]; *Asparagus* [1]; *Asteraceae* [1]; *Aubergines* [1]; *Banana* [1]; *Beans without pods (Fresh)* [1]; *Beetroot* [1]; *Broccoli* [1]; *Cabbages* [1]; *Chillies* [1]; *Chives* [1]; *Clovers* [1]; *Cotton* [1]; *Cucumbers* [1]; *Cyclamen* [1]; *Freesias* [1]; *Garlic* [1]; *Geraniums* [1]; *Gladioli* [1]; *Grapevines* [1]; *Hibiscus trionum* [1]; *Honeysuckle* [1]; *Lettuce* [1]; *Lilies* [1]; *Lupins* [1]; *Maize* [1]; *Marrows* [1]; *Melons* [1]; *Morning Glory* [1]; *Onions* [1]; *Orchids* [1]; *Ornamental plant production* [1]; *Peas* [1]; *Pelargoniums* [1]; *Peppers* [1]; *Poinsettias* [1]; *Potatoes* [1]; *Primulas* [1]; *Protected ornamentals* [1]; *Prunus* [1]; *Pumpkins* [1]; *Roses* [1]; *Spinach* [1]; *Squashes* [1]; *Strawberries* [1]; *Sugar beet* [1]; *Sunflowers* [1]; *Swedes* [1]; *Sweet potato* [1]; *Tomatoes* [1]; *Turnips* [1];
- Insect control in *Agricultural premises* [2];
- Mealybugs in *All edible crops (outdoor and protected)* [1]; *Apples* [1]; *Asparagus* [1]; *Asteraceae* [1]; *Aubergines* [1]; *Banana* [1]; *Beans without pods (Fresh)* [1]; *Beetroot* [1]; *Broccoli* [1]; *Cabbages* [1]; *Chillies* [1]; *Chives* [1]; *Clovers* [1]; *Cotton* [1]; *Cucumbers* [1]; *Cyclamen* [1]; *Freesias* [1]; *Garlic* [1]; *Geraniums* [1]; *Gladioli* [1]; *Grapevines* [1]; *Hibiscus trionum* [1]; *Honeysuckle* [1]; *Lettuce* [1]; *Lilies* [1]; *Lupins* [1]; *Maize* [1]; *Marrows* [1]; *Melons* [1]; *Morning Glory* [1]; *Onions* [1]; *Orchids* [1]; *Ornamental plant production* [1]; *Peas* [1]; *Pelargoniums* [1]; *Peppers* [1]; *Poinsettias* [1]; *Potatoes* [1]; *Primulas* [1]; *Protected ornamentals* [1]; *Prunus* [1]; *Pumpkins* [1]; *Roses* [1]; *Spinach* [1]; *Squashes* [1]; *Strawberries* [1]; *Sugar beet* [1]; *Sunflowers* [1]; *Swedes* [1]; *Sweet potato* [1]; *Tomatoes* [1]; *Turnips* [1];
- Powdery mildew in *All edible crops (outdoor and protected)* [1]; *Apples* [1]; *Asparagus* [1]; *Asteraceae* [1]; *Aubergines* [1]; *Banana* [1]; *Beans without pods (Fresh)* [1]; *Beetroot* [1]; *Broccoli* [1]; *Cabbages* [1]; *Chillies* [1]; *Chives* [1]; *Clovers* [1]; *Cotton* [1]; *Cucumbers* [1]; *Cyclamen* [1]; *Freesias* [1]; *Garlic* [1]; *Geraniums* [1]; *Gladioli* [1]; *Grapevines* [1]; *Hibiscus trionum* [1]; *Honeysuckle* [1]; *Lettuce* [1]; *Lilies* [1]; *Lupins* [1]; *Maize* [1]; *Marrows* [1]; *Melons* [1]; *Morning Glory* [1]; *Onions* [1]; *Orchids* [1]; *Ornamental plant production* [1]; *Peas* [1]; *Pelargoniums* [1]; *Peppers* [1]; *Poinsettias* [1]; *Potatoes* [1]; *Primulas* [1]; *Protected ornamentals* [1]; *Prunus* [1]; *Pumpkins* [1]; *Roses* [1]; *Spinach* [1]; *Squashes* [1]; *Strawberries* [1]; *Sugar beet* [1]; *Sunflowers* [1]; *Swedes* [1]; *Sweet potato* [1]; *Tomatoes* [1]; *Turnips* [1];
- Scale insects in *All edible crops (outdoor and protected)* [1]; *Apples* [1]; *Asparagus* [1]; *Asteraceae* [1]; *Aubergines* [1]; *Banana* [1]; *Beans without pods (Fresh)* [1]; *Beetroot* [1]; *Broccoli* [1]; *Cabbages* [1]; *Chillies* [1]; *Chives* [1]; *Clovers* [1]; *Cotton* [1]; *Cucumbers* [1]; *Cyclamen* [1]; *Freesias* [1]; *Garlic* [1]; *Geraniums* [1]; *Gladioli* [1]; *Grapevines* [1]; *Hibiscus trionum* [1]; *Honeysuckle* [1]; *Lettuce* [1]; *Lilies* [1]; *Lupins* [1]; *Maize* [1]; *Marrows* [1]; *Melons* [1]; *Morning Glory* [1]; *Onions* [1]; *Orchids* [1]; *Ornamental plant production* [1]; *Peas* [1]; *Pelargoniums* [1]; *Peppers* [1]; *Poinsettias* [1]; *Potatoes* [1]; *Primulas* [1]; *Protected ornamentals* [1]; *Prunus* [1]; *Pumpkins* [1]; *Roses* [1]; *Spinach* [1]; *Squashes* [1]; *Strawberries* [1]; *Sugar beet* [1]; *Sunflowers* [1]; *Swedes* [1]; *Sweet potato* [1]; *Tomatoes* [1]; *Turnips* [1];
- Spider mites in *All edible crops (outdoor and protected)* [1]; *Apples* [1]; *Asparagus* [1]; *Asteraceae* [1]; *Aubergines* [1]; *Banana* [1]; *Beans without pods (Fresh)* [1]; *Beetroot* [1]; *Broccoli* [1]; *Cabbages* [1]; *Chillies* [1]; *Chives* [1]; *Clovers* [1]; *Cotton* [1]; *Cucumbers* [1]; *Cyclamen* [1]; *Freesias* [1]; *Garlic* [1]; *Geraniums* [1]; *Gladioli* [1]; *Grapevines* [1]; *Hibiscus trionum* [1]; *Honeysuckle* [1]; *Lettuce* [1]; *Lilies* [1]; *Lupins* [1]; *Maize* [1]; *Marrows* [1]; *Melons* [1]; *Morning Glory* [1]; *Onions* [1]; *Orchids* [1]; *Ornamental plant production* [1]; *Peas* [1]; *Pelargoniums* [1]; *Peppers* [1]; *Poinsettias* [1]; *Potatoes* [1]; *Primulas* [1]; *Protected ornamentals* [1]; *Prunus* [1]; *Pumpkins* [1]; *Roses* [1]; *Spinach* [1]; *Squashes* [1];

Strawberries [1]; *Sugar beet* [1]; *Sunflowers* [1]; *Swedes* [1]; *Sweet potato* [1]; *Tomatoes* [1]; *Turnips* [1];
- Suckers in *All edible crops (outdoor and protected)* [1]; *Apples* [1]; *Asparagus* [1]; *Asteraceae* [1]; *Aubergines* [1]; *Banana* [1]; *Beans without pods (Fresh)* [1]; *Beetroot* [1]; *Broccoli* [1]; *Cabbages* [1]; *Chillies* [1]; *Chives* [1]; *Clovers* [1]; *Cotton* [1]; *Cucumbers* [1]; *Cyclamen* [1]; *Freesias* [1]; *Garlic* [1]; *Geraniums* [1]; *Gladioli* [1]; *Grapevines* [1]; *Hibiscus trionum* [1]; *Honeysuckle* [1]; *Lettuce* [1]; *Lilies* [1]; *Lupins* [1]; *Maize* [1]; *Marrows* [1]; *Melons* [1]; *Morning Glory* [1]; *Onions* [1]; *Orchids* [1]; *Ornamental plant production* [1]; *Peas* [1]; *Pelargoniums* [1]; *Peppers* [1]; *Poinsettias* [1]; *Potatoes* [1]; *Primulas* [1]; *Protected ornamentals* [1]; *Prunus* [1]; *Pumpkins* [1]; *Roses* [1]; *Spinach* [1]; *Squashes* [1]; *Strawberries* [1]; *Sugar beet* [1]; *Sunflowers* [1]; *Swedes* [1]; *Sweet potato* [1]; *Tomatoes* [1]; *Turnips* [1];
- Whitefly in *All edible crops (outdoor and protected)* [1]; *Apples* [1]; *Asparagus* [1]; *Asteraceae* [1]; *Aubergines* [1]; *Banana* [1]; *Beans without pods (Fresh)* [1]; *Beetroot* [1]; *Broccoli* [1]; *Cabbages* [1]; *Chillies* [1]; *Chives* [1]; *Clovers* [1]; *Cotton* [1]; *Cucumbers* [1]; *Cyclamen* [1]; *Freesias* [1]; *Garlic* [1]; *Geraniums* [1]; *Gladioli* [1]; *Grapevines* [1]; *Hibiscus trionum* [1]; *Honeysuckle* [1]; *Lettuce* [1]; *Lilies* [1]; *Lupins* [1]; *Maize* [1]; *Marrows* [1]; *Melons* [1]; *Morning Glory* [1]; *Onions* [1]; *Orchids* [1]; *Ornamental plant production* [1]; *Peas* [1]; *Pelargoniums* [1]; *Peppers* [1]; *Poinsettias* [1]; *Potatoes* [1]; *Primulas* [1]; *Protected ornamentals* [1]; *Prunus* [1]; *Pumpkins* [1]; *Roses* [1]; *Spinach* [1]; *Squashes* [1]; *Strawberries* [1]; *Sugar beet* [1]; *Sunflowers* [1]; *Swedes* [1]; *Sweet potato* [1]; *Tomatoes* [1]; *Turnips* [1];

Approval information
- Products included in this profile are not subject to the Control of Pesticides Regulations/Plant Protection Products Regulations because they act by physical means only

Efficacy guidance
- Products act by physical means following direct contact with spray. They may therefore be used at any time of year on all pest growth stages
- Ensure thorough spray coverage of plant, paying special attention to growing points and the underside of leaves
- Treat as soon as target pests are seen and repeat as often as necessary

Restrictions
- No limit on the number of treatments and no minimum interval between applications
- Before large scale use on a new crop, treat a few plants to check for crop safety
- Do not treat ornamental crops when in flower

Crop-specific information
- HI zero

Environmental safety
- May be used in conjunction with biological control agents. Spray 24 h before they are introduced

Hazard classification and safety precautions
Hazard Irritant [1], Highly flammable
UN Number N/C
Risk phrases H319 [2], H336 [2]
Operator protection U10 [1], U14 [1], U15 [1]; A, H
Storage and disposal D01 [1], D02 [1], D05 [1], D06h [1]

296   pinoxaden

A phenylpyrazoline grass weed herbicide for cereals
HRAC mode of action code: 1 (A)

Products

1	Axial Pro	Syngenta	60 g/l	EC	19010
2	Kanaster	Syngenta	60 g/l	EC	20008

SEE SECTION 3 FOR PRODUCTS ALSO REGISTERED

Uses

- Blackgrass in *Spring barley* [1-2]; *Winter barley* [1-2];
- Italian ryegrass in *Spring barley* [1-2]; *Spring wheat* [1-2]; *Winter barley* [1-2]; *Winter wheat* [1-2];
- Perennial ryegrass in *Spring barley* [1-2]; *Spring wheat* [1-2]; *Winter barley* [1-2]; *Winter wheat* [1-2];
- Wild oats in *Spring barley* [1-2]; *Spring wheat* [1-2]; *Winter barley* [1-2]; *Winter wheat* [1-2];

Approval information

- Pinoxaden included in Annex 1 under EC Regulation 1107/2009

Efficacy guidance

- Best results obtained from treatment when all grass weeds have emerged. There is no residual activity
- Broad-leaved weeds are not controlled
- Treat before emerged weed competition reduces yield
- Blackgrass in winter and spring barley is also controlled when used as part of an integrated control strategy. Product is not recommended for blackgrass control in winter wheat
- Grass weed control may be reduced if rain falls within 1 hr of application
- Pinoxaden is an ACCase inhibitor herbicide. To avoid the build up of resistance do not apply products containing an ACCase inhibitor herbicide more than twice to any crop. In addition do not use any product containing pinoxaden in mixture or sequence with any other product containing the same ingredient
- Use these products as part of a resistance management strategy that includes cultural methods of control and does not use ACCase inhibitors as the sole chemical method of grass weed control
- Applying a second product containing an ACCase inhibitor to a crop will increase the risk of resistance development; only use a second ACCase inhibitor to control different weeds at a different timing
- Always follow WRAG guidelines for preventing and managing herbicide resistant weeds. See Section 5 for more information

Restrictions

- Maximum number of treatments 1 per crop
- Do not spray crops under stress or suffering from waterlogging, pest attack, disease or frost damage
- Do not spray crops undersown with grass mixtures
- Avoid the use of hormone-containing herbicides in mixture or in sequence. Allow 21 d after, or 7 d before, hormone application
- To avoid the build-up of resistance do not apply products containing an ACCase inhibitor herbicide more than twice to any crop
- This product may only be applied after 1st February in the year of harvest

Crop-specific information

- Latest use: before flag leaf extending stage (GS 41)
- Spray cereals in autumn, winter or spring from the 2-leaf stage (GS 12)

Following crops guidance

- There are no restrictions on succeeding crops in a normal rotation
- In the event of failure of a treated crop ryegrass, maize, oats or any broad-leaved crop may be planted after a minimum interval of 4 wk from application

Environmental safety

- Dangerous for the environment
- Toxic to aquatic organisms
- Do not allow spray to drift onto neighbouring crops of oats, ryegrass or maize

Hazard classification and safety precautions

Hazard Irritant, Dangerous for the environment
Transport code 9 [1-2]
Packaging group III
UN Number 3082

FOR FULL CONDITIONS OF USE ALWAYS READ THE PRODUCT LABEL

Risk phrases H315, H317, H361 [2]
Operator protection U05a, U09a, U20b; A, C, H
Environmental protection E15b, E38, H411
Storage and disposal D01, D02, D05, D09a, D10c, D11a, D12a

297   pirimicarb

A carbamate insecticide for aphid control
IRAC mode of action code: 1A

Products

1	Aphox	Adama	50% w/w	WG	18562
2	Clayton Pirimicarb	Clayton	50% w/w	WG	19635
3	Jaspin	Adama	50% w/w	WG	19685

Uses

- Aphids in **Broad beans** [1-3]; **Celeriac** *(off-label)* [1]; **Celery (outdoor)** *(off-label)* [1]; **Chillies** *(off-label)* [1]; **Combining peas** [1-3]; **Dwarf beans** *(off-label)* [1]; **Edible podded peas** *(off-label)* [1]; **Florence fennel** *(off-label)* [1]; **French beans** *(off-label)* [1]; **Horseradish** *(off-label)* [1]; **Ornamental plant production** *(off-label)* [1]; **Parsley root** *(off-label)* [1]; **Peppers** *(off-label)* [1]; **Protected aubergines** *(off-label)* [1]; **Protected baby leaf crops** *(off-label)* [1]; **Protected chilli peppers** *(off-label)* [1]; **Protected courgettes** *(off-label)* [1]; **Protected cress** *(off-label)* [1]; **Protected cucumbers** *(off-label)* [1]; **Protected edible flowers** *(off-label)* [1]; **Protected endives** *(off-label)* [1]; **Protected gherkins** *(off-label)* [1]; **Protected herbs (see appendix 6)** *(off-label)* [1]; **Protected lamb's lettuce** *(off-label)* [1]; **Protected land cress** *(off-label)* [1]; **Protected lettuce** *(off-label)* [1]; **Protected peppers** *(off-label)* [1]; **Protected pumpkins** *(off-label)* [1]; **Protected red mustard** *(off-label)* [1]; **Protected rocket** *(off-label)* [1]; **Protected summer squash** *(off-label)* [1]; **Protected tomatoes** *(off-label)* [1]; **Radishes** *(off-label)* [1]; **Red beet** *(off-label)* [1]; **Rhubarb** *(off-label)* [1]; **Runner beans** *(off-label)* [1]; **Spring field beans** [1-3]; **Vining peas** [1-3]; **Winter field beans** [1-3];
- Bean aphid in **Dwarf beans** *(off-label)* [1]; **Edible podded peas** *(off-label)* [1]; **French beans** *(off-label)* [1]; **Runner beans** *(off-label)* [1];
- Cotton aphids in **Celeriac** *(off-label)* [1]; **Protected chilli peppers** *(off-label)* [1]; **Protected peppers** *(off-label)* [1];
- Foxglove aphid in **Celeriac** *(off-label)* [1]; **Protected chilli peppers** *(off-label)* [1]; **Protected peppers** *(off-label)* [1];
- Pea aphid in **Dwarf beans** *(off-label)* [1]; **Edible podded peas** *(off-label)* [1]; **French beans** *(off-label)* [1]; **Runner beans** *(off-label)* [1];
- Peach-potato aphid in **Celeriac** *(off-label)* [1]; **Protected chilli peppers** *(off-label)* [1]; **Protected peppers** *(off-label)* [1];

Extension of Authorisation for Minor Use (EAMUs)

- **Celeriac** *20182151* [1]
- **Celery (outdoor)** *20182135* [1]
- **Chillies** *20182149* [1]
- **Dwarf beans** *20182152* [1]
- **Edible podded peas** *20182152* [1]
- **Florence fennel** *20182153* [1]
- **French beans** *20182152* [1]
- **Horseradish** *20182136* [1]
- **Ornamental plant production** *20194196* [1]
- **Parsley root** *20182136* [1]
- **Peppers** *20182149* [1]
- **Protected aubergines** *20191393* [1]
- **Protected baby leaf crops** *20191178* [1]
- **Protected chilli peppers** *20182150* [1]
- **Protected courgettes** *20182154* [1]
- **Protected cress** *20191178* [1]
- **Protected cucumbers** *20191395* [1]

SEE SECTION 3 FOR PRODUCTS ALSO REGISTERED

- ***Protected edible flowers*** *20191178* [1]
- ***Protected endives*** *20191178* [1]
- ***Protected gherkins*** *20182154* [1]
- ***Protected herbs (see appendix 6)*** *20191178* [1]
- ***Protected lamb's lettuce*** *20191178* [1]
- ***Protected land cress*** *20191178* [1]
- ***Protected lettuce*** *20191178* [1]
- ***Protected peppers*** *20182150* [1]
- ***Protected pumpkins*** *20191391* [1]
- ***Protected red mustard*** *20191178* [1]
- ***Protected rocket*** *20191178* [1]
- ***Protected summer squash*** *20182154* [1]
- ***Protected tomatoes*** *20191393* [1]
- ***Radishes*** *20182136* [1]
- ***Red beet*** *20182136* [1]
- ***Rhubarb*** *20182135* [1]
- ***Runner beans*** *20182152* [1]

Approval information
- Pirimicarb included in Annex I under EC Regulation 1107/2009
- In 2006 CRD required that all products containing this active ingredient should carry the following warning in the main area of the container label: "Pirimicarb is an anticholinesterase carbamate. Handle with care"

Efficacy guidance
- Chemical has contact, fumigant and translaminar activity
- Best results achieved under warm, calm conditions when plants not wilting and spray does not dry too rapidly. Little vapour activity at temperatures below 15°C
- Apply as soon as aphids seen or warning issued and repeat as necessary
- Addition of non-ionic wetter recommended for use on brassicas
- On cucumbers and tomatoes a root drench is preferable to spraying when using predators in an integrated control programme
- Where aphids resistant to pirimicarb occur control is unlikely to be satisfactory

Restrictions
- Contains an anticholinesterase carbamate compound. Do not use if under medical advice not to work with such compounds
- Maximum number of treatments not specified in some cases but normally 2-6 depending on crop
- Must only be applied between 1 May and 31 August

Crop-specific information
- Latest use in accordance with harvest intervals below
- HI protected courgettes and gherkins 24 h; other edible crops 3 d;

Environmental safety
- Dangerous for the environment
- Very toxic to aquatic organisms
- Dangerous to fish or other aquatic life. Do not contaminate surface waters or ditches with chemical or used container
- Chemical has little effect on bees, ladybirds and other insects and is suitable for use in integrated control programmes on apples and pears
- Keep all livestock out of treated areas for at least 7 d. Bury or remove spillages
- LERAP Category B [1-3]

Hazard classification and safety precautions
  Hazard Toxic, Dangerous for the environment, Toxic if swallowed, Harmful if inhaled
  Transport code 6.1 [1-3]
  Packaging group III
  UN Number 2757
  Risk phrases H319
  Operator protection U05a, U08, U19a, U20b; A, C, D, H, J, M

FOR FULL CONDITIONS OF USE ALWAYS READ THE PRODUCT LABEL

Environmental protection E06c (7 d), E15b, E16a, E34, E38, H410
Storage and disposal D01, D02, D09a, D11a, D12a
Medical advice M02, M03, M04a

298   pirimiphos-methyl

A contact, fumigant and translaminar organophosphorus insecticide
IRAC mode of action code: 1B

Products

1 Actellic 50 EC	Syngenta	500 g/l	EC	19325
2 Actellic Smoke Generator No. 20	Syngenta	22.5% w/w	FU	15739

Uses

- Flour beetle in **Crop handling & storage structures** [1]; **Stored grain** [1];
- Flour moth in **Crop handling & storage structures** [1]; **Stored grain** [1];
- Grain beetle in **Crop handling & storage structures** [1]; **Stored grain** [1];
- Grain storage mite in **Crop handling & storage structures** [1]; **Stored grain** [1];
- Grain storage pests in **Grain stores** [2];
- Grain weevil in **Crop handling & storage structures** [1]; **Stored grain** [1];
- Insect pests in **Grain stores** [2];
- Mites in **Grain stores** [2];
- Warehouse moth in **Crop handling & storage structures** [1]; **Stored grain** [1];

Approval information

- Pirimiphos-methyl included in Annex I under EC Regulation 1107/2009
- In 2006 CRD required that all products containing this active ingredient should carry the following warning in the main area of the container label: "Pirimiphos-methyl is an anticholinesterase organophosphate. Handle with care"
- Accepted by BBPA for use in stores for malting barley

Efficacy guidance

- Chemical acts rapidly and has short persistence in plants, but persists for long periods on inert surfaces
- Best results for protection of stored grain achieved by cleaning store thoroughly before use and employing a combination of pre-harvest and grain/seed treatments [2, 1]
- Best results for admixture treatment obtained when grain stored at 15% moisture or less. Dry and cool moist grain coming into store but then treat as soon as possible, ideally as it is loaded [1]
- Surface admixture can be highly effective on localised surface infestations but should not be relied on for long-term control unless application can be made to the full depth of the infestation [1]
- Where insect pests resistant to pirimiphos-methyl occur control is unlikely to be satisfactory

Restrictions

- Contains an organophosphorus anticholinesterase compound. Do not use if under medical advice not to work with such compounds
- Maximum number of treatments 2 per grain store [2, 1]; 1 per batch of stored grain [1]
- Do not apply surface admixture or complete admixture using hand held equipment [1]
- Surface admixture recommended only where the conditions for store preparation, treatment and storage detailed in the label can be met [1]
- Treated structural areas must remain locked for at least 24 hours, must carry clear warning of treatment and must be ventilated thoroughly before entry

Crop-specific information

- Latest use: before storing grain [2, 1]
- Disinfect empty grain stores by spraying surfaces and/or fumigation and treat grain by full or surface admixture. Treat well before harvest in late spring or early summer and repeat 6 wk later or just before harvest if heavily infested. See label for details of treatment and suitable application machinery

SEE SECTION 3 FOR PRODUCTS ALSO REGISTERED

- Treatment volumes on structural surfaces should be adjusted according to surface porosity. See label for guidelines
- Treat inaccessible areas with smoke generating product used in conjunction with spray treatment of remainder of store
- Predatory mites (*Cheyletus*) found in stored grain feeding on infestations of grain mites may survive treatment and can lead to rejection by buyers. They do not remain for long but grain should be inspected carefully before selling [2, 1]

Environmental safety
- Dangerous for the environment
- Very toxic to aquatic organisms [1]
- Toxic to aquatic organisms [2]
- Highly flammable [2]
- Flammable [1]
- Ventilate fumigated or fogged spaces thoroughly before re-entry
- Unprotected persons must be kept out of fumigated areas within 3 h of ignition and for 4 h after treatment
- Wildlife must be excluded from buildings during treatment [1]
- Keep away from combustible materials [2]

Hazard classification and safety precautions
Hazard Harmful, Highly flammable [2], Flammable [1], Dangerous for the environment, Flammable liquid and vapour [1], Harmful if swallowed, Very toxic to aquatic organisms
Transport code 3 [1], 9 [2]
Packaging group III
UN Number 1993, 3077
Risk phrases H304 [1], H317 [1], H318 [1], H335 [1], H336 [1], H372 [1]
Operator protection U04a [1], U05a, U09a [1], U14 [2], U16a [1], U19a, U20b, U23a [1], U24 [1]; A, C, D, H, J, M
Environmental protection E02a (4 h) [2], E15a, E34, E38, H410
Consumer protection C09 [1], C12
Storage and disposal D01, D02, D05 [1], D07 [2], D09a, D10c [1], D11a [2], D12a
Treated seed S06a [1]
Medical advice M01, M03, M05b [1]

299   potassium bicarbonate (commodity substance)

An inorganic fungicide for use in horticulture approved until 31/8/2025

Products

1	potassium bicarbonate	various	100% w/w	AP	00000

Uses
- Disease control in *All edible crops (outdoor and protected)* [1]; *All non-edible crops (outdoor)* [1]; *All protected non-edible crops* [1];

Approval information
- Approval for the use of potassium bicarbonate as a commodity substance was granted on 26 July 2005 by Ministers under regulation 5 of the Control of Pesticides Regulations 1986
- Potassium bicarbonate is being supported for review in the fourth stage of the EC Review Programme under EC Regulation 1107/2009. It was included in Annex 1 under the alternative name of potassium hydrogen carbonate and now products containing it will need to gain approval in the normal way if they carry label claims for pesticidal activity
- Accepted by BBPA for use on hops

Efficacy guidance
- Adjuvants authorised by CRD may be used in conjunction with potassium bicarbonate
- Crop phytotoxicity can occur

Restrictions
- Food grade potassium bicarbonate must only be used

FOR FULL CONDITIONS OF USE ALWAYS READ THE PRODUCT LABEL

Hazard classification and safety precautions
UN Number N/C

300 potassium hydrogen carbonate (commodity substance)

An insecticide for use in a range of crops

Products

1	Atilla	Certis Belchim B V	85% w/w	SP	16026
2	Karma	Certis Belchim B V	85.42% w/w	WP	16363

Uses

- Fungus diseases in **Apples** [2]; **Nursery fruit trees and bushes** [2];
- Insect pests in **Pears** [1];
- Pear sucker in **Pears** *(reduction)* [1];
- Sooty moulds in **Apples** [2]; **Nursery fruit trees and bushes** [2]; **Pears** *(reduction)* [1];

Approval information

- Potassium hydrogen carbonate included in Annex 1 under EC Regulation 1107/2009
- Accepted by BBPA for use on hops

Restrictions

- Product must not be applied prior to the beginning of flowering (BBCH 61) if used on cereal crops
- A minimum interval of 7 days must be observed between applications

Hazard classification and safety precautions
UN Number N/C
Operator protection U05a, U19a [2]; C
Environmental protection E15a
Storage and disposal D09b

301 potassium phosphonates

Inorganic fungicide (formerly potassium phosphite) for disease control in grapes
FRAC mode of action code: P07

See also dithianon + potassium phosphonates

Products

1	Frutogard	Certis Belchim B V	342 g/l	SL	19105
2	Soriale	BASF	755 g/l	SC	19166

Uses

- Disease control in **Table grapes** [1]; **Wine grapes** [1];
- Phytophthora in **Table grapes** [1]; **Wine grapes** [1];
- Pythium in **Table grapes** [1]; **Wine grapes** [1];
- Scab in **Apples** [2]; **Pears** [2];

Approval information

- Potassium phosphonate included in Annex I under EC Regulation 1107/2009

Restrictions

- A minimum interval of 7 days must be observed between applications.
- Do not handle treated foliage until 24 days after treatment.

Environmental safety

- Buffer zone requirement 15 m
- LERAP Category B [1-2]

Hazard classification and safety precautions
Hazard Very toxic to aquatic organisms [2]
UN Number N/C

SEE SECTION 3 FOR PRODUCTS ALSO REGISTERED

SECTION 2

Risk phrases H317 [2], H319 [2], H351 [2]
Operator protection U19a [2], U20a [2]; A, C, H
Environmental protection E16a, E17a (30 m) [2], H410 [2]
Storage and disposal D01, D02, D05

302   potassium salts of fatty acids

An insecticide for use on protected crops
IRAC mode of action code: Unknown

Products

1	Tec-bom	Fargro	41.62% w/v	SL	19547

Uses
- Aphids in **Protected aubergines** *(some activity)* [1]; **Protected chilli peppers** *(some activity)* [1]; **Protected peppers** *(some activity)* [1]; **Protected tomatoes** *(some activity)* [1];
- Spider mites in **Protected aubergines** *(some activity)* [1]; **Protected chilli peppers** *(some activity)* [1]; **Protected peppers** *(some activity)* [1]; **Protected tomatoes** *(some activity)* [1];
- Whitefly in **Protected aubergines** *(moderate control)* [1]; **Protected chilli peppers** *(moderate control)* [1]; **Protected peppers** *(moderate control)* [1]; **Protected tomatoes** *(moderate control)* [1];

Approval information
- Fatty acids C7-C18 and C18 unsaturated potassium salts (CAS 67701-09-1) included in Annex 1 under EC Regulation 1107/2009

Efficacy guidance
- Although no phytotoxicity has been seen, test on a small area before use on new crops or varieties
- This is a contact insecticide so it is important that the spray contacts the pest. Efforts should be made to thoroughly reach all plant parts, including the underside of leaves
- Clean application equipment thoroughly before use as TEC-BOM may release residues from the equipment of products applied earlier

Restrictions
- Consult processor before use on crops grown for processing
- Do not mix with acidic products

Hazard classification and safety precautions
Hazard Irritant
UN Number N/C
Risk phrases H319, H335
Operator protection U11, U20a; A, C, H
Environmental protection E12h, E15b, E34, H412
Storage and disposal D01, D09a, D10b
Medical advice M03, M04a

303   prochloraz

A broad-spectrum protectant and eradicant imidazole fungicide
FRAC mode of action code: 3

See also prochloraz + tebuconazole

Products

1	Mirage (GB only)	Adama	450 g/l	EC	18361

Uses
- Anthracnose in **Managed amenity turf** *(off-label)* [1];
- Eyespot in **Rye** *(moderate control)* [1]; **Spring wheat** *(moderate control)* [1]; **Triticale** *(moderate control)* [1]; **Winter wheat** *(moderate control)* [1];

FOR FULL CONDITIONS OF USE ALWAYS READ THE PRODUCT LABEL

- Fusarium patch in **Managed amenity turf** *(off-label)* [1];
- Glume blotch in **Spring wheat** [1]; **Winter wheat** [1];
- Leaf spot in **Managed amenity turf** *(off-label)* [1]; **Spring wheat** [1]; **Winter wheat** [1];
- Powdery mildew in **Rye** [1];
- Rhynchosporium in **Rye** *(moderate control)* [1]; **Triticale** *(moderate control)* [1];

Extension of Authorisation for Minor Use (EAMUs)
- **Managed amenity turf** *20220408* [1]

Approval information
- Prochloraz included in Annex 1 under EC Regulation 1107/2009

Efficacy guidance
- Spray cereals at first signs of disease. Protection of winter crops through season usually requires at least 2 treatments. See label for details of rates and timing. Treatment active against strains of eyespot resistant to benzimidazole fungicides
- Tank mixes with other fungicides recommended to improve control of rusts in wheat. See label for details
- A period of at least 3 h without rain should follow spraying
- Prochloraz is a DMI fungicide. Resistance to some DMI fungicides has been identified in Septoria leaf blotch which may seriously affect performance of some products. For further advice contact a specialist advisor and visit the Fungicide Resistance Action Group (FRAG)-UK website

Restrictions
- No application is permitted to any crop before BBCH 30, i.e. only apply after stem elongation has begun [1]

Crop-specific information
- Application with other fungicides may cause cereal crop scorch

Environmental safety
- Dangerous for the environment
- Very toxic to aquatic organisms
- LERAP Category B [1]

Hazard classification and safety precautions
Hazard Very toxic to aquatic organisms
Transport code 9 [1]
Packaging group III
UN Number 3082
Operator protection A, H
Environmental protection E16a, H410

304 prochloraz + tebuconazole

A broad spectrum systemic fungicide mixture for cereals, oilseed rape and amenity use
FRAC mode of action code: 3 + 3

See also tebuconazole

Products
1	Agate EW	Nufarm UK	267:133 g/l	EW	19000
2	Agate EW (GB only)	Nufarm UK	267:133 g/l	EW	19000
3	Monkey	Nufarm UK	267:133 g/l	EW	18997
4	Monkey (GB only)	Nufarm UK	267:133 g/l	EW	18997
5	Orius P	Nufarm UK	267:133 g/l	EW	18994
6	Orius P (GB only)	Nufarm UK	267:133 g/l	EW	18994

Uses
- Brown rust in **Spring rye** [4,6]; **Spring wheat** [2,4,6]; **Triticale** [2,4,6]; **Winter rye** [2,4,6]; **Winter wheat** [2,4,6];

SEE SECTION 3 FOR PRODUCTS ALSO REGISTERED

- Eyespot in *Spring rye* [2,4,6]; *Spring wheat* [2,4,6]; *Triticale* [2,4,6]; *Winter rye* [2,4,6]; *Winter wheat* [2,4,6];
- Glume blotch in *Spring wheat* [2,4,6]; *Triticale* [2,4,6]; *Winter wheat* [2,4,6];
- Late ear diseases in *Spring wheat* [2,4,6]; *Triticale* [2,4,6]; *Winter wheat* [2,4,6];
- Powdery mildew in *Spring rye* [2,4,6]; *Spring wheat* [2,4,6]; *Triticale* [2,4,6]; *Winter rye* [2,4,6]; *Winter wheat* [2,4,6];
- Rhynchosporium in *Spring rye* [2,4,6]; *Winter rye* [2,4,6];
- Septoria leaf blotch in *Spring wheat* [2,4,6]; *Triticale* [2,4,6]; *Winter wheat* [2,4,6];
- Yellow rust in *Spring rye* [2,4,6]; *Spring wheat* [2,4,6]; *Triticale* [2,4,6]; *Winter rye* [2,4,6]; *Winter wheat* [2,4,6];

Approval information
- Prochloraz and tebuconazole included in Annex I under EC Regulation 1107/2009

Efficacy guidance
- Optimum application timing is normally when disease first seen but varies with main target disease - see labels
- Prochloraz and tebuconazole are DMI fungicides. Resistance to some DMI fungicides has been identified in Septoria leaf blotch which may seriously affect performance of some products. For further advice contact a specialist advisor and visit the Fungicide Resistance Action Group (FRAG)-UK website

Restrictions
- Maximum total dose on cereals equivalent to two full dose treatments
- Do not exceed 450 g prochloraz/hectare per application for outdoor uses
- To protect non target insects/arthropods respect an unsprayed buffer zone of 5 m to non-crop land

Crop-specific information
- Latest use: Up to and including flowering (anthesis) just complete (GS 69)
- HI 6 wk for cereals
- Occasionally transient leaf speckling may occur after treating wheat. Yield responses should not be affected

Environmental safety
- Dangerous for the environment
- Very toxic to aquatic organisms
- To protect non target insects/arthropods respect an unsprayed buffer zone of 5 m to non-crop land
- LERAP Category B [1-6]

Hazard classification and safety precautions
Hazard Harmful, Dangerous for the environment, Harmful if swallowed, Very toxic to aquatic organisms [4,6]
Transport code 9
Packaging group III
UN Number 3082
Risk phrases H319, H361
Operator protection U05a, U09b, U11 [2], U14, U20a; A, C, H
Environmental protection E13b, E16a, E34, E38, H410
Storage and disposal D01, D02, D05, D09a, D10c, D12a
Medical advice M03

305 prohexadione-calcium

A cyclohexanecarboxylate growth regulator for use in apples

See also mepiquat chloride + prohexadione-calcium
mepiquat chloride + prohexadione-calcium + pyraclostrobin

Products

1	Attraxor	BASF	10% w/w	WG	18939
2	Bolster	De Sangosse	50 g/l	OD	18773
3	Flagstaff	De Sangosse	50 g/l	OD	18586
4	Kopis	Pan Amenity	10% w/w	WG	19572
5	Kudos	Fine	10% w/w	WG	17578
6	Regalis Plus	BASF	10% w/w	WG	16485
7	Shortcut	Pan Amenity	10% w/w	WG	19557

Uses

- Botrytis in **Protected ornamentals** *(off-label)* [6]; **Table grapes** *(off-label)* [6]; **Wine grapes** *(off-label)* [6];
- Control of shoot growth in **Apples** [5-6]; **Hops in propagation** *(off-label)* [6]; **Nursery fruit trees** *(off-label)* [6]; **Ornamental plant production** *(off-label)* [6]; **Pears** *(off-label)* [6]; **Plums** *(off-label)* [6];
- Disease control in **Protected ornamentals** *(off-label)* [6];
- Growth regulation in **Amenity grassland** [1]; **Managed amenity turf** [1,4,7]; **Rye** [2-3]; **Spelt** [2-3]; **Spring barley** [2-3]; **Spring oats** [2-3]; **Spring wheat** [2-3]; **Triticale** [2-3]; **Winter barley** [2-3]; **Winter oats** [2-3]; **Winter wheat** [2-3];

Extension of Authorisation for Minor Use (EAMUs)

- **Hops in propagation** *20150175* [6]
- **Nursery fruit trees** *20150175* [6]
- **Ornamental plant production** *20150175* [6]
- **Pears** *20150182* [6]
- **Plums** *20171174* [6]
- **Protected ornamentals** *20192153* [6]
- **Table grapes** *20150180* [6]
- **Wine grapes** *20150180* [6]

Approval information

- Prohexadione-calcium included in Annex I under EC Regulation 1107/2009

Efficacy guidance

- For a standard orchard treat at the start of active growth and again after 3-5 wk depending on growth conditions
- Alternatively follow the first application with four reduced rate applications at two wk intervals
- The maximum concentration for apple and pear must not exceed 0.2 kg of product per 100 litre water. The maximum concentration for cherry must not exceed 0.125 kg of product per 100 litre water.

Restrictions

- Maximum total dose equivalent to two full dose treatments
- Do not apply in conjunction with calcium based foliar fertilisers
- Do not apply by hand-held equipment

Crop-specific information

- HI 55 d for apples

Environmental safety

- Harmful to aquatic organisms
- Avoid spray drift onto neighbouring crops and other non-target plants
- Product does not harm natural insect predators when used as directed

---

SEE SECTION 3 FOR PRODUCTS ALSO REGISTERED

Hazard classification and safety precautions
UN Number N/C
Risk phrases H317 [1,4,6-7]
Operator protection U02a [5]; U05a [1,4-7]; U08 [5]; U19a [5]; U20b [5]; A, H
Environmental protection E15b [5]; E38 [1,4-7]; H412 [5]
Storage and disposal D01 [1,4-7], D02 [1,4-7], D09a [1,4-7], D10b [5], D10c [1,4,6-7], D12a [1,4-7]
Medical advice M05a [5]

306 prohexadione-calcium + trinexapac-ethyl

A plant growth regulator for use in cereal crops
FRAC mode of action code: 11

See also trinexapac-ethyl

Products

1	Medax Max	BASF	0.5:0.75% w/w	SG	17263

Uses
- Growth regulation in **Durum wheat** [1]; **Grass seed crops** *(off-label)* [1]; **Spring barley** [1]; **Spring oats** [1]; **Spring rye** *(off-label)* [1]; **Spring wheat** [1]; **Triticale** [1]; **Winter barley** [1]; **Winter oats** [1]; **Winter rye** [1]; **Winter wheat** [1];

Extension of Authorisation for Minor Use (EAMUs)
- **Grass seed crops** *20201502* [1]
- **Spring rye** *20220660* [1]

Approval information
- Prohexadione-calcium and trinexapac-ethyl included in Annex I under EC Regulation 1107/2009

Efficacy guidance
- Do not use on stressed crops

Following crops guidance
- Any crop can follow a normally harvested or failed cereal crop. Ploughing is not necessary before sowing a following crop.

Hazard classification and safety precautions
Transport code 9 [1]
Packaging group III
UN Number 3077
Operator protection U08, U19a, U20a; A
Environmental protection E15b, E34, H412
Storage and disposal D01, D02, D05, D09a, D10c
Medical advice M03

307 propamocarb hydrochloride

A translocated protectant carbamate fungicide
FRAC mode of action code: 28

See also cymoxanil + propamocarb
dimethomorph + propamocarb
fluopicolide + propamocarb hydrochloride
fosetyl-aluminium + propamocarb hydrochloride

Products

1	Promess	Arysta	722 g/l	SL	16008
2	Proplant	Arysta	722 g/l	SL	15422

Uses

- Damping off in **Protected broccoli** [1-2]; **Protected Brussels sprouts** [1-2]; **Protected calabrese** [1-2]; **Protected cauliflowers** [1-2]; **Protected tomatoes** *(off-label)* [2];
- Downy mildew in **Forest nurseries** *(off-label)* [1]; **Ornamental plant production** *(off-label)* [1]; **Protected broccoli** [1-2]; **Protected Brussels sprouts** [1-2]; **Protected bulbs** *(off-label)* [1]; **Protected calabrese** [1-2]; **Protected cauliflowers** [1-2]; **Protected ornamentals** *(off-label)* [1]; **Protected watercress** *(off-label)* [1-2]; **Radishes** *(off-label)* [1-2];
- Peronospora spp in **Forest nurseries** *(off-label)* [1-2]; **Ornamental plant production** *(off-label)* [1-2]; **Protected broccoli** [1-2]; **Protected Brussels sprouts** [1-2]; **Protected bulbs** *(off-label)* [1]; **Protected calabrese** [1-2]; **Protected cauliflowers** [1-2]; **Protected ornamentals** *(off-label)* [1-2];
- Phytophthora in **Forest nurseries** *(off-label)* [1-2]; **Ornamental plant production** *(off-label)* [1-2]; **Protected broccoli** [1-2]; **Protected Brussels sprouts** [1-2]; **Protected bulbs** *(off-label)* [1]; **Protected calabrese** [1-2]; **Protected cauliflowers** [1-2]; **Protected ornamentals** *(off-label)* [1-2]; **Protected tomatoes** *(off-label)* [1-2]; **Protected watercress** *(off-label)* [1-2];
- Pythium in **Forest nurseries** *(off-label)* [1-2]; **Ornamental plant production** *(off-label)* [1-2]; **Protected broccoli** [1-2]; **Protected Brussels sprouts** [1-2]; **Protected bulbs** *(off-label)* [1]; **Protected calabrese** [1-2]; **Protected cauliflowers** [1-2]; **Protected ornamentals** *(off-label)* [1-2]; **Protected watercress** *(off-label)* [1-2];
- Root rot in **Protected broccoli** [1-2]; **Protected Brussels sprouts** [1-2]; **Protected calabrese** [1-2]; **Protected cauliflowers** [1-2]; **Protected tomatoes** *(off-label)* [1-2]; **Protected watercress** *(off-label)* [1];

Extension of Authorisation for Minor Use (EAMUs)

- **Forest nurseries** *20160796* [1], *20160785* [2]
- **Ornamental plant production** *20160796* [1], *20160785* [2]
- **Protected bulbs** *20160796* [1]
- **Protected ornamentals** *20160796* [1], *20160785* [2]
- **Protected tomatoes** *20160797* [1], *20160784* [2]
- **Protected watercress** *20160792* [1], *20160783* [2]
- **Radishes** *20160795* [1], *20160782* [2]

Approval information

- Propamocarb hydrochloride included in Annex I under EC Regulation 1107/2009

Efficacy guidance

- Chemical is absorbed through roots and translocated throughout plant
- Incorporate in compost before use or drench moist compost or soil before sowing, pricking out, striking cuttings or potting up
- Concentrated solution is corrosive to all metals other than stainless steel

Restrictions

- Maximum number of treatments 1 per crop for listed brassicas;
- When applied over established seedlings rinse off foliage with water and do not apply under hot, dry conditions
- Do not apply in a recirculating irrigation/drip system
- Store away from seeds and fertilizers
- Consult processor before use on crops grown for processing

Crop-specific information

- Latest use: before transplanting for brassicas

Hazard classification and safety precautions

Hazard Irritant
UN Number N/C
Risk phrases H317
Operator protection U02a, U05a, U08, U19a, U20b; A, H, M
Environmental protection E15a, E34
Storage and disposal D01, D02, D05, D09a, D10c

SEE SECTION 3 FOR PRODUCTS ALSO REGISTERED

SECTION 2

308 propaquizafop

A phenoxy alkanoic acid foliar acting herbicide for grass weeds in a range of crops
HRAC mode of action code: 1 (A)

Products

1	Clayton Satchmo	Clayton	100 g/l	EC	17061
2	Falcon	Adama	100 g/l	EC	16459
3	Falcon	Adama	100 g/l	EC	20010
4	Longhorn	AgChem Access	100 g/l	EC	16951
5	Shogun	Adama	100 g/l	EC	16527
6	Tithe	Adama	100 g/l	EC	19515

Uses

*   Annual grasses in *All edible seed crops grown outdoors* (off-label) [2,5]; *All non-edible seed crops grown outdoors* (off-label) [2,5]; *Broad beans* [1-6]; *Bulb onions* [1-6]; *Carrots* [1-6]; *Combining peas* [1-6]; *Dwarf beans* [1-6]; *Early potatoes* [1-6]; *Fodder beet* [1-6]; *Forest* (off-label) [2]; *Forest nurseries* (off-label) [2]; *French beans* [1-6]; *Game cover* (off-label) [2,5]; *Garlic* (off-label) [5]; *Linseed* [1-6]; *Lupins* (off-label) [5]; *Maincrop potatoes* [1-6]; *Mustard* (off-label) [2,5]; *Poppies* (off-label) [2]; *Poppies for morphine production* (off-label) [2,5]; *Poppies grown for seed production* (off-label) [2]; *Red beet* (off-label) [5]; *Shallots* (off-label) [5]; *Spring field beans* [1-6]; *Spring oilseed rape* [1-6]; *Sugar beet* [1-6]; *Swedes* [1-6]; *Turnips* [1-6]; *Winter field beans* [1-6]; *Winter oilseed rape* [1-6];
*   Annual meadow grass in *All edible seed crops grown outdoors* (off-label) [5]; *All non-edible seed crops grown outdoors* (off-label) [5]; *Game cover* (off-label) [5]; *Garlic* (off-label) [2,5]; *Lupins* (off-label) [2,5]; *Poppies* (off-label) [2]; *Poppies for morphine production* (off-label) [5]; *Poppies grown for seed production* (off-label) [2]; *Red beet* (off-label) [5]; *Shallots* (off-label) [2,5];
*   Couch in *Mustard* (off-label) [2];
*   Grass weeds in *Garlic* (off-label) [2]; *Lupins* (off-label) [2]; *Shallots* (off-label) [2];
*   Perennial grasses in *All edible seed crops grown outdoors* (off-label) [2,5]; *All non-edible seed crops grown outdoors* (off-label) [2,5]; *Broad beans* [1-6]; *Bulb onions* [1-6]; *Carrots* [1-6]; *Combining peas* [1-6]; *Dwarf beans* [1-6]; *Early potatoes* [1-6]; *Fodder beet* [1-6]; *Forest* (off-label) [2]; *Forest nurseries* (off-label) [2]; *French beans* [1-6]; *Game cover* (off-label) [2,5]; *Garlic* (off-label) [5]; *Linseed* [1-6]; *Lupins* (off-label) [5]; *Maincrop potatoes* [1-6]; *Mustard* (off-label) [5]; *Poppies* (off-label) [2]; *Poppies for morphine production* (off-label) [2,5]; *Poppies grown for seed production* (off-label) [2]; *Red beet* (off-label) [5]; *Shallots* (off-label) [5]; *Spring field beans* [1-6]; *Spring oilseed rape* [1-6]; *Sugar beet* [1-6]; *Swedes* [1-6]; *Turnips* [1-6]; *Winter field beans* [1-6]; *Winter oilseed rape* [1-6];
*   Volunteer cereals in *Mustard* (off-label) [2,5]; *Poppies* (off-label) [2]; *Poppies for morphine production* (off-label) [2,5]; *Poppies grown for seed production* (off-label) [2]; *Red beet* (off-label) [2,5];
*   Volunteer rye in *Red beet* (off-label) [2];

Extension of Authorisation for Minor Use (EAMUs)

*   *All edible seed crops grown outdoors* 20141283 [2], 20141927 [5], 20142509 [5]
*   *All non-edible seed crops grown outdoors* 20141283 [2], 20141927 [5], 20142509 [5]
*   *Forest* 20163458 [2]
*   *Forest nurseries* 20163458 [2]
*   *Game cover* 20141283 [2], 20141927 [5], 20142509 [5]
*   *Garlic* 20141281 [2], 20141926 [5]
*   *Lupins* 20141282 [2], 20141928 [5]
*   *Mustard* 20172803 [2], 20180165 [5]
*   *Poppies* 20180090 [2]
*   *Poppies for morphine production* 20152056 [2], 20160285 [5]
*   *Poppies grown for seed production* 20180090 [2]
*   *Red beet* 20151827 [2], 20160286 [5]
*   *Shallots* 20141281 [2], 20141926 [5]

Approval information
- Propaquizafop included in Annex 1 under EC Regulation 1107/2009

Efficacy guidance
- Apply to emerged weeds when they are growing actively with adequate soil moisture
- Activity is slower under cool conditions
- Broad-leaved weeds and any weeds germinating after treatment are not controlled
- Annual meadow grass up to 3 leaves checked at low doses and severely checked at highest dose
- Spray barley cover crops when risk of wind blow has passed and before there is serious competition with the crop
- Various tank mixtures and sequences recommended for broader spectrum weed control in oilseed rape, peas and sugar beet. See label for details
- Severe couch infestations may require a second application at reduced dose when regrowth has 3-4 leaves unfolded
- Products contain surfactants. Tank mixing with adjuvants not required or recommended
- Propaquizafop is an ACCase inhibitor herbicide. To avoid the build up of resistance do not apply products containing an ACCase inhibitor herbicide more than twice to any crop. In addition do not use any product containing propaquizafop in mixture or sequence with any other product containing the same ingredient
- Use these products as part of a resistance management strategy that includes cultural methods of control and does not use ACCase inhibitors as the sole chemical method of grass weed control
- Applying a second product containing an ACCase inhibitor to a crop will increase the risk of resistance development; only use a second ACCase inhibitor to control different weeds at a different timing
- Always follow WRAG guidelines for preventing and managing herbicide resistant weeds. See Section 5 for more information

Restrictions
- Maximum number of treatments 1 or 2 per crop or yr. See label for details
- See label for list of tolerant tree species
- Do not treat seed potatoes

Crop-specific information
- Latest use: before crop flower buds visible for winter oilseed rape, linseed, field beans; before 8 fully expanded leaf stage for spring oilseed rape; before weeds are covered by the crop for potatoes, sugar beet, fodder beet, swedes, turnips; when flower buds visible for peas
- HI onions, carrots, early potatoes, parsnips 4 wk; combining peas, early potatoes 7 wk; sugar beet, fodder beet, maincrop potatoes, swedes, turnips 8 wk; field beans 14 wk
- Application in high temperatures and/or low soil moisture content may cause chlorotic spotting especially on combining peas and field beans
- Overlaps at the highest dose can cause damage from early applications to carrots and parsnips

Following crops guidance
- In the event of a failed treated crop an interval of 2 wk must elapse between the last application and redrilling with winter wheat or winter barley. 4 wk must elapse before sowing oilseed rape, peas or field beans, and 16 wk before sowing ryegrass or oats

Environmental safety
- Dangerous for the environment
- Toxic to aquatic organisms
- Risk to certain non-target insects or other arthropods. See directions for use
- LERAP Category B [1-6]

Hazard classification and safety precautions
Hazard Harmful, Dangerous for the environment
Transport code 9 [1-6]
Packaging group III
UN Number 3082
Risk phrases H304, H319
Operator protection U02a, U05a, U08, U11, U13, U14, U15, U19a, U20b; A, C, H, M

SEE SECTION 3 FOR PRODUCTS ALSO REGISTERED

SECTION 2

Environmental protection E15a, E16a, E22b, E34, E38, H411
Storage and disposal D01, D02, D05, D09a, D10b [1-3], D12a, D12b, D14
Medical advice M05b

309   propyzamide

A residual benzamide herbicide for use in a wide range of crops
HRAC mode of action code: 3 (K1)

See also aminopyralid + propyzamide

Products

1	Barclay Propyz	Barclay	400 g/l	SC	15083
2	Cleancrop Forward	Corteva	400 g/l	SC	18145
3	Cleancrop Rumble	Agrii	400 g/l	SC	17600
4	Kerb Flo	Corteva	400 g/l	SC	13716
5	Kerb Flo 500	Corteva	500 g/l	SC	15586
6	Levada	Certis Belchim B V	400 g/l	SC	15743
7	Relva Granules	BelCrop	4% w/w	GR	16744
8	Setanta Flo	Certis Belchim B V	400 g/l	SC	15791
9	Solitaire	Certis Belchim B V	400 g/l	SC	15792
10	Stroller	AgChem Access	400 g/l	SC	14603

Uses

*   Annual and perennial weeds in *Amenity vegetation* [6,8-9];
*   Annual dicotyledons in *All edible seed crops grown outdoors* (off-label) [4]; *All non-edible seed crops grown outdoors* (off-label) [4]; *Almonds* (off-label) [4]; *Amenity vegetation* [1-4,7,10]; *Apple orchards* [1-4,6,8-10]; *Bilberries* (off-label) [4]; *Blackberries* [1-4,6,8-10]; *Blackcurrants* [1-4,6,8-10]; *Blueberries* (off-label) [4]; *Bog myrtle* (off-label) [4]; *Borage* (off-label) [4]; *Broccoli* (off-label) [4]; *Calabrese* (off-label) [4]; *Cauliflowers* (off-label) [4]; *Cherries* (off-label) [4]; *Chestnuts* (off-label) [4]; *Chicory root* (off-label) [4]; *Clover seed crops* [1-4,6,8-10]; *Cob nuts* (off-label) [4]; *Corn Gromwell* (off-label) [4]; *Courgettes* (off-label) [4]; *Cranberries* (off-label) [4]; *Cress* (off-label) [4]; *Endives* (off-label) [4]; *Farm forestry* [1-4,6-10]; *Fodder rape seed crops* [1-4,6,8-10]; *Forest* [1-4,6-10]; *Forest nurseries* [1-4,6,8-10]; *Frise* (off-label) [4]; *Game cover* (off-label) [4]; *Gooseberries* [1-4,6,8-10]; *Hazel nuts* (off-label) [4]; *Hedges* [1-4,6,8-10]; *Herbs (see appendix 6)* (off-label) [4]; *Hops* (off-label) [4]; *Kale seed crops* [1-4,6,8-10]; *Lamb's lettuce* (off-label) [4]; *Leaf brassicas* (off-label) [4]; *Lettuce* [2-4,6,8-10]; *Loganberries* [1-4,6,8-10]; *Lucerne* [2,4,6,8-10]; *Marrows* (off-label) [4]; *Mirabelles* (off-label) [4]; *Ornamental plant production* [2,7]; *Pear orchards* [1-4,6,8-10]; *Plums* [1-4,6,8-10]; *Protected endives* (off-label) [4]; *Protected forest nurseries* (off-label) [4]; *Protected herbs (see appendix 6)* (off-label) [4]; *Protected lettuce* (off-label) [4]; *Pumpkins* (off-label) [4]; *Quinces* (off-label) [4]; *Radicchio* (off-label) [4]; *Raspberries* (England only) [1-4,6,8-10]; *Redcurrants* [1-4,6,8-10]; *Rhubarb* [1-4,6,8-10]; *Rubus hybrids* (England only) [3]; *Salad brassicas* (off-label) [4]; *Scarole* (off-label) [4]; *Squashes* (off-label) [4]; *Strawberries* [1-4,6,8-10]; *Sugar beet seed crops* [1-4,6,8-10]; *Table grapes* (off-label) [4]; *Turnip seed crops* [1-4,6,8-10]; *Walnuts* (off-label) [4]; *Wine grapes* (off-label) [4]; *Winter field beans* [2-6,8-10]; *Winter oilseed rape* [1-6,8-10];
*   Annual grasses in *All edible seed crops grown outdoors* (off-label) [4]; *All non-edible seed crops grown outdoors* (off-label) [4]; *Almonds* (off-label) [4]; *Amenity vegetation* [1-4,7,10]; *Apple orchards* [1-4,6,8-10]; *Bilberries* (off-label) [4]; *Blackberries* [1-4,6,8-10]; *Blackcurrants* [1-4,6,8-10]; *Blueberries* (off-label) [4]; *Bog myrtle* (off-label) [4]; *Borage* (off-label) [4]; *Broccoli* (off-label) [4]; *Calabrese* (off-label) [4]; *Cauliflowers* (off-label) [4]; *Cherries* (off-label) [4]; *Chestnuts* (off-label) [4]; *Chicory root* (off-label) [4]; *Clover seed crops* [1-4,6,8-10]; *Cob nuts* (off-label) [4]; *Corn Gromwell* (off-label) [4]; *Courgettes* (off-label) [4]; *Cranberries* (off-label) [4]; *Cress* (off-label) [4]; *Endives* (off-label) [4]; *Farm forestry* [1-4,6-10]; *Fodder rape seed crops* [1-4,6,8-10]; *Forest* [1-4,6-10]; *Forest nurseries* [1-4,6,8-10]; *Frise* (off-label) [4]; *Game cover* (off-label) [4]; *Gooseberries* [1-4,6,8-10]; *Hazel nuts* (off-label) [4]; *Hedges* [1-4,6,8-10]; *Herbs (see appendix 6)* (off-label) [4]; *Hops* (off-label) [4]; *Kale seed crops* [1-4,6,8-10]; *Lamb's lettuce* (off-label) [4]; *Leaf brassicas* (off-label) [4]; *Lettuce* [2-4,6,8-10]; *Loganberries* [1-4,6,8-10]; *Lucerne*

[2,4,6,8-10]; *Marrows (off-label)* [4]; *Mirabelles (off-label)* [4]; **Ornamental plant production** *(off-label)* [2,4,7]; **Pear orchards** [1-4,6,8-10]; **Plums** [1-4,6,8-10]; **Protected endives** *(off-label)* [4]; **Protected forest nurseries** *(off-label)* [4]; **Protected herbs (see appendix 6)** *(off-label)* [4]; **Protected lettuce** *(off-label)* [4]; **Pumpkins** *(off-label)* [4]; **Quinces** *(off-label)* [4]; **Radicchio** *(off-label)* [4]; **Raspberries** *(England only)* [1-4,6,8-10]; **Redcurrants** [1-4,6,8-10]; **Rhubarb** [1-4,6,8-10]; **Rubus hybrids** *(England only)* [3]; **Salad brassicas** *(off-label)* [4]; **Scarole** *(off-label)* [4]; **Squashes** *(off-label)* [4]; **Strawberries** [1-4,6,8-10]; **Sugar beet seed crops** [1-4,6,8-10]; **Table grapes** *(off-label)* [4]; **Turnip seed crops** [1-4,6,8-10]; **Walnuts** *(off-label)* [4]; **Wine grapes** *(off-label)* [4]; **Winter field beans** [2-4,6,8-10]; **Winter oilseed rape** [1-4,6,8-10];

- Annual meadow grass in **Borage** *(off-label)* [4]; **Corn Gromwell** *(off-label)* [4]; **Winter field beans** [5]; **Winter oilseed rape** [5];
- Blackgrass in **Winter field beans** [5]; **Winter oilseed rape** [5];
- Fat hen in **Ornamental plant production** *(off-label)* [4];
- Perennial grasses in **All edible seed crops grown outdoors** *(off-label)* [4]; **All non-edible seed crops grown outdoors** *(off-label)* [4]; **Almonds** *(off-label)* [4]; **Amenity vegetation** [1-4,7,10]; **Apple orchards** [1-4,6,8-10]; **Bilberries** *(off-label)* [4]; **Blackberries** [1-4,6,8-10]; **Blackcurrants** [1-4,6,8-10]; **Blueberries** *(off-label)* [4]; **Bog myrtle** *(off-label)* [4]; **Broccoli** *(off-label)* [4]; **Calabrese** *(off-label)* [4]; **Cauliflowers** *(off-label)* [4]; **Cherries** *(off-label)* [4]; **Chestnuts** *(off-label)* [4]; **Chicory root** *(off-label)* [4]; **Clover seed crops** [1-4,6,8-10]; **Cob nuts** *(off-label)* [4]; **Corn Gromwell** *(off-label)* [4]; **Courgettes** *(off-label)* [4]; **Cranberries** *(off-label)* [4]; **Cress** *(off-label)* [4]; **Endives** *(off-label)* [4]; **Farm forestry** [1-4,6-10]; **Fodder rape seed crops** [1-4,6,8-10]; **Forest** [1-4,6-10]; **Forest nurseries** [1-4,6,8-10]; **Frise** *(off-label)* [4]; **Game cover** *(off-label)* [4]; **Gooseberries** [1-4,6,8-10]; **Hazel nuts** *(off-label)* [4]; **Hedges** [1-4,6,8-10]; **Herbs (see appendix 6)** *(off-label)* [4]; **Hops** *(off-label)* [4]; **Kale seed crops** [1-4,6,8-10]; **Lamb's lettuce** *(off-label)* [4]; **Leaf brassicas** *(off-label)* [4]; **Lettuce** [2-4,6,8-10]; **Loganberries** [1-4,6,8-10]; **Lucerne** [2,4,6,8-10]; **Marrows** *(off-label)* [4]; **Mirabelles** *(off-label)* [4]; **Ornamental plant production** [2,7]; **Pear orchards** [1-4,6,8-10]; **Plums** [1-4,6,8-10]; **Protected endives** *(off-label)* [4]; **Protected forest nurseries** *(off-label)* [4]; **Protected herbs (see appendix 6)** *(off-label)* [4]; **Protected lettuce** *(off-label)* [4]; **Pumpkins** *(off-label)* [4]; **Quinces** *(off-label)* [4]; **Radicchio** *(off-label)* [4]; **Raspberries** *(England only)* [1-4,6,8-10]; **Redcurrants** [1-4,6,8-10]; **Rhubarb** [1-4,6,8-10]; **Rubus hybrids** *(England only)* [3]; **Salad brassicas** *(off-label)* [4]; **Scarole** *(off-label)* [4]; **Squashes** *(off-label)* [4]; **Strawberries** [1-4,6,8-10]; **Sugar beet seed crops** [1-4,6,8-10]; **Table grapes** *(off-label)* [4]; **Turnip seed crops** [1-4,6,8-10]; **Walnuts** *(off-label)* [4]; **Wine grapes** *(off-label)* [4]; **Winter field beans** [2-4,6,8-10]; **Winter oilseed rape** [1-4,6,8-10];
- Volunteer cereals in **Winter field beans** [5]; **Winter oilseed rape** [5];
- Wild oats in **Winter field beans** [5]; **Winter oilseed rape** [5];

Extension of Authorisation for Minor Use (EAMUs)

- **All edible seed crops grown outdoors** *20082942* [4]
- **All non-edible seed crops grown outdoors** *20082942* [4]
- **Almonds** *20082420* [4]
- **Bilberries** *20082419* [4]
- **Blueberries** *20082419* [4]
- **Bog myrtle** *20120509* [4]
- **Borage** *20194355* [4]
- **Broccoli** *20091902* [4]
- **Calabrese** *20091902* [4]
- **Cauliflowers** *20091902* [4]
- **Cherries** *20082418* [4]
- **Chestnuts** *20082420* [4]
- **Chicory root** *20091530* [4]
- **Cob nuts** *20082420* [4]
- **Corn Gromwell** *20122395* [4], *20194355* [4]
- **Courgettes** *20082416* [4]
- **Cranberries** *20082419* [4]
- **Cress** *20082410* [4]
- **Endives** *20082410* [4]
- **Frise** *20082411* [4]
- **Game cover** *20082942* [4]

- *Hazel nuts* 20082420 [4]
- *Herbs (see appendix 6)* 20082412 [4]
- *Hops* 20082414 [4]
- *Lamb's lettuce* 20082410 [4]
- *Leaf brassicas* 20082410 [4]
- *Marrows* 20082416 [4]
- *Mirabelles* 20082418 [4]
- *Ornamental plant production* 20130207 [4]
- *Protected endives* 20082415 [4]
- *Protected forest nurseries* 20082942 [4]
- *Protected herbs (see appendix 6)* 20082412 [4]
- *Protected lettuce* 20082415 [4]
- *Pumpkins* 20082416 [4]
- *Quinces* 20082413 [4]
- *Radicchio* 20082411 [4]
- *Salad brassicas* 20082410 [4]
- *Scarole* 20082411 [4]
- *Squashes* 20082416 [4]
- *Table grapes* 20082417 [4]
- *Walnuts* 20082420 [4]
- *Wine grapes* 20082417 [4]

Approval information
- Propyzamide included in Annex I under EC Regulation 1107/2009
- Some products may be applied through CDA equipment. See labels for details
- Accepted by BBPA for use on hops

Efficacy guidance
- Active via root uptake. Weeds controlled from germination to young seedling stage, some species (including many grasses) also when established
- Best results achieved by winter application to fine, firm, moist soil. Rain is required after application if soil dry
- Uptake is slow and and may take up to 12 wk
- Excessive organic debris or ploughed-up turf may reduce efficacy
- For heavy couch infestations a repeat application may be needed in following winter
- Always follow WRAG guidelines for preventing and managing herbicide resistant weeds. See Section 5 for more information

Restrictions
- Maximum number of treatments 1 per crop or yr
- Maximum total dose equivalent to one full dose treatment for all crops
- Do not treat protected crops
- Apply to listed edible crops only between 1 Oct and the date specified as the latest time of application (except lettuce)
- Do not apply in windy weather and avoid drift onto non-target crops
- Do not use on soils with more than 10% organic matter except in forestry

Crop-specific information
- Latest use: labels vary but normally before 31 Dec in year before harvest for rhubarb, lucerne, strawberries and winter field beans; before 31 Jan for other crops
- HI: 6 wk for edible crops
- Apply as soon as possible after 3-true leaf stage of oilseed rape (GS 1,3) and seed brassicas, after 4-leaf stage of sugar beet for seed, within 7 d after sowing but before emergence for field beans, after perennial crops established for at least 1 season, strawberries after 1 yr
- Only apply to strawberries on heavy soils. Do not use on matted row crops
- Only apply to field beans on medium and heavy soils
- Only apply to established lucerne not less than 7 d after last cut
- In lettuce lightly incorporate in top 25 mm pre-drilling or irrigate on dry soil
- See label for lists of ornamental and forest species which may be treated

FOR FULL CONDITIONS OF USE ALWAYS READ THE PRODUCT LABEL

Following crops guidance

- Following an application between 1 Apr and 31 Jul at any dose the following minimum intervals must be observed before sowing the next crop: lettuce 0 wk; broad beans, chicory, clover, field beans, lucerne, radishes, peas 5 wk; brassicas, celery, leeks, oilseed rape, onions, parsley, parsnips 10 wk
- Following an application between 1 Aug and 31 Mar at any dose the following minimum intervals must be observed before sowing the next crop: lettuce 0 wk; broad beans, chicory, clover, field beans, lucerne, radishes, peas 10 wk; brassicas, celery, leeks, oilseed rape, onions, parsley, parsnips 25 wk or after 15 Jun, whichever occurs sooner
- Cereals or grasses or other crops not listed may be sown 30 wk after treatment up to 840 g.a.i/ha between 1 Aug and 31 Mar or 40 wk after treatment at higher doses at any time and after mouldboard ploughing to at least 15 cm
- A period of at least 9 mth must elapse between applications of propyzamide to the same land

Environmental safety

- Dangerous for the environment
- Very toxic to aquatic organisms
- Some pesticides pose a greater threat of contamination of water than others and propyzamide is one of these pesticides. Take special care when applying propyzamide near water and do not apply if heavy rain is forecast

Hazard classification and safety precautions

Hazard Harmful, Dangerous for the environment, Very toxic to aquatic organisms [6]
Transport code 9 [1-10]
Packaging group III
UN Number 3077, 3082
Risk phrases H351 [1-7,9], R40 [8,10], R50 [8,10], R53a [8,10]
Operator protection U05a [7], U20c [1-6,8-10]; A, D, H, M
Environmental protection E15a [1-5,7,10], E15b [6,8-9], E38, E39 [6,8-9], H410 [2,4,6,9], H411 [1,3,5-7,9]
Consumer protection C02a (6 wk) [1-4,10]
Storage and disposal D01 [7], D02 [7], D05 [1-5,10], D09a [1-5,7,10], D11a [1-5,10], D12a [1-5,7,10]
Medical advice M03 [7]

310 proquinazid

A quinazolinone fungicide for powdery mildew control in cereals
FRAC mode of action code: U7

Products

1	Justice	Corteva	200 g/l	EC	12835
2	Talius	Corteva	200 g/l	EC	12752

Uses

- Powdery mildew in **Apples** *(off-label)* [1-2]; **Blackcurrants** *(off-label)* [2]; **Courgettes** *(off-label)* [2]; **Cucumbers** *(off-label)* [2]; **Forest nurseries** *(off-label)* [2]; **Gooseberries** *(off-label)* [2]; **Grapevines** *(off-label)* [1-2]; **Marrows** *(off-label)* [2]; **Pears** *(off-label)* [1-2]; **Protected aubergines** *(off-label)* [2]; **Protected courgettes** *(off-label)* [2]; **Protected cucumbers** *(off-label)* [2]; **Protected marrows** *(off-label)* [2]; **Protected pumpkins** *(off-label)* [2]; **Protected squashes** *(off-label)* [2]; **Protected strawberries** *(off-label)* [1-2]; **Protected tomatoes** *(off-label)* [2]; **Pumpkins** *(off-label)* [2]; **Redcurrants** *(off-label)* [2]; **Soft fruit** *(off-label)* [2]; **Spring barley** [1-2]; **Spring oats** [1-2]; **Spring rye** [1-2]; **Spring wheat** [1-2]; **Squashes** *(off-label)* [2]; **Top fruit** *(off-label)* [2]; **Triticale** [1-2]; **Winter barley** [1-2]; **Winter oats** [1-2]; **Winter rye** [1-2]; **Winter wheat** [1-2];

Extension of Authorisation for Minor Use (EAMUs)

- **Apples** *20170726* [1], *20170725* [2]
- **Blackcurrants** *20200741* [2]
- **Courgettes** *20152627* [2]

SEE SECTION 3 FOR PRODUCTS ALSO REGISTERED

- **Cucumbers** *20152627* [2]
- **Forest nurseries** *20090420* [2]
- **Gooseberries** *20200741* [2]
- **Grapevines** *20111763* [1], *20222806* [2]
- **Marrows** *20152627* [2]
- **Pears** *20170726* [1], *20170725* [2]
- **Protected aubergines** *20152627* [2]
- **Protected courgettes** *20152627* [2]
- **Protected cucumbers** *20152627* [2]
- **Protected marrows** *20152627* [2]
- **Protected pumpkins** *20152627* [2]
- **Protected squashes** *20152627* [2]
- **Protected strawberries** *20172436* [1], *20170210* [2]
- **Protected tomatoes** *20152627* [2]
- **Pumpkins** *20152627* [2]
- **Redcurrants** *20200741* [2]
- **Soft fruit** *20090420* [2]
- **Squashes** *20152627* [2]
- **Top fruit** *20090420* [2]

Approval information
- Proquinazid included in Annex 1 under EC Regulation 1107/2009
- Accepted by BBPA on malting barley

Efficacy guidance
- Best results obtained from preventive treatment before disease is established in the crop
- Where mildew has already spread to new growth a tank mix with a curative fungicide with an alternative mode of action should be used
- Use as part of an integrated crop management (ICM) strategy incorporating other methods of control or fungicides with different modes of action

Restrictions
- Maximum number of treatments 2 per crop
- Do not apply to any crop suffering from stress from any cause
- Avoid application in either frosty or hot, sunny conditions
- Must not be applied by hand-held equipment

Crop-specific information
- Latest use: before beginning of heading (GS 49) for barley, oats, rye, triticale; before full flowering (GS 65) for wheat

Environmental safety
- Dangerous for the environment
- Toxic to aquatic organisms
- Dangerous to fish or other aquatic life. Do not contaminate surface waters or ditches with chemical or used container
- Aquatic buffer zone 15 m in redcurrants, blackcurrants and Gooseberries [2]
- LERAP Category B [1-2]

Hazard classification and safety precautions
Hazard Harmful, Dangerous for the environment
Transport code 9 [1-2]
Packaging group III
UN Number 3082
Risk phrases H315, H318, H351
Operator protection U05a, U11; A, C, H
Environmental protection E13b, E16a, E16b, E34, E38, H410
Storage and disposal D01, D02, D05, D09a, D10b, D11a, D12a
Medical advice M03, M05a

FOR FULL CONDITIONS OF USE ALWAYS READ THE PRODUCT LABEL

311   prosulfocarb

A thiocarbamate herbicide for grass and broad-leaved weed control in cereals and potatoes
HRAC mode of action code: 15 (N)

See also clodinafop-propargyl + prosulfocarb

Products

1	Clayton Comply	Clayton	800 g/l	EC	17702
2	Clayton Obey	Clayton	800 g/l	EC	16626
3	Cleancrop Carburettor	Agrii	800 g/l	EC	18772
4	Crozier	Barclay	800 g/l	EC	19052
5	Defy	Syngenta	800 g/l	EC	16202
6	Fantasia	Albaugh UK	800 g/l	EC	19689
7	Fulbourn	Agrii	800 g/l	EC	19735
8	Jade	Syngenta	800 g/l	EC	16203
9	Lees	Rotam	800 g/l	EC	18769
10	Quidam	UPL Europe	800 g/l	EC	17595
11	Spinnaker	Adama	800 g/l	EC	18855
12	Topsail	Adama	800 g/l	EC	19024
13	Wicket	Syngenta	800 g/l	EC	17555

Uses

- Annual dicotyledons in **Almonds** *(off-label)* [1,11]; **Apples** *(off-label)* [1,11]; **Apricots** *(off-label)* [1,11]; **Bulb onion sets** *(off-label)* [11,13]; **Bulb onions** *(off-label)* [1]; **Carrots** *(off-label)* [1,5,11]; **Celeriac** *(off-label)* [1,5]; **Celery (outdoor)** *(off-label)* [1,5,11]; **Cherries** *(off-label)* [1,11]; **Chestnuts** *(off-label)* [1,11]; **Forest nurseries** *(off-label)* [1,5,11]; **Game cover** *(off-label)* [1,5,11]; **Garlic** *(off-label)* [1,5,11,13]; **Hazel nuts** *(off-label)* [1,11]; **Herbs (see appendix 6)** *(off-label)* [1]; **Horseradish** *(off-label)* [1,5,11]; **Leeks** *(off-label)* [1,5,11,13]; **Medlar** *(off-label)* [1,11]; **Miscanthus** *(off-label)* [1,5,11]; **Nectarines** *(off-label)* [1,11]; **Onion sets** *(off-label)* [5]; **Ornamental plant production** *(off-label)* [1,5,11]; **Parsley root** *(off-label)* [1,5,11]; **Parsnips** *(off-label)* [1,5,11]; **Peaches** *(off-label)* [1,11]; **Pears** *(off-label)* [1,11]; **Plums** *(off-label)* [1,11]; **Poppies for morphine production** *(off-label)* [1,5,11]; **Potatoes** [1-13]; **Quinces** *(off-label)* [1,11]; **Rye** *(off-label)* [1,5,11]; **Salsify** *(off-label)* [1,5,11]; **Shallots** *(off-label)* [1,5,11,13]; **Spring barley** *(off-label)* [5,8,11]; **Spring field beans** *(off-label)* [1,5,8,11]; **Spring onions** *(off-label)* [1,5,11,13]; **Top fruit** *(off-label)* [5]; **Triticale** *(off-label)* [1,5]; **Walnuts** *(off-label)* [1,11]; **Winter barley** [1-13]; **Winter field beans** *(off-label)* [1,5,8,11]; **Winter linseed** *(off-label)* [1,5,11]; **Winter wheat** [1-13];

- Annual grasses in **Almonds** *(off-label)* [1,11]; **Apples** *(off-label)* [1,11]; **Apricots** *(off-label)* [1,11]; **Bulb onion sets** *(off-label)* [11,13]; **Bulb onions** *(off-label)* [1]; **Carrots** *(off-label)* [1,5,11]; **Celeriac** *(off-label)* [1,5]; **Celery (outdoor)** *(off-label)* [1,5,11]; **Cherries** *(off-label)* [1,11]; **Chestnuts** *(off-label)* [1,11]; **Forest nurseries** *(off-label)* [1,5,11]; **Game cover** *(off-label)* [1,5,11]; **Garlic** *(off-label)* [1,5,11,13]; **Hazel nuts** *(off-label)* [1,11]; **Herbs (see appendix 6)** *(off-label)* [1]; **Horseradish** *(off-label)* [1,5,11]; **Leeks** *(off-label)* [1,5,11,13]; **Medlar** *(off-label)* [1,11]; **Miscanthus** *(off-label)* [1,5,11]; **Nectarines** *(off-label)* [1,11]; **Onion sets** *(off-label)* [5]; **Ornamental plant production** *(off-label)* [1,5,11]; **Parsley root** *(off-label)* [1,5,11]; **Parsnips** *(off-label)* [1,5,11]; **Peaches** *(off-label)* [1,11]; **Pears** *(off-label)* [1,11]; **Plums** *(off-label)* [1,11]; **Poppies for morphine production** *(off-label)* [1,11]; **Quinces** *(off-label)* [1,11]; **Rye** *(off-label)* [1,5,11]; **Salsify** *(off-label)* [1,5,11]; **Shallots** *(off-label)* [1,5,11,13]; **Spring barley** *(off-label)* [5,8,11]; **Spring field beans** *(off-label)* [5,8]; **Spring onions** *(off-label)* [1,5,11,13]; **Top fruit** *(off-label)* [5]; **Triticale** *(off-label)* [1,5]; **Walnuts** *(off-label)* [1,11]; **Winter field beans** *(off-label)* [5,8]; **Winter linseed** *(off-label)* [1,5,11];

- Annual meadow grass in **Bulb onions** *(off-label)* [10]; **Carrots** *(off-label)* [5]; **Celeriac** *(off-label)* [5]; **Celery (outdoor)** *(off-label)* [5]; **Forest nurseries** *(off-label)* [5]; **Game cover** *(off-label)* [5]; **Garlic** *(off-label)* [10]; **Herbs (see appendix 6)** *(off-label)* [10]; **Horseradish** *(off-label)* [5]; **Leeks** *(off-label)* [10]; **Miscanthus** *(off-label)* [5]; **Ornamental plant production** *(off-label)* [5]; **Parsley root** *(off-label)* [5]; **Parsnips** *(off-label)* [5]; **Poppies** *(off-label)* [5,11]; **Poppies for morphine production** *(off-label)* [5]; **Poppies grown for seed production** *(off-label)* [5,11]; **Potatoes** [1-13]; **Rye** *(off-label)* [5]; **Salsify** *(off-label)* [5]; **Shallots** *(off-label)* [10]; **Spring barley** *(off-label)* [11]; **Spring field beans** *(off-label)* [11]; **Spring onions** *(off-label)* [10]; **Top fruit** *(off-label)* [5];

Triticale *(off-label)* [5]; **Winter barley** [1-13]; **Winter field beans** *(off-label)* [11]; **Winter linseed** *(off-label)* [5,10]; **Winter wheat** [1-13];

- Black nightshade in **Broad beans** *(off-label)* [5,11]; **Bulb onions** *(off-label)* [10]; **Edible flowers** *(off-label)* [11]; **Edible podded peas** *(off-label)* [5,11]; **Garlic** *(off-label)* [10]; **Herbs (see appendix 6)** *(off-label)* [1,5,10-11]; **Leeks** *(off-label)* [10]; **Poppies** *(off-label)* [5,11]; **Poppies grown for seed production** *(off-label)* [5,11]; **Shallots** *(off-label)* [10]; **Spring onions** *(off-label)* [10]; **Vining peas** *(off-label)* [5,11]; **Winter linseed** *(off-label)* [10];
- Blackgrass in **Rye** *(off-label)* [11];
- Chickweed in **Broad beans** *(off-label)* [5,11]; **Bulb onions** *(off-label)* [10]; **Celeriac** *(off-label)* [1,5,11]; **Edible podded peas** *(off-label)* [5,11]; **Garlic** *(off-label)* [10]; **Herbs (see appendix 6)** *(off-label)* [10]; **Leeks** *(off-label)* [10]; **Poppies** *(off-label)* [5,11]; **Poppies grown for seed production** *(off-label)* [5,11]; **Potatoes** [1-13]; **Shallots** *(off-label)* [10]; **Spring onions** *(off-label)* [10]; **Vining peas** *(off-label)* [5,11]; **Winter barley** [1-13]; **Winter linseed** *(off-label)* [10]; **Winter wheat** [1-13];
- Cleavers in **Bulb onions** *(off-label)* [10]; **Carrots** *(off-label)* [1,5,11]; **Edible flowers** *(off-label)* [11]; **Garlic** *(off-label)* [10]; **Herbs (see appendix 6)** *(off-label)* [1,5,10-11]; **Horseradish** *(off-label)* [1,5,11]; **Leeks** *(off-label)* [10]; **Parsley root** *(off-label)* [1,5,11]; **Parsnips** *(off-label)* [1,5,11]; **Poppies** *(off-label)* [5,11]; **Poppies grown for seed production** *(off-label)* [5,11]; **Potatoes** [1-13]; **Salsify** *(off-label)* [1,5,11]; **Shallots** *(off-label)* [10]; **Spring field beans** *(off-label)* [1,11]; **Spring onions** *(off-label)* [10]; **Winter barley** [1-13]; **Winter field beans** *(off-label)* [1,11]; **Winter linseed** *(off-label)* [10]; **Winter wheat** [1-13];
- Crane's-bill in **Broad beans** *(off-label)* [5,11]; **Bulb onions** *(off-label)* [10]; **Edible podded peas** *(off-label)* [5,11]; **Garlic** *(off-label)* [10]; **Herbs (see appendix 6)** *(off-label)* [10]; **Leeks** *(off-label)* [10]; **Poppies** *(off-label)* [5,11]; **Poppies grown for seed production** *(off-label)* [5,11]; **Shallots** *(off-label)* [10]; **Spring onions** *(off-label)* [10]; **Vining peas** *(off-label)* [5,11]; **Winter linseed** *(off-label)* [10];
- Fat hen in **Carrots** *(off-label)* [1,5,11]; **Celeriac** *(off-label)* [1,5,11]; **Edible flowers** *(off-label)* [11]; **Herbs (see appendix 6)** *(off-label)* [1,5,11]; **Horseradish** *(off-label)* [1,5,11]; **Parsley root** *(off-label)* [1,5,11]; **Parsnips** *(off-label)* [1,5,11]; **Salsify** *(off-label)* [1,5,11];
- Field speedwell in **Potatoes** [1-13]; **Winter barley** [1-13]; **Winter wheat** [1-13];
- Forget-me-not in **Broad beans** *(off-label)* [5,11]; **Bulb onions** *(off-label)* [10]; **Edible podded peas** *(off-label)* [5,11]; **Garlic** *(off-label)* [10]; **Herbs (see appendix 6)** *(off-label)* [10]; **Leeks** *(off-label)* [10]; **Poppies** *(off-label)* [5,11]; **Poppies grown for seed production** *(off-label)* [5,11]; **Shallots** *(off-label)* [10]; **Spring onions** *(off-label)* [10]; **Vining peas** *(off-label)* [5,11]; **Winter linseed** *(off-label)* [10];
- Fumitory in **Bulb onion sets** *(off-label)* [11,13]; **Edible flowers** *(off-label)* [11]; **Garlic** *(off-label)* [11,13]; **Herbs (see appendix 6)** *(off-label)* [1,5,11]; **Leeks** *(off-label)* [11,13]; **Shallots** *(off-label)* [11,13]; **Spring onions** *(off-label)* [11,13];
- Grass weeds in **Celeriac** *(off-label)* [11];
- Ivy-leaved speedwell in **Poppies** *(off-label)* [5,11]; **Poppies grown for seed production** *(off-label)* [5,11]; **Potatoes** [1-13]; **Winter barley** [1-13]; **Winter wheat** [1-13];
- Knotgrass in **Carrots** *(off-label)* [1,5,11]; **Edible flowers** *(off-label)* [11]; **Herbs (see appendix 6)** *(off-label)* [1,5,11]; **Horseradish** *(off-label)* [1,5,11]; **Parsley root** *(off-label)* [1,5,11]; **Parsnips** *(off-label)* [1,5,11]; **Salsify** *(off-label)* [1,5,11];
- Loose silky bent in **Bulb onions** *(off-label)* [10]; **Garlic** *(off-label)* [10]; **Herbs (see appendix 6)** *(off-label)* [10]; **Leeks** *(off-label)* [10]; **Poppies** *(off-label)* [5,11]; **Poppies grown for seed production** *(off-label)* [5,11]; **Potatoes** [1-13]; **Shallots** *(off-label)* [10]; **Spring onions** *(off-label)* [10]; **Winter barley** [1-13]; **Winter linseed** *(off-label)* [10]; **Winter wheat** [1-13];
- Mayweeds in **Carrots** *(off-label)* [1,5,11]; **Celeriac** *(off-label)* [1,5,11]; **Edible flowers** *(off-label)* [11]; **Herbs (see appendix 6)** *(off-label)* [1,5,11]; **Horseradish** *(off-label)* [1,5,11]; **Parsley root** *(off-label)* [1,5,11]; **Parsnips** *(off-label)* [1,5,11]; **Salsify** *(off-label)* [1,5,11];
- Nettles in **Celeriac** *(off-label)* [11];
- Polygonums in **Bulb onion sets** *(off-label)* [11,13]; **Celeriac** *(off-label)* [1,5,11]; **Garlic** *(off-label)* [11,13]; **Leeks** *(off-label)* [11,13]; **Shallots** *(off-label)* [11,13]; **Spring onions** *(off-label)* [11,13];
- Red dead-nettle in **Broad beans** *(off-label)* [5,11]; **Edible podded peas** *(off-label)* [5,11]; **Poppies** *(off-label)* [5,11]; **Poppies grown for seed production** *(off-label)* [5,11]; **Vining peas** *(off-label)* [5,11];
- Rough-stalked meadow grass in **Bulb onions** *(off-label)* [10]; **Garlic** *(off-label)* [10]; **Herbs (see appendix 6)** *(off-label)* [10]; **Leeks** *(off-label)* [10]; **Poppies** *(off-label)* [5,11]; **Poppies grown for**

FOR FULL CONDITIONS OF USE ALWAYS READ THE PRODUCT LABEL

seed production (off-label) [5,11]; **Potatoes** [1-13]; **Shallots** (off-label) [10]; **Spring onions** (off-label) [10]; **Winter barley** [1-13]; **Winter linseed** (off-label) [10]; **Winter wheat** [1-13];
- Ryegrass in **Rye** (off-label) [11];
- Speedwells in **Broad beans** (off-label) [5,11]; **Bulb onions** (off-label) [10]; **Edible podded peas** (off-label) [5,11]; **Garlic** (off-label) [10]; **Herbs (see appendix 6)** (off-label) [10]; **Leeks** (off-label) [10]; **Poppies** (off-label) [5,11]; **Poppies grown for seed production** (off-label) [5,11]; **Shallots** (off-label) [10]; **Spring onions** (off-label) [10]; **Vining peas** (off-label) [5,11]; **Winter linseed** (off-label) [10];
- Thistles in **Celeriac** (off-label) [1,5,11];
- Volunteer oilseed rape in **Bulb onion sets** (off-label) [13]; **Garlic** (off-label) [13]; **Leeks** (off-label) [13]; **Shallots** (off-label) [13]; **Spring onions** (off-label) [13];

Extension of Authorisation for Minor Use (EAMUs)
- **Almonds** *20162930* [1], *20191742* [11]
- **Apples** *20162930* [1], *20191742* [11]
- **Apricots** *20162930* [1], *20191742* [11]
- **Broad beans** *20191041* [5], *20191749* [11]
- **Bulb onion sets** *20191740* [11], *20161818* [13]
- **Bulb onions** *20162925* [1], *20181690* [10]
- **Carrots** *20162922* [1], *20131354* [5], *20191747* [11]
- **Celeriac** *20162933* [1], *20131355* [5], *20191737* [11]
- **Celery (outdoor)** *20162924* [1], *20131353* [5], *20191751* [11]
- **Cherries** *20162930* [1], *20191742* [11]
- **Chestnuts** *20162930* [1], *20191742* [11]
- **Edible flowers** *20191750* [11]
- **Edible podded peas** *20191041* [5], *20191749* [11]
- **Forest nurseries** *20162926* [1], *20131432* [5], *20191741* [11]
- **Game cover** *20162931* [1], *20131430* [5], *20191739* [11]
- **Garlic** *20162925* [1], *20131357* [5], *20181690* [10], *20191740* [11], *20161818* [13]
- **Hazel nuts** *20162930* [1], *20191742* [11]
- **Herbs (see appendix 6)** *20162923* [1], *20152809* [5], *20181691* [10], *20191750* [11]
- **Horseradish** *20162922* [1], *20131354* [5], *20191747* [11]
- **Leeks** *20162925* [1], *20131357* [5], *20181690* [10], *20191740* [11], *20161818* [13]
- **Medlar** *20162930* [1], *20191742* [11]
- **Miscanthus** *20162931* [1], *20131430* [5], *20191739* [11]
- **Nectarines** *20162930* [1], *20191742* [11]
- **Onion sets** *20131357* [5]
- **Ornamental plant production** *20162927* [1], *20131431* [5], *20191746* [11]
- **Parsley root** *20162922* [1], *20131354* [5], *20191747* [11]
- **Parsnips** *20162922* [1], *20131354* [5], *20191747* [11]
- **Peaches** *20162930* [1], *20191742* [11]
- **Pears** *20162930* [1], *20191742* [11]
- **Plums** *20162930* [1], *20191742* [11]
- **Poppies** *20180615* [5], *20191743* [11]
- **Poppies for morphine production** *20162934* [1], *20131385* [5], *20191744* [11]
- **Poppies grown for seed production** *20180615* [5], *20191743* [11]
- **Quinces** *20162930* [1], *20191742* [11]
- **Rye** *20162929* [1], *20131429* [5], *20191754* [11]
- **Salsify** *20162922* [1], *20131354* [5], *20191747* [11]
- **Shallots** *20162925* [1], *20131357* [5], *20181690* [10], *20191740* [11], *20161818* [13]
- **Spring barley** *20212794* [5], *20212796* [8], *20212797* [11]
- **Spring field beans** *20212926* [1], *20212937* [5], *20212940* [8], *20212941* [11]
- **Spring onions** *20162925* [1], *20131357* [5], *20181690* [10], *20191740* [11], *20161818* [13]
- **Top fruit** *20131433* [5]
- **Triticale** *20162929* [1], *20131429* [5]
- **Vining peas** *20191041* [5], *20191749* [11]
- **Walnuts** *20162930* [1], *20191742* [11]
- **Winter field beans** *20212926* [1], *20212937* [5], *20212940* [8], *20212941* [11]
- **Winter linseed** *20162928* [1], *20131428* [5], *20181692* [10], *20191752* [11]

SEE SECTION 3 FOR PRODUCTS ALSO REGISTERED

SECTION 2

Approval information
- Prosulfocarb included in Annex I under EC Regulation 1107/2009
- Accepted by BBPA for use on malting barley

Efficacy guidance
- Best results obtained in cereals from treatment of crops in a firm moist seedbed, free from clods
- Pre-emergence use will reduce blackgrass populations but the product should only be used against this weed as part of a management strategy involving sequences with products of alternative modes of action
- Always follow WRAG guidelines for preventing and managing herbicide resistant weeds. See Section 5 for more information

Restrictions
- Maximum number of treatments 1 per crop for winter barley, winter wheat and potatoes
- Do not apply to crops under stress from any cause. Transient yellowing can occur from which recovery is complete
- Winter cereals must be covered by 3 cm of settled soil
- Do not apply by hand held equipment

Crop-specific information
- Latest use: at emergence (soil rising over emerging potato shoots) for potatoes, up to and including early tillering (GS 21) for winter barley, winter wheat
- When applied pre-emergence to cereals crop emergence may occasionally be slowed down but yield is not affected
- For potatoes, complete ridge formation before application and do not disturb treated soil afterwards
- Peas are severely damaged or killed.

Following crops guidance
- Do not sow field or broad beans within 12 mth of treatment
- In the event of failure of a treated cereal crop, winter wheat, winter barley or transplanted brassicas may be re-sown immediately. In the following spring sunflowers, maize, flax, spring cereals, peas, oilseed rape or soya beans may be sown without ploughing, and carrots, lettuce, onions, sugar beet or potatoes may be sown or planted after ploughing

Environmental safety
- Dangerous for the environment
- Very toxic to aquatic organisms
- LERAP Category B [1-13]

Hazard classification and safety precautions
Hazard Irritant, Dangerous for the environment
Transport code 9 [1-13]
Packaging group III
UN Number 3082
Risk phrases H304 [1-8,10-13], H315 [1-8,10-13], H317 [1-8,10-13], H319 [1-8,10-13], R38 [9], R43 [9], R50 [9], R53a [9]
Operator protection U02a, U05a, U08, U14, U15, U20b, U23a [6-7]; A, C, H
Environmental protection E15b, E16a, E16b [1-5,8-13], E38, H410 [1-8,10-13]
Storage and disposal D01, D02, D05, D09a, D10c, D12a
Medical advice M05a

312   prosulfuron

A contact and residual sulfonyl urea herbicide for use in maize crops
HRAC mode of action code: 2 (B)

See also dicamba + prosulfuron

Products
1	Clayton Kibo	Clayton	75% w/w	WG	15822

Products – continued

2 Peak	Syngenta	75% w/w	WG	15521

Uses
- Annual dicotyledons in **Forage maize** [1-2]; **Game cover** *(off-label)* [2]; **Grain maize** [1-2]; **Millet** *(off-label)* [2];
- Black bindweed in **Forage maize** [1-2]; **Grain maize** [1-2];
- Chickweed in **Forage maize** [1-2]; **Grain maize** [1-2];
- Fumitory in **Forage maize** [1-2]; **Grain maize** [1-2];
- Groundsel in **Forage maize** [1-2]; **Grain maize** [1-2];
- Knotgrass in **Forage maize** [1-2]; **Grain maize** [1-2];
- Mayweeds in **Forage maize** [1-2]; **Grain maize** [1-2];
- Redshank in **Forage maize** [1-2]; **Grain maize** [1-2];
- Scarlet pimpernel in **Forage maize** [1-2]; **Grain maize** [1-2];
- Shepherd's purse in **Forage maize** [1-2]; **Grain maize** [1-2];
- Sowthistle in **Forage maize** [1-2]; **Grain maize** [1-2];

Extension of Authorisation for Minor Use (EAMUs)
- **Game cover** *20120906* [2]
- **Millet** *20172107* [2]

Approval information
- Prosulfuron included in Annex I under EC Regulation 1107/2009

Efficacy guidance
- For optimum efficacy, apply when weeds are at the 2 - 4 leaf stage

Restrictions
- Do not apply to forage maize or grain maize grown for seed production
- Do not apply with organo-phosphate insecticides
- Do not apply by hand held equipment or in water volumes less than recommended
- A maximum total dose of 15g of prosulfuron per hectare may only be applied every third year on the same field

Following crops guidance
- In the event of crop failure, wait 4 weeks after treatment and then re-sow
- After normal harvest wheat, barley and winter beans may be sown as a following crop in the autumn once the soil has been ploughed to 15 cms. In spring, wheat, barley, peas or beans may be sown but do not sow any other crop at this time.

Environmental safety
- LERAP Category B [1-2]

Hazard classification and safety precautions
> **Hazard** Harmful, Dangerous for the environment, Harmful if swallowed, Very toxic to aquatic organisms
> **Transport code** 9 [1-2]
> **Packaging group** III
> **UN Number** 3077
> **Operator protection** U05a, U14, U20c; A, H
> **Environmental protection** E15b, E16a, E38, H410
> **Storage and disposal** D01, D02, D05, D09b, D10c, D12a

SEE SECTION 3 FOR PRODUCTS ALSO REGISTERED

313 prothioconazole

A systemic, protectant and curative triazole fungicide
FRAC mode of action code: 3

See also benzovindiflupyr + prothioconazole
bixafen + fluopyram + prothioconazole
bixafen + fluoxastrobin + prothioconazole
bixafen + prothioconazole
bixafen + prothioconazole + spiroxamine
bixafen + prothioconazole + tebuconazole
fenpicoxamid + prothioconazole
fluopyram + prothioconazole
fluopyram + prothioconazole + tebuconazole
fluoxastrobin + prothioconazole
fluoxastrobin + prothioconazole + trifloxystrobin

Products

1	Aurelia	Life Scientific	250 g/l	EC	19350
2	Clayton Tuskar	Clayton	250 g/l	EC	19320
3	Croton Flex	Bayer CropScience	250 g/l	EC	19823
4	Decoy 250EC	BASF	250 g/l	EC	18639
5	Ecana	Bayer CropScience	250 g/l	EC	19818
6	Era 250	Syngenta	250 g/l	EC	19734
7	Euskatel	Rotam	250 g/l	EC	19658
8	Glacis	Bayer CropScience	250 g/l	EC	19817
9	Olbran	Bayer CropScience	250 g/l	EC	19812
10	Profound	Bayer CropScience	250 g/l	EC	19816
11	Proline 275	Bayer CropScience	275 g/l	EC	14790
12	Protendo 300 EC	Certis Belchim B V	300 g/l	EC	19612
13	Prothiolux 250 EC	Agform	250 g/l	EC	19642
14	Rudis	Bayer CropScience	480 g/l	SC	14122
15	Soratel	Adama	250 g/l	EC	19636
16	Tokyo	UPL Europe	250 g/l	EC	19591
17	Trinity 250 EC	Barclay	250 g/l	EC	19732

Uses

- Alternaria in **Broccoli** [14]; **Brussels sprouts** [14]; **Cabbages** [14]; **Calabrese** [14]; **Cauliflowers** [14]; **Ornamental plant production** *(off-label)* [11]; **Red beet** *(off-label)* [14];
- Alternaria blight in **Carrots** [14]; **Parsnips** [14]; **Swedes** [14]; **Turnips** [14];
- Brown rust in **Durum wheat** [1-2,4-5,7,13,16-17]; **Spring barley** [1-13,15-17]; **Spring rye** [12]; **Spring triticale** [12]; **Spring wheat** [1-13,15-17]; **Winter barley** [1-13,15-17]; **Winter rye** [1-13,15-17]; **Winter triticale** [12]; **Winter wheat** [1-13,15-17];
- Cercospora leaf spot in **Red beet** *(off-label)* [14];
- Crown rust in **Spring oats** [1-11,13,15-17]; **Winter oats** [1-11,13,15-17];
- Disease control in **Linseed** *(off-label)* [11]; **Mustard** *(off-label)* [11];
- Eyespot in **Durum wheat** [1-2,4-5,7,13,16-17]; **Spring barley** [1-13,15-17]; **Spring oats** [1,3-4,6,8-11,13,15-17]; **Spring rye** [12]; **Spring triticale** [12]; **Spring wheat** [1-13,15-17]; **Winter barley** [1-13,15-17]; **Winter oats** [1,3-4,6,8-11,13,15-17]; **Winter rye** [1-13,15-17]; **Winter triticale** [12]; **Winter wheat** [1-13,15-17];
- Glume blotch in **Durum wheat** [1-2,4-5,7,13,16-17]; **Spring triticale** [12]; **Spring wheat** [1-13,15-17]; **Winter triticale** [12]; **Winter wheat** [1-13,15-17];
- Kabatiella lini in **Linseed** *(off-label)* [11];
- Late ear diseases in **Durum wheat** [1-2,4-5,7,13,16-17]; **Spring barley** [1-13,15-17]; **Spring triticale** [12]; **Spring wheat** [1-13,15-17]; **Winter barley** [1-13,15-17]; **Winter triticale** [12]; **Winter wheat** [1-13,15-17];
- Leaf blotch in **Leeks** *(useful reduction)* [14];
- Leaf spot in **Ornamental plant production** *(off-label)* [11];
- Light leaf spot in **Broccoli** [14]; **Brussels sprouts** [14]; **Cabbages** [14]; **Calabrese** [14]; **Cauliflowers** [14]; **Winter oilseed rape** [11];

- Net blotch in **Spring barley** [1-13,15-17]; **Winter barley** [1-13,15-17];
- Phoma in **Borage** *(off-label)* [11]; **Broccoli** [14]; **Brussels sprouts** [14]; **Cabbages** [14]; **Calabrese** [14]; **Cauliflowers** [14]; **Ornamental plant production** *(off-label)* [11]; **Red beet** *(off-label)* [14]; **Spring corn gromwell** *(off-label)* [11]; **Winter corn gromwell** *(off-label)* [11]; **Winter oilseed rape** [11];
- Powdery mildew in **Borage** *(off-label)* [11]; **Broccoli** [14]; **Brussels sprouts** [14]; **Cabbages** [14]; **Calabrese** [14]; **Carrots** [14]; **Cauliflowers** [14]; **Durum wheat** [1-2,4-5,7,13,16-17]; **Ornamental plant production** *(off-label)* [11]; **Parsnips** [14]; **Red beet** *(off-label)* [14]; **Spring barley** [1-13,15-17]; **Spring corn gromwell** *(off-label)* [11]; **Spring oats** [1-11,13,15-17]; **Spring rye** [12]; **Spring triticale** [12]; **Spring wheat** [1-13,15-17]; **Swedes** [14]; **Turnips** [14]; **Winter barley** [1-13,15-17]; **Winter corn gromwell** *(off-label)* [11]; **Winter oats** [1-11,13,15-17]; **Winter rye** [1-13,15-17]; **Winter triticale** [12]; **Winter wheat** [1-13,15-17];
- Purple blotch in **Leeks** [14];
- Rhynchosporium in **Spring barley** [1-13,15-17]; **Spring rye** [12]; **Winter barley** [1-13,15-17]; **Winter rye** [1-13,15-17]; **Winter triticale** [12];
- Ring spot in **Broccoli** [14]; **Brussels sprouts** [14]; **Cabbages** [14]; **Calabrese** [14]; **Cauliflowers** [14];
- Rust in **Leeks** [14];
- Sclerotinia in **Borage** *(off-label)* [11]; **Ornamental plant production** *(off-label)* [11]; **Spring corn gromwell** *(off-label)* [11]; **Winter corn gromwell** *(off-label)* [11];
- Sclerotinia rot in **Carrots** [14]; **Parsnips** [14];
- Sclerotinia stem rot in **Spring oilseed rape** [12]; **Winter oilseed rape** [1-13,15-17];
- Septoria leaf blotch in **Durum wheat** [1-2,4-5,7,13,16-17]; **Spring triticale** [12]; **Spring wheat** [1-13,15-17]; **Winter triticale** [12]; **Winter wheat** [1-13,15-17];
- Septoria linicola in **Linseed** *(off-label)* [11];
- Stem canker in **Winter oilseed rape** [11];
- Stemphylium in **Leeks** *(useful reduction)* [14];
- Tan spot in **Durum wheat** [1-2,4-5,7,13,16-17]; **Spring triticale** [12]; **Spring wheat** [1-13,15-17]; **Winter triticale** [12]; **Winter wheat** [1-13,15-17];
- Yellow rust in **Durum wheat** [1-2,4-5,7,13,16-17]; **Spring triticale** [12]; **Spring wheat** [1-13,15-17]; **Winter barley** [1-13,15-17]; **Winter triticale** [12]; **Winter wheat** [1-13,15-17];

Extension of Authorisation for Minor Use (EAMUs)
- **Borage** *20193125* [11]
- **Linseed** *20162863* [11]
- **Mustard** *20111003* [11]
- **Ornamental plant production** *20220726* [11]
- **Red beet** *20193126* [14]
- **Spring corn gromwell** *20193125* [11]
- **Winter corn gromwell** *20193125* [11]

Approval information
- Accepted by BBPA for use on malting barley
- Prothioconazole included in Annex I under EC Regulation 1107/2009

Efficacy guidance
- Best results on cereal foliar diseases obtained from treatment at early stages of disease development. Further treatment may be needed if disease attack is prolonged
- Foliar applications to established infections of any disease are likely to be less effective
- Best control of cereal ear diseases obtained by treatment during ear emergence
- Treat oilseed rape at early to full flower
- Prothioconazole is a DMI fungicide. Resistance to some DMI fungicides has been identified in Septoria leaf blotch which may seriously affect performance of some products. For further advice contact a specialist advisor and visit the Fungicide Resistance Action Group (FRAG)-UK website

Restrictions
- Maximum total dose equivalent to two full dose treatments on barley, oats and oilseed rape, three full dose treatments on wheat and rye

SEE SECTION 3 FOR PRODUCTS ALSO REGISTERED

Crop-specific information
- Latest use: before grain milky ripe for foliar sprays on winter rye, winter wheat; beginning of flowering for barley, oats
- HI 56 d for winter oilseed rape

Environmental safety
- Dangerous for the environment
- Toxic to aquatic organisms
- LERAP Category B [1-17]

Hazard classification and safety precautions
  Hazard Irritant [1-13,15-17], Dangerous for the environment, Harmful if swallowed [6,12,15], Very toxic to aquatic organisms [1-2,4-7,11,13-17]
  Transport code 9 [1-17]
  Packaging group III
  UN Number 3082
  Risk phrases H314 [6,15], H315 [12], H317 [6,15], H318 [12], H319 [1-2,4-5,7,11,13,16-17], H335 [1-2,5,7,11,13,17], R36 [3,8-10], R51 [3,8-10], R53a [3,8-10]
  Operator protection U05a, U09b [1-13,15-17], U20b; A, C, H
  Environmental protection E15a [1-13,15-17], E15b [14], E16a, E34, E38, H410 [1-2,4-7,11,13-17], H411 [12]
  Storage and disposal D01, D02, D05 [1-13,15-17], D09a, D10b, D12a
  Medical advice M03

---

314   prothioconazole + spiroxamine

A broad spectrum fungicide mixture for cereals
FRAC mode of action code: 3 + 5

See also spiroxamine

Products

1	Helix	Bayer CropScience	160:300 g/l	EC	18289

Uses
- Brown rust in *Durum wheat* [1]; *Spring barley* [1]; *Spring rye* [1]; *Spring wheat* [1]; *Triticale* [1]; *Winter barley* [1]; *Winter rye* [1]; *Winter wheat* [1];
- Crown rust in *Spring oats* [1]; *Winter oats* [1];
- Eyespot in *Durum wheat* [1]; *Spring barley* [1]; *Spring oats* [1]; *Spring rye* [1]; *Spring wheat* [1]; *Triticale* [1]; *Winter barley* [1]; *Winter oats* [1]; *Winter rye* [1]; *Winter wheat* [1];
- Glume blotch in *Durum wheat* [1]; *Spring wheat* [1]; *Triticale* [1]; *Winter wheat* [1];
- Late ear diseases in *Durum wheat* [1]; *Spring rye* [1]; *Spring wheat* [1]; *Triticale* [1]; *Winter rye* [1]; *Winter wheat* [1];
- Net blotch in *Spring barley* [1]; *Winter barley* [1];
- Powdery mildew in *Durum wheat* [1]; *Spring barley* [1]; *Spring oats* [1]; *Spring rye* [1]; *Spring wheat* [1]; *Triticale* [1]; *Winter barley* [1]; *Winter oats* [1]; *Winter rye* [1]; *Winter wheat* [1];
- Rhynchosporium in *Spring barley* [1]; *Spring rye* [1]; *Winter barley* [1]; *Winter rye* [1];
- Septoria leaf blotch in *Durum wheat* [1]; *Spring wheat* [1]; *Triticale* [1]; *Winter wheat* [1];
- Tan spot in *Durum wheat* [1]; *Spring wheat* [1]; *Triticale* [1]; *Winter wheat* [1];
- Yellow rust in *Durum wheat* [1]; *Spring barley* [1]; *Spring wheat* [1]; *Triticale* [1]; *Winter barley* [1]; *Winter wheat* [1];

Approval information
- Prothioconazole and spiroxamine included in Annex I under EC Regulation 1107/2009
- Accepted by BBPA for use on malting barley

Efficacy guidance
- Best results obtained from treatment at early stages of disease development. Further treatment may be needed if disease attack is prolonged
- Applications to established infections of any disease are likely to be less effective

FOR FULL CONDITIONS OF USE ALWAYS READ THE PRODUCT LABEL

- Best control of cereal ear diseases obtained by treatment during ear emergence
- Prothioconazole is a DMI fungicide. Resistance to some DMI fungicides has been identified in Septoria leaf blotch which may seriously affect performance of some products. For further advice contact a specialist advisor and visit the Fungicide Resistance Action Group (FRAG)-UK website

Restrictions
- A minimum interval of 21 days must be observed between applications
- Only one application may be made before 30 April followed by a second application after 1 May. Alternatively two applications can be made after 1 May.
- Horizontal boom sprayers must be fitted with three star drift reduction technology for all uses
- Low drift spraying equipment must be operated according to the specific conditions stated in the official three star rating for that equipment as published on HSE Chemicals Regulation Division?s website. These operating conditions must be maintained until the operator is 30m from the top of the bank of any surface water bodies

Crop-specific information
- Latest use: before grain watery ripe for rye, oats, winter wheat; up to beginning of anthesis for barley

Environmental safety
- Dangerous for the environment
- Very toxic to aquatic organisms
- Buffer zone requirement 6m
- LERAP Category B [1]

Hazard classification and safety precautions
Hazard Harmful, Dangerous for the environment, Harmful if swallowed, Harmful if inhaled, Very toxic to aquatic organisms
Transport code 9 [1]
Packaging group III
UN Number 3082
Risk phrases H315, H319, H335, H361, H373
Operator protection U05a, U09b, U11, U19a, U20b; A, C, H
Environmental protection E15a, E16a, E34, E38, H410
Storage and disposal D01, D02, D05, D09a, D10b, D12a
Medical advice M03

315 prothioconazole + spiroxamine + tebuconazole

A broad spectrum fungicide mixture for cereals
FRAC mode of action code: 3 + 5 + 3

See also spiroxamine
*   tebuconazole*

Products

1	Cello	Bayer CropScience	100:250:100 g/l	EC	18290

Uses
- Brown rust in **Durum wheat** [1]; **Spring barley** [1]; **Spring wheat** [1]; **Triticale** [1]; **Winter barley** [1]; **Winter rye** [1]; **Winter wheat** [1];
- Crown rust in **Spring oats** [1]; **Winter oats** [1];
- Eyespot in **Durum wheat** *(reduction)* [1]; **Spring barley** *(reduction)* [1]; **Spring oats** [1]; **Spring wheat** *(reduction)* [1]; **Triticale** *(reduction)* [1]; **Winter barley** *(reduction)* [1]; **Winter oats** [1]; **Winter rye** *(reduction)* [1]; **Winter wheat** *(reduction)* [1];
- Glume blotch in **Durum wheat** [1]; **Spring wheat** [1]; **Triticale** [1]; **Winter wheat** [1];
- Late ear diseases in **Durum wheat** [1]; **Spring barley** [1]; **Spring wheat** [1]; **Triticale** [1]; **Winter barley** [1]; **Winter wheat** [1];
- Net blotch in **Spring barley** [1]; **Winter barley** [1];

SEE SECTION 3 FOR PRODUCTS ALSO REGISTERED

- Powdery mildew in *Durum wheat* [1]; *Spring barley* [1]; *Spring oats* [1]; *Spring wheat* [1]; *Triticale* [1]; *Winter barley* [1]; *Winter oats* [1]; *Winter rye* [1]; *Winter wheat* [1];
- Rhynchosporium in *Spring barley* [1]; *Winter barley* [1]; *Winter rye* [1];
- Septoria leaf blotch in *Durum wheat* [1]; *Spring wheat* [1]; *Triticale* [1]; *Winter wheat* [1];
- Yellow rust in *Durum wheat* [1]; *Spring barley* [1]; *Spring wheat* [1]; *Triticale* [1]; *Winter barley* [1]; *Winter wheat* [1];

Approval information
- Prothioconazole, spiroxamine and tebuconazole included in Annex I under EC Regulation 1107/2009

Efficacy guidance
- Best results obtained from treatment at early stages of disease development. Further treatment may be needed if disease attack is prolonged
- Applications to established infections of any disease are likely to be less effective
- Best control of cereal ear diseases obtained by treatment during ear emergence
- Prothioconazole and tebuconazole are DMI fungicides. Resistance to some DMI fungicides has been identified in Septoria leaf blotch which may seriously affect performance of some products. For further advice contact a specialist advisor and visit the Fungicide Resistance Action Group (FRAG)-UK website

Restrictions
- Maximum total dose equivalent to two full dose treatments
- Newer authorisations for tebuconazole products require application to cereals only after GS 30 and applications to oilseed rape and linseed after GS 20 - check label
- Maintain an interval of at least 21 days between applications

Crop-specific information
- Latest use: before grain milky ripe stage for rye, wheat; up to beginning of anthesis for barley and oats

Environmental safety
- Dangerous for the environment
- Very toxic to aquatic organisms
- LERAP Category B [1]

Hazard classification and safety precautions
Hazard Harmful, Dangerous for the environment, Harmful if inhaled
Transport code 9 [1]
Packaging group III
UN Number 3082
Risk phrases H315, H317, H319, H361
Operator protection U05a, U09b, U11, U14, U19a, U20b; A, C, H
Environmental protection E15a, E16a, E34, E38, H410
Storage and disposal D01, D02, D05, D09a, D10b, D12a
Medical advice M03

316   prothioconazole + tebuconazole

A triazole fungicide mixture for cereals
FRAC mode of action code: 3 + 3

See also tebuconazole

Products

1	Clayton Navaro	Clayton	125:125 g/l	EC	19360
2	Clayton Pontoon	Clayton	125:125 g/l	EC	19972
3	Clayton Trellis	Clayton	160:80 g/l	EC	19706
4	Corinth	Bayer CropScience	80:160 g/l	EC	16742
5	Esker	Life Scientific	160:80 g/l	EC	19303
6	Kestrel	Bayer CropScience	160:80 g/l	EC	16751

FOR FULL CONDITIONS OF USE ALWAYS READ THE PRODUCT LABEL

Products – continued

7	Oraso Pro	Life Scientific	125:125 g/l	EC	19321
8	Prosaro	Bayer CropScience	125:125 g/l	EC	16732
9	Redigo Pro	Bayer CropScience	150:20 g/l	FS	15145
10	Stockholm EC	UPL Europe	125:125 g/l	EC	19965

Uses

- Black stem rust in **Grass seed crops** *(off-label)* [8];
- Blue mould in **Durum wheat** [9]; **Rye** [9]; **Spring oats** [9]; **Spring wheat** [9]; **Triticale** [9]; **Winter oats** [9]; **Winter wheat** [9];
- Brown rust in **Spring barley** [1-3,5-8,10]; **Spring wheat** [1-3,5-8,10]; **Winter barley** [1-3,5-8,10]; **Winter rye** [1-3,5-8,10]; **Winter wheat** [1-3,5-8,10];
- Bunt in **Durum wheat** [9]; **Rye** [9]; **Spring oats** [9]; **Spring wheat** [9]; **Triticale** [9]; **Winter oats** [9]; **Winter wheat** [9];
- Covered smut in **Spring barley** [9]; **Spring oats** [9]; **Winter barley** [9]; **Winter oats** [9];
- Crown rust in **Spring oats** [1-3,5-8,10]; **Winter oats** [1-3,5-8,10];
- Ergot in **Rye** *(qualified minor use)* [9];
- Eyespot in **Spring barley** *(reduction)* [1-3,5-8,10]; **Spring oats** [1-3,5-8,10]; **Spring wheat** *(reduction)* [1-3,5-8,10]; **Winter barley** *(reduction)* [1-3,5-8,10]; **Winter oats** [1-3,5-8,10]; **Winter rye** *(reduction)* [1-3,5-8,10]; **Winter wheat** *(reduction)* [1-3,5-8,10];
- Fusarium ear blight in **Spring barley** [3,5-6]; **Spring wheat** [3,5-6]; **Winter barley** [3,5-6]; **Winter wheat** [3,5-6];
- Fusarium seedling blight in **Durum wheat** [9]; **Rye** [9]; **Spring barley** [9]; **Spring oats** [9]; **Spring wheat** [9]; **Triticale** [9]; **Winter barley** [9]; **Winter oats** [9]; **Winter wheat** [9];
- Glume blotch in **Spring wheat** [1-3,5-8,10]; **Winter wheat** [1-3,5-8,10];
- Late ear diseases in **Spring barley** [1-2,7-8,10]; **Spring wheat** [1-2,7-8,10]; **Winter barley** [1-2,7-8,10]; **Winter wheat** [1-2,7-8,10];
- Leaf spot in **Grass seed crops** *(off-label)* [8];
- Leaf stripe in **Spring barley** [9]; **Winter barley** [9];
- Light leaf spot in **Oilseed rape** *(moderate control only)* [3-6]; **Spring oilseed rape** [1-2,7-8,10]; **Winter oilseed rape** *(moderate control only)* [1-2,4,7-8,10];
- Loose smut in **Durum wheat** [9]; **Rye** [9]; **Spring barley** [9]; **Spring oats** [9]; **Spring wheat** [9]; **Triticale** [9]; **Winter barley** [9]; **Winter oats** [9]; **Winter wheat** [9];
- Net blotch in **Spring barley** [1-3,5-8,10]; **Winter barley** [1-3,5-8,10];
- Phoma in **Spring oilseed rape** [1-2,7-8,10]; **Winter oilseed rape** [1-2,7-8,10];
- Phoma leaf spot in **Oilseed rape** [3-6]; **Winter oilseed rape** [4];
- Powdery mildew in **Grass seed crops** *(off-label)* [8]; **Spring barley** [1-3,5-8,10]; **Spring oats** [1-3,5-8,10]; **Spring wheat** [1-3,5-8,10]; **Winter barley** [1-3,5-8,10]; **Winter oats** [1-3,5-8,10]; **Winter rye** [1-3,5-8,10]; **Winter wheat** [1-3,5-8,10];
- Rhynchosporium in **Grass seed crops** *(off-label)* [8]; **Spring barley** [1-3,5-8,10]; **Winter barley** [1-3,5-8,10]; **Winter rye** [1-3,5-8,10];
- Sclerotinia stem rot in **Oilseed rape** [3-6]; **Spring oilseed rape** [1-2,7-8,10]; **Winter oilseed rape** [1-2,4,7-8,10];
- Septoria leaf blotch in **Spring wheat** [1-3,5-8,10]; **Winter wheat** [1-3,5-8,10];
- Sooty moulds in **Spring barley** [3,5-6]; **Spring wheat** [3,5-6]; **Winter barley** [3,5-6]; **Winter wheat** [3,5-6];
- Stem canker in **Oilseed rape** [4]; **Winter oilseed rape** [4];
- Tan spot in **Spring wheat** [1-3,5-8,10]; **Winter wheat** [1-3,5-8,10];
- Yellow rust in **Spring barley** [1-3,5-8,10]; **Spring wheat** [1-3,5-8,10]; **Winter barley** [1-3,5-8,10]; **Winter wheat** [1-3,5-8,10];

Extension of Authorisation for Minor Use (EAMUs)

- **Grass seed crops** *20160083* [8]

Approval information

- Prothioconazole and tebuconazole included in Annex I under EC Regulation 1107/2009
- Accepted by BBPA for use on malting barley

Efficacy guidance

- Best results on cereal foliar diseases obtained from treatment at early stages of disease development. Further treatment may be needed if disease attack is prolonged

SEE SECTION 3 FOR PRODUCTS ALSO REGISTERED

SECTION 2

- On oilseed rape apply a protective treatment in autumn/winter for Phoma followed by a further spray in early spring from the onset of stem elongation, if necessary. For control of Sclerotinia apply at early to full flower
- Applications to established infections of any disease are likely to be less effective
- Best control of cereal ear diseases obtained by treatment during ear emergence
- Prothioconazole and tebuconazole are DMI fungicides. Resistance to some DMI fungicides has been identified in Septoria leaf blotch which may seriously affect performance of some products. For further advice contact a specialist advisor and visit the Fungicide Resistance Action Group (FRAG)-UK website
- Treated seed should be drilled to a depth of 40mm. Ensure no seed is left on the soil surface . After drilling if conditions allow, field should be harrowed and then rolled to ensure good incorporation [9]

Restrictions
- Maximum total dose equivalent to two full dose treatments on barley, oats, oilseed rape; three full dose treatments on wheat and rye
- Newer authorisations for tebuconazole products require application to cereals only after GS 30 and applications to oilseed rape and linseed after GS20 - check label
- Treated seed should not be left on the soil surface. Bury or remove spillages [9]
- A minimum interval of 21 days must be observed between applications

Crop-specific information
- Latest use: before grain milky ripe for rye, wheat; beginning of flowering for barley, oats
- HI 56 d for oilseed rape

Environmental safety
- Dangerous for the environment
- Toxic to aquatic organisms
- LERAP Category B [1-8,10]

Hazard classification and safety precautions
Hazard Harmful [3-6], Irritant [1-2,7-8,10], Dangerous for the environment [1-8,10], Harmful if inhaled [4], Very toxic to aquatic organisms [1-3,7-8]
Transport code 9 [1-10]
Packaging group III
UN Number 3082
Risk phrases H315 [1-8,10], H317 [1-6,10], H319 [1-5,7-8,10], H335 [1-3,5,7-8,10], H361 [1-8,10]
Operator protection U05a, U09b [1-3,5-8,10], U11 [4], U19a [1-2,4,7-8,10], U20a [9], U20b [1-3,5-8,10], U20c [4], U20e [9]; A, C, D, H
Environmental protection E15a [1-8,10], E15b [9], E16a [1-8,10], E22b [4], E22c [1-3,5-8,10], E34, E38, H410 [1-3,5,7-9], H411 [4,6,10]
Storage and disposal D01 [1-8,10], D02 [1-8,10], D05, D09a, D10b [1-8,10], D10d [9], D12a, D21 [9]
Treated seed S01 [9], S02 [9], S03 [9], S04a [9], S04b [9], S05 [9], S07 [9], S08 [9]
Medical advice M03 [1-8,10]

---

317   prothioconazole + trifloxystrobin

A triazole and strobilurin fungicide mixture for cereals
FRAC mode of action code: 3 + 11

See also trifloxystrobin

Products

1 Jager	Bayer CropScience	175:150 g/l	SC	18942
2 Mobius	Bayer CropScience	175:150 g/l	SC	13395
3 Zephyr	Bayer CropScience	175:88 g/l	SC	13174

Uses
- Brown rust in **Durum wheat** [1-3]; **Rye** [1-3]; **Spring barley** [1-3]; **Spring wheat** [1-3]; **Triticale** [1-3]; **Winter barley** [1-3]; **Winter wheat** [1-3];

FOR FULL CONDITIONS OF USE ALWAYS READ THE PRODUCT LABEL

- Ear diseases in **Durum wheat** [1-2]; **Rye** [1-2]; **Spring wheat** [1-2]; **Triticale** [1-2]; **Winter wheat** [1-2];
- Eyespot in **Durum wheat** [1-3]; **Rye** [1-3]; **Spring barley** *(reduction in severity)* [1-3]; **Spring wheat** [1-3]; **Triticale** [1-3]; **Winter barley** *(reduction in severity)* [1-3]; **Winter wheat** [1-3];
- Glume blotch in **Durum wheat** [1-3]; **Rye** [1-3]; **Spring wheat** [1-3]; **Triticale** [1-3]; **Winter wheat** [1-3];
- Net blotch in **Spring barley** [1-3]; **Winter barley** [1-3];
- Powdery mildew in **Durum wheat** [1-3]; **Rye** [1-3]; **Spring barley** [1-3]; **Spring wheat** [1-3]; **Triticale** [1-3]; **Winter barley** [1-3]; **Winter wheat** [1-3];
- Rhynchosporium in **Spring barley** [1-3]; **Winter barley** [1-3];
- Septoria leaf blotch in **Durum wheat** [1-3]; **Rye** [1-3]; **Spring wheat** [1-3]; **Triticale** [1-3]; **Winter wheat** [1-3];
- Yellow rust in **Durum wheat** [1-3]; **Rye** [1-3]; **Spring barley** [1-3]; **Spring wheat** [1-3]; **Triticale** [1-3]; **Winter barley** [1-3]; **Winter wheat** [1-3];

Approval information
- Prothioconazole and trifloxystrobin included in Annex I under EC Regulation 1107/2009
- Accepted by BBPA for use on malting barley

Efficacy guidance
- Best results obtained from treatment at early stages of disease development. Further treatment may be needed if disease attack is prolonged
- Applications to established infections of any disease are likely to be less effective
- Best control of cereal ear diseases obtained by treatment during ear emergence
- Prothioconazole is a DMI fungicide. Resistance to some DMI fungicides has been identified in Septoria leaf blotch which may seriously affect performance of some products. For further advice contact a specialist advisor and visit the Fungicide Resistance Action Group (FRAG)-UK website
- Trifloxystrobin is a member of the QoI cross resistance group. Product should be used preventatively and not relied on for its curative potential
- Use product as part of an Integrated Crop Management strategy incorporating other methods of control, including where appropriate other fungicides with a different mode of action. Do not apply more than two foliar applications of QoI containing products to any cereal crop
- There is a significant risk of widespread resistance occurring in *Septoria tritici* populations in UK. Failure to follow resistance management action may result in reduced levels of disease control
- Strains of wheat and barley powdery mildew resistant to QoIs are common in the UK. Control of wheat mildew can only be relied on from the triazole component
- Where specific control of wheat mildew is required this should be achieved through a programme of measures including products recommended for the control of mildew that contain a fungicide from a different cross-resistance group and applied at a dose that will give robust control

Restrictions
- Maximum total dose equivalent to two full dose treatments
- To avoid the build-up of resistance do not apply products containing QoI fungicides more than twice per cereal crop

Crop-specific information
- Latest use: before grain milky ripe for wheat; beginning of flowering for barley

Environmental safety
- Dangerous for the environment
- Very toxic to aquatic organisms
- LERAP Category B [1-3]

Hazard classification and safety precautions
Hazard Irritant, Dangerous for the environment, Very toxic to aquatic organisms
Transport code 9 [1-3]
Packaging group III
UN Number 3082
Risk phrases H317

SEE SECTION 3 FOR PRODUCTS ALSO REGISTERED

Operator protection U05a, U09b, U19a, U20b [3], U20c [1-2]; A, C, H
Environmental protection E15a, E16a, E34 [3], E38, H410
Storage and disposal D01, D02, D09a, D10c, D12a
Medical advice M03

318   Pseudomonas chlororaphis MA 342

A bacterial fungicide for control of seed-borne diseases in rye, triticale and wheat
FRAC mode of action code: F6

Products
1	Cerall	Koppert	200 g/l	FS	19001

Uses
- Bunt in *Rye* [1]; *Spring wheat* [1]; *Triticale* [1]; *Winter wheat* [1];
- Fusarium in *Rye* [1]; *Spring wheat* [1]; *Triticale* [1]; *Winter wheat* [1];
- Septoria leaf blotch in *Rye* [1]; *Spring wheat* [1]; *Triticale* [1]; *Winter wheat* [1];

Approval information
- Pseudomonas chlororaphis included in Annex 1 under EC Regulation 1107/2009

Efficacy guidance
- Must be used within 12 weeks of the date of manufacture.

Restrictions
- Treated seed must not be used for food or feed.
- Sacks containing treated seed must not be re-used for food or feed.
- Treated seed must not be applied from the air.

Hazard classification and safety precautions
 UN Number N/C
 Operator protection A, D, H
 Storage and disposal D07

319   Pseudomonas SP (DSMZ 13134)

For use on potato seed

Products
1	Proradix	Fargro	0.535% w/w	WP	18115

Uses
- Rhizoctonia in *Potatoes* *(reduction)* [1];

Efficacy guidance
- For the reduction of root rot caused by Rhizoctonia solani in potato crops. It may be applied as spray during sowing of potato or as seed treatment prior to sowing
- Can be stored for 12 months if stored deep frozen or 3 months if stored at room temperature

Restrictions
- Consult processors before using on crops destined for processing

Hazard classification and safety precautions
 UN Number N/C
 Operator protection U05a, U19a, U20b; A, H
 Storage and disposal D01, D12c
 Medical advice M04a

320   pyraclostrobin

A protectant and curative strobilurin fungicide for cereals
FRAC mode of action code: 11

See also boscalid + pyraclostrobin
fluxapyroxad + pyraclostrobin
mefentrifluconazole + pyraclostrobin
mepiquat chloride + prohexadione-calcium + pyraclostrobin

Products

1	Comet 200	BASF	200 g/l	EC	12639
2	Eland	Rigby Taylor	20% w/w	WG	14549
3	Flyer 200	BASF	200 g/l	EC	17293
4	Halley	AgChem Access	200 g/l	EC	15916
5	Insignia	BASF	20% w/w	WG	19403
6	Lybro 200	BASF	200 g/l	EC	19553
7	Retengo 200	BASF	200 g/l	EC	19551
8	Tucana	BASF	250 g/l	EC	10899
9	Vanguard	Pan Amenity	20% w/w	WG	19533
10	Vivid 200	BASF	200 g/l	EC	17295

Uses

* Brown rust in **Spring barley** [1,3-4,6-8,10]; **Spring wheat** [1,3-4,6-8,10]; **Winter barley** [1,3-4,6-8,10]; **Winter wheat** [1,3-4,6-8,10];
* Crown rust in **Spring oats** [1,3-4,6-8,10]; **Winter oats** [1,3-4,6-8,10];
* Disease control in **Forage maize** [3,7,10]; **Grain maize** [3,7,10];
* Dollar spot in **Managed amenity turf** *(useful reduction)* [2,5,9];
* Eyespot in **Forage maize** [1,3,6-7]; **Grain maize** [1,3,6-7]; **Sweetcorn** *(off-label)* [1];
* Fusarium patch in **Managed amenity turf** *(moderate control)* [2,5,9];
* Glume blotch in **Spring wheat** [1,3-4,6-8,10]; **Winter wheat** [1,3-4,6-8,10];
* Net blotch in **Spring barley** [1,3-4,6-8,10]; **Winter barley** [1,3-4,6-8,10];
* Northern leaf blight in **Forage maize** *(moderate control)* [1,3,6-7]; **Grain maize** *(moderate control)* [1,3,6-7]; **Sweetcorn** *(off-label)* [1];
* Red thread in **Managed amenity turf** [2,5,9];
* Rhynchosporium in **Spring barley** *(moderate control)* [1,3-4,6-8,10]; **Winter barley** *(moderate control)* [1,3-4,6-8,10];
* Rust in **Grass seed crops** *(off-label)* [8]; **Rye** *(off-label)* [8]; **Triticale** *(off-label)* [8];
* Septoria leaf blotch in **Spring wheat** [1,3-4,6-8,10]; **Winter wheat** [1,3-4,6-8,10];
* Yellow rust in **Spring barley** [1,3-4,6-8,10]; **Spring wheat** [1,3-4,6-8,10]; **Winter barley** [1,3-4,6-8,10]; **Winter wheat** [1,3-4,6-8,10];

Extension of Authorisation for Minor Use (EAMUs)

* **Grass seed crops** *20193266* [8]
* **Rye** *20193266* [8]
* **Sweetcorn** *20211339* [1]
* **Triticale** *20193266* [8]

Approval information

* Pyraclostrobin included in Annex I under EC Regulation 1107/2009
* Accepted by BBPA for use on malting barley (before ear emergence only) and on hops

Efficacy guidance

* For best results apply at the start of disease attack on cereals [1, 10, 8]
* For Fusarium Patch treat early as severe damage to turf can occur once the disease is established [2]
* Regular turf aeration, appropriate scarification and judicious use of nitrogenous fertiliser will assist the control of Fusarium Patch [2]
* Best results on Septoria glume blotch achieved when used as a protective treatment and against Septoria leaf blotch when treated in the latent phase [1, 10, 8]
* Yield response may be obtained in the absence of visual disease symptoms [1, 10, 8]

SEE SECTION 3 FOR PRODUCTS ALSO REGISTERED

- Pyraclostrobin is a member of the QoI cross resistance group. Product should be used preventatively and not relied on for its curative potential
- Use product as part of an Integrated Crop Management strategy incorporating other methods of control, including where appropriate other fungicides with a different mode of action. Do not apply more than two foliar applications of QoI containing products to any cereal crop or to grass
- There is a significant risk of widespread resistance occurring in *Septoria tritici* populations in UK. Failure to follow resistance management action may result in reduced levels of disease control [1, 10, 8]
- On cereal crops product must always be used in mixture with another product, recommended for control of the same target disease, that contains a fungicide from a different cross resistance group and is applied at a dose that will give robust control [1, 10, 8]
- Late application to maize crops can give a worthwhile increase in biomass

Restrictions
- Maximum number of treatments 2 per crop on cereals [10, 8, 1]
- Maximum total dose on turf equivalent to two full dose treatments [2]
- Do not apply during drought conditions or to frozen turf [2]
- The maximum concentration must not exceed 1.25 litre of product per 200 litres water

Crop-specific information
- Latest use: before grain watery ripe (GS 71) for wheat; up to and including emergence of ear just complete (GS 59) for barley and oats [1, 10, 8]
- Avoid applying to turf immediately after cutting or 48 h before mowing [2]

Environmental safety
- Dangerous for the environment
- Very toxic to aquatic organisms
- LERAP Category B [1-10]

Hazard classification and safety precautions
Hazard Harmful, Dangerous for the environment, Harmful if swallowed [1,3-4,6-8,10], Harmful if inhaled, Very toxic to aquatic organisms
Transport code 6.1 [1,3-4,6-8,10], 9 [2,5,9]
Packaging group III
UN Number 2902, 3077
Risk phrases H304 [1,3-4,6-7,10], H315 [1,3-4,6-8,10], H317 [1,3-4,6-8,10], H319 [1,3-4,6-7,10], H335 [1-3,5-10]
Operator protection U05a [1,3-4,6-8,10], U14 [1,3-4,6-8,10], U20b [1,3-4,6-8,10]; A, C, D, H
Environmental protection E15a [5,9], E15b [1-4,6-8,10], E16a, E16b, E34 [1,3-4,6-8,10], E38, H410
Storage and disposal D01 [1,3-4,6-8,10], D02 [1,3-4,6-8,10], D08, D09a, D10c, D12a
Medical advice M03 [1,3-4,6-8,10], M05a [1,3-4,6-8,10]

321   pyraflufen-ethyl

A phenylpyrazole herbicide for potatoes
HRAC mode of action code: 14 (E)

See also glyphosate + pyraflufen-ethyl

Products

1 Albis	Certis Belchim B V	26.5 g/l	EC	19410
2 Gozai	Certis Belchim B V	26.5 g/l	EC	17381
3 Kabuki	Certis Belchim B V	26.5 g/l	EC	18187

Uses
- Annual dicotyledons in **Hops** *(off-label)* [2]; **Potatoes** [1-3];
- Desiccation in **Hops** *(off-label)* [2]; **Potatoes** [1-3];

Extension of Authorisation for Minor Use (EAMUs)
- **Hops** *20171023* [2]

FOR FULL CONDITIONS OF USE ALWAYS READ THE PRODUCT LABEL

Approval information
- Pyraflufen-ethyl included in Annex I under EC Regulation 1107/2009

Efficacy guidance
- Use with a methylated vegetable oil adjuvant

Following crops guidance
- After cultivation to at least 20 cms winter cereals can be sown as a following crop but the safety to following broad-leaved crops has not yet been established.

Environmental safety
- Buffer zone requirement 20 m in hops [2]
- LERAP Category B [1-3]

Hazard classification and safety precautions

Hazard Harmful, Dangerous for the environment, Harmful if inhaled, Very toxic to aquatic organisms

Transport code 9 [1-3]

Packaging group III

UN Number 3082

Risk phrases H304, H315, H317, H318

Operator protection U05a, U09a, U11, U14, U15; A, C, H

Environmental protection E15b, E16a, E34, H410

Storage and disposal D01, D02, D05, D09a, D10a, D12a, D12b

Medical advice M03, M05b

322   pyrethrins

A non-persistent, contact acting insecticide extracted from Pyrethrum
IRAC mode of action code: 3

Products

1	Dairy Fly Spray	B H & B	0.75 g/l	AL	H5579
2	Pyrethrum 5 EC	PelGar	50 g/l	EC	18532
3	Spruzit	Certis Belchim B V	4.59 g/l	EC	18434

Uses
- Aphids in *Asparagus* [3]; *Beans without pods (Fresh)* [3]; *Bilberries* (off-label) [3]; *Blackberries* (off-label) [3]; *Blackcurrants* (off-label) [3]; *Blueberries* (off-label) [3]; *Broad beans* [3]; *Bulb onions* [3]; *Cabbages* [3]; *Carrots* [3]; *Celeriac* [3]; *Cranberries* (off-label) [3]; *Dwarf beans* [3]; *Edible podded peas* [3]; *French beans* [3]; *Garlic* [3]; *Gooseberries* (off-label) [3]; *Herbs (see appendix 6)* [3]; *Horseradish* [3]; *Jerusalem artichokes* [3]; *Kohlrabi* [3]; *Leeks* [3]; *Lettuce* [3]; *Loganberries* (off-label) [3]; *Parsley root* [3]; *Parsnips* [3]; *Protected aubergines* (off-label) [3]; *Protected cabbages* [3]; *Protected courgettes* (off-label) [3]; *Protected lettuce* [3]; *Protected marrows* (off-label) [3]; *Protected ornamentals* [3]; *Protected summer squash* (off-label) [3]; *Protected tomatoes* [3]; *Radishes* [3]; *Raspberries* (off-label) [3]; *Red beet* [3]; *Redcurrants* (off-label) [3]; *Rose hips* (off-label) [3]; *Rubus hybrids* (off-label) [3]; *Runner beans* [3]; *Salsify* [3]; *Shallots* [3]; *Soya beans* [3]; *Spinach* [3]; *Swedes* [3]; *Turnips* [3]; *Vining peas* [3]; *Watercress* (off-label) [2];
- Caterpillars in *Asparagus* [3]; *Beans without pods (Fresh)* [3]; *Bilberries* (off-label) [3]; *Blackberries* (off-label) [3]; *Blackcurrants* (off-label) [3]; *Blueberries* (off-label) [3]; *Broad beans* [3]; *Bulb onions* [3]; *Cabbages* [3]; *Carrots* [3]; *Celeriac* [3]; *Cranberries* (off-label) [3]; *Dwarf beans* [3]; *Edible podded peas* [3]; *French beans* [3]; *Garlic* [3]; *Gooseberries* (off-label) [3]; *Herbs (see appendix 6)* [3]; *Horseradish* [3]; *Jerusalem artichokes* [3]; *Kohlrabi* [3]; *Leeks* [3]; *Lettuce* [3]; *Loganberries* (off-label) [3]; *Parsley root* [3]; *Parsnips* [3]; *Protected aubergines* (off-label) [3]; *Protected cabbages* [3]; *Protected courgettes* (off-label) [3]; *Protected lettuce* [3]; *Protected marrows* (off-label) [3]; *Protected ornamentals* [3]; *Protected summer squash* (off-label) [3]; *Protected tomatoes* [3]; *Radishes* [3]; *Raspberries* (off-label) [3]; *Red beet* [3]; *Redcurrants* (off-label) [3]; *Rose hips* (off-label) [3]; *Rubus hybrids* (off-label) [3]; *Runner beans* [3]; *Salsify* [3]; *Shallots* [3]; *Soya beans* [3]; *Spinach* [3]; *Swedes* [3]; *Turnips* [3]; *Vining peas* [3];

SEE SECTION 3 FOR PRODUCTS ALSO REGISTERED

- Common green capsid in *Bilberries* (off-label) [3]; *Blackberries* (off-label) [3]; *Blackcurrants* (off-label) [3]; *Blueberries* (off-label) [3]; *Cranberries* (off-label) [3]; *Gooseberries* (off-label) [3]; *Loganberries* (off-label) [3]; *Raspberries* (off-label) [3]; *Redcurrants* (off-label) [3]; *Rose hips* (off-label) [3]; *Rubus hybrids* (off-label) [3];
- Flea beetle in *Watercress* (off-label) [2];
- Flies in *Dairies* [1]; *Farm buildings* [1]; *Livestock houses* [1]; *Poultry houses* [1];
- Flying ants in *Dairies* [1]; *Farm buildings* [1]; *Livestock houses* [1]; *Poultry houses* [1];
- Gnats in *Dairies* [1]; *Farm buildings* [1]; *Livestock houses* [1]; *Poultry houses* [1];
- Insect pests in *Protected aubergines* [2]; *Protected chilli peppers* [2]; *Protected courgettes* [2]; *Protected cucumbers* [2]; *Protected peppers* [2]; *Protected summer squash* [2]; *Protected tomatoes* [2];
- Midges in *Dairies* [1]; *Farm buildings* [1]; *Livestock houses* [1]; *Poultry houses* [1];
- Mosquitoes in *Dairies* [1]; *Farm buildings* [1]; *Livestock houses* [1]; *Poultry houses* [1];
- Mustard beetle in *Watercress* (off-label) [2];
- Southern green stink bug in *Protected aubergines* (off-label) [3]; *Protected courgettes* (off-label) [3]; *Protected marrows* (off-label) [3]; *Protected summer squash* (off-label) [3];
- Spider mites in *Asparagus* [3]; *Beans without pods (Fresh)* [3]; *Broad beans* [3]; *Bulb onions* [3]; *Cabbages* [3]; *Carrots* [3]; *Celeriac* [3]; *Dwarf beans* [3]; *Edible podded peas* [3]; *French beans* [3]; *Garlic* [3]; *Herbs (see appendix 6)* [3]; *Horseradish* [3]; *Jerusalem artichokes* [3]; *Kohlrabi* [3]; *Leeks* [3]; *Lettuce* [3]; *Parsley root* [3]; *Parsnips* [3]; *Protected aubergines* (off-label) [3]; *Protected cabbages* [3]; *Protected courgettes* (off-label) [3]; *Protected lettuce* [3]; *Protected marrows* (off-label) [3]; *Protected ornamentals* [3]; *Protected summer squash* (off-label) [3]; *Protected tomatoes* [3]; *Radishes* [3]; *Red beet* [3]; *Runner beans* [3]; *Salsify* [3]; *Shallots* [3]; *Soya beans* [3]; *Spinach* [3]; *Swedes* [3]; *Turnips* [3]; *Vining peas* [3];
- Spotted wing drosophila in *Bilberries* (off-label) [3]; *Blackberries* (off-label) [3]; *Blackcurrants* (off-label) [3]; *Blueberries* (off-label) [3]; *Cranberries* (off-label) [3]; *Gooseberries* (off-label) [3]; *Loganberries* (off-label) [3]; *Raspberries* (off-label) [3]; *Redcurrants* (off-label) [3]; *Rose hips* (off-label) [3]; *Rubus hybrids* (off-label) [3];
- Thrips in *Asparagus* [3]; *Beans without pods (Fresh)* [3]; *Broad beans* [3]; *Bulb onions* [3]; *Cabbages* [3]; *Carrots* [3]; *Celeriac* [3]; *Dwarf beans* [3]; *Edible podded peas* [3]; *French beans* [3]; *Garlic* [3]; *Herbs (see appendix 6)* [3]; *Horseradish* [3]; *Jerusalem artichokes* [3]; *Kohlrabi* [3]; *Leeks* [3]; *Lettuce* [3]; *Parsley root* [3]; *Parsnips* [3]; *Protected aubergines* (off-label) [3]; *Protected cabbages* [3]; *Protected courgettes* (off-label) [3]; *Protected lettuce* [3]; *Protected marrows* (off-label) [3]; *Protected ornamentals* [3]; *Protected summer squash* (off-label) [3]; *Protected tomatoes* [3]; *Radishes* [3]; *Red beet* [3]; *Runner beans* [3]; *Salsify* [3]; *Shallots* [3]; *Soya beans* [3]; *Spinach* [3]; *Swedes* [3]; *Turnips* [3]; *Vining peas* [3];
- Two-spotted spider mite in *Bilberries* (off-label) [3]; *Blackberries* (off-label) [3]; *Blackcurrants* (off-label) [3]; *Blueberries* (off-label) [3]; *Cranberries* (off-label) [3]; *Gooseberries* (off-label) [3]; *Loganberries* (off-label) [3]; *Raspberries* (off-label) [3]; *Redcurrants* (off-label) [3]; *Rose hips* (off-label) [3]; *Rubus hybrids* (off-label) [3];
- Wasps in *Dairies* [1]; *Farm buildings* [1]; *Livestock houses* [1]; *Poultry houses* [1];
- Whitefly in *Protected aubergines* (off-label) [3]; *Protected courgettes* (off-label) [3]; *Protected marrows* (off-label) [3]; *Protected summer squash* (off-label) [3];

Extension of Authorisation for Minor Use (EAMUs)

- *Bilberries* *20201258* [3]
- *Blackberries* *20210674* [3]
- *Blackcurrants* *20201258* [3]
- *Blueberries* *20201258* [3]
- *Cranberries* *20201258* [3]
- *Gooseberries* *20201258* [3]
- *Loganberries* *20210674* [3]
- *Protected aubergines* *20210528* [3]
- *Protected courgettes* *20210528* [3]
- *Protected marrows* *20210528* [3]
- *Protected summer squash* *20210528* [3]
- *Raspberries* *20210674* [3]
- *Redcurrants* *20201258* [3]
- *Rose hips* *20201258* [3]

FOR FULL CONDITIONS OF USE ALWAYS READ THE PRODUCT LABEL

- **Rubus hybrids** *20210674* [3]
- **Watercress** *20201366* [2]

Approval information
- Pyrethrins have been included in Annex 1 under EC Regulation 1107/2009
- Some products formulated for ULV application [1]. May be applied through a fogging machine or sprayer - see label for details
- Accepted by BBPA for use in empty grain stores, malting barley and hops

Efficacy guidance
- For indoor fly control close doors and windows and spray or apply fog as appropriate [1]
- For best fly control outdoors spray during early morning or late afternoon and evening when conditions are still [1]
- For all uses ensure good spray coverage of the target area or plants by increasing spray volume where necessary
- Best results on outdoor or protected crops achieved from treatment at first signs of pest attack in early morning or evening
- To avoid possibility of development of resistance do not spray more frequently than once per week

Restrictions
- For use only by professional operators [1]
- Do not allow spray to contact open food products or food preparing equipment or utensils [1]
- Remove exposed milk and collect eggs before application [1]
- Do not treat plants [1]
- Avoid direct application to open flowers
- Do not use space sprays containing pyrethrins or pyrethroid more than once per week in intensive or controlled environment animal houses in order to avoid development of resistance. If necessary, use a different control method or product [1]
- Store away from strong sunlight

Crop-specific information
- Some plant species including *Ageratum*, ferns, *Ficus, Lantana, Poinsettia, Petroselium crispum*, and some strawberries may be sensitive especially where more than one application is made. Test before large scale treatment

Environmental safety
- Dangerous for the environment
- Very toxic to aquatic organisms [1]
- High risk to bees. Do not apply to crops in flower or to those in which bees are actively foraging. Do not apply when flowering weeds are present
- Risk to non-target insects or other arthropods
- Do not apply directly to livestock and exclude all persons and animals during treatment [1]
- Wash spray equipment thoroughly after use to avoid traces of pyrethrum causing damage to susceptible crops sprayed later
- Aquatic buffer zone of 20m on blackberry, bilberry, blackcurrant and redcurrant, blueberry, cranberry, gooseberry, loganberry, raspberry,rose hips and rubus hybrids [3]
- LERAP Category B [3]

Hazard classification and safety precautions
Hazard Harmful [1], Irritant [1], Dangerous for the environment [3]
Transport code 9 [1-3]
Packaging group III
UN Number 3082
Risk phrases H304 [2], H315 [2], H318 [2], R22a [1], R36 [1], R38 [1]
Operator protection U05a [1-2], U09a [1], U19a, U20b [1], U20d [2-3]; A, B, E, H
Environmental protection E05a [1], E12a [2-3], E12c [3], E13c [1], E15b [3], E16a [3], E19b [3], E38 [3], H410 [2-3]
Consumer protection C04 [1], C06 [1], C07 [1], C08 [1], C09 [1], C10 [1], C11 [1]
Storage and disposal D01, D02, D05 [2-3], D09a [1,3], D11a [1,3], D12a [3]

SEE SECTION 3 FOR PRODUCTS ALSO REGISTERED

323   pyridate

A contact phenylpyridazine herbicide for bulb onions and brassicas
HRAC mode of action code: 6 (C3)

See also mesotrione + pyridate

Products

1	Diva	Certis Belchim B V	600 g/l	EC	17774
2	Gyo	Certis Belchim B V	600 g/l	EC	17875
3	Lentagran WP	Certis Belchim B V	45% w/w	WP	19474

Uses

- Annual dicotyledons in **Forage maize** [1-2]; **Garlic** *(off-label)* [3]; **Grain maize** [1-2]; **Leeks** *(off-label)* [3]; **Shallots** *(off-label)* [3];
- Annual mercury in **Sweetcorn** *(off-label)* [1-2];
- Black bindweed in **Sweetcorn** *(off-label)* [1-2];
- Black nightshade in **Asparagus** *(off-label)* [3]; **Broccoli** *(off-label)* [3]; **Brussels sprouts** [3]; **Bulb onions** [3]; **Cabbages** [3]; **Calabrese** *(off-label)* [3]; **Cauliflowers** *(off-label)* [3]; **Choi sum** *(off-label)* [3]; **Collards** *(off-label)* [3]; **Forage maize** [1-2]; **Grain maize** [1-2]; **Green cover on land temporarily removed from production** *(off-label)* [3]; **Kale** *(off-label)* [3]; **Salad onions** *(off-label)* [3]; **Spring greens** *(off-label)* [3]; **Sweetcorn** *(off-label)* [1-2];
- Chickweed in **Sweetcorn** *(off-label)* [1-2];
- Cleavers in **Asparagus** *(off-label)* [3]; **Broccoli** *(off-label)* [3]; **Brussels sprouts** [3]; **Bulb onions** [3]; **Cabbages** [3]; **Calabrese** *(off-label)* [3]; **Cauliflowers** *(off-label)* [3]; **Choi sum** *(off-label)* [3]; **Collards** *(off-label)* [3]; **Forage maize** [1-2]; **Grain maize** [1-2]; **Green cover on land temporarily removed from production** *(off-label)* [3]; **Kale** *(off-label)* [3]; **Salad onions** *(off-label)* [3]; **Spring greens** *(off-label)* [3]; **Sweetcorn** *(off-label)* [1-2];
- Common amaranth in **Sweetcorn** *(off-label)* [1-2];
- Common orache in **Sweetcorn** *(off-label)* [1-2];
- Fat hen in **Asparagus** *(off-label)* [3]; **Broccoli** *(off-label)* [3]; **Brussels sprouts** [3]; **Bulb onions** [3]; **Cabbages** [3]; **Calabrese** *(off-label)* [3]; **Cauliflowers** *(off-label)* [3]; **Choi sum** *(off-label)* [3]; **Collards** *(off-label)* [3]; **Forage maize** [1-2]; **Grain maize** [1-2]; **Green cover on land temporarily removed from production** *(off-label)* [3]; **Kale** *(off-label)* [3]; **Salad onions** *(off-label)* [3]; **Spring greens** *(off-label)* [3]; **Sweetcorn** *(off-label)* [1-2];
- Fumitory in **Forage maize** [1-2]; **Grain maize** [1-2]; **Salad onions** *(off-label)* [3];
- Groundsel in **Forage maize** [1-2]; **Grain maize** [1-2]; **Salad onions** *(off-label)* [3];
- Ivy-leaved speedwell in **Sweetcorn** *(off-label)* [1-2];
- Red dead-nettle in **Sweetcorn** *(off-label)* [1-2];

Extension of Authorisation for Minor Use (EAMUs)

- **Asparagus** *20220491* [3]
- **Broccoli** *20220482* [3]
- **Calabrese** *20220482* [3]
- **Cauliflowers** *20220482* [3]
- **Choi sum** *20220482* [3]
- **Collards** *20220482* [3]
- **Garlic** *20201511* [3]
- **Green cover on land temporarily removed from production** *20220494* [3]
- **Kale** *20220482* [3]
- **Leeks** *20201510* [3]
- **Salad onions** *20201512* [3]
- **Shallots** *20201511* [3]
- **Spring greens** *20220482* [3]
- **Sweetcorn** *20211360* [1], *20211427* [2]

Approval information

- Pyridate included in Annex I under EC Regulation 1107/2009

Efficacy guidance
- Best results achieved by application to actively growing weeds at 6-8 leaf stage when temperatures are above 8°C before crop foliage forms canopy [3]
- Apply when maize is between 2 and 8 leaves for optimum activity [1, 2]

Restrictions
- Do not apply in mixture with or within 14 d of any other product which may result in dewaxing of crop foliage
- Do not use on crops suffering stress from frost, drought, disease or pest attack
- Horizontal boom sprayers must be fitted with three star drift reduction technology for all uses [1, 2]
- Must not be applied via hand-held equipment
- Low drift spraying equipment must be used for 30 m from the top of the bank of any surface water bodies [1, 2]
- Applications must not be made before 15th May [1, 2]
- Extreme care must be taken to avoid spray drift onto non-crop plants outside of the target area [3]

Crop-specific information
- Latest use: before 6 true leaf stage of Brussels sprouts and cabbage; before 7 true leaves for onions; before flower buds visible for oilseed rape
- Apply to cabbages and Brussels sprouts after 4 fully expanded leaf stage. Allow 2 wk after transplanting before treating

Environmental safety
- Dangerous for the environment
- Toxic to aquatic organisms
- Buffer zone requirement 12 m [1, 2]
- LERAP Category B [1-3]

Hazard classification and safety precautions
Hazard Irritant [3], Dangerous for the environment [3], Flammable liquid and vapour [1-2], Very toxic to aquatic organisms [1-2]
Transport code 9 [1-3]
Packaging group III
UN Number 3077, 3082
Risk phrases H315 [1-2], H317, H319 [1-2], H370 [1-2]
Operator protection U05a, U14, U22a; A, C, H
Environmental protection E16a, E38, H410
Storage and disposal D01, D02, D09a, D10c, D12a

324 pyrimethanil

An anilinopyrimidine fungicide for apples and strawberries
FRAC mode of action code: 9

See also fludioxonil + pyrimethanil

Products
1	Clayton Patriot	Clayton	400 g/l	SL	18418
2	Pyrus 400 SC	Arysta	400 g/l	SL	16254
3	Scala	BASF	400 g/l	SC	15222
4	Triple SC	Pan Agriculture	400 g/l	SL	18669

Uses
- Botrytis in **Bilberries** *(off-label)* [3]; **Blackberries** *(off-label)* [1-4]; **Blackcurrants** *(off-label)* [3]; **Blueberries** *(off-label)* [3]; **Cranberries** *(off-label)* [3]; **Forest nurseries** *(off-label)* [3]; **Gooseberries** *(off-label)* [3]; **Grapevines** *(off-label)* [3]; **Ornamental plant production** *(off-label)* [3]; **Pears** *(off-label)* [3]; **Protected aubergines** *(off-label)* [3]; **Protected baby leaf crops** *(off-label)* [3]; **Protected blackberries** *(off-label)* [3]; **Protected herbs (see appendix 6)** *(off-label)* [3];

Protected lettuce (off-label) [3]; *Protected raspberries* (off-label) [3]; *Protected strawberries* [3]; *Protected tomatoes* (off-label) [3]; *Quinces* (off-label) [3]; *Raspberries* (off-label) [1-4]; *Redcurrants* (off-label) [3]; *Strawberries* [1-4]; *Top fruit* (off-label) [3]; *Vaccinium spp.* (off-label) [3]; *Whitecurrants* (off-label) [3];

- Disease control in *Forest nurseries* (off-label) [3]; *Top fruit* (off-label) [3];
- Grey mould in *Protected strawberries* (moderate control) [3]; *Strawberries* [3];
- Scab in *Apples* [3]; *Pears* (off-label) [3]; *Quinces* (off-label) [3];

Extension of Authorisation for Minor Use (EAMUs)
- *Bilberries* 20110291 [3]
- *Blackberries* 20110293 [3]
- *Blackcurrants* 20110291 [3]
- *Blueberries* 20110291 [3]
- *Cranberries* 20110291 [3]
- *Forest nurseries* 20122054 [3]
- *Gooseberries* 20110291 [3]
- *Grapevines* 20110283 [3]
- *Ornamental plant production* 20111315 [3]
- *Pears* 20110295 [3]
- *Protected aubergines* 20110282 [3]
- *Protected baby leaf crops* 20110287 [3]
- *Protected blackberries* 20110292 [3]
- *Protected herbs (see appendix 6)* 20110287 [3]
- *Protected lettuce* 20110287 [3]
- *Protected raspberries* 20110292 [3]
- *Protected tomatoes* 20110282 [3]
- *Quinces* 20110295 [3]
- *Raspberries* 20110293 [3]
- *Redcurrants* 20110291 [3]
- *Top fruit* 20122054 [3]
- *Vaccinium spp.* 20110291 [3]
- *Whitecurrants* 20110291 [3]

Approval information
- Pyrimethanil included in Annex I under EC Regulation 1107/2009.

Efficacy guidance
- On apples a programme of sprays will give early season control of scab. Season long control can be achieved by continuing programme with other approved fungicides
- In strawberries product should be used as part of a programme of disease control treatments which should alternate with other materials to prevent or limit development of less sensitive strains of grey mould

Restrictions
- Maximum number of treatments 4 per yr for apples and pears; 2 per yr for cane fruit, bush fruit, strawberries, tomatoes
- Product does not taint apples. Processors should be consulted before use on strawberries

Crop-specific information
- Latest use: before end of flowering for apples
- HI protected cane fruit, strawberries 1 d; aubergines, tomatoes 3 d; protected lettuce 14 d; bush fruit, grapevines 21 d
- All varieties of apples and strawberries may be treated
- Treat apples from bud burst at 10-14 d intervals
- In strawberries start treatments at white bud to give maximum protection of flowers against grey mould and treat every 7-10 d. Product should not be used more than once in a 3 or 4 spray programme

Environmental safety
- Harmful to aquatic organisms

FOR FULL CONDITIONS OF USE ALWAYS READ THE PRODUCT LABEL

- Product has negligible effect on hoverflies and lacewings. Limited evidence indicates some margin of safety to *Typhlodromus pyri*
- Broadcast air-assisted LERAP [3] (20 m); LERAP Category B [3]

Hazard classification and safety precautions
    Hazard Dangerous for the environment [1-2,4]
    Transport code 9 [2,4]
    Packaging group III
    UN Number 3082, N/C
    Risk phrases R51 [1-2,4], R53a [1-2,4]
    Operator protection U05a [1-2,4], U08 [3], U20b [3]; A, H, M
    Environmental protection E15a [3], E16a [3], E17b (20 m) [3], E38 [1-2,4], H411 [3]
    Storage and disposal D01, D02, D05 [3], D08 [3], D09a [3], D10b [3], D12a [3], D12b [1-2,4], D19 [1-2,4]

325   pyriproxyfen

An insect hormone analog which prevents larvae developing into adults
IRAC mode of action code: 7C

Products

1 Harpun	Clayton	100 g/l	EC	19383
2 Kratox	Pan Agriculture	100 g/l	EC	19616

Uses
- Codling moth in **Apples** *(reduction)* [1-2];
- Pear sucker in **Pears** *(reduction)* [1-2];
- Whitefly in **Protected cucumbers** *(reduction)* [1-2]; **Protected tomatoes** *(reduction)* [1-2];

Approval information
- Pyriproxyfen included in Annex I under EC Regulation 1107/2009

Efficacy guidance
- Mimics juvenile hormone and thereby interferes with insect maturation by disrupting and preventing metamorphosis so effect on target insects is a reduction in populations.

Restrictions
- Where beehives are used for pollination in glasshouses it is recommend to close hives prior to application
- Crops should not be re-entered until spray residues are dry
- Where a crop is destined for processing, consult your processor before treating
- Crop safety, including russeting, on ?Cox? , ?Gala? , ?Bramley? and ?Conference? has not been established

Environmental safety
- Broadcast air-assisted LERAP [1-2] (30 m)

Hazard classification and safety precautions
    Transport code 9 [1-2]
    Packaging group III
    UN Number 3082
    Risk phrases H304, H318, H336
    Operator protection A, D, H
    Environmental protection E17b (30 m), H410

326   pyroxsulam

A triazolopyrimidine herbicide for use in cereals
HRAC mode of action code: 2 (B)

See also florasulam + pyroxsulam

Products

1 Avocet	Corteva	7.5% w/w	WG	14829

Uses
- Annual dicotyledons in *Triticale* [1]; *Winter rye* [1]; *Winter wheat* [1];
- Blackgrass in *Triticale* [1]; *Winter rye* [1]; *Winter wheat* [1];
- Cleavers in *Triticale* [1]; *Winter rye* [1]; *Winter wheat* [1];
- Ryegrass in *Triticale* [1]; *Winter rye* [1]; *Winter wheat* [1];
- Volunteer oilseed rape in *Triticale* [1]; *Winter rye* [1]; *Winter wheat* [1];
- Wild oats in *Triticale* [1]; *Winter rye* [1]; *Winter wheat* [1];

Approval information
- Pyroxsulam included in Annex 1 under EC Regulation 1107/2009

Efficacy guidance
- Avoid overlapping of spray swaths.
- Rainfast in one hour.
- Do not spray when crops are under stress from cold, drought, water logging, pest damage, nutrient deficiency etc.
- Do not roll or harrow 7 days before or after application.
- Take extreme care to avoid drift onto susceptible crops, non-target plants or waterways. Do not apply directly to, or allow spray drift to come into contact with agricultural or horticultural crops, amenity plantings, gardens, ponds, lakes or watercourses.
- To avoid crop injury do not apply in tank mixture with plant growth regulators. A minimum interval of 7 days between applications should be left.
- To avoid crop injury do not apply in tank mix with an organophosphate insecticide. A minimum interval of 14 days between applications should be left.

Restrictions
- To avoid the build-up of resistance, do not apply this or any other product containing an ALS inhibitor herbicide with claims for control of grass-weeds more than once to any crop
- To avoid subsequent injury to crops other than cereals, all spraying equipment must be thoroughly cleaned both inside and out, using All Clear Extra spray cleaner

Crop-specific information
- Winter wheat, winter barley, winter oats, triticale, winter rye, oilseed rape, grass, winter beans or brassicas as transplants may be sown as a following crop in the autumn.
- Following a sequence of pyroxsulam and metsulfuron-methyl, treated ground should be ploughed to a depth of 15 cm before establishing winter oilseed rape
- In spring, spring wheat, spring barley, spring oats, triticale, rye, spring oilseed rape, sugar beet, potatoes, grass, clover (as part of a grass/clover ley), beans, peas, maize, linseed may follow a normally harvested treated crop.

Following crops guidance
- In the event of a crop failure after a crop has been treated before February 1st only the following crops may be planted after ploughing to a depth of at least 15 cm; Spring wheat, spring barley, grass or maize. At least 6 weeks must elapse between treatment and re-drilling.
- Where crop failure occurs following an application after February 1st plough and then allow 6 weeks before planting grass or maize.

Environmental safety
- LERAP Category B [1]

Hazard classification and safety precautions
  Hazard Very toxic to aquatic organisms

FOR FULL CONDITIONS OF USE ALWAYS READ THE PRODUCT LABEL

Transport code 9 [1]
Packaging group III
UN Number 3077
Operator protection U05a, U09b, U14, U15; A, H
Environmental protection E16a, E34, H410
Storage and disposal D01, D02, D09a, D12a

327 quinmerac

A residual herbicide available only in mixtures
HRAC mode of action code: 4 (O)

See also dimethenamid-p + metazachlor + quinmerac
dimethenamid-p + quinmerac
imazamox + quinmerac
metamitron + quinmerac
metazachlor + quinmerac

328 quizalofop-P-ethyl

An aryl phenoxypropionic acid post-emergence herbicide for grass weed control
HRAC mode of action code: 1 (A)

Products

1	Clayton Kwiz	Clayton	50 g/l	EC	19467
2	Pilot Ultra	Gowan	50 g/l	SC	17136
3	Targa Max	Gowan	100 g/l	EC	17135
4	Targa Super	Gowan	50 g/l	EC	17134

Uses
- Annual grasses in **Combining peas** [1-2]; **Fodder beet** [1-4]; **Linseed** [1-2]; **Mangels** [1-4]; **Potatoes** [1-4]; **Red beet** [1-4]; **Spring field beans** [1-2]; **Spring oilseed rape** [1-4]; **Sugar beet** [1-4]; **Vining peas** [1-2]; **Winter field beans** [1-2]; **Winter oilseed rape** [1-4];
- Couch in **Combining peas** [1-2]; **Fodder beet** [1-4]; **Game cover** *(off-label)* [2]; **Linseed** [1-2]; **Mangels** [1-4]; **Potatoes** [1-4]; **Red beet** [1-4]; **Spring field beans** [1-2]; **Spring oilseed rape** [1-4]; **Sugar beet** [1-4]; **Vining peas** [1-2]; **Winter field beans** [1-2]; **Winter oilseed rape** [1-4];
- Italian ryegrass in **Game cover** *(off-label)* [2];
- Perennial grasses in **Combining peas** [1-2]; **Fodder beet** [1-4]; **Linseed** [1-2]; **Mangels** [1-4]; **Potatoes** [1-4]; **Red beet** [1-4]; **Spring field beans** [1-2]; **Spring oilseed rape** [1-4]; **Sugar beet** [1-4]; **Vining peas** [1-2]; **Winter field beans** [1-2]; **Winter oilseed rape** [1-4];
- Sterile brome in **Game cover** *(off-label)* [2];
- Volunteer cereals in **Combining peas** [1-2]; **Fodder beet** [1-4]; **Game cover** *(off-label)* [2]; **Linseed** [1-2]; **Mangels** [1-4]; **Potatoes** [1-4]; **Red beet** [1-4]; **Spring field beans** [1-2]; **Spring oilseed rape** [1-4]; **Sugar beet** [1-4]; **Vining peas** [1-2]; **Winter field beans** [1-2]; **Winter oilseed rape** [1-4];
- Wild oats in **Game cover** *(off-label)* [2];

Extension of Authorisation for Minor Use (EAMUs)
- **Game cover** *20193869* [2]

Approval information
- Quizalofop-P-ethyl has been included in Annex 1 under EC Regulation 1107/2009

Efficacy guidance
- Best results achieved by application to emerged weeds growing actively in warm conditions with adequate soil moisture
- Weed control may be reduced under conditions such as drought that limit uptake and translocation
- Annual meadow-grass is not controlled

SEE SECTION 3 FOR PRODUCTS ALSO REGISTERED

- For effective couch control do not hoe beet crops within 21 d after spraying
- At least 2 h without rain should follow application otherwise results may be reduced
- Quizalofop-P-ethyl is an ACCase inhibitor herbicide. To avoid the build up of resistance do not apply products containing an ACCase inhibitor herbicide more than twice to any crop. In addition do not use any product containing quizalofop-P-ethyl in mixture or sequence with any other product containing the same ingredient
- Use these products as part of a resistance management strategy that includes cultural methods of control and does not use ACCase inhibitors as the sole chemical method of grass weed control
- Applying a second product containing an ACCase inhibitor to a crop will increase the risk of resistance development; only use a second ACCase inhibitor to control different weeds at a different timing
- Always follow WRAG guidelines for preventing and managing herbicide resistant weeds. See Section 5 for more information

Restrictions
- Maximum number of treatments 1 on all recommended crops
- Do not spray crops under stress from any cause or in frosty weather
- Consult processor before use on crops recommended for processing
- An interval of at least 3 d must elapse between treatment and use of another herbicide on beet crops, 14 d on oilseed rape, 21 d on linseed and other recommended crops
- Avoid drift onto neighbouring crops

Crop-specific information
- HI 16 wk for beet crops; 11 wk for oilseed rape, linseed; 8 wk for field beans; 5 wk for peas
- In some situations treatment can cause yellow patches on foliage of peas, especially vining varieties. Symptoms usually rapidly and completely outgrown
- May cause taint in peas

Following crops guidance
- In the event of failure of a treated crop broad-leaved crops may be resown after a minimum interval of 2 wks, and cereals after 2 - 6 wks depending on dose applied
- Onions, leeks and maize are not recommended to follow a failed treated crop

Environmental safety
- Dangerous for the environment
- Toxic to aquatic organisms
- Flammable

Hazard classification and safety precautions
Hazard Harmful [1], Flammable [1], Dangerous for the environment, Harmful if inhaled [4], Very toxic to aquatic organisms [4]
Transport code 3 [1], 9 [3-4]
Packaging group III
UN Number 1993, 3082, N/C
Risk phrases H304 [1,4], H315 [1], H317 [4], H318 [3-4], H319 [1], H336 [1]
Operator protection U05a [1], U09a, U11 [1], U14 [1], U15 [1], U19a [1], U20b; A, C, H
Environmental protection E13b, E15a [1], E34, E38, H410 [4], H411 [1]
Storage and disposal D01 [1], D02 [1], D05 [1], D09a, D10b [1], D10c, D12a
Medical advice M03 [1], M05b [1]

329　quizalofop-P-tefuryl

An aryloxyphenoxypropionate herbicide for grass weed control
HRAC mode of action code: 1 (A)

Products

1	Panarex	UPL Europe	40 g/l	EC	19587
2	Rango	UPL Europe	40 g/l	EC	19588

FOR FULL CONDITIONS OF USE ALWAYS READ THE PRODUCT LABEL

Uses

- Blackgrass in **Combining peas** [1-2]; **Fodder beet** [1-2]; **Potatoes** [1-2]; **Spring field beans** [1-2]; **Spring linseed** [1-2]; **Spring oilseed rape** [1-2]; **Sugar beet** [1-2]; **Winter field beans** [1-2]; **Winter linseed** [1-2]; **Winter oilseed rape** [1-2];
- Couch in **Combining peas** [1-2]; **Fodder beet** [1-2]; **Potatoes** [1-2]; **Spring field beans** [1-2]; **Spring linseed** [1-2]; **Spring oilseed rape** [1-2]; **Sugar beet** [1-2]; **Winter field beans** [1-2]; **Winter linseed** [1-2]; **Winter oilseed rape** [1-2];
- Italian ryegrass in **Combining peas** [1-2]; **Fodder beet** [1-2]; **Potatoes** [1-2]; **Spring field beans** [1-2]; **Spring linseed** [1-2]; **Spring oilseed rape** [1-2]; **Sugar beet** [1-2]; **Winter field beans** [1-2]; **Winter linseed** [1-2]; **Winter oilseed rape** [1-2];
- Perennial ryegrass in **Combining peas** [1-2]; **Fodder beet** [1-2]; **Potatoes** [1-2]; **Spring field beans** [1-2]; **Spring linseed** [1-2]; **Spring oilseed rape** [1-2]; **Sugar beet** [1-2]; **Winter field beans** [1-2]; **Winter linseed** [1-2]; **Winter oilseed rape** [1-2];
- Volunteer cereals in **Combining peas** [1-2]; **Fodder beet** [1-2]; **Potatoes** [1-2]; **Spring field beans** [1-2]; **Spring linseed** [1-2]; **Spring oilseed rape** [1-2]; **Sugar beet** [1-2]; **Winter field beans** [1-2]; **Winter linseed** [1-2]; **Winter oilseed rape** [1-2];
- Wild oats in **Combining peas** [1-2]; **Fodder beet** [1-2]; **Potatoes** [1-2]; **Spring field beans** [1-2]; **Spring linseed** [1-2]; **Spring oilseed rape** [1-2]; **Sugar beet** [1-2]; **Winter field beans** [1-2]; **Winter linseed** [1-2]; **Winter oilseed rape** [1-2];

Approval information

- Quizalofop-P-tefuryl has been included in Annex 1 under EC Regulation 1107/2009

Efficacy guidance

- Best results on annual grass weeds when growing actively and treated from 2 leaves to the start of tillering
- Treat cover crops when they have served their purpose and the threat of wind blow has passed
- Best results on couch achieved when the weed is growing actively and commencing new rhizome growth
- Grass weeds germinating after treatment will not be controlled
- Treatment quickly stops growth and visible colour changes to the leaf tips appear after about 7 d. Complete kill takes 3-4 wk under good growing conditions
- Quizalofop-P-tefuryl is an ACCase inhibitor herbicide. To avoid the build up of resistance do not apply products containing an ACCase inhibitor herbicide more than twice to any crop. In addition do not use any product containing quizalofop-P-tefuryl in mixture or sequence with any other product containing the same ingredient
- Use these products as part of a resistance management strategy that includes cultural methods of control and does not use ACCase inhibitors as the sole chemical method of grass weed control
- Applying a second product containing an ACCase inhibitor to a crop will increase the risk of resistance development; only use a second ACCase inhibitor to control different weeds at a different timing
- Always follow WRAG guidelines for preventing and managing herbicide resistant weeds. See Section 5 for more information

Restrictions

- Maximum number of treatments 1 per crop
- Consult processors before use on peas or potatoes for processing
- Do not treat crops and weeds growing under stress from any cause

Crop-specific information

- HI 60 d for all crops
- Treat oilseed rape from the fully expanded cotyledon stage
- Treat linseed, peas and field beans from 2-3 unfolded leaves
- Treat sugar beet from 2 unfolded leaves

Following crops guidance

- In the event of failure of a treated crop any broad-leaved crop may be planted at any time. Cereals may be drilled from 4 wk after treatment

Environmental safety

- Dangerous for the environment

SEE SECTION 3 FOR PRODUCTS ALSO REGISTERED

- Very toxic to aquatic organisms
- Risk to certain non-target insects or other arthropods. For advice on risk management and use in Integrated Pest Management (IPM) see directions for use
- Avoid spraying within 6 m of the field boundary to reduce effects on non-target insects and other arthropods

Hazard classification and safety precautions

Hazard Toxic [2], Irritant [1], Dangerous for the environment, Very toxic to aquatic organisms
Transport code 9 [1-2]
Packaging group III
UN Number 3082
Risk phrases H304, H315, H317, H319, H336, H341, H360
Operator protection U02a, U04a, U05a, U10, U11, U14, U15, U19a, U20b; A, C, H
Environmental protection E15a, E22b, E38, H410, H411 [1]
Storage and disposal D01, D02, D09a, D10b, D12a
Medical advice M04a, M05b

330   rimsulfuron

A selective systemic sulfonylurea herbicide
HRAC mode of action code: 2 (B)

Products

1	Caesar	AgChem Access	25% w/w	SG	15272
2	Clayton Bramble	Clayton	25% w/w	SG	15281
3	Clayton Nero	Clayton	25% w/w	SG	15511
4	Titus	Corteva	25% w/w	SG	19270

Uses

- Annual dicotyledons in *Forage maize* [1-4]; *Potatoes* [1-4];
- Charlock in *Forage maize* [4]; *Potatoes* [4];
- Chickweed in *Forage maize* [4]; *Potatoes* [4];
- Cleavers in *Forage maize* [4]; *Potatoes* [4];
- Hemp-nettle in *Forage maize* [4]; *Potatoes* [4];
- Red dead-nettle in *Forage maize* [4]; *Potatoes* [4];
- Redshank in *Forage maize* [4]; *Potatoes* [4];
- Scentless mayweed in *Forage maize* [4]; *Potatoes* [4];
- Small nettle in *Forage maize* [4]; *Potatoes* [4];
- Volunteer oilseed rape in *Forage maize* [1-4]; *Potatoes* [1-4];

Approval information

- Rimsulfuron included in Annex I under EC Regulation 1107/2009

Efficacy guidance

- Product should be used with a suitable adjuvant or a suitable herbicide tank-mix partner. See label for details
- Product acts by foliar action. Best results obtained from good spray cover of small actively growing weeds. Effectiveness is reduced in very dry conditions
- Weed spectrum can be broadened by tank mixture with other herbicides. See label for details
- Susceptible weeds cease growth immediately and symptoms can be seen 10 d later
- Rimsulfuron is a member of the ALS-inhibitor group of herbicides

Restrictions

- Maximum number of treatments 1 per crop
- Do not treat maize previously treated with organophosphorus insecticides
- Do not apply to potatoes grown for certified seed
- Consult processor before use on crops grown for processing
- Avoid high light intensity (full sunlight) and high temperatures on the day of spraying
- Do not treat during periods of substantial diurnal temperature fluctuation or when frost anticipated

FOR FULL CONDITIONS OF USE ALWAYS READ THE PRODUCT LABEL

- Do not apply to any crop stressed by drought, water-logging, low temperatures, pest or disease attack, nutrient or lime deficiency
- Do not apply to forage maize treated with organophosphate insecticides
- Do not apply to forage maize undersown with grass or clover

Crop-specific information
- Latest use: before most advanced potato plants are 25 cm high; before 4-collar stage of fodder maize
- All varieties of ware potatoes may be treated, but variety restrictions of any tank-mix partner must be observed
- Only certain named varieties of forage maize may be treated. See label

Following crops guidance
- Only winter wheat should follow a treated crop in the same calendar yr
- Only barley, wheat or maize should be sown in the spring of the yr following treatment
- In the second autumn after treatment any crop except brassicas or oilseed rape may be drilled

Environmental safety
- Dangerous for the environment
- Toxic to aquatic organisms
- Extremely dangerous to fish or other aquatic life. Do not contaminate surface waters or ditches with chemical or used container
- Herbicide is very active. Take particular care to avoid drift onto plants outside the target area
- Spraying equipment should not be drained or flushed onto land planted, or to be planted, with trees or crops other than potatoes or forage maize and should be thoroughly cleansed after use - see label for instructions

Hazard classification and safety precautions
Hazard Dangerous for the environment, Very toxic to aquatic organisms
Transport code 9 [1-4]
Packaging group III
UN Number 3077
Operator protection U08, U19a, U20b; A
Environmental protection E13a, E34, E38, H410
Storage and disposal D01, D09a, D11a, D12a

331 silthiofam

A thiophene carboxamide fungicide seed dressing for cereals
FRAC mode of action code: 38

Products
1 Latitude	Certis Belchim B V	125 g/l	FS	18036

Uses
- Seed-borne diseases in **Rye** *(off-label)* [1]; **Triticale** *(off-label)* [1];
- Take-all in **Rye** *(off-label)* [1]; **Spring wheat** *(seed treatment)* [1]; **Triticale** *(off-label)* [1]; **Winter barley** *(seed treatment)* [1]; **Winter wheat** *(seed treatment)* [1];

Extension of Authorisation for Minor Use (EAMUs)
- **Rye** *20193262* [1]
- **Triticale** *20193262* [1]

Approval information
- Silthiofam included in Annex I under EC Regulation 1107/2009
- Accepted by BBPA for use on malting barley

Efficacy guidance
- Apply using approved seed treatment equipment which has been accurately calibrated
- Apply simultaneously (i.e. not in mixture) with a standard seed treatment using a recommended seed treatment machine

SEE SECTION 3 FOR PRODUCTS ALSO REGISTERED

- Drill treated seed in the season of purchase. The viability of treated seed and fungicide activity may be reduced by physical storage
- Drill treated seed at 2.5-4 cm into a well prepared firm seedbed
- As precaution against possible development of disease resistance do not treat more than three consecutive susceptible cereal crops in any one rotation

Restrictions
- Maximum number of treatments one per batch of seed
- Do not use on seed with more than 16% moisture content, or on sprouted, cracked or skinned seed
- Test germination of all seed batches before treatment

Crop-specific information
- Latest use: immediately prior to drilling

Hazard classification and safety precautions
　UN Number N/C
　Operator protection U20b; A, H
　Environmental protection E03, E15a, E34
　Storage and disposal D05, D09a, D10a
　Treated seed S01, S02, S04b, S05, S06a, S07

332　S-metolachlor

A chloroacetamide residual herbicide for use in forage maize and grain maize
HRAC mode of action code: 15 (K3)

See also mesotrione + S-metolachlor

Products

1	Clayton Smelter	Clayton	960 g/l	EC	14937
2	Dual Gold	Syngenta	960 g/l	EC	14649

Uses
- Annual dicotyledons in **Baby leaf crops** *(off-label)* [2]; **Forest nurseries** *(off-label)* [2]; **Garlic** *(off-label)* [2]; **Herbs (see appendix 6)** *(off-label)* [2]; **Lettuce** *(off-label)* [2]; **Onions** *(off-label)* [2]; **Ornamental plant production** *(off-label)* [2]; **Red beet** *(off-label)* [2]; **Shallots** *(off-label)* [2]; **Swedes** *(off-label)* [2]; **Turnips** *(off-label)* [2];
- Annual grasses in **Begonias** *(off-label)* [2]; **Broccoli** *(off-label)* [2]; **Brussels sprouts** *(off-label)* [2]; **Cabbages** *(off-label)* [2]; **Calabrese** *(off-label)* [2]; **Cauliflowers** *(off-label)* [2]; **Chicory** *(off-label)* [1-2]; **Chinese cabbage** *(off-label)* [2]; **Collards** *(off-label)* [2]; **Dwarf beans** *(off-label)* [1-2]; **Endives** *(off-label)* [1-2]; **French beans** *(off-label)* [1-2]; **Kale** *(off-label)* [2]; **Runner beans** *(off-label)* [1-2]; **Strawberries** *(off-label)* [1-2];
- Annual meadow grass in **Baby leaf crops** *(off-label)* [2]; **Edible podded peas** *(off-label)* [2]; **Forage maize** [1-2]; **Forest nurseries** *(off-label)* [2]; **Garlic** *(off-label)* [2]; **Grain maize** [1-2]; **Herbs (see appendix 6)** *(off-label)* [2]; **Lettuce** *(off-label)* [2]; **Onions** *(off-label)* [2]; **Ornamental plant production** *(off-label)* [2]; **Red beet** *(off-label)* [2]; **Shallots** *(off-label)* [2]; **Spring field beans** *(off-label)* [2]; **Swedes** *(off-label)* [2]; **Sweetcorn** *(off-label)* [2]; **Turnips** *(off-label)* [2]; **Vining peas** *(off-label)* [2];
- Barnyard grass in **Sweetcorn** *(off-label)* [2];
- Black nightshade in **Chicory** *(off-label)* [1-2]; **Endives** *(off-label)* [1-2];
- Chickweed in **Forage maize** *(moderately susceptible)* [1-2]; **Grain maize** *(moderately susceptible)* [1-2]; **Spring field beans** *(off-label)* [2];
- Common field speedwell in **Spring field beans** *(off-label)* [2];
- Crane's-bill in **Sweetcorn** *(off-label)* [2];
- Fat hen in **Forage maize** *(moderately susceptible)* [1-2]; **Grain maize** *(moderately susceptible)* [1-2]; **Spring field beans** *(off-label)* [2];
- Field speedwell in **Forage maize** [1-2]; **Grain maize** [1-2];
- Groundsel in **Sweetcorn** *(off-label)* [2];

- Mayweeds in **Forage maize** [1-2]; **Grain maize** [1-2]; **Spring field beans** *(off-label)* [2]; **Sweetcorn** *(off-label)* [2];
- Red dead-nettle in **Forage maize** [1-2]; **Grain maize** [1-2];
- Sowthistle in **Forage maize** [1-2]; **Grain maize** [1-2]; **Sweetcorn** *(off-label)* [2];

Extension of Authorisation for Minor Use (EAMUs)
- **Baby leaf crops** *20120594* [2]
- **Begonias** *20132574* [2]
- **Broccoli** *20111258* [2]
- **Brussels sprouts** *20111259* [2]
- **Cabbages** *20111259* [2]
- **Calabrese** *20111258* [2]
- **Cauliflowers** *20111258* [2]
- **Chicory** *20111262* [1], *20111257* [2]
- **Chinese cabbage** *20111260* [2]
- **Collards** *20111260* [2]
- **Dwarf beans** *20103103* [1], *20132572* [2]
- **Edible podded peas** *20140164* [2]
- **Endives** *20111262* [1], *20111257* [2]
- **Forest nurseries** *20120501* [2]
- **French beans** *20103103* [1], *20132572* [2]
- **Garlic** *20110840* [2]
- **Herbs (see appendix 6)** *20120594* [2]
- **Kale** *20111260* [2]
- **Lettuce** *20120594* [2]
- **Onions** *20110840* [2]
- **Ornamental plant production** *20120501* [2]
- **Red beet** *20111006* [2]
- **Runner beans** *20103103* [1], *20132572* [2]
- **Shallots** *20110840* [2]
- **Spring field beans** *20212938* [2]
- **Strawberries** *20103104* [1], *20132573* [2]
- **Swedes** *20111006* [2]
- **Sweetcorn** *20172834* [2]
- **Turnips** *20111006* [2]
- **Vining peas** *20140164* [2]

Approval information
- S-metolachlor included in Annex I under EC Regulation 1107/2009

Environmental safety
- LERAP Category B [1-2]

Hazard classification and safety precautions
Hazard Irritant, Dangerous for the environment, Very toxic to aquatic organisms
Transport code 9 [1-2]
Packaging group III
UN Number 3082
Risk phrases H317, H319
Operator protection U05a, U09a, U20b; A, C, H
Environmental protection E15b, E16a, E16b, E38, H410
Storage and disposal D01, D02, D09a, D10c, D11a, D12a

333 sodium silver thiosulphate

A compound for post-harvest use in cut flowers

Products

1	Chrysal AVB	Chrysal	40 g/l	SL	19041

SEE SECTION 3 FOR PRODUCTS ALSO REGISTERED

SECTION 2

Uses

- Flower life prolongation in *Ornamental plant production* [1];

Approval information

- Sodium silver thiosulphate included in Annex I under EC Regulation 1107/2009

Efficacy guidance

- Acts by inhibiting production of ethylene
- Spray flowering pot plants to run off 8-14 d before shipment from glasshouse or at 3 wk intervals. Best time for spraying is late afternoon
- Vase life of cut flowers extended by dip treatment immediately after cutting

Restrictions

- Spraying pot plants just before shipping may cause damage

Environmental safety

- Not to be used on food crops

Hazard classification and safety precautions

Operator protection A
Environmental protection H410

334   spearmint oil

A hot fogging treatment for post-harvest sprout suppression in potatoes

Products

1 Biox-M	Juno	100% w/w	HN	16021

Uses

- Sprout suppression in *Potatoes* [1];

Approval information

- Not listed in Annex 1 under EC Regulation 1107/2009 since it is a natural product.

Hazard classification and safety precautions

Hazard Harmful, Dangerous for the environment, Very toxic to aquatic organisms
UN Number N/C
Risk phrases H304, H317
Operator protection A, C, D, H, J, M
Storage and disposal D12a

335   spinosad

A selective insecticide derived from naturally occurring soil fungi (naturalyte)
IRAC mode of action code: 5

Products

1 Clayton Spinoff	Clayton	480 g/l	SC	19465
2 Conserve	Fargro	120 g/l	SC	12058
3 Spindle	Pan Agriculture	120 g/l	SC	18949
4 Tracer	Landseer	480 g/l	SC	12438

Uses

- Asparagus beetle in *Asparagus* (off-label) [4];
- Blackcurrant sawfly in *Blackcurrants* (off-label) [4]; *Redcurrants* (off-label) [4];
- Blastobasis lacticoella in *Protected strawberries* (off-label) [4];
- Cabbage moth in *Broccoli* [1,4]; *Brussels sprouts* [1,4]; *Cabbages* [1,4]; *Calabrese* [1,4]; *Cauliflowers* [1,4]; *Chinese cabbage* [1,4]; *Collards* (off-label) [4]; *Kale* (off-label) [4]; *Kohlrabi* (off-label) [4]; *Oriental cabbage* (off-label) [4];
- Cabbage root fly in *Collards* (off-label) [4]; *Kale* (off-label) [4]; *Kohlrabi* (off-label) [4]; *Oriental cabbage* (off-label) [4];

FOR FULL CONDITIONS OF USE ALWAYS READ THE PRODUCT LABEL

- Cabbage white butterfly in *Broccoli* [1,4]; *Brussels sprouts* [1,4]; *Cabbages* [1,4]; *Calabrese* [1,4]; *Cauliflowers* [1,4]; *Chinese cabbage* [1,4]; *Collards* *(off-label)* [4]; *Kale* *(off-label)* [4]; *Kohlrabi* *(off-label)* [4]; *Oriental cabbage* *(off-label)* [4];
- Caterpillars in *Celeriac* *(off-label)* [4]; *Celery (outdoor)* *(off-label)* [4]; *Celery leaves* *(off-label)* [4]; *Chard* *(off-label)* [4]; *Chicory* *(off-label)* [4]; *Collards* *(off-label)* [4]; *Courgettes* *(off-label)* [4]; *Cress* *(off-label)* [4]; *Dwarf French beans* *(off-label)* [4]; *Endives* *(off-label)* [4]; *Fennel* *(off-label)* [4]; *Frise* *(off-label)* [4]; *Grapevines* *(off-label)* [4]; *Herbs (see appendix 6)* *(off-label)* [4]; *Kale* *(off-label)* [4]; *Kohlrabi* *(off-label)* [4]; *Lamb's lettuce* *(off-label)* [4]; *Leaf brassicas* *(off-label)* [4]; *Lettuce* *(off-label)* [4]; *Oriental cabbage* *(off-label)* [4]; *Pattisons* *(off-label)* [4]; *Protected herbs (see appendix 6)* *(off-label)* [4]; *Protected melons* *(off-label)* [4]; *Protected pumpkins* *(off-label)* [4]; *Radicchio* *(off-label)* [4]; *Red mustard* *(off-label)* [4]; *Rocket* *(off-label)* [4]; *Scarole* *(off-label)* [4]; *Spinach* *(off-label)* [4]; *Spinach beet* *(off-label)* [4]; *Swedes* *(off-label)* [4]; *Turnips* *(off-label)* [4]; *Witloof* *(off-label)* [4];
- Celery fly in *Celery (outdoor)* *(off-label)* [4]; *Fennel* *(off-label)* [4];
- Celery leaf miner in *Celery (outdoor)* *(off-label)* [4];
- Codling moth in *Apples* [1,4];
- Diamond-back moth in *Broccoli* [1,4]; *Brussels sprouts* [1,4]; *Cabbages* [1,4]; *Calabrese* [1,4]; *Cauliflowers* [1,4]; *Chinese cabbage* [1,4]; *Collards* *(off-label)* [4]; *Kale* *(off-label)* [4]; *Kohlrabi* *(off-label)* [4]; *Oriental cabbage* *(off-label)* [4];
- Insect pests in *Bush fruit* *(off-label)* [4]; *Cane fruit* *(off-label)* [4]; *Hops* *(off-label)* [4]; *Nursery fruit trees* *(off-label)* [4]; *Ornamental plant production* *(off-label)* [4]; *Protected strawberries* [1,4]; *Strawberries* *(off-label)* [4]; *Whitecurrants* *(off-label)* [4];
- Large white butterfly in *Collards* *(off-label)* [4]; *Kale* *(off-label)* [4]; *Kohlrabi* *(off-label)* [4]; *Oriental cabbage* *(off-label)* [4];
- Leaf miner in *Celeriac* *(off-label)* [4]; *Chicory* *(off-label)* [4]; *Courgettes* *(off-label)* [4]; *Fennel* *(off-label)* [4]; *Pattisons* *(off-label)* [4]; *Protected melons* *(off-label)* [4]; *Protected pumpkins* *(off-label)* [4]; *Swedes* *(off-label)* [4]; *Turnips* *(off-label)* [4]; *Witloof* *(off-label)* [4];
- Light brown apple moth in *Blueberries* *(off-label)* [4]; *Cranberries* *(off-label)* [4]; *Gooseberries* *(off-label)* [4]; *Haskap* *(off-label)* [4]; *Protected blackcurrants* *(off-label)* [4]; *Protected blueberries* *(off-label)* [4]; *Protected cranberries* *(off-label)* [4]; *Protected gooseberries* *(off-label)* [4]; *Protected haskap* *(off-label)* [4]; *Protected redcurrants* *(off-label)* [4];
- Onion thrips in *Celery leaves* *(off-label)* [4]; *Chard* *(off-label)* [4]; *Cress* *(off-label)* [4]; *Endives* *(off-label)* [4]; *Frise* *(off-label)* [4]; *Herbs (see appendix 6)* *(off-label)* [4]; *Lamb's lettuce* *(off-label)* [4]; *Leaf brassicas* *(off-label)* [4]; *Lettuce* *(off-label)* [4]; *Protected herbs (see appendix 6)* *(off-label)* [4]; *Radicchio* *(off-label)* [4]; *Red mustard* *(off-label)* [4]; *Rocket* *(off-label)* [4]; *Scarole* *(off-label)* [4]; *Spinach* *(off-label)* [4]; *Spinach beet* *(off-label)* [4];
- Sawflies in *Whitecurrants* *(off-label)* [4];
- Silver Y moth in *Celery (outdoor)* *(off-label)* [4]; *Celery leaves* *(off-label)* [4]; *Chard* *(off-label)* [4]; *Cress* *(off-label)* [4]; *Dwarf French beans* *(off-label)* [4]; *Endives* *(off-label)* [4]; *Fennel* *(off-label)* [4]; *Frise* *(off-label)* [4]; *Herbs (see appendix 6)* *(off-label)* [4]; *Lamb's lettuce* *(off-label)* [4]; *Leaf brassicas* *(off-label)* [4]; *Lettuce* *(off-label)* [4]; *Protected herbs (see appendix 6)* *(off-label)* [4]; *Radicchio* *(off-label)* [4]; *Red mustard* *(off-label)* [4]; *Rocket* *(off-label)* [4]; *Scarole* *(off-label)* [4]; *Spinach* *(off-label)* [4]; *Spinach beet* *(off-label)* [4];
- Small white butterfly in *Broccoli* [1,4]; *Brussels sprouts* [1,4]; *Cabbages* [1,4]; *Calabrese* [1,4]; *Cauliflowers* [1,4]; *Chinese cabbage* [1,4]; *Collards* *(off-label)* [4]; *Kale* *(off-label)* [4]; *Kohlrabi* *(off-label)* [4]; *Oriental cabbage* *(off-label)* [4];
- Spotted wing drosophila in *Blueberries* *(off-label)* [4]; *Cranberries* *(off-label)* [4]; *Gooseberries* *(off-label)* [4]; *Haskap* *(off-label)* [4]; *Loganberries* *(off-label)* [4]; *Protected blackcurrants* *(off-label)* [4]; *Protected blueberries* *(off-label)* [4]; *Protected cranberries* *(off-label)* [4]; *Protected gooseberries* *(off-label)* [4]; *Protected haskap* *(off-label)* [4]; *Protected redcurrants* *(off-label)* [4]; *Rubus hybrids* *(off-label)* [4]; *Strawberries* *(off-label)* [4];
- Thrips in *Asparagus* *(off-label)* [4]; *Blackberries* *(off-label)* [4]; *Bulb onions* [1,4]; *Bush fruit* *(off-label)* [4]; *Cane fruit* *(off-label)* [4]; *Celeriac* *(off-label)* [4]; *Celery (outdoor)* *(off-label)* [4]; *Chicory* *(off-label)* [4]; *Courgettes* *(off-label)* [4]; *Fennel* *(off-label)* [4]; *Garlic* [1,4]; *Hops* *(off-label)* [4]; *Leeks* [1,4]; *Nursery fruit trees* *(off-label)* [4]; *Ornamental plant production* *(off-label)* [4]; *Pattisons* *(off-label)* [4]; *Protected blackberries* *(off-label)* [4]; *Protected melons* *(off-label)* [4]; *Protected pumpkins* *(off-label)* [4]; *Protected raspberries* *(off-label)* [4]; *Protected strawberries* [1,4]; *Raspberries* *(off-label)* [4]; *Salad onions* [1,4]; *Shallots* [1,4]; *Strawberries* *(off-label)* [4]; *Swedes* *(off-label)* [4]; *Turnips* *(off-label)* [4]; *Witloof* *(off-label)* [4];

SEE SECTION 3 FOR PRODUCTS ALSO REGISTERED

- Tomato leaf miner in **Protected tomatoes** *(off-label)* [2];
- Tortrix moths in **Apples** [1,4];
- Western flower thrips in **Protected aubergines** [2-3]; **Protected cucumbers** [2-3]; **Protected ornamentals** [2-3]; **Protected peppers** [2-3]; **Protected tomatoes** *(off-label)* [2-3];
- Winter moth in **Blackcurrants** *(off-label)* [4]; **Redcurrants** *(off-label)* [4]; **Whitecurrants** *(off-label)* [4];

Extension of Authorisation for Minor Use (EAMUs)

- **Asparagus** *20210733* [4]
- **Blackberries** *20210735* [4]
- **Blackcurrants** *20210737* [4]
- **Blueberries** *20210740* [4]
- **Bush fruit** *20210734* [4]
- **Cane fruit** *20210734* [4]
- **Celeriac** *20210743* [4]
- **Celery (outdoor)** *20210745* [4]
- **Celery leaves** *20210746* [4]
- **Chard** *20210746* [4]
- **Chicory** *20210743* [4]
- **Collards** *20210741* [4]
- **Courgettes** *20210743* [4]
- **Cranberries** *20210740* [4]
- **Cress** *20210746* [4]
- **Dwarf French beans** *20210744* [4]
- **Endives** *20210746* [4]
- **Fennel** *20181206* [4]
- **Frise** *20181205* [4], *20210746* [4]
- **Gooseberries** *20210740* [4]
- **Grapevines** *20210732* [4]
- **Haskap** *20210740* [4]
- **Herbs (see appendix 6)** *20210746* [4]
- **Hops** *20210734* [4]
- **Kale** *20210741* [4]
- **Kohlrabi** *20210741* [4]
- **Lamb's lettuce** *20210746* [4]
- **Leaf brassicas** *20210746* [4]
- **Lettuce** *20210746* [4]
- **Loganberries** *20210736* [4]
- **Nursery fruit trees** *20210734* [4]
- **Oriental cabbage** *20210741* [4]
- **Ornamental plant production** *20210734* [4]
- **Pattisons** *20210743* [4]
- **Protected blackberries** *20210738* [4]
- **Protected blackcurrants** *20210740* [4]
- **Protected blueberries** *20210740* [4]
- **Protected cranberries** *20210740* [4]
- **Protected gooseberries** *20210740* [4]
- **Protected haskap** *20210740* [4]
- **Protected herbs (see appendix 6)** *20210746* [4]
- **Protected melons** *20210743* [4]
- **Protected pumpkins** *20210743* [4]
- **Protected raspberries** *20210738* [4]
- **Protected redcurrants** *20210740* [4]
- **Protected strawberries** *20210739* [4]
- **Protected tomatoes** *20181208* [2], *20210721* [2]
- **Radicchio** *20210746* [4]
- **Raspberries** *20210735* [4]
- **Red mustard** *20210746* [4]
- **Redcurrants** *20210737* [4]

FOR FULL CONDITIONS OF USE ALWAYS READ THE PRODUCT LABEL

- **Rocket** *20210746* [4]
- **Rubus hybrids** *20210736* [4]
- **Scarole** *20210746* [4]
- **Spinach** *20210746* [4]
- **Spinach beet** *20210746* [4]
- **Strawberries** *20210734* [4], *20210742* [4]
- **Swedes** *20210743* [4]
- **Turnips** *20210743* [4]
- **Whitecurrants** *20113223* [4]
- **Witloof** *20210743* [4]

Approval information
- Spinosad included in Annex I under EC Regulation 1107/2009

Efficacy guidance
- Product enters insects by contact from a treated surface or ingestion of treated plant material therefore good spray coverage is essential
- Some plants, for example Fuchsia flowers, can provide effective refuges from spray deposits and control of western flower thrips may be reduced [2]
- Apply to protected crops when western flower thrip nymphs or adults are first seen [2]
- Monitor western flower thrip development carefully to see whether further applications are necessary. A 2 spray programme at 5-7 d intervals may be needed when conditions favour rapid pest development [2]
- Treat top fruit and field crops when pests are first seen or at very first signs of crop damage [4]
- Ensure a rain-free period of 12 h after treatment before applying irrigation [4]
- For susceptible predators (parasitic hymenoptera) re-introduction is possible after 7 days following application (with perhaps 14 days in winter months). For most other predators introduction is possible 24 hours after application. Re-introduction of Orius laevigatus is advised one week later [2]
- To reduce possibility of development of resistance, adopt resistance management measures. See label and Section 5 for more information

Restrictions
- Maximum number of treatments 4 per crop for brassicas, onions, leeks [4]; 1 pre-blossom and 3 post blossom for apples, pears [4]
- Apply no more than 2 consecutive sprays. Rotate with another insecticide with a different mode of action or use no further treatment after applying the maximum number of treatments
- Establish whether any incoming plants have been treated and apply no more than 2 consecutive sprays. Maximum number of treatments 6 per structure per yr
- Avoid application in bright sunlight or into open flowers [2]
- In protected situations the total number of applications of any spinosad containing product must not exceed 6 per glasshouse/protected structure in a 12 month period, regardless of the crop being treated (including ornamentals)

Crop-specific information
- HI 3 d for protected cucumbers, brassicas; 7 d for apples, pears, onions, leeks
- Test for tolerance on a small number of ornamentals or cucumbers before large scale treatment
- Some spotting of african violet flowers may occur

Environmental safety
- Dangerous for the environment
- Very toxic to aquatic organisms
- Whenever possible use an Integrated Pest Management system. Spinosad presents low risk to beneficial arthropods
- Products have a low impact on many insect and mite predators but is harmful to adults of most parasitic wasps. Most beneficials may be introduced to treated plants when spray deposits are dry but an interval of 2 wk should elapse before introduction of parasitic wasps. See label for details
- Treatment may cause temporary reduction in abundance of insect and mite predators if present at application

SEE SECTION 3 FOR PRODUCTS ALSO REGISTERED

- Exposure to direct spray is harmful to bees but dry spray deposits are harmless. Treatment of field crops should not be made in the heat of the day when bees are actively foraging [4]
- Buffer zone requirement 10 m for melon and pumpkin, 5 m for celeriac, swede and turnip
- Broadcast air-assisted LERAP [1,4] (40 m); LERAP Category B [1,4]

Hazard classification and safety precautions

Hazard Dangerous for the environment, Very toxic to aquatic organisms [1]
Transport code 9 [1-4]
Packaging group III
UN Number 3082
Operator protection U05a [2-3], U08 [1,4], U20b [1,4]; A, H
Environmental protection E15a [1,4], E15b [2-3], E16a [1,4], E17b (40 m) [1,4], E34, E38, H410
Storage and disposal D01 [2-3], D02, D05 [2-3], D10b, D12a

336   spirotetramat

Tetramic acid derivative that acts on lipid synthesis
IRAC mode of action code: 23

Products

1	Batavia	Bayer CropScience	100 g/l	SC	18449
2	Clayton Occupy	Clayton	150 g/l	OD	18924
3	Cleancrop Sierra	Agrii	150 g/l	OD	19005
4	Movento	Bayer CropScience	150 g/l	OD	18435
5	Spiropan	Pan Amenity	100 g/l	SC	19556

Uses

- Aphids in *Apples* [1,5]; *Apricots* *(off-label)* [1]; *Blackcurrants* *(off-label)* [1]; *Blueberries* *(off-label)* [1]; *Broccoli* [2-4]; *Brussels sprouts* [2-4]; *Bulb onions* [2-4]; *Cabbages* [2-4]; *Calabrese* [2-4]; *Carrots* [2-4]; *Cauliflowers* [2-4]; *Cherries* [1,5]; *Collards* [2-4]; *Gooseberries* *(off-label)* [1]; *Kale* [2-4]; *Lettuce* [2-4]; *Ornamental plant production* *(off-label)* [1]; *Parsnips* [2-4]; *Peaches* *(off-label)* [1]; *Pears* [1,5]; *Plums* [1,5]; *Potatoes* [2-4]; *Protected strawberries* [1,5]; *Redcurrants* *(off-label)* [1]; *Shallots* [2-4]; *Strawberries* [1,5]; *Swedes* [2-4]; *Sweetcorn* *(off-label)* [1]; *Turnips* [2-4]; *Whitecurrants* *(off-label)* [1];
- Bird-cherry aphid in *Sweetcorn* *(off-label)* [1];
- Blackcurrant gall mite in *Blackcurrants* *(off-label)* [1]; *Blueberries* *(off-label)* [1]; *Gooseberries* *(off-label)* [1]; *Redcurrants* *(off-label)* [1]; *Whitecurrants* *(off-label)* [1];
- Carrot willow aphid in *Celery (outdoor)* *(off-label)* [1];
- Damson-hop aphid in *Hops* *(off-label)* [1];
- Glasshouse whitefly in *Protected ornamentals* *(off-label)* [1];
- Grain aphid in *Sweetcorn* *(off-label)* [1];
- Mussel scale in *Apples* [1,5]; *Pears* [1,5];
- Peach-potato aphid in *Celery (outdoor)* *(off-label)* [1]; *Protected ornamentals* *(off-label)* [1];
- Rose-grain aphid in *Sweetcorn* *(off-label)* [1];
- Scale insects in *Wine grapes* *(off-label)* [1];
- Soft scales Coccidae in *Wine grapes* *(off-label)* [1];
- Tarsonemid mites in *Protected strawberries* [1,5]; *Strawberries* [1,5];
- Vine phylloxera in *Wine grapes* *(off-label)* [1];
- Western flower thrips in *Protected ornamentals* *(off-label)* [1];
- Whitefly in *Broccoli* [2-4]; *Brussels sprouts* [2-4]; *Bulb onions* [2-4]; *Cabbages* [2-4]; *Calabrese* [2-4]; *Cauliflowers* [2-4]; *Collards* [2-4]; *Kale* [2-4]; *Shallots* [2-4]; *Swedes* [2-4]; *Turnips* [2-4];
- Woolly aphid in *Apples* *(off-label)* [1];

Extension of Authorisation for Minor Use (EAMUs)

- *Apples* 20192594 [1], 20221261 [1]
- *Apricots* 20191055 [1], 20221269 [1]
- *Blackcurrants* 20210680 [1], 20221263 [1]
- *Blueberries* 20210680 [1], 20221263 [1]
- *Celery (outdoor)* 20201115 [1], 20221260 [1]

FOR FULL CONDITIONS OF USE ALWAYS READ THE PRODUCT LABEL

- **Gooseberries** *20210680* [1], *20221263* [1]
- **Hops** *20191056* [1], *20221270* [1]
- **Ornamental plant production** *20191058* [1], *20221267* [1]
- **Peaches** *20191055* [1], *20221269* [1]
- **Protected ornamentals** *20192597* [1], *20221268* [1]
- **Redcurrants** *20210680* [1], *20221263* [1]
- **Sweetcorn** *20193152* [1], *20221264* [1]
- **Whitecurrants** *20210680* [1], *20221263* [1]
- **Wine grapes** *20191057* [1], *20221262* [1]

Approval information
- Spirotetramat is included in Annex 1 under EC Regulation 1107/2009
- Accepted by BBPA for use on hops

Restrictions
- Application to baby leaf crops, cress, endive, lamb?s lettuce, land cress, rocket and spinach must only be made between growth stages BBCH 12 and BBCH 49
- Applications to blackcurrants, redcurrants, blueberries and gooseberries must be made between March and July and only in a year where there is no harvest of the crop
- DO NOT apply to the following ornamental plant species due to a high risk of phytotoxicity (plant damage): Alstroemeria spp., Begonia spp., Cyclamen spp., Euphorbia spp., Ficus spp., Fuchsia spp., Hedera spp., Hydrangea spp., Impatiens spp., Pelargonium spp., Populus spp., Salix spp., Saintpaulia spp., Tilia spp., Quercus frainetto [1]

Environmental safety
- Aquatic buffer zone 15 m in apples [1]

Hazard classification and safety precautions
Hazard Irritant, Dangerous for the environment [2-4]
Transport code 9 [1-5]
Packaging group III
UN Number 3082
Risk phrases H317, H319 [2-4], H361
Operator protection U05a, U14 [2-4], U20b; A, C, H
Environmental protection E12c, E15b, E17a (10 m) [1,5], E22c [2-4], E34, E38 [2-4], H411
Storage and disposal D02, D09a, D10c, D12a
Medical advice M03

337   starch, protein, oil and water

Natural plant extracts for control of moss, liverworts and algae

Products

1 MossKade	Fargro		n/a g/l	SL	00000

Uses
- Algae in **Hard surfaces** [1];
- Liverworts in **Hard surfaces** [1];
- Moss in **Hard surfaces** [1];

Approval information
- No listing required on Annex 1

Efficacy guidance
- Treat surfaces at first sign of moss, lichen, liverworts or algae since young growth is more susceptible.
- Must dry on the treated surface before it begins to work
- Summer growth is less susceptible than growth developing in autumn, winter or spring
- Repeat applications may be necessary for full efficacy

SEE SECTION 3 FOR PRODUCTS ALSO REGISTERED

Hazard classification and safety precautions
  UN Number N/C
  Operator protection U08, U09a, U19a; A, H
  Storage and disposal D06c, D07

338   sulfosulfuron

A sulfonylurea herbicide for grass and broad-leaved weed control in winter wheat
HRAC mode of action code: 2 (B)

See also glyphosate + sulfosulfuron

Products

1	Monitor	Sumitomo	80% w/w	WG	18887

Uses
- Brome grasses in **Winter wheat** *(moderate control of barren brome)* [1];
- Chickweed in **Winter wheat** [1];
- Cleavers in **Winter wheat** [1];
- Couch in **Winter wheat** *(moderate control only)* [1];
- Loose silky bent in **Winter wheat** [1];
- Mayweeds in **Winter wheat** [1];
- Onion couch in **Winter wheat** *(moderate control only)* [1];

Approval information
- Sulfosulfuron included in Annex I under EC Regulation 1107/2009

Efficacy guidance
- For best results treat in early spring when annual weeds are small and growing actively. Avoid treatment when weeds are dormant for any reason
- Best control of onion couch is achieved when the weed has more than two leaves. Effects on bulbils or on growth in the following yr have not been examined
- An extended period of dry weather before or after treatment may result in reduced control
- The addition of a recommended surfactant is essential for full activity
- Specific follow-up treatments may be needed for complete control of some weeds
- Use only where competitively damaging weed populations have emerged otherwise yield may be reduced
- Sulfosulfuron is a member of the ALS-inhibitor group of herbicides. To avoid the build up of resistance do not use any product containing an ALS-inhibitor herbicide with claims for control of grass weeds more than once on any crop
- Use these products as part of a resistance management strategy that includes cultural methods of control and does not use ALS inhibitors as the sole chemical method of grass weed control

Restrictions
- Maximum total dose equivalent to one treatment at full dose
- Do not treat crops under stress
- Apply only after 1 Feb and from 3 expanded leaf stage
- Do not treat durum wheat or any undersown wheat crop
- Do not use in mixture or in sequence with any other sulfonyl urea herbicide on the same crop

Crop-specific information
- Latest use: flag leaf ligule just visible for winter wheat (GS 39)

Following crops guidance
- In the autumn following treatment winter wheat, winter rye, winter oats, triticale, winter oilseed rape, winter peas or winter field beans may be sown on any soil, and winter barley on soils with less than 60% sand
- In the next spring following application crops of wheat, barley, oats, maize, peas, beans, linseed, oilseed rape, potatoes or grass may be sown
- In the second autumn following application winter linseed may be sown, and winter barley on soils with more than 60% sand

FOR FULL CONDITIONS OF USE ALWAYS READ THE PRODUCT LABEL

- Sugar beet or any other crop not mentioned above must not be drilled until the second spring following application
- Where winter oilseed rape is to be sown in the autumn following treatment soil cultivation to a minimum of 10 cm is recommended

Environmental safety
- Dangerous for the environment
- Very toxic to aquatic organisms
- Extremely dangerous to fish or other aquatic life. Do not contaminate surface waters or ditches with chemical or used container
- Take extreme care to avoid drift onto broad leaved plants or other crops, or onto ponds, waterways or ditches.
- Follow detailed label instructions for cleaning the sprayer to avoid damage to sensitive crops during subsequent use
- LERAP Category B [1]

Hazard classification and safety precautions
Hazard Dangerous for the environment, Very toxic to aquatic organisms
UN Number N/C
Operator protection U20b
Environmental protection E15a, E16a, E16b, E34, E38, H410
Storage and disposal D09a, D11a, D12a

339   sulfoxaflor

An insecticide for use on protected crops
IRAC mode of action code: 4C

Products

1	Sequoia	Fargro	120 g/l	SC	18938

Uses
- Aphids in **Protected angelica** *(off-label)* [1]; **Protected aubergines** [1]; **Protected caraway** *(off-label)* [1]; **Protected celery** *(off-label)* [1]; **Protected chilli peppers** [1]; **Protected coriander** *(off-label)* [1]; **Protected courgettes** [1]; **Protected cucumbers** [1]; **Protected dill** *(off-label)* [1]; **Protected fennel** *(off-label)* [1]; **Protected fenugreek** *(off-label)* [1]; **Protected gherkins** [1]; **Protected lettuce** *(off-label)* [1]; **Protected lovage** *(off-label)* [1]; **Protected melons** [1]; **Protected ornamentals** [1]; **Protected peppers** [1]; **Protected pumpkins** [1]; **Protected salad burnet** *(off-label)* [1]; **Protected sorrel** *(off-label)* [1]; **Protected spinach** *(off-label)* [1]; **Protected sweet cicerly** *(off-label)* [1]; **Protected tomatoes** [1]; **Protected watermelon** [1];
- Cotton aphids in **Protected blackberries** *(off-label)* [1]; **Protected loganberries** *(off-label)* [1]; **Protected raspberries** *(off-label)* [1]; **Protected Rubus hybrids** *(off-label)* [1];
- Large raspberry aphid in **Protected blackberries** *(off-label)* [1]; **Protected loganberries** *(off-label)* [1]; **Protected raspberries** *(off-label)* [1]; **Protected Rubus hybrids** *(off-label)* [1];
- Melon aphid in **Protected blackberries** *(off-label)* [1]; **Protected loganberries** *(off-label)* [1]; **Protected raspberries** *(off-label)* [1]; **Protected Rubus hybrids** *(off-label)* [1];
- Potato aphid in **Protected blackberries** *(off-label)* [1]; **Protected loganberries** *(off-label)* [1]; **Protected raspberries** *(off-label)* [1]; **Protected Rubus hybrids** *(off-label)* [1];
- Whitefly in **Protected aubergines** [1]; **Protected chilli peppers** [1]; **Protected courgettes** [1]; **Protected cucumbers** [1]; **Protected gherkins** [1]; **Protected melons** [1]; **Protected ornamentals** [1]; **Protected peppers** [1]; **Protected pumpkins** [1]; **Protected tomatoes** [1]; **Protected watermelon** [1];

Extension of Authorisation for Minor Use (EAMUs)
- **Protected angelica** *20220259* [1]
- **Protected blackberries** *20221028* [1]
- **Protected caraway** *20220259* [1]
- **Protected celery** *20220259* [1]
- **Protected coriander** *20220259* [1]
- **Protected dill** *20220259* [1]

SEE SECTION 3 FOR PRODUCTS ALSO REGISTERED

- ***Protected fennel*** *20220259* [1]
- ***Protected fenugreek*** *20220259* [1]
- ***Protected lettuce*** *20220259* [1]
- ***Protected loganberries*** *20221028* [1]
- ***Protected lovage*** *20220259* [1]
- ***Protected raspberries*** *20221028* [1]
- ***Protected Rubus hybrids*** *20221028* [1]
- ***Protected salad burnet*** *20220259* [1]
- ***Protected sorrel*** *20220259* [1]
- ***Protected spinach*** *20220259* [1]
- ***Protected sweet cicerly*** *20220259* [1]

Approval information
- Sulfoxaflor included in Annex I under EC Regulation 1107/2009

Efficacy guidance
- Has both systemic and translaminar activity and works by both contact and ingestion against aphids and whitefly. Use 400 ml/ha dose for whitefly control, 200 ml/ha for aphid control
- Acts on the insect nicotinic-acetylcholine site at a unique target receptor. Symptoms appear almost immediately and complete mortality occurs within a few hours
- Test the product on a small number of plants before treating the whole crop, especially when treating sensitive species
- May adversely impact on beneficial insects used in IPM as well as bees used for pollination services. Cover or remove bee colonies in glasshouses until the spray droplets have dried - usually 24 hours after application
- For further information and the latest advice on beneficial insects and mites and their integrated use, consult Corteva or Fargro Ltd

Restrictions
- See label for details of recommended ornamentals
- For EAMU 2022-0259 use is restricted to 200 ml/ha per protected crop with a maximum of two applications of 200 ml product/ha (24 g a.i./ha) per year per growing structure

Following crops guidance
- Edible plants have a 1 day harvest interval. Do not handle ornamentals for 8 days after application

Hazard classification and safety precautions
Transport code 9 [1]
Packaging group III
UN Number 3082
Operator protection U09a, U20b; A, H
Environmental protection E15b, E34, H411
Storage and disposal D01, D02, D10b, D11b, D22

340   sulphur

A broad-spectrum inorganic protectant fungicide, foliar feed and acaricide
FRAC mode of action code: M2

Products

1	Kumulus DF	BASF	80% w/w	WG	19420
2	Microthiol Special	UPL Europe	80% w/w	WG	19419
3	Thiopron	UPL Europe	82.5% g/l	SC	19147
4	Vertipin	Sipcam	700 g/l	SC	19508

Uses
- Black spot in ***Forest nurseries*** *(off-label)* [2]; ***Ornamental plant production*** *(off-label)* [2]; ***Protected ornamentals*** *(off-label)* [2];

- Foliar feed in *Rye* [3]; *Spring barley* ; *Spring oats* [2-3]; *Spring wheat* [2-3]; *Sugar beet* [2-3]; *Swedes* [2-3]; *Triticale* [3]; *Turnips* [2-3]; *Winter barley* [2-3]; *Winter oats* [2-3]; *Winter wheat* [2-3];
- Gall mite in *Blackcurrants* [1]; *Forest nurseries* *(off-label)* [1-2]; *Ornamental plant production* *(off-label)* [1-2]; *Protected ornamentals* *(off-label)* [2]; *Redcurrants* [1];
- Powdery mildew in *Apples* *(off-label)* [2]; *Blackcurrants* [2-3]; *Combining peas* *(off-label)* [3]; *Edible podded peas* *(off-label)* [3]; *Forest nurseries* *(off-label)* [1-2]; *Gooseberries* [1-3]; *Hops* *(off-label)* [2]; *Mange-tout peas* *(off-label)* [3]; *Ornamental plant production* *(off-label)* [1-2]; *Pears* *(off-label)* [2]; *Protected aubergines* *(off-label)* [2-3]; *Protected cayenne peppers* *(off-label)* [3]; *Protected chilli peppers* *(off-label)* [2-3]; *Protected courgettes* *(off-label)* [3]; *Protected cucumbers* *(off-label)* [2-3]; *Protected ornamentals* *(off-label)* [2]; *Protected peppers* *(off-label)* [2-3]; *Protected tomatoes* *(off-label)* [2-3]; *Redcurrants* [2-3]; *Rye* [3]; *Spring barley* [2-4]; *Spring oats* [2-4]; *Spring wheat* [2-4]; *Strawberries* [1-3]; *Sugar beet* [1-3]; *Sugar snap peas* *(off-label)* [3]; *Swedes* [2-3]; *Table grapes* [1-3]; *Triticale* [3]; *Turnips* [2-3]; *Vining peas* *(off-label)* [3]; *Wine grapes* [1-3]; *Winter barley* [2-4]; *Winter oats* [2-4]; *Winter wheat* [2-4];
- Rust in *Forest nurseries* *(off-label)* [2]; *Ornamental plant production* *(off-label)* [2]; *Protected ornamentals* *(off-label)* [2];
- Rust mite in *Protected tomatoes* *(off-label)* [2];
- Scab in *Apples* *(off-label)* [2]; *Forest nurseries* *(off-label)* [2]; *Ornamental plant production* *(off-label)* [2]; *Pears* *(off-label)* [2]; *Protected ornamentals* *(off-label)* [2];
- Tomato russet mite in *Protected aubergines* *(off-label)* [3]; *Protected cayenne peppers* *(off-label)* [3]; *Protected chilli peppers* *(off-label)* [3]; *Protected courgettes* *(off-label)* [3]; *Protected cucumbers* *(off-label)* [3]; *Protected peppers* *(off-label)* [3]; *Protected tomatoes* *(off-label)* [3];

Extension of Authorisation for Minor Use (EAMUs)
- *Apples* *20220412* [2]
- *Combining peas* *20211305* [3]
- *Edible podded peas* *20211305* [3]
- *Forest nurseries* *20201814* [1], *20220428* [2]
- *Hops* *20220412* [2]
- *Mange-tout peas* *20211305* [3]
- *Ornamental plant production* *20201814* [1], *20220428* [2]
- *Pears* *20220412* [2]
- *Protected aubergines* *20202702* [2], *20211970* [3]
- *Protected cayenne peppers* *20211970* [3]
- *Protected chilli peppers* *20202702* [2], *20211970* [3]
- *Protected courgettes* *20211970* [3]
- *Protected cucumbers* *20202702* [2], *20211970* [3]
- *Protected ornamentals* *20220428* [2]
- *Protected peppers* *20202702* [2], *20211970* [3]
- *Protected tomatoes* *20202049* [2], *20211970* [3]
- *Sugar snap peas* *20211305* [3]
- *Vining peas* *20211305* [3]

Approval information
- Sulphur is included in Annex 1 under EC Regulation 1107/2009
- Accepted by BBPA for use on malting barley and hops (before burr)

Efficacy guidance
- Apply when disease first appears and repeat 2-3 wk later. Details of application rates and timing vary with crop, disease and product. See label for information
- Sulphur acts as foliar feed as well as fungicide and with some crops product labels vary in whether treatment recommended for disease control or growth promotion
- In grassland best results obtained at least 2 wk before cutting for hay or silage, 3 wk before grazing
- Treatment unlikely to be effective if disease already established in crop

SEE SECTION 3 FOR PRODUCTS ALSO REGISTERED

Restrictions
- Maximum number of treatments normally 2 per crop for grassland, sugar beet, parsnips, swedes, hops, protected herbs; 3 per yr for blackcurrants, gooseberries; 4 per crop on apples, pears but labels vary
- Do not use on sulphur-shy apples (Beauty of Bath, Belle de Boskoop, Cox's Orange Pippin, Lanes Prince Albert, Lord Derby, Newton Wonder, Rival, Stirling Castle) or pears (Doyenne du Comice)
- Do not use on gooseberry cultivars Careless, Early Sulphur, Golden Drop, Leveller, Lord Derby, Roaring Lion, or Yellow Rough
- Do not use on apples or gooseberries when young, under stress or if frost imminent
- Do not use on fruit for processing, on grapevines during flowering or near harvest on grapes for wine-making
- Do not use on hops at or after burr stage
- Do not spray top or soft fruit with oil or within 30 d of an oil-containing spray
- A minimum interval of 10 days should be observed between applications

Crop-specific information
- Latest use: before burr stage for hops; before end Sep for parsnips, swedes, sugar beet; fruit swell for gooseberries; milky ripe stage for cereals
- HI cutting grass for hay or silage 2 wk; grazing grassland 3 wk

Environmental safety
- Sulphur products are attractive to livestock and must be kept out of their reach
- Do not empty into drains
- To protect non target insects/arthropods respect an unsprayed buffer zone of 5 m to non-crop land
- LERAP Category B [4]

Hazard classification and safety precautions
Hazard Irritant [4]
Transport code 9 [3]
Packaging group III
UN Number 3082, N/C
Risk phrases H317 [3], H318 [4], H319 [3]
Operator protection U05a [4], U19a [4], U20b [2-3], U20c [1,4]; A, H
Environmental protection E15a, E16a [4], E19b [4]
Storage and disposal D01 [2-4], D02 [4], D09a, D11a [1-3]

341   tau-fluvalinate

A contact pyrethroid insecticide for cereals and oilseed rape
IRAC mode of action code: 3

Products

| 1 | Mavrik | Adama | 240 g/l | EW | 10612 |

Uses
- Aphids in **Borage for oilseed production** *(off-label)* [1]; **Evening primrose** *(off-label)* [1]; **Grass seed crops** *(off-label)* [1]; **Honesty** *(off-label)* [1]; **Linseed** *(off-label)* [1]; **Mustard** *(off-label)* [1]; **Rye** *(off-label)* [1]; **Spring barley** [1]; **Spring oilseed rape** [1]; **Spring wheat** [1]; **Triticale** *(off-label)* [1]; **Winter barley** [1]; **Winter oilseed rape** [1]; **Winter wheat** [1];
- Barley yellow dwarf virus vectors in **Winter barley** [1]; **Winter wheat** [1];
- Cabbage stem flea beetle in **Winter oilseed rape** [1];
- Pollen beetle in **Spring oilseed rape** [1]; **Winter oilseed rape** [1];

Extension of Authorisation for Minor Use (EAMUs)
- **Borage for oilseed production** *20141329* [1]
- **Evening primrose** *20141329* [1]
- **Grass seed crops** *20193244* [1]
- **Honesty** *20141329* [1]
- **Linseed** *20141329* [1]
- **Mustard** *20141329* [1]

- **Rye** *20193244* [1]
- **Triticale** *20193244* [1]

Approval information
- Tau-fluvalinate included in Annex 1 under EC Regulation 1107/2009
- Accepted by BBPA for use on malting barley

Efficacy guidance
- For BYDV control on winter cereals follow local warnings or spray high risk crops in mid-Oct and make repeat application in late autumn/early winter if aphid activity persists
- For summer aphid control on cereals spray once when aphids present on two thirds of ears and increasing
- On oilseed rape treat peach potato aphids in autumn in response to local warning and repeat if necessary
- Best control of pollen beetle in oilseed rape obtained from treatment at green to yellow bud stage and repeat if necessary
- Good spray cover of target essential for best results
- Do not mix with boron or products containing boron

Restrictions
- Maximum total dose equivalent to two full dose treatments on oilseed rape. See label for dose rates on cereals
- Must not be applied to a cereal crop if any other product containing a pyrethroid has been applied to that crop after the start of ear emergence
- A minimum of 14 d must elapse between applications to cereals

Crop-specific information
- Latest use: before caryopsis watery ripe (GS 71) for barley; before flowering for oilseed rape; before kernel medium milk (GS 75) for wheat

Environmental safety
- Dangerous for the environment
- Very toxic to aquatic organisms
- High risk to non-target insects or other arthropods. Do not spray within 6 m of the field boundary [1]
- Avoid spraying oilseed rape within 6 m of field boundary to reduce effects on certain non-target species or other arthropods
- Must not be applied to cereals if any product containing a pyrethroid insecticide or dimethoate has been sprayed after the start of ear emergence (GS 51)
- LERAP Category A [1]

Hazard classification and safety precautions
Hazard Dangerous for the environment, Very toxic to aquatic organisms
Transport code 9 [1]
Packaging group III
UN Number 3082
Operator protection U05a, U10, U11, U19a, U20a; A, C, H
Environmental protection E15a, E16c, E16d, E22a, H410
Storage and disposal D01, D02, D05, D09a, D10c

SEE SECTION 3 FOR PRODUCTS ALSO REGISTERED

342   tebuconazole

A systemic triazole fungicide for cereals and other field crops
FRAC mode of action code: 3

See also azoxystrobin + tebuconazole
bixafen + prothioconazole + tebuconazole
bromuconazole + tebuconazole
fludioxonil + tebuconazole
fluopyram + prothioconazole + tebuconazole
fluoxastrobin + tebuconazole
prochloraz + tebuconazole
prothioconazole + spiroxamine + tebuconazole
prothioconazole + tebuconazole

Products

1	Clayton Ohio	Clayton	430 g/l	SC	19390
2	Clayton Tebbit	Clayton	250 g/l	EW	19610
3	Clayton Tebucon 250 EW	Clayton	250 g/l	EW	17823
4	Cleancrop Teboo	Agrii	250 g/l	EW	18111
5	Deacon	Nufarm UK	200 g/l	EW	19003
6	Fezan Extra	Sipcam	250 g/l	EW	19696
7	Hail	Albaugh UK	250 g/l	EW	19327
8	Orius	Nufarm UK	200 g/l	EW	18992
9	Surform	Agform	250 g/l	EW	19548
10	Tebucur 250	Certis Belchim B V	250 g/l	EW	13975
11	Tebuzol 250	UPL Europe	250 g/l	EW	17417
12	Teson	Certis Belchim B V	250 g/l	EW	17128
13	Tesoro	Certis Belchim B V	250 g/l	EW	17222
14	Toledo	Rotam	430 g/l	SC	18298
15	Tubosan	Certis Belchim B V	250 g/l	EW	17127
16	Ulysses	Rotam	430 g/l	SC	16052
17	Zonor	Life Scientific	250 g/l	EW	18650

Uses

- Alternaria in **Cabbages** [1,13-14,16]; **Carrots** [1,13-14,16]; **Celeriac** *(off-label)* [14]; **Horseradish** [1,13-14,16]; **Parsley root** *(off-label)* [14]; **Salsify** *(off-label)* [14]; **Spring oilseed rape** [1-5,7-17]; **Spring wheat** [1,14,16]; **Winter oilseed rape** [1-5,7-17]; **Winter wheat** [1,14,16];
- Blight in **Protected ornamentals** *(off-label)* [14];
- Botrytis in **Blackberries** *(off-label)* [10,12-15]; **Linseed** *(reduction)* [13]; **Loganberries** *(off-label)* [10,12-15]; **Protected blackberries** *(off-label)* [14]; **Protected loganberries** *(off-label)* [14]; **Protected raspberries** *(off-label)* [14]; **Protected Rubus hybrids** *(off-label)* [14]; **Raspberries** *(off-label)* [10,12-15]; **Rubus hybrids** *(off-label)* [10,12-15];
- Brown rust in **Rye** [1,14,16]; **Spring barley** [1-17]; **Spring rye** [5,8,10,12-13,15]; **Spring wheat** [1-17]; **Triticale** [1,14,16]; **Winter barley** [1-17]; **Winter rye** [2-13,15,17]; **Winter wheat** [1-17];
- Cane blight in **Blackberries** *(off-label)* [10,12-15]; **Loganberries** *(off-label)* [10,12-15]; **Protected blackberries** *(off-label)* [14]; **Protected loganberries** *(off-label)* [14]; **Protected raspberries** *(off-label)* [14]; **Protected Rubus hybrids** *(off-label)* [14]; **Raspberries** *(off-label)* [10,12-15]; **Rubus hybrids** *(off-label)* [10,12-15];
- Canker in **Apples** *(off-label)* [5,8,10,12-13,15]; **Chestnuts** *(off-label)* [10,12-13,15]; **Crab apples** *(off-label)* [5,8]; **Hazel nuts** *(off-label)* [10,12-13,15]; **Pears** *(off-label)* [5,8,10,12-13,15]; **Protected ornamentals** *(off-label)* [14]; **Quinces** *(off-label)* [5,8,10,12-13,15]; **Walnuts** *(off-label)* [10,12-13,15];
- Chocolate spot in **Spring field beans** *(moderate control)* [1,4-5,7-11,13-14,16-17]; **Winter field beans** *(moderate control)* [1,4-5,7-11,13-14,16-17];
- Cladosporium in **Spring wheat** [1,14,16]; **Winter wheat** [1,14,16];
- Crown rust in **Spring oats** *(reduction)* [1-7,9-17]; **Winter oats** *(reduction)* [1-17];
- Dark leaf spot in **Spring oilseed rape** [1,14,16]; **Winter oilseed rape** [1,14,16];
- Disease control in **Borage for oilseed production** *(off-label)* [10]; **Canary flower (Echium spp.)** *(off-label)* [10]; **Evening primrose** *(off-label)* [10]; **Grass seed crops** *(off-label)* [10]; **Honesty** *(off-*

label) [10]; **Linseed** [2-4,7,15,17]; **Mallow (Althaea spp.)** *(off-label)* [10]; **Mustard** *(off-label)* [10]; **Parsley root** *(off-label)* [10]; **Spring field beans** [2-3,15]; **Triticale** *(off-label)* [10]; **Winter field beans** [2-3,15];

- Ear diseases in **Spring wheat** [1,14,16]; **Winter wheat** [1,14,16];
- Fusarium in **Spring wheat** [1,14,16]; **Winter wheat** [1,14,16];
- Fusarium ear blight in **Spring wheat** [2-5,7-13,15,17]; **Winter wheat** [2-5,7-13,15,17];
- Glume blotch in **Spring wheat** [1-17]; **Triticale** [1,14,16]; **Winter wheat** [1-17];
- Leaf spot in **Protected ornamentals** *(off-label)* [14];
- Light leaf spot in **Cabbages** [1,13-14,16]; **Spring oilseed rape** [1-17]; **Winter oilseed rape** [1-17];
- Lodging control in **Spring oilseed rape** [2-4,7-8,11-13,15,17]; **Winter oilseed rape** [2-4,7-8,11-13,15,17];
- Net blotch in **Spring barley** [1-4,6-17]; **Winter barley** [1-4,6-17];
- Phoma in **Apples** *(off-label)* [10,12-13,15]; **Chestnuts** *(off-label)* [10,12-13,15]; **Hazel nuts** *(off-label)* [10,12-13,15]; **Pears** *(off-label)* [10,12-13,15]; **Quinces** *(off-label)* [10,12-13,15]; **Spring oilseed rape** [2-4,6-13,15,17]; **Walnuts** *(off-label)* [10,12-13,15]; **Winter oilseed rape** [2-4,6-13,15,17];
- Phoma leaf spot in **Spring oilseed rape** [1,14,16]; **Winter oilseed rape** [1,14,16];
- Powdery mildew in **Apples** *(off-label)* [10,12-13,15]; **Blackberries** *(off-label)* [10,12-15]; **Cabbages** [1,13-14,16]; **Carrots** [1,13-14,16]; **Celeriac** *(off-label)* [14]; **Chestnuts** *(off-label)* [10,12-13,15]; **Hazel nuts** *(off-label)* [10,12-13,15]; **Linseed** [6,9-10,12-13]; **Loganberries** *(off-label)* [10,12-15]; **Parsley root** *(off-label)* [14]; **Parsnips** [1,13-14,16]; **Pears** *(off-label)* [10,12-13,15]; **Protected blackberries** *(off-label)* [14]; **Protected loganberries** *(off-label)* [14]; **Protected ornamentals** *(off-label)* [14]; **Protected raspberries** *(off-label)* [14]; **Protected Rubus hybrids** *(off-label)* [14]; **Quinces** *(off-label)* [10,12-13,15]; **Raspberries** *(off-label)* [10,12-15]; **Rubus hybrids** *(off-label)* [10,12-15]; **Rye** [1,14,16]; **Salsify** *(off-label)* [14]; **Spring barley** *(moderate control)* [1-17]; **Spring oats** [1-17]; **Spring rye** [5,8,10,12-13,15]; **Spring wheat** *(moderate control)* [1-17]; **Swedes** [1,13-14,16]; **Turnips** [1,13-14,16]; **Walnuts** *(off-label)* [10,12-13,15]; **Winter barley** *(moderate control)* [1-17]; **Winter oats** [1-17]; **Winter rye** [2-13,15,17]; **Winter wheat** *(moderate control)* [1-17];
- Rhynchosporium in **Rye** [1,14,16]; **Spring barley** *(moderate control)* [1-17]; **Spring rye** *(moderate control)* [5,8,10,12-13,15]; **Winter barley** *(moderate control)* [1-17]; **Winter rye** *(moderate control)* [2-13,15,17];
- Ring spot in **Cabbages** [1,13-14,16]; **Spring oilseed rape** *(reduction)* [1-4,7-8,11-17]; **Winter oilseed rape** *(reduction)* [1-4,7-8,11-17];
- Rust in **Blackberries** *(off-label)* [12]; **Borage for oilseed production** *(off-label)* [10]; **Broad beans** *(off-label)* [10,12-13,15]; **Canary flower (Echium spp.)** *(off-label)* [10]; **Dwarf French beans** *(off-label)* [10,12-13,15]; **Evening primrose** *(off-label)* [10]; **French beans** *(off-label)* [10,12-13,15]; **Grass seed crops** *(off-label)* [10]; **Honesty** *(off-label)* [10]; **Leeks** [13]; **Linseed** [2-4,6-7,17]; **Loganberries** *(off-label)* [12]; **Mallow (Althaea spp.)** *(off-label)* [10]; **Mustard** *(off-label)* [10]; **Parsley root** *(off-label)* [10]; **Raspberries** *(off-label)* [12]; **Rubus hybrids** *(off-label)* [12]; **Runner beans** *(off-label)* [10,12-13,15]; **Spring field beans** [1-4,6-7,9-11,13-17]; **Triticale** *(off-label)* [10]; **Winter field beans** [1-4,6-7,9-11,13-17];
- Sclerotinia in **Carrots** [1,13-14,16]; **Spring oilseed rape** [1,5-6,14,16]; **Winter oilseed rape** [1,5-6,14,16];
- Sclerotinia stem rot in **Spring oilseed rape** [2-4,7-13,15,17]; **Winter oilseed rape** [2-4,7-13,15,17];
- Septoria leaf blotch in **Spring wheat** [1-4,6-17]; **Triticale** [1,14,16]; **Winter wheat** [1-4,6-17];
- Septoria seedling blight in **Celeriac** *(off-label)* [14]; **Parsley root** *(off-label)* [14]; **Salsify** *(off-label)* [14];
- Sooty moulds in **Spring wheat** [2-5,7-13,15,17]; **Winter wheat** [2-5,7-13,15,17];
- Stem canker in **Spring oilseed rape** [1-4,7-17]; **Winter oilseed rape** [1-4,7-17];
- White rot in **Bulb onion sets** *(off-label)* [10,12-15]; **Bulb onions** *(off-label)* [14]; **Chives** *(off-label)* [14]; **Garlic** *(off-label)* [14]; **Salad onions** *(off-label)* [14]; **Shallots** *(off-label)* [14];
- Yellow rust in **Rye** [1,14,16]; **Spring barley** [1-17]; **Spring rye** [5,8,10,12-13,15]; **Spring wheat** [1-17]; **Winter barley** [1-17]; **Winter rye** [2-13,15,17]; **Winter wheat** [1-17];

SEE SECTION 3 FOR PRODUCTS ALSO REGISTERED

SECTION 2

Extension of Authorisation for Minor Use (EAMUs)

- **Apples** *20201159* [5], *20201160* [8], *20211814* [10], *20211989* [12], *20211994* [13], *20211986* [15]
- **Blackberries** *20211815* [10], *20211992* [12], *20211993* [13], *20210052* [14], *20211987* [15]
- **Borage for oilseed production** *20122270* [10]
- **Broad beans** *20211813* [10], *20211990* [12], *20211996* [13], *20211988* [15]
- **Bulb onion sets** *20211816* [10], *20211991* [12], *20211995* [13], *20212215* [14], *20211985* [15]
- **Bulb onions** *20212215* [14]
- **Canary flower (Echium spp.)** *20122270* [10]
- **Celeriac** *20220439* [14]
- **Chestnuts** *20211814* [10], *20211989* [12], *20211994* [13], *20211986* [15]
- **Chives** *20212215* [14]
- **Crab apples** *20201159* [5], *20201160* [8]
- **Dwarf French beans** *20211813* [10], *20211990* [12], *20211996* [13], *20211988* [15]
- **Evening primrose** *20122270* [10]
- **French beans** *20211813* [10], *20211990* [12], *20211996* [13], *20211988* [15]
- **Garlic** *20212215* [14]
- **Grass seed crops** *20122273* [10]
- **Hazel nuts** *20211814* [10], *20211989* [12], *20211994* [13], *20211986* [15]
- **Honesty** *20122270* [10]
- **Loganberries** *20211815* [10], *20211992* [12], *20211993* [13], *20210052* [14], *20211987* [15]
- **Mallow (Althaea spp.)** *20122272* [10]
- **Mustard** *20122270* [10]
- **Parsley root** *20122272* [10], *20220439* [14]
- **Pears** *20201159* [5], *20201160* [8], *20211814* [10], *20211989* [12], *20211994* [13], *20211986* [15]
- **Protected blackberries** *20210052* [14]
- **Protected loganberries** *20210052* [14]
- **Protected ornamentals** *20220366* [14]
- **Protected raspberries** *20210052* [14]
- **Protected Rubus hybrids** *20210052* [14]
- **Quinces** *20201159* [5], *20201160* [8], *20211814* [10], *20211989* [12], *20211994* [13], *20211986* [15]
- **Raspberries** *20211815* [10], *20211992* [12], *20211993* [13], *20210052* [14], *20211987* [15]
- **Rubus hybrids** *20211815* [10], *20211992* [12], *20211993* [13], *20210052* [14], *20211987* [15]
- **Runner beans** *20211813* [10], *20211990* [12], *20211996* [13], *20211988* [15]
- **Salad onions** *20212215* [14]
- **Salsify** *20220439* [14]
- **Shallots** *20212215* [14]
- **Triticale** *20122271* [10]
- **Walnuts** *20211814* [10], *20211989* [12], *20211994* [13], *20211986* [15]

Approval information

- Tebuconazole is included in Annex 1 under EC Regulation 1107/2009
- Accepted by BBPA for use on malting barley

Efficacy guidance

- For best results apply at an early stage of disease development before infection spreads to new crop growth
- To protect flag leaf and ear from Septoria diseases apply from flag leaf emergence to ear fully emerged (GS 37-59). Earlier application may be necessary where there is a high risk of infection
- For light leaf spot control in oilseed rape apply in autumn/winter with a follow-up spray in spring/summer if required
- For control of most other diseases spray at first signs of infection with a follow-up spray 2-4 wk later if necessary. See label for details
- For disease control in cabbages a 3-spray programme at 21-28 d intervals will give good control
- Tebuconazole is a DMI fungicide. Resistance to some DMI fungicides has been identified in Septoria leaf blotch which may seriously affect performance of some products. For further advice contact a specialist advisor and visit the Fungicide Resistance Action Group (FRAG)-UK website

FOR FULL CONDITIONS OF USE ALWAYS READ THE PRODUCT LABEL

Restrictions

- Maximum total dose equivalent to 1 full dose treatment on linseed; 2 full dose treatments on wheat, barley, rye, oats, field beans, swedes, turnips, onions; 2.25 full dose treatments on cabbages; 2.5 full dose treatments on oilseed rape; 3 full dose treatments on leeks, parsnips, carrots
- Do not treat durum wheat
- Apply only to listed oat varieties (see label) and do not apply to oats in tank mixture
- Do not apply before swedes and turnips have a root diameter of 2.5 cm, or before heart formation in cabbages, or before button formation in Brussels sprouts
- Consult processor before use on crops for processing
- Newer authorisations for tebuconazole products require application to cereals only after GS 30 and applications to oilseed rape and linseed after GS 20 - check label
- Must not be applied via hand-held equipment [14]
- Applications to linseed must be made after BBCH 20, applications to field bean must be made after BBCH 40 [3]
- For use on cereals a maximum individual dose of 1.25 l/ha applies after BBCH 30 and before early boot stage (BBCH39). A further maximum individual dose of 1.25 l/ha cannot be applied until after BBCH 40 stage [8]
- Applications must only be made in the Spring [8]
- For use on oilseed rape a single application must only be made after BBCH 20 [8]
- For use on oilseed rape a maximum total dose of 0.5 litres of product /ha can be applied between growth stages BBCH 14 and BBCH 19; a maximum total dose of 1.0 litre of product /ha can be applied between growth stages BBCH 20 and BBCH 69 but if an application is made before BBCH 19 then no further applications are allowed on the crop

Crop-specific information

- Latest use: before grain milky-ripe for cereals (GS 71); when most seed green-brown mottled for oilseed rape (GS 6,3); before brown capsule for linseed
- HI field beans, swedes, turnips, linseed, winter oats 35 d; market brassicas, carrots, horseradish, parsnips 21 d; leeks 14 d
- Some transient leaf speckling on wheat or leaf reddening/scorch on oats may occur but this has not been shown to reduce yield response to disease control

Environmental safety

- Dangerous for the environment
- Toxic to aquatic organisms
- LERAP Category B [1-17]

Hazard classification and safety precautions

Hazard Harmful [1-4,7,9-17], Irritant [5-6,8], Dangerous for the environment [1-4,7,9-17], Harmful if swallowed [2-4,6-7,9,11,17], Harmful if inhaled [2-4,7,9,11,17], Very toxic to aquatic organisms [2-4,7,9,17]

Transport code 9 [1-17]

Packaging group III

UN Number 3082

Risk phrases H315 [6], H318 [2-4,7,9-13,15,17], H319 [5-6,8], H335 [3,6,10,12-13,15], H361

Operator protection U05a, U09a [1,14,16], U11 [2-13,15,17], U12 [1,14,16], U14 [1,5-6,8,14,16], U15 [1,5-6,8,14,16], U19a [9-10], U20a; A, C, H

Environmental protection E15a [1,4-12,14,16-17], E16a, E34, E38 [1-4,7,11-17], H410 [1-9,11,14,16-17], H411 [10,12-13,15]

Storage and disposal D01, D02, D05 [2-4,7,9-13,15,17], D09a, D10b [2-4,7,9-13,15,17], D10c [1,14,16], D12a [1-4,7,11-17]

Medical advice M03 [1-4,7,9-17], M05a [1,4,7,11,14,16-17]

SEE SECTION 3 FOR PRODUCTS ALSO REGISTERED

343   tebuconazole + trifloxystrobin

A conazole and stobilurin fungicide mixture for wheat and vegetable crops
FRAC mode of action code: 3 + 11

See also trifloxystrobin

Products

1	Clayton Bestow	Clayton	200:100 g/l	SC	17185
2	Dedicate	Bayer CropScience	200:100 g/l	SC	17003
3	Fortify	Pan Amenity	200:100 g/l	SC	17718
4	Fusion	Pan Amenity	200:100 g/l	SC	17225
5	Nativo 75 WG	Bayer CropScience	50:25% w/w	WG	16867

Uses

*   Alternaria in *Broccoli* [5]; *Brussels sprouts* [5]; *Cabbages* [5]; *Calabrese* [5]; *Carrots* [5]; *Cauliflowers* [5]; *Celeriac* (off-label) [5]; *Swedes* (off-label) [5]; *Turnips* (off-label) [5];
*   Anthracnose in *Amenity grassland* [4]; *Managed amenity turf* [4];
*   Black blight in *Celeriac* (off-label) [5]; *Celery (outdoor)* (off-label) [5]; *Swedes* (off-label) [5]; *Turnips* (off-label) [5];
*   Black root rot in *Ornamental plant production* (off-label) [5]; *Protected ornamentals* (off-label) [5];
*   Dollar spot in *Amenity grassland* [1-4]; *Managed amenity turf* [1-4];
*   Fusarium in *Amenity grassland* [1-3]; *Managed amenity turf* [1-3];
*   Fusarium patch in *Amenity grassland* [4]; *Managed amenity turf* [4];
*   Leaf spot in *Horseradish* (off-label) [5]; *Parsley root* (off-label) [5]; *Parsnips* (off-label) [5]; *Salsify* (off-label) [5];
*   Light leaf spot in *Broccoli* [5]; *Brussels sprouts* [5]; *Cabbages* [5]; *Calabrese* [5]; *Cauliflowers* [5];
*   Melting out in *Amenity grassland* [1-3]; *Managed amenity turf* [1-3];
*   Mycosphaerella in *Celeriac* (off-label) [5]; *Swedes* (off-label) [5]; *Turnips* (off-label) [5];
*   Phoma leaf spot in *Broccoli* [5]; *Brussels sprouts* [5]; *Cabbages* [5]; *Calabrese* [5]; *Cauliflowers* [5];
*   Powdery mildew in *Broccoli* [5]; *Brussels sprouts* [5]; *Cabbages* [5]; *Calabrese* [5]; *Carrots* [5]; *Cauliflowers* [5]; *Celeriac* (off-label) [5]; *Horseradish* (off-label) [5]; *Parsley root* (off-label) [5]; *Parsnips* (off-label) [5]; *Salsify* (off-label) [5]; *Swedes* (off-label) [5]; *Turnips* (off-label) [5]; *Wine grapes* (off-label) [5];
*   Red thread in *Amenity grassland* [1-4]; *Managed amenity turf* [1-4];
*   Ring spot in *Broccoli* [5]; *Brussels sprouts* [5]; *Cabbages* [5]; *Calabrese* [5]; *Cauliflowers* [5];
*   Rust in *Amenity grassland* [1-3]; *Leeks* (off-label) [5]; *Managed amenity turf* [1-3]; *Salad onions* (off-label) [5];
*   Sclerotinia in *Horseradish* (off-label) [5]; *Parsley root* (off-label) [5]; *Parsnips* (off-label) [5]; *Salsify* (off-label) [5];
*   Sclerotinia rot in *Carrots* [5];
*   White blister in *Broccoli* [5]; *Brussels sprouts* [5]; *Cabbages* [5]; *Calabrese* [5]; *Cauliflowers* [5];
*   White rust in *Ornamental plant production* (off-label) [5]; *Protected ornamentals* (off-label) [5];

Extension of Authorisation for Minor Use (EAMUs)

*   *Celeriac* 20182969 [5]
*   *Celery (outdoor)* 20182970 [5]
*   *Horseradish* 20151959 [5]
*   *Leeks* 20182971 [5]
*   *Ornamental plant production* 20192513 [5]
*   *Parsley root* 20151959 [5]
*   *Parsnips* 20151959 [5]
*   *Protected ornamentals* 20192513 [5]
*   *Salad onions* 20182971 [5]
*   *Salsify* 20151959 [5]

FOR FULL CONDITIONS OF USE ALWAYS READ THE PRODUCT LABEL

- **Swedes** *20182969* [5]
- **Turnips** *20182969* [5]
- **Wine grapes** *20182972* [5]

Approval information
- Tebuconazole and trifloxystrobin included in Annex I under EC Regulation 1107/2009

Efficacy guidance
- Best results obtained from treatment at early stages of disease development. Further treatment may be needed if disease attack is prolonged
- Applications to established infections of any disease are likely to be less effective
- Treatment may give some control of *Stemphylium botryosum* and leaf blotch in leeks
- Tebuconazole is a DMI fungicide. Resistance to some DMI fungicides has been identified in Septoria leaf blotch which may seriously affect performance of some products. For further advice contact a specialist advisor and visit the Fungicide Resistance Action Group (FRAG)-UK website
- Trifloxystrobin is a member of the QoI cross resistance group. Product should be used preventatively and not relied on for its curative potential
- Use product as part of an Integrated Crop Management strategy incorporating other methods of control, including where appropriate other fungicides with a different mode of action. Do not apply more than two foliar applications of QoI containing products to any cereal crop, broccoli, calabrese or cauliflower. Do not apply more than three applications to Brussels sprouts, cabbage, carrots or leeks
- Do not apply to grass during drought conditions nor to frozen turf

Restrictions
- Maximum number of treatments 2 per crop for broccoli, calabrese, cauliflowers; 3 per crop for Brussels sprouts, cabbages, carrots and 1 for leeks
- In addition to the maximum number of treatments per crop a maximum of 3 applications may be applied to the same ground in one calendar year
- Consult processor before use on vegetable crops for processing
- Horizontal boom sprayers must be fitted with three star drift reduction technology for all uses [5]
- Low drift spraying equipment must be operated according to the specific conditions stated in the official three star rating for that equipment as published on HSE Chemicals Regulation Division?s website. These operating conditions must be maintained until 30m from the top of the bank of any surface water bodies [5]

Crop-specific information
- Latest use: before milky ripe stage on winter wheat
- HI 35 d for wheat; 21 d for all other crops
- Performance against leaf spot diseases of brassicas, *Alternaria* leaf blight of carrots and rust in leeks may be improved by mixing with an approved sticker/wetter

Environmental safety
- Dangerous for the environment
- Very toxic to aquatic organisms
- Buffer zone requirement 12 m, 18 m on protected and outdoor ornamentals [5]
- LERAP Category B [1-4]

Hazard classification and safety precautions
Hazard Harmful, Dangerous for the environment, Very toxic to aquatic organisms [1-4]
Transport code 9 [1-5]
Packaging group III
UN Number 3077, 3082
Risk phrases H317 [5], H319 [5], H335 [4], H361
Operator protection U05a [1-3,5], U09a [1-3], U09b [4-5], U19a [1-4], U20b; A, H, P
Environmental protection E15a, E16a [1-4], E34, E38, H410
Storage and disposal D01, D02, D05 [4], D09a, D10b, D12a
Medical advice M03

344   tebufenpyrad

A pyrazole mitochondrial electron transport inhibitor (METI) aphicide and acaricide
IRAC mode of action code: 21

Products

1	Clayton Bonsai	Clayton	20% w/w	WB	18666

Uses
- Red spider mites in *Apples* [1]; *Pears* [1];

Approval information
- Tebufenpyrad included in Annex 1 under EC Regulation 1107/2009

Efficacy guidance
- Acts on eggs (except winter eggs) and all motile stages of spider mites up to adults
- Treat spider mites from 80% egg hatch but before mites become established
- For effective control total spray cover of the crop is required
- Product can be used in a programme to give season-long control of damson-hop aphids coupled with mite control
- Where aphids resistant to tebufenpyrad occur in hops control is unlikely to be satisfactory and repeat treatments may result in lower levels of control. Where possible use different active ingredients in a programme

Restrictions
- Maximum total dose equivalent to one full dose treatment on apples and pears
- Other mitochondrial electron transport inhibitor (METI) acaricides should not be applied to the same crop in the same calendar yr either separately or in mixture
- Do not treat apples before 90% petal fall
- Inner liner of container must not be removed
- Must not be applied via hand-held equipment

Crop-specific information
- HI apples & pears 7d;
- Product has no effect on fruit quality or finish

Environmental safety
- Dangerous for the environment
- Very toxic to aquatic organisms
- High risk to bees. Do not apply to crops in flower or to those in which bees are actively foraging. Do not apply when flowering weeds are present
- Broadcast air-assisted LERAP [1] (30 m); LERAP Category B [1]

Hazard classification and safety precautions
  Hazard Harmful, Dangerous for the environment, Harmful if swallowed, Harmful if inhaled, Very toxic to aquatic organisms
  Transport code 9 [1]
  Packaging group III
  UN Number 3077
  Risk phrases H319, H335
  Operator protection U02a, U05a, U09a, U13, U14, U20b; A
  Environmental protection E12a, E12e, E15a, E16a, E16b, E17b (30 m), E38, H410
  Storage and disposal D01, D02, D09a, D11a, D12a
  Medical advice M05a

345   tefluthrin

A soil acting pyrethroid insecticide seed treatment
IRAC mode of action code: 3

Products

| 1 | Force ST | Syngenta | 200 g/l | CF | 19042 |

Uses

- Millipedes in **Fodder beet** *(seed treatment)* [1]; **Sugar beet** *(seed treatment)* [1];
- Pygmy beetle in **Fodder beet** *(seed treatment)* [1]; **Sugar beet** *(seed treatment)* [1];
- Springtails in **Fodder beet** *(seed treatment)* [1]; **Sugar beet** *(seed treatment)* [1];
- Symphylids in **Fodder beet** *(seed treatment)* [1]; **Sugar beet** *(seed treatment)* [1];

Approval information

- Tefluthrin included in Annex 1 under EC Regulation 1107/2009

Efficacy guidance

- Apply during process of pelleting beet seed. Consult manufacturer for details of specialist equipment required
- Micro-capsule formulation allows slow release to provide a protection zone around treated seed during establishment

Restrictions

- Maximum number of treatments 1 per batch of seed
- Sow treated seed as soon as possible. Do not store treated seed from one drilling season to next
- If used in areas where soil erosion by wind or water likely, measures must be taken to prevent this happening
- Can cause a transient tingling or numbing sensation to exposed skin. Avoid skin contact with product, treated seed and dust throughout all operations in the seed treatment plant and at drilling
- To protect birds the product must be entirely incorporated in the soil; ensure that the product is fully incorporated at the end of rows. Remove spillages

Crop-specific information

- Latest use: before drilling seed
- Treated seed must be drilled within the season of treatment

Environmental safety

- Dangerous for the environment
- Very toxic to aquatic organisms
- Keep treated seed secure from people, domestic stock/pets and wildlife at all times during storage and use
- Treated seed harmful to game and wild life. Bury spillages
- In the event of seed spillage clean up as much as possible into the related seed sack and bury the remainder completely
- Do not apply treated seed from the air
- Keep livestock out of areas drilled with treated seed for at least 80 d

Hazard classification and safety precautions

Hazard Harmful, Dangerous for the environment, Harmful if inhaled, Very toxic to aquatic organisms
Transport code 9 [1]
Packaging group III
UN Number 3082
Risk phrases H317
Operator protection U02a, U04a, U05a, U08, U20b; A, D, E, H
Environmental protection E06a (80 d), E15a, E34, E38, H410
Storage and disposal D01, D02, D05, D09a, D11a, D12a
Treated seed S02, S03, S04b, S05, S07
Medical advice M05b

SEE SECTION 3 FOR PRODUCTS ALSO REGISTERED

346 tetradecadienyl + trienyl acetate

A pheromone insecticide for use in glass and greenhouses

Products
1 Isonet T   Fargro   8.3:72.0% w/w   VP   17929

Uses
* Tomato leaf miner (Tuta absoluta) in **Protected aubergines** [1]; **Protected chilli peppers** [1]; **Protected peppers** [1]; **Protected tomatoes** [1];

Approval information
* Tetradecadienyl acetate included in Annex I under EC Regulation 1107/2009

Restrictions
* For use in greenhouses only
* Do not contaminate water with the product or its container

Hazard classification and safety precautions
  Hazard Irritant, Dangerous for the environment
  Transport code 9 [1]
  Packaging group III
  UN Number 3082
  Risk phrases H315
  Operator protection U05a, U09c; A, H
  Environmental protection H410
  Storage and disposal D01, D12a, D12b, D22

347 thiabendazole

A systemic, curative and protectant benzimidazole (MBC) fungicide
FRAC mode of action code: 1

See also imazalil + thiabendazole
       metalaxyl + thiabendazole

Products
1 Storite Excel   Syngenta   500 g/l   SC   12705

Uses
* Dry rot in **Seed potatoes** *(tuber treatment - post-harvest)* [1]; **Ware potatoes** *(tuber treatment - post-harvest)* [1];
* Gangrene in **Seed potatoes** *(tuber treatment - post-harvest)* [1]; **Ware potatoes** *(tuber treatment - post-harvest)* [1];
* Silver scurf in **Seed potatoes** *(tuber treatment - post-harvest)* [1]; **Ware potatoes** *(tuber treatment - post-harvest)* [1];
* Skin spot in **Seed potatoes** *(tuber treatment - post-harvest)* [1]; **Ware potatoes** *(tuber treatment - post-harvest)* [1];

Approval information
* Thiabendazole included in Annex I under EC Regulation 1107/2009
* Use of thiabendazole products on ware potatoes requires that the discharge of thiabendazole to receiving water from washing plants is kept within emission limits set by the UK monitoring authority

Efficacy guidance
* For best results tuber treatments should be applied as soon as possible after lifting and always within 24 hr
* Dust treatments should be applied evenly over the whole tuber surface
* Thiabendazole should only be used on ware potatoes where there is a likely risk of disease during the storage period and in combination with good storage hygiene and maintenance

FOR FULL CONDITIONS OF USE ALWAYS READ THE PRODUCT LABEL

- Benzimidazole tolerant strains of silver scurf and skin spot are common in UK and tolerant strains of dry rot have been reported. To reduce the chance of such strains increasing benzimidazole based products should not be used more than once in the cropping cycle

Restrictions
- Maximum number of treatments 1 per batch for ware or seed potato tuber treatments
- Treated seed potatoes must not be used for food or feed
- Do not remove treated potatoes from store for sale, processing or consumption for at least 21 d after application
- Do not mix with any other product

Crop-specific information
- Latest use: 21 d before removal from store for sale, processing or consumption for ware potatoes
- Apply to potatoes as soon as possible after harvest using suitable equipment and always within 2 wk of lifting provided the skins are set. See label for details
- Potatoes should only be treated by systems that provide an accurate dose to tubers not carrying excessive quantities of soil

Environmental safety
- Dangerous for the environment
- Toxic to aquatic organisms

Hazard classification and safety precautions
Hazard Irritant, Dangerous for the environment
Transport code 9 [1]
Packaging group III
UN Number 3082
Risk phrases H317
Operator protection U05a, U14, U19a, U20c; A, C, D, H
Environmental protection E15b, E38, H410
Storage and disposal D01, D02, D05, D09a, D10c, D12a

348   thifensulfuron-methyl

A translocated sulfonylurea herbicide
HRAC mode of action code: 2 (B)

See also fluroxypyr + metsulfuron-methyl + thifensulfuron-methyl
fluroxypyr + thifensulfuron-methyl
metsulfuron-methyl + thifensulfuron-methyl
thifensulfuron-methyl + tribenuron-methyl

Products
1 Pinnacle	FMC Agro	50% w/w	SG	18781
2 Prospect SX	FMC Agro	50% w/w	SG	18777

Uses
- Annual dicotyledons in **Soya beans** *(off-label)* [1]; **Spring barley** [1-2]; **Spring wheat** [1-2]; **Winter barley** [1-2]; **Winter wheat** [1-2];
- Creeping buttercup in **Lucerne** *(off-label)* [1];
- Docks in **Grassland** [1-2]; **Lucerne** *(off-label)* [1];
- Green cover in **Land temporarily removed from production** [1-2];
- Groundsel in **Lucerne** *(off-label)* [1];

Extension of Authorisation for Minor Use (EAMUs)
- **Lucerne** *20212074* [1]
- **Soya beans** *20190886* [1]

Approval information
- Thifensulfuron-methyl included in Annex I under EC Regulation 1107/2009
- Accepted by BBPA for use on malting barley

SEE SECTION 3 FOR PRODUCTS ALSO REGISTERED

SECTION 2

Efficacy guidance

- Best results achieved from application to small emerged weeds when growing actively. Broad-leaved docks are susceptible during the rosette stage up to onset of stem extension
- Ensure good spray coverage and apply to dry foliage
- Susceptible weeds stop growing almost immediately but symptoms may not be visible for about 2 wk
- Only broad-leaved docks (*Rumex obtusifolius*) are controlled; curled docks (*Rumex crispus*) are resistant
- Docks with developing or mature seed heads should be topped and the regrowth treated later
- Established docks with large tap roots may require follow-up treatment
- High populations of docks in grassland will require further treatment in following yr
- Thifensulfuron-methyl is a member of the ALS-inhibitor group of herbicides and products should be used in a planned Resistance Management strategy. See Section 5 for more information

Restrictions

- Maximum number of treatments 1 per year for grassland. Must only be applied from 1 Feb in year of harvest
- Do not treat new leys in year of sowing
- Do not treat where nutrient imbalances, drought, waterlogging, low temperatures, lime deficiency, pest or disease attack have reduced crop or sward vigour
- Do not roll or harrow within 7 d of spraying
- Do not graze grass crops within 7 d of spraying
- Specific restrictions apply to use in sequence or tank mixture with other sulfonylurea or ALS-inhibiting herbicides. See label for details
- Only one application of a sulfonylurea product may be applied per calendar yr to grassland and green cover on land temporarily removed from production

Crop-specific information

- Latest use: before 1 Aug on grass
- On grass apply 7-10 d before grazing and do not graze for 7 d afterwards
- Product may cause a check to both sward and clover which is usually outgrown

Following crops guidance

- Only grass or cereals may be sown within 4 wk of application to grassland or setaside, or in the event of failure of any treated crop
- No restrictions apply after normal harvest of a treated cereal crop

Environmental safety

- Dangerous for the environment
- Very toxic to aquatic organisms
- Keep livestock out of treated areas for at least 7 d following treatment
- Take extreme care to avoid drift onto broad-leaved plants outside the target area or onto surface waters or ditches, or land intended for cropping
- Spraying equipment should not be drained or flushed onto land planted, or to be planted, with trees or crops other than cereals and should be thoroughly cleansed after use - see label for instructions

Hazard classification and safety precautions

Hazard Dangerous for the environment, Very toxic to aquatic organisms
Transport code 9 [1-2]
Packaging group III
UN Number 3077
Operator protection U19a, U20b
Environmental protection E06a (7 d), E15b, E38, H410
Storage and disposal D09a, D11a, D12a

349   thifensulfuron-methyl + tribenuron-methyl

A mixture of two sulfonylurea herbicides for cereals
HRAC mode of action code: 2 + 2 (B + B)

See also tribenuron-methyl

SECTION 2

Products

1	Counter SX	FMC Agro	40:10% w/w	SG	18976
2	Hiatus	Rotam	40:15% w/w	WG	16059
3	Inka SX	FMC Agro	25:25% w/w	SG	18760
4	Nautius	Rotam	40:15% w/w	WG	18838
5	Ratio SX	FMC Agro	40:10% w/w	SG	18786

Uses

- Annual dicotyledons in *Game cover (off-label)* [3]; *Spring barley* [1-5]; *Spring oats* [3]; *Spring rye* [3]; *Spring wheat* [1-5]; *Triticale* [2-4]; *Winter barley* [1-5]; *Winter oats* [1,3,5]; *Winter rye* [2-4]; *Winter wheat* [1-5];
- Black bindweed in *Spring barley* [2,4]; *Spring wheat* [2,4]; *Triticale* [2,4]; *Winter barley* [2,4]; *Winter rye* [2,4]; *Winter wheat* [2,4];
- Charlock in *Spring barley* [1,3,5]; *Spring oats* [3]; *Spring rye* [3]; *Spring wheat* [1,3,5]; *Triticale* [3]; *Winter barley* [1,3,5]; *Winter oats* [1,3,5]; *Winter rye* [3]; *Winter wheat* [1,3,5];
- Chickweed in *Spring barley* [1-5]; *Spring oats* [3]; *Spring rye* [3]; *Spring wheat* [1-5]; *Triticale* [2-4]; *Winter barley* [1-5]; *Winter oats* [1,3,5]; *Winter rye* [2-4]; *Winter wheat* [1-5];
- Docks in *Spring barley* [3]; *Spring oats* [3]; *Spring rye* [3]; *Spring wheat* [3]; *Triticale* [3]; *Winter barley* [3]; *Winter oats* [3]; *Winter rye* [3]; *Winter wheat* [3];
- Fat hen in *Spring barley* [2,4]; *Spring wheat* [2,4]; *Triticale* [2,4]; *Winter barley* [2,4]; *Winter rye* [2,4]; *Winter wheat* [2,4];
- Mayweeds in *Spring barley* [1-5]; *Spring oats* [3]; *Spring rye* [3]; *Spring wheat* [1-5]; *Triticale* [2-4]; *Winter barley* [1-5]; *Winter oats* [1,3,5]; *Winter rye* [2-4]; *Winter wheat* [1-5];
- Poppies in *Spring barley* [1-2,4]; *Spring wheat* [1-2,4]; *Triticale* [2,4]; *Winter barley* [1-2,4]; *Winter oats* [1]; *Winter rye* [2,4]; *Winter wheat* [1-2,4];
- Redshank in *Spring barley* [1]; *Spring wheat* [1]; *Winter barley* [1]; *Winter oats* [1]; *Winter wheat* [1];

Extension of Authorisation for Minor Use (EAMUs)

- *Game cover* 20190897 [3]

Approval information

- Thifensulfuron-methyl and tribenuron-methyl included in Annex I under EC Regulation 1107/2009
- Accepted by BBPA for use on malting barley

Efficacy guidance

- Apply when weeds are small and actively growing
- Apply after end of Feb in year of harvest
- Ensure good spray cover of the weeds
- Apply in a volume of 200 - 400 l/ha [2]
- Susceptible weeds cease growth almost immediately after application and symptoms become evident 2 wk later
- Effectiveness reduced by rain within 4 h of treatment and in very dry conditions
- Various tank mixtures recommended to broaden weed control spectrum
- Thifensulfuron-methyl and tribenuron-methyl are members of the ALS-inhibitor group of herbicides and products should be used in a planned Resistance Management strategy. See Section 5 for more information

Restrictions

- Maximum number of treatments 1 per crop
- Do not apply to cereals undersown with grass, clover or other legumes, or any other broad-leaved crop
- Do not apply within 7 d of rolling

SEE SECTION 3 FOR PRODUCTS ALSO REGISTERED

- Specific restrictions apply to use in sequence or tank mixture with other sulfonylurea or ALS-inhibiting herbicides. See label for details
- Do not apply to any crop suffering from stress
- When treating oats, do not mix with other products [1]
- Consult contract agents before use on crops grown for seed
- Horizontal boom sprayers must be fitted with three star drift reduction technology for all uses [2]
- Low drift spraying equipment must be operated according to the specific conditions stated in the official three star rating for that equipment as published on HSE Chemicals Regulation Division?s website. These operating conditions must be maintained until the operator is 30m from the top of the bank of any surface water bodies [2]

Crop-specific information
- Latest use: before flag leaf ligule first visible (GS 39) for all crops

Following crops guidance
- Only cereals, field beans, grass or oilseed rape may be sown in the same calendar year as harvest of a treated crop.
- In the event of failure of a treated crop sow only a cereal crop within 3 mth of product application and after ploughing and cultivating to at least 15 cm. After 3 mth field beans or oilseed rape may also be sown
- Only cereals, oilseed rape or field beans may be sown in the same calendar year after harvest. In the following spring only cereals, oilseed rape or sugar beet may be sown. In case of crop failure for any reason, sow only spring small grain cereal. Before sowing, soil should be ploughed and cultivated to a depth of at least 15 cm [2]

Environmental safety
- Dangerous for the environment
- Very toxic to aquatic organisms
- Buffer zone requirement 6m [2]
- Spraying equipment should not be drained or flushed onto land planted, or to be planted, with trees or crops other than cereals and should be thoroughly cleansed after use - see label for instructions
- Take particular care to avoid damage by drift onto broad-leaved plants outside the target area or onto surface waters or ditches
- LERAP Category B [1-5]

Hazard classification and safety precautions
Hazard Irritant [1,5], Dangerous for the environment
Transport code 9 [1-5]
Packaging group III
UN Number 3077
Operator protection U05a [1,5], U08 [1,5], U14 [1,5], U19a [2,4], U20a [1,5], U20c [3]; A, H
Environmental protection E15a [1,5], E15b [2-4], E16a, E38, H410
Storage and disposal D01 [1-2,4-5], D02 [1-2,4-5], D09a, D11a, D12a

350   tri-allate

A soil-acting thiocarbamate herbicide for grass weed control
HRAC mode of action code: 15 (N)

Products

1	Avadex Excel 15G	Gowan	15% w/w	GR	17872
2	Avadex Factor	Gowan	450 g/l	CS	17877

Uses
- Annual grasses in *Canary seed* (off-label) [1-2]; *Linseed* (off-label) [1]; *Miscanthus* (off-label) [1]; *Spring barley* [1-2]; *Spring wheat* [1]; *Triticale* (off-label) [1]; *Winter barley* [1-2]; *Winter linseed* (off-label) [1]; *Winter rye* (off-label) [1]; *Winter wheat* [1-2];
- Annual meadow grass in *Canary seed* (off-label) [1]; *Linseed* (off-label) [1]; *Miscanthus* (off-label) [1]; *Nursery fruit trees* (off-label) [1]; *Winter linseed* (off-label) [1];

FOR FULL CONDITIONS OF USE ALWAYS READ THE PRODUCT LABEL

- Blackgrass in **Canary seed** *(off-label)* [1-2]; **Corn Gromwell** *(off-label)* [2]; **Linseed** *(off-label)* [1]; **Miscanthus** *(off-label)* [1]; **Nursery fruit trees** *(off-label)* [1]; **Spring barley** [1-2]; **Spring linseed** *(off-label)* [2]; **Spring wheat** [1]; **Triticale** *(off-label)* [1]; **Winter barley** [1-2]; **Winter linseed** *(off-label)* [1-2]; **Winter rye** *(off-label)* [1]; **Winter wheat** [1-2];
- Italian ryegrass in **Canary seed** *(off-label)* [2]; **Corn Gromwell** *(off-label)* [2]; **Spring linseed** *(off-label)* [2]; **Winter linseed** *(off-label)* [2];
- Loose silky bent in **Canary seed** *(off-label)* [2]; **Corn Gromwell** *(off-label)* [2]; **Spring linseed** *(off-label)* [2]; **Winter linseed** *(off-label)* [2];
- Meadow grasses in **Spring barley** [1]; **Spring wheat** [1]; **Triticale** *(off-label)* [1]; **Winter barley** [1]; **Winter rye** *(off-label)* [1]; **Winter wheat** [1];
- Rough-stalked meadow grass in **Nursery fruit trees** *(off-label)* [1];
- Wild oats in **Canary seed** *(off-label)* [1]; **Corn Gromwell** *(off-label)* [2]; **Linseed** *(off-label)* [1]; **Miscanthus** *(off-label)* [1]; **Nursery fruit trees** *(off-label)* [1]; **Spring barley** [1]; **Spring wheat** [1]; **Triticale** *(off-label)* [1]; **Winter barley** [1]; **Winter linseed** *(off-label)* [1]; **Winter rye** *(off-label)* [1]; **Winter wheat** [1];

Extension of Authorisation for Minor Use (EAMUs)
- **Canary seed** *20170467* [1], *20210552* [2]
- **Corn Gromwell** *20210750* [2]
- **Linseed** *20170468* [1]
- **Miscanthus** *20170466* [1]
- **Nursery fruit trees** *20200359* [1]
- **Spring linseed** *20191053* [2]
- **Triticale** *20171361* [1]
- **Winter linseed** *20170468* [1], *20191053* [2]
- **Winter rye** *20171361* [1]

Approval information
- Tri-allate included in Annex 1 under EC Regulation 1107/2009
- Accepted by BBPA for use on malting barley

Efficacy guidance
- Apply to soil surface pre-emergence
- For maximum activity apply to well-prepared moist seedbeds
- Do not use on soils with more than 10% organic matter
- Wild oats controlled up to 2-leaf stage
- Do not apply with spinning disc granule applicator; see label for suitable types [1]
- Do not apply to cloddy seedbeds
- Use sequential treatments to improve control of barren brome and annual dicotyledons (see label for details)

Restrictions
- Maximum number of treatments 1 per crop
- Drill wheat well below treated layer of soil (see label for safe drilling depths)
- Do not apply to shallow-drilled wheat crops
- Do not undersow grass species. It is safe to undersow with clover and other legumes
- Do not sow oats or grasses within 1 yr of treatment
- Low drift spraying equipment must be operated according to the specific conditions stated in the official three star rating for that equipment as published on HSE Chemicals Regulation Division?s website. These operating conditions must be maintained until the operator is 30m from the top of the bank of any surface water bodies [2]

Crop-specific information
- Latest use: pre-drilling for beet crops; before crop emergence for field beans, spring barley, peas, forage legumes; before first node detectable stage (GS 31) for winter wheat, winter barley, durum wheat, triticale, winter rye

Following crops guidance
- Do not sow oat or grass crops within a year of application

Environmental safety
- Irritating to eyes and skin

SEE SECTION 3 FOR PRODUCTS ALSO REGISTERED

- May cause sensitization by skin contact
- Harmful to fish or other aquatic life. Do not contaminate surface waters or ditches with chemical or used container
- Buffer zone requirement 10 m [1] and 6 m [2]
- Avoid drift on to non-target plants outside the treated area
- LERAP Category B [1-2]

Hazard classification and safety precautions
Hazard Harmful [1], Dangerous for the environment [1], Very toxic to aquatic organisms [1]
Transport code 9 [1-2]
Packaging group III
UN Number 3077, 3082
Risk phrases H317 [2], H373
Operator protection U02a [1], U05a [1], U19a [1], U20a [1]; A, C, H
Environmental protection E16a, H410 [2], H411
Storage and disposal D01 [1], D02 [1], D09a [1], D11a [1], D19 [1]
Medical advice M05a [1]

351   tribenuron-methyl

A foliar acting sulfonylurea herbicide with some root activity for use in cereals
HRAC mode of action code: 2 (B)

See also florasulam + tribenuron-methyl
metsulfuron-methyl + tribenuron-methyl
thifensulfuron-methyl + tribenuron-methyl

Products

1 Cameo SX	FMC Agro	50% w/w	SG	18990
2 Flame	Albaugh UK	50% w/w	WG	17842
3 Nuance	Nufarm UK	75% w/w	WG	18616
4 Taxi	FMC Agro	50% w/w	WG	18995

Uses
- Annual dicotyledons in *Durum wheat* [1,4]; *Green cover on land temporarily removed from production* [1,4]; *Spring barley* [1-4]; *Spring oats* [1,3-4]; *Spring wheat* [1,3-4]; *Triticale* [1,3-4]; *Winter barley* [1-4]; *Winter oats* [1,3-4]; *Winter rye* [1,4]; *Winter wheat* [1-4];

Approval information
- Tribenuron-methyl included in Annex I under EC Regulation 1107/2009
- Accepted by BBPA for use on malting barley

Efficacy guidance
- Best control achieved when weeds small and actively growing
- Good spray cover must be achieved since larger weeds often become less susceptible
- Susceptible weeds cease growth almost immediately after treatment and symptoms can be seen in about 2 wk
- Weed control may be reduced when conditions very dry
- Tribenuron-methyl is a member of the ALS-inhibitor group of herbicides and products should be used in a planned Resistance Management strategy. See Section 5 for more information

Restrictions
- Maximum number of treatments 1 per crop
- Specific restrictions apply to use in sequence or tank mixture with other sulfonylurea or ALS-inhibiting herbicides. See label for details
- Do not apply to crops undersown with grass, clover or other broad-leaved crops
- Do not apply to any crop suffering stress from any cause or not actively growing
- Do not apply within 7 d of rolling
- Must not be applied before end of February in the year of harvest

FOR FULL CONDITIONS OF USE ALWAYS READ THE PRODUCT LABEL

Crop-specific information
- Latest use: up to and including flag leaf ligule/collar just visible (GS 39)
- Apply in autumn or in spring from 3 leaf stage of crop

Following crops guidance
- Only cereals, field beans or oilseed rape may be sown in the same calendar yr as harvest of a treated crop
- In the event of crop failure sow only a cereal within 3 mth of application. After 3 mth field beans or oilseed rape may also be sown

Environmental safety
- Dangerous for the environment
- Very toxic to aquatic organisms
- Take extreme care to avoid drift onto broad-leaved plants outside the target area or onto surface waters or ditches, or land intended for cropping
- Spraying equipment should not be drained or flushed onto land planted, or to be planted, with trees or crops other than cereals and should be thoroughly cleansed after use - see label for instructions

Hazard classification and safety precautions
Hazard Irritant, Dangerous for the environment, Very toxic to aquatic organisms
Transport code 9 [1-4]
Packaging group III
UN Number 3077
Risk phrases H317 [1-2], H319 [2], R43 [4], R50 [4], R53a [4], R70 [4]
Operator protection U05a, U08, U20b; A, C, H
Environmental protection E15a, E38, H410 [1-3]
Storage and disposal D01, D02, D09a, D11a, D12a

352   Trichoderma afroharzianum (Strain T22)

A biological control agent

Products
1	Trianum G	Koppert	1% w/w	GR	16740
2	Trianum P	Koppert	1% w/w	GR	16741

Uses
- Fusarium in *Protected aubergines* [1-2]; *Protected baby leaf crops* [1-2]; *Protected bilberries* (off-label) [2]; *Protected blackberries* (off-label) [2]; *Protected blackcurrants* (off-label) [2]; *Protected blueberries* (off-label) [2]; *Protected broccoli* (off-label) [1]; *Protected Brussels sprouts* (off-label) [1]; *Protected cabbages* [1]; *Protected calabrese* [1]; *Protected cauliflowers* [1]; *Protected celery* (off-label) [1]; *Protected chilli peppers* [1-2]; *Protected chives* (off-label) [1]; *Protected choi sum* (off-label) [1]; *Protected collards* (off-label) [1]; *Protected cranberries* (off-label) [2]; *Protected cress* (off-label) [1]; *Protected cucumbers* [1-2]; *Protected dwarf French beans* [1]; *Protected edible flowers* (off-label) [1]; *Protected elderberries* (off-label) [2]; *Protected endives* (off-label) [1]; *Protected gooseberries* (off-label) [2]; *Protected herbs (see appendix 6)* (off-label) [1]; *Protected kale* [1]; *Protected kohlrabi* [1]; *Protected lamb's lettuce* (off-label) [1]; *Protected lettuce* [1]; *Protected loganberries* (off-label) [2]; *Protected mulberry* (off-label) [2]; *Protected oriental cabbage* [1]; *Protected ornamentals* [1-2]; *Protected parsley* (off-label) [1]; *Protected peppers* [1-2]; *Protected purslane* (off-label) [1]; *Protected raspberries* (off-label) [2]; *Protected red mustard* (off-label) [1]; *Protected redcurrants* (off-label) [2]; *Protected rose hips* (off-label) [2]; *Protected Rubus hybrids* (off-label) [2]; *Protected spinach* (off-label) [1]; *Protected spinach beet* (off-label) [1]; *Protected strawberries* (off-label) [2]; *Protected sweet cicerly* (off-label) [1]; *Protected table grapes* (off-label) [2]; *Protected tomatoes* [1-2]; *Protected wine grapes* (off-label) [2];
- Pythium in *Protected aubergines* [1-2]; *Protected baby leaf crops* [1-2]; *Protected bilberries* (off-label) [2]; *Protected blackberries* (off-label) [2]; *Protected blackcurrants* (off-label) [2]; *Protected blueberries* (off-label) [2]; *Protected broccoli* (off-label) [1]; *Protected Brussels sprouts* (off-label) [1]; *Protected cabbages* [1]; *Protected calabrese* [1]; *Protected*

SEE SECTION 3 FOR PRODUCTS ALSO REGISTERED

cauliflowers [1]; *Protected celery* (off-label) [1]; *Protected chilli peppers* [1-2]; *Protected chives* (off-label) [1]; *Protected choi sum* (off-label) [1]; *Protected collards* (off-label) [1]; *Protected cranberries* (off-label) [2]; *Protected cress* (off-label) [1]; *Protected cucumbers* [1-2]; *Protected dwarf French beans* [1]; *Protected edible flowers* (off-label) [1]; *Protected elderberries* (off-label) [2]; *Protected endives* (off-label) [1]; *Protected gooseberries* (off-label) [2]; *Protected herbs (see appendix 6)* (off-label) [1]; *Protected kale* [1]; *Protected kohlrabi* [1]; *Protected lamb's lettuce* (off-label) [1]; *Protected lettuce* [1]; *Protected loganberries* (off-label) [2]; *Protected mulberry* (off-label) [2]; *Protected oriental cabbage* [1]; *Protected ornamentals* [1-2]; *Protected parsley* (off-label) [1]; *Protected peppers* [1-2]; *Protected purslane* (off-label) [1]; *Protected raspberries* (off-label) [2]; *Protected red mustard* (off-label) [1]; *Protected redcurrants* (off-label) [2]; *Protected rose hips* (off-label) [2]; *Protected Rubus hybrids* (off-label) [2]; *Protected spinach* (off-label) [1]; *Protected spinach beet* (off-label) [1]; *Protected strawberries* (off-label) [2]; *Protected sweet cicerly* (off-label) [1]; *Protected table grapes* (off-label) [2]; *Protected tomatoes* [1-2]; *Protected wine grapes* (off-label) [2];

- Rhizoctonia in *Protected aubergines* [1-2]; *Protected baby leaf crops* [1-2]; *Protected bilberries* (off-label) [2]; *Protected blackberries* (off-label) [2]; *Protected blackcurrants* (off-label) [2]; *Protected blueberries* (off-label) [2]; *Protected broccoli* (off-label) [1]; *Protected Brussels sprouts* (off-label) [1]; *Protected cabbages* [1]; *Protected calabrese* [1]; *Protected cauliflowers* [1]; *Protected celery* (off-label) [1]; *Protected chilli peppers* [1-2]; *Protected chives* (off-label) [1]; *Protected choi sum* (off-label) [1]; *Protected collards* (off-label) [1]; *Protected cranberries* (off-label) [2]; *Protected cress* (off-label) [1]; *Protected cucumbers* [1-2]; *Protected dwarf French beans* [1]; *Protected edible flowers* (off-label) [1]; *Protected elderberries* (off-label) [2]; *Protected endives* (off-label) [1]; *Protected gooseberries* (off-label) [2]; *Protected herbs (see appendix 6)* (off-label) [1]; *Protected kale* [1]; *Protected kohlrabi* [1]; *Protected lamb's lettuce* (off-label) [1]; *Protected lettuce* [1]; *Protected loganberries* (off-label) [2]; *Protected mulberry* (off-label) [2]; *Protected oriental cabbage* [1]; *Protected ornamentals* [1-2]; *Protected parsley* (off-label) [1]; *Protected peppers* [1-2]; *Protected purslane* (off-label) [1]; *Protected raspberries* (off-label) [2]; *Protected red mustard* (off-label) [1]; *Protected redcurrants* (off-label) [2]; *Protected rose hips* (off-label) [2]; *Protected Rubus hybrids* (off-label) [2]; *Protected spinach* (off-label) [1]; *Protected spinach beet* (off-label) [1]; *Protected strawberries* (off-label) [2]; *Protected sweet cicerly* (off-label) [1]; *Protected table grapes* (off-label) [2]; *Protected tomatoes* [1-2]; *Protected wine grapes* (off-label) [2];
- Sclerotinia in *Protected aubergines* [1-2]; *Protected baby leaf crops* [1-2]; *Protected cabbages* [1]; *Protected calabrese* [1]; *Protected cauliflowers* [1]; *Protected chilli peppers* [1-2]; *Protected cucumbers* [1-2]; *Protected dwarf French beans* [1]; *Protected kale* [1]; *Protected kohlrabi* [1]; *Protected lettuce* [1]; *Protected oriental cabbage* [1]; *Protected ornamentals* [1-2]; *Protected peppers* [1-2]; *Protected tomatoes* [1-2];

Extension of Authorisation for Minor Use (EAMUs)

- *Protected bilberries* 20201296 [2]
- *Protected blackberries* 20201299 [2]
- *Protected blackcurrants* 20201296 [2]
- *Protected blueberries* 20201296 [2]
- *Protected broccoli* 20220019 [1]
- *Protected Brussels sprouts* 20220019 [1]
- *Protected celery* 20220176 [1]
- *Protected chives* 20220176 [1]
- *Protected choi sum* 20220176 [1]
- *Protected collards* 20220019 [1]
- *Protected cranberries* 20201296 [2]
- *Protected cress* 20220176 [1]
- *Protected edible flowers* 20220176 [1]
- *Protected elderberries* 20201296 [2]
- *Protected endives* 20220176 [1]
- *Protected gooseberries* 20201296 [2]
- *Protected herbs (see appendix 6)* 20220176 [1]
- *Protected lamb's lettuce* 20220176 [1]
- *Protected loganberries* 20201299 [2]
- *Protected mulberry* 20201296 [2]

- *Protected parsley* *20220176* [1]
- *Protected purslane* *20220176* [1]
- *Protected raspberries* *20201299* [2]
- *Protected red mustard* *20220176* [1]
- *Protected redcurrants* *20201296* [2]
- *Protected rose hips* *20201296* [2]
- *Protected Rubus hybrids* *20201299* [2]
- *Protected spinach* *20220176* [1]
- *Protected spinach beet* *20220176* [1]
- *Protected strawberries* *20201297* [2]
- *Protected sweet cicerly* *20220176* [1]
- *Protected table grapes* *20201298* [2]
- *Protected wine grapes* *20201298* [2]

Approval information
- Trichoderma atroharzianum (Strain T22) included in Annex 1 under EC Regulation 1107/2009

Efficacy guidance
- Can be applied via drench application, drip-irrigation systems or by spraying while sowing to the cultivation medium. Apply as early as possible to the crop for optimal effect

Restrictions
- Store between 4-8 °C; Keep away from direct sunlight

Hazard classification and safety precautions
UN Number N/C
Operator protection U11, U15, U19a, U20b; A, C, D, H
Storage and disposal D01, D02, D09a, D12a

353   Trichoderma asperellum (Strain T34)

A biological control agent

Products

1	T34 Biocontrol	Fargro	12% w/w	WP	17290

Uses
- Fusarium in *Protected baby leaf crops* (off-label) [1]; *Protected bilberries* (off-label) [1]; *Protected blackberries* (off-label) [1]; *Protected blackcurrants* (off-label) [1]; *Protected blueberry* (off-label) [1]; *Protected broccoli* (off-label) [1]; *Protected Brussels sprouts* (off-label) [1]; *Protected cabbages* (off-label) [1]; *Protected calabrese* (off-label) [1]; *Protected cauliflowers* (off-label) [1]; *Protected choi sum* (off-label) [1]; *Protected collards* (off-label) [1]; *Protected cranberries* (off-label) [1]; *Protected cress* (off-label) [1]; *Protected elderberries* (off-label) [1]; *Protected endives* (off-label) [1]; *Protected forest nurseries* (off-label) [1]; *Protected gooseberries* (off-label) [1]; *Protected herbs (see appendix 6)* (off-label) [1]; *Protected kale* (off-label) [1]; *Protected kohlrabi* (off-label) [1]; *Protected lamb's lettuce* (off-label) [1]; *Protected lettuce* (off-label) [1]; *Protected loganberries* (off-label) [1]; *Protected mulberry* (off-label) [1]; *Protected oriental cabbage* (off-label) [1]; *Protected ornamentals* (off-label) [1]; *Protected raspberries* (off-label) [1]; *Protected redcurrants* (off-label) [1]; *Protected rocket* (off-label) [1]; *Protected rose hips* (off-label) [1]; *Protected Rubus hybrids* (off-label) [1]; *Protected spinach* (off-label) [1]; *Protected spinach beet* (off-label) [1]; *Protected strawberries* (off-label) [1]; *Protected watercress* (off-label) [1];
- Fusarium foot rot and seedling blight in *Aubergines* [1]; *Chillies* [1]; *Peppers* [1]; *Protected carnations* [1]; *Tomatoes* [1];
- Pythium in *Protected baby leaf crops* (off-label) [1]; *Protected bilberries* (off-label) [1]; *Protected blackberries* (off-label) [1]; *Protected blackcurrants* (off-label) [1]; *Protected blueberry* (off-label) [1]; *Protected broccoli* (off-label) [1]; *Protected Brussels sprouts* (off-label) [1]; *Protected cabbages* (off-label) [1]; *Protected calabrese* (off-label) [1]; *Protected cauliflowers* (off-label) [1]; *Protected choi sum* (off-label) [1]; *Protected collards* (off-label) [1]; *Protected cranberries* (off-label) [1]; *Protected cress* (off-label) [1]; *Protected elderberries* (off-

label) [1]; ***Protected endives*** *(off-label)* [1]; ***Protected forest nurseries*** *(off-label)* [1]; ***Protected gooseberries*** *(off-label)* [1]; ***Protected herbs (see appendix 6)*** *(off-label)* [1]; ***Protected kale*** *(off-label)* [1]; ***Protected kohlrabi*** *(off-label)* [1]; ***Protected lamb's lettuce*** *(off-label)* [1]; ***Protected lettuce*** *(off-label)* [1]; ***Protected loganberries*** *(off-label)* [1]; ***Protected mulberry*** *(off-label)* [1]; ***Protected oriental cabbage*** *(off-label)* [1]; ***Protected ornamentals*** *(off-label)* [1]; ***Protected raspberries*** *(off-label)* [1]; ***Protected redcurrants*** *(off-label)* [1]; ***Protected rocket*** *(off-label)* [1]; ***Protected rose hips*** *(off-label)* [1]; ***Protected Rubus hybrids*** *(off-label)* [1]; ***Protected spinach*** *(off-label)* [1]; ***Protected spinach beet*** *(off-label)* [1]; ***Protected strawberries*** *(off-label)* [1]; ***Protected watercress*** *(off-label)* [1];

Extension of Authorisation for Minor Use (EAMUs)

- ***Protected baby leaf crops*** *20222337* [1]
- ***Protected bilberries*** *20222339* [1]
- ***Protected blackberries*** *20222339* [1]
- ***Protected blackcurrants*** *20222339* [1]
- ***Protected blueberry*** *20222339* [1]
- ***Protected broccoli*** *20222337* [1]
- ***Protected Brussels sprouts*** *20222337* [1]
- ***Protected cabbages*** *20222337* [1]
- ***Protected calabrese*** *20222337* [1]
- ***Protected cauliflowers*** *20222337* [1]
- ***Protected choi sum*** *20222337* [1]
- ***Protected collards*** *20222337* [1]
- ***Protected cranberries*** *20222339* [1]
- ***Protected cress*** *20222337* [1]
- ***Protected elderberries*** *20222339* [1]
- ***Protected endives*** *20222337* [1]
- ***Protected forest nurseries*** *20222340* [1]
- ***Protected gooseberries*** *20222339* [1]
- ***Protected herbs (see appendix 6)*** *20222337* [1]
- ***Protected kale*** *20222337* [1]
- ***Protected kohlrabi*** *20222337* [1]
- ***Protected lamb's lettuce*** *20222337* [1]
- ***Protected lettuce*** *20222337* [1]
- ***Protected loganberries*** *20222339* [1]
- ***Protected mulberry*** *20222339* [1]
- ***Protected oriental cabbage*** *20222337* [1]
- ***Protected ornamentals*** *20222340* [1]
- ***Protected raspberries*** *20222339* [1]
- ***Protected redcurrants*** *20222339* [1]
- ***Protected rocket*** *20222337* [1]
- ***Protected rose hips*** *20222339* [1]
- ***Protected Rubus hybrids*** *20222339* [1]
- ***Protected spinach*** *20222337* [1]
- ***Protected spinach beet*** *20222337* [1]
- ***Protected strawberries*** *20222339* [1]
- ***Protected watercress*** *20222337* [1]

Approval information

- Trichoderma asperellum (Strain T34) included in Annex 1 under EC Regulation 1107/2009

Efficacy guidance

- May be applied by spraying, through an irrigation system or by root dipping.

Restrictions

- Efficacy has been demonstrated on peat and coir composts but should be checked on the more unusual composts before large scale use.

Crop-specific information

- Crop safety has been demonstrated on a range of carnation species. It is advisable to check the safety to other species on a sample of the population before large-scale use.

FOR FULL CONDITIONS OF USE ALWAYS READ THE PRODUCT LABEL

Hazard classification and safety precautions
 UN Number N/C
 Risk phrases H317
 Operator protection U05a, U09a, U19a, U20b; A, D, H
 Environmental protection E15a
 Storage and disposal D01, D02, D05, D09a, D09b, D10c
 Medical advice M03, M04a

354 Trichoderma atroviride strain SC1

A biological control agent
FRAC mode of action code: BM 02

Products

1	Vintec	Certis Belchim B V	1 x 10x13 spores per gram	WG	19606

Uses
 * Phaeoacremonium in *Wine grapes* [1];
 * Phaeomoniella in *Wine grapes* [1];
 * Togninia minima in *Wine grapes* [1];

Approval information
 * Trichoderma astroviride (Strain SC1) included in Annex 1 under EC Regulation 1107/2009

Efficacy guidance
 * Store and transport between 4°C and 20°C only
 * Apply immediately to cuts after pruning, ideally when temperatures are above 10°C and humidity levels are high (+70%)
 * Avoid rain or frost within 48 hours of application

Restrictions
 * Maintain an interval of at least 7 days between applications

Hazard classification and safety precautions
 UN Number N/C
 Operator protection U05a, U19a; A, D, H
 Environmental protection E15b, E34
 Storage and disposal D01, D02, D05, D09a
 Medical advice M03

355 trifloxystrobin

A protectant strobilurin fungicide for cereals, apples and pears and managed amenity turf
FRAC mode of action code: 11

See also fluopyram + trifloxystrobin
fluoxastrobin + prothioconazole + trifloxystrobin
prothioconazole + trifloxystrobin
tebuconazole + trifloxystrobin

Products

1	Flint	Bayer CropScience	50% w/w	WG	11259
2	Swift SC	Bayer CropScience	500 g/l	SC	11227

Uses
 * Brown rust in *Spring barley* [2]; *Winter barley* [2]; *Winter wheat* [2];
 * Disease control in *Apples* [1]; *Pears* [1];
 * Foliar disease control in *Ornamental specimens* (off-label) [2];
 * Glume blotch in *Winter wheat* [2];
 * Net blotch in *Spring barley* [2]; *Winter barley* [2];

SEE SECTION 3 FOR PRODUCTS ALSO REGISTERED

SECTION 2

- Rhynchosporium in **Spring barley** [2]; **Winter barley** [2];
- Rust in **Grass seed crops** *(off-label)* [2]; **Rye** *(off-label)* [2]; **Triticale** *(off-label)* [2];
- Septoria leaf blotch in **Winter wheat** [2];

Extension of Authorisation for Minor Use (EAMUs)
- **Grass seed crops** *20193265* [2]
- **Ornamental specimens** *20082882* [2]
- **Rye** *20193265* [2]
- **Triticale** *20193265* [2]

Approval information
- Trifloxystrobin included in Annex I under EC Regulation 1107/2009
- Accepted by BBPA for use on malting barley

Efficacy guidance
- Should be used protectively before disease is established in crop. Further treatment may be necessary if disease attack prolonged
- Trifloxystrobin is a member of the QoI cross resistance group. Product should be used preventatively and not relied on for its curative potential
- Use product as part of an Integrated Crop Management strategy incorporating other methods of control, including where appropriate other fungicides with a different mode of action. Do not apply more than two foliar applications of QoI containing products to any cereal crop
- There is a significant risk of widespread resistance occurring in *Septoria tritici* populations in UK. Failure to follow resistance management action may result in reduced levels of disease control
- On cereal crops product must always be used in mixture with another product, recommended for control of the same target disease, that contains a fungicide from a different cross resistance group and is applied at a dose that will give robust control
- Strains of barley powdery mildew resistant to QoIs are common in the UK

Restrictions
- Maximum number of treatments 2 per crop per yr
- Do not apply to turf during dought conditions or to frozen turf

Crop-specific information
- HI barley, wheat 35 d [2]

Environmental safety
- Dangerous for the environment
- Very toxic to aquatic organisms

Hazard classification and safety precautions
Hazard Dangerous for the environment [2], Very toxic to aquatic organisms
Transport code 9 [1-2]
Packaging group III
UN Number 3077, 3082
Risk phrases H317 [1]
Operator protection U02a [2], U05a, U09a [2], U11 [1], U13 [1], U14 [1], U15 [1], U19a [2], U20b; A, H
Environmental protection E13b [2], E15b [1], E16b [1], E17a (15 m) [1], E34 [1], E38, H410
Consumer protection C02a (35 d) [2]
Storage and disposal D01, D02, D09a, D10c, D12a
Medical advice M03 [1]

356   triflusulfuron-methyl

A sulfonyl urea herbicide for beet crops
HRAC mode of action code: 2 (B)

See also lenacil + triflusulfuron-methyl

Products

1	Debut	FMC Agro	50% w/w	WG	18766
2	Shiro	UPL Europe	50% w/w	WG	17439
3	Tricle	UPL Europe	50% w/w	WG	17448
4	Upbeet	FMC Agro	50% w/w	WG	18794

Uses

* Annual dicotyledons in **Chicory root** *(off-label)* [1]; **Endives** *(off-label)* [1]; **Fodder beet** [1-4]; **Sugar beet** [1-4];
* Black bindweed in **Red beet** *(off-label)* [2];
* Charlock in **Ornamental plant production** *(off-label)* [1]; **Red beet** *(off-label)* [1-2];
* Chickweed in **Red beet** *(off-label)* [2];
* Cleavers in **Ornamental plant production** *(off-label)* [1]; **Red beet** *(off-label)* [1-2];
* Fat hen in **Red beet** *(off-label)* [2];
* Field pansy in **Red beet** *(off-label)* [2];
* Flixweed in **Ornamental plant production** *(off-label)* [1]; **Red beet** *(off-label)* [1];
* Fool's parsley in **Ornamental plant production** *(off-label)* [1]; **Red beet** *(off-label)* [1-2];
* Fumitory in **Red beet** *(off-label)* [2];
* Knotgrass in **Red beet** *(off-label)* [2];
* Nipplewort in **Ornamental plant production** *(off-label)* [1]; **Red beet** *(off-label)* [1];
* Red dead-nettle in **Red beet** *(off-label)* [2];
* Redshank in **Red beet** *(off-label)* [2];
* Scentless mayweed in **Red beet** *(off-label)* [2];
* Small nettle in **Red beet** *(off-label)* [2];
* Volunteer oilseed rape in **Red beet** *(off-label)* [2];
* Wild chrysanthemum in **Red beet** *(off-label)* [1];

Extension of Authorisation for Minor Use (EAMUs)

* **Chicory root** *20183483* [1]
* **Endives** *20183483* [1]
* **Ornamental plant production** *20194268* [1]
* **Red beet** *20183484* [1], *20191915* [2]

Approval information

* Triflusulfuron-methyl included in Annex I under EC Regulation 1107/2009

Efficacy guidance

* Product should be used with a recommended adjuvant or a suitable herbicide tank-mix partner - see label for details
* Product acts by foliar action. Best results obtained from good spray cover of small actively growing weeds
* Susceptible weeds cease growth immediately and symptoms can be seen 5-10 d later
* Best results achieved from a programme of up to 4 treatments starting when first weeds have emerged with subsequent applications every 5-14 d when new weed flushes at cotyledon stage
* Weed spectrum can be broadened by tank mixture with other herbicides. See label for details
* Product may be applied overall or via band sprayer
* Triflusulfuron-methyl is a member of the ALS-inhibitor group of herbicides

Restrictions

* Maximum number of treatments 4 per crop
* Do not apply to any crop stressed by drought, water-logging, low temperatures, pest or disease attack, nutrient or lime deficiency

Crop-specific information

* Latest use: before crop leaves meet between rows

SEE SECTION 3 FOR PRODUCTS ALSO REGISTERED

SECTION 2

- HI 4 wk for red beet
- All varieties of sugar beet and fodder beet may be treated from early cotyledon stage until the leaves begin to meet between the rows

Following crops guidance
- Only winter cereals should follow a treated crop in the same calendar yr. Any crop may be sown in the next calendar yr
- After failure of a treated crop, sow only spring barley, linseed or sugar beet within 4 mth of spraying unless prohibited by tank-mix partner

Environmental safety
- Dangerous for the environment
- Very toxic to aquatic organisms
- Extremely dangerous to fish or other aquatic life. Do not contaminate surface waters or ditches with chemical or used container
- Take extreme care to avoid drift onto broad-leaved plants outside the target area or onto surface waters or ditches, or land intended for cropping
- Spraying equipment should not be drained or flushed onto land planted, or to be planted, with trees or crops other than sugar beet and should be thoroughly cleansed after use - see label for instructions
- LERAP Category B [1-4]

Hazard classification and safety precautions
Hazard Irritant, Dangerous for the environment, Very toxic to aquatic organisms
Transport code 9 [1-4]
Packaging group III
UN Number 3077
Risk phrases H351 [2-3]
Operator protection U05a, U08, U19a, U20a; A
Environmental protection E13a, E15a, E16a, E16b, E38, H410
Storage and disposal D01, D02, D05, D09a, D11a, D12a

357   trinexapac-ethyl

A novel cyclohexanecarboxylate plant growth regulator for cereals, turf and amenity grassland

See also prohexadione-calcium + trinexapac-ethyl

Products

1	Axe	Agform	250 g/l	EC	19594
2	Circle	Syngenta	250 g/l	ME	18381
3	Clayton Trinket	Clayton	250 g/l	EC	19394
4	Cleancrop Alatrin	Agrii	250 g/l	EC	15196
5	Cleancrop Alatrin Evo	Agrii	250 g/l	DC	17763
6	Cleancrop Cutlass	Nufarm UK	175 g/l	EC	18820
7	Cleancrop Operandi	Agrii	250 g/l	EC	19413
8	Confine	Zantra	250 g/l	EC	16109
9	Curve	Agrii	250 g/l	EC	19770
10	Freeze NT	FMC Agro	250 g/l	EC	18732
11	Limitar	BelCrop	250 g/l	EC	16301
12	Moddus	Syngenta	250 g/l	EC	15151
13	Moxa	Globachem	250 g/l	EC	16105
14	Optimus	Nufarm UK	175 g/l	EC	18817
15	Palisade	Syngenta	250 g/l	EC	17860
16	Pan Tepee	Pan Agriculture	250 g/l	EC	15867
17	Primo Maxx II	Syngenta	121 g/l	SL	17509
18	Profi Trinex 250 SC	Agrovista	250 g/l	EC	19405
19	Scitec	Syngenta	250 g/l	EC	15588
20	Shrink	AgChem Access	250 g/l	EC	15660
21	Sonis	Syngenta	250 g/l	EC	16891

Products – continued

22	Sudo Mor	Life Scientific	250 g/l	EC	18704
23	Tempo	Syngenta	250 g/l	EC	15170
24	Tridus	Globachem	250 g/l	EC	16938
25	Trimaxx	Nufarm UK	175 g/l	EC	19322
26	Trinexis	Arysta	250 g/l	EC	16429

Uses

- Growth regulation in *Canary seed* (off-label) [6,14]; *Durum wheat* [11,20,22]; *Forest nurseries* (off-label) [12,17]; *Grass seed crops* (off-label) [11-12,22]; *Ornamental plant production* (off-label) [12,17]; *Red clover* (off-label) [12]; *Rye* [11]; *Ryegrass seed crops* [20]; *Spring barley* [11,20,22]; *Spring oats* [11,20,22]; *Spring oilseed rape* [2]; *spring red wheat* (off-label) [8]; *Spring rye* [20,22]; *Spring wheat* [11,22]; *Triticale* [11,20,22]; *Winter barley* [11,20,22]; *Winter oats* [11,20,22]; *Winter oilseed rape* [2]; *Winter rye* [20,22]; *Winter wheat* [11,20,22];
- Growth retardation in *Amenity grassland* [17]; *Managed amenity turf* [17]; *spring red wheat* (off-label) [23];
- Lodging control in *Durum wheat* [1,3-10,12-16,18-19,21,23-26]; *Grass seed crops* (off-label) [1,3-4,6-10,12-14,18,24]; *Red clover* (off-label) [12]; *Rye* [9-10,13,24]; *Ryegrass seed crops* [12,15-16,19,21,23,25-26]; *Spring barley* [1,3-10,12-16,18-19,21,23-26]; *Spring oats* [1,3-10,12-16,18-19,21,23-26]; *spring red wheat* (off-label) [8,12,23]; *Spring rye* [1,3-8,12,14-16,18-19,21,23,25-26]; *Spring wheat* [1,3-5,7,9-10,12-14,18-19,24-25]; *Triticale* [1,3-10,12-16,18-19,21,23-26]; *Winter barley* [1,3-10,12-16,18-19,21,23-26]; *Winter oats* [1,3-10,12-16,18-19,21,23-26]; *Winter rye* [1,3-8,12,14-16,18-19,21,23,25-26]; *Winter wheat* [1,3-10,12-16,18-19,21,23-26];

Extension of Authorisation for Minor Use (EAMUs)

- *Canary seed* *20191183* [6], *20192154* [14]
- *Forest nurseries* *20122055* [12], *20180621* [17]
- *Grass seed crops* *20171035* [12]
- *Ornamental plant production* *20103062* [12], *20180621* [17]
- *Red clover* *20171035* [12]
- *spring red wheat* *20130322* [8], *20121385* [12], *20113201* [23]

Approval information

- Trinexapac-ethyl included in Annex I under EC Regulation 1107/2009
- Accepted by BBPA for use on malting barley

Efficacy guidance

- Best results on cereals and ryegrass seed crops obtained from treatment from the leaf sheath erect stage
- Best results on turf achieved from application to actively growing weed free turf grass that is adequately fertilized and watered and is not under stress. Adequate soil moisture is essential
- Turf should be dry and weed free before application
- Environmental conditions, management and cultural practices that affect turf growth and vigour will influence effectiveness of treatment
- Repeat treatments on turf up to the maximum approved dose may be made as soon as growth resumes

Restrictions

- Maximum total dose equivalent to one full dose on cereals and ryegrass seed crops
- Do not apply if rain or frost expected or if crop wet. Products are rainfast after 12 h
- Only use on crops at risk of lodging
- Do not apply within 12 h of mowing turf
- Do not treat newly sown turf
- Not to be used on food crops
- Do not compost or mulch grass clippings
- Some products stipulate that application should not be made before GS30
- Must not be used on grass seed crops that will be grazed by livestock or cut for fodder
- No more than 28 applications must be carried out on managed amenity turf and amenity grassland per year [17]

SEE SECTION 3 FOR PRODUCTS ALSO REGISTERED

SECTION 2

Crop-specific information

- Latest use: Up to and including flag leaf just visible (GS37) for oats and spring barley, up to and including flag leaf ligule just visible (GS39) for wheat, rye, triticale and winter barley.
- On wheat apply as single treatment between leaf sheath erect stage (GS 30) and flag leaf fully emerged (GS 39)
- On barley, rye, triticale and durum wheat apply as single treatment between leaf sheath erect stage (GS 30) and second node detectable (GS 32), or on winter barley at higher dose between flag leaf just visible (GS 37) and flag leaf fully emerged (GS 39)
- On oats and ryegrass seed crops apply between leaf sheath erect stage (GS 30) and first node detectable stage (GS 32)
- Treatment may cause ears of cereals to remain erect through to harvest
- Turf under stress when treated may show signs of damage
- Any weed control in turf must be carried out before application of the growth regulator

Environmental safety

- Dangerous for the environment
- Toxic to aquatic organisms

Hazard classification and safety precautions

Hazard Irritant [1-4,6-10,12-16,18-26], Dangerous for the environment [1-4,6-10,12-16,18-26], Flammable liquid and vapour [2,13,24], Harmful if inhaled [2,13,17,24]

Transport code 9 [1-5,7-10,12-13,15-24,26]

Packaging group III

UN Number 3082, N/C

Risk phrases H315 [6,25], H317 [1,3,6-7,9-12,14-20,22-23,25-26], H319 [2,5-6,10-11,13-14,24-25], H320 [2], H335 [2,11,13,24], H361 [11], H373 [11,13,17,24], R43 [4,21], R51 [4,21], R53a [4,21]

Operator protection U05a [2,6,8-10,12-17,19-26], U08 [11], U09c [11], U11 [11], U15 [2,6,8-10,12-16,19-26], U19a [6,11], U20a [11], U20b [17], U20c [2,6,8-10,12-16,19-26]; A, C, H, K

Environmental protection E15a [2,6,8-10,12-16,19-26], E15b [11], E34 [11], E38 [2,6,8-16,19-26], H410 [1,3,5,7,12-13,15,18-19,24], H411 [2,8-9,11,16,20,22-23,26], H412 [6,10,14,17,25]

Consumer protection C01 [17]

Storage and disposal D01 [1-3,6,8-17,19-26], D02 [1-3,6,8-10,12-17,19-26], D05 [1-3,6,8-10,12-17,19-26], D09a [2,6,8-16,19-26], D10c [2,6,8-10,12-17,19-26], D11a [11], D12a [2,6,8-16,19-26], D12b [11]

Medical advice M03 [11]

---

358   urea

Commodity substance for fungicide treatment of cut tree stumps approved until 31/8/2025

---

359   zoxamide

A substituted benzamide fungicide available only in mixtures
FRAC mode of action code: 22

See also cymoxanil + zoxamide
        dimethomorph + zoxamide

SECTION 3
PRODUCTS ALSO REGISTERED

Products also Registered

Products listed in the table below have not been notified for inclusion in Section 2 of this edition of the *Guide*. These products may legally be stored and used in accordance with their label until their approval expires, but they may not still be available for purchase.

Product	Approval holder	MAPP No.	Expiry Date
1 (Z)-11-hexadecenal			
Box T Pro Press	M2I Biocontrol	19684	28 Feb 2025
Box T Pro Press	Syngenta	20039	29 Mar 2025
2 1,4-dimethylnaphthalene			
1,4Sight (N I only)	Dormfresh	20361	03 Aug 2026
3 abamectin			
Amec	Nurture	18691	31 Oct 2023
Clayton Abba (N I only)	Clayton	19858	31 Oct 2024
Killermite	6Science	19300	30 Jun 2024
Killermite (N I only)	6Science	20274	31 Oct 2024
Smitten (N I only)	Pan Agriculture	20287	31 Oct 2025
5 acetamiprid			
Acetamex 20 SP	MAC	15888	30 Jun 2024
Antelope	Gemini	18041	30 Jun 2024
Ariel SG	Servem	19514	30 Jun 2024
Clayton Vault (N I only)	Clayton	19917	09 Sep 2099
Gazelle	Certis Belchim B V	12909	09 Sep 2099
Gazelle WSB	Certis Belchim B V	19694	09 Sep 2099
Insyst	Certis Belchim B V	13414	30 Jun 2023
Pan Vulcan SG (N I only)	Pan Agriculture	20281	09 Sep 2099
Persist	RAAT	17633	09 Sep 2099
Pure Ace	Pure Amenity	19053	30 Jun 2024
Symiprid	Simagro	19363	30 Jun 2024
Vulcan (N I only)	Certis Belchim B V	20291	09 Sep 2099
Vulcan SP	Certis Belchim B V	20292	09 Sep 2099
6 acetic acid			
New-way Weed Spray	Sipcam	20148	28 Feb 2026
OWK	Sipcam	20153	28 Feb 2026
OWK	UK Organic	15363	28 Feb 2026
Spot On Pro Weed and Moss Killer	Sipcam	20184	28 Feb 2026
9 adoxophyes orana gv			
Capex	Andermatt	18258	31 Jul 2024
Capex (GB only)	Andermatt	19948	31 Jul 2026
10 alpha-cypermethrin			
Alert (GB only)	BASF	16785	30 Apr 2029
Contest (GB only)	BASF	16764	30 Apr 2029
Eribea (GB only)	Certis Belchim B V	17270	30 Apr 2029
Fastac (GB only)	BASF	16761	30 Apr 2029
Fastac ME (GB only)	BASF	17686	30 Apr 2029
Fasthrin 10EC (GB only)	Sharda	18320	30 Apr 2029
Hi-Aubin (GB only)	Hockley	19408	30 Apr 2029
11 aluminium ammonium sulphate			
Asbo	Sphere	19186	28 Feb 2026
Guardsman	Chiltern	19150	28 Feb 2026

Product	Approval holder	MAPP No.	Expiry Date
12 aluminium phosphide			
Degesch Fumigation Tablets	Rentokil	17035	31 Dec 2023
Detia Gas Ex-P	Rentokil	17036	31 Dec 2023
Detia Gas-Ex-T	Rentokil	17034	28 Feb 2026
Phostoxin Bag	Rentokil	19333	28 Feb 2026
Phostoxin Pellet	Rentokil	19331	28 Feb 2026
Phostoxin Tablet	Rentokil	19330	28 Feb 2026
Quickphos Pellets 56% GE	UPL Europe	16987	28 Feb 2026
15 ametoctradin + dimethomorph			
Resplend	BASF	19702	31 Jan 2025
16 ametoctradin + mancozeb			
Diablo	BASF	16084	30 Jul 2026
17 amidosulfuron			
Eagle	Sumitomo	16490	30 Jun 2026
Squire Ultra	Sumitomo	16491	30 Jun 2026
18 amidosulfuron + iodosulfuron-methyl-sodium			
Sekator OD	Sumitomo	16494	09 Sep 2099
Sekator OD	Sumitomo	18905	09 Sep 2099
21 aminopyralid + fluroxypyr			
Forefront Pro	Corteva	19177	04 Jun 2024
Halcyon	Corteva	18561	04 Jun 2024
22 aminopyralid + halauxifen-methyl			
Trezac	Corteva	18253	30 Jun 2027
24 aminopyralid + propyzamide			
Amino-Pro	HMpG GMBH	18948	30 Jun 2024
Clayton Propel Plus	Clayton	17227	30 Jun 2024
Clayton Propel Plus (N I only)	Clayton	19905	09 Sep 2099
Galactic Pro	Chem-Wise	17180	30 Jun 2024
Galactic Pro (N I only)	Chem-Wise	20059	09 Sep 2099
Milepost	Terrechem	17537	30 Jun 2024
Milestone	PSI	17494	30 Jun 2024
Pyzamid Universe	RealChemie	18169	30 Jun 2024
25 aminopyralid + triclopyr			
Garlon Ultra	Nomix Enviro	19172	31 Oct 2024
26 amisulbrom			
Leimay	Syngenta	19280	31 Mar 2027
Sanblight	Nissan	19273	31 Mar 2027
29 Ampelomyces quisqualis (Strain AQ10)			
AQ 10	Fargro	17102	09 Sep 2099
31 Aureobasidium pullulans strain DSM 14940 and DSM 14941			
Blossom Protect	Masstock	20051	31 Jul 2027
Boni Protect	Nufarm UK	19233	31 Jul 2026
Botector	Nufarm UK	19443	31 May 2026
Botector	Masstock	20000	31 Jul 2027
33 azoxystrobin			
5504	Syngenta	12351	30 Jun 2027
Arixa	Nurture	19034	30 Jun 2024
Azaka	FMC Agro	18731	30 Jun 2027
Azofin	Finchimica	18193	30 Jun 2027

Product	Approval holder	MAPP No.	Expiry Date
Azoshy	Sharda	18072	30 Jun 2027
Azure	Synergy	19448	30 Jun 2027
Chamane	UPL Europe	15922	30 Jun 2027
Clayton Belfry	Clayton	18154	30 Jun 2024
Clayton Belfry N I only)	Clayton	19864	30 Jun 2027
Cleancrop Cortina	Agrii	18506	30 Jun 2027
Conclude	Certis Belchim B V	16905	30 Jun 2027
Floozy	Gemini	19740	30 Jun 2027
Globaztar AZT 250SC	Certis Belchim B V	18441	30 Jun 2027
Globaztar SC	Zantra	15575	30 Jun 2027
Hill-strobin	Hillfield	19450	30 Jun 2027
Hi-Nelson	Hockley	19377	30 Jun 2027
Legado	Jebagro	18087	30 Jun 2027
Legate	Novastar	18531	30 Jun 2027
Mirador 250SC	Adama	18292	30 Jun 2027
Opal (N I only)	Pan Agriculture	20277	30 Jun 2027
Ortiva	Syngenta	10542	30 Jun 2027
Pantha	Terrechem	18619	30 Jun 2027
Phloem	Capital CP	17754	30 Jun 2024
Phloem (N I only)	Capital CP	20071	30 Jun 2027
Priori	Syngenta	10543	30 Jun 2027
Promesa	Galenika	18840	30 Jun 2027
Puma	Ascot Pro-G	17678	30 Jun 2024
Puma (N I only)	Ascot Pro-G	20082	30 Jun 2027
Pure Azoxy	Pure Amenity	18065	30 Jun 2024
Storoso	Syngenta	19742	30 Jun 2027
Toran	Becesane	18239	30 Jun 2024
Vertaza	Syngenta	19795	30 Jun 2027
Weaver	Syngenta	19667	30 Jun 2027
Zakeo 250 SC	Adama	18379	30 Jun 2027

39 azoxystrobin + isopyrazam

Symetra Flex (GB only)	Adama	19220	30 Jun 2027

45 Bacillus firmus I - 1582

Flocter	Bayer CropScience	16480	31 Mar 2026
Votivo	BASF	19087	31 Mar 2026

47 Bacillus subtilis

Solani	Russell	16585	31 Oct 2023

49 Bacillus thuringiensis aizawai GC-91

Agree 50 WG	Certis Belchim B V	20177	31 Oct 2025

52 Bacillus thuringiensis kurstaki

Biocure	Russell	18413	31 Oct 2023
Bruco	Progreen	17919	31 Oct 2023
Delfin WG	Andermatt	18452	31 Oct 2025

63 benthiavalicarb + oxathiapiprolin

Zorvec Endavia	Corteva	20370	24 Feb 2024

65 benthiavalicarb-isopropyl + mancozeb

En-Garde	Certis Belchim B V	14901	31 Jul 2026
En-Garde	Certis Belchim B V	20205	31 Jul 2026
Valbon	Certis Belchim B V	20250	30 Jul 2026
Valbon	Certis Belchim B V	14868	30 Jul 2026

66 benzoic acid

MENNO Florades	Brinkman	15091	09 Sep 2099

SECTION 3

Product	Approval holder	MAPP No.	Expiry Date
Menno Florades	Brinkman	20001	21 Apr 2026
67 benzovindiflupyr			
Aprovia Plus	Syngenta	19174	02 Sep 2025
Kavatur Plus	Syngenta	19566	02 Sep 2025
68 benzovindiflupyr + difenoconazole			
Ascernity	Syngenta	19544	02 Sep 2025
69 benzovindiflupyr + prothioconazole			
Elate	Generica	18091	31 Jan 2024
Levee	Syngenta	19848	02 Sep 2025
Lizard	Syngenta	19855	02 Sep 2025
Pro-Benzo	HMpG GMBH	19058	31 Jan 2024
70 benzyladenine			
MaxCel	Sumitomo	15708	31 Jan 2024
71 benzyladenine + gibberellin			
Promalin	Sumitomo	18325	28 Feb 2023
75 bifenazate			
Floramite 240 SC	Arysta	17958	31 Jan 2024
Inter Bifenazate 240 SC	Iticon	14543	31 Jan 2024
Wopro-Bifenazate 24% SC (N I Only)	Simonis	20305	31 Jan 2025
76 bifenox			
Cleancrop Diode	Agrii	14620	31 Dec 2023
Wolf	RAAT	17378	31 Dec 2023
81 bixafen			
Bixafen EC125	Bayer CropScience	15951	30 Nov 2027
Inception	Bayer CropScience	19962	30 Nov 2027
85 bixafen + prothioconazole			
BAYF869	Bayer CropScience	19315	31 Jan 2026
87 bixafen + prothioconazole + tebuconazole			
BAYF326	Bayer CropScience	19395	31 Jan 2026
89 boscalid			
Bonafide	Sharda	18775	31 Jan 2026
Boscler	Life Scientific	19113	08 Jul 2024
Coli	Gemini	17616	31 Jan 2023
Fulmar (N I only)	AgChem Access	20015	31 Jan 2026
Rasput	Globachem	19252	31 Jan 2026
93 boscalid + metconazole			
Corvus	Stefes	19477	31 Oct 2023
94 boscalid + pyraclostrobin			
Daisy	RAAT	17492	31 Jul 2023
Pyrabos	HMpG GMBH	19080	31 Jan 2024
109 bromuconazole + tebuconazole			
Djembe	Sumitomo	18323	28 Feb 2025
Djembe	Syngenta	19342	28 Feb 2025
Sakura	Sumitomo	18327	28 Feb 2025
Soleil	Sumitomo	16869	28 Feb 2025

Product	Approval holder	MAPP No.	Expiry Date
111 buprofezin			
Applaud 25 WP	Certis Belchim B V	17197	31 Jul 2025
112 Candida oleophila Strain 0			
Nexy 1	BioNext	16961	31 Jan 2024
Nexy 1	Agrauxine	19347	31 Mar 2026
113 captan			
Akotan 80 WG	Aako	16307	31 Jan 2026
Clayton Core (N I only)	Clayton	19879	31 Jan 2025
Malvin WG	Arysta	16308	31 Jan 2026
Orthocide WG	Arysta	16309	31 Jan 2026
Ratan 80 WG	RAAT	17252	31 Jan 2024
127 carfentrazone-ethyl			
Carfen 50	HMpG GMBH	19480	30 Jun 2024
Carfen 50 II	HMpG GMBH	19482	30 Jun 2024
Carfen 50 II	HMpG GMBH	19583	30 Jun 2024
Harrier	FMC Agro	18803	09 Sep 2099
129 carfentrazone-ethyl + mecoprop-P			
Jewel	FMC Agro	18780	31 Jul 2025
139 chlormequat			
Adjust	Taminco	17141	31 May 2026
Barleyquat B	Taminco	17204	31 May 2026
BEC CCC 720	Becesane	19009	31 May 2024
BEC CCC 750	Becesane	17930	31 May 2024
Belcocel	Taminco	17773	31 May 2026
Belcocel 750	Taminco	19125	31 May 2026
Bettequat B	Taminco	17208	31 May 2026
CCC 720	Nurture	19008	31 May 2024
Clayton Everest	Clayton	19389	31 May 2024
Clayton Everest (N I only)	Clayton	19886	31 May 2025
Clayton Everest (N I only)	Clayton	19887	31 May 2025
Clayton Squat	Clayton	19578	31 May 2026
Cleancrop Conicen	Agrii	18986	31 May 2026
Eddystone	Terrechem	18345	31 May 2026
Jadex Plus	Clayton	17524	31 May 2026
Jadex-o-720	SFP Europe	16284	31 May 2026
K2	Taminco	17206	31 May 2026
Manipulator	Taminco	17207	31 May 2026
Palermo	SFP Europe	17900	31 May 2026
Plectrum	Gemini	18317	31 May 2026
Selon	Taminco	17209	31 May 2026
Silk 750 SL	Terrechem	17496	31 May 2024
140 chlormequat + ethephon			
BOGOTA UK	Adama	16318	31 Jan 2023
Ormet Plus	SFP Europe	17816	31 Jan 2023
Ormet Plus	Adama	19479	31 Jan 2026
Spatial Plus	Sumitomo	16665	31 Jan 2026
Vivax	Sumitomo	16682	31 Jan 2026
145 chlormequat + trinexapac-ethyl			
Completto	Adama	18448	09 Jun 2024
179 cholecalciferol			
Replexa	BASF	UK20-1257	27 Apr 2025

Product	Approval holder	MAPP No.	Expiry Date
183 clethodim			
Cavalier	Top Crop	19438	09 Nov 2023
Chellist	Chem-Wise	19439	09 Nov 2023
Chellist	Chem-Wise	19463	09 Nov 2023
Destrier	Top Crop	19382	09 Nov 2023
V-Dim 240 EC	VextaChem	18530	30 Nov 2025
Vextadim 240 EC	Clayton	18465	31 Jul 2024
184 clodinafop-propargyl			
Double	Nufarm UK	20302	31 Oct 2025
185 clodinafop-propargyl + cloquintocet-mexyl			
Ciclope	Industrias Afrasa	17226	31 Oct 2024
188 clodinafop-propargyl + prosulfocarb			
Auxiliary	BASF	14576	30 Apr 2025
Grapple	Syngenta	16523	30 Apr 2025
190 clofentezine			
Acaristop 500 SC	Aako	19819	30 Jun 2026
Ariane	RAAT	17296	30 Jun 2024
191 clomazone			
Angelus	Rotam	18451	30 Apr 2026
Clayton Chrome 360	Clayton	19359	30 Apr 2026
Clayton Chrome CS	Clayton	16383	30 Apr 2024
Clozone	UPL Europe	18979	30 Apr 2026
Colone	Life Scientific	20348	30 Apr 2026
Czar	Terrechem	17423	30 Apr 2024
Evea	Rotam	18511	30 Apr 2026
Hobby	Ascot Pro-G	17667	30 Apr 2024
Notion	AgChem Access	16773	30 Apr 2024
Pegasus	Terrechem	18658	30 Apr 2024
Standon Soulmate	Enviroscience	18499	30 Apr 2024
Token	Becesane	18574	30 Apr 2023
Zone 360 CS	HMpG GMBH	18424	30 Apr 2024
Zone 360 CS-II	HMpG GMBH	18437	30 Apr 2024
Zulkon	Unisem	17610	30 Apr 2024
194 clomazone + metazachlor			
Circuit Synctec	FMC Agro	17118	31 Jan 2026
Nimbus CS	BASF	16573	31 Jan 2026
195 clomazone + metazachlor + napropamide			
Colzor SyncTec	FMC Agro	18716	20 Apr 2024
197 clomazone + napropamide			
Altiplano DAMtec	FMC Agro	18717	30 Apr 2026
199 clopyralid			
Bariloche	Proplan-Plant	17577	31 Oct 2024
Cliophar 600 SL	Arysta	16735	31 Oct 2024
Dow Shield 400	Corteva	14984	31 Oct 2023
GF-2895	Corteva	16821	31 Oct 2023
Glopyr 400	Globachem	15009	31 Oct 2024
Leash	Life Scientific	17969	31 Oct 2024
Leash (GB only)	Life Scientific	20006	31 Oct 2026
Lontrel 600	Corteva	19690	31 Oct 2023
Lontrel 600	Corteva	19878	31 Oct 2024
Lontrel 72SG	Corteva	19877	31 Oct 2024

Product	Approval holder	MAPP No.	Expiry Date
Lontrel 72SG	Corteva	15235	31 Oct 2023
Shield Pro	Corteva	20156	09 Sep 2099

203 clopyralid + florasulam + fluroxypyr

Dingo	Corteva	18412	31 Oct 2023
Greenor Gold	Origin Amenity	19827	31 Oct 2023
Leystar	Corteva	17921	31 Oct 2023
Mogul (N I only)	Pan Amenity	20293	31 Oct 2024
Praxys	Corteva	18953	31 Oct 2023

204 clopyralid + fluroxypyr + MCPA

Brittas	Barclay	18587	30 Apr 2025

207 clopyralid + picloram

Barca 334 SL	Pestila	19121	30 Jun 2026
Chaco	Globachem	18783	30 Jun 2026
Legara (N I only)	AgChem Access	20021	30 Jun 2025

209 clopyralid + triclopyr

Blaster Pro	Corteva	18074	31 Oct 2023
Grazon Pro	Corteva	15785	31 Oct 2023
Headland Flail 2	FMC Agro	16007	31 Jul 2023
Prevail	Corteva	17395	31 Oct 2023
Thistlex	Corteva	16123	31 Oct 2023
Tor	Corteva	17777	31 Oct 2023
Tor	Corteva	19958	31 Oct 2024

214 Coniothyrium minitans

Contans WG	Bayer CropScience	17985	31 Jan 2035
Lalstop Contans WG	Danstar	19974	31 Jan 2035

223 cyantraniliprole

Mainspring	Syngenta	19198	14 Mar 2029

224 cyazofamid

Karitsu (GB only)	Certis Belchim B V	19996	31 Oct 2026
Karitsu (GB only)	Certis Belchim B V	20352	31 Jan 2027
Livarti (GB only)	Certis Belchim B V	20303	31 Jan 2027
Livarti (GB only)	Certis Belchim B V	19995	31 Oct 2026
Ranman Top (GB only)	Certis Belchim B V	14753	31 Oct 2026
Ranman Top (GB only)	Certis Belchim B V	20353	31 Jan 2027
Ranman Top (N I only)	Certis Belchim B V	20354	09 Sep 2099
Roman (GB only)	Cropthetics	19783	31 Jan 2027
RouteOne Roazafod	Albaugh UK	15271	31 Jan 2023
Sugoi	ISK Biosciences	19122	09 Sep 2099
Sugoi (GB only)	ISK Biosciences	19122	31 Jan 2027
Tiberius	Cropthetics	19925	31 Jan 2027

228 Cydia pomonella GV

Cyd-X	Certis Belchim B V	17019	31 Oct 2025
Cyd-X Duo	Certis Belchim B V	16779	31 Oct 2025
Cyd-X Xtra	Certis Belchim B V	17020	31 Oct 2025
Madex Top	Andermatt	19213	31 Oct 2025

229 cyflufenamid

Clayton Midas (N I only)	Clayton	19896	30 Sep 2025
Clayton Roulette (N I only)	Clayton	19908	30 Sep 2025
Cyflufen EW	HMpG GMBH	18425	30 Jun 2024
Diego	Star	16980	30 Jun 2024
Laquna (N I only)	Pan Agriculture	20275	30 Sep 2025

SECTION 3

Product	Approval holder	MAPP No.	Expiry Date
NF-149 SC	Certis Belchim B V	15835	30 Sep 2025
Sydly	Terrechem	18307	30 Jun 2024

230 cyflufenamid + spiroxamine
Cyflamid Plus	Certis Belchim B V	19845	04 Aug 2025

232 cymoxanil
Controlla	Cropthetics	19478	28 Feb 2026
Curago 45 WG	Stefes	19970	28 Feb 2026
Curzate 60 WG	DuPont (Corteva)	19603	28 Feb 2026
Cymbal 45	Certis Belchim B V	20203	28 Feb 2026
Cymbal Flow	Certis Belchim B V	20217	28 Feb 2026
Cymbal Flow	Certis Belchim B V	17843	28 Feb 2026
Cymostraight 45	Certis Belchim B V	17437	28 Feb 2026
Danso Flow	Certis Belchim B V	17963	28 Feb 2026
Dauphin 45	SFP Europe	16434	29 Feb 2024
Dauphin 45	Adama	19487	28 Feb 2026
Drum	Certis Belchim B V	20220	28 Feb 2026
Drum Flow	Certis Belchim B V	20229	28 Feb 2026
Drum Flow	Certis Belchim B V	17906	28 Feb 2026
Itwin 45	Indofil	19924	28 Feb 2026
Krug Flow	Certis Belchim B V	17964	28 Feb 2026
Krug Flow	Certis Belchim B V	20232	28 Feb 2026

234 cymoxanil + fluazinam
Grecale	Sipcam	17002	31 Aug 2025
Shirlan Forte	Syngenta	15542	31 Aug 2025
Tezuma	Certis Belchim B V	17396	31 Aug 2025

236 cymoxanil + mancozeb
Cynthia	Spring	19527	31 Jul 2023
Nautile WP	UPL Europe	16468	30 Jul 2026
Palmas WP (GB only)	Adama	19491	30 Jul 2026
Solace Max	Nufarm UK	16697	30 Jul 2026
Solution	Corteva	19078	30 Jul 2026
Zetanil WG	Sipcam	15488	30 Jul 2026

237 cymoxanil + mandipropamid
Bedrock	Syngenta	18426	31 Jan 2026

238 cymoxanil + propamocarb
Omix Duo	Agria SA	18978	31 Jan 2025
Propicarb Duo	Ascot Pro-G	19883	31 Jan 2025
Rival Duo	Agria SA	18442	31 Jan 2025
Simpro	Globachem	19989	31 Jan 2026

239 cymoxanil + zoxamide
Reboot	Gowan	18202	09 Sep 2099

240 cypermethrin
Afrisect 500 EC	Arysta	17137	30 Apr 2025
Cyper 500	Nurture	18670	30 Apr 2024
Langis 300 ES	Arysta	18178	30 Apr 2024
Langis 300 ES	UPL Europe	19641	30 Apr 2025
Permasect 500 EC	Arysta	17132	30 Apr 2025
Phobi Grain EC	Lodi UK	19691	30 Apr 2025
Protect Ultra	Crop Health	19554	30 Apr 2024
Supasect 500 EC	Arysta	17131	30 Apr 2025

Product	Approval holder	MAPP No.	Expiry Date
249 cyprodinil			
Coracle (N I only)	AgChem Access	20014	31 Oct 2025
250 cyprodinil + fludioxonil			
Botrefin	Clayton	18537	31 Oct 2025
Clayton Creed	Clayton	19461	31 Oct 2025
Clayton Gear	Clayton	17169	31 Oct 2023
Clayton Gear (N I only)	Clayton	20331	30 Apr 2025
Lever	HMpG GMBH	18467	31 Oct 2023
Ludo	Pan Agriculture	18314	30 Apr 2024
Ludo (N I only)	Pan Agriculture	20276	30 Apr 2025
Modif	Life Scientific	18583	31 Oct 2025
Reversal	RAAT	17359	31 Oct 2023
Shift	Clayton	18881	31 Oct 2025
Symchanger WG	Simagro	19432	31 Oct 2023
251 cyprodinil + isopyrazam			
Cebara (GB only)	Adama	18569	31 Oct 2026
253 2,4-D			
Dioweed 50	UPL Europe	13197	30 Sep 2023
Headland Staff 500	FMC Agro	13196	30 Sep 2023
Herboxone	FMC Agro	13958	30 Sep 2023
Herboxone 60	FMC Agro	14080	09 Sep 2099
Maton	FMC Agro	13234	30 Sep 2023
258 2,4-D + dicamba + iron sulphate + mecoprop-P			
Elliott's Professional Feed Weed & Moss Killer	Elliott	19205	31 Jul 2025
Feed 'N' Weed Extra	Vitax	19103	31 Jul 2025
GTHMF Pro	GTHMF	18443	31 Jul 2025
Professional Turf Feed, Weed And Moss Killer	Proctor	19106	31 Jul 2025
Restore Plus	Coburn	19102	31 Jul 2025
Turfmaster Complete Feed, Weed & Mosskiller	Hygeia	18985	31 Jul 2025
259 2,4-D + dicamba + MCPA + mecoprop-P			
Dicophar	Arysta	16460	31 Jul 2025
263 2,4-D + glyphosate			
Diamond	Agrigem	19846	09 Sep 2099
Kurtail Evo	Progreen	19975	09 Sep 2099
Rasto	Nufarm UK	16795	09 Sep 2099
264 2,4-D + MCPA			
Agroxone Combi	Nufarm UK	14907	09 Sep 2099
267 2,4-D + triclopyr			
Genoxone ZX EC	Arysta	17016	09 Sep 2099
268 daminozide			
B-Nine SG	Arysta	14434	30 Apr 2024
Stature	Pure Amenity	18064	30 Apr 2024
273 deltamethrin			
CMI Delta 2.5 EC	CMI	16695	30 Apr 2023
Decis Forte	Bayer CropScience	16110	30 Apr 2026
Deltason-D	DAPT	17735	30 Apr 2023
Deserve	Sharda	19504	30 Apr 2026
GAT Decline 2.5 EC	FMC Agro	18715	30 Apr 2026

SECTION 3

Product	Approval holder	MAPP No.	Expiry Date
Grain-Tect ULV	Barrettine	18076	30 Apr 2024
K-obiol EC 25 (GB only)	Environmental Science	20254	30 Apr 2027
K-obiol EC 25 (N I only)	Environmental Science	20253	30 Apr 2026
K-obiol Ulv6 (GB only)	Environmental Science	20174	30 Apr 2027
K-obiol Ulv6 (N I only)	Environmental Science	20310	30 Apr 2026
Tecsis	HMpG GMBH	18439	30 Apr 2024

279 dicamba
Antuco	Ventura	17409	30 Jun 2026

282 dicamba + MCPA + mecoprop-P
Premier Amenity Selective	Hygeia	19860	31 Jul 2025

283 dicamba + mecoprop-P
Dimeco XL	UPL Europe	17190	30 Sep 2023
Foundation	Nufarm UK	19600	30 Sep 2023
GreenForce Professional	Agrichem	19786	31 Jul 2025
Hygrass P	Agrichem	16802	30 Jun 2023
Mircam	Nufarm UK	11707	31 Jul 2025
Optica Forte	Nufarm UK	14845	31 Jul 2025
Quickfire	Headland Amenity	19787	31 Jul 2025
Quickfire (N I only)	Headland Amenity	20116	30 Jun 2025

285 dicamba + nicosulfuron + prosulfuron
Diniro	FMC Agro	19339	30 Jun 2026
Spandis	Syngenta	19276	30 Jun 2026

286 dicamba + prosulfuron
Casper	Syngenta	15573	09 Sep 2099
Clayton Spook	Clayton	15846	30 Jun 2024
Clayton Spook (N I only)	Clayton	19912	09 Sep 2099
Symmaize	Simagro	19434	30 Jun 2024

292 dichlorprop-P + MCPA + mecoprop-P
Hymec Triple	Agrichem	15753	31 Jul 2025
Optica Trio	FMC Agro	16113	30 Sep 2023

295 difenacoum
Neokil	BASF	UK12-0298	19 Feb 2023
Neosorexa Bait Blocks	BASF	UK12-0360	10 May 2023
Neosorexa Gold	BASF	UK12-0304	20 Feb 2023
Neosorexa Gold Ratpacks	BASF	UK12-0304	20 Feb 2023
Neosorexa Pasta Bait	BASF	UK12-0365	09 Sep 2023
Ratak Cut Wheat	BASF	UK12-0313	22 Feb 2023
Romax D Rat & Mouse Killer	Rentokil	UK14-0808	19 Mar 2023
Sorexa D	BASF	UK12-0319	14 Feb 2023
Sorexa D Mouse Killer	BASF	UK12-0319	14 Feb 2023

297 difenoconazole
Bogard	Syngenta	17310	30 Jun 2026
Brace	Terrechem	19197	30 Jun 2026
Change	RAAT	17400	30 Jun 2024
Difend	Q-Chem	16316	30 Jun 2026
Difenofin	Finchimica	19665	30 Jun 2026
Fen	Synergy	19445	30 Jun 2026
Griffin	Ascot Pro-G	19421	30 Jun 2026
Mavita 250EC	Adama	18293	30 Jun 2026
Stratus	B&W Farms	18624	30 Jun 2026

Product	Approval holder	MAPP No.	Expiry Date
300 difenoconazole + fludioxonil + tebuconazole			
Celest Trio	Syngenta	15510	28 Feb 2026
306 diflufenican			
Banjo	Ascot Pro-G	19442	30 Jun 2026
Beluga	Ascot Pro-G	17806	30 Jun 2024
Beluga (N I only)	Ascot Pro-G	20081	30 Jun 2025
Deflect	Sharda	18236	30 Jun 2026
Denican	Synergy	19457	30 Jun 2026
Dican	Albaugh UK	16922	30 Jun 2026
Diecot	Cropthetics	19243	30 Jun 2026
Diflanil 500 SC	Certis Belchim B V	17024	30 Jun 2026
Diflufenican GL 500	Globachem	17084	30 Jun 2026
Dina 50	Globachem	17083	30 Jun 2026
Flash	Sharda	17943	30 Jun 2026
Flyflo	Cropthetics	19242	30 Jun 2026
Flyflo	Gemini	19481	30 Jun 2026
Ossetia	Rotam	17741	30 Jun 2026
Overlord	MAUK	13521	30 Jun 2026
Prefect	Novastar	17713	30 Jun 2026
Sempra	UPL Europe	16967	30 Jun 2026
Solo 500 SC	Sipcam	17622	30 Jun 2026
Solo D500	Sipcam	16098	30 Jun 2024
Terrier	Terrechem	17761	30 Jun 2026
Turnpike	Rotam	18631	30 Jun 2026
308 diflufenican + flufenacet			
Adept	Albaugh UK	19648	30 Apr 2026
Amaranth	PSI	16949	30 Apr 2024
Aspect	Clayton	19530	30 Apr 2026
Beamer	Agrimar	20235	30 Apr 2026
Bomber	Terrechem	20003	30 Apr 2026
Centaur	Agrimar	20050	30 Apr 2026
Clayton Aspect	Clayton	18550	30 Apr 2026
Clayton Facet	Clayton	18265	30 Apr 2026
Clayton Sabre	Clayton	18535	30 Apr 2026
Clayton Vista	Clayton	16705	30 Apr 2024
Cleancrop Marauder	Agrii	15933	30 Apr 2024
Coliseum	Bayer CropScience	19722	30 Apr 2026
Cougar	Bayer CropScience	19719	30 Apr 2026
Dephend Delta	FMC Agro	19505	30 Apr 2026
Diflufenastar	Life Scientific	17611	30 Apr 2026
Duke	Chem-Wise	16738	30 Apr 2024
Falcetto Turbo	Cropthetics	19826	30 Apr 2026
Feud	Gemini	18384	30 Apr 2026
Firebird	Bayer CropScience	14826	30 Apr 2026
Fludual	Hockley	18554	30 Apr 2026
Flunican	Becesane	18305	30 Apr 2024
Halifax	Terrechem	19688	30 Apr 2026
Hampden	Terrechem	19720	30 Apr 2026
Lockdown	Chem-Wise	19793	30 Apr 2026
Marauder	Agrii	20065	30 Apr 2026
Naceto	Certis Belchim B V	18063	30 Apr 2026
Navigate	FMC Agro	18950	30 Apr 2026
Parachute	Origin	20234	30 Apr 2026
Saviour	Chem-Wise	16702	30 Apr 2024
Sharp Turbo	Cropthetics	19283	30 Apr 2026
Sharp XD	Cropthetics	18431	30 Apr 2026
Shergill	Cropthetics	19494	30 Apr 2026

SECTION 3

Product	Approval holder	MAPP No.	Expiry Date
Tamomi	Stefes	19625	30 Apr 2026
UPL DFF + FFT	UPL Europe	18503	30 Apr 2026

314 diflufenican + iodosulfuron-methyl-sodium

Lockstar (GB only)	Environmental Science	20246	18 Sep 2025
Lockstar (N I only)	Environmental Science	20243	18 Sep 2025
Valdor Expert	Bayer CropScience	20009	18 Sep 2025
Valdor Expert (GB only)	Bayer CropScience	20313	18 Sep 2025
Valdor Expert (N I only)	Environmental Science	20312	18 Sep 2025
Valdor Flex	Bayer CropScience	19033	30 Jun 2025
Valdor Flex	Environmental Science	20240	18 Sep 2025

315 diflufenican + iodosulfuron-methyl-sodium + mesosulfuron-methyl

Kalenkoa	Bayer CropScience	17733	09 Sep 2099

318 diflufenican + metribuzin

Tavas	Adama	18213	31 Jan 2026

321 dimethachlor

Clayton Dimethachlor	Clayton	16188	30 Jun 2024
Clayton Dimethachlor (N I only)	Clayton	19880	30 Jun 2025

323 dimethenamid-p + metazachlor

Caribou	RealChemie	18195	30 Apr 2024
Caribou	RealChemie	18263	30 Jun 2024
Muntjac	BASF	16893	09 Sep 2099
Sika	HMpG GMBH	18428	30 Jun 2024

324 dimethenamid-p + metazachlor + quinmerac

Elk	BASF	16920	09 Sep 2099

325 dimethenamid-p + pendimethalin

Clayton Launch (N I only)	Clayton	19894	09 Sep 2099
Pendi P	HMpG GMBH	19040	30 Jun 2024
Pendi P	HMpG GMBH	19092	30 Jun 2024

326 dimethenamid-p + quinmerac

Butisan Pro	BASF	18726	09 Sep 2099

328 dimethomorph

Morph	Adama	15121	31 Jan 2026
Navio	FMC Agro	17827	31 Jan 2026
Rigel WP	Servem	17347	31 Jan 2024

330 dimethomorph + mancozeb

Filder 69 WG	Arysta	19353	30 Jul 2026
Invader	BASF	15223	30 Jul 2026
Saracen	BASF	15250	30 Jul 2026

332 dimethomorph + pyraclostrobin

Optimo Tech	BASF	16455	31 Jan 2025

333 dimethomorph + zoxamide

Presidium	Gowan	18119	09 Sep 2099

338 dithianon

Dithianon WG	BASF	17018	28 Feb 2027
Mulan 700 WG	Cropthetics	19528	28 Feb 2027
Mullomo 700 WG	Cropthetics	19801	28 Feb 2027
Rathianon 70 WG	RAAT	17471	30 Jun 2024

Product	Approval holder	MAPP No.	Expiry Date
339 dithianon + potassium phosphonates			
Astro Gold (N I Only)	Servem	19806	28 Feb 2027
Clayton Prolan (N I only)	Clayton	19904	28 Feb 2027
340 dithianon + pyraclostrobin			
Maccani	BASF	18546	31 Jul 2025
346 dodine			
Radspor 400	Agriphar	16001	31 Dec 2024
362 esfenvalerate			
Barclay Alphasect	Barclay	16053	30 Jun 2024
Barclay Alphasect (N I only)	Barclay	20030	30 Jun 2025
Clayton Cajole	Clayton	14995	30 Jun 2024
Clayton Cajole (N I only)	Clayton	19869	09 Sep 2099
Clayton Slalom	Clayton	15054	30 Jun 2024
Clayton Slalom (N I only)	Clayton	20345	09 Sep 2099
Clayton Vindicate	Clayton	15055	30 Jun 2024
Clayton Vindicate (N I only)	Clayton	19918	09 Sep 2099
Sumi-Alpha	Sumitomo	14023	09 Sep 2099
Sven	Sumitomo	14859	09 Sep 2099
364 ethanol			
EMC2	Ethylene Co	19513	28 Feb 2025
Ethy-Gen II	Ripe Rite	15839	31 Oct 2024
Restrain Fuel	Restrain	14520	28 Feb 2025
365 ethephon			
Cerone	Nufarm UK	15944	31 Jan 2026
Coryx	Adama	19401	31 Jan 2025
Coryx	SFP Europe	17620	31 Jan 2023
Coupon	Globachem	19123	31 Jan 2026
Coupon 480	Globachem	18865	31 Jan 2026
Ethe 480	Agroquimicos	16887	31 Jan 2024
Floralife Tulipa	Oasis	17996	31 Jan 2026
Fonic	Sharda	20233	31 Jan 2026
Hi-Phone 48	Hockley	15506	31 Jan 2024
Ipanema	SFP Europe	15961	31 Jan 2023
Padawan	SFP Europe	15577	31 Jan 2023
Tephon	Sharda	19455	31 Jan 2024
366 ethephon + mepiquat chloride			
Clayton Proud (N I only)	Clayton	20344	31 Aug 2025
Gunbar	Terrechem	17460	30 Jun 2024
Mepicame	Unisem	17541	30 Jun 2024
Riggid	Gemini	17522	30 Jun 2024
Terpal	PSI	17440	30 Jun 2024
367 ethofumesate			
Nortron Flo	Bayer CropScience	18526	30 Apr 2034
368 ethofumesate + metamitron			
Goltix Plus	Aako	18665	28 Feb 2026
Metafol Super	UPL Europe	18925	28 Feb 2025
370 ethofumesate + phenmedipham			
Brigida	Agroquimicos	15658	30 Jun 2024
373 etofenprox			
Trebon 30 EC	Certis Belchim B V	16666	30 Jun 2026

SECTION 3

Product	Approval holder	MAPP No.	Expiry Date
374 etoxazole			
Borneo (GB only)	Sumitomo	18873	31 Jan 2027
Clayton Java (GB only)	Clayton	15155	30 Jun 2024
377 fatty acids			
Flipper	Fargro	17767	28 Feb 2023
NEU 1170 H	Sinclair	15754	28 Feb 2026
379 fatty acids + pelargonic acid + maleic hydrazide			
Weed Killer Xtra	Hygeia	19682	09 Sep 2099
385 fenazaquin			
Matador 200 SC	Gowan	19626	30 Nov 2025
389 fenhexamid			
Agrovista Fenamid	Agrovista	13733	30 Jun 2024
Agrovista Fenhexamid (N I Only)	Agrovista	20080	09 Sep 2099
RouteOne Fenhex 50	Albaugh UK	15282	30 Jun 2024
391 fenoxaprop-P-ethyl			
Polecat	Sumitomo	16112	30 Jun 2026
Polecat	Sumitomo	18896	30 Jun 2026
394 fenpicoxamid + prothioconazole			
Univoq	Corteva	19680	31 Jan 2024
397 fenpropidin + prochloraz + tebuconazole			
Artemis	Adama	18533	30 Jun 2023
Artemis (GB only)	Adama	18533	30 Jun 2029
407 fenpyrazamine			
Clayton Vitis	Clayton	16974	30 Jun 2024
Prolectus	Sumitomo	18891	30 Jun 2026
Prolectus	Sumitomo	16607	31 Jan 2024
409 ferric phosphate			
Derrex	Certis Belchim B V	15351	30 Jun 2033
Fennec	Doff Portland	19984	16 Mar 2026
Fenomenal	Doff Portland	19990	16 Mar 2026
Firescale	Doff Portland	19985	16 Mar 2026
Ironclad	Doff Portland	19388	30 Jun 2033
Ironclad	Gemini	18970	15 Jan 2023
Ironmax Pro	De Sangosse	17122	31 Oct 2023
Limafer	Frunol Delicia	19145	30 Jun 2033
Magnetite	Doff Portland	19991	16 Mar 2026
NEU 1181 M	Neudorff	14355	30 Jun 2033
Regiment	Certis Belchim B V	19065	30 Jun 2033
TurboDisque	Frunol Delicia	19144	30 Jun 2033
TurboPads	Frunol Delicia	19146	30 Jun 2033
410 ferrous sulphate			
Landscaper Pro Moss Control + Fertiliser	ICL (Everris) Ltd	16723	28 Feb 2026
LSTF Pro	LSTF	16992	28 Feb 2026
Maxicrop Professional Moss Killer & Conditioner	Valagro	18514	28 Feb 2026
Moss-Kill Pro	Bioservices	17029	28 Feb 2026
Sinclair Lawn Sand	Westland Horticulture	17562	28 Feb 2026
413 flazasulfuron			
Chikara	Nomix Enviro	13775	09 Sep 2099

Product	Approval holder	MAPP No.	Expiry Date
Chikara Weed Control	Certis Belchim B V	20185	09 Sep 2099
Clayton Apt (N I only)	Clayton	19861	09 Sep 2099
Flaza 250	HMpG GMBH	18470	30 Jun 2024
Flazaraat	RAAT	17912	30 Jun 2024
Hinoki	Certis Belchim B V	19134	09 Sep 2099
Katana	Certis Belchim B V	20199	31 Jan 2035
Paradise (N I only)	Pan Agriculture	20282	09 Sep 2099
Paradise (N I only)	Pan Agriculture	20283	09 Sep 2099
Paradise (N I only)	Pan Agriculture	20284	09 Sep 2099
Railtrax	PSI	15139	30 Jun 2024
Utopia	Pure Amenity	19356	30 Jun 2024
Valdor Solo	Bayer CropScience	19567	09 Sep 2099

414 flocoumafen

Storm Mini Bits	BASF	UK15-0915	08 Mar 2023
Storm Secure	BASF	UK15-0850	08 Mar 2023
Storm Ultra	BASF	UK18-1164	07 Dec 2023
Storm Ultra Bait Blocks	BASF	UK18-1164	07 Dec 2023
Storm Ultra Secure	BASF	UK18-1164	07 Dec 2023

415 flonicamid

Afinto	Syngenta	19622	30 Apr 2024
Hinode	ISK Biosciences	19135	30 Apr 2024
Mainman	Certis Belchim B V	20214	28 Feb 2026
Pekitek	HMpG GMBH	19029	30 Apr 2023
Pekitek	HMpG GMBH	19070	30 Apr 2023
Primeman	RAAT	17656	30 Jun 2024
Teppeki	Certis Belchim B V	20213	30 Apr 2024

416 florasulam

Globus	Globachem	18834	30 Jun 2033
Rassel 100 SC	Innvigo	20207	30 Jun 2033
Upton	Rotam	19361	20 Jun 2025

417 florasulam + fluroxypyr

Cabadex	Headland Amenity	13948	28 Feb 2023
Celadon	ICL (Everris) Ltd	19364	30 Apr 2023
Envymax	Corteva	19645	30 Jun 2027
Flatline	Aremie	17422	30 Jun 2024
Gal-Gone XL	Agrovista	18918	30 Jun 2027
GF 184	Corteva	19744	30 Jun 2027
Hunter	Corteva	12836	31 Mar 2023
Nevada	Corteva	17349	09 Sep 2099
Orpen	Barclay	19677	01 Feb 2025
Scimitar	Corteva	19646	30 Jun 2027
Scimitar LT	Corteva	19675	30 Jun 2027
Slalom	Corteva	13772	31 Mar 2023
Spitfire	Corteva	15101	31 Jul 2023
Starane XL	Corteva	10921	31 Mar 2023
Valentia	Barclay	19595	01 Feb 2025

418 florasulam + halauxifen-methyl

Mattera	Corteva	19133	05 Feb 2028
Renitar	Corteva	19129	05 Feb 2028

419 florasulam + pinoxaden

Axial One	Syngenta	18837	31 Dec 2028
Duozym	Syngenta	19941	31 Dec 2028

SECTION 3

Product	Approval holder	MAPP No.	Expiry Date
420 florasulam + pyroxsulam			
Gyga	Corteva	19143	31 Oct 2027
421 florasulam + tribenuron-methyl			
Clayton Brazen (N I only)	Clayton	19693	09 Sep 2099
424 fluazinam			
Boyano	Certis Belchim B V	20162	31 Aug 2025
Boyano	Certis Belchim B V	16621	31 Aug 2025
Dalimo	ISK Biosciences	19132	31 Aug 2025
Deltic Pro	FMC Agro	18733	31 Aug 2025
Float	FMC Agro	18810	31 Aug 2025
Fluazifin	Clayton	18549	31 Aug 2025
Fluazinam 500 SC	Finchimica	19261	31 Aug 2025
Flunam 500 III	HMpG GMBH	18480	30 Jun 2024
FOLY 500 SC	Aako	16486	31 Aug 2025
Frowncide	Certis Belchim B V	18392	31 Aug 2025
Frowncide	Certis Belchim B V	20188	31 Aug 2025
Ibiza 500	Certis Belchim B V	20189	31 Aug 2025
Ibiza 500	Certis Belchim B V	16620	31 Aug 2025
Kelsos	Cropthetics	19800	31 Aug 2025
Legacy	Certis Belchim B V	20202	31 Aug 2025
Legacy	Certis Belchim B V	18389	31 Aug 2025
Magic	Cropthetics	18861	31 Aug 2025
Ohayo	Certis Belchim B V	18391	31 Aug 2025
Ohayo	Certis Belchim B V	20197	31 Aug 2025
Shirlan	Certis Belchim B V	20198	31 Aug 2025
Shirlan Programme	Certis Belchim B V	20194	31 Aug 2025
Shirlan Programme	Certis Belchim B V	18407	31 Aug 2025
Smash	RAAT	17488	30 Jun 2024
Winby	Certis Belchim B V	18390	31 Aug 2025
Winby	Certis Belchim B V	20221	31 Aug 2025
426 fludioxonil			
Medaille SC	Glo-Quimicos	19623	30 Apr 2024
Protect	Crop Health	19285	30 Apr 2024
433 fludioxonil + sedaxane			
A20078F	Syngenta	17945	30 Apr 2026
434 fludioxonil + tebuconazole			
Protect Plus	Crop Health	19286	29 Feb 2024
436 flufenacet			
Foxton	Terrechem	18595	30 Apr 2024
Gorgon	Sipcam	17685	30 Apr 2026
Preto	Nurture	18594	30 Apr 2024
Staket	Top Crop	18662	30 Apr 2024
440 flufenacet + pendimethalin			
Clayton Glacier	Clayton	15920	30 Jun 2024
Clayton Glacier (N I only)	Clayton	19888	09 Sep 2099
Cleancrop Hector	Agrii	15932	30 Jun 2024
Crystal S	BASF	19586	30 Jun 2024
Ice	BASF	13930	30 Apr 2026
Kruos	Chem-Wise	15880	30 Jun 2024
Kruos (N I only)	Chem-Wise	20060	09 Sep 2099
Standon Diwana	Standon	17221	09 Sep 2099
Standon Diwana	Enviroscience	18485	30 Jun 2024

Product	Approval holder	MAPP No.	Expiry Date
441 flufenacet + picolinafen			
Equs	BASF	17736	09 Sep 2099
Kudu	BASF	17804	09 Sep 2099
Lantern	BASF	19085	09 Sep 2099
442 flumioxazin			
Digital	Sumitomo	18885	31 Dec 2024
Guillotine	Sumitomo	18876	31 Dec 2024
Sumimax	Sumitomo	18884	31 Dec 2024
446 fluopyram + prothioconazole			
Clayton Repel	Clayton	18037	31 Jan 2024
Prolense	HMpG GMBH	18560	31 Jan 2024
448 fluopyram + trifloxystrobin			
Clayton Stellar	Clayton	18417	30 Apr 2024
Clayton Stellar (N I only)	Clayton	19914	09 Sep 2099
Exteris Stressgard (GB only)	Environmental Science	20242	09 Sep 2099
Exteris Stressgard (N I only)	Environmental Science	20241	09 Sep 2099
Robin (N I only)	Pan Agriculture	20285	09 Sep 2099
Robin (N I only)	Pan Agriculture	20286	09 Sep 2099
449 fluoxastrobin			
Bayer UK 831	Bayer CropScience	12091	31 Jan 2026
450 fluoxastrobin + prothioconazole			
Fandango	Bayer CropScience	19966	31 Jan 2026
Firefly	Bayer CropScience	13692	31 Jan 2026
Maestro	Bayer CropScience	19988	31 Jan 2026
Prostrob 20	Chem-Wise	15698	31 Jan 2024
461 fluroxypyr			
Awac	Globachem	17501	30 Jun 2027
Bandon	Barclay	19955	30 Jun 2027
Casino	Certis Belchim B V	20181	30 Jun 2027
Casino	Certis Belchim B V	17592	31 Aug 2026
Cererish	YC Agro	20036	30 Jun 2027
Clean Crop Gallifrey 200	Globachem	17503	30 Jun 2027
Crescent	Certis Belchim B V	20182	30 Jun 2027
Crescent	Certis Belchim B V	17589	31 Aug 2026
Decathlon	Agrii	19746	30 Jun 2027
Filomena	Synergy	19992	30 Jun 2027
GF-1784	Corteva	16850	30 Jun 2027
Hatchet Xtra	Certis Belchim B V	17593	31 Aug 2026
Hatchet Xtra	Certis Belchim B V	20183	30 Jun 2027
Havoc	Gemini	19057	30 Jun 2027
Hyflux GB	Hygeia	19015	30 Jun 2027
Luxor	Synergy	19449	30 Jun 2027
Norgett	Rotam	19475	30 Jun 2027
Paksava	Leeds	19521	30 Jun 2027
Raptor	Ascot Pro-G	19400	30 Jun 2027
Stefes Fluroxy	Stefes	19687	30 Jun 2027
Taipan	Agro Trade	17063	30 Jun 2027
Thalia	YC Agro	19922	30 Jun 2027
Tomahawk 2	Adama	17468	30 Jun 2027
Toska EC	Novastar	17709	30 Jun 2027
Tumu	Terrechem	19032	30 Jun 2027
Weedclear 200 EC	Rainbow	19857	30 Jun 2027

SECTION 3

Product	Approval holder	MAPP No.	Expiry Date
465 fluroxypyr + metsulfuron-methyl			
Croupier OD	Certis Belchim B V	20178	30 Sep 2025
467 fluroxypyr + thifensulfuron-methyl			
Sentrallas	FMC Agro	19385	28 Sep 2025
469 fluroxypyr + triclopyr			
Crossfire	Corteva	19651	31 Oct 2024
Pas	Corteva	17772	31 Oct 2024
472 flutolanil			
NNF-136	Nichino	14302	31 Aug 2025
Rhino DS	Certis Belchim B V	12763	31 Aug 2025
474 fluxapyroxad			
Clayton Paradox	Clayton	19569	30 Jun 2024
Clayton Paradox (N I only)	Clayton	19900	30 Nov 2027
Flux	Chem-Wise	17602	30 Jun 2024
Pioli	Adama	20007	30 Nov 2027
476 fluxapyroxad + metconazole			
Clayton Tardis (N I only)	Clayton	19915	31 Oct 2024
Librax	BASF	17107	31 Oct 2025
478 folpet			
Clayton Canyon IN I only)	Clayton	19870	31 Jan 2025
Lamast	Syngenta	19140	31 Jan 2026
Mirror	Syngenta	18866	31 Jan 2026
Sesto	Certis Belchim B V	19148	31 Jan 2023
Sesto	Adama	19219	31 Jan 2026
480 foramsulfuron + iodosulfuron-methyl-sodium			
Maister WG	Bayer CropScience	16116	31 Dec 2023
481 foramsulfuron + thiencarbazone-methyl			
Conviso One	Bayer CropScience	20035	31 Mar 2027
482 fosetyl-aluminium			
ESP742	Bayer CropScience	19194	31 May 2024
ESP742 (GB only)	Environmental Science	20249	31 May 2024
ESP742 (N I only)	Environmental Science	20248	31 May 2024
Plant Trust	ICL (Everris) Ltd	15779	31 Oct 2025
483 fosetyl-aluminium + propamocarb hydrochloride			
Avatar	ICL (Everris) Ltd	16608	31 Oct 2025
Pan Cradle	Pan Agriculture	15923	30 Jun 2024
Pan Cradle (N I only)	Pan Agriculture	20278	31 Oct 2025
488 garlic extract			
Grenadier	Certis Belchim B V	17617	29 Feb 2024
Nemguard DE (GB Only)	Certis Belchim B V	19851	28 Feb 2027
NEMguard granules	ECOspray	15254	29 Feb 2024
Nemguard SC (GB only)	ECOspray	19755	30 Apr 2025
Pitcher GR	ECOspray	18126	28 Feb 2024
Pitcher GR (GB only)	ICL (Everris) Ltd	19729	28 Feb 2027
Pitcher SC	ECOspray	18125	29 Feb 2024
490 gibberellins			
Smartgrass	Sumitomo	15628	28 Feb 2023

Product	Approval holder	MAPP No.	Expiry Date
491 Gliocladium catenulatum			
Prestop	Danstar	19458	09 Sep 2099
Prestop Mix	Fargro	15104	09 Sep 2099
493 glyphosate			
Alekto Plus TF	Certis Belchim B V	17143	09 Sep 2099
Alekto Plus TF	Helm	19942	09 Sep 2099
Asteroid	FMC Agro	18958	09 Sep 2099
Asteroid Pro	FMC Agro	18754	09 Sep 2099
Asteroid Pro 450	FMC Agro	18788	09 Sep 2099
Cleancrop Bronco	Certis Belchim B V	17056	09 Sep 2099
Cleancrop Bronco	Helm	19932	09 Sep 2099
Cleancrop Tungsten	Agrii	17553	09 Sep 2099
Clinic Grade	Nufarm UK	19028	09 Sep 2099
Clinic Hi-Load	Agrigem	19910	09 Sep 2099
Clinic TF	Nufarm UK	16716	09 Sep 2099
Conan 360	Top Crop	19531	30 Jun 2024
Egret	Monsanto	17510	09 Sep 2099
Envision	FMC Agro	18770	09 Sep 2099
Glister Ultra	Sinon EU	15209	09 Sep 2099
Glister Ultramax	Sinon EU	18021	09 Sep 2099
Glyder 450 TF	Certis Belchim B V	16520	09 Sep 2099
Glyder 450 TF	Helm	19933	09 Sep 2099
Glyfer	Nouvelle Tec	19200	09 Sep 2099
Glyfos Dakar	FMC Agro	13054	09 Sep 2099
Glyfos Dakar Pro	FMC Agro	18812	09 Sep 2099
Glyfos Gold	FMC Agro	18790	09 Sep 2099
Glyfos Gold ECO	FMC Agro	18730	09 Sep 2099
Glyfos Monte	FMC Agro	18807	09 Sep 2099
Glyfos Supreme XL	FMC Agro	18695	09 Sep 2099
Glyphotec	Aako	19882	09 Sep 2099
Habitat	Barclay	17662	30 Apr 2024
Helosate 450 TF	Helm	19934	09 Sep 2099
Helosate 450 TF	Helm	16326	09 Sep 2099
Master Gly	Ventura	19202	09 Sep 2099
MON 76473	Monsanto	15348	09 Sep 2099
MON 79351	Monsanto	12860	09 Sep 2099
MON 79376	Monsanto	12664	09 Sep 2099
MON 79545	Monsanto	12663	09 Sep 2099
MON 79991	Monsanto	16300	09 Sep 2099
Monosate G	Monsanto	18651	09 Sep 2099
NASA	Agria SA	18170	09 Sep 2099
Nomix G	Nomix Enviro	18359	09 Sep 2099
Nomix Prolite	Nomix Enviro	18346	09 Sep 2099
Ormond	Barclay	17712	09 Sep 2099
Pitch	Nufarm UK	15485	09 Sep 2099
Proliance Quattro	Syngenta	13670	09 Sep 2099
Rootblast Pro	Barclay	20236	09 Sep 2099
Rosate Green	Albaugh UK	15122	09 Sep 2099
Roundup Klik	Monsanto	12866	09 Sep 2099
Roundup Pro Biactive	Monsanto	10330	09 Sep 2099
Surrender T/F	Agrovista	19203	09 Sep 2099
Tangent	FMC Agro	18826	09 Sep 2099
Typhoon 360	Adama	14817	09 Sep 2099
Vesuvius Green	Ventura	19201	09 Sep 2099
495 glyphosate + sulfosulfuron			
Nomix Blade	Nomix Enviro	18358	09 Sep 2099
Nomix Duplex	Nomix Enviro	18350	09 Sep 2099

Product	Approval holder	MAPP No.	Expiry Date
499 halauxifen-methyl			
GF-2573	Corteva	17540	05 Feb 2028
500 halauxifen-methyl + picloram			
Kelbar	HMpG GMBH	19543	30 Jun 2024
Kelbar II	HMpG GMBH	19576	30 Jun 2024
504 imazalil + ipconazole			
Conima	Arysta	18966	31 Dec 2025
Conima	UPL Europe	19843	31 May 2027
Rancona i-MIX	Arysta	15574	31 Dec 2025
509 imazamox + metazachlor			
Cleranda	BASF	15036	09 Sep 2099
510 imazamox + metazachlor + quinmerac			
Clesima	BASF	16275	31 Aug 2023
512 imazamox + quinmerac			
Clentiga	BASF	18121	31 Jan 2027
520 indoxacarb			
Clayton Purser	Clayton	18283	30 Apr 2023
522 iodosulfuron-methyl-sodium + mesosulfuron-methyl			
Atlantis Pro	PSI	18402	30 Jun 2024
Clayton Metropolis (N I only)	Clayton	19895	09 Sep 2099
Lanista	Life Scientific	18118	09 Sep 2099
Mesiodos OD	HMpG GMBH	19071	30 Jun 2024
Mesiothyl OD	HMpG GMBH	19638	30 Jun 2024
526 ipconazole			
Rancona 15 ME	Arysta	16136	28 Feb 2027
530 isofetamid			
Kenja	Certis Belchim B V	19329	30 Sep 2026
Kenja	Certis Belchim B V	20200	31 Mar 2029
Kryor	Certis Belchim B V	20201	15 Mar 2029
Kryor	Certis Belchim B V	19335	30 Sep 2026
Zenby	Certis Belchim B V	19336	31 Aug 2026
Zenby	Certis Belchim B V	20171	15 Mar 2029
533 isopyrazam			
A15149W	Adama	18571	30 Sep 2028
Zulu	Adama	18575	30 Sep 2028
534 isopyrazam + prothioconazole			
Polarize (GB only)	Adama	18980	31 Jan 2027
Prizm (GB only)	Adama	18940	31 Jan 2027
535 isoxaben			
Pan Isoxaben 500 (N I only)	Pan Amenity	20299	28 Feb 2027
538 kresoxim-methyl			
Strong WG	RAAT	17470	30 Jun 2024
539 lambda-cyhalothrin			
Clayton Lambada	Clayton	13231	30 Jun 2024
Clayton Lambada (N I only)	Clayton	19892	09 Sep 2099
Clayton Lanark	Clayton	12942	30 Jun 2024
Clayton Lanark (N I only)	Clayton	19893	09 Sep 2099

Product	Approval holder	MAPP No.	Expiry Date
CM Lambaz 50 EC	CMI	16241	30 Jun 2024
Cyclone	Stefes	19747	30 Sep 2025
Eminentos 10 CS	FMC Agro	18712	09 Sep 2099
Hockley Lambda 5EC	Hockley	15516	09 Sep 2099
Laidir 10 CS	FMC Agro	18705	09 Sep 2099
Lambda-C 100	Becesane	15987	30 Jun 2024
Lambda-C 50	Becesane	15986	30 Jun 2024
MAC-Lambda-Cyhalothrin 50 EC	MAC	14941	30 Jun 2024
Markate 50	Certis Belchim B V	13529	09 Sep 2099
Reparto	Sipcam	15452	30 Jun 2024
RouteOne Lambda C	Albaugh UK	13663	30 Jun 2024
Seal Z	AgChem Access	20027	30 Sep 2025
Triumph CS	Capital CP	18235	09 Sep 2099
Triumph CS (N I only)	Capital CP	20072	30 Jun 2024

543 lenacil

Product	Approval holder	MAPP No.	Expiry Date
Venzar 80 WP	FMC Agro	18757	30 Jun 2026

544 lenacil + triflusulfuron-methyl

Product	Approval holder	MAPP No.	Expiry Date
Safari Duo Active	FMC Agro	19960	30 Jun 2026
Upbeet Lite	FMC Agro	18795	30 Jun 2026

546 magnesium phosphide

Product	Approval holder	MAPP No.	Expiry Date
Degesch Plate	Rentokil	19332	28 Feb 2026
Degesch Plates	Rentokil	17105	31 Dec 2023
Magtoxin Pellet	Rentokil	19334	28 Feb 2026
Magtoxin Pellets	Rentokil	17376	31 Dec 2023
Magtoxin Tablet	Rentokil	19344	28 Feb 2026

547 maleic hydrazide

Product	Approval holder	MAPP No.	Expiry Date
Fazor	Arysta	19354	30 Apr 2035
Himalaya XL	Arysta	19426	30 Apr 2035
Himalaya XL	UPL Europe	19589	30 Apr 2035
Itcan 270 SL	Kreglinger	19807	30 Apr 2035
Itcan 270 SL	Gemini	19459	30 Nov 2025

548 maleic hydrazide + pelargonic acid

Product	Approval holder	MAPP No.	Expiry Date
Finalsan Plus	Certis Belchim B V	17928	09 Sep 2099

549 maltodextrin

Product	Approval holder	MAPP No.	Expiry Date
Eradicoat	Certis Belchim B V	17121	11 May 2024
Terminus	Certis Belchim B V	17319	11 May 2024

550 mancozeb

Product	Approval holder	MAPP No.	Expiry Date
Agria Mancozeb 80WP	Agria SA	16636	30 Jul 2026
Avtar 75 NT	Syngenta	18909	30 Jul 2026
Clayton Zebo	Clayton	18460	31 Jul 2023
Cleancrop Mandrake	Agrii	14792	30 Jul 2026
Cleancrop Mandrake Plus WG	Agrii	18508	30 Jul 2026
Dithane 945	Sumitomo	18900	30 Jul 2026
Dithane NT Dry Flowable	Sumitomo	18889	30 Jul 2026
Fortuna	Hockley	18620	30 Jul 2026
Fortuna Globe	Hockley	18612	30 Jul 2026
Karamate Dry Flo Neotec	Landseer	14632	30 Jul 2026
Laminator 75 WG	Sumitomo	15667	31 Jul 2026
Laminator Plus WG	Sumitomo	18907	30 Jul 2026
Malvi	AgChem Access	14981	31 Jul 2023
Manfil 75 WG	Indofil	15958	31 May 2026
Manfil 75 WG	Indofil	20002	30 Jul 2026
Manfil 80 WP	Indofil	15956	30 Jul 2026

Product	Approval holder	MAPP No.	Expiry Date
Manfil 80 WP	Indofil	20005	31 Jul 2026
Mewati	AgChem Access	15763	31 Jul 2023
Tridex	UPL Europe	18196	30 Jul 2026
Trimanoc 75 DG	UPL Europe	18822	30 Jul 2026
Trimanzone	UPL Europe	16759	30 Jul 2026
Zebra WDG	FMC Agro	14875	30 Jul 2026

551 mancozeb + metalaxyl-M

Product	Approval holder	MAPP No.	Expiry Date
Clayton Mohawk	Clayton	18173	30 Jun 2024

553 mancozeb + valifenalate

Product	Approval holder	MAPP No.	Expiry Date
Valis M	Certis Belchim B V	19368	30 Jul 2026
Valis M	Certis Belchim B V	20237	31 Jul 2026

554 mancozeb + zoxamide

Product	Approval holder	MAPP No.	Expiry Date
Electis 75WG	Gowan	17871	09 Sep 2099
Roxam 75WG	Gowan	17869	09 Sep 2099
Unikat 75WG	Gowan	17868	09 Sep 2099

555 mandipropamid

Product	Approval holder	MAPP No.	Expiry Date
Evagio	Syngenta	18279	31 Jan 2026
Mandoprid	Generica	18182	30 Jun 2024
Manpro	HMpG GMBH	18468	30 Jun 2024
Pergardo Uni	Fargro	17513	31 Jan 2026
Standon Mandor	Standon	17536	31 Jan 2026
Standon Mandor	Enviroscience	18463	30 Jun 2024
Standon Mandor	Enviroscience	18496	30 Jun 2024

557 MCPA

Product	Approval holder	MAPP No.	Expiry Date
Agrichem MCPA 500	UPL Europe	16964	30 Apr 2026
Agritox 50	Nufarm UK	14814	30 Apr 2026
Agroxone	FMC Agro	14909	30 Sep 2023
Clayton Haksar	Clayton	18380	30 Apr 2026
Dow MCPA Amine 50	Corteva	14911	30 Apr 2026
Go-Low Power	DuPont (Corteva)	13812	30 Apr 2026
Haksar 500 SL	Ciech	18446	30 Apr 2026
Headland Spear	FMC Agro	14910	30 Sep 2023
Lagoon	Gemini	17856	30 Apr 2026
MCPA 25%	Nufarm UK	14893	30 Apr 2026
MCPA 50	UPL Europe	14908	30 Sep 2023
MCPA 500	Nufarm UK	18415	30 Apr 2026
Nufarm MCPA 750	Nufarm UK	14892	30 Apr 2026
POL-MCPA 500 SL	Clayton	17808	30 Apr 2026
Tasker 75	FMC Agro	14913	30 Sep 2023

559 MCPA + mecoprop-P

Product	Approval holder	MAPP No.	Expiry Date
Greenmaster Extra	ICL (Everris) Ltd	15817	31 Jul 2025

560 MCPB

Product	Approval holder	MAPP No.	Expiry Date
Bellmac Straight	UPL Europe	18636	30 Sep 2023
Butoxone	FMC Agro	18640	30 Sep 2023

561 mecoprop-P

Product	Approval holder	MAPP No.	Expiry Date
Charge	FMC Agro	18802	31 Jul 2025
Clenecorn Super	Nufarm UK	18378	31 Jul 2025
Isomec	Nufarm UK	18376	31 Jul 2025
Optica	FMC Agro	19609	31 Jul 2025

568 mepiquat chloride + metconazole

Product	Approval holder	MAPP No.	Expiry Date
Regulate OSR	HMpG GMBH	18495	31 Aug 2023

Product	Approval holder	MAPP No.	Expiry Date
569 mepiquat chloride + prohexadione-calcium			
Basf01p	BASF	19949	31 Aug 2025
Medax Top	BASF	16574	31 Aug 2025
573 mesosulfuron-methyl + propoxycarbazone-sodium			
Monolith	Bayer CropScience	17687	30 Sep 2023
574 mesotrione			
A12739A	Syngenta	16158	29 Feb 2024
A12739A	Syngenta	20164	30 Nov 2034
Basilico	Life Scientific	19927	08 Jun 2026
Callisto	Syngenta	12323	31 Jan 2024
Callisto	PSI	17512	30 Jun 2024
Clayton Goldcob	Clayton	16299	30 Jun 2024
Clayton Goldcob (N I only)	Clayton	19889	09 Sep 2099
Clayton Goldcob (N I only)	Clayton	19890	09 Sep 2099
Cuter	Clayton	17668	09 Sep 2099
Hockley Mesotrione 10	Hockley	15519	30 Jun 2024
Kalypstowe	AgChem Access	13844	30 Jun 2024
Kalypstowe (N I only)	AgChem Access	20019	30 Nov 2034
Kideka	Nufarm UK	18033	09 Sep 2099
Mabua	Helm	20066	09 Sep 2099
Meristo	Syngenta	16865	29 Feb 2024
Meruba	Helm	17961	28 Feb 2026
Meruba	Helm	19946	09 Sep 2099
Minerva	Terrechem	17514	30 Jun 2024
Osorno	Certis Belchim B V	18250	09 Sep 2099
RouteOne Trione 10	Albaugh UK	14765	30 Jun 2024
Standon Cobmajor	Standon	16721	09 Sep 2099
Standon Cobmajor	Enviroscience	18489	30 Jun 2024
575 mesotrione + nicosulfuron			
Clayton Aspen	Clayton	16114	30 Jun 2024
Clayton Aspen (N I only)	Clayton	19862	09 Sep 2099
576 mesotrione + pyridate			
Botiga	Certis Belchim B V	20304	30 Jun 2033
Onyx extra	Certis Belchim B V	20319	30 Jun 2033
Onyx Extra	Certis Belchim B V	19774	31 Oct 2026
Sansa	Certis Belchim B V	19773	31 Oct 2026
Sansa	Certis Belchim B V	20318	30 Jun 2033
580 metalaxyl-M			
Clayton Tine (N I only)	Clayton	19916	09 Sep 2099
Fongarid Gold	Syngenta	12547	31 Dec 2022
Hobsen (N I only)	AgChem Access	20018	30 Nov 2037
Hobson	AgChem Access	16529	30 Jun 2024
Lenomax	ProKlass	18395	30 Jun 2024
Shams 465	DAPT	16107	30 Jun 2024
Shams 465 (N I only)	DAPT	20095	30 Jun 2024
582 metamitron			
Beetron (N I only)	AgChem Access	20012	28 Feb 2026
Belta	Gemini	19598	01 Dec 2024
Betra SC	Novastar	17244	28 Feb 2026
Bettix Flo SC	UPL Europe	19451	28 Feb 2026
Celmitron 70% WDG	UPL Europe	19307	28 Feb 2026
Clayton Devoid	Clayton	18252	28 Feb 2026
Defiant	UPL Europe	19454	28 Feb 2026
Defiant WG	UPL Europe	16544	28 Feb 2026

SECTION 3

Product	Approval holder	MAPP No.	Expiry Date
Devoid	Cropthetics	18512	28 Feb 2026
Exodus	Ascot Pro-G	19660	28 Feb 2026
Glotron 700 SC	Certis Belchim B V	17308	28 Feb 2026
Goltix 90	Adama	16497	28 Feb 2026
Goltix Compact	Aako	16545	28 Feb 2026
Goltix Gold	Adama	19993	28 Feb 2026
Inter Metron 700 SC	UPL Europe	17389	28 Feb 2026
Metrix 700 SC	Novastar	19602	28 Feb 2025
Metsar	Hemani	20041	01 Dec 2024
Mitron BX SC	Certiplant NV	19279	28 Feb 2025
MTM 700	Terrechem	19084	28 Feb 2025
Synergy Generics Metamitron	Synergy	18277	28 Feb 2025
Target SC	UPL Europe	19323	28 Feb 2025
Tronix 700 SC	Certis Belchim B V	18137	28 Feb 2025
Vallim	Rotam	20042	01 Dec 2024

583 metamitron + quinmerac

Product	Approval holder	MAPP No.	Expiry Date
Kezuro	BASF	17988	28 Feb 2025

585 Metarhizium anisopliae

Product	Approval holder	MAPP No.	Expiry Date
Laiguard M52 GR	Danstar	19841	31 Oct 2024
Laiguard M52 OD	Danstar	19824	31 Oct 2024
Lycomax	Russell	18860	31 Oct 2023
Met52 granular bioinsecticide	Fargro	15168	31 Oct 2024
Met52 OD	Fargro	17367	31 Oct 2024

586 metazachlor

Product	Approval holder	MAPP No.	Expiry Date
Clayton Buzz	Clayton	16928	31 Jan 2024
Clayton Buzz (N I only)	Clayton	19868	31 Jan 2025
Fuego	Adama	16679	31 Jan 2026
Makila 500 SC	Novastar	18607	31 Jan 2026
Rapsan 500 SC	Certis Belchim B V	16592	31 Jan 2026
Standon New Metazachlor	Enviroscience	18498	31 Jan 2024
Taza 500	Becesane	17284	31 Jan 2024

587 metazachlor + napropamide + quinmerac

Product	Approval holder	MAPP No.	Expiry Date
Torso	Certis Belchim B V	19523	31 Jan 2026

588 metazachlor + quinmerac

Product	Approval holder	MAPP No.	Expiry Date
Metamerac II	RealChemie	18146	31 Jan 2024
Metamerac-II	RealChemie	18171	31 Jan 2024
Metazerac	Becesane	17744	30 Jun 2024
Metazerac (N I only)	Becesane	20075	31 Jan 2026
Parsan Extra	Certis Belchim B V	17050	31 Jan 2026
Rocket	Adama	17093	31 Jan 2026

589 metconazole

Product	Approval holder	MAPP No.	Expiry Date
Ambarac	Life Scientific	17971	31 Oct 2025
Aptrell 90	Corteva	19452	31 Oct 2025
Caramba	BASF	15337	31 Oct 2025
Caramba 90	BASF	15524	31 Oct 2025
Gringo (N I only)	AgChem Access	20016	31 Oct 2025
Life Scientific Metconazole	Life Scientific	16022	31 Oct 2025
Metal	Arysta	17265	31 Oct 2025
Metal 60	Arysta	17074	31 Oct 2025
Metfin	Clayton	18660	31 Oct 2025
Plexeo 60	Syngenta	18281	31 Oct 2025
Plexeo 90	Globachem	18681	31 Oct 2025
Quickstep	Syngenta	18831	31 Oct 2025
Redstar (N I only)	AgChem Access	20026	31 Oct 2025

Product	Approval holder	MAPP No.	Expiry Date
Remocco 90	UPL Europe	20150	31 Oct 2025
Saloon	Globachem	18923	31 Oct 2025
Turret 90	Globachem	19025	31 Oct 2022
Turret 90	Corteva	19295	31 Oct 2025

595 1-methylcyclopropene

EthylBloc Tabs	Oasis	18288	09 Sep 2099
Ethylene Buster	Chrysal	16222	09 Sep 2099
Fysium	Janssen	17584	28 Feb 2025
Ripelock Tabs 2.0	AgroFresh	17486	09 Sep 2099

597 metobromuron

Inigo	Certis Belchim B V	20219	30 Jun 2027
Lianto	Certis Belchim B V	18394	30 Sep 2026
Lianto	Certis Belchim B V	20224	30 Jun 2027
Praxim	Certis Belchim B V	20209	30 Jun 2027
Soleto	Certis Belchim B V	20225	30 Jun 2027

599 metrafenone

Attenzo	BASF	11917	31 Oct 2025
Lexi	AgChem Access	14890	31 Oct 2023
Lexi (N I only)	AgChem Access	20022	31 Oct 2025

600 metribuzin

Acorix 70 WG	Globachem	18755	31 Jan 2026
Clayton Tribute	Clayton	18363	31 Jan 2024
Cleancrop Frizbee	Adama	16642	31 Jan 2025
Clopes	YC Agro	19954	31 Jan 2025
Discus	RAAT	17552	31 Jan 2024
Gribuzz	Simagro	19429	31 Jan 2024
Gribuzz	Origin	19661	31 Jan 2026
Lauran 70 WG	Stefes	19928	31 Jan 2026
Masterly	Spring	19509	31 Jan 2024
Matecor New	Clayton	18315	31 Jan 2026
Metricam	Proplan-Plant	18630	31 Jan 2026
Petrichor	YC Agro	19844	31 Jan 2025
Python	Feinchemie	16010	31 Jan 2024
Python	Adama	19570	31 Jan 2025

601 metsulfuron-methyl

Ally SX	FMC Agro	18675	30 Sep 2025
Clayton Rally CX	Clayton	19805	30 Sep 2025
Laya	Life Scientific	19016	30 Sep 2025
Laya 200 SG	Life Scientific	19355	11 Feb 2024
Ricorso Premium	Rotam	18527	30 Sep 2025
Savvy	Rotam	14266	30 Sep 2025
Snicket	Top Crop	19507	30 Sep 2025
Tatler	Terrechem	19926	30 Sep 2025

602 metsulfuron-methyl + thifensulfuron-methyl

Racing TF	Nufarm UK	19577	09 Sep 2099

603 metsulfuron-methyl + tribenuron-methyl

Ali Macx	Terrachem	18408	30 Apr 2024
Equator SX	FMC Agro	18455	09 Sep 2099

604 Mild Pepino Mosaic Virus isolate VC1 + VX1

V10	Valto	18969	31 May 2023

SECTION 3

Product	Approval holder	MAPP No.	Expiry Date
608 napropamide			
AC 650	UPL Europe	17966	29 Feb 2024
Colzamid	UPL Europe	17967	29 Feb 2024
Colzamid	UPL Europe	19398	12 Jan 2025
Devrinol	UPL Europe	17853	31 Jan 2024
Naprop 450	Certis Belchim B V	18682	30 Jun 2026
609 napropamide + quinmerac			
Brando	Globachem	19979	26 Apr 2026
Brando	Syngenta	20040	26 Apr 2026
610 nicosulfuron			
Accent	Corteva	20373	30 Jun 2026
Alaris	Novastar	18482	30 Jun 2026
Arigo WG	DuPont (Corteva)	19564	30 Jun 2026
Arigo WG	Corteva	20349	30 Jun 2026
Fornet 4 SC	Certis Belchim B V	16082	30 Jun 2025
Fornet 4 SC	Certis Belchim B V	20186	30 Jun 2026
Fornet 6 OD	Certis Belchim B V	20187	30 Jun 2026
Fornet 6 OD	Certis Belchim B V	15891	30 Jun 2023
Hill-Agro 40	Stefes	17654	30 Jun 2026
Ikanos	Nufarm UK	18641	30 Jun 2026
Milagro 40 OD	Certis Belchim B V	19573	30 Jun 2026
Nico Pro 4 SC	Certis Belchim B V	20193	30 Jun 2026
Nicole	Synergy	19464	30 Jun 2026
Nicoron	Stefes	18025	30 Jun 2026
Nicosh 4% OD	Sharda	19044	17 Apr 2023
Pampa 4 SC	Certis Belchim B V	20196	30 Jun 2026
Pampa 4 SC	Certis Belchim B V	17521	30 Jun 2025
Samson	Certis Belchim B V	16425	30 Jun 2025
Samson	Certis Belchim B V	20195	30 Jun 2026
Samson Extra 6%	Certis Belchim B V	20191	30 Jun 2026
Samson Extra 6%	Certis Belchim B V	16418	30 Jun 2023
Scape	Terrechem	18863	30 Jun 2026
Standon Frontrunner Super 6	Enviroscience	18486	30 Jun 2023
Stretch	Zenith	19162	30 Jun 2026
Victus	DuPont (Corteva)	19565	30 Jun 2026
Victus	Corteva	20364	30 Jun 2026
611 nicosulfuron + thifensulfuron-methyl			
Collage	FMC Agro	16820	30 Nov 2026
Collage	FMC Agro	20350	09 Sep 2099
612 orange oil			
Prev-Gold	Oro Agri	19673	25 May 2025
615 oxathiapiprolin			
DP 717	DuPont (Corteva)	18667	19 Jul 2025
DP 717	Corteva	20355	19 Jul 2025
Hesper	Ceres	18814	30 Jun 2024
Orondis Plus	Syngenta	19305	19 Jul 2025
Orondis Plus	Corteva	20367	19 Jul 2025
Zorvec Enicade	Corteva	20366	03 Sep 2029
616 paclobutrazol			
A10784A	Syngenta	17172	30 Nov 2025
617 Paecilomyces fumosoroseus			
Futureco Nofly WP	Futureco	19797	30 Jun 2027

Product	Approval holder	MAPP No.	Expiry Date
619 pelargonic acid			
Kalipe	Certis Belchim B V	18746	28 Feb 2026
Kalipe	Certis Belchim B V	20169	28 Feb 2026
Katoun Gold	Certis Belchim B V	20173	28 Feb 2026
Redialo	Certis Belchim B V	20172	28 Feb 2026
Redialo	Certis Belchim B V	18747	28 Feb 2026
620 penconazole			
Clayton Castile	Clayton	18310	30 Jun 2024
Clayton Castile (N I only)	Clayton	19871	30 Jun 2025
Pan Penco (N I only)	Pan Agriculture	20279	30 Jun 2025
Rapas	RAAT	17740	30 Jun 2024
Topenco	Certis Belchim B V	17539	30 Jun 2024
622 pendimethalin			
Domitrel	Adama	19165	09 Sep 2099
Eximus II	Gemini	18483	09 Sep 2099
Pendifin	Finchimica	18164	09 Sep 2099
Pendinova	Finchimica	19157	09 Sep 2099
Penfox	Sharda	18342	09 Sep 2099
Penta	Sharda	17680	09 Sep 2099
Quarry	Gemini	17789	09 Sep 2099
Sombrero (N I only)	Nurture	19617	30 Jun 2024
Sovereign	BASF	13442	09 Sep 2099
Stomp 400 SC	BASF	13405	09 Sep 2099
623 pendimethalin + picolinafen			
Flight	BASF	12534	09 Sep 2099
627 penthiopyrad			
Avella	Certis Belchim B V	19256	30 Nov 2027
Celica	Certis Belchim B V	19260	30 Nov 2027
DP 747	DuPont (Corteva)	17530	30 Nov 2027
Fontelis	Corteva	20356	30 Nov 2027
Intellis	DuPont (Corteva)	17520	30 Sep 2023
Sentenza	Terrachem	17529	30 Jun 2024
Vertisan	DuPont (Corteva)	17485	30 Sep 2023
630 pepino mosaic virus strain CH2 isolate 1906			
PMV-01	De Ceuster	17587	07 Feb 2033
635 phenmedipham			
Agrichem Phenmedipham EC	UPL Europe	16224	31 Jan 2026
Agrichem PMP EC	UPL Europe	16963	31 Jan 2026
Agrichem PMP SE	UPL Europe	16971	31 Jan 2026
Betanal Flow	Bayer CropScience	13893	31 Jan 2026
Corzal	UPL Europe	16952	31 Jan 2026
Dancer Flow	Sipcam	14395	31 Jan 2026
Herbasan Flow	Nufarm UK	13894	31 Jan 2026
Mandolin Flow	Nufarm UK	13895	31 Jan 2026
Parish	Adama	17464	31 Jan 2026
641 picolinafen			
AC 900001	BASF	10714	09 Sep 2099
Vixen	BASF	13621	09 Sep 2099
643 pinoxaden			
A13814D	Syngenta	19006	31 Dec 2028
Acolnis	Syngenta	19850	31 Dec 2028
Glodiac	Syngenta	19854	31 Dec 2028

SECTION 3

Product	Approval holder	MAPP No.	Expiry Date
Lutimor	Syngenta	19856	31 Dec 2028
Olupal	Syngenta	19849	31 Dec 2028
Preado	Syngenta	19853	31 Dec 2028
Scintilla	Syngenta	19852	31 Dec 2028

644 pirimicarb
Clayton Pirimicarb (N I only)	Clayton	19903	30 Oct 2024
Pomona	Spring	19545	30 Jun 2024

645 pirimiphos-methyl
Phobi Smoke Pro90	Lodi UK	17117	31 Jan 2026

647 potassium hydrogen carbonate (commodity substance)
Karbicure	Agronaturalis	19417	09 Mar 2024

648 potassium phosphonates
Alginure Bio Schutz	Tilco-Alginure	18659	31 Jul 2028

649 potassium salts of fatty acids
Tec-bom	Fargro	17455	29 Feb 2024

650 prochloraz
Mirage	Adama	18361	30 Jun 2023
Sporgon 50 WP	Sylvan	17700	30 Jun 2023
Sporgon 50 WP (GB only)	Sylvan	17700	30 Jun 2029

653 prochloraz + tebuconazole
Amplitude	HMpG GMBH	18488	30 Jun 2023
Amplitude (GB only)	HMpG GMBH	18488	30 Jun 2024
Clayton Prius	Clayton	18946	30 Jun 2023
Clayton Prius (GB only)	Clayton	18946	30 Jun 2029

656 prohexadione-calcium
Cutless	Rigby Taylor	19613	30 Jun 2024
Cutless (N I only)	Rigby Taylor	20295	30 Jun 2026
Kopis (N I only)	Pan Amenity	20298	30 Jun 2025

658 propamocarb hydrochloride
Edipro	Arysta	15564	31 Jan 2026
Propamex-I 604 SL	MAC	15901	31 Jan 2024
Propicarb	Ascot Pro-G	19884	31 Jan 2025
Rival	Agria SA	18177	31 Jan 2025
Sporax	Globachem	20004	31 Jan 2026

659 propaquizafop
Chitral	Terrechem	17655	31 May 2024
Clayton Enigma	Clayton	17391	31 May 2024
Clayton Enigma	Clayton	17435	31 May 2024
Clayton Enigma (N I only)	Clayton	19881	31 May 2025
Clayton Enigma (N I only)	Clayton	19885	31 May 2025
Cleancrop Peregrine	Clayton	18999	30 Jun 2024
Cleancrop Peregrine (N I only)	Clayton	20347	30 Nov 2025
Flanoc	RealChemie	18155	31 May 2024
Profop 100	HMpG GMBH	18429	31 May 2024
Shogun	Adama	20056	31 May 2026
Tithe	Adama	20047	31 May 2026
Zetrola	Adama	20045	31 May 2026
Zetrola	Syngenta	18272	31 May 2025
Zetrola	Syngenta	20045	31 May 2025

Product	Approval holder	MAPP No.	Expiry Date
664 propoxycarbazone-sodium			
Attribut	Sumitomo	18894	09 Sep 2099
665 propyzamide			
Gem Granules	Agrigem	19182	09 Sep 2099
Gemstone Granules	Agrigem	17898	31 Jul 2023
Hockley Propyzamide 40	Hockley	15449	30 Jun 2024
KeMiChem - Propyzamide 400 SC	KeMiChem	14706	30 Jun 2024
MAC-Propyzamide 400 SC	MAC	14786	30 Jun 2024
PureFlo	Pure Amenity	15138	30 Jun 2024
RouteOne Zamide Flo	Albaugh UK	14559	30 Jun 2024
Stymie	Becesane	16328	30 Jun 2024
Verdah 400	DAPT	14680	30 Jun 2024
Verdah 400 (N I only)	DAPT	20096	09 Sep 2099
666 proquinazid			
Justice	Corteva	20359	31 Jan 2026
Quin Pro	HMpG GMBH	18469	30 Jun 2024
Talius	PSI	16912	30 Jun 2024
Zorkem	Unisem	16906	30 Jun 2024
667 prosulfocarb			
A8545G	Syngenta	16204	30 Apr 2026
Atlas Pro 800	Cropthetics	19258	30 Apr 2026
Bokken	Syngenta	17557	30 Apr 2026
Ca-Pro	CAC Nantong	19472	30 Apr 2024
Cheetah	Ascot Pro-G	19444	30 Apr 2026
Clayton Heed	Clayton	17701	30 Apr 2026
Clayton Obey	Clayton	16405	30 Apr 2026
Clayton Obey N I only)	Clayton	19898	30 Apr 2025
Dogo	Proplan-Plant	18628	30 Apr 2026
Elude	Chem-Wise	17195	30 Apr 2024
Fade	Life Scientific	17982	30 Apr 2026
Fidox	Syngenta	16209	30 Apr 2026
Fidox 800 EC	Certis Belchim B V	17904	30 Apr 2026
ICI574	Syngenta	16205	30 Apr 2026
Jolt	Hemani	20032	30 Apr 2026
Krum 800	Proplan-Plant	18420	30 Apr 2026
Moose 800 EC	Certis Belchim B V	17968	30 Apr 2026
Nuron	Gemini	19518	30 Apr 2026
Nuron	Cropthetics	19241	30 Apr 2026
NYX	Terrechem	17737	30 Apr 2024
Peloton	Barclay	19049	30 Apr 2026
Penzo	Proplan-Plant	18629	30 Apr 2026
Prose	Sharda	19196	30 Apr 2026
Pro-Star	Cropthetics	19238	30 Apr 2026
Prosulfix EC	Novastar	18774	30 Apr 2026
PSC Pro	CropStar	18957	30 Apr 2026
Roxy 800 EC	Certis Belchim B V	17859	30 Apr 2026
Stefes Prosulfocarb	Stefes	19539	30 Apr 2026
Surf	Synergy	19447	30 Apr 2026
Zen	Terrechem	19115	30 Apr 2026
668 prosulfuron			
Clayton Kibo (N I only)	Clayton	19891	09 Sep 2099
669 prothioconazole			
Abran 250	Barclay	19737	31 Jan 2026
Alana Star	Life Scientific	19326	31 Jan 2026
Aquila	BASF	19971	31 Jan 2026

Product	Approval holder	MAPP No.	Expiry Date
Aragon	Nuvaros	19277	31 Jan 2023
Aragon	Novastar	19425	31 Jan 2026
Artio	Gemini	19562	31 Jan 2026
Audis	Qingdao	19764	31 Jan 2026
Banguy	PSI	15140	31 Jan 2024
Bouncer	Agrimar	20157	31 Jan 2026
Cac Tai	CAC Nantong	20064	31 Jan 2026
Careline	CAC Nantong	20208	31 Jan 2026
Clayton Corlis	Clayton	20211	31 Jan 2026
Cleancrop Protocol	Agrii	20375	31 Jan 2025
Cobra	Novastar	19808	31 Jan 2026
Dagda	Barclay	19803	31 Jan 2026
Era	Syngenta	19792	31 Jan 2026
Foxline	CAC Nantong	20218	31 Jan 2026
Helsinki	Cropthetics	19653	31 Jan 2026
Innox	BASF	19676	31 Jan 2025
Magan	Adama	19715	25 Jan 2025
Pabi 300 EC	Certis Belchim B V	19810	26 Nov 2024
Patel 250 EC	Certis Belchim B V	19582	31 Jan 2026
Patel 300 EC	Certis Belchim B V	19778	26 Nov 2024
Pecari 250 EC	Syngenta	19581	31 Jan 2026
Pecari 300 EC	Syngenta	19637	11 Jan 2025
Pemba	Terrechem	19409	31 Jan 2026
Pride	Sharda	19705	31 Jan 2026
Procer 300 EC	Helm	19831	31 Jan 2026
Proline	Bayer CropScience	12084	31 Jan 2026
Promino 300 EC	Helm	19832	31 Jan 2026
Protendo 250 EC	Globachem	19175	31 Jan 2026
Prothio	ChemSource	14945	31 Jan 2022
Prothious	Chem-Wise	19663	31 Jan 2026
Province	Clayton	19493	31 Jan 2026
Redigo	Bayer CropScience	12085	31 Jan 2026
Skea	Life Scientific	19351	31 Jan 2026
Tartaros 300 EC	Helm	19703	31 Jan 2025
Teko 250	Barclay	19804	31 Jan 2026
Tokyo EC	Cropthetics	19560	31 Jan 2024
Traciafin Plus	Clayton	19212	31 Jan 2026
Ultraline	Cropthetics	19397	31 Jan 2026

670 prothioconazole + spiroxamine

Product	Approval holder	MAPP No.	Expiry Date
Flexure	Life Scientific	19311	31 Jan 2026
Hint	Finchimica	19727	31 Jan 2026
Pro-Spirox	HMpG GMBH	19062	31 Jan 2024
Torsion	HMpG GMBH	18476	31 Jan 2024
Torsion II	HMpG GMBH	19061	31 Jan 2024

672 prothioconazole + tebuconazole

Product	Approval holder	MAPP No.	Expiry Date
Aragon Duo	Novastar	19627	31 Jan 2026
Cathay	Terrechem	19655	31 Jan 2026
Clayton Finch	Clayton	20210	31 Jan 2026
Clayton Orinoco	Clayton	19574	31 Jan 2026
Clayton Proteb	Clayton	20212	31 Jan 2026
Clayton Zorro Pro	Clayton	17073	31 Jan 2024
Dolla	Terrechem	19656	31 Jan 2026
Euskatel Plus	Albaugh UK	20147	31 Jan 2025
Guardian	Chem-Wise	19678	31 Jan 2026
Neptune	Novastar	19779	31 Jan 2026
Noric	Life Scientific	19298	31 Jan 2026
Pro Fit 25	Chem-Wise	17040	30 Jun 2023

Product	Approval holder	MAPP No.	Expiry Date
Pro Fit 25 (N I only)	Chem-Wise	20061	31 Jan 2026
Protect Extra	Crop Health	19497	31 Jan 2024
Protefin	Clayton	19126	31 Jan 2026
Protendo Extra	Certis Belchim B V	19630	31 Jan 2026
Rimtil	HMpG GMBH	18466	31 Jan 2024
Romtil	RealChemie	18117	31 Jan 2024
Sentinel	Chem-Wise	19670	31 Jan 2026
Sequana	Ascot Pro-G	17818	30 Jun 2024
Sequana (N I only)	Ascot Pro-G	20083	30 Jun 2027
Standon Cumer	Enviroscience	18484	31 Jan 2024
Teborp	Gemini	20238	31 Jan 2026
UltraLegend	Cropthetics	19825	31 Jan 2026

674 prothioconazole + trifloxystrobin

Infinity	HMpG GMBH	18475	30 Jun 2024
Symovitan	Simagro	19367	30 Jun 2024

678 pyraclostrobin

BAS 500 06	BASF	12338	31 Jul 2025
BASF Insignia	BASF	11900	31 Jul 2025
Comet	BASF	10875	31 Jul 2025
Eland	Rigby Taylor	19468	31 Jul 2023
Flyer	BASF	12654	31 Jul 2025
Halley (N I only)	AgChem Access	20017	31 Jul 2025
Inter Pyrastrobin 200	BASF	19039	31 Jul 2025
Modem 200	BASF	19552	31 Jul 2025
Orbit TL	Capital CP	19555	30 Jun 2024
Orbit TL (N I only)	Capital CP	20070	31 Jul 2025
Platoon	BASF	12325	31 Jul 2025
Platoon 250	BASF	12640	31 Jul 2025
Tucana 200	BASF	17294	31 Jul 2025
Vanguard	Sherriff Amenity	13838	31 Jan 2024
Vivid	BASF	10898	31 Jul 2025

679 pyraflufen-ethyl

Albis	Certis Belchim B V	20160	09 Sep 2099
Gozai	Certis Belchim B V	20168	09 Sep 2099
Kabuki	Certis Belchim B V	20159	09 Sep 2099

681 pyridate

Clayton Viva	Clayton	18264	30 Jun 2024
Clayton Viva (N I only)	Clayton	19919	30 Jun 2033
Diva	Certis Belchim B V	20166	30 Jun 2033
Gyo	Certis Belchim B V	20167	30 Jun 2033
Lentagran WP	Certis Belchim B V	20176	30 Jun 2033

682 pyrimethanil

Clayton Patriot (N I only)	Clayton	19901	30 Oct 2024
Penbotec 400 SC	Janssen	17384	31 Oct 2025
Pirim	Sharda	19348	31 Oct 2025
Precious	Emerald	18134	30 Jun 2024
Pro-Pyr	RAAT	19757	31 Oct 2024
Pyrimala	Servem	17325	30 Jun 2024
Spectrum	RAAT	17469	28 Feb 2023
Triple SC (N I only)	Pan Agriculture	20290	31 Oct 2025

683 pyriofenone

Diopyr	Certis Belchim B V	19131	31 Aug 2026
Diopyr	Certis Belchim B V	20129	31 Jul 2027
Property 180SC	Certis Belchim B V	20190	31 Jul 2027

SECTION 3

Product	Approval holder	MAPP No.	Expiry Date
Property 180SC	Certis Belchim B V	17666	30 Sep 2026
686 Pythium oligandrum M1			
Polyversum	De Sangosse	17456	31 Oct 2025
690 quizalofop-P-ethyl			
Clayton Kwiz (N I only)	Clayton	20343	31 May 2025
Leopard 5EC	Adama	16915	31 May 2025
691 quizalofop-P-tefuryl			
Panarex	Arysta	17960	31 May 2024
Rango	Arysta	17959	31 May 2024
692 rimsulfuron			
Caesar (N I only)	AgChem Access	20013	31 Oct 2025
Clayton Rasp	Clayton	15952	30 Jun 2024
Titus	Corteva	20363	31 Oct 2025
694 S-abscisic acid			
Conshape	HD2412	19686	31 Mar 2027
696 Sheep fat			
Trico	Kwizda	18149	28 Feb 2023
697 silthiofam			
Latitude XL	Certis Belchim B V	18030	09 Sep 2099
698 S-metolachlor			
Efica 960EC	Adama	18344	31 Jan 2026
700 sodium silver thiosulphate			
Florissant 100	Dejex	19158	31 Jan 2027
702 spinosad			
Clayton Spinoff (N I only)	Clayton	20332	31 Oct 2025
Spindle (N I only)	Pan Agriculture	20288	31 Oct 2025
Wopro Spinosad 480 SC (NI only)	Simonis	19839	31 Oct 2024
705 spirotetramat			
Clayton Occupy (N I only)	Clayton	19899	31 Jan 2027
Movento (GB only)	Bayer CropScience	18435	31 Jan 2027
Spiropan (N I only)	Pan Amenity	20289	31 Oct 2025
Spirotet	HMpG GMBH	18450	30 Jun 2024
Tetramat	Progreen	18947	30 Jun 2024
709 Streptomyces griseoviridis strain K61			
Lalstop K61 WP	Danstar	19571	09 Sep 2099
Mycostop	Danstar	18613	31 Oct 2023
710 sulfosulfuron			
Monitor	Sumitomo	17695	09 Sep 2099
711 sulfoxaflor			
Sequoia	Fargro	18677	30 Sep 2023
712 sulfuryl fluoride			
ProFume	Douglas	17309	22 Apr 2023
713 sulphur			
Acoidal WG	Quimetal	19109	30 Jun 2026
POL-Sulphur 80 WG	Ciech	17613	30 Jun 2026
POL-Sulphur 800 SC	Ciech	17614	30 Jun 2026

Product	Approval holder	MAPP No.	Expiry Date
Solfa WG	Nufarm UK	19460	30 Jun 2026
714 sulphur + tebuconazole			
Unicorn DF	Sulphur Mills	18109	28 Feb 2026
715 tau-fluvalinate			
Clayton Malin	Clayton	19245	31 Dec 2023
Clayton Spirit	Clayton	19190	31 Dec 2023
Evure	Syngenta	18284	31 Dec 2024
Klartan	Adama	11074	31 Dec 2024
Pylon	Adama	19524	31 Dec 2024
Revolt	Adama	13383	31 Dec 2024
Vespa	Adama	19529	31 Dec 2024
716 tebuconazole			
Amala	Agrimar	20151	28 Feb 2026
Buzz Ultra DF	Arysta	18328	28 Feb 2025
Buzz Ultra DF	UPL Europe	19708	28 Feb 2025
Carpetus	Cropthetics	19802	28 Feb 2025
Cezix	Rotam	18056	28 Feb 2025
Clayton Ohio (N I only)	Clayton	20334	28 Feb 2026
Clayton Tebbit	Clayton	20371	30 Nov 2025
Clayton Tebucon 250 EW	Clayton	20372	30 Nov 2025
Erase	FMC Agro	18741	28 Feb 2025
Erasmus	Rotam	18057	28 Feb 2025
Fathom	FMC Agro	18737	28 Feb 2026
Fezan	Sipcam	17082	28 Feb 2025
Folicur	Bayer CropScience	16731	28 Feb 2025
Hail	Agform	19262	28 Feb 2023
Hi-Tebura	Hockley	18421	28 Feb 2025
Jackal	Gemini	18912	28 Feb 2025
Legend	Cropthetics	18432	28 Feb 2025
Missile	Chem-Wise	19668	28 Feb 2025
Mitre	Nufarm UK	19002	28 Feb 2025
Mystic	Nufarm UK	19156	28 Feb 2025
Odin	Rotam	18016	28 Feb 2025
Odin 250 EW	Rotam	19784	28 Feb 2025
Orius	Nufarm UK	17414	28 Feb 2022
Quarta	Terrechem	17844	28 Feb 2026
Santal	Novastar	17955	28 Feb 2025
Savannah	Rotam	18373	28 Feb 2025
Spekfree	Rotam	16004	28 Feb 2025
Standon Beamer	Standon	17350	29 Feb 2024
Starpro	Rotam	18367	28 Feb 2025
Tebu 430	HMpG GMBH	19485	29 Feb 2024
Tebu-I 430	HMpG GMBH	19470	29 Feb 2024
Tebusha 25 EW	Sharda	18034	28 Feb 2025
Tercero	Albaugh UK	19337	28 Feb 2025
Trident	Ascot Pro-G	18858	28 Feb 2025
Ventura TEB 250	Ventura	19654	28 Feb 2026
Zonal	Synergy	19453	28 Feb 2025
719 tebuconazole + trifloxystrobin			
Clayton Bestow (N I only)	Clayton	19865	09 Sep 2099
Dedicate (GB only)	Environmental Science	20307	28 Feb 2027
Dedicate (N I only)	Environmental Science	20306	09 Sep 2099
Defusa	Rigby Taylor	18232	25 Feb 2026
Defusa (G B only)	Rigby Taylor	20309	28 Feb 2027
Defusa (N I only)	Rigby Taylor	20308	09 Sep 2099
Dualitas	ProKlass	18000	30 Jun 2024

Product	Approval holder	MAPP No.	Expiry Date
Fortify (N I only)	Pan Amenity	20296	09 Sep 2099
Fortitude	Capital CP	19522	30 Jun 2024
Fortitude (N I only)	Capital CP	20069	09 Sep 2099
Fusion (N I only)	Pan Amenity	20297	09 Sep 2099
Inter Tebloxy	Iticon	17311	30 Jun 2024
720 tebufenpyrad			
Clayton Bonsai (N I only)	Clayton	19866	30 Apr 2025
Masai	BASF	18287	30 Apr 2025
Shirudo	Certis Belchim B V	19575	30 Apr 2026
722 tefluthrin			
A13219F	Syngenta	19083	30 Jun 2027
723 tembotrione			
Laudis	Bayer CropScience	17302	31 Jan 2027
726 terpenoid Blend (QRD 460)			
Requiem Prime	Bayer CropScience	19112	27 Jun 2023
727 tetraconazole			
Eminent 125 ME	Certis Belchim B V	18864	30 Jun 2026
728 tetradecadienyl + trienyl acetate			
Tutatec	SEDQ	19503	28 Feb 2025
734 thifensulfuron-methyl			
Harmony SX	FMC Agro	18761	09 Sep 2099
735 thifensulfuron-methyl + tribenuron-methyl			
Calibre SX	Certis Belchim B V	15032	09 Sep 2099
Clayton Pause	Clayton	18937	30 Jun 2024
Clayton Pause (N I only)	Clayton	19902	09 Sep 2099
Clayton Rift	Clayton	18945	30 Jun 2024
Clayton Rift (N I only)	Clayton	19907	09 Sep 2099
Parana	Certis Belchim B V	16258	09 Sep 2099
Seduce	Certis Belchim B V	16261	09 Sep 2099
745 tribenuron-methyl			
Corida	Zenith	18247	09 Sep 2099
Helmstar 75 WG	Certis Belchim B V	17077	09 Sep 2099
Helmstar 75 WG	Helm	19935	09 Sep 2099
Quantum SX	FMC Agro	20357	30 Jul 2036
Quantum SX	FMC Agro	18787	31 May 2024
TBM 75 WG	Sharda	17326	09 Sep 2099
Thunder	Ascot Pro-G	19961	09 Sep 2099
Toscana	Proplan-Plant	17726	09 Sep 2099
Trailer	Industrias Afrasa	17903	09 Sep 2099
Triad SX	FMC Agro	20377	30 Jul 2036
Tribun 75 WG	Certis Belchim B V	17090	09 Sep 2099
Tribun 75 WG	Helm	19953	09 Sep 2099
Trimeo 75 WG	Certis Belchim B V	17076	09 Sep 2099
Trimeo 75 WG	Helm	19944	09 Sep 2099
748 Trichoderma atroviride strain SC1			
Vintec	Certis Belchim B V	20311	06 Jan 2034
749 triclopyr			
Topper	Arysta	15719	31 Oct 2024

Product	Approval holder	MAPP No.	Expiry Date
750 trifloxystrobin			
Martinet	AgChem Access	14614	30 Jun 2024
Martinet	AgChem Access	20024	31 Jan 2036
752 triflusulfuron-methyl			
Kaskad	Life Scientific	19761	30 Jun 2026
Safari	FMC Agro	18796	30 Jun 2026
Zareba	Terrechem	17987	30 Jun 2024
753 trinexapac-ethyl			
A17600C	Syngenta	17548	31 Oct 2024
Clipless NT	Headland Amenity	18729	24 Nov 2023
Confine NT	Zantra	17770	31 Oct 2024
Cratus	Gemini	19628	31 Oct 2024
Cutaway	Syngenta	14445	31 Oct 2024
Gyro	UPL Europe	17426	31 Oct 2024
Iceni	UPL Europe	16835	31 Oct 2024
Jaguar	Ascot Pro-G	18882	31 Oct 2024
Maintain NT	FMC Agro	18736	24 Nov 2023
Modan	Certis Belchim B V	17060	31 Oct 2024
Modan	Helm	19950	31 Oct 2024
Moddus ME	Syngenta	17179	31 Oct 2024
Moddus ME	Syngenta	19763	31 Oct 2024
Moxa 250 EC	Certis Belchim B V	16176	31 Oct 2025
Munia	Syngenta	19561	31 Oct 2024
Next	Sharda	17797	31 Oct 2025
Paket 250 EC	Arysta	17436	31 Oct 2024
Pan Tepee (N I only)	Pan Agriculture	20280	31 Oct 2024
Ponytail	YC Agro	20216	31 Oct 2025
Prop	Terrechem	18605	31 Oct 2024
Sabre	Ascot Pro-G	19309	31 Oct 2024
Seize	FMC Agro	16780	31 Oct 2024
Seize NT	FMC Agro	18735	31 Oct 2024
Shorten	Becesane	15377	30 Jun 2024
Shrink (N I only)	AgChem Access	20028	31 Oct 2025
Sudo	Life Scientific	17979	31 Oct 2024
Tempest	Ascot Pro-G	18027	30 Jun 2024
Tempest (N I only)	Ascot Pro-G	20084	31 Oct 2024
Terplex	Globachem	18405	31 Oct 2025
TP 100	Certis Belchim B V	17605	31 Oct 2025
Tramontane	Synergy	19446	31 Oct 2024
Trexstar	Helm	19951	31 Oct 2024
Trexxus	Helm	19952	31 Oct 2024
Trexxus	Certis Belchim B V	17059	31 Oct 2024
Tribune	Novastar	18647	31 Oct 2024
UPL Trinexapac	UPL Europe	16645	31 Oct 2024
Zira	Terrechem	17467	30 Jun 2024
756 Verticillium alobo-atrum			
Dutch Trig	BTL Bomendienst	17481	09 Sep 2099
757 warfarin			
Grey Squirrel Bait	Killgerm	UK19-1169	01 Jan 2024

SECTION 3

SECTION 4
ADJUVANTS

Adjuvants

Adjuvants are not themselves classed as pesticides and there is considerable misunderstanding over the extent to which they are legally controlled under the Food and Environment Protection Act. An adjuvant is a substance other than water which enhances the effectiveness of a pesticide with which it is mixed. Consent C(i)5 under the Control of Pesticides Regulations allows that an adjuvant can be used with a pesticide only if that adjuvant is authorised and on a list published on the HSE adjuvant database: https://secure.pesticides. gov.uk/adjuvants/search.asp. An authorised adjuvant has an *adjuvant number* and may have specific requirements about the circumstances in which it may be used.

Adjuvant product labels must be consulted for full details of authorised use, but the table below provides a summary of the label information to indicate the area of use of the adjuvant. Label precautions refer to the keys given in Appendix 4, and may include warnings about products harmful or dangerous to fish. The table includes all adjuvants notified by suppliers as available in 2023.

Product	Supplier	Adj. No.	Type
Abacus	De Sangosse	A0543	vegetable oil
Contains	53.43 % w/w oil (rapeseed fatty acid esters), 20.0 % w/w alkoxylated alcohols and 9.0 % w/w oil (tall oil fatty acids)		
Use with	All approved pesticides on all edible crops when used at half their recommended dose or less, and on all non-edible crops up to their full recommended dose. Also at a maximum concentration of 0.1% with approved pesticides on listed crops up to specified growth stages		
Protective clothing	A, C, H		
Precautions	R36, U05a, U11, U14, U19a, U20b, E15a, E19b, E34, D01, D02, D05, D10a, D12a, H04		
Activator 90	De Sangosse	A0547	non-ionic surfactant/wetter
Contains	375 g/kg alkoxylated alcohols, 375 g/kg alkoxylated alcohols, 150 g/kg oil (tall oil fatty acids)		
Use with	All approved pesticides on all edible crops when used at half their recommended dose or less, and on all non-edible crops up to their full recommended dose. Also at a maximum concentration of 0.1% with approved pesticides on listed crops up to specified growth stages		
Protective clothing	A, C, H		
Precautions	R36, R38, R53a, U02a, U04a, U05a, U10, U11, U20b, E15a, E19b, D01, D02, D05, D10a, H04		
Addit	Koppert	A0910	spreader/sticker/wetter
Contains	780.2 g/l oil (rapeseed triglycerides)		
Use with	Mycotal at 0.25% spray solution and all approved pesticides at half or less than half the approved pesticide rate.		
Protective clothing	A, F, H		
Precautions	R20, R21, R36, R51, R53a		
Adhere	Intracrop	A0912	adjuvant/anti-drift agent/sticker
Contains	747.8 g/l oil (rapeseed fatty acid esters) and 99.64 g/l rosin and resin esters		
Use with	All residual herbicides pre- or post emergence at the timing recommended by the herbicide manufacturer at a maximum spray concentration of 1.0 % V/V		
Protective clothing	A, C, H		
Precautions	U11, E15a, D01, D10a		

SECTION 4

Product	Supplier	Adj. No.	Type
Adigor	Syngenta	A0522	wetter
Contains	47 % w/w methylated rapeseed oil		
Use with	All approved plant protection products at 1% of spray volume and with Evolya in grain and forage maize at 1.5% spray volume		
Protective clothing	A, C, H		
Precautions	R43, R51, R53a, U02a, U05a, U09a, U20b, E15b, E38, D01, D02, D05, D09a, D10c, D12a, H04, H11		
AdjiFe	Amega	A0797	wetter
Contains	345 g/l ammonium iron (III) citrate, 300 g/l alkyl polyglycosides and 104.5 g/l ethylene oxide-propylene oxide copolymers		
Use with	All authorised plant protection herbicides and fungicides		
Protective clothing	A		
Precautions	U02a, U04a, U05a, U14, U15, U19a, E13b, E34, D01, D02, D09a, D10a, M03		
AdjiMin	Amega	A0812	adjuvant
Contains	920 g/kg oil (petroleum oils)		
Use with	All approved pesticides at half the approved maximum dose on edible crops and at the full approved dose on non-edible crops and non-crop production at 1% spray volume.		
Protective clothing	A		
Precautions	U02a, U04a, U05a, U14, U15, U19a, E13b, E34, D01, D02, D09a, D10a, M03		
AdjiSil	Amega	A0763	spreader/wetter
Contains	83% w/w trisiloxane organosilicone copolymers		
Use with	All approved pesticides at half the approved maximum dose on edible crops and at the full approved dose on non-edible crops and non-crop production at 0.15% spray volume.		
Protective clothing	A, C, H		
Precautions	U02a, U04a, U05a, U14, U15, U19a, E13b, E34, D01, D02, D09a, D10a, M03, H11		
AdjiVeg	Amega	A0756	adjuvant
Contains	90% w/w fatty acid esters		
Use with	All approved pesticides at half the approved maximum dose on edible crops and at the full approved dose on non-edible crops and non-crop production at 1% spray volume.		
Protective clothing	A, C, H		
Precautions	U02a, U04a, U05a, U11, U14, U15, E15a, E34, D01, D02, D09a, D10a, M03, H04		
Arma	Interagro	A0306	penetrant
Contains	500 g/l alkoxylated fatty amine + 500 g/l polyoxyethylene monolaurate		
Use with	Cereal growth regulators, cereal herbicides, cereal fungicides, oilseed rape fungicides and a wide range of other pesticides on specified crops		
Protective clothing	A, C		
Precautions	R51, R58, U05a, E15a, E34, E37, D01, D02, D05, D09a, D10a, H11		
Asu-Flex	Greenaway	A0677	spreader/sticker/wetter
Contains	10.0% w/w rapeseed oil		
Use with	All authorised formulations of glyphosate and asulam at 70% of spray volume		
Protective clothing	A, C		
Precautions	U11, U12, U15, U20b, E15a, E34, D02		

Product	Supplier	Adj. No.	Type
Attach	Intracrop	A0917	adjuvant/anti-drift agent/sticker
Contains	747.8 g/l oil (rapeseed fatty acid esters) and 99.64 g/l rosin and resin esters		
Use with	All residual herbicides pre- or post-emergence at the timing recommended by the manufacturer for the herbicide at a maximum spray concentration of 1.0 % v/v		
Protective clothing	A, C, H		
Precautions	U11, E15a, D01, D10a		
BackRow	Interagro	A0850	mineral oil
Contains	60% w/w refined paraffinic petroleum oil		
Use with	Pre-emergence herbicides. Refer to label or contact supplier for further details		
Protective clothing	A		
Precautions	R22b, R38, U02a, U05a, U08, U20b, E15a, D01, D02, D05, D09a, D10a, M05b, H03		
Backrow Max	Interagro	A0889	adjuvant
Contains	82.6% w/w refined paraffinic petroleum oil		
Use with	All authorised plant protection products at half or less than half the authorised plant protection product rate on edible crops and all authorised plant protection products on non-edible crops and non crop uses. Max concentration 3.5% spray volume.		
Protective clothing	A		
Precautions	R22b, R38, U02a, U05a, U08, U20b, E15a, D01, D02, D05, D09a, D10a, M05b, H03		
Binder	Amega	A0598	spreader/wetter
Contains	30.0 % w/w alkoxylated alcohols		
Use with	All approved pesticides at half or less than half the approved pesticide rate and all approved formulations of glyphosate		
Protective clothing	A		
Precautions	U02a, U04a, U05a, U14, U15, U19a, E13b, E34, D01, D02, D09a, D10a, M03		
Bio Syl	Intracrop	A0773	spreader/sticker/wetter
Contains	32.67% w/w alkoxylated alcohols and 1.0% w/w trisiloxane organosilicone copolymers		
Use with	Recommended rates of approved pesticides on non-crop and non-edible crops; when used on edible crops the dose of the approved pesticide must be half or less than half the approved dose rate		
Protective clothing	A, C		
Precautions	R36, R38, U05a, U08, U20c, E15a, D01, D02, D10a, H04		
BioPower	Bayer CropScience	A0617	wetter
Contains	276.5 g/l alkyl ether sulphates, sodium salts		
Use with	All approved plant protection products in cereals and all approved herbicides on Natural Surfaces Not Intended To Bear Vegetation, Permeable Surfaces Overlying Soil, Industrial And Amenity Areas, Ornamental Plant Production		
Protective clothing	A, C		
Precautions	R36, R38, U02a, U05a, U08, U13, U19a, U20b, E13c, E34, D01, D02, D05, D09a, D10a, H04		

SECTION 4

Product	Supplier	Adj. No.	Type
Bond	De Sangosse	A0556	extender/sticker/wetter
Contains	45% w/w styrene-butadiene copolymers, 10% w/w alkoxylated alcohols		
Use with	All approved potato blight fungicides. Also with all approved pesticides on all edible crops when used at half their recommended dose or less, and on all non-edible crops up to their full recommended dose. Also at a maximum concentration of 0.14% with approved pesticides on listed crops up to specified growth stages		
Protective clothing	A, C, H		
Precautions	R36, R38, U11, U14, U16b, U19a, E15a, E19b, D01, D02, D05, D10a, D12a, H04		
Boost	Intracrop	A0899	spreader/sticker/wetter
Contains	32.67 % w/w alkoxylated alcohols and 1.0% w/w trisiloxane organosilicone copolymers		
Use with	All approved pesticides at half or less than half the approved pesticide rate on edible crops and all approved pesticides on non-edible crops		
Protective clothing	A, C		
Precautions	R43, R51, R53a, U02a, U05a, U09a, U20b, E15b, E38, D01, D02, D05, D09a, D10c, D12a, H04, H11		
Buzz	De Sangosse	A0520	sticker/wetter
Contains	375 g/l alkoxylated alcohols, 250 g/l alkoxylated coconut amines 50 g/l oil (tall oil fatty acids)		
Use with	All approved pesticides at half or less than half the approved pesticide rate at 0.1% spray solution on edible crops.		
Protective clothing	A, C, H		
Precautions	R36, R38, R53a, U02a, U04a, U05a, U10, U11, U20b, E15a, E19b, D01, D02, D05, D10a, H04		
Byo-Flex	Greenaway	A0545	sticker/wetter
Contains	10.0% w/w rapeseed oil		
Use with	All approved 490 g/l formulations of glyphosate on Natural Surfaces Not Intended To Bear Vegetation, Permeable Surfaces Overlying Soil, Hard Surfaces, Amenity Grassland, Amenity Vegetation, Managed Amenity Turf, Forest, Farm Forestry		
Protective clothing	A, C		
Precautions	U11, U12, U15, U20b, E34		
Byte	De Sangosse	A0847	vegetable oil/wetter
Contains	52.4% w/w oil (rape seed fatty acid esters)		
Use with	All approved pesticides at half or less than half the approved pesticide rate at 1% spray solution on edible crops.		
Protective clothing	A, C, H		
Precautions	R36, U05a, U11, U14, U19a, U20b, E15a, E19b, E34, D01, D02, D05, D10a, D12a, H04		
C-Cure	Interagro	A0851	mineral oil
Contains	60% w/w refined mineral oil		
Use with	Pre-emergence herbicides		
Protective clothing	A		
Precautions	R22b, R38, U02a, U05a, U08, U20b, E15a, D01, D02, D05, D09a, D10a, M05b, H03		

Product	Supplier	Adj. No.	Type
Clayton Bower	Clayton	A0861	wetter
Contains	20.1 % w/w 3,6-dioxaoctadecylsulphate, sodium salt and 6.7 % w/w 3,6-dioxaeicosylsulphate, sodium salt		
Use with	All approved products on cereals and all approved herbicides on natural surfaces not intended to bear vegetation, permeable surfaces overlying soil, industrial and amenity areas and ornamental plant production.		
Protective clothing	A, H		
Precautions	H315, H320, U02a, U05a, U08, U13, U19a, U20b, E13c, E34, D01, D02, D05, D09a, D10a, H04		
Codacide Oil	Microcide	A0629	vegetable oil
Contains	95% w/w oil (rapeseed triglycerides)		
Use with	All approved pesticides and tank mixes. See label for details		
Protective clothing	A, C		
Precautions	U20b, D09a, D10b		
Cogent	Intracrop	A0902	spreader/sticker/wetter
Contains	32.67 % w/w alkoxylated alcohols and 1.0% w/w trisiloxane organosilicone copolymers		
Use with	All approved pesticides at half or less than half the approved pesticide rate on edible crops and all approved pesticides on non-edible crops		
Protective clothing	A, C, H		
Precautions	H318, U05a, U11, U20a, H412, M05a, H302		
Companion Gold	Agrovista	A0723	acidifier/buffering agent/drift retardant/extender
Contains	16% w/w ammonium sulphate + 0.95% w/w polyacrylamide		
Use with	Diquat or glyphosate on oilseed rape at 0.5% solution; glyphosate on wheat, rye or triticale at 0.5% solution and with all approved pesticides on non-edible crops at 1% solution or all approved pesticides at 50% dose rate or less on edible crops		
Protective clothing	A, C, H		
Precautions	U05a, U14, U15, U19a, U19c, U20b, E15a, D02, D05, D09a, D10a		
Dash HC	BASF	A0729	non-ionic surfactant/wetter
Contains	348.75 g/l oil (fatty acid esters) and 209.25 g/l alkoxylated alcohols-phosphate esters		
Use with	May be used at a max. concentration of 1.0% spray solution, with all authorised plant protection products up to their full authorised rate, before edible parts form.		
Protective clothing	A, C, H		
Precautions	U05a, U12, U14, D01, D02, D10c, M05b, H03		
Deploy	De Sangosse	A0801	wetter
Contains	91.04 % w/w rapeseed fatty acid esters, 4.16 % w/w alkoxylated alcohols and 2.07 % w/w oil (tall oil fatty acids)		
Use with	All approved pesticides on all edible crops when used at half their recommended dose or less, and on all non-edible crops up to their full recommended dose. Also at a maximum concentration of 0.5% with approved pesticides on listed crops up to specified growth stages		
Protective clothing	A, H		
Precautions	U14, U15, H411, D01, D09a		

SECTION 4

Product	Supplier	Adj. No.	Type
Designer	De Sangosse	A0660	drift retardant/extender/sticker/wetter
Contains	25% w/w styrene-butadiene copolymers, 7.1% w/w trisiloxane organosilicone copolymers		
Use with	A wide range of fungicides, insecticides and trace elements for cereals and specified agricultural and horticultural crops		
Protective clothing	A, C, H		
Precautions	R36, R38, R52, U02a, U11, U14, U15, U19a, E15a, E37, D01, D02, D05, D09a, D10a, H04		
Drill	De Sangosse	A0544	adjuvant
Contains	63.34% w/w oil (rapeseed fatty acid esters), 15% w/w alkoxylated alcohols and 7.5% w/w oil (tall oil fatty acids)		
Use with	All approved pesticides on all edible crops when used at half their recommended dose or less, and on all non-edible crops up to their full recommended dose. Also at a maximum concentration of 0.1% with approved pesticides on listed crops up to specified growth stages		
Protective clothing	A, C, H		
Precautions	R36, U05a, U11, U14, U19a, U20b, E15a, E19b, E34, D01, D02, D05, D10a, D12a, H04		
Dynetic	De Sangosse	A0743	wetter
Contains	80.0% w/w trisiloxane organosilicone copolymers		
Use with	All approved pesticides on all edible crops when used at half their recommended dose or less, and on all non-edible crops up to their full recommended dose at 0.15% spray solution		
Protective clothing			
Precautions	R20, R21, R22a, R36, R43, R48, R51, R58, U11, U15, U19a, E15a, E19b, D01, D02, D05, D10a, D12a, H03, H11		
Eco-flex	Greenaway	A0696	sticker/wetter
Contains	10% w/w refined rapeseed oil		
Use with	Approved formulations of glyphosate, 2,4-D		
Protective clothing	A, C		
Precautions	U11, U12, U15, U20b, E34		
Eco-Tac	Global Adjuvants	A0867	sticker/wetter
Contains	36.4 % w/w paraffin wax		
Use with	All fungicides and insecticides at 0.2% spray solution		
Protective clothing	A, C, H		
Precautions			
Elan Xtra	Intracrop	A0735	spreader/wetter
Contains	530 g/l ethylene oxide-propylene oxide copolymers and 415 g/l trisiloxane organosilicone copolymers		
Use with	All approved pesticides on non-edible crops and non-crop production and with half dose of all approved pesticides on edible crops		
Protective clothing	A		
Precautions	U19a, U20c, E13c, E37, D09a, D11a		

Product	Supplier	Adj. No.	Type
Firebrand	Barclay	-	fertiliser/water conditioner
Contains	500 g/l ammonium sulphate		
Use with	Use at 0.5% v/v in the spray solution with glyphosate		
Protective clothing	A, C		
Precautions	U08, U20a, E15a, D09a, D10b		
Fortune	FMC Agro	A0878	penetrant/spreader/vegetable oil/wetter
Contains	75% w/w mixed methylated fatty acid esters of seed oil and N-butanol		
Use with	Herbicides and fungicides in a wide range of crops. See label for details		
Protective clothing	A, C		
Precautions	R43, U02a, U05a, U14, U20a, E15a, E34, E38, D01, D02, D05, D09a, D10b, H04		
Gateway	FMC Agro	A0881	extender/sticker/wetter
Contains	14.0 % w/v styrene-butadiene copolymers and 7.5 % w/v trisiloxane organosilicone copolymers		
Use with	All authorised plant protection products on non-edible crops and all authorised plant protection products at half rate or less on edible crops. Max concentration 0.125% spray solution		
Protective clothing	A, C		
Precautions	U11, U15, E38, D01		
Gly-Flex	Greenaway	A0588	spreader/sticker/wetter
Contains	95% w/w refined rapeseed oil		
Use with	All approved 490g/l glyphosate products on Natural Surfaces Not Intended To Bear Vegetation, Hard Surfaces, Permeable Surfaces Overlaying Soil, Forest		
Protective clothing	A, C		
Precautions	R36, U11, U12, U15, U20b, E13c, E34, E37, H04		
Ground Force	Global Adjuvants	A0890	adjuvant/sticker
Contains	725 g/l oil petroleum oils		
Use with	Soil-applied residual herbicides to improve coverage and speed penetration to the first 5 cm of soil		
Protective clothing	A, C, H		
Precautions	U05a, U11, U20c, E15b, E34, D01, D02, D10a, M03, M04a		
Herbi-AKtiv	Global Adjuvants	A0839	spreader/sticker/wetter
Contains	57.0 % w/w ethylene oxide-propylene oxide copolymers		
Use with	All authorised plant protection products at half or less than half the authorised plant protection product rate. Adj No updated due to expiry of A0821 July 2018		
Protective clothing	A, C, H, M		
Precautions	U05a, U20c, E34, D01, D02, D10a, M04a		
Holster	Greenaway	A0682	sticker/wetter
Contains	10.0 % w/w oil (rapeseed triglycerides)		
Use with	All approved herbicides on non-edible crops and on non-crop production and with all approved herbicides at half their approved dose on edible crops		
Protective clothing	A, C, H		
Precautions	U11, U12, U15, U20b, E15a, E34, D02		

SECTION 4

Product	Supplier	Adj. No.	Type
Innovator	Intracrop	A0935	adjuvant
Contains	521.5 g/l ethylene oxide-propylene oxide copolymers or 521.5 g/l ethylene oxide-propylene oxide copolymers and 8.6 g/l trisiloxane organosilicone copolymers		
Use with	All authorised plant protection products on all Non Edible Crops and Non Crop Production at 1% spray solution		
Protective clothing	A, C, H, M		
Precautions	U12, D12a		
Intake	FMC Agro	A0879	penetrant
Contains	450 g/l propionic acid		
Use with	All approved pesticides on any crop not intended for human or animal consumption, and with all approved pesticides on beans, peas, edible podded peas, oilseed rape, linseed, sugar beet, cereals (except triazole fungicides), maize, Brussels sprouts, potatoes, cauliflowers. See label for detailed advice on timing on these crops		
Protective clothing	A, C		
Precautions	R34, U02a, U05a, U10, U11, U14, U15, U19a, E34, D01, D02, D05, D09b, D10b, M04a, H05		
Intracrop Agwet GTX	Intracrop	A0646	spreader/wetter
Contains	500 g/l alkoxylated alcohols		
Use with	All approved pesticides on non-edible crops, all approved pesticides on edible crops at half or less than half the pesticide rate and all approved pesticides with a recommendation for use with a non-ionic or wetting agent.		
Protective clothing	A, C, H		
Precautions	H318, U05a, U11, U14, U15, U19a, U20a, M04c, H226, H302		
Intracrop BLA	Intracrop	A0655	anti-drift agent/anti-transpirant/extender/sticker/UV screen
Contains	22.0 % w/w styrene-butadiene copolymers		
Use with	All potato blight fungicides. Also with recommended rates of approved pesticides in certain non-crop situations and up to the growth stage indicated for specified crops, and with half or less than the recommended rate on these crops after the stated growth stages. Also with pesticides in grassland at half or less their recommended rate		
Protective clothing	A, C		
Precautions	U08, U20b, E13c, E34, E37, D01, D05, D09a, D10a		
Intracrop Dictate	Intracrop	A0673	adjuvant
Contains	91.0 % w/w oil (rapeseed fatty acid esters)		
Use with	All approved pesticides on non-edible crops; all approved pesticides applied at half or less than half dose on edible crops		
Protective clothing	A, C		
Precautions	U19a, U20b, D05, D09a, D10b		

Product	Supplier	Adj. No.	Type
Intracrop Evoque	Intracrop	A0754	spreader/wetter
Contains	652 g/l oil (rapeseed fatty acid esters), 112 g/l trisiloxane organosilicone copolymers, 22.5 g/l alkoxylated alcohols and 19.1 g/l alkoxylated alcohols		
Use with	All authorised plant protection products on non-edible crops and all authorised plant protection products at half or less than half the authorised plant protection product rate on edible crops. Max conc is 0.2% of spray volume.		
Protective clothing	A, C		
Precautions	U08, U20b, E13c, E34, E37, D01, D05, D09a, D10a		
Intracrop F16	Intracrop	A0752	spreader/wetter
Contains	652 g/l oil (rapeseed fatty acid esters), 112 g/l trisiloxane organosilicone copolymers, 22.5 g/l alkoxylated alcohols and 19.1 g/l alkoxylated alcohols		
Use with	All authorised plant protection products on non-edible crops and all authorised plant protection products at half or less than half the authorised plant protection product rate on edible crops. Max conc is 0.2% of spray volume.		
Protective clothing	A, C		
Precautions	U08, U20b, E13c, E34, E37, D01, D05, D09a, D10a		
Intracrop Mica AF	Intracrop	A0808	spreader/sticker/wetter
Contains	250 g/l ethylene oxide-propylene oxide copolymers, 70.2 g/l styrene-butadiene copolymers and 41.5 g/l trisiloxane organosilicone copolymers		
Use with	All approved pesticides at half or less than half the approved rate		
Protective clothing	A, C		
Precautions	U05a, U08, U20c, E13e, D01, D02, D09a, D10a		
Intracrop Predict	Intracrop	A0503	adjuvant/vegetable oil
Contains	91.0 % w/w oil (rapeseed fatty acid esters)		
Use with	All approved pesticides on non-edible crops, and all pesticides approved for use on growing edible crops when used at half recommended dose or less. On specified crops product may be used at a maximum spray concentration of 1% with approved pesticides at their full approved rate up to the growth stages shown in the label		
Protective clothing	A, C		
Precautions	U19a, U20b, D05, D09a, D10b		
Intracrop Quad	Intracrop	A0753	spreader/wetter
Contains	652 g/l oil (rapeseed fatty acid esters), 112 g/l trisiloxane organosilicone copolymers, 22.5 g/l alkoxylated alcohols and 19.1 g/l alkoxylated alcohols		
Use with	All approved pesticides on non-edible crops, and all pesticides approved for use on growing edible crops when used at half recommended dose or less.		
Protective clothing	A, C		
Precautions	U08, U20b, E13c, E34, E37, D01, D05, D09a, D10a		
Intracrop Rigger	Intracrop	A0783	adjuvant/vegetable oil
Contains	840 g/l oil (rapeseed fatty acid esters)		
Use with	All approved pesticides on non-edible crops, and all pesticides approved for use on growing edible crops when used at half recommended dose or less. On specified crops product may be used at a maximum spray concentration of 1.78% with approved pesticides at their full approved rate up to the growth stages shown in the label		
Protective clothing	A, C		
Precautions	U19a, U20b, D05, D09a, D10b		

SECTION 4

Product	Supplier	Adj. No.	Type
Intracrop Stay-Put	Intracrop	A0507	vegetable oil

Contains — 91.0 % w/w oil (rapeseed fatty acid esters)

Use with — Recommended rates of approved pesticides for non-crop uses and with recommended rates of approved pesticides up to the growth stage indicated for specified crops, and with half or less than the recommended rate on these crops after the stated growth stages

Protective clothing — A, C

Precautions — U19a, U20b, D05, D09a, D10b

Product	Supplier	Adj. No.	Type
Intracrop Super Rapeze MSO	Intracrop	A0782	adjuvant/vegetable oil

Contains — 840 g/l oil (rapeseed fatty acid esters)

Use with — All approved pesticides on non-edible crops, and all pesticides approved for use on growing edible crops when used at half recommended dose or less.   On specified crops product may be used at a maximum spray concentration of 1.78% with approved pesticides at their full approved rate up to the growth stages shown in the label

Protective clothing — A, C

Precautions — U19a, U20b, D05, D09a, D10b

Product	Supplier	Adj. No.	Type
Intracrop Vantage	Intracrop	A0719	spreader/wetter

Contains — 30.04 % w/w alkoxylated alcohols

Use with — All approved formulations of glyphosate and all approved plant protection products at half or less than half the approved rate on edible crops

Protective clothing — A, C, H

Precautions — H318, U05a, U11, U20b, E15a, H412, D01, D10a, M03, H302

Product	Supplier	Adj. No.	Type
Intracrop Zenith AF	Intracrop	A0811	spreader/sticker/wetter

Contains — 250 g/l ethylene oxide-propylene oxide copolymers, 70.2 g/l styrene-butadiene copolymers and 41.5 g/l trisiloxane organosilicone copolymers.

Use with — All approved pesticides on non-edible crops, all approved pesticides on edible crops at half or less than half the pesticide rate and all approved pesticides with a recommendation for use on non crop production.

Protective clothing — A, C, H

Precautions — H318, U05a, U11, U14, U15, U19a, U20a, M04c, H226, H302

Product	Supplier	Adj. No.	Type
Intracrop Zodiac	Intracrop	A0755	spreader/wetter

Contains — 652 g/l oil (rapeseed fatty acid esters), 112 g/l trisiloxane organosilicone copolymers), 22.5 g/l alkoxylated alcohols and 19.1 g/l alkoxylated alcohols

Use with — All authorised plant protection products on all non edible crops and all authorised plant protection products on edible crops at half or less than half the authorised plant protection product rate

Protective clothing — A, C

Precautions — U05a, U08, U20c, E13e, E40c, D01, D02, D09a, D10a

---

Product	Supplier	Adj. No.	Type
Kantor	Interagro	A0623	spreader/wetter

Contains — 790 g/l alkoxylated triglycerides

Use with — All approved pesticides

Protective clothing — A, C

Precautions — U05a, U20b, E15b, E19b, E34, D01, D02, D09a, D10a, D12a

Product	Supplier	Adj. No.	Type
Katalyst	Interagro	A0450	penetrant/water conditioner
Contains	90% w/w alkoxylated fatty amine		
Use with	Glyphosate and a wide range of other pesticides on specified crops. Refer to label or contact supplier for further details		
Protective clothing	A, C		
Precautions	R22a, R36, R38, R50, R58, U02a, U05a, U08, U19a, U20b, E15a, E34, E37, D01, D02, D05, D09a, D10a, M03, H03, H11		
Kotek	Agrovista	A0746	spreader/wetter
Contains	83.0% w/w trisiloxane organosilicone copolymers		
Use with	All approved pesticides at half or less than half the approved rate on edible crops; all approved pesticides on non-edible crops or non-production targets. Max conc 0.15% of spray volume		
Protective clothing	A, H		
Precautions	R20, R41, R51, R53a, U11		
Logic	Microcide	A0288	vegetable oil
Contains	95% w/w oil (rapeseed triglycerides)		
Use with	All approved pesticides on edible and non-edible crops for ground or aerial application		
Protective clothing	A, C		
Precautions	U19a, U20b, D09a, D10b		
Logic Oil	Microcide	A0630	vegetable oil
Contains	95% w/w oil (rapeseed triglycerides)		
Use with	All approved pesticides on edible and non-edible crops for ground or aerial application		
Protective clothing	A, C		
Precautions	U19a, U20b, D09a, D10b		
Master Wett	Global Adjuvants	A0826	spreader/sticker/wetter
Contains	84% trisiloxane organosilicone copolymers		
Use with	All approved pesticide products at a dose of 0.25% spray solution.		
Protective clothing	A, C, H		
Precautions	H319, E34, H411, D01, D02, D09a, D10a		
Meco-Flex	Greenaway	A0678	spreader/sticker/wetter
Contains	10.0% w/w oil (rapeseed triglycerides)		
Use with	2,4-D formulations on grassland and all authorised 490g/l glyphosate products on Natural Surfaces Not Intended To Bear Vegetation, Hard Surfaces, Permeable Surfaces Overlaying Soil and Forest		
Protective clothing	A, C		
Precautions	U11, U12, U15, U20b, E15a, E34, D02		
Mero	Bayer CropScience	A0818	wetter
Contains	81.4% w/w rapeseed fatty acid esters		
Use with	All approved cereal and maize herbicides and all approved brassica insecticides		
Protective clothing	A, C, H		
Precautions	R38, U05a, U09a, U13, U20b, E34, D01, D02, D09a, D10a, M05a		

SECTION 4

Product	Supplier	Adj. No.	Type
Mixture B NF	Amega	A0570	non-ionic surfactant/spreader/wetter
Contains	37% w/w alkoxylated alcohols and 41% w/w alkoxylated alcohols and isopropanol		
Use with	All approved pesticides on non-edible crops at their full recommended rate, all approved pesticides on edible crops at half their approved dose rate at 2% spray volume and on non-crop production situations at 0.2% spray volume		
Protective clothing	A, C, H		
Precautions	U02a, U04a, U05a, U11, U14, U15, E13a, E34, D01, D02, D09a, D10a, M03, H03, H11		
Nelson	Agrovista	A0796	spreader/wetter
Contains	300 g/l alkoxylated alcohols and 300 g/l alkoxylated alcohols		
Use with	All approved pesticides at half the approved maximum dose on edible crops and at the full approved dose on non-edible crops and non-crop production at 0.5% spray volume.		
Protective clothing	A, C, H		
Precautions	H318, U02a, U05a, U08, U09c, U11, U13, U14, U15, U20d, M03a, M05b, H302		
Newman Cropspray 11E	De Sangosse	A0863	adjuvant
Contains	99% w/w oil (petroleum oils)		
Use with	All approved pesticides on all edible crops when used at half their recommended dose or less, and on all non-edible crops up to their full recommended dose. Also with listed herbicides on a range of specified crops and with all pesticides on a range of specified crops up to specified growth stages.		
Protective clothing	A, C		
Precautions	R22a, U10, U16b, U19a, E15a, E19b, E34, E37, D01, D02, D05, D10a, D12a, M05b, H03		
Newman's C-80	De Sangosse	A0886	spreader/wetter
Contains	80% w/w alkoxylated coconut amines		
Use with	All approved pesticides on all edible crops when used at half their recommended dose or less, and on all non-edible crops up to their full recommended dose. Also with all authorised formulations of asulam applied to moorland, rough grazing and stubbles via aerial application		
Protective clothing	A, C, H		
Precautions	R36, U05a, U11, U14, U19a, U20b, E15a, E19b, E34, D01, D02, D05, D10a, D12a, H04		
Nu Film P	Intracrop	A0635	anti-drift agent/anti-transpirant/sticker/ UV screen/wetter
Contains	96.0 % w/w pinene oligomers		
Use with	Glyphosate and many other pesticides and growth regulators for which a protectant is recommended. Do not use in mixture with adjuvant oils or surfactants		
Protective clothing	A, C		
Precautions	U19a, U20b, E13c, D09a, D10a		

Product	Supplier	Adj. No.	Type
Pan Oasis	Pan Agriculture	A0411	vegetable oil
Contains	95% w/w methylated rapeseed oil		
Use with	A wide range of pesticides that have a label recommendation for use with authorised adjuvant oils. Contact distributor for further details		
Protective clothing	A		
Precautions	R36, U02a, U05a, U08, U20b, E15a, D01, D02, D05, D09a, D10a, H04		
Phase II	De Sangosse	A0622	vegetable oil
Contains	95.2% w/w oil (rapeseed fatty acid esters)		
Use with	Pesticides approved for use in sugar beet, oilseed rape, cereals (for grass weed control), and other specified agricultural and horticultural crops		
Protective clothing	A, C, H		
Precautions	U19a, U20b, E15a, E19b, E34, D01, D02, D05, D10a, D12a		
Pillion	Interagro	A0882	spreader/wetter
Contains	197 g/l alkoxylated alcohols		
Use with	All approved pesticides on edible crops at half their recommended dose and all approved pesticides on non-edible crops at their full dose		
Protective clothing	A, C, H		
Precautions	H315, H318		
Predict Plus	Intracrop	A0894	adjuvant/sticker
Contains	747.8 g/l oil (rapeseed fatty acid esters) and 99.64 g/l rosin and resin esters		
Use with	All approved pesticides on non-edible crops, and all pesticides approved for use on growing edible crops when used at half recommended dose or less.		
Protective clothing	A, C, H		
Precautions	U04b, U05a, U11, U13, U14, U15, U19a, E15a, D01, D09a, M03a		
Probe	Life Scientific	A0874	wetter
Contains	276.5 g/l alkyl ether sulphates, sodium salts		
Use with	All authorised plant protection products on cereals at 1% solution and all authorised herbicides at 1% solution on Ornamental Plant Production, Amenity Vegetation, Natural Surfaces Not Intended To Bear Vegetation, Permeable Surfaces Overlying Soil, Industrial And Amenity Areas		
Protective clothing	A, H		
Precautions	H318, U05a, U11, U14, U15, U19a, U20a, M04c, H226, H302		
Profit	Microcide	A0289	sticker/wetter
Contains	95.0 % w/w oil (rapeseed triglycerides)		
Use with	All approved pesticides		
Protective clothing	A, H		
Precautions	U19a, U20b, D09a, D10b		
Profit Oil	Microcide	A0631	extender/sticker/wetter
Contains	95% w/w oil (rapeseed triglycerides)		
Use with	All approved pesticides		
Protective clothing	A, C		
Precautions	U19a, U20b, D09a, D10b		

SECTION 4

Product	Supplier	Adj. No.	Type
Pryz-Flex	Greenaway	A0679	spreader/sticker/wetter
Contains	10.0% w/w oil (rapeseed triglycerides)		
Use with	Kerb Flo (MAPP 13716 or 15586) and all approved 490 g/l glyphosate products on permeable surfaces overlying soil		
Protective clothing	A, C		
Precautions	U11, U12, U15, U20b, E34, E37		
Quartz	Intracrop	A0898	spreader/sticker/wetter
Contains	32.67 % w/w alkoxylated alcohols and 1.0% w/w trisiloxane organosilicone copolymers		
Use with	All approved pesticides on non-edible crops, and all pesticides approved for use on growing edible crops when used at half recommended dose or less.		
Protective clothing	A, C		
Precautions	U08, U20b, E13c, E34, E37, D01, D05, D09a, D10a		
Remix	Agrovista	A0916	wetter
Contains	732 g/l petroleum (paraffin) oil		
Use with	All approved pesticides at half or less than half the approved rate on edible crops; all approved pesticides on non-edible crops or non-production targets. Can also be used with all approved pesticides pre-emergence on bulb and stem vegetables, legumes, carrots, parsnips, swedes and turnips. Max conc for all uses is 1% of spray solution.		
Protective clothing	A, C, H		
Precautions	U02a, U05a, U19a, E15a, D01, D02, D05, D09a, D10a, D12a		
Renu	AgriMax	A0627	spreader/wetter
Contains	300 g/l alkoxylated alcohols		
Use with	All plant protection products on all edible crops and all authorised herbicides on managed amenity turf at a maximum concentration of 0.5% spray volume		
Protective clothing			
Precautions			
Renu DS	AgriMax	A0926	spreader/wetter
Contains	600 g/l alkoxylated alcohols		
Use with	All edible crops, all non-edible crops and all non-crop productions at a maximum concentration of 0.5% spray solution		
Protective clothing	A, E		
Precautions	H315, H318, U05a, U14, U15, U19a, U20c, E15a, H412, D01, D09b, M05a, H05, H302, H303		
Retainer	Intracrop	A0711	wetter
Contains	91.0 % w/w oil (rapeseed fatty acid esters)		
Use with	All approved pesticides on non-edible crops, and all pesticides approved for use on growing edible crops when used at half recommended dose or less at 1% spray solution		
Protective clothing	A, C, H, M		
Precautions	H318, H412		
Reward Oil	Microcide	A0632	extender/sticker/wetter
Contains	95% w/w oil (rapeseed triglycerides)		
Use with	All approved pesticides		
Protective clothing	A, C		
Precautions	U19a, U20b, D09a, D10b		

Product	Supplier	Adj. No.	Type
Rheus	FMC Agro	A0883	wetter
Contains	85% w/w polyalkylene oxide modified heptamethyl siloxane		
Use with	Any herbicide, systemic fungicide, systemic insecticide or plant growth regulator (except any product applied in or near water) where the use of a wetting, spreading and penetrating surfactant is recommended to improve foliar coverage		
Protective clothing	A, C		
Precautions	R21, R22a, R38, R41, R43, R51, R58, U02a, U05a, U08, U11, U14, U19a, U20b, E13c, E34, D01, D02, D05, D09a, D10b, M03, H03, H11		
Roller	Agrovista	A0748	spreader/wetter
Contains	832 g/l ethylene oxide-propylene oxide copolymers and 169.3 g/l trisiloxane organosilicone copolymers		
Use with	All approved pesticides up to their maximum dose at a max conc of 0.2% v/v.		
Protective clothing	A, C, H		
Precautions	R36, R38, R41, R52, R53a, U02a, U05a, U08, U11, U20b, E15a, E34, E40c, D01, D02, D05, D09a, D10a, D12b, H04		
SAS 90	Intracrop	A0740	spreader/wetter
Contains	83% w/w trisiloxane organosilicone copolymers		
Use with	All approved herbicides at a maximum concentration of 0.025% spray volume and all approved fungicides and insecticides at a maximum concentration of 0.05% spray volume		
Protective clothing	A, C		
Precautions	U02a, U08, U19a, U20b, E13c, E34, D09a, D10a		
Saturn Plus	Intracrop	A0813	spreader/wetter
Contains	70.0% w/w ethylene oxide-propylene oxide copolymers		
Use with	All approved pesticides on non-edible crops, and all pesticides approved for use on growing edible crops when used at half recommended dose or less.		
Protective clothing	A, C		
Precautions	U19a, U20c, E13c, C02a, D09a, D11a		
Silwet L-77	De Sangosse	A0640	drift retardant/spreader/wetter
Contains	80% w/w trisiloxane organosilicone copolymers		
Use with	All approved fungicides on winter and spring sown cereals; all approved pesticides applied at 50% or less of their full approved dose. A wide range of other uses. See label or contact supplier for details		
Protective clothing	A, C, H		
Precautions	R20, R21, R22a, R36, R43, R48, R51, R58, U11, U15, U19a, E15a, E19b, D01, D02, D05, D10a, D12a, H03, H11		
Slither	De Sangosse	A0458	wetter
Contains	80.0% w/w trisiloxane organosilicone copolymers		
Use with	All approved pesticides on all edible crops when used at half their recommended dose or less, and on all non-edible crops up to their full recommended dose at 0.15% spray solution		
Protective clothing	A, C, H		
Precautions	R20, R21, R22a, R36, R43, R48, R51, R58, U11, U15, U19a, E15a, E19b, D01, D02, D05, D10a, D12a, H03, H11		

SECTION 4

Product	Supplier	Adj. No.	Type
Solar Plus	Intracrop	A0802	spreader/wetter
Contains	70.0% w/w ethylene oxide-propylene oxide copolymers		
Use with	All approved pesticides on non-edible crops, and all pesticides approved for use on growing edible crops when used at half recommended dose or less.		
Protective clothing	A, C		
Precautions	R36, R38, U04a, U05a, U08, U20a, E13c, E34, D01, D02, D09a, D10b, D11a, M03, H04		
Sorrento	Interagro	A0885	wetter
Contains	50.0 % w/w alkoxylated alcohols		
Use with	All approved pesticides on edible crops at half their recommended dose and all approved pesticides on non-edible crops at their full dose		
Protective clothing	A, C, H		
Precautions	H315, H318		
SP058	De Sangosse	A0793	wetter
Contains	80.0% w/w trisiloxane organosilicone copolymers		
Use with	All approved pesticides on all edible crops when used at half their recommended dose or less, and on all non-edible crops up to their full recommended dose at 0.2% spray solution		
Protective clothing	A, C, H		
Precautions	R20, R21, R22a, R36, R43, R48, R51, R58, U11, U15, U19a, E15a, E19b, D01, D02, D05, D10a, D12a, H03, H11		
Speedway Total	ICL (Everris) Ltd	A0820	extender/spreader/wetter
Contains	41.0 % w/w alkoxylated alcohols and 37.0 % w/w alkoxylated alcohols		
Use with	All authorised plant protection products on non-edible crops and non-crop production uses; all authorised plant protection products at half or less than half the authorised plant protection product rate on edible crops.		
Protective clothing	A		
Precautions	U02a, U04a, U05a, U14, U15, U19a, E13b, E34, D01, D02, D09a, D10a, M03		
Spray-fix	De Sangosse	A0559	extender/sticker/wetter
Contains	45% w/w styrene-butadiene copolymers and 10% w/w alkoxylated alcohols		
Use with	All approved potato blight fungicides. Also with all approved pesticides on all edible crops when used at half their recommended dose or less, and on all non-edible crops up their full recommended dose. Also at a maximum concentration of 0.14% with approved pesticides on listed crops up to specified growth stages		
Protective clothing	A, C, H		
Precautions	R36, R38, U11, U14, U16b, U19a, E15a, E19b, D01, D02, D05, D10a, D12a, H04		
Spraymac	De Sangosse	A0549	acidifier/non-ionic surfactant
Contains	100 g/l alkoxylated alcohols and 350 g/l propionic acid		
Use with	All approved pesticides on all edible crops when used at half their recommended dose or less, and on all non-edible crops up to their full recommended dose. Also at a maximum concentration of 0.5% with approved pesticides on listed crops up to specified growth stages		
Protective clothing	A, C, H		
Precautions	R34, U04a, U10, U11, U14, U19a, U20b, E15a, E19b, E34, D01, D02, D05, D10a, D12a, M04a, H05		

Product	Supplier	Adj. No.	Type
Spraymate	De Sangosse	A0621	
Contains	92.5% w/w oil(rapeseed fatty acid esters)		
Use with	All approved pesticides at half or less than half the approved pesticide rate at 1.78% spray solution on edible crops; 0.75% spray solution on potatoes in mix with pyraflufen-ethyl (H I 14 days)		
Protective clothing	A, C, H		
Precautions	D01, D02, D05, D10a, D12a, E15a, E19b, E34, H04, R36, U05a, U11, U14, U19a, U20b		
Spryte Aqua	Amega	A0857	spreader/wetter
Contains	300 g/l alkoxylated alcohols		
Use with	All edible crops, all non-edible crops and all non-crop productions at a maximum concentration of 0.5% spray solution		
Protective clothing	A, E		
Precautions	H315, U05a, U14, U15, U19a, U20c, E15a, D01, D09b, M05a, H05, H302, H303		
Spryte Aqua CF	AgriMax	A0859	spreader/wetter
Contains	600 g/l alkoxylated alcohols		
Use with	All edible crops, all non-edible crops and all non-crop productions at a maximum concentration of 0.25% spray solution		
Protective clothing	A, E		
Precautions	H315, H318, U05a, U14, U15, U19a, U20c, E15a, H412, D01, D09b, M05a, H05, H302, H303		
Squadron	Intracrop	A0913	spreader/wetter
Contains	70.0 % w/w ethylene oxide-propylene oxide copolymers		
Use with	All authorised plant protection products on edible crops at half the recommended dose rate and with all authorised plant protection products on non-edible crops and non-crop situations at 0.5% v/v		
Protective clothing	A, C, H		
Precautions	H318		
Stika	De Sangosse	A0557	extender/sticker/wetter
Contains	22.5% w/w styrene-butadione copolymers and 10% w/w alkoxylated alcohols		
Use with	All approved potato blight fungicides. Also with all approved pesticides on all edible crops when used at half their recommended dose or less, and on all non-edible crops up to their full recommended dose. Also at a maximum concentration of 0.14% with approved pesticides on listed crops up to specified growth stages		
Protective clothing	A, C, H		
Precautions	R36, R38, U11, U14, U16b, U19a, E15a, E19b, D01, D02, D05, D10a, D12a, H04		
SU Wett	Global Adjuvants	A0840	spreader/wetter
Contains	57.0 % w/w ethylene oxide-propylene oxide copolymers		
Use with	All authorised plant protection products at half or less than half the authorised plant protection product rate. Adj No updated due to expiry of A0814 July 2018		
Protective clothing	A, C, H, M		
Precautions	U05a, U20c, E34, D01, D02, D10a, M04a		

SECTION 4

Product	Supplier	Adj. No.	Type
Surfer	Dow (Corteva)	A0800	wetter
Contains	72.0 % w/w alkoxylated alcohols		
Use with	All approved pesticides on all edible crops when used at half their recommended dose or less, and on all non-edible crops up to their full recommended dose.		
Protective clothing	A, C, H		
Precautions	H318, U05a, U11, U14, U15, U20a, E15a, E34, D01, D02, D09a, D10a		
SW7	Omex	A0875	spreader/sticker
Contains	18.0 % w/w alkoxylated alcohols and 8.0 % w/w alkoxylated alcohols		
Use with	All authorised plant protection products at half or less than half the authorised product rate on edible crops and up to full rate on non-edible crops and non-crop situations		
Protective clothing	A, C, H		
Precautions	H315, U11, U12, H410, D06f, M04c, H226, H332		
Tempest	Intracrop	A0887	spreader/wetter
Contains	301 g/l alkoxylated alcohols, 299 g/l alkoxylated alcohols and 18.36 g/l trisiloxane organosilicone copolymers		
Use with	All approved pesticides at a maximum concentration of 0.5% v/v		
Protective clothing	A, C, H		
Precautions	H318, U05a, U09b, U20d, D12b, M03, M05b, H302		
TM 1008	Intracrop	A0675	wetter
Contains	750 g/l ethylene oxide-propylene oxide copolymers		
Use with	All approved pesticides on non-edible crops and non crop production uses; All approved pesticides at half or less than half the approved pesticide rate on edible crops		
Protective clothing	A, C		
Precautions	R36, R38, U04a, U05a, U08, U19a, U20b, E13b, E34, D01, D02, D09a, D10a, M03, M05a, H04		
Toil	Interagro	A0907	vegetable oil
Contains	95% w/w methylated rapeseed oil		
Use with	Sugar beet herbicides, oilseed rape herbicides, cereal graminicides and a wide range of other pesticides that have a label recommendation for use with authorised adjuvant oils on specified crops. Refer to label or contact supplier for further details. Adj No updated June 2021		
Protective clothing	A		
Precautions	U05a, U20b, E15a, D01, D02, D05, D09a, D10a		
Tonto	Intracrop	A0900	spreader/sticker/wetter
Contains	32.67 % w/w alkoxylated alcohols and 1.0% w/w trisiloxane organosilicone copolymers		
Use with	All approved pesticides at half or less than half the approved pesticide rate on edible crops and all approved pesticides on non-edible crops		
Protective clothing	A, C		
Precautions	U05a, U08, U20c, E13e, E40c, D01, D02, D09a, D10a		

Product	Supplier	Adj. No.	Type
Torpedo II	De Sangosse	A0892	spreader/wetter
Contains	21.0 % w/w alkoxylated oleyl amines, 19.0 % w/w alkoxylated alcohols, 19.0 % w/w alkoxylated alcohols and 7.5 % w/w oil (tall oil fatty acids)		
Use with	All approved pesticides on all edible crops when used at half their recommended dose or less, and on all non-edible crops up to their full recommended dose. Also at a maximum concentration of 0.1% with approved pesticides on listed crops up to specified growth stages		
Protective clothing	A, C, H		
Precautions	R36, R38, R53a, U02a, U04a, U05a, U10, U11, U20b, E15a, E19b, D01, D02, D05, D10a, H04		
Transact	Agrii	A0880	penetrant
Contains	40% w/w propionic acid		
Use with	Plant growth regulators in cereals, all approved pesticides in non-edible crops, all approved pesticides on edible crops when used at half or less their recommended rate, and with all pesticides on listed crops up to specified growth stages		
Protective clothing	A, C		
Precautions	R34, U10, U11, U14, U15, U19a, D01, D05, D09b, M04a, H05		
Validate	De Sangosse	A0500	adjuvant/wetter
Contains	50% w/w soybean phospholipids, 25% w/w alkoxylated alcohols and 25% w/w oil (soybean fatty acid esters)		
Use with	Approved pesticides on non-edible crops where the addition of a wetter/spreader or adjuvant oil is recommended on the pesticide label, and pesticides approved for use on growing edible crops when used at half recommended dose or less. On specified crops product may be used at a maximum spray concentration of 0.5% with approved pesticides at their full approved rate up to the growth stages shown in the label		
Protective clothing	A, C, H		
Precautions	R51, R53a, U05a, U11, U14, U19a, U20b, E15a, E19b, E34, D01, D02, D10a, D12a, H11		
Velocity	Agrovista	A0697	adjuvant
Contains	745g/l oil (rapeseed fatty acid esters) and 103g/l trisiloxane organosilicone copolymers.		
Use with	All approved pesticides up to their full approved rate at a maximum conc of 0.5% vol/vol.		
Protective clothing	A, C		
Precautions	R41, R52, R53a, U02a, U05a, U08, U11, U14, U15, U20b, E13c, E15a, E34, E40c, D01, D02, D09a, D10a, H04		
Velomax	De Sangosse	A0831	wetter
Contains	86.8% w/w oil (rapeseed fatty acid esters), 5.2% w/w alkoxylated alcohols and 2.5% w/w oil (tall oil fatty acids)		
Use with	All approved pesticides on all edible crops when used at half their recommended dose or less, and on all non-edible crops up to their full recommended dose at 1% spray solution		
Protective clothing	A, C, H		
Precautions	R20, R21, R22a, R36, R43, R48, R51, R58, U11, U15, U19a, E15a, E19b, D01, D02, D05, D10a, D12a, H03, H11		

SECTION 4

Product	Supplier	Adj. No.	Type
Velvet	Global Adjuvants	A0814	spreader/sticker/wetter
Contains	83% trisiloxane organosilicone copolymers		
Use with	All approved pesticides on all edible crops when used at half their recommended dose or less, and on all non-edible crops up to their full recommended dose.		
Protective clothing	A, C, H		
Precautions	H319, E34, H411, D01, D02, D09a, D10a		
Verdant	Agrovista	A0714	wetter
Contains	8% w/w alkoxylated alcohol, 63% w/w alkoxylated alkyl polyglucoside and 10% w/w alkyl polyglucoside		
Use with	All approved pesticides at 50% or less on edible crops and with all approved pesticides on non-edible crops, non-crop production and stubbles of edible crops at 0.25% spray sloution.		
Protective clothing	A, C, H		
Precautions	R38, R41, U02a, U05a, U11, U14, U15, U20b, E15a, E34, D02, D05, D09a, D10a, H04		
Vindicator	Intracrop	A0914	buffering agent/spreader/water conditioner/wetter
Contains	301 g/l alkoxylated alcohols, 299 g/l alkoxylated alcohols and 18.36 g/l trisiloxane organosilicone copolymers		
Use with	All authorised plant protection products on edible crops at half the recommended dose rate and with all authorised plant protection products on non-edible crops and non-crop situations at 0.5% v/v		
Protective clothing	A, C, H		
Precautions	H318, U02a, U11, E15a, D01, M03, H302		
Wetcit	Oro Agri	A0893	penetrant/wetter
Contains	8.15% w/w alcohol ethoxylate		
Use with	All approved pesticides on all edible crops when used at half their recommended dose or less, and on all non-edible crops up to their full recommended dose. Also at a maximum concentration of 0.25% with approved pesticides on listed crops up to specified growth stages at full rate, and at half rate thereafter		
Protective clothing	A, C, H		
Precautions	R22a, R38, R41, U08, U11, U19a, U20b, E13c, E15a, E34, E37, D05, D08, D09a, D10b, H03		
Zarado	De Sangosse	A0516	vegetable oil
Contains	70% w/w oil (rapeseed oil fatty acid esters)		
Use with	All approved pesticides on edible and non-edible crops when used at half their approved dose or less. Also with approved pesticides on specified crops, up to specified growth stages, at up to their full approved dose		
Protective clothing	A, C, H		
Precautions	R43, R52, U05a, U13, U19a, U20b, E15a, E19b, E34, D01, D02, D05, D10a, M05a, H04		

Product	Supplier	Adj. No.	Type
Zeal	Interagro	A0685	water conditioner/wetter
Contains	60.0 % w/w alkoxylated tallow amines and 37.0 % w/w alkoxylated sorbitan esters		
Use with	Trace elements such as calcium, copper, manganese and sulphur and with macronutrients such as phosphates		
Protective clothing	A, C, H		
Precautions	R22a, R36, R38, R51, R53a, U05a, U09a, U11, U14, U15, U19a, E15b, E34, D01, D02, D09a, D10a, M03, H03, H11		
Zinzan	Certis Belchim	A0600	extender/spreader/wetter
Contains	70% w/w 1,2 bis (2-ethylhexyloxycarbonyl) ethanesulphonate		
Use with	All approved pesticides on non-edible crops, and all pesticides approved for use on growing edible crops when used at half recommended dose or less. On specified edible crops product may be used at a maximum spray concentration of 0.075% with approved pesticides at their full approved rate up to the growth stages shown in the label		
Protective clothing	A, C, P		
Precautions	R38, R41, U11, U14, U20c, D01, D09a, D10a, H04		

SECTION 4

SECTION 5
USEFUL INFORMATION

Pesticide Legislation

Anyone who advertises, sells, supplies, stores or uses a pesticide is bound by legislation, including those who use pesticides in their own homes, gardens or allotments. As the UK has left the EU so European Directives no longer apply to the UK but CRD has issued a statement that EU legislation on pesticides will be mirrored in new UK legislation. The details are not yet available but what applied under EU law will apply under future UK laws. The current EU and UK laws are described below.

Regulation 1107/2009 — The Replacement for EU 91/414

Regulation 1107/2009 entered into force within the EU on 14 Dec 2009 and was applied to new approval applications from 14th June 2011. Since that date all pesticides that held a current approval under 91/414 have been deemed to be approved under 1107/2009 and the new criteria for approval will only be applied when the active substance comes up for review. The new regulation largely mirrors the previous regulation but requires additional approval criteria to be met to weed out the more hazardous chemicals. These include assessing the effect on vulnerable groups, taking into account known synergistic and cumulative effects, considering the effect on coastal and estuarine waters, effects on the behaviour of non-target organisms, biodiversity and the ecosystem. However, these effects will only be evaluated when scientific methods approved by EFSA (European Food Safety Authority) have been developed.

The additional criteria will require that no CMR's (substances which are Carcinogens, Mutagens or toxic to Reproduction), no endocrine disruptors and no POP (Persistent Organic Pollutants), PBT (chemicals that are Persistent, Bioaccumulative or Toxic) or vPVB (very Persistent and very Bioaccumulative) will gain approval unless the exposure is negligible. However, note that for EFSA the term 'negligible' has yet to receive a precise definition. A list of 66 chemicals that pose an endocrine disruptor risk has been produced. It contained maneb and vinclozolin and both actives have now been withdrawn.

If an active substance passes these hurdles, the first approval for basic substances will be granted for 10 years (15 years for 'low risk' substances, 7 years for candidates for substitution – see below) and then at each subsequent renewal a further 15 years will be granted. Actives will be divided between 5 categories – low risk substances (e.g. pheromones, semiochemicals, micro-organisms and natural plant extracts), basic substances, candidates for substitution and then safeners/synergists and co-formulants. Candidates for substitution will be those chemicals which have just scraped past the approval thresholds but are deemed to still carry some degree of hazard. They will be withdrawn when significantly safer alternatives are available. This includes physical methods of control but any alternative must not have significant economic or practical disadvantages, must minimise the risk of the development of resistance and the consequences for any 'minor uses' of the original product must also be considered.

The Water Framework Directive

This became law in the UK in 2003 but is being introduced over a number of years to allow systems to acclimatise to the new rules and required member states of the EU to achieve 'good status' in ground and surface water bodies by 22 December 2015 with reviews every six years thereafter. To achieve 'good status' as far as pesticides are concerned the concentration of any individual pesticide must not exceed 0.1 micrograms per litre and the total quantity of all pesticides must not exceed 0.5 micrograms per litre. These levels are not based on the toxicity of the pesticides but on the level of detection that can be reliably achieved. Failure to meet these limits could jeopardise the approval of the problem active ingredients and so every effort must be made to avoid contamination of surface and ground water bodies with pesticides. The pesticides that are currently being found in water samples above the 0.1 ppb limit are metaldehyde, carbetamide, propyzamide, MCPA, mecoprop, glyphosate, bentazone, clopyralid, metazachlor, chlorotoluron and ethofumesate. These actives are identified as a risk to water in the pesticide profiles in section 2 and great care should be taken when applying them, particularly metaldehyde which is almost impossible to remove from water at treatment plants.

Classification, Labelling and Packaging (CLP) Regulation

The CLP Regulation (EC No 1272/2008) governs the classification of substances and mixtures and is intended to introduce the United Nations Globally Harmonised System (GHS) of classification to Europe. This new system and the hazard icons that appear on pesticide labels is very similar to the old system with the most obvious difference being that the hazard icons have a white background rather than an orange one. The new system of classification became mandatory for new supplies from 1st June 2015 in the UK but for products already in the supply chain the compliance date was deferred to 1st June 2017. Since that date all products placed in the market MUST comply with the CLP Regulation. The biggest change was the replacement of the old HARMFUL icon **'X'** with a new one **'!'** to notify that caution is required when handling or using a product. This brings Europe and the UK in line with the rest of the world. Details of the new phrases are given in Appendix 4.

The Food and Environment Protection Act 1986 (FEPA)

FEPA introduced statutory powers to control pesticides with the aims of protecting human beings, creatures and plants, safeguarding the environment, ensuring safe, effective and humane methods of controlling pests and making pesticide information available to the public. This was supplemented by Control of Pesticides Regulations 1986 (COPR) which has now been replaced by EC Regulation 1107/2009 – see above. Details are given on the websites of the Chemicals Regulation Division (CRD) and the Health and Safety Executive (HSE). Together these regulations mean that:

- Only approved products may be sold, supplied, stored, advertised or used.
- No advertisement may contain any claim for safety beyond that which is permitted in the approved label text.
- Only approved products may be sold, supplied, stored, advertised or used.
- No advertisement may contain any claim for safety beyond that which is permitted in the approved label text.
- Only products specifically approved for the purpose may be applied from the air.
- A recognised Storeman's Certificate of Competence is required by anyone who stores for sale or supply pesticides approved for agricultural use.
- A recognised Certificate of Competence is required by anyone who gives advice when selling or supplying pesticides approved for agricultural use.
- Users of pesticides must comply with the Conditions of Approval relating to use.
- A recognised Certificate of Competence is required for all contractors and persons applying pesticides approved for agricultural use (unless working under direct supervision of a certificate holder).
- Only those adjuvants authorised by CRD may be used.
- For tank-mixes of convenience, when the purpose is to reduce the number of passes in the crop, no efficacy data is required but physical and chemical compatibility data are required in line with EU guidance. For mixtures that claim positive benefits efficacy data remains necessary in addition to the physical and chemical compatibility data. Note however that ALS herbicides (HRAC mode of action code 'B') may only be mixed or applied in sequence with other ALS herbicides when a label statement permits this. Mixtures of anticholinesterase products are not permitted unless CRD has fully assessed the toxicological effects.

Dangerous Preparations Directive (1999/45/EC)

The Dangerous Preparations Directive (DPD) came into force in the UK for pesticide and biocidal products on 30 July 2004. Its aim is to achieve a uniform approach across all Member States to the classification, packaging and labelling of most dangerous preparations, including crop protection products. The Directive is implemented in the UK under the Chemicals (Hazard Information and Packaging for Supply) Regulations 2002, often referred to under the acronym CHIP3. In most cases the Regulations have led to additional hazard symbols, and associated risk and safety phrases relating to environmental and health hazards, appearing on the label. All products affected by the DPD entering the supply chain from the implementation date above must be so labelled. The new environmental hazard classifications and risk phrases are included.

Under the DPD, the labels of all plant protection products now state 'To avoid risks to man and the environment, comply with the instructions for use'. As the instruction applies to all products, it has not been repeated in each fact sheet.

The Review Programme

The review programme of existing active substances will continue under EC Directive 1107/2009. The programme is designed to ensure that all available plant protection products are supported by up-to-date information on safety and efficacy. As this will now be under 1107/2009, the new standards for approval will be applied at the due date and any that are deemed to be hazardous chemicals will be withdrawn. Any that are just within the safety standards yet to be defined will be called 'candidates for substitution' and granted only a seven-year renewal to allow time for significantly safer alternatives to be developed. If these are available at the time of renewal, the approval will be withdrawn. Chemicals deemed to be low risk, such as natural plant extracts, microorganisms, semiochemicals and pheromones, will be renewed for 15 years, while most active substances will be renewed for a further 10 years.

Control of Substances Hazardous to Health Regulations 1988 (COSHH)

The COSHH regulations, which came into force on 1 October 1989, were made under the Health and Safety at Work Act 1974, and are also important as a means of regulating the use of pesticides. The regulations cover virtually all substances hazardous to health, including those pesticides classed as Very Toxic, Toxic, Harmful, Irritant or Corrosive, other chemicals used in farming or industry, and substances with occupational exposure limits. They also cover harmful micro-organisms, dusts and any other material, mixture, or compound used at work which can harm people's health.

The original Regulations, together with all subsequent amendments, have been consolidated into a single set of regulations: The Control of Substances Hazardous to Health Regulations 1994 (COSHH 1994).

The basic principle underlying the COSHH regulations is that the risks associated with the use of any substance hazardous to health must be assessed before it is used, and the appropriate measures taken to control the risk. The emphasis is changed from that pertaining under the Poisonous Substances in Agriculture Regulations 1984 (now repealed) – whereby the principal method of ensuring safety was the use of protective clothing – to the prevention or control of exposure to hazardous substances by a combination of measures. In order of preference the measures should be:

(a) substitution with a less hazardous chemical or product
(b) technical or engineering controls (e.g. the use of closed handling systems, etc.)
(c) operational controls (e.g. operators located in cabs fitted with air-filtration systems, etc.)
(d) use of personal protective equipment (PPE), which includes protective clothing.

Consideration must be given as to whether it is necessary to use a pesticide at all in a given situation and, if so, the product posing the least risk to humans, animals and the environment must be selected. Where other measures do not provide adequate control of exposure and the use of PPE is necessary, the items stipulated on the product label must be used as a minimum. It is essential that equipment is properly maintained and the correct procedures adopted. Where necessary, the exposure of workers must be monitored, health checks carried out, and employees instructed and trained in precautionary techniques. Adequate records of all operations involving pesticide application must be made and retained for at least 3 years.

Biocidal Products Regulations

These regulations concern disinfectants, preservatives and pest control products but only the latter are listed in this book. The regulation aims to harmonise the European market for biocidal products, to provide a high level of safety for humans, animals and the environment and to ensure that the products are effective against the target organisms. Products that have been reviewed and included under these regulations are issued with a new number of the format UK-2012-xxxx and these are summarised in this book as UK??-xxxx.

SECTION 5

Certificates of Competence – the roles of BASIS and NPTC

COPR, COSHH and other legislation places certain obligations on those who handle and use pesticides. Minimum standards are laid down for the transport, storage and use of pesticides, and the law requires those who act as storekeepers, sellers and advisors to hold recognised Certificates of Competence.

BASIS

BASIS is an independent Registration Scheme for the pesticide industry. It is responsible for organising training courses and examinations to enable such staff to obtain a Certificate of Competence.

In addition, BASIS undertakes an annual assessment of pesticide supply stores, enabling distributors, contractors and seedsmen to meet their obligations under the Code of Practice for Suppliers of Pesticides. Further information can be obtained from BASIS. Since 26 November 2015 all businesses must have sufficient number of staff with BASIS certificates to advise customers at the time of sale where previously it was just at the point of sale.

Certificates of Competence

Storage

- BASIS Certificate of Competence in the Storage and Handling of Crop Protection Products

Sale and supply

- BASIS Certificate in Crop Protection (Agriculture)
- BASIS Certificate in Crop Protection (Commercial Horticulture)
- BASIS Certificate in Crop Protection (Amenity Horticulture)
- BASIS Certificate in Crop Protection (Forestry)
- BASIS Certificate in Crop Protection (Seed Treatment)
- BASIS Certificate in Crop Protection (Seed Sellers)
- BASIS Certificate in Crop Protection (Field Vegetables)
- BASIS Certificate in Crop Protection (Potatoes)
- BASIS Certificate in Crop Protection (Aquatic)
- BASIS Certificate in Crop Protection (Indoor Landscaping)
- BASIS Certificate in Crop Protection (Grassland and Forage Crops)
- BASIS/LEAF ICM Certificate
- BASIS Advanced Certificate
- FACTS (Fertiliser Advisers Certification and Training Scheme) Certificate

NPTC

Certain spray operators also require certificates of competence under the Control of Pesticides Regulations. NPTC's Pesticides Award is a recognised Certificate of Competence under COPR and it is aimed principally at those people who use pesticide products approved for use in agriculture, horticulture (including amenity horticulture) and forestry. All contractors must possess a certificate if they spray such products unless they are working under the direct supervision of a certificate holder.

Because the required spraying skills vary widely among the uses listed above, candidates are assessed under one (or more) modules that are most appropriate for their professional work. Assessments are carried out by an approved NPTC or Scottish Skills Testing Service Assessor. All candidates must first complete a foundation module (PA1), for which a certificate is not issued, before taking one of the specialist modules. Certificate holders who change their work so that a different specialist module becomes more appropriate may need to obtain a new certificate under the new module.

Holders are required to produce on demand their Certificate of Competence for inspection to any authorised person. Further information can be obtained from NPTC.

Certificates of Competence for spray operators

- PA1 Foundation Module
- PA2 Ground Crop Sprayers – mounted or trailed
- PA3 Broadcast Air Blast Sprayer. Variable Geometry Boom Air Assisted Sprayer
- PA4 Granule Applicator – Mounted or Trailed
- PA5 Boat Mounted Applicators
- PA6 Hand Held Applicators
- PA7 Aerial Application
- PA8 Mixer/Loader
- PA9 Fogging, Misting and Smokes
- PA10 Dipping Bulbs, Corms, Plant Material or Containers
- PA11 Seed Treating Equipment
- PA12 Application of Pesticides to Material as a Continuous or Batch Process

Maximum Residue Levels

A small number of pesticides are liable to leave residues in foodstuffs, even when used correctly. Where residues can occur, statutory limits, known as maximum residue levels (MRLs), have been established. MRLs provide a check that products have been used as directed; **they are not safety limits**. However, they do take account of consumer safety because they are set at levels that ensure normal dietary intake of residues presents no risk to health. Wide safety margins are built in, and eating food containing residues above the MRL does not automatically imply a risk to health. Nevertheless, it is an offence to put into circulation any produce where the MRL is exceeded.

The surrounding legislation is complex. MRLs may be specified by several different bodies. The UK has set statutory MRLs since 1988. The European Union intends eventually to introduce MRLs for all pesticide/commodity combinations. These are being introduced initially by a series of priority lists but will subsequently be covered by the review programme under EC Regulation 1107/2009. However, in cases where no information is available, EU Regulation 2000/42/EC requires many MRLs to be set at the limit of determination (LOD). As a result, certain approvals could leave residues above the MRL set in the Directive.

MRLs apply to imported as well as home-produced foodstuffs. Details of those that have been set have been published in *The Pesticides (Maximum Residue Levels in Crops, Food and Feeding Stuffs) Regulations 1994*, and successive amendments to these Regulations. These Statutory Instruments are available from The Stationery Office (www.tsoshop.co.uk).

MRLs are set for many chemicals not currently marketed in Britain. Because of this and the ever-changing information on MRLs, direct access is provided to the comprehensive MRL databases on the Chemicals Regulation Division website (www.hse.gov.uk/crd/index.htm). This online database sets out in table form the levels specified by UK Regulations, EC Directives, and the Codex Alimentarius for each commodity.

REACH (Registration, Evaluation, Authorisation and restriction of CHemicals)

REACH is a European regulation that came into force on 1 June 2007 to produce a centralized database of all substances made or imported into the EU in quantities greater than 1 tonne per annum. Its aim was to enhance the protection of human health and the environment, to allow free movement of substances within the EU and to promote the use of alternatives to hazardous materials where possible. The data on these substances is held by the European Chemicals Agency (ECHA) and in the UK the competent authority is Health and Safety Executive (HSE). Full details and background information are available of the HSE website at www.hse.gov.uk/reach/index.htm.

SECTION 5

Approval (On-label and Off-label)

Only officially approved pesticides can be marketed and used in the UK. Approvals are granted by UK Government Ministers in response to applications that are supported by satisfactory data on safety, efficacy and, where relevant, humaneness. The Chemicals Regulation Division (CRD), (www.hse.gov.uk/crd/index.htm.uk) comes under the Health and Safety Executive (HSE), (www.hse.gov.uk), and is the UK Government Agency for regulating agricultural pesticides and plant protection products. The HSE currently fulfils the same role for other pesticides, with the two organisations now merged. The main focus of the regulatory process in both bodies is the protection of human health and the environment.

Statutory Conditions of Use

Approvals are normally granted only in relation to individual products and for specified uses. It is an offence to use non-approved products or to use approved products in a manner that does not comply with the statutory conditions of use, except where the crop or situation is the subject of an off-label extension of use (see below).

Statutory conditions have been laid down for the use of individual products and may include:

- field of use (e.g. agriculture, horticulture etc.)
- crop or situations for which treatment is permitted
- maximum individual dose
- maximum number of treatments or maximum total dose
- maximum area or quantity which may be treated
- latest time of application or harvest interval
- operator protection or training requirements
- environmental protection
- any other specific restrictions relating to particular pesticides.

Products must display these statutory conditions in a boxed area on the label entitled 'Important Information', or words to that effect. At the bottom of the boxed area must be shown a bold text statement: **'Read the label before use. Using this product in a manner that is inconsistent with the label may be an offence. Follow the Code of Practice for Using Plant Protection Products'.** This requirement came into effect in October 2006, and replaced the previous 'Statutory Box'.

Types of Approval

Where there were once three levels of approval (full; provisional; experimental permit), there will now be just two levels (authorisation for use; limited approval for research and development). Provisional approval used to be granted while further data were generated to justify label claims, but under the new regulations, decisions on authorisation will be reached much more quickly, and this 'half-way stage' is deemed to be unnecessary. The official list of approved products, excluding those approved for research and development only, are shown on the websites of CRD and the Health and Safety Executive (HSE).

Withdrawal of Approval

Product approvals may be reviewed, amended, suspended or revoked at any time. Revocation may occur for various reasons, such as commercial withdrawal, or failure by the approval holder to meet data requirements.

From September 2007, where an approval is being revoked for purely administrative, 'housekeeping' reasons, the existing approval will be revoked and in its place will be issued:

- an approval for advertisement, sale and supply by any person for 24 months, and
- an approval for storage and use by any person for 48 months.

Where an approval is replaced by a newer approval, such that there are no safety concerns with the previous approval, but the newer updated approval is more appropriate in terms of reflecting the latest regulatory standard, the existing approval will be revoked and in its place will be issued:

- an approval for advertisement, sale and supply by any person for 12 months, and
- an approval for storage and use by any person for 24 months.

Where there is a need for tighter control of the withdrawal of the product from the supply chain, for example failure to meet data submission deadlines, the current timelines will be retained:

- immediate revocation for advertisement, sale or supply for the approval holder
- approval for 6 months for advertisement, sale and supply by 'others', and
- approval for storage and use by any person for 18 months.

Immediate revocation and product withdrawal remain an option where serious concerns are identified, with immediate revocation of all approvals and approval for storage only by anyone for 3 months, to allow for disposal of product.

The expiry date shown in the product fact sheet is the final date of legal use of the product.

Off-label Extension of Use

Products may legally be used in a manner not covered by the printed label in several ways:

- In accordance with an Extension of Authorisation for Minor Use (EAMU). EAMUs are uses for which individuals or organisations other than the manufacturers have sought approval. The Notices of Approval are published by CRD and are widely available from ADAS or NFU offices. Users of EAMUs must first obtain a copy of the relevant Notice of Approval and comply strictly with the conditions laid down therein. Users of this Guide will find details of extant EAMUs and direct links to the CRD website, where EAMU notices can also be accessed.
- In tank mixture with other approved pesticides in accordance with Consent C(i) made under FEPA. Full details of Consent C(i) are given in Annex A of Guide to Pesticides on the CRD website, but there are two essential requirements for tank mixes. First, all the conditions of approval of all the components of a mixture must be complied with. Second, no person may mix or combine pesticides that are cholinesterase compounds unless allowed by the label of at least one of the pesticides in the mixture.
- In conjunction with authorised adjuvants.
- In reduced spray volume under certain conditions.
- In the use of certain herbicides on specified set-aside areas subject to restrictions, which differ between Scotland and the rest of the UK.
- By mutual recognition of a use fully approved in another Member State of the European Union and authorised by CRD.

Although approved, off-label uses are not endorsed by manufacturers and such treatments are made entirely at the risk of the user.

SECTION 5

Using Crop Protection Chemicals

Use of Herbicides In or Near Water

Products in this Guide approved for use in or near water are listed in Table 5.1. Before use of any product in or near water, the appropriate water regulatory body (Environment Agency/Local Rivers Purification Authority; or in Scotland the Scottish Environmental Protection Agency) must be consulted. Guidance and definitions of the situation covered by approved labels are given in the Defra publication Guidelines for the Use of Herbicides on Weeds in or near Watercourses and Lakes. Always read the label before use.

Table 5.1 Products approved for use in or near water

Chemical	Product
glyphosate	Barclay Gallup Biograde 360, Barclay Gallup Biograde Amenity, Barclay Gallup Hi-Aktiv, Barclay Glyde 144, Cleancrop Tungsten Ultra, Discman Biograde, Gallup Hi-Aktiv Amenity, Glypho-Rapid 450, Mascot Hi-Aktiv Amenity, Roundup Biactive, Roundup Biactive GL, Roundup POWERMAX, Trustee Amenity,

Use of Pesticides in Forestry

Table 5.2 Products in this *Guide* approved for use in forestry

Chemical	Product	Use
aluminium ammonium sulphate	Curb Liquid Crop Spray, Curb Liquid Crop Spray,	Animal deterrent/repellent
cycloxydim	Laser,	Herbicide
cypermethrin	Forester,	Insecticide
ferric phosphate	Chicane, Daxxos, Epitaph, Ferrex, Ironflexx, Iroxx, Luminare, Menorexx, Minixx, Sluxx HP, Spinner,	Molluscicide
fluazifop-P-butyl	Clayton Maximus, Fusilade Forte, Fusilade Max,	Herbicide
glyphosate	Amega Duo, Azural, Barclay Gallup Biograde 360, Barclay Gallup Biograde Amenity, Barclay Gallup Hi-Aktiv, Barclay Glyde 144, Cleancrop Tungsten Ultra, Clinic UP, Credit, Crestler, Discman Biograde, Ecoplug Max, Gallup Hi-Aktiv Amenity, Giddyup, Glypho-Rapid 450, Glypho-Rapid 450, Hilite, Intercept, Liaison, Mascot Hi-Aktiv Amenity, Mentor, Monsanto Amenity Glyphosate, Monsanto Amenity Glyphosate XL, Motif, Nomix Conqueror Amenity, Ovation, Rattler, Rodeo, Samurai, Scorpion, Snapper, Tanker, Trustee Amenity,	Herbicide
metazachlor	Stalwart, Sultan 50 SC,	Herbicide
Phlebiopsis gigantea	PG Suspension,	Biological agent
propaquizafop	Clayton Satchmo, Falcon, Falcon, Longhorn, Shogun, Tithe,	Herbicide
propyzamide	Barclay Propyz, Cleancrop Rumble, Relva Granules,	Herbicide
pyrethrins	Spruzit,	Insecticide

Pesticides Used as Seed Treatments

Information on the target for these products can be found in the relevant pesticide profile in Section 2.

Table 5.3 Products used as seed treatments (including treatments on seed potatoes)

Chemical	Product	Formulation	Crop(s)
Bacillus amyloliquefaciens strain MBI600	Integral Pro		Angelica, Baby leaf crops, Balm, Basil, Bay, Caraway, Celery leaves, Chervil, Chives, Choi sum, Coriander, Cress, Curry plant, Dill, Edible flowers, Endives, Fennel leaves, Fenugreek, Hyssop, Lamb's lettuce, Land cress, Lavender, Lettuce, Linseed, Lovage, Marjoram, Mint, Mustard, Nasturtium, Nettle, Oregano, Oriental cabbage, Parsley, Plantain, Protected angelica, Purslane, Rocket, Rosemary, Sage, Salad burnet, Savory, Sorrel, Spinach, Spinach beet, Sweet ciceley, Tarragon, Tatsoi, Thyme, Winter oilseed rape
cypermethrin	Signal 300 ES	ES	Spring barley, Spring wheat, Winter barley, Winter wheat
difenoconazole + fludioxonil	Celest Extra	FS	Winter oats, Winter rye, Winter wheat
	Difend Extra	FS	Winter wheat
fludioxonil	Beret Gold	FS	Rye, Spring barley, Spring oats, Spring wheat, Triticale, Winter barley, Winter oats, Winter wheat
	Maxim 100FS	FS	Potatoes, Seed potatoes
	Maxim 480FS	FS	Baby leaf crops, Broccoli, Brussels sprouts, Bulb onions, Cabbages, Carrots, Cauliflowers, Celeriac, Collards, Fennel, Kale, Kohlrabi, Oriental cabbage, Parsley root, Parsnips, Protected broccoli, Protected Brussels sprouts, Protected cauliflowers, Protected celeriac, Protected collards, Protected fennel, Protected kale, Protected kohlrabi, Protected oriental cabbage, Protected parsley root, Protected parsnips, Protected radishes, Radishes, Red beet, Salad onions, Shallots, Spinach

Chemical	Product	Formulation	Crop(s)
	Prepper	FS	Rye, Spring barley, Spring oats, Spring wheat, Triticale, Winter barley, Winter oats, Winter wheat
fludioxonil + fluxapyroxad + triticonazole	Kinto Plus	FS	Winter barley, Winter oats, Winter rye, Winter triticale, Winter wheat
fludioxonil + sedaxane	Vibrance Duo	FS	Spring barley, Spring oats, Spring wheat, Triticale, Winter barley, Winter rye, Winter wheat
fludioxonil + tebuconazole	Fountain	FS	Triticale, Winter barley, Winter oats, Winter rye, Winter wheat
fluopyram + prothioconazole + tebuconazole	Raxil Star	FS	Winter barley
flutolanil	Rhino DSG	DS	Potatoes
metalaxyl-M	Apron XL	ES	Beetroot, Brassica leaves and sprouts, Broccoli, Brussels sprouts, Bulb onions, Cabbages, Calabrese, Cauliflowers, Chard, Choi sum, Collards, Herbs (see appendix 6), Kale, Kohlrabi, Oriental cabbage, Ornamental plant production, Radishes, Shallots, Spinach
prothioconazole + tebuconazole	Redigo Pro	FS	Durum wheat, Rye, Spring barley, Spring oats, Spring wheat, Triticale, Winter barley, Winter oats, Winter wheat
Pseudomonas chlororaphis MA 342	Cerall	FS	Rye, Spring wheat, Triticale, Winter wheat

Aerial Application of Pesticides

New aerial spraying permit arrangements came into force in June 2012 and those undertaking aerial applications must ensure that the spraying is done in line with an Approved Application Plan that is subject to approval by CRD. Plans will only be approved where there is no viable alternative method of application or where aerial application results in reduced impact on human health and/or the environment compared to land-based application. Thus applications to bracken, forestry and possibly blight sprays to potatoes may meet these requirements but few others are likely to do so. Template application plans are available from the CRD web site and once all the necessary details have been received by CRD they undertake to approve or reject the plan within 10 working days.

There are currently no products approved for aerial application

Resistance Management

Pest species are, by definition, adaptable organisms. The development of resistance to some crop protection chemicals is just one example of this adaptability. Repeated use of products with the same mode of action will clearly favour those individuals in the pest population able to tolerate the treatment. This leads to a situation where the tolerant (or resistant) individuals can dominate the population and the product becomes ineffective. In general, the more rapidly the pest species reproduces and the more mobile it is, the faster is the emergence of resistant populations, although some weeds seem able to evolve resistance more quickly than would be expected. In the UK, key independent research organisations, chemical manufacturers and other organisations have collaborated to share knowledge and expertise on resistance issues through three action groups. Participants include **ADAS**, the **Chemicals Regulation Division (CRD)**, universities, colleges and the **Agriculture and Horticulture Development Board (AHDB)** The groups have a common aim of monitoring resistance in the UK and devising and publishing management strategies designed to combat it where it occurs. The groups are:

- **The Weed Resistance Action Group (WRAG)**, formed in 1989 (Secretary: Richard Hull, Rothamsted Research, Harpenden, Herts AL5 2JQ *Tel: 01954 268219*)
  Email: Richard.hull@rothamsted.ac.uk Web: https://ahdb.org.uk/knowledge-library/the-weed-resistance-action-group-wrag
- **The Fungicide Resistance Action Group (FRAG)**, formed in 1995 (Secretary: Mr Paul Ashby, HSE, Email: paul.ashby@hse.gsi.gov.uk). Web: https://ahdb.org.uk/knowledge-library/frag-uk
- **The Insecticide Resistance Action Group (IRAG)**, formed in 1997 (Secretary: Bethan Shaw, NIAB EMR, New Road, East Malling, Kent, ME19 6BJ. Email: bethan.shaw@emr.ac.uk. Web: https://ahdb.org.uk/knowledge-library/the-insecticide-resistance-action-group-irag

The above groups publish detailed advice on resistance management relevant for each sector and, in some cases, specific to a pest problem. This information, together with further details about the function of each group, can be obtained from the **Chemicals Regulation Division** website www.hse.gov.uk/crd/index.htm.

The speed of appearance of resistance depends on the mode of action of the crop protection chemicals, as well as the manner in which they are used. Resistance among insects and fungal diseases has been evident for much longer than weed resistance to herbicides, but examples in all three categories are now widespread and increasing. This has created a need for agreement on the advice given for the use of crop protection chemicals in order to reduce the likelihood of the development of resistance and to avoid the loss of potentially valuable products in the chemical armoury. Mixing or alternating modes of action is one of the guiding principles of resistance management. To assist appropriate product choices, mode of action codes, published by the international Resistance Action Committees (see below), are shown in the respective active ingredient fact sheets. The product label and/or a professional advisor should always be consulted before making decisions. The general guidelines for resistance management are similar for all three problem areas.

Preparation in advance

- Be aware of the factors that favour the development of resistance, such as repeated annual use of the same product, and assess the risk.
- Plan ahead and aim to integrate all possible means of control.
- Use cultural measures such as rotations, stubble hygiene, variety selection and, for fungicides, removal of primary inoculum sources, to reduce reliance on chemical control.
- Monitor crops regularly.
- Keep aware of local resistance problems.
- Monitor effectiveness of actions taken and take professional advice, especially in cases of unexplained poor control.

Using crop protection products

- Optimise product efficacy by using it as directed, at the right time, in good conditions.
- Treat pest problems early
- Mix or alternate chemicals with different modes of action.

SECTION 5

- Avoid repeated applications of very low doses.
- Keep accurate field records.

Label guidance depends on the appropriate strategy for the product. Most frequently it consists of a warning of the possibility of poor performance due to resistance, and a restriction on the number of treatments that should be applied in order to minimise the development of resistance. This information is summarised in the profiles in the active ingredient fact sheets, but detailed guidance must always be obtained by reading the label itself before use.

International Action Committees

Resistance to crop protection products is an international problem. Agrochemical industry collaboration on a global scale is via three action committees whose aims are to support a coordinated industry approach to the management of resistance worldwide. In particular they produce lists of crop protection chemicals classified according to their mode of action. These lists and other information can be obtained from the respective websites:

Herbicide Resistance Action Committee (HRAC) – www.hracglobal.com

Fungicide Resistance Action Committee (FRAC) – www.frac.info

Insecticide Resistance Action Committee (IRAC) – www.irac-online.org

Poisons and Poisoning

Chemicals Subject to the Poison Law

Certain products are subject to the provisions of the Poisons Act 1972. The Poisons Rules 1982 have been revoked under the Deregulation Act 2015 (Poisons and Explosive Precursors) and replaced by The Control of Poisons and Explosive Precursors Regulations 2015 with appropriate amendments to The Poisons Act 1972. These Rules include provisions for the storage and sale and supply of listed non-medicine poisons. Details can be accessed on the HSE website. The nature of the formulation and the concentration of the active ingredient allow some products to be exempted from the Rules, while others with the same active ingredient are included (see below). The chemicals approved for use in the UK are specified under Parts I and II of the Poisons List as follows.

Part I Poisons (sale restricted to registered retail pharmacists and to registered non-pharmacy businesses provided sales do not take place on retail premises):

- aluminium phosphide
- magnesium phosphide.

Part II Poisons (sale restricted to registered retail pharmacists and listed sellers registered with a local authority):

- formaldehyde

Occupational Exposure Limits

A fundamental requirement of the COSHH Regulations is that exposure of employees to substances hazardous to health should be prevented or adequately controlled. Exposure by inhalation is usually the main hazard, and in order to measure the adequacy of control of exposure by this route various substances have been assigned occupational exposure limits.

There are two types of occupational exposure limits defined under COSHH: Occupational Exposure Standards (OES) and Maximum Exposure Limits (MEL). The key difference is that an OES is set at a level at which there is no indication of risk to health; for a MEL a residual risk may exist and the level takes socio-economic factors into account. In practice, MELs have been most often allocated to carcinogens and to other substances for which no threshold of effect can be identified and for which there is no doubt about the seriousness of the effects of exposure.

OESs and MELs are set on the recommendation of the Advisory Committee on Toxic Substances (ACTS). Full details are published by HSE in EH 40/2005 available at:

http://www.hse.gov.uk/pubns/books/eh40.htm

As far as pesticides are concerned, OESs and MELs have been set for relatively few active ingredients. This is because pesticide products usually contain other substances in their formulation, including solvents, which may have their own OES/MEL. In practice inhalation of solvent may be at least, or more, important than that of the active ingredient. These factors are taken into account by the regulators when approving a pesticide product under the Control of Pesticides Regulations. This indicates one of the reasons why a change of pesticide formulation usually necessitates a new approval assessment under COPR.

First Aid Measures

If pesticides are handled in accordance with the required safety precautions, as given on the container label, poisoning should not occur. It is difficult, however, to guard completely against the occasional accidental exposure. Thus, if a person handling, or exposed to, pesticides becomes ill, it is a wise precaution to apply first aid measures appropriate to pesticide poisoning even though the cause of illness may eventually prove to have been quite different. An employer has a legal duty to make adequate first aid provision for employees. Regular pesticide users should consider appointing a trained first aider even if numbers of employees are not large, since there is a specific hazard.

The first essential in a case of suspected poisoning is for the person involved to stop work, to be moved away from any area of possible contamination and for a doctor to be called at once. If no doctor is available the patient should be taken to hospital as quickly as possible. In either event it is most important that the name of the chemical being used should be recorded and preferably the whole product label or leaflet should be shown to the doctor or hospital concerned.

Some pesticides, which are unlikely to cause poisoning in normal use, are extremely toxic if swallowed accidentally or deliberately. In such cases get the patient to hospital as quickly as possible, with all the information you have. Some labels now include Material Safety Data Sheets and these contain valuable information for both the first aider and medical staff. If not included on the label, MSDS are available from company websites.

General Measures

Measures appropriate in all cases of suspected poisoning include the following:

- Remove any protective or other contaminated clothing (taking care to avoid personal contamination).
- Wash any contaminated areas carefully with water or with soap and water if available.
- In cases of eye contamination, flush with plenty of clean water for at least 15 minutes.
- Lay the patient down, keep at rest and under shelter. Cover with one clean blanket or coat, etc. Avoid overheating.
- Monitor level of consciousness, breathing and pulse rate.
- If consciousness is lost, place the casualty in the recovery position (on his/her side with head down and tongue forward to prevent inhalation of vomit).

Reporting of Pesticide Poisoning

Any cases of poisoning by pesticides must be reported without delay to an HM Agricultural Inspector of the Health and Safety Executive. In addition any cases of poisoning by substances named in schedule 2 of The Reporting of Injuries, Diseases and Dangerous Occurrences Regulations 1985, must also be reported to HM Agricultural Inspectorate (this includes organophosphorus chemicals, mercury and some fumigants).

Cases of pesticide poisoning should also be reported to the manufacturer concerned.

Additional Information

General advice on the safe use of pesticides is given in a range of Health and Safety Executive leaflets available from HSE Books. The major agrochemical companies are able to provide authoritative medical advice about their own pesticide products. Useful information is now available from the Material Safety Data Sheet available from the manufacturer or the manufacturer's website.

New arrangements for the provision of information to doctors about poisons and the management of poisonings have been introduced as part of the modernisation of the National Poisons Information Service (NPIS). The Service provides a year-round, 24-hour-a-day service for healthcare staff on the diagnosis, treatment and management of patients who may have been poisoned. The new arrangements are aimed at moving away from the telephone as the first point of contact for poisons information, to the use by doctors of an online database, supported by a second-tier, consultant-led information service for more complex clinical advice.

NPIS no longer provides direct information on poisoning to members of the public. **Anyone suspecting poisoning by pesticides should seek professional medical help immediately via their GP or NHS Direct (www.nhs.uk) or NHS 24 (www.nhs24.scot)**.

Environmental Protection

Environmental Land Management: 8 simple steps for arable farmers

Choosing the right measures, putting them in the right place, and managing them in the right way will make all the difference to your farm environment. The general principles given here should be considered in conjunction with local priorities for soil and water protection and wildlife conservation. This approach complements best practice in soil, crop, fertiliser and pesticide management.

An improved farm environment can be achieved by good management of around 4% of the arable area where high quality habitats are maintained or created. However, the actual area required on your farm will depend on factors such as the area of vulnerable soils and length of watercourses.

If you need further advice then consult a competent environmental adviser.

What you can do

It is important to have a balance of environmental measures that contribute to each of the relevant points below to achieve improved environmental benefits.

1. **Look after established wildlife habitats**
   Start by assessing what you already have on the farm! Maintaining, or where necessary, restoring any existing wildlife habitats, such as woodland, ponds, flower-rich grassland or field margins, is critical to the survival of much of the wildlife on the farm, and may count towards some of the following measures without the need to create new habitats. Unproductive land can be used to create new habitats to complement what you already have.
2. **Maximise the environmental value of field boundaries**
   Hedgerow management and ditch management on a 2–3 years rotation boosts flowers, fruit and refuges for wildlife. This is most suited to hedges dominated by hawthorn and blackthorn, and ditches where rotational management will not compromise the drainage function. Establish new hedgerow trees to maintain or restore former numbers within the landscape.
3. **Create a network of grass margins**
   The highest priority is to buffer watercourses, ideally with a minimum of 5 m buffer strips. Grass margins can also be used to boost beneficial insects and small mammals, and buffer hedges, ponds and other environmental features. Beetle banks can be used to reduce soil erosion and run-off on slopes greater than 1:20 and boost beneficial insects in fields greater than 20 ha.
4. **Establish flower rich habitats**
   Evidence suggests that a network of flower-rich margins on 1% of arable land will support beneficial insects and a wealth of wildlife that feeds on insects. Assess whether this is best done by allowing arable plants in the seed bank to germinate, establishing perennial margins with a grass and wildflower mix, or using nectar flower mixtures. Improving the linkages between these features on the farm will also help wildlife move across the landscape.
5. **Provide winter food for birds with weedy over-wintered stubbles or wild bird cover**
   Provision of seed for wildlife is best achieved by leaving over-wintered stubbles unsprayed and uncultivated until at least mid-February on at least 5% of arable land, or growing seed-rich crops as wild bird cover on 2% of arable land.
6. **Use of spring cropping or in-field measures to help ground-nesting birds**
   Spring crops provide better habitat for a range of plants and insects, and birds such as lapwings and skylarks. Use rotational fallows, skylark plots in winter cereals or (if breeding lapwings occur) fallow plots to support ground-nesting birds where spring cropping forms less than 25% of the arable area. Fallow plots should not be created on land liable to runoff or erosion. Evidence suggests that at least 20 skylark plots or a 1 ha fallow plot per 100 ha would support ground-nesting birds.

7. **Use winter cover crops to protect water**
    You should consider whether a winter cover crop (e.g. mustard) is necessary to capture residual nitrogen on cultivated land left fallow through the winter. This is not necessary if the stubble is retained until at least mid-February and forms a green cover.
8. **Establish in-field grass areas to reduce soil erosion and run-off**
    Land liable to act as channels for soil erosion or run-off (e.g. steep slopes or field corners) should be converted to in-field grass areas.

Remember

- Right measures
- Right place
- Right management

This guidance has been produced by The Voluntary Initiative and the Campaign for the Farmed Environment in conjunction with RSPB, Natural England, GWCT, Plantlife, Butterfly Conservation, BCPC and Buglife

Protection of Bees

Honey bees

Honey bees are a source of income for their owners and important to farmers and growers as pollinators of their crops. It is irresponsible and unnecessary to use pesticides in such a way that may endanger them. Pesticides vary in their toxicity to bees, but those that present a special hazard carry a specific warning in the precautions section of the label. They are indicated in this Guide in the hazard classification and safety precautions section of the pesticide profile.

Product labels indicate the necessary environmental precautions to take, but where use of an insecticide on a flowering crop is contemplated the British Beekeepers Association have produced the following guidelines for growers:

- Target insect pests with the most appropriate product.
- Choose a product that will cause minimal harm to beneficial species.
- Follow the manufacturer's instructions carefully.
- Inspect and monitor crops regularly.
- Avoid spraying crops in flower or where bees are actively foraging.
- Keep down flowering weeds.
- Spray late in the day, in still conditions.
- Avoid excessive spray volume and run-off.
- Adjust sprayer pressure to reduce production of fine droplets and drift.
- Give local beekeepers as much warning of your intention as possible.

Wild bees

Wild bees also play an important role. Bumblebees are useful pollinators of spring flowering crops and fruit trees because they forage in cool, dull weather when honey bees are inactive. They play a particularly important part in pollinating field beans, red and white clover, lucerne and borage. Bumblebees nest and overwinter in field margins and woodland edges. Avoidance of direct or indirect spray contamination of these areas, in addition to the creation of hedgerows and field margins, and late cutting or grazing of meadows, all help the survival of these valuable insects.

BeeConnected

It is best practice to notify local beekeepers before using certain crop protection products where there is a risk to bees. The online BeeConnected tool (www.beeconnected.org.uk) can help you do this.

Accidental and Illegal Poisoning of Wildlife

The Campaign Against Accidental or Illegal Poisoning (CAIP) of Wildlife, was launched back in March 1991 and has now been replaced by **The Wildlife Incident Investigation Scheme**.

This has two objectives:

SECTION 5

- To provide information to HSE on hazards to wildlife, pets and beneficial invertebrates:
- To enforce the correct use of pesticides, identifying and penalizing those who deliberately or recklessly misuse and abuse pesticides.

Full details can be found on the HSE website at www.hse.gov.uk/pesticides/topics/reducing-environmental-impact/wildlife/index.htm

Water Quality

Even when diluted, some pesticides are potentially dangerous to fish and other aquatic life. Not only this, but many watercourses and groundwaters are sources of drinking water, and it requires only a tiny amount of contamination to breach the stringent European Union water quality standards. The EEC Drinking Water Directive sets a maximum admissible level in drinking water for any pesticide, regardless of its toxicity, at 1 part in 10,000 million. As little as 250 grams could be enough to cause the daily supply to a city the size of London to exceed the permitted levels (although it would be very unlikely to present a health hazard to consumers).

The protection of groundwater quality is therefore vital. The Food and Environment Protection Act 1985 (FEPA) places a special obligation on users of pesticides to "safeguard the environment and in particular avoid the pollution of water". Under the Water Resources Act 1991 it is an offence to pollute any controlled waters (watercourses or groundwater), either deliberately or accidentally. Protection of controlled waters from pollution is the responsibility of the Environment Agency (in England and Wales) and the Scottish Environmental Protection Agency. Addresses for both organisations can be found under Useful Contacts.

Users of pesticides therefore have a duty to adopt responsible working practices and, unless they are applying herbicides in or near water, to prevent them getting into water. Guidance on how to achieve this is given in the Defra Code of Good Agricultural Practice for the Protection of Water.

The duty of care covers not only the way in which a pesticide is sprayed, but also its storage, preparation and disposal of surplus, sprayer washings and the container. Products that are a major hazard to fish, other aquatic life or aquatic higher plants carry one of several specific label precautions in their fact sheet, depending on the assessed hazard level.

Where to get information

For general enquiries telephone 03708 506 506 (Environment Agency) or 0131 449 7296 (SEPA). Both Agencies operate a 24-hour emergency hotline for reporting all environmental incidents relating to air, land and water:

0800 80 70 60 (for environmental incidents in England, Wales and Scotland).

In addition, printed literature concerning the protection of water is available from the Environment Agency and CropLife UK (see Useful Contacts).

Protecting surface waters

Surface waters are particularly vulnerable to contamination. One of the best ways of preventing those pesticides that carry the greatest risk to aquatic wildlife from reaching surface waters is to prohibit their application within a boundary adjacent to the water. Such areas are known as no-spray, or buffer, zones. Certain products are restricted in this way and have a legally binding label precaution to make sure the potential exposure of aquatic organisms to pesticides that might harm them is minimised.

Before 1999 the protected zones were measured from the edge of the water. The distances were 2 metres for hand-held or knapsack sprayers, 6 metres for ground crop sprayers, and a variable distance (but often 18 metres) for broadcast air-assisted applications, such as in orchards. The introduction of LERAPs (see below) has changed the method of measuring buffer zones.

'Surface water' includes lakes, ponds, reservoirs, streams, rivers and watercourses (natural or artificial). It also includes temporarily or seasonally dry ditches, which have the potential to carry water at different times of the year. Buffer zone restrictions do not necessarily apply to all products containing the same active ingredient. Those in formulations that are not likely to contaminate surface water through spray drift do not pose the same risk to aquatic life and are not subject to the restrictions.

Local Environmental Risk Assessment for Pesticides (LERAPs)

Local Environment Risk Assessments for Pesticides (LERAPs) were introduced in March 1999, and revised guidelines were issued in January 2002. They give users of most products currently subject to a buffer zone restriction the option of continuing to comply with the existing buffer zone restriction (using the new method of measurement), or carrying out a LERAP and possibly reducing the size of the buffer zone as a result. In either case, there is a new legal obligation for the user to record his decision, including the results of the LERAP.

The scheme has changed the method of measuring the buffer zone. Previously the zone was measured from the edge of the water, but it is now the distance from the top of the bank of the watercourse to the edge of the spray area. In 2013 new rules were introduced which allowed LERAP buffer zones greater than the normal 5 metres for LERAP B pesticides to avoid the loss of some important actives. The products concerned are identified in the Environmental safety section of the affected pesticide profiles.

The LERAP provides a mechanism for taking into account other factors that may reduce the risk, such as dose reduction, the use of low drift nozzles, and whether the watercourse is dry or flowing. The previous arrangements applied the same restriction regardless of whether there was actually water present. Now there is a standard zone of 1 m from the top of a dry ditch bank.

Other factors to include in a LERAP that may allow a reduction in the buffer zone are:

- The size of the watercourse, because the wider it is, the greater the dilution factor, and the lower the risk of serious pollution.
- The dose applied. The lower the dose, the less is the risk.
- The application equipment. Sprayers and nozzles are star-rated according to their ability to reduce spray drift fallout. Equipment offering the greatest reductions achieves the highest rating of three stars. The scheme was originally restricted to ground crop sprayers; new, more flexible rules introduced in February 2002 included broadcast air-assisted orchard and hop sprayers.
- Other changes introduced in 2002 allow the reduction of a buffer zone if there is an appropriate living windbreak between the sprayed area and a watercourse.

Not all products that had a label buffer zone restriction are included in the LERAP scheme. The option to reduce the buffer zone does not apply to organophosphorus or synthetic pyrethroid insecticides. This group are classified as Category A products. All other products that had a label buffer zone restriction are classified as Category B. In addition some products have a 'Broadcast air-assisted LERAP' classification. The wording of the buffer zone label precautions has been amended for all products to take account of the new method of measurement, and whether or not the particular product qualifies for inclusion in the LERAP scheme.

Products in this Guide that are in Category A or B, or have a broadcast air-assisted LERAP, are identified with an appropriate icon on the product fact sheet. Updates to the list are published regularly by CRD and details can be obtained from its website at www.pesticides.gov.uk.

The introduction of LERAPs was an important step forward because it demonstrated a willingness to reduce the impact of regulation on users of pesticides by allowing flexibility where local conditions make it safe to do so. This places a legal responsibility on the user to ensure the risk assessment is done either by himself or by the spray operator or by a professional consultant or advisor. It is compulsory to record the LERAP and make it available for inspection by enforcement authorities. In 2013 new rules were introduced which allowed LERAP buffer zones greater than 5 m to avoid the loss of some important actives. These products are identified in the section of Environmental safety for the pesticide profiles affected.

More details of the LERAP arrangements and guidance on how to carry out assessments are published in the Ministry booklet PB5621 *Local Environmental Risk Assessment for Pesticides – Horizontal Boom Sprayers*, and booklet PB6533 *Local Environment Risk Assessment for*

SECTION 5

Pesticides – Broadcast Air-Assisted Sprayers. Additional booklets, PB2088 Keeping Pesticides Out of Water, and PB3160 Is Your Sprayer Fit for Work give general practical guidance.

Groundwater Regulations

Groundwater Regulations were introduced in 1999 to complete the implementation in the UK of the EU Groundwater Directive (Protection of Groundwater Against Pollution Caused by Certain Dangerous Substances – 80/68/EEC). These Regulations help prevent the pollution of groundwater by controlling discharges or disposal of certain substances, including all pesticides.

Groundwater is defined under the Regulations as any water contained in the ground below the water table. Pesticides must not enter groundwater unless it is deemed by the appropriate Agency to be permanently unsuitable for other uses. The Agricultural Waste Regulations were introduced in May 2006 and apply to the disposal of all farm wastes, including pesticides. With certain exemptions farmers are required to obtain a waste management licence for most waste disposal activities.

As far as pesticides are concerned the new Regulations made little difference to existing controls, except for the disposal of empty containers and the disposal of sprayer washings and rinsings. It remains an offence to dispose of pesticides onto land without official authorisation. Normal use of a pesticide in accordance with product approval does not require authorisation. This includes spraying the washings and rinsings back on the crop provided that, in so doing, the maximum approved dose for that product on that crop is not exceeded. However, those wishing to use a lined biobed for this purpose need to obtain an exemption from the Agency.

In practice the best advice to farmers and growers is to plan to use all diluted spray within the crop and to dispose of all washings via the same route making sure that they stay within the conditions of approval of the product. The enforcing agencies for these Regulations are the Environment Agency (in England and Wales) and the Scottish Environment Protection Agency. The Agencies can serve notice at any time to modify the conditions of an authorisation where necessary to prevent pollution of groundwater.

Integrated Farm Management (IFM)

Integrated farm management is a method of farming that balances the requirements of running a profitable farming business with the adoption of responsible and sensitive environmental management. It is a whole-farm, long-term strategy that combines the best of modern technology with some basic principles of good farming practice. It is a realistic way forward that addresses the justifiable concerns of the environmental impact of modern farming practices, at the same time as ensuring that the industry remains viable and continues to provide wholesome, affordable food. IFM embraces arable and livestock management. The phrase 'integrated crop management' is sometimes used where a farm is wholly arable.

Pest control is essential in any management system. Much can be done to minimise the incidence and impact of pests, but their presence is almost inevitable. IFM ensures that, where a pest problem needs to be contained, the action taken is the best combination of all available options. Pesticides are one of these options, and form an essential, but by no means exclusive, part of pest control strategy.

Where chemicals are to be used, the choice of product should be made not only with the pest problem in mind, but with an awareness of the environmental and social risks that might accompany its use. The aim should be to use as much as necessary, but as little as possible. A major part of the approval process aims to safeguard the environment, so that any approved product, when used as directed, will not cause long-term harm to wildlife or the environment. This is achieved by specifying on the label detailed rules for the way in which a product may be used.

The skill in implementing an IFM pest control strategy is the decision on how easily these rules may be complied with in the particular situation where use is contemplated. Although many chemical options may be available, some are likely to be more suitable than others. This Guide lists, under the heading Special precautions/Environmental safety, the key label precautions that need to be considered in this context, although the actual product label must always be read before a product is used.

Campaign for the Farmed Environment

With the demise of set-aside, it was realised that the environmental benefits achieved by this scheme could be lost. The Campaign for the Farmed Environment (CFE) scheme seeks to retain or even exceed the environmental benefits achieved by set-aside by voluntary measures.

It has three main themes:

- farming for cleaner water and a healthier soil
- helping farmland birds thrive
- improving the environment for farm wildlife.

The scheme encourages farmers to sign up for Entry Level Stewardship (ELS) schemes and to target their already extensive knowledge of habitat management to greater effect. If successful, it will help to fend off the proposal that farmers of cultivated land in England will have to adopt environmental management options on up to 6% of their land.

SECTION 5

SECTION 6
APPENDICES

Appendix 1
Suppliers of Pesticides and Adjuvants

Adama:
Adama Agricultural Solutions UK Ltd
  Third Floor East
  1410 Arlington Business Park
  Theale
  Reading
  Berks.
  RG7 4SA
  Tel: 01635 860555
  Fax: 01635 861555
  Email: ukenquiries@adama.com
  Web: www.adama.com

AgChem Access: AgChemAccess Limited
  Cedar House,
  41 Thorpe Road
  Norwich
  Norfolk
  NR1 1ES
  Tel: 0845 459 9413
  Fax: 0207 149 9815
  Email: enquiries@agchemaccess.com
  Web: www.agchemaccess.com

Agform: Agform Ltd
  Unit One, Concorde Park
  Concorde Way
  Segensworth North
  Fareham
  Hampshire
  PO15 5FG
  Tel: 02382 122379
  Email: sales@agform.com
  Web: www.agform.com

Agrauxine: Agrauxine
  137, rue Gabriel Péri
  59700 Marcq-en-Baroeul
  France
  Web: www.agrauxine.com

Agria SA: Agria SA
  Asenovgradsko Shose,
  4009 Plodiv
  Bulgaria

Agrichem: Agrichem (International) Ltd
  Arus Hygeia
  Oranmore
  Co Galway
  Ireland
  Tel: 01733 204019
  Email: info@agrichem.co.uk
  Web: www.agrichem.co.uk

Agrii: United Agri Products Ltd
  Throws Farm
  Stebbing
  Great Dunmow
  Essex
  DM6 3AQ
  Web: www.agrii.co.uk

AgriMax: AgriMax Ltd
  Yew Tree Cottage
  Crickley Hill
  Gloucester
  GL3 4UQ
  Tel: 01452 862696
  Fax: 01452 862096

Agrovista: Agrovista UK Ltd
  Rutherford House
  Nottingham Science and Technology
  Park
  University Bulevard
  Nottingham
  NG7 2PZ
  Tel: (0115) 939 0202
  Fax: (0115) 921 8498
  Email: enquiries@agrovista.co.uk
  Web: www.agrovista.co.uk

Albaugh UK: Albaugh UK Ltd
  1 Northumberland Avenue
  Trafalgar Square
  London
  WC2N 5BW
  Tel: 0800 078 9649
  Email: info@albaugh.eu
  Web: www.albaugh.eu

Amega: Amega Sciences
  Lanchester Way
  Royal Oak Industrial Estate
  Daventry
  Northants.
  NN11 8PH
  Tel: (01327) 704444
  Fax: (01327) 71154
  Email: admin@amega-sciences.com
  Web: www.amega-sciences.com

Arysta: Arysta Life Sciences
  Route d'Artix, BP80
  64150
  Nogueres
  France
  Email: info.uk@upl-td.com
  Web: www.upl-ltd.com/uk

B H & B: Battle Hayward & Bower Ltd
Victoria Chemical Works
Crofton Drive
Allenby Road Industrial Estate
Lincoln
LN3 4NP
Tel: (01522) 529206
Fax: (01522) 538960
Email: orders@battles.co.uk
Web: www.battles.co.uk

Barclay:
Barclay Chemicals Manufacturing Ltd
Damastown Way
Damastown Industrial Park
Mulhuddart
Dublin 15
Ireland
Tel: (+353) 1 811 2900
Fax: (+353) 1 822 4678
Email: info@barclay.ie
Web: www.barclay.ie

Barrier: Barrier BioTech Ltd
36/37 Haverscroft Industrial Estate
New Road
Attleborough
Norfolk
NR17 1YE
Tel: (01953) 456363
Fax: (01953) 455594
Email: sales@barrier-biotech.com
Web: www.barrier-biotech.com

BASF: BASF plc
Agricultural Divison
PO Box 4, Earl Road
Cheadle Hulme
Cheshire
SK8 6QG
Tel: (0845) 602 2553
Fax: (0161) 485 2229
Web: www.agricentre.basf.co.uk

Bayer CropScience:
Bayer CropScience Limited
230 Cambridge Science Park
Milton Road
Cambridge
CB4 0WB
Tel: (01223) 226500
Email: ukcropsupport@bayer.com
Web: www.cropscience.bayer.co.uk

BelCrop: BelCrop BV
Tiensestraat 300
B-3400 Landen
Belgium
Tel: +32 (0) 11 59 83 60
Fax: +32 (0) 11 59 83 61
Email: info@belcrop.com
Web: www.BELCROP.com

Certis Belchim B V: Certis UK & Ireland
Suite 5, 3 Riverside
Granta Park
Great Abington
Cambs
CB21 6AD
Tel: (0845) 370305
Email: infocertisuk@certiseurope.com
Web: www.certiseurope.co.uk/

Chrysal : Chrysal UK Limited
Ardsley Mills
Common Lane
East Ardsley
Wakefield
West Yorkshire
WF3 2DW
Web: www.chrysal.com

Clayton: Clayton Plant Protection Ltd
Bracetown Business Park
Clonee
Dublin 15
Ireland
Tel: (+353) 1 821 0127
Fax: (+353) 81 841 1084
Email: info@cpp.ag
Web: www.cpp.ag

Corteva: Corteva Agriscience UK Limited
CPC2 Capital Park
Fulbourn
Cambridge
CB21 5XE
Tel: 0800 689 8899
Email: ukhotline@corteva.com
Web: www.corteva.co.uk

Cropco Ltd: Cropco Limited
Northeys Farm, Yeldham Rd
Belchamp Walter
Sudbury
CO10 7BB
Tel: 01787 238200
Fax: 01787 238222
Email: info@cropco.co.uk
Web: www.cropco.co.uk

De Sangosse: De Sangosse Ltd
De Sangosse House
Goodwin Business Park
Willie Snaith Road
Newmarket
Cambridge
CB8 7 SQ
Tel: (01223) 811215
Email: info@desangosse.co.uk
Web: www.desangosse.co.uk

Fargro: Fargro Ltd
Vinery Fields
Arundel Road
Poling
Arundel
West Sussex
BN18 9PY
Tel: (01903) 721591
Fax: (01903) 883303
Email: info@fargro.co.uk
Web: www.fargro.co.uk

Fine: Fine Agrochemicals Ltd
Hill End House
Whittington
Worcester
WR5 2RQ
Tel: (01905) 361800
Fax: (01905) 361810
Email: enquire@fine.eu
Web: www.fine.eu

FMC Agro: FMC Agro Ltd
Rectors Lane
Pentre
Deeside
Flintshire
CH5 2DH
Tel: 01423 205011
Fax: 01244 532097
Web: www.fmc-agro.co.uk

Forest Research: Forest Research
Alice Holt Lodge
Farnham
Surrey
GU10 4LH
Tel: 01420 22255
Fax: 01420 23653
Email:
Katherine.tubby@forestry.gsi.gov.uk
Web: www.forestry.gov.uk

Gemini: Gemini Agriculture Ltd
8 Mythop Road
Lytham St. Annes
Lancs
FY8 4JD

Globachem: Globachem NV
Brustem Industriepark
Lichtenberglaan 2019
3800 Sint-Truiden
Belgium
Tel: (+32) 11 69 01 73
Fax: (+32) 11 68 15 65
Email: globachem@globachem.com
Web: www.globachem.com

Global Adjuvants :
Global Adjuvants Company Ltd
20-22 Wenlock Road
London
N1 7GU
Email: office@global-adjuvants.com
Web: www.global-adjuvants.com

Gowan: Gowan Crop Protection Ltd
Danial Hall Building
Rothamsted Research
West Common
Harpenden
Herts
AL5 2JQ
Tel: 01582 280390
Email: ukmarketing@gowanco.com
Web: www.avadex.co.uk

Greenaway: Greenaway Amenity Ltd
7 Browntoft Lane
Donington
Spalding
Lincs.
PE11 4TQ
Tel: (01775) 821031
Fax: (01775) 821034
Email: info@greenawayamenity.co.uk
Web: www.greenawaycda.co.uk

Greencrop: Greencrop Technology Ltd
c/o Clayton Plant Protection Ltd
Bracetown Business Park
Clonee
Dublin 15
Ireland
Tel: (+353) 1 821 0127
Fax: (+353) 81 841 1084
Email: info@cpp.ag
Web: www.cpp.ag

Headland Amenity:
Origin Amenity Solutions Limited
1 - 3 Freeman Court
Jarman Way
Royston
Herts.
SG8 5HW
Tel: 01763 255550
Email: info@headlandamenity.com
Web: www.headlandamenity.com

ICL (Everris) Ltd: Everris Limited
Epsilon House
West Road
Ipswich
Suffolk
IP3 9FJ
Tel: 01473 237111
Fax: 01473 237150
Email: prof.sales@icl-group.com
Web: www.icl-group.com

SECTION 6

Interagro: Interagro (UK) Ltd
Thorley Wash Barn
London Road
Thorley
Bishop's Stortford
Hertfordshire
CM23 4AT
Tel: (01279) 714970
Fax: (01279) 758227
Email: info@interagro.co.uk
Web: www.interagro.co.uk

Intracrop: Intracrop
Unit 21, Raleigh Hall Industrial Estate
Eccleshall
Staffordshire
SL21 6JL
Tel: (01926) 634801
Fax: (01926) 634798
Email: admin@intracrop.co.uk
Web: www.intracrop.co.uk

Juno : Juno (Plant Protection) Ltd
Little Mill Farm
Underlyn Lane
Marden
Kent
TN12 9AT
Tel: 01580 765079
Email: nick@junopp.com

Killgerm: Killgerm Chemicals Ltd
115 Wakefield Road
Flushdyke
Ossett
W. Yorks.
WF5 9AR
Tel: (01924) 268400
Fax: (01924) 264757
Email: info@killgerm.com
Web: www.killgerm.com

Koppert: Koppert (UK) Ltd
Unit 8, Tudor Rose Court
53 Hollands Road
Haverhill
Suffolk
CB9 8PJ
Tel: (01440) 704488
Fax: (01440) 704487
Email: info@koppert.co.uk
Web: www.koppert.co.uk

Landseer: Landseer Ltd
Lodge Farm
Goat Hall Lane
Galleywood
Chelmsford
Essex
CM2 8PH
Tel: (01245) 357109
Web: www.lanfruit.co.uk

Life Scientific: Life Scientific Limited
Block 4
Belfield Office Park
Beech Hill Road
Dublin 4
Ireland
Tel: 00353 1 283 2024
Web: www.lifescientific.com

Microcide: Microcide Ltd
Shepherds Grove
Stanton
Bury St. Edmunds
Suffolk
IP31 2AR
Tel: (01359) 251077
Fax: (01359) 251545
Email: microcide@microcide co.uk
Web: www.microcide.co.uk

Mitsui: Mitsui Chemicals Agro, Inc.
Tel: +81-3-5290-2700
Fax: +81-3-3231-1171

Monsanto: Bayer CropScience Limited
230 Cambridge Science Park
Milton Road
Cambridge
CB4 0WB
Tel: 01223 226500
Email: ukcropsupport@bayer.com
Web: www.cropscience.bayer.co.uk

Nomix Enviro: Nomix Enviro Ltd.
The Grain Silos
Weyhill Road
Andover
Hampshire
SP10 3NT
Tel: 01264 388050
Fax: 01522 866176
Email: nomixenviro@frontierag.co.uk
Web: www.nomix.co.uk

Novastar: Novastar Link Limited
17 St Ann's Square,
Manchester,
M2 7PW
UK

Nufarm UK: Nufarm UK Ltd
Wyke Lane
Wyke
West Yorkshire
BD12 9EJ
Tel: 01274 69 1234
Fax: 01274 69 1176
Email: infouk@uk.nufarm.com
Web: www.nufarm.co.uk

Omex: Omex Agriculture Ltd
Bardney Airfield
Tupholme
Lincoln
LN3 5TP
Tel: (01526) 396000
Fax: (01526) 396001
Email: enquire@omex.com
Web: www.omex.co.uk

Pan Agriculture: Pan Agriculture Ltd
8 Cromwell Mews
Station Road
St Ives
Huntingdon
Cambs.
PE27 5HJ
Tel: (01480) 467790
Fax: (01480) 467041
Email: info@panagriculture.co.uk

Pan Amenity: Pan Amenity Ltd
8 Cromwell Mews
Station Road
St Ives
Cambs.
PE27 5HJ
Tel: 01480 467790
Fax: 01480 467041

Pangaea : Pangaea Agrochemicals Limited
St Johns Innovation Centre, Cowley
Road
Cambridge
CB4 0WS
Tel: 01603 883736
Web:
www.pangaeaagrochemicals.co.uk

PelGar: PelGar International
13 Newman Lane
Alton
GU34 2QR
Tel: 01420 80744
Web: www.pelgar.co.uk

ProKlass: ProKlass Products Limited
145-157 St John Street
London
EC1V 4PW
UK
Tel: 01480 810137
Email: office@proklass-products.com

Rentokil: Rentokil Initial 1927 plc
The Power Centre
Units A1 & A2
Link 10 Napier Way
Crawley
W. Sussex
RH10 9RA
Tel: 01293 858306
Email: dawn.kirby@rentokil-initial.com
Web: www.rentokil.com

Resource Chemicals:
Resource Chemicals Ltd
Resource House
76 High Street
Brackley
Northants
NN13 7DS
Tel: 01280 843800

Rigby Taylor: Rigby Taylor Ltd
Rigby Taylor House
Crown Lane
Horwich
Bolton
Lancs.
BL6 5HP
Tel: (01204) 677777
Fax: (01204) 677715
Email: info@rigbytaylor.com
Web: www.rigbytaylor.com

Rotam : Rotam Europe Ltd
Hamilton House
Mabledon Place
London
WC1H 9BB
Tel: 0207 953 0447
Web: www.rotam.com/uk

SFP Europe: SFP Europe SA
11 Boulevard de la Grand Thumine
Parc d'Ariane - Bat B
13090 Aix en Provence
France

Sherriff Amenity:
Sherriff Amenity Services
The Pines
Fordham Road
Newmarket
Cambs
CB8 7LG
Tel: (01638) 721888
Fax: (01638) 721815
Web: www.sherriffamenity.com

SECTION 6

Sipcam: Sipcam UK Ltd
4C Archway House
The Lanterns
Melbourn Street
Royston
Herts.
SG8 7BX
Tel: (01763) 212100
Fax: (01763) 212101
Email: stewart@sipcamuk.co.uk
Web: sipcamuk.co.uk

Sphere: Sphere Laboratories (London) Ltd
c/o Mainswood
Putley Common
Ledbury
Herefordshire
HR8 2RF
Tel: 01684 899306
Fax: 01684 893322
Email:
Fiona.Homes@Sphere-London.co.uk

Stefes: Stefes GMBH
Wendenstrasse 21b
20097 Hamburg
Germany
Tel: +49 (0)40 533083-0
Fax: +49 (0)40 5330833-29
Email: info@stefes.eu
Web: www.stefes.eu

Sumi Agro: Sumi Agro Europe Limited
Vintners' Place
68 Upper Thames Street
London
EC4V 3BJ
Tel: (020) 7246 3697
Fax: (020) 7246 3799
Email: summit@summit-agro.com

Sumitomo : Sumitomo Chemical Plc
Hythe House
200 Shepherds Bush Road
London
W6 7NL
Tel: 0203 538 3099
Web: www.sumitomo-chemical.co.uk/

Synergy : Synergy Generics Limited
Market House
Church Street
Harlston
Norfolk
IP20 9BB
Tel: 07884 435 282
Email: thomas@agchemaccess.com

Syngenta: Syngenta UK Limited
CPC4
Capital Park
Fulbourn
Cambridge
CB21 5XE
Tel: 0800 1696058
Email:
customer.services@syngenta.com
Web: www.syngenta.co.uk

UPL Europe: UPL Europe Ltd
Engine Rooms
Birchwood Park
Warrington
Cheshire
WA3 6YN
Tel: (01925) 819999
Fax: (01925) 856075
Email: info.uk@upl-td.com
Web: www.upl-ltd.com/uk

Valto: Valto BV
Leehove 81
2678 MB De Lier
The Netherlands
Tel: +31 174 51 45 19
Email: info@valto.nl
Web: www.valto.nl

Zantra: Zantra Limited
Westwood
Rattar Mains
Scarfskerry
Thurso
Caithness
KW14 8XW
Tel: 01480 861066
Fax: 01480 861099

Appendix 2
Useful Contacts

Agriculture Industries Confederation Ltd
First Floor, Unit 4
The Forum
Minerva Business Park
Lynch Wood
Peterborough PE2 6FT
Tel: (01733) 385230
Web: www.agindustries.org.uk

AHDB Horticulture
Stoneleigh Park
Kenilworth
Warwickshire CV8 2TL
Tel: (024) 7669 2051
Web: www.ahdb.org.uk/horticulture

BASIS Ltd
St Monica's House
39 Windmill Lane
Ashbourne
Derbyshire DE6 1EY
Tel: (01335) 343945
Web: www.basis-reg.co.uk

British Beekeepers' Association
National Beekeeping Centre
Stoneleigh Park
Kenilworth
Warwickshire CV8 2LG
Tel: (024) 7669 6679
Web: www.bbka.org.uk

British Beer & Pub Association (BBPA)
Ground Floor
61 Queen Street
London EC4R 1EB
Tel: (0207) 627 9173
Web: www.beerandpub.com

British Crop Production Council (BCPC)
93 Lawrence Weaver Road
Cambridge, CB3 0LE
Tel: (01223) 342495
Web: www.bcpc.org

British Pest Control Association (BPCA)
4A Mallard Way, Pride Park
Derby DE24 8GX
Tel: (01332) 294288
Web: www.bpca.org.uk

Chemicals Regulation Division
Room 1A, Mallard House
King's Pool
3 Peasholme Green
York YO1 7PX
Tel: (01904) 640500 or (03459) 335577
Fax: (01904) 455733
Web: www.hse.gov.uk/crd

CropLife UK (formerly Crop Protection Association)
Stuart House
St John's Street
Peterborough PE1 5DD
Tel: (01733) 355370
Web: www.croplife.uk

CropLife International
326 Avenue Louise
Box 35 B-1050 Brussels
Belgium
Tel: (+32) 2 542 0410
Web: www.croplife.org

Department of Agriculture and Rural Development (Northern Ireland)
Dundonald House
Upper Newtownards Road
Ballymiscaw
Belfast BT4 3SB
Tel: (028) 9052 4704 or (0300) 200 7852
Web: www.daera-ni.gov.uk

Department of Environment, Food and Rural Affairs (Defra)
Seacole Building
2 Marsham Street
London SW1P 4DF
Tel: (020) 7238 6000 / 03459 335577
Web: www.defra.gov.uk

Environment Agency
National Customer Contact Centre
PO Box 544
Rotherham S60 1BY
Tel: (03708) 506 506
Email: enquiries@environment-agency.gov.uk
Web: www.environment-agency.gov.uk

CropLife Europe (formerly ECPA)
9 rue Guimard
B-1040 Brussels
Belgium
Tel: (+32) 2 663 1550
Web: www.croplifeeurope.eu

European Chemicals Agency (ECHA)
Mailing address:
PO Box 400
00121 Helsinki, Finland
Tel: +358 9 686180
Web: www.echa.europa.eu

Farmers' Union of Wales
Llys Amaeth
Plas Gogerddan
Aberystwyth
Ceredigion SY23 3BT
Tel: (01970) 820820
Web: www.fuw.org.uk

Forestry Commission
620 Bristol Business Pk
Coldharbour Lane
Bristol BS16 1EJ
Tel: (0300) 067 4000
Web: www.forestry.gov.uk

Game and Wildlife Conservation Trust (GWCT)
Burgate Manor
Fordingbridge
Hampshire SP6 1EF
Tel: 01425 652381
Email: info@gwct.org.uk
Web: www.gwct.org.uk

Health and Safety Executive
Product Approvals
Chemical Regulation
Redgrave Court
Merton Road Bootle
Merseyside L20 7HS
Tel: (0151) 951 4000
Web: www.hse.gov.uk

Health and Safety Executive – Books
TSO Customer Services
PO Box 29
Norwich NR3 1GN
Tel: 0333 202 5070

Lantra
National Agricultural Centre
Lantra House
Stoneleigh
Kenilworth
Warwickshire CV8 2LG
Tel: (024) 7669 6996
Web: www.lantra.co.uk

National Association of Agricultural Contractors (NAAC)
The Old Cart Shed
Easton Lodge Farms
Wansford,
Peterborough PE8 6NP
Tel: (01780) 784631
Web: www.naac.co.uk

National Chemicals Emergency Centre (NCEC)
The Gemini Building
Fermi Avenue
Harwell
Didcot
Oxfordshire OX11 0RG
Tel: (01235) 753654
Web: www.the-ncec.com

National Farmers' Union
Agriculture House
Stoneleigh Park Stoneleigh
Warwickshire CV8 2TZ
Tel: (024) 7685 8500
Web: www.nfu.org.uk

National Poisons Information Service (NPIS)
(Birmingham Unit)
City Hospital, Dudley Road
Birmingham B18 7QH
Tel: (0121) 507 4123
Email: sds.npis@nhs.net
Web: www.npis.org

Natural England (enquiries)
County Hall
Spetchley Road
Worcester
WR5 2NP
Tel: 0300 060 3900
Suspected wildlife/pesticide poisoning
Tel: 0800 321 600
Email: enquiries@naturalengland.org.uk
Web: www.gov.uk/government/organisations/natural-england

NIAB
93 Lawrence Weaver Road,
Cambridge, Cambs
CB3 0LE
Tel: (01223) 342200
Email: info@niab.com
Web: www.niab.com

Processors and Growers Research Organisation (PGRO)
The Research Station
Great North Road, Thornhaugh
Peterborough, Cambs. PE8 6HJ
Tel: (01780) 782585
Web: www.pgro.org

**Scottish Beekeepers' Association
Secretary**
Ormiston Hill
Kirknewton EH27 8DQ
*Email: secretary@scottishbeekeepers.
org.uk*
Web: www.scottishbeekeepers.org.uk

**Scottish Environmental Protection
Agency (SEPA)**
Edinburgh Office, Silvan House
SEPA 3rd Floor
231 Corstorphine Road
Edinburgh EH12 7AT
Tel: (0131) 449 7296
Web: www.sepa.org.uk

TSO (The Stationery Office)
18 Central Avenue
St Andrews Business Park
Norwich NR7 0HR
Tel: (0333) 202 5070
Email:customer.services@tso.co.uk
Web: www.tsoshop.co.uk

UK Flour Millers (formerly NABIM)
21 Arlington Street
London SW1A 1RN
Tel: (020) 7493 2521
Web: www.ukflourmillers.org

SECTION 6

Appendix 3
Keys to Crop and Weed Growth Stages

Decimal Code for the Growth Stages of Cereals
Illustrations of these growth stages can be found in the reference indicated below and in some company product manuals.

0 Germination

00	Dryseed
03	Imbibition complete
05	Radicle emerged from caryopsis
07	Coleoptile emerged from caryopsis
09	Leaf at coleoptile tip

1 Seedling growth

10	First leaf through coleoptile
11	First leaf unfolded
12	2 leaves unfolded
13	3 leaves unfolded
14	4 leaves unfolded
15	5 leaves unfolded
16	6 leaves unfolded
17	7 leaves unfolded
18	8 leaves unfolded
19	9 or more leaves unfolded

2 Tillering

20	Main shoot only
21	Main shoot and 1 tiller
22	Main shoot and 2 tillers
23	Main shoot and 3 tillers
24	Main shoot and 4 tillers
25	Main shoot and 5 tillers
26	Main shoot and 6 tillers
27	Main shoot and 7 tillers
28	Main shoot and 8 tillers
29	Main shoot and 9 or more tillers

3 Stem elongation

30	Ear at 1 cm
31	1st node detectable
32	2nd node detectable
33	3rd node detectable
34	4th node detectable
35	5th node detectable
36	6th node detectable
37	Flag leaf just visible
39	Flag leaf ligule/collar just visible

4 Booting

41	Flag leaf sheath extending
43	Boots just visibly swollen
45	Boots swollen
47	Flag leaf sheath opening
49	First awns visible

5 Inflorescence

51	First spikelet of inflorescence just visible
52	1/4 of inflorescence emerged
55	1/2 of inflorescence emerged
57	3/4 of inflorescence emerged
59	Emergence of inflorescence completed

6 Anthesis

60	
61	Beginning of anthesis
64	
65	Anthesis half way
68	
69	Anthesis complete

7 Milk development

71	Caryopsis watery ripe
73	Early milk
75	Medium milk
77	Late milk

8 Dough development

83	Early dough
85	Soft dough
87	Hard dough

9 Ripening

91	Caryopsis hard (difficult to divide by thumb-nail)
92	Caryopsis hard (can no longer be dented by thumb-nail)
93	Caryopsis loosening in daytime

(From Tottman, 1987. *Annals of Applied Biology*, **110**, 441–454)

Stages in Development of Oilseed Rape

Illustrations of these growth stages can be found in the reference indicated below and in some company product manuals.

0 Germination and emergence

1 Leaf production

1,0 Both cotyledons unfolded and green
1,1 First true leaf
1,2 Second true leaf
1,3 Third true leaf
1,4 Fourth true leaf
1,5 Fifth true leaf
1,10 About tenth true leaf
1,15 About fifteenth true leaf

2 Stem extension

2,0 No internodes ('rosette')
2,5 About five internodes

3 Flower bud development

3,0 Only leaf buds present
3,1 Flower buds present but enclosed by leaves
3,3 Flower buds visible from above ('green bud')
3,5 Flower buds raised above leaves
3,6 First flower stalks extending
3,7 First flower buds yellow ('yellow bud')

4 Flowering

4,0 First flower opened
4,1 10% all buds opened
4,3 30% all buds opened
4,5 50% all buds opened

5 Pod development

5,3 30% potential pods
5,5 50% potential pods
5,7 70% potential pods
5,9 All potential pods

6 Seed development

6,1 Seeds expanding
6,2 Most seeds translucent but full size
6,3 Most seeds green
6,4 Most seeds green-brown mottled
6,5 Most seeds brown
6,6 Most seeds dark brown
6,7 Most seeds black but soft
6,8 Most seeds black and hard
6,9 All seeds black and hard

7 Leaf senescence

8 Stem senescence

8,1 Most stem green
8,5 Half stem green
8,9 Little stem green

9 Pod senescence

9,1 Most pods green
9,5 Half pods green
9,9 Few pods green

(From Sylvester-Bradley, 1985. *Aspects of Applied Biology*, **10**, 395–400)

Stages in Development of Peas

Illustrations of these growth stages can be found in the reference indicated below and in some company product manuals.

0   Germination and emergence

000   Dry seed
001   Imbibed seed
002   Radicle apparent
003   Plumule and radicle apparent
004   Emergence

1   Vegetative stage

101   First node (leaf with one pair leaflets, no tendril)
102   Second node (leaf with one pair leaflets, simple tendril)
103   Third node (leaf with one pair leaflets, complex tendril)
•
l0x   X nodes (leaf with more than one pair leaflets, complex tendril)
•
•
10n   Last recorded node

2   Reproductive stage (main stem)

201   Enclosed buds
202   Visible buds
203   First open flower
204   Pod set (small immature pod)
205   Flat pod
206   Pod swell (seeds small, immature)
207   Podfill
208   Pod green, wrinkled
209   Pod yellow, wrinkled (seeds rubbery)
210   Dry seed

3   Senescence stage

301   Desiccant application stage. Lower pods dry and brown, middle yellow, upper green. Overall moisture content of seed less than 45%
302   Pre-harvest stage. Lower and middle pods dry and brown, upper yellow. Overall moisture content of seed less than 30%
303   Dry harvest stage. All pods dry and brown, seed dry

(From Knott, 1987. *Annals of Applied Biology*, **111**, 233–244)

Stages in Development of Faba Beans

Illustrations of these growth stages can be found in the reference indicated below and in some company product manuals.

0   Germination and emergence

000   Dry seed
001   Imbibed seed
002   Radicle apparent
003   Plumule and radicle apparent
004   Emergence
005   First leaf unfolding
006   First leaf unfolded

1   Vegetative stage

101   First node
102   Second node
103   Third node
•
•
l0x   X nodes
•
•
10n   N, last recorded node

2   Reproductive stage (main stem)

201   Flower buds visible
203   First open flowers
204   First pod set
205   Pods fully formed, green
207   Pod fill, pods green
209   Seed rubbery, pods pliable, turning black
210   Seed dry and hard, pods dry and black

3   Pod senescence

301   10% pods dry and black
•
•
305   50% pods dry and black
•
•
308   80% pods dry and black, some upper pods green
309   90% pods dry and black, most seed dry. Desiccation stage.
310   All pods dry and black, seed hard. Pre-harvest (glyphosate application stage)

4   Stem senescence

401   10% stem brown/black
•
•
405   50% stem brown/black
•
•
409   90% stem brown/black
410   All stems brown/black. All pods dry and black, seed hard.

(From Knott, 1990. *Annals of Applied Biology*, **116**, 391–404)

SECTION 6

Stages in Development of Potato

Illustrations of these growth stages can be found in the reference indicated below and in some company product manuals.

0   Seed germination and seedling emergence

000   Dry seed
001   Imbibed seed
002   Radicle apparent
003   Elongation of hypocotyl
004   Seedling emergence
005   Cotyledons unfolded

1   Tuber dormancy

100   Innate dormancy (no sprout development under favourable conditions)
150   Enforced dormancy (sprout development inhibited by environmental conditions)

2   Tuber sprouting

200   Dormancy break, sprout development visible
21x   Sprout with 1 node
22x   Sprout with 2 nodes
•
•
29x   Sprout with 9 nodes
21x(2) Second generation sprout with 1 node
22x(2) Second generation sprout with 2 nodes
•
•
29x(2) Second generation sprout with 9 nodes

Where x = 1, sprout >2 mm;
2, 2-5 mm; 3, 5-20 mm;
4, 20-30 mm; 5, 50-100 mm;
6, 100-150 mm long

3   Emergence and shoot expansion

300   Main stem emergence
301   Node 1
302   Node 2
•
•
319   Node 19
Second order branch
321   Node 1
•
•
Nth   order branch
3N1   Node 1
•
•
3N9   Node 9

4   Flowering

Primary flower
400   No flowers
410   Appearance of flower bud
420   Flower unopen
430   Flower open
440   Flower closed
450   Berry swelling
460   Mature berry
Second order flowers
410(2) Appearance of flower bud
420(2) Flower unopen
430(2) Flower open
440(2) Flower closed
450(2) Berry swelling
460(2) Mature berry

5   Tuber development

500   No stolons
510   Stolon initials
520   Stolon elongation
530   Tuber initiation
540   Tuber bulking (>10 mm diam)
550   Skin set
560   Stolon development

6   Senescence

600   Onset of yellowing
650   Half leaves yellow
670   Yellowing of stems
690   Completely dead

(From Jefferies & Lawson, 1991. *Annals of Applied Biology*, **119**, 387–389)

Stages in Development of Linseed
Illustrations of these growth stages can be found in the reference indicated below and in some company product manuals.

0   Germination and emergence

00   Dry seed
01   Imbibed seed
02   Radicle apparent
04   Hypocotyl extending
05   Emergence
07   Cotyledon unfolding from seed case
09   Cotyledons unfolded and fully expanded

1   Vegetative stage (of main stem)

10   True leaves visible
12   First pair of true leaves fully expanded
13   Third pair of true leaves fully expanded
1n   n leaf fully expanded

2   Basal branching

21   One branch
22   Two branches
23   Three branches
2n   n branches

3   Flower bud development (on main stem)

31   Enclosed bud visible in leaf axils
33   Bud extending from axil
35   Corymb formed
37   Buds enclosed but petals visible
39   First flower open

4   Flowering (whole plant)

41   10% of flowers open
43   30% of flowers open
45   50% of flowers open
49   End of flowering

5   Capsule formation (whole plant)

51   10% of capsules formed
53   30% Of capsules formed
55   50% of capsules formed
59   End of capsule formation

6   Capsule senescence (on most advanced plant)

61   Capsules expanding
63   Capsules green and full size
65   Capsules turning yellow
67   Capsules all yellow brown but soft
69   Capsules brown, dry and senesced

7   Stem senescence (whole plant)

71   Stems mostly green below panicle
73   Most stems 30% brown
75   Most stems 50% brown
77   Stems 75% brown
79   Stems completely brown

8   Stems rotting (retting)

81   Outer tissue rotting
85   Vascular tissue easily removed
89   Stems completely collapsed

9   Seed development (whole plant)

91   Seeds expanding
92   Seeds white but full size
93   Most seeds turning ivory yellow
94   Most seeds turning brown
95   All seeds brown and hard
98   Some seeds shed from capsule
99   Most seeds shed from capsule

(From Jefferies & Lawson, 1991. *(From Freer, 1991. Aspects of Applied Biology,* **28***, 33–40)*

Stages in Development of Annual Grass Weeds

Illustrations of these growth stages can be found in the reference indicated below and in some company product manuals.

0 Germination and emergence

00	Dry seed
01	Start of imbibition
03	Imbibition complete
05	Radicle emerged from caryopsis
07	Coleoptile emerged from caryopsis
09	Leaf just at coleoptile tip

1 Seedling growth

10	First leaf through coleoptile
11	First leaf unfolded
12	2 leaves unfolded
13	3 leaves unfolded
14	4 leaves unfolded
15	5 leaves unfolded
16	6 leaves unfolded
17	7 leaves unfolded
18	8 leaves unfolded
19	9 or more leaves unfolded

2 Tillering

20	Main shoot only
21	Main shoot and 1 tiller
22	Main shoot and 2 tillers
23	Main shoot and 3 tillers
24	Main shoot and 4 tillers
25	Main shoot and 5 tillers
26	Main shoot and 6 tillers
27	Main shoot and 7 tillers
28	Main shoot and 8 tillers
29	Main shoot and 9 or more tillers

3 Stem elongation

31	First node detectable
32	2nd node detectable
33	3rd node detectable
34	4th node detectable
35	5th node detectable
36	6th node detectable
37	Flag leaf just visible
39	Flag leaf ligule just visible

4 Booting

41	Flag leaf sheath extending
43	Boots just visibly swollen
45	Boots swollen
47	Flag leaf sheath opening
49	First awns visible

5 Inflorescence emergence

51	First spikelet of inflorescence just visible
53	1/4 of inflorescence emerged
55	1/2 of inflorescence emerged
57	3/4 of inflorescence emerged
59	Emergence of inflorescence completed

6 Anthesis

61	Beginning of anthesis
65	Anthesis half-way
69	Anthesis complete

(From Lawson & Read, 1992. *Annals of Applied Biology*, **12**, 211–214)

Growth Stages of Annual Broad-leaved Weeds

Preferred Descriptive Phrases
Illustrations of these growth stages can be found in the reference indicated below and in some company product manuals.

Pre-emergence	Plants up to 50 mm across/high
Early cotyledons	Plants up to 100 mm across/high
Expanded cotyledons	Plants up to 150 mm across/high
One expanded true leaf	Plants up to 250 mm across/high
Two expanded true leaves	Flower buds visible
Four expanded true leaves	Plant flowering
Six expanded true leaves	Plant senescent
Plants up to 25 mm across/high	

(From Lutman & Tucker, 1987. *Annals of Applied Biology*, **110**, 683–687)

SECTION 6

Appendix 4
Key to Hazard Classifications and Safety Precautions

Every product label contains information to warn users of the risks from using the product, together with precautions that must be followed in order to minimise the risks. A hazard classification (if any) and the symbol are shown with associated risk phrases, followed by a series of safety numbers under the heading **Hazard classification and safety precautions**.

The codes are defined below, under the same sub-headings as they appear in the pesticide profiles.

Where a product label specifies the use of personal protective equipment (PPE), the requirements are listed under the sub-heading **Operator protection**, using letter codes to denote the protective items, e.g. handling the concentrate, cleaning equipment etc., but it is not possible to list them separately. The lists of PPE are therefore an indication of what the user may need to have available to use the product in different ways. **When making a COSHH assessment it is therefore essential that the product label is consulted for information on the particular use that is being assessed.**

Where the generalised wording includes a phrase such as '... for *xx* days', the specific requirement for each pesticide is shown in brackets after the code.

Hazard
H01	Very toxic
H02	Toxic
H03	Harmful
H04	Irritant
H05	Corrosive
H06	Extremely flammable
H07	Highly flammable
H08	Flammable
H09	Oxidising agent
H10	Explosive
H11	Dangerous for the environment
H220	Extremely flammable gas
H225	Highly flammable liquid and vapour
H226	Flammable liquid and vapour
H228	Flammable solid
H260	In contact with water releases flammable gases which may ignite spontaneously
H300	Fatal if swallowed
H301	Toxic if swallowed
H302	Harmful if swallowed
H303	May be harmful if swallowed
H311	Toxic in contact with skin
H330	Fatal if inhaled
H331	Toxic if inhaled
H332	Harmful if inhaled
H400	Very toxic to aquatic organisms
H401	Toxic to aquatic organisms

Risk phrases
H280	Contains gas under pressure; may explode if heated
H290	May be corrosive to metals
H304	May be fatal if swallowed and enters airways
H310	Fatal in contact with skin

H312	Harmful in contact with skin
H314	Causes severe skin burns and eye damage
H315	Causes skin irritation
H317	May cause an allergic skin reaction
H318	Causes serious eye damage
H319	Causes serious eye irritation
H320	Causes eye irritation
H334	May cause allergy or asthma symptoms or breathing difficulties if inhaled
H335	May cause respiratory irritation
H336	May cause drowsiness or dizziness
H340	May cause genetic defects
H341	Suspected of causing genetic defects
H351	Suspected of causing cancer
H360	May damage fertility or the unborn child
H361	Suspected of damaging fertility or the unborn child
H362	May cause harm to breast-fed children
H370	Causes damage to organs
H371	May cause damage to organs
H372	Causes damage to organs through prolonged or repeated exposure
H373	May cause damage to organs through prolonged or repeated exposure
R08	Contact with combustible material may cause fire
R09	Explosive when mixed with combustible material
R15	Contact with water liberates extremely flammable gases
R16	Explosive when mixed with oxidising substances
R19	Fatal if swallowed
R20	Harmful by inhalation
R21	Harmful in contact with skin
R22a	Harmful if swallowed
R22b	May cause lung damage if swallowed
R22c	May be fatal if swallowed and enters airways
R23a	Toxic by inhalation
R23b	Fatal if inhaled
R24	Toxic in contact with skin
R25	Toxic if swallowed
R25b	Toxic; Danger of serious damage to health by prolonged exposure if swallowed
R26	Very toxic by inhalation
R27	Very toxic in contact with skin
R28	Very toxic if swallowed
R31	Contact with acids liberates toxic gas
R34	Causes burns
R35	Causes severe burns
R36	Irritating to eyes
R37	Irritating to respiratory system
R38	Irritating to skin
R39	Danger of very serious irreversible effects
R40	Limited evidence of a carcinogenic effect
R40b	Suspected of causing cancer
R41	Risk of serious damage to eyes
R42	May cause sensitization by inhalation
R43	May cause sensitization by skin contact
R45	May cause cancer
R46	May cause heritable genetic damage
R48	Danger of serious damage to health by prolonged exposure
R50	Very toxic to aquatic organisms
R51	Toxic to aquatic organisms
R52	Harmful to aquatic organisms
R53a	May cause long-term adverse effects in the aquatic environment
R53b	Dangerous to aquatic organisms
R54	Toxic to flora
R55	Toxic to fauna
R56	Toxic to soil organisms

SECTION 6

R57   Toxic to bees
R58   May cause long-term adverse effects in the environment
R60   May impair fertility
R60b   May cause heritable genetic damage
R61   May cause harm to the unborn child
R62   Possible risk of impaired fertility
R63   Possible risk of harm to the unborn child
R64   May cause harm to breast-fed babies
R65   Harmful: May cause lung damage if swallowed
R66   Repeated exposure may cause skin dryness or cracking
R66b   May cause damage to organs through prolonged or repeated exposure
R67   Vapours may cause drowsiness and dizziness
R68   Possible risk of irreversible effects
R69   Danger of serious damage to health by prolonged oral exposure.
R70   May produce an allergic reaction

Operator protection

A   Suitable protective gloves (the product label should be consulted for any specific requirements about the material of which the gloves should be made)
B   Rubber gauntlet gloves
C   Face-shield
D   Approved respiratory protective equipment
E   Goggles
F   Dust mask
G   Full face-piece respirator
H   Coverall
J   Hood
K   Apron/Rubber apron
L   Waterproof coat
M   Rubber boots
N   Waterproof jacket and trousers
P   Suitable protective clothing
U01   To be used only by operators instructed or trained in the use of chemical/product/type of produce and familiar with the precautionary measures to be observed
U02a Wash all protective clothing thoroughly after use, especially the inside of gloves
U02b Avoid excessive contamination of coveralls and launder regularly
U02c Remove and wash contaminated gloves immediately
U03   Wash splashes off gloves immediately
U04a Take off immediately all contaminated clothing
U04b Take off immediately all contaminated clothing and wash underlying skin. Wash clothes before re-use
U04c Wash clothes before re-use
U05a When using do not eat, drink or smoke
U05b When using do not eat, drink, smoke or use naked lights
U06   Handle with care and mix only in a closed container
U07   Open the container only as directed (returnable containers only)
U08   Wash concentrate/dust from skin or eyes immediately
U09a Wash any contamination/splashes/dust/powder/concentrate from skin or eyes immediately
U09b Wash any contamination/splashes/dust/powder/concentrate from eyes immediately
U09c If on skin, wash with plenty of soap and water
U10   After contact with skin or eyes wash immediately with plenty of water
U11   In case of contact with eyes rinse immediately with plenty of water and seek medical advice
U12   In case of contact with skin rinse immediately with plenty of water and seek medical advice
U12b After contact with skin, take off immediately all contaminated clothing and wash immediately with plenty of water
U13   Avoid contact by mouth
U14   Avoid contact with skin
U15   Avoid contact with eyes

U16a	Ensure adequate ventilation in confined spaces
U16b	Use in a well ventilated area
U18	Extinguish all naked flames, including pilot lights, when applying the fumigant/dust/liquid/product
U19a	Do not breathe dust/fog/fumes/gas/smoke/spray mist/vapour. Avoid working in spray mist
U19b	Do not work in confined spaces or enter spaces in which high concentrations of vapour are present. Where this precaution cannot be observed distance breathing or self-contained breathing apparatus must be worn, and the work should be done by trained operators
U19c	In case of insufficient ventilation, wear suitable respiratory equipment
U19d	Wear suitable respiratory equipment during bagging and stacking of treated seed
U19e	During fumigation/spraying wear suitable respiratory equipment
U19f	In case of accident by inhalation: remove casualty to fresh air and keep at rest
U20a	Wash hands and exposed skin before eating, drinking or smoking and after work
U20b	Wash hands and exposed skin before eating and drinking and after work
U20c	Wash hands before eating and drinking and after work
U20d	Wash hands after use
U20e	Wash hands and exposed skin after cleaning and re-calibrating equipment
U21	Before entering treated crops, cover exposed skin areas, particularly arms and legs
U22a	Do not touch sachet with wet hands or gloves/Do not touch water soluble bag directly
U22b	Protect sachets from rain or water
U23a	Do not apply by knapsack sprayer/hand-held equipment
U23b	Do not apply through hand held rotary atomisers
U23c	Do not apply via tractor mounted horizontal boom sprayers
U24	Do not handle grain unnecessarily
U24a	Do not handle treated crops for at least 4 days after treatment
U25	Open the container only as directed
U26	Keep unprotected workers out of treated areas for at least 5 days after treatment.

Environmental protection

E02a	Keep unprotected persons/animals out of treated/fumigation areas for at least xx hours/days
E02b	Prevent access by livestock, pets and other non-target mammals and birds to buildings under fumigation and ventilation
E02c	Vacate treatment areas before application
E02d	Exclude all persons and animals during treatment
E03	Label treated seed with the appropriate precautions, using the printed sacks, labels or bag tags supplied
E05a	Do not apply directly to livestock/poultry
E05b	Keep poultry out of treated areas for at least xx days/weeks
E05c	Do not apply directly to animals
E06a	Keep livestock out of treated areas for at least xx days/weeks after treatment
E06b	Dangerous to livestock. Keep all livestock out of treated areas/away from treated water for at least xx days/weeks. Bury or remove spillages
E06c	Harmful to livestock. Keep all livestock out of treated areas/away from treated water for at least xx days/weeks. Bury or remove spillages
E06d	Exclude livestock from treated fields. Livestock may not graze or be fed treated forage nor may it be used for hay, silage or bedding
E07a	Keep livestock out of treated areas for up to two weeks following treatment and until poisonous weeds, such as ragwort, have died down and become unpalatable
E07b	Dangerous to livestock. Keep livestock out of treated areas/away from treated water for at least xx weeks and until foliage of any poisonous weeds, such as ragwort, has died and become unpalatable
E07c	Harmful to livestock. Keep livestock out of treated areas/away from treated water for at least xx days/weeks and until foliage of any poisonous weeds such as ragwort has died and become unpalatable
E07d	Keep livestock out of treated areas for up to 4-6 weeks following treatment and until poisonous weeds, such as ragwort, have died down and become unpalatable
E07e	Do not take grass crops for hay or silage for at least 21 days after application

SECTION 6

E07f	Keep livestock out of treated areas for up to 3 days following treatment and until poisonous weeds, such as ragwort, have died
E07g	Keep livestock out of treated areas for at least 50 days following treatment
E08a	Do not feed treated straw or haulm to livestock within xx days/weeks of spraying
E08b	Do not use on grassland if the crop is to be used as animal feed or bedding
E09	Do not use straw or haulm from treated crops as animal feed or bedding for at least xx days after last application
E10a	Dangerous to game, wild birds and animals
E10b	Harmful to game, wild birds and animals
E10c	Dangerous to game, wild birds and animals. All spillages must be buried or removed
E11	Paraquat can be harmful to hares; spray stubbles early in the day
E12a	High risk to bees
E12b	Extremely dangerous to bees
E12c	Dangerous to bees
E12d	Harmful to bees
E12e	Do not apply to crops in flower or to those in which bees are actively foraging. Do not apply when flowering weeds are present
E12f	Do not apply to crops in flower, or to those in which bees are actively foraging, except as directed on [crop]. Do not apply when flowering weeds are present
E12g	Apply away from bees
E12h	Reasonable precautions must be taken to prevent access of birds, wild mammals and honey bees to treated crops
E13a	Extremely dangerous to fish or other aquatic life. Do not contaminate surface waters or ditches with chemical or used container
E13b	Dangerous to fish or other aquatic life. Do not contaminate surface waters or ditches with chemical or used container
E13c	Harmful to fish or other aquatic life. Do not contaminate surface waters or ditches with chemical or used container
E13d	Apply away from fish
E13e	Harmful to fish or other aquatic life. The maximum concentration of active ingredient in treated water must not exceed XX ppm or such lower concentration as the appropriate water regulatory body may require.
E14a	Extremely dangerous to aquatic higher plants. Do not contaminate surface waters or ditches with chemical or used container
E14b	Dangerous to aquatic higher plants. Do not contaminate surface waters or ditches with chemical or used container
E15a	Do not contaminate surface waters or ditches with chemical or used container
E15b	Do not contaminate water with product or its container. Do not clean application equipment near surface water. Avoid contamination via drains from farmyards or roads
E15c	To protect groundwater, do not apply to grass leys less than 1 year old
E16a	Do not allow direct spray from horizontal boom sprayers to fall within 5 m of the top of the bank of a static or flowing waterbody, unless a Local Environment Risk Assessment for Pesticides (LERAP) permits a narrower buffer zone, or within 1 m of the top of a ditch which is dry at the time of application. Aim spray away from water
E16b	Do not allow direct spray from hand-held sprayers to fall within 1 m of the top of the bank of a static or flowing waterbody. Aim spray away from water
E16c	Do not allow direct spray from horizontal boom sprayers to fall within 5 m of the top of the bank of a static or flowing waterbody, or within 1m of the top of a ditch which is dry at the time of application. Aim spray away from water. This product is not eligible for buffer zone reduction under the LERAP horizontal boom sprayers scheme.
E16d	Do not allow direct spray from hand-held sprayers to fall within 1 m of the top of the bank of a static or flowing waterbody. Aim spray away from water. This product is not eligible for buffer zone reduction under the LERAP horizontal boom sprayers scheme.

E16e	Do not allow direct spray from horizontal boom sprayers to fall within 5 m of the top of the bank of a static or flowing water body or within 1 m from the top of any ditch which is dry at the time of application. Spray from hand held sprayers must not in any case be allowed to fall within 1 m of the top of the bank of a static or flowing water body. Always direct spray away from water. The LERAP scheme does not extend to adjuvants. This product is therefore not eligible for a reduced buffer zone under the LERAP scheme.
E16f	Do not allow direct spray/granule applications from vehicle mounted/drawn hydraulic sprayers/applicators to fall within 6 m of surface waters or ditches/Do not allow direct spray/granule applications from hand-held sprayers/applicators to fall within 2 m of surface waters or ditches. Direct spray/applications away from water
E16g	Do not allow direct spray from train sprayers to fall within 5 m of the top of the bank of any static or flowing waterbody.
E16h	Do not spray cereals after 31st March in the year of harvest within 6 metres of the edge of the growing crop
E16i	Do not allow direct spray from horizontal boom sprayers to fall within 12 metres of the top of the bank of a static or flowing water body.
E16j	Do not allow direct spray from horizontal boom sprayers to fall within the specified distance of the top of the bank of a static or flowing waterbody,
E17a	Do not allow direct spray from broadcast air-assisted sprayers to fall within xx m of surface waters or ditches. Direct spray away from water
E17b	Do not allow direct spray from broadcast air-assisted sprayers to fall within xx m of the top of the bank of a static or flowing waterbody, unless a Local Environmental Risk Assessment for Pesticides (LERAP) permits a narrower buffer zone, or within 5 m of the top of a ditch which is dry at the time of application. Aim spray away from water
E18	Do not spray from the air within 250 m horizontal distance of surface waters or ditches
E19a	Do not dump surplus herbicide in water or ditch bottoms
E19b	Do not empty into drains
E20	Prevent any surface run-off from entering storm drains
E21	Do not use treated water for irrigation purposes within xx days/weeks of treatment
E22a	High risk to non-target insects or other arthropods. Do not spray within 6 m of the field boundary
E22b	Risk to certain non-target insects or other arthropods. For advice on risk management and use in Integrated Pest Management (IPM) see directions for use
E22c	Risk to non-target insects or other arthropods
E23	Avoid damage by drift onto susceptible crops or water courses
E34	Do not re-use container for any purpose/Do not re-use container for any other purpose
E35	Do not burn this container
E36a	Do not rinse out container (returnable containers only)
E36b	Do not open or rinse out container (returnable containers only)
E37	Do not use with any pesticide which is to be applied in or near water
E38	Use appropriate containment to avoid environmental contamination
E39	Extreme care must be taken to avoid spray drift onto non-crop plants outside the target area
E40	To protect groundwater/soil organisms the maximum total dose of this or other products containing ethofumesate MUST NOT exceed 1.0 kg ethofumesate per hectare in any three year period
E40b	To protect aquatic organisms respect an unsprayed buffer zone to surface waters in line with LERAP requirements
E40c	This product must not be used with any pesticide to be applied in or near water
E40d	To protect non-target arthropods respect an untreated buffer zone of 5m to non crop land
E40e	To protect aquatic organisms respect an unsprayed buffer zone of 20 metres to surface water bodies
E41	Hay for silage must not be cut from treated crops for at least 21 days after treatment
H410	Very toxic to aquatic life with long-lasting effects
H411	Toxic to aquatic life with long lasting effects
H412	Harmful to aquatic life with long lasting effects

SECTION 6

H413   May cause long lasting harmful effects to aquatic life

Consumer protection
C01   Do not use on food crops
C02a   Do not harvest for human or animal consumption for at least xx days/weeks after last application
C02b   Do not remove from store for sale or processing for at least 21 days after application
C02c   Do not remove from store for sale or processing for at least 2 days after application
C04   Do not apply to surfaces on which food/feed is stored, prepared or eaten
C05   Remove/cover all foodstuffs before application
C06   Remove exposed milk before application
C07   Collect eggs before application
C08   Protect food preparing equipment and eating utensils from contamination during application
C09   Cover water storage tanks before application
C10   Protect exposed water/feed/milk machinery/milk containers from contamination
C11   Remove all pets/livestock/fish tanks before treatment/spraying
C12   Ventilate treated areas thoroughly when smoke has cleared/Ventilate treated rooms thoroughly before occupying

Storage and disposal
D01   Keep out of reach of children
D02   Keep away from food, drink and animal feeding-stuffs
D03   Store away from seeds, fertilizers, fungicides and insecticides
D04   Store well away from corms, bulbs, tubers and seeds
D05   Protect from frost
D06a   Store away from heat
D06b   Do not store near heat or open flame
D06c   Do not store in direct sunlight
D06d   Do not store above 30/35 °C
D06e   Store in a well-ventilated place. Keep container tightly closed
D06f   Keep away from sources of ignition - No smoking
D06g   Take precautionary measures against static discharges
D06h   Store at 10 - 25 °C
D07   Store under cool, dry conditions
D08   Store in a safe, dry, frost-free place designated as an agrochemical store
D09a   Keep in original container, tightly closed, in a safe place
D09b   Keep in original container, tightly closed, in a safe place, under lock and key
D09c   Store unused sachets in a safe place. Do not store half-used sachets
D10a   Wash out container thoroughly and dispose of safely
D10b   Wash out container thoroughly, empty washings into spray tank and dispose of safely
D10c   Rinse container thoroughly by using an integrated pressure rinsing device or manually rinsing three times. Add washings to sprayer at time of filling and dispose of container safely
D10d   Do not rinse out container
D11a   Empty container completely and dispose of safely/Dispose of used generator safely
D11b   Empty container completely and dispose of it in the specified manner
D11c   Ventilate empty containers until no phosphine is detected and then dispose of as hazardous waste via an authorised waste-disposal contractor
D12a   This material (and its container) must be disposed of in a safe way
D12b   This material and its container must be disposed of as hazardous waste
D12c   Dispose of contents/container to a licensed hazardous waste disposal contractor or collection site except for empty, clean containers which can be disposed of normally
D13   Treat used container as if it contained pesticide
D14   Return empty container as instructed by supplier (returnable containers only)
D15   Store container in purpose built chemical store until returned to supplier for refilling (returnable containers only)
D16   Do not store below 4 degrees Centigrade
D17   Use immediately on removal of foil
D18   Place the tablets whole into the spray tank - do not break or crumble the tablets

D19   Do not empty into drains
D20   Clean all equipment after use
D21   Open the container only as directed
D22   Collect spillage

Treated Seed

S01   Do not handle treated seed unnecessarily
S02   Do not use treated seed as food or feed
S03   Keep treated seed secure from people, domestic stock/pets and wildlife at all times during storage and use
S04a   Bury or remove spillages
S04b   Harmful to birds/game and wildlife. Treated seed should not be left on the soil surface. Bury or remove spillages
S04c   Dangerous to birds/game and wildlife. Treated seed should not be left on the soil surface. Bury or remove spillages
S04d   To protect birds/wild animals, treated seed should not be left on the soil surface. Bury or remove spillages
S05   Do not reuse sacks or containers that have been used for treated seed for food or feed
S06a   Wash hands and exposed skin before meals and after work
S06b   Wash hands and exposed skin after cleaning and re-calibrating equipment
S07   Do not apply treated seed from the air
S08   Treated seed should not be broadcast
S08a   Treated seed should not be stored longer than 1 month before planting
S09   Label treated seed with the appropriate precautions using printed sacks, labels or bag tags supplied
S10   Only use automated equipment for planting treated seed potatoes

Vertebrate/Rodent control products

V01a   Prevent access to baits/powder by children, birds and other animals, particularly cats, dogs, pigs and poultry
V01b   Prevent access to bait/gel/dust by children, birds and non-target animals, particularly dogs, cats, pigs, poultry
V02   Do not prepare/use/lay baits/dust/spray where food/feed/water could become contaminated
V03a   Remove all remains of bait, tracking powder or bait containers after use and burn or bury
V03b   Remove all remains of bait and bait containers/exposed dust/after treatment (except where used in sewers) and dispose of safely (e.g. burn/bury). Do not dispose of in refuse sacks or on open rubbish tips.
V04a   Search for and burn or bury all rodent bodies. Do not place in refuse bins or on rubbish tips
V04b   Search for rodent bodies (except where used in sewers) and dispose of safely (e.g. burn/bury). Do not dispose of in refuse sacks or on open rubbish tips
V04c   Dispose of safely any rodent bodies and remains of bait and bait containers that are recovered after treatment (e.g. burn/bury). Do not dispose of in refuse sacks or on open rubbish tips
V05   Use bait containers clearly marked POISON at all surface baiting points

Medical advice

M01   This product contains an anticholinesterase organophosphorus compound. DO NOT USE if under medical advice NOT to work with such compounds
M02   This product contains an anticholinesterase carbamate compound. DO NOT USE if under medical advice NOT to work with such compounds
M03   If you feel unwell, seek medical advice immediately (show the label where possible)
M03a   If exposed or concerned, get medical advice or attention
M04a   In case of accident or if you feel unwell, seek medical advice immediately (show the label where possible)
M04b   In case of accident by inhalation, remove casualty to fresh air and keep at rest

SECTION 6

M04c   If inhaled, remove victim to fresh air and keep at rest in a position comfortable for breathing. IMMEDIATELY call Poison Centre or doctor/physician
M05a   If swallowed, seek medical advice immediately and show this container or label
M05b   If swallowed, do not induce vomiting: seek medical advice immediately and show this container or label
M05c   If swallowed induce vomiting if not already occurring and take patient to hospital immediately
M05d   If swallowed, rinse the mouth with water but only if the person is conscious
M06   This product contains an anticholinesterase carbamoyl triazole compound. DO NOT USE if under medical advice NOT to work with such compounds

Appendix 5
Key to Abbreviations and Acronyms

The abbreviations of formulation types in the following list are used in Section 2 (Pesticide Profiles) and are derived from the *Catalogue of Pesticide Formulation Types and International Coding System* (CropLife International Technical Monograph 2, 5th edn, March 2002).

1 Formulation Types

AB	Grain bait
AE	Aerosol generator
AL	Other liquids to be applied undiluted
AP	Any other powder
AS	Aqeous suspension
BB	Block bait
BR	Briquette
CB	Bait concentrate
CC	Capsule suspension in a suspension concentrate
CF	Capsule suspension for seed treatment
CG	Encapsulated granule (controlled release)
CL	Contact liquid or gel (for direct application)
CP	Contact powder (for direct application)
CR	Crystals
CS	Capsule suspension
DC	Dispersible concentrate
DP	Dustable powder
DS	Powder for dry seed treatment
EC	Emulsifiable concentrate
EG	Emulsifiable granule
EO	Water in oil emulsion
ES	Emulsion for seed treatment
EW	Oil in water emulsion
FG	Fine granules
FP	Smoke cartridge
FS	Flowable concentrate for seed treatment
FT	Smoke tablet
FU	Smoke generator
FW	Smoke pellets
GA	Gas
GB	Granular bait
GE	Gas-generating product
GG	Macrogranules
GL	Emulsifiable gel
GP	Flo-dust (for pneumatic application)
GR	Granules
GS	Grease
GW	Water soluble gel
HN	Hot fogging concentrate
KK	Combi-pack (solid/liquid)
KL	Combi-pack (liquid/liquid)
KN	Cold-fogging concentrate
KP	Combi-pack (solid/solid)
LA	Lacquer
LI	Liquid, unspecified
LS	Solution for seed treatment
ME	Microemulsion
MG	Microgranules
MS	Microemulsion for seed treatment

OD	Oil dispersion
OL	Oil miscible liquid
PA	Paste
PC	Gel or paste concentrate
PO	Powder
PS	Seed coated with a pesticide
PT	Pellet
RB	Ready-to-use bait
RC	Ready-to-use low volume CDA sprayer
RH	Ready-to-use emulsion
RH	Ready-to-use solution
RH	Ready-to-use spray in hand-operated sprayer
SA	Sand
SC	Suspension concentrate (= flowable)
SE	Suspo-emulsion
SG	Water soluble granules
SL	Soluble concentrate
SL	Soluble concentrate; Water-soluble granule
SP	Water soluble powder
SS	Water soluble powder for seed treatment
ST	Water soluble tablet
SU	Ultra low-volume suspension
TB	Tablets
TC	Technical material
TP	Tracking powder
UL	Ultra low-volume liquid
VP	Vapour releasing product
WB	Water soluble bags
WG	Water dispersible granules
WP	Wettable powder
WS	Water dispersible powder for slurry treatment of seed
WT	Water dispersible tablet
XX	Other formulations
ZZ	Not Applicable

2 Other Abbreviations and Acronyms

ACP	Advisory Committee on Pesticides
ACTS	Advisory Committee on Toxic Substances ADAS Agricultural Development and Advisory Service
a.i.	active ingredient
AIC	Agriculture Industries Confederation
BBPA	British Beer and Pub Association
CDA	Controlled droplet application
CPA	Crop Protection Association
cm	centimetre(s)
COPR	Control of Pesticides Regulations 1986
COSHH	Control of Substances Hazardous to Health Regulations
CRD	Chemicals Regulation Directorate
d	day(s)
Defra	Department for Environment, Food and Rural Affairs
EA	Environment Agency
EBDC	ethylene-bis-dithiocarbamate fungicide
EAMU	Extension of Authorisation for Minor Use
FEPA	Food and Environment Protection Act 1985
g	gram(s)
g.a.e.	grams acid equivalent
GS	growth stage (unless in formulation column)
h	hour(s)
ha	hectare(s)
HBN	hydroxybenzonitrile herbicide
HI	harvest interval

HSE	Health and Safety Executive
ICM	integrated crop management
IPM	integrated pest management
kg	kilogram(s)
l	litre(s)
LERAP	Local Environmental Risk Assessments for Pesticides
m	metre(s)
MBC	methyl benzimidazole carbamate fungicide
MEL	maximum exposure limit
min	minute(s)
mm	millimetre(s)
MRL	maximum residue level
mth	month(s)
NA	Notice of Approval
NFU	National Farmers' Union
OES	Occupational Exposure Standard
OLA	off-label approval
PPE	personal protective equipment
PPPR	Plant Protection Products Regulations
SOLA	specific off-label approval
ULV	ultra-low volume
VI	Voluntary Initiative
w/v	weight/volume
w/w	weight/weight
wk	week(s)
yr	year(s)

3 UN Number details

UN No.	Substance	EAC	APP	Hazards Class	Sub risks	HIN
1692	Strychnine or Strychnine salts	2X		6.1		66
1760	Corrosive Liquid, N.O.S., Packing groups II & III	2X	B	8		88
1830	Sulphuric acid, with more than 51% acid	2P		8		80
1993	Flammable Liquid, N.O.S., Packing group III	•3Y		3		30
2011	Magnesium Phosphide	4W[1]		4.3	6.1	
2588	Pesticide, Solid, TOXIC, N.O.S.	2X		6.1		66/60
2757	Carbamate Pesticide, Solid, Toxic	2X		6.1		66/60
2783	Organophosphorus Pesticide, Solid, TOXIC	2X		6.1		66/60
2902	Pesticide, Liquid, TOXIC, N.O.S., Packing Groups I, II & III	2X	B	6.1		66/60
3077	Environmentally hazardous substance, Solid, N.O.S.	2Z		9		90
3082	Environmentally hazardous substance, Solid, N.O.S.	•3Z		9		90
3265	Corrosive Liquid, Acidic, Organic, N.O.S., Packing group I, II or III	2X	B	8		80/88
3351	Pyrethroid Pesticide, Liquid, Toxic, Flammable, Flash Point 23°C or more	•3W	A(fl)	6.1	3	663/63
3352	Pyrethroid Pesticide, Liquid, TOXIC, Packing groups I, II & III	2X	B	6.1		66/60

Ref: Dangerous Goods Emergency Action Code List 2009 (TSO).
[1] Not applicable to the carriage of dangerous goods under RID or ADR.

SECTION 6

Appendix 6
Definitions

The descriptions used in this *Guide* for the crops or situations in which products are approved for use are those used on the approved product labels. These are now standardised in a Crop Hierarchy published by the Chemicals Regulation Division in which definitions are given. To assist users of this *Guide* the definitions of some of the terminology where misunderstandings can occur are reproduced below.

Rotational grass: Short-term grass crops grown on land that is likely to be growing different crops in future years (*e.g. short-term intensively managed leys for one to three years that may include clover*)

Permanent grassland: Grazed areas that are intended to be permanent in nature (*e.g. permanent pasture and moorland that can be grazed*).

Ornamental Plant Production: All ornamental plants that are grown for sale or are produced for replanting into their final growing position (*e.g. flowers, house plants, nursery stock, bulbs grown in containers or in the ground*).

Managed Amenity Turf: Areas of frequently mown, intensively managed, turf that is not intended to flower and set seed. It includes areas that may be for intensive public use (*e.g. all types of sports turf*).

Amenity Grassland: Areas of semi-natural or planted grassland subject to minimal management. It includes areas that may be accessed by the public (*e.g. railway and motorway embankments, airfields, and grassland nature reserves*). These areas may be managed for their botanical interest, and the relevant authority should be contacted before using pesticides in such locations.

Amenity Vegetation: Areas of semi-natural or ornamental vegetation, including trees, or bare soil around ornamental plants, or soil intended for ornamental planting. It includes areas to which the public have access. It does NOT include hedgerows around arable fields.

Natural surfaces not intended to bear vegetation: Areas of soil or natural outcroppings of rock that are not intended to bear vegetation, including areas such as sterile strips around fields. It may include areas to which the public have access. It does not include the land between rows of crops.

Hard surfaces: Man-made impermeable surfaces that are not intended to bear vegetation (*e.g. pavements, tennis courts, industrial areas, railway ballast*).

Permeable surfaces overlying soil: Any man-made permeable surface (excluding railway ballast) such as gravel that overlies soil and is not intended to bear vegetation

Green Cover on Land Temporarily Removed from Production: Includes fields covered by natural regeneration or by a planted green cover crop that will not be harvested (*e.g. green cover on setaside*). It does NOT include industrial crops.

Forest Nursery: Areas where young trees are raised outside for subsequent forest planting.

Forest: Groups of trees being grown in their final positions. Covers all woodland grown for whatever objective, including commercial timber production, amenity and recreation, conservation and landscaping, ancient traditional coppice and farm forestry, and trees from natural regeneration, colonisation or coppicing. Also includes restocking of established woodlands and new planting on both improved and unimproved land.

Farm forestry: Groups of trees established on arable land or improved grassland including those planted for short rotation coppicing. It includes mature hedgerows around arable fields.

Indoors (for rodenticide use): Situations where the bait is placed within a building or other enclosed structure, and where the target is living or feeding predominantly within that building or structure.

Herbs: Reference to Herbs or Protected Herbs when used in Section 2 may include any or all of the following. The particular label or EAMU Notice will indicate which species are included in the approval.

Agastache spp.
Angelica
Applemint
Balm
Basil
Bay
Borage (except when grown for oilseed)
Camomile
Caraway
Catnip
Chervil
Clary
Clary sage
Coriander
Curry plant
Dill
Dragonhead
English chamomile
Fennel
Fenugreek
Feverfew
French lavender
Gingermint
Hyssop
Korean mint
Land cress
Lavandin
Lavender
Lemon balm
Lemon peppermint

Lemon thyme
Lemon verbena
Lovage
Marigold
Marjoram
Mint
Mother of thyme
Nasturtium
Nettle
Oregano
Origanum heracleoticum
Parsley root
Peppermint
Pineapplemint
Rocket
Rosemary
Rue
Sage
Salad burnet
Savory
Sorrel
Spearmint
Spike lavender
Tarragon
Thyme
Thymus camphoratus
Violet
Winter savory
Woodruff

Herbs for Medicinal Uses: Reference to Herbs for Medicinal Uses when used in Section 2 may include any or all of the following. The particular label or EAMU Notice will indicate which species are included in the approval

Black cohosh
Burdock
Dandelion
Echinacea
Ginseng

Goldenseal
Liquorice
Nettle
Valerian

SECTION 6

Appendix 7
References

The information given in *The UK Pesticide Guide* provides some of the answers needed to assess health risks, including the hazard classification and the level of operator protection required. However, the Guide cannot provide all the details needed for a complete hazard assessment, which must be based on the product label itself and, where necessary, the Health and Safety Data Sheet and other official literature.

Detailed guidance on how to comply with the Regulations is available from several sources.

Pesticides: Code of Practice

The sale, supply, storage and use of pesticides are strictly controlled by EU law. The key acts, Regulations and Codes of Practice setting out the duties of employers, supervisors and operators are:

* The *Plant Protection Products Regulations 2011*;
* The *Plant Protection Products (Sustainable Use) Regulations 2012*;
* The *Control of Substances Hazardous to Health Regulations 2002* (COSHH) (as amended);
* Pesticides: *Code of Practice for Using Plant Protection Products 2006*, (incorporating the former 'Green and Orange Codes') with specific Codes of Practice for Scotland and Northern Ireland;
* Classification, Labelling and Packaging (CLP) Regulations.

The Code is currently being updated (as of May 2016) and may be replaced by a series of guidance notes as opposed to a Code in future. The current Code should be read in conjunction with guidance published on the HSE website on how new pesticide legislation has affected the responsibilities of the regulated community.

Further information and details of legislation relating to water, environment protection, waste management and transportation can be found in the UK National action Plan for the Sustainable Use of Pesticides (www.gov.uk/government/publications/pesticides-uk-national-action-plan).

Other Codes of Practice

The Food and Environment Protection Act and its regulations aim to protect the health of human beings, creatures and plants, safeguard the environment and secure safe, efficient and humane methods of controlling pests.

Code of Practice for Suppliers of Pesticides to Agriculture, Horticulture and Forestry (the 'Yellow Code') (Defra Booklet PB 0091)

Code of Good Agricultural Practice for the Protection of Soil (Defra Booklet PB 0617)

Code of Good Agricultural Practice for the Protection of Water (Defra Booklet PB 0587)

Code of Good Agricultural Practice for the Protection of Air (Defra Booklet PB 0618)

Approved Code of Practice for the Control of Substances Hazardous to Health in Fumigation Operations. Health and Safety Commission (ISBN 0-717611-95-7)

Code of Best Practice: Safe use of Sulphuric Acid as an Agricultural Desiccant. National Association of Agricultural Contractors, 2002. (Also available at www.naac.co.uk/?Codes/acidcode.asp)

Safe Use of Pesticides for Non-agricultural Purposes. HSE Approved Code of Practice. HSE L21 (ISBN 0-717624-88-9)

Other Guidance and Practical Advice

HSE (by mail order from HSE Books – See Appendix 2)

COSHH – A brief guide to the Regulations 2003 (INDG136)

A Step by Step Guide to COSHH Assessment, 2004 (HSG97) (ISBN 0-717627-85-3)

Defra (from The Stationery Office – see Appendix 2)

Local Environment Risk Assessments for Pesticides (LERAP): Horizontal Boom Sprayers.

Local Environment Risk Assessments for Pesticides (LERAP): Broadcast Air-assisted Sprayers.

Crop Protection Association and the Voluntary Initiative (see Appendix 2)

Every Drop Counts: Keeping Water Clean

Best Practice Guides. A range of leaflets giving guidance on best practice when dealing with pesticides before, during and after application.

H2OK? Best Practice Advice and Decision Trees – July 2009/10

BCPC (British Crop Production Council – see Appendix 2)

UK Pesticide Guide online (ukpesticideguide.co.uk) – subscription resource

The Pesticide Manual (19th edition) (ISBN 978-1-9998966-5-2)

The Pesticide Manual online (https://www.bcpc.org/product/bcpc-online-pesticide-manual-latest-version) – subscription resource

The GM Crop Manual (ISBN 978-1-901396-20-1)

Compendium of Pesticide Common Names (http://www.bcpcpesticidecompendium.org/) – free resource

The Manual of Biocontrol Agents online (https://www.bcpc.org/product/manual-of-biocontrol-agents-online) – subscription resource.

BCPC Knowledge Bank (https://www.bcpc.org/open-access/knowledge-bank) – free resource

IdentiPest (www.identipest.co.uk) – free resource

Small Scale Spraying (ISBN 1-901396-07-X)

The Environment Agency (see Appendix 2)

Best Farming Practices: Profiting from a Good Environment

Use of Herbicides in or Near Water

SECTION 7
INDEX

Index of Proprietary Names of Products

The references are to entry numbers, not to pages. Adjuvant names are referred to as 'Adj' and are listed separately in Section 4. Products also registered, i.e. not actively marketed, are entered as *PAR* and are listed by their active ingredient in Section 3.

REFERENCES ARE TO ENTRY NUMBERS NOT PAGES

REFERENCES ARE TO ENTRY NUMBERS NOT PAGES

REFERENCES ARE TO ENTRY NUMBERS NOT PAGES

REFERENCES ARE TO ENTRY NUMBERS NOT PAGES

REFERENCES ARE TO ENTRY NUMBERS NOT PAGES

REFERENCES ARE TO ENTRY NUMBERS NOT PAGES

NOTES

NOTES

THE UK PESTICIDE GUIDE 2023

RE-ORDERS

☐ Please send me _____ more copies of *The UK Pesticide Guide 2023* at £63.50 each

Postage: single copy £4.95; orders up to £200 £9.95... orders over £200 £12.50

Outside the UK please add £10 per order for delivery.

Name _____ Position_____

Institution _____ Department_____

Address _____

City _____ Region _____ Postcode_____ Country_____

Tel_____ Fax_____ E-mail_____

EU countries except UK – VAT No:_____

Payment (pre-payment is required):

☐ I enclose a cheque/draft for £_____ payable to BCPC. Please send me a receipt.

☐ I wish to pay by credit card: ☐ Visa   ☐ Mastercard   ☐ Amex   ☐ Switch

Credit Card Payments
If you wish to pay by credit card the membership team will contact you to take your payment details.

Please photocopy and return to:
BCPC BCPC Publications Sales, 93 Lawrence Weaver Road, Cambridge CB3 0LE.
Tel: 01223 342495, Email: publications@bcpc.org, Web: www.bcpc.org

FUTURE EDITIONS

☐ I wish to take out an annual order for _____ copies of each new edition of *The UK Pesticide Guide*

☐ Please send me advance price details for the 2024 edition of *The UK Pesticide Guide* when available

Order by phone: 01223 342495 or online: www.bcpc.org

Bulk discount:
100+ copies   Contact BCPC
50–99   10%
10–49   5%
List price £63.50

Thank you for your order